P9-DYI-984

Human Biochemistry

Human Biochemistry

James M. Orten, Ph.D.

Emeritus Professor of Biochemistry and
Director of Graduate Programs,
Wayne State University School of Medicine,
Detroit, Michigan; Fellow, American Institute of Nutrition

Otto W. Neuhaus, Ph.D.

Professor and Chairman of the Division of Biochemistry,
Physiology, and Pharmacology,
University of South Dakota, School of Medicine,
Vermillion, South Dakota

Tenth edition with 1100 illustrations and 1 color plate

The C. V. Mosby Company

ST. LOUIS • TORONTO • LONDON 1982

A TRADITION OF PUBLISHING EXCELLENCE

Editor: John E. Lotz
Manuscript editors: Marjorie L. Sanson and Alan Sorkowitz
Design: James A. Buddenbaum
Production: Jeanne Bush

Tenth edition

Printed in the United States of America

The C.V. Mosby Company
11830 Westline Industrial Drive, St. Louis, Missouri 63141

Library of Congress Cataloging in Publication Data

Orten, James M., 1904-
Human biochemistry.

Includes bibliographies and index.
1. Biological chemistry. I. Neuhaus, Otto Wilhelm, 1922- . II. Title. [DNLM: 1. Biochemistry. QU 4 078b]
QP514.2.077 1982 612'.015 81-14089
ISBN 0-8016-3730-9 AACR2

TS/CB/B 9 8 7 6 5 4 3 2 1 02/C/229

To our mentors

Reuben Gilbert Gustavson*
Robert Curtis Lewis*
Lafayette Benedict Mendel*

Howard Bishop Lewis*
Lila Miller
Arthur Henry Smith*

*In Memoriam

Preface

With the completion of the tenth edition of *Human Biochemistry,* this textbook is now approaching its fortieth anniversary, the first edition appearing in August 1945 under the authorship of the late Israel S. Kleiner. Thus it now is the senior current American textbook of biochemistry. The present edition is strikingly different from its original ancestor, both in size and in content. Many of the topics in the first edition that were incompletely understood or were vague possibilities are now well-established facts. For example, the glycolytic pathway was just being elucidated, and the citric acid cycle had been postulated only a few years earlier. DNA and RNA were scarcely mentioned, and then only casually as "nucleic acids." The entire subject of purine-pyrimidine metabolism—structures, biosynthesis, and functions—occupied only six pages! The expansion of knowledge in this as well as in the entire field indeed has been phenomenal during the past four decades—in fact, *five* decades for the senior author (JMO), this being his "golden anniversary" in biochemistry.

The continued dramatic progress in biochemistry, even since the publication of the ninth edition of this textbook, has made necessary some rather drastic changes in its organization to maintain a logical sequence of presentation. Also, the reorganization has been focused on a central theme, *the dynamic state of metabolism,* with special emphasis on application to the *human* organism. A considerable amount of older material has been deleted or condensed to make room for newer concepts and to limit the text to a practical size. References to the literature have been restricted largely to selected review-type articles, monographs, and books.

To facilitate an orderly, logical presentation, this edition has been divided into seven major sections. The first, an introductory section, deals with the nature and scope of biochemistry, the biochemical morphology of the cell, and the dynamic state of metabolism. The second section is concerned with basic principles of metabolism and its regulation, including the properties and functions of proteins, enzymes, and coenzymes, mechanisms of energy transduction, and the genetic control of metabolism. The third major section, almost completely rewritten and carefully updated, considers the metabolism of carbohydrates, lipids, amino acids, purines, and pyrimidines. This portion is followed by sections on specialized tissues and body fluids and on the chemical communication between tissues (the hormones, transport mechanisms, and metabolic interrelationships). Section six deals with some applications of human biochemistry and is composed of a reorganized and rewritten chapter on nutrition and a new chapter on molecular aspects of disease. The final section is designed for review purposes in applicable areas of organic and physical chemistry. There are three chapters devoted to the chemistry of the major cellular constituents, selected physicochemical topics, and analytical techniques frequently used in biochemistry.

Several new chapters have been added. One on coenzymes brings together coenzyme chemistry and principal functions in a way that does not detract from the flow of discussions in the chapters on metabolism and nutrition. Bioenergetics and biological oxidations also are treated in a separate chapter for the sake of emphasis and clarity. Likewise, a full chapter is devoted to transport mechanisms to discuss more advantageously recent advances in this active area and their importance in the control of metabolism. The subject of nutrition has been consolidated into a single chapter, including a concise discussion of energy metabolism as well as newer developments in this rapidly expanding area of biochemistry. The new (1980) recommended dietary allowances of the Food and Nutrition Board (N.R.C.) and the rationale for their formulation are thoroughly considered. The material on digestion and absorption has been transferred into appropriate chapters on metabolism. A new chapter on molecular aspects of disease has been added to localize and emphasize this subject of increasing medical significance.

This edition has been prepared with the needs of students in the medical and other health sciences uppermost in mind. Excessive detail, redundancy, and verbosity have been minimized as far as possible without detracting from clarity of expression and accuracy of meaning. At the same time, we have tried to avoid oversimplification and to challenge the curiosity and imagination of the more thoughtful students of biochemistry. The reorganization and revision of the text, we believe, will facilitate the study and understanding of this dynamic and challenging discipline.

In the preparation of much of the manuscript for this edition we were fortunate in having the collaboration of Dr. Francis C. Neuhaus of Northwestern University as a consultant. His critical, detailed reading and thoughtful suggestions on most of the chapters added greatly to their clarity, accuracy, and completeness. Likewise, we are indebted to Dr. Aline Underhill Orten for numerous helpful suggestions and comments, and especially for critical reviews of the chapters on nutrition and molecular aspects of disease.

We also wish to express our sincere appreciation to a number of colleagues and friends who have offered constructive comments or materials for the manuscript. Special words of gratitude are due Drs. Ray K. Brown, Samuel C. Brooks, Robert J. Peanasky, Danica Dabich, Gary D. Small, Walter H. Seegers, John A. Thomas, William W. Winder, Bennie Zak, David F. Hastings, and Paul Loach.

Finally, we thank the editors and staff of The C.V. Mosby Company for their cooperation and patience during the preparation and publication of the manuscript, and our respective departmental staffs for their generous help and encouragement over the many months required for the completion of this edition.

James M. Orten
Otto W. Neuhaus

Contents

Human Biochemistry

Section One

Introduction: some perceptions of human biochemistry

Chapter 1

Nature and scope of biochemistry

Nature of biochemistry

Biochemistry, in broad terms, is the study of the chemical composition of living matter and the biochemical processes that underlie life activities during growth and maintenance. Such a general concept obviously includes plant as well as animal life. However, this book concerns primarily the biochemistry of animal life, and especially biochemistry as it applies to humans.

Early inquiries into the biochemistry of living matter by pioneering scientists such as Priestly and Laviosier in the late eighteenth century and Liebig, Berzelius, and Bernard in the nineteenth century were concerned with the whole animal and gross tissue samples. Their work was more descriptive, seeking answers to the "what" types of questions.

With the modern refinement of biochemical techniques and the development of more sophisticated and sensitive instrumentation, particularly during the last 25 years, it has become possible to delve much more deeply into these questions and to seek answers to "how" and "why" questions. For example, the development of the ultracentrifuge by Svedberg made it possible to investigate molecular problems at a cellular level and indeed even their component structural parts, i. e., the membranes, mitochondria, nuclei, endoplasmic reticulum, and other cellular parts (Chapter 2). It became possible to explore the chemical mechanisms involved in the development and differentiation of cells and what factors shape their ultimate form: how and why, for example, a primitive erythropoietic stem cell differentiates into a mature erythrocyte capable of transporting oxygen rather than into a muscle cell or a nerve cell with entirely different functions. Modern biochemistry has provided many answers to basic questions such as these.

Another exciting example is the biochemical elucidation of the structures of deoxyribonucleic acid (DNA) and ribonucleic acid (RNA). This has had far-reaching effects in biology and genetics on the understanding of the structure of genes and the transmission of genetic information and, in turn, on the development and differentiation of cells. Other examples are the chemical transformations by which glucose, amino acids, and lipids derived from foods are transformed into essential body components (anabolism) by way of so-called metabolic pathways or are used for energy (catabolism) in cells. Insight into the precise mechanisms involved in the control of these metabolic reactions in cells is another monumental accomplishment of modern biochemistry. The potential practical importance of this information is emphasized by the fact that a failure of normal metabolic control mechanisms resulting in wild growth is a unique biochemical characteristic of cancer cells. The control of the cancerous processes thus may depend on the reestablishment of normal cellular regulatory mechanisms.

Other recent accomplishments in biochemistry include the chemical isolation and determination of the molecular structures, three dimensional in many instances, of those catalysts of metabolic reactions, the various enzymes, hormones, and vitamins to be considered in later sections. In some cases even visualization of changes in molecular conformation, by x-ray diffraction techniques, in relation to the functional activity of the molecule has been achieved. Likewise of great importance has been the use of isotopic tracers in unraveling the complexities of metabolic pathways, of chromatographic procedures in isolating the minute amounts of intermediates involved in metabolic reactions, and of radioimmunoassay techniques in determining even picomolar quantities of certain vital biomolecules, such as insulin and other hormones in living tissues and body fluids. These procedures also have proven invaluable in the diagnosis of a number of human diseases. Nor should one omit mention of the current evidence linking proper nutrition with the prevention of human "killer diseases" such as ischemic heart disease, stroke, certain forms of cancer, diabetes mellitus, and possibly even hypertension. The foregoing, randomly selected examples emphasize not only the great progress made in biochemistry itself during recent years but also the tremendous impact of these advances on all other life sciences. Biochemistry has thus become the center around which studies of all life sciences must revolve.

Development of biochemistry

Biochemistry, as such, is a relatively young science; the term itself was not introduced until 1903 by the eminent German chemist Carl Neuberg. However, investigations of the chemical composition of plant and animal tissues as early as the mideighteenth century perhaps may be regarded as the beginning of biochemistry as a separate discipline. This was followed in the later eighteenth and early nineteenth centuries by chemical studies of respiration, fermentation, and the quantitative analysis of naturally occurring substances. The nature of many of these earlier discoveries and the names of the investigators involved are referred to later. From these rather fragmentary beginnings biochemistry matured into a distinct entity in the later nineteenth century, although it was then termed *physiological chemistry* or by some *pathological chemistry*. The first journal devoted exclusively to the new discipline, Hoppe-Seyler's *Zeitschrift für Physiologische Chemie,* began in 1877; its American counterpart, the *Journal of Biological Chemistry,* began publication in 1906 as the official journal of the newly formed American Society of Biological Chemists.

The period of greatest progress in biochemistry began in the 1920s with the American biochemists playing an increasingly prominent role in its development, previously accomplished primarily by German, French, English, and Swedish biochemists. This period included monumental discoveries such as the isolation, determination of the chemical structure, and synthesis of a number of the vitamins and certain hormones, the concepts of the essential amino acids and the citric acid cycle, and the beginning elucidation of metabolic pathways by means of isotopes as tracers. The post–World War II era, however, beginning in the early 1950s, was the most remarkable period of progress in biochemistry, with knowledge in the field nearly doubling every 8 years. This was due in part to the development of the new, improved research techniques and equipment and in part to the availability of more adequate funds for research, largely from federal agencies.

Biochemistry thus has developed into one of the most, if not the most, dynamic and exciting areas of human endeavor. Many of the outstanding achievements have been made by American biochemists, a number of whom have received Nobel Prizes and other prestigious awards in recognition of the quality and significance of their work.

Chemical composition of living matter

Living matter is composed of a variety of nitrogenous and nonnitrogenous organic compounds, a number of inorganic elements, and water. Water is the most abundant constituent of animal organisms, composing 75% to 85% of the weight of most tissues, with the exception, of course, of bones, teeth, and a few others.

Water

The water of the tissue and body fluids is mostly in the free state, i.e., substances may be dissolved in it and it may pass back and forth from blood to tissues, in and out of cells. A small fraction of the water is believed to be bound. In other words, some of the water in hydrophilic colloid systems is combined so that the activity of the water molecules is reduced considerably. Free water varies according to diet and physiological activity, whereas bound water is a rather constant constituent of the tissues.

Studies using deuterated water (D_2O) in dogs have shown that the average water content of the body as a whole is 61% of body weight, with a range of 55% to 67%. The water content of the human body apparently has about the same range, being less than average in fat individuals and somewhat greater in thin persons. The water content of individual tissues also varies considerably, as will be discussed later. There are several mechanisms for maintaining and controlling the water content of tissues (Chapter 15).

Water is needed for various reasons. It is a solvent, the vehicle that enables water-soluble, water-miscible, or emulsifiable substances to be transferred in the body, not only in the blood but also intercellularly and intracellularly. Ionization takes place in water, and ionization is a prerequisite to most biochemical reactions.

Organic constituents

The organic compounds found in all living matter include a wide variety of substances indispensable to life processes. Quantitatively, three major groups of organic compounds predominate: the carbohydrates, lipids, and proteins. Some of these are present partly as highly complex macromolecules, as discussed in some detail later. Various other essential biomolecules are present in smaller amounts. These include the nucleic acids (derivatives of purines and pyrimidines), many hormones, vitamins, and a number of other nitrogenous and nonnitrogenous compounds. These are discussed in subsequent chapters.

Sources

Most animal organisms, including humans, depend on plants and certain microorganisms for a preformed supply of many of the aforementioned nitrogenous and nonnitrogenous substances. Therefore they have been categorized as essential nutrients (Chapter 21). Plants and microorganisms, in turn, form them from simpler molecules, such as CO_2, N_2 or NH_3, SO_4, and H_2O, by way of photosynthesis, using energy (photons) from the sun to drive the synthetic reactions. Important constituents of animals cells such as glucose, the essential amino acids, the essential unsaturated fatty acids, and the vitamins (Chapter 21) thus are derived from food sources.

Interesting recent investigations have demonstrated that many of the previously mentioned compounds also were synthesized in prebiotic times, probably several billion years ago, from simple molecules (CO_2, CH_4, NH_3, H_2, SCN, and H_2O) using radiant energy from sources such as the sun, cosmic rays, and lightning. Indeed, a variety of organic molecules of biological importance, including glucose, a number of amino acids, adenine, ribose, malic acid, and even proteinoids, have been formed in the laboratory using these simple molecules, an electric arc for energy, and suitable temperature and pressure. Supporting evidence has come from other sources. Some 22 different amino acids have been identified in samples of pre-Cambrian sedimentary rocks that are at least 3.1 billion years old. Trace amounts of some 11 amino acids have been found in lunar rocks and other materials brought back from the surface of the moon by astronauts. Human and other contamination as a source of the amino acids apparently was ruled out. Likewise, evidence from optical spectra and radiotelescopic studies shows that porphyrin derivatives and some 31 other molecular species exist in interstellar space.

Thus the prebiotic formation of organic compounds needed for converting the sun's radiant energy into chemical energy in the form of essential substances by primordial organisms, then by photosynthetic microorganisms and plants, made possible their ultimate use as constituents of all living matter, including that of humans.

Inorganic constituents

At least 60 of the 102 or more elements known to be present in the universe occur in biological matter. Only about 22 to 26 of these are found consistently in human tissue, however, and some are present only in extremely minute amounts. About 1% of the total weight of an average soft tissue is ash, or inorganic salts, chiefly of the cations Na^+, K^+, Mg^{++}, Ca^{++}, and NH_4^+, and the anions Cl^-, $H_2PO_4^-$, $HPO_4^=$, HCO_3^-, and $SO_4^=$. Some of these may be present in organic molecules, as is also the case for the trace elements iron (Fe), iodine (I), copper (Cu), zinc (Zn), and manganese (Mn). Other trace elements consistently found in nearly all forms of living matter include boron (B), chromium (Cr), cobalt (Co), fluorine (F), molybdenum (Mo), selenium (Se), and silicon (Si). Biochemical functions of Co, F, and probably Cr, Mo, and Se, are now known, as is discussed in Chapter 21. Other elements are found in small amounts in some species, but as yet no definite function for them has been established. These include silver (Ag), aluminum (Al), arsenic (As), barium (Ba), beryllium (Be), bromine (Br), cadmium (Cd), cesium (Cs), germanium (Ge), lithium (Li), nickel (Ni), lead (Pb), rubidium (Rb), tin (Sn), strontium (Sr), titanium (Ti), and vanadium (V). A few other elements that are regarded as contaminants or accidental constituents may be found in living matter. These include argon (Ar), gold (Au), bismuth (Bi), helium (He), mercury (Hg), and thallium (Tl).

A number of elements occur in living matter as mixtures of the more common form with varying amounts of other forms of the same element. These have slightly different atomic structure and atomic weight from the more common form and are called *isotopes*. Thus ordinary chlorine, with an atomic weight of 35.457, has been found to be a mixture of two isotopes, the first and more abundant one having an atomic mass of approximately 35, and the second less abundant one an atomic mass of 37. Since isotopes in general have the same chemical and biological properties as the more abundant form, they have proved extremely valuable as tracers in biochemical research. Metabolites labeled with isotopic atoms can be followed through an organism, and metabolic pathways can thus be determined. Also, it is possible by determining the amount of an isotope, e.g., ^{14}C, in a specimen of wood or fossil to accurately estimate the age of the specimen by the isotope dating technique. The amount of the isotope present is determined by either its radioactivity, as in the case of ^{14}C or ^{3}H (tritium), or by its mass (in a mass spectrometer), as in the case of ^{13}C or ^{2}H (deuterium).

Gases

Relatively small amounts of several free gases, including oxygen, carbon dioxide, nitrogen, and traces of others, are found in animal tissues and body fluids. Oxygen and carbon dioxide have extremely important functions and are considered in Chapters 16 and 17. Larger amounts of oxygen, carbon, and nitrogen are found, of course, as constituents of a wide variety of organic and inorganic compounds.

Dynamic state of metabolism

Since biochemistry is concerned with the chemical components of and chemical reactions in living matter, it is logical at this point to inquire into the characteristics that distinguish living matter from the nonliving, inanimate molecules of which it is composed. Although knowledge of this fundamental question is far from complete, one such characteristic is the high degree of *organization* of the unit cells of which living matter is composed (Chapter 2). Each cellular subunit has specific, purposeful functions to perform in establishing and maintaining the life of the cell. Some are concerned with generating energy for use in vital cell functions such as mobility, the absorption of essential nutrients from the environment, or the replication of new molecules to replace those lost or rendered nonfunctional by the wear and tear of life processes. Also, specialization of cellular functions is common in living organisms, especially in higher forms, some functions being emphasized genetically and some being repressed (Chapter 3). Familiar examples in humans are muscle cells for locomotion, red blood cells for oxygen transport, glandular epithelial cells for special secretory products (enzymes, hormones, etc.), and nerve cells for intercellular coordination. However, all cells have in common two major general functions: energy generation and energy utilization for growth and/or maintenance. These may be termed *metabolic reactions* or more simply *metabolism*. These metabolic reactions must be maintained and regulated continuously to sustain life. Therefore they are in a *dynamic state*.

Metabolic processes in normal cells are not random but are rigidly controlled by a number of interrelated intrinsic and extrinsic checks and balances. The regulation of the dynamic state of metabolism is effected by genetic (coarse control) and fine control mechanisms (modulation of cellular enzyme activities) (Chapter 3). The dynamic state of metabolism with its highly organized regulatory mechanisms is thus a vital biochemical characteristic of "normal" living matter and is considered

Figure 1-1 The dynamic nature of metabolism with its interrelationships to the major areas of biochemistry.

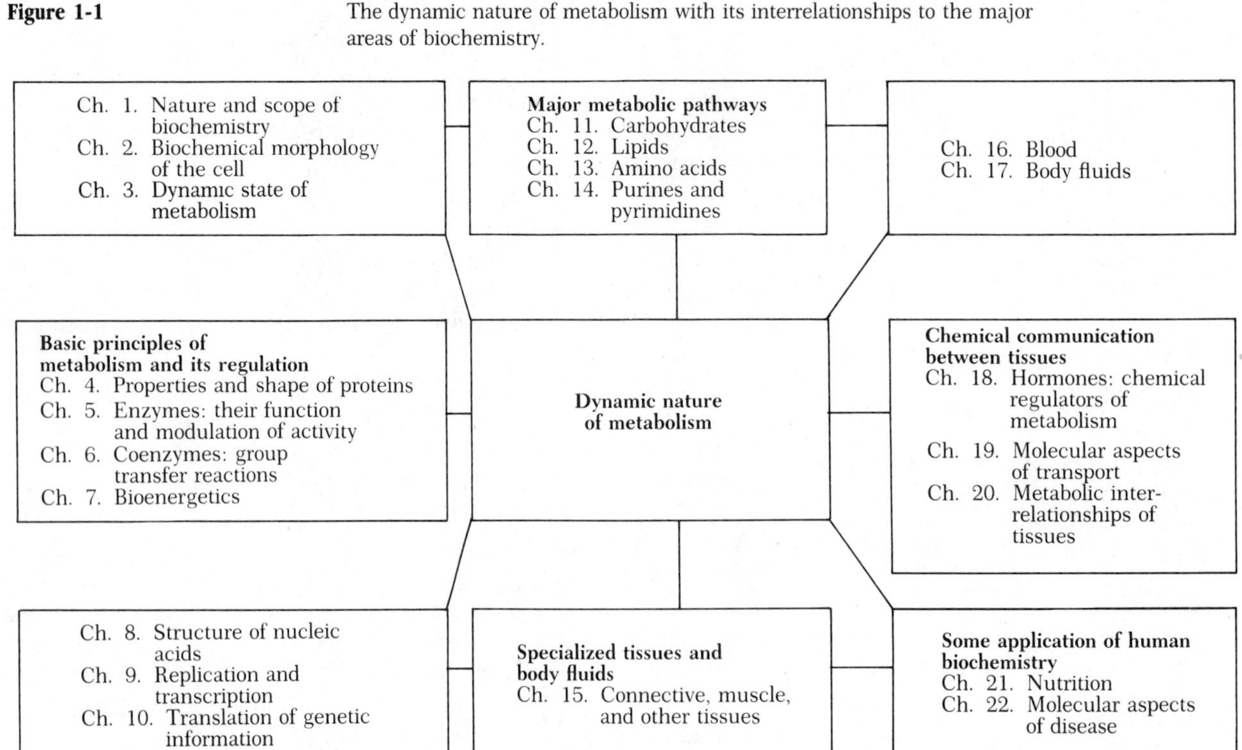

to be a central theme of the present discussion of human biochemistry. The ensuing chapters of this book are in fact subtopics of this guiding theme, arranged in logical progression, as depicted schematically in Fig. 1-1.

Molecular aspects of disease

Unfortunately perhaps, animal cells, especially human, are unable to accomplish complete metabolism because of an inability to synthesize all their required components. Thus humans are dependent, nutritionally speaking, on other living organisms to supply these substances: plants for glucose as carbohydrates by photosynthesis, bacteria for nitrogen fixation, and microorganisms for the synthesis of certain amino acids and vitamins. But animal and human cellular systems themselves are not always perfect. Defective enzymes may be synthesized, or an insufficient production of an enzyme may lead to an insufficient supply of an essential intermediate or product (metabolite). A hereditary disease of metabolism thus may result from an incorrect expression of an enzyme. The molecular aspects of hereditary metabolic diseases, one of the most important and frequently occurring groups of diseases afflicting mankind, are discussed in Chapter 22.

Modern biochemistry is thus a multifaceted discipline, closely interrelated with other life sciences including the medical sciences. Consequently it encompasses a wide variety of life scientists who use chemical, biological, physical, immunological, and nutritional techniques to probe life processes—molecular events in cells, cellular organelles, tissues and body fluids, and even whole organisms. Biochemistry has become a unique blend of contributions from classical biochemists, cell biologists, physiologists, immunologists, nutritionists, medical scientists, and others. In turn, it has become a foundation for and is enriched by a wide variety of life science disciplines. Biochemistry is now a basic language for all biology and is essential to the expression and understanding of biological and medical phenomena.

Suggested readings

Anders, E., Hayatsu, R., and Studier, M.H.: Organic compounds in meteorites, Science **182:**781, 1973.

Anfinsen, C.B.: The molecular basis of evolution, New York, 1959, John Wiley & Sons, Inc.

Bronowski, J.: The ascent of man, Boston, 1974, Little, Brown & Co., Inc.

Buvet, R., and Ponnamperuma, C., editors: Chemical evolution and the origin of life, New York, 1971, American Elsevier Publishing Co., Inc.

Calvin, M.: Chemical evolution, New York, 1969, Oxford University Press, Inc.

Chittenden, R.H.: The development of physiological chemistry in the United States, American Chemical Society Monograph No. 54, New York, 1930, The Chemical Catalog Co., Inc.

FASEB Conference: Genetics and biological evolution, Fed. Proc. **35:**2077, 1976.

Florkin, M., and Stotz, E., editors: Comprehensive biochemistry, New York, 1962-1967, American Elsevier Publishing Co., Inc.

Fox, S.W., and Dose, K.: Molecular evolution and the origin of life, San Francisco, 1972, W.H. Freeman & Co., Publishers.

Johnson, F.M.: Porphyrins in space, Chem. Eng. News **49:**5, 1971.

Lewis, J.S.: The chemistry of the solar system, Sci. Am. **230:**51, 1974.

Lunar knowledge from moon samples, Chem. Eng. News **50:**35, 1973.

Nobel Foundation: Nobel lectures (1901-1962, 1963-1970): Chemistry, 4 vols.; Physiology and Medicine, 4 vols., New York, 1966 and 1972, American Elsevier Publishing Co., Inc.

Oparin, A.I.: Life: its nature, origin, and development, New York, 1962, Academic Press, Inc.

Ponnamperuma, C., et al.: Non-protein amino acids in the Murchison meteorite, Proc. Nat. Acad. Sci. **69:**486, 1971.

Rohlfing, D.L., and Oparin, A.I., editors: Molecular evolution—prebiological and biological, New York, 1973, Plenum Publishing Corp.

Rose, W.C.: Recollections of personalities involved in the early history of American biochemistry, J. Chem. Educ. **46:**759, 1969.

Wilson, A.C., Carlson, S.S., and White, T.J.: Biochemical evolution, Annu. Rev. Biochem. **46:**573, 1977.

Chapter 2

Biochemical morphology of the cell

The human body is composed of a multitude of specialized tissues which in turn consist of vast clusters of cells that have differentiated into specialized chemical factories. Each human begins as a single cell, a zygote, possessing a nucleus formed by the fusion of a sperm and an egg cell. Thus all the genetic information needed to produce a person (about 10^{13} cells) is represented in the original, fertilized egg cell. With growth and development cells differentiate and become highly specialized. Even though all the cells in the body contain the same genetic information, much of it has become repressed and is silent or not expressed.

In our consideration of human biochemistry it is first appropriate to describe the nature of the unit cell and its morphological substructures. In time we must appreciate the relation of substructure to cellular function and to the function of the tissue type of the individual cell. Substructures will eventually be seen as sites where specific molecular activities occur. This view of structure related to biochemical function is *biochemical morphology*.

Concept of the cell

The cell is a living, reproducing, structural unit that contains substructures or *organelles*. Two general types of cells are recognized in nature from a biochemical-morphological point of view. These are *prokaryotes* and *eukaryotes*. As the names imply, the difference lies primarily in the presence or absence of a specific organelle, namely the nucleus.

Prokaryotes are relatively simple cells representing most bacteria, as well as blue-green algae, mycoplasmas, spirochetes, and other types of microorganisms. They have no discrete nucleus and no nuclear membrane. Eukaryotic cells possess a distinct nucleus and are present in all higher organisms, both plant and animal, as well as in fungi, protozoa, and most algae.

Eukaryotic cells

In a human, eukaryotic cells exist in a variety of sizes and shapes that allow adaptation to their special functions (Fig. 2-1). Despite their diversity they have certain features in common. The nucleus is always a structure in which the DNA-containing chromosomes (Chapter 8) are separated from the rest of the cell's contents by an envelope called the perinuclear membrane. There are also other structures or subcellular organelles that involve membranes and have special functions. These are the mitochondria, the endoplasmic reticulum, the Golgi apparatus, and various kinds of vacuoles. Therefore eukaryotic cells possess a variety of internal membrane systems.

The schematic diagrams in Fig. 2-1 show that the cell units are separated from their surroundings by a *plasma membrane*. It is apparent that this membrane per-

Figure 2-1
The diversity of human cells.

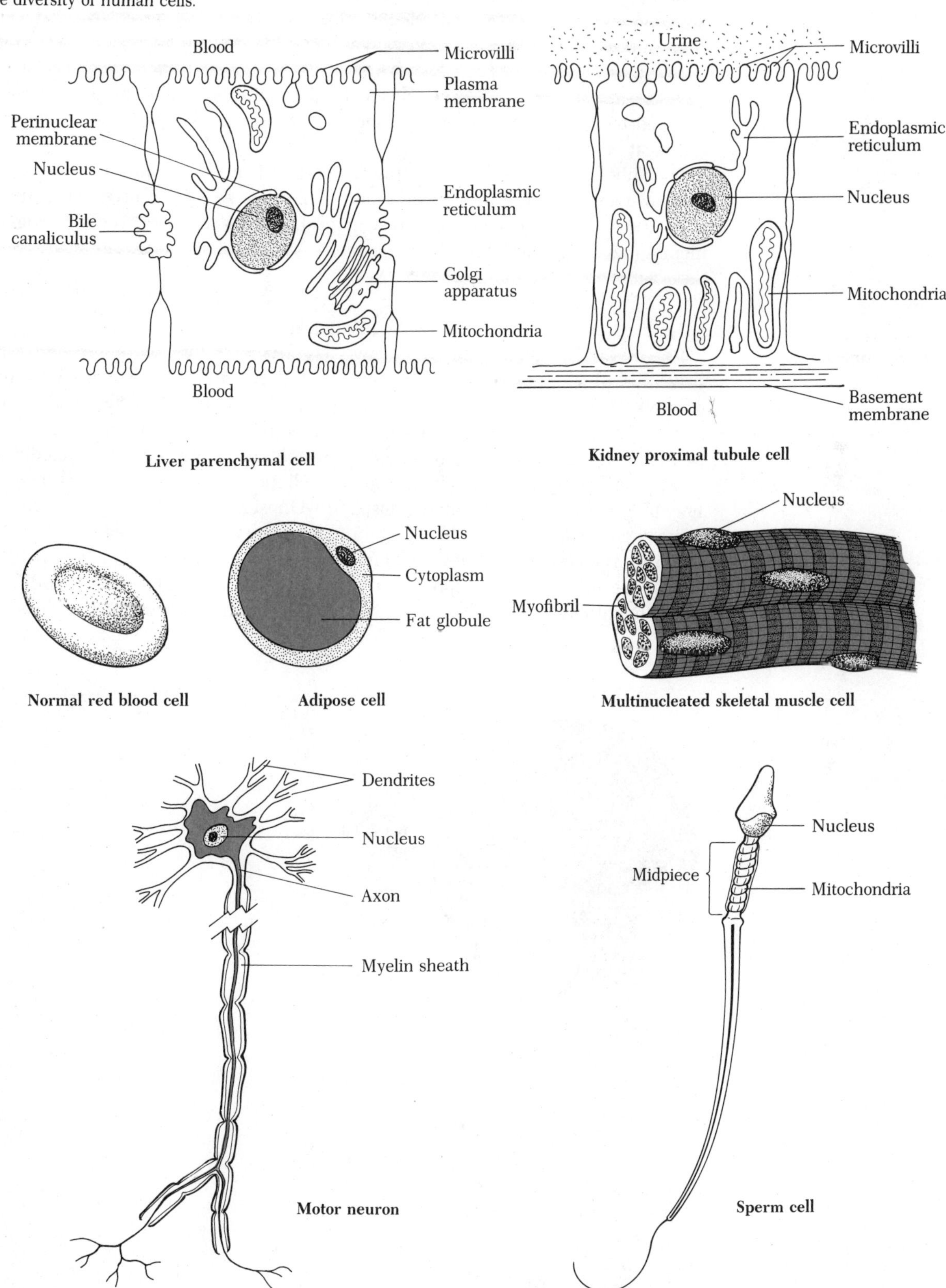

forms a number of important functions: it holds the cell together; it serves as a selective barrier to the outside, permitting (and even enhancing) the entrance of essential nutrients while at the same time preventing the loss of needed substances; it secretes waste products; it keeps out toxic materials; and it binds certain regulatory substances, particularly hormones, in a locale most advantageous to the performance of their function. Obviously, a unique type of structure must be required to serve diverse functions such as mechanical support and a selective chemical barrier.

The subcellular organelles illustrated in Fig. 2-1 are also composed, in part at least, of membranes. Although internal membranes perform a diversity of functions, they all seem to share certain structural features. In other words, biological membranes are remarkably alike in structure. As a least common denominator, all membranes contain *phospholipids*.

Membrane structure

Phospholipids (Chapter 23) are distinguished by a common structure, the backbone of which is glycerol. As is evident in Fig. 2-2, two of the three hydroxyl groups of glycerol serve as points of attachment for long-chain carboxylic acids (fatty acids). To the third glycerol carbon is attached a phosphate group linking compounds such as choline, ethanolamine, inositol, and serine via a phosphodiester bond. The principal phospholipids found in cell membranes are phosphatidyl choline, phosphatidyl ethanolamine, phosphatidyl inositol, and phosphatidyl serine. In mammalian cells there are also certain neutral lipids such as cholesterol. As an example, the composition of various membranes of rat liver cells is given in Table 2-1. Since phospholipids represent many individual molecules distinguished by their constituent fatty acids (mostly 16- and 18-carbon saturated fatty acids or the

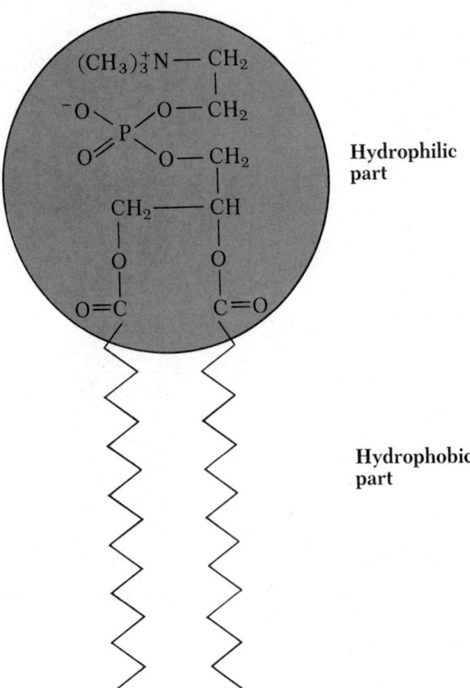

Figure 2-2
Structure of a phospholipid.

Table 2-1
Relative distribution of lipids in membrane fractions of rat liver

Lipids	Endoplasmic reticulum*	Golgi apparatus*	Plasma membrane*
Phospholipids			
Phosphatidylcholine	61	45	40
Lysophosphatidylcholine	5	6	7
Phosphatidylethanolamine	19	17	18
Lysophosphatidylethanolamine	—	6	6
Phosphatidylserine	3	4	4
Phosphatidylinositol	9	9	6
Sphingomyelin	4	12	19
TOTAL PHOSPHOLIPIDS†	85	54	62
Neutral lipids			
Triglycerides	25	35	22
Free fatty acids	41	39	35
Cholesterol esters	10	10	8
Cholesterol	25	17	35
TOTAL NEUTRAL LIPIDS†	15	46	38

Modified from Keenan, T.W., and Morré, D.J.: Biochem. **9**:19, 1970.
*Percent distribution of individual phospholipid; percent distribution of individual neutral lipids.
†Percent of total lipids.

18-carbon monounsaturated oleic acid), the diversity of membrane composition is extensive and varies from tissue to tissue. It has been estimated that the membranes of red blood cells contain 150 to 200 chemically different lipid molecules.

Phospholipids contain both charged and uncharged groups. In phosphatidyl choline, for example, a positively charged quaternary ammonium ion contributes to the molecule's *hydrophilic*, or water-loving, properties (represented by the colored area in Fig. 2-2). The two fatty acid "tails," on the other hand, provide this molecule with distinctly *hydrophobic*, or water-hating, characteristics. A molecule exhibiting hydrophobic characteristics at one end and hydrophilic properties at the other is called an *amphipathic* molecule.

Model membranes

When a solution of phospholipid is pipetted onto an aqueous surface, the lipid molecules spontaneously orient themselves at the interface between the air and water (Fig. 2-3, *A*). The aliphatic side chains are oriented into the air, and the hydrophilic groups are in the aqueous medium. If, however, such a solution of phospholipids is very carefully injected into the aqueous medium, a globule or *micelle* forms in which the lipid molecules are oriented with their aliphatic side chains facing inward to produce an hydrophobic core (Fig. 2-3, *B*). The hydrophilic groups form the outside surface of the sphere in contact with the aqueous medium. This spontaneous orientation is a consequence of the amphipathic nature of the phospholipid molecules. The nonpolar side chains will come together to the exclusion of any water molecules, producing a hydrophobic core. The hydrophilic groups orient themselves so as to interact with the water molecules. When larger amounts of phospholipid solution are injected into the aqueous medium, the lipid molecules may spontaneously achieve a more complex structure. The droplets now consist of phospholipid bilayers; such globules are called *vesicles* (Fig. 2-3, *C*). Even more complex globules composed of multiple phospholipid bilayers may form. These are called *liposomes*. In lipid bilayers the phospholipids orient themselves so that the aliphatic side chains face each other to form a hydrophobic domain (Fig. 2-3, *D*). The hydrophilic groups are at each aqueous surface of the layer. Lipid bilayers can

Figure 2-3

Spontaneous orientation of phospholipid molecules.
A As a monolayer at air-water interface.
B As a micelle with a hydrophobic core.
C As a vesicle composed of a lipid bilayer.
D Orientation of phospholipid molecules in a typical bilayer arrangement.

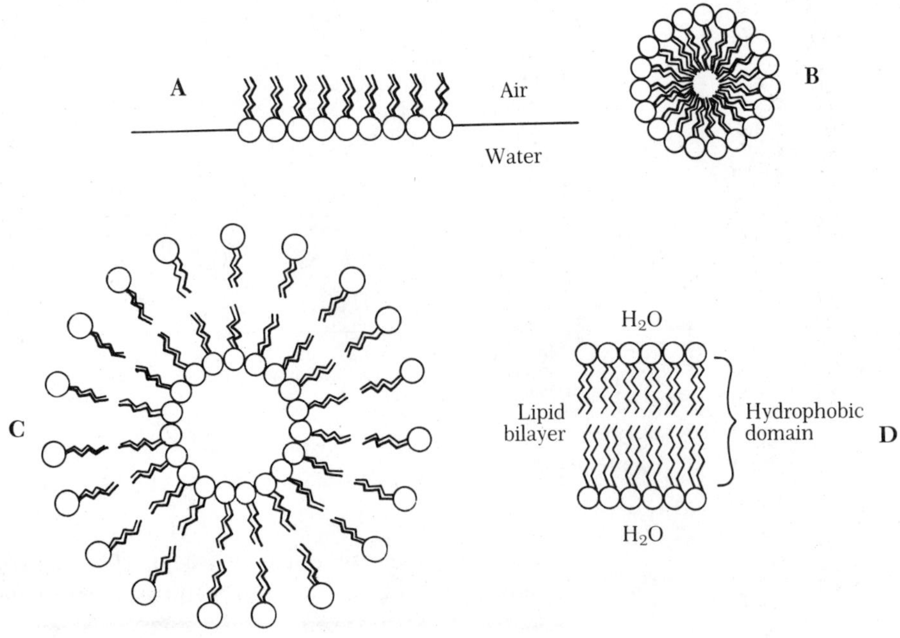

Figure 2-4
Lipid-protein mosaic model of a cell membrane. Two classes of membrane protein are illustrated. "Extrinsic" proteins lie at or near either membrane surface, as illustrated by *A*. "Intrinsic" proteins are either partially or totally buried in the lipid bilayer, as shown by *B* and *C*. In the case of *C*, proteins serve to bridge the bilayer, either singly or in pairs. The branched structures associated with the upper proteins represent polysaccharide side chains.

Modified from Capaldi, R.A.: A dynamic model of cell membranes, Sci. Am. **230**:27, 1974.

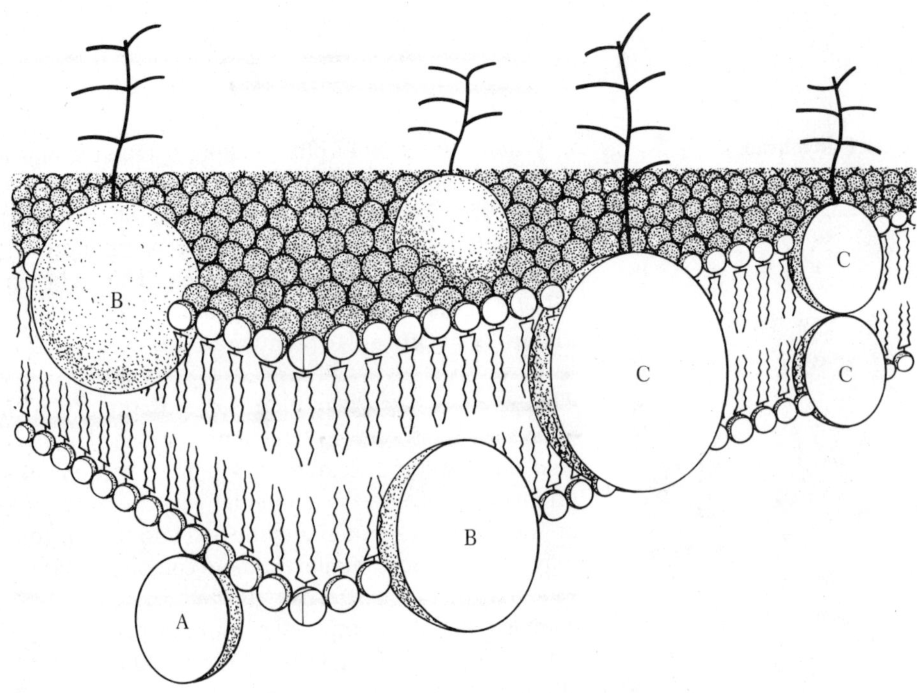

Table 2-2
Composition of membranes from human erythrocyte

Component	Percent of dry mass
Protein	49.2
Lipid (total)	43.6
Phospholipid	32.5
Cholesterol	11.1
Carbohydrate (total)	7.2

only be considered to be models of biological membranes; they do not have all the properties exhibited by true cell membranes.

As seen in Table 2-2, a cell membrane such as that derived from red blood cells is composed of 49% protein, 44% total lipid, and 7% carbohydrate. Furthermore, electron microscopy has shown that the typical plasma membrane, which is 7 to 10 nm thick, is a lipid bilayer in which is dispersed a considerable amount of protein. The model membrane is considered to be composed of many irregularly shaped protein molecules floating in a lipid sea (Fig. 2-4). There exist two kinds of proteins: integral (or intrinsic) and extrinsic (or peripheral). *Integral* proteins are either partially or totally immersed in the lipid bilayer and are difficult to remove by any means other than the disruption of the membrane with a detergent. These proteins have regions rich in amino acids with nonpolar side chains (Chapter 4). Parts of the protein molecules that are rich in polar amino acids are oriented to the outside or into the aqueous environment adjacent to the membrane. Fig. 2-4 also shows that some proteins have carbohydrate side chains that contribute to their hydrophilic nature. Such glycoproteins are arranged with the carbohydrate oriented into the external aqueous environment. *Extrinsic* protein only interacts with the hydrophilic groups and therefore is readily removed by extraction with an aqueous medium.

Red blood cells possess a plasma membrane with a supportive latticework composed of a glycoprotein called *spectrin*. This protein occupies the place indicated by *A* in Fig. 2-4. Actually, the spectrin molecules are relatively long and threadlike. The fibers interlace to provide the membrane with the properties of a rigid lattice.

Both protein and lipid molecules have freedom of motion within the framework of the membrane, made possible by the fluid nature of the bilayer. Movement is mainly in a lateral direction. This important property of fluidity facilitates conformational changes that occur in the protein molecules of the membrane during their functional activity.

The nucleus

The nucleus, the "information center" of the eukaryotic cell, is its most prominent feature. Approximately 90% of the cell's DNA (Chapter 8) is found concentrated in this organelle as part of structures called *chromosomes*. Human cells contain genetic information distributed among 46 such chromosomes. These structures are composed of a nucleoprotein known as *chromatin,* which consists approximately of equal quantities of DNA and protein; some RNA is also found.

The nucleus is surrounded by two concentric membranes punctuated by discontinuities, or nuclear pores (Fig. 2-5, *A*). Each of the membranes appears to be of the typical bilayer type composed of phospholipids and cholesterol. They are approximately 8 nm thick and are separated from each other by a space, the perinuclear cisterna, which is 10 to 50 nm thick. The two membranes fuse periodically to produce the pores that permit the exchange of materials between the nucleus and

Figure 2-5

Structure of perinuclear membrane.

A Electron micrograph showing a nuclear membrane (*N*) punctuated by numerous pores (arrows).

B Parts of a nuclear envelope and its relation to endoplasmic reticulum.

Courtesy D. Branton, Harvard University.

A

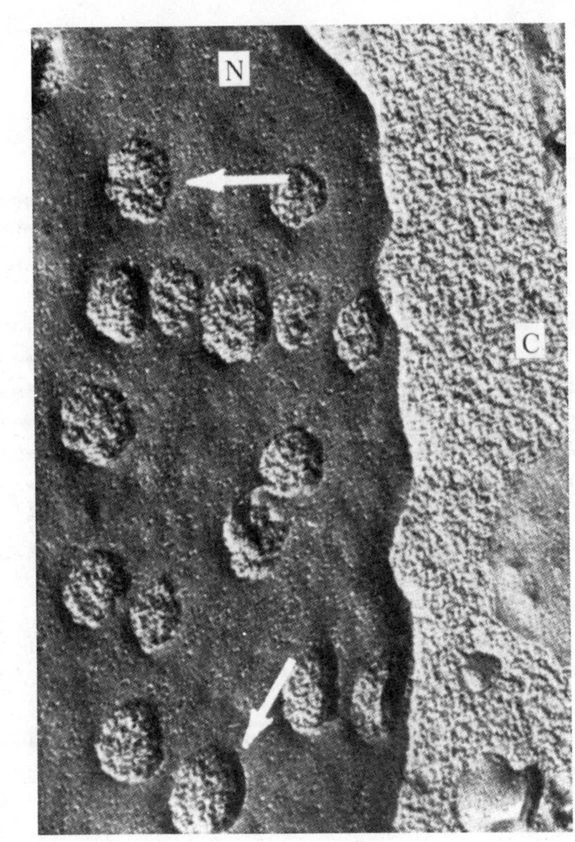

B

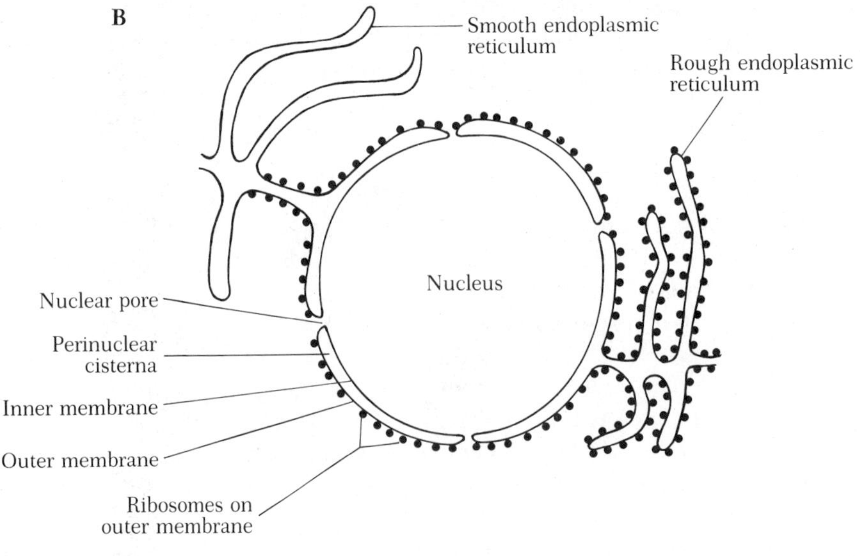

the cytoplasm. The outer membrane is usually covered with ribosomes and is continuous with the endoplasmic reticulum (Fig. 2-5, *B*).

Nucleolus

Nuclei contain one or more small, dense bodies known as the *nucleolus(i)*. Their actual number varies from one type of cell to another. They are not encompassed by a membrane but are in direct contact with the matrix of the nucleus or *nucleoplasm*. Approximately 80% of the nucleolus is protein; the remainder is about 11% RNA and 8% DNA. The nucleolar RNA is chiefly ribosomal RNA and appears to be synthesized in this body prior to its movement into the cytoplasm by way of the nuclear pores.

Cytomembranes

Many eukaryotic cells possess an extensive network of internal membranes pervading the cytoplasm. They are given the general designation of *cytomembranes* and are usually subdivided into the *endoplasmic reticulum* and the *Golgi apparatus*.

Rough endoplasmic reticulum

The *endoplasmic reticulum* consists of a complicated array of vesicular spaces separated from the cytoplasmic fluid or cytosol by a system of membranes. These have the same lipid bilayer structure as described previously, although they are usually somewhat thinner than the plasma membrane (approximately 7 nm). The endoplasmic reticulum is separated into rough and smooth aspects (Fig. 2-6). The rough is lined on the cytoplasmic surface with a multitude of particles called *ribosomes* or, if in clusters of many, *polyribosomes* or simply *polysomes* (Chapter 10).

Figure 2-6
Various types of cytomembranes in a liver cell.

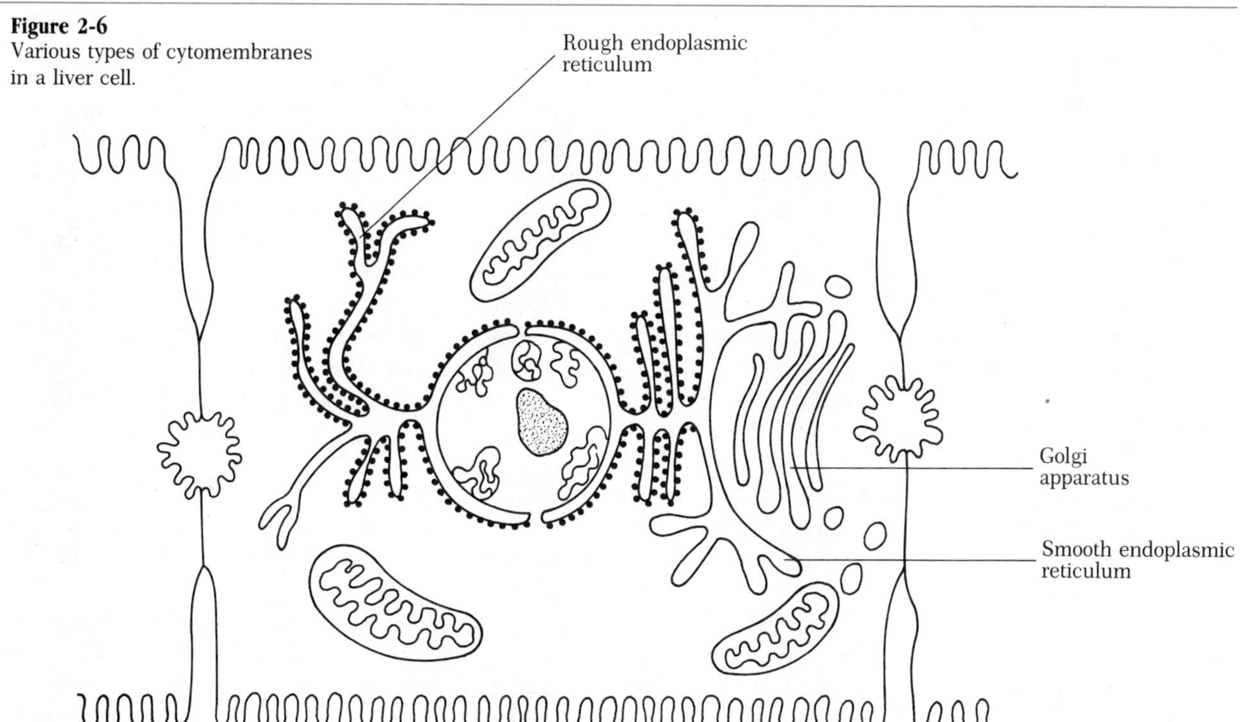

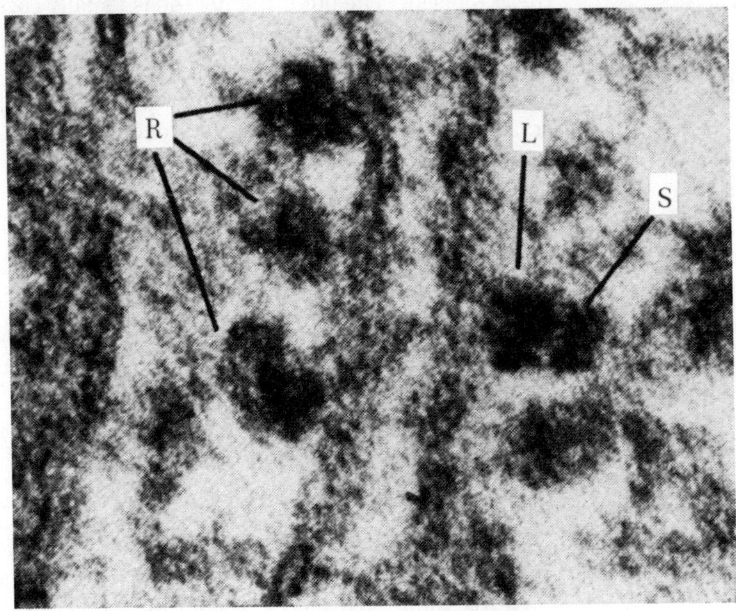

Figure 2-7
Ribosomes (*R*) of guinea pig hepatocyte showing the large (*L*) and small (*S*) subunits. The large ones apparently attach the ribosomes to the membrane.

Courtesy D.D. Sabatini, Y. Tahiro, and G.E. Palade.

These are responsible for the rough appearance of this type of endoplasmic reticulum in electron micrographs. The rough endoplasmic reticulum (RER) is also distinguished by its seeming continuity with the outer perinuclear membrane (Figs. 2-5, *B*, and 2-6).

Ribosomes

The RER and its ribosomal particles are the site of protein synthesis (Chapter 10). A careful scrutiny of electron micrographs of ribosomes shows that they consist of two subunits, a larger and a smaller unit (Fig. 2-7). The function of these subunits is discussed in detail in Chapter 10. As indicated, ribosomes are usually connected by a single strand of messenger RNA into clusters called polysomes.

It appears, however, that not all polysomes are membrane bound (RER). For example, immature red blood cells (reticulocytes) that synthesize hemoglobin, a protein that remains within the confines of the cell, do not possess an endoplasmic reticulum. Rather, all the polysomes appear to be free in the cytoplasm. On the other hand, cells that produce proteins for secretion or export, such as the pancreatic acinar cells and the intestinal goblet cells, are particularly rich in membrane-bound ribosomes. The cytomembrane system, that is, the endoplasmic reticulum, is apparently essential for the storage and secretion of the protein products outside the cell.

Smooth endoplasmic reticulum

Continuous with the rough endoplasmic reticulum is the *smooth (agranular)* membrane system (SER). It is distinguishable from the rough by the absence of ribosomes. However, the smooth membranes also differ in the enzymes present and the function of the system.

In the absorptive cells of the intestinal mucosa the SER functions in the synthesis of triglycerides (Chapter 12). In the steroid-secreting cells of the adrenal cortex important enzymes involved in the biosynthesis of steroid hormones are found. In the liver these membranes are the locale of cytochrome P_{450} involved in the metabolism of fatty acids, steroids, and various drugs. In many cells the SER is also the

Figure 2-8
Electron micrograph showing formation of lysosomal vesicles *(V)* from Golgi apparatus *(G)* of an immature white cell of a rabbit.

From Nichols, B., Branton, D., and Farquhar, J.: J. Cell Biol. **50:**498, 1971.

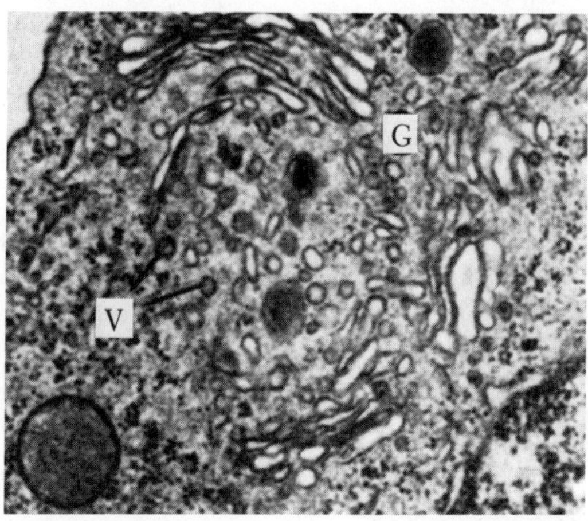

location of glycoside transferases, enzymes required for the addition of carbohydrates to the polypeptide chains in the biosynthesis of glycoproteins.

Golgi apparatus

Another kind of smooth membrane system is the Golgi apparatus (GA), which appears in electron micrographs as flattened sacs or vesicles, often continuous with the endoplasmic reticulum (Fig. 2-8). They are organelles to which the newly synthesized proteins are transported and temporarily stored (e.g., zymogen granules found in pancreatic acinar cells prior to the release of the protein from the cells). They represent the "packaging stations" and are particularly abundant in cells that produce proteins to be secreted.

Therefore it appears that secreted proteins are synthesized on membrane-bound polyribosomes. They must pass successively through each of the cytomembrane systems.

RER → SER → GA → Vacuoles → Environment

Secretion may involve fusion of the vacuoles with the plasma membrane followed by a discharge of the contents into the extracellular space. This process is called *exocytosis.*

Mitochondria

Mitochondria are the second largest organelles in the cell, measuring from 0.5 to 1 μm in diameter and from 1 to 10 μm in length. There are usually 1 to 1000 mitochondria per cell, depending on the cell's nature and function. Although usually ellipsoidal, they are highly plastic and may assume a variety of shapes; some are filamentous, others are spherical. In some cells the mitochondria remain fixed in place, whereas in others they move or stream throughout the cytosol. These organelles possess an outer and an inner membrane. The outer is typical of most and is composed of a phospholipid and cholesterol-rich bilayer and an equal amount of protein and functions as the limiting mitochondrial membrane. The inner mem-

Figure 2-9

Electron micrographs of mitochondrial structure.
A Cardiac muscle (24,000 ×).
B Skeletal muscle (24,500 ×).
C Substructure of heart muscle mitochondria (360,000 ×).
Middle circle is an artist's concept of a mitochondrion, whereas at bottom is a diagram of the elementary functional particles such as seen in **C** above. Each elementary particle consists of a spherical headpiece, a connecting stalk, and a basepiece.

A and **B** courtesy E.B. Scott, formerly of Department of Anatomy, University of South Dakota, School of Medicine, Vermillion, S. Dak. **C** courtesy H. Fernandez-Moran, J. Cell. Biol. **22**:63, 1964.

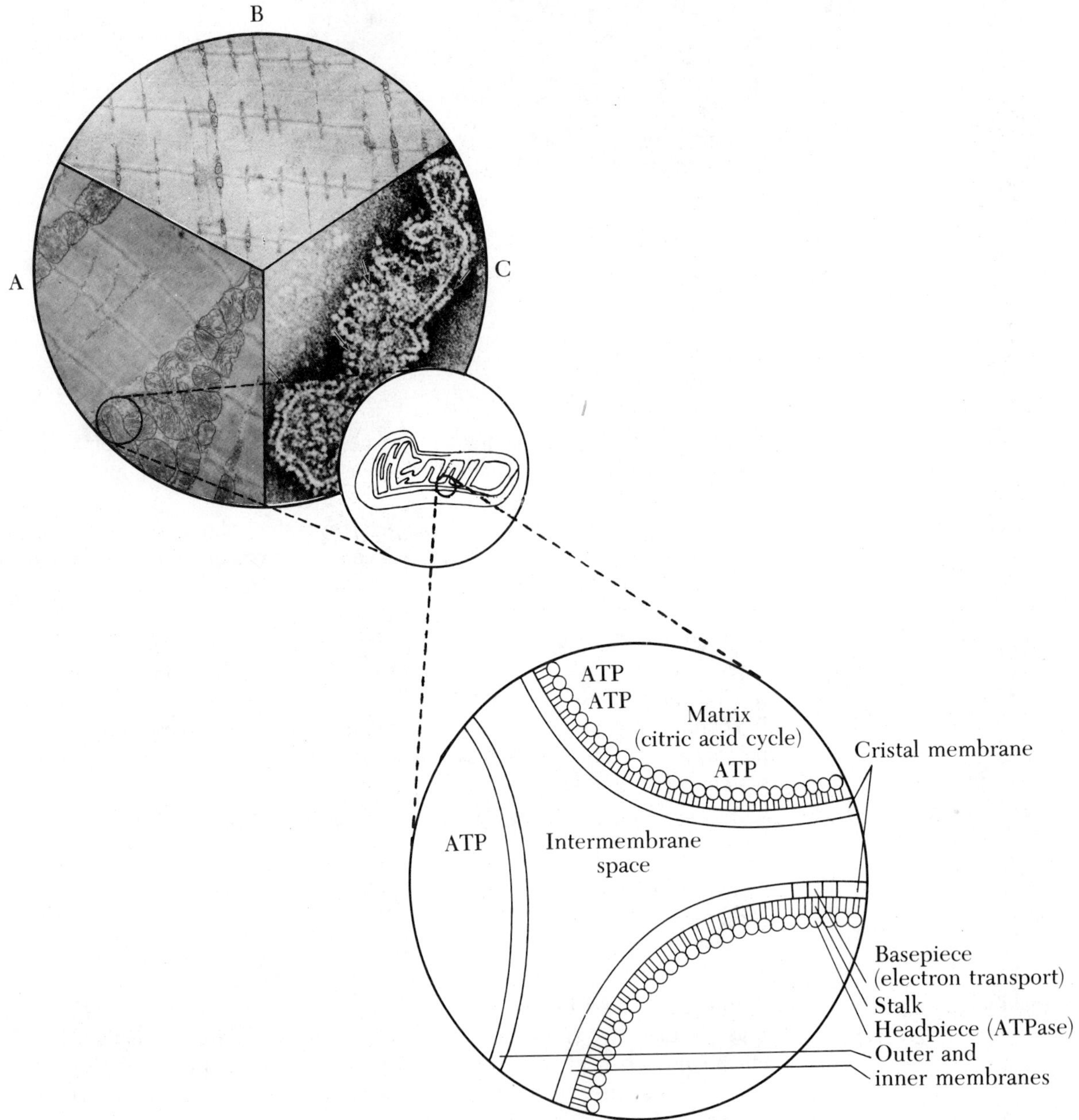

brane is of somewhat different composition and is convoluted to form numerous invaginations, shelves or *cristae* extending into the matrix of the organelle (Fig. 2-9).

Mitochondria are often termed the *powerhouse* of the cell because they are the site of respiration and the concomitant production of adenosine triphosphate (ATP). As discussed in Chapter 7, ATP supplies the energy for vital cellular functions such as motility, active transport of nutrients, and certain energy-requiring metabolic reactions. The number and size of mitochondria may reflect the need for energy and the particular nature of metabolic activity occurring in the tissue. A good example is Fig. 2-9 which shows the mitochondria of cardiac versus skeletal muscle cells. In cardiac muscle, where metabolism is essentially aerobic or oxygen dependent, the mitochondria are large and numerous, whereas in skeletal muscle cells, which can function anaerobically or in the absence of oxygen, the mitochondria are smaller and fewer.

Enzymes involved in electron transport and oxidative phosphorylation (Chapter 7) are part of the inner membrane and cristae. These cristae vary considerably in number, shape, and continuity from one tissue to another. In cardiac muscle the number of cristae is large, whereas in the liver there are fewer.

Close inspection of the inner membrane shows the existence of even more complexities in structure. On electron micrographs one can see that the cristae are covered with knoblike structures. These are thought to be composed of a spherical headpiece or knob (8 or 9 nm in diameter), a stalk, and a basepiece (Fig. 2-9). There may be as many as 50,000 such subunits, which are involved in performing the functions of electron transport and oxidative phosphorylation. These functions are the result of the appropriate arrangement of enzyme molecules within these structures, an excellent example of biochemical morphology (Chapter 7).

The inner membrane delineates the lumen, or inner chamber, of the mitochondrion, which is filled with a viscous or semisolid medium called the matrix. This medium also contains fine granules 4 to 5 nm in diameter. The enzymes involved in the citric acid cycle are found in this matrix.

Mitochondria contain a certain amount of DNA as well as ribosomes and are capable of synthesizing some of the protein for this organelle.

Lysosomes

The cytoplasm also contains a variety of vesicles surrounded by membranes that are 0.4 to 0.9 μm in diameter. These are called *lysosomes* and contain a variety of hydrolytic and degradative enzymes. Their function is directly or indirectly associated with intracellular digestion. This may affect substances either produced intracellularly or of extracellular origin. Lysosomes appear to be derived from the Golgi apparatus (Fig. 2-8). These bodies, termed *primary lysosomes,* are membrane-covered packages of at least six different classes of enzymes (i.e., proteases, nucleases, glycosidases, aryl sulfatase, lipases, and phospholipases and phosphatases). The primary lysosomes fuse with membrane-bound vesicles containing substances such as proteins, destined to be partially or wholly digested (Fig. 2-10). The fused, or *digestion, lysosome* is referred to as the *secondary lysosome*. Digestion products such as amino acids leave the secondary lysosome and enter the bloodstream. When undigested or indigestible substances accumulate, they are called *residual bodies*. The cells of the proximal tubules are a good example of this function. They absorb proteins from the urine by *pinocytosis*. The protein-containing pinocytotic vesicles fuse with the primary lysosomes, and the absorbed proteins are then totally digested to amino acids by the lysosomal proteases.

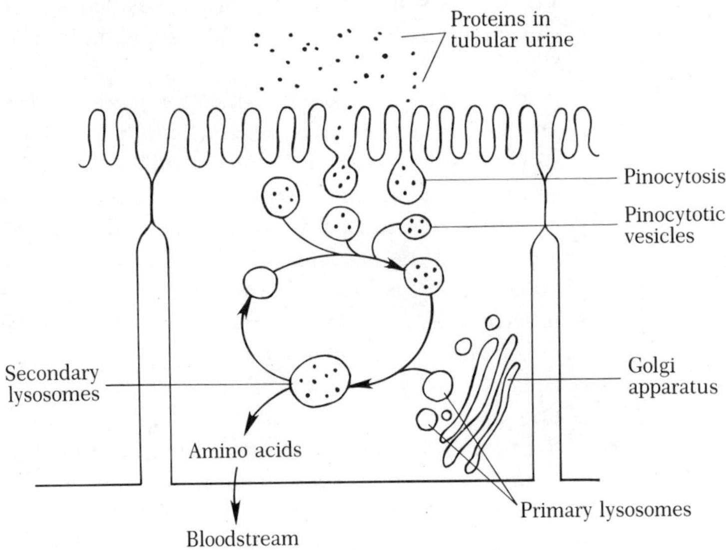

Figure 2-10
Kidney tubule cell showing lysosomal action in metabolizing proteins.

Peroxisomes

In addition to the lysosomes, almost all cells possess membrane-bound vesicles that have oxidative functions and are called *peroxisomes*. They are spherical, 0.3 to 1.5 μm in diameter, and intimately associated with the endoplasmic reticulum. Although these vesicles have no clearly defined morphological characteristics, they all are distinguishable by their content of catalase, an enzyme that degrades hydro-

$$2\ H_2O_2 \rightarrow 2\ H_2O + O_2$$

gen peroxide. In addition, peroxisomes possess certain other oxidases, such as urate oxidase, D- and L-amino acid oxidases, and α-hydroxy acid oxidase. These oxidative reactions involve the conversion of molecular oxygen to hydrogen peroxide, a product that is very toxic to the cell. As shown above, hydrogen peroxide is degraded by the high concentration of catalase (about 40% of the total peroxisomal protein). Uric acid metabolism (urate oxidase) of these organelles is important in most animals except humans and the higher apes.

Cytosol

The cytoplasm has been described as all of the extranuclear cell content that possesses both soluble and insoluble constituents. Once the insoluble components are removed, only an aqueous matrix remains. This continuum in which the organelles are bathed is called the cell sap, or *cytosol*. The amount of cytosol is suggested by the abundance of water in the cell, making up 70% to 75% of its weight (Table 2-3). Many proteins, multitudes of enzymes, and numerous other solutes are found in this medium. Some important metabolic processes that occur in the cytosol are glycolysis, much of gluconeogenesis, the hexose monophosphate pathway, the activation of amino acids, and fatty acid synthesis. The latter process occurs via a multienzyme system (fatty acyl synthetase) in which molecules of each constituent enzyme are associated as a tightly bound aggregate, a unique example of biochemical morphology. In such a complex the product of each enzymic reaction is passed on progressively to the next enzyme without actually diffusing into the cytosol. However, in some of the other systems mentioned the enzymes are in solution and

Table 2-3
Composition of rat liver and skeletal muscle

	Liver (%)	Skeletal muscle (%)
Water	69	75
Protein	16	7
Glycogen	3	4
Phospholipids	3	2
Neutral lipids	2	9

function independently; their substrates must diffuse to the enzyme molecules before a reaction can occur.

Properties of water

Obviously a most important factor in the cytosol is the water itself. It is well known that the heat of vaporization of water is considerably higher than that of various other common liquids. This is because the water molecules form hydrogen bonds with neighboring water molecules. Oxygen attracts the single electrons of the two hydrogen atoms, leaving each of these atoms with a partial positive charge (designated δ^+); the oxygen atom in turn becomes negatively charged ($\delta^=$). Therefore neutral water molecules actually are electric dipoles. As seen at left, a total of five water molecules can interact to form a tetrahedral structure. In the cytosol, however, the water molecules are also attracted to the charged protein molecules (Fig. 2-11). Water molecules associate with or hydrate the ionic side chains of protein molecules; they also hydrate other polar side chains. They do not associate with the *nonpolar* side chains. This hydration process slows the motion of water molecules largely because of the very slow motion of the proteins.

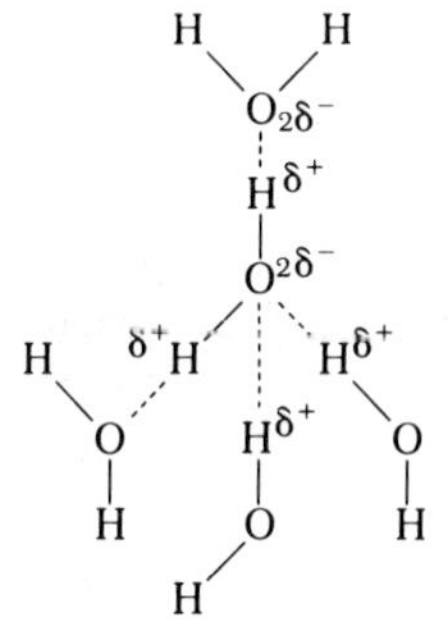

Actually, the true state of water in the cell is poorly understood. It is apparent, however, that the cytosol is not free flowing. Its fluidity is restricted not only by the process of hydration of protein molecules but also by the presence of protein aggregates or structures called *microtubules* and *microfilaments*. The microfilaments, 4 to 6 nm in diameter, are often found next to the plasma membrane, where they probably participate in the movement of membrane proteins. Microtubules are composed of a glycoprotein called *tubulin*, which aggregates in a helical fashion to form hollow cylinders approximately 25 nm in diameter. They function in the movement of cytosol from one part of the cell to another (intracellular transport) and in the maintenance and control of cell shape.

Fractionation of organelles

Much of our understanding of cellular function and metabolism has been obtained by studying the various subcellular organelles. To obtain purified preparations of organelles, the tissue is first carefully broken up in a homogenizing apparatus in a suitable medium, usually 0.25M sucrose. In this way the cell membranes are ruptured and the contents are uniformly distributed in the medium without undue mechanical damage to their organelles. The endoplasmic reticulum, however, is broken into pieces that spontaneously form vesicles called *microsomes*. The homogenate is subjected to increasing centrifugal forces in a preparative ultracentrifuge (Fig. 2-12). The dense nuclei are sedimented first, followed by the mitochondria, and finally the microsomal fraction at the highest forces. After all the particulate matter has been removed, the soluble remnant is the cytosol. These fractions have been extensively studied for their content of various constituents such as DNA, RNA, protein, or various specific enzymes, as well as for their biochemical function in metabolism.

Figure 2-11
Hydration of polar amino acid side chains on the surface of a protein molecule.

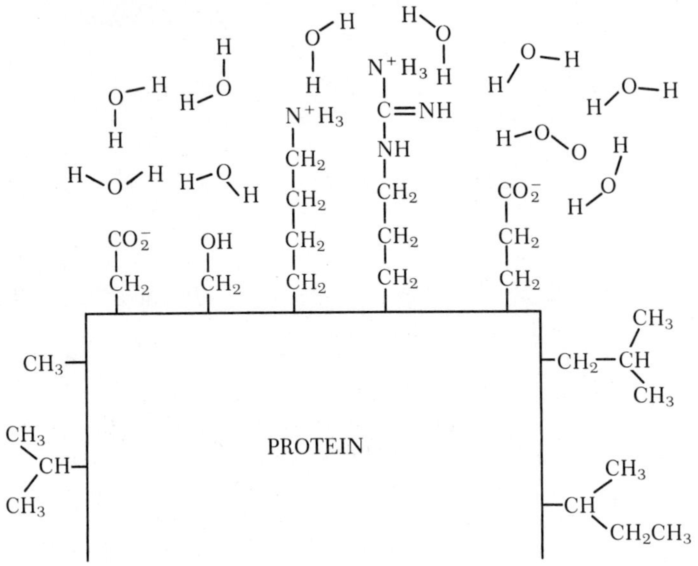

PROTEIN

Recapitulation of biochemical morphology

As the individual biochemical function of the various cellular organelles became known, their special roles and interrelationships could be delineated. For example, the human organism requires the production of a special chemical regulator—insulin. This protein is uniquely produced by cells located in the pancreas. The genetic information needed to produce insulin is retained in the DNA-containing chromosomes of the cell's nucleus. Since the DNA cannot leave this organelle, the information must first be transcribed or copied as another molecule known as messenger RNA. It represents a blueprint of the final insulin molecule. The reading of the blueprint requires the activities of the ribosomes of the rough endoplasmic reticulum. The product, however, is not yet ready to be exported, it must first be

Figure 2-12

Separation of principal organelles of animal tissue cells by gradient ultracentrifugation in sucrose solution. Microsomal fraction contains ribosomes, Golgi body membranes, and fragmented endoplasmic reticulum. Cytosol contains numerous cellular enzymes, e.g., those for glycogenesis and glycogenolysis and the glycolytic system. Lysosomes are sedimented at about 20,000 g.

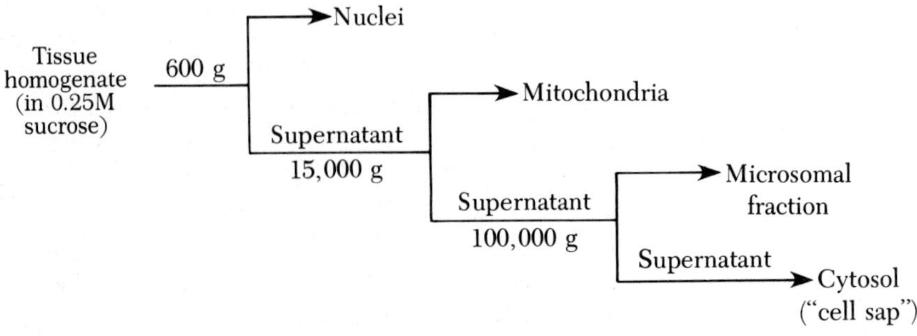

chemically modified by certain enzymes retained in the Golgi apparatus and the storage vacuoles. The active insulin is then stored in these vacuoles until needed. Protein synthesis is an energy-requiring process. ATP must be produced by the breakdown of a nutrient such as glucose. The first stage of glucose metabolism utilizes enzymes found in the cytosol, whereas the second phase needs enzymes found in the mitochondria. The glucose is ultimately converted into CO_2 and H_2O; the energy released is harvested as ATP. In other words, the specialized functions of the mitochondria and the cytosol contribute to the synthesis of the insulin in the endoplasmic reticulum by providing the necessary ATP. This kind of functional interrelationship is true for all cells.

Suggested readings

Bangham, A.D.: Lipid bilayers and biomembranes, Annu. Rev. Biochem. **41:**753, 1972.

Beck, F., and Lloyd, J.B.: The cell in medical science, vol. 1, The cell and its organelles, New York, 1974, Academic Press, Inc.

Bolis, L., Keynes, R.D., and Willbrandt, W.: Role of membranes in secretory processes, New York, 1973, American Elsevier Publishing Co.

Brachet, J., and Mirsky, A.E., editors: The cell: biochemistry, physiology, and morphology, New York, 1959-1961, Academic Press, Inc.

Bretscher, M.S.: Membrane structure: some general principles, Science **181:**622, 1973.

Capaldi, R.A.: A dynamic model of cell membranes, Sci. Am. **230:**27, 1974.

Chapnick, P.: The skin of our cells, The Sciences (N.Y. Acad. Sci.) **13:**20, 1973.

Coleman, R.: Membrane-bound enzymes and membrane ultrastructure, Biochim. Biophys. Acta **300:**1, 1973.

Dyson, R.D.: Cell biology, a molecular approach, Boston, 1978, Allyn & Bacon, Inc.

Fawcett, D.W.: The cell, its organelles and inclusions, Philadelphia, 1966, W.B. Saunders Co.

Gran, F.C., editor: Structure and function of the endoplasmic reticulum in animal cells, New York, 1968, Academic Press, Inc.

Green, D.E., editor: Membrane structure and its biological applications, Ann. N.Y. Acad. Sci. **195:**5, 1972.

Guidotti, G.: Membrane proteins, Annu. Rev. Biochem. **41:**731, 1972.

Lehninger, A.L.: The mitochondrion, New York, 1965, W.A. Benjamin, Inc.

Lodish, H.F., and Rothman, J.E.: The assembly of cell membranes, Sci. Am. **240:**48, 1979.

Loewy, A.G., and Siekevitz, P.: Cell structure and function, ed. 2, New York, 1969, Holt, Rinehart & Winston, Inc.

Manson, L.A., editor: Biomembranes, New York, 1973, Plenum Publishing Corp.

Marchesi, V.T., et al.: Molecular features of the major glycoprotein of the human erythrocyte membrane, Fed. Proc. **32:**1833, 1973.

Nichols, B.A., Bainton, D.F., and Farquhar, M.G.: Differentiation of monocytes: origin, nature, and fate of their azurophil granules, J. Cell Biol. **50:**498, 1971.

Nicholson, G.L.: Transmembrane control of the receptors on normal and tumor cells. I. Cytoplasmic influence over cell surface components, Biochim. Biophys. Acta **457:**57, 1976.

Northcote, D.H.: The Golgi apparatus, Endeavour **30:**26, 1971.

Novikoff, A.B., and Holtzman, E.: Cells and organelles, New York, 1970, Holt, Rhinehart & Winston, Inc.

Oseroff, A.R., Robbins, P.W., and Burger, M.M.: The cell surface membrane: biochemical aspects and biophysical probes, Annu. Rev. Biochem. **42:**647, 1973.

Palade, G.E.: The organization of living matter, Proc. Nat. Acad. Sci. **52:**613, 1964.

Porter, K.R., and Bonneville, M.A.: The fine structure of cells and tissues, Philadelphia, 1972, Lea & Febiger.

Schönheimer, R.: The dynamic state of body constituents, Cambridge, 1946, Harvard University Press.

Singer, S.J., and Nicolson, G.L.: The fluid mosaic model of the structure of cell membranes, Science **175:**720, 1972.

Steck, T.L.: The organization of proteins in the human red blood cell membrane, J. Cell. Biol. **62:**1, 1974.

Threadgold, L.T.: The ultrastructure of the animal cell, New York, 1968, Pergamon Press, Inc.

Wallach, D.F.H.: The plasma membrane: dynamic perspectives, genetics and pathology, vol. 18, London, 1972, Heidelberg Science Library, The English Universities Press Ltd.

Chapter 3

Dynamic state of metabolism

Two of the most striking characteristics of all living matter are the dynamic state of its biochemical processes, its *metabolism,* both at subcellular and molecular levels (Chapter 1), and the high degree of its structural organization (Chapter 2). Evolutionary processes have selected and preserved those reactions and structural features which are most useful and efficient in enabling the organism to cope with and adapt to its environment. Indeed, the sequence of chemical reactions used in converting raw materials, nutrients, into the organism's vitally important constituents invariably is the best and most energy-efficient possible mechanism for their biosynthesis.

With the explosive growth of knowledge in the fields of enzymology (Chapter 5) and metabolism (Chapters 11-14), the latter eventually was perceived as a series of linked reactions, termed *metabolic pathways.* Each step of the pathways is catalyzed by a specific enzyme, frequently aided by ancillary substances called coenzymes and cofactors (Chapter 6). These reactions may take place either in the cytosol, the subcellular organelles, or both (Chapter 2). Furthermore, the coupling of reactions through group transfer (Chapter 7) forms the basis for much of our present understanding of cellular metabolism, its integration and control.

The dynamic state of metabolism as used in this text (Chapter 1) describes the regulation and interrelationships of the metabolic processes in the human body. It is apparent that in a healthy adult the body's overall composition remains relatively unchanging, yet there is a constant input of fuel (food) to provide the energy needed for both synthetic and mechanical activities. Furthermore, food intake is required to replace those constituents broken down in the course of time. Physiological demand and stresses as well as disease may alter these requirements.

Dynamic equilibrium between anabolism-catabolism

Assimilation and breakdown of nutrients are in a constant state of dynamic flux so that in the adult there is a delicately regulated balance between anabolic (biosynthetic) and catabolic (degradative) processes. In the growing child, however, input of nutrients and anabolism must exceed catabolism so that growth of tissues may occur. In the aging process or in the case of wasting diseases the reverse is true; catabolic processes exceed anabolic ones.

Food is not provided in a continuous stream but in a fashion dependent on the dietary pattern or habits of the person consuming it; in America this usually represents two or three meals per day. An intermittent supply means that fuel elements must be stored when provided and remobilized when needed. Storage and mobilization are processes that involve the interrelationship of tissues having those functions. The balance of input and output processes is here referred to as the *dynamic state of metabolism.*

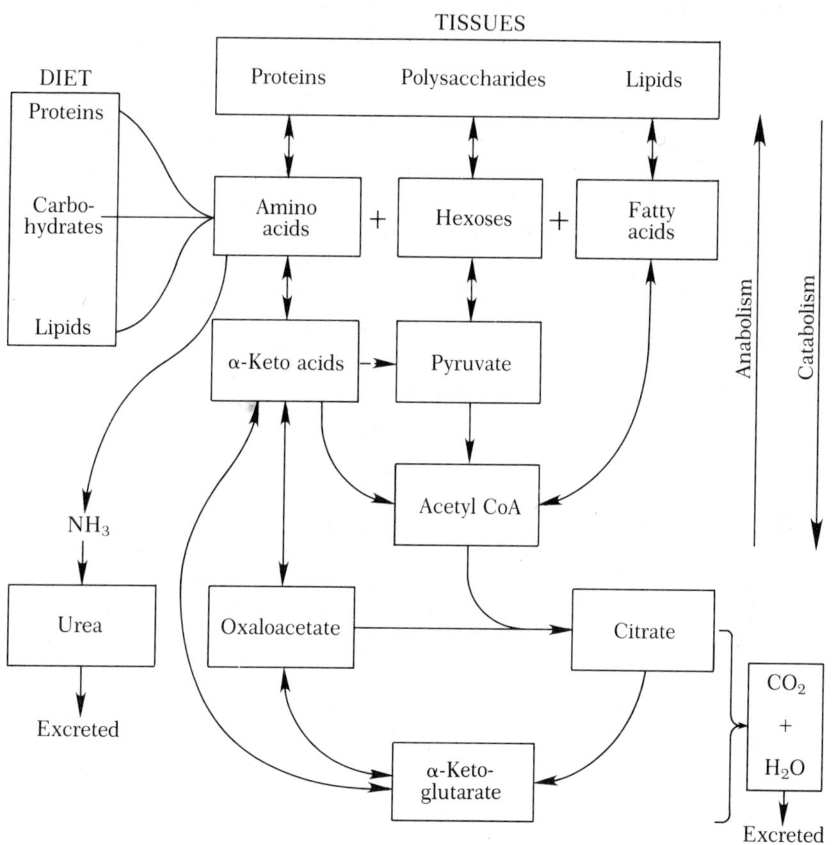

Figure 3-1
Dynamic interrelationships between various anabolic and catabolic processes.

Metabolism must be considered at two levels. First, we need to know about those processes occurring in specific tissue cells. Second, we need to understand how the actions of specialized tissues are integrated into a unified human system. Fig. 3-1 provides a perspective of the human organism as a metabolic machine. It illustrates the interrelationships of metabolism of three groups of nutrients: proteins, carbohydrates, and fats (lipids). These are first digested to their subunits, the amino acids, hexoses, and fatty acids (Fig. 3-1). They in turn may be converted by anabolism to protein, polysaccharides, and lipids for the use of the various tissue cells during growth and replacement. Or they may be degraded progressively to CO_2, H_2O and NH_3 by catabolic reactions for the purpose of supplying energy to the system. Glucose and fatty acids are continually used to meet the energy requirements of the body and therefore must be replaced by the diet. They are also derived by the catabolism of tissue components and from storage depots (e.g., adipose tissue).

Fig. 3-1 places emphasis on another important feature regarding metabolism: the interrelationships between these processes. The amino acids, hexoses, and fatty acids can be catabolized to a common chemical substance called acetyl coenzyme A (acetyl CoA), which in turn is degraded to CO_2 and H_2O. On the other hand, the carbon atoms of acetyl CoA, originally derived from hexoses, can be converted into fatty acids. This means that dietary carbohydrates provided in excess of need can be used to synthesize fatty acids and can be stored to meet future requirements for fuel.

Concept of metabolic pathways

Metabolism generally occurs by orderly, stepwise sequences that are dependent on the basic laws of thermodynamics. These metabolic pathways consist of a series of progressive, individual chemical reactions. Such pathways assume a variety of complexities, the simplest obviously being a single step, as in the conversion of a precursor (A) to a given product (B).

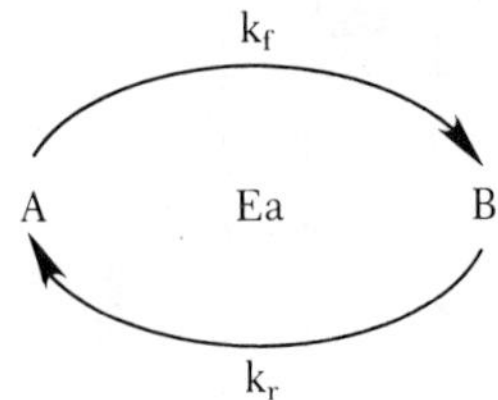

This reaction is catalyzed by a protein molecule, E_a, called an enzyme (Chapter 5). As seen at left, an enzymatic reaction is reversible. The forward reaction is described by a rate constant, k_f, whereas the reverse reaction is described by k_r. Reversible reactions may achieve an equilibrium state in which $k_f = k_r$. In most metabolic pathways, however, the product B of this reaction is in turn the substrate for a second reaction yielding product C and catalyzed by a second enzyme, E_b. This second reaction may deplete the supply of B, thereby promoting a continued formation of more B from A.

$$B \xleftrightarrow{E_b} C$$

Linear sequences

Such a sequence may continue for numerous individual steps so that the original substrate, A, is converted to the ultimate product, P. This is illustrated by Fig. 3-2 as a typical linear metabolic pathway. Whereas A is the original precursor or nutrient, it forms P only by passing sequentially through B, C, D, and so forth. These intermediate substances are called *metabolites* and we speak of the overall process as *intermediary metabolism*.

Branched sequences

Metabolic pathways may begin as a linear sequence of steps and then branch into two or more separate pathways (Fig. 3-3). In this way the precursor or nutrient may

Figure 3-2 Typical metabolic pathway, showing initial starting substance or precursor (A), intermediate compounds formed $(B$ to $D)$, and the final product (P). The specific enzymes required for each step are E_a, E_b, and E_c. Cofactors, if required, are x_a, x_b, and x_c.

Precursor **Intermediates** **Product**

$$A \xrightarrow[x_a]{E_a} B \xrightarrow[x_b]{E_b} C \xrightarrow[x_c]{E_c} D \longrightarrow \rightarrow \rightarrow P$$

Figure 3-3 Branched metabolic pathway, using B as a common starting point. Deficiency of an enzyme $(E_b$, E_c, etc.) or a cofactor $(x_b$, x_c, etc.) required for any of the pathways shown may result in a diversion of B into one or both of the other pathways.

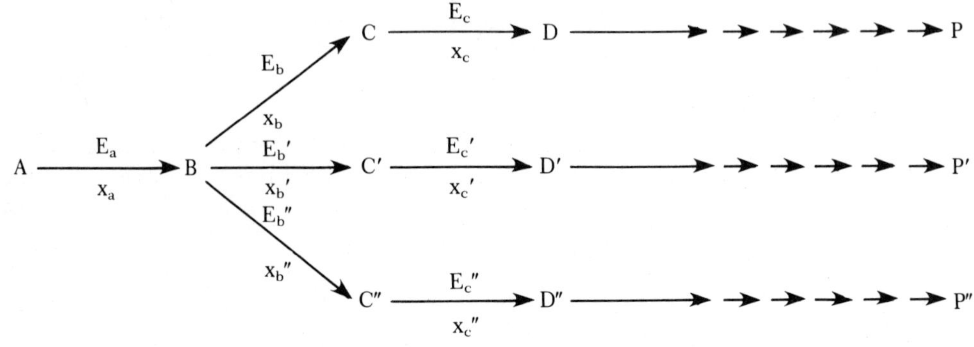

lead to the production of several products. Usually the first reaction beyond the common metabolite, B, is called the committed step. Thus $B \rightarrow C$ is the committed step in the production of P; $B \rightarrow C''$ is the committed step in the production of P''.

Reversibility of metabolic sequences

To achieve the necessary control of metabolism, it is frequently desirable for a given step in a pathway to be irreversible. The reaction $A \rightarrow B$ occurs by a given route catalyzed by the enzyme E_a (Fig. 3-4). Although all enzymatic reactions are in principle reversible, in some steps the change in free energy may be sufficiently large as to make the reverse reaction improbable. Furthermore, the concentration of product is too low for the reverse reaction to be appreciable. Under these circumstances the reaction is said to be physiologically irreversible. For the metabolic pathway to proceed from $B \rightarrow A$, a second process catalyzed by another enzyme, E_a', is required. The two enzyme systems, E_a and E_a', may be oppositely controlled, i.e., a given metabolite may activate E_a and at the same time inhibit E_a'. This prevents E_a from catalyzing the conversion of $A \rightarrow B$ only to have B restored to A by E_a'. Such a process is called a *futile cycle* and would involve a wasteful consumption of energy.

Some potential futile cycles may be viewed as proceeding from a precursor to a product by way of a number of intervening steps and then back to the precursor by an entirely different series. This is illustrated at left. The enzymes that catalyze precursor $\rightarrow$ product are controlled independently of those leading in the reverse direction. Furthermore, the two sequences may occur simultaneously, the net achievement in either direction being dependent on the metabolic needs. An example is the biosynthesis and degradation of glycogen. Glycogenesis (synthesis) and glycogenolysis (degradation) occur by independent pathways, each requiring its own special enzymes. Were this not the case, then the precursor glucose 1-phosphate would yield glycogen that would immediately return to glucose 1-phosphate, the only accomplishment being a wasteful expenditure of energy.

PRODUCT

Independent control

PRECURSOR

Role of cellular compartmentalization

Anabolic and catabolic pathways may occur simultaneously not only because they involve different enzyme sequences but also because they may be separated in their location within the cell. For example, the degradation of fatty acids to acetyl CoA (Fig. 3-1) occurs by way of enzymes located within the mitochondria, whereas the anabolic conversion of acetyl CoA to fatty acids occurs in the cytoplasm by way of a separate system. Therefore the anabolic biosynthesis of fatty acids and their catabolic degradation may occur simultaneously and in separate cellular compartments; the process predominating again depends on the metabolic needs.

Figure 3-4

Typical nonreversible metabolic pathway, showing point of irreversibility (A to B). For reversal to occur, different enzymes (E_a, and E_a') and usually different cofactors (x_a, and x_a') are required for the forward and backward reactions, respectively.

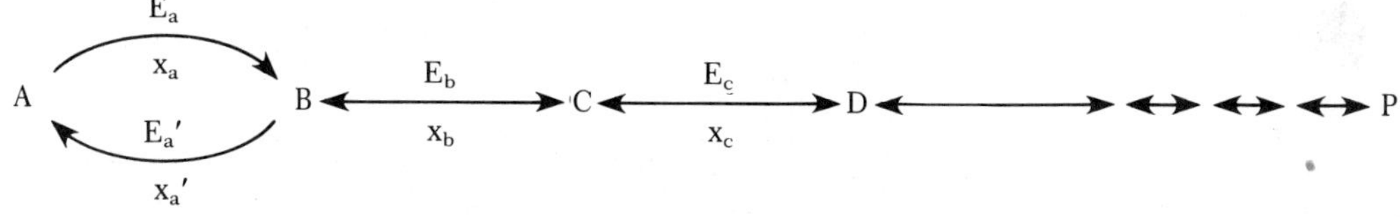

Nutrients and energy needs

Energy needed for anabolic processes as well as for mechanical work and other chemical processes is derived from nutrients such as glucose, fatty acids, and amino acids. Glucose is catabolized to CO_2 and H_2O, accompanied by a release of a large amount of chemical energy. The total energy in calories available to do work provided by the oxidation of 1 mole of glucose is 686,000 calories (686 kcal).

$$C_6H_{12}O_6 + 6\ O_2 \rightarrow 6\ CO_2 + 6\ H_2O + 686{,}000\ \text{calories/mole}$$

This energy would be totally lost to the cell or the human organism were it not for the fact that it is conserved in the simultaneous biosynthesis of ATP. In the cell, however, glucose is not oxidized to carbon dioxide and water in a single step; rather, a sequential metabolic pathway is involved, requiring many steps.

In a multistep pathway such as the metabolism of glucose, we might expect each step to proceed spontaneously and to achieve its lowest free-energy state. But this is not the case, as is evident by the first reaction in the metabolism of glucose. This step involves a phosphorylation to produce a molecule of glucose 6-phosphate. Thus:

$$\text{Glucose} + \text{Phosphate} \rightarrow \text{Glucose 6-phosphate} + H_2O$$

However, this reaction does not go spontaneously in the direction written; the formation of glucose 6-phosphate is thermodynamically an unfavorable reaction. The reverse reaction, or the hydrolysis of glucose 6-phosphate, is spontaneous and involves the release of a certain amount of free energy.

The energy released in a general reaction $A \rightarrow B$ depends on the nature of the system and the concentrations of reactants (A) and products (B). It is described as the free-energy change ΔG, or:

$$\Delta G = \Delta G^\circ + 2.303\ \text{RT}\ \log \frac{[B]}{[A]}$$

where R is the gas constant, T the absolute temperature, and $[A]$ and $[B]$ are concentrations of A and B in moles per liter. The constant ΔG°, the standard free-energy change, depends on the chemical makeup of A and B and may be calculated from the equilibrium constant for $A \leftrightarrow B$. At the equilibrium the free-energy change is zero ($\Delta G = 0$), and therefore: $\Delta G^\circ = -2.303\ \text{RT} \log [B]/[A]$. Since the ratio $[B]/[A]$ describes the equilibrium constant (K_{eq}), the above equation becomes:

$$\Delta G^\circ = -2.303\ \text{RT}\ \log K_{eq}$$

In biological systems the enzyme reactions occur at or near pH 7.0. Therefore the extra designation of $\Delta G^{\circ\prime}$ is made for the standard free-energy change at this pH. For the hydrolysis of glucose 6-phosphate, $\Delta G^{\circ\prime} = -3300$ calories/mole. Other examples of standard free energies of hydrolysis are illustrated in Table 7-2. When the reaction is spontaneous and yields energy, then $\Delta G^{\circ\prime}$ has a negative value and the reaction is said to be *exergonic*. However, we are concerned currently with the phosphorylation of glucose, that is:

$$\text{Glucose} + \text{Phosphate} \rightarrow \text{Glucose 6-phosphate} + H_2O$$

Since this is the reverse of hydrolysis, it is an *endergonic* reaction and requires an input of an amount of energy equal to the free energy of hydrolysis. In an endergonic reaction, $\Delta G^{\circ\prime}$ has a positive value and would be $+3300$ calories/mole for the above reaction.

In metabolic pathways reactions frequently proceed by way of thermodynamically unfavorable steps. The question arises, How are endergonic reactions driven in the desired direction?

Coupled reactions

To drive endergonic reactions in the cell, a second spontaneous reaction, usually the highly exergonic hydrolysis of ATP, is coupled with it. The standard free energy of hydrolysis of ATP is −7300 calories/mole (Table 3-1). For example, in the phosphorylation of glucose the coupled system may be written as the sum of the two component reactions. The final free-energy change would be the sum of the individual free-energy changes.

1.	Glucose + Phosphate	→ Glucose 6-phosphate + H_2O	$+3300\ \Delta G^{\circ\prime}$
2.	ATP + H_2O	→ ADP + Phosphate	$-7300\ \Delta G^{\circ\prime}$
SUM:	Glucose + ATP	→ Glucose 6-phosphate + ADP	$-4000\ \Delta G^{\circ\prime}$

Since the sum of the two reactions is negative (−4000 calories/mole), the overall coupled process is highly exergonic. The reaction as written, however, is only a formal one; it does not describe the actual mechanism as it occurs. A coupled reaction in a metabolic pathway may be written as follows:

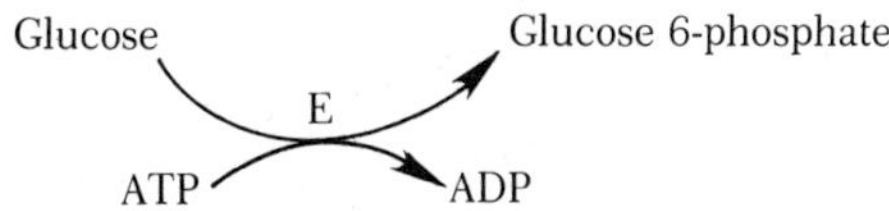

where *E* is the specific enzyme, e.g., hexokinase or glucokinase I (Chapter 11), that catalyzes this phosphorylation.

Regulation of metabolic pathways

The various metabolic pathways of living cells and in multicellular or multitissue systems must be closely regulated so that there is available, when needed, sufficient amounts of the essential products or intermediates of the pathway but not an oversupply. Excessive accumulation of metabolites would be not only uneconomical from an overall energetic point of view but even destructive, in that some intermediates may have toxic effects on the cell of origin or on other tissues.

The point or step of a metabolic pathway at which regulation usually occurs is the committed step. In the branched pathway suggested in Fig. 3-3, regulation might be expected to occur at the step catalyzed by E_b'. Thus the accumulation of product P' might inhibit this enzyme and not E_a, E_b, or E_b''. In other words the product P' may inhibit only that branch that produces it and not the others. This is known as *feedback inhibition*. Other modes of regulation also exist, e.g., where two products (P_1 and P_2) are derived from a common precursor, P_2 can accelerate the branch producing P_1 so as to maintain an equal balance in the amounts of the two products. This is seen in the effects of ATP and guanosine triphosphate (GTP) on each other's formation from inosine monophosphate (Chapter 14). Not all regulatory actions are inhibitory; some are *feed-forward* activations, which increase the rate of flow through a pathway thereby preventing the accumulation of intermediates. This is seen in the effect of fructose 1,6-bisphosphate on pyruvate kinase (Chapter 11).

Since feedback inhibition usually affects the committed step to that branch of the pathway, it prevents the formation of the remaining intermediates and the ultimate product of that branch. This being the case, it is now no longer necessary for the intervening enzymes to exist. Therefore, in the interest of economy, the cell decreases the further synthesis of the unnecessary enzymes, a control mechanism that occurs at the genetic or nuclear level. Once feedback inhibition is lifted and a need for more enzymes of the pathway exists, then their selective synthesis is restored. These basic principles of regulation are suggested by Fig. 3-5.

It is evident from the figure that there are two general levels of metabolic control. One level is exemplified by the feedback and feed-forward processes. These repre-

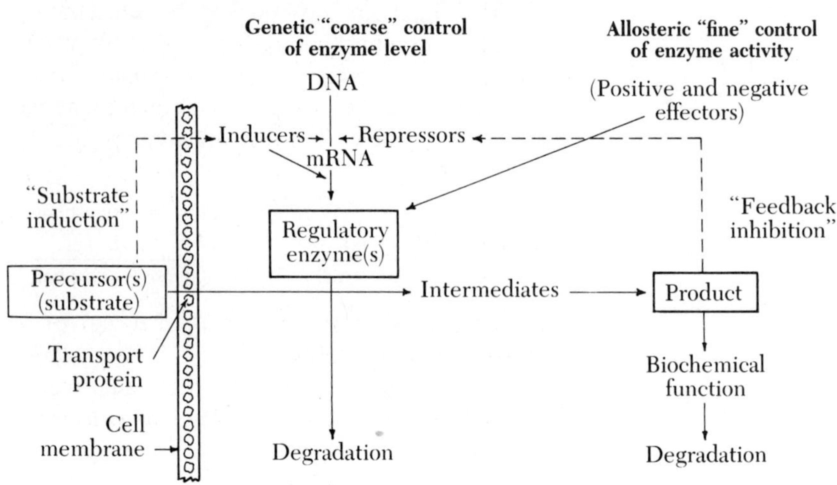

Figure 3-5
Principal factors concerned in regulation of metabolic pathways. Details are described in the text.

sent rapid or instant-to-instant regulations. Therefore in Fig. 3-5 we suggest calling this level "fine" control. The second level of metabolic control are the long-range, longer lasting effects. These are called "coarse" control (Fig. 3-5) and are represented by processes involving changes in the rates of enzyme biosynthesis.

There are a number of factors that regulate metabolic pathways:

1. Precursor (substrate) concentration
2. Enzyme level
3. Enzyme activity
4. Cofactor availability
5. End-product concentration

Precursor concentration

Before a nutrient can be assimilated into its metabolic pathways, it must first enter the cell. The passage of glucose and amino acids through the plasma membrane requires special transport proteins and often also an input of energy (Chapter 19). These processes may be regulated by hormones as is, for example, the transport of glucose into muscle cells by insulin.

Cellular enzyme levels

The amounts and kinds of enzymes that are present within a cell at any given time are determined by a balance between the biosynthesis of the enzymes on the one hand and their removal from the cell or degradation on the other. This represents a long-range coarse control of metabolism. In other words, metabolism is controllable by the amounts and specific kinds of enzymes that are present within the cell at any given time. The production of enzymes through biosynthesis may reflect a stimulation of protein synthesis either by increasing concentrations of nutrients or by a regulatory factor such as a hormone. When the level of amino acids in the cytoplasm increases, certain essential enzymes will be synthesized in response. The enzyme tyrosine amino transferase (Chapter 20) is a specific example. A hormone such as insulin will stimulate the production of glucokinase in liver cells so as to promote the phosphorylation of glucose.

Genetic control of enzyme synthesis may also involve a reduction in the biosynthesis of a specific protein. As shown in Chapter 10, this kind of genetic control is usually called *repression* and is typically involved in part of the feedback-inhibition process (Fig. 3-5). Since it is a slow, protracted response, involving as it does nuclear and biosynthetic processes, it also is a coarse control mechanism.

Cellular enzyme activity

Modulation of the activities of enzymes already present within the cell is fundamental to the fine control of metabolic pathways. This involves activation or inhibition in enzyme activities in response to the accumulation of substrate, products, or other metabolites.

Regulatory enzymes often possess special sites for binding small molecules that may be a product of the pathway, precursors, or even of metabolites not directly related to the pathway in which the enzyme functions. Such molecules are called *allosteric effectors* and may be positive if the effect is to increase the enzymatic activity or negative if the effect is to decrease it. Binding to the special allosteric site results in a change in the shape or conformation of the protein so as to influence its activity. An example of allosteric modulation is the effect of ATP on the glycolytic enzyme phosphofructokinase (Chapter 11). ATP acts both as a substrate and an allosteric regulator. As ATP levels increase, they have an inhibitory effect on phosphofructokinase, thereby reducing glycolysis. Another allosteric inhibitor of this enzyme is citrate, which also serves to slow down the glycolytic pathway.

Fine control can involve the chemical modification of enzymes. Covalent modification may alter the catalytic efficiency of the enzyme (V_{max}) or the affinity for its substrate (K_m) (Chapter 5). A variety of different types of chemical modification occur, the most well known being phosphorylation. The classic example is the conversion of inactive phosphorylase b to active phosphorylase a by the enzyme phosphorylase kinase in the presence of ATP (Chapter 11). Phosphorylase a catalyzes a key reaction in the process of mobilizing glucose from its storage carbohydrate, glycogen. The activation is triggered by the need for glucose and is stimulated by the hormone glucagon. The activation by phosphorylation is reversed in turn by a dephosphorylation catalyzed by a corresponding phosphatase. The control of glycogen phosphorylase by phosphorylation/dephosphorylation is illustrated in the following diagram.

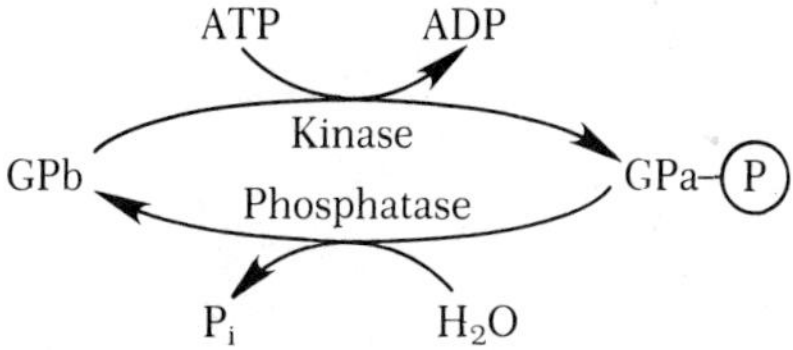

Cofactor availability

The availability and the particular state of certain cofactors may regulate some enzyme systems and thereby affect the metabolic pathways. An example is the competition between several pathways for the cofactor NAD^+. The metabolism of ethanol, for example, has a relatively high requirement for NAD^+, converting it to NADH. This fact may inhibit reactions requiring NAD^+. High ratios of NADH/NAD^+ will slow the metabolism of glucose, for example, by inhibiting pyruvate and isocitrate dehydrogenases, which are important metabolites in the complete oxidation to CO_2 and H_2O.

End-product level

The need for a particular product of a metabolic pathway may serve as a signal regulating that pathway. Thus the physiological requirement for blood glucose stimulates its production and release by increasing such hepatic metabolic processes as glycogenolysis and gluconeogenesis (anabolic production of glucose). Excessive levels of glucose after a meal will in turn stimulate glycogenesis and glycolysis while inhibiting glycogenolysis and gluconeogenesis.

Interrelation and integration of tissue metabolism

The foregoing discussion emphasizes metabolism within tissue cells, the dynamic nature of metabolic pathways, and the regulation of these pathways. Rapid shifts from one pathway to another are important, depending on cellular needs and supplies of nutrients. We need to appreciate how these metabolic processes, occurring as they do to varying degrees in specialized tissues and located remotely from each other, are integrated into a whole. An important illustration of metabolic integration is afforded by the dynamic regulation of blood glucose levels, called *homeostasis*. Immediately following a meal the systemic blood glucose level rises because of the increased input. The blood from the intestinal tract first passes directly through the liver by way of the portal system before it joins the systemic blood supply. During this passage almost two-thirds of the dietary glucose is removed and converted to hepatic glycogen (Chapter 11), which is the first-line storage facility for glucose. The increased systemic blood glucose is sensed by the pancreatic β-cells of the islets of Langerhans, which release the hormone insulin into the bloodstream. Insulin in turn increases the uptake (transport) of glucose by certain of the tissues, especially the skeletal muscle and the adipose tissue. It also acts to stimulate the glycogenic and glycolytic pathways in the liver. The increased uptake of glucose by the liver and adipose tissue results in its conversion into acetyl CoA and then into fatty acids and lipids. (Fig. 3-1). This means that excess glucose is converted to fats, which are not only a readily available source of fuel but are efficiently stored for future use. Once the blood glucose level declines below normal (2 to 3 hours after the meal), the pancreatic α-cells of the islets of Langerhans release the hormone glucagon and the secretion of insulin is suppressed. Glucagon stimulates the pathways of glucose *production* by the liver, using as the first source the polysaccharide glycogen. Later glucagon stimulates the lipolysis (lipid breakdown) of the fat stored in the adipose tissues. The fatty acids released are carried by the blood to various

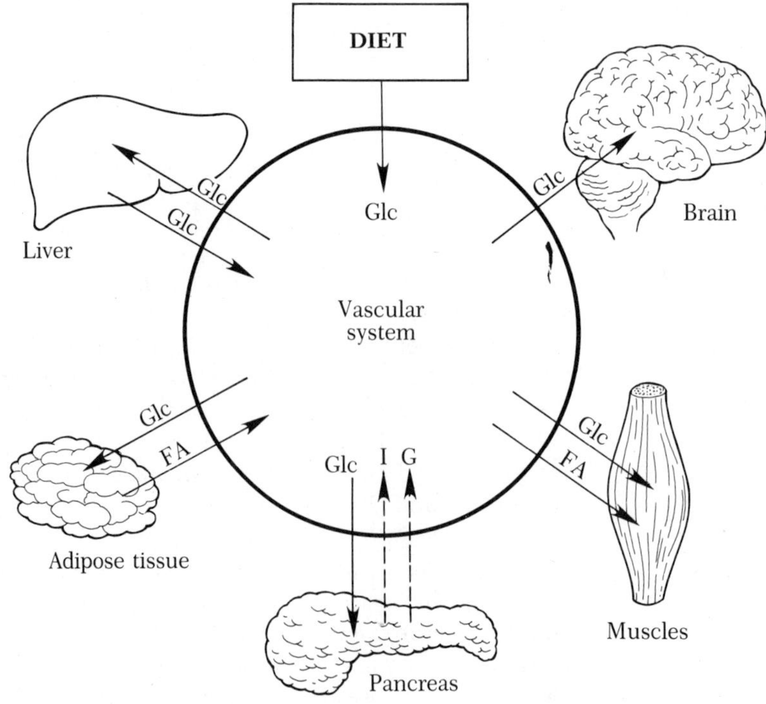

Figure 3-6
Metabolic interrelationship of certain tissues with regard to glucose metabolism. The hormone insulin favors storage of fuel (energy), whereas glucagon favors mobilization. (*Glc*, Glucose; *FA*, fatty acids; *I*, insulin; *G*, glucagon.)

tissues, e.g., skeletal and cardiac muscle, where they are used to provide the necessary energy for mechanical work. It is important to appreciate that the central nervous system (brain) has the highest priority for glucose use. This is so because normally it does not have the metabolic option of using other nutrients; it must have glucose to function.

Using blood glucose as an example, it is now apparent that metabolic pathways function to use a dietary nutrient not only for ultimate metabolism to provide energy but also for storage, both as a carbohydrate (glycogen) and as a lipid, for use as a fuel source. This overall process serves to illustrate the integration of metabolism in the liver, skeletal muscle, and adipose tissues. It also illustrates the interrelationship of metabolic pathways. Fuel is stored when present in excess following a meal and is again released by a process of mobilization when it is needed. Furthermore, each different tissue, e.g., the liver and adipose tissue, serves in its own way to supply the body as a whole with the necessary sources of fuel between meals (Fig. 3-6).

Suggested readings

Altman, P.L., and Dittmer, D.S., editors: Biological handbook: metabolism, Bethesda, Md., 1968, Federation of the American Societies for Experimental Biology.

Bondy, P.K., and Rosenberg, L.E., editors: Metabolic control and diseases, ed. 8, Philadelphia, 1980, W.B. Saunders Co.

Florkin, M., and Stotz, E., editors: Comprehensive biochemistry, section IV, Metabolism, vols. 17 to 21, New York, 1968-1971, American Elsevier Publishing Co., Inc.

Greenberg, D.M., editor: Metabolic pathways, ed. 3, New York, 1968, Academic Press, Inc.

Hammes, G.G., and Wu, C.W.: Regulation of enzyme activity, Science **172:**1205, 1971.

Koshland, D.E., Jr., and Neet, K.E.: The catalytic and regulatory properties of enzymes, Annu. Rev. Biochem. **37:**359, 1968.

Krebs, H.A., and Johnson, W.A.: The role of citric acid in intermediate metabolism in animal tissues, Enzymologia **4:**148, 1937.

Kun, E., and Grisolia, S., editors: Biochemical regulatory mechanisms in eukaryotic cells, New York, 1972, John Wiley & Sons, Inc.

Larner, J.: Intermediary metabolism and its regulation, Englewoods Cliffs, N.J., 1971, Prentice-Hall, Inc.

Rechcigl, M., Jr., editor: Enzyme synthesis and degradation in mammalian systems, Basel, 1971, S. Karger, AG.

Schimke, R.T.: Control of enzyme levels in mammalian tissues, Adv. Enzymol. **37:**135, 1973.

Schimke, R.T., and Doyle, D.: Control of enzyme levels in animal tissues, Annu. Rev. Biochem. **39:**929, 1970.

Segal, H.L.: Enzymatic interconversion of active and inactive forms of enzymes, Science **180:**25, 1973.

Vogel, H.J., editor: Metabolic pathways, vol 5, Metabolic regulation, New York, 1971, Academic Press, Inc.

Weber, G., and Weber, C.E.F., editors: Advances in enzyme regulation, vols. 1 to 17, New York, 1963-1979, Pergamon Press, Inc.

Section Two

Basic principles of metabolism and its regulation

Chapter 4

Properties and shape of proteins

Proteins are polymers in which the monomeric units are the amino acids (Chapter 23). These are held together by a specific kind of covalent linkage, namely the peptide bond, to form a long polypeptide chain. The exact sequence of amino acids in the chain is called the *primary structure*. The primary structure of a protein molecule determines how the chain is folded into higher levels of conformation or shape. For example, a polypeptide chain may be folded first into helical and pleated sheet regions, which are termed the *secondary structure*. This is folded further into the final three-dimensional conformation known as the *tertiary structure*. These three aspects of structure determine the nature and function of the protein molecules. In enzymes, for example, the three-dimensional shape establishes the exact spatial arrangement or proximity of specific amino acid residues essential to their biological activity. This precise positioning of the amino acids by the primary, secondary, and tertiary conformation creates the enzyme's catalytically active site (Chapter 5). Furthermore, it must be appreciated that the three-dimensional shape is flexible and subject to changes due to the binding of a variety of substances such as the allosteric factors mentioned in Chapter 3. Some proteins have yet a higher level of molecular architecture, called the *quaternary structure*. In this case several chains aggregate and function together as a unit. Aggregation is important in many enzymes that exhibit regulatory functions in metabolism.

There are, however, many biologically active proteins other than enzymes. The fundamental importance of proteins to the living cell and to the human organism has long been appreciated; in fact, the very word *protein* was derived from the Greek *proteos,* meaning "holding first place." Proteins are important (1) in the structure of membranes, musculature, or connective tissues; (2) for transport of oxygen by hemoglobin or electrons by the cytochromes; (3) for maintenance of fluid balance by serum albumin; (4) in genetics as nucleoproteins; (5) in defense mechanisms involving the γ-globulins of blood and in blood clotting by prothrombin and fibrinogen; and (6) as enzymes in the catalysis of thousands of reactions and in the regulation of metabolism by hormones such as glucagon and insulin.

Primary structure

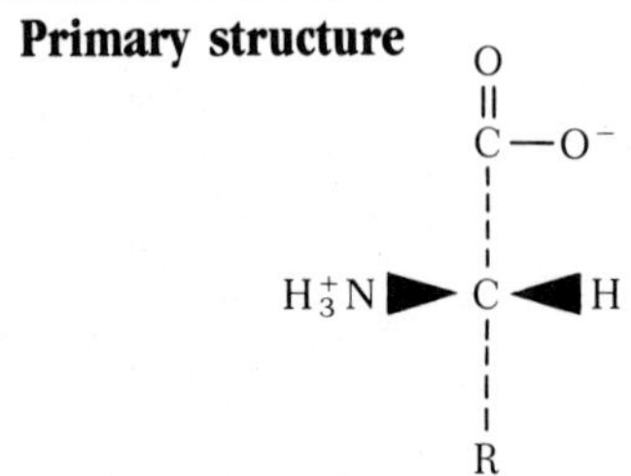

As just indicated, all proteins are linear polymers consisting of the L-α-*amino acids* (Chapter 23). The general structure of an L-α-amino acid is shown at left. (The *R* side chain is more fully elaborated on on p. 38. These are joined together in a specific order or sequence (primary structure) that is unique for every protein. The bond that links them together is actually a substituted amide called the *peptide bond*. It is formed when two amino acids are joined by the union of a carboxyl of one with the amino group of the other; a molecule of water is eliminated in the process. As an example, when alanine is joined to glycine, the product is the *dipeptide* alanylglycine.

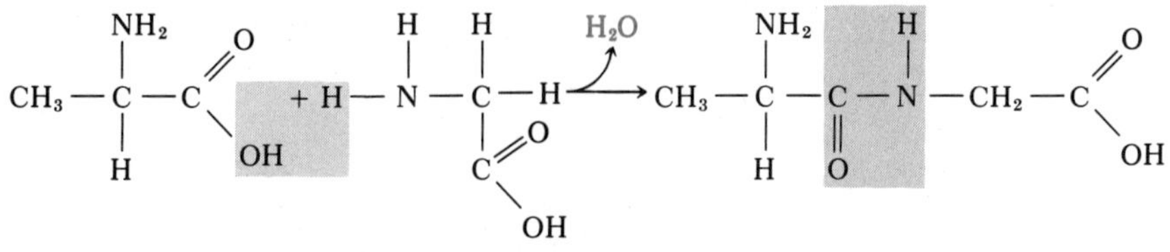

Alanine **Glycine** **Alanylglycine**

A dipeptide has two amino acids bound by a single peptide bond; a *tripeptide* is composed of three amino acids joined by two peptide bonds; a *polypeptide* is one in which any number (n) amino acids, or (amino acids)$_n$, are linked together by $n - 1$ peptide bonds.

The primary structure of a protein is determined genetically, that is, the necessary information is derived from the chromosomes in the cell nucleus. Specific genetic information exists for 20 different amino acids (see Codons, Chapter 10). Those amino acids used in the actual biosynthesis of proteins are presented below. They are arranged so that their side chains are grouped according to common physicochemical properties: nonpolar (hydrophobic), polar neutral (hydrophilic), and polar charged (hydrophilic). Those with nonpolar side chains (R groups) contribute hydrophobic properties to the protein; they tend to stabilize the tertiary structure. Amino acids contributing hydrophilic properties are divided into those with uncharged side chains and those with charged side chains. The neutral polar residues frequently participate in hydrogen bonds that also stabilize protein conformation. Charged polar residues are the dicarboxylic or acidic amino acids and those having cationic groups (ϵ-ammonium, guanidinium, and imidazolium) or basic amino acids. These two groups can exist in charged or uncharged states, depending on the pH of the surrounding medium. Thus at low pH the protein will have a positive charge because the aspartic acid and glutamic acids will be uncharged, whereas the basic amino acids will be protonated. In a medium of high pH the protein will have a negative charge because the basic amino acids will be uncharged and the acidic ones will have a negative charge. Obviously the net charge of a protein in a physiological medium will depend on the relative amounts of the acidic and basic amino acids.

When a protein is heated at 110° C for 18 hours in an acid (5.7 N HCl) or basic ($Ba(OH)_2$ or NaOH) medium, then the peptide bonds are broken and the constituent amino acids are released. Hydrolysis may also be accomplished with suitable digestive or proteolytic enzymes. When the amino acid content of such protein hydrolysates is determined, certain modified amino acids will be found. Three commonly found are cystine, hydroxyproline, and hydroxylysine. Their structures are shown below. It is evident from the structure of cystine that it was produced by the oxidative joining of two sulfhydryl groups from cysteine residues.

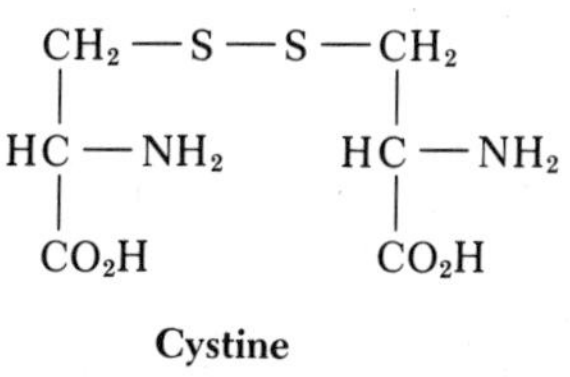

Cystine 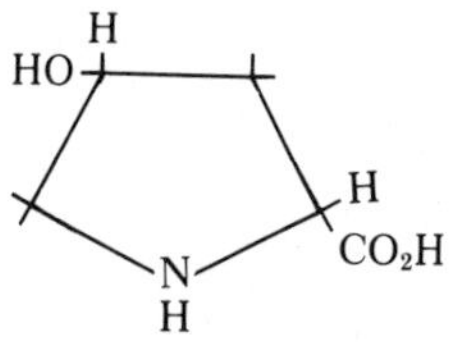 **Hydroxyproline** 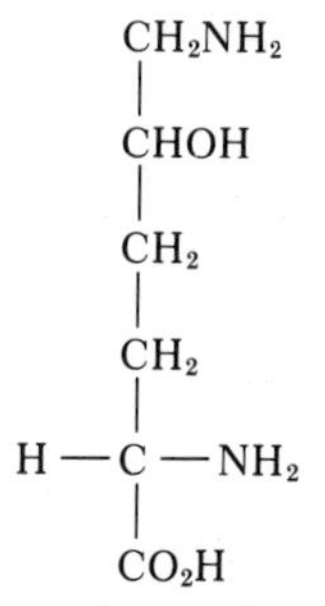 **Hydroxylysine**

A CLASSIFICATION OF SIDE CHAINS (R-GROUPS) FOR THE 20 NATURALLY OCCURRING L-α-AMINO ACIDS

	R =		
	$—H$ **Glycine (Gly)**	$—CH_3$ **Alanine (Ala)**	CH_3 / $—CH$ / CH_3 **Valine (Val)**
NON POLAR-HYDROPHOBIC	CH_3 / $—CH_2CH$ / CH_3 **Leucine (Leu)**	CH_2CH_3 / $—CH$ / CH_3 **Isoleucine (Ile)**	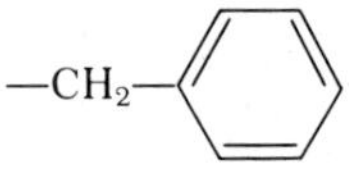 $—CH_2—$ **Phenylalanine (Phe)**
	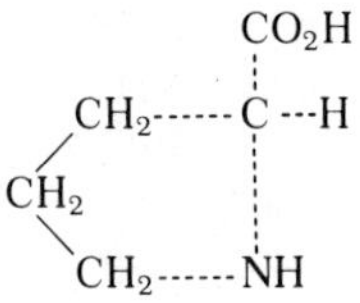 CO_2H / CH_2---C---H / CH_2 / CH_2---NH **Proline (Pro)**	$—CH_2CH_2—S—CH_3$ **Methionine (Met)**	
NEUTRAL HYDROPHILIC (POLAR)	$—CH_2OH$ **Serine (Ser)**	$—CH—OH$ / CH_3 **Threonine (Thr)**	$—CH_2C(=O)—NH_2$ **Asparagine (Asn)**
	$—CH_2CH_2C(=O)—NH_2$ **Glutamine (Gln)**	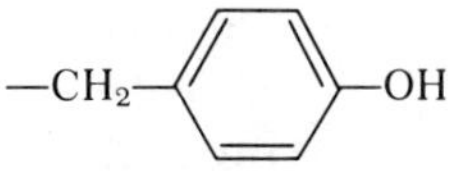 $—CH_2—$ … $—OH$ **Tyrosine (Tyr)**	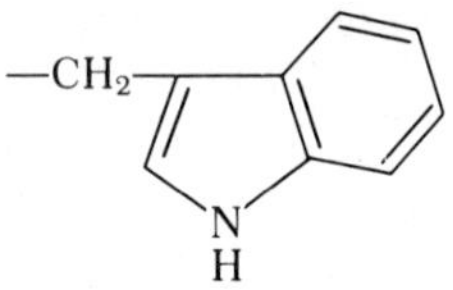 $—CH_2—$ … N / H **Tryptophan (Trp)**
	$—CH_2—SH$ **Cysteine (Cys)**		
ACIDIC HYDROPHILIC (POLAR)	$—CH_2—C(=O)—OH$ **Aspartic acid (Asp)**	$—CH_2CH_2—C(=O)—OH$ **Glutamic acid (Glu)**	
BASIC HYDROPHILIC (POLAR)	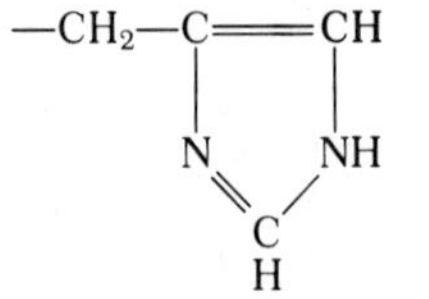 $—CH_2—C{=}CH$ / N / NH / C / H **Histidine (His)**	$—CH_2CH_2CH_2CH_2NH_2$ **Lysine (Lys)**	$—CH_2CH_2CH_2—NH—C(=NH)—NH_2$ **Arginine (Arg)**

This modification to form a disulfide linkage occurs after the polypeptide chain has been synthesized and cysteine has been inserted. Cystine itself is not an amino acid that is introduced into a polypeptide chain during synthesis; there is no codon for cystine. Likewise, hydroxyproline and hydroxylysine are the consequence of post-synthetic modification processes; the precursors are actually proline and lysine.

Convention for naming polypeptides

The convention used in naming *peptides* or *polypeptides* is shown in the following example. Note that the amino acid contributing the free amino group (NH_2 terminal or, for convenience, N-terminal) is named first and placed to the left in the sequence. The free carboxyl (COOH terminal, or C-terminal) is placed on the right side, and its amino acid is named last.

Alanyl-leucyl-cysteinyl-tyrosyl-glycine

Composition of proteins

The wide diversity of amino acid composition found in nature is suggested by the data for a few selected proteins in Table 4-1. For example, protamine, which is a part of the nucleoproteins of sperm, is 66% arginine. Collagen, which represents almost 30% of protein in the human body, contains 27% proline and hydroxyproline as well as 27% glycine. Other proteins of diverse properties with unusual amino acid distributions are gliadin, the protein of wheat flour, with almost 50% glutamic acid; insulin, a hormone, with 18% glutamic acid, 12% cystine, and 13% tyrosine; and pepsin, a digestive enzyme with 12% glutamic acid and 9% tyrosine.

To understand the various physical and chemical properties of proteins as well as their biological function, it is necessary to establish the primary structure. A first step in this direction is the determination of the overall amino acid composition. This is accomplished by hydrolyzing the protein with hydrochloric acid (5.7 N), alkali ($Ba(OH)_2$), or proteolytic enzymes (pepsin, trypsin, chymotrypsin).

The hydrolysates are then separated into the individual amino acids by ion-exchange chromatography (Chapter 25). A sulfonated polystyrene resin is used in a continuous and automated process whereby all amino acids can be analyzed quantitatively on a single sample. Tryptophan is destroyed by acid hydrolysis and must be determined on a sample following alkaline hydrolysis.

Sequence determination

Insulin is the first protein for which a complete amino acid sequence was established (by Sanger in 1956). This hormone consists of two polypeptide chains, an A-chain having 21 and a B-chain having 30 amino acid residues (Fig. 4-1). The first enzyme to be sequenced was ribonuclease (by Hirs, Moore, and Stein in 1960), having 124 amino acid residues in a single polypeptide chain (Fig. 4-2). The largest single chain to be sequenced to date is the enzyme β-galactosidase, with 1170 residues (*Escherichia coli*).

Table 4-1
Amino acid composition of some selected proteins

	Albumin (human)†	Casein‡	Myoglobin (bovine)*	Collagen (bovine)	Cytochrome c*	Insulin‡	Ribonuclease (bovine)*	Gliadin‡	Trypsin inhibitor (*Ascaris*)*	Protamine†	Histone IV*	Pepsin†
Ala	—	3.0	10.1	10.3	7.9	4.5	6.1	2.1	7.6	3.6	6.9	4.5
Arg	6.2	4.1	1.9	8.2	2.2	3.1	4.0	2.7	4.5	66.2	13.7	1.0
Asn	—	—	1.7	—	4.8	—	8.0	—	3.0	0.0	2.0	—
Asp	10.4	7.1	4.7	7.0	3.6	6.8	4.5	1.7	3.0	0.0	2.9	16.6
Cys	5.6	0.4	0.0	—	2.0	12.5	6.4	2.6	15.2	0.0	0.0	1.5
Gln	—	—	2.8	—	3.3	—	6.4	—	6.1	0.0	2.0	—
Glu	17.4	23.3	8.3	11.2	5.9	18.6	3.7	45.7	7.6	0.0	3.9	11.3
Gly	1.6	2.7	8.7	26.2	12.2	4.3	4.0	< 0.5	9.1	4.1	16.7	8.1
His	3.5	3.1	7.5	0.7	2.5	4.9	3.7	2.3	0.0	0.0	2.0	0.5
Hyp	—	—	—	12.8	—	—	—	—	—	—	—	—
Ile	1.7	6.1	5.2	1.9	4.8	2.8	3.2	5.4	4.5	0.9	5.9	10.0
Leu	11.9	9.2	11.4	3.7	6.2	13.2	1.9	6.5	0.0	0.0	7.8	10.4
Lys	12.3	8.2	12.8	4.0	15.2	2.5	7.7	1.1	10.6	0.0	10.8	0.4
Met	1.3	3.4	1.5	1.0	1.7	0.3	3.5	1.7	0.0	0.0	1.0	2.1
Phe	7.8	5.0	4.7	2.4	3.8	8.1	2.4	6.9	3.0	0.0	2.0	6.7
Pro	5.1	11.3	2.5	14.4	4.2	2.5	4.8	13.4	9.1	8.1	1.0	4.9
Ser	3.7	7.7	3.9	4.3	3.0	3.2	12.3	4.9	1.5	10.8	2.0	13.2
Thr	5.0	4.9	3.4	2.3	7.2	2.1	6.9	2.1	6.1	1.4	6.9	9.5
Trp	0.2	1.2	1.3	—	1.0	0.3	0.0	0.6	1.5	0.0	0.0	3.5
Tyr	4.7	6.3	1.3	1.0	4.4	13.0	3.7	3.2	0.0	0.0	3.9	9.4
Val	7.7	7.2	4.6	2.5	3.6	7.8	6.7	2.7	3.0	5.0	8.8	7.1

*From Dayhoff, M.O.: Atlas of protein sequence and structure, vol. 5, Silver Spring, Md., 1972, National Biomedical Research Foundation.
†From Tristram, G.R., and Smith, R.H. In Neurath, H., editor: The proteins, ed. 2, vol. 1, New York, 1963, Academic Press Inc.
‡From Block, R.J.: Personal communication.

Table 4-2
Amino acid content of cytochrome c (hippopotamus)

	Residues/molecule	
Residues	**From analysis**	**From amino acid sequence**
Aspartic acid	8.19	8
Threonine	7.98	8
Serine	2.43	2
Glutamic acid	12.59	12
Proline	N.D.	4
Glycine	14.10	14
Alanine	6.46	6
½-Cystine	N.D.	2
Valine	3.03	3
Methionine	1.67	2
Isoleucine	5.91	6
Leucine	6.22	6
Tyrosine	4.06	4
Phenylalanine	4.08	4
Histidine	2.89	3
Lysine	16.95	17
Arginine	2.12	2
Tryptophan	N.D.	1

From Thompson, R.B., et al.: J. Biol. Chem. **253**:8957, 1978.

Figure 4-1 Primary structure of beef insulin.

From Sanger, F: Science **129:**1340, 1959.

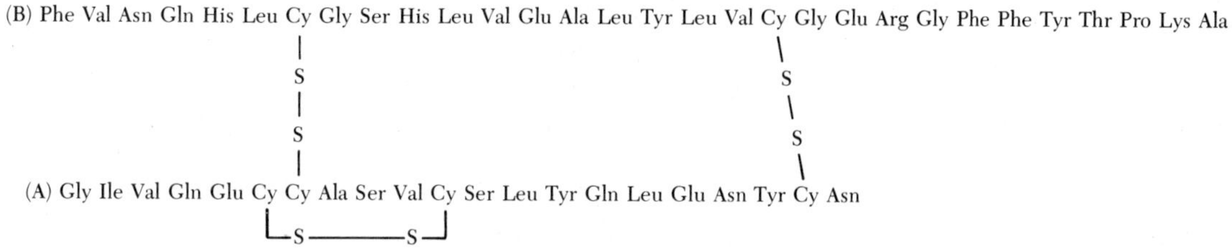

```
(B) Phe Val Asn Gln His Leu Cy Gly Ser His Leu Val Glu Ala Leu Tyr Leu Val Cy Gly Glu Arg Gly Phe Phe Tyr Thr Pro Lys Ala
                            |                                              \
                            S                                               S
                            |                                                \
                            S                                                 S
                            |                                                  \
(A) Gly Ile Val Gln Glu Cy Cy Ala Ser Val Cy Ser Leu Tyr Gln Leu Glu Asn Tyr Cy Asn
                        └─S───────────S─┘
```

Figure 4-2 Sequence of amino acid residues in bovine pancreatic ribonuclease A, based on experiments of Hirs and associates, Spackman and associates, and Smyth and associates.

Modified from Smyth, D.G., Stein, W.H., and Moore, S.: J. Biol. Chem **238:**227, 1963.

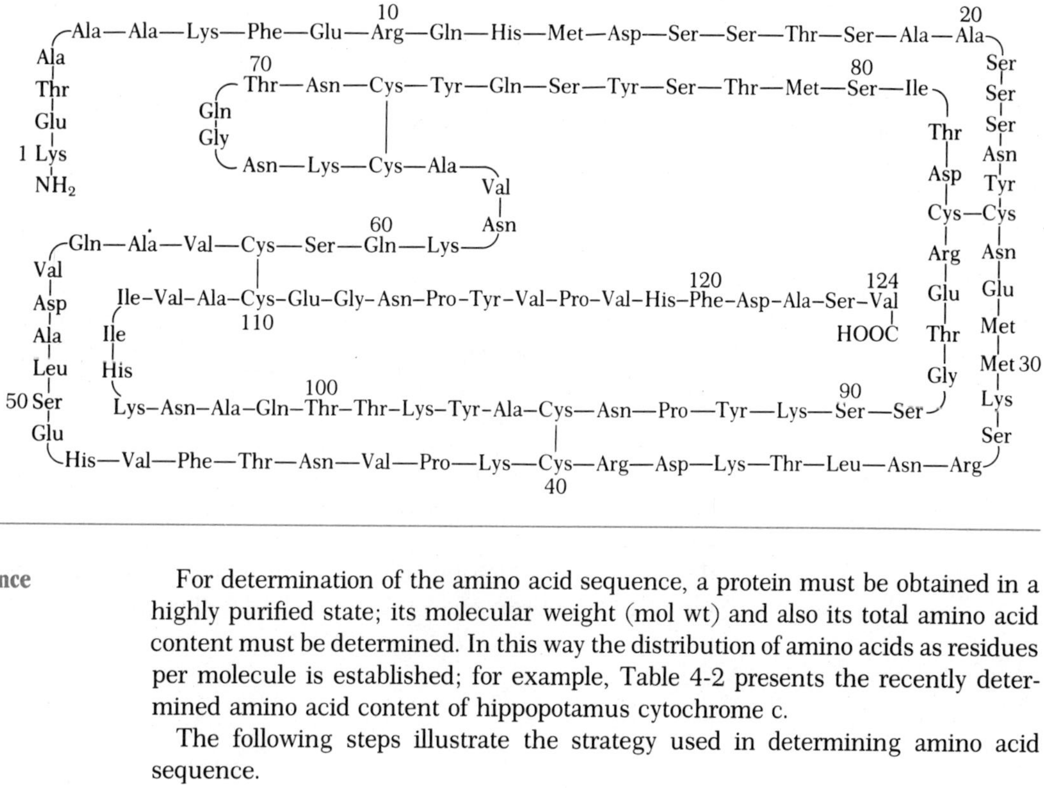

```
                                10                                              20
 ┌Ala—Ala—Lys—Phe—Glu—Arg—Gln—His—Met—Asp—Ser—Ser—Thr—Ser—Ala—Ala┐
 Ala              70                                        80                Ser
 Thr           ┌Thr—Asn—Cys—Tyr—Gln—Ser—Tyr—Ser—Thr—Met—Ser—Ile┐         Ser
 Glu          Gln          |                                          |         Ser
 1 Lys        Gly          |                                          Thr      Asn
 NH2           └Asn—Lys—Cys—Ala┐                                     Asp      Tyr
                                  Val                                 Cys—Cys
                        60        Asn                                 Arg      Asn
 ┌Gln—Ala—Val—Cys—Ser—Gln—Lys┘                                                 
 Val               |                       120                   124  Glu      Glu
 Asp     Ile-Val-Ala-Cys-Glu-Gly-Asn-Pro-Tyr-Val-Pro-Val-His-Phe-Asp-Ala-Ser-Val   Met
 Ala     Ile          110                                         HOOC  Thr      Met 30
 Leu     His                                                           Gly      Lys
 50 Ser  Lys-Asn-Ala-Gln-Thr-Thr-Lys-Tyr-Ala-Cys-Asn—Pro—Tyr—Lys—Ser—Ser┘         Ser
 Glu          100                          |                     90
 └His—Val—Phe—Thr—Asn—Val—Pro—Lys—Cys—Arg—Asp—Lys—Thr—Leu—Asn—Arg┘
                                      40
```

Strategy for sequence determination

For determination of the amino acid sequence, a protein must be obtained in a highly purified state; its molecular weight (mol wt) and also its total amino acid content must be determined. In this way the distribution of amino acids as residues per molecule is established; for example, Table 4-2 presents the recently determined amino acid content of hippopotamus cytochrome c.

The following steps illustrate the strategy used in determining amino acid sequence.

1. The N- and C-terminal residues are determined.
2. If the sample protein is composed of more than one chain held together by interchain covalent bonds, e.g., disulfide bonds in insulin, it is necessary to break these bonds and isolate the constituent chains by chromatography.
3. Each polypeptide chain is hydrolyzed into small peptides using proteolytic enzymes or chemical means. The smaller peptides are isolated chromatographically.
4. The amino acid sequences of the small peptides are determined using the Edman degradation.
5. Amino acid sequence of the parent polypeptide is deduced by proper arrangement of the small peptides.
6. The location of interchain disulfide bonds is established.

Step 1: N- and C-terminal amino acids The N-terminal amino acid of a polypeptide chain is identified by treating the sample with fluorodinitrobenzene (FDNB). This reagent reacts with free N-terminal amino groups to form a DNP-peptide derivative. Acid hydrolysis then yields a mixture of all the amino acids except for the N-terminal residue, which is stable to this treatment and may be isolated as the DNP derivative. The identity of the DNP amino acid is established by adsorption chromatography. Another reagent often used is dimethylaminonaphthalene-5-sulfonyl chloride (dansyl chloride); the procedure is much the same as that using FDNB. A covalent bond is formed between the free N-terminal amino group. The dansylated protein is hydrolyzed with acid, and the dansylated amino acid is separated and identified by chromatography. These fluorescent derivatives make this technique highly sensitive and economical of sample.

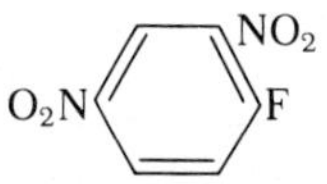

O_2N, NO_2, F

FDNB

CH_3 CH_3 N

$O \leftarrow S \rightarrow O$, Cl

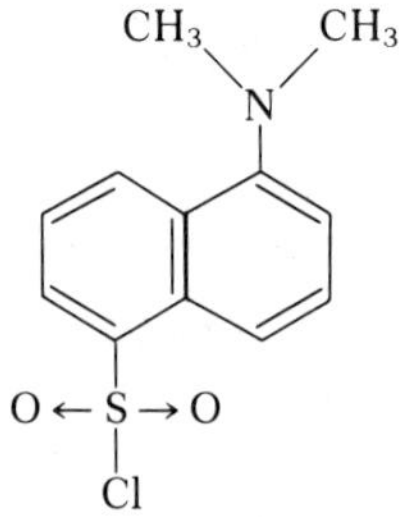

Dansyl chloride

C-terminal residues are usually identified with the enzyme carboxypeptidase, which attacks only the peptide bond joining the last residue with a free α-carboxyl group to the polypeptide chain. Amino acids released are identified by chromatography. Pancreatic carboxypeptidase consists of carboxypeptidase A, which functions until proline contributes the carboxyl group of the last peptide bond or lysine or arginine becomes the terminal residue. Also present is carboxypeptidase B, which releases the lysine and arginine from the C-terminal end of polypeptides.

Alternatively, the polypeptide may be treated with anhydrous hydrazine, which splits peptide bonds forming hydrazides with the carbonyl carbons. As seen below, the C-terminal residue does not form a hydrazide because its carboxyl is free. After removal of the hydrazides, this amino acid is then identified chromatographically.

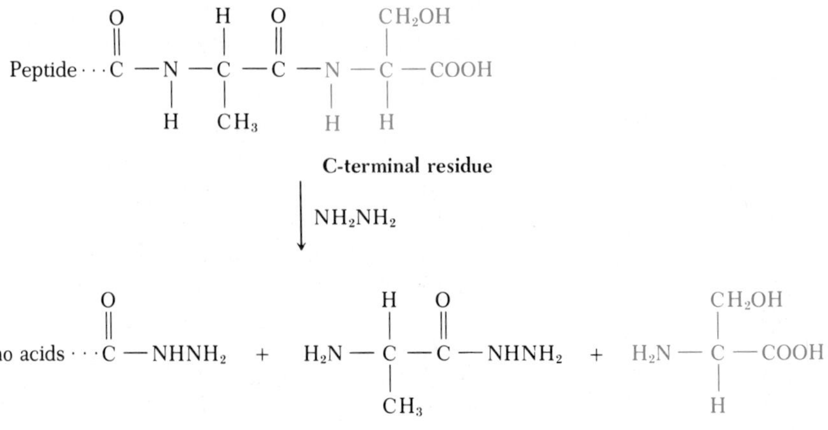

$$\text{Peptide}\cdots\overset{\displaystyle O}{\overset{\|}{C}}-\underset{\underset{\displaystyle H}{|}}{N}-\overset{\displaystyle H}{\overset{|}{\underset{\underset{\displaystyle CH_3}{|}}{C}}}-\overset{\displaystyle O}{\overset{\|}{C}}-\underset{\underset{\displaystyle H}{|}}{N}-\overset{\displaystyle CH_2OH}{\overset{|}{\underset{\underset{\displaystyle H}{|}}{C}}}-COOH$$

C-terminal residue

$$\downarrow NH_2NH_2$$

$$\text{Amino acids}\cdots\overset{\displaystyle O}{\overset{\|}{C}}-NHNH_2 + H_2N-\overset{\displaystyle H}{\overset{|}{\underset{\underset{\displaystyle CH_3}{|}}{C}}}-\overset{\displaystyle O}{\overset{\|}{C}}-NHNH_2 + H_2N-\overset{\displaystyle CH_2OH}{\overset{|}{\underset{\underset{\displaystyle H}{|}}{C}}}-COOH$$

C-terminal amino acid

Step 2: separating constituent polypeptides When a protein is composed of two polypeptide chains held together by interchain disulfide bonds, it is necessary to break these bonds and isolate the individual chains. This is accomplished by treating the protein with performic acid, which oxidizes the disulfide bonds into two sulfonic acid groups. The two chains are separated and isolated by chromatography.

```
—NHCHCO—                     —NHCHCO—
    |                            |
   CH2                          CH2
    |                            |
    S         H·COOOH           SO3H
    |       ---------->
    S                           SO3H
    |                            |
   CH2                          CH2
    |                            |
—NHCHCO—                     —NHCHCO—
```

Disulfide bonds can also be cleaved by reduction to the sulfhydryl groups using mercaptoethanol. To prevent reoxidation, the sulfhydryls are blocked by alkylation with iodoacetate.

```
—NHCHCO—                        —NHCHCO—                      —NHCHCO—
    |                               |                             |
   CH2                             CH2                           CH2
    |                               |                             |
    S        HSCH2CH2OH            SH       ICH2CO2H              S—CH2CO2H
    |       ------------>                  ---------->
    S                              SH                             S—CH2CO2H
    |                               |                             |
   CH2                             CH2                           CH2
    |                               |                             |
—NHCHCO—                        —NHCHCO—                      —NHCHCO—
```

Step 3: enzymatic and chemical cleavage to smaller peptides The polypeptide chain for which the C- and N-terminal residues have been established is now hydrolyzed into peptides by treatment with a proteolytic enzyme. Usually enzymes such as trypsin and chymotrypsin are employed; they have a unique specificity for the peptide bond split. Trypsin hydrolyzes peptide bonds in which the carbonyl groups are contributed by lysine or arginine. It is used to produce smaller peptides having C-terminal lysine or arginine residues.

Usually, it is necessary to hydrolyze the parent polypeptide into an entirely different set of smaller peptides using either another enzyme or a chemical method of differing specificity. The enzyme chymotrypsin is frequently used because it splits those peptide bonds in which the carbonyl groups are contributed by aromatic amino acids. The chemical method most frequently used employs cyanogen bromide (CNBr), a reagent that attacks methionyl residues, resulting in the cleavage at the carbonyl of the peptide bond contributed by this amino acid. Two peptides are formed; one is terminated with homoserine lactone and the other begins with a newly created N-terminal amino acid. Regardless of the processes used, all the smaller peptides produced are separated chromatographically and their amino acid sequences are determined.

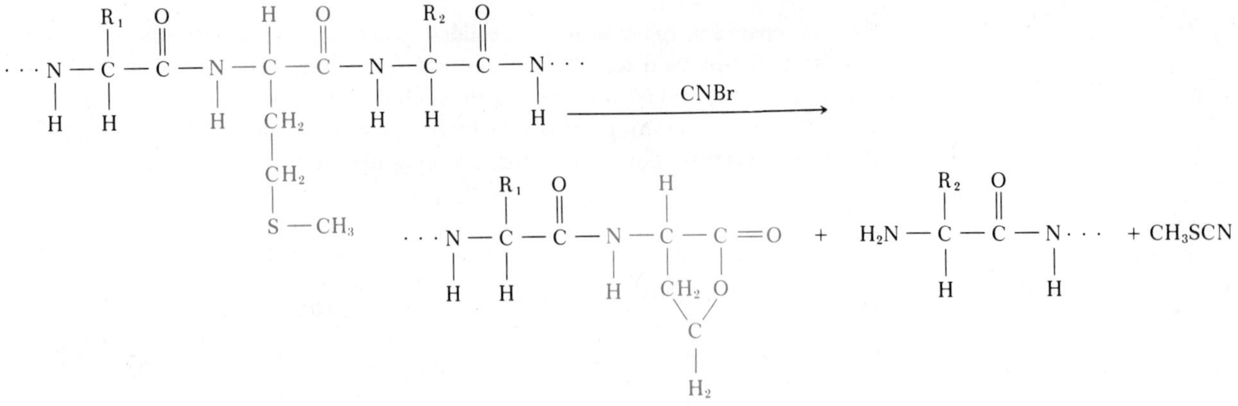

$$\cdots N(H)-C(H)(R_1)-C(=O)-N(H)-C(H)(CH_2CH_2SCH_3)-C(=O)-N(H)-C(H)(R_2)-C(=O)-N(H)\cdots \xrightarrow{CNBr}$$

$$\cdots N(H)-C(H)(R_1)-C(=O)-N(H)-C(H)-C{=}O \text{ (homoserine lactone ring: } CH_2-CH_2-O) + H_2N-C(H)(R_2)-C(=O)-N(H)\cdots + CH_3SCN$$

Homoserine lactone

Step 4: sequence by Edman degradation All the small peptides that are produced are sequenced by using the Edman degradation technique. The sample is treated with phenylisothiocyanate or Edman's reagent. The product is the phenylthiocarbamyl derivative of the peptide, which cyclizes in an acid medium, thereby breaking the first peptide bond and yielding the phenylthiohydantoin (PTH) of the N-terminal amino acid. Since this is a nonhydrolytic process, the remainder of the peptide is intact but shorter by one residue. The reaction with phenylisothiocyanate is then repeated on the shortened peptide. The PTH amino acids are extracted from the reaction mixtures and identified chromatographically. The amino acid sequences are thus determined from the N-terminal end of the peptide. The process has been automated in an apparatus called the Protein Sequenator.* With this apparatus it has been possible to establish the sequence of as many as 60 amino acids in a given polypeptide.

*Beckman Instrument Co., Fullerton, Calif.

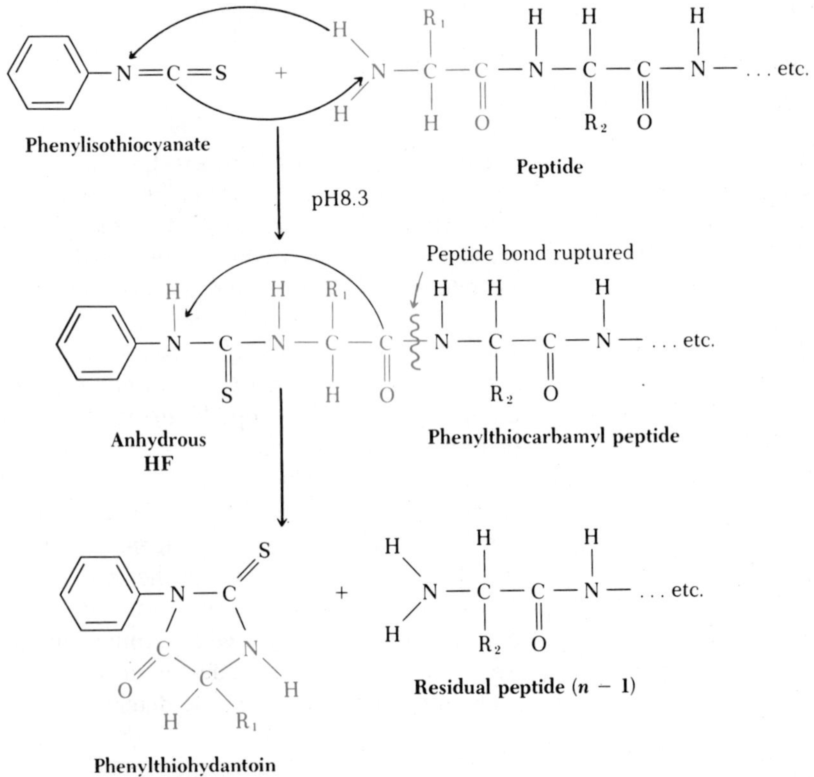

$$C_6H_5-N{=}C{=}S + H_2N-C(H)(R_1)-C(=O)-N(H)-C(H)(R_2)-C(=O)-N(H)-\ldots \text{etc.}$$

Phenylisothiocyanate — Peptide

pH8.3

$$C_6H_5-N(H)-C(=S)-N(H)-C(H)(R_1)-C(=O)\,\wr\,N(H)-C(H)(R_2)-C(=O)-N(H)-\ldots \text{etc.}$$

Peptide bond ruptured

Phenylthiocarbamyl peptide

Anhydrous HF

Phenylthiohydantoin (PTH) + $H_2N-C(H)(R_2)-C(=O)-N(H)-\ldots$ etc.

Phenylthiohydantoin (PTH)

Residual peptide (n − 1)

Figure 4-3 Principle of sequence determination. The positions of peptides *1* and *5* are established because the N- and C-terminal amino acids are the same as those of the parent polypeptide (likewise, peptides *A* and *D*). Note that peptide *A* overlaps peptide *1* and its neighbor. This establishes that 2 must follow *1*.

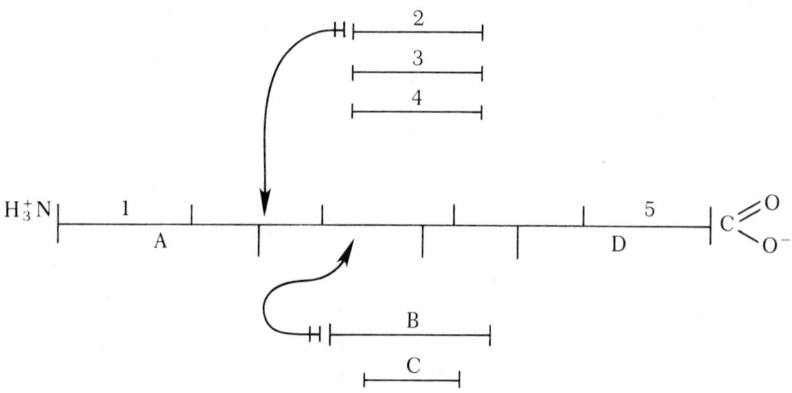

Step 5: deduction of the amino acid sequence If a hypothetical polypeptide were to yield a number of smaller peptides, e.g. five with trypsin, one could predict which of the five should be first because it would have the same N-terminal amino acid residue as the parent polypeptide (Fig. 4-3). Likewise, it should be possible to predict which would be last, since it would have to provide the C-terminal residue. However, the arrangement of *2*, *3*, and *4* would not be possible. To overcome this problem, a second and different series of peptides are produced by the action of chymotrypsin, e.g., *A*, *B*, *C*, and *D*. If peptide *A* is longer than peptide *1*, as suggested in Fig. 4-3, then it will overlap peptides *1* and 2. Peptide *A* will have at its C-terminal end the same sequence as peptide 2 has at its N-terminal. In this way it is established that *1* and 2 are neighbors and not *1* and *3* or *1* and *4*. As seen in Fig. 4-3, it is possible to arrange the two sets of peptides in their proper order. The combined sequence of peptides *1*-*2* will overlap and establish the proper sequence *A*-*B* as opposed to *A*-*C*. Thus the principle of deducing the actual amino acid sequences depends on the appropriate arrangement of overlapping sequences of two sets of small peptides produced by two methods of differing specificity.

An actual example in which this principle was employed is provided in the recent determination of the amino acid sequence of cytochrome c derived from hippopotamus heart (Fig. 4-4).

Step 6: disulfide bridges Once the amino acid sequences of constituent polypeptide chains have been established, the protein with disulfide bonds intact is studied to determine which of the cysteine residues participate in disulfide bridges. For this purpose the protein is often hydrolyzed with pepsin at pH 2 to prevent any disulfide bond exchanges from taking place. The cystine-containing peptides are identified and separated chromatographically from non–cystine-containing ones and then are treated with performic acid. This process produces two peptides, each containing half the cystine molecule oxidized to sulfonic acid. Once the two half–cystine-containing peptides have been separated, their terminal residues and amino acid composition are determined. Since the overall amino acid sequence of the parent protein is known, the position of each half-cystine peptide can be established by the amino acids around it and hence the location of the disulfide bridges determined.

Figure 4-4 Principle whereby the primary structure of an unknown polypetide chain (hippopotamus cytochrome c) is deduced from the overlapping peptide fragments produced by treating the original with two enzymes of differing specificity.

Data from Thompson, R.B., et al.: J. Biol. Chem **253**:8957, 1978.

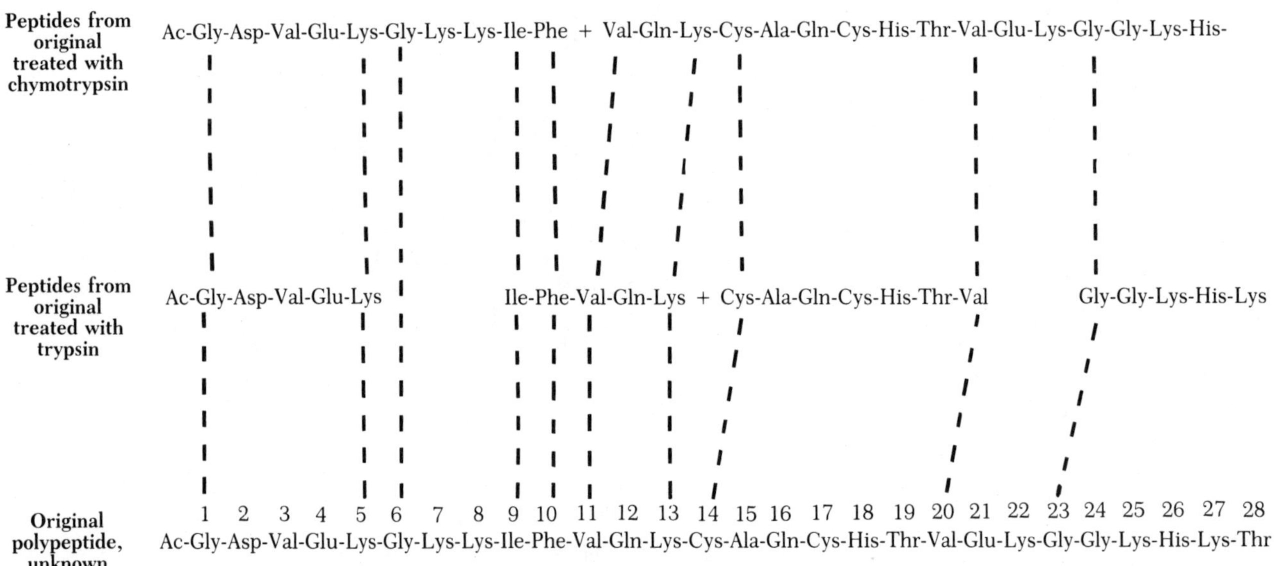

Synthesis of peptides

Proteins for which the primary structure has been established may be synthesized by purely chemical means. An automated procedure that has been used to synthesize peptide hormones such as bradykinin, angiotensin, and oxytocin as well as the antibiotics gramicidin S and tyrocidin is available. The protein insulin has also been synthesized. It is now possible to produce biological analogues that differ from the parent proteins in precisely known ways. Chemical synthesis can help produce hormones with greater or more prolonged activity than the natural ones have. Studies of other protein analogues will make it possible to learn how the sequence of amino acids determines overall protein structure and hence biological activity.

The synthesis of proteins is usually accomplished by a method introduced by Merrifield and his group in 1964. The enzyme ribonuclease was synthesized using an automatic *solid-phase* system. The peptide chain in this procedure is assembled one amino acid at a time beginning with the C-terminal residue. The C-terminal amino acid is first attached to small polystyrene beads by a means of benzyl ester bonds (Fig. 4-5). A second amino acid, with its amino group protected by a tertiary butyloxycarbonyl group, is attached to the amino group of the first amino acid residue. In Fig. 4-5 the reagent dicyclohexylcarbodiimide (diimide) is used as the coupling agent. At the end of the synthesis of the desired polypeptide chain the resin is suspended in anhydrous trifluoroacetic acid, and dry hydrogen bromide is used to split the peptide from the polystyrene beads. This also removes many of the protecting groups. The entire procedure may be accomplished in an automated apparatus. One cycle of the synthesis requires 4 hours.

Figure 4-5
General reaction scheme for solid-phase peptide synthesis.

From Merrifield, R.N.: Recent Progr. Horm. Res. **23**:451, 1967.

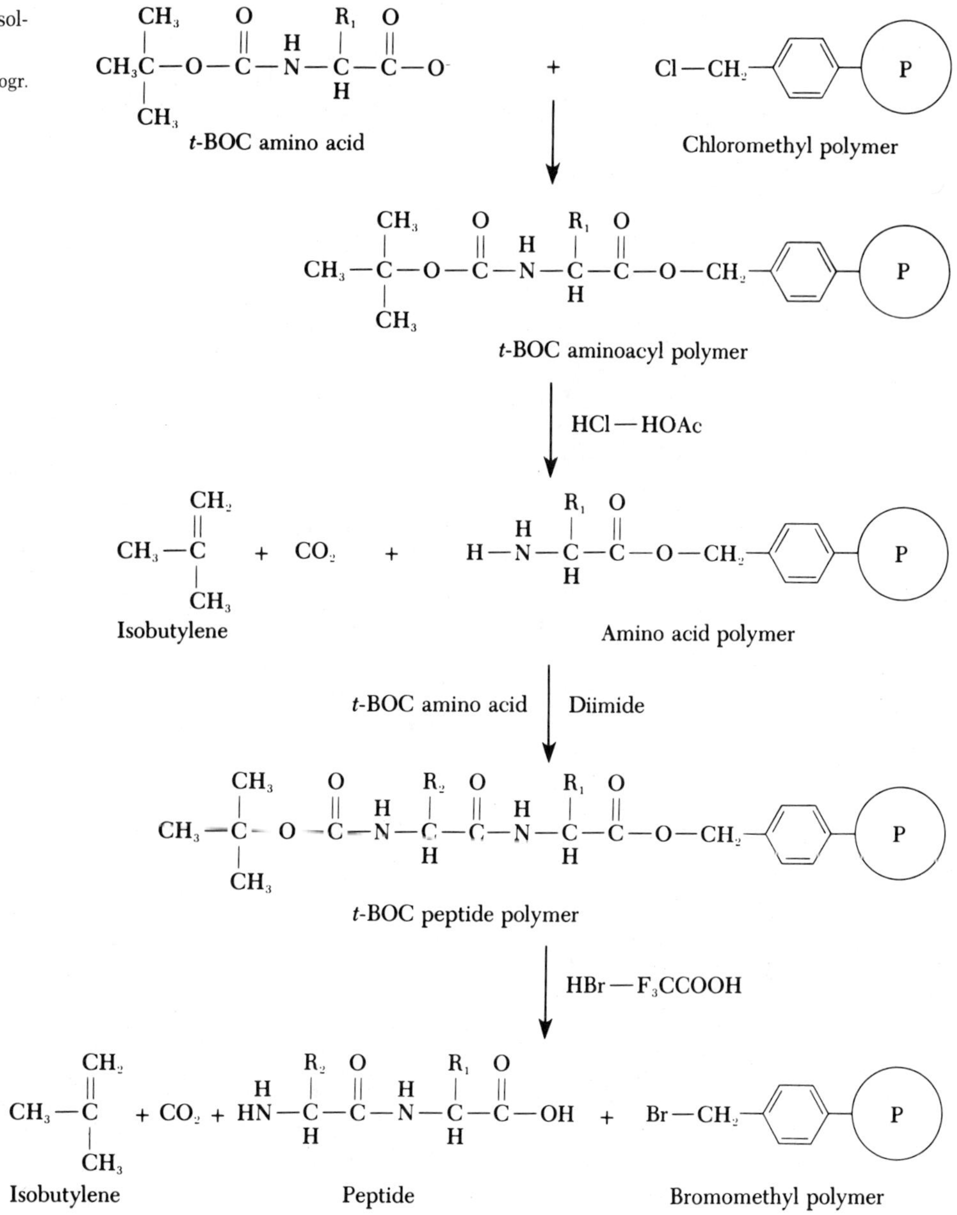

$$(CH_3)_3C{-}O{-}\overset{O}{\overset{\|}{C}}{-}\overset{H}{N}{-}\underset{H}{\overset{R_1}{C}}{-}\overset{O}{\overset{\|}{C}}{-}O^- \quad + \quad Cl{-}CH_2{-}C_6H_4{-}(P)$$

t-BOC amino acid + Chloromethyl polymer

$$\downarrow$$

$$CH_3{-}C(CH_3)_2{-}O{-}\overset{O}{\overset{\|}{C}}{-}\overset{H}{N}{-}\underset{H}{\overset{R_1}{C}}{-}\overset{O}{\overset{\|}{C}}{-}O{-}CH_2{-}C_6H_4{-}(P)$$

t-BOC aminoacyl polymer

$$\downarrow \ HCl{-}HOAc$$

$$CH_3{-}\overset{CH_2}{\overset{\|}{C}}{-}CH_3 \quad + \quad CO_2 \quad + \quad H{-}\overset{H}{N}{-}\underset{H}{\overset{R_1}{C}}{-}\overset{O}{\overset{\|}{C}}{-}O{-}CH_2{-}C_6H_4{-}(P)$$

Isobutylene + Amino acid polymer

$$\downarrow \ t\text{-BOC amino acid} \mid \text{Diimide}$$

$$CH_3{-}C(CH_3)_2{-}O{-}\overset{O}{\overset{\|}{C}}{-}\overset{H}{N}{-}\underset{H}{\overset{R_2}{C}}{-}\overset{O}{\overset{\|}{C}}{-}\overset{H}{N}{-}\underset{H}{\overset{R_1}{C}}{-}\overset{O}{\overset{\|}{C}}{-}O{-}CH_2{-}C_6H_4{-}(P)$$

t-BOC peptide polymer

$$\downarrow \ HBr{-}F_3CCOOH$$

$$CH_3{-}\overset{CH_2}{\overset{\|}{C}}{-}CH_3 + CO_2 + \overset{H}{HN}{-}\underset{H}{\overset{R_2}{C}}{-}\overset{O}{\overset{\|}{C}}{-}\overset{H}{N}{-}\underset{H}{\overset{R_1}{C}}{-}\overset{O}{\overset{\|}{C}}{-}OH \quad + \quad Br{-}CH_2{-}C_6H_4{-}(P)$$

Isobutylene + Peptide + Bromomethyl polymer

Secondary structure

Although one first needs to determine the amino acid sequence of a protein, i.e., its primary structure, this is not sufficient information to understand the basis for its chemical and physical properties or its biological activity. A protein is not simply a series of amino acids "strung out" in a long chain; it has a particular three-dimensional shape or *conformation*. This is usually described in terms of higher orders of structure, i.e., the secondary and tertiary. The arrangement of the polypeptide chain into helical, pleated sheet, and random coil configurations is called the *secondary structure*.

First, however, we must understand a basic characteristic of the peptide bond, i.e., its planar nature. Studies of model peptides using the x-ray diffraction method made it possible to measure the various interatomic distances, as seen in the diagram below. Thus the distance between the carbonyl C and the amide N was found to be 0.132 nm, which is intermediate between that for a double bond (as C═O), 0.124 nm, and the single C_α—N bond distance of 0.147 nm. This indicates that the peptide bond exists in a state of resonance between the double and single bonds.

O
0.124 nm
C 0.147 nm N 0.132 nm C 0.153 nm C
H
O −
C C
+ N C
H

A direct consequence of this resonance and the double bond characteristic of the peptide bond is that carbonyl carbon and the amide nitrogen lie in a plane accompanied by the oxygen and the amide hydrogen. This planar unit is common to protein structures. Furthermore, as seen in the diagram below, the orientation of the carbonyl oxygen and the amide hydrogen about the peptide bond is trans, as is that of the α-carbons. This places the R groups alternately on either side of the polypeptide backbone.

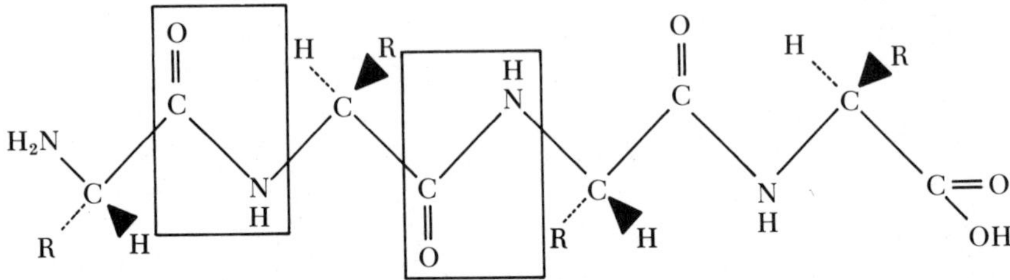

As is apparent from this diagram, each of two neighboring planar units are joined by an α-carbon. Since these single bonds, C_α—N and C_α—carbonyl C (C′), can rotate freely, a large number of angular relationships may conceivably exist between planar units. The relative positions of any two neighboring planes joined by an α-carbon are described by torsion angles, namely ϕ for the C_α—N angle and ψ for the C_α—C′ (Fig. 4-6). The magnitude of these two angles defines the precise spatial relationship between planes. Although the combination of torsion angles may be unlimited, the size and nature of the R groups exert a degree of constraint on these angles. Studies of various model polypeptides have shown the existence of limited combinations of values for the two torsion angles.

The first, regular secondary structure proposed for proteins was that of a right-handed helix in which the angles ϕ and ψ are −57 degrees and −47 degrees, respectively. This helix has 3.6 amino acid residues per turn and has a pitch or rise per turn of 0.54 nm. Such a conformation is now usually called an α-helix. Although a left-handed α-helix is also possible, the right-handed one is the more stable structure.

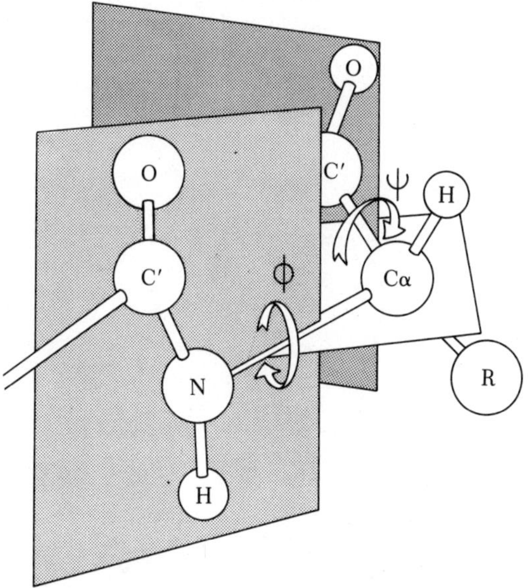

Figure 4-6
Two planar peptide groups linked by an α-carbon, showing a segment of a right-handed helix of L-amino acids.

Modified from Schellman, J.A., and Schellman, C. In Neurath, H., editor: The proteins: composition, structure, and function, vol. 2, New York, 1964, Academic Press, Inc.

A hypothetical polypeptide chain in the α-helical conformation is illustrated in Fig. 4-7, *A*. Two features are noteworthy. First, the peptide carbonyl groups are oriented parallel to the helical axis and are directed at amide groups to which they are joined by hydrogen bonds. Thus the helix is stabilized by hydrogen bonds between an amide in one planar group and the carbonyl oxygen three planar units, or 13 atoms, further along the polypeptide chain (Fig. 4-7, *B*). Second, the amino acid side chains, or R groups, point outward or away from the helical axis.

Another stable helical form that has been found to a limited degree, for example, in the enzyme lysozyme, is the 3_{10} helix. This form has three residues per turn and a pitch of 0.6 nm. Hydrogen bonding occurs between amide hydrogens and carbonyl oxygens 10 atoms away.

A second, regular type of protein structure that has been defined is the β-pleated sheet or the β-conformation. In this case neighboring planes exist in an alternating right-left arrangement. The ϕ and ψ are twisted in opposite (negative versus positive values) directions. For example, a model polypeptide composed of alanine residues has values of ϕ and ψ that are −139 degrees and +135 degrees respectively. Fig. 4-8, *A*, illustrates this alternating relationship between planar peptide units.

The arrangement of polypeptide chains in the pleated sheet conformation can occur in two ways. In one, called the *parallel* pleated sheet (Fig. 4-8, *B*), the polypeptides are side by side and lie in the same direction so that their N-terminal residues are at the same end. Fig. 4-8, *B*, also shows that the pleated sheet conformation is stabilized by hydrogen bonding. Here the bonds are formed between an amide hydrogen of a peptide plane in one chain and the carbonyl oxygen in a planar unit of a neighboring chain. In the second kind of β-conformation the polypeptide chains lie in opposing directions such that the N-terminal end of one is next to the C-terminal of the other. (Fig. 4-8, *C*). This type of β-conformation is called the *antiparallel* pleated sheet. It is also stabilized by interchain hydrogen bonding.

A third type of secondary structure is the so-called random coil conformation in which there is no consistent relationship between planes; consequently there are no regularly distributed stabilizing hydrogen bonds.

Figure 4-7 Representation of the α-helical structure of proteins.
A Scheme illustrating the relative arrangement of planar peptide units.
B Intrahelical hydrogen bonds formed between amide hydrogens and carbonyl oxygens.

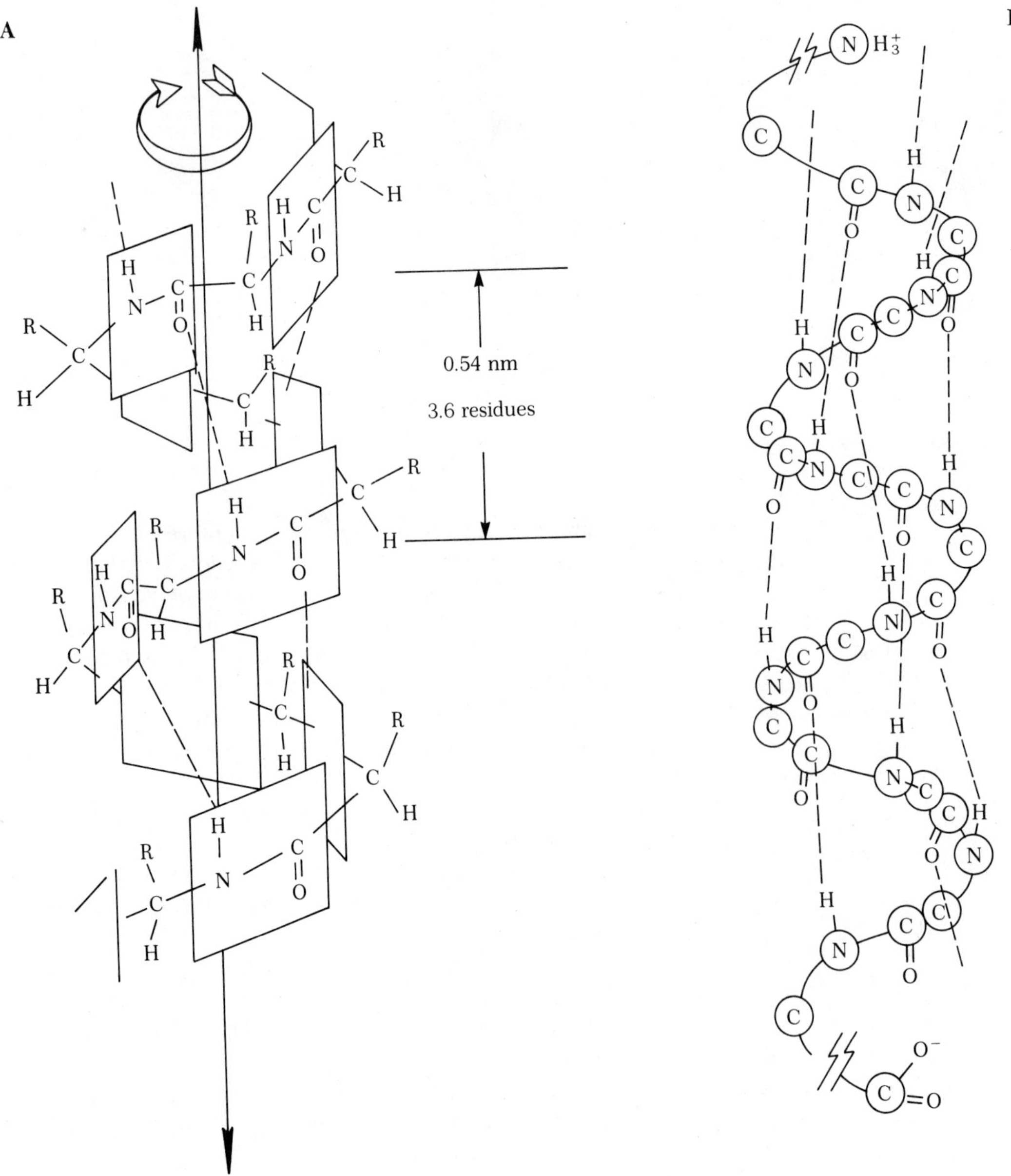

Figure 4-8
Some aspects of the β-pleated sheet or the β-conformation.
A Alternating relationship between planar peptide units.
B Parallel arrangement of polypeptide chains stabilized by hydrogen bonding.
C Antiparallel arrangement of polypeptide chains stabilized by hydrogen bonding.

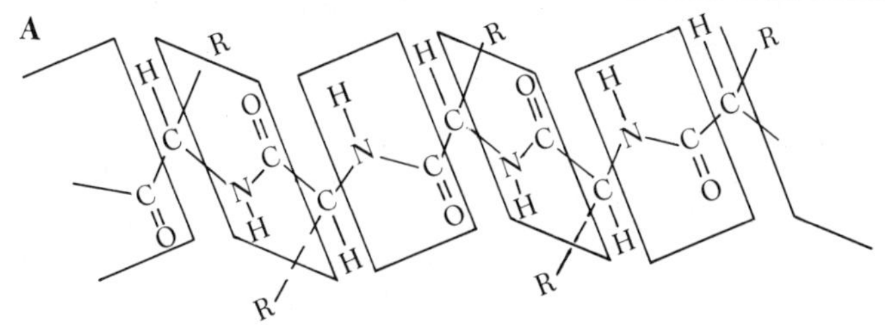

X-ray diffraction analysis coupled with a knowledge of the amino acid sequence has made it possible to know the exact structure of a great many proteins. It is evident that globular proteins are composed to varying degrees of the types of secondary structure described. For example, as seen in Table 4-3, the presence of α-helix varies from protein to protein. Myoglobin and the α- and β-subunits of hemoglobin consist of α-helical regions held together without any pleated sheet structure. At the other extreme is α-chymotrypsin, which has extensive β-pleated sheets with only two small regions of α-helix.

Why certain regions of a protein are helical and others are not is difficult to understand. It reflects properties of the amino acid side chains in the primary sequence. Certain amino acids have a tendency to form the α-helical secondary structure, e.g., glutamic acid, alanine, and leucine are strong α-helix formers, whereas methionine, valine, isoleucine, and aspartic acid are formers of the β-conformation. Proline and hydroxyproline are incompatible with the α-helix; therefore departure from α-helical structure occurs when these amino acids appear in the sequence. The nitrogens of the two heterocyclic amino acids are part of a ring

Table 4-3
The α-helical content of a group of selected proteins

Protein	Percent of α-helix
α-Chymotrypsin	5
Ribonuclease	15
Carboxypeptidase	30
Insulin	66
Myoglobin	77

structure that makes them incapable of forming a resonating peptide bond; furthermore, the rings do not fit into the constraints of the α-helix.

Proline and hydroxyproline are major constituents of collagen, a protein composed of three polypeptide chains, each wound in a special *left-handed* helix. This conformation, which represents yet another form of secondary structure, has three residues per turn compared with 3.6 for the right-handed α-helix. The polypeptides of collagen contain 27% proline and hydroxyproline and 33% glycine. Typically the primary structure is Gly-X-Pro, Gly-X-Hypro, and Gly-Pro-Hypro. (Here X represents another amino acid residue.) The three left-handed helices are bound together by interchain hydrogen bonds; two of every three amide and carbonyl groups of one chain bond with carbonyls and amides in one of the other two chains. Then the three polypeptides together are given a right-handed twist to form a threefold *superhelix* (Chapter 15). This structure gives collagen its unusual strength; it is said to have the tensile strength of steel.

Tertiary structure

Biological function of protein molecules is contingent on their ability to form a unique three-dimensional structure or shape. For example, the enzyme ribonuclease, which hydrolyzes RNA, depends on two histidine residues for its catalytic activity. Yet one of these histidines appears as residue 12, far removed from the other at 119. Folding of the molecules must occur so as to bring these into their proper juxtaposition. Most proteins are distinguished by their ability to fold spontaneously in their biological environment to yield their native, biologically active form. This ability to fold depends on the primary structure and the nature and interaction of the amino acid side chains.

In the cell the newly synthesized protein must be able to fold with extreme rapidity and accuracy to achieve its biologically active conformation. Therefore folding must occur along relatively limited pathways. In general, the folding process results in a closely packed structure that serves to bury nonpolar side chains internally so as to yield a nonpolar or hydrophobic core. Hydrophilic polar groups are predominantly exposed to interact with water molecules of the aqueous environment. These interactions between nonpolar groups and between polar groups and water are the forces that drive the folding process. The tertiary structure achieved is usually the most stable, i.e., the structure of least free energy content.

Proteins may be classified as follows:

1. The *α-group,* having exclusively α-helical structure
2. The *β-group,* having only pleated sheet
3. *α + β group,* containing both α-helical and β-sheets; however, they are not interspersed
4. *α/β group,* where the α and β structures are mixed or alternate
5. The *coil group,* containing almost no regular secondary structure

It has been suggested that intracellular proteins are usually of the α or α/β type, whereas most extracellular proteins are of the β or α + β type.

Plate 1
Three-dimensional model of myoglobin molecule showing cleft formed for heme prosthetic group. Heme is attached to the α-helical regions (*E* and *F*) by the histidine residues indicated. The back of the cleft is formed by the helices *G* and *H*. *His,* Histidine; *Phe,* phenylalanine.

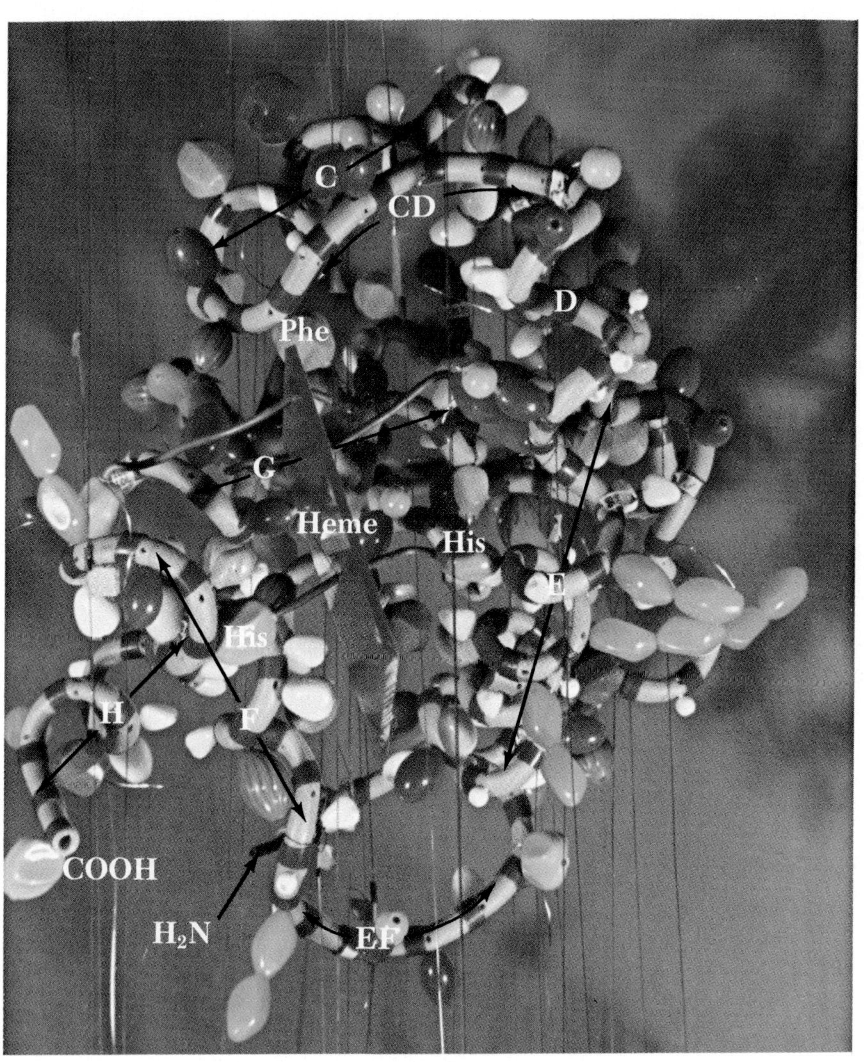

Myoglobin, an α-group protein, serves as a useful and relatively simple example of the importance of tertiary folding in achieving the final, functional protein molecule. It has eight α-helical regions. If these were simply stretched out in a single line, the molecule would have a tubular shape, some 20 times as long as its diameter. X-ray diffraction studies have shown this not to be the case at all; myoglobin is a football-shaped molecule with dimensions of 4.3 × 3.5 × 2.5 nm. Plate 1 shows the tertiary structure and the relative positions of the eight helical regions, usually designated by the letters *A* to *H*. The interhelical sequences are assigned letters corresponding to the two helices joined, e.g., region *CD* joins helices *C* and *D*. These regions are random coils. The helices arrange themselves to form a pocket or cleft into which the heme prosthetic group is inserted. This cleft is formed by helices *E* and *F*, which provide the walls, whereas *B*, *G*, and *H* form the floor. *CD*, a nonhelical portion, closes the top of the pocket.

Most of the hydrophobic residues of myoglobin are turned inward. The attraction between them is a most important force stabilizing its tertiary conformation and is caused by an exclusion of solvent molecules and by Van der Waals forces (Fig. 4-9). Thus the molecule is compact, having a hydrophobic core in which no more than five water molecules are trapped. Nearly all the polar groups are on the surface, where they bind water molecules. Some of the hydrophobic side chains line the heme pocket to provide the prosthetic group with an environment in which it can function in its oxygen-storing capacity. The tertiary structure also brings two histidines close enough together so that they may bind to either side (coordination bond to the iron) of the heme (Plate 1).

The tertiary conformation and packing of α-helical and β-sheets for an important intracellular enzyme (α/β), lactic dehydrogenase, is illustrated in Fig. 4-10. This diagram illustrates the arrangement of helices (cylinders) and β-sheets (flat arrows) in space.

In extracellular proteins, i.e., group α + β, disulfide bonds usually stabilize the tertiary structure. Pancreatic ribonuclease, which functions in the intestines, is a well-known example in which four disulfide bonds serve to maintain its active conformation.

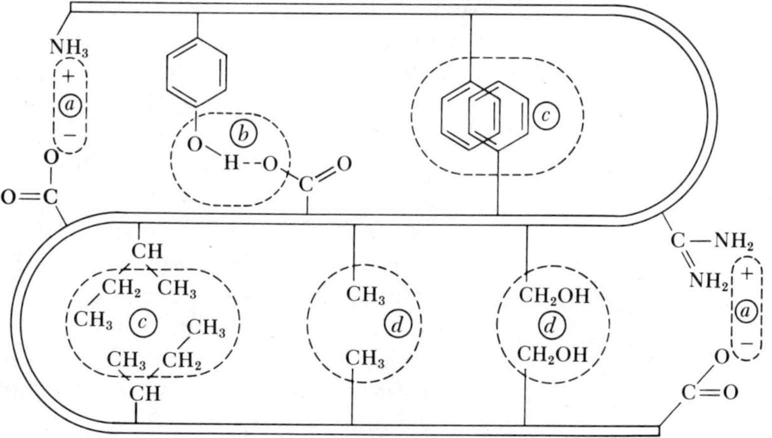

Figure 4-9
Some noncovalent bonds that stabilize proteins. *a*, Electrostatic interaction; *b*, hydrogen bonding between tyrosine residues and carboxylate groups; *c*, hydrophobic interaction; *d*, Van derWaals interactions.

From Anfinsen, C.B.: The molecular basis of evolution, New York, 1959, John Wiley & Sons, Inc.

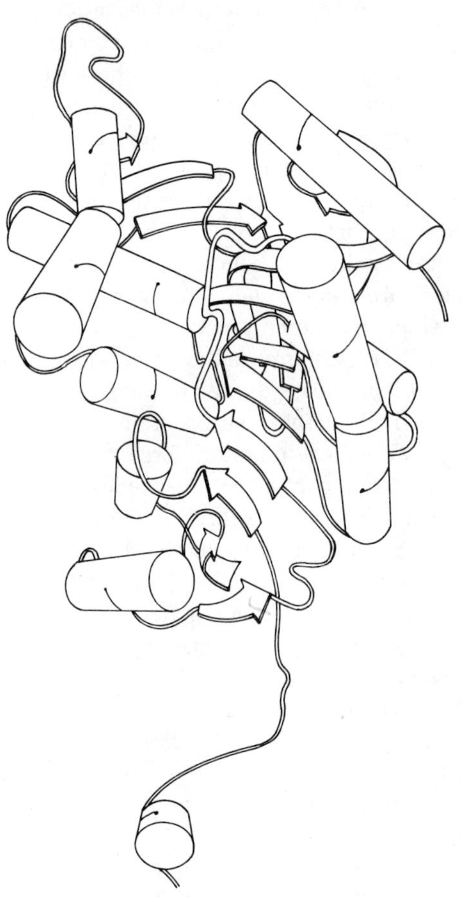

Figure 4-10
Folding of α-helices (cylinders) and β-sheets (flat arrows) of lactic dehydrogenase.

From Rossmann, M.G.: Lactic dehydrogenase. In Boyer, P.D., editor: The enzymes, vol. 11, New York, 1975, Academic Press, Inc.

Quaternary structure

As seen from Table 4-4, proteins generally have very large molecular weights ranging from 3000 to several million. Molecular weights are usually determined by ultracentrifugation, measurements of osmotic pressure, molecular sieving, and by polyacrylamide gel electrophoresis in the presence of a detergent sodium dodecyl sulfate (Chapter 25). By ultracentrifugation the molecular weight of hemoglobin was established as 64,500. If the molecular weight is calculated from its iron content (0.34%), then a value of only 16,400 (minimal molecular weight) is obtained. In this calculation it is assumed that each molecule of protein has at least one atom of iron, i.e., 55.8 gm/mol wt. These two values (64,500 and 16,400) allow the conclusion that hemoglobin actually is composed of four subunits, each having an atom of iron and a molecular weight of 16,400. Such a grouping or aggregate of subunits is called the *quaternary structure* of the protein.

Hemoglobin actually possesses two kinds of subunits, called the α- and β-chains, each having a distinctive structure. The α-chain has 141 residues; valine is the amino N-terminal and arginine is at the carboxyl end. The β-chain has 146 residues, with valine and histidine at the N- and C-terminal ends, respectively. Although the primary structures of the α- and β-chains differ significantly from that of myoglobin, their secondary and tertiary conformations are remarkably similar. They have a tetrahedral shape with most of the nonpolar amino acid residues

Table 4-4
Quaternary structure of selected proteins

Protein	Molecular weight*	Subunits Number	Subunits Molecular weight
Insulin	11,466	2	5,733
Alcohol dehydrogenase (liver)	80,000	4	20,000
Ceruloplasmin	151,000	8	18,000
Histidine decarboxylase	109,000	10	19,000
Fumarase	194,000	4	48,500
Glucose 6-phosphate dehydrogenase	240,000	6	43,000
Apoferritin	480,000	20	24,000
Urease	483,000	6	83,000
Glutamine synthetase	592,000	12	48,500
Glutamate dehydrogenase	2,000,000	8	250,000
Acetyl CoA carboxylase (liver)	8,300,000	2	4,100,000
Tobacco mosaic virus	40,000,000	2130	17,500

From Klotz, I.M., and Darnall, D.W.: Science **169:**127, 1969.
*At the present time many biologists are using the term *dalton* as a unit of mass, synonymous with *molecular weight.* Such usage is questionable at this time because no commission of any international scientific union or indeed any other appropriate body has officially recognized this unit for such a purpose. Molecular weight, or perhaps more clearly termed *relative molecular mass,* is a pure number and dimensionless and is the *ratio* of the *mass* of one molecule of a substance to one twelfth the mass of an atom of ^{12}C (IUPAC definition). The dalton is a unit of mass and is numerically identical with *molar mass* in gram-mole^{-1}, but it is clearly imcompatible with the above IUPAC definition of molecular weight. Thus subtle distinctions are involved. The term *dalton* may be useful, however, for example, to specify the mass of crystals of protein that contain water and salts as well as the molecule of primary interest, or of ribosomes or other cellular organelles containing many different kinds of protein and nucleic acid molecules. Until the nomenclature committees of the International Union of Biochemistry, the American Society of Biological Chemists, and the IUPAC clarify matters, the term molecular weight will continue to be used in this book as a dimensionless number, a ratio in the IUPAC sense. (See Edsall, J.T.: Definitions of molecular weight, Nature **228:**888, 1970; Paul, M.A.: International system of units (S.I.)—development and progress, J. Chem. Documentation **11:**3, 1971; National Bureau of Standards Policy on Usage of S.I. units, J. Chem. Educ. **48:**569, 1971.)

buried in the interior of the molecule, whereas the polar chains and those which are ionized at neutral pH are near the surface. In a schematic representation of the two subunits as cones (white for α-subunits and black for β-subunits [Fig. 4-11]) the quaternary structure may be represented as a nesting of the two pairs of cones. The two oppositely oriented white erect cones are nested with two upside-down, black cones. By this arrangement the positively charged N-terminal ammonium group of one α-chain is brought close to the negatively charged C-terminal carboxylate of the second α-chain (Fig. 4-12). Electrostatic attraction is critical to stabilizing the quaternary structure of hemoglobin. Figs. 4-11 and 4-12 show that there can be little surface contact between like cones, i.e., between α- or between β-subunits, whereas there is considerable contact between unlike ones, i.e., between α- and β-subunits. The majority of contact interactions are hydrophobic in nature.

Increasing evidence points to a functional role of the quaternary structure in the biological activity of proteins. The classic examples are afforded by the oxygen binding of hemoglobin and feedback control of *E. coli* aspartyl transcarbamoylase. Other examples are provided in Table 4-4.

Figure 4-11
Model illustrating the nesting of two α-chains (white) and two β-chains (black) to form a molecule of hemoglobin.

Figure 4-12

Hemoglobin molecule, as deduced from x-ray diffraction studies, shown from above (top) and side (bottom). Drawings follow scheme used in three-dimensional models built by Perutz and his co-workers. Irregular blocks represent electron-density patterns at various levels in the hemoglobin molecule. The molecule is built up from four subunits: two identical α-chains (light blocks) and two identical β-chains (dark blocks). *N* in the top view identifies the amino ends of the two α-chains; *C* identifies the carboxyl ends. Each chain enfolds a heme group (white disks), the iron-containing structure that binds oxygen to the molecule.

From Perutz, M.F.: The hemoglobin molecule, Sci. Am. **211**:64, 1964.

Denaturation and renaturation

It is apparent from the preceding description that the primary, secondary, and tertiary structures are the basis for the chemical, physical, and biological properties of proteins. When normal conformation is altered but not accompanied by a rupture of peptide bonds, the protein is *denatured* and its distinctive properties and biological activity also may be changed.

Denaturation can take many forms; the tertiary structure may be unfolded, the secondary structure may assume varying degrees of random coil, or disulfide bonds may be cleaved. The most sensitive index of denaturation is a loss of biological activity. Changes in solubility occur with more extreme changes, the coagulation of egg white being an obvious example. Factors that denature proteins are temperature (50° to 60° C), extremes in pH (below 4 or above 10), high concentration of organic solvents such as alcohols, acetone, or ether, and urea or guanidine HCl. Denaturation may also result from exposure to ultraviolet light and ionizing radiations as well as from mechanical agitation, heavy metals, and detergents.

Urea and guanidine HCl have been used to denature proteins in the process of studying their structure. Many are unfolded by 8M urea or 5M guanidine HCl; those having a quaternary structure are disaggregated into subunits.

A detergent commonly used as a denaturing agent is sodium dodecyl sulfate (SDS). SDS complexes with proteins to produce negatively charged, rod-shaped particles. The amount of SDS (1.4 gm) that binds to each gram of protein is constant for different proteins. This high level of binding "swamps out" the existing charge on most. These negatively charged, SDS-protein complexes mutually repel one another and therefore prevent aggregation, which accounts for the solubilizing effect of the detergent. This property is used in the determination of molecular weights in conjunction with acrylamide gel electrophoresis (Chapter 25). The electrophoretic mobility of the SDS-protein complex is molecular weight dependent.

Denaturation may be either reversible or irreversible. Obviously, coagulated egg white is an extreme example and represents an irreversible change. Where only limited unfolding and refolding are encountered, the denaturation is usually reversible. Reversal of denaturation is called *renaturation* and was first extensively studied using ribonuclease (Fig. 4-13). Denaturation was accomplished in 8M urea with β-mercaptoethanol in a mildly alkaline medium. On reduction of the four disulfides the protein assumed a random coil structure. Dialysis to remove urea, in the presence of oxygen, resulted in the reformation of the proper disulfide bonds, the native conformation, and a restoration of 95% to 100% of the enzymatic activity. If reforming the proper disulfide bonds were a completely random process, then

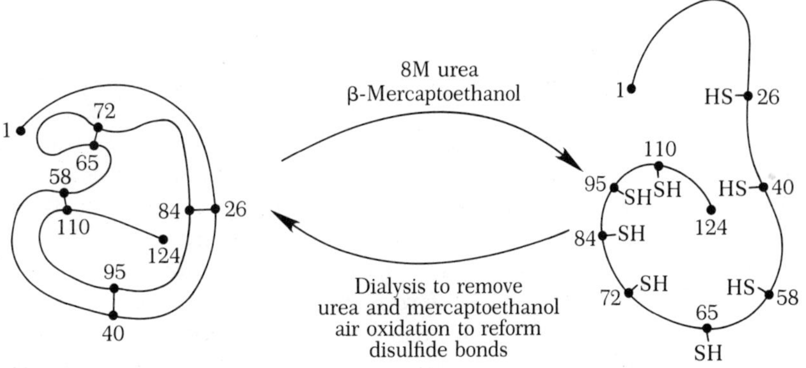

Figure 4-13
Reversible denaturation and renaturation of ribonuclease. This illustrates the spontaneous reformation of the enzymatically active conformation and reformation of the appropriate disulfide bonds.

only about 1% of the activity should have been recovered. This experiment shows that the native conformation forms spontaneously; it automatically achieves a state of lowest free energy and of greatest stability. This is the native and biologically active form of the protein and is determined by the primary structure.

Denaturation-renaturation experiments have been conducted on proteins exhibiting quaternary structure. For example, when aldolase, a tetramer, was treated with 4M urea, the protein disaggregated and the individual subunits assumed a random coil-like structure. Dialysis against a suitable buffer resulted in a recovery of 70% of the enzyme activity. This shows that both the reformation of individual subunit structure and the proper state of aggregation occurs spontaneously.

The fact that denatured proteins may spontaneously be restored to their native and biologically active state provides an important insight into the manner in which proteins are assembled in the cell. Once the primary structure has been completed by the cell's protein-synthesizing process (Chapter 10), the polypeptide chain spontaneously assumes its biologically active three-dimensional conformation. It is a mistake, however, to assume that a protein achieves an active shape and is held that way like a golf ball. Rather, proteins exist in a dynamic state of conformation in which changes of shape may occur under the influence of various factors in the environment of the molecule. This concept is especially pertinent to the activity and regulation of enzymes.

Ionization

A factor of great importance to the properties and function of proteins is their ionization or net charge. Since they contain both acidic and basic amino acids and since these are located primarily at the surface of the compact molecule, proteins are amphoteric and are capable of donating and accepting protons. In an acid environment containing excess protons the protein molecules assume a net positive charge, whereas in an alkaline medium the protein molecules have a net negative charge. As with the amino acids themselves, there exists a pH at which the protein does not migrate in an electric field; its net charge is zero. This is the *isoelectric point,* illustrated in the following diagram:

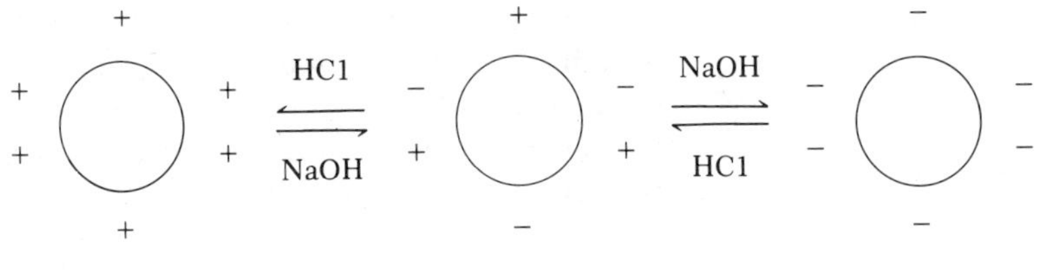

HC1 ⇌ NaOH   NaOH ⇌ HC1

Although a protein is usually in solution at its isoelectric point, it is least soluble at this pH and consequently is most easily precipitated. Proteins differ greatly in their isoelectric points. Most proteins have isoelectric points near neutrality, reflecting similar numbers of acidic and basic residues. Salmine is one extreme example, in which 66% of the amino acids are arginine, and therefore this protein has an isoelectric point at 12.1. At the other extreme is the gastric protease pepsin, which has a predominance of aspartic and glutamic acids and an isoelectric point of 1.0.

A protein may be titrated just as amino acids. Because of the large numbers of titratable groups, however, interpretation of the curve tends to be complex. In general, titration curves for proteins are divided into three pH regions: the pH range from 1.5 to 6 corresponds to the ionization of carboxyl groups; pH 6 to 8.5 is attributed to histidine residues and terminal α-amino groups (Table 4-5); and from pH 8.5 up are the ϵ-amino groups of lysine, the phenolic group of tyrosine, and sulfhydryl group of cysteine. Usually the guanidine group of arginine has a pK too high for observation in the titration curve. The pK value of ionizable groups in proteins

Table 4-5
pK values for various groups in selected proteins

Group	pK
α-Carboxyl	
Insulin	3.6
Side-chain carboxyl	
Serum albumin	4.0
Insulin	4.7
β-Lactoglobulin (49 of 51 groups)	4.8
β-Lactoglobulin (2 of 51 groups)	7.3
Imidazole	
Insulin	6.4
Serum albumin	6.9
Phenolic	
Insulin	9.6
Serum albumin	10.4
ϵ-Amino	
Serum albumin	9.8
Ribonuclease	10.2
Guanidyl	
Insulin	11.9

From Tanford, C.: Adv. Protein Chem. **17**:69, 1962.

may be modified by factors such as the ionization of other groups in the protein, the proximity of hydrophobic residues, and the participation in hydrogen bonding. Consequently the pK values for the side-chain carboxyls vary from 3.5 to 4.8, but, as an extreme case, two carboxyls of β-lactoglobulin exhibit pKs as high as 7.3; one carboxyl of pepsin has a pK of 1.0 (Table 4-5). It is suggested that these carboxyls are buried in the hydrophobic interior of the molecule.

In a physiological environment proteins are in the pH range of 6 to 8.5 in the titration curve. This means that the only amino acid residues that donate or accept protons are histidine (imidazole) and the N-terminal α-amino groups. With the exception of a few buried carboxyls, all side-chain carboxyl groups would be ionized at this pH, and all ϵ-amino groups of lysine and the guanidyls of arginine would be protonated. Since there are relatively few N-terminal α-amino groups, it is the imidazole of histidine that exerts the buffering action of most proteins in this region.

Since proteins exist in ionic forms on either side of their isoelectric point, they may form salts with various anions or cations. For example, in acid solution they may form soluble hydrochlorides or insoluble salts with anions such as salicylates, trichloroacetate, perchlorate, tungstate, picrate, and tannate. In an alkaline solution proteins exist as sodium salts and are precipitated by cations such as Ag^{+}, Ca^{++}, Zn^{++}, Fe^{+++}, Cu^{++}, and Pb^{++}.

The insolubility of the various protein salts on either side of the isoelectric point is used in a number of practical ways. For example, picric acid hardens the skin around minor burns, tannic acid prepares and preserves leather, silver nitrate cauterizes wounds, and trichloroacetic acid and phosphotungstic acid are often employed for the purpose of deproteinizing blood samples for certain clinical measurements. Proteins such as those in milk and egg white are frequently recommended as antidotes in metal poisoning.

Classification of proteins

The diversity of physical and chemical properties of proteins prevents assigning them to rigorous categories or classes. Two broad types of proteins are still recognized: (1) *simple proteins,* composed of amino acids only, and (2) *conjugated* or *compound proteins,* which contain components other than amino acids. These non–amino acid components are called the *prosthetic group(s)*. The nucleic acid of nucleoproteins is the prosthetic group. Likewise, in glycoproteins the prosthetic group is the carbohydrate and in lipoproteins it is the lipids. Other important examples include the metalloproteins, to which various metals are attached, and phosphoproteins, which contain phosphate groups.

Suggested readings

Anfinsen, C.B.: The molecular basis of evolution, New York, 1959, John Wiley & Sons, Inc.

Anfinsen, C.B.: Principles that govern the folding of protein chains, Science **181:**223, 1973.

Anfinsen, C.B., Jr., Edsall, J.T., and Richards, F.M.: Advances in protein chemistry, vols. 1-26, New York, 1944-1972, Academic Press, Inc.

Bailey, J.L.: Techniques in protein chemistry, ed. 2, Amsterdam, 1967, Elsevier Publishing Co.

Dickerson, R.E., and Geis, I.: The structure and action of proteins, New York, 1969, Harper & Row, Publishers.

Florkin, M., and Stotz, E.H., editors: Comprehensive biochemistry: proteins, part I, vol. 7, Amsterdam, 1963, Elsevier Publishing Co.

Greenstein, J.P., and Winitz, M.: Chemistry of the amino acids, New York, 1961, John Wiley & Sons, Inc.

Haschemeyer, R.H., and Haschemeyer, A.E.V.: Proteins: a guide to study by physical and chemical methods, New York, 1973, John Wiley & Sons, Inc.

Hirs, C.H.W., editor: Enzyme structure. In Colowick, S.P., and Kaplan, N.O.: Methods of enzymology, vol. 11, New York, 1967, Academic Press, Inc.

Hirs, C.H.W., and Timasheff, S.N., editors: Enzyme structure, part B. In Methods in enzymology, vol. 25, New York, 1972, Academic Press, Inc.

Kendrew, J.C.: The three-dimensional structure of a protein molecule, Sci. Am. **205:**96, 1961.

Konigsberg, W.H., and Steinman, H.M.: Strategy and methods of sequence analysis. In Neurath, H., and Hill, R.L., editors: The proteins, ed. 3, vol. 3, New York, 1977, Academic Press, Inc.

Means, G.E., and Feeney, R.E.: Chemical modification of proteins, San Francisco, 1971, Holden-Day, Inc.

Merrifield, R.B.: Solid-phase peptide synthesis, Adv. Enzymol. **32:**221, 1969.

Moore, S., and Stein, W.H.: Chemical structure of pancreatic ribonuclease and deoxyribonuclease, Science **180:**458, 1973.

Neurath, H., editor: The proteins—composition, structure and function, ed. 2, vols. 1-5, New York, 1963-1968, Academic Press, Inc.

Neurath, H., and Hill, R.L., editors: The proteins, ed. 3, vols. 1-4, New York, 1975-1979, Academic Press, Inc.

Perutz, M.F.: The hemoglobin molecule, Sci. Am. **211:**64, 1964.

Perutz, M.F.: Proteins and nucleic acids: structure and function, New York, 1963, American Elsevier Publishing Co., Inc.

Ramachandran, G.N., editor: Aspects of protein structure, New York, 1963, Academic Press, Inc.

Schroeder, W.E.: The primary structure of proteins, New York, 1968, Harper & Row, Publishers.

Sheraga, H.A.: Protein structure, New York, 1961, Academic Press, Inc.

Sternberg, M.J.E., and Thornton, J.M.: Prediction of protein structure from amino acid sequence, Nature **271:**15, 1978.

Tanford, C.: Protein denaturation, Adv. Protein Chem. **24:**2, 1970.

Thompson, R.B., Borden, D., Tarr, G.E., and Margoliash, E.: Heterogeneity of amino acid sequence in hippopotamus cytochrome c, J. Biol. Chem. **253:**8957, 1978.

Chapter 5

Enzymes: their function and modulation of activity

The metabolism of nutrients and their assimilation into the tissues of the body depend on the activity of thousands of enzymes. Enzymes are protein catalysts produced by the living cell yet not dependent on the intact cell for their activity. Some, the digestive enzymes for example, function outside their cells of origin; normally they act in the stomach or intestinal lumen. The nutrients that result are used by cells throughout the body for metabolism (Chapter 3). The metabolic pathways shown in Fig. 3-1 require the catalytic action of intracellular enzymes. It is important to appreciate fully that all enzymes are dependent on their structure and not on the existence of an intact cell for activity. As suggested in Chapter 4, a protein's biological action as well as its chemical and physical properties is dependent on its primary, secondary, and tertiary conformation. In enzymes it is the specific spatial relationships or proximity of amino acids that confer on it the particular catalytic activity. Also, as suggested in Chapter 3, the activity of some enzymes can be controlled by changing the shape and therefore the spatial arrangement of amino acids.

Characteristics of enzymes

Enzymes are proteins

Like all proteins, enzymes have a unique sequence of amino acids and three-dimensional structure. Yet a detailed knowledge of this structure does not explain how an enzyme works. What is there about pepsin, trypsin, or chymotrypsin that makes each capable of hydrolyzing other proteins also having unique primary and three-dimensional structures? Enzymes such as pepsin, trypsin, and chymotrypsin are composed only of amino acids. Many others, however, are conjugated proteins containing, in addition to amino acids, a nonprotein organic component, a metal ion, or both. For example, an important digestive enzyme of pancreatic origin is carboxypeptidase, which contains an atom of zinc. Other examples are provided in Table 5-1. In general such enzymes are composed of a heat-labile, nondialyzable protein called the *apoprotein* and either a heat-stable, organic group called the *coenzymes* or one of several metal ions. The coenzymes are well-defined organic compounds and in many instances have structures related to the vitamins. Because of their great importance to metabolism and nutrition, they are the subject of a separate chapter (Chapter 6). Neither the apoenzyme nor the coenzyme functions independently, but together they form the catalytically active unit called the *holoenzyme*. Thus:

$$\text{Apoenzyme} + \text{Coenzyme} = \text{Holoenzyme}$$

Enzymes are catalysts

Enzymes are protein catalysts and therefore are capable of accelerating the rate of a chemical reaction; yet they neither initiate the reaction nor alter the extent to

Table 5-1
Examples of enzymes containing metal ions or coenzymes

Enzyme	Metal ion	Coenzyme
Cytochrome oxidase	Fe, Cu	Porphyrin
Catalase	Fe	Porphyrin
Succinic dehydrogenase	Fe	FAD
Tyrosinase	Cu	
Ascorbic acid oxidase	Cu	
Carbonic anhydrase	Zn	
Carboxypeptidase	Zn	
Pyruvate carboxylase	Zn, Mn	Biotin
Glucose oxidase		FAD
Glyceraldehyde 3-phosphate dehydrogenase		NAD^+
Glucose 6-phosphate dehydrogenase		$NADP^+$
Glutamic oxaloacetic transaminase		Pyridoxal phosphate

which it occurs. As described in Chapter 3, all chemical reactions are at least theoretically reversible. In the following reaction, reactant *A* is converted to product *B* at a rate dependent on the rate constant k_1, whereas the reverse is dependent on k_{-1}. These two reactions proceed until a state of equilibrium exists, at which point

$$A \underset{k_{-1}}{\overset{k_1}{\rightleftharpoons}} B$$

there will be a characteristic mixture of *A* and *B*. If the concentration of *B* molecules increases, then a compensatory increase in the formation of *A* takes place so as to maintain the state of equilibrium. The ratio of *B* to *A* is dependent on k_1/k_{-1}. Therefore, the equilibrium state is defined by a constant, $K_{eq} = [B]/[A]$, or k_1/k_{-1}. Remember that $[A]$ and $[B]$ are the molar concentrations of reactant and product. An enzyme, like all catalysts, accelerates the velocities of both the forward and reverse reactions; it does not affect the equilibrium state (the ratio $[B]/[A]$, or K_{eq}).

When k_1 is very much larger than k_{-1}, the reverse reaction is exceedingly slow compared with the forward reaction. For all practical purposes, the reaction is irreversible and *A* is very unstable. Likewise, if k_1 is very much smaller than k_{-1}, then the forward reaction is exceedingly slow and never reaches equilibrium. Substance *A* would then be stable.

Normally the uncatalyzed hydrolysis of a protein, the forward reaction involving the hydrolysis of peptide bonds, would not be measurable in a reasonable amount of time. It would be practically impossible to meet our nutritional needs were it not for the presence of digestive enzymes (pepsin, trypsin, etc.). These catalysts accelerate the hydrolysis of peptide bonds by factors ranging in the millions.

The rate at which a chemical reaction proceeds in the forward direction depends on the amount of energy that must be provided before the reactant can yield products. In the conversion of $A \rightarrow B$, the reaction occurs with a loss in free energy (ΔG), as illustrated in Fig. 5-1. However, the figure also shows that the free-energy change must first pass through a transition state (A^*) that has a higher free-energy content than either *A* or *B*. In other words, for $A \rightarrow B$, there must first be an input of energy equivalent to ΔE_a. The rate of the reaction depends on the number of *A* molecules having sufficient energy to achieve A^*. This difference in the free energy, that is, ΔE_a (calories/mole), is called the *free energy of activation*. One way for the chemist to increase the proportion of *A* molecules having ΔE_a is to increase the temperature. Obviously, in the body it would not be possible to aid the digestion of protein or any other biological reaction by increasing the body temperature,

Figure 5-1
Equilibrium for the simple reaction $A \underset{k_{-1}}{\overset{k_1}{\rightleftharpoons}} B$ will favor the formation of *B* because it is at a lower energy level than that of initial state *A*. For this to happen, the reaction must first pass through an activated transition state A^*. If the energy difference $A^* - A$ (ΔE_a) is small, the reaction $A \rightarrow B$ will occur readily and k_1 will be large. However, if ΔE_a is large, then the rate will be slow and k_1 will be small. The enzyme *E* combines with A and lowers the ΔE_{aE} activation energy so that A will more readily change to B (k_1 will be large).

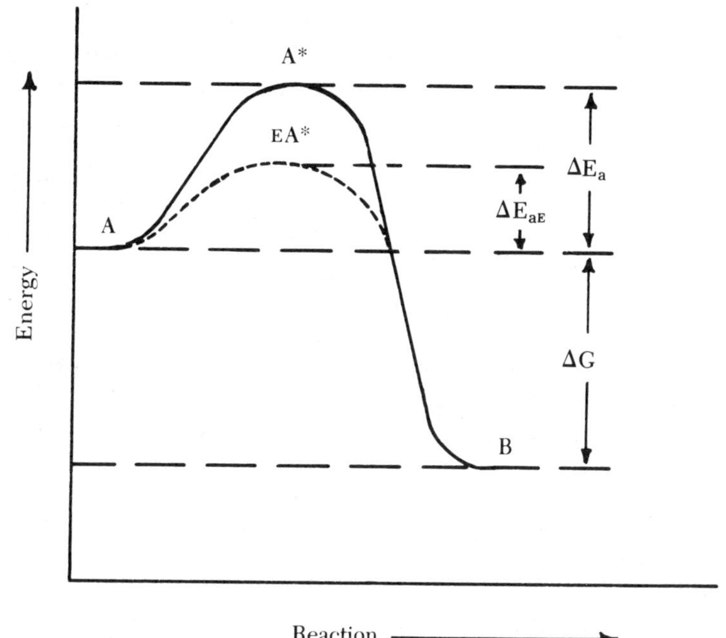

which normally remains at 37° C. Rather, the catalytic action of enzymes acts to reduce the activation energy (ΔE_{aE}), thereby lowering the energy barrier to the reaction. Table 5-2 illustrates the extent to which some enzymes lower the activation energy. Noteworthy is the greater reduction of activation energies by enzymes as compared with that of some other catalysts.

The activation energy in proteolysis is the amount of energy required to cleave the peptide bond and to overcome the forces of attraction in a molecule of water so that a hydroxide ion and a proton may be added to the carbonyl and amide component of the peptide. The final products are the free carboxyl and amino groups. It is this energy hurdle that is lowered by the enzyme so as to increase the rate of the reaction. When, in the presence of the enzyme, the ΔE_a is reduced to ΔE_{aE}, then many more molecules will have the necessary activation energy, and the rate of hydrolysis of the peptide bond will be enhanced enormously.

Table 5-2
Examples of activation energies of various catalyzed reactions

Reaction	Catalyst	Energy of activation (calories/mole)
Hydrogen peroxide decomposition	None	18,000
	Colloidal Pt	11,700
	Liver catalase	< 2000
Sucrose hydrolysis	H^+	26,000
	Yeast invertase	11,500
Casein hydrolysis	H^+	20,600
	Trypsin	12,000
	Chymotrypsin	12,000
Ethyl butyrate hydrolysis	H^+	16,800
	OH^-	10,200
	Lipase	4500

The fact that the enzyme lowers the free energy of activation suggests that the enzyme participates in the reaction in some tangible fashion. Enzymes, like all catalysts, enter into the reaction but remain unchanged. Consider the following general statement of an enzymatic reacton:

$$E + S \leftrightarrow ES \leftrightarrow E + P$$

Here the enzyme (E) enters into the overall reaction by combining with the substrate or reactant (S) to form ES. This intermediary complex then dissociates into the product (P) and restores the unchanged enzyme to the system. The enzyme is then free to repeat the process. Consequently, enzymes are effective in extremely small quantities.

Concept of active site

The formation of the intermediary complex is evident from substrate-velocity curves, which are typically rectangular hyperbolas (Fig. 5-11). As seen from Fig. 5-11, increasing the substrate concentration leads to a velocity beyond which further increases of substrate have no apparent effect. This indicates that the capacity of the enzyme to function can be saturated. The action of an enzyme can be saturated because it has only a finite number of sites at which catalysis can occur. These are called the *active sites*, specific regions on the protein molecules responsible for the catalytic activity. It is at the active site that the substrate molecule is bound.

Although the nature of this site varies with the enzyme and the kind of reaction catalyzed, it is a special sequence and unique grouping of amino acids. In principle it has two functional parts: the *binding site(s)* and the *catalytic site*. The substrate is bound at the binding site and thereby is oriented so that the susceptible bond can be attacked at the catalytic site. This brings about the formation of the product(s).

To identify the amino acid residues that are a functional part of the active site, a reagent or probe is used that reacts specifically with the responsible amino acid residues. Such a reagent is *diisopropylfluorophosphate* (DFP). This extremely poisonous compound, developed during World War II as a nerve gas, reacts with the hydroxyl group of serine. Therefore, it inactivates many enzymes that are dependent on a particular serine for their activity. Examples of such serine-dependent proteases are trypsin and chymotrypsin; others are provided in Table 5-3. In chymotrypsin, only one of 28 serine residues reacts with DFP thereby yielding an inactive enzyme. This is serine 195, which is located at the active site. The nature of the reaction of DFP with the serine residue of a polypeptide chain is illustrated below:

```
             CH3   CH3                                   CH3   CH3
               \   /                                       \   /
                CH                                          CH
                |                                           |
    CH3         O                                CH3        O
       \        |                                   \       |
        CH—O—P=O  +  Seryl     →                   CH—O—P=O  +  HF
       /        |      residue                      /       |
    CH3         F                                CH3        O
                                                            |
                                                            CH2   O
                                                            |     ||
                                                       —N—CH—C—
                                                        |
                                                        H
```

Table 5-3
Sequences of serine-containing active sites

Enzyme	Amino acid sequence
Proteolytic	
Bovine trypsin	··· AsnSerCysGlnGlyAsp**Ser**GlyGlyProValVal ···
Bovine chymotrypsin	··· SerSerCysMetGlyAsp**Ser**GlyGlyProLeuVal ···
Porcine elastase	··· SerGlyCysGlnGlyAsp**Ser**GlyGlyProLeuHis ···
Porcine thrombin	··· GluGlyAsp**Ser**GlyGly ···
Acetylcholine esterase	··· Glu**Ser**Ala ···
Subtilisin (Carlsberg)	··· AsnGlyThr**Ser**MetAla ···
Phosphorylating enzymes	
Alkaline phosphatase	Asp**Ser**Ala
Phosphoglucomutase (*E. coli*)	ThrAla**Ser**HisAsp
Phosphorylase a (rabbit)	LysGlnIle**Ser**ValArg

From Dayhoff, M.O., and Eck, R.V.: Atlas of protein sequence and structure, Washington, D.C., 1967-1968, National Biomedical Research Foundation.

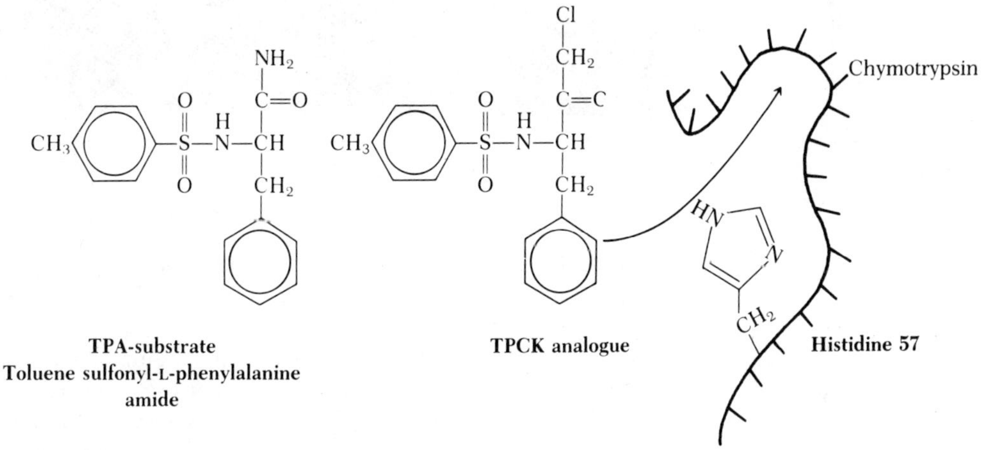

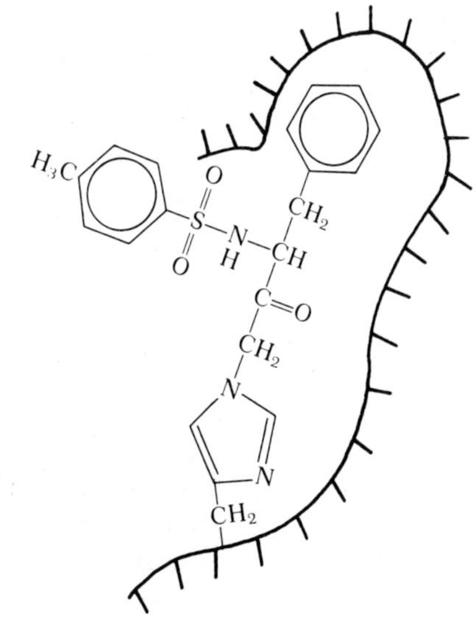

Figure 5-2
Reaction of toluene sulfonyl-L-phenylalanine chloromethyl ketone (TPCK) with the active site of chymotrypsin. TPCK is an analogue of the synthetic substrate toluene sulfonyl-L-phenylalanine amide (TPA).

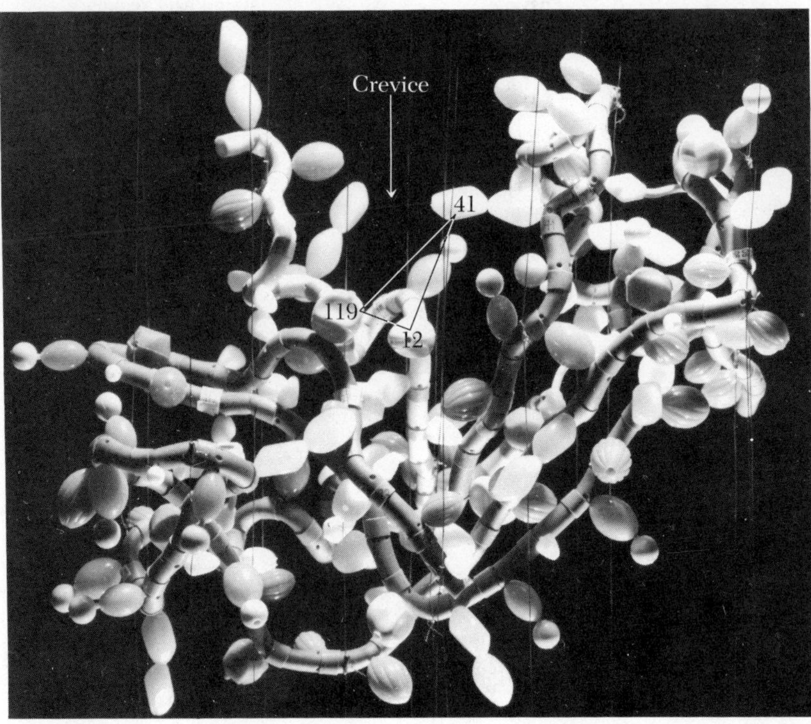

Figure 5-3
Model of ribonuclease showing three-dimensional conformation. Residues are labeled to show proximity of histidines 12 and 119 and lysine 41. Also to be noted is the crevice formed by the β-pleated sheet structure.

The amino acid sequences about the functional residue have been determined for numerous serine-dependent enzymes. As seen in Table 5-3, many share a common aspartyl-seryl-glycyl sequence at the active site.

Another method of identifying functional amino acid residues is called *affinity labeling* and uses a group-specific reagent that is also an analogue of a substrate. This means that the analogue binds in precisely the same manner as the substrate but then reacts irreversibly with an amino acid side chain that is functional in catalysis. An example of such a compound is toluene sulfonyl-L-phenylalanine chloromethyl ketone, or TPCK. This is an analogue of toluene sulfonyl-L-phenylalanine amide, or TPA (Fig. 5-2), a simple, synthetic substrate for chymotrypsin. The enzyme hydrolyzes the amide group of TPA. The phenylalanine side chain is held in an hydrophobic pocket so as to orient the reactive chloromethyl group in the same position that the amide would have been in had the synthetic substrate been used (Fig. 5-2). The chloromethyl group then reacts with histidine 57; the resulting derivative of chymotrypsin is enzymatically inactive. Therefore it is concluded that the imidazole of histidine 57 is essential to the action of this enzyme.

In the primary sequence of chymotrypsin, histidine 57 is far from serine 195, as is another group needed for activity, the aspartyl residue 102. To bring these residues into a unique orientation requires the folding of the polypeptide into the specific three-dimensional shape of the active molecule. Fig. 5-3 illustrates the importance of three-dimensional shape to the enzymatic activity of ribonuclease. This protein is a single chain of 124 amino acid residues of which histidine 12 and 119 and lysine 41 are required for catalytic activity. As seen in the figure, the three residues are brought together in space by the conformation of the protein. Four disulfide bonds stabilize the enzymatically active molecule. Fig. 5-3 illustrates the importance of

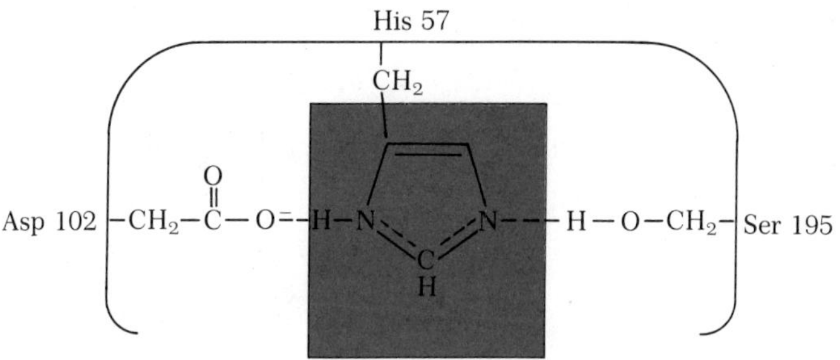

Figure 5-4
Diagram illustrating a possible arrangement of the catalytically important amino acids required for the action of chymotrypsin. This shows the functional role of histidine as a proton donor and acceptor and also the formation of a nucleophilic serine residue.

the protein's topology, for the active site is located in a crevice formed by the folding of the β-pleated sheets that make up much of the ribonuclease. The polynucleotide substrate (RNA) fits into this crevice.

Mechanism of action

It should now be clear that the catalytic action of an enzyme requires that the protein interact with its substrate at a special, active site having a unique orientation of amino acid residues. But what is the reaction pathway that takes place in forming the intermediate enzyme-substrate complex and the ultimate products? What is the sequence of steps that lowers the activation energy and thereby accelerates the reaction? In practically all enzyme-catalyzed reactions, the pathway somehow involves the addition and removal of protons from the substrate. *Acid-base* catalysis may be a general mechanism in enzymatic action.

Acid-base catalysis may be illustrated using chymotrypsin as an example. Fig. 5-4 shows a possible arrangement of aspartic acid 102, histidine 57, and serine 195. Adjacent to this active grouping of amino acids is a hydrophobic pocket into which is tucked a bulky aromatic side chain of the substrate polypeptide. In this way a susceptible peptide bond is properly oriented with respect to the catalytic site. The sequence of events leading to hydrolysis is illustrated in Fig. 5-5. In Step I aspartic acid is thought to draw away a proton from serine 195, using histidine 57 as the intermediate. This creates a negative charge on the serine, making it strongly nucleophilic. In Step II the susceptible peptide interacts with the active site, the peptide nitrogen is protonated by the imidazole of histidine 57. Thus the histidine acts as an acid or proton-donating group. The nucleophile then attacks the carbonyl of the susceptible peptide, resulting in the cleavage of the substrate peptide bond and the formation of an acylated enzyme molecule. In Steps II and III the "leaving" group, $R'NH_2$, is replaced by a molecule of water. Finally in Step III the water attacks the carbonyl of the acyl group assisted by histidine 57, which now acts as a proton acceptor or base. This step results in a release of the second product (RCOOH) and regenerates the original active grouping of amino acids as in Step IV. It is noteworthy that a covalent intermediate or acylated enzyme is formed in the reaction pathway. This shows that the reaction is more complicated than is suggested by the general statement:

$$E + S \rightleftharpoons ES \rightleftharpoons E + P$$

Rather it might be more accurate to express the process as:

$$E + S \rightleftharpoons ES \rightleftharpoons \underset{\displaystyle P_1}{\overset{}{EP_2 \atop +}} \rightleftharpoons E + P_2$$

P_1 and P_2 are the two products produced in the hydrolysis of the peptide bond.

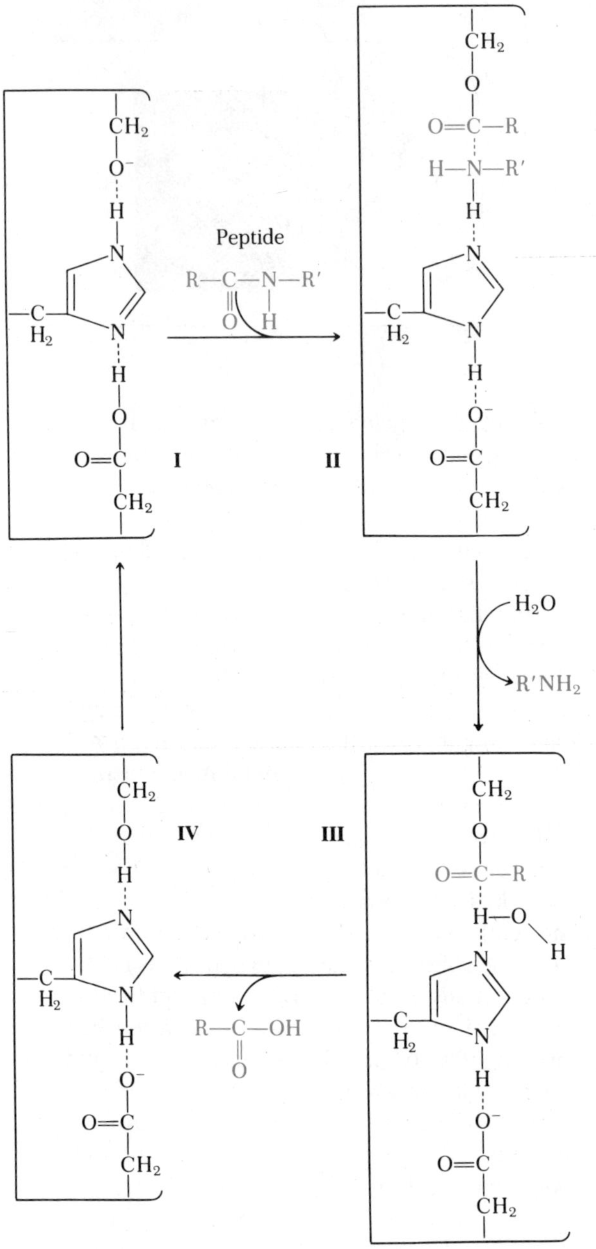

Figure 5-5
Stepwise pathway illustrating the hydrolysis of peptides as catalyzed by chymotrypsin. This sequence shows the formation of an intermediate covalent acylated molecule *(step III)*.

There is evidence that the triad of Step I forms only when a bulky group is bound in the hydrophobic pocket. Therefore, the actual formation of the active site and the change of serine 195 to a highly nucleophilic group is contingent on binding the substrate.

Another valuable example of an enzymatic reaction pathway is pancreatic carboxypeptidase A. As previously mentioned, this enzyme acts specifically on the peptide bond that binds the C-terminal amino acid residue to a polypeptide chain. Carboxypeptidase has a total of 307 amino acids and is a compact ellipsoid molecule

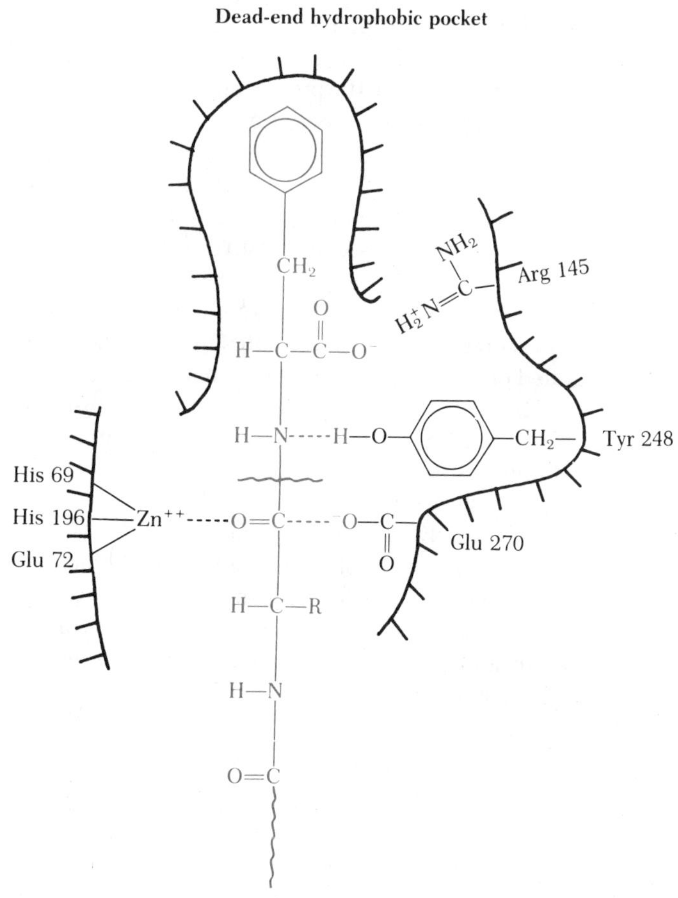
Dead-end hydrophobic pocket
CH2
O
H—C—C—O⁻
NH2
H2+N=C—
Arg 145
H—N-----H—O—
—CH2—
Tyr 248
His 69
His 196
Glu 72
Zn++-----O=C----- ⁻O—C—
O
Glu 270
H—C—R
H—N
O=C

Figure 5-6
Diagram showing the binding of the C-terminal end of a polypeptide substrate to the active site of carboxypeptidase.

having a deep cleft in its structure. Each molecule has a single atom of zinc bound at the bottom of the cleft. As seen from the schematic diagram (Fig. 5-6), the first step in the catalytic pathway involves an electrostatic attraction between the carboxylate anion of the substrate and the cationic guanidinium group of arginine 145 in the enzyme. To form this ionic bond requires the displacement of the arginine by about 0.2 nm. The side chain of the substrate's terminal amino acid must be tucked into a pocket of the carboxypeptidase. The zinc ion (2^+) is bound to the enzyme by histidine 69, histidine 196, and glutamic acid 72. It interacts with the carbonyl oxygen of the substrate's susceptible peptide bond. The effect of the zinc is to make the carbonyl group more susceptible to nucleophilic attack. In this process the zinc must move about 0.1 nm. The phenolic hydroxyl of tyrosine 248 then moves approximately 1.2 nm to form a hydrogen bond with the amide nitrogen. Glutamic acid 270 also moves a distance of about 0.2 nm. All of these changes in the shape of the protein molecule are required to bring the groups into an active catalytic team. Once the groups are in their proper proximity, it is thought that tyrosine 248 provides the proton to the leaving group, the C-terminal amino acid (Fig. 5-6). One interpretation of these events is that the carboxylate of glutamic acid 270 acts as a general base, removing a proton from a water molecule as it attacks the carbonyl group of the susceptible peptide bond.

Induced-fit concept

Proteins and hence enzymes are not rigid molecules whose shape is unalterable. As described in the preceding section, enzyme molecules such as carboxypeptidase are changed even when they are binding their substrates. Only the appropriate substrate can cause the precise alignment of side chains essential for catalysis. This is the *induced-fit* concept of enzyme action.

Enzyme activation

Another phenomenon that is dependent on the rearrangement of enzyme molecules is the process of *activation*. Many of the proteolytic enzymes that function in the gastric and intestinal digestive process are first produced in the cells as inactive precursors called *zymogens*. To avoid self destruction the protease-producing cells synthesize these proteins in an inactive form. Proteases therefore are synthesized and secreted into the gastrointestinal tract in the form of zymogens. They are activated only on entering the intestinal lumen. Activation involves limited proteolysis, which allows the realignment of amino acid side chains to form the active site.

Again, chymotrypsin is a convenient example. This protease is synthesized and secreted as its zymogen, called chymotrypsinogen (suffix *-ogen* denotes *zymogen*). Fig. 5-7 shows the location of five disulfide bonds in this protein: between residues 1 and 122, 42 and 58, 136 and 201, 168 and 182, and 191 and 220. Activation involves the hydrolysis by trypsin of a single peptide bond between arginine 15 and isoleucine 16. This yields the most active species, known as π-chymotrypsin. A new N-terminal is created, with isoleucine the terminal amino acid. The free α-ammonium group is attracted to the carboxylate anion of aspartate 194. A number of changes in conformation are triggered that ultimately bring serine 195, histidine 57, and aspartate 102 into proper proximity. Also formed is the hydrophobic pocket for binding the substrate. Further hydrolysis by chymotrypsin releases the dipeptide serylarginine to form δ-chymotrypsin; finally, hydrolysis of bonds 146-147 and 148-149 releases threonylasparagine to produce α-chymotrypsin. Therefore α-chy-

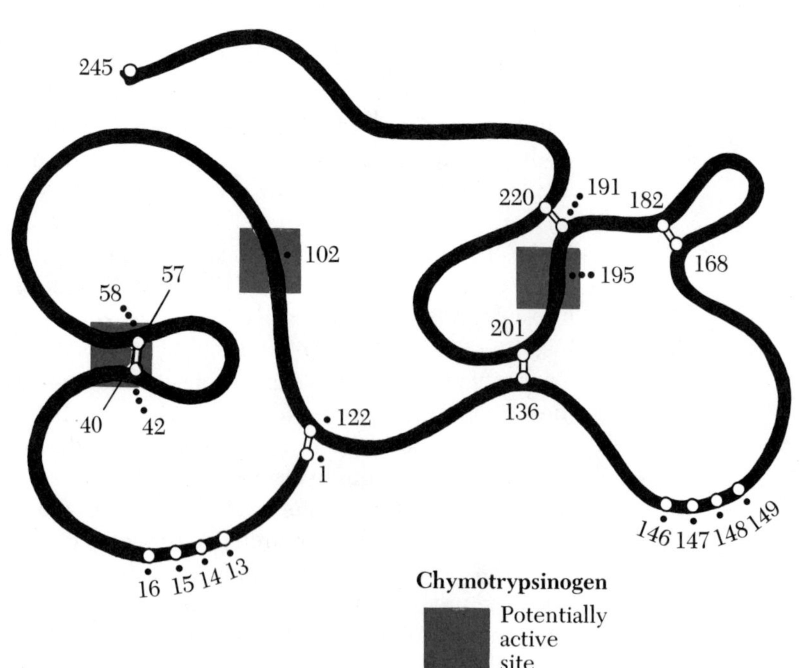

Figure 5-7
Model of structure of chymotrypsinogen. Activation to chymotrypsin minimally requires hydrolysis between residues 15 (Arg) and 16 (Ile). Subsequent conformational changes bring residues 195 (Ser), 102 (Asp), and 57 (His) into proximity. Other residues labeled are 14 (Ser), 13 (Leu), 1 (Cys), 40 (His), 42 and 58 (Cys), 146 (Tyr), 147 (Thr), 148 (Asn), 149 (Ala), 191 (Cys), and 245 (Asn).

motrypsin consists of three chains, the A-chain (residues 1-13), the B-chain (16-146), and the C-chain (149-245). All are joined by disulfide bonds (Fig. 5-7).

Activation of trypsinogen, having 229 residues, is accomplished by removal of a N-terminal hexapeptide, Val-Asp-Asp-Asp-Asp-Lys, by enterokinase (Chapter 13) as well as autocatalytically by trypsin itself. Enterokinase hydrolyzes the peptide bond between lysine 6 and isoleucine 7, thereby releasing the negatively charged hexapeptide. The trypsin molecule then undergoes a change in conformation in which the serine 183 and histidine 29 are brought into proximity.

Pepsinogen, the zymogen of pepsin, has 362 amino acids and is activated by the hydrogen ion concentration of the gastric juice and autocatalytically by pepsin. This involves the loss of 42 amino acids from the N-terminal portion of the zymogen.

Enzyme specificity

Enzymes are unique as catalysts partly because of their specificity of action. Some act on a particular kind of covalent bond in closely related substrates and therefore exhibit a *relative* specificity; others appear to act only on one substrate, exhibiting *absolute* specificity. Proteolytic enzymes hydrolyze the peptide bonds of polypeptides and therefore exhibit relative specificity. Likewise, glycosidases act on glycosidic bonds of carbohydrates, and esterases such as lipases split the ester bonds of lipids. Examples of absolute specificity are more restricted: urease, which acts only on urea; carbonic anhydrase, which acts only on carbonic acid; and fumarase, which acts only on fumaric acid. In addition, many enzymes exhibit stereospecificity. Arginase acts only on L-arginine, not on the D-isomer. Glucose oxidase converts only the β-D-anomer of glucose to gluconic acid, not the α-D-anomer.

Examples of specificity in proteolytic enzymes are shown on Fig. 5-8 for pepsin, chymotrypsin, trypsin, and carboxypeptidases A and B. All hydrolyze peptide bonds, yet each has certain preferences regarding the participating amino acid(s).

Although pepsin is not highly specific, it prefers those peptide bonds between hydrophobic amino acids. Since the bonds attacked are located in the interior of the protein substrate, pepsin is called an *endopeptidase*. Trypsin likewise is an endopeptidase but hydrolyzes bonds in which the carbonyl group is contributed by the basic amino acids lysine or arginine. Chymotrypsin, also an endopeptidase, preferentially splits peptide bonds in which the carbonyl group is from an aromatic amino acid.

Figure 5-8
Examples of specificity of some proteolytic enzymes. These five enzymes are important in gastric and intestinal digestion of proteins.

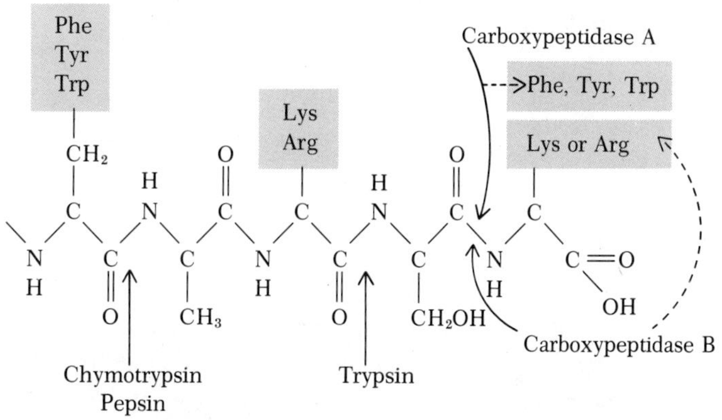

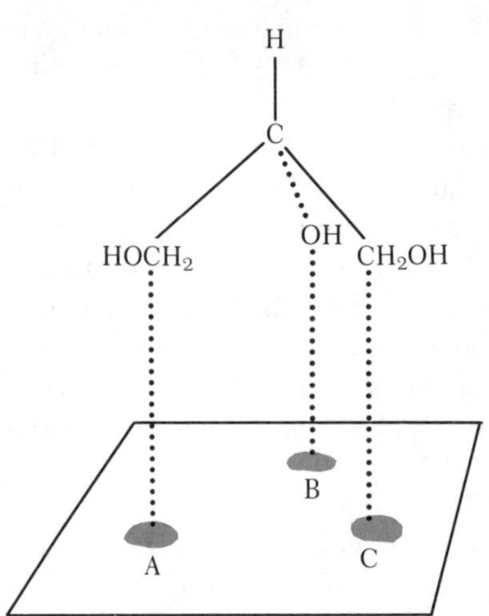

Figure 5-9
Model showing three-point attachment of substrate glycerol to the enzyme "surface." There is only one way in which the glycerol can bind in precisely this manner. Therefore only one of the CH_2OH groups is phosphorylated by glycerokinase. *A* is the active phosphorylating site; *B* binds only the $-OH$ group; *C* binds one $-CH_2-OH$ group.

The carboxypeptidases, on the other hand, are *exopeptidases* and therefore act on peptide bonds that bind the C-terminal amino acids to the polypeptide chains. Carboxypeptidase A acts preferentially but not exclusively on peptide bonds adjacent to C-terminal aromatic residues and continues to remove C-terminal amino acids until it reaches terminal lysine or arginine residues or until the terminal residue is adjacent to a peptide bond in which proline contributes the carbonyl group. Carboxypeptidase B (porcine) acts exclusively on bonds adjacent to C-terminal arginine or lysine.

At one time specificity was explained by analogy to the opening of a lock by its key (lock and key mechanism). Only the proper substrate can fit into a complementary enzyme surface as a key fits into its lock. Now we can appreciate that specificity depends on the shape of the protein molecule as well as on the binding of the substrate to the enzyme's binding sites. As we have already seen, it is also essential for the susceptible bond to be in a proper proximity to the active amino acid residues of the catalytic site.

Another form of specificity that is explained by the linking of the substrate to multiple binding sites is *optical specificity*, or *stereospecificity*. An example of this type is the enzyme glycerokinase, which acts on the symmetrical molecule glycerol to form only L-glycerol 3-phosphate

$$\begin{array}{c} CH_2OH \\ | \\ HO{-}C{-}H \\ | \\ CH_2OH \end{array} \xrightarrow{ATP \quad ADP} \begin{array}{c} CH_2OH \\ | \\ HO{-}C{-}H \quad O^- \\ | \qquad\quad | \\ CH_2O{-}P{=}O \\ | \\ O^- \end{array}$$

Glycerol **L-Glycerol 3-phosphate**

Glycerol does not possess an asymmetric carbon atom, therefore it is not readily evident how the enzyme distinguishes one hydroxymethyl group from the other. If the enzyme were not perceptive of these two groups, an equal mixture of D- and L-isomers would result. However, if there are three specific points of attachment, then the glycerol will bind to the enzyme only in one way (Fig. 5-9). Three-point

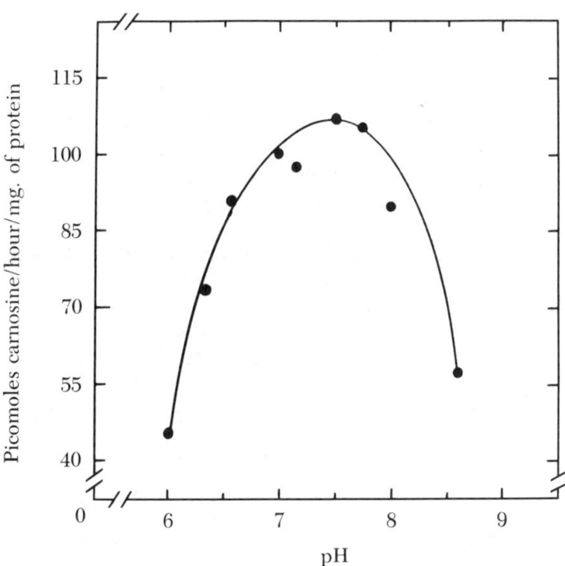

Figure 5-10
Effect of pH on enzyme activity. Example shown is of action of homocarnosine-carnosine synthetase of rat brain. The ordinate (velocity) is the picomoles of carnosine produced per hour per milligram of protein.

Courtesy Skaper, S.D., Das, S., and Marshall, F.D.: J. Neurochem. **21:**1429, 1973.

attachment explains how only one of the two hydroxymethyl groups can be positioned on the active site and be phosphorylated.

Factors influencing enzyme action

Since the enzyme protein is usually present in biological systems in amounts too small to measure quantitatively, their activity is determined instead. A *unit of activity* (International unit) is defined as the amount of enzyme that will produce 1 micromole of product per minute at 25° C under standardized conditions. Frequently one speaks of the *specific activity* of the enzyme, meaning the number of enzyme units per mg of protein. Another mode of expressing enzyme activity is the *turnover number,* which is the number of moles of substrate transformed (or moles of product formed) per mole of enzyme per unit of time. To make this mode of expression possible, one needs to know the molecular weight of the enzyme. Turnover numbers may vary from 1150 (per minute) for succinate dehydrogenase to 36,000,000 for carbonic anhydrase.

Factors that influence the velocity of enzyme-catalyzed reactions are:

1. Hydrogen ion concentration
2. Temperature
3. Substrate concentration
4. Enzyme concentration
5. Time
6. Products of the reaction

To determine the effects of these factors on enzyme reactions, one usually studies the velocity of the reaction under different conditions. It is possible to measure either the accumulation of product or the disappearance of substrate per unit of time. In any event, the activity of the enzyme is expressed in terms of velocity (v).

Hydrogen ion concentration

Enzyme reactions are influenced by varying the hydrogen ion concentration (Fig. 5-10). The *optimum pH* is that at which a certain enzyme causes a reaction to progress most rapidly. On either side of the optimum, the rate of reaction is lower,

Table 5-4
Michaelis-Menten constants of some enzymes and their optimum pHs*

Enzyme	Source	Substrate	K_m (molar)	Optimum pH
Sucrase	Intestine	Sucrose	0.016-0.04	6.2
Ribonuclease	Pancreas	(3′-5′)-Cytidylyladenine	1.4×10^{-3}	7.0
α-Glucosidase	Yeast	Methyl-α-D-glucoside	0.037-0.075	5.4
β-Glucosidase	Almond	O-Nitrophenyl β-glucoside	6×10^{-3}	4.5
Acetylcholinesterase	Erythrocytes	Acetylcholine	2.68×10^{-4}	7.5
Enolase	Rabbit muscle	2-Phospho-D-glycerate	7×10^{-5}	6.8
Arginase	Beef liver	L-Arginine	1.16×10^{-2}	8.4-9.7
Pepsin	Gastric mucosa	Acetyl-L-phenylalanyl-L-phenylalanine	4.3×10^{-4}	1.5-2.5
Trypsin	Pancreas	Benzoyl-L-arginine ethylester	5×10^{-5}	7.8-8.0
Urease	Soybean	Urea	0.025	6.4-6.9
Alkaline phosphatase	Bone	*p*-Nitrophenyl phosphate	2.5×10^{-4}	9.5
Catalase	Liver	Hydrogen peroxide	1.17	7.0-7.6
Xanthine oxidase	Milk	Xanthine	1.7×10^{-6}	7.8
Succinic dehydrogenase	Beef heart	Succinate	1.3×10^{-3}	7.6

*These data have been obtained from a number of sources. Most of the enzymes are not pure preparations.

and at certain pHs an enzyme may be inactivated or even destroyed. Although ideally the pH activity curve is bell-shaped, not all enzymes exhibit the ideal. As shown in Table 5-4, the optimum pH for pepsin, the enzyme of the gastric juice, is in the region of 1.5 to 2.5. This means that the pH curve is skewed well to the low range of pH. Cholinesterase yields a curve with reduced activity at pHs below 7 and reaching a maximum that is not affected by increasing the pH to levels higher than 7.

The bell-shaped pH-versus-activity curve may represent an effect on the ionization of groups both in the enzyme and the substrate. The pH would conceivably affect the binding of enzyme to substrate and also the functioning of the active site. Therefore in enzyme studies buffers are used to keep the enzyme at an optimum or at least a favorable hydrogen ion concentration. The optimum pH is dependent on various conditions; the kind of buffer, the particular substrate, and the source of the enzyme may all have an influence. Moreover, Table 5-4 shows the wide range of optimum hydrogen ion concentrations for various enzymes.

Influence of temperature

Enzyme-catalyzed reactions are accelerated by increasing the temperature until an optimum value is reached; thereafter the reaction velocity declines. An *optimum temperature* is usually reached at 40° to 50° C for animal enzymes, whereas for plant enzymes it is higher, usually 50° to 60° C. Above this, the rate decreases, because the enzyme is denatured at a rate faster than the increase in the reaction. Most enzymes are denatured above 60° C. Obviously the time of exposure is a factor also; an enzyme may withstand higher temperatures for short periods of time.

Substrate concentration

For a given quantity of enzyme, the velocity of the reaction increases as the concentration of the substrate is increased. At first, this relationship is almost linear; but later the reaction curve assumes a hyperbolic shape (Fig. 5-11). This means that the reaction kinetics tend to be essentially biphasic. The reaction in the initial, nearly linear phase of the curve (I) follows *first-order kinetics*, that is to say

Figure 5-11 Effect of substrate concentration on enzyme activity. Typical relationship obtained with increasing substrate concentration at constant enzyme concentration. The plot is a rectangular hyperbola.

Figure 5-12 Effect of enzyme concentration on enzyme activity. Amount of enzyme was increased. Substrate level was kept constant.

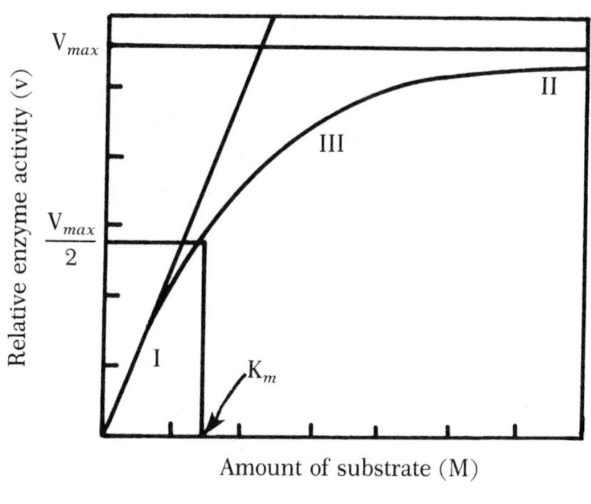

Figure 5-11

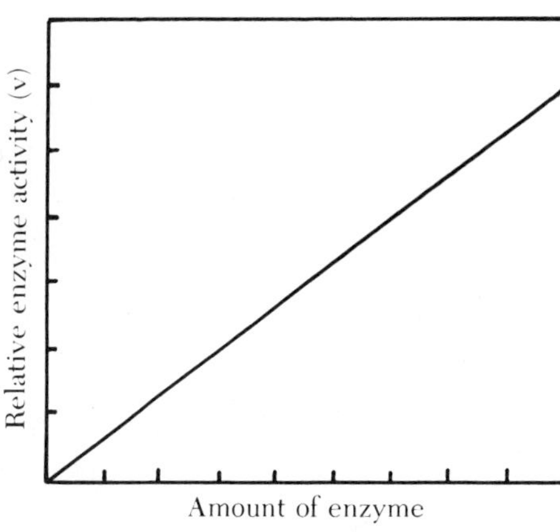

Figure 5-12

$v = k\,[E]\,[S]$ or for a given amount of enzyme the velocity is directly proportional to the initial substrate concentration, or $[S]$. In the second phase (II) a plateau is reached in which the velocity is constant, and the reaction is said to exhibit *zero-order kinetics*. This occurs at substrate concentrations such that further increases no longer affect the apparent velocity. Here all of the enzyme molecules are saturated with substrate and therefore exist only as the enzyme-substrate (*ES*) complex. Since the velocity-substrate curve is hyperbolic, the zero-order portion approaches a limiting, or maximum, velocity, usually termed V_{max}. At V_{max}, the velocity is constant for any given enzyme concentration ($v = k\,[E]$) and independent of the substrate concentration. Strictly speaking V_{max} is reached only as S approaches infinity. V_{max} is an important characteristic of a given enzyme-substrate system and is dependent only on the concentration of active enzyme. The region between these two extremes (III) is considered to be a mix of both first- and zero-order kinetics.

Enzyme concentration

In addition to the effect of substrate concentration, enzyme-catalyzed reactions are also dependent on the amount of enzyme. Thus, the initial velocity of an enzyme reaction is directly proportional to the concentration of enzyme (provided that the substrate is present in excess), and the reaction exhibits zero-order kinetics (Fig. 5-12).

Effect of time and the accumulation of products

In the presence of an excess of substrate the amount of product produced with time is initially linear, then it declines and eventually reaches a plateau value. The decrease is partly the result of a simultaneous reduction in the amount of substrate and the accumulation of product. The latter by mass action increases the rate of the reverse reaction. Furthermore, some products may actually have a deleterious effect by denaturing the enzyme. In biological systems, however, the product is usually removed, for example it becomes a substrate for a succeeding enzyme in a metabolic pathway.

Kinetics of enzyme reactions

The simplest expression of an enzyme-catalyzed reaction involving a single substrate, S, and requiring the formation of an intermediate *ES* complex is:

$$E + S \underset{k_{-1}}{\overset{k_1}{\rightleftharpoons}} ES \underset{k_{-2}}{\overset{k_2}{\rightleftharpoons}} E + P$$

The combination of *E* and S to yield *ES*, a rapid and reversible process, is followed by the conversion of the substrate into the product and by the regeneration of the enzyme. As seen in the above equation, k_1, k_{-1}, k_2 and k_{-2} are the rate constants of the individual steps for the overall process. The amount of *ES* complex depends on the amount of free enzyme and substrates; that is, when *ES* is converted to product it is replaced by the formation of new *ES*. This is called the steady state and is expressed as $d[ES]/dt = 0$.

Michaelis-Menten equation

If all measurements of the velocity are made very soon after the enzyme and substrate have been mixed, (i.e., initial velocities), then only very small amounts of product are formed. Also it is assumed that the concentration of substrate is much larger than the enzyme concentration. Under these conditions the extent of the conversion of product to *ES* is negligible and k_{-2} may be dismissed. This assumption greatly simplifies the derivation of an equation that describes the kinetics of the reaction (by Michaelis and Menten in 1913). To illustrate the derivation of this equation for a rectangular hyperbola consider again the simplified enzyme-catalyzed reaction:

$$E + S \underset{k_{-1}}{\overset{k_1}{\rightleftharpoons}} ES \xrightarrow{k_2} E + P$$

The velocity of *ES*-complex formation is:

$$v_1 = k_1[E][S]$$

where [*E*] and [S] are the molar concentrations of enzyme and substrate, respectively. The velocity of *ES* breakdown is:

$$v_{-1} = k_{-1}[ES] \text{ for the regeneration of } E \text{ and S and}$$
$$v_2 = k_2[ES] \text{ for the formation of } P \text{ and } E$$

Under steady-state conditions the velocity of *ES* complex formation must equal the sum of the velocities of breakdown.

$$v_1 = v_{-1} + v_2$$

then

$$k_1[E][S] = k_{-1}[ES] + k_2[ES]$$

$$\frac{[E][S]}{[ES]} = \frac{k_{-1} + k_2}{k_1} = K_m \text{ (Michaelis-Menten constant)} \quad \textbf{(1)}$$

The ratio of rate constants is called the Michaelis constant and is represented by K_m.

It is seldom possible to measure the concentration of free enzyme (*E*) or of the enzyme-substrate complex [*ES*]. Therefore, *E* and *ES* must be eliminated from the equation. Recognizing that $[E_f]$ is the concentration of free enzyme whereas the total enzyme $[E_t] = [E_f] + [ES]$ or $[E_f] = [E_t] - [ES]$, then:

$$K_m = \frac{([E_t] - [ES])[S]}{[ES]}$$

or

$$K_m[ES] = [E_t][S] - [S][ES]$$
$$K_m[ES] + [S][ES] = [E_t][S]$$
$$[ES](K_m + [S]) = [E_t][S]$$
$$[ES] = \frac{[E_t][S]}{K_m + [S]}$$

The velocity of the forward reaction, v_2, is proportional to $[ES]$:

$$v_2 = k_2[ES] \quad (2)$$

Then substituting v_2/k_2 for $[ES]$ in equation 2:

$$v_2 = \frac{k_2[E_t][S]}{K_m + [S]} \quad (3)$$

When the system is saturated with substrate, all the enzyme, $[E_t]$, will exist as $[ES]$ and the velocity will be maximal, V_{max}. Thus:

$$V_{max} = k_2[E_t]$$

Substituting in equation 3:

$$v = \frac{V_{max}[S]}{K_m + [S]} \quad (4)$$

Equation 4 is the usual form of the Michaelis-Menten equation.

If, in equation 4, the velocity (v) is set to $V_{max}/2$, then it will be apparent that K_m may be expressed in terms of the substrate concentration. Or K_m may be defined as the substrate concentration in moles per liter when the initial velocity is one half of V_{max}.

$$\frac{V_{max}}{2} = \frac{V_{max}[S]}{K_m + [S]}$$

$$[S] = K_m$$

This is true for an enzyme-catalyzed reaction involving a single substrate. As indicated in Table 5-4, the K_m is a characteristic of enzymes and is independent of the enzyme concentration itself. The relation of K_m to $V_{max}/2$ is apparent from Fig. 5-11.

Since $v = V_{max}$ only when S is infinite, it is not possible to determine the value of V_{max} precisely from a hyperbolic curve. To obtain a value for K_m, it is preferable to transpose the Michaelis-Menten relationship into the equation for a straight line. Lineweaver and Burk made this transformation by taking the reciprocal form of equation 4, as shown below.

$$\frac{1}{v} = \frac{K_m + [S]}{V_{max}[S]}$$

$$\frac{1}{v} = \frac{K_m}{V_{max}} \cdot \frac{1}{[S]} + \frac{1}{V_{max}}$$

Since this fits the equation for a straight line ($y = mx + b$), a linear graph is obtained when $1/v$ is plotted against $1/[S]$ (Fig. 5-13, *A*). The straight line must have a slope of K_m/V_{max} and a y-intercept of $1/V_{max}$. If the line is extrapolated until it crosses the x axis, then the intercept equals $-1/K_m$. For an alternate linear plot, equation 4 may be rearranged such that $v = V_{max} - K_m v/[S]$. This plot of v versus $v/[S]$ is called the Eadie-Hofstee plot. As seen in Fig. 5-13, *B*, the intercept with the y axis is V_{max}, that of the x axis is V_{max}/K_m, and the slope is $-K_m$.

The Lineweaver-Burk plot has the disadvantage of compressing the data at the higher substrate concentrations into a small region and unduly emphasizing those of the lower concentrations. The Eadie-Hofstee plot allows the data to be uniformly distributed and therefore is considered to be the more accurate method of expressing the data.

The Michaelis-Menten equation is of great value in the study of enzymes, and the constants of many enzymes have been determined (Table 5-4). The range of values is from 1×10^{-8} to 1M or higher.

Figure 5-13 Two examples of linear evaluation of kinetic data for an enzyme-catalyzed reaction.
A The Lineweaver-Burk plot.
B The Eadie-Hofstee plot.

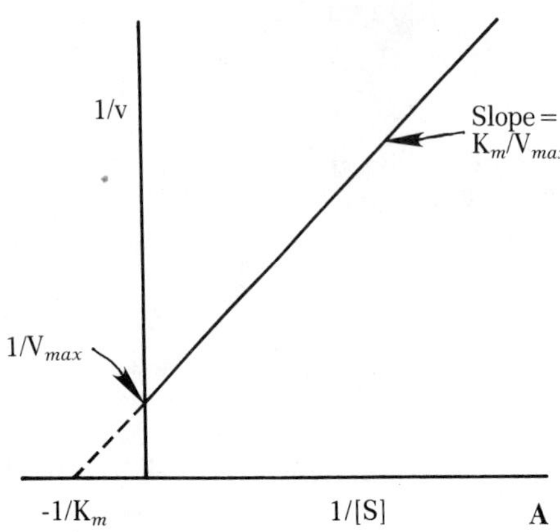

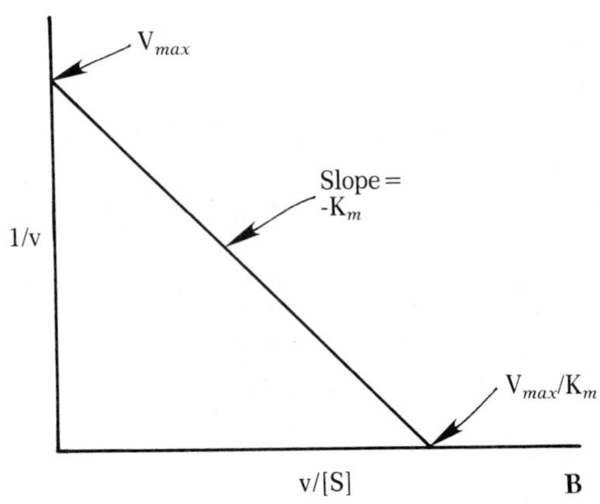

A knowledge of the K_m permits an approximation regarding the order of the enzymatic reaction at a given substrate concentration. For example, if $[S] > K_m$ by a factor of 100, then the equation

$$v = \frac{V_{max}[S]}{K_m + [S]}$$

will simplify to

$$v = V_{max} = k_2[E_t]$$

Thus V_{max} is a constant and is independent of the substrate concentration. This is zero-order kinetics and is represented by the plateau of the curve (Fig. 5-11). If $[S] < K_m$ by a factor of 100, then the equation

$$v = \frac{V_{max}[S]}{K_m + [S]}$$

will become

$$\frac{V_{max}}{K_m} \cdot [S]$$

or

$$\frac{k_2 E_t}{K_m} \cdot [S]$$

Thus, the velocity of the reaction is directly proportional to the substrate concentration. The reaction exhibits first-order kinetics.

The kinetic considerations made so far are for a single-substrate system. Actually most enzymatic reactions involve more than one substrate. They are of the type:

$$A + B \xrightarrow{E} P + Q$$

Kinetics of multisubstrate reactions

Bisubstrate reactions usually obey Michaelis-Menten kinetics when one substrate saturates the enzyme and the other is varied. Thus the initial velocities may be a function of substrate *A* when *B* saturates the enzyme. The same is true using *B*

Figure 5-14

A Lineweaver-Burk plot of a bisubstrate reaction and exhibiting a ternary complex–forming mechanism.
B Same plot, but illustrating a reaction occurring by means of ping-pong mechanism.

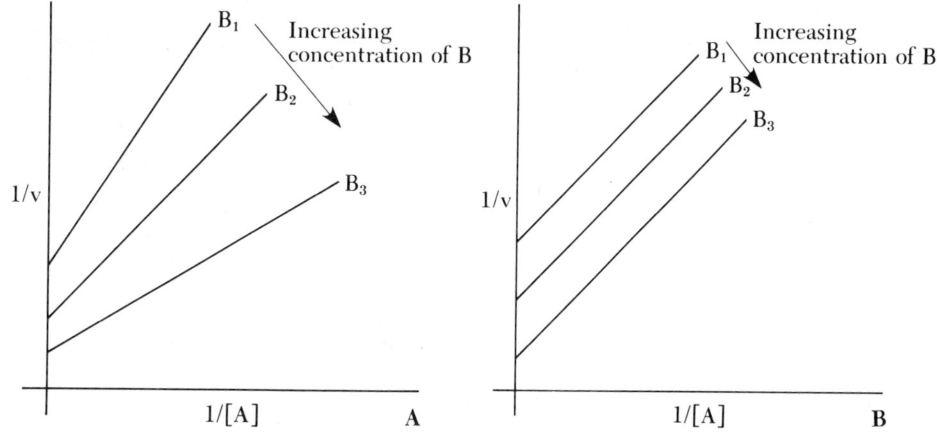

as a variable and saturating amounts of *A*. Therefore a characteristic K_m value can be obtained for each substrate.

Basically there are two kinds of enzymatic multisubstrate reactions. One involves the addition of all the substrates to the enzyme before the first product is formed and released. This is called a single displacement, or sequential, mechanism. In the second kind, termed the double displacement, or ping pong, mechanism, one or more products are released before all of the substrates are added.

Single displacement or sequential mechanism

The steps involved in this process are illustrated below.

$$\begin{array}{ccccccccc} & A & & B & & P & & Q & \\ & \downarrow & & \downarrow & & \uparrow & & \uparrow & \\ \hline E & & AE & & (AEB, QEP) & & QE & & E \end{array}$$

Here both substrates *A* and *B* bind to *E* to form the ternary complex *AEB*. Actually the diagram illustrates one of two possible sequential processes. The one shown is *ordered*, meaning that the substrates *A* and *B* bind to *E* while *P* and *Q*, the products, dissociate in an obligatory sequence. Alternatively no obligatory order could be involved and the process would be a *random* one.

If the initial velocities for a sequential, bisubstrate process are plotted on a double reciprocal graph, a family of straight lines having differing slopes is obtained (Fig. 5-14, *A*). Typical enzymes yielding such kinetics and proceeding by way of ternary complex formation in a sequential fashion are the NAD^+- and $NADP^+$-requiring dehydrogenases. In these cases the coenzyme actually behaves as an obligate substrate. The coenzyme (substrate *A*) reacts to yield *AE*, which then combines with metabolite *B* to form the ternary complex *AEB*. Dehydrogenation leads to the reduction of NAD^+ (NADH) or product *Q* as well as the oxidized metabolite *P* (hence *QEP*). These two products are released in an ordered fashion, first *P* then *Q*. This ordered process is illustrated below:

$$\begin{array}{lcl} NAD^+ + E & \rightleftharpoons & NAD^+ \cdot E\ (AE) \\ NAD^+ \cdot E + MH_2 & \rightleftharpoons & NAD^+ \cdot E \cdot MH_2\ (AEB) \\ NAD^+ \cdot E \cdot MH_2 & \rightleftharpoons & NADH(H^+) \cdot E \cdot M\ (QEB) \\ NADH(H^+) \cdot E \cdot M & \rightleftharpoons & NADH(H^+) \cdot E + M\ (P) \\ NADH(H^+) \cdot E & \rightleftharpoons & E + NADH(H^+)\ (Q) \end{array}$$

An enzyme that catalyzes an ordered bisubstrate reaction is malate dehydrogenase (Chapter 11). The overall reaction is the following:

$$\text{L-Malate} + \text{NAD}^+ \rightleftharpoons \text{Oxaloacetate} + \text{NADH(H}^+\text{)}$$

As indicated above, when no obligatory order is followed, then the enzymatic process is a random one. This is suggested by the following equations:

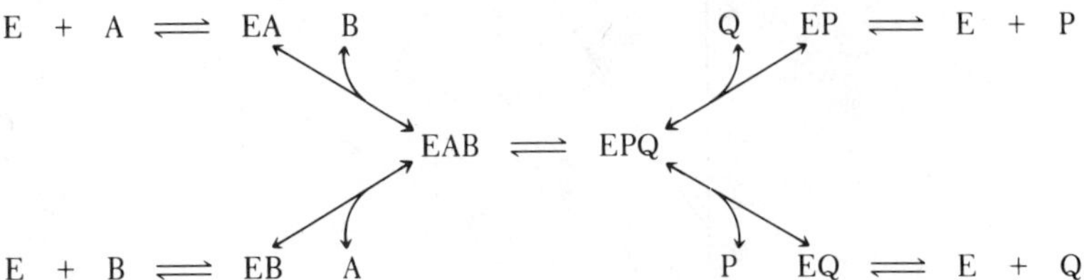

These steps are frequently symbolized as:

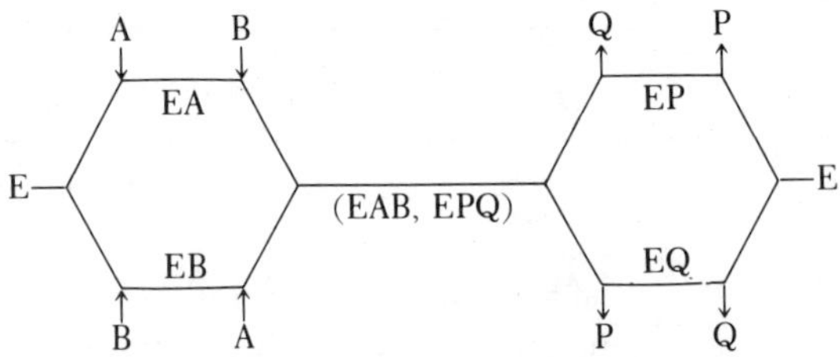

Double displacement, or ping-pong, mechanism

The sequence of steps in a double displacement mechanism is represented by the following:

	$A \downarrow$		$P \uparrow$		$B \downarrow$		$Q \uparrow$	
E		$(EA, E'P)$		E'		$(E'B, EQ)$		E

The kinetics of these reactions are illustrated by the Lineweaver-Burk plots of Fig. 5-14, *B*. Here the plot of initial velocities yields a family of straight lines all having the same slope.

A typical enzyme system yielding ping-pong kinetics is the aminotransferases. The diagram shows that the enzyme first binds an amino acid (substrate *A*) to yield the intermediate *EA*. The corresponding α-keto acid (product *P*) is then released, leaving an altered state of the enzyme, E′ (containing pyridoxamine instead of pyridoxal). E′ then reacts with substrate *B* (an α-keto acid) and the corresponding amino acid, Q, is released. The individual steps involved in this transfer of the amino group are explained elsewhere (Chapter 6).

$$E + A \rightleftharpoons EA$$
$$EA \rightleftharpoons E' + P$$
$$E' + B \rightleftharpoons E'B$$
$$E'B \rightleftharpoons EQ$$
$$EQ \rightleftharpoons E + Q$$

A specific enzyme that functions by the ping pong mechanism is glutamic oxaloacetic transaminase (aminotransferase). The overall reaction catalyzed is:

$$\text{L-Glutamate} + \text{Oxaloacetate} \rightleftharpoons \alpha\text{-Ketoglutarate} + \text{L-Aspartate}$$

Presence of inhibitors

Much can be learned about the nature of active sites and the reactions involved in *ES*-complex formation by studying the interaction of enzymes with substrate analogues and other inhibitors. Not only is this information of theoretical importance, but it also has practical applications in medicine.

Inhibition is either *reversible* or *irreversible*. The former involves interactions of a noncovalent nature, whereas the latter involves chemical modification of the enzyme molecule. These may be distinguished by kinetic analysis using Lineweaver-Burk plots. In general there are three kinds of reversible inhibitions: competitive, noncompetitive, and uncompetitive.

Reversible inhibition

Competitive inhibition In competitive inhibition an inhibitor (I) whose structure resembles a substrate or a product of the reaction binds with the enzyme at its active site. Therefore, it competes with the substrate for the enzyme. As suggested below, both EI and ES are formed. Unlike the ES complex, EI cannot yield product;

$$\begin{array}{l} E + S \rightleftharpoons ES \rightleftharpoons E + P \\ + \\ I \rightleftharpoons EI \end{array}$$

nor can it bind S. Likewise, ES cannot bind inhibitor. Since the presence of the inhibitor results in the formation of a nonproductive complex, the overall rate of the reaction is retarded. However, EI dissociates to regenerate the enzyme, which then can bind to S if this is present. The more substrate available, the more enzyme will be bound to S and the more product will be formed. In this process the actual conversion of S to P at infinite substrate concentration remains unchanged, so that V_{max} is the same. However, the amount of S needed to reach V_{max} is greater than normal. Therefore the K_m, or the amount of substrate needed to achieve $V_{max}/2$, is also greater than normal.

The association of enzyme and inhibitor (that is, $E + I \rightleftharpoons EI$) allows the definition of a dissociation constant, which is designated K_I ($K_I = [E][I]/[EI]$). If we now keep in mind that the total enzyme $E_t = ES + EI + E$, then a modified Michaelis-Menten equation involving K_I can be derived.

$$v = \frac{V_{max}[S]}{K_m\left(1 + \frac{[I]}{K_I}\right) + [S]}$$

The K_m determined in the presence of inhibitor is not the same as that for the normal enzyme-substrate system. It is therefore customary to designate the K_m in the presence of inhibitor as the *apparent Michaelis constant*, or $K_{m(app)}$.

$$K_{m(app)} = K_m(1 + [I]/K_I)$$

Figure 5-15

A Linear double reciprocal Lineweaver-Burk plots showing competitive and noncompetitive inhibition.
B Lineweaver-Burk plot showing uncompetitive inhibition.

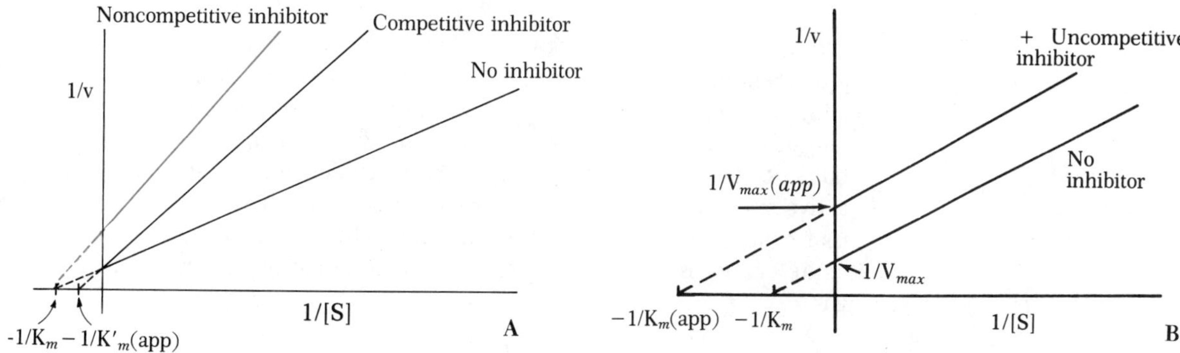

The reciprocal form of the Michaelis-Menten equation is:

$$\frac{1}{v} = \frac{K_m\left(1 + \frac{[I]}{K_I}\right)}{V_{max}} \cdot \frac{1}{[S]} = \frac{1}{V_{max}}$$

A Lineweaver-Burk plot (Fig. 5-15, *A*) has a slope greater than the normal by a factor of $(1 + [I]/K_I)$. As seen in the figure, the *y*-intercept is $1/V_{max}$, just as in the absence of inhibitor.

A classic example of competitive inhibition involves the enzyme succinic dehydrogenase (Chapter 11). This enzyme acts on its substrate, succinic acid, by removing two hydrogen atoms to form the product fumaric acid.

$$\begin{matrix} COOH \\ | \\ CH_2 \\ | \\ CH_2 \\ | \\ COOH \end{matrix} \xrightarrow{} \begin{matrix} COOH \\ | \\ CH \\ \| \\ CH \\ | \\ COOH \end{matrix} + 2H$$

Malonic acid is also a dicarboxylic acid, but it has one less carbon atom. It behaves as an analogue of succinic acid and therefore binds to succinic dehydrogenase. When malonic acid is added to the reaction mixture, it competes with succinic acid for the active site of the enzyme. However, as shown below, two hydrogens cannot be removed from malonic acid.

$$\begin{matrix} COOH \\ | \\ CH_2 \\ | \\ COOH \end{matrix} \not\rightarrow \text{Product}$$

Noncompetitive inhibition Noncompetitive inhibition is another kind of reversible inhibition. The inhibitor, however, does not bind at the same site on the enzyme as the substrate, i.e., the binding of substrate and inhibitor is not mutually exclusive or competitive for the same site. Since *I* and *S* bind at different sites, the inhibitor may react both with free *E* and with *ES* (see diagram below).

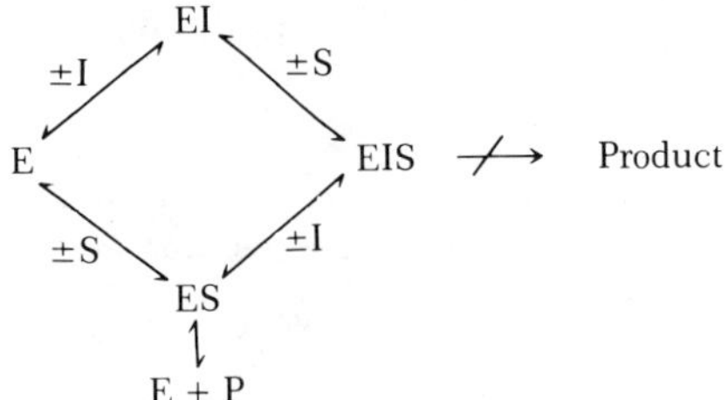

If the complex *EIS* is inactive and does not yield a product, then the overall rate of product formation is suppressed. For this the Lineweaver-Burk plot (Fig. 5-15, *A*) is a line with reduced V_{max} but with the same K_m as the uninhibited reaction. This tells us that the inhibitor does not affect the affinity of enzyme for its true substrate, but since *EIS* does not yield product the inhibition actually reduces the amount of active enzyme. This may be the consequence of a change in molecular shape so that catalysis does not occur. Unlike competitive inhibition, increasing the substrate cannot reverse the inhibition. The equation for noncompetitive inhibition is:

$$\frac{1}{v} = \left(1 + \frac{[I]}{K_I}\right)\left(\frac{K_m}{V_{max}} \cdot \frac{1}{[S]} + \frac{1}{V_{max}}\right)$$

This shows that both the slope and intercept are increased by the factor $(1 = [I]/K_I)$. When the reciprocal plot is extrapolated to the x-intercept $(1/v = 0)$, a value for $-1/K_m$ is obtained that is the same as that for the uninhibited reaction. The y-intercept is increased by $(1 + [I]/K_I)$ (Fig. 5-15, *A*).

$$V_{m(app)} = V_{max}/(1 + [I]/K_I)$$

Uncompetitive inhibition Another form of reversible inhibition, which is neither competitive nor noncompetitive, is called uncompetitive. In this case the inhibitor binds only to *ES* (see diagram on p. 82). The reciprocal equation for uncompetitive inhibition is:

$$\frac{1}{v} = \frac{K_m}{V_{max}} \cdot \frac{1}{[S]} + \frac{1}{V_{max}}\left(1 + \frac{[I]}{K_I}\right)$$

The double reciprocal plot yields a family of parallel lines; both V_{max} and K_m are altered (Fig. 5-15, *B*). Therefore $V_{m(app)} = V_{max}/(1 + [I]/K_I)$ and $K_{m(app)} = K_m/(1 + [I]/K_I)$.

Irreversible inhibition Irreversible inhibition occurs when the inhibitor forms a covalent derivative with the enzyme. Reagents such as DFP or iodoacetate irreversibly react with the seryl and cysteinyl residues respectively. Ths inactivates some of the enzyme. Kinetic studies reflect a decrease in V_{max} and an unchanged K_m.

Regulation of metabolic pathways

Metabolic pathways are regulated by the modifying effect of various factors within the cell on the catalytic activity of certain enzymes within the pathway. As described in Chapter 3, regulation usually occurs at the first or committed steps in a pathway. These enzymes serve as valves controlling the flow of molecules in the metabolic scheme. The regulatory enzyme responds to signals which may be a substrate, a product of the pathway, or some completely unrelated compound such as a cyclic nucleotide (cAMP). Regulation may be negative, i.e., inhibitory in nature, or it may be positive and stimulatory. When the product of a metabolic pathway is not consumed and accumulates, it may inhibit the enzyme catalyzing the first or committed step by *negative feedback inhibition*. In this way the entire sequence of reactions is suppressed, thereby conserving metabolites. In general there are three kinds of regulatory enzyme systems: allosteric systems, isozymic systems, and regulation by covalent modification.

Allosteric systems

Enzymes are allosteric when the regulating factor, called the effector, binds to the enzyme at a site that is separate from the catalytically active site. In general allosteric enzymes have four important characteristics.

1. The proteins are oligomers, consisting of multiple subunits that may or may not be the same.
2. The enzyme may exist as an equilibrium mixture of two conformational states representing active and inactive (or less active) forms.
3. The allosteric binding site and the catalytic sites are distinct and separated spatially on the enzyme. The two sites may occur on different subunits, although this is not an essential feature.
4. In some cases the substrate and effector are the same compounds; in others they are different. The former is often called an homotropic system; the latter is called a heterotropic system. Many allosteric enzymes, however, exhibit both homotropic and heterotropic allosterism.

Michaelis-Menten kinetics do not apply to allosteric enzymes. The usual velocity-substrate concentration curves are not rectangular hyperbolas but rather are sig-

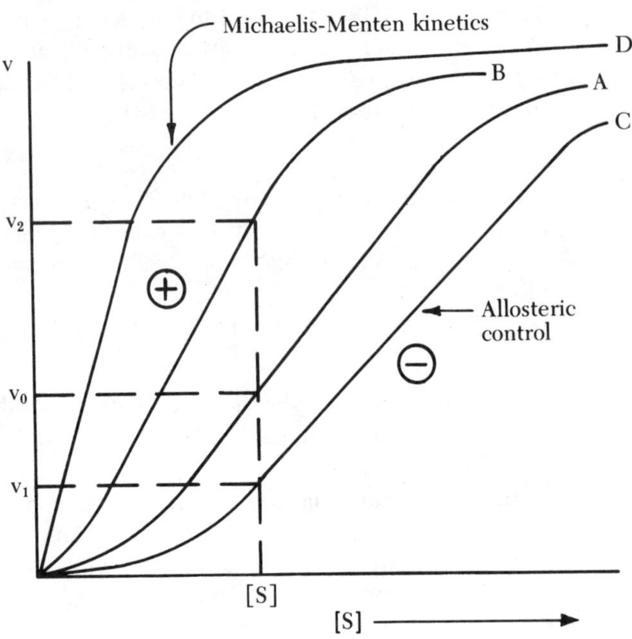

Figure 5-16
Relationship between velocity and substrate concentration of an allosteric enzyme showing the effect of heterotrophic effectors. Curve *A:* in the absence of an effector, the velocity is V_0; *B:* V_2 is the velocity in the presence of a positive effector; *C:* V_1 is the velocity in the presence of a negative effector; *D:* a nonallosteric enzyme exhibiting Michaelis-Menten kinetics.

moidal, or S-shaped. This is illustrated by curve *A* of Fig. 5-16, which represents the kinetics for an allosteric enzyme in the presence of substrate only.

The reason for the sigmoidal shape first became apparent from a nonenzymatic system, i.e., the oxygenation of hemoglobin. As already described (Chapter 4) hemoglobin is a tetramer made up of two different kinds of subunits, α and β. Each polypeptide is capable of binding a molecule of oxygen. When hemoglobin devoid of oxygen (deoxyhemoglobin) binds a molecule of oxygen, alteration occurs in the tertiary and quaternary structures of the subunits, resulting in a progressive increase in the affinity for oxygen of the other binding sites. Thus the binding of oxygen is sequential, as indicated below.

$$\text{Hb} \xrightleftharpoons{O_2} \text{HbO}_2 \xrightleftharpoons{O_2} \text{Hb(O}_2)_2 \xrightleftharpoons{O_2} \text{Hb(O}_2)_3 \xrightleftharpoons{O_2} \text{Hb(O}_2)_4$$

The affinity of the fourth subunit for oxygen may be 100 to 1000 times as great as that for the first oxygen. The progressively increasing affinity for oxygen is an example of *positive cooperativeness*. The sigmoidal shape of the oxygen saturation curve for hemoglobin is described in a later chapter (see Fig. 16-18). On the other hand, myoglobin, an oxygen-binding protein of muscle cells, is a single polypeptide that does not exhibit a quaternary structure. Therefore, the binding of oxygen by myoglobin is not cooperative and does not have a sigmoidal oxygen saturation curve. The oxygen saturation curve for myoglobin is a rectangular hyperbola (see Fig. 16-18).

In like manner an allosteric enzyme may also exhibit positive cooperativeness. The affinity for the substrate molecules becomes increasingly strong as the shape of the subunits is altered. Fig. 5-16 compares the sigmoidal kinetic curve for an allosteric enzyme *(A)* with the curve for a nonallosteric enzyme *(D)* exhibiting Michaelis-Menten kinetics.

Figure 5-17
Feedback inhibition of bacterial aspartyl transcarbamoylase by cytidine triphosphate (CTP) and activation by adenosine triphosphate (ATP).

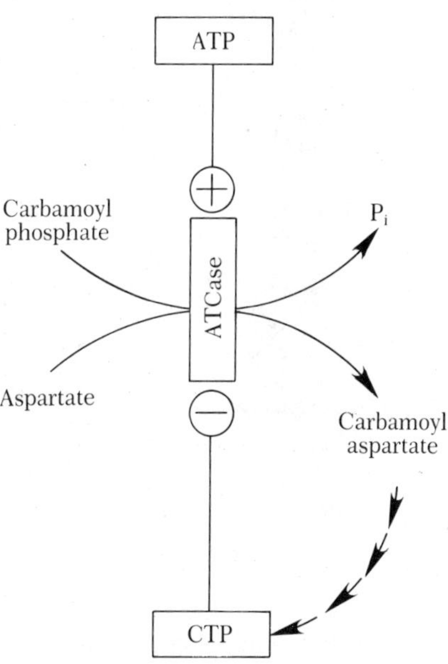

It is apparent from curve *D*, Fig. 5-16, that the velocity of a nonallosteric enzyme increases rapidly in the presence of very low amounts of substrate. On the other hand, with an allosteric enzyme (curve *A*) the velocity may remain relatively small at low substrate concentrations. There is a range of concentration at which the velocity rapidly increases approaching V_{max}. Therefore an allosteric enzyme, exhibiting sigmoidal kinetics, allows the accumulation in the cell of some metabolite; above this level, it is rapidly converted to product. In this way the enzyme may respond sensitively to changes in substrate concentration and therefore serve an important regulatory function in metabolism.

A classic example of metabolic control by an allosteric enzyme is that of the aspartyl transcarbamoylase (ATCase) from *E. coli* (Fig. 5-17). This enzyme catalyzes the combination of carbamoyl phosphate with aspartic acid to form carbamoylaspartate and ultimately cytidine triphosphate (CTP). When the ultimate product, CTP, is not sufficiently used by the cell and accumulates, it binds to the ATCase and inhibits its activity. This is an example of negative feedback control. As suggested on p. 86, ATCase has a quaternary structure consisting of two trimeric sets of catalytic subunits (Cat) and three dimeric regulatory subunits [Reg]. The overall molecular weight of the aggregate is 310,000. One molecule of CTP can bind to each of the six regulatory chains, causing a change in their conformation. This change in shape of the regulatory subunits is transmitted to the flexible catalytic polypeptides, causing a reduced affinity for aspartate and thereby inhibiting the initial step of pyrimidine biosynthesis. Therefore CTP is a negative effector that allosterically inhibits the enzyme. On the other hand, ATP acts as a positive effector; it competes with CTP for the same allosteric binding site. In the presence of ATP, CTP is displaced from the molecule.

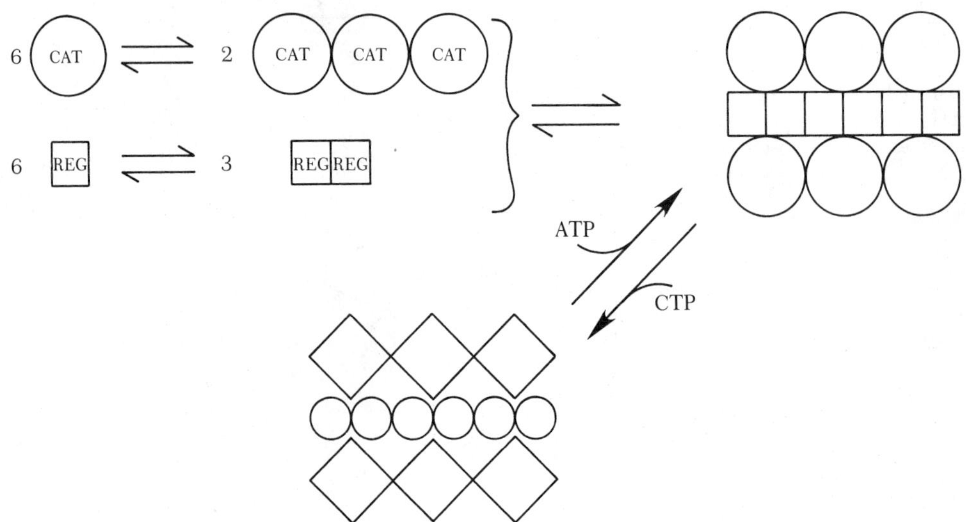

ATCase exhibits a sigmoidal reaction curve when the concentration of aspartate is varied and carbamoyl phosphate is present in saturating amounts (Fig. 5-16, curve *A*). In other words, in the absence of CTP it shows a cooperative effect for aspartate binding to the six catalytic sites. Binding of aspartate to one catalytic site leads to an increased affinity of the other catalytic subunits for the substrate and hence to an increased rate of catalysis. However, if ATCase is treated with mercurial compounds, its quaternary structure is disrupted and the catalytic and regulatory subunits dissociate and can be separated. The catalytic polypeptides in the absence of the regulatory polypeptides are still active, but they now exhibit typical Michaelis-Menten kinetics with respect to aspartate (Fig. 5-16, curve *D*).

The addition of the negative effector CTP shifts the sigmoidal curve for aspartate to the right (Fig. 5-16, curve C). In other words in the presence of the negative effector, larger amounts of aspartate are needed to reach a given velocity. ATP, on the other hand, shifts the sigmoidal curve to the left (Fig. 5-16, curve *B*) so that less aspartate would be required to achieve the same velocity. It is important to appreciate that in cases such as exemplified by the ATCase from *E. coli* the negative effector exerts its action not by influencing the amount of active catalytic sites but by decreasing the affinity for the substrate. Thus, control is realized by raising the K_m, thereby increasing the amount of substrate required to achieve a given velocity. In these cases the V_{max} remains essentially unaltered.

Sometimes, for example with pyruvate kinase and isocitrate dehydrogenase, positive effectors result in a total change in the enzyme kinetics. For these two enzymes the sigmoidal curve is converted to a rectangular hyperbola, i.e., in the presence of positive effectors the enzyme now exhibits Michaelis-Menten kinetics. Therefore in these cases it abolishes the positive cooperativeness normally exhibited.

Isozymes

Another phenomenon that may play a role in control mechanisms and that depends on the quaternary structure of the enzyme protein is the formation of *isozymes*. Isozymes, or isoenzymes, are enzyme variants, i.e., they are enzymes which perform the same catalytic action but which are present in multiple molecular forms within the same animal species. It is possible for an enzyme to be produced by a single gene and yet be modified after synthesis by a variety of factors that would influence net charge or conformation. Isozymes are usually separated

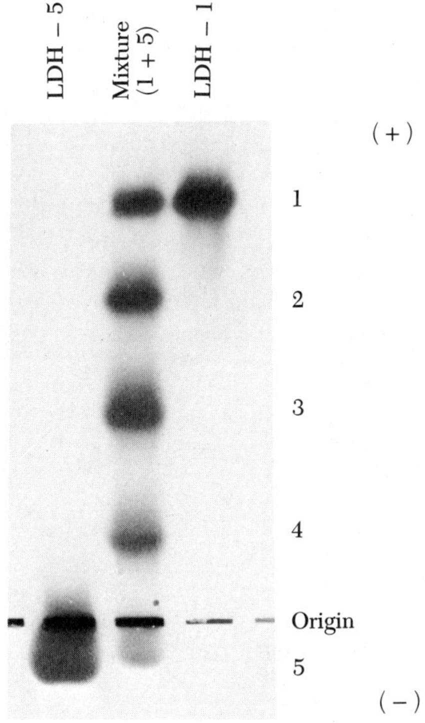

Figure 5-18
Electrophoretic migration of purified ox heart LDH-1 and LDH-5. The two isozymes were separated by processes of chromatography and electrophoresis. The separation obtained in the central slot is that of an equal mixture of LDH-1 and LDH-5.

From Markett, C.L.: Ann. N.Y. Acad. Sci. **151:**14, 1968.

electrophoretically and hence exhibit variations in charge. However, isozymes may also result from more than one gene. For example, two distinct polypeptides, each exhibiting the same catalytic activity, may stem from individual genes. The classic example of this type of isozyme involves the enzyme *lactate dehydrogenase* (LDH), which is actually a tetramer consisting of two distinct subunit molecules. One polypeptide subunit is found predominantly in heart muscle (H subunit); the other is found in skeletal muscle (M subunit). These subunits are produced by separate genes, yet each has the same enzymatic activity. The two proteins have an amino acid composition sufficiently different that they are readily separated in an electrophoretic system and do not cross-react in immunological tests. When the two subunits are mixed in equal proportions, a sequence of five bands is obtained by electrophoresis (Fig. 5-18). These bands represent a regular distribution of each subunit in the quaternary structure. Thus in Fig. 5-18, band *1* is a tetramer of *LDH-1* (H) only; band 2 contains three *LDH-1* (H) and one *LDH-5* (M); band *3* contains two *LDH-1* (H) and two *LDH-5* (M); band *4* contains one *LDH-1* (H) and three *LDH-5* (M); and band 5 contains only *LDH-5* (M).

The two subunits of LDH differ in their kinetic characteristics despite the fact that they both catalyze the reversible reduction of pyruvate to lactate. The LDH-5, or M_4, isozyme from skeletal muscle has a high V_{max} for pyruvate. Skeletal muscle is capable of anaerobic metabolism of glucose and readily converts pyruvate to lactate. By comparison the LDH-1, or H_4, isozyme found in heart muscle has a relatively low V_{max} for pyruvate, so it does not favor the production of lactate. Furthermore, excess pyruvate inhibits the H_4 isozyme. Heart muscle, which is essentially an aerobic tissue, metabolizes glucose to pyruvate and then to CO_2 and H_2O; it produces very little lactate. However, in situations of oxygen deficit, pyruvate may be converted to lactate as an emergency measure. Thus the distinct kinetic characteristics of the two isozymes play a role in determining the kind of metabolism in the tissue in which they occur.

Covalent modification of enzymes

The activity of some enzymes may be regulated by a reversible covalent modification. This results either in an activation or an inhibition of enzymatic activity. Frequently modification takes the form of the phosphorylation of specific hydroxyl amino acids such as serine. From the following illustration, it is evident that the phosphorylation requires ATP and is catalyzed by a second enzyme called a *protein kinase*. Once modified, the enzyme protein can be reconverted to its former state by hydrolysis of the phosphoester bond catalyzed by a third enzyme, *phosphoprotein phosphatase*. These opposing processes are illustrated below:

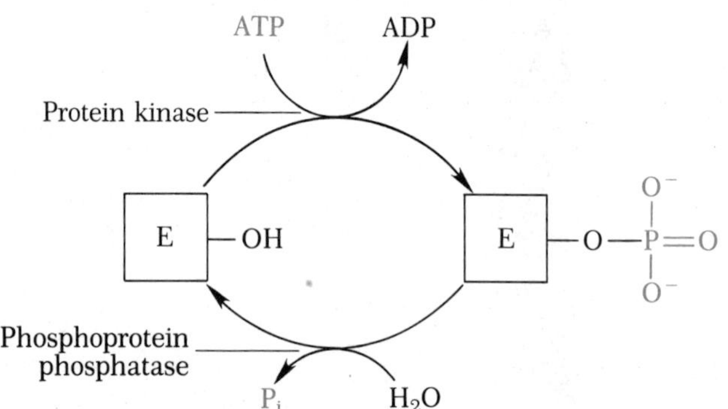

Covalent modification stabilizes a particular protein conformation. In some cases the stabilized conformation is that of the active enzyme; in others the modified protein is inactive. This process is illustrated by Fig. 5-19 and suggests the existence of a small amount of the unstabilized but active conformation in equilibrium with the inactive enzyme. Phosphorylation stabilizes the active form, resulting in a shift toward the active conformation. For example, in the metabolism of the polysaccharide glycogen (Chapter 11), the breakdown of the polymer (glycogenolysis) requires the activation of *glycogen phosphorylase*. Normally it is a tetrameric enzyme that exists in an inactive (b) conformation. Phosphorylation of seryl hydroxyl groups in the b form stabilizes the active conformation, or phosphorylase a. The biosynthesis of glycogen requires the action of *glycogen synthase* (Chapter 11); phosphorylation of this enzyme leads to a loss of activity.

In glycogenolysis, the phosphorylase kinase also exists in inactive (unphosphorylated) and active (phosphorylated) states. Therefore, glycogenolysis is controlled by a sequence of two covalently regulated enzymes. Fig. 5-20 illustrates in a general fashion the sequential action of two interconvertible enzymes dependent on phosphorylation-dephosphorylation processes. Such a sequential process is called a *cascade mechanism*. The initial activation of enzyme E-1 may require the action of a small number of kinase molecules (k) (Fig. 5-20). The effect of k is multiplied greatly, since many E-1 molecules are phosphorylated. The active E-1 then catalyzes the phosphorylation of many more E-2 molecules, which finally catalyze the reaction $A \rightarrow B$. Therefore a cascade represents an amplification process. The initial, small effect literally leads to an avalanche of product in the final step of $A \rightarrow B$. It is important not to lose sight of the reconversion of phosphorylated to dephosphorylated forms; a balance is achieved between the active and inactive enzymes. Cascade processes provide metabolism with a process that exists under steady-state conditions and is exquisitely tuned to changes in the levels of various metabolites.

Figure 5-19 Stabilization of the active conformation of an enzyme regulated by phosphorylation-dephosphorylation.

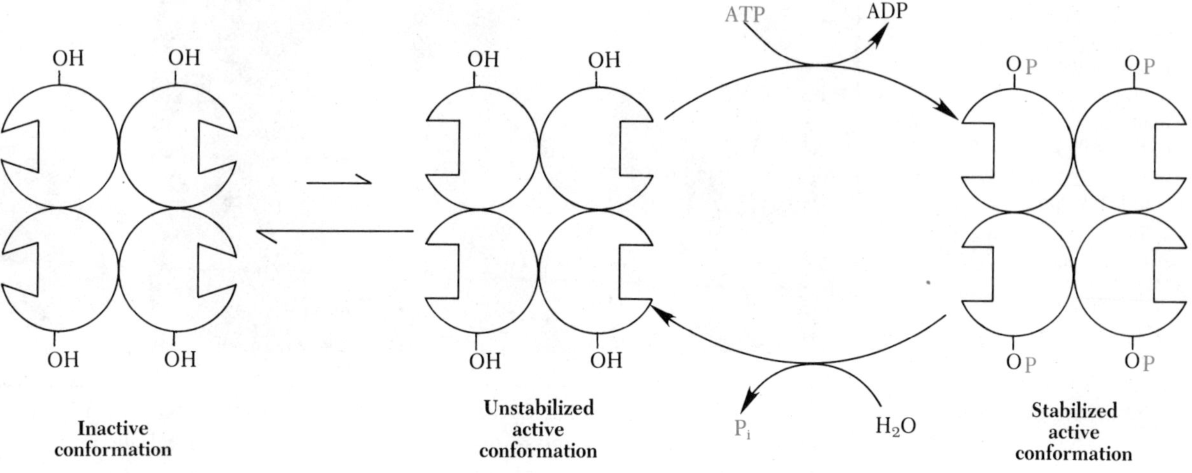

Figure 5-20 Sequential action of two covalently modified enzymes that function together to control the conversion of metabolite *A* to product *B*. This diagram illustrates the cascade phenomenon.

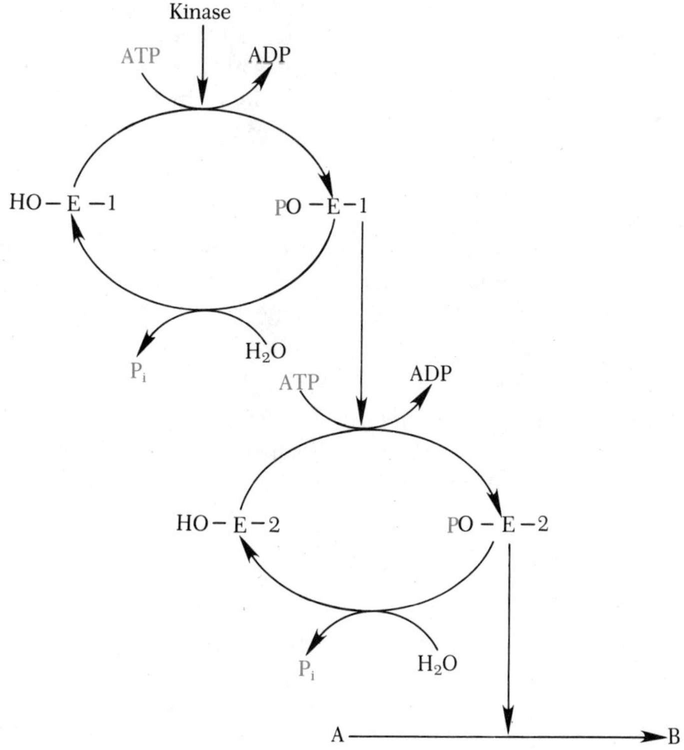

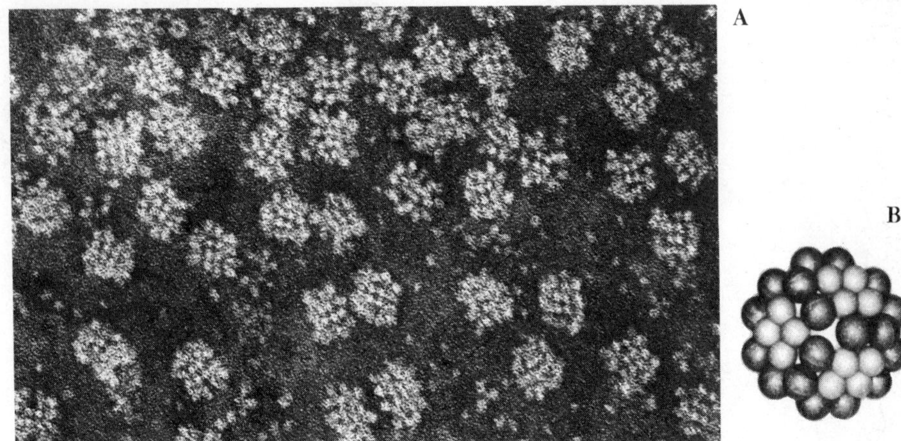

Figure 5-21
A Electron micrograph of the pyruvate dehydrogenase complex of *E. coli,* negatively stained with methylamine tungstate (×250,000).
B Interpretive model of the complex showing 24 dihydrolipoyl dehydrogenase subunits (white spheres) and 24 molecules of pyruvate decarboxylase (black spheres).

From Reed, L.J., and Oliver, R.M.: Brookhaven Symp. Biol. **21:**397, 1968.

Multienzyme systems

Another level of enzyme structure that is important in metabolic regulatory processes is the highly organized multienzyme complex. Some metabolic enzyme systems are known to exist, not as independent molecules but in a state of aggregation or architecture involving several different enzymes. The pyruvic acid dehydrogenase complex of *E. coli* is one that has received considerable attention. This complex (mol wt = 4.8 million) consists of three different enzymes, 24 molecules of pyruvate decarboxylase (mol wt = 90,000), 24 molecules of dihydrolipoic dehydrogenase (mol wt = 55,000), and 8 molecules of lipoyl reductase transacetylase (mol wt = 120,000). Fig. 5-21 shows an electron micrograph of this enzyme as well as a tentative model. The functions of the various coenzymes associated with this system are described in greater detail in Chapters 6 and 11. Molecular organization provides a mosaic of enzymes in which each component is arranged so as to afford an efficient coupling of the individual reactions catalyzed by these enzymes. In other words, the product of the first enzyme becomes the substrate of the second and so on.

Classification

Enzymes often have been named for their substrate, the type of reaction, and sometimes on other bases. Consequently one enzyme may have several names and occasionally the same name has been given to more than one enzyme. Today enzymes are named according to the following categories established by the International Union for Biochemistry (IUB).

1. *Oxidoreductases.* Oxidoreductases catalyze oxidation and reduction processes. Their substrate is hydrogen or an electron donor. The name recommended for them is dehydrogenases; oxidase is used only when oxygen is the electron acceptor.

REACTION TYPE

$$A_{ox} + B_{red} \rightleftharpoons A_{red} + B_{ox}$$

SPECIFIC EXAMPLE

$$\begin{array}{c} CH_3 \\ | \\ H{-}C{-}OH \\ | \\ C{-}O^- \\ \| \\ O \end{array} \xrightarrow[NAD^+ \;\curvearrowright\; NADH + H^+]{\text{Lactate dehydrogenase}} \begin{array}{c} CH_3 \\ | \\ C{=}O \\ | \\ C{-}O^- \\ \| \\ O \end{array}$$

2. *Transferases.* Transferases catalyze the transfer of a group from the donor compound to the acceptor. In many cases, the donor is a coenzyme carrying the group to be transferred.

REACTION TYPE

$AB + C \rightleftharpoons A + BC$

SPECIFIC EXAMPLE

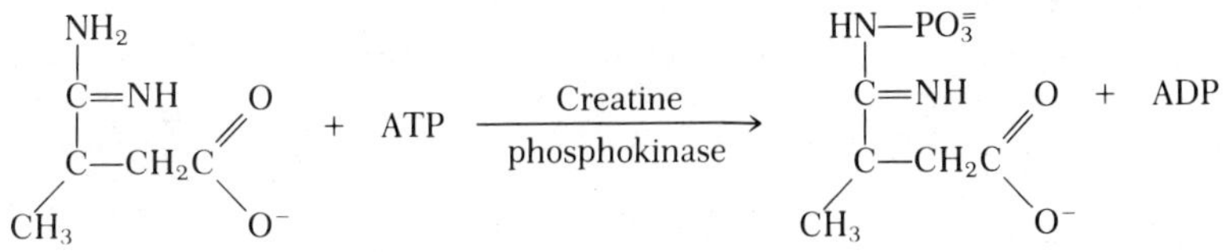

$$\text{Creatine} + \text{ATP} \xrightarrow[\text{phosphokinase}]{\text{Creatine}} \text{Phosphocreatine} + \text{ADP}$$

3. *Hydrolases*. Hydrolases catalyze the hydrolysis of various bonds.

REACTION TYPE

$AB + H_2O \rightleftharpoons AOH + BH$

SPECIFIC EXAMPLE

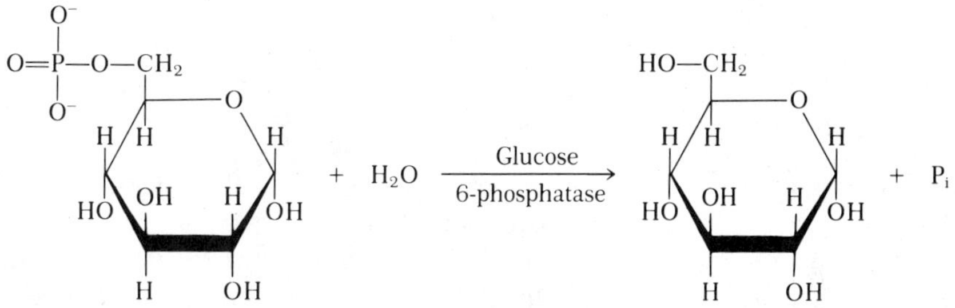

$$\text{Glucose 6-phosphate} + H_2O \xrightarrow[\text{6-phosphatase}]{\text{Glucose}} \text{Glucose} + P_i$$

4. *Lyases*. Lyases catalyze the cleavage of C—C, C—O, and C—N bonds by means other than hydrolysis or oxidation.

REACTION TYPE

$AB \rightleftharpoons A + B$

SPECIFIC EXAMPLE

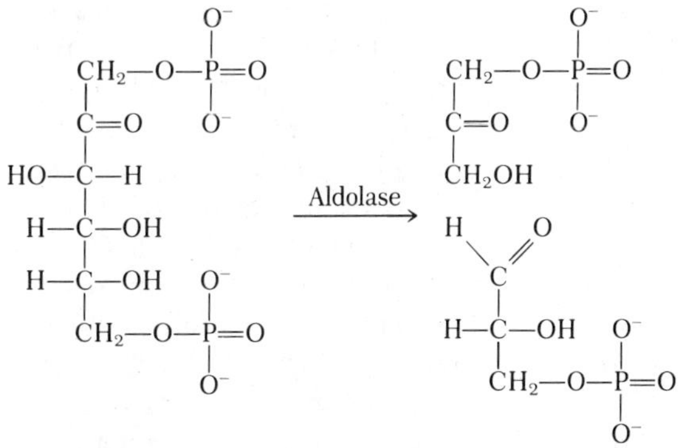

$$\text{Fructose 1,6-bisphosphate} \xrightarrow{\text{Aldolase}} \text{Dihydroxyacetone phosphate} + \text{Glyceraldehyde 3-phosphate}$$

5. *Isomerases*. Isomerases catalyze intramolecular rearrangement.

REACTION TYPE

$ABC \rightleftharpoons ACB$

SPECIFIC EXAMPLE

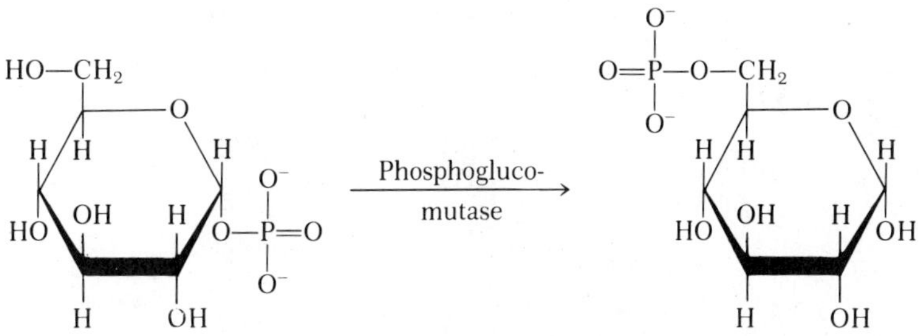

$$\text{Glucose 1-phosphate} \xrightarrow{\text{Phosphoglucomutase}} \text{Glucose 6-phosphate}$$

6. *Ligases (synthetases).* Ligases catalyze the joining of two molecules coupled with the hydrolysis of a pyrophosphate bond in ATP or similar triphosphate. They may form C—O, C—S, C—N, C—C.

REACTION TYPE

$$A + B + ATP \rightleftharpoons AB + ADP + P_i$$

SPECIFIC EXAMPLE

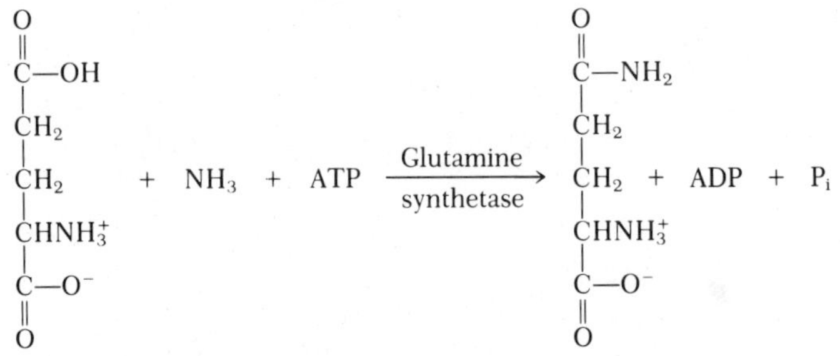

$$\begin{array}{c} O \\ \| \\ C{-}OH \\ | \\ CH_2 \\ | \\ CH_2 \\ | \\ CHNH_3^+ \\ | \\ C{-}O^- \\ \| \\ O \end{array} + NH_3 + ATP \xrightarrow[\text{synthetase}]{\text{Glutamine}} \begin{array}{c} O \\ \| \\ C{-}NH_2 \\ | \\ CH_2 \\ | \\ CH_2 \\ | \\ CHNH_3^+ \\ | \\ C{-}O^- \\ \| \\ O \end{array} + ADP + P_i$$

Clinical enzymology

Enzymology has been considered for the purpose of understanding the metabolic processes of the body. However, enzymology also is useful in the diagnosis of disease and to some extent in the treatment of clinical problems.

In the clinical laboratory, enzymes are used either as specific reagents for the quantitative determination of various constituents in tissue samples or for the measurement of enzyme levels that serve as an index in the diagnosis of disease processes.

Enzymes as reagents

Enzymes as reagents have distinct advantages over other chemical reagents; they are highly specific for the substrate, extremely sensitive, and rapid in action. Generally enzyme technology requires a strict adherence to optimum environmental conditions. Where the purpose is to determine a substrate concentration, an excess of enzyme should be used. In other words, the enzyme concentration must be such that it is not saturated with substrate. Under these conditions the amount of product produced or the amount of substrate consumed is proportional to the initial substrate concentration.

Sometimes the substrate or the products of a reaction are not easily measured. In that event a second enzyme reaction may be used to measure the product produced by the first. If the secondary reaction requires NAD^+ or $NADP^+$, then the changes in coenzyme are readily determined in a spectrophotometer. Fig. 5-22 shows the absorption of light at 340 nm by NADH and NADPH compared with that of their oxidized forms. It is therefore possible to measure an enzymic reaction by determining the change in light absorbance either as NAD^+ is reduced (an increase in optical density) or as NADH is oxidized (a decrease in optical density).

Blood urea nitrogen Coupling an NAD^+-dependent enzyme to the determination of a tissue constituent is illustrated by the measurement of blood urea nitrogen (BUN). This assay consists of two reactions, of which the first is the conversion of urea to ammonia and CO_2 by the enzyme urease.

$$1.\ \text{Urea} + H_2O \xrightarrow{\text{Urease}} 2\ NH_3 + CO_2$$

To measure the ammonia liberated, the NAD^+-dependent glutamate dehydrogenase (GDH) is added.

$$2.\ NH_3 + \alpha\text{-Ketoglutarate} \xrightarrow[\text{NADH} \to \text{NAD}^+]{\text{GDH}} \text{Glutamate} + H_2O$$

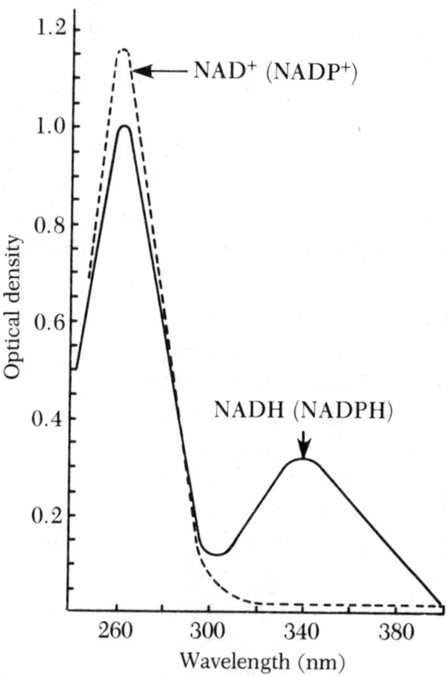

Figure 5-22
Absorption spectra for oxidized (dotted line) and reduced (solid line) forms of nicotinamide adenine dinucleotide (NAD^+).

The amination of α-ketoglutarate requires the oxidation of NADH. Therefore the progress of the reaction can be followed by the decrease in optical density. Normal values of BUN are 8 to 19 mg/100 ml.

A rapid micromethod uses a filter-paper strip impregnated with urease. The ammonia released is absorbed in an area containing bromcresol green. The extent of color change is a measure of the concentration of urea.

Blood glucose Glucose is oxidized to gluconic acid and hydrogen peroxide by glucose oxidase. Methods involving this procedure include the enzyme peroxidase, which converts the hydrogen peroxide into water and oxygen (nascent). Usually a dye such as O-dianisidine, which is oxidized to a colored compound, is also included. Color production is proportional to glucose concentration.

1. $\text{D-Glucose} + O_2 + H_2O \xrightarrow[\text{oxidase}]{\text{Glucose}} \text{Gluconic acid} + H_2O_2$

2. $H_2O_2 \xrightarrow{\text{Peroxidase}} H_2O + \frac{1}{2}\,O_2$

This procedure is also available in qualitative and semiquantitative form as filter-paper strip tests.

Alcohol Alcohol is converted to acetaldehyde by alcohol dehydrogenase (ADH).

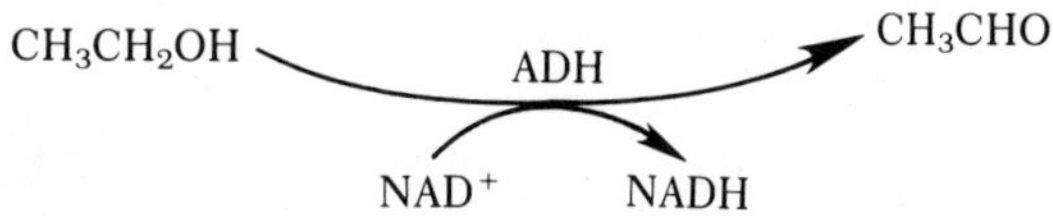

The increase in absorbance at 340 nm is measured and related to concentration. There is also a colorimetric method in which a blue dye is reduced to its colorless form. Again change in absorbance is related to concentration of alcohol.

Enzyme levels as an index of disease

Enzymes in the blood serum are either normal or abnormal constituents. Those normally present are of hepatic origin; the liver functions to maintain a constant blood level. Liver disease or genetic defects may result in a reduction in serum enzyme levels. Enzymes not normally found in serum are derived from tissue damage. Actually such enzymes are present at minimal levels as a result of the turnover of tissue cells, i.e., the "wear and tear" process. Tissue damage (as in myocardial infarction) and infection (as in hepatitis) cause release of tissue enzymes into the bloodstream. Proliferative processes, as in osteosarcoma or prostatic carcinoma, lead to increased production of enzymes and spillage into the blood. Liver disease leading to biliary obstruction causes an increase in some serum enzymes by blocking their elimination into the bile.

Levels of normal serum enzymes Two enzymes whose primary function is in the blood serum are ceruloplasmin (transferrin) and cholinesterase.

Ceruloplasmin Ceruloplasmin is a plasma protein with ferroxidase activity. It is measured using the rate of oxidation of *p*-phenylenediamine. The oxidation product is purple and has an absorption peak of 520 to 530 nm. Normal values are 280 to 570 units. This enzyme is used to establish the presence of Wilson's disease (cirrhotic liver), in which the levels are markedly decreased from normal.

Cholinesterase The cholinesterase found in blood serum is usually called Type II cholinesterase, or pseudocholinesterase, to distinguish it from Type I, or true, cholinesterase of the nervous system. Measurement of the serum enzyme uses acetylthiocholine as substrate. The product, thiocholine, reacts with dithiobisnitrobenzoic acid to yield a colored product that is measured at 405 nm. Normal levels of serum cholinesterase are 1900 to 3800 mU per ml of serum. Serum cholinesterase levels are useful diagnostic tests for liver disease when the level is decreased from normal.

Levels of abnormal serum enzymes (nonplasma, specific enzymes) These enzymes may be divided into two groups: those that are secretory enzymes and function outside the body as in digestion and those that are intracellular enzymes that perform their roles in metabolism.

Secretory enzymes

α-Amylase α-Amylase is produced by the pancreas; hence its appearance in the blood serum is indicative of pancreatic disorders. The action of serum, plasma, or urine on amylopectin labeled with a blue dye (amylopectin azure) is measured as the amylase hydrolyzes and amylopectin azure to form soluble colored products. The intensity of blue color released is proportional to the amount of enzyme. The normal serum level is increased in acute pancreatitis during the first 72 hours after the onset of symptoms.

Intracellular enzymes

Alkaline phosphatase The method often used to detect alkaline phosphatases is that of Bessey-Lowry-Brock. It involves the hydrolysis of *p*-nitrophenylphosphate to *p*-nitrophenol at a pH of 8.5 to 10. Levels of serum enzymes are elevated in osteological diseases (e.g., osteogenic sarcoma, rickets) or biliary disorders (obstructive jaundice cirrhosis).

Acid phosphatase The method for the detection of acid phosphatase is similar to that for alkaline phosphatase except that the optimum pH is 5. Levels are elevated in carcinoma of the prostrate.

Creatine phosphokinase (CPK) This enzyme, found primarily in skeletal and cardiac muscle, is elevated in blood after myocardial infarction. The assay involves the following reaction:

$$\text{Creatine phosphate} \xrightarrow[\text{ADP} \;\curvearrowright\; \text{ATP}]{\text{CPK}} \text{Creatine}$$

$$\text{ATP} + \text{Glucose} \xrightarrow{\text{Hexokinase}} \text{Glucose 6-phosphate} + \text{ADP}$$

$$\text{Glucose 6-phosphate} \xrightarrow[\text{NADP}^+ \;\curvearrowright\; \text{NADPH}]{\text{G6P DH}} \text{6-Phosphogluconate}$$

Three enzymes, of which the first (CPK) is the actual unknown, are involved. The increase in optical density at 340 nm is converted to units per milliliter of serum.

CPK is a dimer consisting of one subunit found in the brain (B) and another in muscle (M). It is found as three isozymes, termed CPK_1 (BB), CPK_2 (MB), and CPK_3 (MM). These isozymes are separated electrophoretically, stained, and scanned. In normal serum, 95% of the CPK activity exists as the CPK_3 species. Following myocardial infarction the serum CPK_2 (MB) species increases specifically.

Serum transaminases Two serum transaminases* are used clinically. These are serum glutamic oxaloacetic transaminase (SGOT) and serum glutamic pyruvic transaminase (SGPT). The principle of transamination reactions is detailed elsewhere.

$$\alpha\text{-Ketoglutarate} + \text{L-Aspartate} \longrightarrow \text{Oxaloacetate} + \text{L-Glutamate}$$

$$\text{Oxaloacetate} \xrightarrow[\text{NADH} \;\curvearrowright\; \text{NAD}^+]{\text{MDH}} \text{L-Malate}$$

The added enzyme, malate dehydrogenase (MDH), catalyzes the reduction of oxaloacetate to malate and the oxidation of NADH. The reduction in optical density at 340 nm may be expressed in International milliunits per milliliter. Normal values are 5 to 20 mU/ml.

$$\alpha\text{-Ketoglutarate} + \text{L-Alanine} \xrightarrow{\text{SGPT}} \text{Pyruvate} + \text{L-Glutamate}$$

$$\text{Pyruvate} \xrightarrow[\text{NADH} \;\curvearrowright\; \text{NAD}^+]{\text{LDH}} \text{L-Lactate}$$

The added enzyme, LDH, catalyzes the reduction of pyruvate and oxidation of NADH. The decrease in absorbance at 340 nm may be converted to International milliunits per milliliter. Normal values are 5 to 24 mU/ml.

Increased serum levels of SGOT are observed after myocardial infarction; it is also elevated in liver disease. On the other hand, SGPT is not elevated after myocardial infarction but is increased in liver disease. Thus the two conditions may be differentiated enzymatically.

Lactate dehydrogenase The procedure frequently employed uses pyruvate and NADH to form lactate. It is referred to as LDH-P and is faster than the older procedure using the forward reaction beginning with lactate.

$$\text{Pyruvate} \xrightarrow[\text{NADH} \;\curvearrowright\; \text{NAD}^+]{\text{LDH}} \text{Lactate}$$

*The new and preferred terms for the *serum aminotransferases* are *serum glutamate oxaloacetate aminotransferase* (SGOT) and *serum glutamate pyruvate aminotransferase* (SGPT).

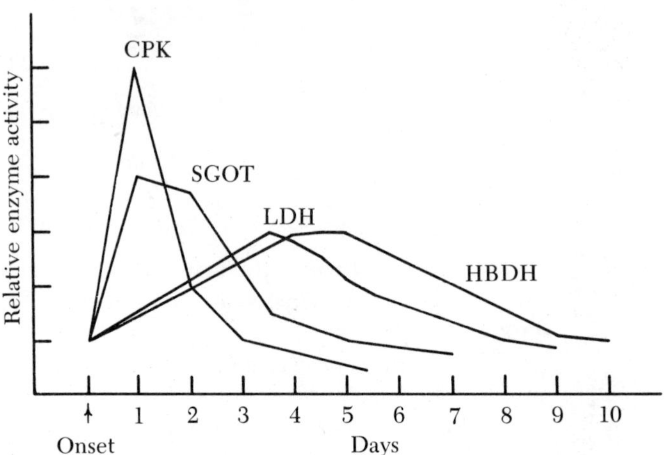

Figure 5-23
Changes in serum activities for CPK, SGOT, LDH, and HBDH after myocardial infarction.

The decrease in absorbance at 340 nm is converted to International milliunits per milliliter. Normal values are 125 to 270 mU/ml. LDH is increased in myocardial infarction but does not reach a peak until 48 to 72 hours after the episode.

As described previously, serum LDH can be separated electrophoretically into 5 isozymes (Fig. 5-18). The electropherograms are stained by the reduction of a tetrazolium salt by NADH to the blue formazan dye. The stained isozymic pattern is then scanned in a recording densitometer. In normal serum LDH-2 (H_3M) is greater than LDH-1 (H_4) so that the ratio of LDH-1/LDH-2 is less than 1. Following myocardial infarction the ratio is inverted (i.e., greater than 1) as a result of the release of LDH-1 into the bloodstream. Disease of the liver and skeletal muscle can be detected as an increased LDH-5 activity. Increases may also be seen in jaundice secondary to drug toxicity, hepatic contusions, and infectious mononucleosis.

α-Hydroxybutyrate dehydrogenase (α-HBDH) The decrease in absorbance is converted to International milliunits per milliliter. For α-HBDH, normal values are 96 to 210 mU/ml. Actually this enzyme appears to be the same as LDH-1; its richest source is heart muscle, kidney, and erythrocytes. In myocardial infarction α-HBDH is usually elevated for a longer time than is total LDH. Elevation will continue for 2 to 3 weeks after infarction.

$$\alpha\text{-Ketobutyrate} \xrightarrow[\text{NADH} \quad \text{NAD}^+]{\alpha\text{-HBDH}} \alpha\text{-Hydroxybutyrate}$$

The temporal variations of four serum enzymes used to confirm the diagnosis of myocardial infarction are illustrated in Fig. 5-23. A few hours after the onset of infarction the activity of CPK increases; it reaches a peak in 24 to 36 hours and then returns to normal after 3 to 6 days. The SGOT activity increases a few hours later, reaches a peak in 1 to 2 days, and returns to normal within 5 to 7 days. LDH is slower than either of the former enzymes, reaching a maximum after 48 to 60 hours and returning to normal after 6 to 7 days. α-HBDH is the slowest to respond, reaching a maximum in 3 to 5 days and returning to normal after 10 days.

Enzymes as therapeutic agents

Enzymes are frequently used therapeutically. For example, fibrinous and purulent material is often removed from infected or necrotic wounds or other lesions by enzymes such as fibrinolysin and deoxyribonuclease. Various proteolytic enzymes,

Table 5-5
Some therapeutic uses for enzymes

Enzyme	Application of enzyme
α-Chymotrypsin, trypsin, bromelain, papain, fibrinolysin; Pancreatic dornase (DNAase), streptodornase (DNAase)	Debridement; treatment of abscesses, infected burns, infected skin ulcers, cervicitis, and vaginitis
Streptokinase plus streptodornase	Dissolves thrombi and fibrinous adhesions
Pepsin, trypsin, peptidases, lipase, amylase, nucleases, elastase, cellulase	Gastrointestinal disorders, chronic pancreatitis, pancreatectomy, enzyme deficiency
Hyaluronidase	Facilitates intradermal injection of fluid; dispersion of accumulated fluids in trauma

especially trypsin, as well as activators of fibrinolysin (streptokinase) are used as treatments of thrombosis. A summary of some of the therapeutic uses is presented in Table 5-5.

Suggested readings

Batsakis, J.G., and Briere, R.O.: Interpretive enzymology, Springfield, Ill., 1967, Charles C Thomas, Publisher.
Bernhard, S.: The structure and function of enzymes, New York, 1968, W.A. Benjamin, Inc.
Boyer, P.D., editor: The enzymes, ed. 3, New York, 1970, Academic Press, Inc.
Colowick, S.P., and Kaplan, N.O., editors: Methods in enzymology, vols. I-LXIII, New York, 1958-1979, Academic Press, Inc.
Criss, W.E.: A review of isozymes in cancer, Cancer Res. **31:**1523, 1971.
Dixon, M., and Webb, E.C.: Enzymes, ed. 2, New York, 1964, Academic Press, Inc.
Fersht, A.: Enzyme structure and mechanism, Reading, England, 1977, W.H. Freeman and Co. Publishers.
Gutfreund, H.: Enzymes: physical principles, New York, 1972, John Wiley & Sons, Inc.
Hammes, G.G., and Wu, C.W.: Regulation of enzyme activity, Science **172:**1205, 1971.
Hess, B.: Enzymes in blood plasma, New York, 1963, Academic Press, Inc.
Holzer, H., and Duntze, W.: Metabolic regulation by chemical modification of enzymes. Annu. Rev. Biochem. **40:**345, 1971.
Jencks, W.P.: Catalysis in chemistry and enzymology, New York, 1969, McGraw-Hill Book Co.
Kirsch, J.F.: Mechanism of enzyme action, Annu. Rev. Biochem. **42:**205, 1973.
Koshland, D.E., Jr..: Conformation changes at the active site during enzyme action, Fed. Proc. **23:**719, 1964.
Koshland, D.E., Jr.: Protein shape and biological control, Sci. Am. **229:**52, 1973.
Koshland, D.E., Jr., and Neet, K.E.: The catalytic and regulatory properties of enzymes, Annu. Rev. Biochem. **37:**359, 1968.
Lienhard, G.E.: Enzymatic catalysis and transition-state theory, Science **180:**149, 1973.
Lowenstein, J.M.: The pyruvate dehydrogenase complex and the citric acid cycle. In Florkin, M., and Stotz, E.H., editors: Comprehensive biochemistry, vol. 18S, Amsterdam, 1971, Elsevier Publishing Co., p. 1.
Mattenheimer, H.: Clinical enzymology, Ann Arbor, Mich., 1971, Ann Arbor Science Publishers, Inc.
Meister, A., editor: Advances in enzymology, vols. 35-48, New York, 1972-1979, John Wiley & Sons, Inc.
Nord, F.F., editor: Advances in enzymology, vols. 1-34, New York, 1941-1971, John Wiley & Sons, Inc.
Phillips, D.C.: The three-dimensional structure of an enzyme molecule. Sci. Am. **215:**78, 1966.
Plowman, K.M.: Enzyme kinetics, New York, 1972, McGraw-Hill Book Co.
Reed, L.J., and Oliver, R.M.: The multienzyme α-keto acid dehydrogenase complexes, Brookhaven Symp. Biol. **21:**397, 1968.
Segal, H.L.: Enzymatic interconversion of active and inactive forms of enzymes, Science **180:**25, 1973.
Segel, I.H.: Enzyme kinetics, New York, 1975, John Wiley & Sons, Inc.
Sigman, D.S., and Mooser, G.: Chemical studies of enzyme active sites, Annu. Rev. Biochem. **44,** 889, 1975

Smellie, R.M.S., editor: Chemical reactivity and biological role of functional groups in enzymes, Biochem. Soc. Symp., no. 31, New York, 1970, Academic Press, Inc.

Stadtman, E.R., and Chock, P.B.: Interconvertible enzyme cascades in metabolic reactions. In Horecker, B.L., and Stadtman, E.R., editors: Current topics in cellular regulation, vol. 13, New York, 1978, Academic Press, Inc., p. 53.

Wang, J.H.: Facilitated proton transfer in enzyme catalysis, Science **161**:328, 1968.

Wimmer, M.J., and Rose, I.A.: Mechanisms of enzyme-catalyzed group transfer reactions, Annu. Rev. Biochem. **47**:1031, 1978.

Wolf, P.L., and William, D.: Practical clinical enzymology: techniques and interpretations, New York, 1973, John Wiley & Sons, Inc.

Chapter 6

Coenzymes: group transfer reactions

Two important factors of human nutrition that play an essential role in the function of certain enzymes and in the control of metabolism are vitamins and minerals. Most of the water-soluble vitamins (Chapter 21) are in turn converted in the body into low–molecular weight, heat-stable, dialyzable, organic compounds called coenzymes. As stated previously (Chapter 5) many intracellular enzymes are a combination of a coenzyme and an apoenzyme; the complete, functional unit being termed the holoenzyme. The need for coenzymes was demonstrated in early experiments using extracts of yeast for the fermentation of glucose. When such yeast extracts were dialyzed against water, a substance was lost without which fermentation could not proceed. Addition of the essential factor to the system restored its capacity for fermentation. The substance was isolated and called Cozymase or Coenzyme I. Minerals are also an important factor in metabolism; approximately one third of the known enzymes require a metal for their activity.

We may group the coenzymes and minerals together under the more general term *cofactor,* which means any nonprotein substance required for the activity of the enzyme. As seen below, the cofactor can then be subdivided into inorganic and organic categories.

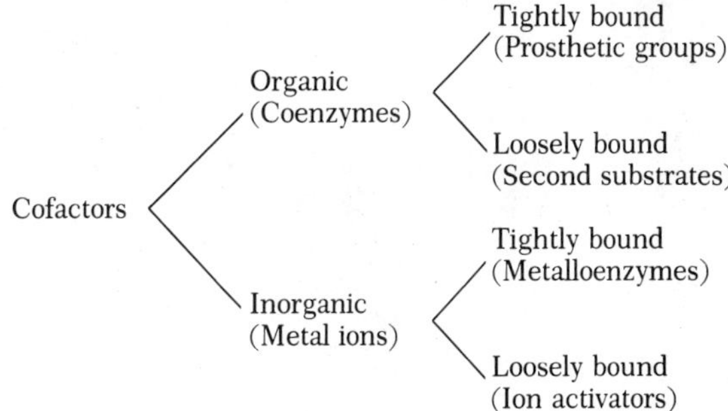

By definition, the organic cofactors are the coenzymes. They may either be an integral or structural element of the enzyme (in such a case they are called prosthetic groups) or they may be loosely bound and serve as a second substrate. Clear distinctions between coenzymes, prosthetic groups, and substrates are often difficult to make. Generally, the water-soluble vitamins, with the exception of vitamin C, are accepted as precursors of coenzymes. Some of these are also prosthetic groups.

Not all prosthetic groups, however, are considered to be coenzymes. For example, the heme of hemoglobin (myoglobin) or the cytochromes are usually not so categorized. Furthermore such substrates of bisubstrate reactions as *S*-adenosyl methionine or ATP are usually not considered as coenzymes despite the fact that they are essential, organic cofactors in a particular reaction.

If we consider under coenzymes those substrates that are present in limited amounts and that therefore must be continually regenerated to maintain metabolism, then such factors as S-adenosyl methionine and ATP also may be included in this category. Furthermore, based on the characteristic of recycling, a prosthetic group such as heme may also be included, provided we consider it in the context of enzymes and not transport proteins such as in hemoglobin. This added qualification to characterize a coenzyme is used in this chapter.

Organic cofactors or coenzymes

Pools

Since some coenzymes serve as a second substrate in a reaction, the corresponding enzyme must possess binding sites specific not only for the primary substrate but also for the coenzyme molecule. In the course of the reaction, the coenzyme acts as an acceptor of some group(s) and therefore is chemically altered. As described in Chapter 5, in the ordered, single displacement reaction:

$$A + \mathrm{E} \rightleftharpoons AE$$
$$AE + \mathrm{B} \rightleftharpoons AEB$$
$$AEB \rightleftharpoons QE + P$$
$$QE \rightleftharpoons \mathrm{E} + \mathrm{Q}$$

B and Q could be the oxidized and reduced states of a coenzyme, whereas *A* and *P* are the reduced and oxidized states of the primary substrate. In the cell, *A* may be continually supplied and *P* removed. On the other hand, the total amount of coenzyme, *B* and Q, remains constant. This coenzyme pool must supply the needs of other enzymes as well. Once a coenzyme is reduced or altered in one reaction, it must be reoxidized or regenerated in another if it is to be reused in the initial reaction. Coenzymes, however, are not all in one state or the other; actually a ratio of reduced molecule to oxidized form exists. Such a cyclical alteration-regeneration process is illustrated below. As seen in the diagram, the substrate MH_2 is oxidized by the enzyme E_M to the product *M*. In the process two hydrogens are removed and transferred to the coenzyme (Co → CoH_2). The reduced coenzyme in turn becomes the substrate for a second reaction catalyzed by E_p in which the hydrogens are transferred to the metabolite *P* to form PH_2 as well as other reactions represented by Q to QH_2. In this way the coenzyme is restored to its original form (CoH_2 → Co) and reused in the initial reaction.

Coenzyme pool

CoH_2 / Co

M, MH_2, E_M, CoH_2, Co

P, PH_2, E_P, CoH_2, Co

Q, QH_2, E_Q, CoH_2, Co

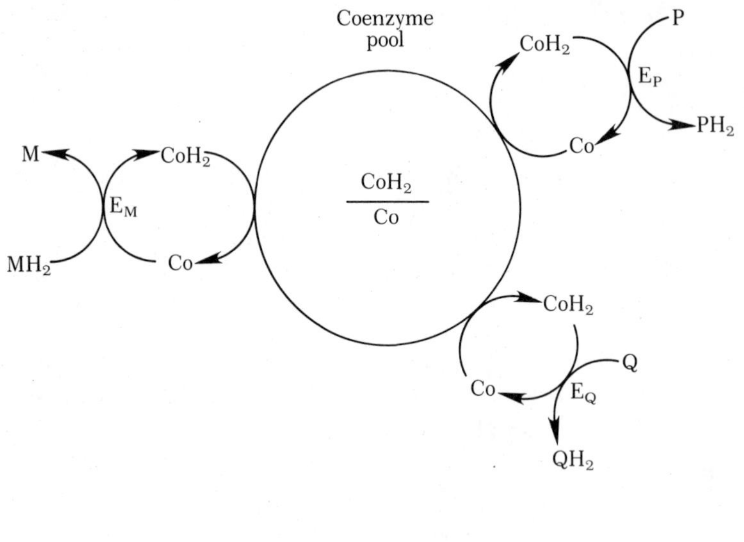

Although we might envision a given coenzyme molecule as oscillating between a number of enzyme systems, actually it is more realistic to consider the two forms of the coenzyme (CoH_2 and Co) as part of a pool. The degree of regeneration may depend entirely on the state of metabolism. Therefore, the ratio CoH_2/Co may serve as a regulatory signal in metabolism.

Other coenzymes are tightly bound to their enzymes and are not part of soluble pools. In such cases the coenzyme is considered to be a prosthetic group and a structural part of the enzyme. These coenzymes must be regenerated while attached to the enzyme molecule. When the coenzyme is altered, the enzyme is inactive; activity is restored when the original state of the coenzyme is regenerated. Therefore the metabolic system would consist of a pool of active and inactive enzyme molecules.

Water-soluble vitamins as precursors of some coenzymes

As described above, the water-soluble vitamins serve as precursors for many coenzymes. The provision of an adequate supply of vitamins is an important aspect of nutrition (Chapter 21). It should be apparent that a vitamin deficiency will lead to a depletion of the coenzyme pool and therefore will have a serious effect on the functioning of many important enzyme systems. Obviously this will adversely influence the metabolic pathways in which these enzymes function. The importance of achieving an adequate supply of vitamins and minerals is discussed in the chapter on nutrition, whereas their molecular functions are described here.

Figure 6-1
Structure of NAD^+ and $NADP^+$.

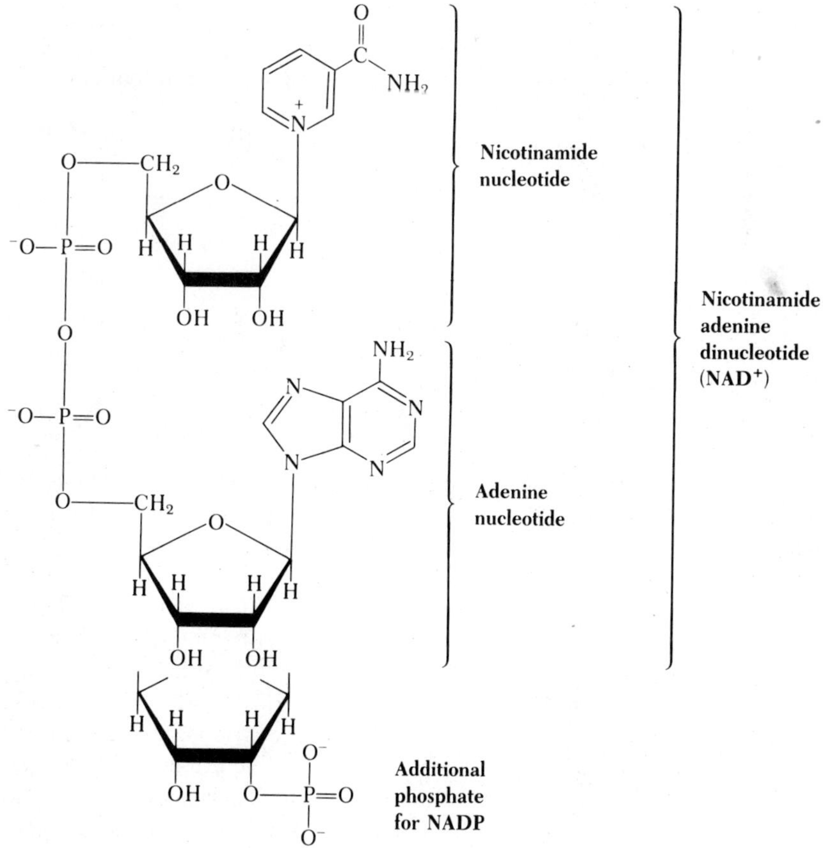

Nicotinamide adenine dinucleotide

The first oxidative step in the metabolism of carbohydrates, lipids and amino acids requires the removal of two hydrogen atoms. The coenzyme that is directly or indirectly involved is NAD^+ (Fig. 6-1). As seen in this figure, NAD^+ consists of two components, nicotinamide nucleotide and adenine nucleotide (adenosine monophosphate, or AMP). It is the nicotinamide nucleotide that is derived from the vitamin niacin, or nicotinic acid. Actually nicotinic acid, a derivative of the pyridine ring, is present as its amide, nicotinamide (Fig. 6-1). The pyridine ring is attached to ribose by a β-*N*-glycosidic bond. Carbon-5 of the ribose has a phosphate group, so the nicotinamide ribose phosphate is a nucleotide (Chapter 23). (Strictly speaking the term *nucleotide* is not accurate because the pyridine ring is not a purine or a pyrimidine.) AMP is attached by its 5′-phosphate group. Therefore, the term *dinucleotide* is used to describe the complete structure. *Nicotinamide adenine dinucleotide* is the correct name, or simple NAD^+. The positive charge refers to the quaternary nitrogen in the pyridine ring. NAD^+ is used in preference to the original name, *Coenzyme I*, or the later but inaccurate term *DPN*, which stands for diphosphopyridine nucleotide.

In the dehydrogenation of a general metabolite, MH_2, two electrons and a proton are transferred to the NAD^+. Therefore the coenzyme acts as an acceptor of a hydrogen atom and an extra electron. This may be viewed as the transfer of a hydride ion (H^-) to carbon-4 of the nicotinamide. The double bonds are rearranged and the quaternary nitrogen becomes a tertiary nitrogen. The second proton enters the environment.

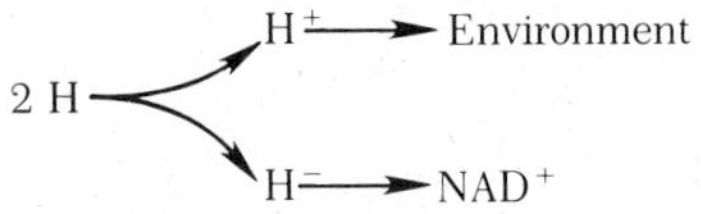

$$2\,H \rightarrow \begin{cases} H^+ \rightarrow \text{Environment} \\ H^- \rightarrow NAD^+ \end{cases}$$

The following diagram illustrates the reduction of the pyridinium ring structure.

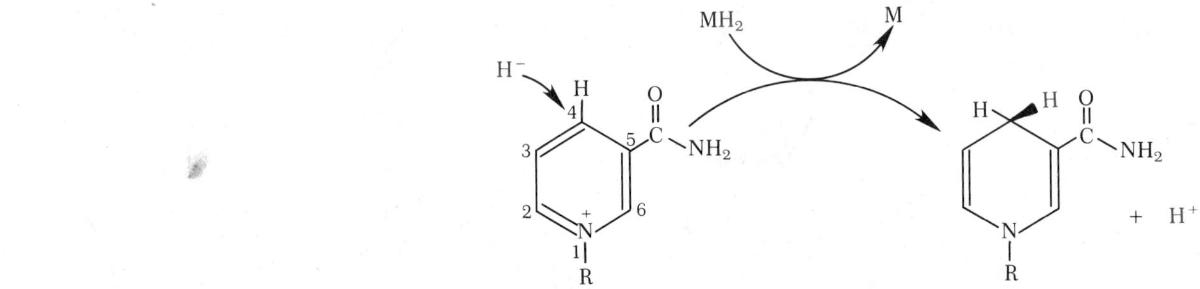

Although by convention the reduced form of NAD^+ is written *NADH*, we must not lose sight of the fact that two electrons and two protons have actually been transferred. It is instructive but cumbersome to write the reduced form as $NADH(H^+)$.

In the illustration above, it is apparent that the two hydrogens on carbon-4 of NADH project in front of and behind the plane of the pyridinium ring. It may be imagined that the substrate hydride ion may approach carbon-4 from either the front or the back of the plane. Which approach occurs depends on the positioning of the pyridinium ring with respect to the substrate molecule at the active site of the dehydrogenase. Therefore the transfer of hydrogen by the enzyme is actually stereospecific. As seen in the diagrams below, the two sides of the pyridine ring are termed the A and B sides. In the case of lactate dehydrogenase the substrate, lactic acid, and the NAD^+ are positioned at the active site so that the hydride ion

Table 6-1
Examples of enzyme systems requiring pyridine nucleotide coenzymes

Coenzyme	Enzyme
NAD^+	L-Malate dehydrogenase
	Mitochondrial isocitrate dehydrogenase
	Lactate dehydrogenase
	Glyceraldehyde 3-phosphate dehydrogenase
	β-Hydroxy acyl CoA dehydrogenase
	β-Ketoisovalerate dehydrogenase
$NADP^+$	Cytoplasmic isocitrate dehydrogenase
	Malic enzyme
	β-Ketoacyl-ACP-reductase
	Dihydrofolate reductase
	HMG-CoA reductase

approaches the pyridinium ring from the A side. All NAD^+ and $NADP^+$-dependent dehydrogenases that have been studied fall either into the A or B class of stereospecificity. Examples are L-lactate, L-malate, and isocitrate dehydrogenases, which are all of the A type. Those showing B stereospecificity are L-glycerol 3-phosphate, 3-phosphoglyceraldehyde, glucose 6-phosphate, and 6-phosphogluconate dehydrogenases.

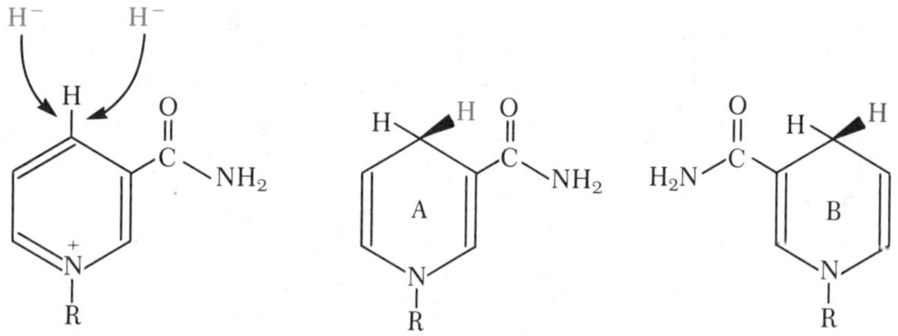

In some oxidations and especially in biosynthetic processes the phosphorylated form of NAD^+ is required. This coenzyme is called $NADP^+$ for *nicotinamide adenine dinucleotide phosphate* (Fig. 6-1). It has a third phosphate group on carbon-2 of the adenosine ribose. $NADP^+$ was formerly called *Coenzyme II* and *TPN* for triphosphopyridine nucleotide. Table 6-1 shows some of the enzymes that require NAD^+ and $NADP^+$ to catalyze their hydrogen-transferring reactions.

NAD^+ and $NADP^+$ oscillate between their oxidized and reduced states, accepting hydrogens from the metabolite and transferring them in turn to an acceptor molecule or the coenzyme of another system. Frequently these enzymes are flavoproteins and have flavin mononucleotide (FMN) and flavin adenine dinucleotide (FAD) as a prosthetic group.

Flavin mononucleotide and flavin adenine dinucleotide

Flavoprotein enzymes possess either FMN or FAD as coenzymes. The functional part of the coenzymes is the isoalloxazine ring of riboflavin, or vitamin B_2. The isoalloxazine ring is combined with the sugar alcohol ribitol and phosphate to yield what is usually termed flavin mononucleotide, or FMN (Fig. 6-2). Actually this terminology is inaccurate, since the sugar attached to the flavin group is not the aldopentose ribose and therefore is not, strictly speaking, a nucleotide. However, the terminology continues to be used. Furthermore, as seen in Fig. 6-2, the combination of FMN with AMP is called flavin adenine dinucleotide, or FAD. There are numerous examples of flavoprotein enzymes, as seen in Table 6-2.

Figure 6-2
Structures of vitamin B_2 (riboflavin) and flavin coenzymes.

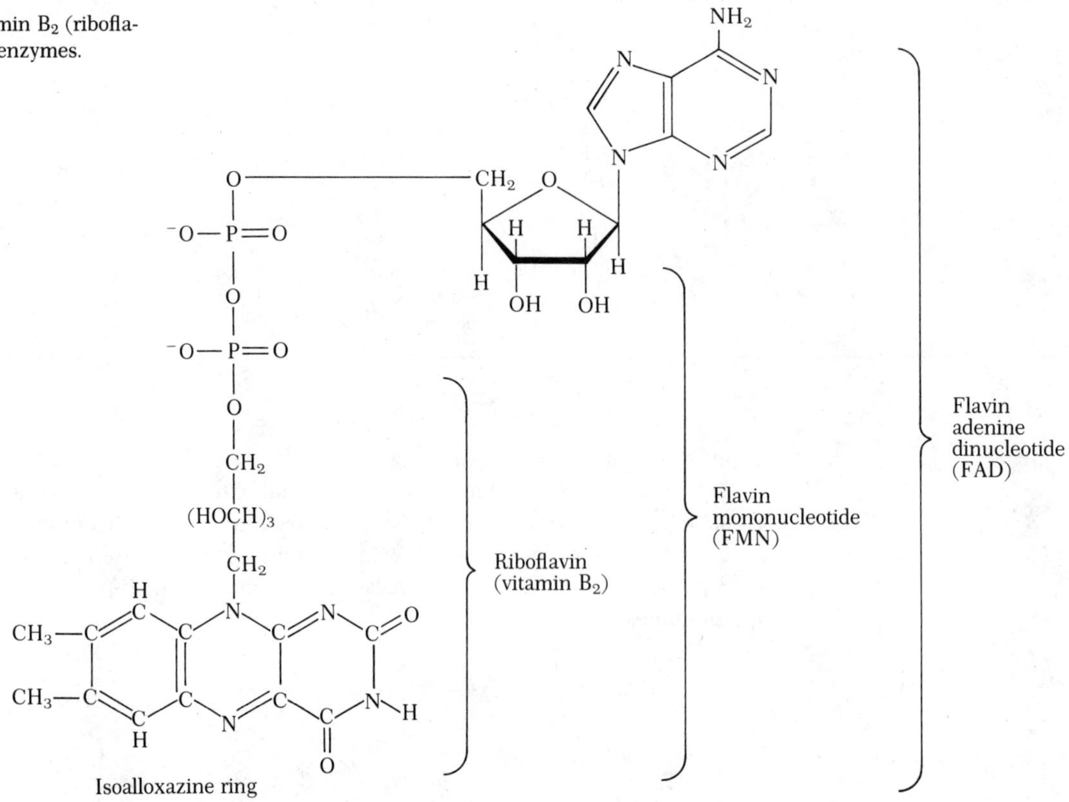

Isoalloxazine ring

Riboflavin (vitamin B_2)

Flavin mononucleotide (FMN)

Flavin adenine dinucleotide (FAD)

The coenzyme components of flavoprotein enzymes are synthesized from riboflavin as follows:

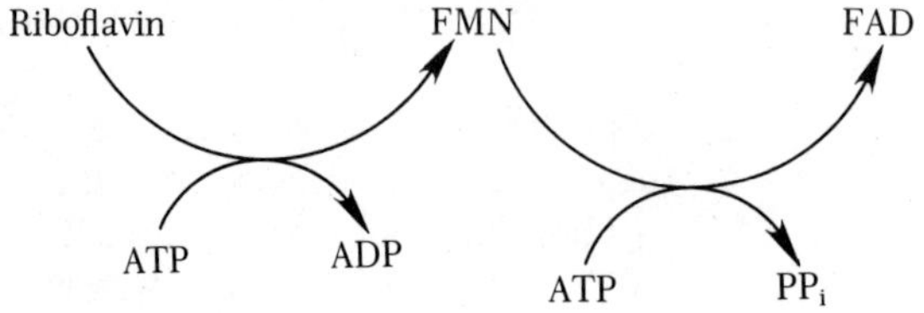

Riboflavin → FMN (ATP → ADP) → FAD (ATP → PP_i)

The flavin moiety is bound to the apoenzyme; in the case of succinate dehydrogenase, for example, it is covalently bound to a histidyl residue.

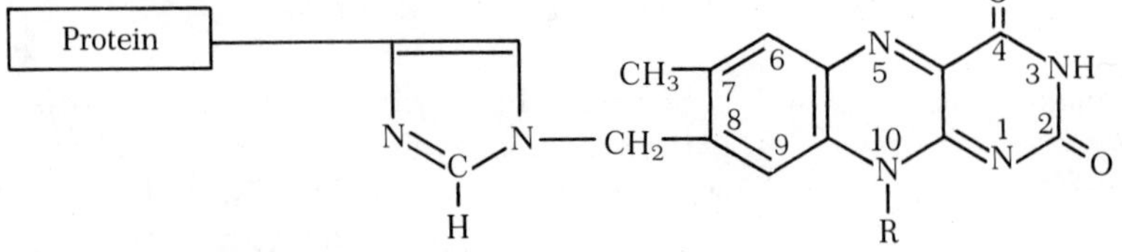

The flavoprotein enzymes function in oxidation-reduction reactions; the flavin moiety is reduced to $FADH_2$ or $FMNH_2$. As seen on p. 105, the transfer of two hydrogens from the substrate is to the isoalloxazine ring. The hydrogens are distributed between the nitrogens 1 and 5, as seen in the colored area.

Table 6-2
Characteristics of some flavoprotein enzymes

Enzyme	Metal	Electron donor	Product	Coenzyme	Electron acceptor	Product
NADH dehydrogenase	Fe	NADH	NAD^+	FMN	CoQ	$CoQH_2$
Succinate dehydrogenase	Fe	Succinate	Fumarate	FAD	CoQ	$CoQH_2$
Electron-transferring flavoprotein (ETF)		Reduced CoA	Oxidized acyl CoA	FAD	CoQ	$CoQH_2$
Dihydrolipoyl dehydrogenase		Dihydrolipoate	Oxidized lipoate	2 FAD	NAD^+	NADH
Dihydroorotate oxidase	4Fe	Dihydroorotate	Orotic acid	2 FMN, 2 FAD	NAD^+	NADH
Glycerol 3-phosphate dehydrogenase	Fe	Glycerol 3-phosphate	Dihydroxy acetone phosphate	FAD	NAD^+	NADH
Acyl CoA dehydrogenase		Fatty acyl CoA		FAD	FAD(ETF)	$FADH_2$(ETF)
D-Amino acid oxidase (kidney)		D-Amino acids	α-Keto acids + NH_3	2 FAD	O_2	H_2O_2
L-Amino acid oxidase		L-Amino acids	α-Keto acids + NH_3	2 FMN	O_2	H_2O_2
Glucose oxidase		Glucose	Gluconic acid	FAD	O_2	H_2O_2
Xanthine oxidase	Fe, Mo	Xanthine	Uric acid	FAD	O_2	O_2^-, H_2O_2

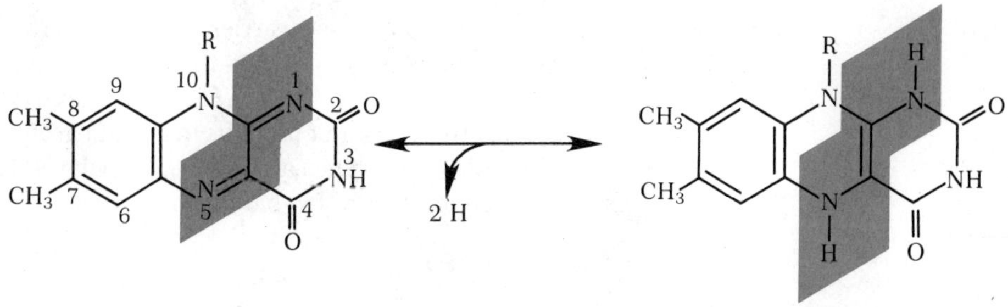

It is apparent from Table 6-2 that the flavin-containing enzymes may be considered in two categories, one in which the reduced flavin is oxidized directly by oxygen and another in which another electron acceptor, e.g., coenzyme Q, is involved. In the former group, enzymes such as D-amino acid oxidase and xanthine oxidase, the final product of the reaction is hydrogen peroxide.

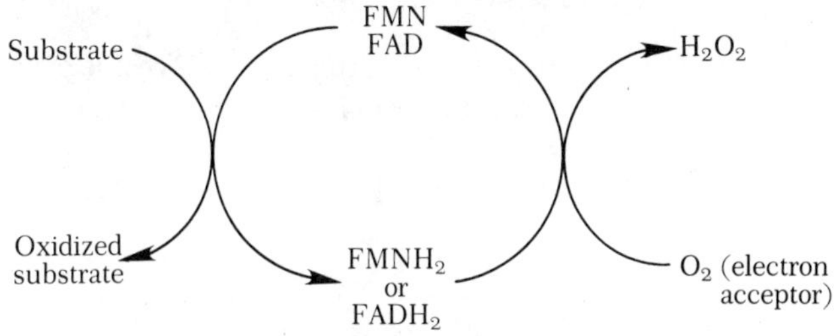

Some of the flavoprotein enzymes are also metalloproteins and require inorganic cofactors for their activity. For example xanthine oxidase has both iron and molybdenum in addition to FAD. The metal ions likewise may function by oscillating between the oxidized and reduced states.

Biotin

Biotin is an essential component of enzymes that catalyze the incorporation of CO_2 into metabolites, i.e., enzymes involved in carboxylation reactions. The vita-

Figure 6-3
Attachment of biotin to the biotin carboxyl carrier protein (*BCCP*) by linkage with the ε-amino group of a lysyl residue.

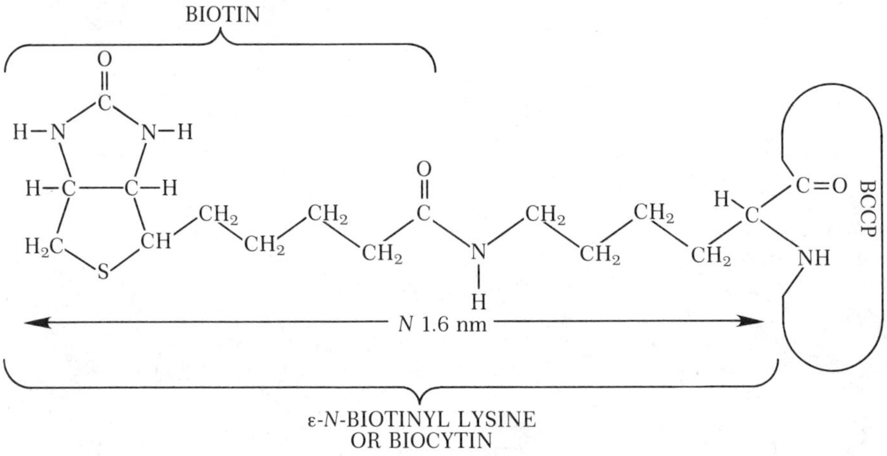

min is covalently linked to the apoenzyme by an amide bond that involves the ε-amino group of a lysyl residue. Therefore, the vitamin acts as a prosthetic group for the enzyme. The coenzyme form is considered to be ε-*N*-biotinyl lysine, also called *biocytin*. As seen in Fig. 6-3, the ring structure is attached to a flexible arm that is some 1.6 nm long.

The coenzyme serves as a carrier of carboxyl groups in the overall carboxylation reaction. Two steps are envisioned that in some cases are known to be catalyzed by two different enzymes. Actually in *E. coli* the carboxylation of acetyl CoA to malonyl CoA involves three distinct proteins: a biotin carboxylase, a biotin carboxyl carrier protein (BCCP), and a transcarboxylase. The biotin is covalently linked to the BCCP. Carboxylation of biotin by biotin carboxylase is an ATP-dependent reaction.

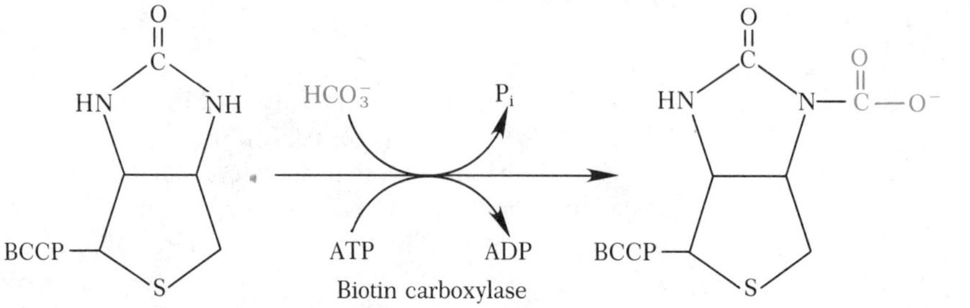

In the second step (see below) the covalently linked carboxyl group is transferred to a substrate such as acetyl CoA to yield the corresponding dicarboxylic or acid in this case its CoA derivative, malonyl CoA. The enzyme catalyzing this particular reaction is acetyl CoA: malonyl CoA transcarboxylase.

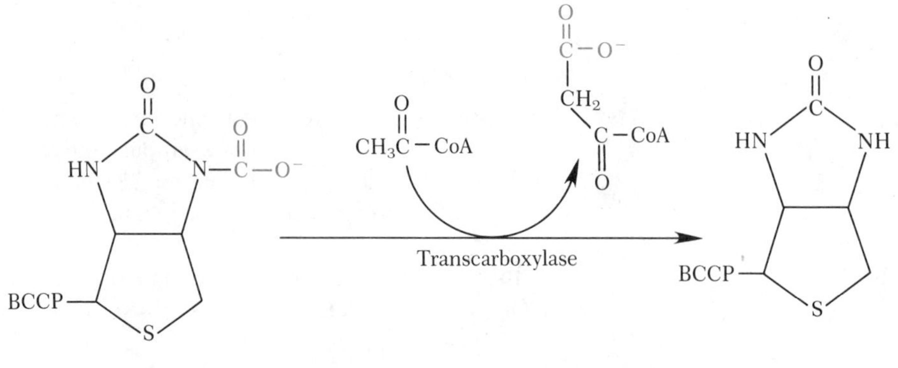

Figure 6-4
Structure of thiamin pyrophosphate.

SUBSTITUTED PYRIMIDINE RING

THIAZOLE RING

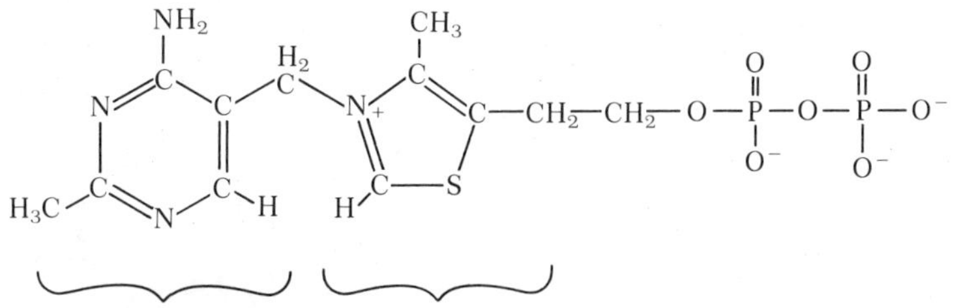

Other enzymes that catalyze carboxylation reactions are the pyruvate, propionyl CoA, and β-methyl crotonyl CoA carboxylases.

Thiamin pyrophosphate

Thiamin, or vitamin B_1, is involved in the intermediary metabolism of carbohydrates in all cells as its coenzyme form, thiamin pyrophosphate (TPP). The vitamin component consists of a substituted pyrimidine ring attached to a thiazole moiety by a methylene group (Fig. 6-4). In the active coenzyme the thiamin is linked to pyrophosphate.

TPP participates in three kinds of enzymatic reactions—nonoxidative decarboxylation, oxidative decarboxylation, and transketolation—all of which depend on the same basic mechanism. In each of the three, a C—C bond adjacent to a carbonyl group is cleaved. This is illustrated in Fig. 6-5 as the cleavage between the carbonyl carbon and the carboxyl carbon of pyruvate. The mechanism involves the addition of the α-keto acid to carbon 2 of the thiazole ring to form product I. As indicated below, the hydrogen of carbon-2 of the thiazole ring dissociates as a proton, leaving a carbanion.

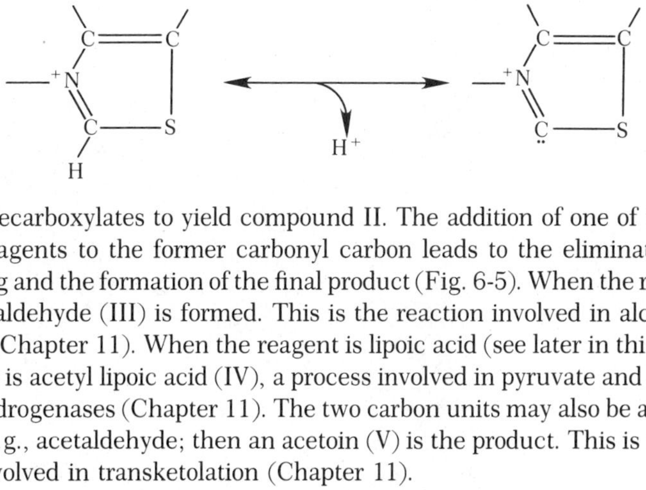

H^+

Product I decarboxylates to yield compound II. The addition of one of three electrophilic reagents to the former carbonyl carbon leads to the elimination of the thiazole ring and the formation of the final product (Fig. 6-5). When the reagent is a proton, an aldehyde (III) is formed. This is the reaction involved in alcoholic fermentation (Chapter 11). When the reagent is lipoic acid (see later in this chapter), the product is acetyl lipoic acid (IV), a process involved in pyruvate and α-ketoglutarate dehydrogenases (Chapter 11). The two carbon units may also be added to an aldehyde, e.g., acetaldehyde; then an acetoin (V) is the product. This is the type of reaction involved in transketolation (Chapter 11).

Pyridoxal phosphate

A variety of specific amino acid transformations require the coenzyme pyridoxal phosphate. This coenzyme is the phosphorylated derivative of one of a family of the B_6 vitamins (Fig. 6-6). As seen from the figure, vitamin B_6 consists of pyridoxine, pyridoxal, and pyridoxamine. Pyridoxal phosphate serves as a coenzyme for such metabolic processes as transaminations (aminotransferases) and specific decarboxylations (decarboxylases).

Figure 6-5
Three types of reactions requiring thiamin pyrophosphate as a coenzyme in carbohydrate metabolism (Chapter 11). *Reaction 1* is the mechanism involved in the production of acetaldehyde during fermentation. *Reaction 2* is involved in pyruvate dehydrogenase. *Reaction 3* is the mechanism in the transketolase step in the pentose pathway.

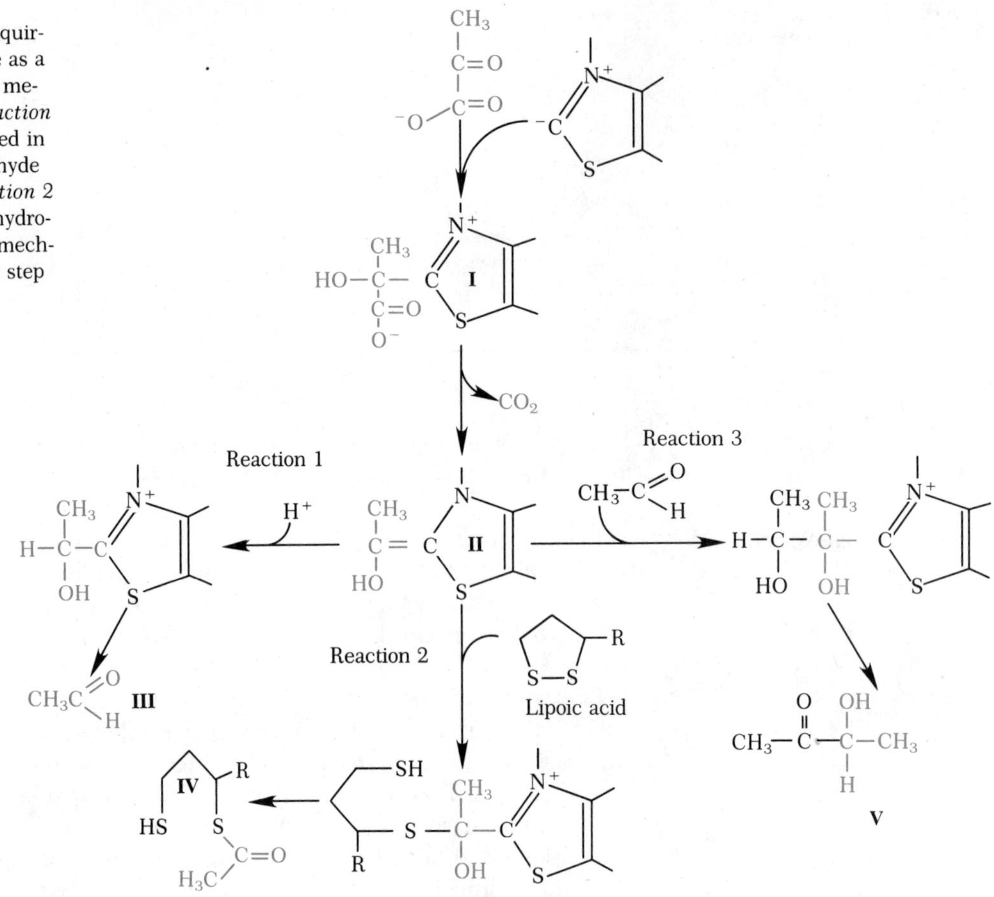

Reaction 1

Reaction 2

Reaction 3

Lipoic acid

Figure 6-6
Structural formulas of vitamin B_6 and phosphorylated forms, which serve as coenzymes.

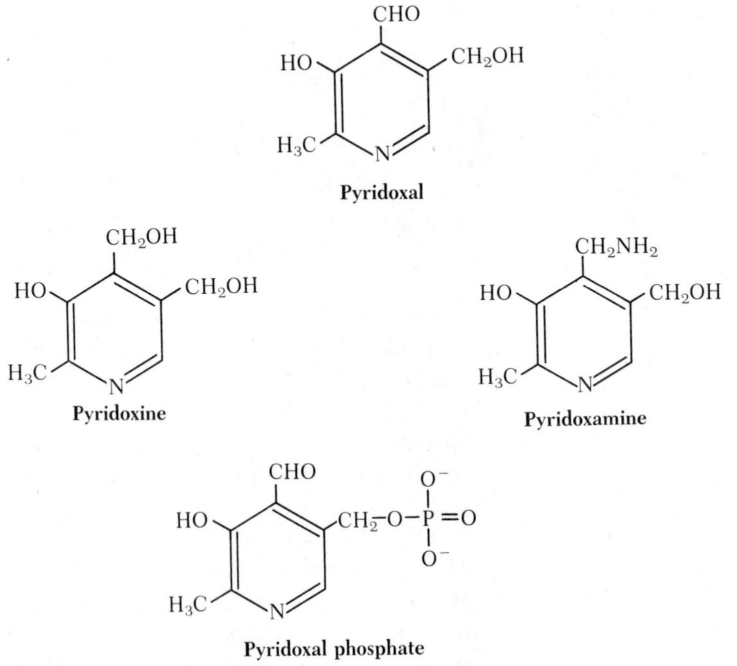

Pyridoxal

Pyridoxine

Pyridoxamine

Pyridoxal phosphate

Figure 6-7 Steps involved in transamination, illustrating the role of pyridoxal phosphate in the function of glutamate-oxaloacetate transaminase.

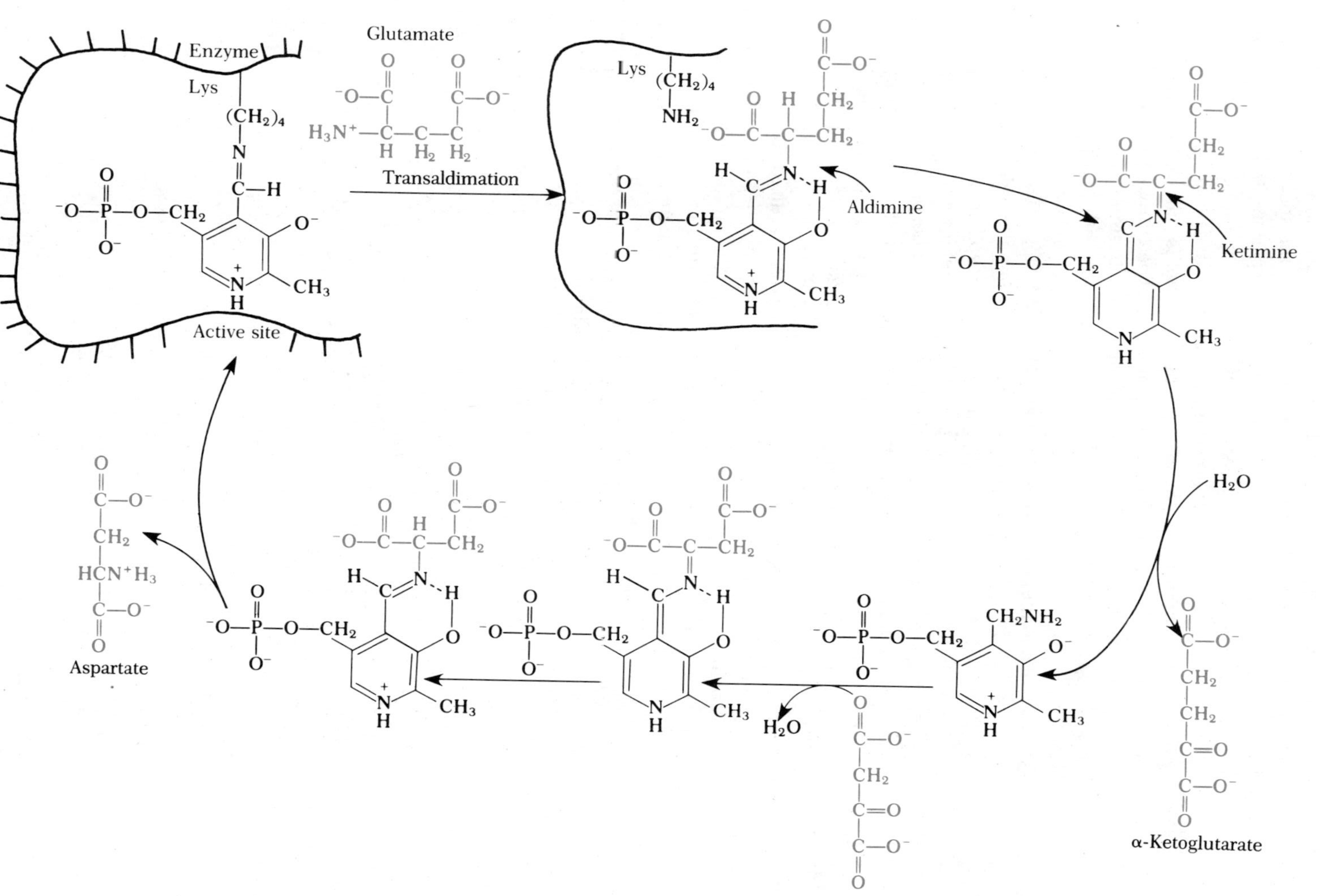

Fig. 6-7 illustrates the stepwise process of transamination. First the coenzyme is covalently linked to the apoenzyme by a Schiff base or aldimine structure. This involves the removal of a molecule of water between the ϵ-amino group of a lysyl residue and the aldehyde of the coenzyme. In the specific case of glutamate-oxaloacetate transaminase (GOT), also called aspartate aminotransferase, the substrate (glutamate) displaces the lysyl nitrogen by transaldimation. The result is a Schiff base between the glutamate and the coenzyme. The double bond of the aldimine then shifts to yield a ketimine. This is followed by hydrolysis and the release of the α-keto acid α-ketoglutarate. The nitrogen of the glutamic acid is transferred to the coenzyme to produce pyridoxamine phosphate. In the second phase the α-keto acid oxaloacetate combines with the pyridoxamine, water is removed, and a ketimine structure is formed. Rearrangement of double bonds results in the restoration of the aldimine, and subsequent hydrolysis yields aspartic acid and regenerates the coenzyme to its original pyridoxal form.

In addition to transamination, pyridoxal phosphate also serves as coenzyme for specific decarboxylases. For example, with the enzyme glutamate decarboxylase, the glutamate combines with the pyridoxal phosphate by transaldimation, as shown in Fig. 6-7 for the aminotransferase. Then the bond between the α-carbon and carboxyl group is cleaved, with the resulting elimination of carbon dioxide and release of the important neural transmitter gamma aminobutyric acid (GABA).

Tetrahydrofolic acid

Tetrahydrofolic acid (FH_4) is the coenzyme required for the transfer of such single carbon units as the formyl, methyl, methylene, and methenyl groups. The coenzyme is actually a reduced form of the vitamin folic acid, whose structure is seen in Fig. 6-8. Folic acid is composed of a substituted pteridine ring covalently bound to *p*-aminobenzoic acid (PABA) and L-glutamate. The vitamin is found in green leaves as a polyglutamyl derivative (Folacin) containing as many as seven glutamyl residues linked to the terminal carboxyl group by γ-glutamyl peptide bonds. In the intestine all of these residues except one are removed, leaving folic acid. To be converted to the coenzyme forms the vitamin must be successively reduced (NADPH), first to the dihydrofolate and then to the tetrahydrofolate.

1. $$\text{Folate} \xrightarrow[\text{Folate reductase}]{\text{NADPH} \quad \text{NADP}^+} FH_2$$

2. $$FH_2 \xrightarrow[\text{Dihydrofolate reductase}]{\text{NADPH} \quad \text{NADP}^+} FH_4$$

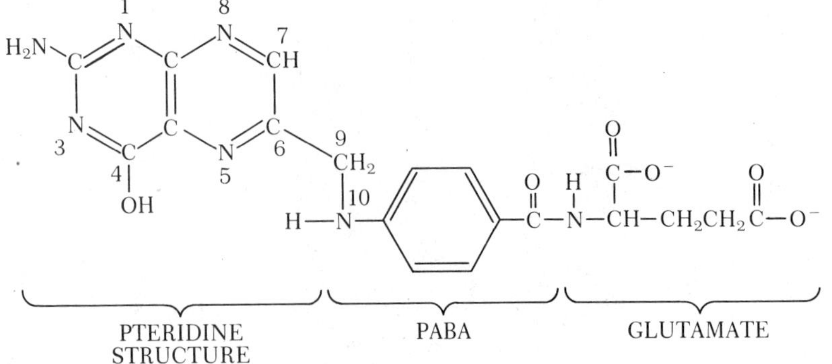

Figure 6-8
Structure of pteroyl-L-glutamate or folic acid.

The distribution of the four hydrogens is shown in color in the accompanying structure.

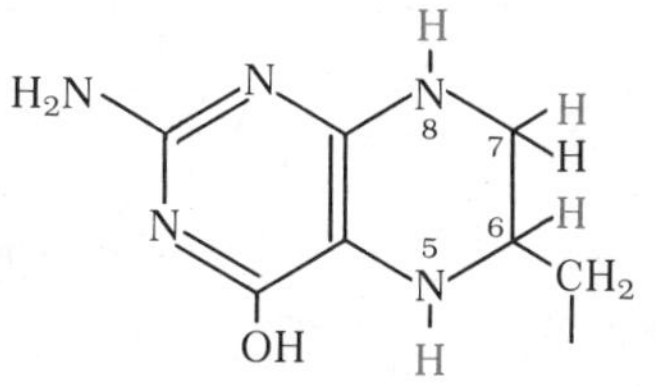

5,6,7,8-Tetrahydropteridine or FH_4

FH_4 functions as a carrier of one-carbon units in the biosynthesis of certain amino acids (Chapter 13) as well as of purines and pyrimidines (Chapter 14). In these processes the important forms of the coenzyme are the N^{10} formyl, N^5, N^{10} methenyl, and N^5, N^{10} methylene derivatives. Formation of the one-carbon derivatives involves a covalent linkage of the group with the N^5 atom of the pteridine ring, with the N^{10} of the PABA, or with both.

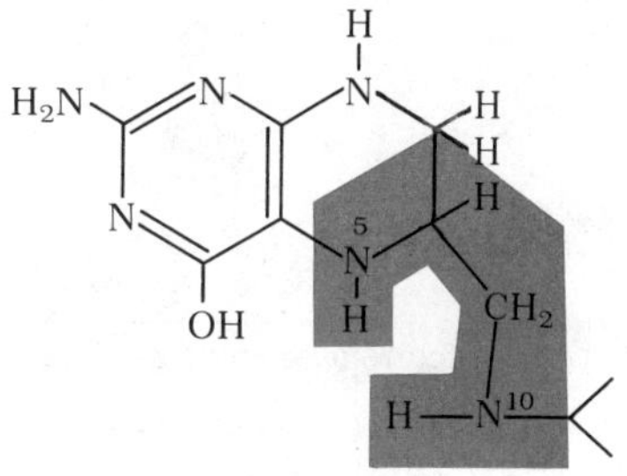

Figure 6-9 Basic role of tetrahydrofolate (FH_4) in the interconversion of one-carbon units. Fig. 13-5 is a complete diagram showing various sources and fates of one-carbon units.

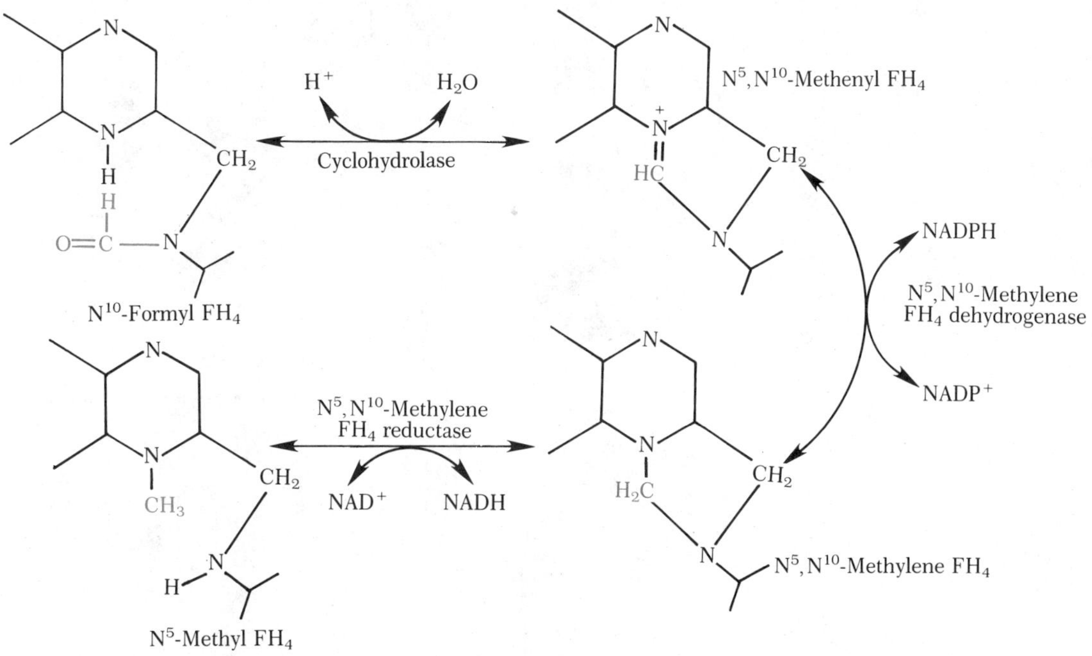

Figure 6-10 Pathway showing the requirements of pyridoxal phosphate (*PLP*) and $N^{5,10}$-methylene FH_4 in the biosynthesis of L-serine from glycine.

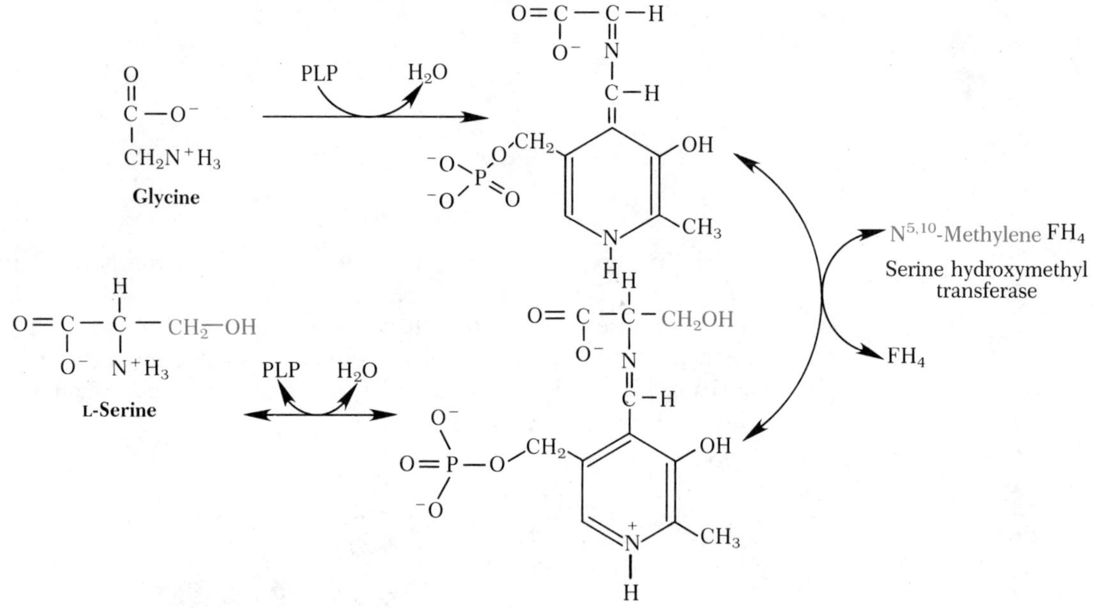

Figure 6-11 Structure of vitamin B_{12} (cyanocobalamin).

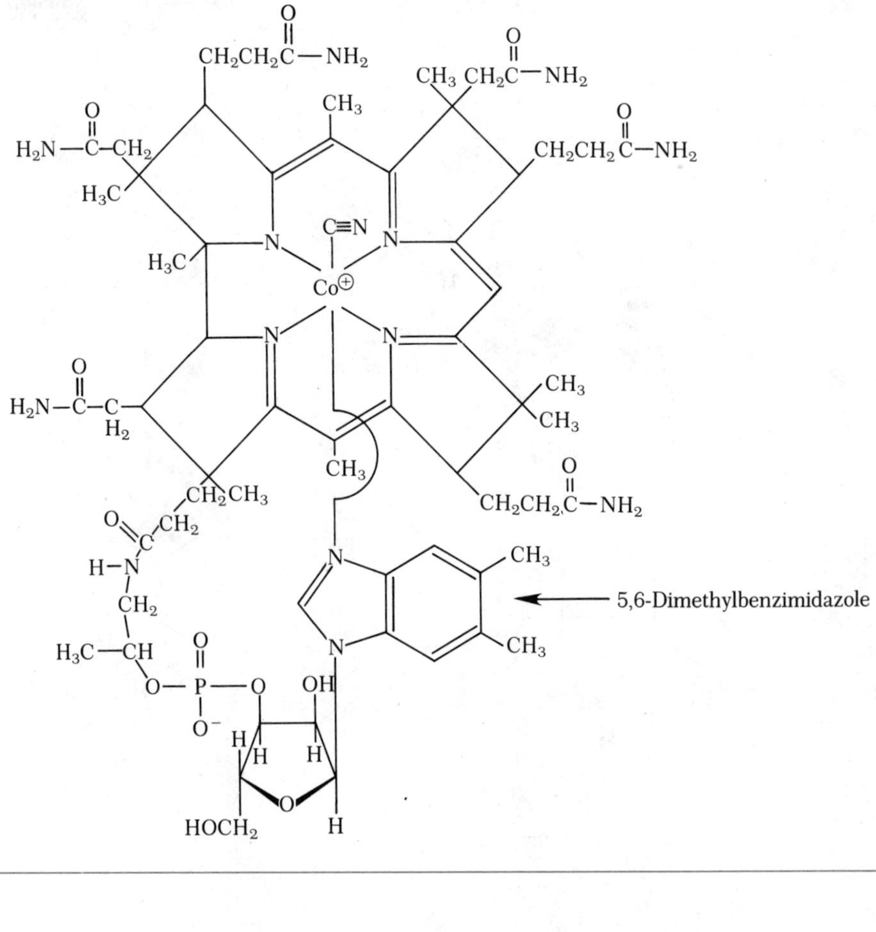

They are interrelated, as seen in Fig. 6-9. In the first step, the N^5, N^{10} methenyl FH_4 is produced by the elimination of a molecule of water from N^{10}-formyl FH_4; the enzyme is called *cyclohydrolase*. The methylene derivative is obtained by reduction with NADPH, catalyzed by *N^5, N^{10}-methylene FH_4 dehydrogenase*. Finally the N^5 methyl derivative is produced by further reduction (NADPH), catalyzed by *N^5, N^{10}-methylene FH_4 reductase*.

The role of FH_4 as a carrier of one-carbon units is exemplified in the biosynthesis of serine from glycine (Fig. 6-10). Here the N^5, N^{10}-methylene FH_4 is the coenzyme for *serine hydroxymethyl transferase*. However, pyridoxal phosphate is also required in this process. Therefore the formation of serine from glycine illustrates not only the role of FH_4 but also the functional interrelationship of two water-soluble vitamins.

Cobamide or B_{12} coenzyme

An unusual cobalt-containing vitamin called cobamide or vitamin B_{12} is part of a coenzyme required in certain group transfer reactions. Vitamin B_{12} consists of a highly substituted ring structure called the *corrin* ring (Fig. 6-11). This nucleus contains an atom of cobalt chelated to four pyrrole nitrogens. In addition a fifth coordinate linkage is to the nitrogen of a 5, 6-dimethylbenzimidazole ring. The vitamin as isolated from liver contains a cyanide ion and therefore is known as *cyanocobalamin* (Fig. 6-11).

In its coenzyme form, the cyanide is replaced by a 5-deoxyadenosyl group, as shown below.

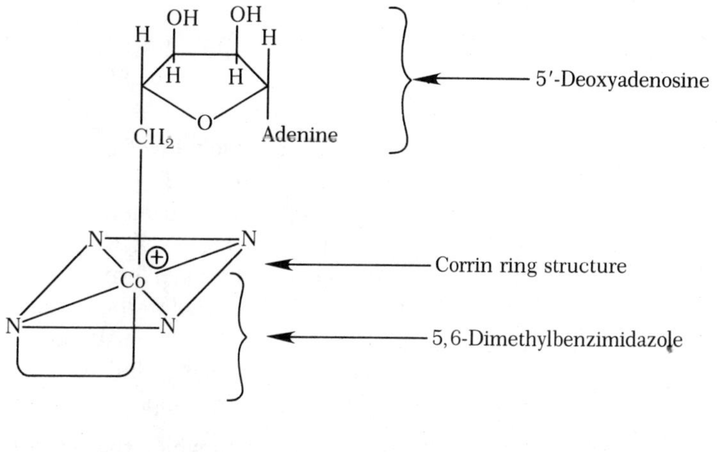

OH OH
H H
H H
O
CH_2
Adenine
5′-Deoxyadenosine
N N
⊕
Co
N N
Corrin ring structure
5,6-Dimethylbenzimidazole

The conversion of vitamin B_{12} to the coenzyme requires *coenzyme B_{12} synthetase* and ATP.

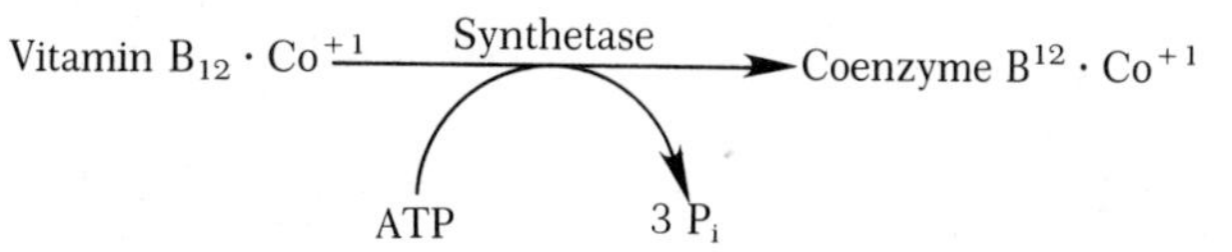

$$\text{Vitamin } B_{12} \cdot Co^{+1} \xrightarrow[\text{ATP} \;\rightarrow\; 3\,P_i]{\text{Synthetase}} \text{Coenzyme } B^{12} \cdot Co^{+1}$$

Two kinds of reactions serve to illustrate the function of coenzyme B_{12}. The first is an intramolecular rearrangement of a type exemplified below in the conversion of L-methylmalonyl CoA to succinyl CoA (Chapters 12 and 13). Here there is a shift of a hydrogen from carbon-1 to 2 and a reverse shift of the thioester from 2 to 1. In general, therefore, coenzyme B_{12} participates in a rearrangement in which there is a 1,2 shift of a hydrogen accompanied by a 2,1 shift of some other group.

$$\text{L-Methylmalonyl CoA} \xrightarrow[\text{B}_{12}]{\text{L-Methylmalonyl CoA mutase}} \text{Succinyl CoA}$$

The second type of reaction involves the transfer of methyl groups in certain transmethylations. In these, the methyl group is coordinated to the cobalt in place of the 5′-deoxyadenosyl group, forming methylcobalamin. This process is exemplified by the methylation of homocysteine to yield methionine (Chapter 13).

$$\text{Homocysteine } (HS{-}CH_2{-}CH_2{-}CHN^+H_3{-}CO_2^-) \xrightarrow{\text{Transmethylase}} \text{L-Methionine } (CH_3{-}S{-}CH_2{-}CH_2{-}CHN^+H_3{-}CO_2^-)$$

($CH_3{-}CoB_{12} \rightarrow CoB_{12}$; $N^5{-}CH_3\ FH_4 \rightarrow FH_4$)

Coenzyme A

Coenzyme A is widely involved in the transfer of acyl groups. This factor, first described by Lipmann in 1947, is a pantothenic acid–containing compound that also possesses β-alanine, β-mercaptoethylamine, two phosphates, and a 3′-phosphoadenosine unit (Fig. 6-12). The pantothenic acid component is the vitamin; coenzyme A is its functional form. The free sulfhydryl group of the β-mercaptoethylamine, which is coupled to the pantothenic acid by a peptide bond, is the functional group of the coenzyme. It serves as the acceptor site for the binding of carboxylic acids; a thioester linkage is involved.

In general acyl CoA compounds participate in four kinds of reactions.

1. Condensations in which the acyl group is attached to a nucleophile, resulting in the release of CoA (Chapter 15).

$$\underset{\text{Choline}}{HO{-}CH_2CH_2N^+(CH_3)_3} \xrightarrow{CH_3C(=O){-}SCoA \ \rightarrow\ CoASH} CH_3C(=O){-}O{-}CH_2CH_2N^+(CH_3)_3$$

2. Reactions involving dehydrogenation and dehydration (Chapter 12).

$$RCH_2CH_2C(=O){-}SCoA \xrightarrow{FAD \ \rightarrow\ FADH_2} RCH{=}CHC(=O){-}SCoA$$

Figure 6-12
Structure of coenzyme A.

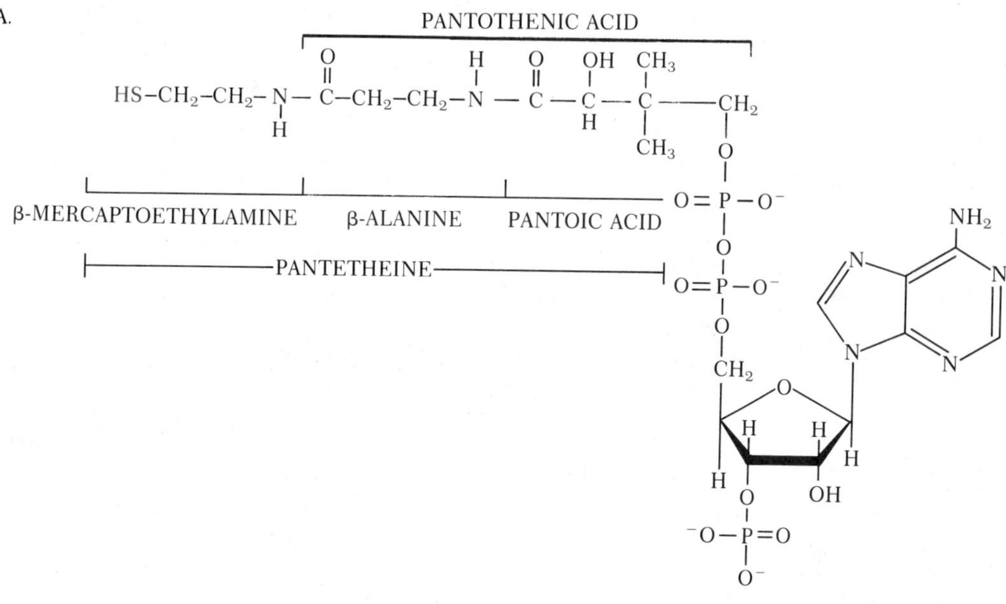

3. Reactions involving the condensation of the α-carbon of the acyl group with another group (Chapter 11).

$$O{=}C(CO_2H){-}CH_2{-}CO_2H + CH_3\overset{\overset{O}{\|}}{C}{-}SCoA \longrightarrow HO{-}C(CH_2CO_2H)(CO_2H){-}CH_2{-}CO_2H + CoASH$$

Citric acid

4. Reactions involving acyl group interchange (Chapter 12).

$$CO_2H{-}CH_2{-}CH_2{-}C{=}O{-}SCoA + CH_3\overset{\overset{O}{\|}}{C}{-}CH_2CO_2H \longrightarrow CH_3\overset{\overset{O}{\|}}{C}{-}CH_2\overset{\overset{O}{\|}}{C}{-}SCoA + CO_2H{-}CH_2{-}CH_2{-}CO_2H$$

Pantothenic acid also is an important constituent of the coenzyme (prosthetic group) in the acyl carrier protein (ACP) of the fatty acyl synthetase complex (Chapter 12). ACP has a 4′-phosphoryl pantetheine structure covalently linked to the hydroxyl group of a serine residue in the protein.

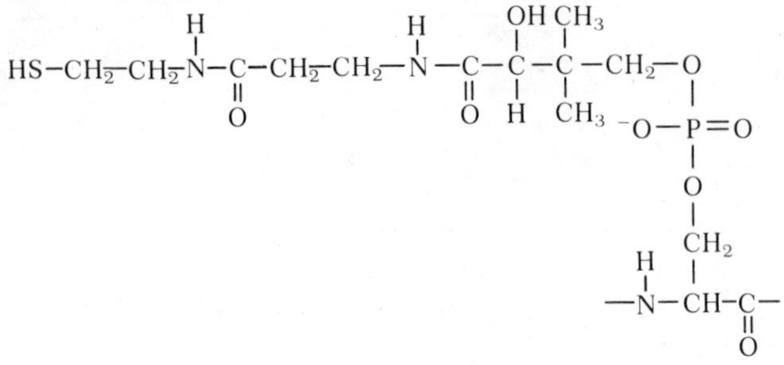

HS–CH$_2$–CH$_2$–N(H)–C(=O)–CH$_2$–CH$_2$–N(H)–C(=O)–C(OH)(H)–C(CH$_3$)$_2$–CH$_2$–O–P(=O)(O$^-$)–O–CH$_2$–CH(–NH–)–C(=O)–

ACYL-CARRIER PROTEIN

As is shown in detail in Chapter 12, the sulfhydryl group of the pantetheine moiety functions in a manner analogous to that of coenzyme A. In other words it forms a thioester with the carboxylic acid and carries the growing fatty acid throughout the biosynthetic sequence.

Coenzymes derived from nonvitamin precursors

Lipoic acid

Lipoic acid is required in the decarboxylation of pyruvate and α-ketoglutarate; it serves as a coenzyme in pyruvate dehydrogenase and α-ketoglutarate dehydrogenase. As seen below, lipoic acid is an eight-carbon, disulfide-containing carboxylic acid that is attached to dihydrolipoyl transacetylase by an amide bond (ε-amino group of lysine).

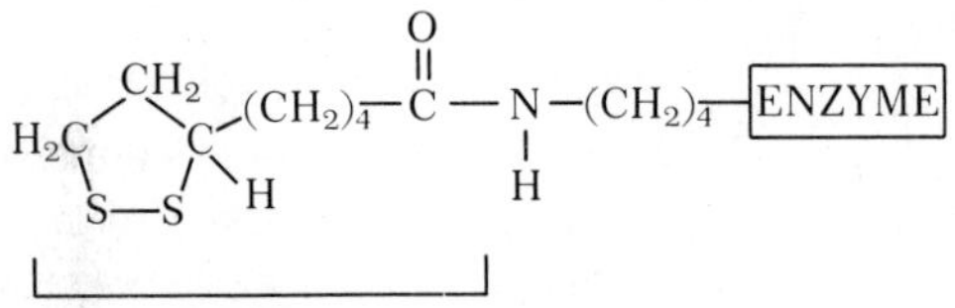

H$_2$C–CH$_2$–C(H)(–S–S–)–(CH$_2$)$_4$–C(=O)–N(H)–(CH$_2$)$_4$–ENZYME

Lipoic acid

In the multienzyme complex, pyruvate dehydrogenase, the lipoic acid functions as a carrier of acyl groups. The acyl group is transferred to lipoic acid from the thiamin pyrophosphate derivative (Fig. 6-5). The disulfide bond is reduced and an acyl thioester linkage is formed, as illustrated in Fig. 6-5. The lipoic acid then transfers the acyl group to the final acceptor molecule, which is coenzyme A. This process is shown below.

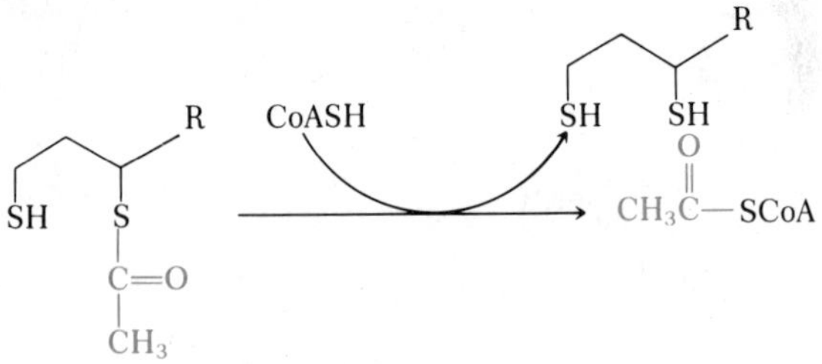

SH, S–C(=O)–CH$_3$ (R) + CoASH ⟶ SH, SH (R) + CH$_3$C(=O)–SCoA

The overall sequence of events beginning with pyruvate and ending in acetyl CoA, including the reoxidation of lipoyl sulfhydryl groups to the oxidized disulfide bond, is shown in Fig. 11-7.

Coenzyme Q

Coenzyme Q, often called ubiquinone because of its ubiquitous occurrence in nature, is a lipid-soluble hydrogen (electron) carrier found in mitochondrial membranes. As is evident from its structure (Fig. 6-13), it is a benzoquinone having an

Figure 6-13
General structure of coenzyme Q, or ubiquinone. Mitochondria from human tissue have coenzyme Q with 10 isoprenoid units.

BENZOQUINONE { CH_3O, CH_3O, CH_3, O, O }
$(CH_2{-}CH{=}C(CH_3){-}CH_2)_n{-}H$
ISOPRENOID UNITS

extended isoprenoid side chain. The size of this chain varies from source to source; the form derived from human mitochondria has 10 repeating units. Other homologs of coenzyme Q contain 6 to 10 units and have been isolated from various microorganisms, chloroplasts of green plants, and the mitochondria of beef and other animal tissues.

This factor functions in mitochondria primarily as a link between flavin-containing dehydrogenases and the cytochromes of the electron transport chain (Chapter 7). As illustrated below, coenzyme Q acts as an acceptor of hydrogens (electrons) from such flavoproteins as NADH dehydrogenase and succinate dehydrogenase, thereby serving to restore the constituent FMN and FAD to their oxidized states. Coenzyme Q is restored to its quinone form by the transfer of two electrons to cytochromes (heme iron) and the simultaneous release of two protons into the environment.

FMN H_2 FMN
FAD H_2 FAD
CH_3O, CH_3O, CH_3, R (quinone, O, O)
CH_3O, CH_3O, CH_3, R (hydroquinone, OH, OH)
2 H^+ 2 Fe^{++} 2 Heme Fe^{+++}

Biopterin

Another pteridine-containing coenzyme is biopterin (Fig. 6-14), which is required for the action of a group of hydroxylases such as the hepatic *phenylalanine hydroxylase* (Chapter 13). As seen below, tetrahydrobiopterin *(TB)* serves as the source of hydrogen in the hydroxylation of the aromatic amino acids. The dihydrobiopterin *(DB)* is reduced to tetrahydrobiopterin by NADPH; the enzyme is *dihydropteridine reductase*.

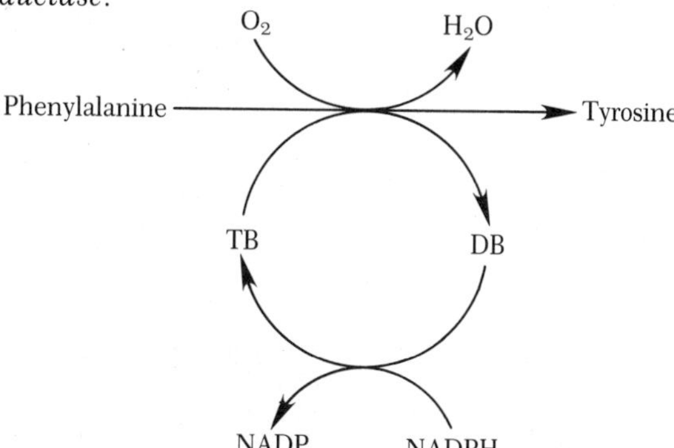

Figure 6-14
Structures of dihydrobiopterin and tetrahydrobiopterin. This compound functions as a coenzyme for phenylalanine hydroxylase.

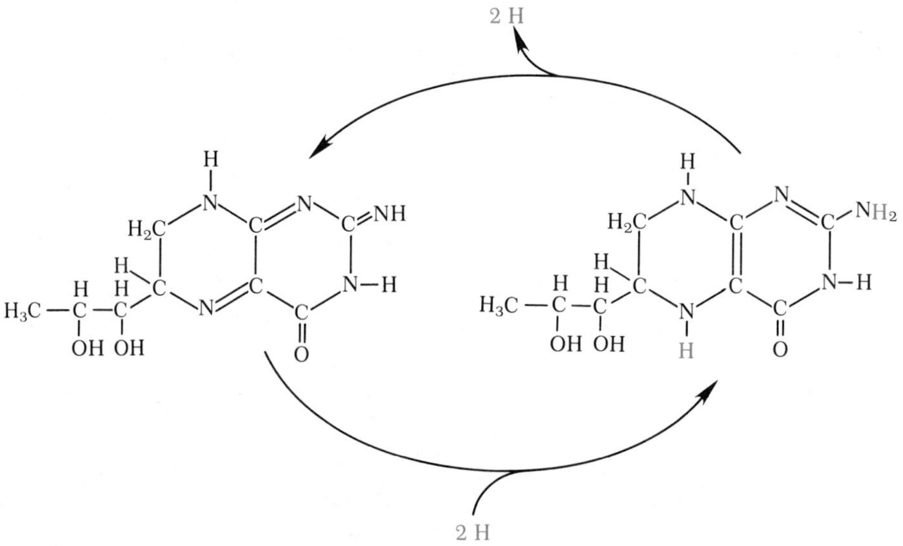

Another important example of a biopterin-dependent enzyme is *tryptophan hydroxylase* (Chapter 13), which produces 5-hydroxytryptophan, the precursor of an important neurotransmitter, 5-hydroxytryptamine.

Nucleoside triphosphates: ATP, UTP, GTP, and CTP

Many metabolic processes are energy requiring. These endergonic reactions are usually coupled to the hydrolysis of ATP (Chapter 7). The energy needed to drive otherwise energetically unfavorable reactions is derived from the hydrolysis of the phosphoanhydride bonds. Nucleoside triphosphates are also used to modify and activate certain enzymes such as glycogen phosphorylase kinase and glycogen

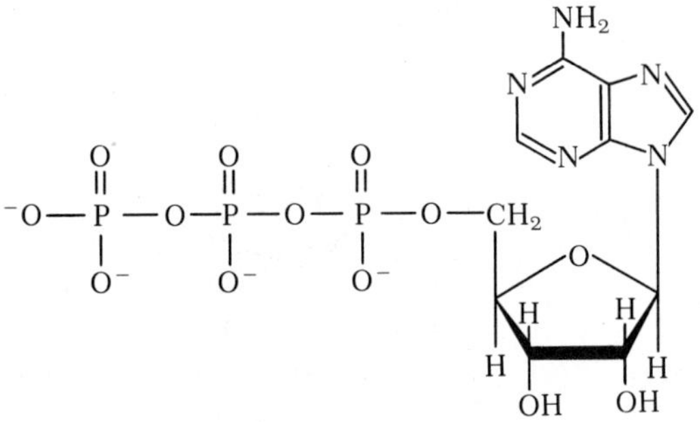

phosphorylase (Chapters 5 and 11). ATP and the other nucleoside triphosphates, uridine triphosphate (UTP), cytidine triphosphate (CTP), and guanosine triphosphate (GTP), are in fact substrates in phosphorylation reactions. Since they must be constantly restored and since their levels serve as metabolic signals, they are also described here as coenzymes. As is shown in Chapter 7, the restoration of ATP from

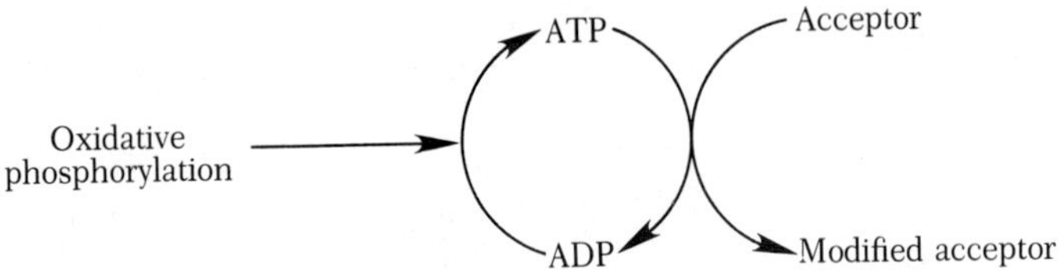

ADP occurs as a consequence of a transfer of electrons from a substrate to molecular oxygen. The energy changes occurring in this process are harvested in the form of the ATP molecules.

The ATP-dependent reactions may be considered to fall into two classes, namely those in which a component of the ATP is transferred to an acceptor and those in which the cleavage of ATP provides the necessary driving force for the overall thermodynamically unfavorable reactions. Many reactions involving ATP require the donation of a group to an acceptor (e.g., phosphoryl, pyrophosphoryl, adenyl, and adenosyl structures), whereas, on the other hand, ligases and synthetases require ATP as an energy source.

The following are general ATP-requiring reactions in the metabolism of nutrients. They are considered in detail in their appropriate chapters.

Lipid metabolism (Chapter 12):

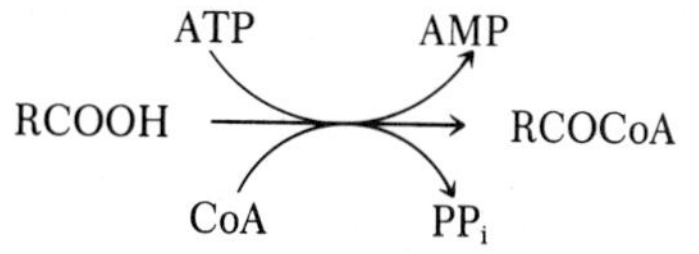

$$\text{RCOOH} \xrightarrow[\text{CoA} \;\to\; \text{PP}_i]{\text{ATP} \;\to\; \text{AMP}} \text{RCOCoA}$$

Carbohydrate metabolism (Chapter 11):

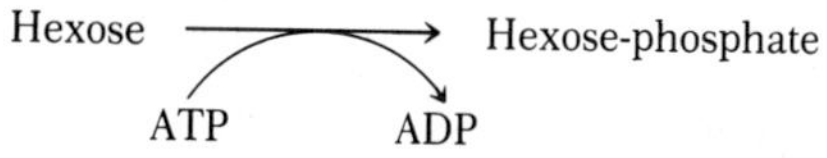

$$\text{Hexose} \xrightarrow[\text{ATP} \;\to\; \text{ADP}]{} \text{Hexose-phosphate}$$

Amino acid metabolism (Chapter 13):

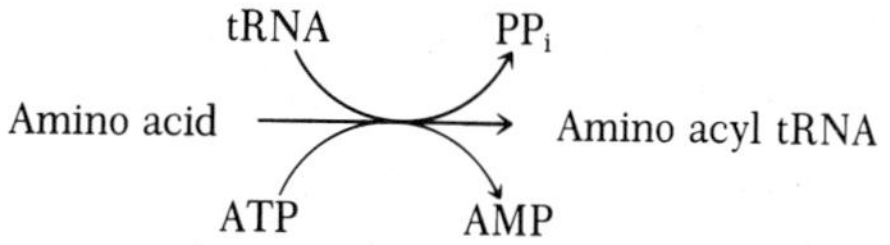

$$\text{Amino acid} \xrightarrow[\text{ATP} \;\to\; \text{AMP}]{\text{tRNA} \;\to\; \text{PP}_i} \text{Amino acyl tRNA}$$

UTP, GTP, and CTP are each important in the metabolism of carbohydrates (Chapter 11). In addition CTP plays an essential role in the biosynthesis of phospholipids (Chapter 12). Each of these metabolic processes results in a depletion of these nucleoside triphosphates. They must be regenerated at the expense of ATP (Chapter 14).

S-Adenosylmethionine

A coenzyme that plays a key role in methylation reactions is S-adenosylmethionine (SAM). As seen below, it is a coenzyme that makes the methyl groups of methionine available for transmethylation reactions. The activating enzyme is *methionyl adenosyl transferase.*

$$CH_3-S-CH_2-CH_2-CHN^+H_3-COO^- \xrightarrow[\text{Methionyl adenosyl transferase}]{\text{ATP} \;\to\; PP_i + P_i} CH_3-S^+(\text{-CH}_2\text{-adenosine})-CH_2-CH_2-CHN^+H_3-COO^-$$

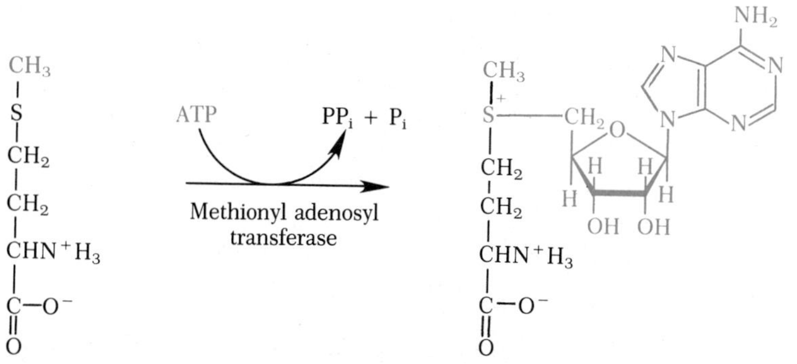

L-Methionine **S-Adenosylmethionine (SAM)**

It is apparent from the above that both inorganic phosphate and pyrophosphate are removed from the ATP. The adenosine unit is attached to the sulfur of methi-

Figure 6-15
A Pyrrole ring structure.
B Protoporphyrin IX structure.
C Heme structure.

A

Pyrrole ring

B

Methine bridges

Protoporphyrin IX

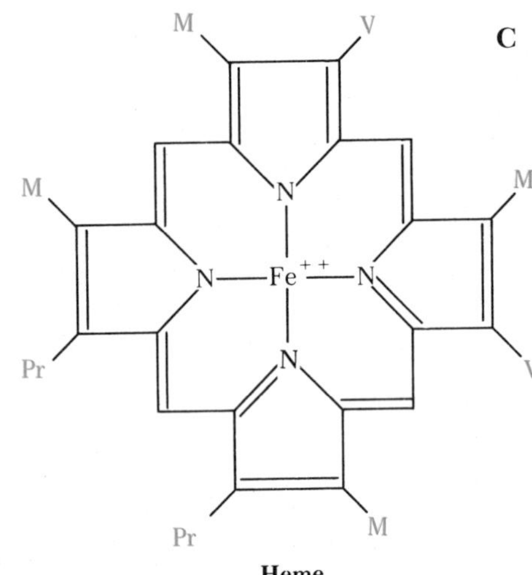

Heme

onine to yield a sulfonium derivative. SAM is a high-energy compound that acts as a donor of methyl groups in reactions producing methylated derivatives. Enzymes requiring SAM as a coenzyme are called *transmethylases*.

$$\text{Acceptor} \xrightarrow[\text{SAM} \quad \text{SAM}]{\text{Transmethylase}} \text{Methylated acceptor}$$

A product of transmethylation is S-adenosylhomocysteine, which is subsequently converted to homocysteine and adenosine. Homocysteine may be restored to methionine by the transfer of a methyl group from N^5-methyl FH_4. This reaction is catalyzed by *homocysteine methyl transferase (methionine synthetase)*, a vitamin B_{12}–containing enzyme. The cyclical metabolic fate of SAM and its biosynthesis are described in Chapter 13.

Heme

Heme is an iron-containing prosthetic group found in a variety of conjugated proteins often called hemoproteins. The most prevalent of these is hemoglobin, the protein of red blood cells that functions in the transport of oxygen. Since hemoglobin is not an enzyme, the heme unit cannot truly be called a coenzyme. It is best termed a functional prosthetic group. There are, however, heme-containing enzymes in the cell so that in these cases, the heme may legitimately be considered a coenzyme.

Heme is a chelate of iron with a particular tetrapyrrole ring structure called protoporphyrin IX. The porphyrin structure is composed of four pyrrole rings (Fig. 6-15, *A* and *B*) joined by methene bridges. In protoporphyrin IX the tetrapyrrole

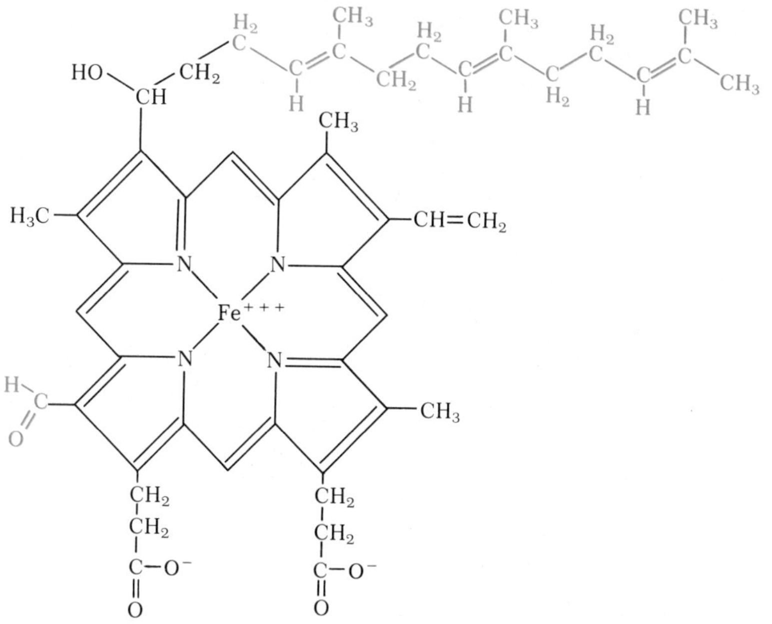

Figure 6-16
Structure of heme A, the prosthetic group of cytochrome oxidase.

bears a specific sequence of side chains, i.e., methyl-vinyl-methyl-vinyl-methyl–propionic acid–propionic acid–methyl (Fig. 6-15, *B*). This structure is planar and has alternating single and double bonds. When iron is introduced to yield heme, it forms coordination bonds with the four pyrrole nitrogens (Fig. 6-15, *C*). Each nitrogen contributes a pair of electrons to the iron. The formation of these bonds displaces two protons from the pyrrole nitrogens, leaving a net negative charge. These two negative charges are balanced by the two positive charges of ferrous iron.

Enzymes that require heme as a coenzyme are *cytochrome oxidase,* hydroxylases, and oxygenases. *Peroxidase* and *catalase* are heme proteins that catalyze reactions with hydrogen peroxide.

As is described in Chapter 7, the electron transport chain consists of a sequence of heme-containing proteins, i.e., cytochromes b, c, and c_1. These hemoproteins are electron-transport proteins. The chain is terminated by an enzyme, cytochrome oxidase. Its optical properties show it to be two proteins, which are designated cytochrome a and cytochrome a_3. Two heme prosthetic groups are present as well as two atoms of copper. Fig. 6-16 shows that the heme of cytochrome oxidase called heme A is different from that shown in Fig. 6-15. In this case the porphyrin ring is unique in having a formyl group instead of a methyl group and a hydrocarbon chain in place of a vinyl group. In the electron transport chain (Chapter 7) cytochrome oxidase transfers electrons from cytochrome c to oxygen (see below).

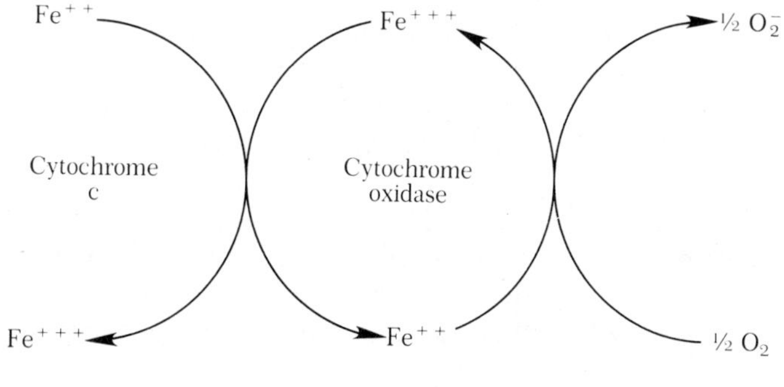

Fe^{++} → Cytochrome c → Fe^{+++}; Fe^{+++} → Cytochrome oxidase → Fe^{++}; ½ O_2 → ½ O_2^-

Figure 6-17
Pathway illustrating the role of cytochrome P_{450} in hydroxylation reactions.

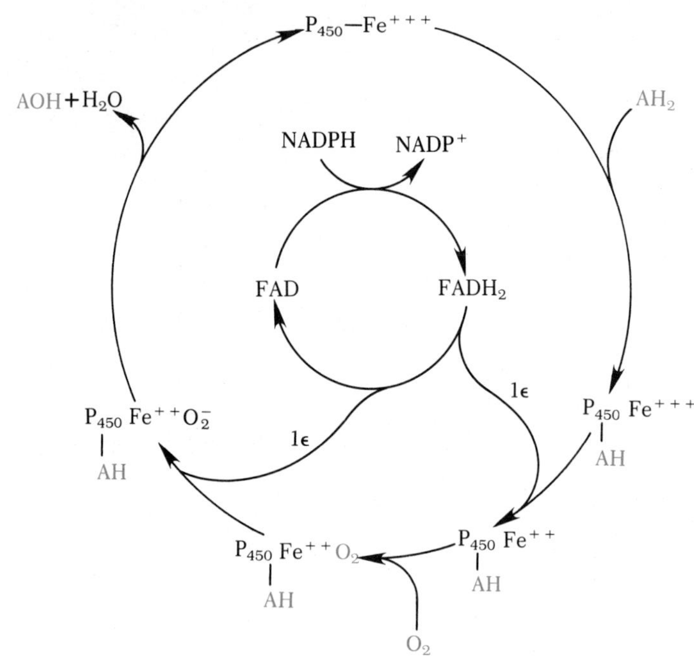

The enzyme peroxidase catalyzes the transfer of an electron from a substrate to H_2O_2.

$$AH_2 + H_2O_2 \rightarrow A + 2\ H_2O$$

In this process there is a transient oxidation of the iron. Polymorphonuclear leukocytes or granulocytes that phagocitize and destroy bacteria do so by the action of *myeloperoxidase* and hydrogen peroxide.

A second heme enzyme of great importance is catalase, which catalyzes the conversion of hydrogen peroxide to water and oxygen.

$$H_2O_2 + H_2O_2 \rightarrow 2\ H_2O + O_2$$

Catalase is found in the subcellular organelles called peroxisomes (Chapter 2). Apparently the function of the enzyme is protective, serving to remove hydrogen peroxide which otherwise would be toxic to the organism.

In the hydroxylation of steroids an important component of the system is the hemoprotein cytochrome P_{450}. It is located in the endoplasmic reticulum of the liver and is also an essential component of enzyme systems that metabolize certain drugs. The role of this cytochrome in hydroxylation reactions is illustrated in Fig. 6-17.

Cytochrome b_5 is yet another hemoprotein associated with the endoplasmic reticulum of the liver and is an essential part of the desaturation of fatty acids (desaturase). The conversion of stearoyl CoA to oleoyl CoA (Chapter 12) may involve an initial hydroxylation in which the heme moiety of the cytochrome b_5 serves as an electron carrier for the desaturase system.

Inorganic cofactors: metal ions

As already indicated many enzymes require metal ions as functional elements. A few examples are presented in Table 5-1 and Table 6-2. From these it is also evident that some enzymes require both organic and inorganic cofactors. Since the only

Table 6-3
Characteristics of metal-activated enzymes and metalloenzymes

Metal-activated enzymes	Metalloenzymes
Metal reversibly bound	Metal firmly bound
Metal to protein ratio is variable	Metal to protein ratio is constant
Metal to enzyme activity ratio is variable	Metal to enzyme activity ratio is constant
Metal is not necessarily unique	Metal is unique
Enzyme activity may exist without metal	No enzyme activity without metal

source of the minerals is the diet (Chapter 21), the adequacy of their consumption, as also is true of the vitamins, is a matter of considerable nutritional concern.

From the scheme presented on p. 99 it is evident that the association of the metal ions may be a tight one ($K_d = 10^{-7}$ to 10^{-8} molar), in which case the term used to describe such proteins is *metalloenzyme*. Other enzymes also may require the presence of metal ions for optimal activity but do not form a stable complex (K_d = greater than 10^{-7} molar). These enzymes are often called metal-activated enzymes. Some characteristics for these two classes are compared in Table 6-3.

Metal ions either function in the enzymatic process or else they serve to stabilize the catalytically active conformation of the protein. Usually metal ions form ternary complexes with the enzyme and substrates. This is especially true for metalloenzymes. Four types of complexes are visualized: (1) a metal bridge between enzyme and substrate, Ⓔ—M—S; (2) a cyclic complex, Ⓔ——M (with both Ⓔ and M bonded to S); (3) a complex in which metal and substrate bind to separate sites, Ⓔ——M (with S bonded to Ⓔ); and (4) a substrate bridge complex, Ⓔ—S—M. In those metalloenzymes that form the third kind of complex, the metal ion may serve primarily to stabilize the catalytically active conformation (active site). The fourth class is composed of those complexes in which the metal is bound to the substrate and not directly to the protein. In these enzymes the metal ion may not be tightly bound.

In the ensuing discussion only a few examples are cited to illustrate the function of metal ions in enzyme catalysis. They are arranged according to the distinction between metalloenzymes and metal-activated enzymes.

Metalloenzymes

This group of enzymes can be further segregated into metalloproteins, in which the metal and protein only are associated; metalloporphyrins, in which the metal ion is associated with the porphyrin ring structure; and metalloflavin enzymes, in which the metal-containing protein also requires a flavin unit for enzymatic activity.

Metalloprotein

Copper Superoxide dismutase is a metalloenzyme having both zinc and copper ions. This enzyme functions in the catalytic process as follows:

$$O_2^- + O_2^- + 2\,H^+ \rightarrow H_2O_2 + O_2$$

The copper functions in the catalytic process as follows:

$$E—Cu^{++} + O_2^- \rightleftharpoons E—Cu^+ + O_2$$
$$E—Cu^+ + O_2^- + 2\,H^+ \rightleftharpoons E—Cu^{++} + H_2O_2$$

The role of the zinc ion is currently unknown.

Zinc Carboxypeptidase A, a pancreatic enzyme already described in Chapter 5 (Fig. 5-6), possesses a single zinc ion that is tightly coordinated to two histidine imidazole rings and the carboxylate group of a glutamate residue. The zinc, as seen below, reacts with the carbonyl group, making it more susceptible to nucleophilic attack by the carboxylate group of glutamate 270 (Fig. 5-6). Therefore as indicated below, the ternary complex formed is of the cyclic type.

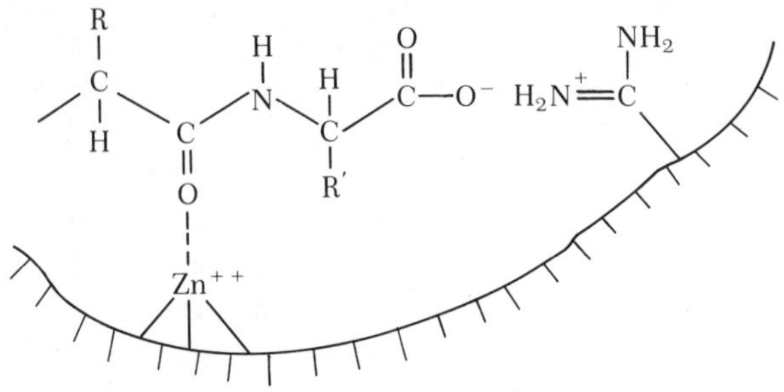

Carboxypeptidase

Another zinc-containing metalloenzyme of great physiological significance is *carbonic anhydrase*. In this enzyme the zinc is tightly bound by coordination to three imidazole rings. A molecule of water serves as the fourth ligand. This Zn^{++}—H_2O dissociates readily (pK of about 7.0) into Zn^{++}—$OH^-$and H^+. As seen below, carbon dioxide is then captured and a bicarbonate ion liganded to the zinc is formed.

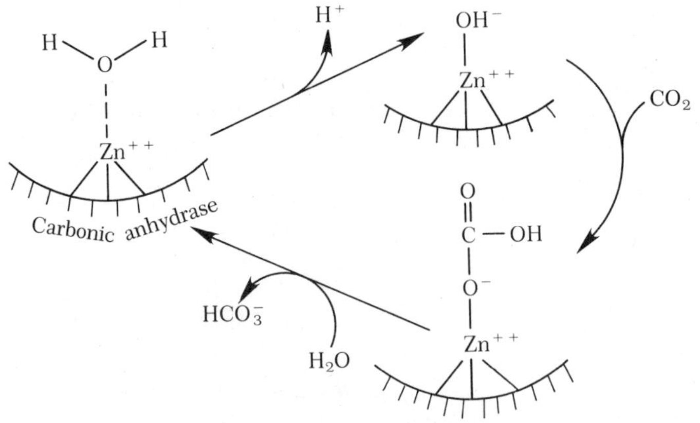

A metalloenzyme possessing both inorganic and organic (NAD^+) cofactors is *alcohol dehydrogenase*. This enzyme has two zinc ions for each NAD^+. One of them ligates a molecule of water, leading again to the formation of Ⓔ—Zn^{++}—OH^-. This species then binds the alcohol (RCH_2OH), as seen below. A hydride ion is transferred to the NAD^+, resulting in the aldehyde liganded to the zinc. The second zinc ion serves to stabilize the active conformation.

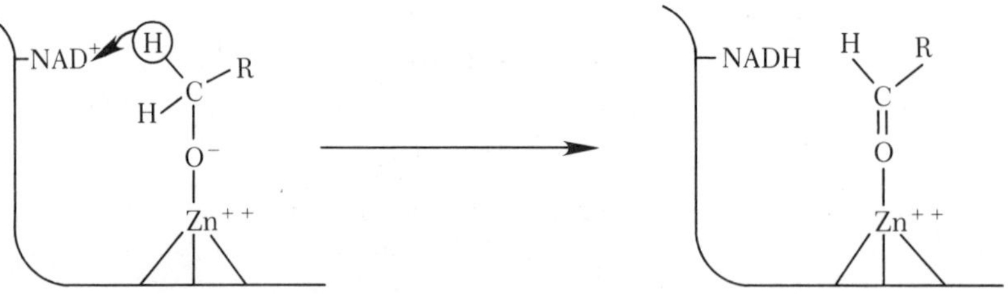

Metalloporphyrin

Iron porphyrin (heme) Cytochrome oxidase is an important example of this type of metalloenzyme. The structure and function of the heme moiety has already been described in this chapter. In addition to the iron-porphyrin ring unit, this enzyme also possesses copper. Both the metal ions oscillate between their oxidized and reduced states. It appears that they function to transmit electrons to molecular oxygen in the mitochondrial electron transport chain (p. 133). Water is the final product.

Metalloflavin enzyme Xanthine oxidase is an enzyme that depends on the presence of two atoms of molybdenum, eight atoms of iron, and two molecules of FAD. In this enzyme the iron is bound directly to the protein; a porphyrin structure is not involved. The enzyme catalyzes the oxidation of xanthine, i.e.:

$$\text{Xanthine} + O_2 + H_2O \rightleftharpoons \text{Uric acid} + H_2O_2$$

The mechanism is complex but appears to include the reduction of Mo^{6+} to Mo^{5+}. A molybdenum-coordinated hydroxyl group is added to the substrate (xanthine). Electrons from the substrate then pass by way of the molybdenum to the FAD and iron and finally to oxygen.

Metal-activated enzymes

Magnesium Many phosphotransferases involve metal ions complexed with two of the phosphoryl groups of ATP. Thus:

$$^{-}O-P(=O)(O^{-})-O-P(=O)(O^{-})-O-P(=O)(O^{-})-O-\boxed{\text{Adenosine}}$$

$$\text{(two } O^{-} \text{ coordinated to } Mg^{++}\text{)}$$

Creatine kinase, for example, which catalyzes the transfer of a phosphoryl group to creatine (Chapter 13), uses the $MeATP^{=}$:

$$MeATP^{=} + \text{Creatine} \rightleftharpoons MeADP^{-} + \text{Phosphorylcreatine} + H^{+}$$

In this case the obligatory divalent metal may be magnesium, manganese, calcium, or cobalt. This illustrates the absence of an absolute uniqueness of the metal requirement for this type of enzyme. The metal ion does not act directly on the enzyme but by way of the nucleotide substrate. Thus the mechanism involves the Ⓔ—S—M type of complex.

Calcium The allosteric activation of an enzyme by calcium is demonstrated by *glycogen phosphorylase kinase.* Calcium is required in the muscle cell to initiate contraction (Chapter 15). At the same time these ions activate glycogen phosphorylase kinase, which then triggers glycogenolysis (Chapter 11).

Calcium is also needed to activate salivary and pancreatic α-amylases, as are chloride ions. Although their function is not certain, the calcium ions probably serve to stabilize the active conformation of the enzymes.

Suggested readings

Bowen, H.J.M.: Trace elements in biochemistry, New York, 1966, Academic Press, Inc.

Brändén, C.I., et al.: Alcohol dehydrogenases. In Boyer, P.D., editor: The enzymes, vol. 11, ed. 3, New York, 1975, Academic Press, Inc., p. 103.

Herbert, V., and Das, K.C.: The role of vitamin B_{12} and folic acid in hemato- and other cell poiesis. In Vitamins and hormones, vol. 34, New York, 1976, Academic Press, Inc., p. 2

McCormick, D.B., and Wright, L.D., Vitamins and coenzymes. In Colowick, S.P., and Kaplan, N.O.: Methods in enzymology, vol. 62, part D, New York, 1979, Academic Press, Inc.; vol. 66, part E; vol. 67, part F, 1980, Academic Press, Inc.

Mildvan, A.S.: Metals in enzyme catalysis. In Boyer, P.D., Lardy, H., and Myrbäck, K., editors: The enzymes, vol. 2, New York, 1970, Academic Press, Inc., p. 472.

Mildvan, A.S.: Mechanism of enzyme action, Annu. Rev. Biochem. **43**:357, 1974.

Prasad, A.S.: Trace elements and iron in human metabolism, New York, 1978, Plenum Medical Book Co.

Scrutton, M.C: Metal enzymes. In Eichehorn, G.L., editor: Inorganic biochemistry, vol. 1, New York, 1973, Elsevier Scientific Publishing Co.

Walsh, C.: Enzymatic reaction mechanisms, San Francisco, 1979, W.H. Freeman & Co., Publishers.

Chapter 7

Bioenergetics and biological oxidation

The complete oxidative degradation of the major metabolic fuels involves numerous dehydrogenases requiring coenzymes such as NAD^+, $NADP^+$, and FAD. As described in Chapter 6, their supply is limited to a cellular pool; the reduced coenzymes must be reoxidized by other enzyme systems. Two alternate processes whereby this may be accomplished are suggested in Fig. 7-1. These are reductive, synthetic processes, which depend on a direct requirement for NADPH, and the electron transport chain, which results in the production of ATP through oxidative phosphorylation. The purpose of this chapter is to describe the second of these alternate pathways: electron transport and the oxidative phosphorylation coupled with it.

As an example of the complete oxidative metabolism of a major fuel, consider glucose. When the chemist burns this sugar under laboratory conditions in an atmosphere of oxygen, a total of 686,000 calories/mole is released.

$$C_6H_{12}O_6 + 6\ O_2 \rightarrow 6\ CO_2 + 6\ H_2O$$

However, in the cell, oxidation of glucose cannot proceed this way; it occurs in a stepwise fashion involving oxidation by dehydrogenation, as in the following:

$$C_6H_{12}O_6 + 6\ H_2O \rightarrow 6\ CO_2 + 24\ H$$

The 24 H that are removed by dehydrogenation at various steps require the reduction of NAD^+ and FAD. Thus the oxidation of a mole of glucose produces 20 H, which reduce 10 NAD^+ to 10 NADH + 10 H^+ and 4 H, which reduce 2 FAD to 2 $FADH_2$. In eukaryotic cells the use of molecular oxygen to oxidize the 24 hydrogen atoms occurs in the mitochondria (Chapter 2).

Most of the NADH produced from metabolic fuels, whether glucose, fatty acids, or amino acids, occurs in the catabolism of the common intermediate *acetyl CoA* (see Fig. 3-1). The oxidation of this compound occurs by way of the citric acid cycle, whose enzymes are located in the mitochondria. Aerobic oxidation of hydrogens also proceeds in this compartment in a stepwise process called *electron transport*. This is coupled to oxidative phosphorylation, a process that yields energy in a usable form, i.e., ATP. For this reason the mitochondria often are called the *power plants* of the cell.

Anatomy of mitochondria

As described in Chapter 2, mitochondria are bounded by outer and inner membranes. Fig. 2-9 shows that the outer membrane is the limiting mitochondrial membrane. It is also evident that the inner one possesses numerous invaginations that cross the lumen of the organelle, giving the interior the appearance of having many baffles. These are usually called *cristae* and vary in number, shape,

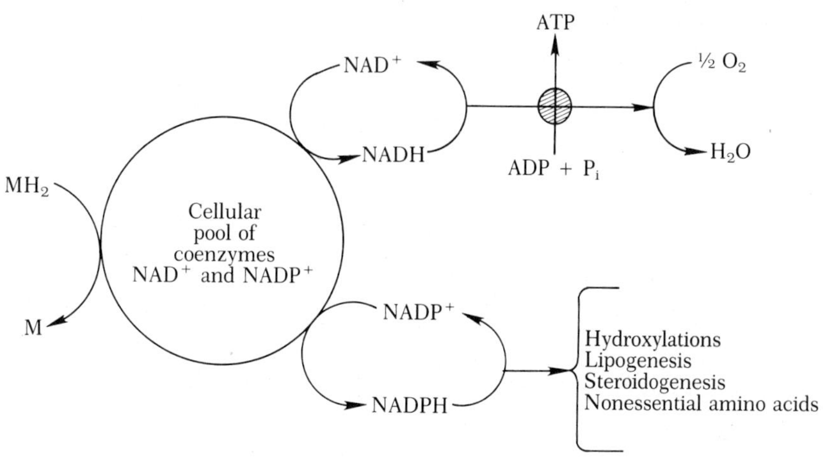

Figure 7-1
Pyridine nucleotides that are reduced by the oxidation of metabolites *(MH_2)*. Reduced coenzymes are then reoxidized by two general processes—electron transport coupled to oxidative phosphorylation (ATP production) and reductive synthetic processes such as hydroxylations, lipogenesis, and steroidogenesis.

and continuity from one tissue to another. In mammalian heart muscle the number of cristae is large; in mitochondria of liver and bronchial cells they are far fewer. In muscle cells the mitochondria function primarily to provide energy by oxidative phosphorylation for contraction. Since the enzyme systems involved are associated with the inner membrane, and therefore the cristae, these mitochondria are particularly rich in the invaginations. The cristae are covered by numerous knoblike structures (see Fig. 2-9). These consist of a spherical headpiece or knob (9 nm in diameter), a stalk, and a base.

The mitochondrial lumen is filled with a viscous or semisolid medium called the matrix. All the processes identified with the citric acid cycle occur here (Table 7-1), whereas the restoration of NADH and $FADH_2$, as well as the production of ATP, involves enzyme systems of the inner mitochondrial membrane.

Some metabolic processes that produce NADH are located outside the mitochondria and in the cytoplasm. For example, of the 24 hydrogens removed in the metabolism of glucose, four are removed during glycolysis (Chapter 11). If these are also to be converted to water with a concomitant production of ATP, they must enter the mitochondria. The mitochondrial membranes, however, have a limited permeability. The accompanying table lists some substances to which mitochondria are permeable and impermeable. By permeability is meant the existence or absence in the membrane of appropriate transport or translocating processes.

Permeable	**Impermeable**
Pyruvate	Acetyl CoA
ATP, ADP	NAD^+, NADH
Ca^{++}	K^+
Citrate and isocitrate	Oxaloacetate
L-Aspartate, L-glutamate	
α-Ketoglutarate	
L-Malate and succinate	
L-Glycerol 3-phosphate	
H^+	

The table shows that NADH cannot carry the hydrogens and their electrons (referred to as reducing equivalents) into the mitochondrion directly; there must be a supplementary process whereby these are transported.

Table 7-1
Location of mitochondrial enzymes

Outer membrane	Intermembranous space	Inner membrane	Matrix
Monoamine oxidase	Adenylate kinase	Respiratory chain	Malate dehydrogenase
Kynurenine hydrolase	Nucleoside diphosphokinase	Cytochromes b,c_1, a-a_3	Isocitrate dehydrogenase (NADP specific)
Acyltransferase	Nucleoside monophosphokinase	Succinate dehydrogenase	Isocitrate dehydrogenase (NAD specific)
Choline phosphotransferase		Succinate–cytochrome c reductase	α-Ketoglutarate dehydrogenase
Phosphatidate phosphatase		Succinate oxidase	Citrate synthase
Nucleoside diphosphokinase		"Rotenone-sensitive" NADH–cytochrome c reductase	Aconitase
Fatty acid elongation system		NADH oxidase	Fumarase
		Choline–cytochrome c reductase	Pyruvate carboxylase
		Cytochrome c oxidase	Phosphopyruvate carboxylase
		Respiratory chain—linked phosphorylation	Aspartate aminotransferase
		β-Hydroxybutyrate dehydrogenase	Ornithine carbamoyltransferase
		Ferrochelatase	Fatty acyl CoA synthetase(s)
		α-Aminolevulinic acid synthetase (?)	Fatty acid oxidation systems (?)
		Carnitine palmityltransferase	Glutamate dehydrogenase
		Fatty acid oxidation system (?)	
		Fatty acid elongation system	

Transport of reducing equivalents via enzymatic shuttle mechanisms

The transport of hydrogens as NADH or NADPH regardless of source occurs by enzymatic processes called shuttle mechanisms. Two such processes that can lead to the transport of reducing equivalents from the cytoplasm into the mitochondria are the *glycerol phosphate shuttle* and the *malate-aspartate shuttle.*

Glycerol phosphate shuttle

In this process dihydroxyacetone phosphate, an intermediate of glycolysis (Chapter 11), is reduced to L-glycerol 3-phosphate by cytoplasmic *glycerol phosphate dehydrogenase* and NADH (Fig. 7-2). The L-glyccerol 3-phosphate in turn diffuses through the outer mitochondrial membrane and comes into contact with the outer surface of the inner membrane. Here is found the mitochondrial glycerol phosphate dehydrogenase, a flavoprotein having FAD as its prosthetic group. Dihydroxyacetone phosphate is regenerated, and the reduced FAD then feeds the reducing equivalents to the electron transport chain. The dihydroxyacetone phosphate returns to the cytosol and can be reused for reduction to L-glycerol 3-phosphate or conversion to D-glyceraldehyde 3-phosphate. The glycerol phosphate shuttle is unidirectional and proceeds only to carry reducing equivalents into the mitochondria.

Malate-aspartate shuttle

The malate-aspartate shuttle also appears to be a unidirectional process and therefore carries reducing equivalents only into the mitochondrion (Fig. 7-3). In this shuttle oxaloacetate is reduced to L-malate by a cytosolic *malate dehydrogenase* and NADH. L-Malate is then carried into the mitochondrion by an exchange process (Chapter 19) in which α-ketoglutarate simultaneously leaves the mitochondrion. Thus L-malate enters the mitochondrion and α-ketoglutarate exits by the

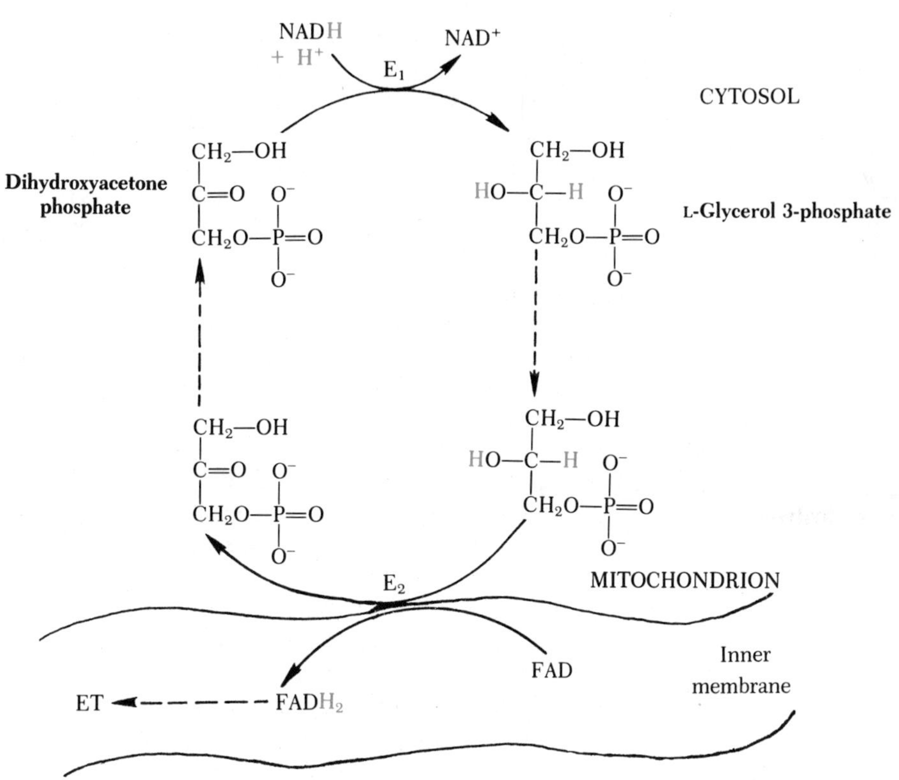

Figure 7-2
Glycerol phosphate shuttle for the transfer of reducing equivalents into the mitochondrion. E_1 is the glycerol phosphate dehydrogenase found in the cytosol; E_2 is the mitochondrial enzyme found in the inner membrane. *ET* suggests the presence of the electron transport chain.

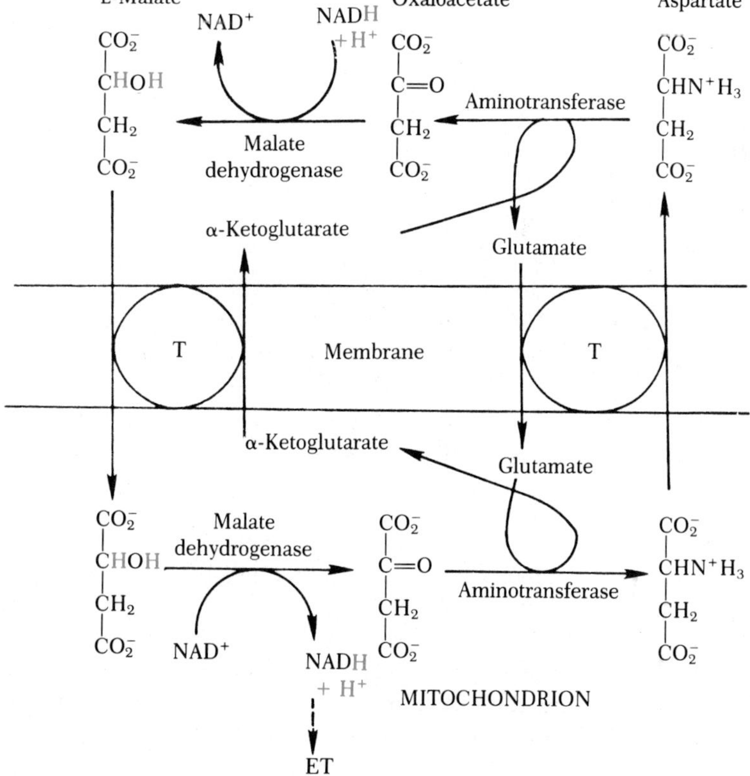

Figure 7-3
Malate-aspartate shuttle mechanism. Enzymes involved are the cytosolic and the mitochondrial malate dehydrogenases and aspartate aminotransferases. *ET* is the electron transport chain; *T* represents membrane transporters. The aspartate-glutamate exchange is driven by the membrane potential.

same carrier or transport protein. Inside the mitochondrion the L-malate is reoxidized to oxaloacetate using the citric acid cycle enzyme, thereby depositing the reducing equivalents within the organelles. The oxaloacetate is then transaminated to L-aspartate by *aspartate aminotransferase*; the glutamate is converted to α-ketoglutarate (Chapter 13). Finally, the aspartate is transported out of the mitochondrion in exchange for glutamate and enters the cytosol. This latter exchange process is unidirectional and is driven by the membrane potential created by the respiratory process (p. 141). The aspartate is converted back to oxaloacetate by transamination using α-ketoglutarate as the nitrogen acceptor. Transamination reactions occur by means of the cytosolic and mitochondrial aspartate aminotransferases.

Transhydrogenase

Although the principal metabolic function of cytoplasmic NADPH is as a reductant in certain anabolic processes, both cytoplasmic and mitochondrial NADPH may also form ATP. However, there is no direct mitochondrial path for electrons from NADPH to oxygen. The oxidation of NADPH requires an intermediary reaction in which a hydride ion is accepted by NAD^+. As seen below, reversible transhydrogenation (1) is catalyzed by *transhydrogenase*. However, in the presence of ATP the transhydrogenation is pushed in the direction of NADPH production (2), as observed in heart and liver mitochondria.

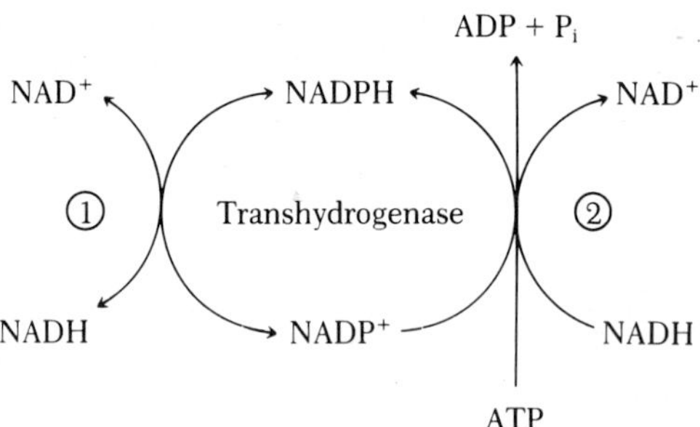

Isocitrate shuttle

The transport of reducing equivalents from the mitochondrion into the cytoplasm for biosynthetic purposes requires the operation of the isocitrate shuttle. NADH reduces $NADP^+$ by the *ATP-requiring transhydrogenase*. The accumulated NADPH then drives the $NADP^+$-dependent *isocitrate dehydrogenase* in the reverse direction, converting α-ketoglutarate into isocitrate. The latter penetrates the mitochondrial membrane by exchange with L-malate and enters the cytoplasm where it acts as a substrate for cytoplasmic $NADP^+$-dependent isocitrate dehydrogenase. The overall scheme is illustrated in Fig. 7-4.

Electron transport mechanism

NADH dehydrogenase

All the reduced coenzymes produced during metabolism may be reoxidized, directly or indirectly, by the action of the electron transport chain. It will be recalled from Chapter 6 that in the dehydrogenation of a metabolite (MH_2) two electrons and a proton are actually transferred to the NAD^+ (specific dehydrogenase); this may be perceived as the transfer of a hydride ion (H^-). The hydride ion is accepted by the nicotinamide moiety of NAD^+ while the second proton is released to the aqueous environment. NADH in turn is oxidized by an FMN-containing enzyme, *NADH dehydrogenase*. The dimethylisoalloxazine ring of the FMN accepts two electrons and a proton from NADH plus one proton from the environment.

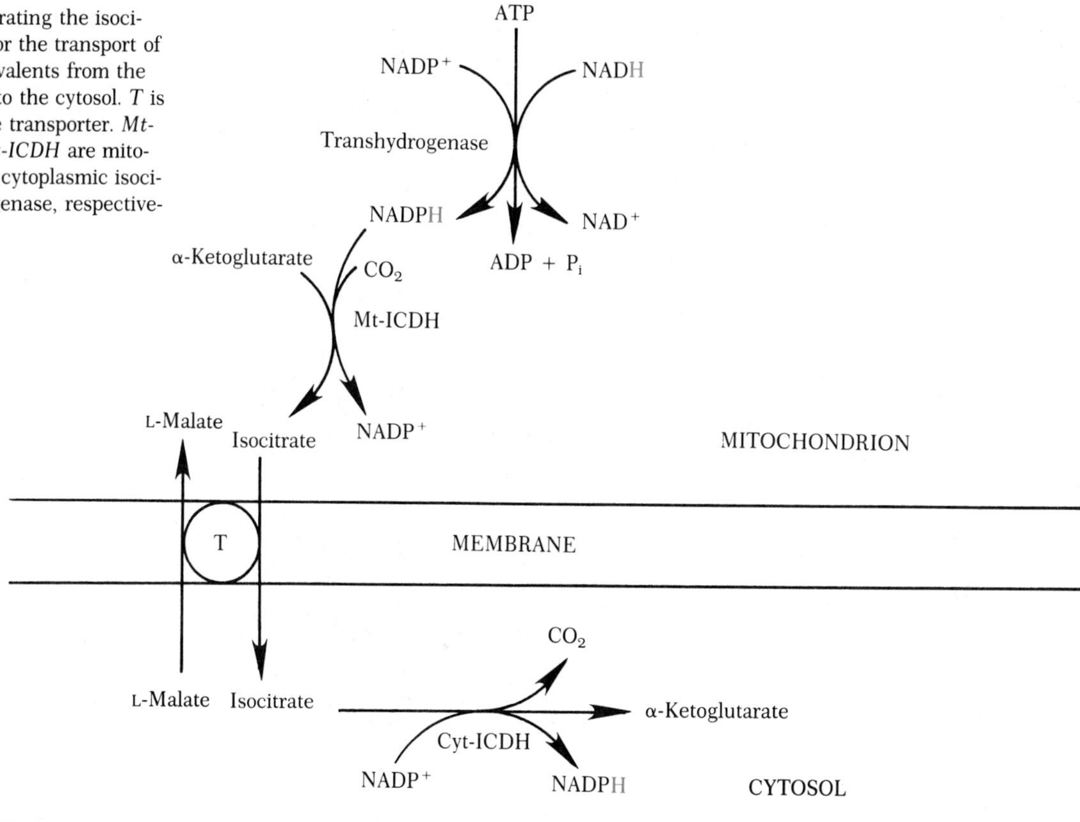

Figure 7-4
Pathway illustrating the isocitrate shuttle for the transport of reducing equivalents from the mitochondria to the cytosol. *T* is the membrane transporter. *Mt-ICDH* and *Cyt-ICDH* are mitochondrial and cytoplasmic isocitrate dehydrogenase, respectively.

NADH dehydrogenase is actually a complex enzyme having, in addition to the flavoprotein, at least four iron-sulfur centers. Although the exact structure of the latter is unknown, they are related to the iron-sulfur proteins from bacteria and plants known as *ferredoxins*. These proteins contain complexes of iron and sulfur atoms. For example, in the ferredoxin of spinach the sulfurs are contributed by

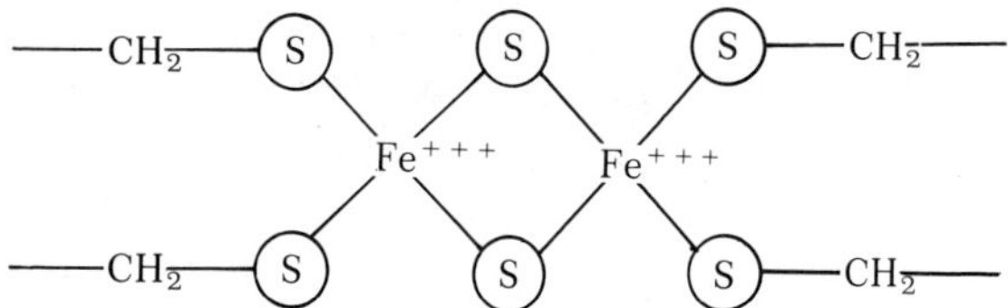

cysteine and inorganic sulfide. The function of the iron-sulfur center appears to be in the transfer of electrons. Thus:

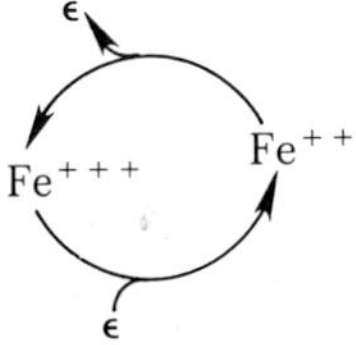

In its completely oxidized state both irons are in the ferric state; when reduced, one iron becomes ferrous.

The acceptor of electrons and protons from the NADH dehydrogenase is coenzyme Q (Chapter 6). As noted previously, two hydrogens from the reduced FMN (two electrons and two protons) are accepted by the quinone to produce hydroquinone. This process also may involve a transfer of electrons by way of iron-sulfur centers. Coenzyme Q functions as an acceptor of electrons and protons from a number of metabolic inputs (see below). All these serve to reoxidize coenzymes or the prosthetic groups of flavoproteins.

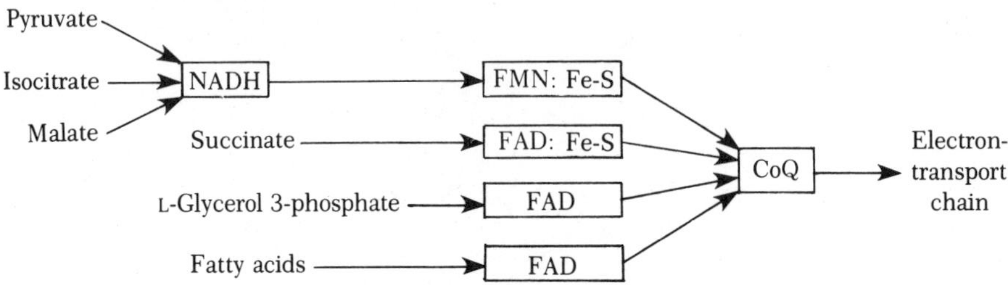

Typical of these processes is the dehydrogenation of succinate to yield fumarate (citric acid cycle) by the flavoprotein *succinate dehydrogenase* (Chapter 11). In this protein the flavin moiety, FAD, is covalently bound to the apoenzyme. Also present are two iron-sulfur centers. Other flavoproteins oxidized by coenzyme Q are glycerol phosphate dehydrogenase and the special electron transferring flavoprotein for the various *fatty acyl CoA dehydrogenases* (Chapter 12).

The electron transport chain itself is visualized as a sequence of individual heme-containing proteins, called *cytochromes*, in addition to iron-sulfur centers that act as electron carriers. The sequential arrangement of cytochromes in mitochondria is as follows:

$$\text{CoQ} \rightarrow \text{Cyto b} \rightarrow \text{Cyto c}_1 \rightarrow \text{Cyto c} \rightarrow \text{Cyto a-a}_3 \rightarrow H_2O \;(\tfrac{1}{2}\,O_2)$$

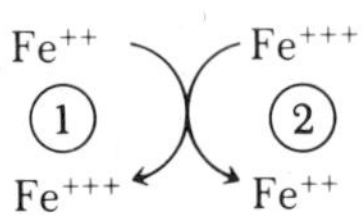

The functional entity or oxidation-reduction center of these proteins is the iron atom. Oxidation-reduction involves the reversible removal (oxidation) or addition (reduction) of an electron from the iron. This may be visualized to occur between two consecutive cytochrome-iron atoms as shown at left. In this scheme it is evident that one electron has moved from protein *1* to protein 2. That is, the iron of cytochrome 1, which is in the ferrous state, yields an electron to the ferric iron in cytochrome 2, thereby reducing it to the ferrous state. The ferrous iron of 1 is simultaneously oxidized to the ferric state.

Cytochromes

The prosthetic group of the cytochrome protein is heme. As noted in Chapter 6, the heme iron is inserted in the center of the protoporphyrin IX ring structure and is bonded to the pyrrole nitrogens. Of the various mitochondrial cytochromes, i.e., a, a_3, b, c_1, and c, cytochrome c is best known as far as structure is concerned. This protein consists of a single polypeptide chain of 104 amino acid residues (mol wt = 12,400). Its three-dimensional structure forms a crevice in which the heme prosthetic group is buried. The heme in cytochrome c is covalently bound to the polypeptide chain by two thioether links (Fig. 7-5). The sulfurs of cysteines 14 and 17 are linked to the vinyl groups of the porphyrin structure. The arrangement of the planar heme in the protein crevice brings the iron in apposition to histidine 18 on one side of the crevice and methionine 80 on the other. Therefore these two residues serve as the fifth and sixth ligands for the iron (iron may have six coordination bonds). Iron is at the center of an octahedron of ligands contributed entirely by the hemoprotein (four pyrrole nitrogens and two amino acid ligands) and thereby dif-

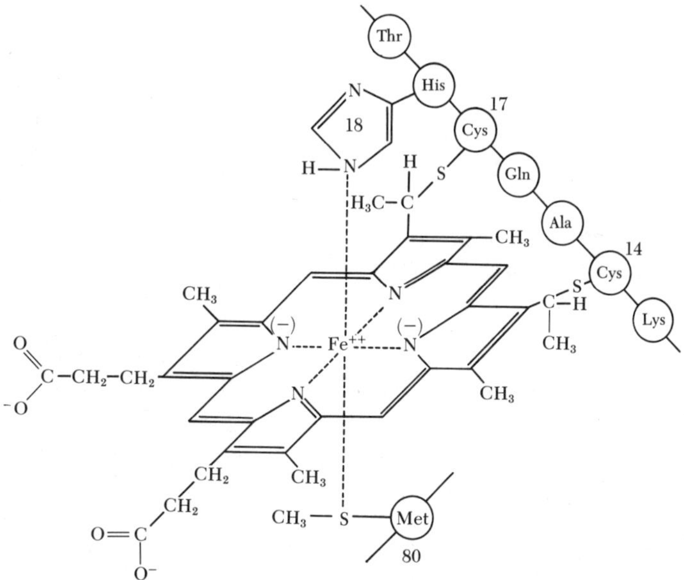

Figure 7-5
Positioning and covalent attachment of the heme prosthetic group in the crevice of cytochrome c. This represents one apsect of its resonating structure.

fers from the oxygen-carrying proteins myoglobin and hemoglobin. In cytochrome c there is no way in which an external ligand such as oxygen can enter the crevice to coordinate with the iron. Although its exact structure is uncertain, cytochrome c_1 is also a heme-containing protein involving iron protoporphyrin IX covalently bound to the protein (mol wt = 31,000). Cytochrome b has the same heme prosthetic groups; however, it is not covalently linked to the polypeptide chain.

Cytochrome oxidase, already described in Chapter 6, is the terminal unit in the electron transport chain. Of the various cytochromes, only the a_3 subunit of cytochrome oxidase is capable of reacting directly with oxygen. The cytochrome a unit probably functions in the transport of an electron by the reversible oxidation and reduction of the ferrous ion. Copper ions ($Cu^{++} \leftrightarrow Cu^{+}$) also may be involved in the transfer of electrons from cytochrome a to a_3.

A perspective of the electron transport chain

However, the cytochromes apparently do not function independently of other factors. For example, as indicated below, an iron-sulfur center links cytochrome b to c_1. Furthermore, copper ions in the cytochrome oxidase serve to transfer electrons from the iron of cytochrome a to that of a_3. The various components of the electron transport chain may be imagined to function as a sequence of cogwheels, each being turned by the preceding one and then turning the succeeding cog.

$CoQH_2$ | Fe^{+++} | Fe^{++} | Fe^{+++} | Fe^{++} | Fe^{+++} | Cu^{+} | Fe^{+++} | ½ $O_2^{=}$ | 2 H
2 Cyto b | 2 Fe-S | 2 Cyto c_1 | 2 Cyto c | 2 Cyto a | 2 Cyto a_3 | H_2O
CoQ | Fe^{++} | Fe^{+++} | Fe^{++} | Fe^{+++} | Fe^{++} | Cu^{++} | Fe^{++} | ½ O_2
2 H^+

Since the original step is the oxidation of a metabolite, two hydrogen atoms, i.e., two electrons, are removed and accepted by NAD^+. Therefore the electron transport chain must cycle twice to pass along the two electrons to an atom of oxygen. Each oxygen atom with two extra electrons accepts two protons from the environment,

Figure 7-6 The four lipid-soluble complexes in the electron transport chain.

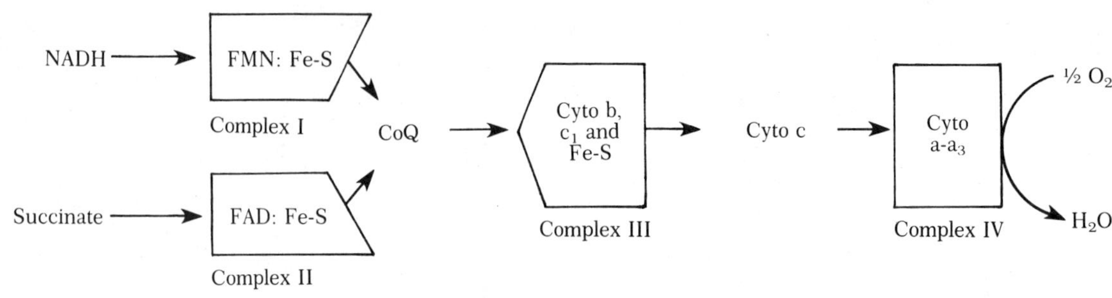

and a molecule of water results. In this entire process only two components are consumed: the metabolite (or fuel) and the oxygen. Each intermediate constituent is considered to oscillate between the oxidized and reduced form, i.e., they are continuously recycled.

The dehydrogenases and cytochromes are an integral part of the mitochondrial inner membrane. Studies using detergents have made it possible to identify four lipid-soluble complexes of the electron transporting constituents. Fig. 7-6 illustrates their functional interrelationship. They aid understanding of the structural organization of the electron transport system and the manner in which substrates feed into it. The NADH dehydrogenase (FMN)-Fe:S protein is indicated as the first complex, and the succinate dehydrogenase (FAD)-Fe:S protein is the second. CoQ is not considered to be fixed in the membrane; rather, it appears to be a mobile component. The third complex consists of cytochromes b and c_1 and an Fe:S protein. Cytochrome c is water soluble and easily diffusible, whereas cytochromes b, c_1 and a-a_3 are lipid soluble and therefore are fixed components of the membrane. Thus complex I catalyzes the oxidation of NADH and the reduction of CoQ. Complex II reduces CoQ at the expense of succinate. Complex III reduces cytochrome c, and complex IV reduces oxygen.

The disposition of the four complexes also indicates that those metabolites directly oxidized by flavin-containing dehydrogenases, e.g., succinate dehydrogenase, bypass the NADH dehydrogenase step. The importance of this consideration is evident on realization that the oxidation of NADH to H_2O yields three molecules of ATP, whereas the oxidation of succinate yields only two ATPs.

Oxidative phosphorylation: energy conservation

The oxidation of NADH and $FADH_2$ is accompanied by the formation of ATP; the energy released in the oxidative process is not dissipated as heat but conserved as the so-called high-energy phosphate bonds of ATP. As suggested by the accompanying diagram, energy conservation occurs by coupling another process, oxidative phosphorylation, to the transport of electrons.

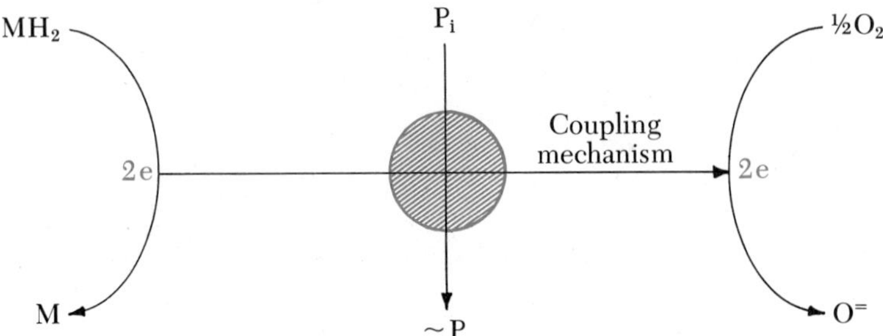

Table 7-2
Free energy of hydrolysis of various biological compounds

Compound hydrolyzed	Bond hydrolyzed	$-\Delta G^{\circ\prime}$ (calories/mole)
AMP	Ester	2200
Glucose 6-phosphate	Ester	3300
ATP $\rightarrow$ ADP + P_i	Anhydride	7300
ADP $\rightarrow$ AMP + P_i	Anhydride	7300
Acetyl CoA	Thioester	8200
Phosphocreatine	N-Phosphate	10,500
1,3-Bisphosphoglycerate	Anhydride	11,800
Phosphoenol pyruvate	Ester: enol phosphate	14,800

The high-energy phosphate bond

The concept of a high-energy bond is a controversial one. At the outset one must recognize that a high-energy bond, usually symbolized by ~ (a tilde), does not mean that this is some unusual type of phosphate bond. Rather, it refers to the fact that the free energy of hydrolysis ($\Delta G^{\circ\prime}$) is higher than for stable phosphates such as phosphate esters. Table 7-2 compares the standard free energy of hydrolysis for a number of different biological phosphate compounds.

For purposes of instruction we may consider phosphate compounds in biological systems in two categories:

1. *Relatively inert phosphate compounds*, such as the ester phosphates shown in Table 7-2. Hydrolysis of glyceraldehyde 3-phosphate or glucose 6-phosphate yields only 3300 calories/mole. This is what is usually considered to be an ordinary phosphate compound.
2. *Active or high-energy phosphate compounds*. This category may be subdivided into (a) the active phosphate carriers or nucleoside triphosphates ATP, UTP, CTP, and GTP and (b) metabolites such as 1,3-bisphosphoglyceric acid, phosphoenolpyruvate, and phosphocreatine. Hydrolysis of these compounds yields approximately 5000 to 15,000 calories/mole, and therefore according to definition they are called high-energy phosphate compounds.
   a. *Active phosphate carriers* include the familiar di- and triphosphates of various nucleosides. In ATP the first (α) phosphate is bound to the 5′-carbon; this represents the usual phosphate ester, which on hydrolysis yields 2200 calories/mole. The next phosphate (β) is attached by the removal of a molecule of water to form a phosphoanhydride bond; its hydrolysis yields 7300 to 7800 calories/mole. Likewise, the third phosphate (γ) is added by the removal of a molecule of water to form a phosphoanhydride bond; it is also a high-energy (~) bond yielding 7300 to 7800 calories/mole on hydrolysis. ATP furnishes the necessary energy in the performance of biological work. This includes biosynthetic functions, muscular contraction, activity of brain and nerve tissues, absorption and transport processes, and myriads of other energy-requiring processes in the cell.

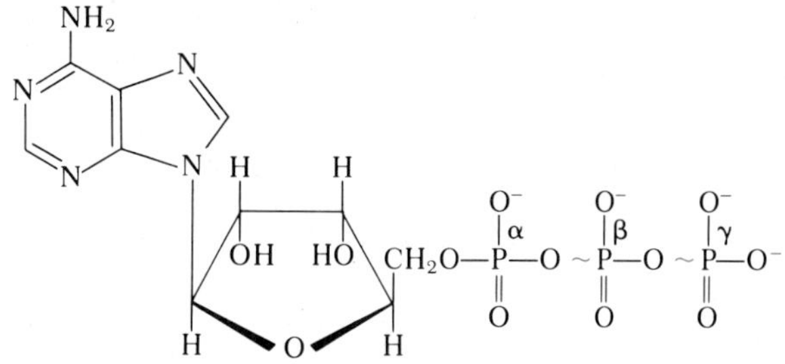

b. *A second group of high-energy compounds* includes esters and mixed anhydrides. As seen in Table 7-2, hydrolysis of the bonds yields 5000 to 15,000 calories/mole. Examples of ester linkages are the high-energy bond involved in aminoacyl tRNAs (Chapter 10), the thioester in CoA derivatives, and the enol phosphate ester of phosphoenolpyruvate. 1,3-Bisphosphoglycerate is an example of a mixed anhydride involving a carboxyl group and phosphate, whereas phosphocreatine is a phosphoguanidine.

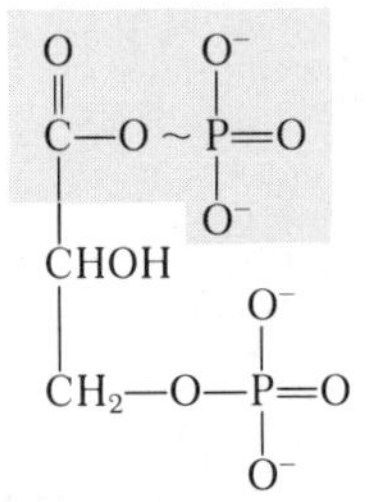

Mixed anhydride (1,3-bisphosphoglycerate) 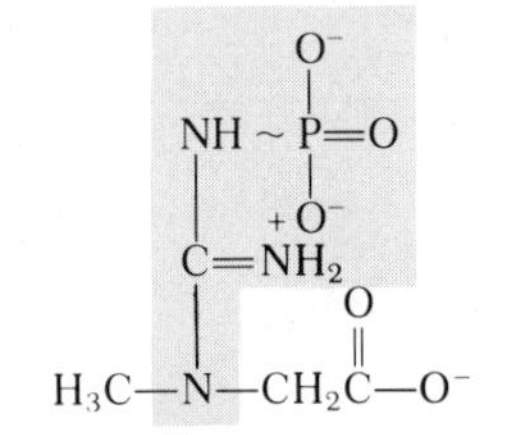 **Guanidinium phosphate (phosphocreatine)**

Bioenergetics

As described above, the electron transport system involves a sequence of individual oxidation-reduction reactions. Each has a characteristic ability to donate or accept electrons. This relative capacity for each component to give and receive electrons is expressed in terms of a standard reduction potential: E_o. These values are established under specific experimental conditions and are related to a common standard value. The standard is the hydrogen electrode whose reduction potential is arbitrarily set at zero volts. Zero potential exists for an H^+-containing solution at pH 0.0 or 1.0M and at a hydrogen gas pressure of 1 atmosphere. Then:

$$2\,H^+ + 2\,\epsilon \rightleftharpoons H_2$$

The standard reduction potential of any other oxidation-reduction system can be conveniently related to that of the hydrogen electrode. Thus it forms half a complete electric cell while the system (or electrode) to be compared forms the other. The standard reduction potential for the hydrogen electrode is usually corrected to a pH of 7.0, a condition that is more biologically meaningful. Under this condition, the reduction potential E_o' equals -0.42 volts. E_o' is used to indicate this extra condition.

The reduction potential of any other substance, such as NAD^+, may be established using the hydrogen electrode as a standard (pH = 7.0). Table 7-3 provides a list of such values for biologically important oxidation-reduction reactions. All the

Table 7-3
Standard reduction potentials for components of the electron transport chain

Half reaction	E_o' (volts)*
$\frac{1}{2}\,O_2 + 2\,H^+ + 2\,\epsilon \rightarrow H_2O$	+0.815
Cytochrome a: $Fe^{+++} + \epsilon \rightarrow Fe^{++}$	+0.29
Cytochrome c: $Fe^{+++} + \epsilon \rightarrow Fe^{++}$	+0.254
Ubiquinone + $2\,H^+ + 2\,\epsilon \rightarrow$ Ubihydroquinone	+0.10
Cytochrome b: $Fe^{+++} + \epsilon \rightarrow Fe^{++}$	+0.075
$NAD^+ + H^+ + 2\,\epsilon \rightarrow NADH$	−0.32
$2\,H^+ + 2\,\epsilon \rightarrow H_2$	−0.420

Reprinted with permission from Loach, P.A. In Hosman, G.D., editor: Handbook of biochemistry and molecular biology, ed. 3, vol. 1, Cleveland, 1976, CRC Press, Inc. Copyright The Chemical Rubber Co., CRC Press, Inc.
*Standard reduction potentials at pH 7.0 and 30° C.

half reactions listed are given as relative reduction reactions, e.g., $NAD^+ + H^+ + 2\ \epsilon \rightarrow NADH$ ($E'_o = -0.315$ volts). The more negative the value for E'_o, the greater is its capacity to yield electrons and the more effective it is as a reducing agent. The hydrogen electrode, $E'_o = -0.42$ volts, is the most potent reducing agent listed in Table 7-3. On the other hand, for the reaction $\frac{1}{2}\ O_2 + 2\ H^+ + 2\ \epsilon \rightarrow H_2O$, $E'_o = +0.82$, indicating oxygen to be the most avid acceptor of the electrons. Oxygen therefore has the greatest tendency to become reduced and therefore is the best oxidizing agent.

In the electron transport chain NADH is the compound to be oxidized while oxygen is reduced. Since NADH is a relatively good reducing agent, its oxidation may be coupled to the reduction of oxygen by way of the electron transport chain. The overall reaction is:

Since E'_o for oxygen $\rightarrow H_2O$ is very positive, the above reaction will proceed as shown; oxygen will be reduced and NADH will be oxidized.

Such a reaction is expressed in more general terms as follows:

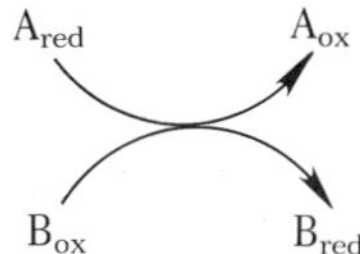

The standard reduction potentials for the two general electrode reactions ($A_{red} \rightleftharpoons A_{ox}$ and $B_{ox} \rightleftharpoons B_{red}$) would be E'^{A}_o and E'^{B}_o. Electrode potentials are expressed quantitatively by the following equation:

$$E = E'_o - 2.303\ \frac{RT}{nF}\ \log \frac{\text{(Reduced species)}}{\text{(Oxidized species)}}$$

Here R is the gas constant, T the absolute temperature, F the faraday constant (23,000 calories absolute volt equivalent), and n the number of electrons transferred in the reaction. E is the observed potential and E'_o is the standard electrode potential.

We can calculate the free-energy change for the overall electron transport chain beginning with NADH. Knowing its magnitude, we may then predict how many molecules of ATP might be synthesized for every two electrons transported.

Consider again the two general electrode reactions: $A_{red} + B_{ox} \rightleftharpoons A_{ox} + B_{red}$. These are expressed as:

$$E_A = E'^{A}_o - 2.303\ \frac{RT}{nF}\ \log \frac{(A_{red})}{(A_{ox})}$$

$$E_B = E'^{B}_o - 2.303\ \frac{RT}{nF}\ \log \frac{(B_{red})}{(B_{ox})}$$

At equilibrium they may be equated ($E_A = E_B$) and rearranged so that:

$$E'^{B}_o - E'^{A}_o = \Delta E'_o = 2.3\ \frac{RT}{nF}\left[\log \frac{(B_{red})}{(B_{ox})} - \log \frac{(A_{red})}{(A_{ox})}\right]$$

$$\text{and } \Delta E'_o = 2.3\ \frac{RT}{nF}\ \log \frac{(A_{ox})(B_{red})}{(A_{red})(B_{ox})}$$

$$= 2.3\ \frac{RT}{nF}\ \log K_{eq}$$

Recalling that $\Delta G^{\circ\prime} = -2.3\ RT \log K_{eq}$ (Chapter 3), it may now be said that

$$\Delta G^{\circ\prime} = -nF\ \Delta E_o'$$

This important relationship makes it possible to calculate the overall change in free energy across the entire electron transport chain beginning with NADH.

Thus for $NADH + H^+ + \frac{1}{2}\ O_2 \rightleftharpoons NAD^+ + H_2O$:

$$\begin{aligned} \Delta E_o' &= E_o'^{O_2} - E_o'^{NADH} \\ &= 0.82 - (-0.32) = +1.14 \text{ volts} \\ \text{Then } \Delta G^{\circ\prime} &= -2\ (23{,}000)\ (1.14) \\ &= -52{,}600 \text{ calories/mole } (-52.6 \text{ kcal/mole}) \end{aligned}$$

This shows that a considerable amount of energy is evolved in the oxidation of NADH. The question now arises, How many moles of ATP can be produced? If we consider that the formation of ATP involves the reversal of its hydrolysis, i.e.,

$$ADP + P_i \rightleftharpoons ATP + H_2O : \Delta G^{\circ\prime} = +7.3 \text{ kcal/mole}$$

then, under standard conditions (1M concentrations of reactants and products) +7.3 kcal are required to drive the incorporation of P_i into ADP to produce 1 mole of ATP. However, these figures require the existence not only of standard conditions but also thermodynamic equilibrium, two situations that are incompatible with the living cell. The ATP/ADP ratio is high within the cell, and the amount of energy required to produce a mole of ATP is considerably greater than 7.3 kcal; the value may be closer to 14 kcal.

Oxidative phosphorylation

Measurements of the number of ATPs produced in the oxidation of NADH were obtained in mitochondria using L-malate as substrate. In this experiment the consumption of oxygen and inorganic phosphate by the mitochondria was measured. The results were expressed as a P:O (phosphorus:oxygen) ratio. With L-malate the P:O ratio was 3, i.e., three molecules of phosphate were converted from inorganic to an organically bound form (ATP) per atom of oxygen consumed. In effect this means that as two hydrogen atoms (one molecule of NADH) or their electron equivalents passed along the electron transport chain, three molecules of ATP were formed per atom of oxygen consumed. When the citric acid cycle intermediate succinate was used, the P:O ratio was 2, indicating that only two molecules of ATP were produced per atom of oxygen consumed.

The location of specific ATP-producing sites in the electron transport chain was inferred from the large free-energy changes calculated from standard reduction potential for the individual steps shown in Table 7-4. The free energy changes in steps *1*, 2, and 5 are sufficient for each to yield 1 mole of ATP, provided that there is an appropriate system for the conversion of inorganic phosphate to the organically bound, high-energy phosphoanhydride of ATP. Each step originally was considered to be coupled in some way with the phosphorylative process at three distinct coupling sites. Thus the flow of two electrons from NADH to $\frac{1}{2}\ O_2$ was visualized as

Table 7-4
Free-energy changes at various stages of electron transport chain

Reaction	$\Delta G^{\circ\prime}$
1. NAD → FMN	12,450
2. FMN → Cyto b	4,150
3. Cyto b → Cyto c	10,150
4. Cyto c → Cyto a	1,380
5. Cyto a → ½ O_2	24,450
TOTAL	52,580

passing through ATP-producing sites, I, II, and III. Succinate, on the other hand, was considered to bypass coupling site I and the two electrons passed through sites II and III. This concept was applied to the so-called *chemical theory* of oxidative phosphorylation, which is described later. Today the preferred explanation of oxidative phosphorylation is the *chemiosmotic theory* of Mitchell. In this hypothesis the overall free-energy change is utilized in the transport of protons from the matrix to the cytoplasmic surface of the inner membrane. Phosphorylation is visualized as a single process driven by the downhill flow of protons back across the membrane.

However, final answers are still not available as to exactly how the oxidative energy is harvested and used for the synthesis of ATP. Therefore the three existing hypotheses will be described that have been and are used to explain the energy transducing process. As previously described, the inner membrane is covered with knoblike structures (see Fig. 2-9). Treatment with sonic vibrations removes the spherical headpieces and simultaneously eliminates the ability to perform oxidative phosphorylation. The remaining membranous material is capable of electron transport. As seen in Fig. 7-7, the headpiece or F_1 unit is attached to the membrane by an aggregate of proteins called the F_o unit, which is embedded in and is thought to penetrate the membrane. The complex of F_1 and F_o units (F_1-F_o) has the capacity to hydrolyze ATP. As indicated before, ATP formation is seen as the reversal of its hydrolysis, i.e., ADP + P_i → ATP + H_2O. Electron transport itself occurs in the membrane, where the various components of the respiratory chain are located. As suggested in Fig. 7-7, there appears to be a distinct orientation or topography of the electron transport proteins. For example, the NADH dehydrogenase, cytochrome b, and cytochrome oxidase (a and a_3) penetrate the membrane, whereas the cytochromes c and c_1 are buried on the cytoplasmic side of the inner membrane.

Chemical theory The oldest of the three hypotheses usually cited to explain oxidative phosphorylation is the *chemical theory* (Fig. 7-8). It proposes the existence of specific carrier proteins (C_1, C_2, and C_3) at each of three distinct ATP-producing sites just described. Another factor (I), common to all three sites, then combines

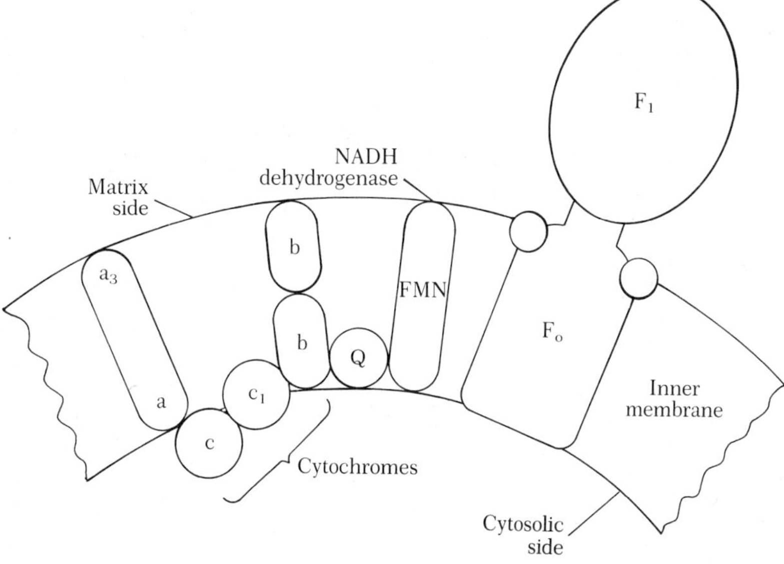

Figure 7-7
Spatial orientation of electron transport proteins and the F_1-F_o-ATPase (knobs) in the inner mitochondrial membrane.

Modified from Racker, E.: Annu. Rev. Biochem. **46:**1006, 1977. Reproduced, with permission, from the Annual Review of Biochemistry. © 1977 by Annual Reviews Inc.

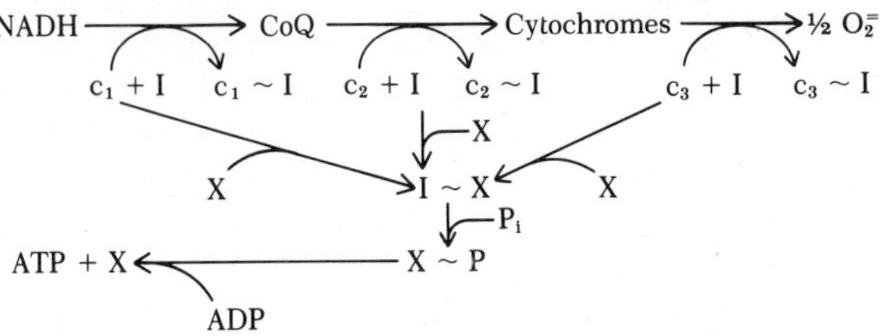

Figure 7-8
Chemical theory of oxidative phosphorylation. c_1, c_2, and c_3 are carrier proteins specific for each of the three ATP-producing sites. Both *I* and *X* are hypothetical intermediates involved in forming the high-energy phosphate *X ~ P*.

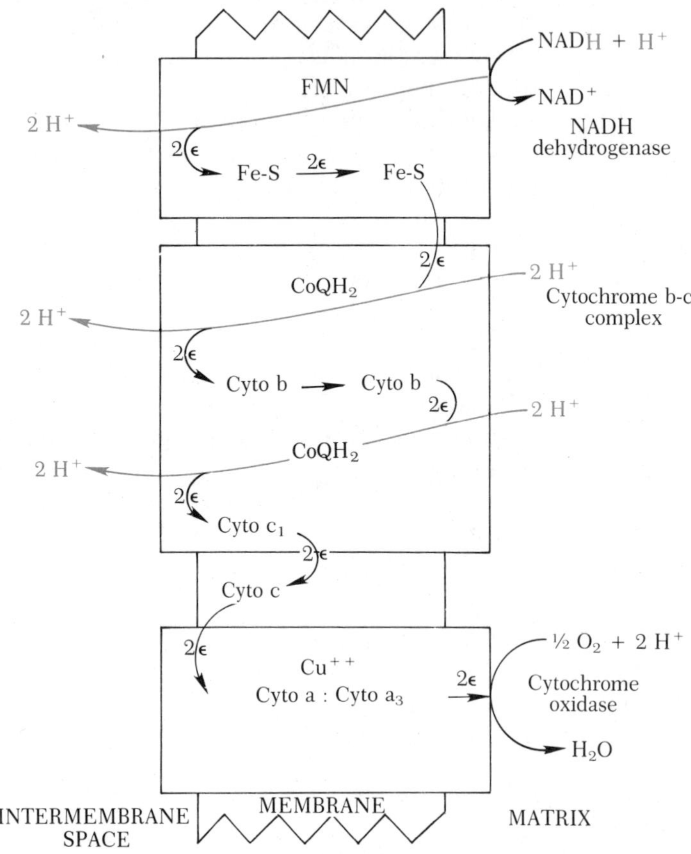

Figure 7-9
Translocation of protons by the spatial arrangement of electron transport components. The formation of a proton gradient is basic to the Mitchell chemiosmotic hypothesis of oxidative phosphorylation. The flow of protons is shown by the colored arrows.

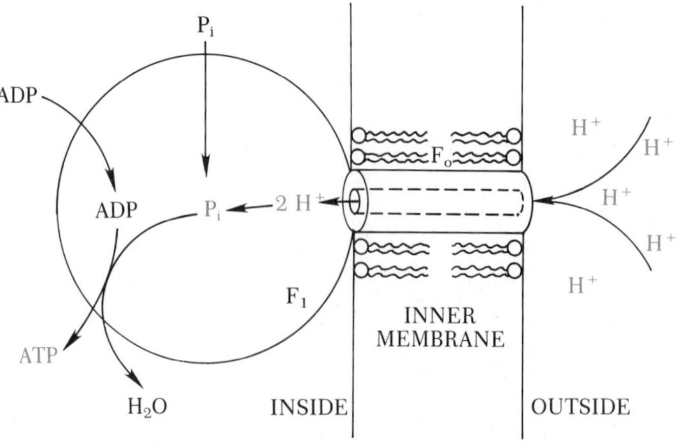

Figure 7-10
Direct role of the proton gradient in the formation of ATP. F_o is considered to be an ion channel for protons.

with the specific carriers; the linkage formed is presumed to be a high-energy bond. In the next stage another factor common to all three steps, factor or enzyme X, is believed to combine with I to form X ~ I; again a high-energy bond is involved. Then the I component is replaced by inorganic phosphate to yield a phosphorylated enzyme (X ~ P). Finally the phosphate is transferred to ADP to yield ATP. The reactions

$$I \sim X + P_i \rightleftharpoons X \sim P$$
$$X \sim P + ADP \rightleftharpoons ATP + X$$

are thought to be catalyzed by the ATPase. Since none of the phosphorylated intermediates has been found, the chemical theory has generally been discarded as a possible mechanism for energy conservation as ATP.

Chemiosmotic theory The mechanism for coupling the synthesis of ATP to electron transport most often cited is the chemiosmotic process of Mitchell. Fundamental to this mechanism is a spatially arranged electron transport system in the inner mitochondrial membrane. In the flow of two electrons from NADH to ½ O_2, protons are transported across the membrane into the space between the inner and outer membranes. This creates an H^+ gradient across the membrane. In the original formulation each pair of electrons originating in NADH results in the outward flow of six protons (Fig. 7-9). In the first step two electrons and a proton from NADH, in addition to one proton from the aqueous matrix, are added to the FMN of the NADH dehydrogenase, reducing the prosthetic group to $FMNH_2$. The carrier protein is pictured here as being embedded in the membrane, extending from the inner (matrix) to the outer surface. Two electrons from the $FMNH_2$ are then accepted by the iron-sulfur centers, which cycle the electrons back across the membrane to the interior. Simultaneously the 2 H^+ are released into the inter membrane space. The two electrons (Fe-S centers) are now transferred to CoQ accompanied by 2 H^+ derived from the aqueous matrix. The reduced CoQ ($CoQH_2$) translocates 2 H^+ to the exterior surface where two electrons are donated to cytochrome b and the protons are released into the medium. Then the electrons (cytochrome b) are returned to the inner surface where, together with 2 H^+ from the matrix, they may again join CoQ. Once the third pair of H^+ has been released, the electrons are donated to cytochromes c_1 and c on the outer membrane surface. Then they are accepted by cytochrome a and cross the membrane for the final cycle. On the matrix side the electrons are transferred by cytochrome a_3 to an atom of oxygen. With the acquisition of two protons a molecule of water is formed. This cyclical process involves the flow of an electric current in the membrane and the establishment of a gradient of H^+ (pH is lower outside than inside). Currently, however, the stoichiometry of protons translocated per electron pair is uncertain; as many as 12 may actually be involved. Once the gradient is established, the H^+ may flow spontaneously back across the membrane in a downhill fashion. However, the protons are not able to traverse the membrane except at certain sites. Such a site is the F_o protein of the F_1-F_o complex. Fig. 7-10 describes one perspective of the events that may occur. Two protons enter the F_o protein, which provides an ion channel through the lipid bilayer and into the headpiece, F_1. F_1 has binding sites for inorganic phosphate near the proton channel. The energy of the proton gradient and also the membrane potential promotes a reaction of two protons with one of the oxygens of the phosphate. Water is eliminated, and the resulting highly reactive phosphate group then joins and ADP to form ATP.

Conformation hypothesis The third hypothesis to explain energy transduction suggests that conformational changes are involved in the proteins (ATPase) of the inner membrane. This concept also depends on the formation of the proton gradient (see Fig. 7-9). As proposed, the energy of the gradient serves to induce a con-

Figure 7-11 Potential function of conformational changes of membrane protein (ATPase) in the biosynthesis of ATP.

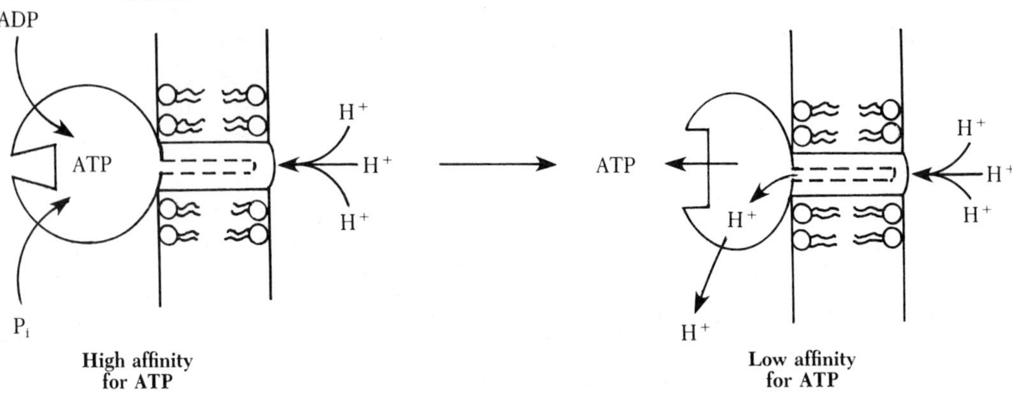

formational change in the ATPase. It is suggested in Fig. 7-11 that ADP and phosphate bind at the ATPase binding site and spontaneously produce ATP. However, the affinity of the active site for the ATP is thought to be very high, and therefore it is not released. The proton-gradient induced conformational change decreases the affinity for ATP, resulting in its release. The ATPase then returns to its original conformation.

Inhibitors of electron transport–oxidative phosphorylation

A variety of substances inhibit biological oxidations in diverse ways. The determination of their particular points of action has been important in developing our understanding of electron transport–oxidative phosphorylation processes. A side benefit from this kind of research has yielded an understanding of the mode of action of certain poisons and drugs. Fig. 7-12 locates the points of action for various inhibitors of electron transport and oxidative phosphorylation.

One can demonstrate the actions of inhibitors by using a mitochondrial preparation, maintained in a closed system containing a known volume of oxygen and pyruvate as substrate (Fig. 7-13). When all the endogenous ADP has been converted to ATP, oxidation (oxygen consumption) slows. Now, if an excess of ADP is introduced, rapid respiration resumes and the oxygen in the system diminishes. If rotenone, a potent fish poison, or barbiturates such as secobarbital or amobarbital are added to this system, oxygen consumption is suppressed. However, the addition of succinate leads to a resumption of respiration. This indicates that the three inhibitors exert their effect somewhere in the electron transport chain between NADH dehydrogenase and CoQ (Fig. 7-12). The addition to the mitochondrial system of oligomycin, an antibiotic from actinomycetes, also suppresses oxygen consumption (Fig. 7-13). Likewise, the plant glucoside called atractyloside inhibits respiration. However, if DNP (dinitrophenol) is now added to these two, respiration resumes. This is because DNP is an uncoupling agent that separates oxidative phosphorylation from electron transport. Since uncoupling eliminates the inhibition of oligomycin and atractyloside, the effects of these two compounds must occur in oxidative phosphorylation and not in electron transport. Oligomycin is believed to interfere with the phosphorylation of ADP, whereas atractyloside inhibits the ATP-translocating mechanism required to transport ATP through the mitochondrial membranes (see Fig. 7-12).

DNP acts as a proton ionophore, i.e., it appears to function by translocating H^+ back across the lipid bilayer, thereby abolishing the gradient. The uncoupling effect

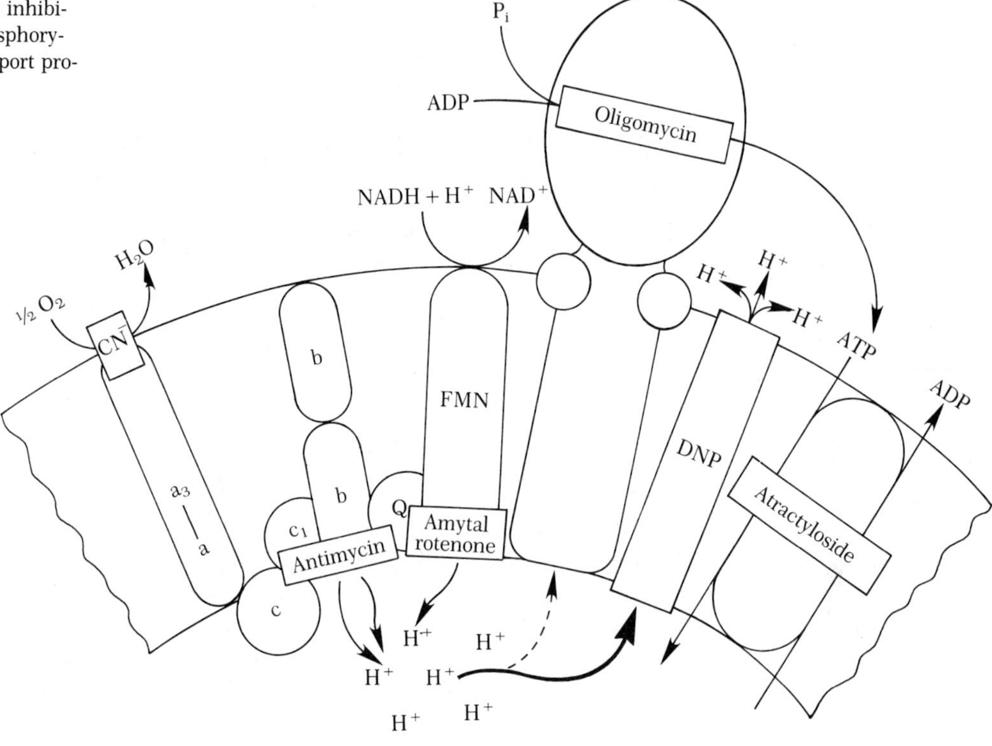

Figure 7-12
Sites of action of various inhibitors of the oxidative phosphorylation and electron transport processes.

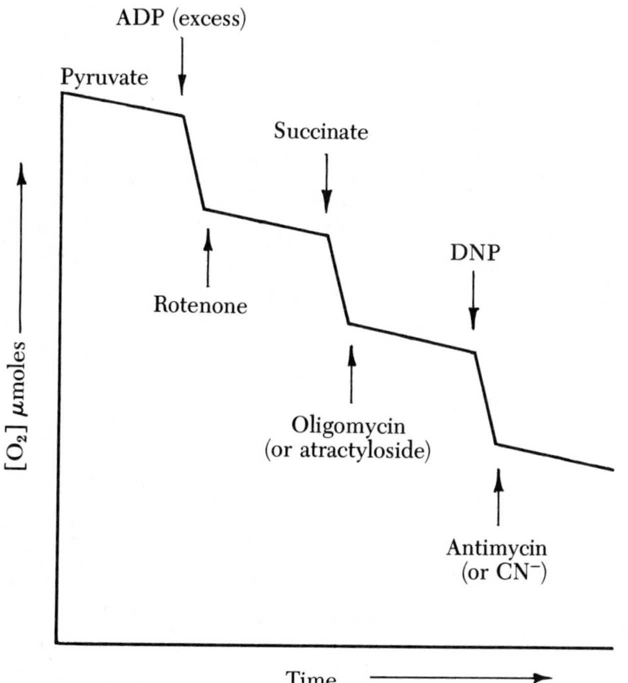

Figure 7-13
Scheme showing effect of various substrates and inhibitors on oxygen consumption in a closed system containing suspension of mitochondria.

of DNP results in an accumulation of ADP. This stimulates electron transport as the cell vainly attempts to supply its needs for ATP. Since the energy normally conserved as ATP is lost as heat, the cell is energetically "spinning its wheels." For this reason, DNP on occasion has been used as an ingredient in reducing pills.

Fig. 7-13 shows that respiration can be inhibited by both antimycin A and cyanide. Antimycin A inhibits the cytochrome c reductase, whereas cyanide complexes cytochrome a-a_3 prevent its interaction with oxygen. Cyanide is a good ligand for iron and probably prevents oxygen from coordinating with the iron in this final step of the electron transport chain.

In summary, inhibitors of mitochondrial action may be classified as follows:

1. Those affecting electron transport: rotenone, amobarbital, antimycin A, and cyanide
2. Those affecting phosphorylation: oligomycin and atractyloside
3. Those which uncouple phosphorylation from electron transport: DNP

Respiratory control

In mitochondria the process of oxidative phosphorylation is normally tightly coupled to electron transport. In the presence of excess substrate and oxygen, respiration continues until all the ADP is converted to ATP. Then the respiration rate, i.e., the utilization of oxygen, reaches a low resting level. Addition of ADP elicits a rapid rate of respiration by the sudden consumption of oxygen. This means that in the presence of adequate oxygen and substrate, ADP becomes rate limiting; it exerts a close control over the entire oxidative phosphorylation process. This control, which depends on the tight coupling of electron transport to phosphorylation, is referred to as *respiratory control.*

Substrate level energy transduction

In addition to oxidative phosphorylation, some ATP is produced in direct association with individual steps of certain metabolic pathways. Thus there is a considerable free-energy change in the oxidation of glyceraldehyde 3-phosphate to the corresponding carboxylic acid. As will be seen in Chapter 11, the loss of energy is prevented by forming 1,3-bisphosphoglycerate instead. The energy is conserved in the mixed anhydride bond of the 1,3-bisphosphoglycerate. In the next reaction of the pathway this energy is utilized to form ATP directly; the enzyme involved is *3-phosphoglycerate kinase.* Formation of ATP directly coupled to a metabolic process is called the production of ATP at the *substrate* level. Other substrate level processes are in the conversion of phosphoenolpyruvate to pyruvate and succinyl CoA to succinate (citric acid cycle).

Storage and release of high-energy phosphate

As is evident from Fig. 2-9, cardiac muscle is rich in large mitochondria; its metabolism is aerobic. Skeletal muscle, which is called on for varying degrees of activity, has comparatively fewer and smaller mitochondria. Under conditions of sudden and extended activity skeletal muscle metabolism may be anaerobic. The amount of ATP available is sufficient only for approximately 2 seconds of work; therefore it is important for the supply of ATP to be continually replenished. In the muscle cell this is accomplished at the expense of phosphocreatine, a compound that serves as a ready source of stored, metabolic energy.

Phosphocreatine (creatine phosphate)

The role of phosphocreatine as a storage compound of high-energy phosphates is suggested by the following reaction:

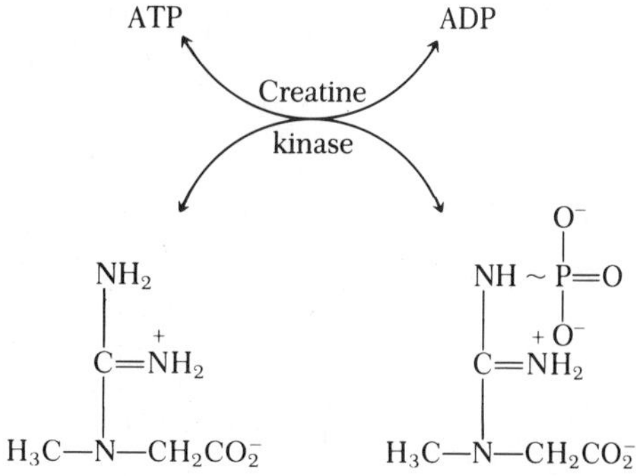

ATP ⇄ ADP (Creatine kinase)

$$H_3C{-}N(C(=\overset{+}{N}H_2)NH_2){-}CH_2CO_2^- \quad\rightleftharpoons\quad H_3C{-}N(C(=\overset{+}{N}H_2)NH\sim PO_3^{2-}){-}CH_2CO_2^-$$

Creatine **Phosphocreatine**

In the resting muscle cell the ATP produced is used for the phosphorylation of creatine, a reaction catalyzed by *creatine phosphotransferase* or *creatine kinase*.

As seen in Table 7-2, the standard free energy of hydrolysis of phosphocreatine is −10.5 kcal/mole. Therefore the reverse of the above reaction (the transfer of phosphate from phosphocreatine to ADP) is thermodynamically favorable and occurs with a free-energy change of −3 kcal/mole. The phosphorylation of creatine by ATP, however, is unfavorable and occurs when the intracellular concentration of ATP is sufficiently high. Muscle contraction is accompanied by a utilization of ATP. The reaction readily reverses, and ATP is reformed at the expense of the phosphocreatine (Chapter 15). Thus the net effect is to maintain the ATP at relatively constant levels.

The biosynthesis of creatine from arginine, glycine, and methionine, as well as the conversion of phosphocreatine to creatinine (a product excreted in the urine), is described in Chapter 13.

Myokinase reaction

Another source of ATP is actually derived from the ADP that accumulates during muscular activity (Chapter 15). As seen by the accompanying diagram, two molecules of ADP may interact to yield one molecule of ATP and one of AMP. This reversible reaction, catalyzed by *adenylate kinase* (myokinase), occurs only during periods of high ATP need. When the demand for ATP is no longer an overriding one, the reaction reverses; AMP is phosphorylated to ADP. It is the ADP, not the AMP, that functions in ATP production in glycolysis, the citric acid cycle, and oxidative phosphorylation. AMP, derived from the reactions involved in the activation of amino acids and fatty acids, must also be converted to ADP. AMP, however, is an important positive effector that acts on phosphofructokinase and thus stimulates glycolysis.

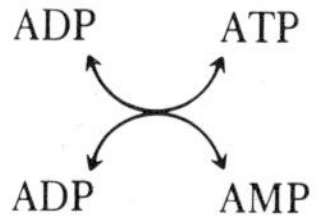

ADP + ADP ⇄ ATP + AMP

Suggested readings

Boyer, P.D.: Coupling mechanisms in capture, transmission, and use of energy, Annu. Rev. Biochem. **46:**957, 1977.

Chance, B.: Electron transfer: pathways, mechanisms, and controls, Annu. Rev. Biochem. **46:**967, 1977.

Ciba Foundation Symposia: Energy transformation in biological systems, Amsterdam, 1975, Association of Scientific Publishers.

Dawson, A.G.: Oxidation of cytosolic NADH formed during aerobic metabolism in mammalian cells, Trends in Biochem. Science **4:**171, 1980.

Ernster, L.: Chemical and chemiosmotic aspects of electron transport-linked phosphorylation, Annu. Rev. Biochem. **46:**981, 1977.

Fernández-Morán, H., Oda, T., Blair, P.V., and Green, D.E.: A macromolecular repeating unit of mitochondrial structure and function, J. Cell Biol. **22:**63, 1964.

Fessenden-Raden, J.M., and Racker, E.: Structural and functional organization of mitochondrial membranes. In Rothfield, L.I., editor: Structure and function of biological membranes, New York, 1971, Academic Press Inc.

Green, D.E.: The conformational model and the structure function relationship in mitochondria. In Manson, L.A., editor: Biomembranes, vol. 2, New York, 1971, Plenum Press.

Hatefi, Y., and Djanadi-Ohaniance, L., editors: The structural basis of membrane function, New York, 1976, Academic Press, Inc.

Hinkle, P.C., and McCarty, R.E.: How cells make ATP, Sci. Am. **238:**104, 1978.

Mehlman, M.A., and Hanson, R.W., editors: Energy metabolism and the regulation of metabolic processes, New York, 1972, Academic Press, Inc.

Mitchell, P.: Vectorial chemiosmotic processes, Annu. Rev. Biochem. **46:**996, 1977.

Racker, E., editor: Membranes of mitochondria and chloroplasts, ACS Monograph no. 165, Princeton, N.J., 1970, D. Van Nostrand Co., Inc.

Racker, E.: A new look at mechanisms in bioenergetics, New York, 1976, Academic Press Inc.

Racker, E.: Mechanism of energy transformation, Annu. Rev. Biochem. **46:**1006, 1977.

Slater, E.C.: Mechanism of oxidative phosphorylation, Annu. Rev. Biochem. **46:**1015, 1977.

Tedeschi, H.: Mitochondria: structure, biogenesis and transducing function, New York, 1976, Springer Verlag.

Tzagoloff, A.: The adenosine triphosphatase complex of mitochondria. In Martonosi, A., editor: The enzymes of biological membranes, vol. 4, New York, 1976, Plenum Press.

Wainio, W.W.: The mammalian mitochondrial respiratory chain, New York, 1970, Academic Press Inc.

Wilson, D.F.: Energy transduction in biological membranes. In Bittar, E.E., editor: Membrane structure and function, vol. 1, New York, 1980, John Wiley & Sons.

Chapter 8

Genetic control of metabolism: structure of nucleic acids

Cellular metabolism and its regulation depend on the presence within the cell of the proper amounts of the enzymes needed. The dynamic control of metabolism in humans, responding to the input of nutrients or diet and physiological needs, is controlled by the selective biosynthesis of enzymes. This response is relatively long term and therefore represents a coarse control of metabolism (Chapter 3).

Since enzymes are proteins (Chapter 5), an understanding of the nature of coarse control must be sought at the level of protein biosynthesis. The information required for a selective synthesis of proteins is inherent in the cell's genetic material, i.e., the nucleic acids (DNA) of the chromosomes. The biosynthesis of enzymes in response to chemical communication either from the nutrients themselves or by certain chemical regulators (the hormones) (Chapter 18), requires first a selective retrieval of information from the chromosomal DNA located exclusively in the cell's nucleus. Information retrieval involves the biosynthesis of another kind of nucleic acid called ribonucleic acid (RNA).

This chapter describes the structural aspects of DNA and the various kinds of RNA. The process of DNA duplication (replication) and the biosynthesis of RNA (transcription) are described in Chapter 9.

Biochemical genetics

DNA: genetic information

That genetic information is inherent in a chemical substance within cells was first appreciated in 1928 by Griffith. This investigator used two strains of *Pneumococcus* in his studies. The normal bacterium produces a protective polysaccharide capsule and is pathogenic to mice. When grown on agar media, it produces smooth and shiny colonies (S strain). A mutant was obtained that had lost the genetic information needed to produce its protective capsule and therefore was nonpathogenic to mice. Growth on agar plates yielded rough-looking colonies; hence they were designated the R strain. Griffith killed the pathogenic bacteria by heating and then injected into mice a combination of dead S and living R bacteria. The mice died; both living S and R bacteria were found present in the dead animals. Therefore the genetic information needed to produce the polysaccharide capsule had been transmitted from the dead, pathogenic S strain to the living but nonpathogenic R strain. The substance involved he called the *transforming factor* because it transformed the nonpathogenic R into the pathogenic S strain. In 1944 Avery, McLeod, and McCarty extracted the transforming factor from the S strain and showed that it was DNA. When the DNA from the pathogenic *Pneumococcus* was combined with living R strain and then injected into mice, the microorganism was found to be lethal. This and other similar studies have verified that processes of genetic transformation involve the chemical substance DNA.

Now it is clear that genetic information is inherent in the structure of the DNA molecules. In eukaryotic cells the DNA is gathered into units called chromosomes, which are located in the cell nucleus. There are roughly 10^{13} cells in the human body, each possessing the same complement of genetic information as the original fertilized, diploid egg cell. In the process of embryonic development the cells divide and the DNA itself is duplicated precisely with each division. Replication of DNA can now be understood in molecular terms involving the Watson and Crick model of DNA and the principle of *base complementarity*. It is possible to visualize how the structure of DNA and the genetic information in each cell are duplicated exactly. This model yields an insight into the process of mutation and the concept of molecular disease.

Nucleic acids

Nucleic acids are polyanionic molecules of high molecular weight composed of a specific sequence of subunits or monomers called nucleotides (Chapter 23); the whole is termed a *polynucleotide*. Nucleic acids are of two main categories: the *deoxyribonucleic acid*, or DNA, and *ribonucleic acid*, or RNA. DNA is found primarily in the chromosomes of the cell nucleus, whereas 90% of the RNA is present in the cell cytoplasm and 10% in the nucleolus. On total hydrolysis the two nucleic acids yield the constituents shown in Table 8-1. They are distinguished primarily on the basis of the 5-carbon sugars or pentose present. Thus DNA contains the β-anomer of D-2-deoxyribose, whereas RNA has β-D-ribose instead. There are also two classes of bases present in both nucleic acids: the purines and the pyrimidines. The purines, adenine and guanine, are found in both polynucleotides. However, RNA is distinguished from DNA by its content of the pyrimidine uracil rather than the thymine present in DNA. Cytosine is found in both nucleic acids. The structure of the purines and pyrimidines are found in Chapter 23.

Deoxyribonucleic acid

Proportions of components

The proportions of the nucleotide bases from complete hydrolysates of DNAs from a variety of sources were first investigated by Chargaff. Some typical results are illustrated in Table 8-2. On the basis of such chemical analyses the following relationships were established:

1. The sum of purines equals the sum of the pyrimidines.
2. The molar proportion of adenine equals that of thymine.
3. The molar proportion of guanine equals that of cytosine.

The importance of these observations to the total structure of DNA was not apparent until suitable x-ray diffraction studies were completed. Thereafter it was possible to visualize the probable location of the constituent atoms in a three-dimensional model.

Double-helical structure

On the basis of the x-ray diffraction studies of Wilkins and the analytical data of Chargaff, Watson and Crick proposed the three-dimensional model of DNA (Fig. 8-1). They suggested that DNA consists of two polynucleotide chains wound about a common axis to form a right-handed double helix having a diameter of 2 nm and a pitch or rise of 3.4 nm.

Fig. 8-1, which shows a space-filling model as well as a diagrammatic representation, illustrates how the two helices are wound so as to yield two interchain spacings, a minor and a major groove. Such a double helix cannot be pulled apart and can be separated only by an unwinding process. They are called *plectonemic coils*.

Table 8-1
Components of DNA and RNA

Type of component	DNA	RNA
Acid moiety	Phosphoric acid	Phosphoric acid
Sugar	D-2-Deoxyribose	D-Ribose
Bases		
Purines	Adenine	Adenine
	Guanine	Guanine
Pyrimidines	Cytosine	Cytosine
	Thymine	Uracil

Table 8-2
Base composition of representative types of DNA (moles per 100 moles)*

	A	G	C	MC	T	A + G/T + C + MC
Human	30.5	19.9	20.6	—	28.9	1.02
Rat (bone marrow)	28.6	21.4	20.4	1.1	28.4	1.00
Rainbow trout (sperm)	29.7	22.2	20.5	—	27.5	1.08
Wheat germ	27.3	22.7	22.8	—	27.1	1.00
Escherichia coli (K-12)	26.0	24.9	25.2	—	23.9	1.08

*A, Adenine; G, guanine; C, cytosine; MC, methylcytosine; T, thymine.

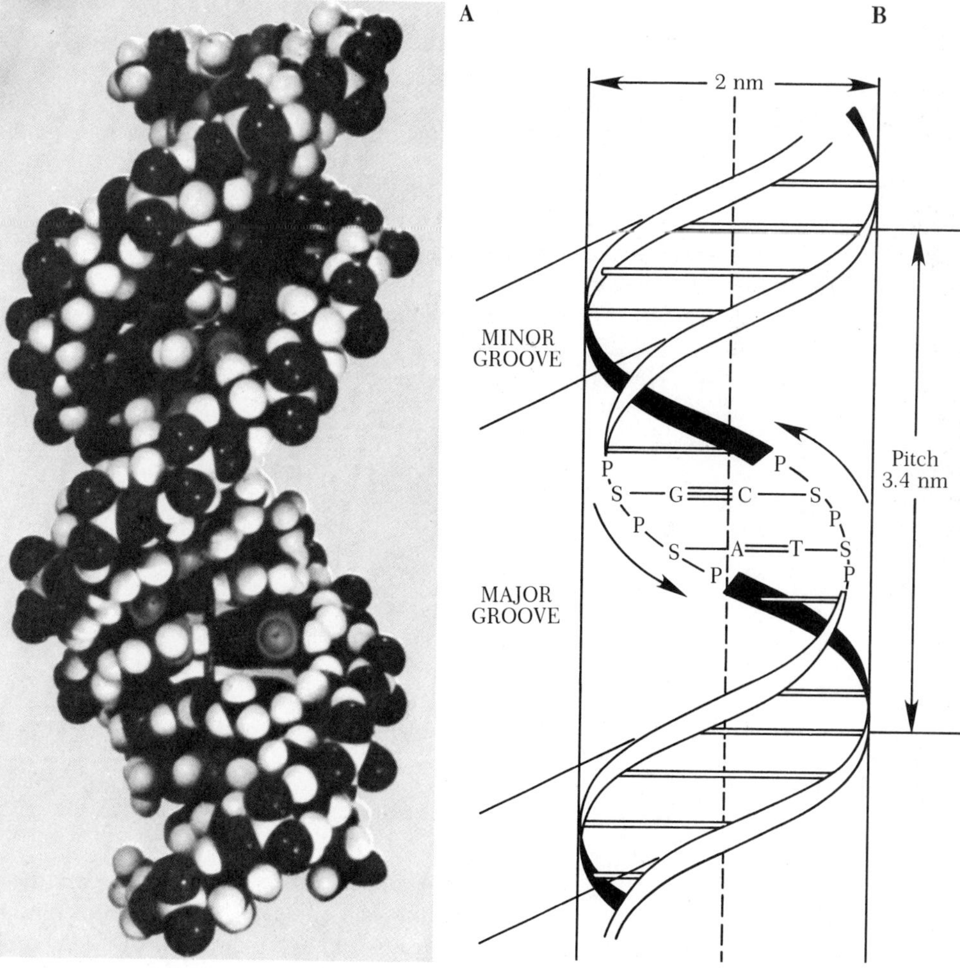

Figure 8-1
A Photograph of Watson-Crick model of DNA produced from space-filling atomic models.
B Diagrammatic representation of double helix. Horizontal lines symbolize hydrogen-bonded base pairs. Also illustrated are the major and minor grooves and the dimensions of the double helix.

A courtesy F. Ferguson, Ealing Corp.

Base complementarity

Each of the two polynucleotide chains in the double helix consists of a specific sequence of nucleotides connected by 3′,5′-phosphodiester bonds. The sequence of bases or primary structure is illustrated below. Nucleotide sequences are arranged conventionally so that the 5′ terminal is at the left and the 3′-OH terminal is to the right. It is customary to arrange a sequence so that the 3′,5′-phosphodiester bonds have a left-right directionality as described by the arrow. One of several symbolic representations is also used to describe the primary structure. For example, the deoxyribose may be symbolized by a straight line and the phosphodiester bonds may be represented as shown in the illustration. The bases are then indicated by an appropriate letter (*A* = adenine, *G* = guanine, *C* = cytosine, and *T* = thymine). The sequence may be further shortened to pGpTpApC or G-T-A-C. The hyphen represents the phosphodiester bond.

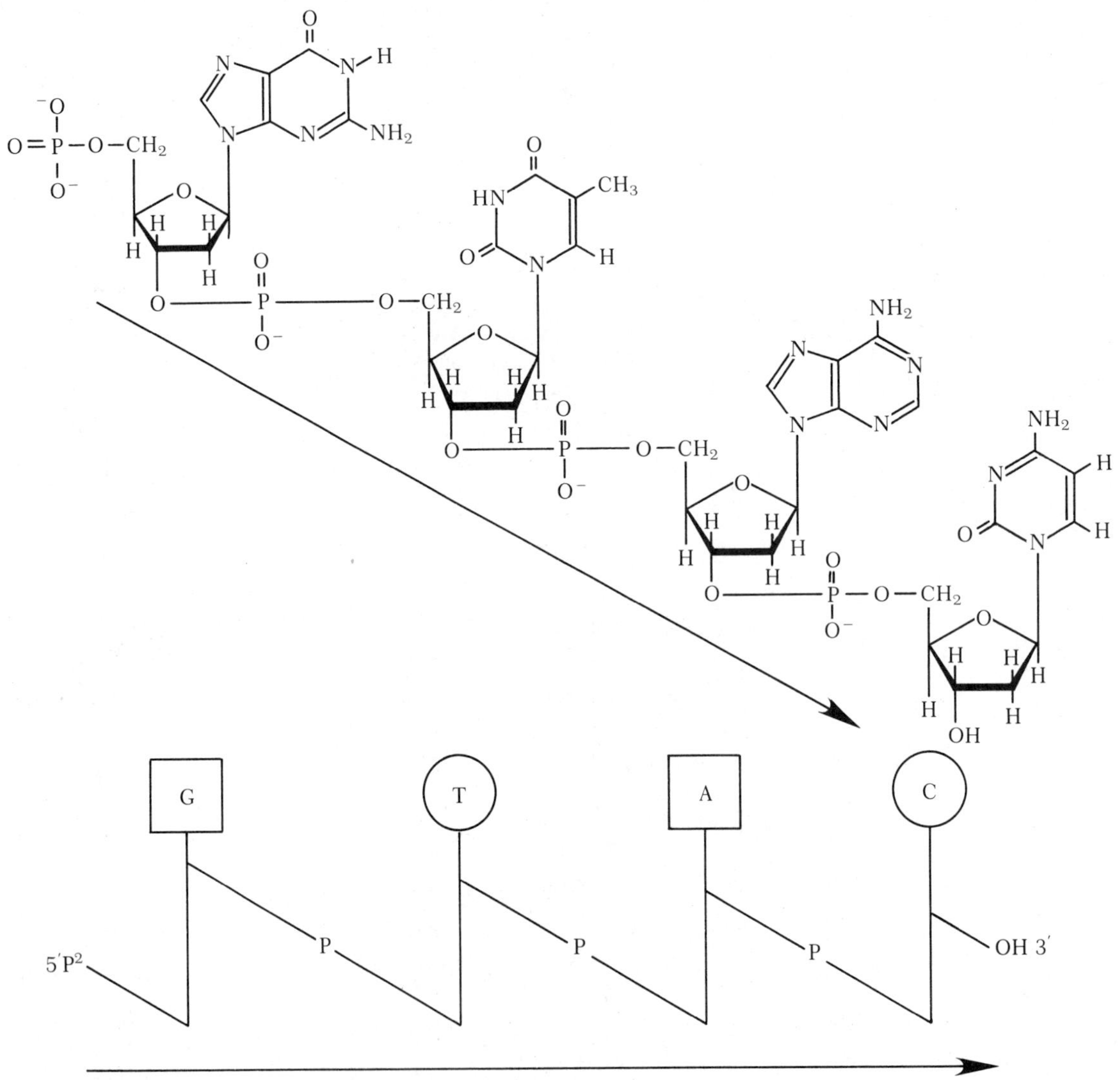

The question now arises: Exactly how are the two chains arranged within the double helix? The cue may be taken from the observations regarding the molar ratios of purines and pyrimidines, i.e., that the proportion of adenine equals that of thymine, and guanine equals that of cytosine. These were interpreted to mean that

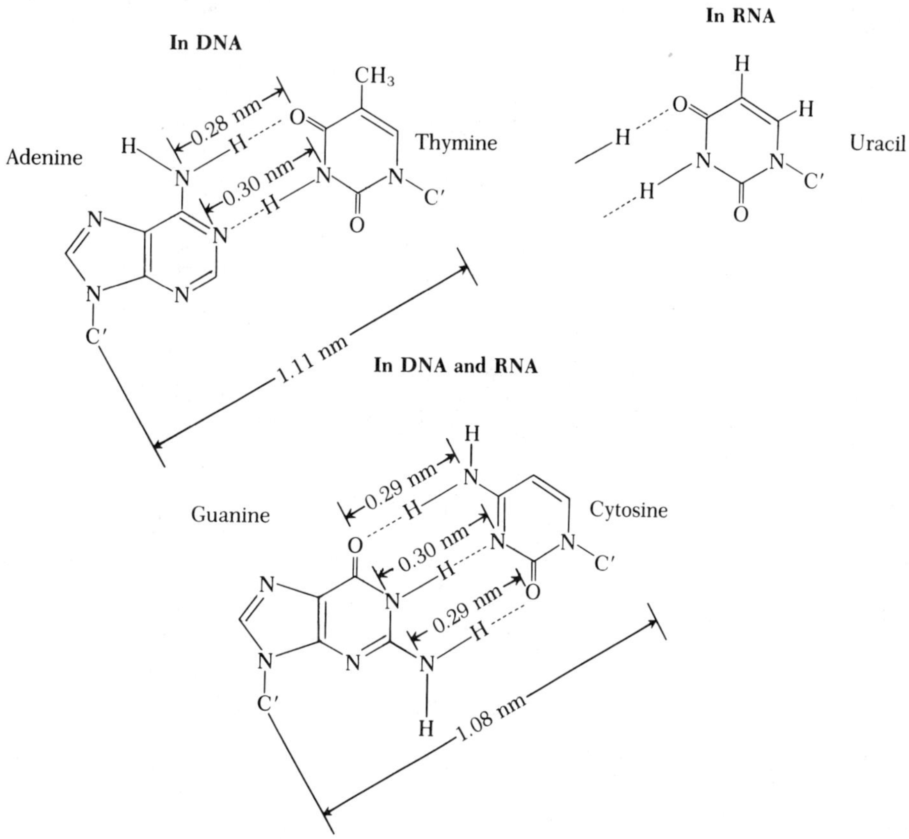

Figure 8-2
Structure of an adenine-thymine or adenine-uracil base pair *(upper left and right)* and guanine-cytosine base pair *(bottom)*.

the side-by-side arrangement of the two polynucleotide chains is such that the adenine in one chain is always opposite a thymine in the companion chain; likewise, guanine in one is always located opposite cytosine in the other. This positioning or pairing of the bases is called *base complementarity*.

The x-ray diffraction data showed that the complementary bases approach each other so that some of the hydrogen and nitrogen atoms are within a distance of 0.3 nm (Fig. 8-2). This means that in some instances the hydrogen atoms may be shared by a nitrogen of one base and the oxygen or nitrogen of another. The hydrogen bonds between adenine and thymine and between guanine and cytosine are illustrated in Fig. 8-2. Although hydrogen bonds taken individually are weak, the large numbers involved in a molecule of DNA ensure that these contribute to the stability of the double helix.

In Fig. 8-1, *B*, the outer lines of the polynucleotide chain symbolize the phosphodiester bonds that hold together the individual nucleotide units. The base pairs in turn are represented by the parallel lines. This indicates that the purine-pyrimidine base pairs are found in the center of the double helix, with the phosphates at the periphery. There are 10 such base pairs per turn of the helix. Furthermore, it is apparent from Fig. 8-1, *A*, that the base pairs are planar structures stacked one above the other at a distance of 0.34 nm (center to center). Therefore another important stabilizing force of the DNA molecule is the hydrophobic interaction between stacked base pairs.

The existence of two hydrogen bonds between adenine and thymine (or uracil) and three between guanine and cytosine is fundamental to the principle of base complementarity. This principle is essential to an understanding not only of DNA and RNA structure but also of the process of DNA replication, the transcription of DNA to yield RNA, and the translation of messenger RNA into protein structure. Thus in DNA adenine and thymine are paired, whereas in RNA the comparable base pair is adenine-uracil. Furthermore, in the process of DNA replication the base pair is adenine-thymine, whereas in transcription and translation the base pair is adenine-uracil. This interchangeability of thymine and uracil in the base pairs is illustrated in Fig. 8-2.

Antiparallel chains

Analysis of the primary structure of the two complementary DNA chains shows that they are arranged to read in opposite directions. From the diagram below it is apparent that the left-hand chain is read in the conventional 5′ to 3′ direction indicated by the arrow. On the other hand, the right-hand chain is arranged so that its 5′, 3′ directionality is read from the top down (note arrow). Two complementary polynucleotide chains that exhibit opposite polarity are said to be *antiparallel*. Accordingly, if the sequence of bases in one chain is TCAG, the sequence in the other must be CTGA when both are read in the 5′ to 3′ direction. The antiparallel nature of the two polynucleotide chains in the DNA helix is indicated in Fig. 8-1, *B*, by the two opposing arrows. As in the case of base complementarity, the principle of antiparallel chains is a general one and is observed not only in DNA structure but also in replication, transcription, and translation.

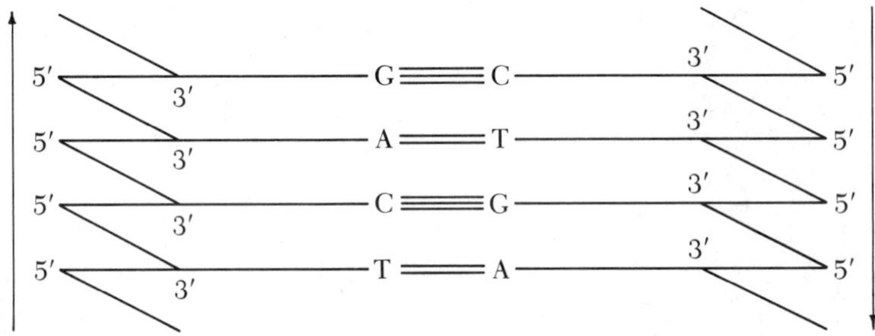

Denaturation and renaturation

Although the DNA double helix is stabilized by hydrogen bonding and the hydrophobic interactions between stacked base pairs, its structure can be disrupted or denatured. Denaturation involves breaking the stabilizing forces and unwinding and separation of the helix into two flexible polynucleotide chains. Conditions and reagents that denature DNA are low and high pH, low ionic strengths, formamide, urea, and elevated temperatures.

The process of unwinding during heat denaturation may be followed as a melting curve by measuring the light absorbance (260 nm) of solutions of nucleic acids. Fig. 8-3 shows melting curves for various kinds of DNA. Nucleic acids absorb light at 260 nm, a property of their purine and pyrimidine bases. Native DNA, however, does not absorb as much light as expected from its base content, probably because of the interaction of the stacked bases. When a solution of native DNA is heated slowly, the light absorbance increases (hyperchromicity), yielding the patterns shown in Fig. 8-3. The unwinding process is not complete until a maximum absorbance is reached. The inflection points of these curves are called the melting temperatures (T_m) and are characteristic of the particular DNAs used. They range between 84° and 96° C. If the completely denatured DNA (maximum absorbance) is slowly cooled to room temperature, the original structure is often restored. This

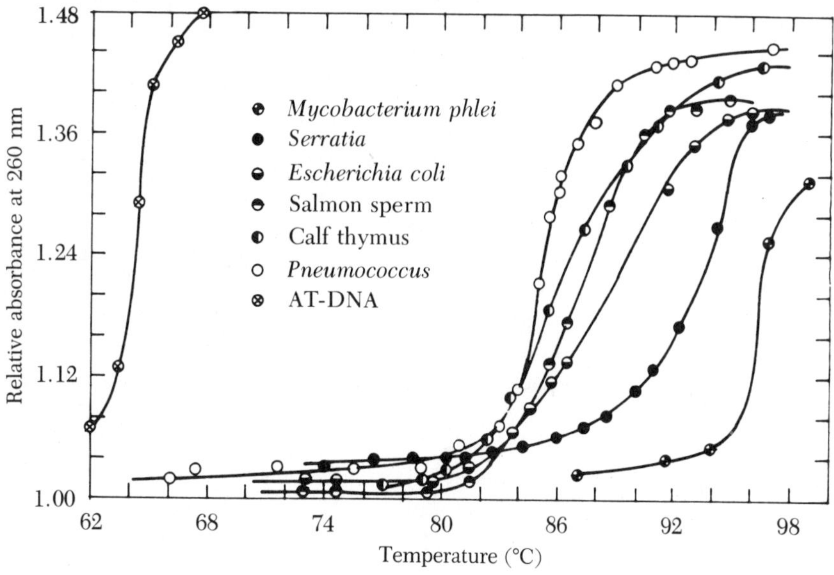

Figure 8-3
Melting curves for representative samples of DNA. Shown is the variation in absorbance of light (260 nm) as the temperatures of the DNA solutions are increased.

From Marmur, J., and Doty, P.: Nature **183:**1427, 1959.

renaturation process, also called annealing, occurs because the complementary chains gradually find one another, base pairs again are hydrogen bonded, and the double helices reform. If, on the other hand, the mixture of denatured DNA is cooled rapidly, then renaturation or annealing may not occur and the strands will remain in their separated state.

Since the guanine-cytosine base pair has three hydrogen bonds, compared with two for adenine-thymine, it follows that DNAs with high concentrations of guanine and cytosine may be more stable and have higher melting temperatures than do those with high concentrations of adenine and thymine. An extreme situation would involve helices of poly-A and poly-T chains. Such helices would have a comparatively low T_m (around 65° C), whereas the DNA of *Mycobacterium phlei*, with 70% of its bases represented by guanine-cytosine, has a T_m greater than 90° C.

Complexity of eukaryotic DNA

Denaturation-renaturation can provide an insight into the complexity of eukaryotic DNA. The rate of annealing depends on the relative amounts of complementary sequences present in the denatured sample. Thus, if there are many regions of highly repetitive sequences, the annealing process may be relatively rapid. This rate of renaturation or annealing is expressed as a C_ot value. Its meaning may be explained by considering the renaturation of a simple mixture of complementary polynucleotide chains, e.g., poly-A and poly T-chains. Helices can be formed under annealing conditions and the time required measured. The more concentrated the polynucleotide solution, the more readily the complementary molecules will find one another and therefore the shorter the time required. This concentration is expressed as C_o (moles of DNA phosphate per liter); t is the time in seconds required for the annealing to occur. The rate of renaturation is described as the product of the two: C_ot. A mixture of complementary poly-A and poly-T chains would have a relatively low C_ot value. Actually, the time required for renaturation to be half complete is a useful characteristic of DNA. Therefore it is customary to measure the half C_ot value ($C_ot_{1/2}$).

To appreciate the complexity of eukaryotic DNA, the extracted material is first fragmented into smaller pieces, which are then separated into fractions of like physical and chemical properties. Such fractions are studied by the denaturation-renaturation (heating and cooling) technique; their $C_ot_{1/2}$ values are established. When mouse DNA was extracted and fragmented, a fraction was obtained having a surprisingly low $C_ot_{1/2}$ value and consisting of relatively simple and highly repetitive sequences. This component, termed *satellite DNA*, has been located in the centromeric region of the mouse chromosomes. Since there are some 10^6 copies of this sequence, the satellite DNA is a significant component of chromosomal DNA; however, its function is still unknown. At the other extreme lies a fraction with the largest $C_ot_{1/2}$ value; it therefore consists of sequences that are practically nonrepetitive. These unique sequences, of which there are only a few copies per cell, require long annealing times before the complementary chains can find one another and reform double helices. The unique or nonrepetitive sequences represent the structural genes. Approximately 50% of the mammalian DNA has an intermediate $C_ot_{1/2}$ value, suggesting the existence of sequences of several thousand nucleotides and repeated some 10^2 to 10^4 times per cell. It is thought that these intermediate or moderately repetitive sequences include the spacer segments that lie between the structural genes. This arrangement is illustrated in the following scheme:

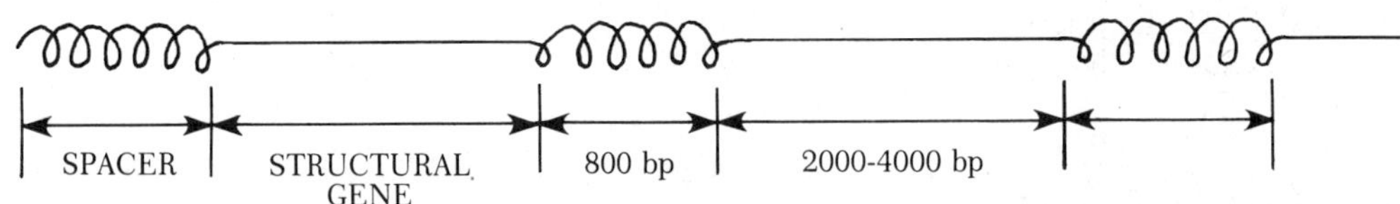

The moderately repetitive sequences also include the genes for ribosomal RNA.

Molecular size and shape

Depending on their source, molecules of DNA may be relatively small or enormously large and complex. As seen in Table 8-3, the DNA from an average single human chromosome may be more than 4 cm long and have a molecular weight in the billions. These molecules are linear and unbranched. The smallest human chromosome contains a linear duplex (double helix) DNA molecule with a molecular weight of 3×10^{10} and some 1.4 cm long; the largest has a molecular weight of 1.5×10^{11} and is 7.3 cm long. Some DNA molecules, as from simian virus 40 (SV40), ϕX174, or *E. coli*, are much smaller, carry less genetic information, and are

Table 8-3
Size and shape of some representative DNA molecules

Organism	Base pairs	Total size of genome: Length (mm)	Total size of genome: Molecular weight	Shape
Simian virus 40 (SV40)	5.1×10^3	1.7×10^{-3}	3.0×10^6	Circular duplex
Bacteriophage ϕX174	5.4×10^3	1.8×10^{-3}	1.7×10^6	Circular single-stranded; circular duplex, replicative form
Bacteriophage T_2	1.44×10^5	4.9×10^{-2}	130×10^6	Linear
E. coli	4×10^6	1.36	2.8×10^9	Circular
Yeast	1.35×10^7 (17)*	4.6		
Drosophila melanogaster	1.65×10^8 (4)*	56	80×10^9	Linear
Human	2.9×10^9 (23)*	990	1.8×10^{12}	Linear

Data from Kornberg, A.: DNA synthesis, San Francisco, 1974, W.H. Freeman. Co.
*Number of chromosomes (haploid).

circular. Furthermore, ϕX174 is a single-stranded DNA having 5375 nucleotides, *E. coli* is double stranded with 4×10^6 base pairs, and human DNA is linear and double stranded and may have 125 million base pairs. The size of the DNA molecules can be seen in Fig. 8-4, which shows the DNA dispersed from a T_2 bacteriophage.

It is now evident that the double helix is wound not about a linear but about a curved axis. As just indicated, some DNA is circular and does not possess free terminal ends. Circular double helical DNA can then be wound in space to yield a new helix: a supercoil (Fig. 8-5). Chromosomal DNA of eukaryotic cells is wound around a core of proteins to yield superhelices (see Fig. 8-10).

Figure 8-4
Electron micrograph of T_2 DNA.

From Kleinschmidt, A.K., Lang, D., and Zahn, R.K.: Biochim. Biophys. Acta **61**:857, 1962.

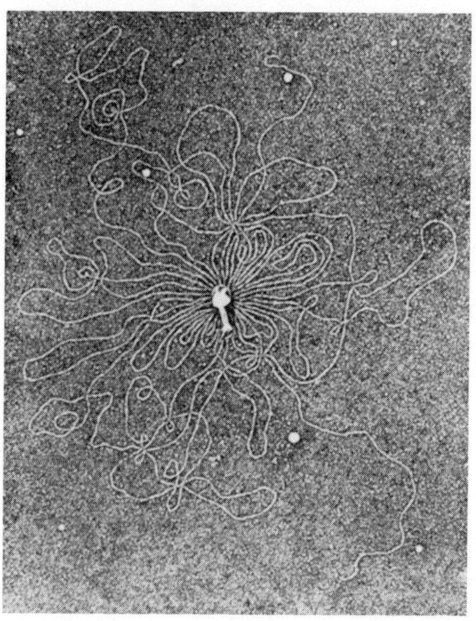

Figure 8-5
A Circular double-helical DNA.
B Circular double-helical DNA exhibiting a superhelix.

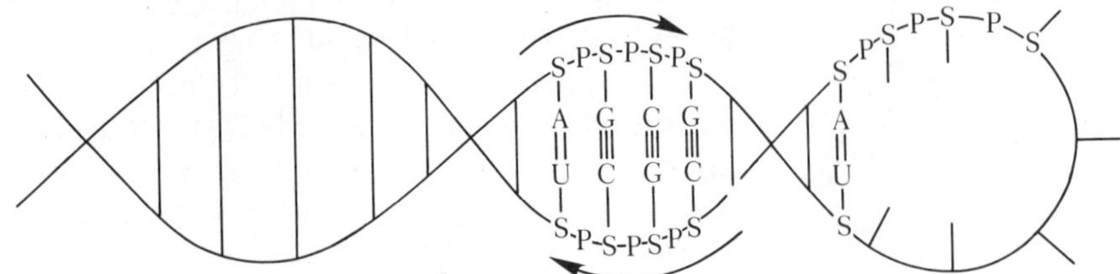

Figure 8-6
Schematic diagram showing a local region of antiparallel double helix in a single strand of RNA.

Ribonucleic acid

Like DNA, RNA consists of polynucleotide chains held together by 3′,5′-phosphodiester bonds. Unlike DNA, however, RNA is single stranded and does not exhibit the equivalence of adenine with uracil, and cytosine with guanine. It does exhibit a melting curve and a fairly sharp T_m, which represents an untwisting of local antiparallel, double-helical regions within the single strand. Such a region, held together by base pairing, is illustrated in Fig. 8-6.

In living cells there are three fundamental classes of RNA. About 75% to 80% of RNAs are structural and functional components of the ribosomes, the particles associated with the rough endoplasmic reticulum (Chapter 2), and are therefore called ribosomal RNA (rRNA). Also associated with the ribosomes, specifically the polysomes, is the messenger RNA (mRNA), representing 5% to 10% of the total. Messenger RNA provides the necessary genetic information for the proper assembly of amino acids into proteins. The third category, representing 10% to 15% of the whole, is called transfer RNA (tRNA).

General characteristics

Ribosomal RNA Ribosomes are dissociable into two subunits (Chapter 2). These, as well as the original particles, are usually described in terms of their sedimentation coefficients, a constant derived from their rates of sedimentation in an ultracentrifuge. Ribosomes from prokaryotic and eukaryotic cells are characterized by having sedimentation coefficients of 70S (Svedberg units) and 80S, respectively. Each of these can be dissociated into two subunits that are characteristic of the original particle. Thus the 70S yields 50S and 30S, whereas the 80S ribosome yields 60S and 40S subunits. Each subunit is composed of one or two strands of rRNA and numerous protein molecules. The manner in which they are combined to produce the final ribosome is described later. Characteristics of rRNA from prokaryotic and eukaryotic cells are given in Table 8-4.

Table 8-4
Characteristics of rRNA from prokaryotic and eukaryotic cells

	Sedimentation coefficient (S)			
	Ribosome	rRNA	Molecular weight	Number of bases
Prokaryote (70S)	30	16	5.6×10^5	1700
	50	23	1.1×10^6	3300
		5	3.6×10^4	120
Eukaryote (80S)	40	18	6.5×10^5	2000
	60	28	1.7×10^6	5000
		5.8	5×10^4	150
		5	4×10^4	120

Messenger RNA Messenger RNA is a heterogeneous mixture of large–molecular weight RNAs (approximately 2×10^6). While in bacteria the mRNA is distinguished by its rapid turnover, which in eukaryotic cells may or may not be so characterized. For example, in reticulocytes the mRNA for hemoglobin appears to be fairly stable. Eukaryotic mRNA is distinguished by its unique terminal residues. A molecule may have approximately 200 adenylyl units at its 3′ terminal and be capped by a 7-methyl guanosine triphosphate at its 5′ end. The guanosine triphosphate "cap" is attached to the 5′-carbon of the terminal ribose, which also is methylated on carbon 2′. These features are illustrated in the accompanying diagram. (*B*, Any of the four bases: *A*, adenine.)

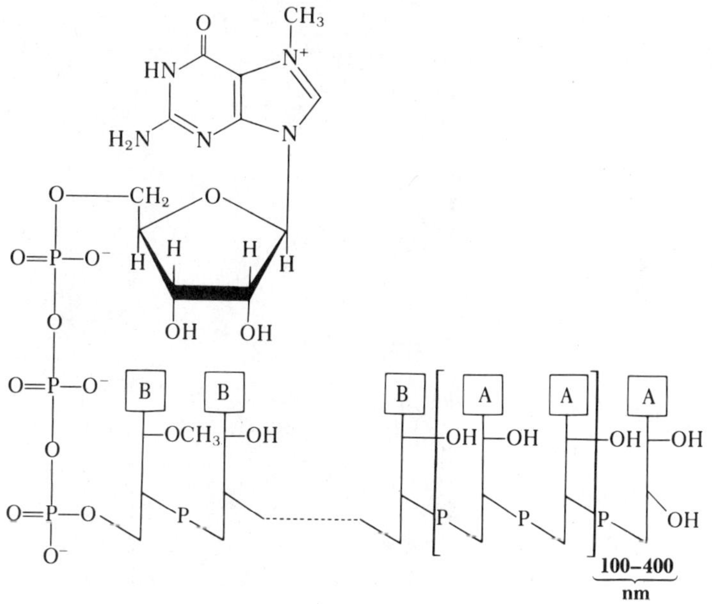

7-Methylguanosine triphosphate cap

Poly A terminus

Transfer RNA Transfer RNA has a small molecular weight: 25,000 (4S). Its function is to bind the specific amino acids in the process of translation; therefore there must be at least one for each amino acid.

It appears that all the tRNAs are of approximately the same length (75 to 90 nucleotides) and all terminate at the 3′-hydroxyl end in the sequence cytosine-cytosine-adenine. Transfer RNAs also are distinguished by their content of unusual and methylated purines and pyrimidines. These are 6-*N*-methyladenosine, 1-methyladenosine, 2-methyladenosine, 6-*N*-isopentenyladenosine, 5-methylcytidine, 4-*N*-acetylcytidine, 2′-*O*-methylcytidine, 3-methylcytidine, 2-thiocytidine, 2-*N*-methylguanosine, 1-methylguanosine, 2-*N*-methylguanosine, 2′-*O*-methylguanosine, 2-*N*-dimethylguanosine, inosine, 1-methylinosine, 4,5-dihydrouridine (DHU), 2′-*O*-methyluridine, 2-thiouridine, and pseudouridine (ψ). The structures for some of these are given in Chapter 23.

The primary structure, i.e., the exact sequence of purines and pyrimidines, has been established for about 100 tRNAs. As is true for all, tRNA is a single chain beginning with a 5′-phospho terminal and ending with a 3′-hydroxyl group. Despite the absence of individual complementary chains, tRNA has regions of antiparallel and complementary base sequences. Therefore the single chain can fold back on itself to form regions of double-helical structure. A typical structure for tRNA has three major helical regions or loops and has the appearance of a cloverleaf (Fig. 8-7, *B*). These loops are stabilized by hydrogen bonds between base pairs in complementary regions or stems. The first (*I*), or D loop, is named after its content of dihydrouridine. Loop *II* is the anticodon loop; its sequence contains the anticodon triplet (as described in Chapter 10). A third arm of variable length is

Figure 8-7
A Space-filling model of a tRNA molecule.
B Primary structure of alanyl tRNA. Alanine is attached to the terminal adenosine residue. *I-G-C* represents the sequence inosine, guanine, cytosine and is the anticodon for alanine (Chapter 10).
C Perspective diagram showing a tRNA molecule. The different shading represents the various loops and stem regions of the tRNA molecule. Note that the TψC and DHU loops come in close contact.

A courtesy Alexander Rich; **B** adapted from Madison, J.T., Everett, G.A. and King, H.K.: Symp. Quant. Biol. **31**:409, 1966; **C** from Rich, A. and Rajbhandary, U.L.: Annu. Rev. Biochem. **45**:805, 1976; reproduced, with permission, from the Annual Review of Biochemistry, volume 45;

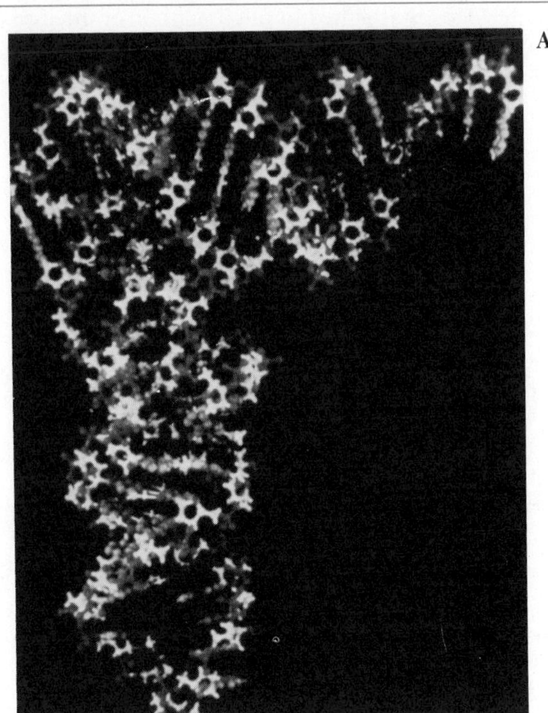

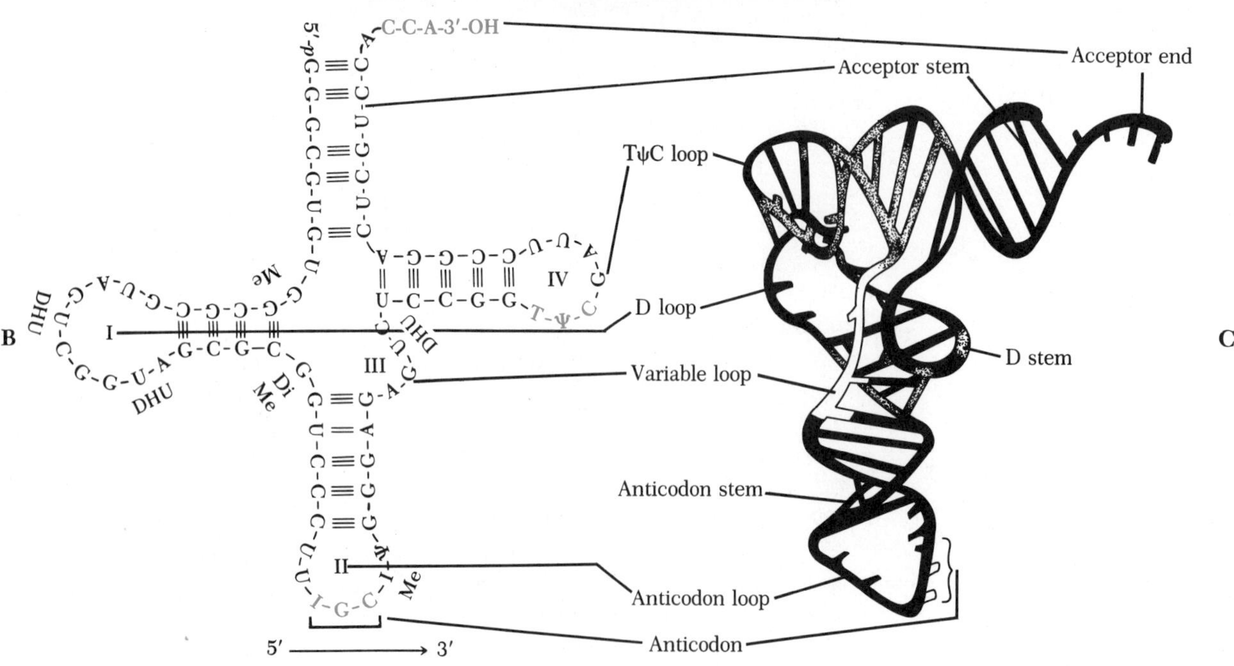

called the extra, or variable, loop. The fourth (*IV*), the ribothymidine-pseudouridine-cytidine or *TψC* loop is named for its content of thymine, a pyrimidine usually found only in DNA and not in RNA other than tRNA.

The tRNA structure terminates at the 3′-hydroxyl end with a cytosine-cytosine-adenine sequence. At this terminal the appropriate amino acid is attached by an ester bond (Chapter 10). Furthermore, the bases of the 3′-hydroxyl end (acceptor) pair with some of the bases of the 5′-phosphate end, forming the acceptor stem. The base pairs of the acceptor stem, as well as the TψC, anticodon, and D stems, appear to be constant for all tRNA species regardless of source, whether from animal, plant, or microorganisms.

Recent evidence shows that the cloverleaf structure is folded into a three-dimensional or tertiary conformaton (Fig. 8-7, *A* and *C*). This folded arrangement consists of two columns of bases at right angles to one another, forming an L shape. The arms of the L-shaped molecule have both helical and nonhelical regions corresponding to the stems and loops of the cloverleaf structure. Each of the two helical regions is composed of about 10 complementary base pairs and forms a single turn of a double helix (Fig. 8-7, *C*). In the nonhelical portions the bases may point inward, providing opportunities for internal hydrogen bonding. Some of these are unconventional bonds because they involve the C-2 hydroxyl group of one of the riboses rather than a nucleotide base. Although tRNA can be denatured by heat, on cooling the molecule snaps back into its native conformation. Also noteworthy is the fact that the C-C-A sequence (amino acid acceptor) terminates one arm of the L-shaped structure, whereas the anticodon loop is located at the other end (Fig. 8-7, *C*).

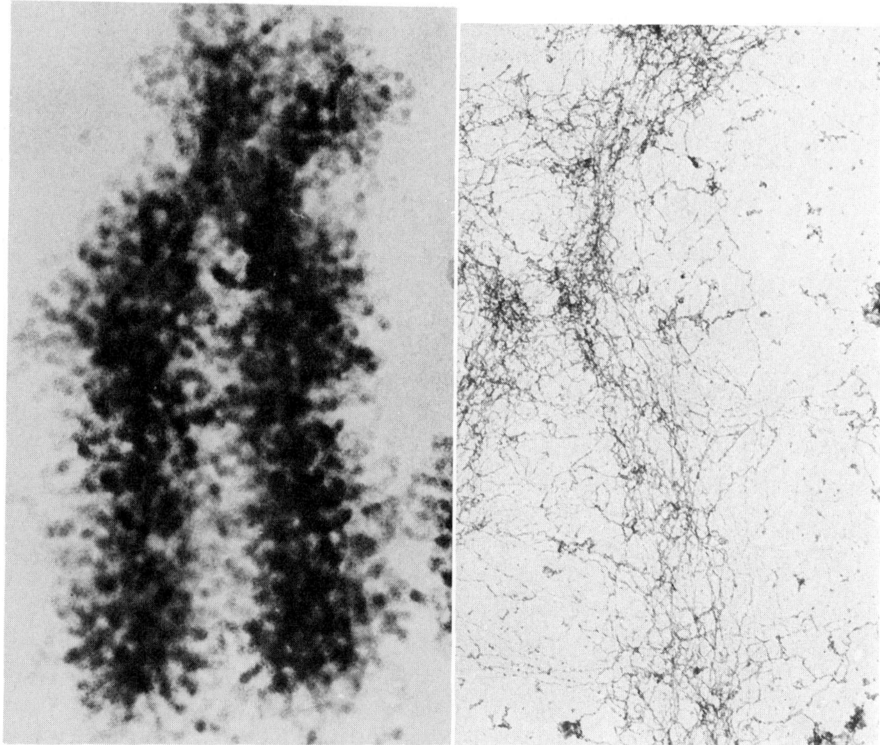

Figure 8-8
A Electron micrograph showing a human chromosome.
B Electron micrograph of a well-dispersed human chromosome.

A from Dupraw, E.J.: Symp. Quant. Biol. **38:**91, 1973; **B** from Comings, D.E., and Okada, T.A.: Symp. Quant. Biol. **38:**145, 1973. Copyright 1973 by Cold Spring Harbor Laboratory.

Nucleoproteins: part of subcellular structures

Most of the nucleic acids in eukaryotic cells are complexed with proteins and make up the subcellular organelles, the chromosomes (DNA), and the ribosomes (RNA).

Chromosomes

Approximately 90% of the cell's DNA exists as a nucleoprotein (nucleic acid–protein complex) called *chromatin*. This is the substance of the 23 pairs of chromosomes in human diploid cells. Fig. 8-8, *A*, shows a single intact chromosome; Fig. 8-8, *B*, shows a chromosome in which the DNA has been dispersed. These figures serve to illustrate the complexity of the packing of the chromatin in the chromosomal structure. Chromatin, as suggested below, is composed equally of the polyanionic DNA molecules combined with polycationic *histone* proteins. The force holding them together is electrostatic attraction.

$$\text{Chromatin}\begin{cases}\text{DNA}\\ \text{Histones and nonhistone proteins}\end{cases}$$

In addition to the histones, chromatin also contains a certain amount of nonhistone proteins.

Histones The highly cationic histones are rich in the basic amino acids lysine and arginine (approximately 25% of the total amino acid content) and have molecular weights in the range of 11,000 to 21,000. They are classified into five major fractions according to their contents of lysine and arginine and other parameters. Most of the basic amino acids are clustered at the amino-terminal end of the polypeptide chain (the first 30 to 40 amino acids); hydrophobic amino acids are at the other end. The nomenclature used to describe the five classes follows.

Histones	Lys/Arg ratio	Molecular weight
H1	20	21,000 (approximately)
H2a	1 to 2.5	14,000
H2b		13,800
H3	0.7 to 0.8	15,300
H4		11,300

Nucleosomes Histone arrangement in the chromosomal structure is a repeating pattern nearly invariant among organisms, cell types, and stages in the cell cycle. Chromatin exists as a regular array of globules; it has the appearance of many beads on a string (Fig. 8-9). Each discrete unit is about 10 nm in diameter and is called a *nucleosome*. The nucleosomes consist of four of the histone types arranged as an octamer, e.g., $(H4)_2$ $(H3)_2$ $(H2a)_2$ $(H2b)_2$. The beads form a core around which the DNA, approximately 140 base pairs long, is wound in two shallow turns (Fig. 8-10). The nucleosomes are interconnected by spacer or linker sequences consisting of 15 to 100 base pairs. The spacer regions are also the points of attachment for the H1 category, which is larger than the rest and more basic, having about 37% basic amino acids. H1 is thought to be important for the formation of higher orders of structure within the chromosomes. It now appears that six nucleosomes may wind about a central channel forming a chromatin fiber. Also, chromosomes may possess "superbeads" consisting of 10 or more nucleosomes.

Histones function not only in folding the DNA into the chromosomal structure but also in regulating gene activity. They may be involved in blocking (repressing) or selecting (derepressing) specific genetic information. Information retrieval by the copying of selected base sequences as RNA (Chapter 9) may require a chemical alteration of the histones by phosphorylation and acetylation. Chemical modification of certain amino acids could serve to decrease the total cationic charge and thereby loosen the interaction of the histone with DNA. It would be expected that this could lead to an increased and selective exposure of the DNA to the process of

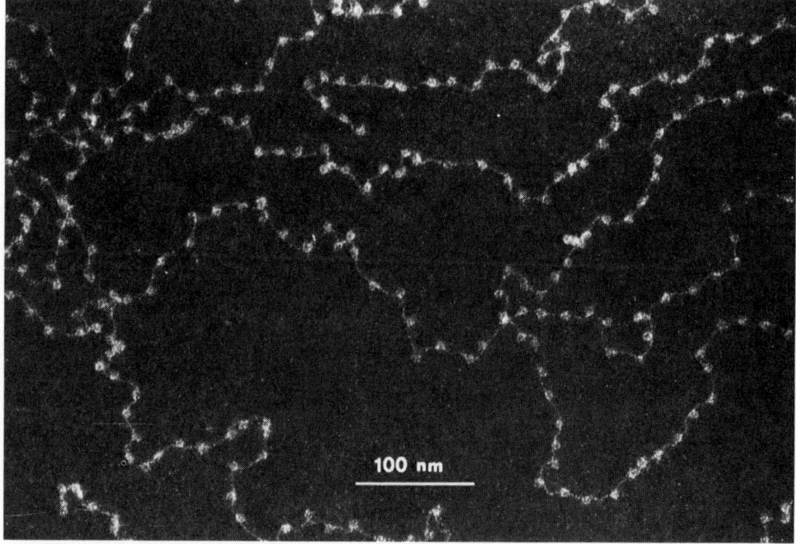

Figure 8-9
Electron micrograph showing "beads" or nucleosomes of chromatin derived from chicken erythrocytes. Also visible are the connecting strands or spacer regions.

From Olins, D.E., and Olins, A.L.: Am. Sci. **65**:707, 1978.

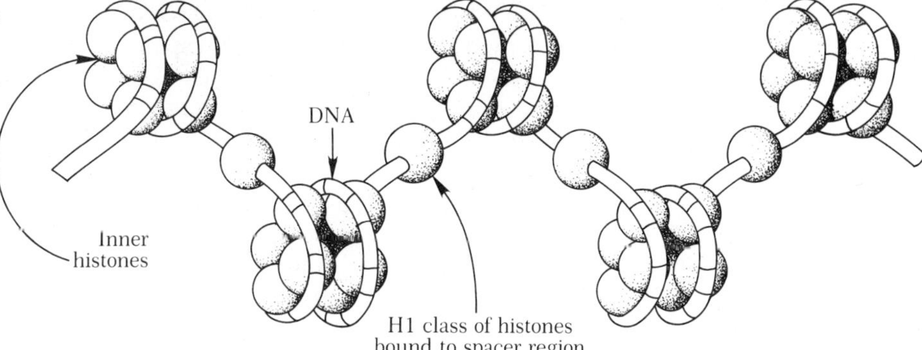

Figure 8-10
Nucleosome model of chromatin structure. Inner histones are close-packed in the nucleosome core, and DNA is wrapped around the outside in two turns.

From Olins, D.E., and Olins, A.L.: Am. Sci. **66**:704, 1978.

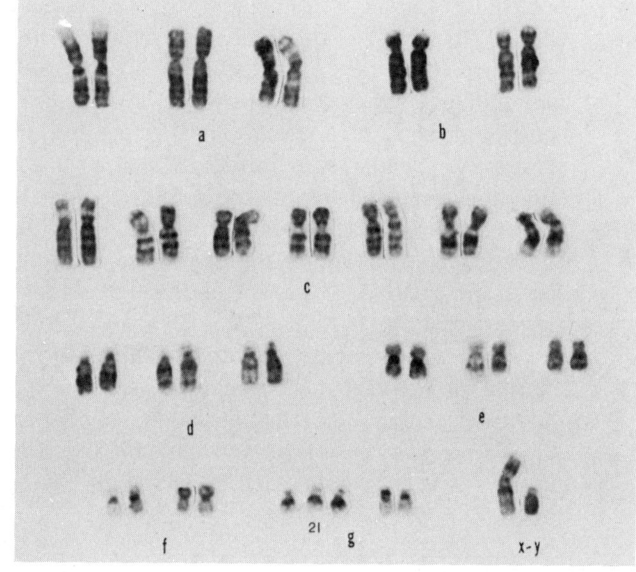

Figure 8-11
Karyotype or chromosomal pattern of a boy with Down's syndrome (mongolism). Cells possess a third representative of chromosome 21 and therefore have a total of 47 chromosomes.

Courtesy V.P. Johnson, M.D., University of South Dakota School of Medicine.

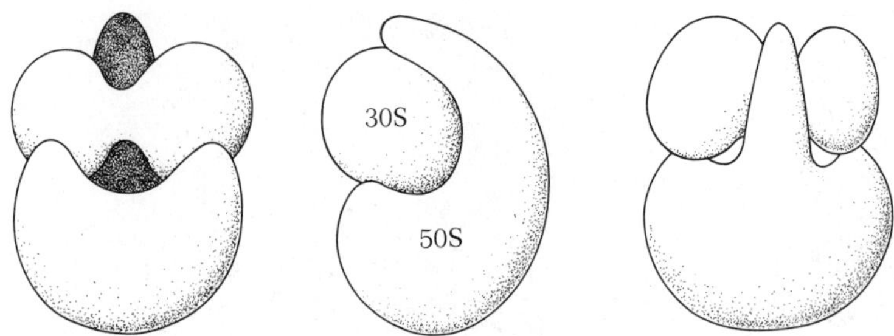

Figure 8-12
Three views of *E. coli* ribosome. These drawings are based on the three-dimensional models of Wittmann.

From Wittmann, H.G.: Eur. J. Biochem. **61**:1, 1976.

transcription. Furthermore, it has been shown that nonhistone proteins also play an important role.

As already indicated, normal human somatic cells typically have 46 chromosomes existing as 23 pairs (diploid cells). These chromosomes can be arranged by pairs in an order moving from the largest to the smallest. This arrangement also takes into consideration shape and location of the centromeric region. Such a profile of chromosomes is called a *karyotype* (Fig. 8-11). A prenatal diagnosis of genetic disease can be made by determining whether the pattern of chromosomal size and shape is normal or if aberrant chromosomes are to be found. Samples of amniotic fluid are withdrawn from the uterus by a hypodermic needle (amniocentesis). Cells originating from fetal skin or respiratory tract are used to study the karyotype of the fetus. Fig. 8-11 is a karyotype of a boy with Down's syndrome (mongolism). Note the presence in the chromosomal pattern of a third representative of chromosome 21; hence the abnormal total number of 47.

Ribosomes

As described previously, 75% to 80% of the total cellular RNA exists as a structural and functional part of the ribosomes. To describe what is known about these subcellular organelles, it is first necessary to discuss extensive progress that has been made using the ribosomes of prokaryotic cells.

In *E. coli* the 30S ribosomal subunit consists of a single strand of 16S rRNA and 21 proteins varying from a molecular weight of 10,700 to 65,000. The process of forming this subunit involves a sequential association of six proteins with the nucleotide sequences that make up the 3′ terminal end of the rRNA. Initial binding changes the shape of the RNA, thereby increasing the association with the other protein molecules (i.e., the assembly is cooperative). The final shape of the 30S ribosomal subunit is shown in Fig. 8-12. Similarly, to form the 50S subunit, the constituent 28S rRNA strand binds cooperatively with approximately 34 proteins varying in molecular weight from 9000 to 28,500. Some of the proteins of the 50S subunit join specifically with the 5S rRNA strand. It is of interest to note that the 5S RNA plus three specific proteins (called the L5, L8, and L25f proteins) bind the oligonucleotides T-ψ-C-G; therefore they may represent the binding site for the T-ψ-C arm of tRNA. The shape of the 50S subunit and the complete 70S ribosome are shown in Fig. 8-12.

Eukaryotic ribosomes are larger (80S) than their prokaryotic counterparts. It appears, however, that the structural aspects are comparable. Thus the 40S subunit consists of an 18S rRNA strand joined to approximately 30 proteins, whereas the 60S contains 28S rRNA and some 50 different proteins. In addition, this subunit contains one strand each of 5.8S and 5S rRNA.

Suggested readings

Pertinent readings are listed at the conclusion of Chapter 9.

Chapter 9

Genetic control of metabolism: replication and transcription

Genetic information is stored in the DNA structure as the specific sequence of bases in the polynucleotide chain. To transfer this genetic information from a parent to a daughter cell during cellular reproduction, the DNA base sequence must be precisely copied. The duplication of DNA is called *replication*. Information retrieval, which occurs when cells and tissues respond or adapt to nutritional and physiological needs, requires that RNA be synthesized from selected genes. In this process, called *transcription,* the genetic information is transcribed as the specific base sequence of the RNA chains. The final manifestation of genetic information as proteins is called *translation*. Here the base sequence of the RNA is translated into the amino acid sequence of specific proteins. This flow of genetic information from the genes (DNA) to the production of enzymes (proteins) may be illustrated as follows:

$$\text{DNA} \xrightarrow{\text{Replication}} \text{DNA} \xrightarrow{\text{Transcription}} \text{RNA} \xrightarrow{\text{Translation}} \text{Protein}$$

Although there are circumstances when an enzyme known as *reverse transcriptase* promotes the synthesis of DNA from RNA, the normal progression of events begins with DNA and ends with proteins.

This chapter considers the molecular events involved in replication and transcription. Principles basic to both have been developed using prokaryotic systems, especially with *E. coli*. Where known, information relevant to human systems will be included. It is assumed generally that those systems functioning in eukaryotic cells are comparable, in principle at least, to those of prokaryotic cells.

Replication

Biosynthesis of DNA

The biosynthesis of DNA is catalyzed by *DNA polymerase,* an enzyme first isolated from *E. coli* by Arthur Kornberg in 1959. It forms a polymer of deoxyribonucleotide linked together by 3′,5′-phosphodiester bonds. The overall reaction involves the incorporation of the four deoxyribonucleoside triphosphates (dNTPs) into the polynucleotide; a molecule of pyrophosphate is eliminated. It is now clear that the polymerase activity of *E. coli* actually represents three enzymes; the original "Kornberg" enzyme is known as polymerase I. A single cell possesses approximately 400 molecules of this Zn^{++}-containing metalloenzyme (mol wt = 109,000).

DNA polymerases do not determine the actual sequence of the bases in the polynucleotide product; the enzyme must have some preexisting DNA that it can copy. In other words, there must be present in the system strands of DNA that serve as a template to establish the primary structure of the newly synthesized chain. The base sequence is determined by base complementarity. DNA polymerase also has a second requirement; polymerization occurs by the sequential addition of nucleoside monophosphates to the free 3′-hydroxyl group of a short piece of preexisting polynucleotide. The enzyme needs a primer for its action. These needs are illustrated in the accompanying diagram, in which template and primer functions are provided by a hypothetical piece of double-stranded DNA to be replicated. As illustrated, the dNTPs are aligned by base complementarity. This is followed by a nucleophilic attack on the innermost or α-phosphate of the dNTP by the 3′-hydroxyl of the primer. Pyrophosphate is eliminated as the 3′,5′-phosphodiester linkage is formed. Therefore as seen in the diagram, the biosynthesis of the polynucleotide segment grows in a 5′ to 3′ direction. In addition the scheme shows that the newly synthesized chain is antiparallel to its template.

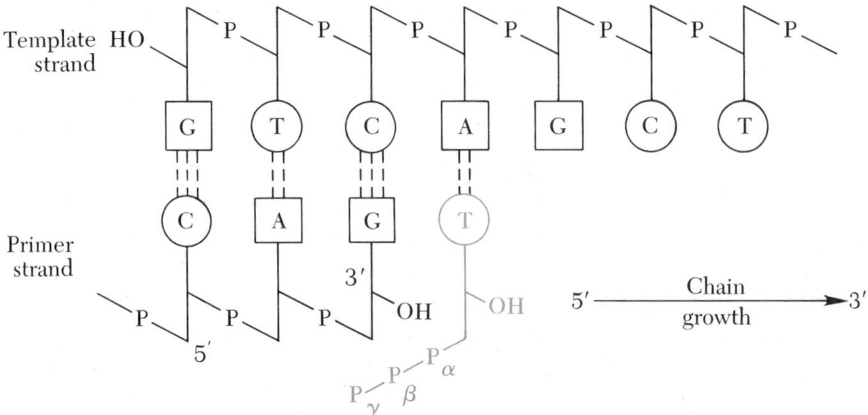

This system was tested by using synthetic polynucleotides as templates and determining the nature of the newly synthesized products. For example, a poly-A chain yielded, in the presence of *E. coli* polymerase and the four dNTPs, a new chain of thymidine only (poly-T). Likewise, an alternating poly-AT chain yielded a complementary poly-TA product. The chemical compositions of a variety of enzymatically synthesized DNAs, produced from different natural templates and the microbial polymerase, are compared in Table 9-1.

To reiterate, replication requires preexistent DNA, of which one strand is the template that establishes base sequence while a small segment or primer is needed to provide the free 3′-hydroxyl for chain growth.

Although DNA polymerase I is highly effective in copying single strands of DNA, it is much less active in the presence of undenatured double-helical DNA. It functions at a rate of incorporation of 1000 nucleotides per minute per mole of enzyme, whereas the estimated in vivo rate is approximately 100 times faster. Furthermore, a mutant of *E. coli* has been found that multiplies normally but contains only 1% or less of polymerase I. These considerations show that DNA polymerase I is not the only polymerase involved in replication in vivo.

DNA polymerases

The enzymatic activity of polymerase I is not restricted to the formation of phosphodiester bonds; it also can hydrolyze these linkages. Actually polymerase I has two nucleolytic activities, one to hydrolyze the DNA chain from the 3′ end and the other to hydrolyze the DNA chain from the 5′ terminal. It can degrade DNA chains by a *3′ to 5′ exonuclease* activity, removing nucleotides, and also by a *5′ to 3′ exonuclease* activity. The former serves to recognize and then remove an erroneous, non–base-paired terminal residue, thereby correcting or “proofreading” the growing DNA chain. The 5′ to 3′ exonuclease removes single nucleotides or even oli-

Table 9-1
Chemical composition of enzymatically synthesized DNA, synthesized with different primers

DNA	A*	T	G	C	$\frac{A+G}{T+C}$	$\frac{A+T}{G+C}$
Mycobacterium phlei						
Primer	0.65	0.66	1.35	1.34	1.01	0.49
Product	0.66	0.65	1.34	1.37	0.99	0.48
Escherichia coli						
Primer	1.00	0.97	0.98	1.05	0.98	0.97
Product	1.04	1.00	0.97	0.98	1.01	1.02
Calf thymus						
Primer	1.14	1.05	0.90	0.85	1.05	1.25
Product	1.12	1.08	0.85	0.85	1.02	1.29
Bacteriophage T_2						
Primer	1.31	1.32	0.67	0.70	0.98	1.92
Product	1.33	1.29	0.69	0.70	1.02	1.90
AT copolymer	1.99	1.93	<0.05	<0.05	1.03	40

From Kornberg, A.: Science **131**:1503, 1960.
*A, Adenine; *T*, thymine; *G*, guanine; *C*, cytosine. The numbers are molar proportions.

gonucleotides from the DNA, beginning at the 5′ terminal end. When the enzyme binds at a nick, i.e., a point at which the phosphodiester linkage either has not been formed or has been hydrolyzed, it can catalyze in a coordinated fashion the addition of new nucleotides to the free 3′-hydroxyl end of the nick while at the same time removing nucleotides from the 5′ end. This process translocates or moves the nick along the chain. Such concerted exonuclease and polymerase activities are illustrated in Fig. 9-6. It is now thought that DNA polymerase I of *E. coli* functions principally to repair damaged DNA or to correct or proofread the growing chain rather than as the primary factor in polymerization. Nonetheless, it does play an important role in the overall replicating process as is described later.

Two other polymerases from *E. coli,* II and III, have been isolated. Although polymerase II synthesizes DNA in the 5′ to 3′ direction, its rate is too slow to account for the in vivo synthesis of DNA. Certain thermosensitive mutants of *E. coli* that are not viable at 42° C have been produced in which DNA replication does not occur at the elevated temperature. In these, polymerases I and II are normal, whereas polymerase III is altered. Therefore it has been concluded that polymerase III is essential to viability, independent of the other two enzymes. It also catalyzes the addition of nucleotides to the 3′-hydroxyl end of the primer (i.e., growth occurs in the 5′ to 3′ direction). A single cell contains approximately 10 molecules (mol wt = 180,000), each capable of incorporating 15,000 nucleotides into DNA per minute. This rate of synthesis observed in vitro is compatible with the rate of synthesis in vivo.

Eukaryotic cells possess similar polymerizing enzymes but they do not exhibit the exonuclease activities of the enzymes found in prokaryotes. Most of the enzymatic activity appears to be associated with large heterogeneous molecules referred to as α-DNA polymerase and having a molecular weight in the range of 200,000 to 230,000. Another nuclear enzyme, β-DNA polymerase, is a smaller basic molecule of molecular weight 40,000. A mitochondrial DNA polymerase (called mt-polymerase) of molecular weight 106,000 has also been described.

Semiconservative replication

The replicative process is semiconservative in nature, i.e., each strand of the parent double-helical DNA molecule is conserved and copied by base pairing. To prove that this is true Meselson and Stahl grew *E. coli* in a medium containing ammonium chloride labeled with the heavy isotope of nitrogen (^{15}N). After many

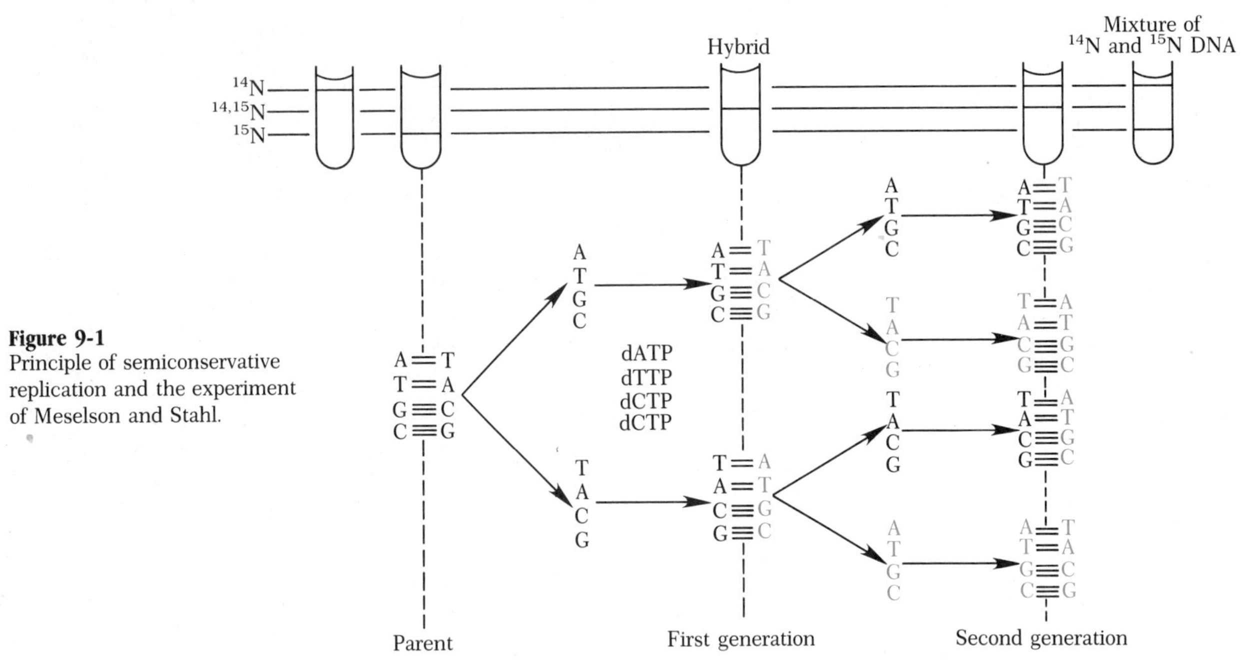

Figure 9-1
Principle of semiconservative replication and the experiment of Meselson and Stahl.

generations the progeny possessed DNA labeled with ^{15}N. DNA from these bacteria was isolated and separated by ultracentrifugation on a cesium chloride density gradient. In the gradient the ^{15}N-DNA sedimented further than ^{14}N-DNA because of its greater density. Consequently it was possible to separate ^{14}N- and ^{15}N-species because of their differences in density. This is illustrated in Fig. 9-1. When *E. coli* saturated with ^{15}N was allowed to grow in the presence of ^{14}N–ammonium chloride for a single generation and this DNA was centrifuged in the cesium chloride gradient, an intermediate layer of hybrid $^{14,15}N$-DNA was observed (Fig. 9-1). Subsequent generations of bacteria grown in the presence of ^{14}N–ammonium chloride yielded a mixture of $^{14,15}N$-DNA and ^{14}N-DNA. The existence of the hybrid at each generation showed the presence of double helices composed of one parent and one newly formed strand. Fig. 9-1 illustrates the manner in which base pairing and the Meselson-Stahl experiment explain semiconservative replication. During reproduction each parent, double-stranded DNA molecule yields two templates to be copied. In the first generation of progeny the cells produce only DNA hybrid molecules consisting of a single parent strand combined with a newly synthesized, complementary chain. Reproduction to yield the second generation of progeny proceeds by the unwinding of the DNA hybrid molecules; again each strand is copied during replication.

Unwinding of helix during replication

As noted previously, the double helix of DNA is a plectonemic coil, i.e., the two polynucleotide chains cannot be pulled apart except by an unwinding process. Therefore to comply with the principle of semiconservative replication, it is necessary for the DNA to unwind as replication proceeds. In Fig. 9-2 this unwinding process is illustrated as occurring at one end of the duplex. Such an unwinding process raises several difficult questions. One is the mechanical restriction of the unwinding itself, i.e., it must proceed within the confines of the cell at a rate approximating 15,000 to 20,000 rpm in order that a cell (*E. coli*) can duplicate itself

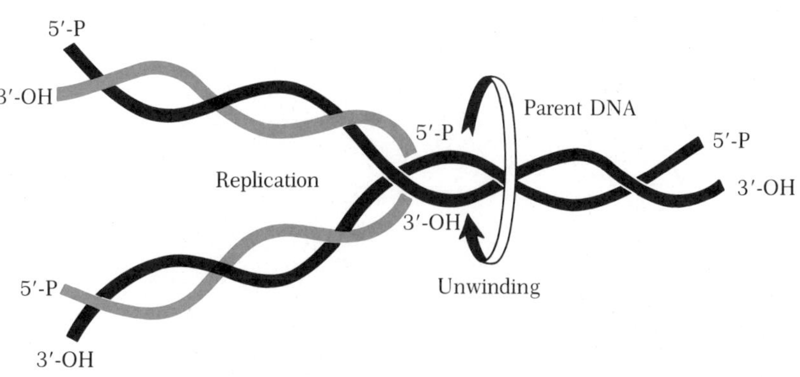

Figure 9-2
Suggested model of DNA replication, showing unwinding of a double helix. Each parent strand *(black)* serves as a template for the synthesis of a new polynucleotide chain *(color)*.

within the required time (20 minutes for *E. coli*). Apart from the mechanical difficulties, the model raises a second problem stemming from the antiparallel nature of the two strands. If they unwind as illustrated, these chains would present both a 3′ terminal and a 5′ terminal to the polymerase system. Since DNA replication occurs only in the 5′ → 3′ direction, one is compelled to suggest the need for two polymerase molecules, each building daughter chains simultaneously but in opposite directions. However, as is apparent from Fig. 9-2, a complementary strand growing continuously from right to left would be incompatible with the net direction of replication, which in the figure is from left to right.

Okazaki fragments

It is now evident that the polymerase that uses the 5′ → 3′ parent chain (upper strand, Fig. 9-2) as a template does not synthesize a continuous, complementary daughter strand but rather produces a series of short pieces. Therefore the polymerases on both parent strands are visualized as moving in the same direction, i.e., with replication. The polymerase using the 3′ → 5′ strand (lower) creates a continuous daughter called the leading strand while another polymerase using the 5′ → 3′ template (upper) produces a discontinuous sequence of chains called the lagging strand. Studies using very brief (5 to 60 sec) exposures of *E. coli* to ^{3}H-thymidine showed that some of the new DNA chains are produced in segments of 400 to 2000 nucleotides all synthesized in a 5′ to 3′ direction. These short pieces of DNA are called *Okazaki fragments*.

Chain initiation

Fig. 9-2 raises yet a third problem, i.e., how is the new chain initiated? What is the actual primer? Rifampicin, an antibiotic known to inhibit specifically RNA polymerase (a DNA-dependent, RNA polymerase), was observed to block DNA replication. Therefore the synthesis of some RNA is a prerequisite to DNA replication. Subsequently the Okazaki fragments were shown to consist not only of the 400 to 2000 nucleotide segments of DNA but also of 50 to 100 nucleotide lengths of RNA. It was concluded that replication is initiated by the formation of a short segment of RNA that serves as the primer for the DNA polymerase. This role of RNA as primer for replication is illustrated in Figs. 9-3 and 9-4.

The DNA-dependent, RNA polymerase required in replication is now called *primase*. It acts much like the DNA polymerase but uses ribonucleotide triphosphates as substrate. Like DNA polymerase it requires a single strand of DNA as template, but unlike DNA polymerase it does not need a primer nucleotide. Instead a single GTP or ATP unit serves for the initiation of RNA synthesis. The RNA polymerase catalyzes the formation of phosphodiester linkages (pyrophosphate is eliminated),

Figure 9-3
Initiation of replication by the formation of an RNA primer.

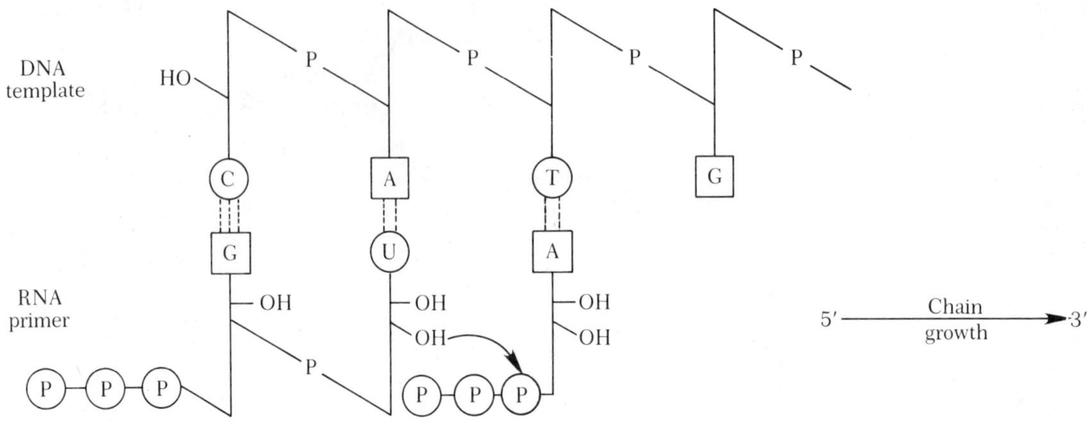

Figure 9-4
Elongation of DNA chain using the primer RNA as the site of initiation. The covalent hybrid of RNA and DNA is known as the Okazaki fragment.

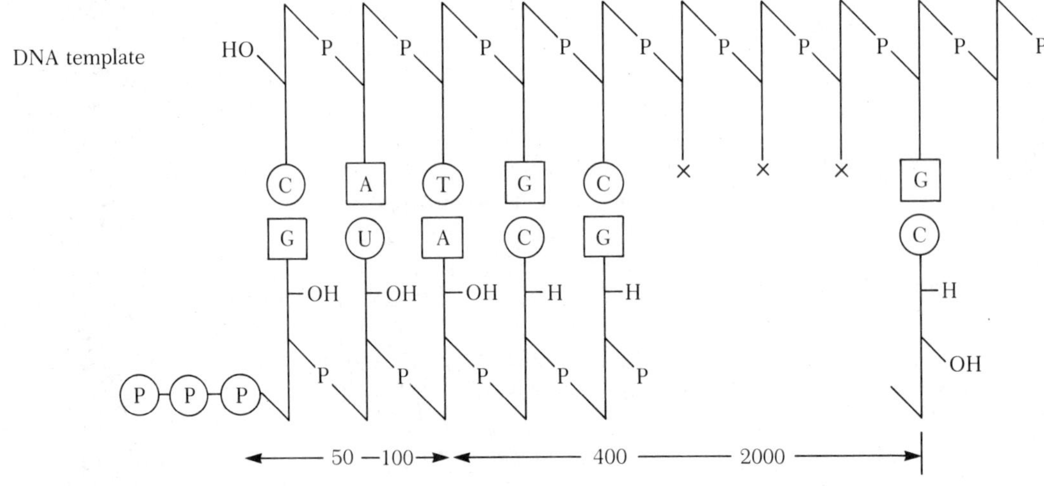

Figure 9-5
Replication of a DNA duplex molecule showing that the continuous (leading strand) forms on a single RNA primer while the discontinuous (lagging strand) forms on multiple RNA segments. The direction of replication or the movement of the growing fork is to the right.

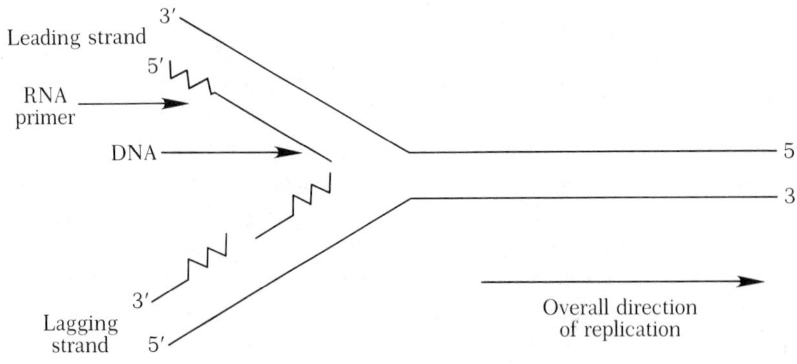

joining a succession of ribonucleotides until a segment of 50 to 100 is produced. This fragment then serves as the point of initiation for the synthesis of a DNA segment. The relationship of the RNA primer to replication is illustrated in Figs. 9-3 and 9-4.

Chain elongation

Once the primer has been completed, the actual synthesis of DNA begins, using the free 3′-hydroxyl terminal of the RNA (Fig. 9-3). This process requires a tetrameric (four subunits) complex of two proteins known as polymerase III★ (III star) and copolymerase III★ as well as ATP. The ATP and copolymerase are required to start the DNA synthesis but not to sustain it. As seen in Fig. 9-4, the polymerase then catalyzes the addition of dNTPs and the formation of the phosphodiester bonds. In *E. coli* approximately 400 to 2000 deoxyribonucleotides are added to complete the Okazaki fragment.

The products of these two processes are illustrated in Fig. 9-5. Thus there are two daughter strands formed at the point of chain growth (i.e., the replicating fork), one of which is the continuous leading strand and the other of which is the discontinuous lagging strand. As is evident from Fig. 9-5, the continuous strand requires only a single RNA primer segment, whereas the discontinuous strand has multiple RNA pieces that serve as primers for the Okazaki fragments. Therefore all replication occurs in the 5′ to 3′ direction and the overall chain growth and the movement of the replicating fork are to the right in the figure.

The appearance of the Okazaki fragments is transitory; they are ultimately joined together to form a continuous daughter DNA strand. This process requires removing the RNA, refilling the gap formed, and finally joining together the neighboring segments of DNA.

Removal of the RNA primer

In *E. coli,* it is thought that the RNA primer segments are removed by the action of the 5′ to 3′ exonuclease activity of polymerase I. Simultaneously the same enzyme may fill the gap by polymerizing the DNA strand, beginning with the neigboring DNA segment (Fig. 9-6). The net effect is to propagate the nick to the right until all of the RNA has been eliminated and refilled with the appropriate deoxyribonucleotide residues. Alternatively the RNA also may be removed by the specific ribonuclease called RNase H (for hybrid).

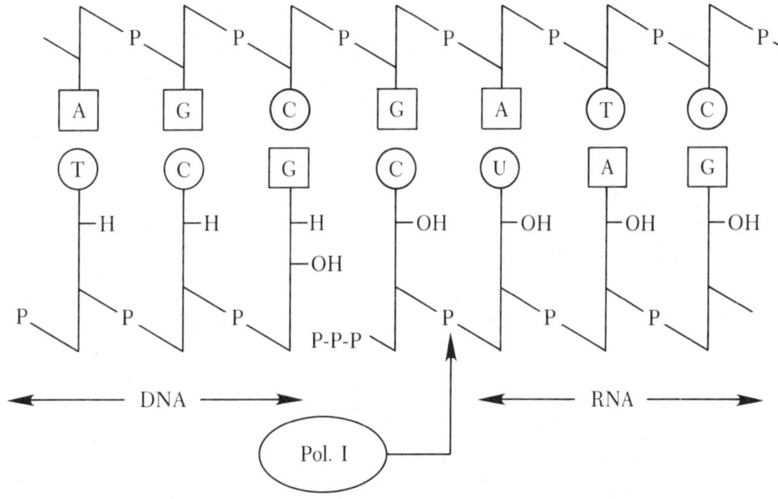

Figure 9-6
The concerted exonuclease and polymerase actions of *E. coli* polymerase I. This process removes the RNA primer and replaces the ribonucleotides with the corresponding deoxyribonucleotides. The exonuclease activity removes the end nucleotide (cytidine) while the polymerase activity fills the gap using dNTP (dCTP).

Figure 9-7 DNA ligase closes a nick by joining neighboring strands. It catalyzes the formation of a 3′,5′ phosphodiester bond using NAD^+.

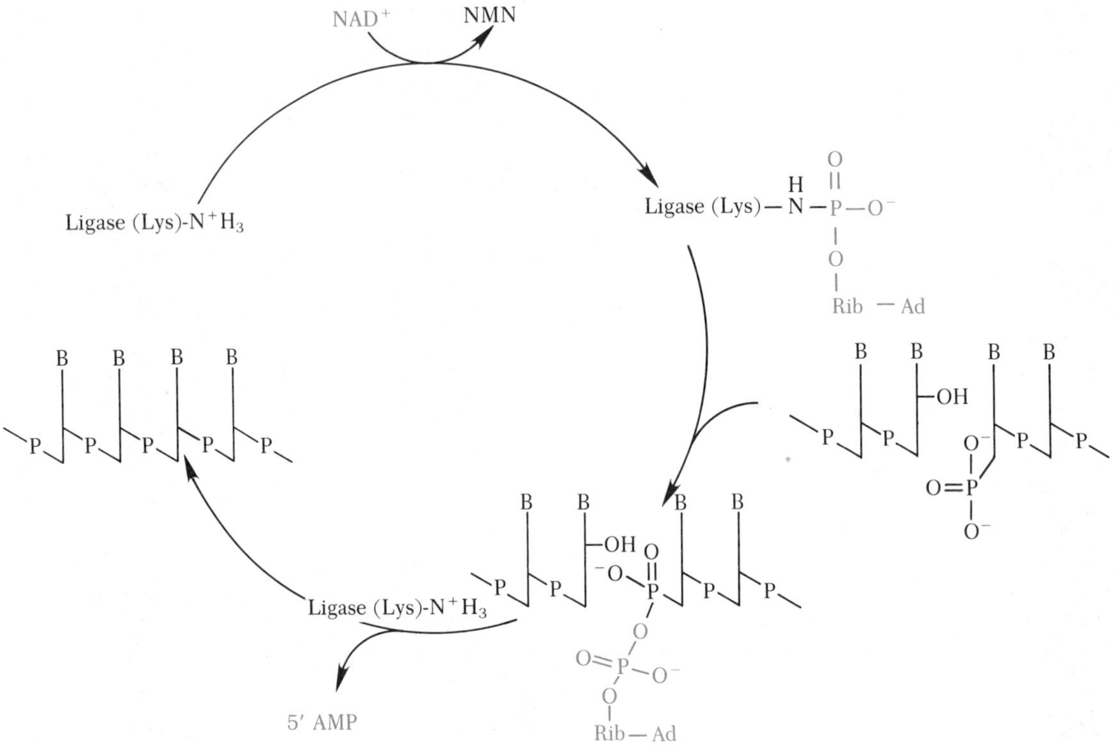

DNA ligase

The DNA fragments are then united by a joining enzyme called *DNA ligase*. The ligase catalyzes the formation of a 3′,5′-phosphodiester bond between a free 5′-phosphate terminal of one polynucleotide fragment and the 3′-hydroxyl of another. In *E. coli* the DNA ligase derives the needed energy for this joining reaction from NAD^+. An activated adenylyl ligase is formed by transferring the adenylyl group onto an ε-amino group of a ligase lysyl residue; nicotinamide mononucleotide is released in the process (Fig. 9-7). In the joining reaction, the adenylyl group is then transferred from the ligase to the 5′-phosphate end of the DNA to produce the pyrophosphoryl derivative. The enzyme then catalyzes the formation of the phosphodiester bond with the concomitant release of AMP. In mammalian systems the adenylyl ligase is formed by the action of ATP instead of NAD^+.

Chromosomal replication

The unwinding of the double helix preparatory to and concomitant with replication is not easy to visualize. There is the problem of size, shape, and packaging of the DNA in the chromosome. A system that has been much studied is the chromosome of *E. coli,* which is approximately 500 times as long as the cell itself. The DNA duplex is a circle when completely "relaxed" (Fig. 9-8, *A*). Normally, however, the chromosome exhibits considerable supercoiling (Fig. 9-8, *B*). Studies of the replication of the circular chromosome of *E. coli* have shown that the DNA duplex unwinds locally in such a way as to form an "eye" or "bubble," as illustrated below. In effect this local unwinding provides two template strands to be copied. Local

Figure 9-8

A Autoradiograph of *E. coli* chromosome labeled with tritiated thymidine for two generations. Figure illustrates replication process as it occurs in a circular, duplex chromosome. *X* and *Y* are the replicating forks.
B Supercoiling that exists in circular duplex chromosome of *E. coli.*

A from Cairns, J.: Symp. Quant. Biol. **28**:43, 1963; **B** with permission from Worcel, A., and Burgi, E.: J. Mol. Biol. **71**:143, 1972. Copyright by Academic Press Inc. (London) Ltd.

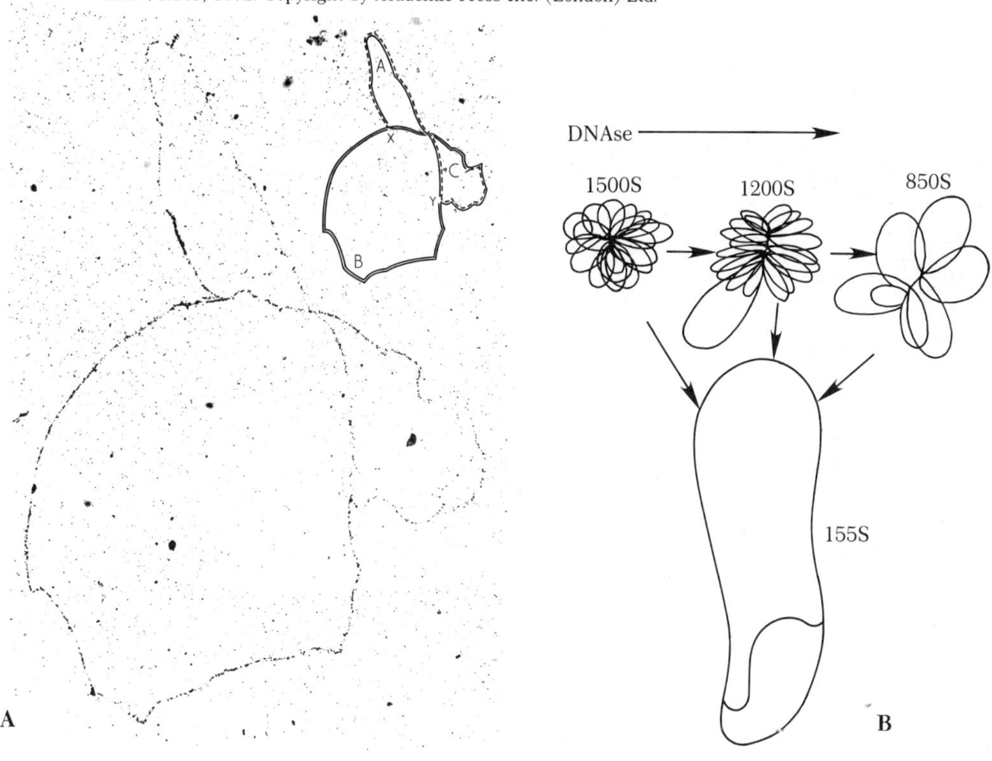

unwinding apparently is not a random process; rather it occurs at a specific site on the circular chromosome. Actually it provides two points at which replication occurs; these are known as the replicating forks. As illustrated below, unwinding and replication occur in opposite directions, i.e., bidirectionally. Replication at the two forks proceeds simultaneously and at the same velocity so that they meet at the opposite side of the chromosome from the point of initiation.

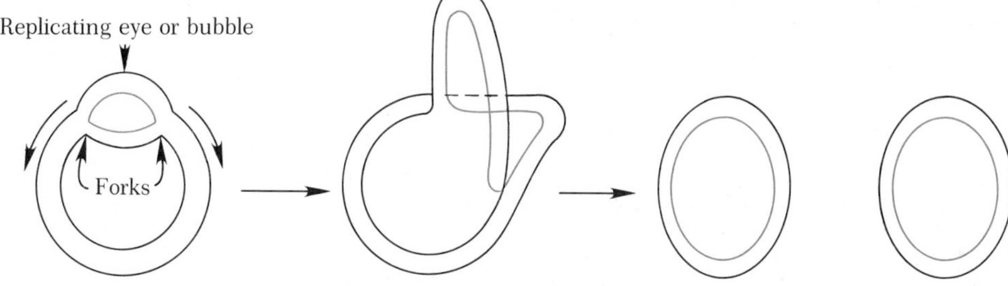

As shown in Fig. 9-8, *B*, the chromosome of *E. coli* does not exist in the cell simply as a relaxed circular duplex; rather it is supercoiled in about 50 tightly twisted loops approximately 20 μm long and organized into a multiple coiled structure.

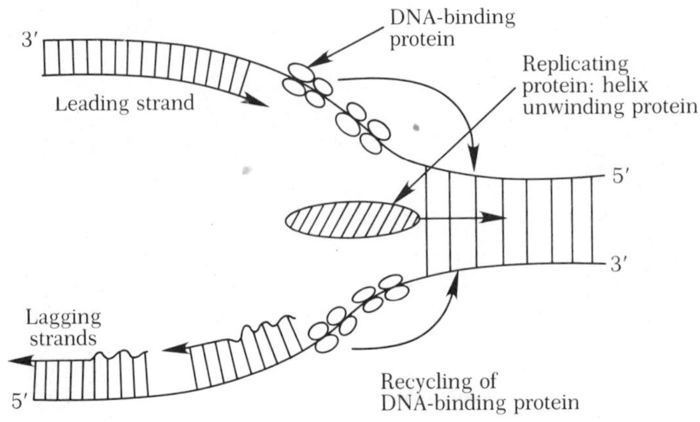

Figure 9-9
Role of the unwinding protein, which ruptures the hydrogen bonds in the duplex. The DNA-binding proteins then stabilize the single strands.

Each cell contains some 10,000 copies of a special DNA binding protein that aids in the unwinding of the duplex. This binding protein has a very high affinity for single-stranded DNA so that the duplex undergoes a transient melting process. The protein has an elongated shape and can bind to DNA stretches encompassing approximately 10 nucleotides. This protein may be recycled as the replicating fork proceeds (Fig. 9-9). Actually two activities are indicated in Fig. 9-9; one is the action of an unwinding protein that ruptures the hydrogen bonds; the other is the action of a DNA-binding protein that stabilizes the single strands before replication. There probably is yet another process required that leads to a relaxation of the supercoils. This protein, called *swivelase,* is thought to catalyze a transient formation and resealing of nicks in the polynucleotide chains, allowing an uncoiling or relaxation of the tightly wound DNA in the chromosome. Relaxation of the supercoils is essential if the replicative bubble is to proceed and involve the complete circular chromosome as suggested above.

As replication continues to completion circular DNA must be wound into the coiled form indicated in Fig. 9-8, *B*. Supercoiling is catalyzed by yet another enzyme, termed *DNA gyrase*.

Although the principles of replication were established largely with the aid of *E. coli* and various viral systems, the basic rules developed are applicable to eukaryotic cells also. For example, replication occurs in short discontinuous spurts involving initiation with the formation of RNA. In eukaryotic cells the Okazaki fragments are smaller and consist of only 10 nucleotides in the RNA segment and 100 to 150 in the DNA. Replication is bidirectional and, as illustrated below, originates from multiple sites, thereby forming many eyes or bubbles. In mammalian cells there are some 6000 forks, or 1 growth fork per 10,000 bases. Cellular nuclei also contain a DNA-unwinding enzyme.

MULTIPLE REPLICATING EYES IN EUKARYOTIC CHROMOSOMES

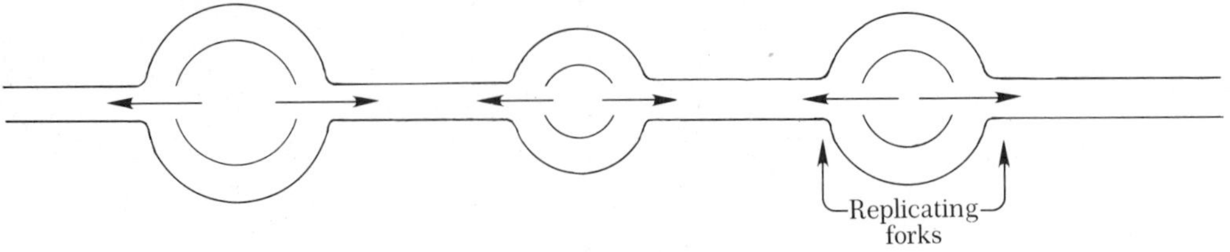

Figure 9-10 Formation of cyclobutane ring between two neighboring thymidines induced by ultraviolet radiation. Thymidine dimer formed creates a distortion or "kink" in the polynucleotide strand.

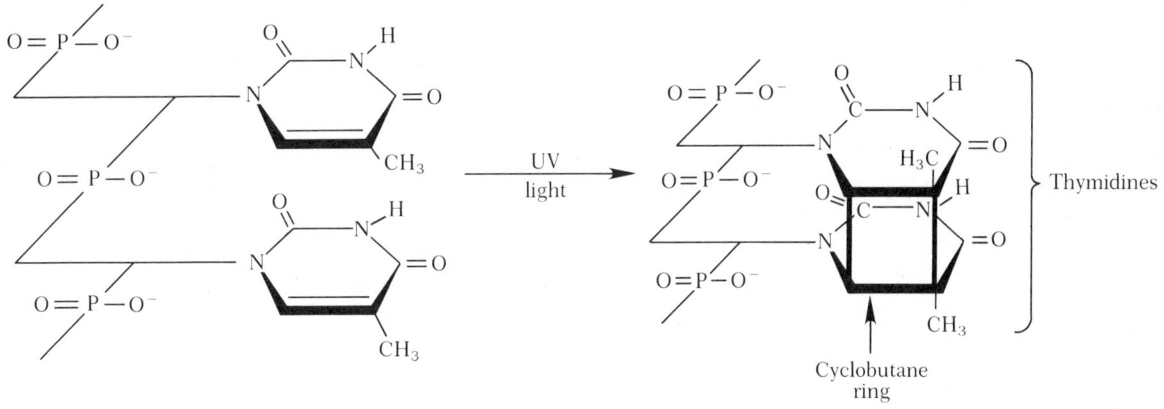

Genetic repair mechanisms

The accuracy of the replicative process is dependent on the exactness of base pairing. A bacterial gene may be duplicated 1×10^8 times before there is a 50:50 chance that even a single gene has been altered. The precision of base pairing may be adversely affected, however, by a variety of environmental factors that may damage the DNA, alter the genetic information, and yield a defective cell or mutant. Mutation or damage to DNA may take the form of several processes, specifically: (1) alteration of single purine or pyrimidine bases, (2) alteration of two bases, (3) a break in the polynucleotide chain, and (4) a joining or cross-linking of the two chains of the helix. To survive a potentially lethal or otherwise damaging effect of the mutation, cells possess mechanisms to repair the DNA. Of the various repair mechanisms, the best known is the one that corrects the two-base alteration induced by exposure to ultraviolet light. Absorption of ultraviolet light energy by two neighboring thymine bases results in the formation of C—C bonds (Fig. 9-10). The resulting cyclobutane ring involving the two thymines creates a distortion or "kink" in the polynucleotide strand that may interfere in subsequent replicaton. Repair may involve either the reversal of this process or the actual removal of the offending dimer.

Photoreactivation

Bacteria damaged by exposure to ultraviolet light can be caused to repair the DNA kink by exposure to an intense source of visible light (blue region). This process, called *photoreactivation,* requires a protein that recognizes the kink and binds to the polynucleotide strand at that point. The protein absorbs light and is activated into a system that splits the cyclobutane ring between the two thymine bases and so restores the DNA to its normal state.

Dark repair

A second mechanism for the repair of DNA is the so-called *dark* or *excision* repair process. Again the aberrant portion of the DNA strand is recognized by a protein that in this case is an endonuclease that nicks or hydrolyzes a phosphodiester bond at the 5′ side of the thymidine dimer (Fig. 9-11). Then an exonuclease removes the dimer from the strand, leaving a gap that must be filled. Reinsertion of appropriate nucleotides into the gap is catalyzed by DNA polymerase I (in *E. coli*) and DNA ligase.

Endonucleases that recognize kinks in DNA appear to be present in all cells. A rare genetic disease exists in which this and other repair enzymes are missing or

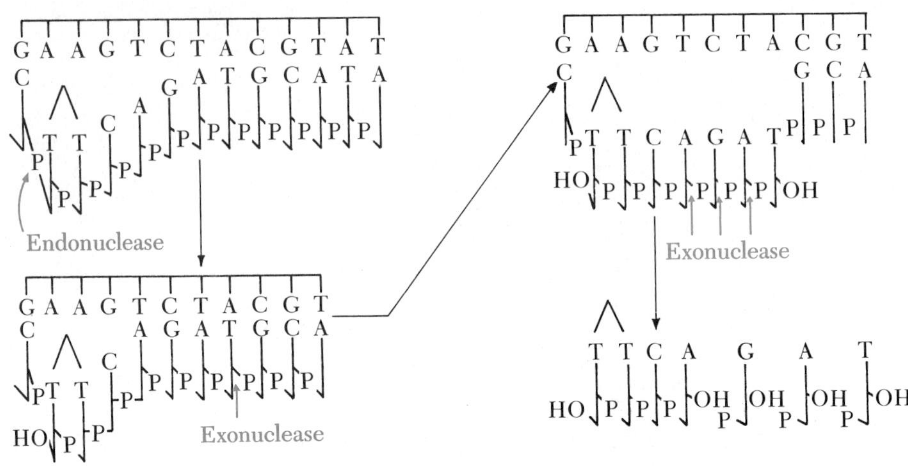

Figure 9-11
Probable sequence of steps in the excision repair of UV light–damaged DNA. Gap resulting from excision is filled by subsequent action of polymerase I *(E. coli)* and DNA ligase.

defective; this affliction is called *xeroderma pigmentosum*. Such individuals are extremely sensitive to sunlight and tend to develop skin cancer following exposure.

The base sequence of DNA

As stated in the introduction to this chapter, genetic information is stored in the chromosomal DNA as the exact base sequence of the polynucleotide chains. The ability to determine this base sequence in polynucleotides represents a major technical advance of the past 5 years. To provide insight into the kind of procedure used for this pupose, the basic principles of the Maxam-Gilbert method are now described.

1. First the unwieldly DNA molecule must be reduced to smaller units whose sequence can be more readily determined. This is accomplished by cleaving the sample with selected restriction endonucleases (p. 190). The fragments are isolated by polyacrylamide gel electrophoresis (PAGE).
2. The double-stranded fragments are labeled with the radioisotope of phosphorus (^{32}P). This is accomplished by treating the fragments with ^{32}P-labeled ATP and a polynucleotide kinase. The labeled double-stranded fragments are denatured (high pH) and the single strands are separated by PAGE. The result is two samples of single-stranded and complementary polynucleotides labeled with ^{32}P at their 5′ ends.
3. The labeled, single-stranded polynucleotide chains are then sequenced. Each is first studied for the location of purines. This is accomplished by methylation with dimethyl sulfate (pH 8.0). Methylation occurs at N-7 for the guanines and at N-3 for the adenines. The glycosidic bonds holding the modified purine to its deoxyribose is labilized. Under the right conditions the purines may be selectively removed. Also the phosphoester bonds at the depurinated sites are ruptured. Conditions are chosen so that depurination and cleavage occur only at single sites. The result is a mixture of fragments of varying lengths (Fig. 9-12). Such partial hydrolysates can be separated by PAGE. The location of fragments is visualized by exposing an x-ray film to the radioactive phosphorus. As seen in Fig. 9-12, a distribution of bands is obtained representing the increasing size of the fragments. The degree of migration is inversely related

Figure 9-12

Hypothetical polynucleotide of alternating pyrimidines (○) and purines (□) is sequenced as illustrated. The polynucleotide, labeled with ^{32}P, is treated with dimethyl sulfate under conditions that methylate the purines. A sample is then treated so as to eliminate only the adenines and cleave the chain at the empty sites. The deadenylated and cleaved mixture is separated on PAGE in track *1*. A regular array of fragments is illustrated. Other samples are similarly treated to cause deguanylation and cleavage (track *2*). Then samples are treated with hydrazine to modify the cytosines and thymines. Track *3* represents fragments from cleavage at the sites where the cytosines had been removed; track *4* shows the fragments from cleavage at both cytosine and thymine sites. The base sequence may be read as shown at the right side of the PAGE profile.

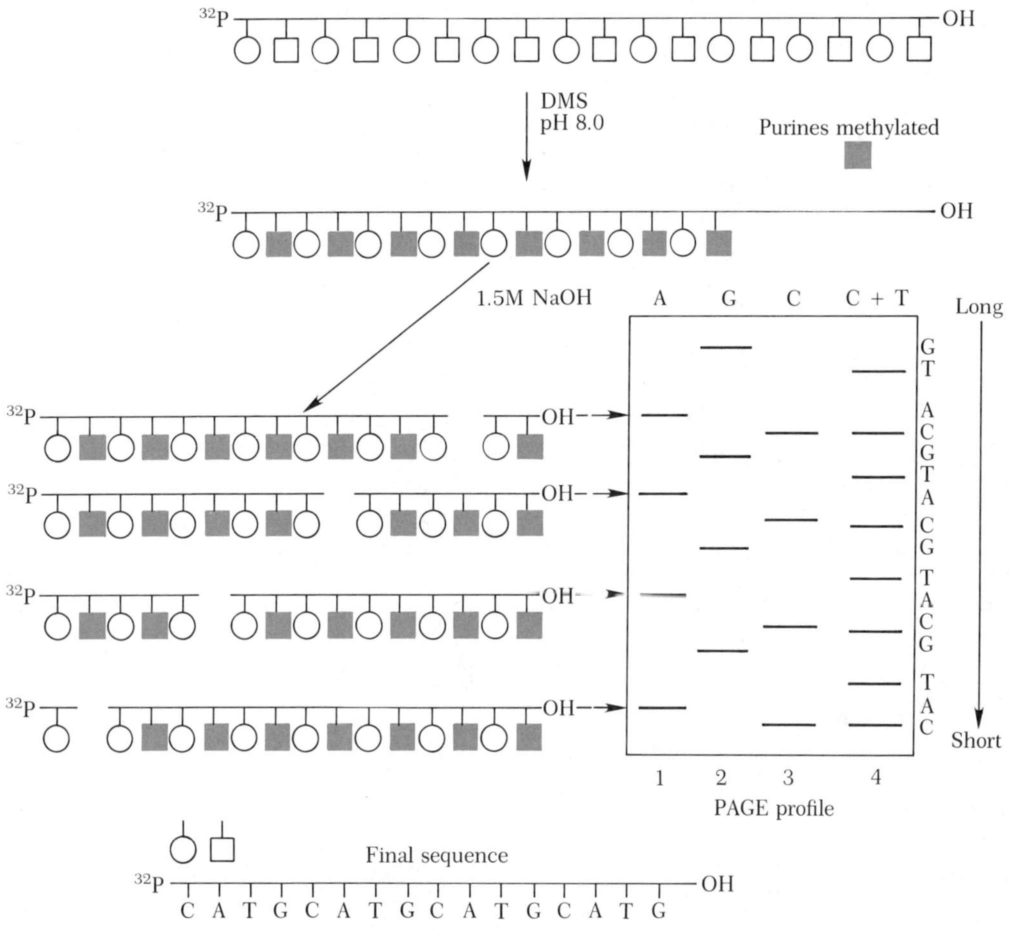

to the length of the chain. Another methylated sample is treated under conditions that result in deguanylation and cleavage. A ladder of bands such as shown in the second track (Fig. 9-12) is obtained. Then a sample of the original ^{32}P-labeled single strand is treated with hydrazine under conditions that eliminate selectively either cytosine or both cytosine and thymine. In this way mixtures of gradually increasing fragments cleaved at cytosine and cytosine plus thymine sites are obtained that can be separated by PAGE (third and fourth tracks). Fig. 9-12 illustrates the determination of base sequence of a hypothetical polynucleotide having alternating purines and pyrimidines. The actual sequence that would be obtained is ^{32}P-C-A-T-G-C-A-T-G-C-A-T-G-C-A-T-G-OH.

A complete base sequence was first obtained for the single-stranded, closed-loop DNA molecule from the bacterial virus ϕX174. This molecule contains 5375 nucleotides and has 9 genes. Sequencing has demonstrated that some of the genes overlap, a phenomenon of biological economy that was heretofore not appreciated. In addition sequencing has made it possible to locate precisely not only the genes themselves but also the various control, start-stop sequences.

Transcription

Biosynthesis of RNA

For the genetic information stored in chromosomal DNA to be used, it must first be transcribed as the base sequence of RNA. Ribosomal and transfer RNAs are synthesized by transcription of appropriate genes. The manifestation of genetic information as protein requires the prior transcription of the genes as messenger RNA.

The general biosynthesis of RNA may be viewed as follows:

$$\text{ATP, GTP, CTP, UTP} \xrightarrow[\substack{\text{2. DNA template} \\ \text{3. Mg}^{++}}]{\text{1. DNA-dependent RNA polymerase}} \text{RNA} + \text{PP}_i$$

Transcription therefore requires a supply of the four ribonucleoside triphosphates, a polymerizing enzyme called DNA-dependent RNA polymerase, and a DNA template that determines the exact sequence of nucleotides in the final RNA transcript.

DNA template

When the RNA polymerase binds to the double-helical DNA a local unwinding of the two strands occurs. Only one of these serves as the template and it therefore is termed the *sense strand*. As visualized in Fig. 9-13, only the sense strand is transcribed. The role of the second, or untranscribed, strand (the *antisense strand*) is unknown; it may be required to release the newly formed RNA molecules.

As seen below, the arrangement of nucleoside triphosphates (NTPs) is determined by the base sequence of the sense strand. Polymerization is similar to that described for replication but does not require a primer polynucleotide. Rather, RNA polymerase may simply begin with an ATP or a GTP. The RNA chain grows in the 5′ to 3′ direction. Since the growing RNA chain is antiparallel to its template, it follows that the DNA strand is read in the opposite, or 3′ to 5′, direction.

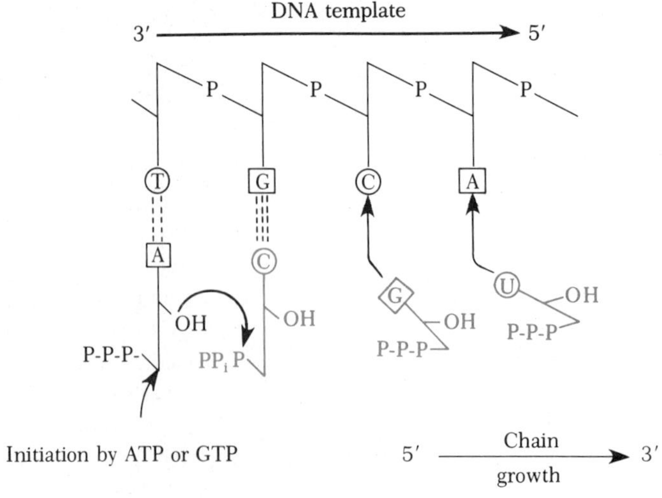

Initiation

In *E. coli* initiation of transcription always occurs with a GTP or ATP. This also requires the presence of a special protein called the initiation, or sigma (σ), factor. Apparently the σ factor has no enzymatic activity but is a part of the RNA polymerase structure and serves to recognize the start signals (a specific sequence of nucleotides) in the DNA molecule.

RNA polymerase from *E. coli* is a complex consisting of a number of subunits. Its overall molecular weight is approximately 500,000, consisting of two α-chains (mol wt = 39,000), one β^1-chain of molecular weight 165,000, one β-chain of molecular weight 155,000, and one σ factor of molecular weight 95,000. The subunits are assembled as follows:

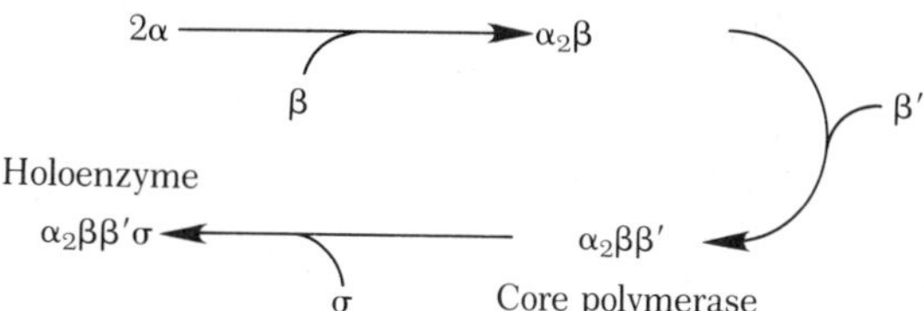

Although the core RNA polymerase is enzymatically active, it binds only randomly to the DNA. As is described later, the transcriptional process begins by a selective binding of polymerase at a point of initiation called the *promotor site*. Sigma reduces the nonspecific binding of the polymerase; it may provide the holoenzyme with a conformation favoring a tight complex specifically in the promoter staging area. After elongation of RNA by approximately 10 nucleotides, σ dissociates and may then become available for further combination with free core polymerase to produce new, additional holoenzyme (Fig. 9-13).

Termination

The completion of the transcriptive process (i.e., when the polymerase reaches the end of the gene) is also signaled. Termination is recognized by another protein (mol wt = 60,000) called the rho (ρ) factor, which causes the release of the nascent RNA molecule (Fig. 9-13). If there is no ρ factor, RNA polymerase simply continues to transcribe the next genetic region. The ρ factor may be part of the polymerase or it may attach directly to the DNA.

Mechanism of transcription

The principal steps of DNA transcription into RNA are summarized in Fig. 9-13. First the RNA polymerase, in conjunction with σ factor, recognizes a special sequence of nucleotides that marks the beginning of the gene (Phase I). Then RNA polymerase progresses along the sense strand, which acts as the template. NTPs arrange themselves in a sequence dictated by base complementarity. The synthesis of RNA occurs at the 3′-hydroxyl and elongation proceeds in the 5′ to 3′ direction (Phase II). As the RNA growth progresses simultaneously with the movement of the polymerase along the DNA template, the two DNA strands are believed to reform the double helix behind the polymerase and thereby cause the RNA to be peeled off. In some way the ρ factor now enters the picture (Phase III), recognizes the end of the gene, and causes the complex system to dissociate, as shown in Phase IV. From this concept it is apparent that all product RNAs should have two characteristics in common: (1) a terminal triphosphate group and (2) a 5′ terminal guanine or adenine base. Since this is not generally the case for isolated RNAs, apparently a postranscriptional modification is involved to complete the functional molecules.

Although our understanding of transcription in eukaryotic cells and hence in mammals is still incomplete, it is based in principle on the process as outlined in *E. coli*. Thus mammalian transcription should require phases of initiation, elongation,

Figure 9-13
Principal phases in the biosynthesis of RNA as determined in *E. coli*.

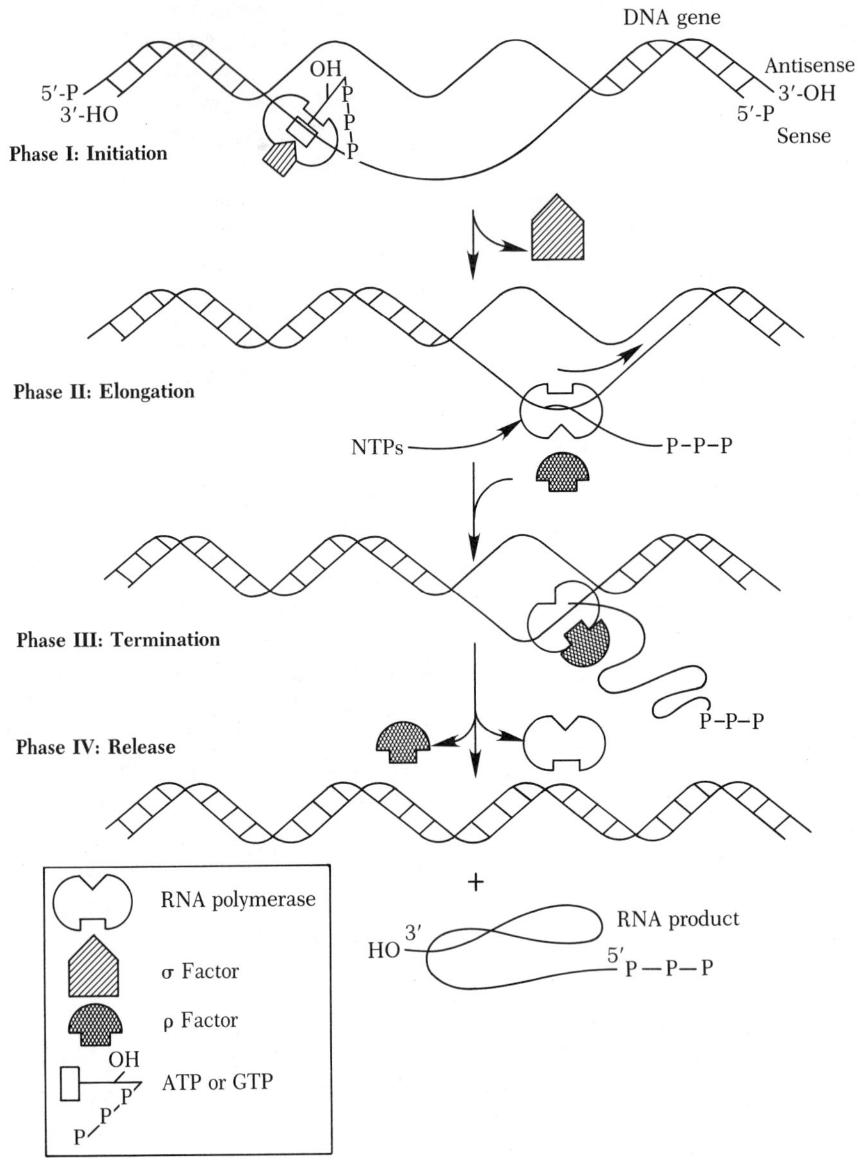

termination, and release. Since transcription occurs within the confines of the nucleus, three RNA polymerases are found there. These are polymerase I, found in the nucleolus, and polymerases II and III, both found in the nucleoplasm. Their functions are described in Table 9-2. These enzymes may be distinguished by their relative sensitivity to the mushroom *(Amanita phalloides)* toxin α-amanitin. Polymerase I is not inhibited by α-amanitin, II is sensitive to low concentration of the toxin (10^{-8}M-10^{-9}M), and III is sensitive to high concentrations (10^{-5}M-10^{-6}M). The eukaryotic polymerases are complex in structure; polymerase I consists of 6 subunits, II of 5, and III of at least 10 polypeptides. A separate RNA polymerase is found in mitochondria; it is a single polypeptide chain of molecular weight 64,000.

Table 9-2
Location and function of eukaryotic RNA polymerases

RNA polymerases	Location in nucleus	RNA produced
I (A)	Nucleolus	rRNA
II (B)	Nucleoplasm	hnRNA, mRNA
III (C)	Nucleoplasm	tRNA, 5S rRNA

Visualization of transcription

The transcriptive process was dramatically visualized by using the giant multistrand or salivary gland chromosomes of the fruit fly, *Drosophila,* or midge, *Chironomus*. Histological studies have demonstrated the formation of local bulges or puffs on the chromosomes of larvae previously injected with the molting hormone ecdysone. These regions of local activity on the chromosome were found to be newly synthesized RNA. When the salivary glands from larvae given ecdysone were exposed to radioactive uridine, the puffs became highly radioactive, as shown in radioautograph *A* of Fig. 9-14.

In a similar experiment, in which the antibiotic actinomycin D was included, very little radioactivity was present in the puff region, showing that transcription had been inhibited. Actinomycin D is known to block transcription by combining noncovalently with the DNA at the sites of guanine-cytosine base pairs. The ring structure of actinomycin D is positioned (intercalated) between two neighboring G≡C pairs and thereby inhibits the process of transcription. Therefore this antibiotic is an important tool to establish the existence of active RNA biosynthesis in a biological system. Other inhibitors are also commonly used; these are listed in the following table.

Inhibitors	Mode of action
Actinomycin D	Binds DNA at guanine and blocks RNA elongation
Rifampicin and rifamycin	Inhibit prokaryotic RNA polymerase by competing for binding of purine triphosphates involved in chain initiation but not in elongation
α-Amanitin	Toxin from toadstool *(Amanita phalloides)* inhibits eukaryotic polymerases II and III but not polymerase I

When algae were grown in the presence of radioactive uridine for a very short time only (pulse labeled), radioautographs showed the location of RNA to be restricted to the nucleus. Continued incubation of the labeled algae in a nonra-

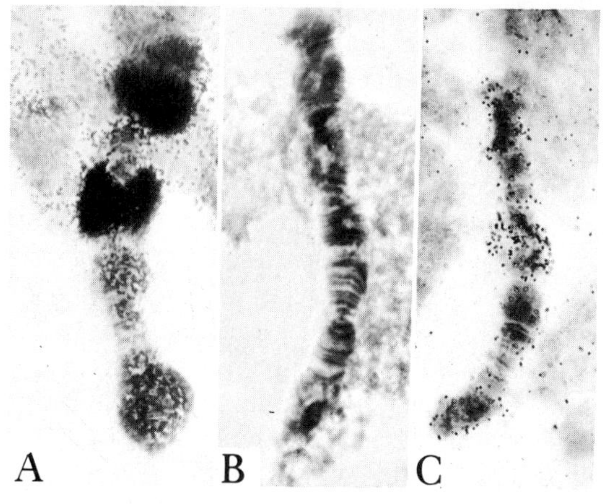

Figure 9-14
A Autoradiographs of salivary gland chromosomes of *Chironomus tentans* after incubation of the glands in a medium containing ^{3}H-uridine for 30 minutes.
B and **C** Larvae were treated with actinomycin D, 10 and 20 hours, respectively, before salivary glands were incubated with ^{3}H-uridine.

From Clever, U.: Science **146:**794, 1964.

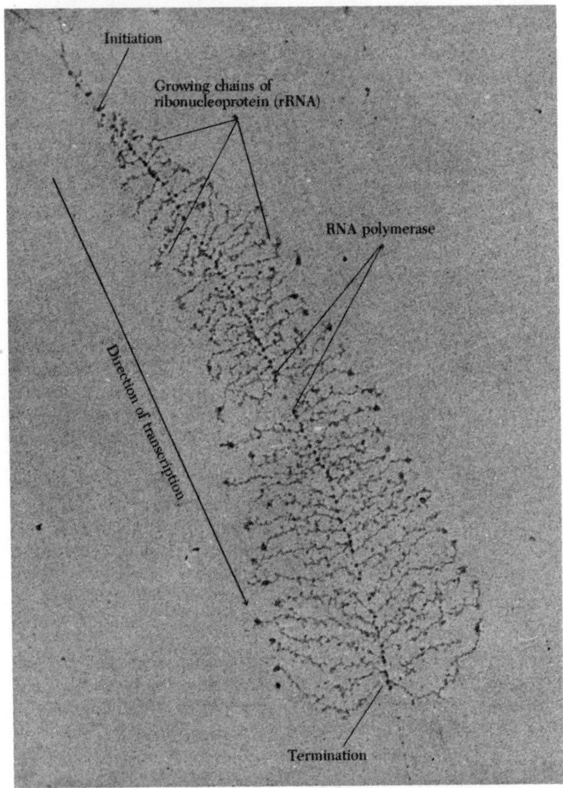

Figure 9-15
Electron micrograph showing a nucleolar gene *(Triturus viridescens)* in the process of active synthesis of rRNA. RNA polymerase molecules are attached to the DNA axis. A polymerase must move from the extreme left of the leaflike figure to the right before it has transcribed the entire gene. The figure shows ribonucleoprotein fibrils in varying degrees of completion.

Courtesy Miller, O.M.: Acta Endocrinol. (supp.) **168:**155, 1972.

dioactive medium led to a migration of the radioactivity from the nucleus to the cytoplasm. Such experiments illustrate the sequence of events in which the genetic information residing in chromosomal DNA is transcribed into RNA and then transmitted to the cytoplasm.

The genes of amphibian oocytes have been visualized during transcription using electron microscopy. Fig. 9-15 shows a nucleolar gene *(Triturus viridescens)* engaged in the active synthesis of rRNA. As the nascent rRNA strands are produced they become associated with protein. The result is a system of hairlike ribonucleoprotein fibrils increasing in length from left to right. A nucleolar gene such as that visualized in Fig. 9-15 has 80 to 100 polymerase molecules that simultaneously transcribe each gene. These enzyme molecules appear as a series of particles attached to the central core (DNA) of the leaflike figure. The polymerase particles at the extreme left have the shortest fibrils and therefore are just beginning the transcriptive process. As the polymerases move along the gene, the fibrils of ribonucleoprotein increase in length until they are terminated at the end of the gene (i.e., the extreme right of the figure). Similarly, nucleolar transcription has been visualized in mammalian cells (HeLa cells and mouse oocytes).

Processing of RNA

Most RNAs undergo modification after the actual transcriptional process is comlete. Structural genes may be transcribed bearing nonfunctional sequences that must be removed by a "trimming process." Finally some RNAs, especially the tRNAs, are highly modified, the unusual bases being formed after the actual transcription is complete. Specific kinds of RNA processing are described in the following discussion.

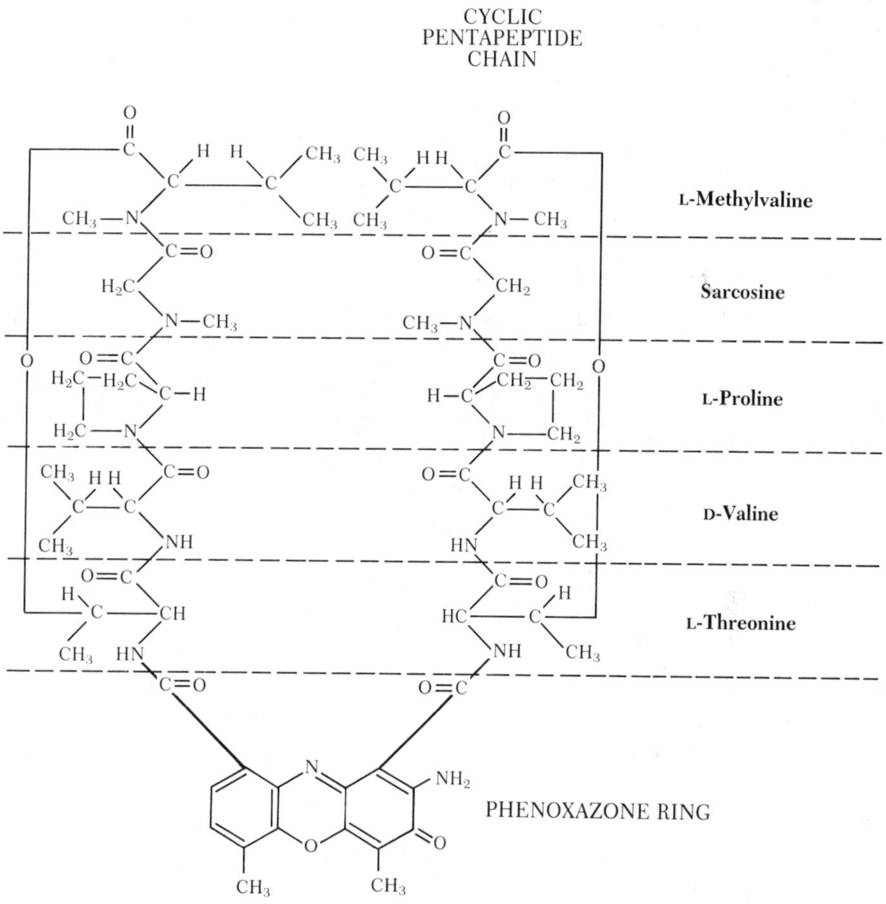

CYCLIC PENTAPEPTIDE CHAIN

L-Methylvaline

Sarcosine

L-Proline

D-Valine

L-Threonine

PHENOXAZONE RING

Actinomycin D

Ribosomal RNA In the *E. coli* chromosome the genes for RNA are arranged in a sequential order specifying the 16S, 23S and 5S rRNA units. RNA polymerase transcribes the three genes in order, as shown in Fig. 9-16, *A*. However, a primary transcript encompassing the three units is not usually observed, although a precursor 30S rRNA can be demonstrated in a mutant strain. This mutant is deficient in an enzyme, ribonuclease III, that normally removes the 16S component before the completion of the overall transcriptive process. The sequence of events as shown in Fig. 9-16, *A*, involves the synthesis and removal of the precursor 16S transcript (mol wt = 0.6×10^6) followed by the precursor 23S (mol wt = 1.2×10^6) transcript. The precursor 5S transcript is released last. A subsequent trimming process removes the spacer nucleotides and yields the final, or "mature," rRNA. It is thought that a special ribonuclease called RNase M or "maturase" has the capability of trimming the precursors to yield the final functional rRNAs.

A definite precursor transcript for rRNAs is produced from the sequence of genes in eukaryotic cells. The primary transcript is the precursor 45S polynucleotide (mol wt = 4.5×10^6). Fig. 9-16, *B*, illustrates a possible stepwise process whereby individual functional rRNAs (18S, 28S, and 5.8S) are removed from the single precursor chain. The 45S transcript is highly modified by methylation; in human cells this RNA has 65 or more methyl groups attached to the C-2 oxygen of the ribose moieties. In addition there are six methylated bases. Methylation of the ribose units

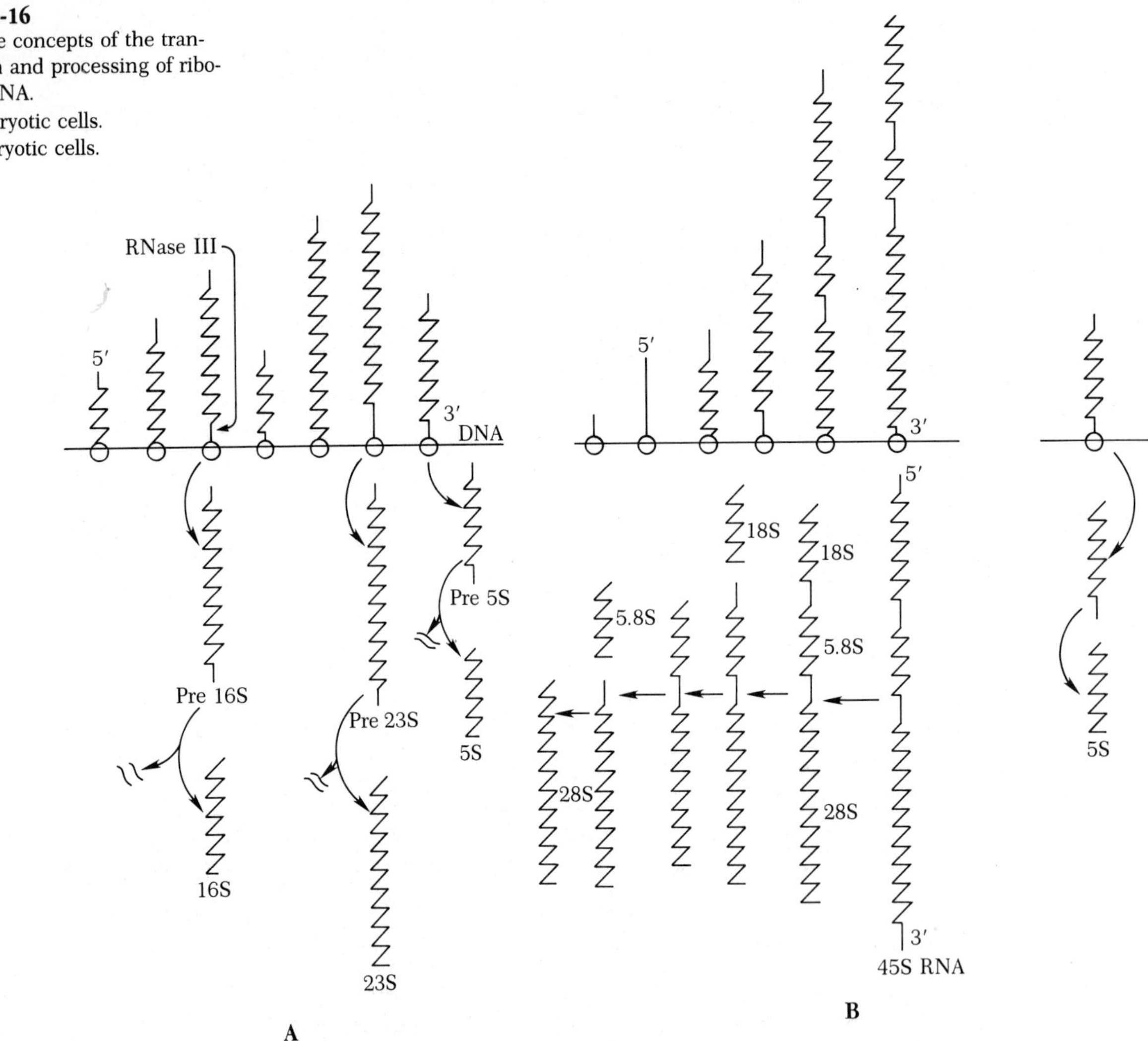

Figure 9-16
Plausible concepts of the transcription and processing of ribosomal RNA.
A Prokaryotic cells.
B Eukaryotic cells.

may be an important factor in controlling the extent of hydrolysis of the RNA by ribonucleases involved in the maturation process. In addition the 5S rRNA is produced from a separate gene.

Transfer RNA In bacterial cells tRNA is produced as precursors that are trimmed by the action of ribonucleases to form the final tRNA species. For some the tRNA precursor is the transcript of a single gene; for others it is a multiple sequence of genes. The general structure for precursor tRNA is as follows:

$$5'\text{p-leader-(tRNA-spacer)}_n\text{-tRNA-trailer } 3'\text{-OH}$$

The leader sequence is removed by RNase P in *E. coli*. Subsequently the trailer sequence is removed by another ribonuclease (RNase Q). RNase P also separates the individual tRNA species (Fig. 9-17). Once the trimmed species have been formed some ($tRNA^{Pro}$, $tRNA^{Ser}$) must have the -CCA (amino acid acceptor end) terminal added. This posttranscriptional process requires the action of a tRNA-CMP-AMP pyrophosphorylase. Other species, such as $tRNA^{Tyr}$ and $tRNA^{Gly}$, possess the -CCA terminal as an integral part of the transcript (Fig. 9-17).

In addition to the trimming process, extensive modifications are required to

Figure 9-17
Primary structure of precursor tRNA for tyrosine. Those bases which are modified in the mature tRNA are so indicated. The points of trimming are also shown.

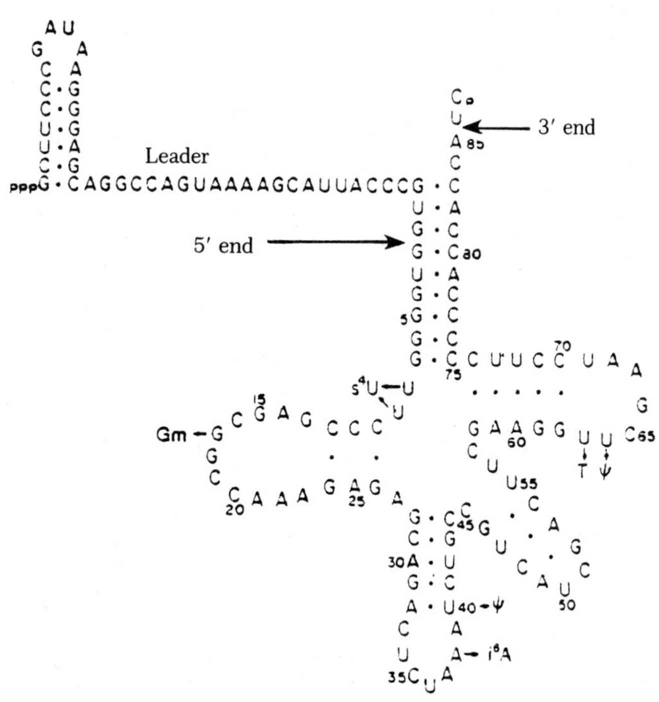

produce the final tRNA. These involve methylation of the 2′-hydroxyl of the ribose and of the 5 position of uridine or cytosine. Methylation also occurs on certain guanines. Reduction of the double bond in uridine yields the dihydrouridine; replacement of the C-4 oxygen with sulfur produces 4-thiouridine. A rearrangement of the uracil ring results in pseudouridine. The nature and diversity of these modifications can best be appreciated by reference to Fig. 9-17.

In eukaryotic cells the transcription and modification occur in the nucleus. The tRNAs appear to be produced as individual precursor molecules. The bulk of the methylases have also been observed in the nuclei.

Messenger RNA The posttranscriptional processing of eukaryotic mRNA is especially involved. It appears that genes for given proteins are not a continuous sequence of bases in the DNA as might have been anticipated but rather consist of regions of coding sequences divided in a regular manner by intervening spacers. Thus the gene for the β-globin chain in the mouse and rabbit hemoglobin consists of three coding sequences called *exons* separated by intervening spacers, or *introns*. With the apparent exception of those for the histones, genes for proteins as a rule may be interrupted in this fashion. To date the gene for conalbumin, an egg-white protein, holds the record with 16 intervening sequences.

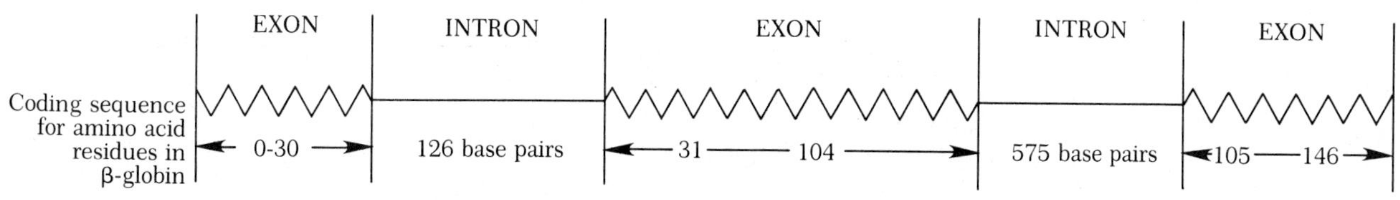

Thus the premessenger transcript is much longer than the final mature mRNA. As a matter of fact it is estimated that approximately 10% of the transcribed RNA eventually leaves the nuclear compartment as mature, functional mRNA; the

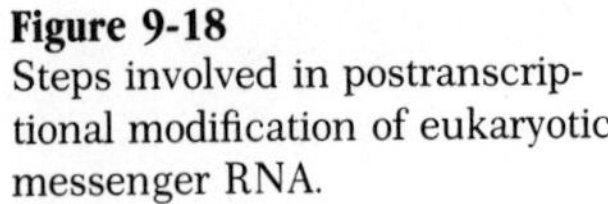

Figure 9-18
Steps involved in postranscriptional modification of eukaryotic messenger RNA.

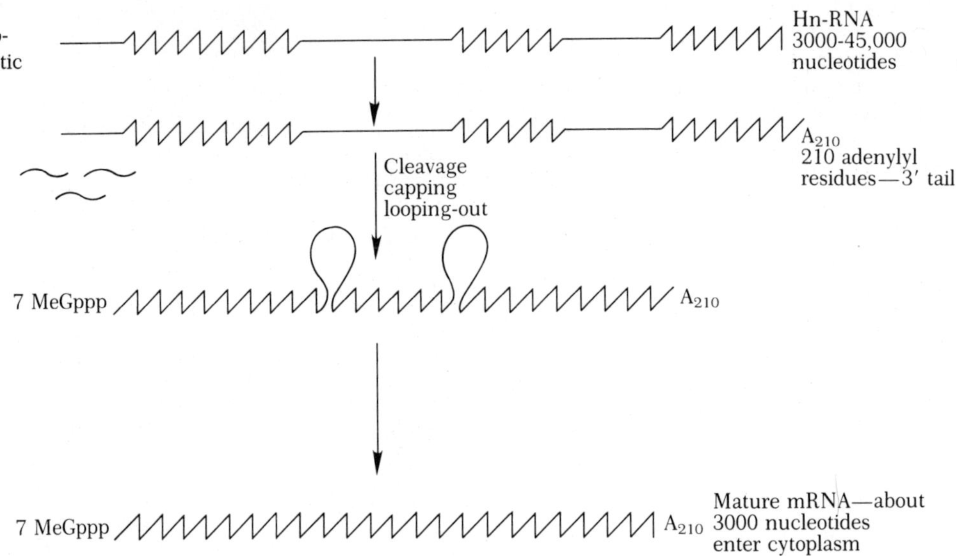

remainder presumably is degraded to mononucleotides. Because of the diversity of gene transcripts and also because of the apparent universal presence of the intron spacers, the nuclear mRNA is extremely heterogeneous and hence is designated as hn-RNA (heterogeneous nuclear RNA). Maturation requires some form of excision-ligation process that involves the removal of the internal segments, or introns, and the precise joining of the coding segments, or exons. Sequences that flank the coding segments are removed by a trimming process.

Maturation also involves the addition of a sequence of adenylyl residues (poly A) to the 3′ end (Fig. 9-18). The synthesis of the poly-A sequence may be catalyzed by a poly-A polymerase that requires ATP. This sequence is fairly homogeneous and has 210 ± 20 nucleotides.

Both the hn- and the cytoplasmic mRNAs are further modified by the addition to the 5′ terminal of a "cap" consisting of 7-methyl guanosine 5′-triphosphate. It should be recalled from the previous discussion of transcription that the 5′ end of the transcript consists of a nucleoside triphosphate. As seen below, capping first involves guanylation with GTP.

$$\text{ppp}^{A}_{G}\text{——RNA} \xrightarrow[\text{Hydrolase}]{H_2O \;\; P_i} \text{pp}^{A}_{G}\text{——RNA}$$

$$\text{pp}^{A}_{G}\text{——RNA} \xrightarrow[\text{Transferase}]{\text{GTP} \;\; \text{pp}_i} \text{Gppp}^{A}_{G}\text{——RNA}$$

$$\text{Gppp}^{A}_{G}\text{——RNA} \xrightarrow[\text{Transferase}]{\text{SAM} \;\; \text{SAH}} \text{7-Methyl Gppp}^{A}_{G}\text{——RNA}$$

$$\text{7-Methyl Gppp}^{A}_{G}\text{——RNA} \xrightarrow[\text{Transferase}]{\text{SAM} \;\; \text{SAH}} \text{7-Methyl Gppp}^{A}_{G}\text{-2′-O-methyl RNA}$$

This is followed by two methylation steps, the first at guanine N-7 and the second at the 2′-hydroxyl of the terminal RNA nucleotide. Specific methylases are required to catalyze these reactions. The coenzyme SAM is the methyl donor.

Once in the cytoplasm, the poly-A sequence may play a role in controlling the functional lifespan of the mRNA by preserving it from the degradative action of cytoplasmic RNase.

Drugs that interfere with replication and transcription

Several types of drugs (Fig. 9-19) interfere in the processes of replication and transcription and therefore are useful in the treatment of cancer and some infectious diseases.

Structural analogues of thymine such as 5-bromodeoxyuridine (BUdR) and 5-iododeoxyuridine, or idoxuridine (IUdR), can be inserted into DNA. Since their presence in the polynucleotide chain leads to errors in base pairing, the resulting DNA is nonfunctional. In vivo these compounds inhibit the replication of herpes simplex virus and are useful in the treatment of herpetic keratitis of the cornea.

Compounds such as actinomycin D (p. 181) bind tenaciously to DNA by intercalation between G≡C base pairs and interfere with transcription as well as replication. Daunorubicin, adriamycin, chromomycin-A_3, and mithramycin are used in cancer chemotherapy. Adriamycin appears to be useful in treating lung, breast, and bladder carcinoma. Quinone, the common antimalarial, can also form a hydrogen-bonded complex with DNA, apparently preventing strand separation and hence replication and transcription.

Some drugs can modify DNA chemically and form cross-links between the two neighboring polynucleotide chains in the helix. Typical of this group is cyclophosphamide, a derivative of bis(chloroethyl) amine. Modification of DNA occurs by alkylation of the N^7 of a guanine in each of the two polynucleotide chains involved. The alkylating drugs are used in the treatment of malignancies of the hemopoietic tissues and neuroblastomas and carcinomas of the lungs, ovaries, and testes. Bleomycin alters DNA by breaking 3′,5′-phosphodiester links. It is active in squamous cell skin cancers, testicular tumors, and most lymphomas.

Bis(chloroethyl) amine

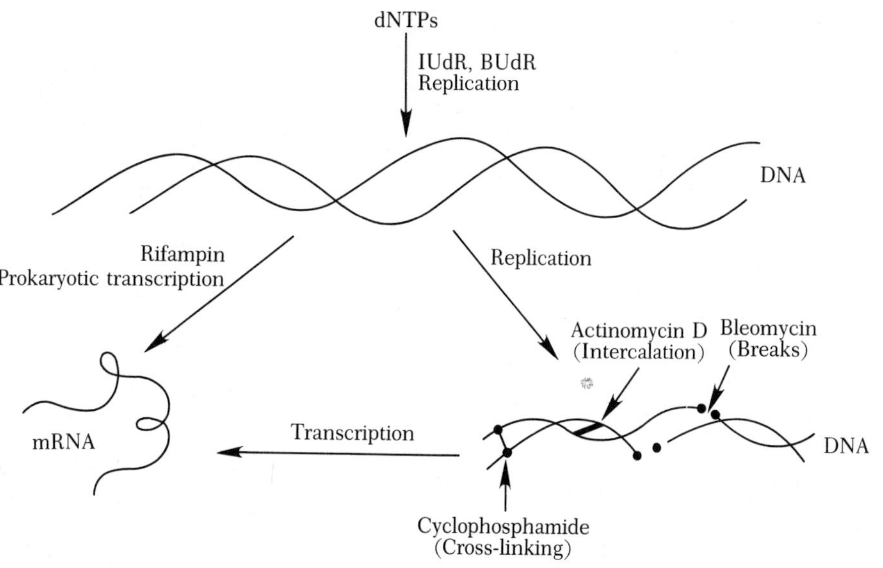

Figure 9-19
Summary of some drugs that act by interfering with the processes of replication and transcription.

Rifampicin is a derivative of rifamycin, an antibiotic produced by *Streptomyces mediterranei*. It binds strongly to DNA-dependent RNA polymerase of prokaryotic but not eukaryotic cells and therefore inhibits bacterial RNA synthesis. This antibiotic finds use in the treatment of tuberculosis.

Figure 9-20
Structure of tobacco mosaic virus. Inner helix represents RNA. Helically assembled ellipsoids represent protein subunits.

From Lauffer, M.A.: Adv. Virus Res. **13**:2, 1968.

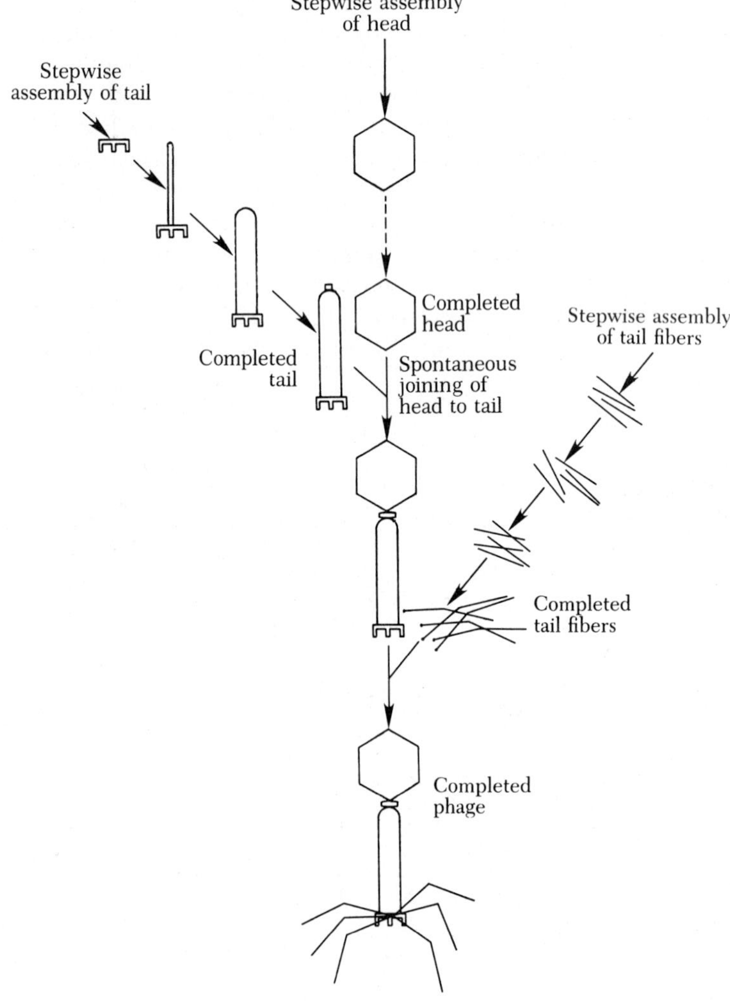

Figure 9-21
Sequential assembly of the T_4 bacteriophage.

Redrawn from Wood, W.B.: Fed. Proc. **27**:1160, 1968.

Viruses and their replication

Replication and transcription play an important role in the life cycle of viruses as well as in cells. Since viruses are the single major cause of human disease, it is appropriate at this point to gain an insight into the viral life cycle and to see how this knowledge may lead to a treatment of viral disease.

A simple virus such as tobacco mosaic virus (TMV) is a nucleoprotein composed of a single strand of RNA surrounded by molecules of a coat protein, or capsomere (Fig. 9-20). Viruses have either RNA or DNA but not both. Furthermore the polynucleotides exist as single-stranded RNA (TMV, polio, rabies) or DNA (ϕX174) or as double-stranded RNA (reovirus) or DNA (T_4, SV40). A single viral particle, or virion, such as the Qβ has only 3 genes, whereas the complex poxvirus has nearly 250 genes. In the case of the simplest virions (e.g., TMV) the assembly process is spontaneous, whereas for the more complex viruses, (e.g., T_4) the assembly requires the action of certain enzymes (Fig. 9-21).

Viral infection

Viruses are characterized by their inability to reproduce apart from a host cell. Thus they must invade the cell and redirect its cellular activity for their reproduction. For example, when the single-stranded RNA chain from the Qβ virus enters its host cells *(E. coli),* it redirects the metabolic activity of the cells to its own needs. The viral strand may act both as mRNA and as the template for replication. Since the base sequence directly dictates the genetic information, the viral RNA is designated RNA(+) to distinguish it from its antisense complementary copy RNA(−). The host, protein-synthesizing machinery translates the three genes into their respective proteins. One of the three is the enzyme *replicase,* which is capable of producing complementary RNA(−) copies of the RNA(+). Actually the active enzyme is a tetrameric moleule consisting of one subunit of viral protein and three host proteins.

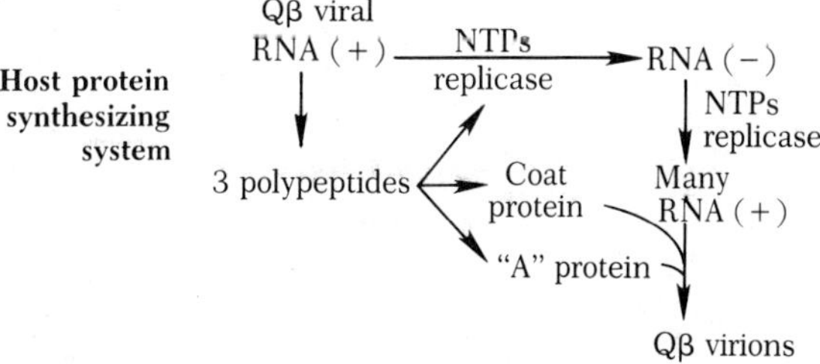

Translation is so timed as to yield the proteins before replication. The coat proteins aggregate with the RNA(+) to yield new virions. The host cell is lysed and the progeny viruses are released into the environment.

Not all single-stranded RNA viruses contain the sense strand; rabies is caused by an RNA(−) virus and has the antisense strand. In this case, the RNA cannot serve directly as the mRNA but must first be replicated to yield the complementary (+) strand. The virions possess RNA polymerase, which enters the host cell accompanied by the polynucleotide.

RNA (−) —NTPs / polymerase→ RNA (+) → RNA (−); RNA (+) → Proteins (Host protein synthesizing system)

The reovirus is an example of an RNA virus in which the polynucleotide is double stranded. Actually the virion contains some 10 different double-stranded molecules and also carries in its coat an RNA polymerase. Once inside the host cell, the polymerase catalyzes the biosynthesis of (+) strands only.

$$\text{RNA}(\pm) \xrightarrow[\text{polymerase}]{\text{NTPs}} \text{RNA}(+) \rightarrow \text{Capsid} + \text{RNA}(-) \rightarrow \text{Reovirus}$$

Of the DNA-containing viruses, the T_4 bacteriophage is best kown. The virion has a double-stranded DNA consisting of 165 genes. It has a head, tail, and six tail fibers. The DNA, packed into the head piece, possesses three sets of genes. These are described according to the timing of their transcription, i.e., immediate-early, delayed-early, and late. Once the phage DNA enters the host cell *(E. coli)* the early genes are transcribed and translated into phage-specific enzymes. One of these hydrolyzes dCTP in the host cell to dCMP to prevent its use for the biosynthesis of new bacterial DNA. A second acts on dCMP to produce the 5-hydroxymethylcytidylate. A third converts this derivative to the triphosphate, which is subsequently used for the biosynthesis of T_4-DNA. Some of these bases are glucosylated by yet another "early" enzyme. The glucosylated T_4-DNA is resistant to deoxyribonuclease, which is also produced by the early genes. This enzyme degrades the host cell DNA, which does not contain the cytidylate derivative, thereby preventing cell reproduction.

5-Hydroxymethyl dCMP **Glucosylated 5-hydroxymethylcytidine**

Spontaneous assembly

Viruses are usually assembled spontaneously from the appropriate nucleic acids and the protein capsomeres. This was first established using TMV, which possesses a central RNA helix containing some 6000 nucleotides. The RNA is arranged inside a shell of 2150 identical protein molecules, each having 158 amino acid residues. Under conditions of abnormal salt concentration and pH the virus disassembles and the RNA and protein may be separated from one another. Subsequently by remixing of TMV-RNA with protein subunits and by readjusting the salt concentration and pH to normal values, the protein units spontaneously arrange themselves about the RNA strand (Fig. 9-20).

Lysogeny

Not all viruses are compelled to reproduce themselves at the expense of the host cell and to cause the lysis of the cell, thereby releasing new generations of viruses. An alternative to this lytic process is the *lysogenic pathway,* in which the virus polynucleotide is incorporated into the host genome. An example is the λ-phage, which is called a temperate virus and does not at first kill its host. The virion DNA is linear, double-stranded, and has a molecular weight of 30.8×10^6, corresponding to 46,500 base pairs. This polynucleotide is characterized by the fact that its 5′ terminal consists of single-stranded sequences of some 12 bases that are complementary. Since the ends can join together (Fig. 9-22), they are called "sticky" or

Figure 9-22
Formation of circular λ-DNA from the linear form.

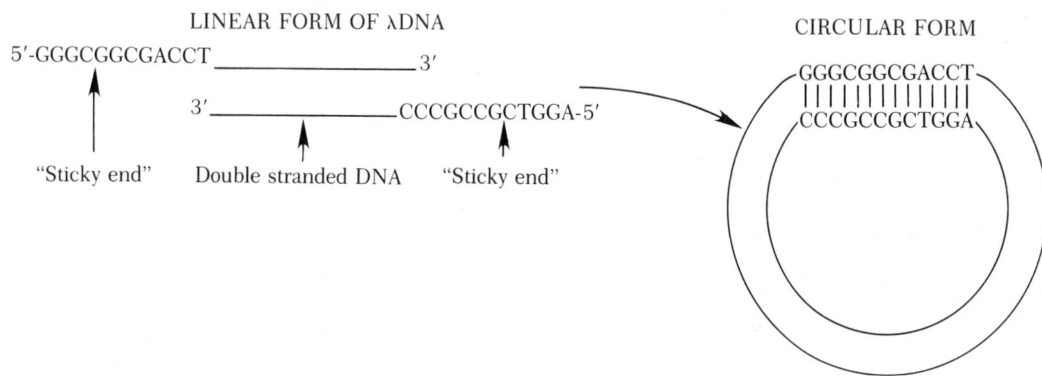

Figure 9-23
Recombination of λ-DNA and the *E. coli* chromosome.

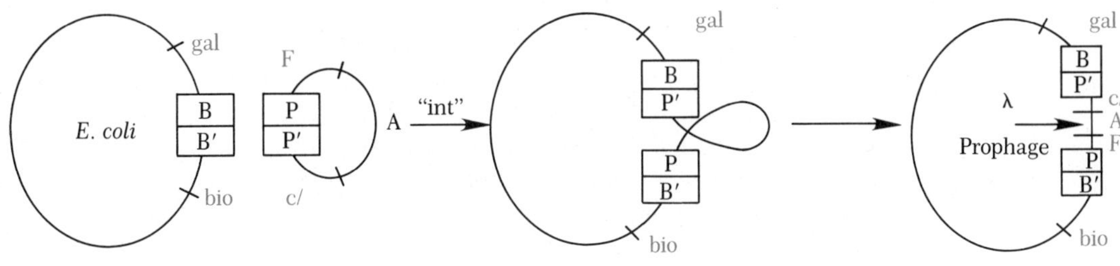

cohesive. Once the linear polynucleotide has entered the host cell a circular structure is formed; host DNA ligase seals the gaps, thereby creating a circular λ DNA molecule. This viral genome is then integrated into the bacterial chromosome by a *recombination* process at specific loci in the λ and *E. coli* DNA. As shown in Fig. 9-23, a specific protein, "int" for *integrase*, recognizes appropriate binding sites, namely B and B′ in the chromosome and P and P′ in the circular λ-DNA and brings about the recombination illustrated. Once recombination has occurred, the λ-DNA is an integral part of the *E. coli* chromosome; the *E. coli* is now a *lysogenic* bacterium. As the bacteria reproduce the λ-DNA is replicated along with the regular chromosome. At some signal another protein, called the "xis" protein for *excise,* is produced in the bacterium. This protein plus the "int," which is also needed, reverses the recombination process. Excision of the λ-DNA protein releases the circular molecules from the remaining *E. coli* chromosome. The λ-DNAs are again replicated in their host cell, RNA is transcribed, and all proteins necessary to produce new virions are translated. Following lysis the viruses are released.

Treatment of viral infections

As mentioned earlier (p. 185), replication of the viral genome may be inhibited by such thymidine analogues as *idoxuridine* (5-iodo-2′-deoxyuridine). This drug has a number of effects, such as the inhibition of thymidylate synthetase and thymidylate phosphorylase thereby suppressing the production of dTTP. However, DNA polymerase cannot distinguish the iodo analogue from dTTP and therefore lethal iododeoxyuridine-containing DNA is produced.

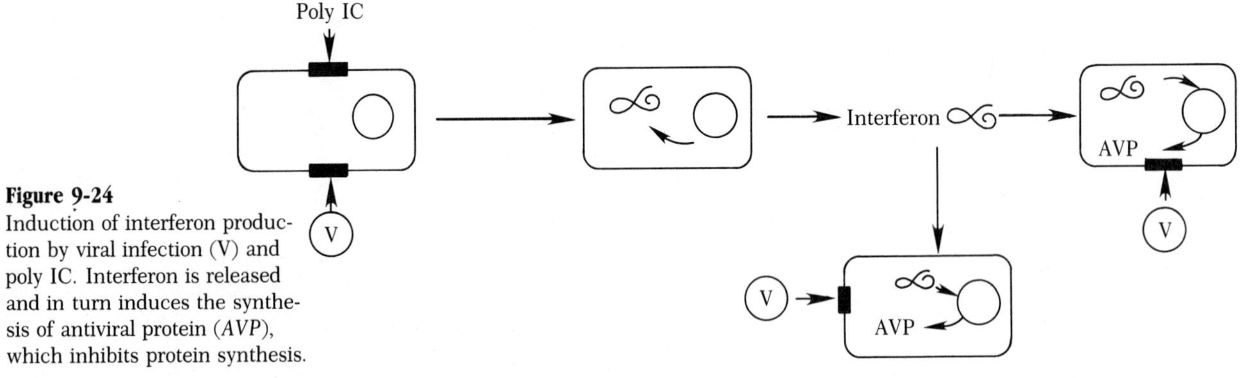

Figure 9-24
Induction of interferon production by viral infection (V) and poly IC. Interferon is released and in turn induces the synthesis of antiviral protein (*AVP*), which inhibits protein synthesis.

Mammalian cells also have their own kind of defense mechanism in the form of a glycoprotein (mol wt = 20,000) called *interferon*. Interferon production is induced following viral infection by double-stranded RNA molecules as well as by the synthetic polynucleotide poly IC (Fig. 9-24). Interferon from the infected cells is expelled and subsequently enters other cells. Although interferon itself is not the antiviral agent, it induces new RNA synthesis and the production of an antiviral protein in the responsive cells. This antiviral protein (AVP) is thought to inhibit the "capping" of mRNA, thereby slowing down protein synthesis. It is apparent from Fig. 9-24 that infection of one cell by the virus can lead to the protection of many because of the release of the interferon. Preparations of human interferon have been tested as a prophylactic treatment during influenza epidemics. Since poly IC is the most potent interferon inducer, it has high activity in the prevention of viral infections in animals.

Modification of host DNA and the role of restriction enzymes

Bacterial cells protect themselves against the invasion of viral or foreign polynucleotides by the action of endogenous endonucleases that selectively hydrolyze foreign DNA. The protective endonucleases are called *restriction* enzymes, and they hydrolyze the duplex DNA molecules at specific regions in which the base pairs exhibit a special symmetrical arrangement called a *palindrome*. This term is used to designate a symmetrical sequence in which the base pairs read the same forward (to the right) or backward (to the left) relative to the axis of symmetry.

3′—C—T—T—A—A↓G—5′
5′—G↑A—A—T—T—C—3′
Axis of symmetry

A restriction enzyme of *E. coli* such as ECo R1 hydrolyzes the duplex DNA molecule at the points shown (arrows) in the above palindrome. Other examples of palindromes and restriction enzymes are illustrated below:

Enzymes	Origin	Palindrome
ECo R1	*E. coli*	G↓A A T T C C T T A A↑G
Hind III	*Haemophilus influenzae*	A↓A G C T T T T C G A↑A
Hpa II	*Haemophilus parainfluenzae*	C↓C G G G G C↑C
Hae III	*Haemophilus aegypticus*	G G↓C C C C↑G G

To distinguish between the endogenous and exogenous DNAs the cell possesses highly selective methylases that modify the host but not the foreign DNA by adding methyl groups to such bases as adenine and cytosine. Methylation of these bases occurs at the palindromes, which thereby become resistant to hydrolysis by the restriction enzyme. Methylation of DNA is catalyzed by the specific methylases, using SAM as the methyl group donor. The products of the methylation of adenine and cytosine are 6-*N*-methyl adenine and 5-methyl cytosine. Unmethylated foreign DNA is hydrolyzed by the host cell's restriction enzymes, whereas the methylated endogenous DNA is not.

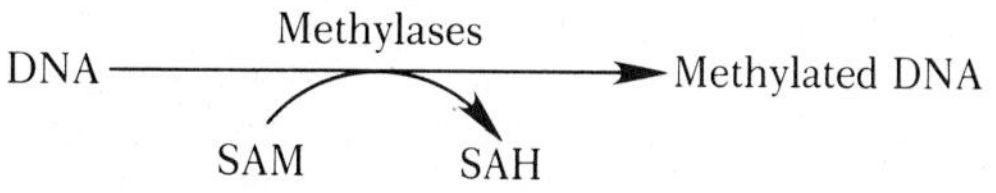

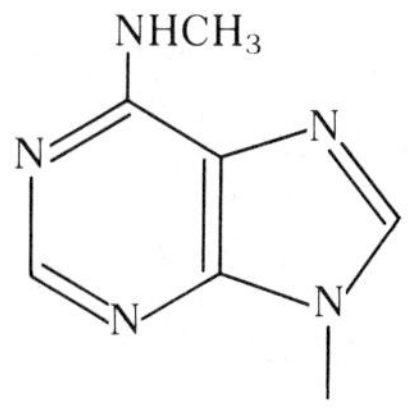

$$\text{DNA} \xrightarrow[\text{SAM} \;\curvearrowright\; \text{SAH}]{\text{Methylases}} \text{Methylated DNA}$$

$NHCH_3$ · NH_2 · CH_3

6-*N*-Methyladenine 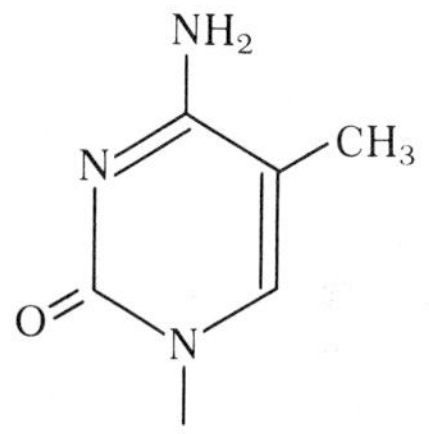 **5-Methylcytosine**

Oncogenic viruses (cancer)

A variety of so-called oncogenic viruses such as leukemia, Rous sarcoma, polyoma, SV40, and papilloma viruses are capable of inducing cancer in the host. Oncogenic viruses may possess either RNA or DNA. Viruses of the papova group (polyoma, SV40, and papilloma) have circular DNA molecules containing 5 genes (5100 base pairs). Invasion of the host cell results in the recombination with the host genome, thereby permanently "infecting" the cell. The cell is said to be *transformed*. These transformed cells differ from the normal in that they have lost their capacity for contact inhibition. This is a process in which normal cells multiply until they come in direct contact with one another. On contact they cease to multiply, thereby limiting growth. Cancerous cells have lost this restriction on growth and continue to multiply in an uncontrolled, unabated fashion. This abnormal characteristic is perpetuated from cell generation to generation.

RNA-containing, oncogenic viruses enter the cells and are "transcribed" to yield DNA. This seemingly backwards process is catalyzed by "reverse transcriptase." The latter viral enzyme forms a DNA(−) chain from the RNA(+). The enzyme can also produce DNA(+) from the DNA(−), thereby yielding a duplex DNA that can then be integrated into the host genome. The infected cell is thus permanently transformed.

Recombinant DNA and bioengineering

The principles involved in replication and transcription have opened the door to an area of research whose potentialities are at once mind boggling and controversial. This is the field of genetic engineering. At this point the primary goal is the commercial production of specific human proteins, e.g., hormones, for therapeutic purposes. The strategy of bioengineering involves first inserting the desired gene into the genome of a suitable bacterial vehicle and then allowing the "host" to reproduce, thereby cloning the desired gene. Furthermore, all conditions being suitable, the growing bacterial vehicle should transcribe and translate the foreign gene into the appropriate polypeptide. This strategy may be exemplified by the

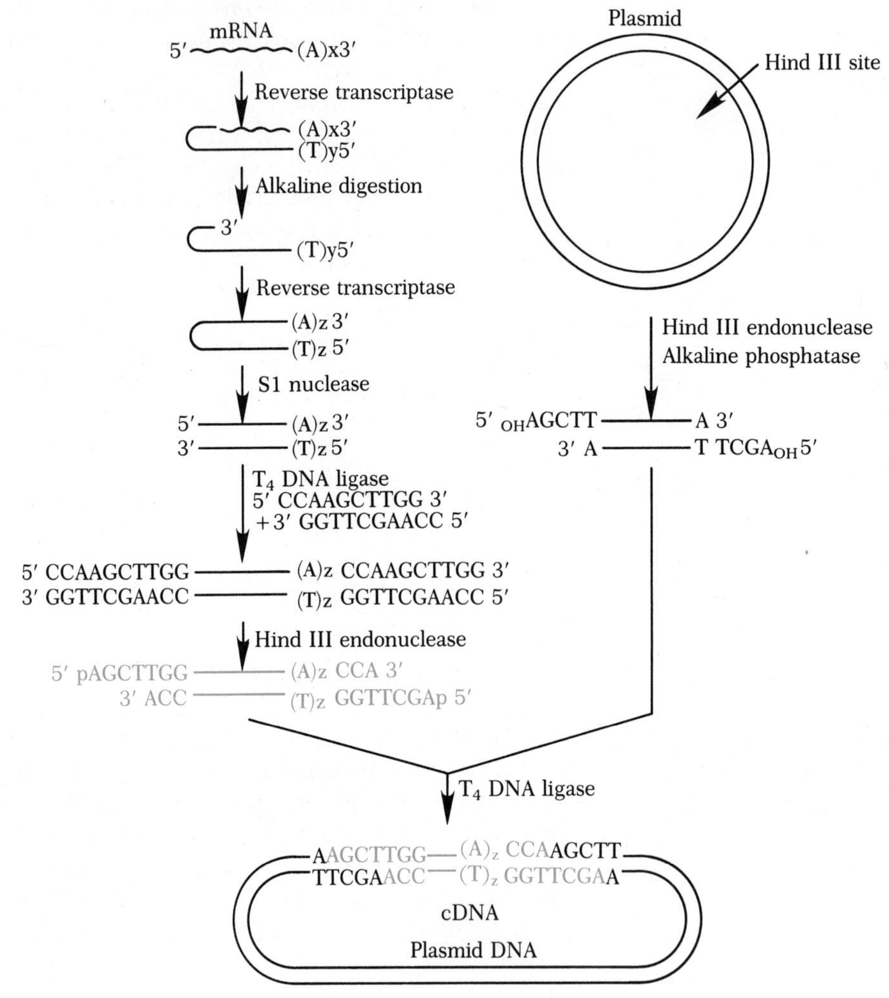

Figure 9-25
Use of the recombinant DNA technique to reproduce (clone) a mammalian gene (rat proinsulin) in a bacterium (*E. coli*).

From Ullrich, A., et al.: Science **196**:1313, 1977. Copyright 1977 by the American Association for the Advancement of Science.

production of the proinsulin gene in *E. coli* (Fig. 9-25). Pure mRNA for rat proinsulin is isolated and used as the template for the formation of complementary DNA (cDNA) by reverse transcriptase. This Zn-containing metalloenzyme (mol wt = 160,000), which is an RNA-dependent polymerase, uses the single strand of RNA as a template to produce a complementary strand of DNA. As seen in Fig. 9-25, the product is an RNA-DNA hybrid held together by a hairpin loop. The RNA component is removed by treatment with potassium hydroxide, leaving the single strand of cDNA. Further action of reverse transcriptase produces a loop of DNA, which may then be treated with S1 nuclease. The product is a duplex of DNA with open ends. "Sticky" or complementary ends are added as shown, using the enzyme DNA ligase and the restriction enzyme Hind III. Plasmid DNA (extra-chromosomal DNA) is isolated from *E. coli*. These circular duplex DNA molecules are also treated with Hind III so as to open the duplex at a palindrome and thereby leave complementary or "sticky" ends. The proinsulin DNA gene is added to the solution of opened plasmid molecules. As seen in the figure, some of the proinsulin DNA will align itself within the plasmid. The nicks are then sealed with DNA-ligase.

This modified plasmid is added to *E. coli* previously treated with calcium chloride so as to increase the permeability of their cell walls. In this way a cell is produced containing a mammalian gene as part of its genome. As the bacteria reproduce the mammalian gene is replicated, or "cloned". Similarly the gene for the hormone somatostatin has been inserted and cloned and the host cells have been induced to produce a human protein.

It is also conceivable that a replicable piece of RNA could act as a carrier for an RNA gene, coding for a specific enzyme that is deficient in an individual having a particular genetic disease, e.g., galactosemia. Such a synthetic virus might enter the host's cells and introduce the gene into the chromosomes by means of an RNA-dependent DNA polymerase. Or a DNA-containing virus such as SV40 could also be used to carry a specific DNA gene into the host cell. Such efforts at gene manipulation may be premature, but an era of "bioengineering" is not unlikely.

Suggested readings

Alberts, B., and Sternglanz, R.: Recent excitement in the DNA replication problem, Nature **269:**655, 1977.

Baron, S., and others: Interferon, Science **180:**779, 1973.

Bouché, J-P., Rowen, L., and Kornberg, A.: The RNA primer synthesized by primase to initiate phage G4 DNA replication, J. Biol. Chem. **253:**765, 1978.

Cozzarelli, N.R.: The mechanism of action of inhibitors of DNA synthesis, Annu. Rev. Biochem. **46:**641, 1977.

Cozzarelli, N.R.: DNA gyrase and the supercoiling of DNA, Science **207:**953, 1980.

Crick, F.: Split genes and RNA splicing, Science **204:**267-271, 1979.

Darnell, J.E., Jelinek, W.R., and Molloy, G.R.: Biogenesis of mRNA: genetic regulation in mammalian cells, Science **181:**1215, 1973.

Davidson, E.H., and Britten, R.J.: Regulation of gene expression: possible role of repetitive sequences, Science **204:**1052, 1979.

Davidson, J.N.: The biochemistry of nucleic acids, New York, 1972, Academic Press, Inc.

Davidson, J.N., and Cohn, W.E., editors: Progress in nucleic acid research and molecular biology, vols. 1-13, New York, 1963-1973, Academic Press, Inc.

DeLange, R.J., and Smith, E.L.: Histones: structure and function, Annu. Rev. Biochem. **40:**279, 1971.

Friedman, T., and Robbin, R.: Gene therapy for human genetic disease, Science **175:**949, 1972.

Gauss, D.H., and others: Recent results in tRNA research, Annu. Rev. Biochem. **40:**1045, 1971.

Gilbert, W.: Why genes in pieces? Nature **271:**501, 1978.

Goulian, M.: Biosynthesis of DNA, Annu. Rev. Biochem. **40:**855, 1970.

Goulian, M., and Hanawalt, P.C., editors: DNA synthesis and its regulation, Menlo Park, Calif., 1975, W.A. Benjamin, Inc.

Goulian, M., Kornberg, A., and Sinsheimer, R.L.: Enzymatic synthesis of DNA, XXIV. Synthesis of infectious phage ϕX174 DNA, Proc. Nat. Acad. Sci. U.S.A. **58:**2321, 1967.

Hamkalo, B.A., and Miller, O.L., Jr.: Electronmicroscopy of genetic activity, Annu. Rev. Biochem. **42:**379, 1973.

Hanawalt, P.C., and Setlow, R.B. editors: Molecular mechanisms for repair of DNA, New York, 1975, Plenum Press.

Harris, M.: Interferon: clinical application of molecular biology, Science **170:**1068, 1970.

Holley, R.W.: The nucleotide sequence of a nucleic acid, Sci. Am. **214:**30, 1966.

Hurwitz, J., and others: On the copying of DNA in the RNA polymerase reaction, Symp. Quant. Biol. **28:**59, 1963.

Ingram, V.M.: Biosynthesis of macromolecules, ed. 2, New York, 1972, W.A. Benjamin, Inc.

Kasamatsu, H., and Vinograd, J.: Replication of circular DNA in eukaryotic cells, Annu. Rev. Biochem. **43:**695, 1974.

Khorana, H.G.: Total synthesis of a gene, Science **203:**614, 1979.

Kim, S.H., and others: Three-dimensional structure of yeast phenylalanine transfer RNA: folding of the polynucleotide chain, Science **179:**285, 1973.

Klein, A., and Bonhoeffer, F.: DNA replication, Annu. Rev. Biochem. **41:**301, 1972.

Kleinschmidt, W.J.: Biochemistry of interferon and its inducers, Annu. Rev. Biochem. **41:**517, 1972.

Kolata, G.B.: Overlapping genes: more than anomalies, Science **196:**1187, 1977.

Kornberg, A.: DNA synthesis, San Francisco, 1980, W.H. Freeman & Co. Publishers.

Kushner, S.R., and others: Enzymatic repair of deoxyribonucleic acid. IV. Mechanism of photoproduct excision. Biochemistry **10:**3325, 1971.

Lehman, I.R., and Uyemura, D.G.: DNA polymerase I: essential replication enzyme, Science **193:**963, 1976.

Lewin, Benjamin M.: Gene expression, vol. 2, eucaryotic chromosomes, New York, 1974, John Wiley & Sons, Inc.

Losick, R.: In vitro transcription, Annu. Rev. Biochem. **41:**409, 1972.

Losick, R., and Chamberlin, M., editors: RNA polymerases, Cold Spring Harbor, N.Y., 1976, Cold Spring Harbor Laboratory.

Maxam, A.M., and Gilbert, W.: A new method for sequencing DNA, Proc. Natl. Acad. Sci. U.S.A. **74:**560, 1977.

McMacken, R., Ueda, K., and Kornberg, A.: Migration of *Escherichia coli* dnaB protein on the template DNA strand as a mechanism in initiating DNA replication, Proc. Natl. Acad. Sci. U.S.A. **74:**4190, 1977.

Miller, O.L., Jr.: The visualization of genes in action, Sci. Am. **228:**34, 1973.

Miller, O.L., Jr., and Bakken, A.H.: Morphological studies of transcription, Acta Endocrinol. [Suppl.] **168:**155, 1972.

Mills, D.R., Kramer, F.R., and Spiegelman, S.: Complete nucleotide sequence of a replicating RNA molecule, Science **180:**816, 1973.

Otto, B., Bonhoeffer, F., and Schaller, H.: Purification and properties of DNA polymerase III, Eur. J. Biochem. **34:**440, 1973.

Perry, R.P. In Davidson, J.N., and Cohn, W.E., editors: Progress in nucleic acid research and molecular biology, vol. 6, New York, 1967, Academic Press, Inc., p. 219.

Rapp, F.: Herpes viruses, venereal disease, and cancer, Am. Scientist **66:**670, 1978.

Rich, A., and RajBhandary, U.L.: Transfer RNA: molecular structure, sequence, and properties, Annu. Rev. Biochem. **45:**805, 1976.

Sinsheimer, R.L.: Recombinant DNA, Annu. Rev. Biochem. **46:**415, 1977.

Sobell, H.M.: How actinomycin binds to DNA, Sci. Am. **231:**82, 1974.

Temin, H.M.: RNA-directed DNA synthesis, Sci. Am. **227:**24, 1972.

Ullrich, A., and others: Rat insulin genes: construction of plasmids containing the coding sequences, Science **196:**1313, 1977.

Worcel, A., and Burgi, E.: On the structure of the folded chromosome of *Escherichia coli*, J. Molec. Biol. **71:**127, 1972.

Chapter 10

Genetic control of metabolism: translation of information

As described in Chapter 8, cellular metabolism and its regulation depend on the presence within the cell of not only the correct enzymes but also appropriate amounts of each. Selective transcription or information retrieval is the mechanism whereby specific mRNA's and therefore the proper enzymes are synthesized. This chapter primarily describes the process by which the genetic information (base sequence) is translated into specific proteins (amino acid sequence). However, the amount of proteins produced is not restricted to the quantity of mRNAs. That is, the overall translational process also depends on a variety of other factors that reflect the diet and the physiological state of the body. For example, a suitable supply of amino acids is needed. This is controlled by an energy-dependent transport process (Chapter 19). Once inside the cell, the amino acids must be activated and joined to their appropriate tRNA molecules; this is also an ATP-consuming process. Furthermore, the actual mechanism for joining the amino acids by means of peptide bonds requires a supply of GTP. In addition, other factors affect translation, such as the production and aggregation of ribosomes into clusters called polysomes.

The accompanying diagram summarizes various factors that influence the overall translational process.

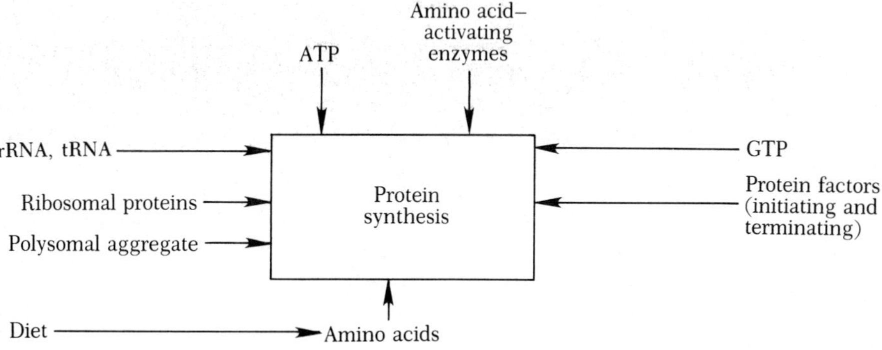

Once the protein chain has been or is being assembled, modifying processes control the formation of the final molecule and its ultimate delivery to the body. The glycoproteins require the addition of oligosaccharide moieties to the polypeptide chain. As with peptide synthesis itself, the formation of these is an energy- and metabolite-requiring process. In proteins that possess prosthetic groups, such as the heme of hemoglobin, the availability of the prosthetic group serves as a controlling factor in protein synthesis. Exocytosis controls the ultimate delivery of some proteins to the outside of the cells of origin and thus to the body. An example is the release of insulin from the β-cells of the islets of Langerhans; the protein is stored within the cell until its release is signalled by the blood glucose level.

Genetic code

It is apparent from the preceding chapters that genetic information is inherent in the structure of DNA. Since DNA is composed of the four bases—adenine, guanine, cytosine, and thymine—it follows that the information is somehow determined by the base sequence. As described in Chapter 9, the base sequence of the mRNA is copied from that of the DNA. In translation the base sequence determines the primary structure of the protein produced. It is now clear that each amino acid is represented in the nucleic acids in the form of a genetic code composed of a combination of three nucleotide bases (triplet code). Since there are four bases available from which to produce the triplets, 4^3 or 64 different combinations of the triplet code are possible. Obviously this is more than enough to represent the 20 available amino acids.

Once it was determined that the genetic code occurs in the form of groupings of three bases, experiments were designed to establish that they were read in an end-to-end, nonoverlapping manner. In studies using T_4 bacteriophage Crick and his co-workers showed that messages in the DNA are read as triplets, in a nonoverlapping sequence beginning from one end of the chain. This discovery was accomplished by using mutant bacteriophage from which one, two, or three bases were either deleted from or added to the DNA by the action of certain acridine dyes. The experiment is instructive and therefore is illustrated in Table 10-1 for a hypothetical segment of bacteriophage DNA. It is apparent from the table that a single deletion would throw the sequential reading of triplets out of phase and therefore would yield nonsense information. Two deletions likewise would yield an out-of-phase transcription of information. Neither of these deleted DNAs (mutants) could infect host cells to produce bacteriophage. However, when three bases were deleted, the mutant DNA was infective again. It is apparent from the table that the gene would be out of phase initially but would later be in phase again so that most of the DNA would be functional. This experiment confirmed that genetic information occurs in the form of triplets of bases, read in an end-to-end sequence; it does not say what the triplets mean in terms of amino acids.

Breaking the code

To decipher this triplet code, Nirenberg and Matthai designed cell-free, protein-synthesizing systems in which synthetic mRNAs could be translated. Such an amino acid–incorporating system had the following components:

1. Ribosomes (*E. coli*)
2. Cell sap (tRNAs, amino acid–activating enzymes)
3. Suitable buffer system, GTP, Mg^{++}
4. ATP-generating system
5. Mixture of amino acids (one radioactively labeled amino acid plus 19 unlabeled)
6. mRNA (natural and synthetic)

In the presense of isolated mRNA the labeled amino acid was incorporated into newly synthesized protein. In the absence of mRNA little amino acid incorporation occurred. When a synthetic polynucleotide was included, such as poly U (a synthetic polynucleotide containing only uracil), only phenylalanine was incorporated into a polypeptide. The peptide produced was polyphenylalanine. This meant that poly U served as an mRNA and was translated by the system as a sequence of uracil triplets: UUU, UUU, UUU. Similarly, it was found that AAA coded for lysine and

Table 10-1
Nucleotide sequence comprising a hypothetical gene

Number of bases deleted	
0	CAT CAT CAT CAT CAT CAT CAT
1	out of phase CAT C TC ATC ATC ATC ATC ATC A
2	out of phase CAT C TC A CA TCA TCA TCA TCA A T
3	out of phase / in phase CAT C TC A CA T AT CAT CAT CAT A T C

Table 10-2
Triplet codons for amino acids

First letter	Second letter: U	C	A	G	Third letter
U	Phe UUU	Ser UCU	Tyr UAU	Cys UGU	U
	UUC	UCC	UAC	UGC	C
	Leu UUA	UCA	Term. UAA*	Term. UGA*	A
	UUG	UCG	Term. UAG*	Trp UGG	G
C	Leu CUU	Pro CCU	His CAU	Arg CGU	U
	CUC	CCC	CAC	CGC	C
	CUA	CCA	Gln CAA	CGA	A
	CUG	CCG	CAG	CGG	G
A	Ile AUU	Thr ACU	Asn AAU	Ser AGU	U
	AUC	ACC	AAC	AGC	C
	AUA	ACA	Lys AAA	Arg AGA	A
	Met AUG*	ACG	AAG	AGG	G
G	Val GUU	Ala GCU	Asp GAU	Gly GGU	U
	GUC	GCC	GAC	GGC	C
	GUA	GCA	Glu GAA	GGA	A
	GUG*	GCG	GAG	GGG	G

**Term.*, Codons can act as signals for chain termination; *AUG* and *GUG* signal chain initiation by coding *N*-formylmethionine (prokaryotes) or methionine (eukaryotes).

that CCC coded for proline. Triplets of bases present in mRNA are now known as *codons*. Subsequent studies with synthetic mRNA polynucleotides allowed a complete designation of 61 codons with their corresponding amino acids. It should be clear that the codons presented in Table 10-2 refer to the sequence of bases in mRNA. Most of the amino acids have two or more codons; only tryptophan and methionine are represented by a single codon each.

The synthesis of the polynucleotides of known triplet sequences needed for the complete determination of the amino acid code was accomplished by Khorana and co-workers.

Further experiments with synthetic polynucleotides established that the reading of mRNAs occurs in a 5′ to 3′ direction.

Basic characteristics of the genetic code

The following statements summarize the basic characteristics of the genetic code:

1. The code is read as sequences of three bases, i.e., as triplet codons in the mRNA.
2. The code is said to be degenerate or redundant, meaning that most amino acids have more than one codon. Reference to Table 9-2 shows that leucine has six; isoleucine, three; phenylalanine, two; methionine and tryptophan have only single codons. However, the genetic code is not ambiguous, meaning that a given codon represents only a single and specific amino acid.
3. The code is universal in nature and, as far as is known, is the same for animals, plants, and microorganisms.
4. Within a given message the codons are read end to end in a nonoverlapping fashion. In some organisms, however, biological conservation has resulted in the reading of a different message overlapping the first. In such cases the precise reading frame determines which message is read.
5. Not all of the 64 possible triplets are codons for amino acids. Three—UAA, UAG, and UGA—serve as terminating signals for the message; they do not code for amino acids. Two codons—AUG and GUG—serve as signals for the initiation of a message in addition to acting as codons for specific amino acids. Therefore these five codons act as punctuation signals.
6. The direction of reading the messenger RNA is always in the 5′ to 3′ direction. Also, peptide synthesis occurs from the amino to the carboxy terminal.

Direction in which mRNA is read

$\longrightarrow$

mRNA	5′	p-G-U-G-G-C-U - - - - - - - - - - -	U-U-C	3′
Polypeptide		H_2N-Met-Ala - - - - - - - - - - - - -	Phe-CO_2H	

$\longrightarrow$

Direction in which polypeptide is synthesized

7. For the sake of nomenclature the triplets of bases are always written and read in the 5′ to 3′ direction.

Translation of the genetic code

The fact that the codon is a sequence of three bases in mRNA does not describe how they "recognize" the individual amino acids. The principle of codon recognition is independent of amino acid structure; rather, it is dependent entirely on base complementarity between the codon and a corresponding antiparallel sequence called the *anticodon*, which is found in Loop II (anticodon loop) of all tRNAs (Fig. 8-7, *B*). Thus the anticodon IGC in phenylalanyl-tRNA forms a base pair with GCC, GCA, or GCU (see following wobble hypothesis) in mRNA, thereby serving to translate these codons into the amino acid alanine. (It is important to remark that by convention of nomenclature all sequences are read in the 5′ to 3′ direction whether for codons or anticodons.)

Wobble hypothesis

Although there are 61 codons for the 20 amino acids, only 40 different tRNAs with as many anticodon sequences appear to exist. According to Crick, there is a degree of biological conservatism whereby a given tRNA may recognize more than one codon. The major limitation to this apparent flexibility is that the first two bases in the codon have an absolute requirement for their standard complementary bases. As seen at the top of p. 199, the third position exhibits a degree of flexibility allowing the formation of nonstandard base pairs. It is thought that some movement or "wobble" occurs at this position, allowing the formation of hydrogen bonds between

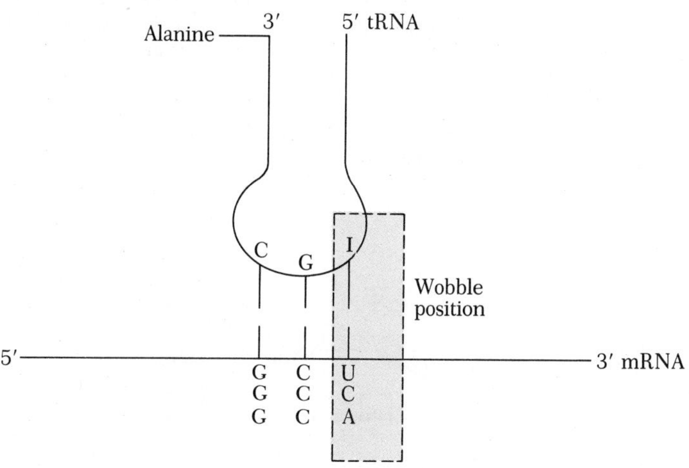

bases other than the standard ones; these deviations are listed in Table 10-3. For example, the anticodon for alanine, IGC, includes the unusual nucleoside inosine (base is hypoxanthine). As seen below, inosine may hydrogen-bond with adenosine, cytidine, and uridine. Therefore, as noted in the preceding diagram, a single alanyl-tRNA having the anticodon IGC can recognize three codons—GCA, GCC, and GCU—all for alanine. Wobble, however, does not confer ambiguity to the code; all the nonstandard base pairs formed involve recognition of codons for the proper amino acid, such as alanine in the example cited.

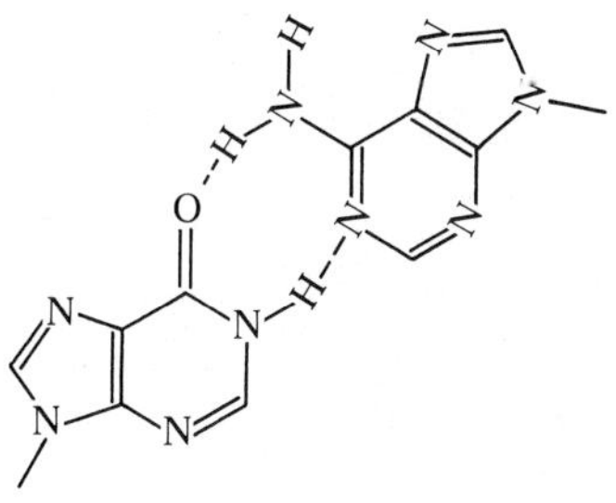

Inosine-adenosine 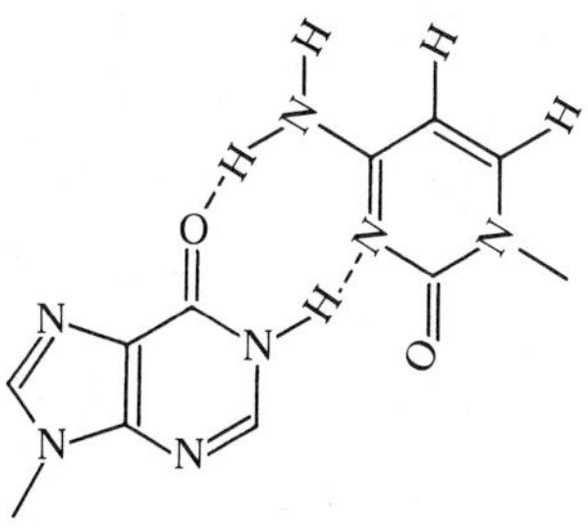 **Inosine-cytidine** 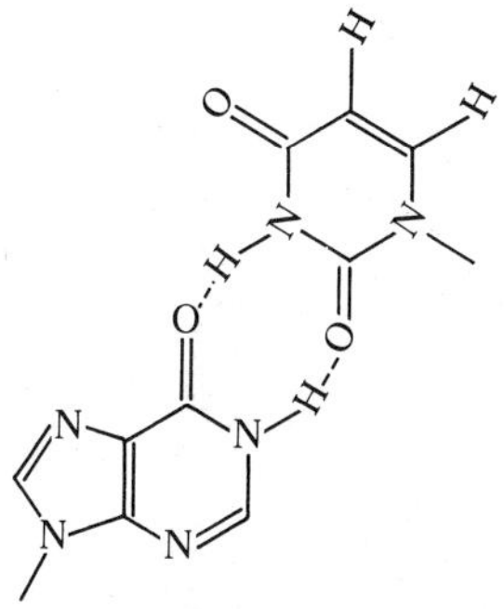 **Inosine-uridine**

Aminoacyl synthetases

The actual process of combining an amino acid with its proper tRNA is the function of the appropriate *aminoacyl tRNA synthetase*. There are 20 individual enzymes corresponding to each amino acid and its appropriate tRNA(s). Some of these enzymes are known to have single polypeptide chains (valine, leucine, and isoleucine), whereas others exhibit a quaternary structure and are aggregates of multiple subunits (glycine, serine, methionine, and tryptophan). The aminoacyl

Table 10-3
Nonstandard base pairs formed by wobble between codon and anticodon

Base in anticodon	Base in codon
G	U or C
C	G
A	U
U	A or G
I	A, U, or C

Figure 10-1

Scheme showing amino acid activation and the biosynthesis of aminoacyl tRNA by aminoacyl synthetase. The enzyme is visualized here as having three binding sites. All bind ATP (site *II*). Site *I* is presumed to recognize the specific amino acid by its *R* group (side chain). Site *III* is considered to be specific for the tRNA, which is unique for the amino acid bound. The synthetase acts as the cell's dictionary because it defines which amino acid will be combined to a particular tRNA.

Modified from Arnstein, H.R.V.: Br. Med. Bull. **21**:217, 1965.

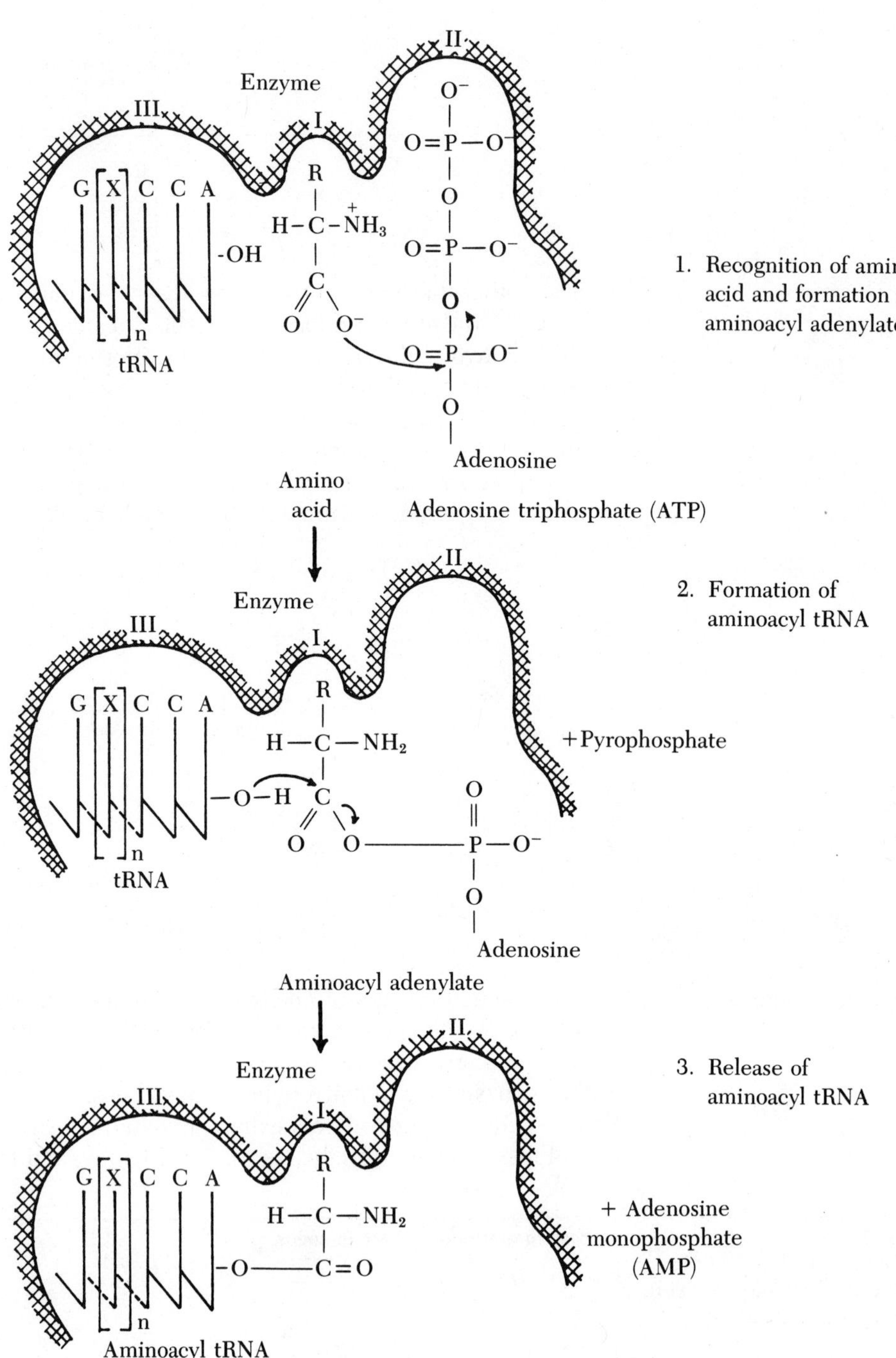

synthetases can be visualized as having three binding sites, one each for the amino acid, tRNA, and ATP (Fig. 10-1). Probably each enzyme must be able to recognize all the tRNAs for a given amino acid.

As seen below, the combination of an amino acid with its specific tRNA is a two-step process.

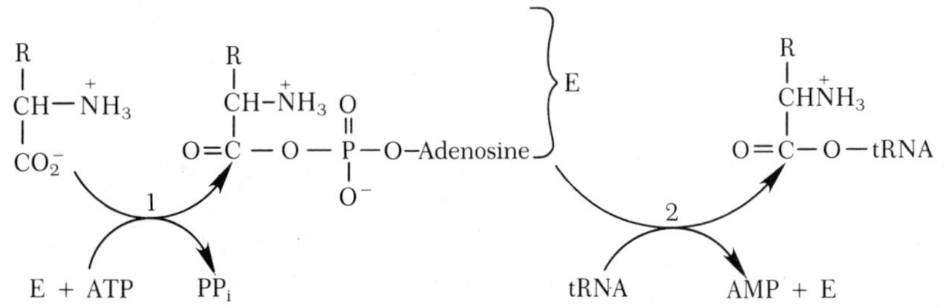

$$\begin{array}{c} R \\ | \\ CH-\overset{+}{N}H_3 \\ | \\ CO_2^- \end{array} \xrightarrow[E + ATP \;\to\; PP_i]{1} \underbrace{\begin{array}{c} R \\ | \\ CH-\overset{+}{N}H_3 \quad O \\ | \qquad\quad \| \\ O=C-O-P-O\text{–Adenosine} \\ | \\ O^- \end{array}}_{E} \xrightarrow[tRNA \;\to\; AMP + E]{2} \begin{array}{c} R \\ | \\ CH\overset{+}{N}H_3 \\ | \\ O=C-O-tRNA \end{array}$$

In the first step the amino acid is activated by reacting with ATP to yield an aminoacyl adenylate. This involves the formation of a high-energy mixed anhydride bond between the carboxyl group of the amino acid and the phosphate of adenylic acid (AMP); pyrophosphate is eliminated. Apparently the adenylate remains attached to the synthetase for the second reaction, when the activated amino acid is transferred to the specific tRNA. As seen below, the amino acid is covalently linked to the 3′-hydroxyl group (either or both the 2′- and 3′-hydroxyl may be involved) of the terminal adenosine moiety of the tRNA by way of an ester bond.

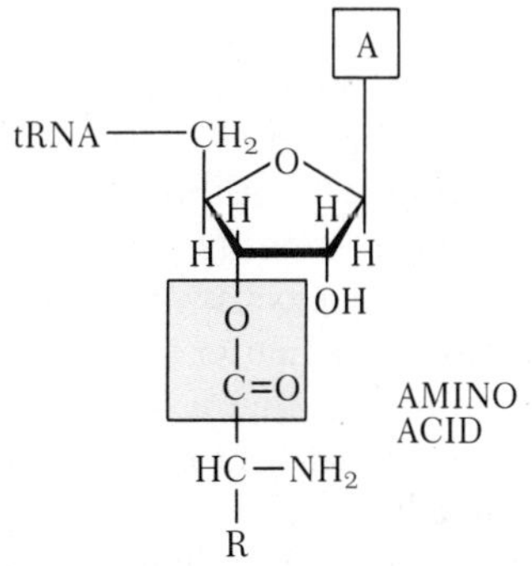

tRNA—CH_2 (ribose ring with O, H, H, H, H; A; OH)

O
|
C=O
|
HC—NH_2
|
R

AMINO ACID

The nature of the overall activation process and the role of the aminoacyl tRNA synthetases are illustrated in Fig. 10-1.

The fact that the amino acid structure itself is not involved in decoding the message has been clearly demonstrated. To prove this point, cysteine linked to its specific tRNA was treated with hydrogen in the presence of a nickel catalyst. In this way the sulfhydryl group of the cysteine was replaced with hydrogen, thereby converting it to alanine. The alanine, however, was still attached to the cysteine-specific tRNA.

$$\text{Cys-tRNA}^{\text{Cys}} \xrightarrow[\text{Ni}]{\text{H}} \text{Ala-tRNA}^{\text{Cys}}$$

The Ala-tRNA^{Cys} was then used in an amino acid–incorporating system including a polynucleotide containing U and G. The poly UG (UGUGUGUGU) supplied codons (UGU) that are necessary for the incorporation of cysteine into peptide chains. Then, however, alanine was included in the polypeptide product instead of cysteine. Yet when Ala-tRNA^{Ala} was used, i.e., alanine coupled to its specific tRNA, alanine was not incorporated by poly UG. This experiment demonstrated that the

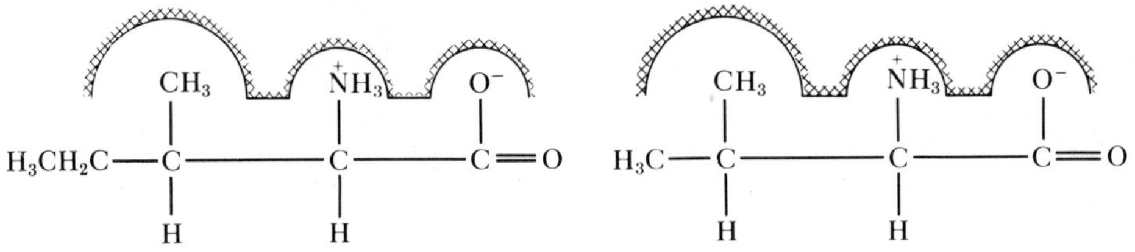

Figure 10-2
Potential error in the recognition of valine by isoleucine-specific aminoacyl synthetase.

translation of genetic messages depends on the specific tRNA and not on the particular structure of the amino acid.

In general, the specificity of each aminoacyl tRNA synthetase is such as to preclude errors. There is the possibility that error in recognition might occur in the particular case of valine by the isoleucine-specific enzyme and isoleucine by the valine enzyme (Fig. 10-2). The erroneous formation of Val-$tRNA^{Ile}$ and the Ile-$tRNA^{Val}$ is prevented by the spontaneous hydrolysis of the hybrid complex.

The aminoacyl tRNA molecules translate the genetic information provided by the mRNA into the appropriate amino acid sequence. The key to this process is the exact joining of the amino acid with its proper tRNA. Since this is the function of the aminoacyl tRNA synthetases, these enzymes act as the "dictionary" of the cell. The site at which this codon-anticodon recognition occurs and the site of peptide bond formation are provided by the ribosomes.

Protein biosynthesis: primary structure

Like transcription, the translation of genetic information into the primary structure of proteins can be considered in the three stages: initiation, elongation, and termination. Although much of the pertinent experimentation has involved prokaryotic systems *(E. coli)*, the principles that have been developed can be applied to eukaryotic cells with appropriate modifications. The major difference between the two is the matter of compartmentalization. In prokaryotes protein synthesis begins on mRNA that is still associated with the chromosome and even before the transcriptional process is complete. In eukaryotic cells transcription and translation are totally independent of one another and occur in separate compartments, the nucleus versus the cytoplasm.

Chain initiation

The initiation of protein synthesis in a prokaryotic system involves first the stepwise aggregation of a 30S ribosomal subunit with *N*-formylmethionyl tRNA, GTP, certain initiation factors (IF-1, IF-2, IF-3), and a strand of mRNA. The source of the 30S subunit as well as the 50S is a cytoplasmic pool. These subunits eventually combine to form a 70S ribosome (see Fig. 8-12).

The first amino acid (amino terminal) of all proteins produced in prokaryotic cells is *N*-formylmethionine (see further). The fMet requires first the combination of methionine itself to a special initiating tRNA. In all cells there are two species of tRNA, both specific for the amino acid methionine. These are designated $tRNA_m^{Met}$ and $tRNA_f^{Met}$. The aminoacyl tRNA synthetase for methionine apparently has a binding site that accepts either of the two species. Once the $tRNA_f^{Met}$ is charged with methionine (methionyl $tRNA_f^{Met}$), it serves in prokaryotes as substrate for a *transformylase*, which catalyzes the transfer of a formyl group from N^{10}-formyl tetrahydrofolate (Chapter 6) to the amino group of the methionine. Formylation occurs only after the addition of methionine to $tRNA_f^{Met}$; preformed fMet will not combine with the tRNA. The primary structure of $tRNA_f^{Met}$ is somewhat different

Figure 10-3
Stepwise assembly of the ribosomal initiation complex. Probable role of the various initiating factors is shown.

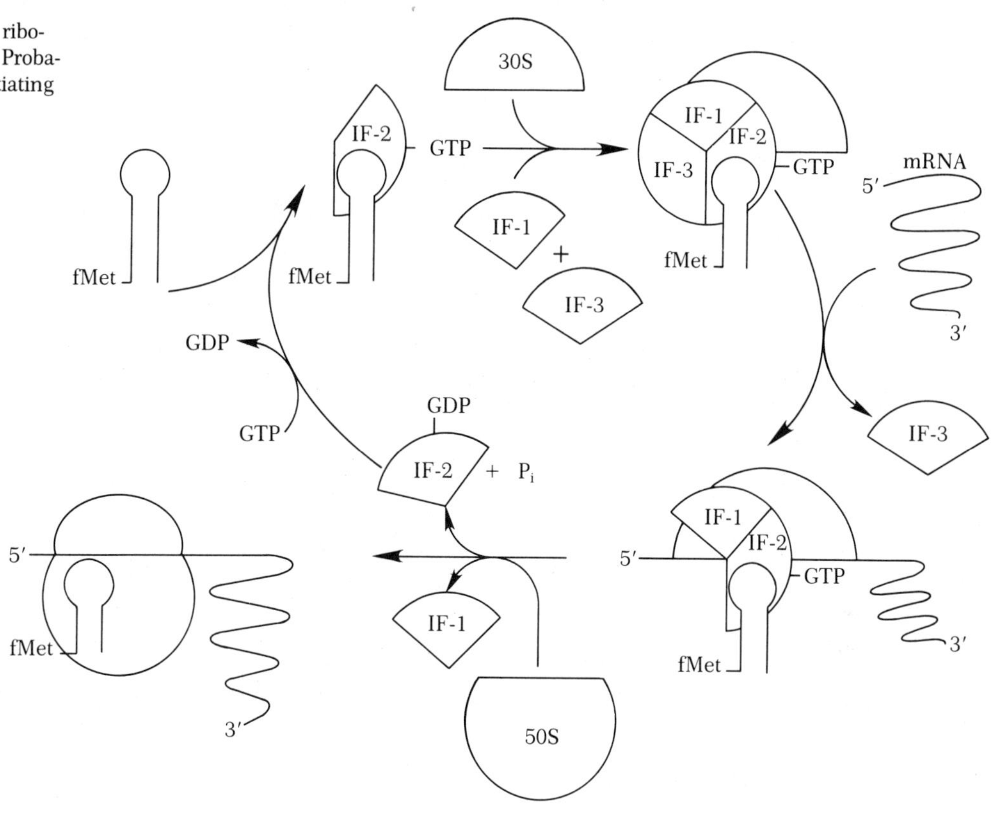

from that of the second species, $tRNA_m^{Met}$; it does not have the T-ψ-C sequence in loop IV. Therefore, unlike all other tRNAs, fMet-$tRNA_f^{Met}$ cannot bind to the 50S subunit at the site where the 5S rRNA surfaces. Rather, this special initiating fMet-$tRNA_f^{Met}$ binds only to the 30S subunit and always serves in prokaryotic cells as the initiating amino acid. On the other hand, Met-$tRNA_m^{Met}$ is not formylated; it binds to the 50S subunit and therefore serves to insert methionine into the growing polypeptide chain.

$$\begin{array}{l} CH_3 \\ | \\ S \\ | \\ CH_2 \\ | \\ CH_2 \qquad\quad O \\ | \qquad\qquad // \\ HC-N-C \\ | \quad\;\; H \quad\;\; \backslash \\ | \qquad\qquad\;\; H \\ CO_2H \end{array}$$

N-**Formylmethionine (fMet)**

Initiation is signaled by the appearance of the codon AUG (or GUG) either at or adjacent to the 5′ end of the mRNA or at the beginning of a cistron, i.e., a unit of information translatable into a polypeptide chain. These codons dictate the presence of the special initiating fMet-$tRNA_f^{Met}$, the initiating factor IF-2, and GTP. The 30S subunit combines with the fMet-$tRNA_f^{Met}$:IF-2:GTP complex as well as with IF-1 and IF-3 (Fig. 10-3). Apparently IF-3 prevents a premature formation of the 70S ribosomal particle prior to the addition of the mRNA. The proper joining of the mRNA to the complex requires IF-1. Following the subsequent release of IF-3 the 50S subunit is added. This formation of the 70S ribosome is accompanied by the hydrolysis of GTP and the dissociation of IF-1. The IF-2:GDP complex dissociates and is recycled for the formation of another initiating complex.

The binding of the mRNA occurs in such a way as to bring the initiating codon (AUG) in direct apposition to the anticodon (CAU) of the fMet-$tRNA_f^{Met}$. The proper registry of mRNA with fMet-$tRNA_f^{Met}$ and therefore the recognition of the beginning of the cistron may be the consequence of a special leading base sequence in the mRNA. In the case of the *lac* gene of *E. coli* the cistron for β-galactosidase is preceded by a sequence of 63 untranslated bases. The cistron begins with AUG, bases 39 to 41. The second codon is ACC and signals the insertion of threonine. Therefore the correct reading frame is established by the AUG following the untranslated leader sequence.

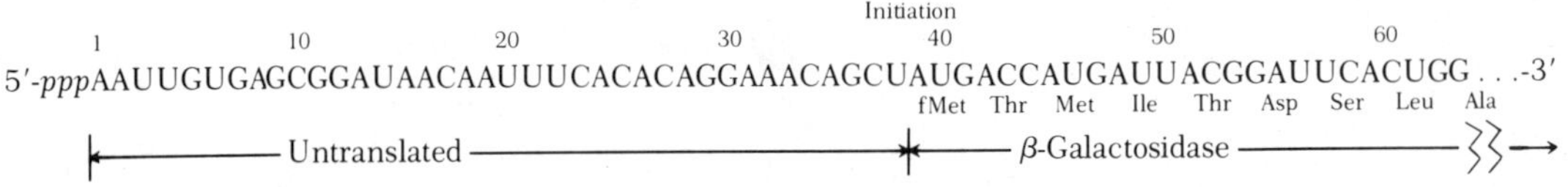

Perhaps the first 21 bases of the untranslated sequence form a double-helical loop, which then serves as a recognition site in the joining of the mRNA to the 30S subunit.

Elongation

Fig. 10-4 illustrates the completed 70S ribosomal mRNA complex. It shows that ribosomes have two binding sites: one is the *A*, or aminoacyl tRNA binding site, and the other is the *P* site, where the growing peptide chain bound to tRNA is attached. The direction of translation of the mRNA is in the 5′ to 3′ direction. As in the case of replication and transcription, the anticodon sequence of the tRNAs is arranged antiparallel to the appropriate codon in the mRNA. The anticodon indicated in Fig. 10-4 for Ala-tRNA is IGC. As described previously, inosine frequently appears in anticodons and, according to the wobble hypothesis, can form a base pair with A, U, and C.

When initiation is complete, the fMet-$tRNA_f^{Met}$ is located in the P site and the second codon in mRNA has moved opposite site A. This codon then determines the identity of the incoming aminoacyl tRNA by the principle of base complementarity. Therefore A may be viewed as the *recognition* or *decoding* site. Binding of the new aminoacyl tRNA requires an additional factor called the elongation factor, or EF-T.

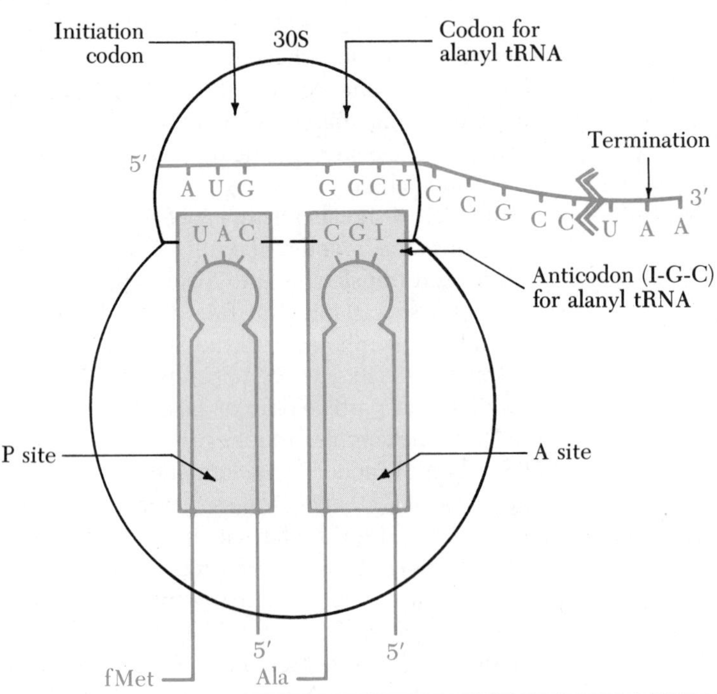

Figure 10-4
Scheme of completed ribosome (70S) showing the two subunits, 50S and 30S. Messenger RNA is represented by mRNA and is found associated only with the 30S subunit. Two binding sites are indicated, the A, or aminoacyl-binding, and the P, or peptidyl-binding, sites.

There are actually two components of EF-T: one is termed *Tu* (for temperature unstable) and the other is *Ts* (for temperature stable). GTP is needed and is cleaved to GDP and P_i during binding. A cyclical process involving the formation of a Tu-GTP aminoacyl tRNA complex has been suggested and is illustrated in Fig. 10-6.

Peptidyl transferase, a constituent of the 50S subunit, catalyzes the formation of a peptide bond between the carboxyl of fMet (P site) and the amino group of the aminoacyl tRNA at the A site (Fig. 10-5). Peptide bond formation occurs by nucleophilic attack of the amino nitrogen of the aminoacyl tRNA on the ester carbonyl of fMet. The energy required is derived from the ester linkage of the aminoacyl tRNA.

Once the peptide bond has been formed, a *translocase,* or the EF-G factor, shifts both the peptidyl tRNA and its codon to the P site. This also means that the mRNA has been translocated with respect to the ribosome, and the third codon appears opposite the A or decoding site. The system then can select and bind the third aminoacyl tRNA; again, GTP is hydrolyzed. During the translocation process the tRNA freed or unloaded of its amino acid (fMet) is released from the ribosome and

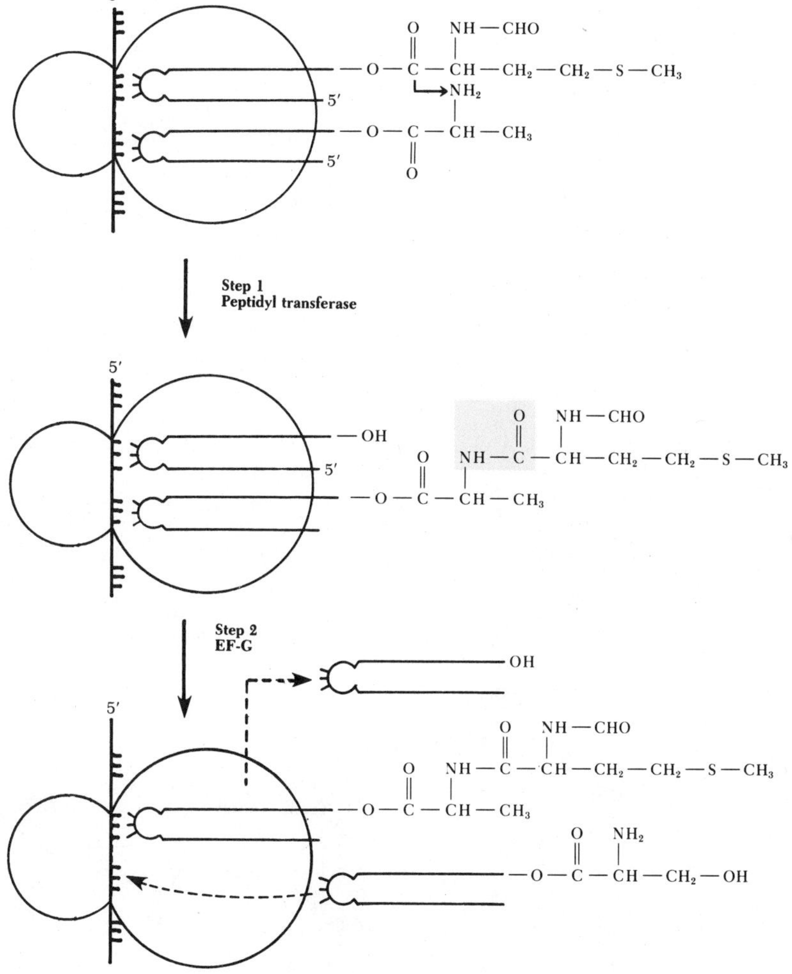

Figure 10-5
Stepwise representation of peptide bond formation. *Step 1:* Peptidyl transferase catalyzes transfer of an amino acid from P site (fMet) to free α-amino group of amino acid (Ala) on A site. Peptide bond formed is shaded in color. Newly formed dipeptide attached to A site via carboxyl group to tRNA. *Step 2:* Translocase (*EF-G*) catalyzes transfer of the peptidyl tRNA from the A to the P sites; GTP is hydrolyzed. At the same time the "unloaded" tRNA is discharged. The A site is then charged with a new aminoacyl tRNA unit.

returns to the cytosol where it may once again engage in amino acid activation. The sequence of events illustrated in Fig. 10-5 is repeated, and the peptide is thereby extended in a stepwise fashion. The growing polypeptide always remains attached to the ribosome by the most recently introduced aminoacyl tRNA.

In general, the growing peptide chain on the P site always contributes its carboxyl to the newly formed peptide bond, whereas the nitrogen is from the free amino group of the most recently bound aminoacyl tRNA.

Chain termination

Ultimately the ribosome reaches the end of the message or cistron. At this point a terminating codon, e.g., UAA, UAG, UGA (Table 10-2), signals the end of the translational process. When such codons arrive at the A site, the ribosome will not bind another aminoacyl tRNA. Instead it accepts a terminating protein (R_1 and R_2), which activates the peptidyl transferase, thereby releasing the polypeptide chain and the ribosomal subunits. The latter are restored to the free subunit pool for reuse.

The various steps and factors involved in the overall synthesis of protein (*E. coli*) are summarized in Fig. 10-6.

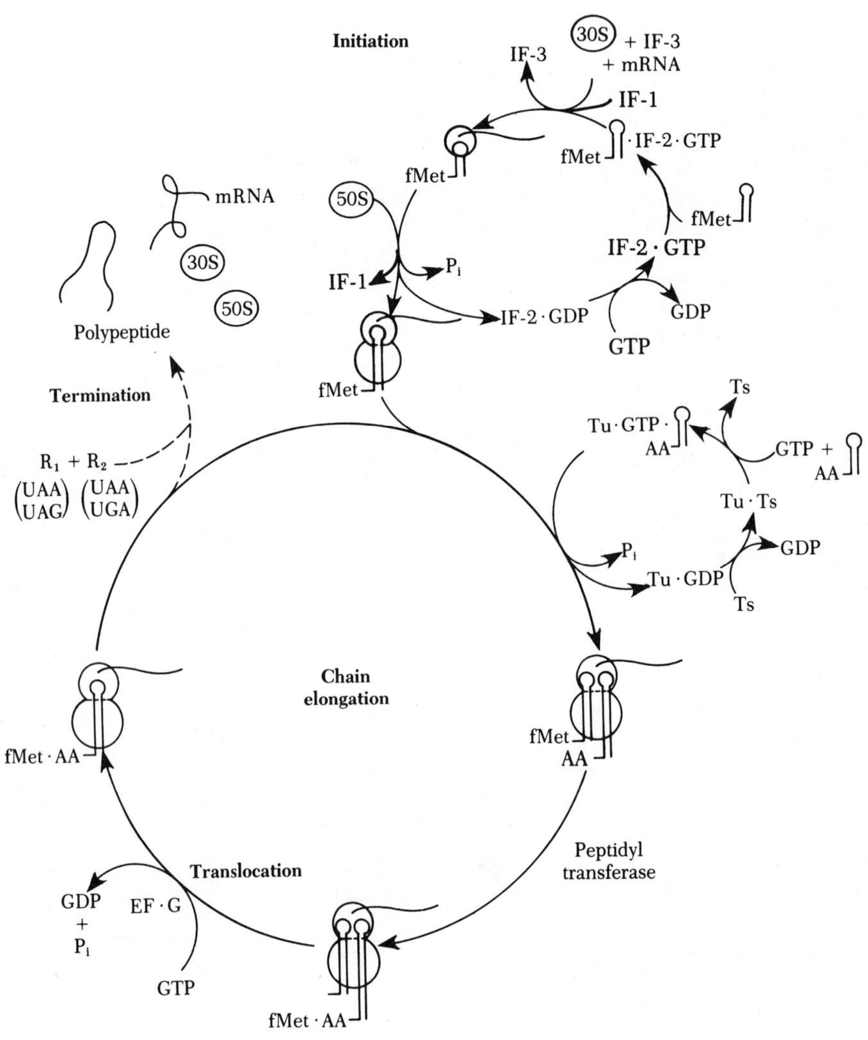

Figure 10-6
Summary of steps involved in processes of protein biosynthesis in *E. coli*. Illustrated are requirements for chain initiation, elongation, and termination.

Although it is clear from the preceding description that polypeptides always begin with *N*-formylmethionine in prokaryotic cells, it is equally clear that not all completed *E. coli* proteins have this amino acid for an N-terminal residue. Therefore at some point during or shortly after peptide synthesis the formyl group may be eliminated by a specific *deformylase*. Furthermore, of the soluble proteins only 45% terminate in methionine, 30% in alanine, and 15% in serine. Therefore the nascent polypeptides in *E. coli* must be further modified by the action of one or more aminopeptidases.

Polysomes

As the peptide chain grows, the first ribosome shifts along the mRNA sufficiently to allow a new initiation to occur and a second polypeptide to be synthesized. This process continues with the acceptance of new ribosomal subunits so that protein synthesis may be viewed in terms of multiple ribosomes, i.e., polyribosomes or simply *polysomes* (Fig. 10-7). As the individual ribosomes advance over the mRNA, they will bear peptide chains of increasing lengths. When a polypeptide is complete, it and the ribosomal subunits dissociate. Fig. 10-7 suggests a cyclical process in which the mRNA is continually translated. Initiation removes ribosomal subunits from the cytoplasmic ribosomal pool; termination restores them.

Protein synthesis in eukaryotic cells

The processes of initiation, elongation, and termination are also applicable to eukaryotic cells. The various factors required in protein synthesis by prokaryotic and eukaryotic cells are compared in Table 10-4. Certain differences, however, do exist not only in the factors but also in the protein synthesis process itself. Eukaryotic cells contain the same species of methionine-specific tRNAs as do the prokaryotic cells. The $tRNA_f^{Met}$ differs from the $tRNA_m^{Met}$ in the absence of the T-ψ-C sequence from loop IV; furthermore, it does not bind to the 60S subunit. However, eukaryotic cells do not possess the transformylase required for the formation of fMet-$tRNA_f^{Met}$. Therefore initiation of all polypeptide chains begins with methionine instead of its formyl derivative.

Another difference is seen in the properties of the translocase. The eukaryotic translocase EF-2 reacts with NAD^+ in the presence of diphtheria toxin. As seen below, the ADP ribosylated EF-2 (an ADP-ribose moiety attached to EF-2) is inac-

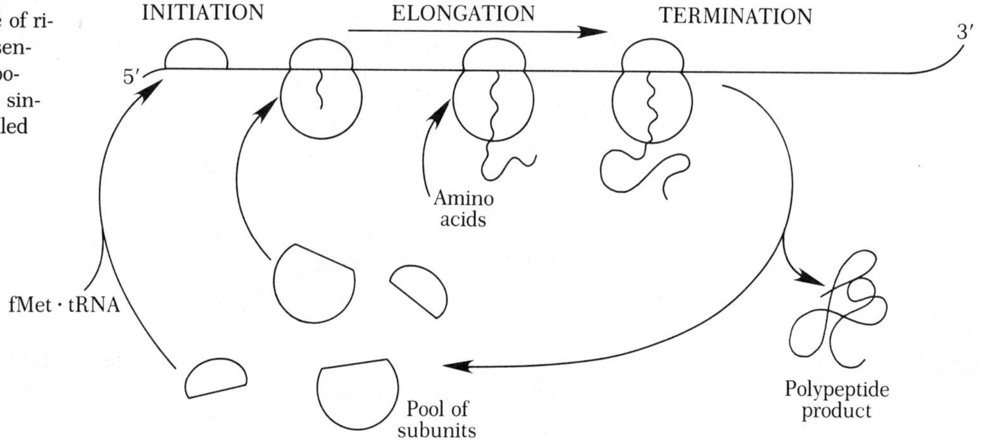

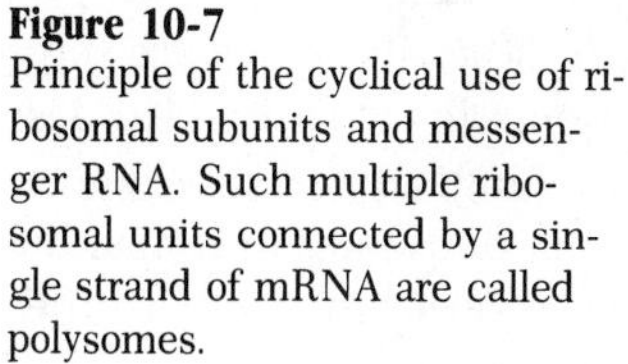
Figure 10-7
Principle of the cyclical use of ribosomal subunits and messenger RNA. Such multiple ribosomal units connected by a single strand of mRNA are called polysomes.

Table 10-4
Comparison of factors required for protein synthesis in prokaryotic and eukaryotic cells

Stages in synthesis	Prokaryotes	Eukaryotes
Initiation	fMet on $tRNA_f^{Met}$	Met on $tRNA_f^{Met}$; cells do not possess transformylase
	AUG and GUG initiation codons IF-1, IF-2, IF-3	AUG and GUG initiation codons eIF-1, eIF-2, eIF-3
	GTP	GTP
Elongation	EF-Tu, EF-Ts	EF-1 (also transferase I)
	GTP	GTP
Peptide bond	Peptidyl transferase in 50S	Peptidyl transferase in 60S
Translocation	EF-G	EF-2 (also transferase II)
	GTP	GTP
Termination	R_1 and R_2 (no GTP)	Releasing factors? Requires GTP

tive. Therefore diphtheria toxin acts by inhibiting the translocation step in protein synthesis of eukaryotic cells only.

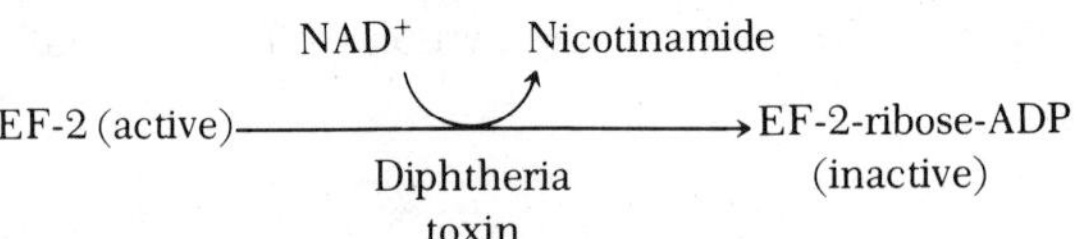

In eukaryotic cells protein synthesis is restricted to the cytoplasm. The mRNA formed in the nucleus as hn-RNA is extensively modified before traversing the perinuclear membrane. As already described, eukaryotic mRNA is distinguished by the presence of a 7-methyl guanosine triphosphate cap at the 5′ terminal and a poly A tail at the 3′ end. Although the purposes for these unusual features are not yet ascertained, it is believed that the length of the poly A tail determines the life span of the mRNA. The cap may be required for the proper binding of the mRNA by the ribosomes.

In the cytoplasm the mRNA is translated by the accession of ribosomal particles. The existence of multiribosomal systems or polysomes is well established. Ultracentrifugation on sucrose gradients followed by electron microscopy has made it possible to visualize polysomes from many sources (Fig. 10-8). A typical example are the polysomes involved in the biosynthesis of the globin molecules of hemoglobin. The α- and β-chains possess 141 and 146 amino acid residues, respectively. This means that their mRNAs must possess minimally 423 nucleotides to supply the necessary codons. Furthermore, it has been found that the ribosomes are separated by about 80 nucleotides. This means that a polysome for globin molecules could accommodate four to six ribosomes; actually the average appears to be five (pentamer). Electron microscopy shows that the individual ribosomes are separated by a thin strand of mRNA (Fig. 10-8, *B*). The polysome varies in size with the protein molecule produced; thus myosin, a protein of large molecular weight, requires an aggregate of 100 or more ribosomes. The number of ribosomes per polysome also may be variable and serve as a level of translational control. When protein synthesis is stimulated, a larger number of ribosomes would make the translation of a single strand of mRNA more efficient.

Rough endoplasmic reticulum

Polysomes appear in the cytoplasmic compartment either in the free state, i.e., in the cytosol, or bound to the outer surface of the endoplasmic reticulum (Chapter 2). It is thought that those proteins which are produced for intracellular purposes are made by free polysomes, whereas proteins for export are synthesized by the membrane-bound polysomes. The endoplasmic reticulum and the Golgi apparatus function in the chemical modification of newly synthesized proteins and for their transport or secretion outside the cell.

It now appears that proteins destined for secretion are synthesized with an initial segment of 15 to 30 largely hydrophobic amino acids at its N-terminal. This "leader" sequence is believed to serve as a signal which determines that the protein and its associated ribosome are to be membrane bound. The largely hydrophobic sequence threads its way through the lipid bilayer (Fig. 10-9); the 60S ribosomal

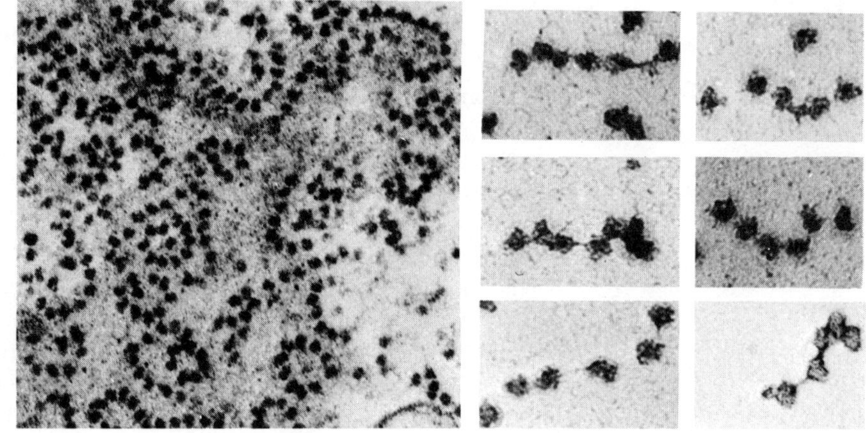

Figure 10-8
A Electron micrograph showing polysomes in root cells of radish (64,000 ×).
B Representative polysomes from rabbit reticulocytes showing the connecting strands of RNA (100,000 ×).

A from Bonnett, H.R., and Newcomb, E.H.: J. Cell Biol. **27**:423, 1965; **B** from Slayter, H.S., Warner, J.R., Rich, A., and Hall, C.E.: J. Molec. Biol. **7**:652, 1963.

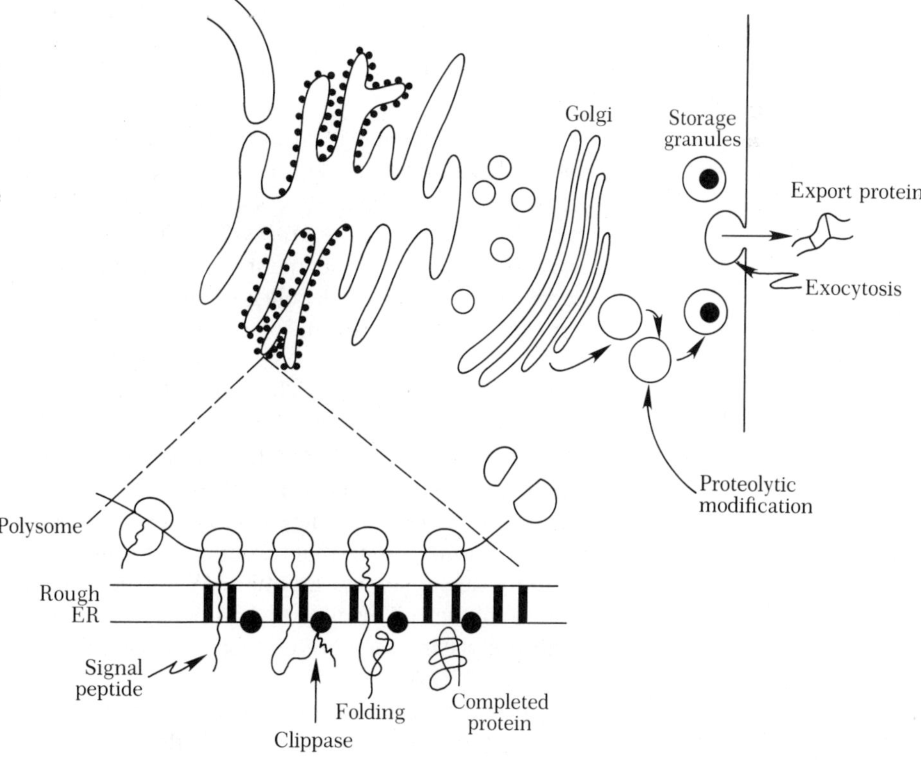

Figure 10-9
Role played by the endoplasmic reticulum with membrane-bound polysomes in protein synthesis. Signal peptide is thought to lead the growing polypeptide chain into and through the lipid bilayer. Signal peptidase (clippase) removes the signal peptide once inside the cisternal cavity of the endoplasmic reticulum. Further covalent modification occurs in smooth endoplasmic reticulum and in the storage vesicles or granules. Complete protein leaves cell by exocytosis.

subunit is bound to the membrane by a special binding site. The leader peptide forms a channel through the lipid, perhaps with the aid of two proteins, ribophorin I and II. As the peptide continues to elongate, the leader chain is removed inside the cisterna of the endoplasmic reticulum by the action of an enzyme appropriately called *clippase* or, signal peptidase (Fig. 10-9). The main polypeptide chain enters the lumen and assumes its proper secondary and tertiary conformation.

Folding of polypeptide

Once the primary structure has been assembled, the polypeptide achieves a three-dimensional structure that in its normal environment (such as pH and salt concentration) is at its lowest free energy. For example, it is known that the renaturation of experimentally denatured myoglobin takes only a few seconds or less. Folding, following biosynthesis, must be rapid and the consequence of a structural "memory," contingent on the genetically determined primary structure. On the basis of this premise, it is believed that the nascent protein reaches its native conformation either during chain growth or immediately thereafter.

Covalent modification

After assembly the folded shape may be stabilized by the formation of disulfide cross-linkages, electrostatic and hydrogen bonds. A protein such as ribonuclease not only achieves a proper conformation but also must form the *correct* disulfide bonds. This protein has eight sulfhydryl groups that are oxidized to four disulfide bonds; therefore many isomeric disulfide-bonded forms are possible. Studies using experimentally reduced pancreatic ribonuclease have shown that oxidation to the enzymic form is slow and not at all compatible with the in vivo synthesis of active ribonuclease. Efforts to resolve this problem led to the discovery of a rearranging enzyme in the endoplasmic reticulum. This enzyme, which requires the tripeptide glutathione (Chapter 13) for its activity, acts as an unscrambling system for the rapid formation of the single active form of ribonuclease.

Other covalent modifications that frequently occur include limited proteolysis and the addition of a variety of prosthetic groups such as carbohydrate side chains. The completed protein is finally transported from the cell by the process of exocytosis.

A valuable example illustrating these steps in protein synthesis is provided by the biosynthesis of insulin. This hormone (Chapter 18) is produced exclusively by pancreatic β-cells in the islets of Langerhans. Insulin is synthesized as a single polypeptide chain, known as *proinsulin* (Fig. 10-9). The actual nascent chain, however, includes a leader segment of 23 additional amino acids attached to the N-terminal. The entire molecule is called *preproinsulin*. As the growing polypeptide is inserted into the cisterna of the endoplasmic reticulum, the signal peptide is removed (signal peptidase) and the proinsulin folds spontaneously into its proper three-dimensional shape. This conformation is stabilized by the formation of three disulfide bonds. Proinsulin is transported through the endoplasmic reticulum into the Golgi apparatus, where it is packaged in membranes by a pinching-off process. Further modification occurs in these vesicles; a trypsin-like protease removes the connecting C-peptide, thereby producing the final active hormone. Insulin aggregates with Zn^{++} to form a hexameric chelate that crystallizes within the vesicles, forming dense storage granules. When needed, the hormone is secreted to the outside of the cell by exocytosis, a process in which the vesicular membrane fuses with the plasma membrane, thereby expelling the contents to the outside.

Glycoproteins

One of the most important and general modification processes in the mammalian system is the attachment of carbohydrate to polypeptide chains to yield glycoproteins. Carbohydrate units are added as the growing chains enter the cisterna of the endoplasmic reticulum. Nearly all secretory proteins are glycoproteins; a notable exception is serum albumin. Much of the addition of carbohydrate occurs in that portion of the endoplasmic reticulum which is devoid of ribosomes, known as the smooth endoplasmic reticulum, as well as in the Golgi apparatus.

Glycoproteins may be divided into two categories, depending primarily on the nature of the linkage between the first sugar residue and the peptide chain. Two distinct kinds of glycosidic linkages are found, those involving oxygen and those using nitrogen as the bridging atom.

N-Acetyl galactosamine **Threonine** **N-Acetyl glucosamine** **Asparagine**

α-O-Glycosidic bond **β-N-Glycosidic bond**

Glycoproteins such as salivary mucin are representative of the first type and involve hydroxy–amino acids and an O-glycosidic bond. Such proteins are glycosylated by a sequential array of enzymes called glycosyl transferases, which catalyze the stepwise addition of carbohydrate units. The sugars themselves must be

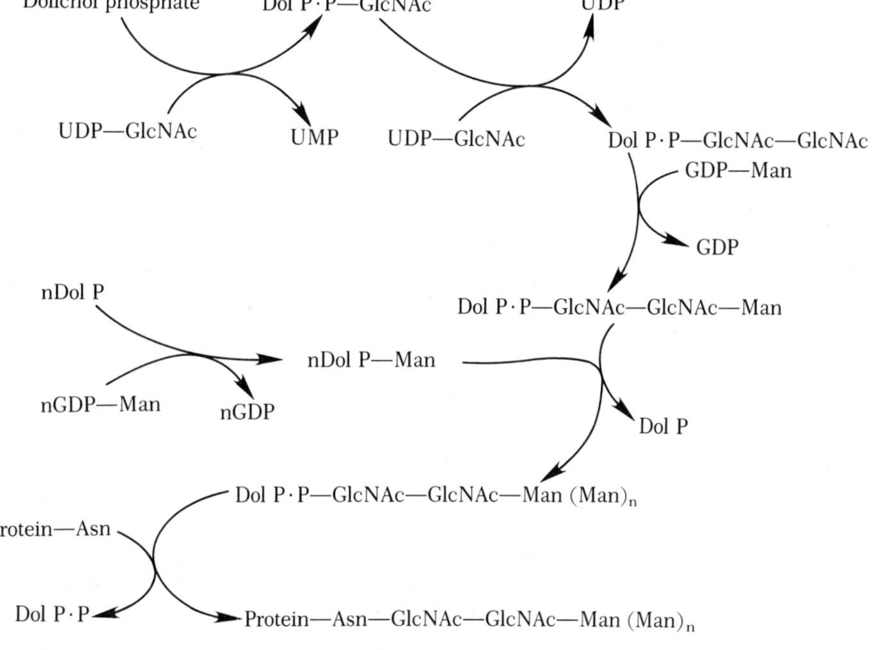

Figure 10-10
Scheme suggesting the sequential assembly of a polysaccharide core on a dolichol phosphate carrier. Each step is thought to be catalyzed by a specific glycosyl transferase. The final addition to the protein backbone involves the formation of an N-glycosidic bond with an asparagine residue.

activated as their nucleoside diphosphates (Chapter 6). Although their sequence is not under direct genetic control, this is determined by the stepwise location of the specific transferases in the membranes of the SER and the Golgi apparatus. The first sugar added is *N*-acetylgalactosamine. Sheep salivary mucin has disaccharide units of *N*-acetylneuraminic acid attached to *N*-acetylgalactosamine. Apparently the polypeptide comes in contact first with *N*-acetylgalactosaminyl transferase (UDP-*N*-acetylgalactosamine) followed by an *N*-acetylneuraminyl transferase (CMP-sialic acid). However, since the biosynthesis of these enzymes is under genetic control, it may be said that the sequence of the carbohydrate in the glycoprotein product is also, albeit indirectly, under genetic control.

In the second category of glycoprotein the first sugar is added to the amino acid asparagine, and the covalent bond formed is an *N*-glycoside (see above). It now appears that these side chains are formed in a quite different fashion from that of the mucin. Instead of a direct accession of sugars, the polysaccharide side chains are preassembled on a lipid carrier molecule. This intermediate carrier is called *dolichol phosphate*. Dolichols are polyisoprene compounds having about 20 isoprene units; the terminal unit is saturated and ends in a hydroxyl group. Dolichol phosphates are amphipathic, i.e., the isoprene end is lipophilic, whereas the phosphate terminal is hydrophilic. A suggested sequence of reactions leading to the assembly of sugars and their final transfer to the polypeptides is indicated in Fig. 10-10.

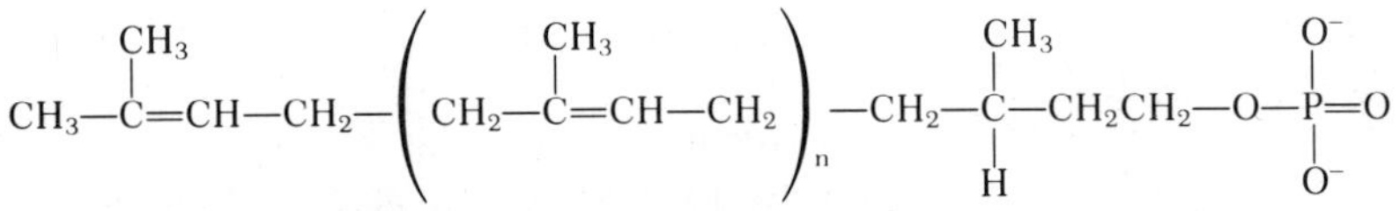

$$CH_3-\overset{\overset{\displaystyle CH_3}{|}}{C}=CH-CH_2-\left(CH_2-\overset{\overset{\displaystyle CH_3}{|}}{C}=CH-CH_2\right)_n-CH_2-\underset{\underset{\displaystyle H}{|}}{\overset{\overset{\displaystyle CH_3}{|}}{C}}-CH_2CH_2-O-\underset{\underset{\displaystyle O^-}{|}}{\overset{\overset{\displaystyle O^-}{|}}{P}}=O$$

Dolichol phosphate

The attachment of carbohydrate chains to protein via the *N*-glycosidic bond occurs in proteins such as ovalbumin, IgG (immunoglobulin), α_1-acid glycoprotein, fetuin, corneal keratin sulfate, ovomucoid, and thyroglobulin. Attachment via the *O*-glycosidic bond to threonine (and serine) where the sugar is *N*-acetylgalactosamine or xylose is found in various mucins, such as from sheep and hog, as well as the glycoproteins of the blood group substances. Xylose is the linking sugar for mucopolysaccharide-protein complexes, i.e., the proteoglycans, and involves specifically the amino acid serine. Another variation of the *O*-glycosidic linkage occurs in collagen. Here galactose is linked to the hydroxyl group of hydroxylysine.

Protein synthesis and the mode of action of certain antibiotics

An understanding of the molecular processes of protein synthesis has paved the way toward a better appreciation of the function of various antibiotics. Many of these serve to inhibit growth by interfering in the process of protein synthesis.

Puromycin

Puromycin is an antibiotic that functions by aborting peptide chain synthesis. As seen from its structure (p. 213), puromycin is a structural analogue for tyrosyl tRNA. It is apparently comparable to the terminal adenosine unit and therefore binds to the A site provided that translocation has shifted the existing peptidyl tRNA to the P site. The peptidyl transferase catalyzes peptide bond formation between the growing peptide chain and the free amino group of puromycin. The peptidyl-puromycin is nonfunctional and is released from the ribosome, thereby terminating chain growth. In this way all nascent or growing chains are released prematurely from the polysomes. Puromycin, however, is unselective and affects all protein synthesis, bacterial and mammalian alike.

Puromycin

Cycloheximide

Cycloheximide is an antibiotic that binds to the 60S ribosomal subunit of eukaryotic cells. It blocks protein synthesis by inhibiting the peptidyl transferase activity.

Antibiotics acting on bacterial ribosomes

The following antibiotics are specific in their ability to inhibit bacterial protein synthesis while leaving mammalian protein synthesis unimpaired. This selectivity arises from the fact that these compounds bind to 50S and 30S ribosomal subunits but not to their eukaryotic counterparts. This is the basis for their therapeutic value. The action of the various antibiotics listed below is summarized in Fig. 10-11.

1. *Streptomycin* is an aminoglycoside. It binds specifically to the 30S subunit at the A site, thereby causing a cessation of chain elongation by preventing the adequate binding of additional aminoacyl tRNA. Its binding also distorts the A site and may result in the insertion of improper amino acids.

Figure 10-11
Site of action of some drugs that act by interfering in bacterial protein synthesis.

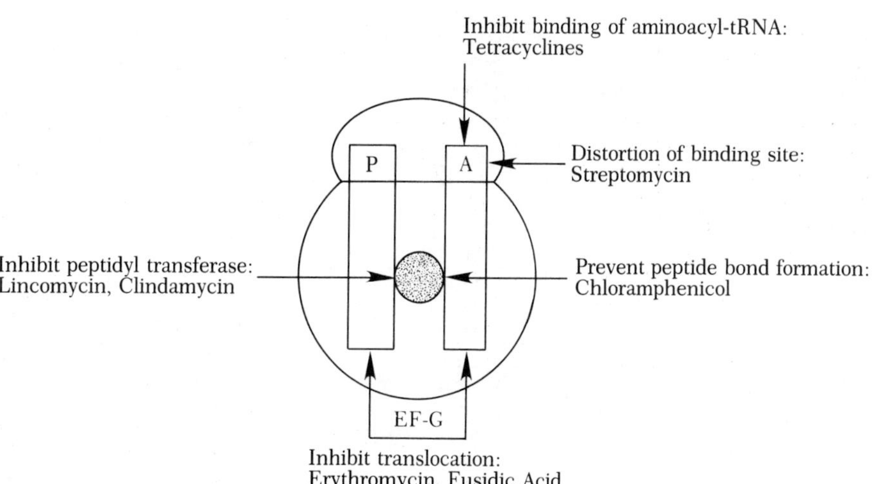

2. *Tetracyclines* are compounds that also bind to the A site, thereby preventing the addition of aminoacyl tRNAs.
3. *Chloramphenicol* binds to the 50S subunit but functions to block the action of the peptidyl transferase. It inhibits, in addition, the attachment of the mRNA-30S complex to the 50S subunit.
4. *Erythromycin* binds to the 50S ribosomal subunit and blocks the aminoacyl translocation process, thereby preventing chain elongation. The same is true for *fusidic acid,* an antibiotic available in Europe but not in the United States.
5. *Lincomycin* and *clindamycin* bind to the 50S subunit and inhibit the peptidyl tansferase, thereby preventing peptide bond formation.

Control of protein synthesis at the genetic level

As described in Chapter 3, the living organism adapts to physiological and metabolic demands by altering the spectrum and quantities of enzymes. It should now be evident that this adaptation involves effects at the level of transcription and translation. Both are subject to internal, local metabolic changes and external demands signalled by hormones (Chapter 18) as well as by the availability of nutrients (Fig. 10-12). Hormones *(H1, H2)* stimulate a selective production of mRNA either directly as a hormone-receptor protein complex *(HR)* or indirectly via the formation of cAMP. These processes are described in detail in Chapter 18. Certain nutrients *(Nu)* also may elicit a selective biosynthesis of mRNA. The mRNA then enters the cytoplasm and is translated into specific enzymes. The amount of protein synthesized is regulated both by the amount of mRNA produced and by the various translational factors discussed earlier in this chapter.

Selective information retrieval in response to a given signal, whether from a hormone or a nutrient, requires the activation of certain genes or groups of genes so that the RNA polymerase will transcribe only that segment of DNA to produce the desired mRNA. Just how this selective process occurs is not altogether clear. However, a model system was suggested in 1961 by Jacob and Monod for *E. coli.* It provides a perspective, at least, of the principles involved. This model was proposed to explain the ability of *E. coli* to adapt itself to a change in nutrients in the growth medium. Genes in *E. coli* can be turned on or off in response to these nutrients. Normally the microorganisms are grown on glucose to provide the source of carbon atoms. However, they can also use lactose, provided that this disaccharide can first

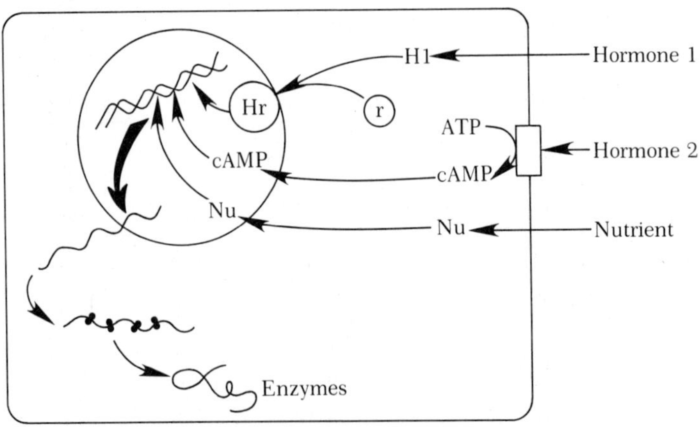

Figure 10-12
Selective production of enzymes under the stimulus of hormones and nutrients. Effect of the signals is to affect the particular kinds of mRNA transcribed as well as the amount produced.

be hydrolyzed to glucose and galactose by β-galactosidase. On the glucose medium *E. coli* organisms ordinarily possess fewer than five molecules of this enzyme per cell and perhaps one or less molecules of its mRNA. However, when glucose is replaced by lactose, then the synthesis of β-galactosidase is stimulated more than a thousandfold. The selective production of a particular protein (enzyme) and its mRNA is called *induction* or *derepression*. The effector causing the induction is termed the *inducer*. Apparently the microorganism possesses the necessary genetic information for β-galactosidase, but it is normally kept in a dormant or *repressed* state. The presence of lactose instead of glucose serves as a signal that brings about the change needed to synthesize the β-galactosidase. This adaptive process illustrates the coarse control of metabolism (Chapter 3).

Regulation by induction

The inductive process observed in the response of *E. coli* to lactose, however, involves more than β-galactosidase. In addition, two other companion proteins are

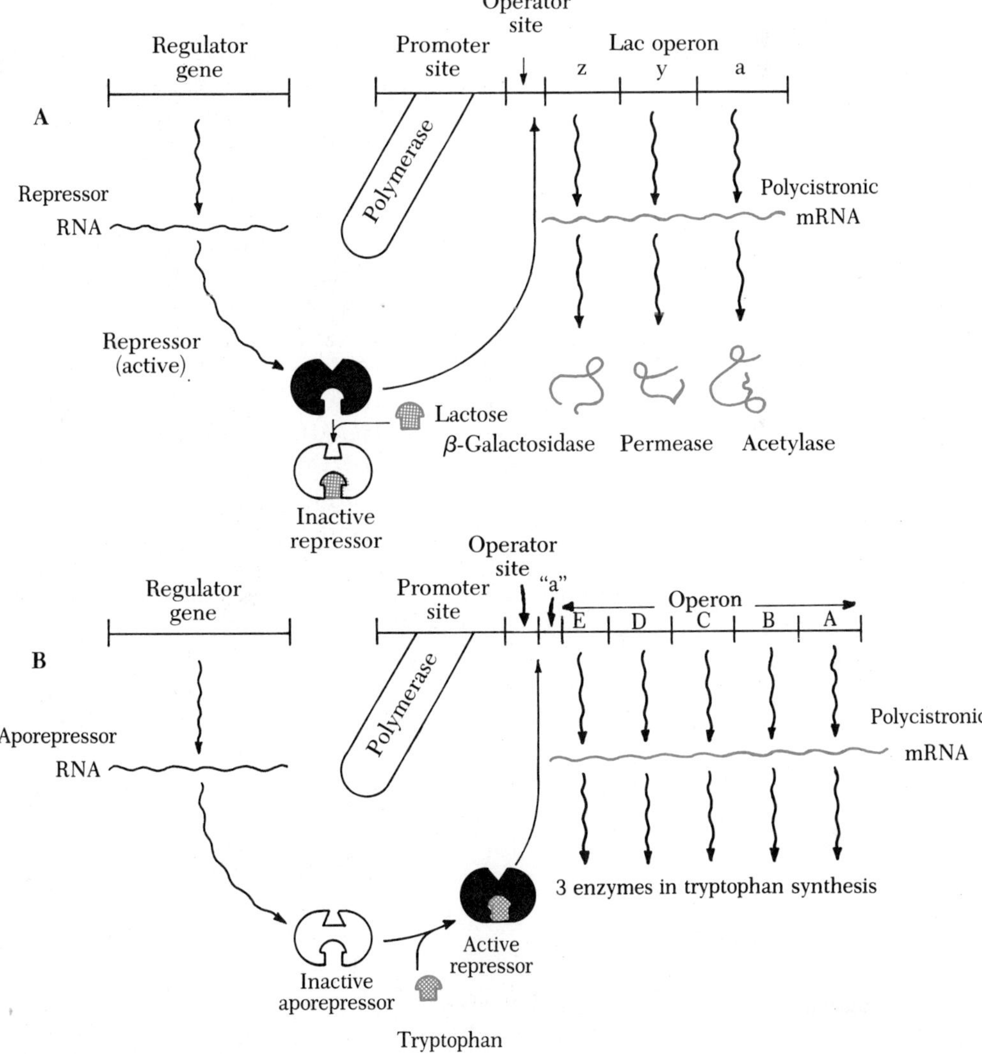

Figure 10-13
Genetic control of protein biosynthesis.

A Induction-derepression. *E. coli lac* operon is normally repressed if bacteria are grown on glucose. Lactose (β-galactoside) as sole source of carbon in growth medium binds to allosteric site on *lac* repressor, yielding an inactive repressor. Consequently the structural genes *z*, *y*, and *a* are transcribed.

B Repression. Tryptophan operon in *E. coli* is normally functional. The repressor is synthesized in an inactive form (aporepressor). The effector, tryptophan, binds to aporepressor and converts it into an active repressor, which then binds to operator gene, thereby repressing the operon and preventing transcription. *"a"* represents the attenuator, a site which results in the termination of the RNA chain biosynthesis when tryptophan accumulates.

produced: *permease* and *galactoside transacetylase*. According to Jacob and Monod, the genes for these three proteins are actually part of a composite system called an *operon*. An operon is a series of structural genes or cistrons for several proteins that are under the common control of a single regulatory site or *operator gene*. The entire operon is controlled by a *regulator gene,* which produces a protein called the *repressor* and which has a specific affinity for the operator gene (Fig. 10-13, *A*). For the *lac* operon the repressor is a tetrameric protein having a molecular weight of 150,000 and is sufficiently large to recognize 12 to 20 base pairs in the operator gene. The DNA-directed RNA polymerase begins transcribing the operon by first combining with a special region called the *promoter site*. The repressor bound at the operator gene can block or prevent the binding of the RNA polymerase.

Repressor proteins appear to possess two allosteric sites, one having a specific affinity for the operator gene and the other for the inducer. In this case the lactose (actually allolactose) is the effector that binds to the appropriate site, thus altering the conformation and rendering it incapable of further binding to the operator.

$$\text{Inducer} + \text{Active repressor} \rightleftharpoons \text{Inactive repressor}$$

In the presence of inactive repressor molecules the operon is no longer repressed or turned off. The RNA polymerase binds to the promoter site and transcribes the operon. In the case of the *lac* operon, mRNA is synthesized for β-galactosidase, permease, and transacetylase; therefore these are called inducible proteins. There is one strand of mRNA bearing the information for all three proteins; it is called a polycistronic mRNA (Fig. 10-13, *A*). In general, mRNAs for prokaryotic systems are polycistronic.

If some glucose is then added to the lactose medium, the production of β-galactosidase is repressed and the available glucose is used preferentially. It appears that some metabolite of glucose, X, serves as a signal that results in a repression (catabolite repression) of the *lac* operon. The preferential use of glucose involves another controllable factor, which has been shown to be cyclic AMP. As suggested below, glucose affects the intracellular level of cAMP. Thus by stimulating the breakdown of cAMP or by inhibiting its synthesis, X—the catabolite—decreases the concentration of cAMP. The level of the cyclic nucleotide is sensed by a specific binding protein called CAP (catabolite activator protein). Normally the CAP:cAMP complex binds specifically with the operon near the promoter site. This alters the promoter so that RNA polymerase can bind tightly and be ready to initiate transcription, provided that its way is not barred by the repressor protein. The cAMP increases the synthesis of mRNA from the *lac* operon by increasing the frequency of chain initiation. A depletion of cAMP serves to minimize the process of transcription. Catabolite repression by glucose is not restricted to the *lac* operon; some 12 systems involving carbohydrate, amino acid, purine, and pyrimidine metabolism are also regulated by cAMP.

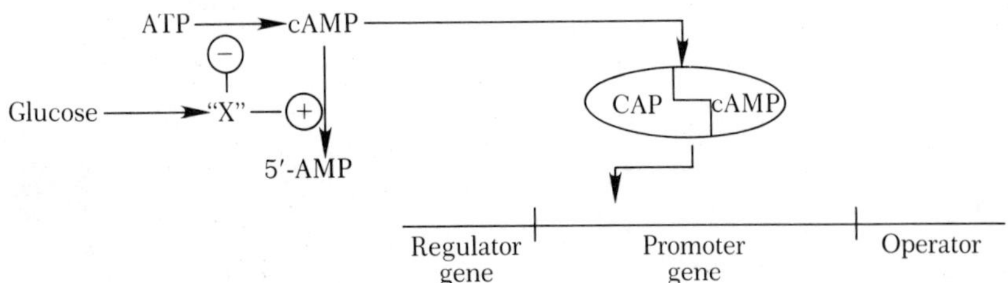

Regulation by repression

Metabolism may also be regulated by a positive suppression of the biosynthesis of certain enzymes. As an example, *E. coli* growing in a glucose medium actively synthesizes tryptophan by a pathway requiring three enzymes. The mRNA for these enzymes is transcribed from five genes that are linked together as the *trp* operon. Normally RNA polymerase binds at the operator-promotor site, and the polycistronic mRNA is transcribed. When tryptophan accumulates and the cell no longer needs to synthesize the amino acid, then it halts its own synthesis by genetic repression. This also may be explained by the model of Jacob and Monod (Fig. 10-13, *B*). The regulator gene then is viewed as producing an inactive repressor or *aporepressor* that requires the addition of a metabolite, the *corepressor*, which in this instance is tryptophan.

$$\text{Corepressor} + \text{Aporepressor} \rightleftharpoons \text{Active repressor}$$

The aporepressor molecule also has two allosteric sites; one is for the corepressor, and the other enables it to bind to the operator site. In the absence of the corepressor the aporepressor has little affinity for the binding site, and therefore the operon is actively transcribed (Fig. 10-13, *B*). When the tryptophan accumulates, the aporepressor is converted to its active form: the repressor-tryptophan complex. This complex then binds to the operon and excludes the binding of RNA polymerase from the promoter binding site. The net effect as suggested is the repression of the transcriptive process.

The *trp* operon also has a second level of regulation that affects transcription. The attenuator site ("a") lies between the promoter-operator and the first structural gene, *trp* E. The attenuator senses changes in the tryptophan level and serves as an operational terminator of transcription. Thus, as the amount of tryptophan increases, the attenuator causes a termination of mRNA synthesis even after transcription has been initiated. This appears to enable a greater sensitivity of control over transcription than is possible from the repression process alone.

The hypothesis of Jacob and Monod therefore serves to explain two opposing regulatory mechanisms: in one the operon is normally repressed and must be induced for expression; in the other the operon is usually functional and must be "turned off."

Regulation of protein turnover in mammalian cells

Genetic level

Although the concept of Jacob and Monod is a useful model, its application to eukaryotes and multitissue organisms is, at the very least, an oversimplification.

In eukaryotic cells the structural genes are not necessarily physically related as composite, polycistronic operons. All the genetic information is distributed among the 23 pairs of chromosomes. For example, α and β subunits of hemoglobin are not physically linked. Therefore the mammalian system must rely on a mechanism far more complex than one depending on a simple contiguity of structural genes, as in the operon. Thus a given effector such as a hormone may enter the nucleus of a target cell and elicit a transcriptive response of a large program. Information retrieval may involve a library of different combinations of structural genes. An inducer such as a hormone may derepress a battery of structural genes. The battery concept might be visualized as a system of various structural genes under the control of sensing or receptor sites (Fig. 10-14). In the scheme illustrated four inducers could elicit four different response profiles, yet some overlapping would occur. All four would signal activation of gene *B*, for example. The diversity of metabolic processes that are controlled by a single hormone can be visualized by the increased responses elicited by estrogens (a steroid effector) in uterine cells. These responses are the following*:

1. Total cell protein
2. Transport of amino acids into cell

*From Britten, R.J., and Davidson, E.H.: Science **165**:349, 1969.

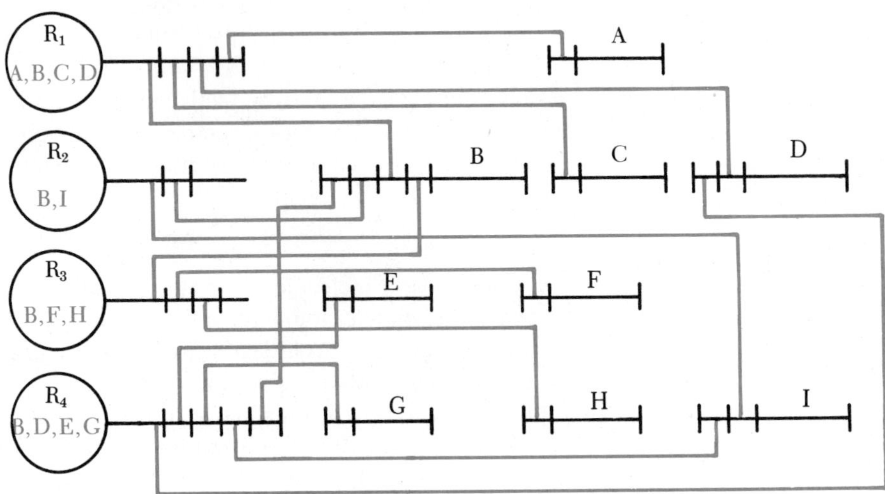

Figure 10-14
Concept of overlapping control of various structural or producer genes by several receptor genes. Activation of receptor gene R_1 releases a battery of producer genes, *A* to *D*, whereas activation of R_4 yields activity at *B, C, E,* and *G*. All four receptor genes would elicit activity from gene *B*.

Adapted from Britten, R.J., and Davidson, E.H.: Science **165**:349, 1969.

3. Protein synthesis activity per unit amount of polyribosomes
4. Synthesis of new ribosomes
5. Amount of polyribosomes per cell
6. Nucleolar mass and number
7. Activity of two RNA polymerases
8. Synthesis of contractile proteins
9. Imbibition of water
10. Synthesis of many phospholipids
11. Synthesis of purines
12. Synthesis of various mucopolysaccharides

Furthermore, tissue differentiation plays a most important role in the control of protein synthesis by various tissues of multitissue organisms. Although all cells possess the same complement of genetic information, differentiation has made the genes selectively transcribable, and many are totally repressed. For example, the enzyme glucose 6-phosphatase is present primarily in the liver; it is not found in muscle tissue. Yet the genetic information needed to produce this enzyme must be present in both. This selective transcribability has an important effect on the control of metabolic activities and the function of various tissues.

Regulation of translation

Once the DNA has been transcribed, the accumulation of mRNA serves as an important level of control. The more mRNA existing at any time, the more protein can be synthesized. The amount of mRNA present in the cell depends on its life span. Although the mRNAs of eukaryotic cells are considered to be longer lived than those of prokaryotes, it does appear that both short- and long-lived species exist. Thus in rat liver the total mRNA has been separated grossly into a short-lived population ($t_{1/2}$ = 3 to 3½ hours) and a long-lived group ($t_{1/2}$ = 80 hours). Here $t_{1/2}$ is the half-life, or the time required for half the mRNA molecules to be either degraded or resynthesized. Those mRNAs which are translated into enzymes subject to control at the level of synthesis are generally among those of short half-life.

When an enzyme is no longer needed, its further synthesis is halted. This occurs at the level of translation; apparently chain initiation is the rate-limiting step. Control of translation depends on the availability of certain essential amino acids and the hormone insulin. A deficiency of essential amino acids halts the binding of Met-tRNA$_f^{Met}$ to the 40S ribosomal subunit. Also affected are the numbers and the state of polysomal aggregation.

Regulation of protein degradation

In prokaryotic cells, elevated levels of enzymes can be reduced simply by cellular reproduction and subsequent dilution. There appears to be little degradation of intracellular proteins. In the mammalian tissue cells, which reproduce slowly or not at all, the synthesis of enzymes must be opposed by an active process of degradation. Were this not the case, control of metabolism via transient changes in enzyme concentration would be impossible.

The level of enzymes present in the cell at any given time is a steady state involving, as illustrated below, the opposing processes of biosynthesis and degradation.

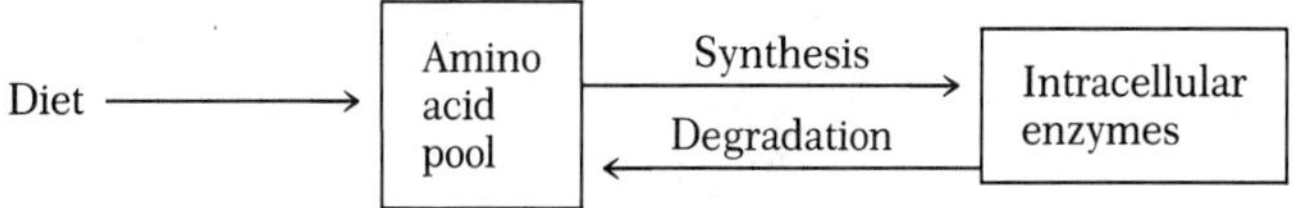

The actual life span of an enzyme varies; the half-life of ornithine decarboxylase is 11 minutes; that for tryptophan dioxygenase is 2 to 4 hours, and for arginase it is 4 to 5 days. Enzymes, whose in vivo reactions are rate limiting in metabolic pathways, are rapidly degraded and have a relatively short half-life.

Although there is little current information regarding the mechanism of intracellular protein degradation, it appears to occur by proteolysis within the lysosomes (Chapter 2). These organelles are known to contain a class of proteolytic enzymes called *cathepsins*. As the intracellular supply of amino acids accumulates, the degradation of protein is suppressed, whereas a depletion of amino acids, as in starvation, results in an increased proteolysis. Insulin, which stimulates protein synthesis, inhibits the lysosomal degradation of proteins. By increasing synthesis and at the same time inhibiting degradation, insulin ensures an accumulation of certain enzymes. These observations show that the action of the lysosomes in the degradation of protein is subject to control.

The general breakdown of proteins in the lysosomes does not, however, explain how a specific enzyme may be selectively degraded or not degraded depending on metabolic need. Some insight into this question is provided by the studies of Schimke on tryptophan dioxygenase (pyrrolase). The administration of both a glucocorticoid hormone and the substrate L-tryptophan increases the tissue level of the enzyme. The hormone stimulates the biosynthesis while degradation is unaffected. L-Tryptophan decreases the rate of degradation; the biosynthesis is unaffected. It appears that the presence of the substrate in some way protects the enzyme from proteolysis. Tryptophan dioxygenase contains a dissociable cofactor (hematin). In the liver the total enzyme includes both the apoenzyme and the holoenzyme. The presence of L-tryptophan promotes the binding of hematin and therefore the conversion of apoenzyme to holoenzyme. Unlike the apoenzyme, the holoenzyme is stable to proteolytic enzymes. Therefore in the presence of tryptophan the holoenzyme is the predominate species in the cell. Tryptophan ensures an adequate level of enzyme by promoting the binding of hematin and consequently the protection of the enzyme from degradation.

It should now be clear that the metabolic adaptations to nutrients and hormones may depend on a selective biosynthesis of an enzyme protein as well as a selective reduction in its degradation.

Suggested readings

Anderson, W.F. et al.: International Symposium on Protein Synthesis, FEBS Lett. **76:**1, 1976.

Anfinsen, C.B.: Principles that govern the folding of protein chains, Science **181:**223, 1973.

Bertrand, K., et al.: New features of the regulation of the tryptophan operon, Science **189:**22, 1975.

Britten, R.J., and Davidson, E.H.: Gene regulation for higher cells: a theory, Science **165:**349, 1973.

Clark, B.F.C., and Marcker, K.A.: How proteins start, Sci. Am. **218**(1):36, 1968.

Frenster, J.H., and Herstein, P.R.: Gene de-repression, N. Engl. J. Med. **288:**1224, 1973.

Goldberg, A.L., and Dice, J.F.: Intracellular protein degradation in mammalian and bacterial cells, Annu. Rev. Biochem. **43:**835, 1974.

Goldberg, A.L., and St. John, A.C.: Intracellular protein degradation in mammalian and bacterial cells, Annu. Rev. Biochem. **45:**747, 1976.

Goulian, M., and Hanawalt, P.C., editors: DNA synthesis and its regulation, Menlo Park, Calif., 1975, W.A. Benjamin, Inc.

Haselkorn, R., and Rothman-Denes, L.B.: Protein synthesis, Annu. Rev. Biochem. **42:**397, 1973.

Holzer, H., and Heinrich, P.C.: Control of proteolysis, Annu. Rev. Biochem. **49:**64, 1980.

Ingram, V.M.: The biosynthesis of macromolecules, ed. 2, Menlo Park, Calif., 1972, W.A. Benjamin, Inc.

Kurland, C.G.: Structure and function of the bacterial ribosome, Annu. Rev. Biochem. **46:**173, 1977.

Leder, P.: Discontinuous genes, N. Engl. J. Med. **298:**1079, 1978.

Lengyel, P. In Taylor, J.H., editor: Molecular genetics, New York, 1967, Academic Press, Inc.

Lengyel, P., and Söll, D.: Mechanism of protein biosynthesis, Bact. Rev. **33:**264, 1969.

Lennarz, W.J., and Scher, M.G.: Metabolism and function of polyisoprenol sugar intermediates in membrane-associated reactions, Biochim. Biophys. Acta **265:**417, 1972.

Lipmann, F.: Polypeptide chain elongation in protein biosynthesis, Science **164:**1024, 1969.

Lucas-Lenard, J., and Lipmann, F.: Protein biosynthesis, Annu. Rev. Biochem. **40:**409, 1971.

Maniatis, T., and Ptashne, M.: A DNA operator-repressor system, Sci. Am. **243**(1):64, 1976.

Marks, P.A., and Rifkind, R.A.: Protein synthesis: its control in erythropoiesis, Science **175:**955, 1972.

Marshall, R.D.: Glycoproteins, Annu. Rev. Biochem. **41:**673, 1972.

Mechanism of protein synthesis, Symp. Quant. Biol. **34:**1,1969.

Munro, H.N.: A general survey of mechanisms regulating protein metabolism in mammals. In Munro, H.N., editor: Mammalian protein metabolism, vol. 4, New York, 1970, Academic Press, Inc.

Noll, H.: Organelle integration and the evolution of ribosome structure and function, Symp. Soc. Exp. Biol. **24:**419, 1970.

Parodi, A.J., et al.: The role of polyprenol-bound saccharides as intermediates in glycoprotein synthesis in liver, Proc. Nat. Acad. Sci. **69:**3268, 1972.

Pastan, I., and de Crombrugghe, B.: Control of expression of genetic information in *E. coli* by cyclic AMP, Fifth Karolinska Symposium on Research Methods in Reproductive Endocrinology, May 29-31, 1972, p. 298.

Ptashne, M., and Gilbert, W.: Genetic repressors, Sci. Am. **222**(6):36, 1970.

Rannels, D.E., McKee, E.E., and Morgan, H.E.: Regulation of protein synthesis and degradation in heart and skeletal muscle. In Litwack, G., editor: Biochemical action of hormones, vol. 4, New York, 1977, Academic Press, Inc.

Schimke, R.T.: Regulation of protein degradation in mammalian tissues. In Munro, H.N., editor: Mammalian protein metabolism, vol. 4, New York, 1970, Academic Press, Inc.

Watson, J.D.: Molecular biology of the gene, ed. 3, Menlo Park, Calif., 1976, W.A. Benjamin, Inc.

Weinberg, R.A.: Nuclear RNA metabolism, Annu. Rev. Biochem. **42:**329, 1973.

Weissbach, H., and Pestka, S., editors: Molecular mechanism of protein biosynthesis, New York, 1977, Academic Press, Inc.

Yanofsky, C.: Gene structure and protein structure, Sci. Am. **216**(5):80, 1967.

Section Three

Major metabolic pathways

Chapter 11

Metabolism of carbohydrates

Carbohydrate serves as the major source of fuel or energy in the average human diet (approximately 46%) (Chapter 21). Of this, much is in the form of polysaccharides, 22% as starches and dextrins, largely derived from cereal grains and vegetables such as corn and potatoes. The remaining 24% is supplied as sucrose, lactose, glucose, maltose, and other sugars. Cellulose is a major food polysaccharide that is not usable by humans and forms most of the nondigestible residue or fibers of foods.

Digestion

Before the dietary carbohydrates can be absorbed by the intestinal epithelium, it is essential that the various polysaccharides, oligosaccharides, and disaccharides are hydrolyzed to their constituent monosaccharide subunits. This cleavage occurs sequentially in different parts of the gastrointestinal tract by a series of hydrolases or *carbohydrases*. Fig. 11-1 illustrates the function of these enzymes, which are secreted first in the saliva (an α-amylase [ptyalin]) and subsequently as another α-amylase (amylopsin) in the pancreatic juice. Digestion is continued and completed in the intestinal tract by enzymes found both in the secretions of the pancreas and in a fluid called the succus entericus, found in the intestinal mucosa.

Both salivary and pancreatic amylases are 1,4-glucosidases and serve to hydrolyze only the 1,4-glucosidic bonds found in starch and glycogen. The specificity of this enzyme is such that it can only hydrolyze alternate glucosidic bonds; also, it cannot affect the 1,6-glucosidic bonds or branch points that are essential to the structure of amylopectin (Chapter 23). Consequently the result of the amylolytic process beginning in the saliva involves the conversion of starches and glycogen into maltose and dextrins. The latter is a mixture of large-chain, branched oligosaccharides or partial digestion products. The pancreatic α-amylase serves to digest further the (1,4-glucosidic bonds) mixture of dextrins into maltose, isomaltose (the α-1,6-disaccharide), glucose, and residual oligosaccharides (Fig. 11-1).

The final digestive process occurs at the mucosal lining and involves the action of an α-dextrinase, an oligo-1,6-glucosidase. This enzyme hydrolyzes the 1,6-glucosidic bonds, resulting in the formation of glucose from oligosaccharides and also from isomaltose. An α-glucosidase or maltase is also present. As a result of the actions of these two enzymes, the oligosaccharides are converted to glucose. Another group of enzymes includes the disaccharidases sucrase, maltase, and lactase, which are found in the succus entericus. The end products of carbohydrate digestion are chiefly glucose, fructose, and galactose, which are readily absorbed through the intestinal mucosal cells into the bloodstream. The removal of the end products by absorption favors the continued digestive process. Thus, in the normal

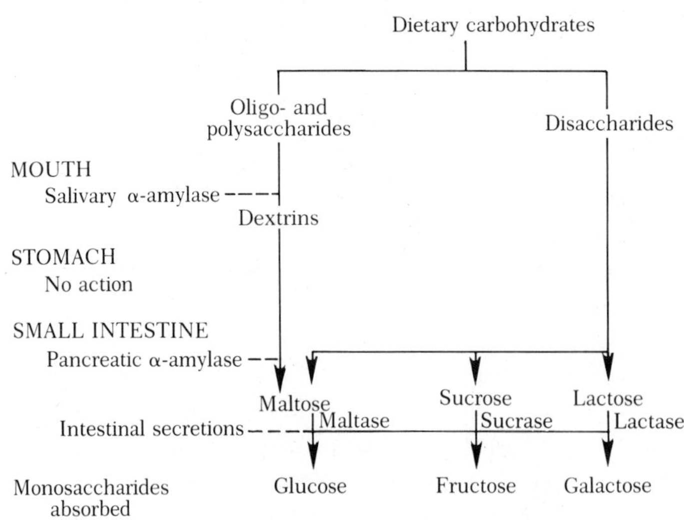

Figure 11-1
Distribution and actions of various carbohydrate-digesting enzymes.

individual, digestion and absorption of usable carbohydrates are 95% or more complete.

Differences in intestinal disaccharidases may exist according to race. Orientals and blacks, for example, are less tolerant to lactose than is the average white person because of a deficiency of the enzyme lactase. Thus it appears that a large portion of the world's population has a low tolerance to the lactose of milk; for them milk is not a satisfactory food. Hereditary deficiencies of the enzyme α-dextrinase, sucrase, and lactase have been described in infants and children with disaccharide intolerance or carbohydrate malabsorption. The failure to digest completely amylopectin, maltose, sucrose, or lactose results in their bacterial decomposition in the lower intestine, with the production of glucose and other monosaccharides, organic acids, low pHs (5.5 or less), and gases. This results in bloating, flatulence, and diarrhea from irritation of the bowel with the ensuing increased motility. The reduction or elimination of the offending carbohydrate from the diet is necessary for the control of this condition.

Absorption of monosaccharides

Only monosaccharides are absorbed from the intestinal lumen by passage through the mucosal epithelial cells into the bloodstream. Although passage of monosaccharides through the cell membranes may occur to some extent by simple diffusion (see Chapter 19), this is not the primary mechanism for the absorption of the hexoses, especially glucose and galactose.

The absorption of various monosaccharides from the intestinal lumen was found to have the following relative rates, based on glucose as 100:

D-Galactose	110
D-Glucose	100
D-Fructose	43
D-Mannose	19
D-Xylose	15
D-Arabinose	9

The transport of glucose and galactose across the brush border membrane of the mucosal cells occurs by an active, energy-requiring process that involves a specific transport protein and the presence of sodium ions. Actually, the influx of glucose requires the cotransport of Na^+ (symport; Chapter 19) by means of the same carrier protein. The accumulation of glucose occurs against a concentration gradient, i.e., from a relatively low concentration in the intestinal lumen to a higher intracellular concentration. The energy needed is derived from the simultaneous inflow of Na^+ from a relatively high concentration in the lumen to a lower one inside the cell. Therefore the trend is for the sodium ions to flow into the cell in a downhill direction, carrying the glucose molecules with them. Naturally, the flow of Na^+ can proceed only so long as the proper gradient is maintained. To do so requires the function of a sodium pump, which transfers the accumulating sodium back out of the cells and into the bloodstream. This uphill process requires energy in the form of ATP. Actually, as Na^+ ions are expelled, K^+ ions enter the cell and ATP is hydrolyzed. For this reason the pump is called a *sodium-potassium ATPase.* (See Fig. 19-7.)

Insulin (Chapter 18), which enhances most glucose transport across cellular membranes, especially of the muscle and adipose tissues, does not influence the intestinal transmucosal uptake of glucose.

Apparently fructose and mannose are absorbed by a facilitated diffusion process (Chapter 19). Their movement across the membranes is more rapid than would be expected from simple diffusion. Facilitated diffusion involves the intermediary formation of a complex with a specific transport or carrier protein. However, the movement of sugar is strictly downhill, going from a higher concentration to a lower one until it reaches an equilibrium. Other sugars, such as the pentoses, are absorbed by simple diffusion through the lipid bilayer of the membrane. The absorption of monosaccharides occurs only in the small intestine.

Fate of dietary carbohydrates

On entering the intestinal blood supply the monosaccharides are carried by the portal circulation directly to the liver. Therefore the immediate fate of the dietary carbohydrate occurs in the liver, where more than 60% is metabolized. Normally the liver contains little free glucose; most of the entering molecules are phosphorylated to glucose 6-phosphate. In this way a downhill gradient of free glucose is maintained so long as the portal carbohydrate level is high. Furthermore, the sugar is trapped within the cell because its plasma membrane is impermeable to the hexose phosphate. The remainder of the glucose passes on into the systemic blood supply. As will be seen later, an essential function of the liver is to control the systemic blood level (glucose homeostasis) between 70 and 90 mg/100 ml of blood. It serves as the primary reservoir for fuel by converting the glucose into the storage polysaccharide glycogen. The liver also converts glucose into lipid, which then is transported to the adipose tissues. Other dietary monosaccharides, fructose and galactose, are phosphorylated and may be converted into glucose.

Glucose as a source of energy

In the liver, as in all tissues, glucose is metabolized for the production of ATP and to provide metabolic intermediates needed in various biosynthetic processes. In addition, a major part of the dietary glucose is converted to glycogen for storage.

As described in Chapter 7, the total oxidation of glucose results in a large free-energy change ($\Delta G^{\circ\prime} = -686{,}000$ calories/mole). This energy is used for the production of ATP.

In the cell glucose is degraded gradually by way of a series of phosphorylated intermediates, principally two metabolic pathways: the Embden-Meyerhof pathway, or glycolysis, and the citric acid cycle (tricarboxylic acid cycle, or the Krebs cycle, after Sir Hans Adolph Krebs) of oxidation. These pathways include exergonic reactions, yielding energy primarily in the form of ATP. ATP then serves as the

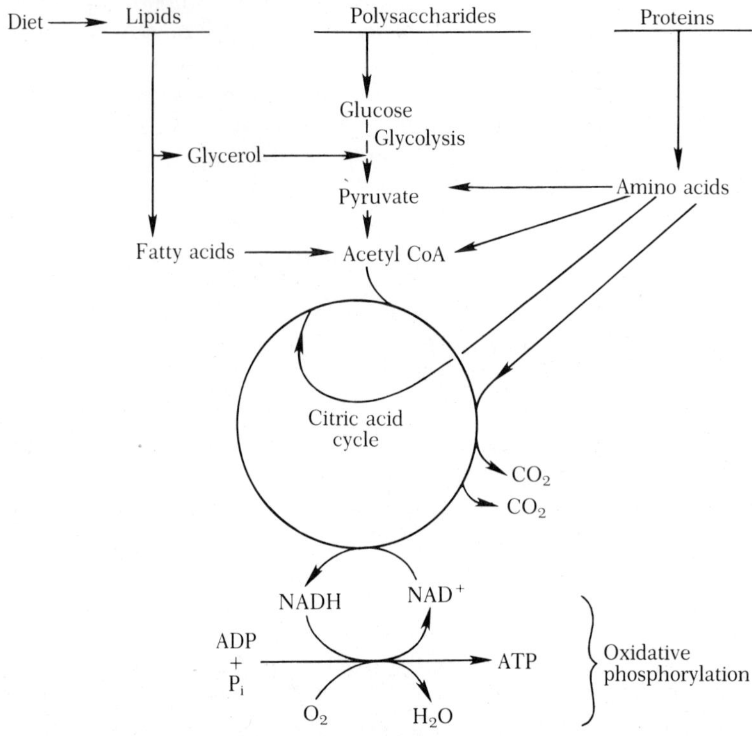

Figure 11-2
Central role of the citric acid cycle in the metabolism of the three classes of nutrients: lipids, carbohydrates, and proteins.

common intermediary for all energy-requiring processes. In the living cell, therefore, the overall reaction for the oxidation of glucose is the following:

$$C_6H_{12}O_6 + 36\text{-}38\ ADP + 36\text{-}38\ P_i + 6\ O_2 \rightarrow 6\ CO_2 + 6\ H_2O + 36\text{-}38\ ATP + \text{Unavailable energy}$$

However, glucose is not the only metabolic fuel for living cells. The liver and to varying degrees other tissues also use dietary fatty acids and amino acids as sources of energy. All three nutrients are metabolized by means of the citric acid cycle, thereby producing ATP in the course of metabolism (Fig. 11-2).

Glycolysis

In all cells carbohydrate metabolism proceeds first to pyruvate and lactate in the cytoplasm (Fig. 11-3). This is accomplished by way of phosphorylated intermediates with the formation of a limited amount of ATP. The pyruvate produced in glycolysis enters the mitochondria and is there oxidized to carbon dioxide and water. This process occurs by way of acetyl CoA and the citric acid cycle with associated oxidative phosphorylation and ATP formation (Fig. 11-2). The major portion of the ATP formed from glucose is produced in the mitochondria.

Formation of glucose 6-phosphate

The first step in the metabolism of glucose is its phosphorylation (actually as an α,β-D-glucopyranose equilibrium mixture) to yield the phosphate ester α-D-glucose 6-phosphate. This reaction consumes 1 mole of ATP and is catalyzed by a group of four distinct enzymes: types I to IV *hexokinases*. Although the hexokinases are found in most cells, their relative distribution of types varies. Thus brain and kidney have chiefly the type I enzyme, skeletal muscle has type II, adipose tissue has both types I and II, and the liver possesses all four. In the liver the principal species of hexokinase is type IV, commonly known as *glucokinase*.

In general, the kinases catalyze the transfer of phosphate from a nucleoside

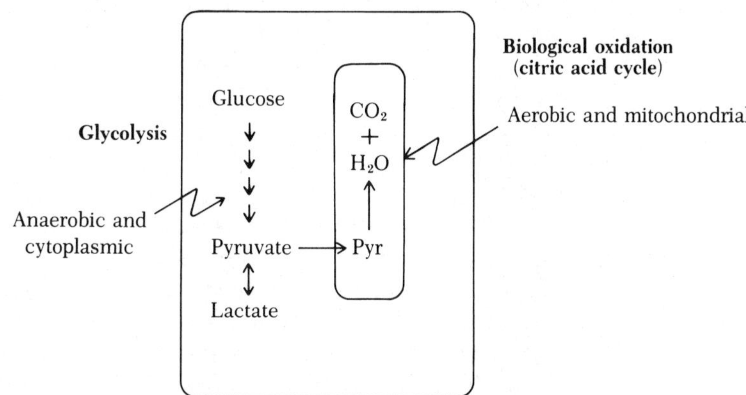

Figure 11-3
Compartmentalization of glycolysis and biological oxidation.

triphosphate, usually ATP, to their substrates. As seen below, the hexokinases catalyze the phosphorylation of D-glucose at the expense of ATP. Mg^{++} ions are required as a cofactor for such reactions (Chapter 6); ATP chelates the metal ion.

α D Glucose + ATP —(Hexokinase, Mg^{++})→ α D Glucose 6 phosphate + ADP

The oxygen on carbon-6 of glucose is involved in a nucleophilic attack (electron donating) on the terminal phosphate of ATP. Actually, the formation of glucose 6-phosphate is formally a reversal of its hydrolysis, an endergonic reaction ($\Delta G^{\circ\prime}$ = +3.3 kcal/mole) that would require an input of energy to proceed. For this purpose it is coupled with the hydrolysis of ATP, a highly exergonic reaction ($\Delta G^{\circ\prime}$ = −7.3 kcal/mole). The coupled reaction is the summation of these two constituent reactions, i.e.:

Reaction	
D-Glucose + P_i → D-Glucose 6-phosphate + H_2O	$\Delta G^{\circ\prime} = +3.3$ kcal
ATP + H_2O → ADP + P_i	$\Delta G^{\circ\prime} = -7.3$ kcal
D-Glucose + ATP → D-Glucose 6-phosphate + ADP	$\Delta G^{\circ\prime} = -4.0$ kcal

Since the sum of the free-energy changes is negative (−4.0 kcal/mole), the overall coupled reaction is exergonic. An equilibrium constant of 4900 (pH = 7.5) clearly favors the forward reaction. Furthermore, the kinetic parameters, the V_{max} and K_m, also favor the phosphorylation of glucose over the reverse reaction. An important characteristic of hexokinases types I to III is their extreme sensitivity to inhibition by the product, glucose 6-phosphate. All these factors combine to make the hexokinase-catalyzed phosphorylation of glucose a unidirectional process under physiological conditions. Therefore, when the liver is called on to produce free glucose, the kinase-catalyzed reaction is not reversed; rather glucose 6-phosphate is hydrolyzed by a specific phosphatase.

It will be seen subsequently that there are actually three such physiologically unidirectional reactions in the glycolytic pathway. These steps are important regulatory stages in glycolysis.

The liver functions as a "glucostat," i.e., when the blood glucose levels are high, glucose is metabolized or stored; when low, it is then remobilized and returned to the circulation. Therefore the liver cells must be able to phosphorylate glucose in response to fluctuations in blood glucose levels. The chief hexokinase in liver parenchymal cells is the type IV enzyme *glucokinase*. As seen in the accompanying table, it has a high K_m compared with that of the type I enzyme. Since the normal blood glucose level is 70 to 100 mg/100 ml (approximately 5 mM), the glucokinase with a K_m of 20 mM functions near first-order kinetics. Recall that the velocity is proportional to the substrate (glucose) concentration at first-order kinetics (Chapter 5). In other words, glucokinase activity varies with the blood glucose levels. Furthermore, it is an inducible enzyme; the amount is under the control of the hormone insulin. On the other hand, in tissues such as the brain the hexokinase predominating is the type I enzyme with a K_m of 0.01 mM (see table). Since its K_m is small compared with that of the blood glucose (5 mM), it functions at or near V_{max}. Phosphorylation, in this case, does not fluctuate with blood glucose levels; thus the brain is ensured a constant input of glucose 6-phosphate.

	K_m (mM)	Glucose concentration when $v = V_{max}/2$
Hexokinase (type I)	0.01	0.18 mg/100 ml
Glucokinase (hexokinase IV)	20	360 mg/100 ml

It is important to reinforce the value of this initial phosphorylation of glucose. As the sugar enters the cell by facilitated diffusion, it is promptly converted to glucose 6-phosphate in the cytoplasm. The phosphate ester cannot diffuse back out of the cell because the membrane is impermeable to it. Therefore the downhill gradient of free glucose is maintained because of the action of the two kinases.

Glucose 6-phosphate is an important central intermediate in several metabolic pathways. As seen below, it represents a branch point in the metabolism of glucose.

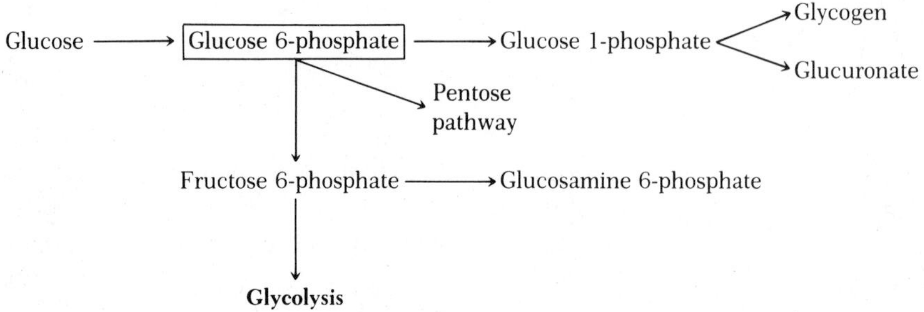

Interconversion of glucose 6-phosphate and fructose 6-phosphate

Glucose 6-phosphate and fructose 6-phosphate are freely interconvertible in the cell. The reaction is catalyzed by an isomerase, *phosphoglucoisomerase*. This enzyme is specific for glucose 6-phosphate. As seen below, the reaction involves the interconversion of an aldose and a ketose (Chapter 23).

D-Glucose 6-phosphate ⟷ (Phosphogluco-isomerase) D-Fructose 6-phosphate

Phosphorylation of fructose 6-phosphate

Fructose 6-phosphate is further phosphorylated to fructose 1,6-bisphosphate. As in the phosphorylation of glucose, this reaction also occurs at the expense of ATP and is catalyzed by a specific kinase, *phosphofructokinase.* The ATP-coupled reaction is exergonic ($\Delta G^{\circ\prime} = -3.4$ kcal), physiologically irreversible, and is a key step in the control of glycolysis. Since the formation of fructose 1,6-bisphosphate is actually the first unique step in glycolysis, it is called the *committed step.*

D-Fructose 6-phosphate + ATP —(Phosphofructokinase)→ D-Fructose 1,6-bisphosphate + ADP

A curve relating the velocity of fructose 1,6-bisphosphate synthesis to ATP concentration at a constant level of fructose 6-phosphate exhibits a biphasic response. The velocity at first increases, reaches a maximum, and then declines (Fig. 11-4, *A*). This shows that the velocity increases with [ATP] up to a certain point; thereafter further increments of ATP serve to inhibit the enzyme. In other words, ATP acts both as a substrate and an inhibitor. Fig. 11-4, *B*, shows that the curve of velocity versus (fructose 6-phosphate) at optimal ATP concentration is sigmoidal. Increasing the amount of ATP above the optimum shifts the sigmoid curve to the right; the shift thereby demonstrates that ATP actually acts allosterically to inhibit phosphofructokinase activity. The allosteric ATP-binding site is separate from the substrate-binding site. Likewise, citrate also behaves as a negative effector. In the presence of physiological concentrations of ATP the enzyme is almost inactive, thereby suppressing the glycolytic pathway. As ATP is consumed in metabolism, AMP and ADP accumulate. AMP and, to a lesser degree, ADP activate phosphofructokinase primarily by relieving the inhibition caused by the ATP.

Since the phosphorylation of fructose 6-phosphate is physiologically irreversible, its reversal in the cell is catalyzed by a separate enzyme. The hydrolysis of fructose

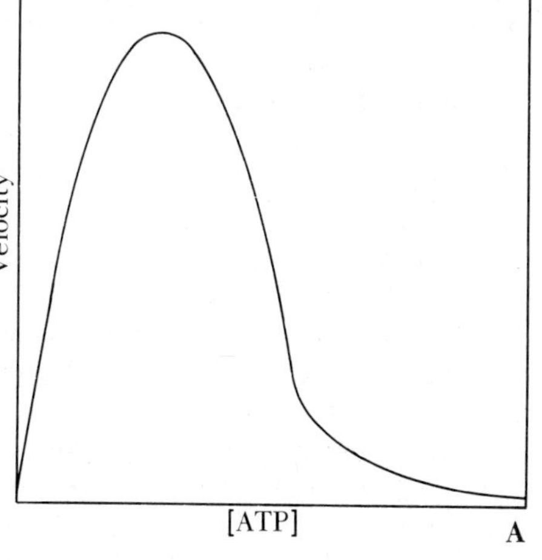

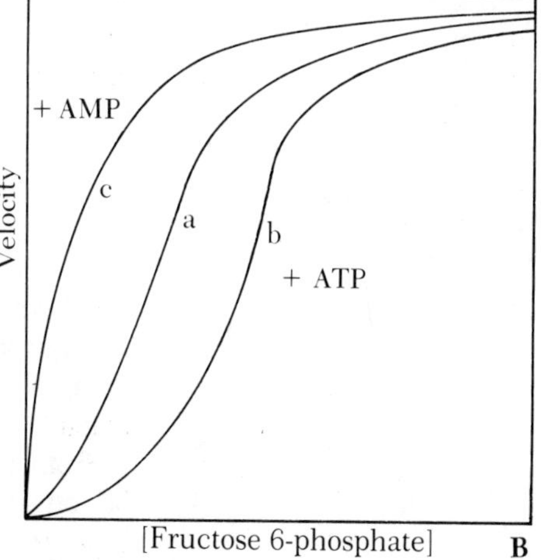

Figure 11-4
A ATP saturation curve for phosphofructokinase.
B Fructose 6-phosphate saturation curves for phosphofructokinase. Curve *a* represents normal data typical of enzyme; *b* is typical of native enzyme in presence of inhibitory ATP concentration, whereas *c* is a typical curve in presence of an activator, AMP.

1,6-bisphosphate to fructose 6-phosphate requires the action of *fructose bisphosphatase.*

Cleavage of D-fructose 1,6-bisphosphate

The reversible cleavage of fructose 1,6-bisphosphate by the enzyme *aldolase* to form D-glyceraldehyde 3-phosphate and dihydroxyacetone phosphate (DHAP) is the next step in glycolysis. The ketose structure of fructose 1,6-bisphosphate combines with a lysyl ϵ-amino group at the active site of aldolase to form a Schiff base. Cleavage of fructose bisphosphate, actually a reverse aldol condensation, results in the breaking of a carbon-carbon bond.

$CH_2O-PO_3^{2-}$ | C=O | HOCH | HCOH | HCOH | $CH_2O-PO_3^{2-}$

D-Fructose 1,6-bisphosphate

Aldolase →

$CH_2O-PO_3^{2-}$ | C=O | CH_2OH **Dihydroxyacetone phosphate (DHAP)**

Triosephosphate isomerase ⇅

CHO | HCOH | $CH_2O-PO_3^{2-}$ **D-Glyceraldehyde 3-phosphate**

Liver and muscle aldolases require no cofactors. In the reverse reaction DHAP forms the Schiff base that facilitates the aldol condensation with D-glyceraldehyde 3-phosphate.

As indicated above, the interconversion of DHAP and D-glyceraldehyde 3-phosphate is catalyzed by *triosephosphate isomerase.* This reaction is analogous to the interconversion of hexose phosphates by phosphoglucoisomerase. The isomerization is strongly in favor of DHAP.

DHAP may be reduced by NADH in the presence of *glycerol phosphate dehydrogenase* to L-glycerol 3-phosphate, which is needed in the biosynthesis of triacylglycerols (Chapter 12).

It is important to reflect at this point on the fact that one molecule of a hexose has now been cleaved to yield two molecules of triose. This point must always be remembered in considering the amount of ATP or other products formed. Also, it is important to note that only D-glyceraldehyde 3-phosphate is directly in the pathway and is the substrate for the next enzyme.

Phosphorylation and oxidation of D-glyceraldehyde 3-phosphate

The aldehyde group of D-glyceraldehyde 3-phosphate is converted to a mixed anhydride between the carboxyl of 3-phosphoglycerate and phosphoric acid; thus the product is 1,3 bisphospho-D-glycerate. For this reaction NAD^+ and P_i are required in addition to the enzyme *glyceraldehyde 3-phosphate dehydrogenase.* The enzyme obtained from muscle is a tetramer, with each monomeric unit containing at the active site a sulfhydryl group and one molecule of tightly bound NAD^+. The reaction sequence is shown in Fig. 11-5.

From a purely formal standpoint the production of 1,3-bisphosphoglycerate may be considered the summation of two distinct steps: the oxidation of the aldehyde to a carboxyl group and its subsequent endergonic phosphorylation to yield the anhydride. The free-energy change involved in the highly exergonic oxidation ($\Delta G^{\circ\prime}$ =

Figure 11-5
Outline of steps involved in the oxidation and phosphorylation of glyceraldehyde 3-phosphate by the enzyme glyceraldehyde 3-phosphate dehydrogenase.

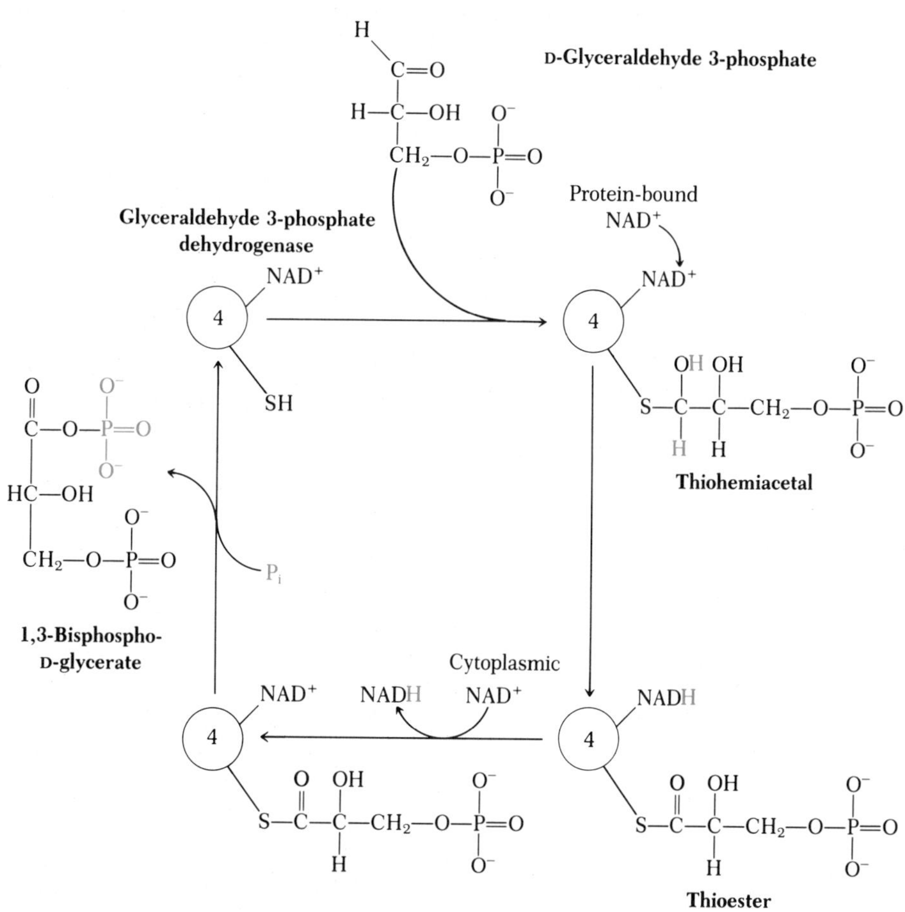

−10.3 kcal) is used in the formation of the mixed anhydride ($\Delta G^{\circ\prime} = +11.8$ kcal) so that the energy of the oxidation is conserved. The $\Delta G^{\circ\prime}$ for the formation of this high-energy phosphate compound is $+11.8 - 10.3 = +1.5$.

The overall reaction requires the stoichiometric participation of NAD^+ and P_i. The mechanism outlined in Fig. 11-5 indicates the formation first of a thiohemiacetal derivative with the enzyme. This occurs only on one of four cysteinyl residues present in each monomeric polypeptide chain. Two hydrogens are removed from the hemiacetal, converting it to the thioester. The resultant reduced NAD^+ is less tightly bound to the enzyme than the original oxidized form, so that it is now displaced from its binding site by cytoplasmic NAD^+. The thioester is released from the enzyme by phosphorolysis, the P_i being added to yield the final product, 1,3-bisphosphoglycerate. At the same time the free enzyme, with −SH group and oxidized NAD^+ restored, is ready again to repeat the process.

Under anaerobic conditions the cytoplasmic NADH is oxidized in the presence of (1) pyruvic acid to yield lactate, (2) dihydroxyacetone phosphate to yield L-glycerol 3-phosphate, or (3) oxaloacetate to yield L-malate. On the other hand, under aerobic conditions the NADH may be oxidized by means of the electron transport chain (Chapter 7), resulting in the formation of 2 or 3 moles of ATP per mole of triosephosphate metabolized.

The importance of the cysteinyl −SH group at the active site to the mechanism was demonstrated by blocking this group with iodoacetate. This reaction, shown below, inhibited the catalytic activity of glyceraldehyde 3-phosphate dehydrogenase and blocked the glycolytic pathway.

$$\text{(4)}\langle^{NAD^+}_{SH} + ICH_2CO_2H \longrightarrow \text{(4)}\langle^{NAD^+}_{S—CH_2CO_2H} + HI$$

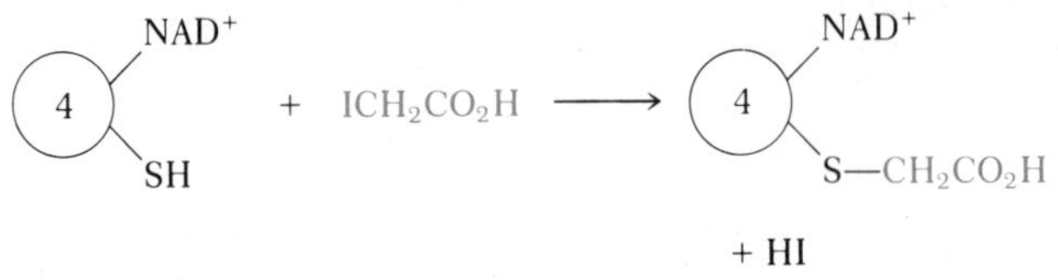

The phosphorolytic process (Fig. 11-5) is inhibited by arsenate ($AsO_4^{\equiv}$), which competes with the inorganic phosphate. The product formed is 1-arseno 3-phosphoglycerate, which hydrolyzes spontaneously to yield 3-phosphoglycerate. The energy usually conserved as the mixed anhydride here is lost as heat.

Formation of ATP at the substrate level

In the presence of *phosphoglycerate kinase* the free-energy change in the conversion of 1,3-bisphosphoglycerate to 3-phosphoglycerate is used to form ATP from ADP. As seen below, the $\Delta G^{\circ\prime}$ of 1,3-bisphosphoglycerate is −11.8 kcal. This is sufficient for the phosphorylation of ADP. In the coupled reaction the net free-energy change is −4.5 kcal.

Reaction	$\Delta G^{\circ\prime}$ *(kcal)*
1. 1,3-Bisphosphoglycerate + H_2O → P_i + 3-Phosphoglycerate	−11.8
2. ADP + P_i → ATP + H_2O	+7.3
3. 1,3-Bisphosphoglycerate + ADP → ATP + 3-Phosphoglycerate	−4.5

The overall free-energy change beginning with glyceraldehyde 3-phosphate and ending with 3-phosphoglycerate is $\Delta G^{\circ\prime} = -3$ kcal. Thermodynamically speaking these reactions are driven in the forward direction.

Conversion of 3-phosphoglycerate to 2-phosphoglycerate

In the next step of the glycolytic pathway the phosphate group is transferred from the carbon-3 to carbon-2 of the glyceric acid, a reaction catalyzed by *phosphoglyceromutase.* A bisphospho intermediate, 2,3-bisphosphoglycerate, is required. As seen below, the phosphate of carbon-3 is transferred to the carbon-2 of the 3-phosphoglycerate. The products therefore are 2-phosphoglycerate and a new 2,3-bisphosphoglycerate molecule (or DPG, formerly 2,3-diphosphoglycerate). In this case we may consider 2,3-bisphosphoglycerate or DPG to be a coenzyme. In red blood corpuscles 2,3-bisphosphoglycerate also plays an important role in oxygen release from hemoglobin.

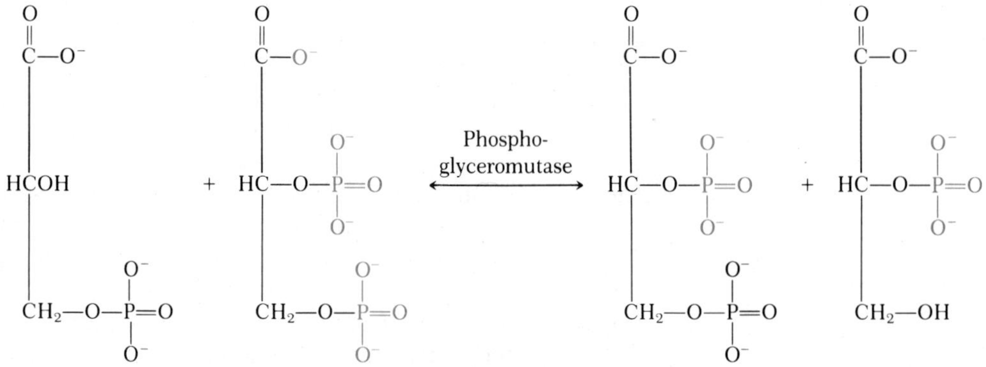

$$\underset{\textbf{3-Phosphoglycerate}}{\begin{array}{l} C(=O)—O^- \\ HCOH \\ CH_2—O—PO_3^{2-} \end{array}} + \underset{\textbf{2,3-Bisphosphoglycerate}}{\begin{array}{l} C(=O)—O^- \\ HC—O—PO_3^{2-} \\ CH_2—O—PO_3^{2-} \end{array}} \xrightleftharpoons{\text{Phospho-glyceromutase}} \underset{\textbf{2,3-Bisphosphoglycerate}}{\begin{array}{l} C(=O)—O^- \\ HC—O—PO_3^{2-} \\ CH_2—O—PO_3^{2-} \end{array}} + \underset{\textbf{2-Phosphoglycerate}}{\begin{array}{l} C(=O)—O^- \\ HC—O—PO_3^{2-} \\ CH_2—OH \end{array}}$$

Formation of phosphoenolpyruvate (PEP)

Dehydration of the 2-phosphoglycerate leads to the formation of a high-energy phosphate compound, *phosphoenolpyruvate (PEP)*. The enzyme that catalyzes the dehydration is called *enolase* (mol wt = 85,000) and requires Mg^{++} or Mn^{++} cofactors. Fluoride combines with magnesium fluorophosphate complex and thereby can inhibit this reaction and, consequently, glycolysis.

$$\underset{\text{2-Phosphoglycerate}}{\begin{array}{l} \text{C(=O)—O}^- \\ | \\ \text{HC—O—PO}_3^{2-} \\ | \\ \text{CH}_2\text{OH} \end{array}} \underset{\text{Mg}^{++}\ \to\ \text{H}_2\text{O}}{\overset{\text{Enolase}}{\rightleftharpoons}} \underset{\text{Phosphoenolpyruvate (PEP)}}{\begin{array}{l} \text{C(=O)—O}^- \\ | \\ \text{C—O—PO}_3^{2-} \\ \| \\ \text{CH}_2 \end{array}}$$

Formation of ATP from PEP

Pyruvate kinase catalyzes the transfer of the high-energy phosphate from PEP to ADP; Mg^{++} and K^+ are essential cofactors. The free-energy change is strongly exergonic ($\Delta G°' = -6.1$ kcal). It represents the third, rate-controlling step encountered in glycolysis. Fructose 1,6-bisphosphate acts as a positive effector for the liver, adipose tissue, kidney, and red blood cells but not the muscle enzyme. It stimulates this reaction, whereas ATP, Ca^{++}, and L-alanine inhibit it. Thus, if metabolic circumstances favor glycolysis and fructose 1,6-bisphosphate accumulates, the pyruvate kinase reaction is accelerated. This is an example of *feed-forward* activation. On the other hand, when the supply of ATP is more than adequate for the needs of the cell, its accumulation will inhibit this enzyme as well as the phosphofructokinase.

$$\underset{\text{PEP}}{\begin{array}{l} \text{C(=O)—O}^- \\ | \\ \text{C—O—PO}_3^{2-} \\ \| \\ \text{CH}_2 \end{array}} \underset{\text{ADP}\ \curvearrowright\ \text{ATP}}{\overset{\text{Pyruvate kinase}}{\xrightarrow{\text{Mg}^{++}}}} \left[\begin{array}{l} \text{C(=O)—O}^- \\ | \\ \text{C—OH} \\ \| \\ \text{CH}_2 \end{array}\right] \xrightarrow{\text{Spontaneous}} \underset{\text{Pyruvate}}{\begin{array}{l} \text{C(=O)—O}^- \\ | \\ \text{C=O} \\ | \\ \text{CH}_3 \end{array}}$$

Formation of lactic acid

Pyruvate is the end product of glycolysis. From here several options are available to the cell, depending on the environment and the particular demands made on the metabolic processes occurring at the time. In the absence of O_2, i.e., under anaerobic conditions, the process of oxidation and phosphorylation of 3-phosphoglyceraldehyde soon consumes the available supply of NAD^+ (Chapter 6). In this event the accumulating NADH can be reoxidized by entering into the following reaction with pyruvate:

$$\underset{\text{Pyruvate}}{\begin{array}{l} \text{C(=O)—O}^- \\ | \\ \text{C=O} \\ | \\ \text{CH}_3 \end{array}} \underset{\text{NADH}\ \curvearrowright\ \text{NAD}^+}{\overset{\text{LDH}}{\rightleftharpoons}} \underset{\text{L-Lactate}}{\begin{array}{l} \quad\ \text{C(=O)—O}^- \\ \quad\ | \\ \text{HO—C—H} \\ \quad\ | \\ \quad\ \text{CH}_3 \end{array}}$$

Lactic dehydrogenase (LDH) is the enzyme that catalyzes this reversible reaction. The free-energy change, however, favors the formation of lactic acid ($\Delta G°' = -6.0$ kcal). This preference may be modified by differences in properties of the

LDH isozymes. As has already been discussed (Chapter 5), LDH consists of varying proportions of two distinct polypeptides, LDH-1 and LDH-5. The K_m's suggest that the affinity for lactate by LDH-1 is somewhat greater than that by LDH-5; the reaction in heart muscle tends away from lactate toward pyruvate. Furthermore, LDH-1 is inhibited by high concentrations of pyruvate (substrate inhibition), thereby further minimizing the reaction pyruvate → lactate. LDH-5, on the other hand, is not inhibited by pyruvate, so lactate accumulates. These effects on LDH are consistent with the fact that in heart muscle carbohydrate metabolism is almost entirely aerobic, whereas anaerobic glycolysis can play an important role in the function of skeletal muscle, especially under conditions of oxygen debt.

Tissue	Isozyme	K_m (lactate)
Heart muscle	LDH-1	0.44 to 1.65 × 10^{-2}
Skeletal muscle	LDH-5	1.96 to 2.82 × 10^{-2}

Energy balance in glycolysis

In summary, glycolysis accomplishes the following production of ATP:

	ATP
1. Glucose → Glucose 6-phosphate	−1
2. Fructose 6-phosphate → Fructose 1,6-bisphosphate	−1
3. Glyceraldehyde 3-phosphate → 3-Phosphoglycerate	+2 (per glucose)
4. PEP → Pyruvate	+2
Net ATP production	+2 (per glucose)

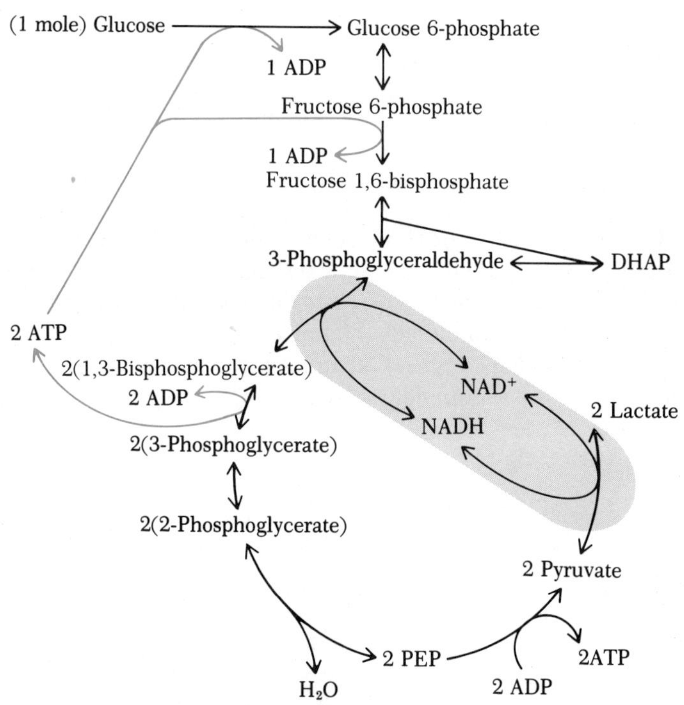

Figure 11-6
Summary of steps in glycolysis showing net synthesis of 2 moles of ATP per mole of glucose *(colored arrows)*. Also indicated is closed system for reoxidizing NADH *(color)*.

Thus in anaerobic glycolysis:

$$\text{Glucose} + 2\ \text{ADP} + 2\ \text{P}_i \rightarrow 2\ \text{Lactate} + 2\ \text{ATP} + 2\ \text{H}_2\text{O}$$

Approximately 15 kcal have been conserved as ATP, compared with a free-energy change of −48 kcal for anaerobic glycolysis. This amount of ATP has been produced strictly at the substrate level; no oxygen has been used.

It is important to reflect again on the concept of coenzyme pools discussed in Chapter 6. The cell does not have an unlimited supply of NAD^+. As the coenzyme is reduced, other enzyme systems must function to reoxidize the NADH. This is essential in anaerobic glycolysis; the glyceraldehyde 3-phosphate dehydrogenase requires NAD^+ and converts it to NADH. To maintain its activity, the pool of NAD^+ must be continuously restored by enzymes such as lactate, glycerolphosphate, and malate dehydrogenases. As already stated, in the muscle the lactate dehydrogenase restores the NADH to NAD^+. This dependence of anaerobic glycolysis on the reduction of pyruvate to lactate is illustrated in Fig. 11-6. Note the "internal" provision of ATP by the pathway for the phosphorylation of glucose and fructose 6-phosphate.

Metabolism of lactate and pyruvate

Of the various pathways that exist in cells for the further metabolism and disposal of pyruvate, oxidative decarboxylation to acetyl CoA is primary. This step, however, occurs only in the mitochondria. Pyruvate readily passes through the membrane and therefore leaves the cytosol and enters the mitochondrial compartment (Fig. 11-3).

Formation of acetyl CoA

The conversion of pyruvate to acetyl CoA (Chapter 6) is a central step linking the glycolytic pathway with the aerobic oxidative pathway, or the citric acid cycle. Acetyl CoA is also an important intermediate in lipid metabolism, cholesterol biosynthesis, and acetylation reactions.

The first loss of carbon in the metabolism of glucose takes place as carbon dioxide in the formation of acetyl CoA from pyruvate. In this process three individual enzymes and five coenzymes are involved. These enzymes are aggregated as a multimolecular complex called *pyruvate dehydrogenase* (Chapter 5). The initial reaction involves the coenzyme thiamin pyrophosphate (TPP) (Chapter 6) and is catalyzed by *pyruvate decarboxylase.* This step involves the decarboxylation of the α-keto acid and the formation of α-hydroxyethyl derivative of thiamin pyrophosphate (see Fig. 6-5).

CH_3–C(=O)–CO_2^- + HC(TPP thiazolium ring, N^+, S) → ($-CO_2$) CH_3–C(OH)(H)–C(thiazolium ring, N^+, S)

Pyruvate **TPP** **α-Hydroxyethyl thiamin pyrophosphate**

In the second step the 2-carbon acetyl group is donated to lipoic acid (Chapter 6). As described previously, this 8-carbon, disulfide-containing carboxylic acid is linked to the enzyme *dihydrolipoate transacetylase* by an amide bond (ε-amino group of lysine). This reaction is reductive; the disulfide is opened, and one sulfur accepts the acetyl while the other is reduced to a sulfhydryl group.

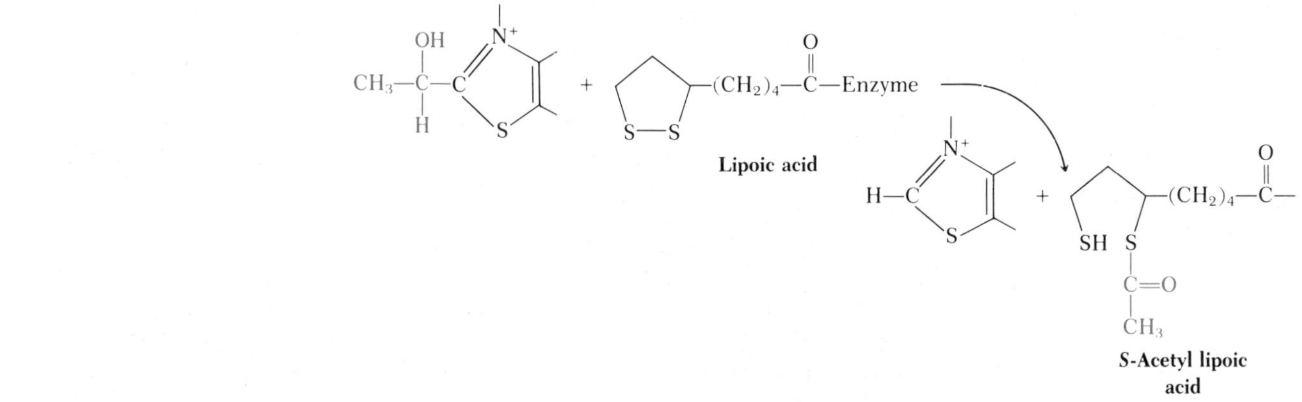

$CH_3-C(OH)(H)-C$(thiazolium) + (dithiolane)$-(CH_2)_4-C(=O)-$Enzyme → $H-C$(thiazolium) + (S-acetyl dihydrolipoyl)$-(CH_2)_4-C(=O)-$

Lipoic acid

S-Acetyl lipoic acid

The same enzyme, dihydrolipoate transacetylase, then catalyzes a third reaction in which CoA is the acceptor molecule. This means that the acetyl group is transferred from the sulfur on lipoate to the sulfur on CoA (Chapter 6). This thioester,

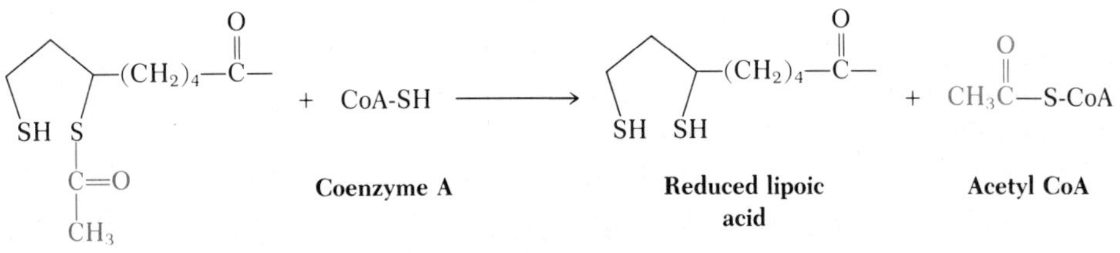

S-Acetyl lipoic acid $-(CH_2)_4-C(=O)-$ + CoA-SH → Reduced lipoic acid $-(CH_2)_4-C(=O)-$ + $CH_3C(=O)-S\text{-}CoA$

Coenzyme A

Reduced lipoic acid

Acetyl CoA

i.e., acetyl CoA, is a high-energy compound. The reduced form of lipoate must be reoxidized to the disulfide by the third enzyme, *dihydrolipoyl dehydrogenase*, an FAD-containing protein. The fourth step then involves the transfer of a pair of hydrogen atoms from the reduced lipoate, thereby reforming the disulfide bond and donating them to the isoalloxazine ring of the FAD (Chapter 6). The reduced FAD is then restored to its oxidized state by the fifth coenzyme, NAD^+; the final product is NADH.

The steps in the oxidative decarboxylation of pyruvate (the following reaction) are summarized in Fig. 11-7.

$$\text{Pyruvate} + NAD^+ + \text{CoA} \rightarrow \text{Acetyl CoA} + \text{NADH} + CO_2$$

The metabolism of pyruvate by this pathway is an example of the general metabolism of α-keto acids. It will be discussed again in the metabolism of α-ketoglutarate in the citric acid cycle and in the degradation of the branched-chain amino acids (Chapter 13).

Recall that the multienzyme complex pyruvate dehydrogenase of mammalian tissue involves the following enzymes:

1. 20 molecules of pyruvate decarboxylase
2. 60 molecules of reductase transacetylase
3. 5 or 6 molecules of dihydrolipoate dehydrogenase

As shown in Fig. 5-21, these enzymes are arranged in a compact structure. Furthermore, the five coenzymes needed are thiamin pyrophosphate, lipoic acid, CoA, FAD, and NAD^+. It should be appreciated that this is the site of action of four water-soluble vitamins: thiamin, riboflavin, nicotinic acid, and pantothenic acid.

It is apparent from Fig. 11-2 that pyruvate and acetyl CoA are important central compounds in metabolism. Acetyl CoA is produced not only from glucose but also from fatty acids and amino acids. There are circumstances in which sufficient or

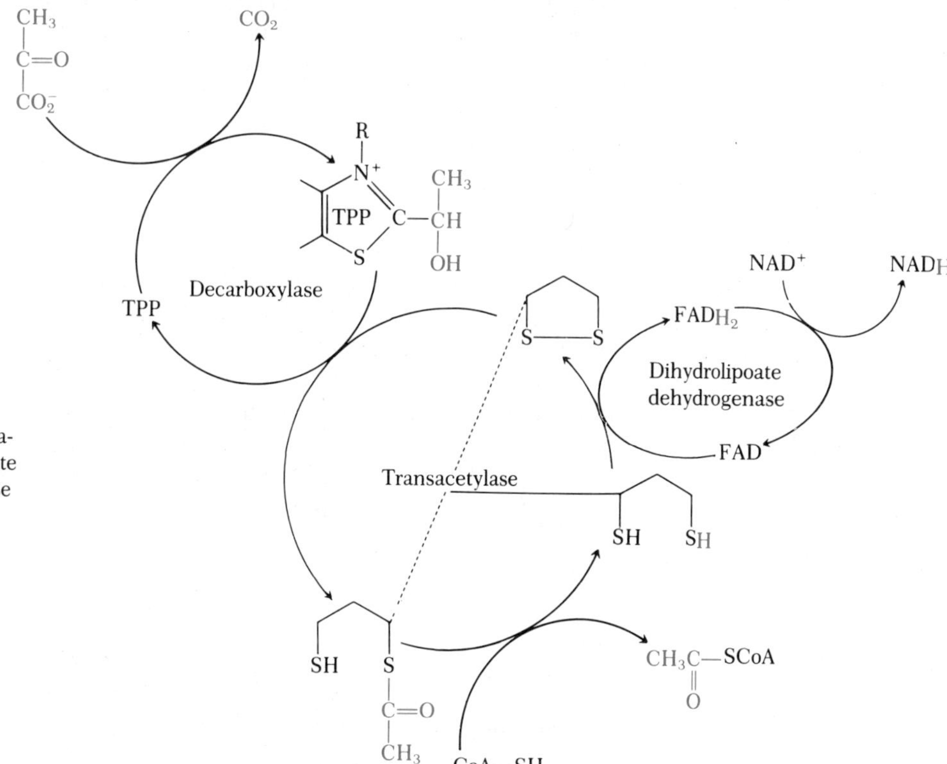

Figure 11-7
Summary of steps in the oxidative decarboxylation of pyruvate by the pyruvate dehydrogenase complex.

even excessive amounts of acetyl CoA are derived from lipids so that the cell does not need to produce more from glucose. For this reason the oxidative decarboxylation of pyruvate is an important metabolic control site. If the input of acetyl CoA from lipid is adequate, then the pyruvate dehydrogenase is inhibited, preventing further input from glucose. The control of pyruvate dehydrogenase occurs via a phosphorylation-dephosphorylation process of the decarboxylase (Chapter 5). Phosphorylation requires ATP and is catalyzed by a protein kinase. The latter is present in an approximate ratio of one molecule of kinase to 10 of decarboxylase. The phosphorylation catalyzed by the kinase leads to a loss of decarboxylase activity by converting the active enzyme to its inactive derivative. Also present in the pyruvate dehydrogenase complex is a phosphoprotein phosphatase that, by removing the phosphates, restores the decarboxylase to its active form. As seen below, the kinase is activated by an increasing ratio of acetyl CoA/CoA and NADH/NAD^+. Furthermore, this activating effect is counteracted by ADP and pyruvate.

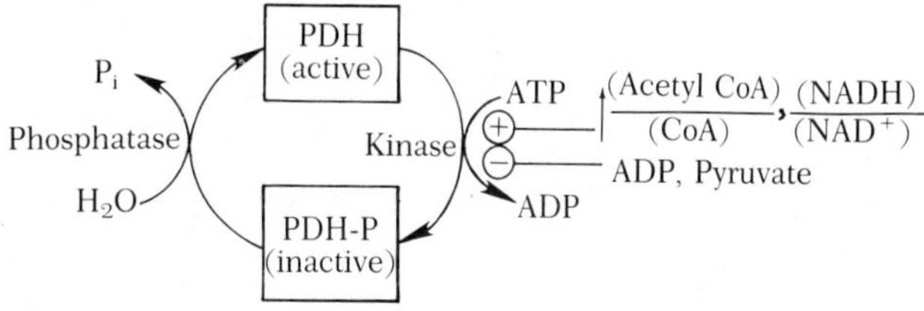

Yeast fermentation

In yeast the breakdown of hexoses proceeds by pathways involving the same chemical reactions as in glycolysis in animal cells. However, the fate of pyruvate is different in that it is converted to ethyl alcohol rather than acetyl CoA. Pyruvate is decarboxylated by a decarboxylase to form the same thiamin-hydroxyethyl intermediate and CO_2 as described on pp. 236 and 237. However, the hydroxyethyl thiamin compound is handled differently; it forms acetaldehyde and is then reduced to ethyl alcohol by means of alcohol dehydrogenase and reduced NAD^+.

$$CH_3-\underset{H}{\overset{OH}{C}}-C(\text{thiazolium}) \xrightarrow{\ \ \text{TPP}\ \ } \underset{\textbf{Acetaldehyde}}{CH_3-\overset{O}{\overset{\|}{C}}-H} \xrightarrow{\text{NADH} \rightarrow \text{NAD}^+} \underset{\textbf{Ethyl alcohol}}{CH_3CH_2OH}$$

Fermentation, in the classic sense, involves the anaerobic formation of ethyl alcohol and CO_2. In effect, the reduction of aldehyde to alcohol plays the same role in yeast as does the conversion of pyruvate to lactate in the mammalian cell. That is to say, the oxidation of NADH restores the supply of NAD^+ and therefore permits a continuation of the oxidation of 3-phosphoglyceraldehyde to 1,3-bisphosphoglycerate. The reduction of aldehyde is the built-in mechanism to continue glycolysis.

If, on the other hand, yeast is grown under aerobic conditions, then the NADH is oxidized by the cytochrome system and the acetaldehyde is oxidized to acetic acid.

Citric acid cycle

Once acetyl CoA is produced, from whatever precursor, it is degraded via the *citric acid cycle*. The cycle may be viewed as a closed system into which are fed the acetyl CoA molecules (see Fig. 11-2). In the course of the function of this cycle the two carbons represented by the acetyl group are lost as CO_2.

To place the system into its proper perspective with regard to glucose metabolism, we must remember that for every glucose molecule two pyruvates are produced at the stage of cytoplasmic glycolysis. These pyruvates are transferred in turn into the mitochondria, where decarboxylation leads to the formation of two acetyl CoA units and two molecules of CO_2. It now remains to be seen how the two acetyl CoAs are further degraded until all six glucose carbons have been released as six molecules of CO_2.

Many of the oxidative processes in the cycle are actually dehydrogenations, in which an enzyme catalyzes the removal of two hydrogens to an acceptor molecule, the coenzyme. The fate of the reduced coenzyme carrier and the ultimate formation of ATP by oxidative phosphorylation is described in Chapter 7.

Condensation to form citrate

As indicated previously, the formation of acetyl CoA is the link between glycolysis and the citric acid cycle. Entrance of the 2-carbon acetyl group into the citric acid cycle occurs by a condensation of acetyl CoA with oxaloacetate to yield citric acid. The *condensing enzyme*, called *citrate synthase*, catalyzes the formation of a carbon-to-carbon bond and the transfer of a hydrogen from the methyl group of the acetyl group to the carbonyl oxygen of the oxaloacetate. The methyl carbon of acetyl CoA attaches to the α-carbon of oxaloacetate as follows:

$$\text{Acetyl CoA}\ CH_3\overset{O}{\overset{\|}{C}}-SCoA + \underset{\textbf{Oxaloacetate}}{O{=}C(-CO_2^-)CH_2CO_2^-} \xrightarrow{\text{Citrate synthase}} \text{CoASH} + \underset{\textbf{Citrate}}{HO-\underset{CH_2CO_2^-}{\overset{CH_2CO_2^-}{C}}-CO_2^-}$$

The products of this condensation are both citric acid and CoA. The enzyme catalyzes the removal of a proton from the acetyl CoA methyl group with the resultant formation of a carbanion. The carbanion then attacks the carbonyl carbon of the oxaloacetate. Simultaneously, the intermediate citryl thioester is hydrolyzed to yield CoA and citric acid. Because of the hydrolysis of the high-energy thioester, the equilibrium of the reaction is far toward citrate formation, and therefore the reaction is physiologically unidirectional.

Interconversion of citrate, *cis*-aconitate, and isocitrate

The enzyme *aconitase* catalyzes the interconversion of citrate, *cis*-aconitate, and isocitrate. This reversible process represents the removal and restoration of a water molecule in these three tricarboxylic acids. The equilibrium mixture at 38° C and pH 6.8 consists of approximately 90% citrate, 4% *cis*-aconitate, and 6% isocitrate. Because the citrate molecule is symmetrical (see below), the enzyme could be expected to remove the hydroxyl group and one of two hydrogens, either from the acetyl portion *(color)* or from the oxaloacetate part of the molecule *(black)*. Aconitase, however, sees citrate as an asymmetrical molecule; dehydration and hydration occur only with respect to the carboxymethyl of the oxaloacetate portion. The manner in which this specificity occurs may be understood in terms of the three-point attachment hypothesis of Ogston (see Fig. 5-9). Furthermore, both the removal and addition of water are *trans* reactions. The unique specificity of aconitase determines that the incoming, or two "new," acetyl carbons are retained in the cycle at least for one complete circuit, whereas two carboxyl groups of the oxaloacetate, i.e., two "old" carbons, are lost as CO_2. The loss of two carbons, therefore, is not a random process but a directed one.

$$\underset{\textbf{Citrate}}{\begin{array}{c} CH_2CO_2^- \\ | \\ HO\text{—}C\text{—}CO_2^- \\ | \\ CH_2CO_2^- \end{array}} \underset{\text{Aconitase}}{\overset{H_2O}{\rightleftharpoons}} \underset{\textbf{\textit{cis}-Aconitate}}{\begin{array}{c} CH_2CO_2^- \\ | \\ C\text{—}CO_2^- \\ \| \\ CHCO_2^- \end{array}} \overset{H_2O}{\rightleftharpoons} \underset{\textbf{Isocitrate}}{\begin{array}{c} CH_2CO_2^- \\ | \\ H\text{—}C\text{—}CO_2^- \\ | \\ HO\text{—}C\text{—}CO_2^- \\ H \end{array}}$$

Citrate itself plays several important roles in metabolism other than in the citric acid cycle. For example, it readily passes through the mitochondrial membrane and then may act, if it should accumulate, as a negative effector on phosphofructokinase. Thus it inhibits this enzyme by allosteric binding and thereby slows down further metabolism of glucose by way of glycolysis. Its cytoplasmic accumulation also results in citrate cleavage (citrate-cleavage enzyme) to form acetyl CoA and oxaloacetate. Furthermore, it stimulates the activity of acetyl CoA carboxylase. These are important in the biosynthesis of fatty acids (Chapter 12).

Conversion of isocitrate to α-ketoglutarate

Isocitrate dehydrogenase catalyzes the dehydrogenation of isocitrate to form an intermediate α-keto acid, oxalosuccinate. This reaction is catalyzed in most animal cells by two distinct dehydrogenases, one requiring $NADP^+$ and the other requiring NAD^+ as coenzyme. In animal tissues approximately 75% to 90% of the $NADP^+$-specific isocitrate dehydrogenase is found in the cytosol, whereas all the NAD^+ enzyme is in the mitochondria. These enzymes require either Mg^{++} or Mn^{++} for activity. Once the oxalosuccinate has been formed, decarboxylation to α-ketoglutarate occurs spontaneously. The oxalosuccinate apparently never leaves the enzyme. In animal cells the NAD^+-specific enzyme requires ADP as an activator. It acts to increase the affinity of the enzyme for isocitrate; in its absence the enzyme rapidly becomes inactive. Furthermore, the enzyme is very sensitive to the accumulation of NADH. Isocitrate dehydrogenase therefore serves as a sensitive regulator of the activity of the citric acid cycle. NADH must be oxidized by the electron transport system for dehydrogenation of isocitrate to continue. Accumulation of ADP, signaling a need for ATP, stimulates the activity of this enzyme. Too much ATP may inhibit the system by chelating the essential metal ions Mn^{++} and Mg^{++}.

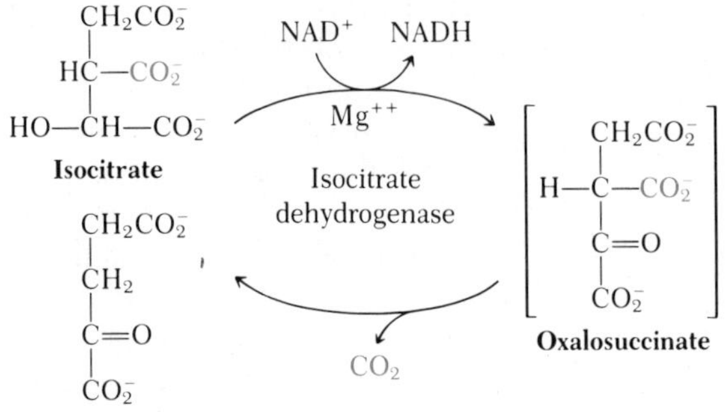

$CH_2CO_2^-$ | $HC—CO_2^-$ | $HO—CH—CO_2^-$

Isocitrate

NAD^+ → NADH; Mg^{++}; Isocitrate dehydrogenase

[$CH_2CO_2^-$ | $H—C—CO_2^-$ | C=O | CO_2^-]

Oxalosuccinate

→ CO_2

$CH_2CO_2^-$ | CH_2 | C=O | CO_2^-

α-Ketoglutarate

It is important to note that the first loss of CO_2 in the citric acid cycle occurs at the step catalyzed by isocitrate dehydrogenase. This reaction appears to be unidirectional; the carboxylation of α-ketoglutarate at the expense of NADH has not been demonstrated. On the other hand, the mitochondrial $NADP^+$-dependent enzyme is capable of reversing the reaction. This is required in the isocitrate shuttle described in Chapter 7, whereby reducing equivalents may be transported out of the mitochondrion.

Conversion of α-ketoglutarate to succinyl CoA

Oxidative decarboxylation of α-ketoglutarate is catalyzed by a multienzyme complex similar to that described for the conversion of pyruvate to acetyl CoA. This system is called *α-ketoglutarate dehydrogenase*. The decarboxylating enzyme (α-ketoglutarate decarboxylase) catalyzes the formation of the succinate semialdehyde with thiamin pyrophosphate (α-hydroxy-γ-carboxyl propylthiamin pyrophosphate), as seen below. Simultaneously CO_2 is evolved. The succinyl group is transferred to the lipoic acid component of *transsuccinylase* to yield the succinyl lipoyl enzyme. Then the succinyl group is joined to CoA, leaving the lipoic acid in its reduced state. The disulfide bond is restored by the action of the flavoprotein *dihydrolipoamide dehydrogenase*. This is the same enzyme found in the pyruvate dehydrogenase. Finally, the riboflavin is reoxidized by NAD^+. The overall process is comparable with that described for pyruvate dehydrogenase (Fig. 11-7), except that the succinyl replaces the acetyl group throughout the reaction.

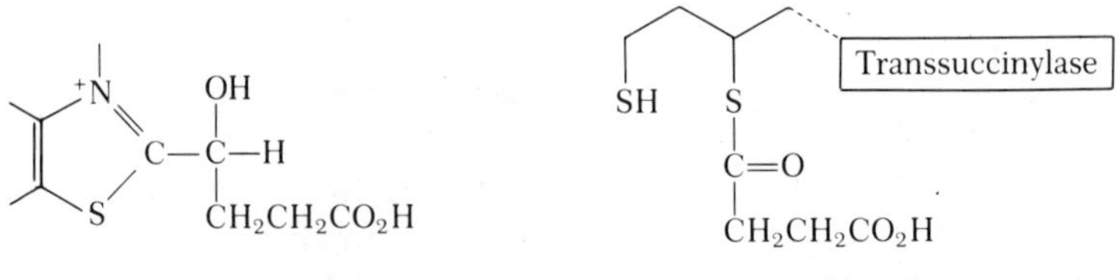

^+N, S (thiazolium ring) C—C(OH)—H | $CH_2CH_2CO_2H$

α-Hydroxy-γ-carboxyl propylthiamin pyrophosphate

SH, S—C=O | $CH_2CH_2CO_2H$ — Transsuccinylase

Succinyl lipoyl transsuccinylase

The CO_2 released in this reaction is the second one liberated in the citric acid cycle. The thioester bond formed first with the lipoyl group and then with CoA are high-energy bonds. Because of the large free-energy change involved in the overall oxidative decarboxylation ($\Delta G^{\circ\prime} = -8.0$ kcal), the reaction is not reversible and is sufficiently large to serve for the subsequent formation of ATP.

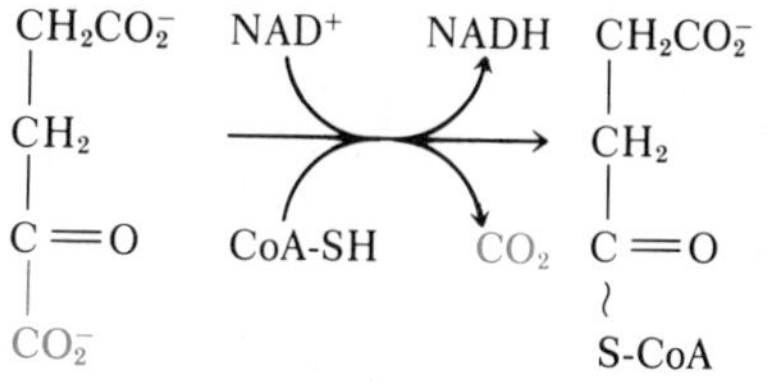

$$\underset{\alpha\text{-Ketoglutarate}}{\begin{array}{c} CH_2CO_2^- \\ | \\ CH_2 \\ | \\ C{=}O \\ | \\ CO_2^- \end{array}} \xrightarrow[\text{CoA-SH} \;\to\; CO_2]{NAD^+ \;\to\; NADH} \underset{\text{Succinyl CoA}}{\begin{array}{c} CH_2CO_2^- \\ | \\ CH_2 \\ | \\ C{=}O \\ \wr \\ \text{S-CoA} \end{array}}$$

α-Ketoglutarate is also important because it serves as a point at which amino acid and carbohydrate metabolism interact. It can be transaminated to glutamate, and conversely glutamate can be converted to α-ketoglutarate, which in turn is used in carbohydrate metabolism.

Formation of succinate

The energy conserved from the previous step as the thioester bond is now liberated by means of the enzyme *succinyl thiokinase*. The enzyme catalyzes the synthesis of GTP or inosine triphosphate (ITP) at the expense of the thioester. It is specific for either guanosine diphosphate or inosine diphosphate; inorganic phosphate is removed from the environment, and GTP (or ITP) is produced. The overall reaction is believed to occur in three steps so as to conserve the energy of the thioester compound.

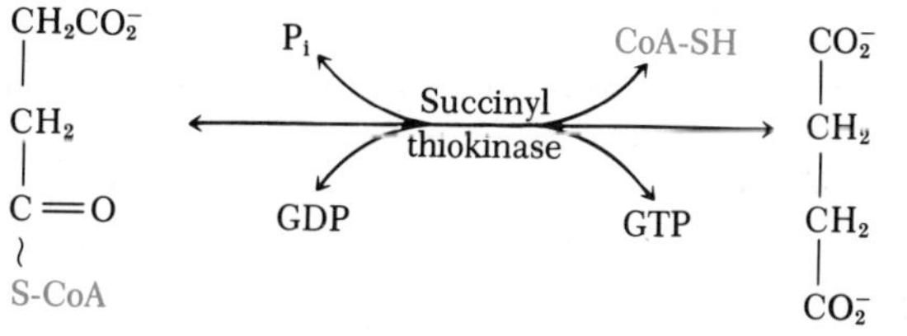

$$\underset{\text{Succinyl CoA}}{\begin{array}{c} CH_2CO_2^- \\ | \\ CH_2 \\ | \\ C{=}O \\ \wr \\ \text{S-CoA} \end{array}} \underset{\text{GDP} \;\to\; \text{GTP}}{\overset{P_i \;\to\; \text{CoA-SH}}{\xleftrightarrow{\text{Succinyl thiokinase}}}} \underset{\text{Succinate}}{\begin{array}{c} CO_2^- \\ | \\ CH_2 \\ | \\ CH_2 \\ | \\ CO_2^- \end{array}}$$

Thus

$$\text{Succinyl} \sim \text{S-CoA} + \text{Enzyme} \longleftrightarrow \text{CoA-S} \sim \text{Enzyme} + \text{Succinate}$$

$$\text{CoA-S} \sim \text{Enzyme} + P_i \xleftrightarrow{Mg^{++}} \text{Phosphoryl} \sim \text{Enzyme} + \text{CoA-SH}$$

$$\text{Phosphoryl} \sim \text{Enzyme} + \text{GDP} \xleftrightarrow{Mg^{++}} \text{Enzyme} + \text{GTP}$$

GTP plays a number of important metabolic roles, e.g., in protein synthesis. It also may be converted into ATP by the action of the enzyme *nucleoside diphosphate kinase*. Therefore this step in the citric acid cycle may be viewed as one producing ATP at the substrate level.

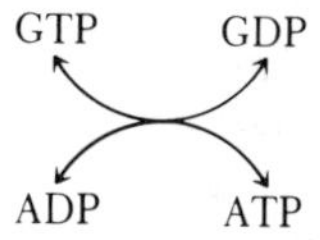

$$\text{GTP} + \text{ADP} \longrightarrow \text{GDP} + \text{ATP}$$

Succinyl CoA derived from the citric acid cycle is also an important intermediate (with glycine) in the biosynthesis of δ-aminolevulinic acid, the precursor of the porphyrin ring structure (Chapter 16).

Dehydrogenation of succinate to fumarate

The flavoprotein enzyme *succinate dehydrogenase* catalyzes a reversible dehydrogenation of succinate to produce fumarate.

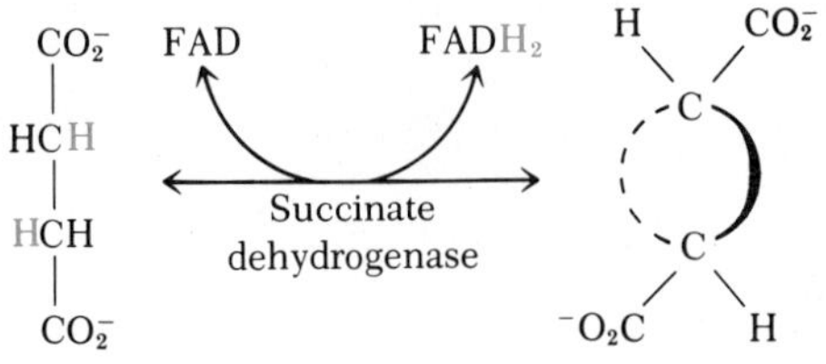

$$\text{Succinate } (CO_2^- - HCH - HCH - CO_2^-) \xrightarrow[\text{Succinate dehydrogenase}]{FAD \rightarrow FADH_2} \text{Fumarate } (H(CO_2^-)C{=}C(H)CO_2^-)$$

Succinate dehydrogenase has been crystallized from heart muscle mitochondria and shown to be a flavoprotein with a molecular weight of about 100,000. It consists of two subunits, a larger (mol wt = 70,000) and a smaller one (mol wt = 27,000). The former has the covalently bound FAD (Chapter 6) as well as nonheme iron atoms. The smaller polypeptide is an iron sulfide–containing protein (Chapter 7) that is closely associated with the flavoprotein. The two in turn are tightly bound to the mitochondrial membrane. Dehydrogenation of succinate occurs by the *trans* elimination of hydrogens to produce fumaric acid only. The *cis* isomer, maleic acid, is not produced and, in fact, is extremely toxic to the organism. Another dicarboxylic acid, malonic acid, is a specific competitive inhibitor of succinate dehydrogenase; its administration results in the accumulation of citrate and other preceding acids of the citric acid cycle.

Succinate also serves as a substrate for the electron transport system. It reduces coenzyme Q (CoQ) in mitochondria (Chapter 7); the overall oxidation to water of hydrogens from a mole of succinate involves the formation of 2 moles of ATP. In this process the FAD of succinate dehydrogenase serves as the first acceptor of the substrate hydrogens; the role of iron-sulfur centers is currently unknown.

Hydration of fumarate

The reversible insertion of water into the double bond of fumarate is a *trans* addition and yields only L-malic acid. *Fumarase,* the enzyme catalyzing this reaction, has been crystallized from pig hearts. It has a molecular weight of about 200,000 and has four identical subunits. Free sulfhydryl groups are required for its activity. The importance of stereospecificity is illustrated in the following diagram.

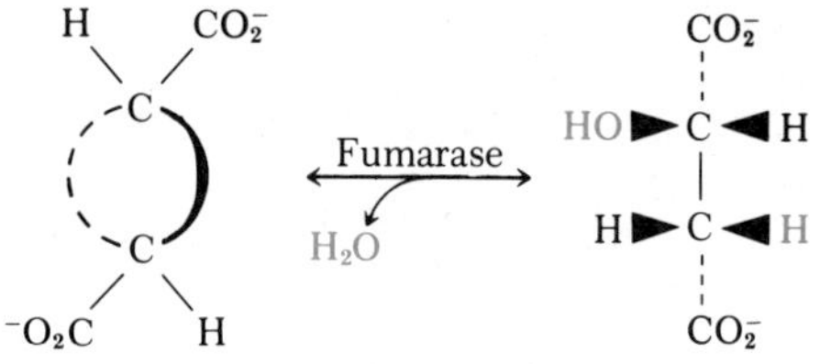

$$\text{Fumarate } (H(CO_2^-)C{=}C(H)CO_2^-) \xrightarrow[H_2O]{\text{Fumarase}} \text{L-Malate } (CO_2^- - C(HO)(H) - C(H)(H) - CO_2^-)$$

L-Malate readily penetrates the mitochondrial membrane so that it can either leave or enter and participate in the citric acid cycle. The importance of this role of malate is considered later as a means of transporting *reducing equivalents* from the cytosol into the mitochondria—the so-called shuttle mechanism (Chapter 7).

Oxidation of L-malate to oxaloacetate

The reversible dehydrogenation of L-malate to oxaloacetate is catalyzed by *malate dehydrogenase* and requires NAD^+. Although the reaction equilibrium is in favor of malate, the low level of oxaloacetate causes the reaction to proceed in the direction of the dehydrogenation.

$$\text{L-Malate} \xrightleftharpoons[\text{Malate dehydrogenase}]{\text{NAD}^+ \;\rightarrow\; \text{NADH}} \text{Oxaloacetate}$$

L-Malate: CO_2^-—CH_2—HO—C—H—CO_2^-

Oxaloacetate: CO_2^-—CH_2—C=O—CO_2^-

Review of the citric acid cycle

Fig. 11-8 provides an overview of the citric acid cycle and emphasizes the location of crucial control points. It is apparent from this figure that high ratios of ATP/ADP, acetyl CoA/CoA, and NADH/NAD$^+$ will serve as signals to inhibit the operation of the cycle. An excess of ATP, NADH, and acetyl CoA occurs when the energy supply is sufficient for the cell. As energy is used, the ratio of ATP/ADP, for example, declines and the inhibition of the cycle is relieved. Increasing ADP concentrations stimulate the isocitrate dehydrogenase and thereby increase the activity of the citric acid cycle.

The following features of the cycle are important to appreciate:

1. Three carbons enter the mitochondrion as pyruvate; one is lost as CO_2 in the conversion to acetyl CoA. Two carbons then enter the citric acid cycle as acetyl CoA and subsequently leave as CO_2, one in the isocitrate dehydrogenase reaction and the other in the α-ketoglutarate dehydrogenase step. A complete turn of the cycle consumes one molecule of oxaloacetate and in turn restores it again. Since for every two carbons that enter two also are released, it is impossible for the cycle to produce a net amount of oxaloacetate from the incoming acetyl groups.

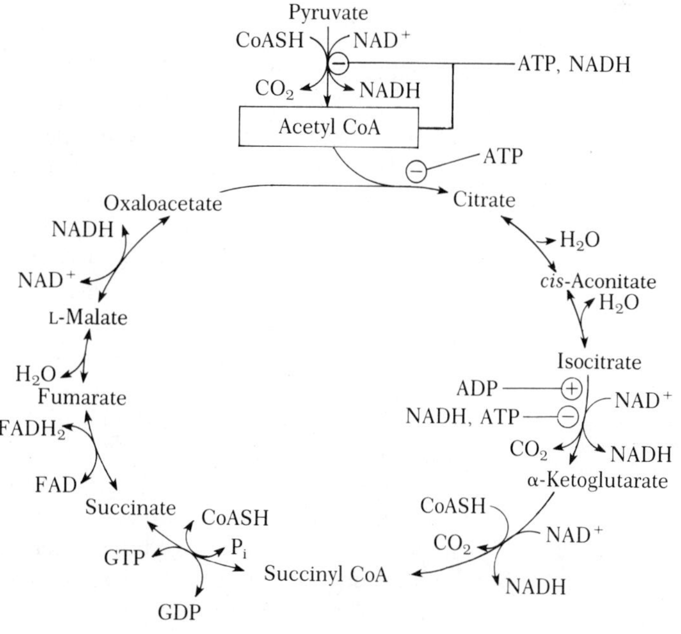

Figure 11-8
Metabolic regulation of the citric acid cycle.

2. The oxidation of acetyl CoA and the resulting oxidative decarboxylation yields a total of 12 ATPs. These are the following:

Isocitrate to α-ketoglutarate	3 ATP
α-Ketoglutarate to succinyl CoA	3 ATP
Succinyl CoA to succinate	1 GTP (ATP)
Succinate to fumarate	2 ATP
Malate to oxaloacetate	3 ATP
Total	12 ATP

A total of 12 ATPs, or 87.6 kcal, is realized per turn of the cycle (15 ATPs are obtained beginning with pyruvate). As discussed in Chapter 7, the oxidation of one NADH by way of oxidative phosphorylation yields three molecules of ATP, whereas the oxidation of $FADH_2$ (succinate dehydrogenase) yields two.

3. Some citric acid cycle constituents serve as points of entry for compounds from other pathways. Some of these intermediates also are involved in biosynthetic pathways.
   a. *Amino acid metabolism* (Chapter 13). Glutamate is reversibly converted to α-ketoglutarate, aspartate to oxaloacetate, and alanine to pyruvate via transamination. Also, α-ketoglutarate is reversibly converted to glutamate by glutamate dehydrogenase.
   b. *Lipid metabolism.* Acetyl CoA and citrate are the starting points for the biosynthesis of fatty acids (Chapter 12). Acetyl CoA is also the product of the breakdown of fatty acids and the starting point of steroid synthesis (Chapter 12).
   c. *Porphyrin biosynthesis.* Succinyl CoA together with glycine is a substrate for the biosynthesis of porphyrins and therefore is essential in the production of hemoglobin, cytochromes, and other hemoproteins.
   d. *Nitrogen metabolism.* Aspartate serves as a source of nitrogen in some biosynthetic pathways such as the production of argininosuccinate and purines. The end product of these processes is fumarate, which then enters the cycle.

Gluconeogenesis

As mentioned previously, an essential function of the liver is the control of the systemic blood glucose level. When needed, the liver not only remobilizes glucose from glycogen but also has the capacity of synthesizing this hexose. The biosynthesis of glucose from noncarbohydrate carbon chains is called *gluconeogenesis*. Gluconeogenesis and glycolysis are opposing metabolic pathways. These pathways share a number of enzymes (Fig. 11-9). In addition, gluconeogenesis has four unique enzymic steps: (1) the carboxylation of pyruvate to oxaloacetate and (2) its subsequent decarboxylation to PEP, (3) the hydrolysis of fructose 1,6-bisphosphate to yield fructose 6-phosphate, and (4) the hydrolysis of glucose 6-phosphate to glucose.

Production of phosphoenolpyruvate (PEP) from pyruvate

Consider the fate of the three carbons of L-alanine. Alanine is first converted to pyruvate by transamination (Chapter 13). The pyruvate, however, cannot be directly converted to PEP by a reversal of the pyruvate kinase step; as mentioned earlier, this step is unidirectional, and its reversal does not occur under physiological conditions. In gluconeogenesis, therefore, pyruvate is first carboxylated in the mitochondria to form oxaloacetate. This serves to replenish oxaloacetate under metabolic conditions that place a particular drain on the available supply of the α-keto acid. Gluconeogenesis is such a situation. The diversion of pyruvate into

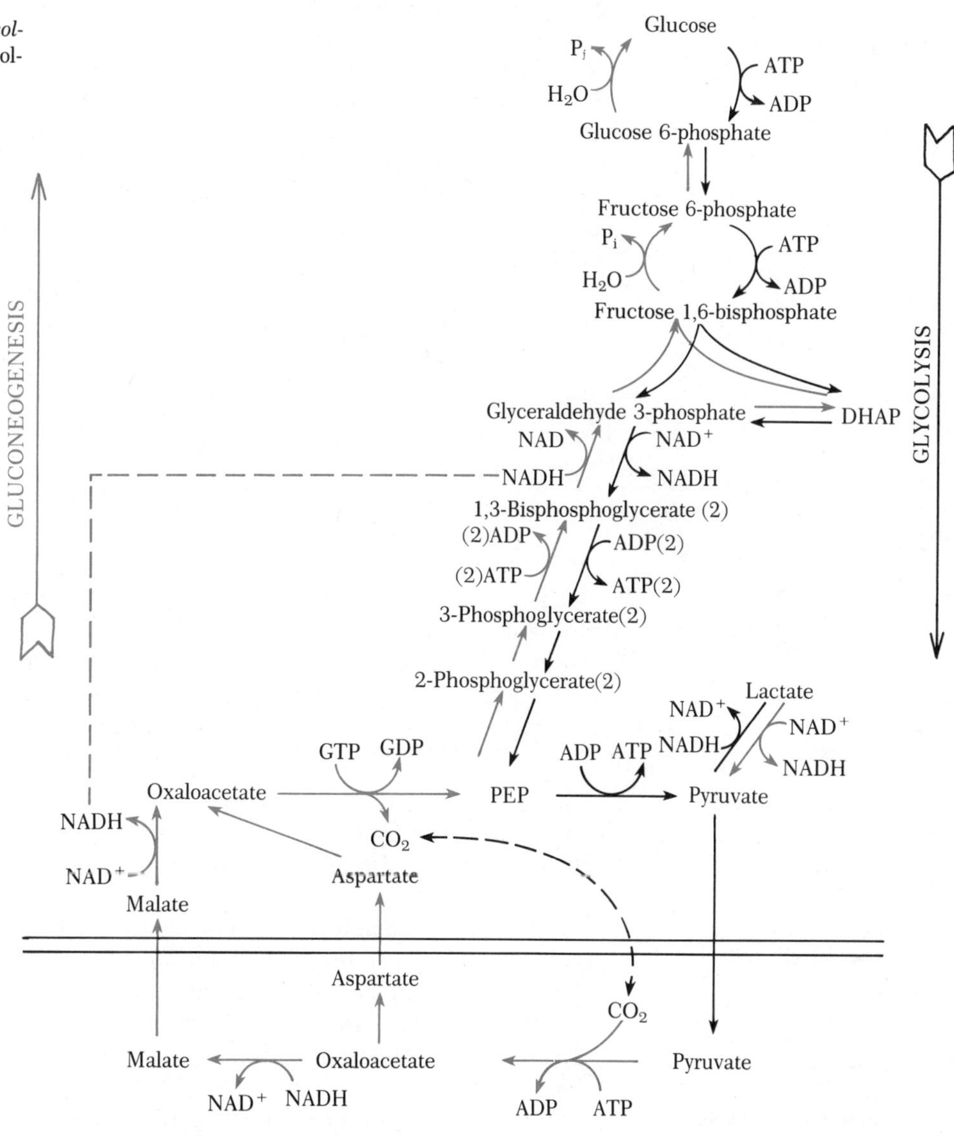

Figure 11-9
Pathway of gluconeogenesis *(color)* compared with that of glycolysis *(black)*.

oxaloacetate is called an *anaplerotic* process; it resupplies the system with oxaloacetate. Carboxylation of pyruvate is accomplished by transferring CO_2 to the methyl group of pyruvate. The enzyme is *pyruvate carboxylase* and requires for its activity the coenzyme biotin (Chapter 6). ATP and magnesium (or manganese) ions as well as K^+ are needed. The reaction also has an absolute requirement for acetyl CoA as a positive effector.

The overall reaction is as follows:

$$\text{Pyruvate} + HCO_3^- + \text{ATP} \rightarrow \text{Oxaloacetate} + \text{ADP} + P_i$$

Biotin is *N*-carboxylated in an initial ATP-dependent step, followed by a second step in which the CO_2 is transferred to the methyl group of pyruvate to yield oxaloacetate (Chapter 6).

Since the conversion of oxaloacetate to PEP involves a cytoplasmic enzyme, *phosphoenolpyruvate carboxykinase,* it is essential for the oxaloacetate to leave the mitochondrion. The mitochondrial membrane is impermeable to the α-keto acid; therefore it must be converted to either L-malate (malate dehydrogenase and NADH) or aspartate (aminotransferase) and glutamate; see Chapter 13). Malate is transported across the inner mitochondrial membrane by a specific antiport system (Chapter 19) in exchange for inorganic phosphate, other dicarboxylic acids, citrate, or α-ketoglutarate. Aspartate is exchanged in an antiport system for glutamate. In the cytoplasm malate is reoxidized to oxaloacetate by a cytoplasmic malate dehydrogenase; aspartate is reconverted to oxaloacetate by aminotransferase.

Oxaloacetate is then decarboxylated to PEP by phosphoenolpyruvate carboxykinase. In animal cells the energy for this reaction is provided by GTP or ITP.

$$\underset{\textbf{Oxaloacetate}}{CO_2^- - CH_2 - C(=O) - CO_2^-} \xrightleftharpoons[GTP \;\; Mn^{++},\ K^+ \;\; GDP]{\text{PEP Carboxykinase} \;\; CO_2} \underset{\textbf{Phosphoenol pyruvate}}{CH_2 = C(CO_2^-) - O - P(=O)(O^-)_2}$$

The combination of the two steps—pyruvate carboxylase and phosphoenolpyruvate carboxykinase—provides a key mechanism whereby the cell circumvents the irreversibility of the pyruvate kinase reaction. The overall process is as follows:

$$\begin{array}{ll} & \text{Pyruvate} + HCO_3^- + ATP \rightarrow OAA + ADP + P_i \\ & OAA + GTP \rightarrow PEP + GDP + HCO_3^- \\ \hline \text{Sum:} & GTP + \text{Pyruvate} + ATP \rightarrow PEP + GDP + P_i + ADP \end{array}$$

This reaction could simply recycle the various products at the expense of ATP. Such a "futile" cycle is prevented by the fact that high levels of ATP, which favor gluconeogenesis, also inhibit the pyruvate kinase.

Conversion of PEP to fructose 1,6-bisphosphate

PEP is converted to fructose 1,6-bisphosphate by using the freely reversible glycolytic enzymes enolase, phosphoglyceromutase, phosphoglycerate kinase, glyceraldehyde 3-phosphate dehydrogenase, triosephosphate isomerase, and aldolase. This requires ATP to form the 1,3-bisphosphoglycerate and NADH to produce the glyceraldehyde 3-phosphate (see Fig. 11-9).

Conversion of fructose 1,6-bisphosphate to fructose 6-phosphate

Since, in the glycolytic pathway, the phosphorylation of fructose 6-phosphate to fructose 1,6-bisphosphate is unidirectional and not reversible under physiological conditions, a special enzyme exists that catalyzes the hydrolysis of the bisphosphate. This enzyme, which is unique to the gluconeogenic pathway, is *fructose 1,6-bisphosphatase.*

$$\textbf{Fructose 1,6-bisphosphate} \xrightarrow[H_2O \;\rightarrow\; P_i]{\text{Fructose 1,6-bisphosphatase}} \textbf{Fructose 6-phosphate}$$

Fructose 1,6-bisphosphatase is an important factor in the regulation of gluconeogenesis and is responsive to several allosteric effectors. High levels of ATP favor its action and thus gluconeogenesis, whereas high levels of AMP and ADP act as negative effectors and inhibit gluconeogenesis. This regulatory enzyme is found only in the liver, kidneys, and intestinal epithelia, i.e., the gluconeogenic organs.

Conversion of glucose 6-phosphate to glucose

Fructose 6-phosphate, obtained as just described, is converted to glucose 6-phosphate by using the freely reversible glycolytic enzyme phosphoglucoisomerase. The glucose 6-phosphate is then hydrolyzed to free glucose, a reaction catalyzed by *glucose 6-phosphatase*. This enzyme is found only in gluconeogenic organs.

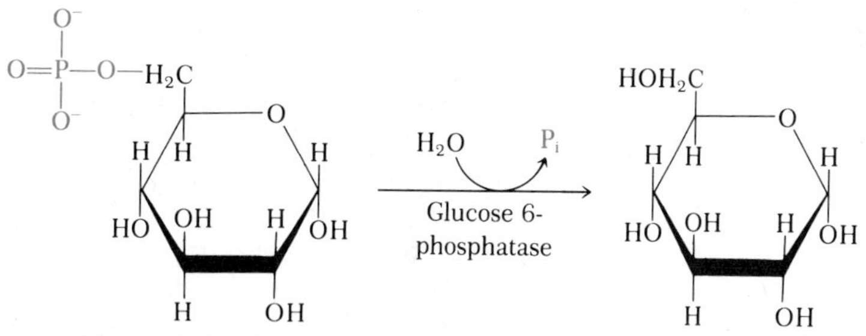

O^- | $O=P-O-H_2C$ | O^- ... $\xrightarrow[\text{Glucose 6-phosphatase}]{H_2O \quad P_i}$ HOH_2C ...

D-Glucose 6-phosphate → D-Glucose

In addition to its importance in gluconeogenesis, glucose 6-phosphatase is essential in the mobilization of glucose from glycogen. This role of the enzyme is described later.

Review of gluconeogenesis

Fig. 11-9 shows that the gluconeogenic pathway shares a number of its steps with that of glycolysis. Four enzymes, however, are unique: pyruvate carboxylase, phosphoenolpyruvate carboxykinase, fructose 1,6-bisphosphatase, and glucose 6-phosphatase. Seven enzymes are shared with glycolysis: phosphoglucoisomerase, aldolase, triosephosphate isomerase, glyceraldehyde 3-phosphate dehydrogenase, phosphoglycerate kinase, phosphoglyceromutase, and enolase. Glycolytic enzymes generally are found in all tissues; the four unique enzymes are only in the gluconeogenic organs.

Although L-alanine was used to illustrate the pathway, numerous substances may provide carbon chains for the production of glucose. Lactate, for example, is a very important precursor, since it is produced during muscle metabolism and is returned to the liver for use in the gluconeogenic process. Other sources are glutamate and aspartate, which by transamination yield α-ketoglutarate and oxaloacetate, respectively. Of the lipids, only the glycerol of the triglyceride molecules and propionyl CoA derived from odd-numbered fatty acids can yield glucose. The carbons of the even-numbered fatty acids are degraded entirely to acetyl CoA (Chapter 12), which cannot yield a net production of glucose.

In retrospect, the gluconeogenic process from lactate involves the following:

$$2 \text{ Lactate} + 4 \text{ ATP} + 2 \text{ GTP} \rightarrow \text{Glucose} + 4 \text{ ADP} + 2 \text{ GDP} + 6 \text{ P}_i$$

Therefore the equivalent of 6 ATPs is required to produce one glucose from two lactates. Comparison of this equation with that on p. 235 shows that gluconeogenesis is not simply a direct reversal of glycolysis.

Storage of carbohydrate as glycogen

As previously stated, the liver serves as the primary reservoir for carbohydrate by converting dietary glucose into the polysaccharide glycogen. This biosynthetic process is called *glycogenesis*. Once produced, glycogen is a ready source of glucose by a mobilizing process: *glycogenolysis*. Glycogenesis and glycogenolysis, it will be seen, are not the reversal of a single metabolic pathway but rather are two different and independently regulatable processes. Glycogen storage occurs in most tissues, especially in the musculature. Here it serves as a means of providing carbohydrate for metabolism and the production of ATP. In this way the muscle cells are ensured of a continuing supply of energy despite intermittent food intake.

Glycogen is a homopolysaccharide (Chapter 23) consisting entirely of glucose subunits. Practically all mammalian tissue cells contain some glycogen, but the liver and muscles are the richest sources. The liver, which weighs approximately 1800 gm in a 70-kg man, usually contains about 108 gm of glycogen (6% by weight), whereas the muscle tissues, which weigh about 35 kg, store 245 gm (0.7% by weight).

In glycogen the glucose units are attached to one another by α-1,4-glucosidic bonds. However, this polysaccharide is distinguished by its highly branched nature. Each individual branch has approximately 7 to 12 glucose units and is separated from its neighbor by at least 3 units. These short chains are connected by α-1,6-glucosidic bonds at the branch points. This results in a treelike pattern (Fig. 11-10). Glycogen does not exist in the form of homogeneous molecules that all have the same molecular weight and possess the same number of monosaccharide units. It is a mixture of molecules of varying size, some with molecular weights as high as

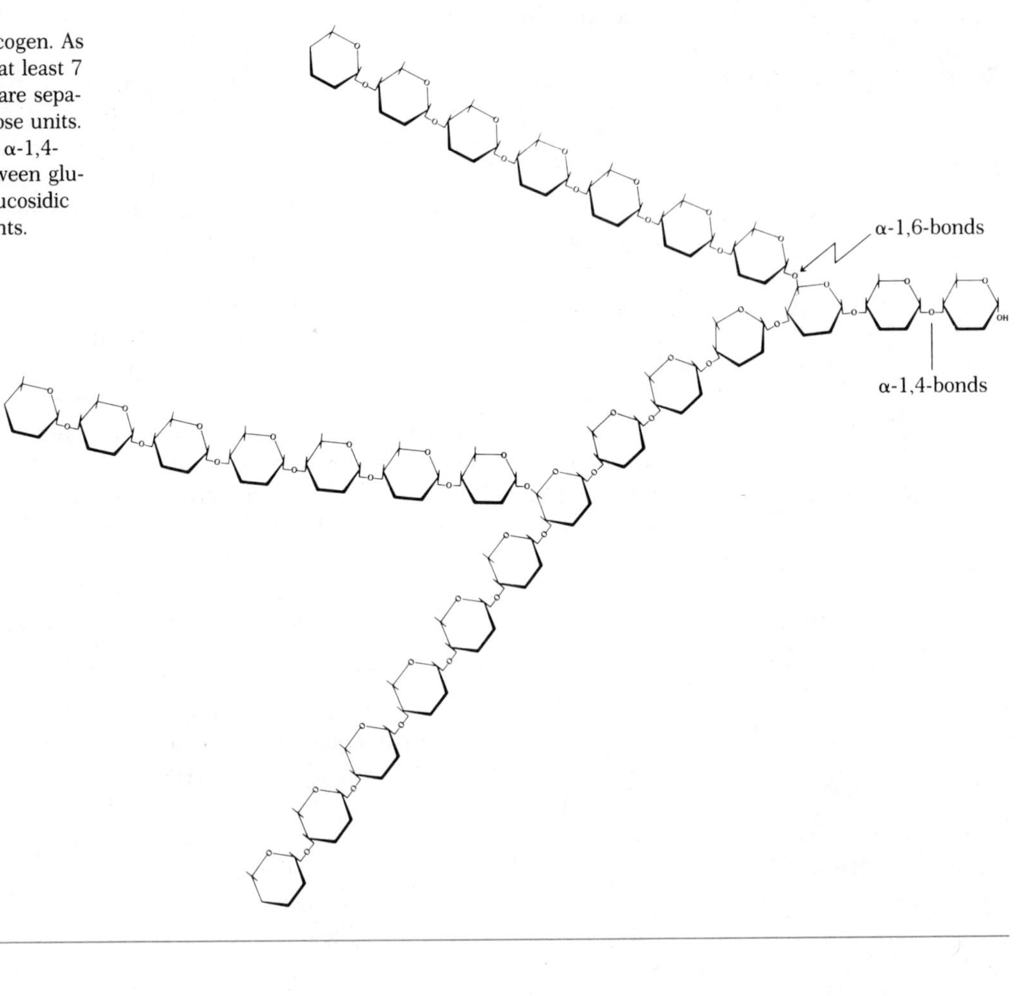

Figure 11-10
Branched nature of glycogen. As seen, the branches are at least 7 glucose units long and are separated by at least 3 glucose units. The bonds involved are α-1,4-glucosidic linkages between glucose units and α-1,6-glucosidic units at the branch points.

900 million. The glycogen molecules aggregate into insoluble granules that deposit in the cell's cytoplasm. This deposition of glycogen as granules is an important factor in storing glucose molecules. As mentioned earlier, the sugar readily diffuses into the liver cell and is phosphorylated, thereby trapping it within the cell. If large amounts were stored as hexose phosphates, it would be accompanied by an osmotic accumulation of water. To prevent the swelling and disruption of the cells, the glucose units are converted to the insoluble polymer glycogen, which does not contribute to the osmotic properties of the cytosol. After a meal heavy in carbohydrates as much as 10% of the liver mass may be glycogen.

As will be seen, glycogen is degraded by *glycogen phosphorylase*, an enzyme closely associated with the glycogen granule. Phosphorylase acts at the nonreducing ends of the polymer. Fig. 11-10 shows that the glycogen molecules have many nonreducing ends and only a single reducing terminal. Therefore the branched structure provides many sites of action for phosphorylase.

Glycogenolysis

Glycogenolysis occurs primarily by phosphorolysis or the insertion of inorganic phosphate into the α-1,4-glucosidic bonds. The products produced are glucose 1-phosphate and the residual glycogen, which is then reduced by one glucose unit (Fig. 11-11). This process is catalyzed by the enzyme *glycogen phosphorylase* and, as just stated, occurs at the nonreducing chain terminals. Phosphorolysis proceeds on all outer chains until approximately four glucose units remain on each side of the branch point. Phosphorolysis cannot continue until the branch is removed. This is accomplished by a *debranching enzyme*, which acts in two distinct steps. First, it acts as an *α-1,4 to α-1,4 glucan transferase* and transfers three of the remaining residues as a trisaccharide unit to the end of another branch. This exposes the glucose unit, which is attached by the α-1,6-glucosidic bond (Fig. 11-12). In the second step the enzyme acts as an *amylo 1,6-glucosidase*, catalyzing the hydrolysis of the α-1,6-glucosidic bond. This step releases free glucose. Further splitting of the glycogen can then proceed by the action of phosphorylase until the neighborhood of

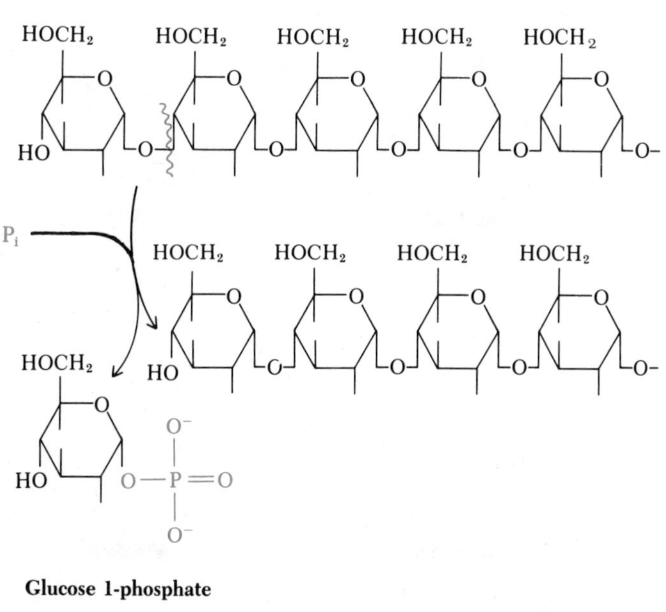

Figure 11-11
Glycogenolysis. Phosphorolytic degradation of glycogen by phosphorylase.

Figure 11-12
Two-step action of the glycogen debranching enzyme. The enzyme is believed to have two active sites, one a 1,4 → α-1,4-glucan transferase and the other an amylo-1,6-glucosidase.

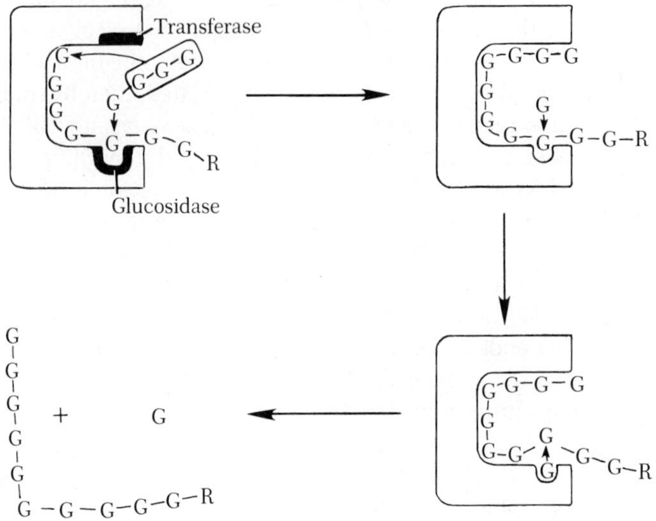

another branch point is reached. The actions of the transferase and debranching enzymes are repeated. The partial or even complete cleavage of the glycogen molecule is accomplished with the formation of glucose 1-phosphate and free glucose from the hydrolytic cleavage of the 1,6-glucosidic bonds. Eventually the glucose 1-phosphate is converted to glucose 6-phosphate by *phosphoglucomutase*. Phosphoglucomutase catalyzes the interconversion of glucose 6-phosphate and glucose 1-phosphate. This reversible reaction requires the presence of a small amount of glucose 1,6-bisphosphate. As seen below, the enzyme catalyzes the transfer of the carbon-6 phosphate of the bisphosphate to the carbon-1 of the glucose 6-phosphate. In this way the products formed are glucose 1-phosphate and a new molecule of glucose 1,6-bisphosphate.

Glucose 6-phosphate + **Glucose 1,6-bisphosphate**

Phosphoglucomutase

Glucose 1,6-bisphosphate + **Glucose 1-phosphate**

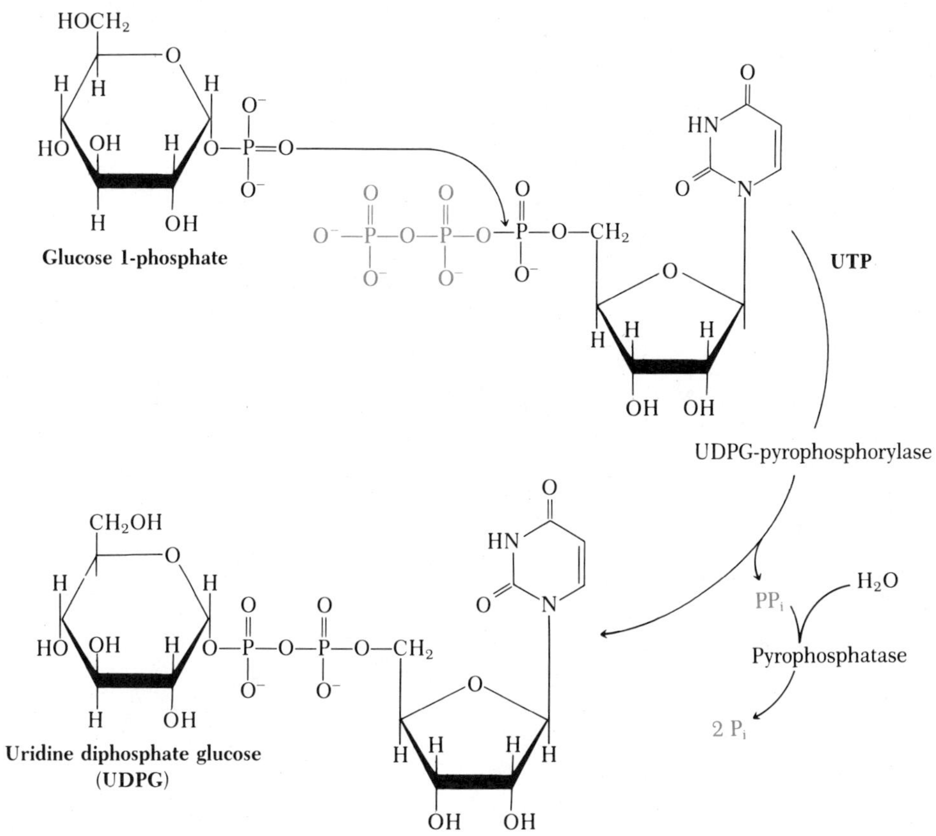

Figure 11-13
Biosynthesis of UDP-glucose from glucose 1-phosphate and UTP. Catalyzed by UDPG-pyrophosphorylase.

In the liver the glucose 6-phosphate may then be hydrolyzed to free glucose by glucose 6-phosphatase.

Glycogenesis

Glycogenesis begins with the intermediate substance that is central to glucose metabolism: glucose 6-phosphate. The first step leading to glycogenesis is the conversion of the glucose 6-phosphate to glucose 1-phosphate, a reaction catalyzed by *phosphoglucomutase*. As previously noted, this process requires Mg^{++} and a small amount of glucose 1,6-bisphosphate.

Glucose 1-phosphate is converted to UDP-glucose. This reaction, catalyzed by *UDPG-pyrophosphorylase*, requires UTP and occurs with the loss of pyrophosphate and the attachment of the glucose 1-phosphate to uridine monophosphate (Fig. 11-13). Since the pyrophosphate produced is hydrolyzed by a pyrophosphatase, the overall reaction is unidirectional.

The synthesis of glycogen from UDPG is catalyzed by *glycogen synthase* (UDPG-glycogen: glycosyl transferase). Glucose units are specifically added to the carbon-4 hydroxyl groups of preexisting glycogen "primer" molecules by α-1,4-glucosidic linkages. UDP is released (Fig. 11-14). In other words, glycogen grows by the sequential addition of glucose units to form a straight chain held together by the α-1,4 bonds. Eventually, a *branching enzyme*, oligo-1,4 to 1,6-glucan transferase, transfers chain segments six or seven units long from the growing end either to another point on the same chain (Fig. 11-15) or to a neighboring chain. The branch is attached by an α-1,6-glucosidic linkage. Each branch may then be extended by

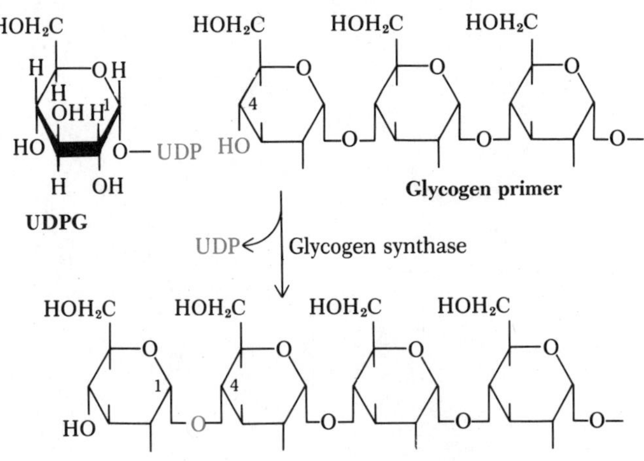

Figure 11-14
Process of glycogenesis. A new 1,4-glucosidic bond (color) is formed between UDPG and the C-4 OH of the primer glycogen.

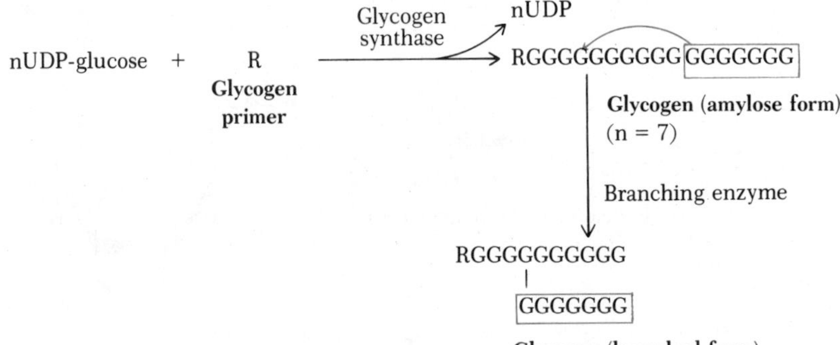

Figure 11-15
Formation of branch points in glycogen.

Figure 11-16
Actions of glycogenesis and glycogenolysis.

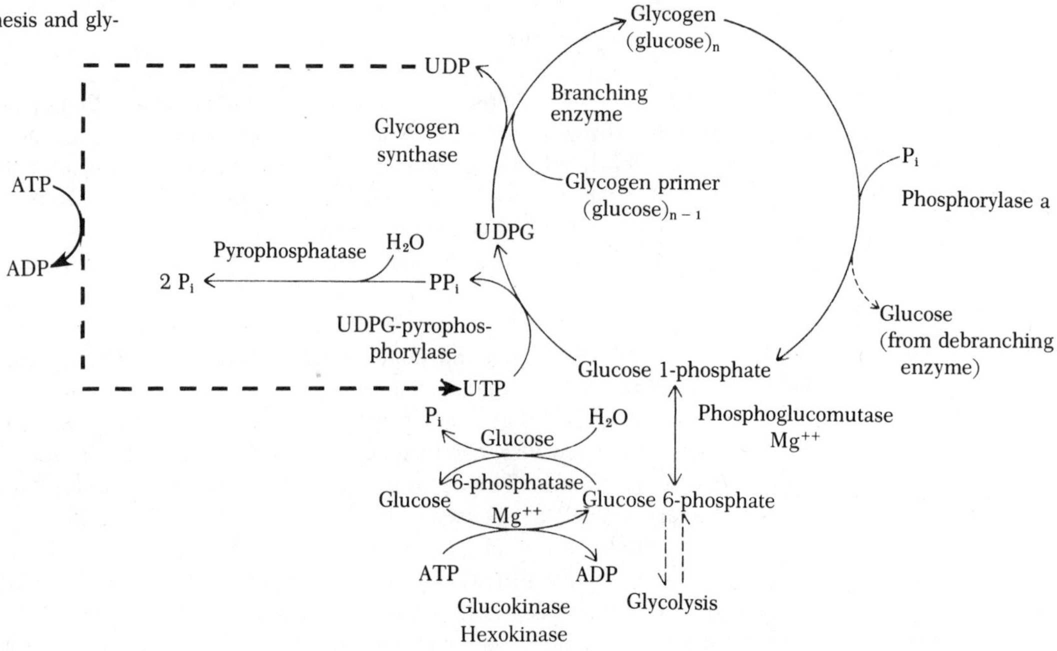

Glycogen $(glucose)_n$
UDP
Branching enzyme
Glycogen synthase
P_i
Phosphorylase a
Glycogen primer $(glucose)_{n-1}$
ATP
ADP
UDPG
Pyrophosphatase H_2O
2 P_i
PP_i
Glucose (from debranching enzyme)
UDPG-pyrophosphorylase
Glucose 1-phosphate
UTP
P_i
Glucose
H_2O
Phosphoglucomutase Mg^{++}
6-phosphatase
Glucose
Mg^{++}
Glucose 6-phosphate
ATP
ADP
Glucokinase
Hexokinase
Glycolysis

the addition of more glucose until enough has been added to transfer other sequences of 6 to 10 units in length. The branches are at least three units, as suggested in Fig. 11-10, or as many as 8 to 12 units apart. This treelike growth continues; the mechanism whereby its ultimate size is limited is not understood.

The opposing pathways for glycogenesis and glycogenolysis are summarized in Fig. 11-16.

Control of glycogenesis-glycogenolysis in the muscle

Glycogen synthase exists in two interconvertible forms that differ by the presence or absence of phosphate on certain serine residues of the enzyme. The phosphorylated form, termed synthase D, is dependent for its activity on the presence of the allosteric effector glucose 6-phosphate. The unphosphorylated enzyme, on the other hand, is not activated by glucose 6-phosphate and therefore is called the I, or independent, form. The following diagram shows that the phosphorylating process is catalyzed by a *cAMP-dependent protein kinase.*

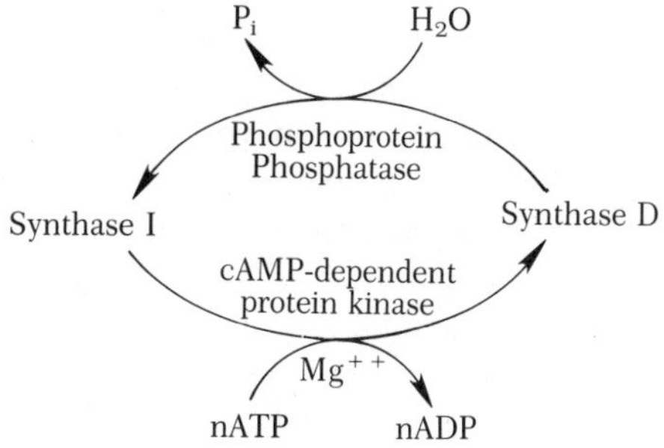

The protein kinase that requires cAMP for its activity is a complex of two subunits, one catalytic and the other inhibitory. In the absence of the cyclic nucleotide the complex is enzymatically inert. However, when cAMP is present, it is bound by the inhibitory peptide, thereby resulting in a dissociation of the complex and the release of the catalytically active subunit. This same enzyme is involved in triggering glycogenolysis.

The conversion of glycogen synthase D to the active I species by hydrolysis of the phosphates is catalyzed by a *phosphoprotein phosphatase* and is accompanied by a liberation of inorganic phosphate (note preceding diagram).

Glycogenolysis likewise is controlled by the phosphorylation-dephosphorylation of an enzyme, in this case *glycogen phosphorylase.* This enzyme exists in muscle cells as phosphorylase b (unphosphorylated) and phosphorylase a (phosphorylated form). The former is a dimer and has a molecular weight of 200,000. Although it exists in an inactive conformation, it can be activated allosterically by binding AMP at a nucleotide-binding site. Activation by AMP is opposed by ATP. Glucose 6-phosphate also binds allosterically to overcome the AMP stimulation. In the phosphorylation process to produce phosphorylase a the enzyme in effect is "frozen" in its catalytically active form. Thus phosphorylase a is not influenced further by the effectors AMP, ATP, or glucose 6-phosphate.

Phosphorylation involves a single serine residue in each subunit of phosphorylase b. When modified in this way, two dimers join to form a tetramer with a molecular weight of 400,000. The following diagram illustrates not only the phosphorylation process but also the reverse, i.e., dephosphorylation with the restoration of the inactive dimeric phosphorylase b. As seen at the top of p. 254, phosphorylation is catalyzed by *phosphorylase b kinase,* and dephosphorylation requires *phosphorylase a phosphatase.*

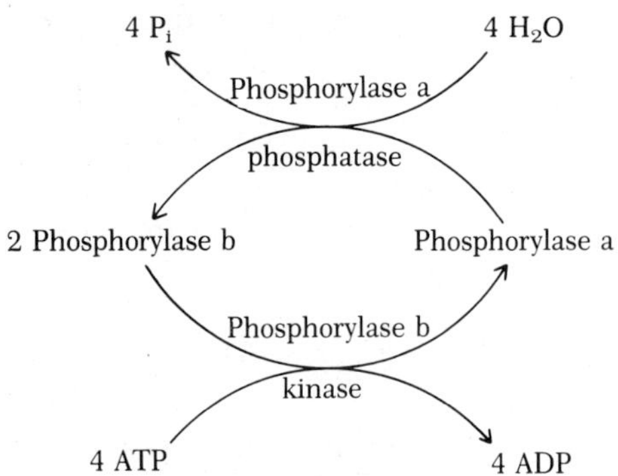

Glycogen phosphorylase contains one molecule of pyridoxal phosphate (Chapter 6) per subunit. Although the pyridoxal is essential to the enzyme's activity, its function is unknown.

Phosphorylase b kinase also exists in an active (phosphorylated) and an inactive (unphosphorylated) form. Its quaternary structure involves three types of subunits: A, B, and C, the probable composition being $A_4B_4C_8$. This kinase is a calcium-dependent enzyme that, in its unphosphorylated state, has a high K_m for the metal ion. At the normal cytosolic Ca^{++} concentration the enzyme is practically inactive. Phosphorylation of the B subunit by the cAMP-dependent protein kinase results in a lowering of the K_m for calcium. In this way the enzyme becomes activated. In addition, stimulation of phosphorylase kinase activity occurs as a result of an increased cytosolic concentration of calcium ions.

Phosphorylation of phosphorylase b kinase is increased not only by the cAMP-dependent protein kinase but also independently by glycogen.

As seen in the following scheme, the phosphorylation of the phosphorylase b kinase is catalyzed by the cAMP-dependent protein kinase. This is also a reversible process; dephosphorylation occurs hydrolytically, catalyzed by a phosphoprotein phosphatase.

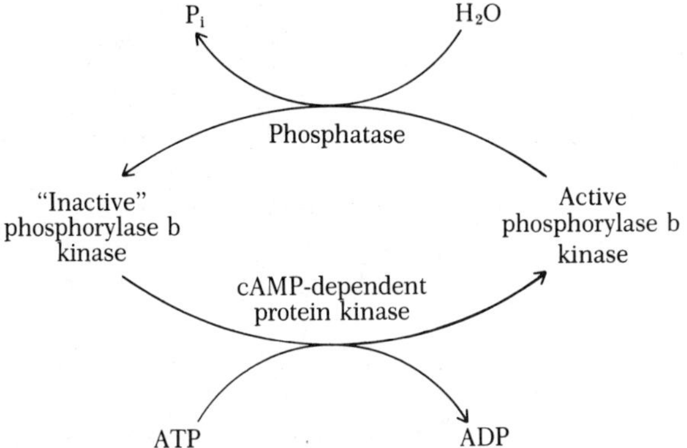

All the foregoing steps described for glycogenolysis can be joined into a single continuous process (Fig. 11-17). A single molecule of cAMP-dependent protein kinase may catalyze the activation of many molecules of phosphorylase b kinase, each of which in turn would catalyze the activation of many molecules of phosphorylase a, etc. Thus there exists a process of molecular amplification that ultimately results in the mobilization of thousands of molecules of glucose 1-phosphate from glycogen. Such an amplification process is called a cascade.

We now need to see how this glycogenolytic process functions to provide the

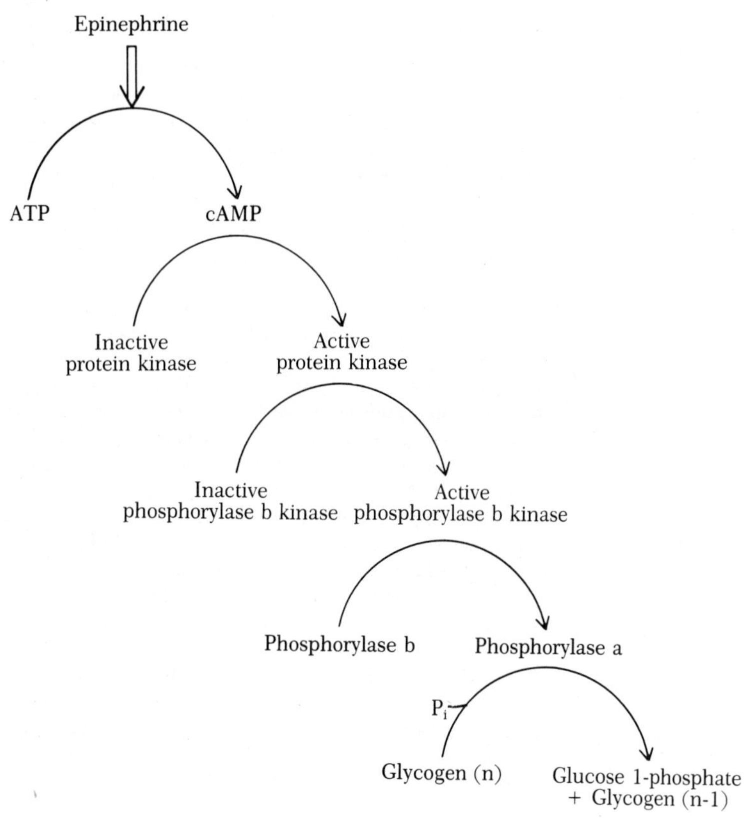

Figure 11-17
Glycogenolysis as a cascade phenomenon.

energy needs of muscle cells. Metabolic regulation of glycogenolysis depends on three factors: (1) intracellular metabolic signals, (2) extracellular hormonal signals, and (3) neural stimulation.

1. *Intracellular metabolic signals.* Muscular contraction results in an accumulation of AMP at the expense of ATP; the ATP/AMP ratio declines. Increases in AMP may do two things: (1) stimulate glycolysis by means of the activation of phosphofructokinase and (2) stimulate glycogenolysis by allosterically activating glycogen phosphorylase b. The negative effects of ATP and glucose 6-phosphate are relieved because of their decreased concentration.
2. *Extracellular hormonal signals.* The stimulus of glycogenolysis that accompanies sudden exercise ("flight or fright") involves the hormone epinephrine (Chapter 18). This hormone activates adenyl cyclase and produces cAMP. As seen in Fig. 11-17, epinephrine stimulates the entire glycogenolytic cascade.
3. *Neural stimulation.* As will be seen in Chapter 15, muscle contraction is triggered when a neuromuscular impulse causes a release of Ca^{++} from the sarcoplasmic reticulum. The increased cytoplasmic Ca^{++} simultaneously activates contraction and phosphorylase b kinase. The result is a total activation of all kinase molecules, whether phosphorylated or not. The activation of the kinase catalyzes the phosphorylation of phosphorylase b to a, and thus glycogenolysis is stimulated.

Figure 11-18

Factors regulating the processes of glycogenesis and glycogenolysis in the muscle cell.
1, Cyclic AMP (cAMP) activates a nonspecific protein kinase.
2, cAMP-stimulated protein kinase catalyzes the phosphorylation of glycogen synthase I to the D form. In the absence of glucose 6-phosphate, this reduces glycogenesis.
3, cAMP-stimulated protein kinase also activates phosphorylase b kinase, which converts phosphorylase b to a.
4, Active phosphorylase a catalyzes the breakdown of glycogen to glucose 1-phosphate, which is converted to glucose 6-phosphate and then is utilized in glycolysis or, in the liver, can be converted to glucose.
5, When stimulation of cAMP production is halted, the action of the various phosphatases reverses the activation processes. This is also likely to be a time of resting conditions when the demand for glucose 6-phosphate is diminished. An accumulation of glucose 6-phosphate activates the D form of glycogen synthase, thereby favoring glycogenesis. Glucose 6-phosphate also serves to retain phosphorylase b in an inactive conformation. It stimulates phosphorylase phosphatase. These actions inhibit glycogenolysis and favor glycogenesis.
6, Glycogen also acts as a controlling factor. For example, this polysaccharide inhibits the glycogen synthase phosphatase, preventing the conversion of the D form to the active I form. This places a damper on further glycogenesis. Glycogen stimulates the phosphorylation of phosphorylase b kinase and increases the activity of this enzyme. Thus glycogen build-up favors glycogenolysis and counteracts glycogenesis.
7, Dephosphorylation of phosphorylase b kinase leads to a dephosphorylation of phosphorylase a to the less active b form. This reduces glycogenolysis and therefore favors a retention of glycogen.
8, The ATP/AMP ratio is an important controlling factor in glycogen turnover. AMP is a positive effector for phosphorylase b, i.e., the less active enzyme form, whereas ATP prevents AMP binding and hence inhibits the b form. As ATP accumulates and AMP diminishes, even the less active phosphorylase is inhibited and the phosphorolysis of glycogen is reduced to a minimum. If the reverse occurs, i.e., the ATP/AMP ratio decreases, a need for more ATP and hence glucose is signaled. Accumulating AMP acts as a positive effector on phosphorylase b, and hence the breakdown of glycogen is increased.

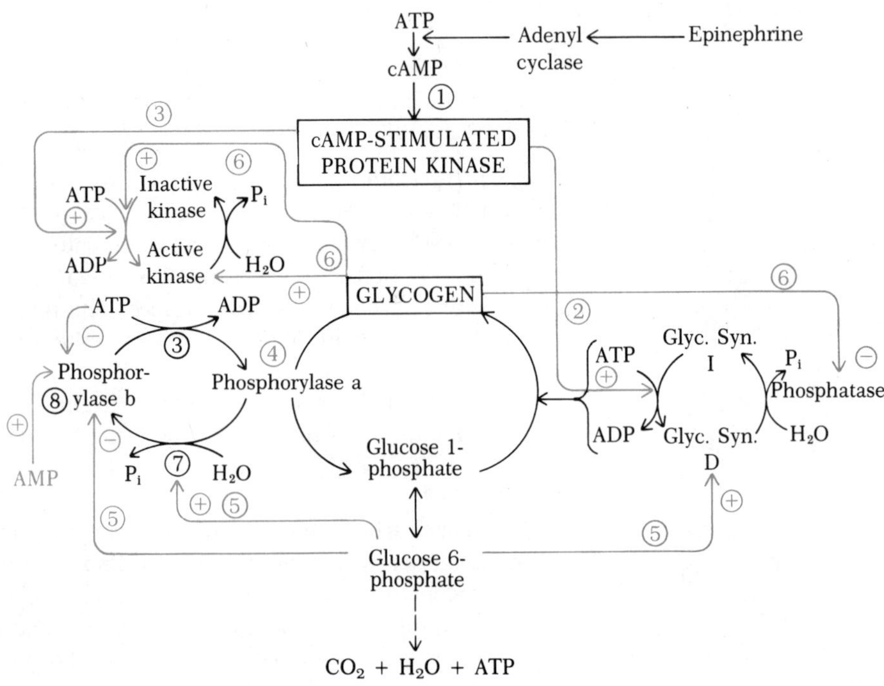

During the stimulation of glycogenolysis, as by exercise, the biosynthesis of glycogen is suppressed. This means that the two opposing processes, glycogenesis and glycogenolysis, are under reciprocal control. Were this not the case, a futile cycle would ensue in which the glucose 1-phosphate produced by glycogenolysis would simply be reconverted to glycogen with a consumption of UTP (ATP). When epinephrine triggers cAMP production, the cAMP-dependent protein kinase is activated. This enzyme has two simultaneous effects (Fig. 11-18). First, it stimulates glycogenolysis by increasing the conversion of phosphorylase b to a, and at the same time it inhibits the synthesis of glycogen by promoting the conversion of glycogen synthase I to D. The phosphorylated synthase D is virtually inactive in the absence of sufficient glucose 6-phosphate. As ATP is used for muscle contraction, the accumulation of AMP allosterically activates any phosphorylase b so as to achieve a maximum glycogenolytic action. With the decline of muscular activity, the utilization of glucose 6-phosphate is reduced. The accumulating hexose phosphate then allosterically activates the glycogen synthase D and promotes glycogenesis. Simultaneously it inhibits the AMP-stimulated phosphorylase b kinase.

In resting muscle most of the glycogen synthase is in the active I form, whereas phosphorylase is primarily in its inactive b form. Under these conditions the metabolic stance of the muscle is glycogen synthesis as opposed to glycogenolysis. As the glycogen accumulates, it serves as its own feedback regulator; it acts as an activator of phosphorylase b kinase and an inhibitor of glycogen synthase phosphatase. It serves to promote glycogenolysis and inhibit its own synthesis by preventing further conversion of the synthase D (inactive) to the I (active) form.

Control of glycogenesis-glycogenolysis in the liver

Liver cells contain the same complement of enzymes for glycogenesis and glycogenolysis as do the muscle cells, although there are important structural differences. Interconvertible phosphorylated and unphosphorylated glycogen synthase and phosphorylase are present. The hormone that serves as the primary signal for the mobilization of glucose from liver glycogen is glucagon (Chapter 18). It is released from the pancreatic islets of Langerhans (α-cells) in response to low blood glucose levels. This hormone functions via a cAMP-stimulated protein kinase system. As in muscle cells, the activated cAMP-dependent protein kinase stimulates glycogenolysis and suppresses glycogenesis. Epinephrine also affects these processes in the liver.

In the liver, glucose itself serves as an important regulator. An increase in glucose concentrations results in the binding of the free sugar to phosphorylase a, causing a partial inhibition. Simultaneously, it activates phosphorylase a phosphatase, thereby increasing dephosphorylation of phosphorylase a. These two effects serve to suppress hepatic glycogenolysis. Furthermore, studies of glycogenesis in the liver show that glycogen synthase activity is observed only in the presence of very low levels of glycogen phosphorylase a. It appears, therefore, that free glucose not only suppresses glycogenolysis but also results in an increased synthesis of glycogen.

Control of carbohydrate metabolism

During the discussions of the various individual reactions in carbohydrate metabolism, attention was repeatedly focused on the enzymes and cofactors required. Their levels and activities as well as the substrate concentrations, on the one hand, and the product concentrations, on the other, serve as key regulators of the various reactions. Control of carbohydrate metabolism may therefore be dependent on (1) the availability of enzymes, (2) the availability of substrate, and (3) environmental factors affecting reaction velocities.

Availability of enzyme

The availability of enzymes will depend on the rate of enzyme synthesis versus its degradation, as well as on the interconversion of active and inactive forms. In the discussion of glycolysis attention was focused on certain physiologically irreversible steps that require separate enzyme systems for reversal, as in gluconeogenesis. These steps are points of metabolic control. For example, if the amount of glucokinase were increased by a selective stimulation of protein synthesis, glycolysis would be favored. It is instructive to divide the various enzymes into three categories: group I—those which favor glycolysis; group II—those which are common to both pathways; and group III—those which are unique to gluconeogenesis. The following table shows the designation of the enzymes into these three groups.

Group I: glycolytic enzymes	**Group II: bifunctional enzymes**	**Group III: gluconeogenic enzymes**
Glucokinase	Phosphoglucoisomerase	Glucose 6-phosphatase
Phosphofructokinase	Aldolase	Fructose bisphosphatase
Pyruvate kinase	Triosephosphate isomerase	PEP carboxykinase
	Glyceraldehyde 3-phosphate dehydrogenase	Pyruvate carboxylase
	Phosphoglycerate kinase	
	Phosphoglyceromutase	
	Enolase	
	Lactate dehydrogenase	

Groups I and III are inducible enzymes, i.e., they are formed by the stimulated transcription of new mRNA under the control of certain hormones. Group II enzymes are not inducible and are produced at a steady rate under the control of constitutive gene systems. Thus glucocorticoids stimulate gluconeogenesis by inducing the formation of the key gluconeogenic enzymes, and insulin may stimulate glycolysis by inducing an increased synthesis of the key glycolytic enzymes.

Control of carbohydrate metabolism by the interconversion of active and inactive forms has been described as part of the turnover of glycogen. The glycogenic and glycogenolytic systems depend on the interconversion of active and inactive forms of existing enzyme molecules. Thus, by phosphorylating muscle phosphorylase b, the active phosphorylase a is produced, thereby favoring the breakdown of glycogen. Likewise, glycogenesis is controlled by the interconversion of glycogen synthase I and D.

Availability of substrate

The available pool of substrate molecules depends on at least two factors. These are the rate at which the molecules are transported through the cell membrane and the rate at which they are used by a subsequent step in the metabolic pathways. In the cells of muscle and adipose tissue, for example, glucose is carried across the plasma membrane by facilitated transport that is controlled by the hormone insulin. This may result from an induced synthesis of carrier protein.

Control of metabolism by the interrelationship of various enzymic reactions is indicated by the relative activities of competing reactions. For example, in liver cells the enzyme glucose 6-phosphatase is antagonistic to glucokinase; the ratio of activities is 1000:150. Therefore dephosphorylation is favored over phosphorylation. At the step converting fructose 6-phosphate to fructose 1,6-bisphosphate, the relative activities of fructose bisphosphatase to phosphofructokinase is 600:150. Again, gluconeogenesis is favored. In the conversion of PEP to pyruvate the ratio of pyruvate kinase to the competing reaction PEP carboxykinase is 5000:50. Here glycolysis is favored by a factor of 100:1. The first two reactions favor gluconeogenesis, whereas the latter drives pyruvate into the citric acid cycle.

Factors affecting reaction velocities

Regulatory factors that control carbohydrate metabolism are the ratio of ATP to AMP, the presence of NADH compared with NAD^+, and the accumulation of metabolites such as citrate, acetyl CoA, free fatty acids, L-alanine, fructose 1,6-bisphosphate, glucose 6-phosphate, L-phenylalanine, and glyceraldehyde 3-phosphate. The effects exerted by these compounds involve allosteric control mechanisms. When ATP is used, ADP plus P_i is formed. As already described, the enzyme adenylate kinase catalyzes the interconversion of the following:

$$ADP + ADP \rightleftharpoons ATP + AMP$$

AMP is also the product of synthetic reactions such as the activation of amino acids and fatty acids. Therefore a controlling signal that regulates carbohydrate metabolism is the ATP/AMP ratio. AMP is a positive effector for two of the three rate-limiting glycolytic enzymes, phosphofructokinase and pyruvate kinase. As noted earlier, AMP discharges the ATP inhibition of phosphofructokinase, thereby increasing the rate of fructose 1,6-bisphosphate production. Increasing levels of fructose bisphosphate should in turn increase its hydrolysis by fructose bisphosphatase. However, AMP simultaneously inhibits the reverse reaction, preventing the development of a "futile cycle," the consequence of which would be the net hydrolysis of ATP.

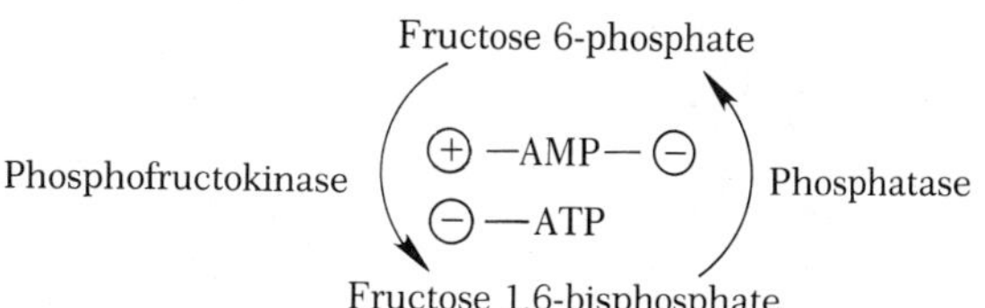

Thus AMP exerts a reciprocal regulatory effect. Increasing the ATP levels inhibits phosphofructokinase, whereas decreasing the AMP releases the fructose bisphosphatase activity from inhibition. The overall effect is to favor gluconeogenesis over glycolysis (Fig. 11-19).

AMP is also a positive effector for glycogen phosphorylase b, thus favoring glycogenolysis. ADP is a positive effector for isocitrate dehydrogenase, thereby favoring aerobic metabolism by the citric acid cycle. On the other hand, ATP is a negative effector for the pyruvate dehydrogenase system, thus reducing the input of pyruvate into the citric acid cycle. Increasing the ATP/AMP ratio reduces glycolysis and favors gluconeogenesis, whereas a decreasing ratio favors glycolysis and limits gluconeogenesis.

Regulators favoring glycolysis are glucose 6-phosphate, fructose, 1,6-bisphosphate, and D-glyceraldehyde 3-phosphate (Fig. 11-19). These modulators stimulate glycolysis by "signaling ahead" to pyruvate kinase. If the level of NADH exceeds its oxidation, the accumulation of reduced coenzyme will inhibit pyruvate kinase and suppress glycolysis. In the mitochondrion, NADH is a negative effector of isocitrate dehydrogenase, so the citrate metabolism is diminished. ATP also inhibits isocitrate dehydrogenase. These effects lead to an accumulation of citrate, which enters the cytoplasm and inhibits phosphofructokinase. The resulting accumulation of glucose 6-phosphate stimulates the D form of glycogen synthase and promotes glycogenesis. Inhibition of isocitrate dehydrogenase also leads to an intramitochondrial accumulation of acetyl CoA. An increased degradation of free fatty acids likewise increases this metabolite. Acetyl CoA is the obligatory activator of pyruvate carboxylase and thus would divert the metabolism of pyruvate away from further acetyl CoA formation and into the anaplerotic step, forming oxaloacetate. An accumulation of acetyl CoA inhibits pyruvate dehydrogenase and therefore limits further

Figure 11-19
Regulatory effects of various metabolites on carbohydrate metabolism.

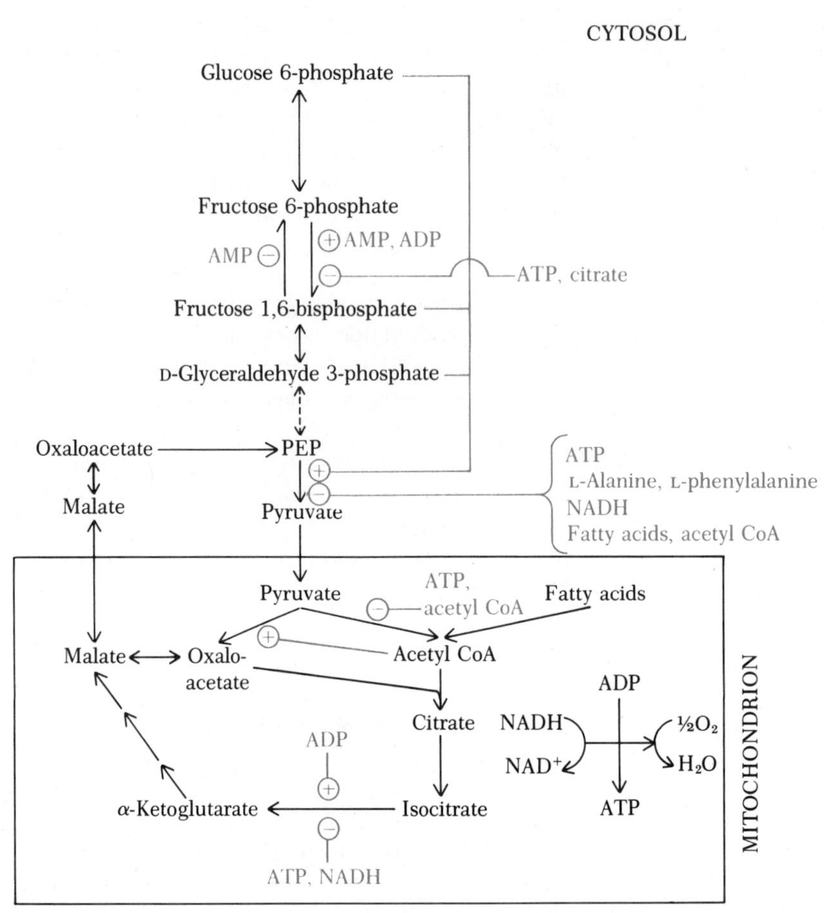

entrance of pyruvate into the citric acid cycle. Cytoplasmic acetyl CoA (citrate cleavage enzyme), as well as free fatty acids, inhibits pyruvate kinase. Therefore acetyl CoA acts as a reciprocal controlling factor by activating pyruvate carboxylase and inhibiting pyruvate kinase. This action again prevents the dual development of a futile cycle in which the net result would only be the hydrolysis of GTP (Fig. 11-19).

The amino acids L-alanine and L-phenylalanine also affect the direction of carbohydrate metabolism. Both are inhibitors of pyruvate kinase and therefore serve to limit glycolysis.

Fig. 11-19 illustrates the impact of various metabolites on the control of carbohydrate metabolism. This is viewed as two opposing directions: glycolysis versus gluconeogenesis. The control exerted by various factors, such as the ATP/AMP ratio, free fatty acids, and amino acids, ensures an appropriate adaptation of the mammalian organism to the impact of nutritional and hormonal factors.

Phosphogluconate or pentose phosphate pathway

As previously described, glucose catabolism proceeds by way of a system of branched pathways in which glucose 6-phosphate is the central intermediate. From this point the sugar may be converted to glucose 1-phosphate and glycogen or diverted into the glycolytic pathway. In some tissues an important alternative fate for the oxidation of glucose is the phosphogluconate pathway (also known as the pentose phosphate pathway, the Warburg-Dickens pathway, and the hexosemonophosphate shunt), in which one molecule of hexose yields 12 of NADPH and 6 of CO_2. Like glycolysis, this pathway is anaerobic and its enzymes are found in the cellular cytoplasm. But whereas glycolysis is ATP producing, the phosphogluconate pathway is the primary source of reduced $NADP^+$, which is essential for the biosynthesis of fatty acids and cholesterol. In addition, it is a pentose (ribose) supply source, which is required in the biosynthesis of nucleotides and nucleic acids (Chapter 14).

To simplify our consideration of the phosphogluconate pathway, the steps will be divided into three phases. Phase I includes only the oxidative process, which begins with glucose 6-phosphate; phases II and III are nonoxidative isomerizations and rearrangements.

Phase I: $NADP^+$-dependent dehydrogenases

Oxidation of glucose 6-phosphate occurs by a reversible dehydrogenation using $NADP^+$ in the presence of Mg^{++} and is catalyzed by *glucose 6-phosphate dehydrogenase.* This reaction is specific for the β-anomer of glucose 6-phosphate. Dehydrogenation yields a lactone, 6-phospho-D-gluconolactone, which in turn is opened by the insertion of a molecule of water, a reaction catalyzed by *6-phosphoglucono-lactonase*, also a Mg^{++}-requiring enzyme. The product is 6-phosphogluconic acid. These two steps are outlined below.

PHASE I

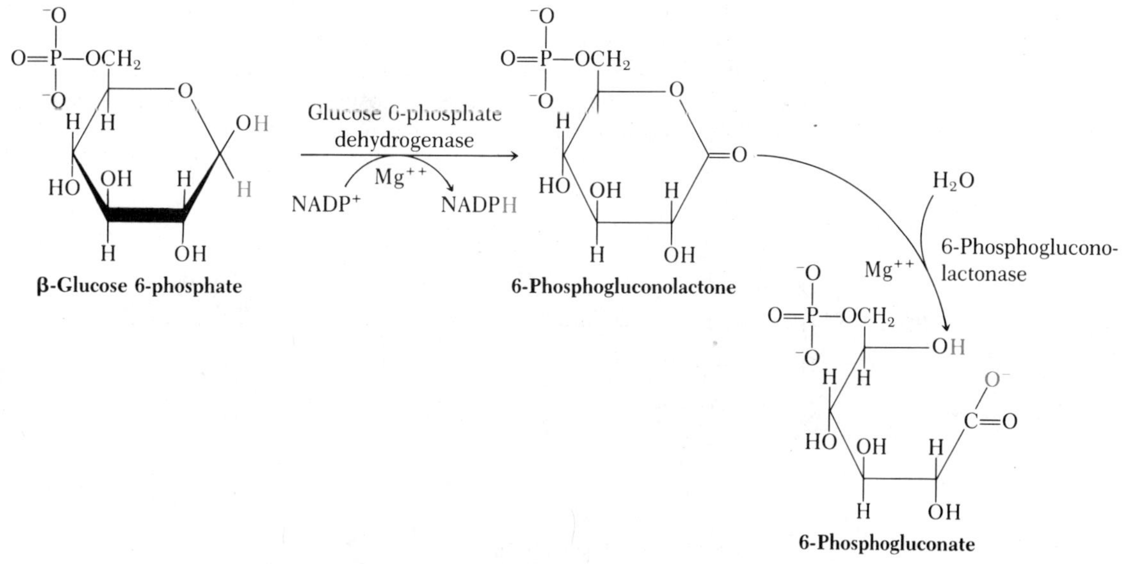

β-Glucose 6-phosphate —(Glucose 6-phosphate dehydrogenase; Mg^{++}; $NADP^+$ → NADPH)→ **6-Phosphogluconolactone** —(H_2O; 6-Phosphoglucono-lactonase; Mg^{++})→ **6-Phosphogluconate**

In the next oxidative step of phase I, 6-phosphogluconate is dehydrogenated by *6-phosphogluconate dehydrogenase*, again using $NADP^+$ as coenzyme; Mn^{++} is also needed. The product, 3-ketophosphogluconate, decarboxylates spontaneously to yield the pentose D-ribulose 5-phosphate (see following diagram). Both these oxidative steps, which are unique to the phosphogluconate pathway, are strongly inhibited by their common product: NADPH. Inhibition is relieved when the reduced NADPH is utilized and reoxidized, e.g., in the biosynthesis of fatty acids. Therefore NADPH is an important metabolic signal, controlling the oxidative steps of the phosphogluconate pathway.

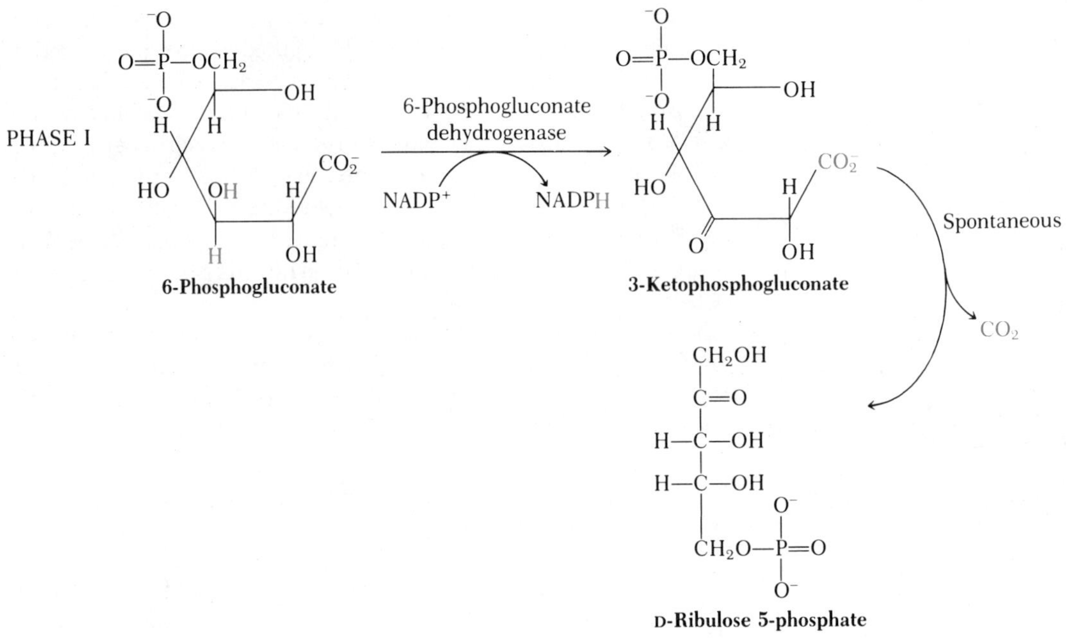

PHASE I

6-Phosphogluconate → (6-Phosphogluconate dehydrogenase; $NADP^+$ → NADPH) → 3-Ketophosphogluconate → (Spontaneous; → CO_2) → D-Ribulose 5-phosphate

6-Phosphogluconate

3-Ketophosphogluconate

D-Ribulose 5-phosphate

Phase II: isomerization steps

The final product of the phase I oxidative processes, D-ribulose 5-phosphate, is the substrate for an epimerase, *ribulose 5-phosphate epimerase.* In this reaction the carbon-3 of D-ribulose 5-phosphate is inverted to yield D-xylulose 5-phosphate (as shown below). Alternatively the ketose D-ribulose 5-phosphate is converted to the corresponding aldose D-ribose 5-phosphate. This isomerization is catalyzed by *ribose 5-phosphate isomerase*, a reaction comparable to the reversible conversion of glucose 6-phosphate to fructose 6-phosphate. These two isomerization reactions are illustrated in the following diagram.

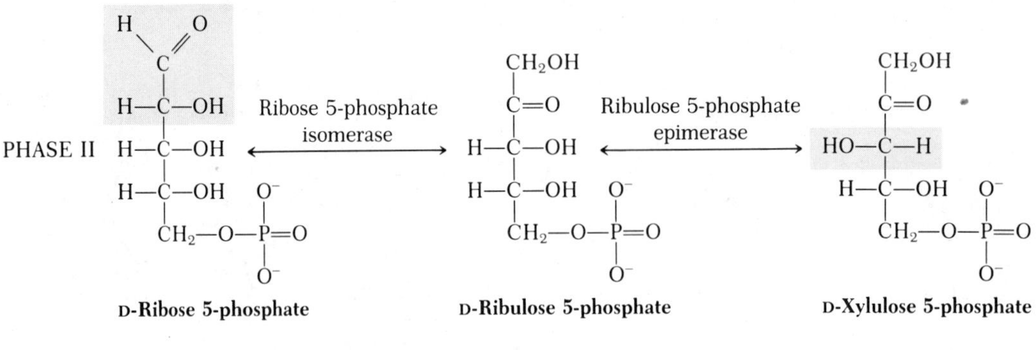

PHASE II

D-Ribose 5-phosphate ⟷ (Ribose 5-phosphate isomerase) ⟷ D-Ribulose 5-phosphate ⟷ (Ribulose 5-phosphate epimerase) ⟷ D-Xylulose 5-phosphate

D-Ribose 5-phosphate **D-Ribulose 5-phosphate** **D-Xylulose 5-phosphate**

Phase III: pentose interconversions

Phase III of the phosphogluconate pathway can be considered a "reshuffling" or rearrangement of carbon atoms in which the overall consequence is the reassembly of pentoses into D-fructose 6-phosphate. Rearrangement is achieved by the transfer of 2- or 3-carbon units to appropriate acceptor molecules. The first of these, i.e., the transfer of a 2-carbon unit, is comparable to a mechanism already considered (see Fig. 6-5). This specifically refers to the transfer of a 2-carbon *active glycolaldehyde* unit to thiamin pyrophosphate (Fig. 11-20).

Specifically, carbon-1 and carbon-2 of D-xylulose 5-phosphate are transferred by an intermediate active glycolaldehyde unit to an acceptor molecule, which is D-ribose 5-phosphate. The products of this reaction are the unusual 7-carbon sugar phosphate sedoheptulose 7-phosphate and the triose phosphate D-glyceraldehyde

Figure 11-20
General mechanism of transketolation reaction showing role of thiamin pyrophosphate.

3-phosphate. The enzyme that catalyzes this reaction is called *transketolase* and requires thiamin pyrophosphate and Mg^{++} for its activity. Transketolase exhibits a stereospecificity such that the hydroxyl on carbon-3 of the sedoheptulose has the L-configuration.

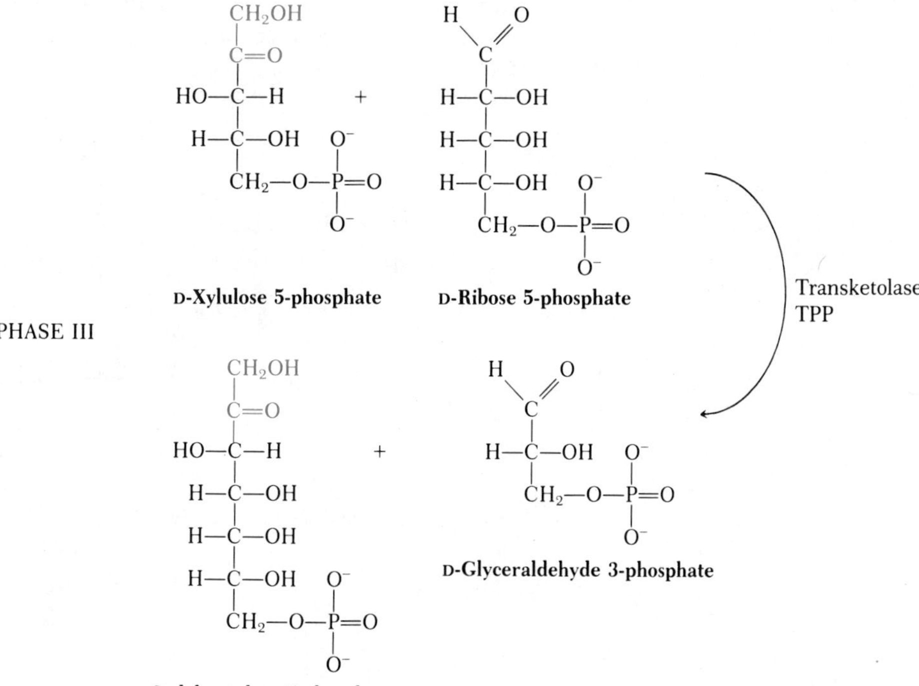

The next step requires that carbons-1, -2 and -3 of the sedoheptulose 7-phosphate are transferred as a single unit to an acceptor molecule, which in this case is D-glyceraldehyde 3-phosphate. The products are D-fructose 6-phosphate and D-erythrose 4-phosphate (see below). This reaction is comparable to the aldolase step in glycolysis; therefore its enzyme is called *transaldolase*. Again, the specificity is such that the hydroxyl group of carbon-3 assumes the L-configuration.

PHASE III

```
CH2OH
|
C=O                                 H     O
|                                    \   //
HO—C—H                                 C
|                                      |
H—C—OH              +             H—C—OH   O⁻
|                                      |       |
H—C—OH                              CH2—O—P=O
|                                              |
H—C—OH   O⁻                                    O⁻
|        |                  D-Glyceraldehyde 3-phosphate
CH2—O—P=O
         |                                          Transaldolase
         O⁻
D-Sedoheptulose 7-phosphate

CH2OH                               H     O
|                                    \   //
C=O                                    C
|                                      |
HO—C—H                            H—C—OH
|                                      |
H—C—OH              +             H—C—OH   O⁻
|                                      |       |
H—C—OH   O⁻                         CH2—O—P=O
|        |                                     |
CH2—O—P=O                                      O⁻
         |                  D-Erythrose 4-phosphate
         O⁻
D-Fructose 6-phosphate
```

Finally, transketolase again catalyzes the transfer of an active glycolaldehyde unit by thiamin pyrophosphate. This time the first 2 carbons of D-xylulose 5-phosphate are added to an acceptor molecule, which is the D-erythrose 4-phosphate from the transaldolase reaction. The products formed are D-fructose 6-phosphate and D-glyceraldehyde 3-phosphate.

PHASE III

```
CH2OH                               H     O
|                                    \   //
C=O                                    C
|                                      |
HO—C—H              +             H—C—OH
|                                      |
H—C—OH   O⁻                       H—C—OH   O⁻
|        |                             |       |
CH2—O—P=O                           CH2—O—P=O
         |                                     |
         O⁻                                    O⁻
D-Xylulose 5-phosphate       D-Erythrose 4-phosphate
                                                    Transketolase
                                                    TPP
CH2OH
|
C=O                                 H     O
|                                    \   //
HO—C—H                                 C
|                                      |
H—C—OH              +             H—C—OH   O⁻
|                                      |       |
H—C—OH   O⁻                         CH2—O—P=O
|        |                                     |
CH2—O—P=O                                      O⁻
         |
         O⁻
D-Fructose 6-phosphate       D-Glyceraldehyde
                             3-phosphate
```

In retrospect, phase III can represent a sequence of three reactions catalyzed by the following enzymes:

Transketolase → Transaldolase → Transketolase

To complete the degradation of glucose by the phosphogluconate pathway, a fourth phase is needed, i.e., glycolysis. It is apparent from the foregoing discussion that the pathway ultimately yields two compounds, both of which are intermediates in glycolysis; these are D-fructose 6-phosphate and D-glyceraldehyde 3-phosphate.

Review of phosphogluconate pathway

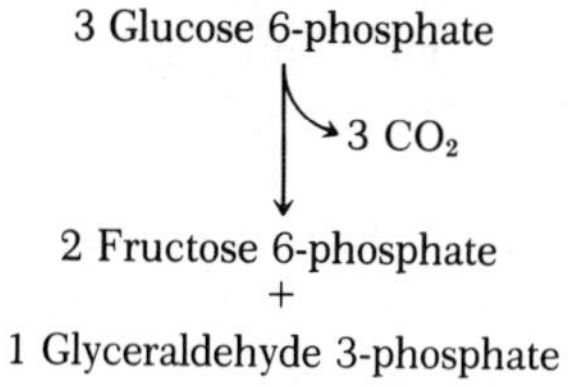

3 Glucose 6-phosphate
↓ → 3 CO_2
2 Fructose 6-phosphate
+
1 Glyceraldehyde 3-phosphate

The phosphogluconate pathway is summarized in Fig. 11-21. Note that the fate of three molecules of glucose ought to be considered simultaneously. Then it becomes clear that three molecules of CO_2 are liberated with the concomitant formation of one molecule of D-ribose 5-phosphate and two of D-xylulose 5-phosphate. One D-ribose 5-phosphate and one of two xylulose 5-phosphate molecules serve as precursors for the eventual formation of one D-fructose 6-phosphate. The second xylulose 5-phosphate and erythrose 4-phosphate yield a second fructose 6-phosphate and one glyceraldehyde 3-phosphate (see diagram at left).

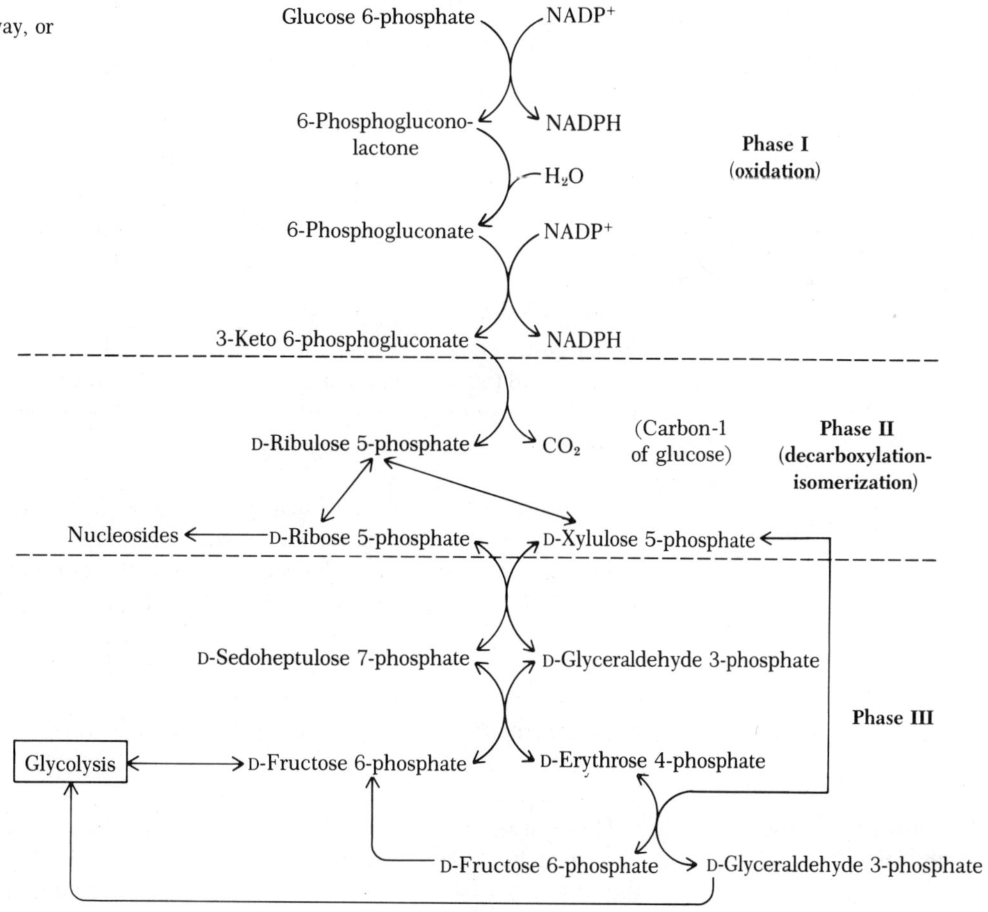

Figure 11-21
Pentose phosphate pathway, or phosphogluconate shunt.

If this scheme is doubled, so that six glucose 6-phosphates are involved, then six molecules of CO_2 will be produced, equivalent to the complete degradation of a hexose unit. Four fructose 6-phosphates and two D-glyceraldehyde 3-phosphates are returned to the glycolytic pathway. Actually, these can yield five molecules of glucose 6-phosphate.

An important feature of the pathway is that no ATP is required for its operation once the glucose 6-phosphate has been formed. This means that the pathway may continue to function under anaerobic conditions. As already indicated, $NADP^+$ is reformed because of the participation of NADPH in biosynthetic processes, such as fatty acid or cholesterol synthesis. Reversal of the pentose pathway allows an anaerobic production of pentoses beginning with fructose 6-phosphate from the glycolytic pathway. This represents an increased pentose synthesis for nucleotides without a dependence on a supply of $NADP^+$.

The question of ATP formation by way of the phosphogluconate pathway is complex. Since the reactions all occur in the cytosol, the NADPH formed must react with NAD^+ to enter the mitochondria. This transfer of hydrogens from NADPH to NAD^+ occurs by a transhydrogenation process. Therefore each mole of NADPH would ultimately yield 3 moles of ATP. In this way the complete oxidation of *glucose* would yield 35 moles of ATP. This amount is nearly as much as that formed by the oxidation of 1 mole of glucose (36 to 38 moles) by way of glycolysis and the citric acid cycle.

$$6 \text{ Glucose 6-phosphate} + 12\ NADP^+ \rightarrow 5 \text{ Glucose 6-phosphate} + 6\ CO_2 + 12\ NADPH + P_i$$

Also

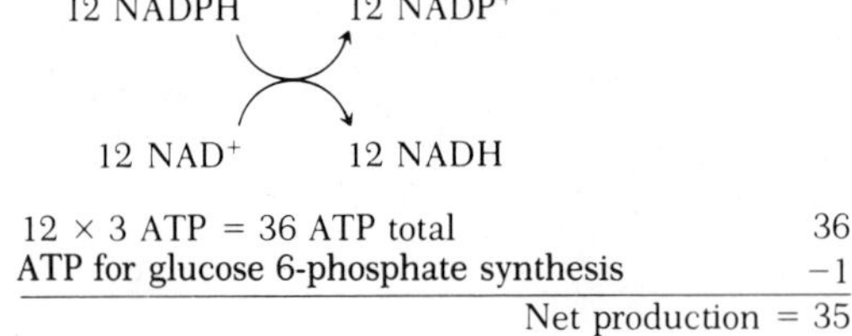

12 × 3 ATP = 36 ATP total	36
ATP for glucose 6-phosphate synthesis	−1
Net production =	35

The phosphogluconate pathway is particularly active in tissues that are sites for the biosynthesis of fatty acids and steroids, e.g., the liver, adipose tissue, lactating mammary glands, adrenal cortex, and testes. However, in the liver approximately 30% of glucose degradation occurs via the pentose phosphate pathway. It is also important in the metabolism of erythrocytes for maintaining membrane structure.

Glucose may be metabolized to a greater degree by way of the pentose pathway in situations resulting from tissue anoxia. The lack of tissue oxygen probably decreases the pyruvate metabolism by way of the citric acid cycle and causes an accumulation of intermediates of the glycolytic pathway. These in turn result in a diversion of glucose 6-phosphate into the pentose pathway. The accumulating NADPH is diverted to fatty acid synthesis, thus accounting for the fatty infiltration of tissues subjected to anoxia for extended periods. This sequence of events has been demonstrated in infarcted myocardial tissue after coronary occlusion.

Metabolism of other carbohydrates

Interconversion of hexoses

There are pathways for the metabolism of glucose other than those already described. For example, glucuronic acid production competes with glycogenesis for the supply of UDPG, and the biosynthesis of hexosamines competes with glycolysis for fructose 6-phosphate. These competitive pathways convert glucose into precursors required in the production of proteoglycans and in detoxication processes. Alternate pathways are required for the metabolism of dietary sugars other than glucose, such as fructose and galactose.

Biosynthesis of uronic acid

In the biosynthesis of D-glucuronic acid, glucose 1-phosphate is converted to UDP-glucose as in glycogenesis. The enzyme is the same, i.e., *UDPG-pyrophosphorylase.* UDPG then serves as a substrate for *UDPG-dehydrogenase,* which requires NAD^+ as coenzyme (see following diagram). Two hydrogens are removed from the primary alcohol group of carbon-6. It is not clear just how the carboxyl group and therefore UDP glucuronic acid is formed, but an intermediate aldehyde group has been proposed. UDP-glucuronic acid is required in detoxication reactions forming glucuronides (Chapter 17) and for the biosynthesis of proteoglycans (Chapter 15).

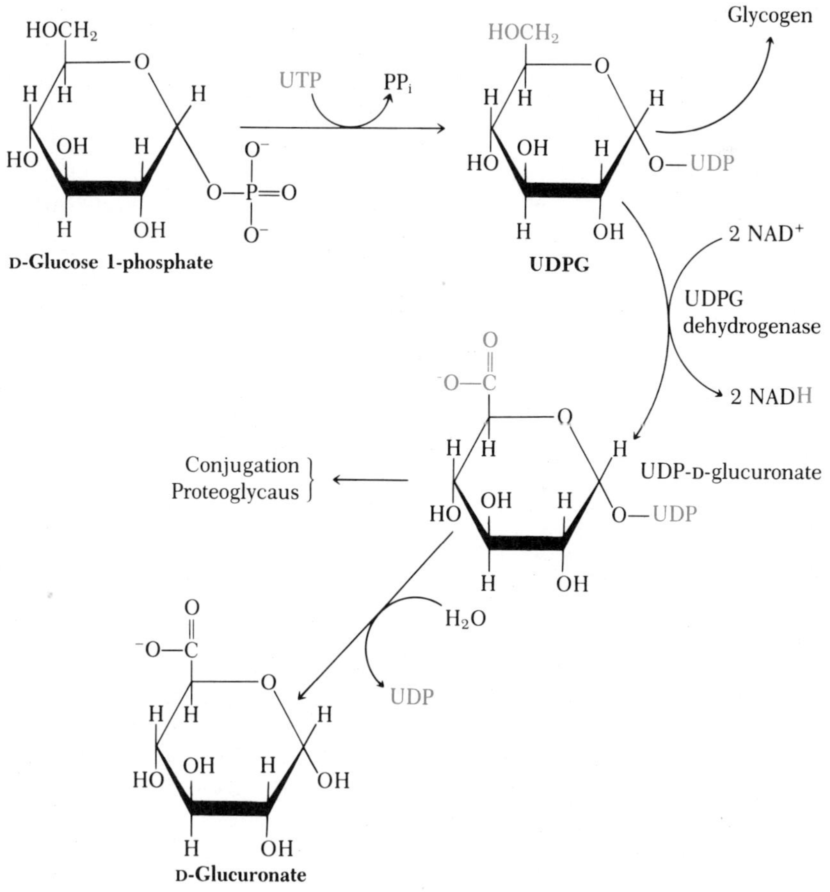

$HOCH_2$ O H H H OH H HO O^- O—P=O H OH O^-
D-Glucose 1-phosphate

UTP PP_i

$HOCH_2$ O H H H OH H HO O—UDP H OH
UDPG

Glycogen

$2\ NAD^+$
UDPG dehydrogenase
2 NADH

O $^-$O—C O H H H OH H HO O—UDP H OH
UDP-D-glucuronate

Conjugation
Proteoglycaus

H_2O
UDP

O $^-$O—C O H H H OH H HO OH H OH
D-Glucuronate

The formation of L-gulonic acid is yet another option for the metabolic utilization of glucose. In this case glucuronic acid is reduced at carbon-1 to yield the corresponding primary alcohol. L-Gulonic acid is actually an aldonic acid, since the carboxyl group, formerly carbon-6 of glucuronic acid (see diagram), is the active group and is called carbon-1. Therefore the sugar acid is L-gulonic acid because the new carbon-5 is of the L-configuration and is related to the hexose L-gulose.

L-Gulonic acid is the precursor of L-ascorbic acid (vitamin C). The biosynthesis of L-ascorbic acid (see diagram on top of p. 268) does not occur in humans.

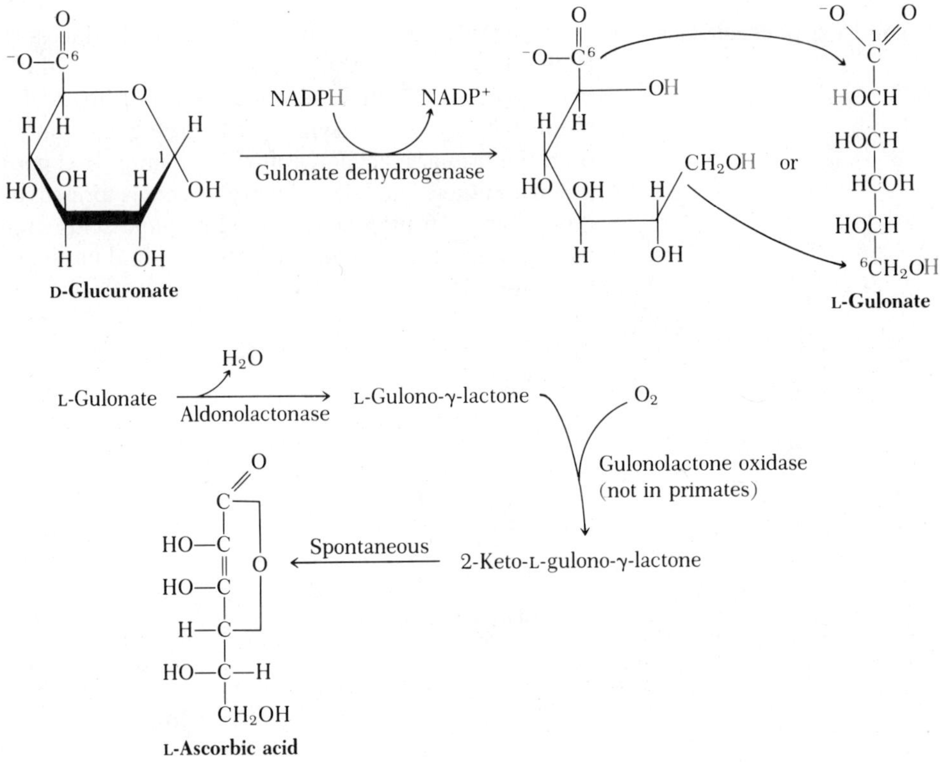

D-Glucuronate

NADPH → $NADP^+$

Gulonate dehydrogenase

CH_2OH or

L-Gulonate

L-Gulonate —(H_2O)→ L-Gulono-γ-lactone

Aldonolactonase

O_2

Gulonolactone oxidase (not in primates)

2-Keto-L-gulono-γ-lactone

Spontaneous

L-Ascorbic acid

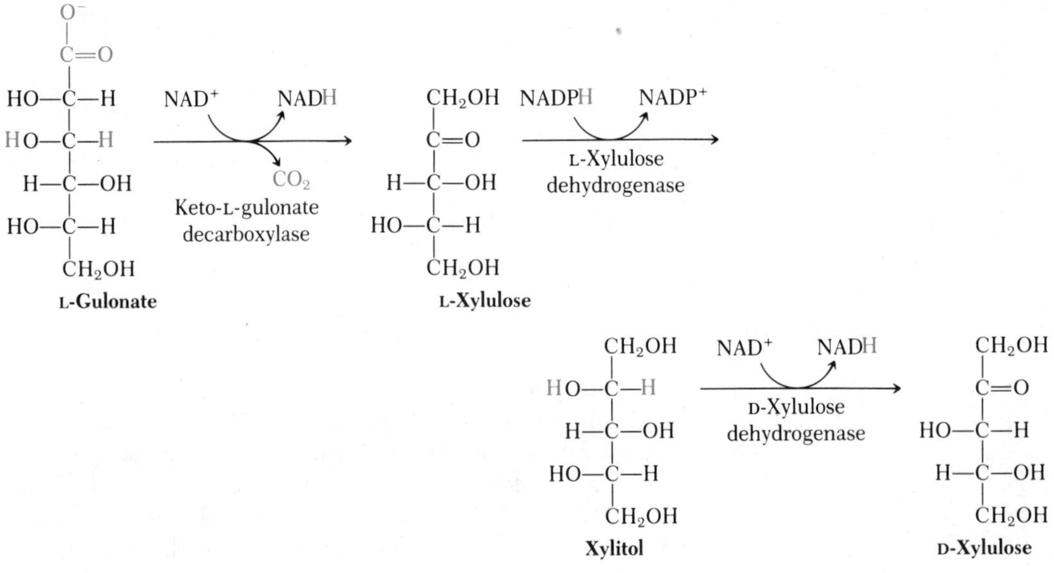

L-Gulonate

NAD^+ → NADH, CO_2

Keto-L-gulonate decarboxylase

L-Xylulose

NADPH → $NADP^+$

L-Xylulose dehydrogenase

Xylitol

NAD^+ → NADH

D-Xylulose dehydrogenase

D-Xylulose

L-Gulonic acid is also decarboxylated and oxidized by the enzyme *keto-L-gulonate decarboxylase*; the product is L-xylulose. L-Xylulose is reduced in turn by the enzyme *L-xylulose dehydrogenase* in the presence of NADPH to the sugar alcohol xylitol. Xylitol, a symmetrical molecule, is also a substrate for the enzyme *D-xylulose dehydrogenase*, which is an NAD^+-requiring enzyme. The product of reoxidation of xylitol to xylulose is stereospecific and yields D-xylulose. D-Xylulose in turn is phosphorylated by ATP in the presence of *xylulokinase* to yield D-xylulose 5-phosphate, which can enter the pentose phosphate pathway and therefore can lead to the production of glucose. (See bottom of p. 268.)

Biosynthesis of hexosamines

Just as glucuronic acid is an alternate fate for UDPG, the biosynthesis of amino sugars represents an optional pathway for the metabolism of fructose 6-phosphate. Glucosamine is derived from fructose 6-phosphate by a process of transamidation. The amino acid glutamine serves as the donor of the amino group. *Transamidase* (L-glutamine:D-fructose 6-phosphate amidotransferase) is the responsible enzyme, and glucosamine 6-phosphate is its product. Further metabolism requires acetylation of the amino group (glucosamine phosphate transacetylase) by acetyl CoA; the product is *N*-acetylglucosamine 6-phosphate. *N*-Acetylglucosamine 6-phosphate is a precursor for the biosynthesis of proteoglycans and the oligosaccharide side chains for glycoproteins.

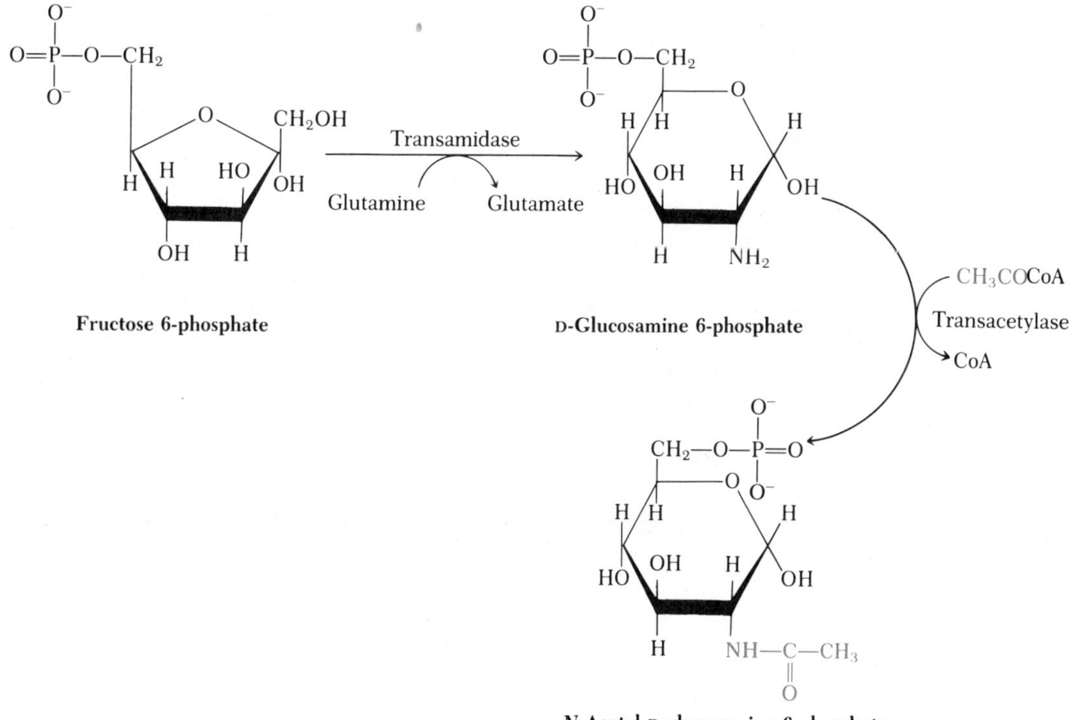

Transamidase: Glutamine → Glutamate

CH_3COCoA → Transacetylase → CoA

Fructose 6-phosphate → **D-Glucosamine 6-phosphate** → ***N*-Acetyl-D-glucosamine 6-phosphate**

The formation of these polysaccharides requires the activity of specific transferase enzymes and the UDP-activated sugars. For glucosamine to participate in such biosynthetic steps, it must first be converted into the corresponding UDP-amino sugar. The first step in UDPG formation is the conversion of glucose 6-phosphate into glucose 1-phosphate. *N*-Acetylglucosamine 6-phosphate is converted to the corresponding 1-phosphate, a reaction requiring *N-acetyl glucosamine phosphomutase* and *N*-acetylglucosamine 1,6-bisphosphate. *N*-Acetylglucosamine 1-phosphate is then converted to the nucleotide UDP-*N*-acetylglucosamine in the presence of *UDP-N-acetylglucosamine*.

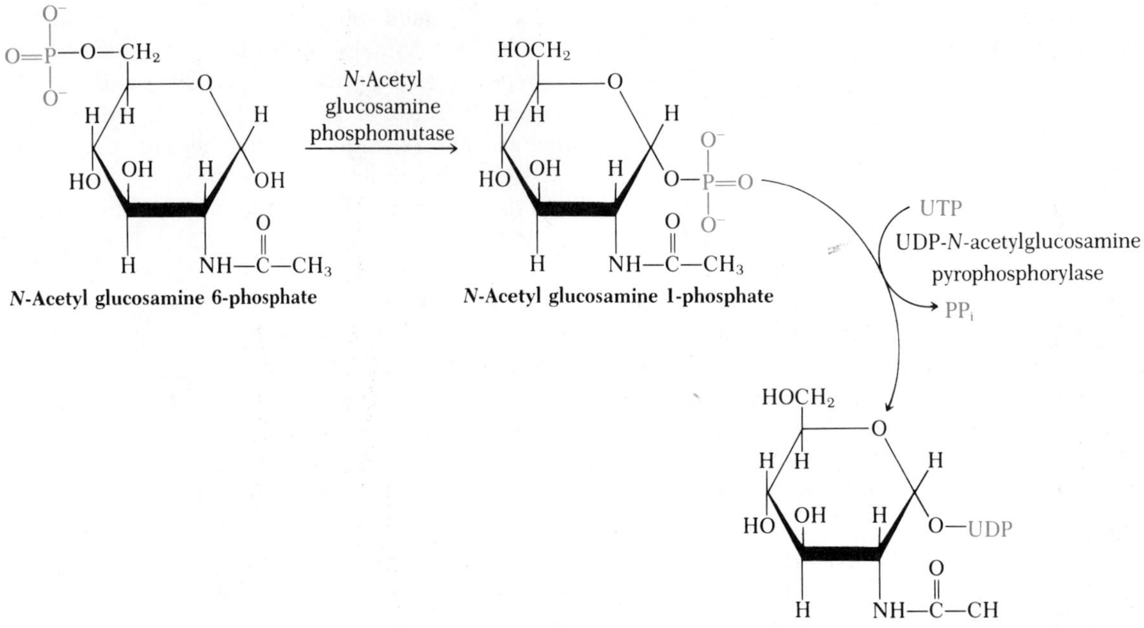

N-Acetyl glucosamine 6-phosphate

N-Acetyl glucosamine phosphomutase

N-Acetyl glucosamine 1-phosphate

UTP
UDP-N-acetylglucosamine pyrophosphorylase
PP_i

UDP-N-acetyl glucosamine

Another amino sugar used in the biosynthesis of proteoglycans and glycoproteins is galactosamine. UDP-N-acetylgalactosamine is derived from UDP-N-acetylglucosamine by the change in configuration around carbon-4 catalyzed by *UDP-N-acetylglucosamine epimerase.*

Biosynthesis of sialic acids

Sialic acids are a group of acetylated derivatives of the parent 9-carbon amino sugar called neuraminic acid. They are usually present as the terminal carbohydrate residue of the oligosaccharide side chains of glycoproteins. Neuraminic acid is made up of pyruvate and N-acetylmannosamine. The first step in its biosynthesis is the formation of N-acetylmannosamine. In mammalian cells this requires the inversion of carbon-2 by an epimerase that changes UDP-N-acetylglucosamine to N-acetylmannosamine with a loss of UDP. The enzyme is *N-acetylmannosamine-2-epimerase.* N-Acetylmannosamine is then phosphorylated to the 6-phosphate by ATP and the appropriate kinase.

Neuraminate is actually produced by condensation with PEP. The enzyme is *enolpyruvyltransferase*, and the product is N-acetylneuraminate 9-phosphate.

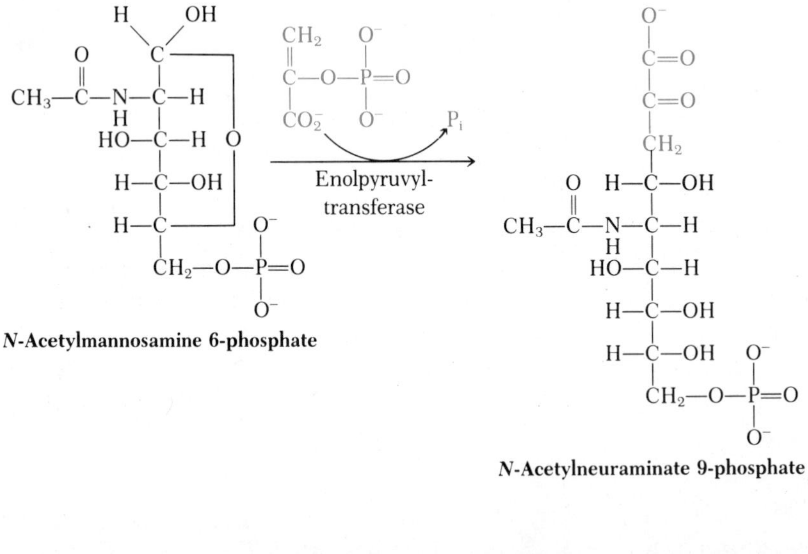

N-Acetylmannosamine 6-phosphate

Enolpyruvyl-transferase

P_i

N-Acetylneuraminate 9-phosphate

Before this compound can participate in the transfer reaction that will place it in an oligosaccharide side chain of a glycoprotein, it must be activated by reaction with a nucleotide. In this particular case the nucleotide is cytosine monophosphate rather than UDP. The responsible enzyme is *N-acetylneuraminate-CMP pyrophosphorylase.*

N-Acetylneuraminate 9-phosphate $\xrightarrow{H_2O \;\; P_i}$ **N-Acetylneuraminate (NANA)**

CTP → PP$_i$ (N-Acetylneuraminate-CMP-pyrophosphorylase)

O^- — C=O — C(—O—P(=O)(O$^-$)—O—Cytidine) — CH_2 — H—C—OH — CH_3—C(=O)—N(H)—C—H — C—H (ring O) — H—C—OH — H—C—OH — CH_2OH

CMP-NANA

Figure 11-22
Metabolic pathway showing feedback control in the biosynthesis of UDP-*N*-acetylglucosamine and CMP-*N*-acetylneuraminic acid (CMP-NANA).

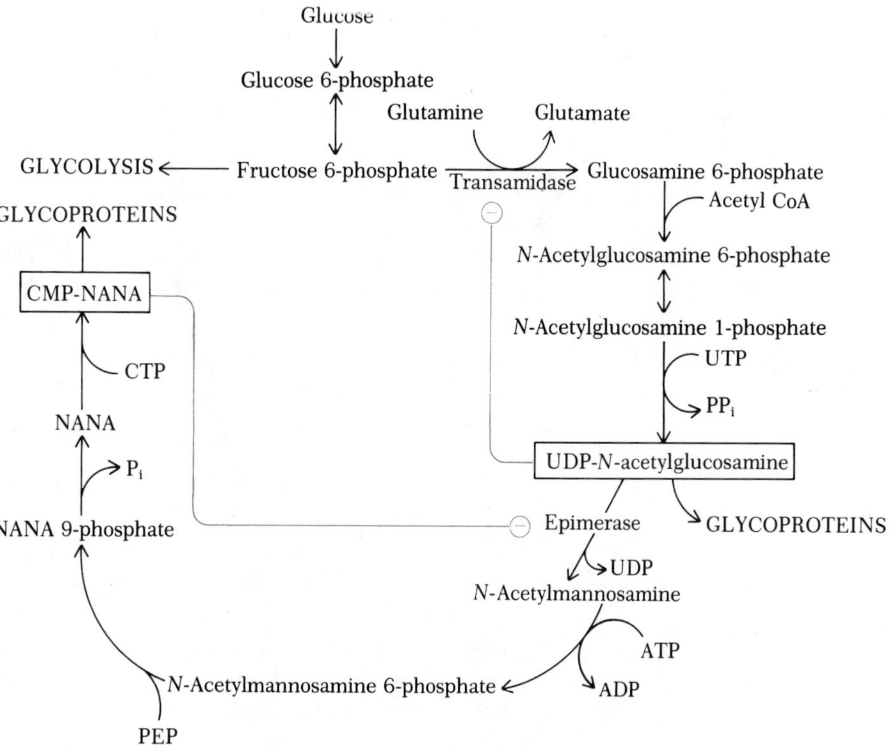

Control of hexosamine biosynthesis

Key enzymes in the control of the biosynthesis of amino sugars are fructose 6-phosphate transamidase and UDP-*N*-acetylglucosamine 2-epimerase. Both are under feedback control, the former by UDP-*N*-acetylglucosamine and the latter by CMP-*N*-acetylneuraminic acid. The control of the transamidase renders the synthesis of *UDP-N*-acetylglucosamine relatively independent of the concentration of fructose 6-phosphate, which fluctuates widely in response to diet and the rate of glycolysis. Control of both key steps enables the animal to adapt easily to the demands for glycoprotein synthesis. This feedback mechanism is summarized in Fig. 11-22.

Biosynthesis of fucose

Another pathway involving the interconversion of hexoses leads to the production of a methylpentose, L-fucose. Beginning with fructose 6-phosphate, isomerization at carbon-2 (phosphomannoisomerase) yields mannose 6-phosphate. A mutase then catalyzes the conversion to mannose 1-phosphate. Activation in the presence of a specific pyrophosphorylase and GTP results in the formation of GDP-D-mannose. A reaction sequence involving dehydration, isomerization, and reduction yields the GDP-L-fucose.

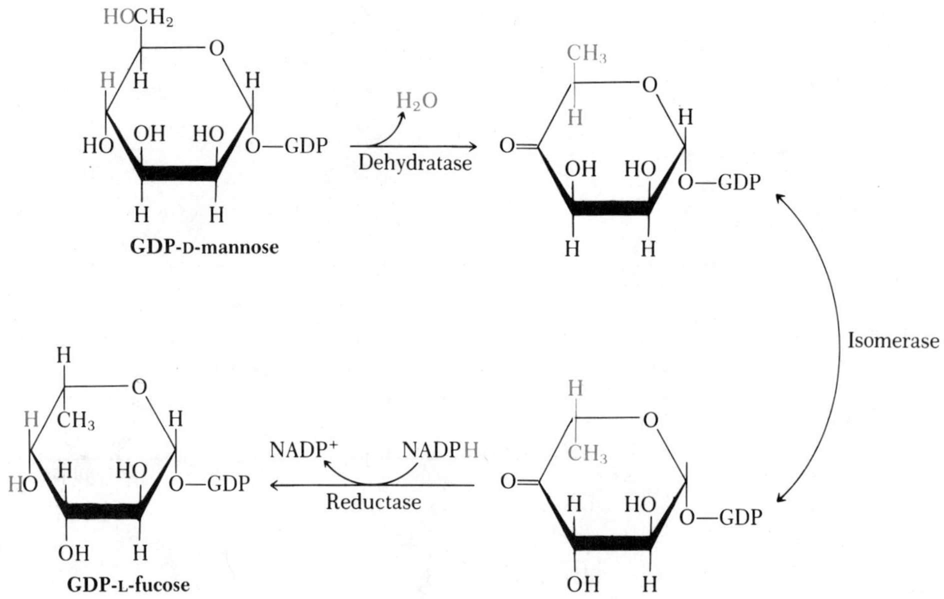

HOCH$_2$; H; H; H; HO; OH; HO; O—GDP; H; H

GDP-D-mannose

$\xrightarrow[\text{Dehydratase}]{H_2O}$

CH$_3$; O; H; H; O=; OH; HO; O—GDP; H; H

Isomerase

H; O; CH$_3$; O=; H; HO; O—GDP; OH; H

$\xleftarrow[\text{Reductase}]{NADP^+ \quad NADPH}$

H; O; H; CH$_3$; H; HO; H; HO; O—GDP; OH; H

GDP-L-fucose

Metabolism of other hexoses

Fructose Fructose is an important factor in the average American diet. As indicated in the introduction to this chapter, nearly half the carbohydrate consumed is sucrose. Since the liver is the main site of fructose metabolism, this represents an important hepatic function.

Fructose does not freely enter the hepatocyte; the process is by facilitated diffusion. There appears to be a steep gradient across the cell membrane. Once inside, fructose is phosphorylated to fructose 1-phosphate by *fructokinase*. There is no mutase present that can convert the 1- to the 6-phosphate, nor is fructose 1-phosphate a substrate for phosphofructokinase. Therefore fructose does not serve as a precursor for fructose 1,6-bisphosphate. However, fructose 1-phosphate is cleaved by liver aldolase (aldolase B), which is equally active on this substrate as on fructose 1,6-bisphosphate. The products formed are dihydroxyacetone phosphate and glyceraldehyde. The latter is phosphorylated by *glyceraldehyde kinase* (or *triokinase*) to D-glyceraldehyde 3-phosphate. These two triose phosphates then enter the glyco-

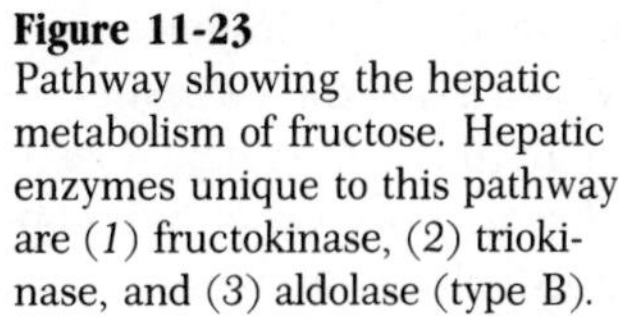

Figure 11-23
Pathway showing the hepatic metabolism of fructose. Hepatic enzymes unique to this pathway are (*1*) fructokinase, (*2*) triokinase, and (*3*) aldolase (type B).

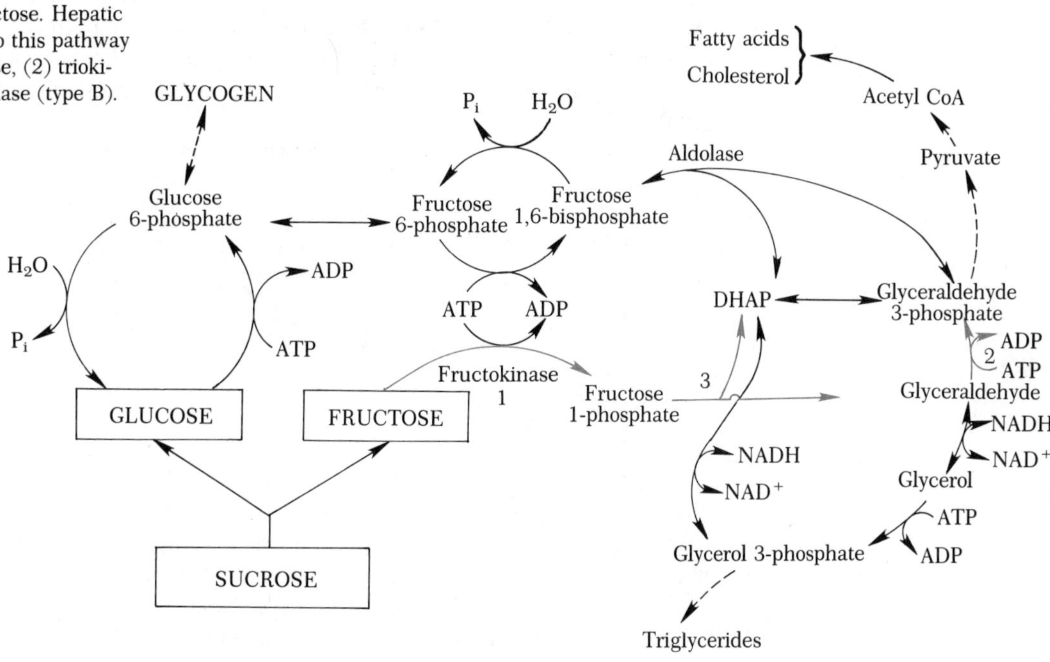

lytic pathway, gluconeogenesis and glycogenesis according to the metabolic posture of the tissue. Actually, fructose is the best gluconeogenic substrate and therefore leads to the production of glucose. The pathway of fructose metabolism is summarized in Fig. 11-23.

Since fructose metabolism bypasses the control of glycolysis at phosphofructokinase, it increases the flux through the latter half of the glycolytic pathway and may provide an uncontrolled production of acetyl CoA. This in turn may be used to produce excessive amounts of fatty acids and cholesterol. Simultaneously, glyceraldehyde may be reduced to glycerol and phosphorylated to glycerol 3-phosphate. The overall result is a stimulation of the biosynthesis of triacylglycerols. Efforts have been made to link atherosclerosis to the high fructose content of contemporary American diets.

In extrahepatic tissues, especially the adipose and muscle tissues, fructose can be phosphorylated by hexokinase to fructose 6-phosphate. This, however, is a slow reaction and occurs only in the presence of high levels of the ketose.

Fructose is present in significant amounts in seminal fluid. Apparently it is synthesized in the prostate gland by the following pathway:

$$\beta\text{-D-Glucose} \xrightarrow[\text{Aldose reductase}]{\text{NADPH} \quad \text{NADP}^+} \text{D-Sorbitol} \xrightarrow[\text{Ketose reductase}]{\text{NAD}^+ \quad \text{NADH}} \beta\text{-D-Fructose}$$

It has been suggested that spermatozoa use fructose more readily than do the epithelial cells of the vagina or uterus, and therefore the necessary carbohydrates for sperm motility are readily retained in the seminal fluid without a competitive usage by other cells.

Galactose Galactose is derived chiefly from lactose of the diet. It is important for the formation of glycolipids and glycoproteins and, of course, for the formation of lactose during lactation. As the first step in its metabolism in the liver galactose is phosphorylated to form galactose 1-phosphate (*galactokinase*), which is then converted to glucose 1-phosphate by the formation of intermediary derivatives with UDP.

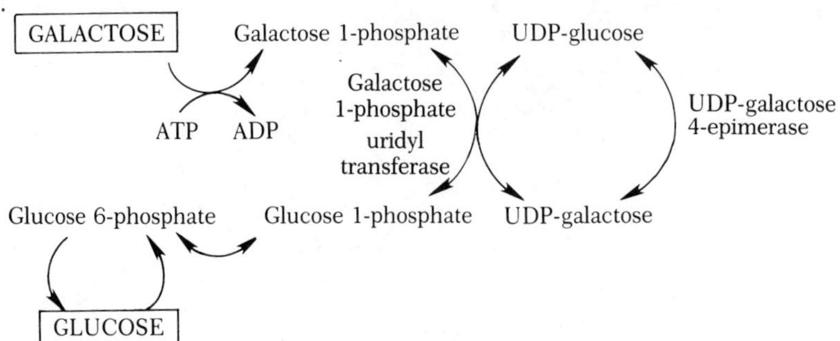

In this way galactose is converted to glucose 1-phosphate, which can yield glucose.

UDP-galactose can be synthesized in the mammary gland from UDP-glucose by using the enzyme UDP-galactose 4-epimerase. *Lactose synthase* then catalyzes a reaction of UDP-galactose with glucose to yield the disaccharide lactose.

Mannose Mannose is of little practical importance, since only small amounts appear in dietary carbohydrates. However, this hexose is well used. It is first phosphorylated to mannose 6-phosphate and then converted to fructose 6-phosphate by the enzyme *phosphomannose isomerase*. Further metabolism occurs by any of the pathways described previously.

Pentoses The pentoses in ingested food are not readily absorbed from the intestinal tract. They are used as well as are the hexoses.

Formation of glycosidic bonds

Glycosides are produced usually by the formation of a 1,4-glycosidic bond between two sugars. In general, the sugar contributing carbon-1 must enter the reaction in the activated form, i.e., as a nucleotide. The following diagram illustrates the reaction using the specific case of lactose biosynthesis catalyzed by *lactose synthase*.

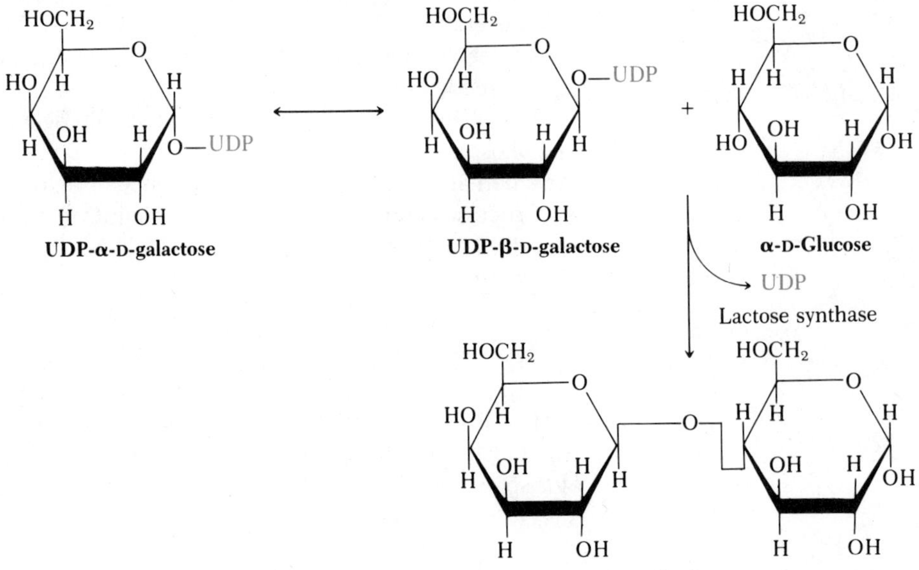

UDP-α-D-galactose ⟷ **UDP-β-D-galactose** + **α-D-Glucose**

→ UDP

Lactose synthase

Lactose

Thus the enzyme *lactose synthase* is a UDP-galactose:D-glucose 1-β-galactosyl transferase.

This enzyme, found in mammary gland tissue, has two specificities depending on the presence or absence of another protein, α-lactalbumin. In these cells there exists a unique control mechanism that determines whether lactose synthase uses *N*-acetylglucosamine or glucose as an acceptor molecule. During pregnancy, in the absence of the milk protein α-lactalbumin, reaction *A* below predominates. The disaccharide produced is *N*-acetyllactosamine, a component of some glycoproteins. After birth and during lactation, when α-lactalbumin is produced, this protein forms a complex with the lactose synthase. The consequence is a drastic reduction in K_m for glucose, perhaps as a result of an allosteric effect. Glucose is now utilized as a primary acceptor molecule so that milk sugar or lactose becomes the important product of the reaction. This is an example of cell economy in a single enzyme serves two purposes, depending on the presence or absence of another protein having in itself no known catalytic activity.

Glucose
UDP-galactose
N-Acetylglucosamine
B
+
α-Lactalbumin
A
Lactose
UDP
N-Acetyllactosamine

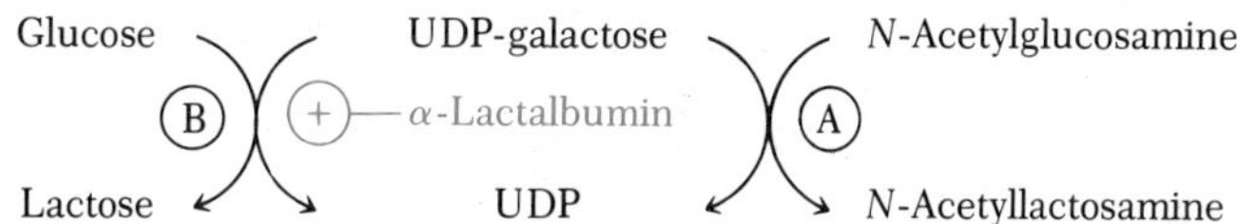

The principal competitive pathways involved in the metabolism of the four major hexose monosaccharides of dietary origin are summarized in Fig. 11-24.

Figure 11-24
Some competitive pathways in carbohydrate metabolism showing the possible fates of four dietary sugars *(color)*.

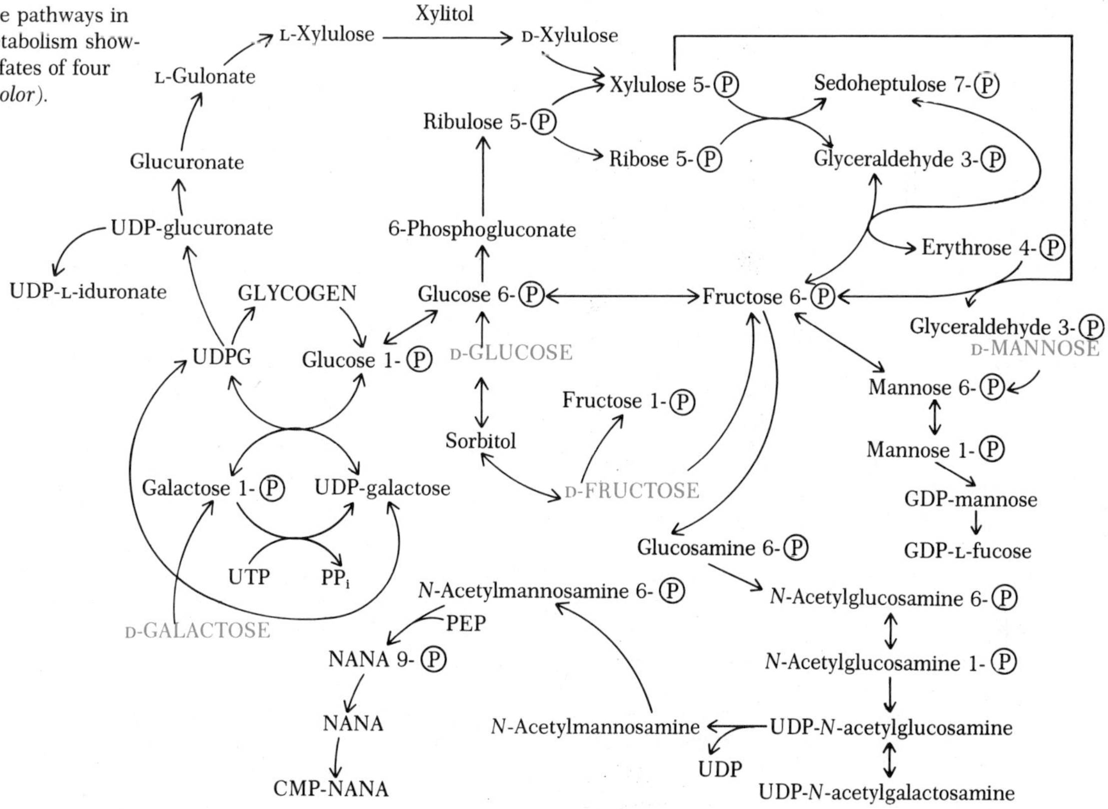

Xylitol
L-Xylulose
D-Xylulose
L-Gulonate
Xylulose 5-(P)
Sedoheptulose 7-(P)
Ribulose 5-(P)
Glucuronate
Ribose 5-(P)
Glyceraldehyde 3-(P)
UDP-glucuronate
6-Phosphogluconate
Erythrose 4-(P)
UDP-L-iduronate
GLYCOGEN
Glucose 6-(P)
Fructose 6-(P)
Glyceraldehyde 3-(P)
D-MANNOSE
UDPG
Glucose 1-(P)
D-GLUCOSE
Mannose 6-(P)
Fructose 1-(P)
Sorbitol
Mannose 1-(P)
Galactose 1-(P)
UDP-galactose
D-FRUCTOSE
GDP-mannose
Glucosamine 6-(P)
GDP-L-fucose
UTP
PP$_i$
N-Acetylmannosamine 6-(P)
N-Acetylglucosamine 6-(P)
D-GALACTOSE
PEP
NANA 9-(P)
N-Acetylglucosamine 1-(P)
NANA
N-Acetylmannosamine
UDP-N-acetylglucosamine
UDP
CMP-NANA
UDP-N-acetylgalactosamine

Suggested readings

Axelrod, B.: Glycolysis. In Greenberg, D.M., editor: Metabolic pathways, ed. 3, vol. 1, New York, 1969, Academic Press, Inc.

Carbohydrate metabolism. In Florkin, M., and Stotz, E.H., editors: Comprehensive biochemistry, vol. 17, Amsterdam, 1969, Elsevier Publishing Co.

Dickens, F., Randle, P.J., and Whelan, W.J., editors: Carbohydrate metabolism and its disorders, vols. 1-3, New York, 1968, Academic Press, Inc.

Hanson, R.W., and Mehlman, M.A., editors: Gluconeogenesis: its regulation in mammalian species, New York, 1976, John Wiley & Co.

Hers, H.G.: The control of glycogen metabolism in the liver, Annu. Rev. Biochem. **45:**167, 1976.

Hess, B., Baiteux, A., and Kruger, J.: Cooperation of glycolytic enzymes, Adv. Enzyme Regul. **7:**149, 1969.

Holzer, H., and Duntze, W.: Metabolic regulation by chemical modification of enzymes, Annu. Rev. Biochem. **40:**345, 1971.

Huijing, F.: Glycogen metabolism and glycogen storage diseases, Physiol. Rev. **55:**609, 1975.

Jost, J.P., and Rickenberg, H.V.: Cyclic AMP, Annu. Rev. Biochem. **40:**741, 1971.

Krebs, E.G.: Protein kinases. In Horecker, B.L., and Stadtman, E.R., editors: Current topics in cellular regulation, vol. 5, New York, 1972, Academic Press, Inc.

Larner, J., and Villar-Palasi, C.: Glycogen synthase and its control. In Horecker, B.L., and Stadtman, E.R., editors: Current topics in cellular regulation, vol. 3, New York, 1971, Academic Press, Inc.

Lowenstein, J.M.: The tricarboxylic cycle. In Greenberg, D.M., editor: Metabolic pathways, ed. 3, vol. 1, New York, 1967, Academic Press, Inc.

Madsen, N.B., et al.: Studies on allosteric phenomena in glycogen phosphorylase b, Mol. Cell. Biochem. **11:**35, 1976

Mansour, T.E.: Phosphofructokinase. In Horecker, B.L., and Stadtman, E.R., editors: Current topics in cellular regulation, vol. 5, New York, 1972, Academic Press, Inc.

Plaut, G.W.E.: DPN-linked isocitrate dehydrogenase of animal tissue. In Horecker, B.L., and Stadtman, E.R., editors: Current topics in cellular regulation, vol. 2, New York, 1970, Academic Press, Inc.

Purich, D.L., Fromm, H.J., and Rudolph, F.G.: The hexokinases: kinetics, physical and regulatory properties, Adv. Enzymol. **39:**249, 1973.

Ramberg, C.F., and Vranic, M.: Glucose recycling and gluconeogenesis, Fed. Proc. **36:**225, 1977.

Reed, L.J.: Pyruvate dehydrogenase complex. In Horecker, B.L., and Stadtman, E.R., editors: Current topics in cellular regulation, vol. 1, New York, 1969, Academic Press, Inc.

Ryman, B.E., and Whelan, W.J.: New aspects of glycogen metabolism, Adv. Enzymol. **34:**285, 1971.

Scrutton, M.C., and Utter, M.F.: Regulation of glycolysis and gluconeogenesis in animal tissues, Annu. Rev. Biochem. **37:**249, 1968.

Seubert, W., and Schmet, W.: The regulation of pyruvate kinase. In Horecker, B.L., and Stadtman, E.R., editors: Current topics in cellular regulation, vol. 3, New York, 1971, Academic Press, Inc.

Stalmans, W.: The role of the liver in homeostasis of blood glucose, Curr. Top. Cell. Regul. **11:**51, 1976

Utter, M.F., and Scrutton, M.C.: Pyruvate carboxylase. In Horecker, B.L., and Stadtman, E.R., editors: Current topics in cellular regulation, vol. 1, New York, 1969, Academic Press, Inc.

Van den Berghe, G.: Metabolic effects of fructose on the liver. In Horecker, B.L., and Stadtman, E.R., editors: Current topics in cell regulation, vol. B, New York, 1978, Academic Press, Inc.

Walsh, D.A., et al.: The receptor protein for cyclic AMP in the control of glycogenolysis. In Greengard, P., and Costa, E., editors: Role of cyclic AMP in cell function, New York, 1970, Raven Press.

Weber, G.: Regulation of pyruvate kinase, Adv. Enzyme Regul. **7:**15, 1969.

Weber, G., Singhal, R.L., and Srivastava, S.K.: Action of glucocorticoid as inducer and insulin as suppressor of biosynthesis of hepatic gluconeogenic enzymes, Adv. Enzyme Regul. **3:**43, 1965.

Wood, W.A., editor: Carbohydrate metabolism. In Colowick, S.P., and Kaplan, N.O., editors: Methods in enzymology, vol. 9, New York, 1966, Academic Press, Inc.

Chapter 12

Lipid metabolism

The lipids present in the human body, like the carbohydrates discussed in the preceding chapter, are in a dynamic state of metabolism, constantly undergoing changes in the various cells of the body. They are continually being oxidized for energy (mainly as ATP), converted to other essential tissue constituents, or stored as reserve fat in adipose tissue. Even stored fat, formerly regarded as an inert mass, is now known to be extremely active in metabolism. Indeed, the half-life ($t_{1/2}$) of the fatty acids in the reserve fat of laboratory mice and rats has been shown by isotope labeling studies to be only some 5 to 8 days.

Lipids include a wide variety of chemical substances, such as "neutral" fat (triacylglycerols,* or triglycerides), fatty acids and their derivatives, phospholipids, glycolipids, sterols, and carotenes. The physical properties and chemical nature of the various lipids are considered in Chapter 23 for review or reference purposes.

In this chapter primary emphasis is placed on the metabolism of the triacylglycerols and the fatty acids, since they are quantitatively the predominant lipid in the human body.

General functions

Lipids constitute over 10% of the body weight of normal adult individuals and approximately 40% of the calories of the average American's daily food intake. In addition to being an important factor in the diet, they are of great concern because of their implication in heart disease, the nation's number one "killer disease." In general, lipids are important as (1) a *ready source of energy,* because lipids supply over half of the energy used in basal metabolism; (2) a *thermal blanket,* because their presence in subcutaneous tissue insulates the body against heat loss; (3) a *structure of cell membranes,* because lipids are an integral part of all cell membranes; (4) a *protective cushion* for many tissues and organs; and (5) a *structure of secondary sex characteristics.*

When considering lipids as a source of energy, one should keep in mind that humans rely on a limited number of meals each day for the necessary energy-producing metabolites. A readily available store of energy is needed because glycogen cannot support the human organism from the evening meal until breakfast. Energy must be supplied from other sources; stored body fat is used for further

*The chemical name for fats is *triglycerides* or perhaps more properly *triacylglycerols,* the terminology approved by the IUB and the International Union of Pure and Applied Chemistry (IUPAC). All three terms will be used interchangeably in this book, however, since all appear in the current scientific and medical literature. For example, the term *triglycerides* continues to appear in such biochemical publications as the *Journal of Biological Chemistry* and the *Annual Review of Biochemistry.*

energy needs. Stored fat is not an inert mass but a dynamic system; the fatty acids are continually turned over. Rittenberg and Schönheimer, using isotopically labeled fatty acids, showed that even small amounts of labeled dietary fatty acids are incorporated into body fat before being oxidized.

After extensive starvation, lipids are still found in the tissues. This is not triglyceride but primarily phospholipid and cholesterol. The basic tissue lipid is a constant constituent of the body and is independent of the diet. It is essential for the life of the cell, comprising structural and functional materials. Lipid that is stored in excess of this is the variable element and represents that stored for energy. When deposited it is called "depot fat" and occurs in the adipose tissue. Depot fat is stored primarily for its fuel value and is composed mostly of triglycerides.

Fats yield twice the amount of energy upon oxidation (9 kcal per gm) as either carbohydrates or proteins (4 kcal per gm). When a mole of a typical fatty acid, e.g., palmitic acid, is completely oxidized to carbon dioxide and water, a total of 2340 kcal is realized under standard conditions.

$$C_{16}H_{32}O_2 + 23\ O_2 \rightarrow 16\ CO_2 + 16\ H_2O + 2340\ \text{kcal}$$

The larger yield of energy stems from the lower state of oxidation of fatty acids compared with carbohydrates. To understand how fatty acids are broken down to carbon dioxide and water and how they act as a ready source of energy, we must appreciate the relationship that lipid metabolism bears to carbohydrate metabolism. Both processes occur simultaneously; one is not subordinate to the other. Furthermore, the metabolism of lipids and carbohydrates can be controlled by the dietary fat, as is discussed later (Chapter 20).

Digestion of lipids

Before food lipids can be used by the body they must be suitably altered (digested) in the intestinal tract and absorbed into the bloodstream for distribution to the various cells.

The digestion of fats and other lipids poses a special problem because of their insolubility in water and because lipolytic enzymes, like other enzymes, are soluble in an aqueous medium. The problem is solved by the emulsification of lipids, particularly by bile salts, as is discussed in Chapter 17. The breaking of large fat or oil globules into fine particles increases the surface exposed to interaction with lipases, and the rate of digestion is proportionally increased.

Little or no lipid digestion occurs in the mouth or stomach since no significant amount of lipase is present in the secretions of these organs and no mechanism for the emulsifying of fatty material exists. There is some evidence for small amounts of a gastric lipase, but its minimal action, if it has any at all, is confined to highly emulsified lipids (e.g., those of milk or egg yolk) or to fats with short-chain fatty acids, since these are somewhat more water soluble. Furthermore, the acid pH of gastric secretions is not conducive to lipid digestion.

Lipids do play one important role in the stomach, however. They delay the rate of emptying of the stomach, presumably by way of the hormone *enterogastrone* (Chapter 18), which inhibits gastric motility and retards the discharge of foods from the stomach. Fats thus have a "high satiety value."

The major site of lipid digestion is the small intestine. This is attributable to the presence of a powerful lipase, *steapsin,* in the pancreatic juice and bile salts (Chapter 17), the latter acting as an effective emulsifying agent for fats. Pancreatic juice and bile enter the upper small intestine, the *duodenum,* by way of the pancreatic and bile ducts respectively. Secretion of pancreatic juice is stimulated by the passage of the acid gastric contents (chyme) into the duodenum by the hormones *secretin* and *pancreozymin,* now known to be identical with the hormone *cholecys-*

tokinin (Chapter 18). Secretin increases the secretion of electrolyte and fluid components of pancreatic juice, whereas pancreozymin stimulates the secretion of pancreatic enzymes. Lipid in the acid chyme stimulates the secretion of cholecystokinin (Chapter 18), which, in turn, causes contraction of the gallbladder and discharge of the bile into the duodenum. These events thus prepare the small intestine for the digestion of lipids.

Pancreatic lipase, steapsin, is an α-lipase specifically attacking the ester linkages at the 1- and 3-positions in the triglycerides, leaving a monoglyceride with the fatty acid esterified at the carbon-2 atom of glycerol. This linkage may then be cleaved by an esterase to release the third fatty acid molecule and glycerol. There is some evidence that steapsin is secreted in an inactive zymogen form, steapsinogen, that is converted to active steapsin rather nonspecifically by calcium salts, soaps, bile salts, etc.

Investigations in which ^{14}C-labeled mono-, di-, and triolein were administered to rats with cannulated thoracic ducts indicated that there was digestive cleavage of *all* the fatty acids in the 1- and 3-positions (α- and α′-) stated above. There was hydrolysis of only 22% of the fatty acids in position two (β), however. Thus about 75% of the fatty acids of the labeled triolein were split and absorbed as free fatty acids. The remainder was absorbed primarily as a β-monoglyceride.

Monoglycerides, along with bile salts, play an important role in stabilizing and further increasing the emulsification of fat in the small intestine. The emulsified fat droplets are reduced in size to *micelles,* 0.1 to 0.5 μm in diameter. This further enhances the digestion of fats and other lipids solubilized in the micellar particle.

Thus 50% to 75% of dietary fat is split by steapsin to free fatty acids, which are absorbed as such. A smaller amount is partially digested to and absorbed as β-monoglycerides, and still less as di- and triglycerides. Some estimates are that the final digestion and absorption of usual dietary fats are about 95% complete in normal individuals.

Several other enzymes secreted in the pancreatic juices are involved in the digestion (hydrolysis) of certain lipids. For example, phospholipases hydrolyze the carbon-2 (β) fatty acid from lecithin, forming *lysolecithin.* Phosphatases and esterases, with α-lipases, complete the hydrolysis. Cholesterol esters are hydrolyzed to cholesterol plus fatty acids by cholesterol esterase.

Absorption of fats and other lipids

Absorption of the digestion products of fats, primarily free fatty acids (70%) and β-monoglycerides (25%), occurs from the micelles in the microvilli *(brush border)* of the epithelial cells of the small intestinal mucosa. Bile salts of the micelle apparently are not absorbed at this point but are redissolved in other emulsoid particles, which solubilize them into micelles for later absorption into the microvilli. They are reabsorbed in the lower part of the small intestine and returned to the liver by the portal vein for resecretion into the bile. This is known as the *enterohepatic circulation* of the bile salts (Chapter 17).

Evidence based on electron microscopic studies indicates that the products of fat digestion, mainly free fatty acids and β-monoglycerides, enter the microvilli and the apical pole of the absorptive mucosal epithelial cell by passive diffusion through the cell membrane. Rapid absorption is assured by the maintenance of a concentration gradient across the cell membrane, accomplished in turn by the rapid use of fatty acids and monoglycerides in the cell's cytoplasm. Pinocytosis does not appear to play a significant role in fat absorption, as was formerly believed. The short-to-medium–chain (6 to 10 carbons) and unsaturated fatty acids are more readily absorbed than the long-chain fatty acids (12 to 18 carbons). Also the short-chain fatty acids appear to enhance the absorption of fats in general, whereas long-chain fatty acids tend to impair the process. Furthermore, the monoglycerides of the less well-absorbed, long-chain fatty acids (e.g., stearic) are better absorbed than are the corresponding free fatty acids.

Differences in the rates of digestion and absorption of the individual fatty acids are reflected in the overall rates of digestion and absorption of the dietary fats from which they are derived. Fats and oils with lower melting points (i.e., below 50° C) are more rapidly and completely hydrolyzed and absorbed than are those with higher melting points. Animal and vegetable triglycerides having similar melting points seem to be equally well digested and absorbed. Likewise, human milk fat is absorbed better than cow's milk fat because it contains a higher percent of unsaturated fatty acids and more palmitate in the carbon-2 position.

The products of digestion next appear to be taken up by the smooth endoplasmic reticulum and resynthesized into triglycerides by enzymes present in the membranes or cavities of the reticulum. The rapid removal of fatty acids and monoglycerides by their resynthesis into triglycerides maintains a sharp gradient of concentration within the mucosal cell that favors the continued, rapid diffusion of free fatty acids and monoglycerides into the cell from the intestinal lumen. There is a merging of the smooth endoplasmic reticulum into rough endoplasmic reticulum (Chapter 2) in which probably enzymes for triglyceride resynthesis are formed as well as the protein component of the chylomicron coat of the minute fat globules. The resynthesized triglycerides, in the form of chylomicrons, are discharged from the mucosal cell into the lymph and thence into the lacteals of the small intestine for transport into the lymphatic system and finally into the bloodstream for metabolic use. These events are shown schematically in Fig. 12-1.

The resynthesis of triglycerides from free fatty acids or monoglycerides in the mucosal cells, shown schematically in Fig. 12-2, is discussed later. Briefly summarized, the steps involved in resynthesis are (1) conversion of the free fatty acid to the

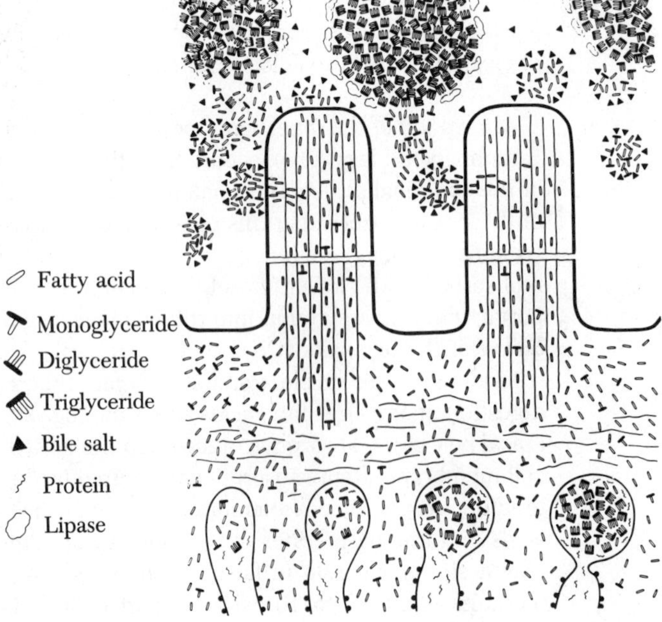

Figure 12-1
Biochemical events in digestion and absorption of fat.

From Porter, K. R.: Fed. Proc. Symp. **28:**35, 1969.

fatty acid CoA derivative by ATP and CoA; (2) conversion of the fatty acid CoA derivative to monoglyceride phosphate in the presence of L-glycerol 3-phosphate; and (3) successive conversions, in the presence of 2 moles of fatty acid CoA, to diglyceride phosphate, diglyceride, and finally triglyceride. Aggregates of triglycerides, plus small amounts of extraneous phospholipids, cholesterol, etc., are then "coated" with protein and secreted from the mucosal cell into the intracellar fluid, thence into the lacteals and lymphatics, and finally into the general circulation for metabolic disposal.

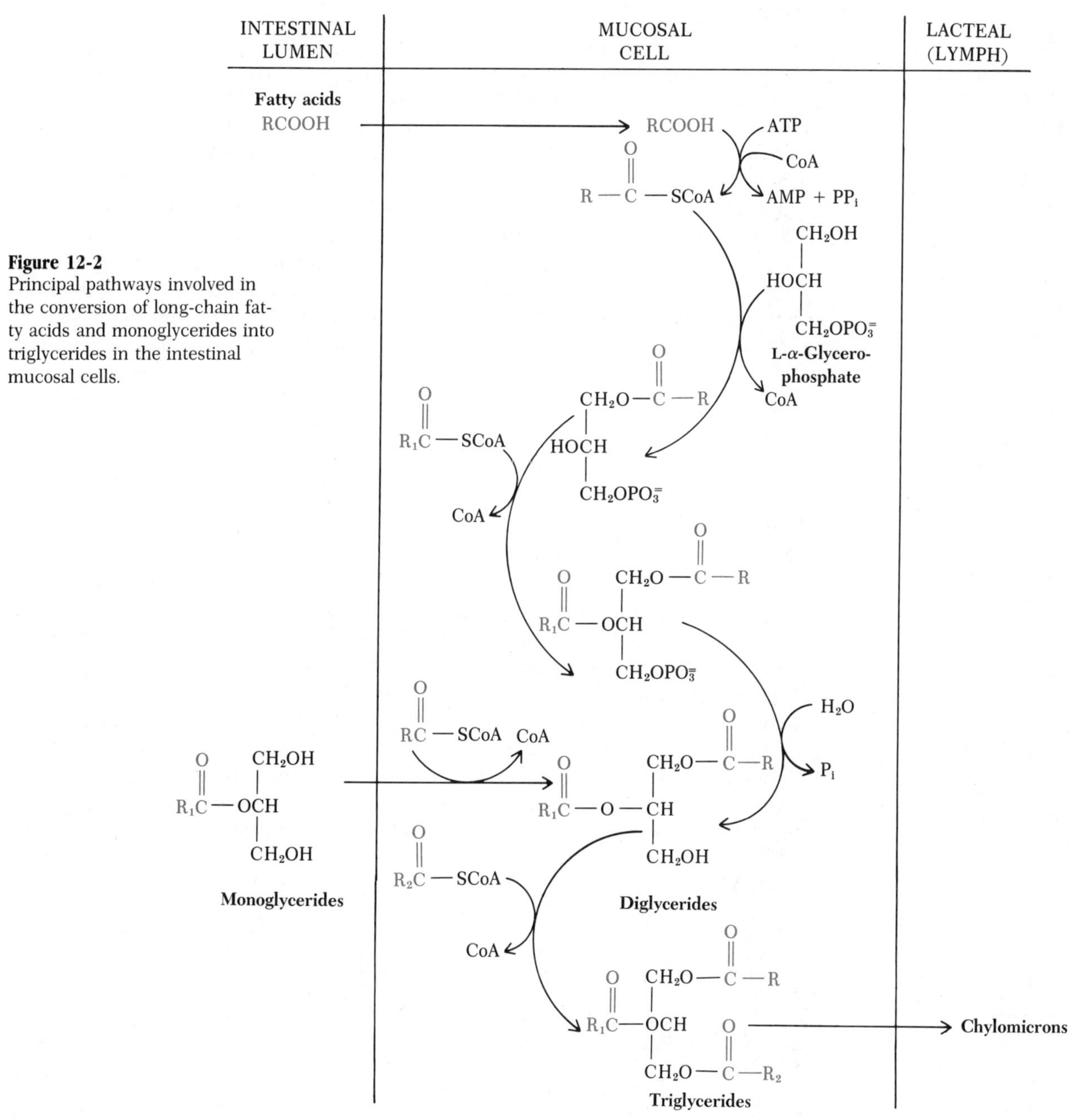

Figure 12-2
Principal pathways involved in the conversion of long-chain fatty acids and monoglycerides into triglycerides in the intestinal mucosal cells.

The foregoing discussion of the absorption of fatty acids and their resynthesis into triglycerides applies to long-chain fatty acids (12 to 18 carbons). The lower molecular weight free fatty acids (6 to 10 carbons), representing less than 30% of the absorbed fat, are distributed, according to Frazer's *partition theory,* bound to plasma albumin, mainly by the blood capillaries. They then pass into the portal blood and thence directly to the liver for oxidation or lengthening into long-chain fatty acids, as is discussed later.

Lipid malabsorption

Impaired intestinal absorption of lipids occurs not infrequently in man, sometimes with serious consequences. Basically this may result from (1) defective lipolysis in the lumen of the small intestine or (2) defective mucosal cell synthesis of the protein moiety of the lipoproteins.

Intestinal resynthesis of triglycerides

The free fatty acids and monoglycerides from digested dietary fat are synthesized into triglycerides by the mucosal cells of the small intestine (Fig. 12-2). The free fatty acids are activated by conversion to the fatty acyl CoA derivatives. The enzyme *fatty acyl CoA thiokinase* catalyzes the reaction with ATP and CoA; the products are AMP and pyrophosphate. Although this reaction is readily reversible, it is driven toward the formation of fatty acyl CoA by the concomitant hydrolysis of the pyrophosphate. *Pyrophosphatase* is universally present in mammalian tissues. The thiokinases exhibit some degree of specificity with regard to the size of the fatty acid substrates. There is a long-chain fatty acyl thiokinase acting on chains having 10 to 20 carbons (saturated and unsaturated), a medium-chain thiokinase acting on chains of 4 to 12 carbons, and acetate thiokinase, which acts on acetate and propionate.

$$\underset{\textbf{Fatty acid}}{R-\overset{\displaystyle O}{\overset{\|}{C}}-OH} \xrightleftharpoons[\text{ATP} \quad \text{AMP}]{\text{CoA} \quad \text{PP}_i}_{\text{Acyl CoA thiokinase}} \underset{\textbf{Fatty acyl CoA}}{R-\overset{\displaystyle O}{\overset{\|}{C}}-SCoA}$$

Once the absorbed fatty acid is activated, it reacts with L-glycerol 3-phosphate to yield the corresponding monoglyceride phosphate (see below). The enzyme catalyzing this reaction is an *acyl CoA transferase.* In general, L-glycerol 3-phosphate is obtained from two potential sources. One is by a reduction of DHAP using L-glycerol 3-phosphate dehydrogenase (Chapter 11). This interrelationship between lipid and carbohydrate metabolism occurs in all tissues. The second source is the phosphorylation of free glycerol by glycerokinase at the expense of ATP. Liver and kidneys contain glycerokinase; it is found to some degree in the intestine, whereas adipose tissue and skeletal muscle do not have this enzyme.

$$\underset{\textbf{L-Glycerol 3-phosphate}}{\begin{array}{l} CH_2OH \\ HO-C-H \\ CH_2-O-PO_3^{2-} \end{array}} + R-\overset{\displaystyle O}{\overset{\|}{C}}-SCoA \xrightarrow{\text{Acyl CoA transferase}} \underset{\textbf{Monoglyceride phosphate}}{\begin{array}{l} CH_2O-\overset{O}{\overset{\|}{C}}-R \\ HO-C-H \\ CH_2-O-PO_3^{2-} \end{array}} + CoA$$

A second residue of fatty acid (by fatty acyl CoA) is added to the monoglyceride phosphate to yield a diglyceride phosphate (Fig. 12-2). The sequential formation of the two esters is catalyzed by *acyl CoA transferases*. The diglyceride phosphate, also called a phosphatidic acid, is then hydrolyzed to the α, β-diglyceride by means of *phosphatidic acid phosphatase*.

Fig. 12-2 shows that absorbed dietary monoglycerides can also be converted to diglycerides by the esterification of a second fatty acid from fatty acyl CoA. This pathway for the use of dietary monoglycerides is important principally in intestinal mucosal cells and serves to maintain the concentration gradient required to favor their rapid absorption.

Finally diglycerides are converted to triglycerides by the incorporation of another fatty acid (fatty acyl CoA), catalyzed by a *diglyceride acyl transferase*.

Aggregates of triglycerides (80%), phospholipids (7%), and cholesterol (9%) are "coated" with proteins to produce particles called *chylomicrons*. These particles, about 1 μm in diameter, are secreted from the mucosal cells into the intracellular fluid, thence into the lacteals and the lymphatics, and finally into the systemic bloodstream. Triglycerides that follow this route have predominantly long-chain fatty acids of more than 10 carbons. Medium-chain fatty acids (10 carbons or less), on the other hand, are usually absorbed directly from the intestinal contents into the portal blood system as free fatty acids bound to plasma albumin. This difference in fate may be the consequence of their greater solubility.

Transport of lipids

After an average meal, there is a transient elevation of blood lipids (alimentary hyperlipemia). The peak level of lipid in blood plasma usually occurs after ½ to 3 hours and returns to normal in 5 to 6 hours. The lipid transported in the blood appears to be in at least three forms: (1) chylomicrons, (2) lipids associated with proteins as lipoproteins, and (3) unesterified fatty acids loosely bound to serum albumin. The chylomicrons are responsible for the turbid or even milky appearance of the plasma after a meal rich in fat. The amounts of various lipids in the blood plasma of normal young adults in the postabsorptive state are given in Table 12-1.

Role of the lipoproteins

Since lipids are insoluble in an aqueous medium, their transport in the blood plasma presents a special problem. This has been solved by associating the more hydrophobic lipids with the more hydrophilic ones, such as phospholipids, and then combining the mixture with cholesterol and protein to form a globular hydrophilic lipoprotein particle. As revealed by electron micrographs, the larger lipoprotein

Table 12-1
Plasma lipids of normal young adults

Lipid	mg/100 ml
Total lipid	590
Neutral fat	150
Total fatty acid	350
Phospholipid fatty acid	130
Cholesterol ester fatty acid	75
Neutral fat fatty acid	150
Free fatty acids	20
Total cholesterol	160
Combined cholesterol	115
Free cholesterol	45
Phospholipid	200

particles, chylomicrons and the very low–density lipoproteins (VLDL), typically consist of a core of nonpolar triglycerides and cholesteryl esters surrounded by amphipathic lipids such as phospholipid and free cholesterol, and specially-synthesized proteins, such as the apoprotein moiety of the lipoprotein complex. The latter three components solubilize the particle in the aqueous plasma medium.

Physical properties and chemical composition of the major lipoproteins Four broad categories of human plasma lipoprotein particles involved in lipid transport have been separated and characterized by high-speed centrifugation and by various types of chromatographic procedures. These four groups, together with the principal types of lipids they transport, are included in Table 12-2.

A fifth category of lipoproteins, termed very high–density lipoproteins (VHDL), is recognized by some authorities. This group has a higher protein content (approximately 57%) and a higher density (>1.21) than the HDL group (Tables 12-3 and 12-4), as would be expected. Its triglyceride, phospholipid, and cholesterol contents are somewhat lower than HDL. The apoproteins are types A-I and A-II, described later. Extensive studies of this group also are in progress.

Some of the outstanding physical properties of the four major categories of human plasma lipoproteins are given in Table 12-3. The data emphasize the relatively large particle size, low density, and high flotation rate (sf) of the chylomicrons, in keeping with their primary function of triglyceride transport. These values shift, as might be expected with the decreasing triglyceride and increasing cholesterol, protein, and phospholipid contents of the other types of lipoproteins, as shown in Table 12-4.

Table 12-2
The major types of human plasma lipoproteins and the principal lipids they transport

Type	Principal lipid transported
Chylomicrons	Triglycerides (intestinal)
Very low–density lipoproteins (VLDL)	Triglycerides (from liver)
Low-density lipoproteins (LDL)	Endogenous cholesterol
High-density lipoproteins (HDL)	Cholesterol

Table 12-3
Some physical properties of the major categories of human plasma lipoproteins

Lipoprotein	Particle size (nM)	Density (range)	Flotation (sf)	Principal lipid
Chylomicron	100-1000	<0.95	>400	Triglyceride (intestine)
VLDL	30-75	0.95-1.006	12-400	Triglyceride (liver)
LDL	20-25	1.019-1.063	0-12	Cholesteryl esters
HDL	10-15	1.063-1.210	0-2	Cholesterol

Table 12-4
Lipid and protein contents of human plasma lipoproteins

		Approximate composition (percent)				
				Cholesterol		
Lipoprotein	Amount (mg/100 ml plasma)	Protein	Triglycerides	Free	Esters	Phospholipids
Chylomicron	100-250	2	85	2	5	7
VLDL	130-200	9	50	8	15	18
LDL	200-400	25	10	9	36	20
HDL	50-130	40	5	5	22	28

The values in Table 12-4 also emphasize the roles of the major lipoproteins in lipid transport. The relatively high content of triglycerides in chylomicrons and VLDL is a reflection of their principal roles in the transport of triglycerides from the intestine to the liver and from the liver to other tissues respectively. There is evidence indicating that LDL is formed from the VLDL fraction by the removal of triglycerides.

The markedly increased cholesterol content of the LDL and HDL categories is in accord with their principal functions in the transport of cholesterol, primarily as cholesteryl esters and especially as the linoleate. Approximately 70% of the plasma cholesterol is contained in the LDL fraction. HDL appears to serve particularly in removing cholesterol from peripheral cells and transporting it back to the liver, where some 95% of the cholesterol excreted from the body is removed. There is some evidence that HDL actually competes for the cellular binding sites of LDL, thus preventing the deposition of cholesteryl esters in cells. The increased protein and phospholipid content of the LDL and HDL fractions also is noteworthy.

Although all categories of lipoproteins contain significant amounts of phospholipid, it appears to play no known role in lipid transport. When ^{14}C-labeled palmitic acid is fed to rats, 96% of the labeled acid can be recovered from the blood in forms other than phospholipid.

It is also significant that little or no free fatty acid is found in any of the lipoprotein complexes. Fatty acids are transported mainly in association with plasma albumin. However, as mentioned previously, some may be transported as cholesteryl esters, mainly as linoleate.

The apoprotein moiety The apoprotein components of the various lipoproteins have been studied extensively in recent years. These components differ among the groups and are mixtures of proteins within a given category. For this reason, the various apoproteins are now designated by the letters A, B, and C, with subgroups given in Roman numerals (I, II, III, etc). Thus the apoproteins of the VLDL group include B, C-I, C-II, and C-III, and an arginine-rich protein (11%) designated as such. The apoprotein of the LDL group is B; those of the HDL category are A-I and A-II, with minor C-I, C-II, and C-III components. "Apo-B" of the LDL type has been reported to have a molecular weight of about 500,000 and to consist of a number of subunits having molecular weights averaging about 25,000. It is a glycoprotein with 5% to 9% carbohydrate. The amino acid composition and sequence of the various apoproteins, as well as any special functions, are being investigated intensively.

The apoprotein of the chylomicrons is synthesized in the epithelial cells of the intestinal mucosa, as previously stated. That of the VLDL is formed in the liver. However, the apoprotein portion of the LDL group is apparently derived from the VLDL apoprotein after the removal of peptide fragments. It has a half-life of approximately 2½ days. It is probably degraded in nonhepatic tissues. The LDL apoprotein migrates with the β_1-plasma proteins (Chapter 16) on electrophoresis and thus sometimes is termed a β-plasma protein. Similarly, the HDL apoproteins are associated with the α_1-plasma proteins.

The apoprotein moieties of several lipoproteins have important functions in addition to solubilizing lipids. For example, that of HDL fraction C-II is an activator for extrahepatic lipoprotein lipase. Another, A-I of HDL, is an activator of lecithin-cholesterol acyl transferase (LCAT). Further investigations undoubtedly will disclose additional ways in which the apoprotein moieties of the lipoproteins perform important functions in lipid metabolism.

The rate-limiting step in the transport of triglycerides is their conversion to chylomicrons in the mucosal cells of the intestine and their subsequent secretion into the lymphatics. This is evident in a condition involving a deficiency of β-lipoproteins, termed *abetalipoproteinemia* (Chapter 22). Chylomicron formation and the transport of lipid into the lymphatics are thus greatly impaired. Triglycerides in particular accumulate in the mucosal cells. The importance of protein synthesis to chylomicron formation and the transport of fat is also illustrated by the accumulation of triglycerides in the intestinal mucosal cells in animals treated with puromycin.

Variations in the plasma levels of human lipoproteins

Variations in the levels of the various lipoproteins in human blood plasma from the average values given in Table 12-4 have become of increasing interest in recent years. As determined by electrophoretic procedures, they have been found to vary widely in a number of physiological and pathological conditions. Their determination provides extremely useful information in the diagnosis and treatment of a number of diseases of lipid metabolism, as is discussed later.

Variations in the pattern of plasma lipoproteins occur under usual physiological activities such as exercise. For example, the level of HDL tends to be increased by jogging and running. This relation may have clinical significance. The level of HDL also tends to increase with age and with the use of estrogens or alcohol in modest amounts. On the other hand, the plasma level of HDL is decreased by the progestin-type of contraceptives, androgens, smoking, obesity, and dietary carbohydrate (sucrose or glucose).

Likewise the plasma level of LDL varies under differing physiological conditions. It is higher in men 20 to 40 years old than in women of the same age. Older men and women have higher levels than younger ones. Furthermore, obese strains of laboratory mice and rats have significantly higher levels of LDL than do control animals. These findings may have clinical implications in the coronary atherosclerosis problem. Indeed, recent studies indicate that high levels of LDL appear to render the human species uniquely susceptible to coronary heart disease, whereas higher levels of HDL appear to afford protection against heart attacks. This important relationship is considered further in Chapter 22.

A number of disorders of plasma lipoprotein metabolism have been described and characterized in recent years. Most of these are rare, inherited abnormalities. They may be grouped into two general categories: those in which there is a deficiency of plasma lipoproteins, *hypolipoproteinemia,* and those in which there is an excess, the *hyperlipoproteinemias*. These are discussed in Chapter 22.

Metabolism of triglycerides

Release of fatty acids in tissues

After the ingestion of fat-containing foods, the turbid hyperlipemia plasma "clears" as it passes through various tissues. This is caused by the hydrolysis of the triglyceride component of lipoproteins by the enzyme *lipoprotein lipase,* also known as the "clearing factor." Lipoprotein lipase is present in most tissues and is found in relatively large amounts in adipose tissue and muscle. The hydrolytic action of the enzyme thus releases free fatty acids primarily for oxidation by the cells of the tissue to form energy as ATP. If not needed for energy, the free fatty acids can associate with plasma albumin for transport to other tissues for use or conversion into other lipids, or they can be resynthesized into triglycerides by pathways previously described and stored in adipose or other tissues for use as required. The processes of fatty acid oxidation and of its storage-mobilization in adipose tissue are described later in this chapter.

The free fatty acid plasma albumin complex has an exceedingly rapid turnover rate, its half-life being only 2 to 3 minutes. Its high metabolic activity is probably the result of its primary use for oxidation to meet energy needs.

The plasma level of free fatty acids is affected by a number of hormones, particularly by epinephrine and norepinephrine. This results from an increased mobilization of fatty acids from adipose tissue and may be another stress effect of the catecholamines analogous to the breakdown of liver glycogen. Insulin, as well as glucose, decreases free fatty acid levels; growth hormones, thyroxine, and ACTH appear to elevate the values. Further details, including the reactions involved, are presented in a later section of this chapter.

Lipid metabolism in the liver

Formerly, the liver was regarded as the major site of lipid metabolism. However, subsequent investigations have demonstrated that most tissues have the ability to oxidize fatty acids completely and to synthesize them at least to a limited extent. Adipose tissue is extremely active in lipid metabolism. Nevertheless, the liver does play an important and unique role in lipid metabolism not only in the oxidation of fatty acids and in their biosynthesis and conversion into triglycerides but also in the synthesis of the apoproteins of VLDL and LDL and possibly of other lipoproteins.

The biosynthesis of triglycerides occurs in the liver and adipose tissue, but as has already been described, the epithelial cells of the intestinal mucosa are also active in this respect. The fatty acids involved are primarily 16- and 18-carbon acids, both saturated and unsaturated. In the liver the L-α-glycerophosphate is derived from either free glycerol, DHAP, or both. The adipose tissues, however, derive L-α-glycerophosphate entirely from DHAP because they do not possess the required glycerokinase needed to phosphorylate glycerol.

Figure 12-3
Pathway showing biosynthesis of triglycerides in mammalian liver.

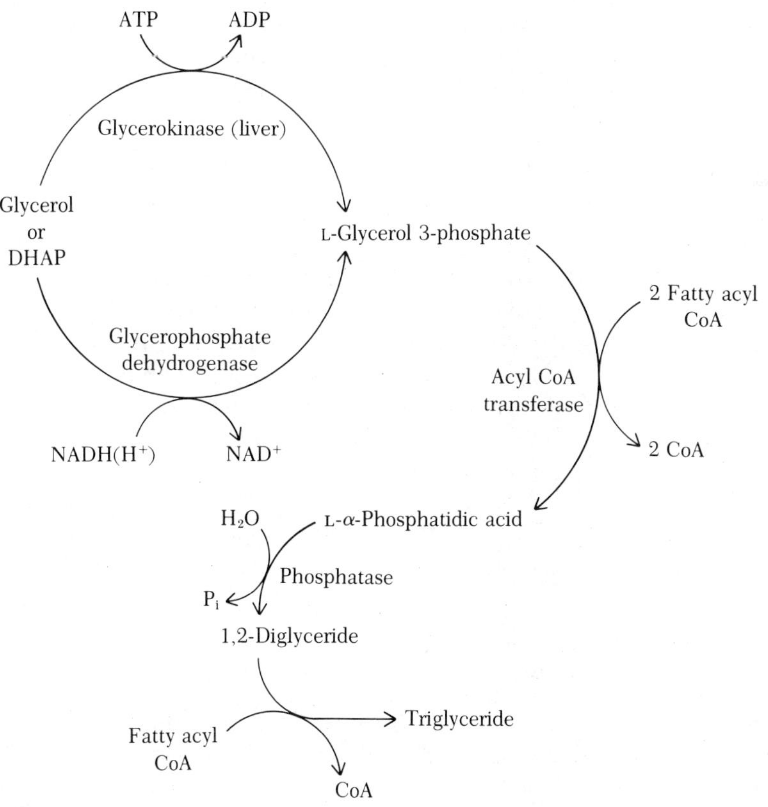

Fig. 12-3 reviews the stepwise synthesis of a triglyceride; this is the same as described for the intestinal mucosal cells (Fig. 12-2). L-α-Phosphatidic acid is the central intermediate for the biosynthesis of both triglycerides and phospholipids. DHAP may serve as a precursor in an alternate pathway.

$$CH_2OH{-}C({=}O){-}CH_2O{-}P({=}O)(O^-)_2 \xrightarrow[\;CoA\;]{R_1{-}C({=}O){-}SCoA} CH_2O{-}C({=}O){-}R_1,\ C{=}O,\ CH_2{-}O{-}P({=}O)(O^-)_2$$

$$\xrightarrow{NADH(H^+)\ \rightarrow\ NAD^+} CH_2O{-}C({=}O){-}R_1,\ HO{-}C{-}H,\ CH_2{-}O{-}P({=}O)(O^-)_2$$

$$\xrightarrow[\;CoA\;]{R_2{-}C({=}O){-}SCoA} CH_2O{-}C({=}O){-}R_1,\ R_2{-}C({=}O){-}O{-}C{-}H,\ CH_2O{-}P({=}O)(O^-)_2$$

Fatty liver syndrome The liver appears to be uniquely vulnerable to the accumulation of fats and other lipids, the so-called fatty liver syndrome. If prolonged, this condition may lead to enlargement of the liver, fibrotic changes, cirrhosis, and severely impaired liver function. Fatty livers may result from a number of stresses or abnormalities in lipid metabolism. One such causative factor may be prolonged increases in the plasma level of free fatty acids. This may result from the excessive mobilization of fatty acids from adipose tissue or from the hydrolysis of plasma lipoproteins or chylomicrons. If the production of the apoprotein for plasma lipoproteins, particularly for VLDL, does not parallel the influx of free fatty acids into the liver, triglycerides in particular will accumulate because of an impaired transport of lipids from the liver, and fatty liver will result. A similar situation may result from the prolonged feeding of a high-fat diet, or during starvation, or in uncontrolled diabetes mellitus, or in ketosis.

Another major cause of fatty livers is an impairment in the biosynthesis of plasma lipoproteins, either in the apoprotein moiety or in the lipid-protein complex itself. Lipid transport is impaired, primarily that of triglycerides, and fat accumulation in the liver ensues. This general type of fatty liver is often associated with a lack of *lipotropic factors,* particularly choline or its precursors (Chapter 21). Choline is required for the formation of a major type of phospholipid, lecithin (phosphatidylcholine), which in turn is an essential component of the lipid component of lipoproteins, as described previously. Much of the choline required for phospholipid formation is synthesized by way of transmethylation from methionine (Chapter 13). Hence, diets low in protein may cause fatty livers. Support for this concept comes from the fact that the VLDL fraction of plasma lipoproteins is low in choline-deficient rats.

Fatty livers also may result from the use of any of a variety of drugs or from certain toxic substances. The list includes such substances as certain antibiotics (e.g., puromycin), alcohol, chloroform, carbon tetrachloride, phosphorus, lead, and arsenic. The mechanism of action of these substances appears to be the inhibition of hepatic protein synthesis, at least in most instances. Deficiencies of certain vitamins (E, B_6, and pantothenic acid) and nutrients (essential fatty acids and selenium) (Chapter 21) have been reported to produce fatty livers in experimental animals.

The mechanism involved in the production of fatty livers by alcohol is unclear. A single intoxicating dose of alcohol administered to rats produces an increase in plasma free fatty acids and triglycerides. Whether this is a result of mobilization of fatty acids from adipose tissue, or an increase in hepatic synthesis, or a decrease in fatty acid oxidation, or some combination of these is as yet uncertain.

Oxidation of fatty acids

β-Oxidation

Fatty acids are oxidized mainly by a process called β-oxidation in which two-carbon units are sequentially removed, beginning from the carboxyl-terminal end. The principle of β-oxidation was first described in 1904 by Knoop. He studied the oxidation of various fatty acids by tagging the acids at the methyl end with a benzene ring. These phenyl derivatives were fed to animals; the compounds containing the benzene ring, isolated from the urine, were found to be derivatives of benzoic and phenylacetic acids. Neither acid is oxidized by the body; both are excreted in combination with glycine, i.e., as hippuric and phenaceturic acids (Chapter 17). By the same token, when benzoic or phenylacetic acids were fed, hippuric or phenaceturic acids were excreted. However, on feeding phenylpropionic acid, the next higher homolog, phenylaceturic acid was not formed; instead hippuric acid was excreted. Phenylbutyric acid was cleaved to phenylacetic acid, which was excreted as phenaceturic acid.

$$C_6H_5-CH_2CH_2\overset{\beta}{\wr}CH_2C(=O)-OH \longrightarrow C_6H_5-CH_2C(=O)-OH + CH_3C(=O)-OH$$

Cleavage

Phenylbutyric acid — Phenylacetic acid — Acetic acid

Compound fed	Compound excreted
C_6H_5COOH	$C_6H_5CONHCH_2COOH$
$C_6H_5CH_2COOH$	$C_6H_5CH_2CONHCH_2COOH$
$C_6H_5CH_2CH_2COOH$	$C_6H_5CONHCH_2COOH$
$C_6H_5CH_2CH_2CH_2COOH$	$C_6H_5CH_2CONHCH_2COOH$

The first and third derivatives (see above) yield the same product, hippuric acid. For this to happen, phenylpropionic acid must lose two carbons. The same relationship exists between the second and fourth compounds. The conclusion reached was that fatty acids lose two carbons at a time, i.e., oxidation must begin with the β-carbon and, when this results in the loss of two carbons, the new β-carbon is oxidized. On the basis of these observations, Knoop formulated the principle of β-oxidation of fatty acids.

β-Oxidation is the major mechanism for the catabolism of fatty acids. However, they first must be converted to their CoA derivatives. This reaction requires cytoplasmic thiokinase and is similar to the one already described for fatty acid absorption and reesterification. Subsequent steps occur in the mitochondria of the liver and other tissue cells; fatty acid degradation thus is compartmentalized. The mitochondrial membrane, however, is impermeable to both fatty acids and their CoA derivatives. Therefore entrance of the fatty acid into the mitochondrion requires the action of a membrane transport system; the carrier molecule is called *carnitine*.

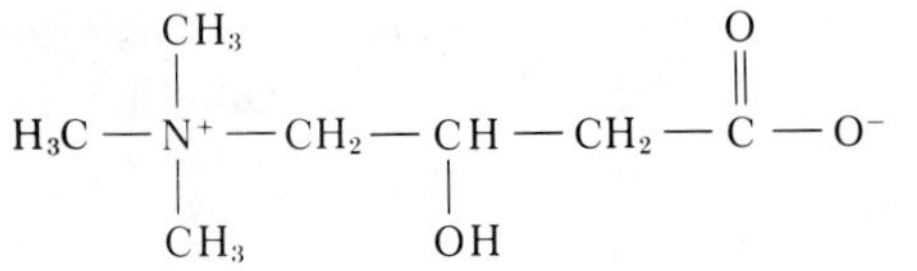

$$H_3C-\overset{\displaystyle CH_3}{\underset{\displaystyle CH_3}{\overset{|}{\underset{|}{N^+}}}}-CH_2-\underset{\displaystyle OH}{\underset{|}{CH}}-CH_2-\overset{\displaystyle O}{\overset{\|}{C}}-O^-$$

Carnitine

Fig. 12-4 shows that the transport of fatty acids across the mitochondrial membrane requires an exchange of the CoA group for carnitine to form an acyl-carnitine derivative. This reaction is catalyzed by *carnitine–acyl CoA transferase*. Reversal of the exchange probably occurs on the inside of the mitochondrial membrane, with the acyl group once again being attached to CoA. However, the CoA now is of intramitochondrial origin. The entire process, beginning with plasma free fatty acids, activation in the cytoplasm, transport into the mitochondrion, followed by β-oxidation, is outlined in Fig. 12-5.

Inside the mitochondrion, the fatty acyl CoA is the substrate for a class of enzymes called *fatty acyl CoA dehydrogenases*, which are flavoproteins having FAD as coenzyme. There are four such dehydrogenases, each specific for a range of chain lengths. One is palmitoyl CoA dehydrogenase; it acts on medium- and long-chain fatty acids. Octanoyl CoA dehydrogenase and two butyryl dehydrogenases

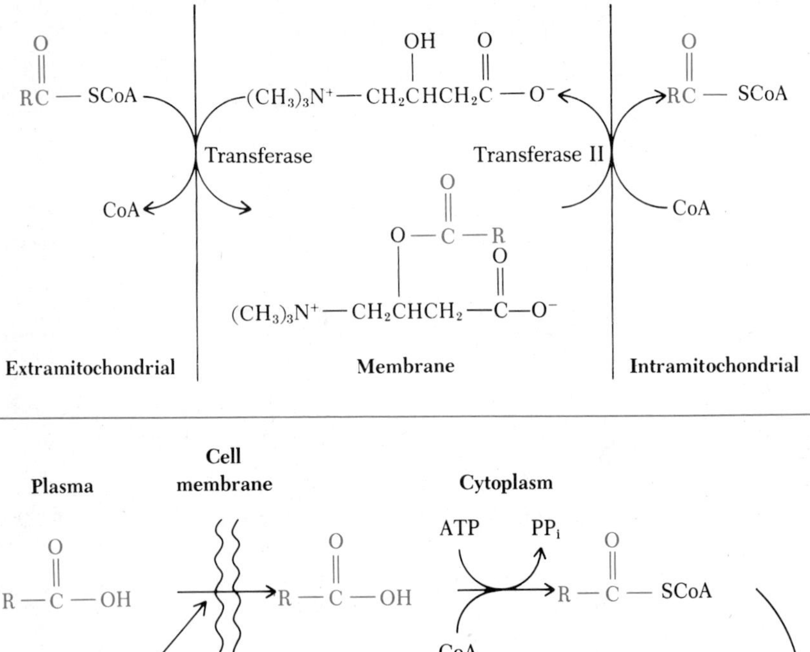

Figure 12-4
Transport of a fatty acid across mitochondrial membrane. Acyl carrier is carnitine; the transport process is catalyzed by carnitine–acyl CoA transferase.

$RC(=O)-SCoA$ — $(CH_3)_3N^+-CH_2CH(OH)CH_2C(=O)-O^-$ — $RC(=O)-SCoA$

Transferase; CoA

Transferase II; CoA

$(CH_3)_3N^+-CH_2CH(O-C(=O)-R)CH_2-C(=O)-O^-$

Extramitochondrial | Membrane | Intramitochondrial

Figure 12-5
Passage of fatty acids through various membranes.

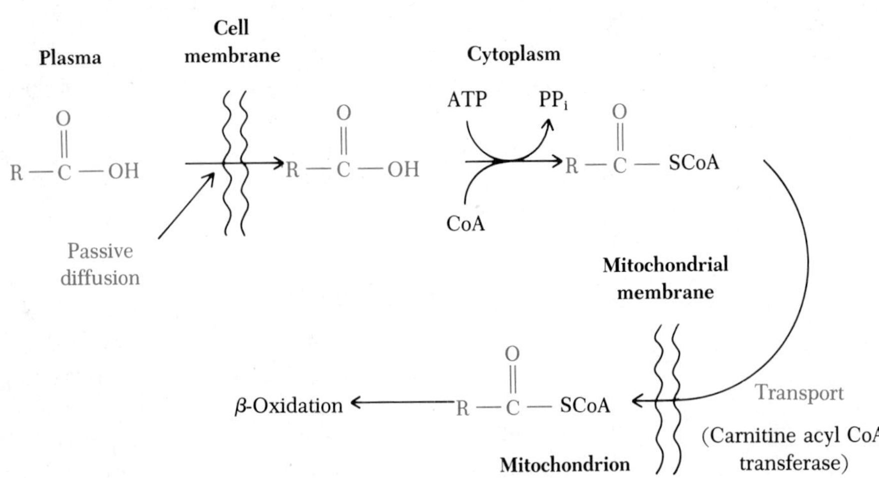

serve to dehydrogenate smaller fatty acids. As in the comparable oxidation of succinate to fumarate, the fatty acyl CoA yields two hydrogens, one each from the α- and β-carbons, to FAD. The unsaturated product also has a *trans* configuration.

$$R-\overset{\beta}{C}H_2-\overset{\alpha}{C}H_2-\overset{O}{\overset{\|}{C}}-SCoA \xrightarrow[\text{Dehydrogenase}]{FAD \;\to\; FADH_2} R-CH=CH-\overset{O}{\overset{\|}{C}}-SCoA$$

Acyl CoA **α,β-Unsaturated acyl CoA**

The enzyme-bound $FADH_2$ is then oxidized by means of another flavoprotein called the *electron-transfer flavoprotein*. This $FADH_2$ yields its two hydrogens to CoQ (Chapter 7) and therefore feeds two electrons into the electron-transport system.

$$FADH_2 \to FAD;\quad FAD \to FADH_2 \quad | \quad FADH_2 \to FAD;\quad CoQ \to CoQ \cdot H_2$$

Electron-transfer flavoprotein

As with succinate dehydrogenase, oxidation of acyl CoA flavoprotein by the electron-transport chain yields two ATPs. The unsaturated acyl CoA then accepts a molecule of water, a reaction catalyzed by *enoyl CoA hydrase,* or *crotonase*. This is

$$R-CH=CH-\overset{O}{\overset{\|}{C}}-SCoA \xrightarrow[\text{Enoyl CoA hydrase}]{H_2O} R-CH(OH)-CH_2-\overset{O}{\overset{\|}{C}}-SCoA$$

α,β-Unsaturated acyl CoA **L-β-Hydroxyacyl CoA**

reminiscent of the hydration of fumarate to L-malate. The product L-β-hydroxyacyl CoA is oxidized by *β-hydroxyacyl CoA dehydrogenase,* a reaction requiring NAD^+. As before (Chapter 7) oxidation of NADH by the electron-transport system leads to the formation of three ATPs.

$$R-CH(OH)-CH_2-\overset{O}{\overset{\|}{C}}-SCoA \xrightarrow[NAD^+ \;\to\; NADH(H^+)]{\beta\text{-Hydroxyacyl CoA dehydrogenase}} R-\overset{O}{\overset{\|}{C}}-CH_2-\overset{O}{\overset{\|}{C}}-SCoA$$

L-β-Hydroxyacyl CoA **β-Ketoacyl CoA**

The final step in β-oxidation is the cleavage of the β-ketoacyl CoA by *β-ketothiolase,* which results in the insertion of CoA and cleavage at the β-carbon. The products are acetyl CoA and a saturated acyl CoA with two fewer carbons than the original substrate.

$$R-\overset{O}{\overset{\|}{C}}-CH_2-\overset{O}{\overset{\|}{C}}-SCoA \xrightarrow[\beta\text{-Ketothiolase}]{CoA} R-\overset{O}{\overset{\|}{C}}-SCoA + CH_3\overset{O}{\overset{\|}{C}}-CoA$$

β-Ketoacyl CoA **Acyl CoA** **Acetyl CoA**

The entire sequence of reactions is repeated, with two carbons being cleaved each time. Finally a four-carbon butyryl CoA is the intermediate; β-oxidation yields acetoacetyl CoA and then two molecules of acetyl CoA.

The cyclical β-oxidation of fatty acids has the following distinguishing features: (1) the fatty acid needs to be activated only once, which costs the cell two high-energy bonds represented by the elimination and hydrolysis of pyrophosphate; (2) all of the enzymes, except thiokinase, are mitochondrial; and (3) the process involves repetitive steps of oxidation, hydration, and oxidation.

Energy considerations

Since the activation of fatty acids actually requires two high-energy bonds, there is a deficit of two ATPs per mole of fatty acid oxidized. Each cleavage of a carbon-carbon bond yields five ATPs. The acetyl CoAs produced are metabolized by the citric acid cycle of CO_2 and water. For each acetyl CoA oxidized, 12 ATPs are realized (Chapter 11).

For the metabolism of a mole of palmitic acid (16 carbons) the yield of ATP would be as follows:

1. 7 cleavages occur	7 × 5 =	35
2. 8 acetyl CoA oxidized by citric acid cycle	8 × 12 =	96
Total ATP produced		131
3. 2 ATPs for activation		−2
Net ATP produced		129

This yield in ATP would represent approximately 980 kcal per mole of palmitic acid oxidized. The complete oxidation yields theoretically 2340 kcal per mole.

Fatty acids having odd numbers of carbon atoms

Although most fatty acids metabolized by the animal organism are composed of an even number of carbon atoms, small amounts of fatty acids containing odd numbers of carbon atoms are also used. β-Oxidation of these chains takes place by the preceding sequence of steps to form acetyl CoA until only a residual propionyl CoA is left. The metabolism of propionyl CoA then occurs by the following scheme:

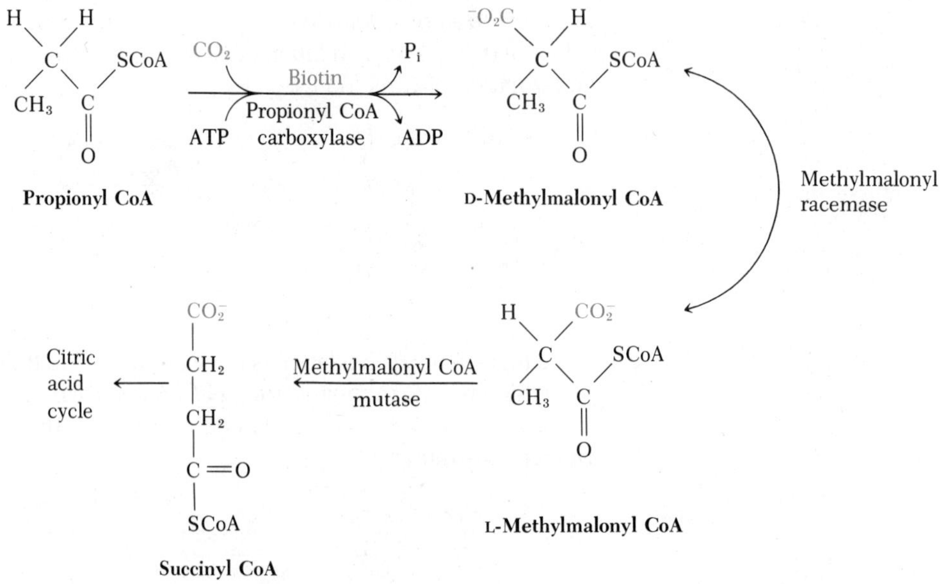

$$CH_3CH_2C(=O)SCoA \xrightarrow[\text{ATP} \to \text{ADP}]{CO_2 \to P_i;\ \text{Biotin; Propionyl CoA carboxylase}} {}^{-}O_2C\text{–}CH(CH_3)\text{–}C(=O)SCoA$$

Propionyl CoA → D-Methylmalonyl CoA

Methylmalonyl racemase

D-Methylmalonyl CoA → L-Methylmalonyl CoA

$$CO_2^-\text{–}CH(CH_3)\text{–}C(=O)SCoA \xrightarrow{\text{Methylmalonyl CoA mutase}} CO_2^-\text{–}CH_2\text{–}CH_2\text{–}C(=O)\text{–}SCoA \longrightarrow \text{Citric acid cycle}$$

L-Methylmalonyl CoA → Succinyl CoA

The succinyl CoA formed is metabolized in the citric acid cycle. *Propionyl CoA carboxylase* contains 4 moles of biotin per mole of enzyme. Thus carboxylation involves the intermediary bindings of CO_2 by biotin (Chapter 6). Propionic acid is also produced during protein metabolism by way of the oxidation of certain branched-chain amino acids (Chapter 13). Of interest also, is the fact that methylmalonyl CoA mutase is a vitamin B_{12}–dependent enzyme (Chapters 6 and 21). In vitamin B_{12} deficiency this reaction is blocked and relatively large amounts of methylmalonyl CoA are excreted in the urine, as would be expected.

Unsaturated fatty acids

Since nearly one half of dietary and body fatty acids are unsaturated, they must provide some share of the energy derived from lipids. Oleic acid is a *cis*-Δ^9 fatty acid (Chapter 23). Oxidation proceeds by the process of β-oxidation; α, β-dehydrogenation, hydration, and oxidation with a subsequent cleavage to acetyl CoA. The oxidation of oleyl CoA results in the formation of the products:

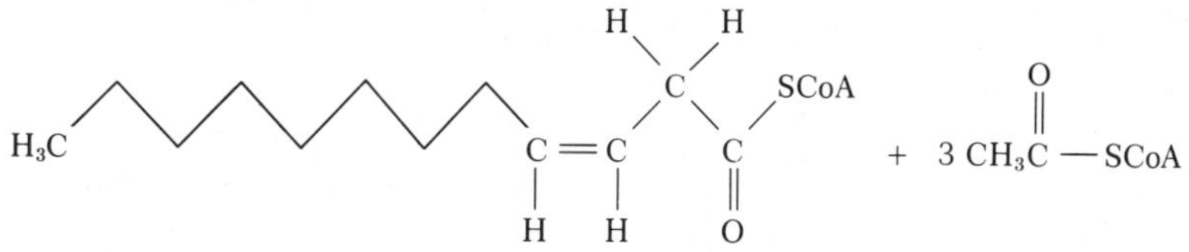

$$H_3C\text{—(zigzag chain)—}CH{=}CH{-}CH_2{-}C({=}O){-}SCoA \quad + \quad 3\ CH_3C({=}O){-}SCoA$$

This compound cannot serve as a substrate either for acyl CoA dehydrogenase, which produces a Δ^2 unsaturated fatty acid, or for enoyl CoA hydratase, which hydrates only the Δ^2 unsaturated fatty acid. The presence of a Δ^3-*cis*-Δ^2-*trans*-*enoyl CoA isomerase* reversibly shifts the double bond from carbons 3 and 4 to 2 and 3, yielding a *trans*-enoyl CoA, and allows the hydratase (crotonase) to catalyze the insertion of a molecule of water.

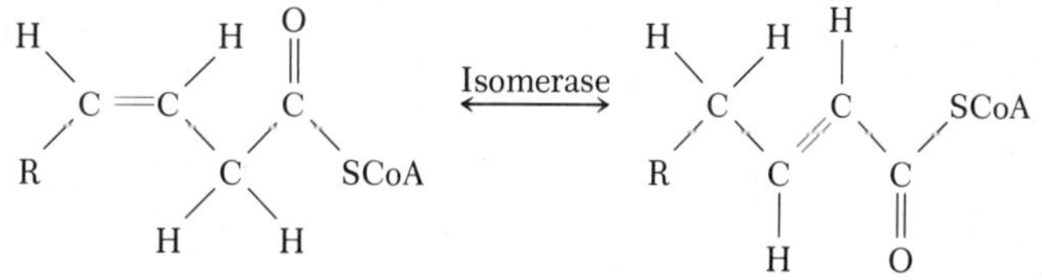

$$R{-}CH{=}CH{-}CH_2{-}C({=}O){-}SCoA \xrightleftharpoons{\text{Isomerase}} R{-}CH_2{-}CH{=}CH{-}C({=}O){-}SCoA$$

cis-Δ^3-Fatty acyl CoA | **trans-Δ^2-Fatty acyl CoA**

Likewise in the metabolism of linolenic acid the *cis*-$\Delta^{9,12,15}$-octadecatrienoic acid is oxidized successively by β-oxidation (see top of p. 294).

The compound that is the product of the last reaction has two double bonds, both in the *cis* configuration. This unsaturated acyl CoA is a substrate for enoyl hydratase; the product, however, is the D-hydroxyl acid and therefore is not a substrate for the β-hydroxyacyl dehydrogenase. Another special enzyme is required at this point, namely, an *epimerase*, which changes the configuration to the L-isomer. In this way the unsaturated fatty acids can be completely metabolized by the β-oxidative process.

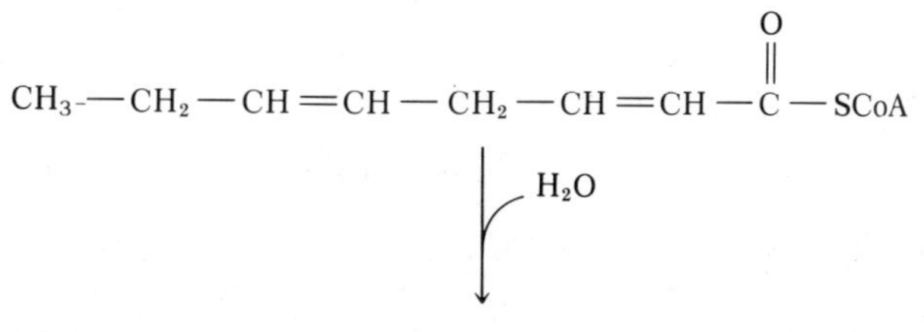

$$CH_3{-}CH_2{-}CH{=}CH{-}CH_2{-}CH{=}CH{-}C({=}O){-}SCoA \qquad \textit{cis-}\Delta^{2,5}$$

$$\Big\downarrow\ H_2O$$

The number of ATP molecules produced by the complete oxidation of unsaturated fatty acids is somewhat less (the amount depending number of double bonds)

$$CH_3-CH_2-CH=CH-CH_2-CH=CH-CH_2-CH=CH-CH_2-\overset{O}{\overset{\|}{C}}-SCoA$$

$$+ \quad 3 \text{ Acetyl CoA}$$

$$\downarrow \text{Enoyl CoA isomerase}$$

$$CH_3-CH_2-CH=CH-CH_2-CH=CH-CH_2-CH_2-CH=CH-\overset{O}{\overset{\|}{C}}-SCoA \qquad \Delta^2 = \textit{trans}$$

$$\downarrow H_2O$$

$$CH_3-CH_2-CH=CH-CH_2-CH=CH-CH_2-CH_2-\overset{OH}{\overset{|}{C}H}-CH-\overset{O}{\overset{\|}{C}}-SCoA \qquad \text{L-}\beta\text{-Hydroxy}$$

$$\downarrow NAD^+ \rightarrow NADH(H^+)$$

$$CH_3-CH_2-CH=CH-CH_2-CH=CH-CH_2-CH_2-\overset{O}{\overset{\|}{C}}-CH_2-\overset{O}{\overset{\|}{C}}-SCoA$$

$$\downarrow CoA$$

$$CH_3-CH_2-CH=CH-CH_2-CH=CH-CH_2-CH_2-\overset{O}{\overset{\|}{C}}-SCoA \quad + \quad CH_3-\overset{O}{\overset{\|}{C}}-SCoA$$

$$\downarrow \beta\text{-Oxidation}$$

$$CH_3-CH_2-CH=CH-CH_2-CH=CH-\overset{O}{\overset{\|}{C}}-SCoA \quad + \quad CH_3-\overset{O}{\overset{\|}{C}}-SCoA$$

$$CH_3-CH_2-CH=CH-CH_2-\underset{OH}{\underset{|}{C}H}-CH_2-\overset{O}{\overset{\|}{C}}-SCoA \qquad \text{D-Isomer}$$

$$\downarrow \text{Epimerase}$$

$$CH_3-CH_2-CH=CH-CH_2-\overset{OH}{\overset{|}{C}H}-CH_2-\overset{O}{\overset{\|}{C}}-SCoA \qquad \text{L-Isomer}$$

$$\downarrow NAD^+ \rightarrow NADH(H^+)$$

$$CH_3-CH_2-CH=CH-CH_2-\overset{O}{\overset{\|}{C}}-CH_2-\overset{O}{\overset{\|}{C}}-SCoA$$

$$\downarrow CoA$$

$$CH_3-CH_2-CH=CH-CH_2-\overset{O}{\overset{\|}{C}}-SCoA \quad + \quad CH_3-\overset{O}{\overset{\|}{C}}-SCoA$$

than that of saturated fatty acids since the latter have more hydrogen atoms and hence more electrons to be transferred by way of the respiratory chain to oxygen (Chapter 7). (See figure at bottom of p. 294.)

Alternate pathways for fatty acid oxidation

Although β-oxidation represents the major pathway for the metabolism of fatty acids, at least two others exist. These are α-oxidation, which functions to introduce an α-hydroxyl group followed by oxidative decarboxylation, and ω-oxidation, which yields dicarboxylic acids.

α-Oxidation A direct hydroxylation of long-chain fatty acids yields α-hydroxy fatty acids, which are constituents of complex lipids of the brain. Oxidative decarboxylation eliminates one carbon and converts the even-numbered fatty acids to odd-numbered chains, which are also constituents of complex lipids. The microsomal fraction of mammalian brain and also liver oxidize the free fatty acid directly; the CoA ester is not involved.

$$RCH_2COOH \xrightarrow[Fe^{++},\ \text{ascorbic acid}]{O_2\ \text{Monoxygenase}} \underset{\text{L-}\alpha\text{-Hydroxy fatty acid}}{R{-}\underset{\substack{|\\ OH}}{CH}{-}COOH} \xrightarrow[NAD^+ \rightarrow NADH]{\text{Dehydrogenase}} R{-}\underset{\substack{\|\\ O}}{C}{-}COOH \xrightarrow[\text{ATP, } NAD^+\text{, ascorbate}]{\text{Decarboxylase}} RCOOH + CO_2$$

A fatty acid that can only be metabolized by α-oxidation is *phytanic acid,* or 3,7,11,15-tetramethylhexanoic acid. This compound, derived from chlorophyll, accumulates in the blood and tissues of individuals who have a rare neurologic disorder called *Refsum's disease*. These patients have a genetically defective α-oxidation system and therefore are unable to metabolize phytanic acid (Chapter 22).

ω-Oxidation Liver microsomal fraction is capable of metabolizing fatty acids by an oxidative attack of the alkyl chain on the carbon furthest (ω) from the carboxyl end. The reaction is as follows:

$$CH_3(CH_2)_nCOOH \xrightarrow[O_2\ \text{Monoxygenase}]{NADPH \rightarrow NADP^+} HOCH_2(CH_2)_n{-}COOH \xrightarrow{\text{Enzyme II}} HOOC{-}(CH_2)_n{-}COOH$$

In this mechanism, as in α-oxidation, the substrate is the free fatty acid, not the CoA ester. The reaction requires cytochrome P_{450} (Chapter 17), molecular oxygen, and reduced $NADP^+$. The enzyme system has been resolved into (1) cytochrome P_{450}, (2) a NADPH-cytochrome P_{450} reductase, and (3) a heat-stable lipid factor. The α-ω-dicarboxylic acid thus formed may be shortened from either end by β-oxidation, as described above. Short-chain dicarboxylic acids, such as pimelic acid (a precursor of biotin; see Chapter 21), are ultimately formed.

Biosynthesis of fatty acids

As with the opposing processes of glycogenesis and glycogenolysis, the biosynthesis and degradation of fatty acids also occur by separate pathways. Furthermore, they occur in different compartments; degradation of β-oxidation is mitochondrial whereas biosynthesis is cytoplasmic. This separation allows them to occur simultaneously and provides for careful control of both. As with the metabolism of glycogen, which begins and ends with glucose 1-phosphate, the biosynthesis and degradation of fatty acids also begins and ends with the same compound, i.e., acetyl CoA.

Acetyl CoA, whether from pyruvate or fatty acids, is produced in the mitochondria. Mitochondrial membranes are impermeable to acetyl CoA. It is possible that the acetyl group may leave the mitochondrion by transport via carnitine; this, however, is not a major mechanism for the passage of acetyl groups into the cytoplasm.

Mitochondrial membranes are permeable, however, to citrate produced in the citric acid cycle from acetyl CoA by condensation with oxaloacetate. Once in the cytoplasm the process is reversed; citrate is cleaved into acetyl CoA and oxaloacetate by the *citrate cleavage enzyme,* or citrate lyase.

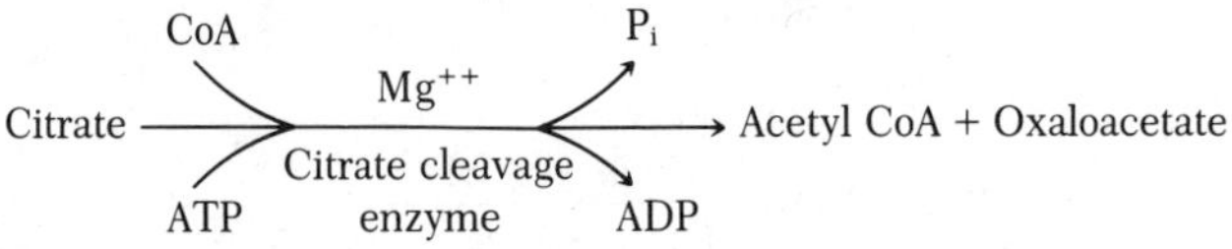

$$\text{Citrate} \xrightarrow[\text{ATP} \rightarrow \text{ADP};\ \text{Citrate cleavage enzyme}]{\text{CoA},\ Mg^{++} \rightarrow P_i} \text{Acetyl CoA} + \text{Oxaloacetate}$$

Acetyl CoA is the substrate for *acetyl CoA carboxylase,* a biotin-containing enzyme. This enzyme first forms the *N*-carboxybiotin complex, a reaction requiring ATP (Chapter 6). It is considered to be the committed step in the synthesis of fatty acids. Then the carboxyl group is transferred to acetyl CoA to form a malonyl CoA.

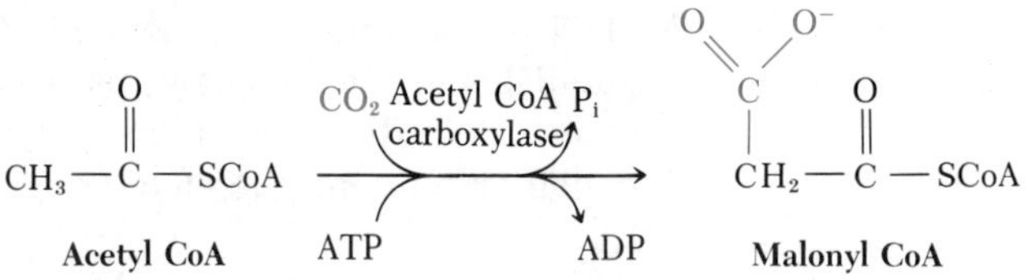

$$\underset{\text{Acetyl CoA}}{CH_3-\overset{O}{\overset{\|}{C}}-SCoA} \xrightarrow[\text{ATP} \rightarrow \text{ADP}]{CO_2,\ \text{Acetyl CoA carboxylase},\ P_i} \underset{\text{Malonyl CoA}}{\overset{^{-}O_2C}{\overset{|}{C}H_2}-\overset{O}{\overset{\|}{C}}-SCoA}$$

Malonyl CoA is the substrate for the multienzyme complex called *fatty acid synthase.* This complex was crystallized from yeast by Lynen and co-workers. It has a molecular weight of 2.3 million and appears to consist of an many as 21 subunits. The enzymes required for fatty acid synthesis are part of this molecular aggregate. Included is a molecule called the *acyl carrier protein* (ACP). ACP is a heat-stable protein consisting of 77 amino acid residues and having a molecular weight of 9500 *(E. coli).* ACP from *E. coli* has been completely characterized as to primary structure; it has been synthesized by the solid-phase technique. As shown in the accompanying diagram, it has a prosthetic group, 4′-phosphopantetheine, attached at residue 36 (serine). As with CoA, ACP also has a mercaptoethylamine group. It is believed that the growing fatty acyl group is transported from enzyme to enzyme in the manner of an assembly line while being attached to ACP as the thioester.

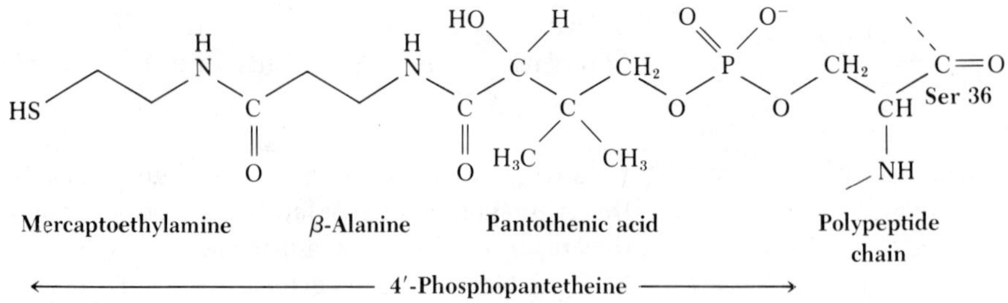

$$HS-CH_2-CH_2-NH-\overset{O}{\overset{\|}{C}}-CH_2-CH_2-NH-\overset{O}{\overset{\|}{C}}-CH(OH)-C(CH_3)_2-CH_2-O-\overset{O}{\overset{\|}{P}}(O^-)-O-CH_2-\underset{\text{Ser 36}}{CH}(NH-)-C{=}O$$

Mercaptoethylamine | β-Alanine | Pantothenic acid | Polypeptide chain

←— 4′-Phosphopantetheine —→

The stepwise biosynthesis of fatty acids by fatty acid synthase begins with acetyl CoA, followed by the repeated addition of malonyl CoA and the simultaneous loss of CO_2. Stoichiometrically the biosynthesis of palmitate would be the following:

$$CH_3CO\text{-}SCoA + 7\ HOOCCH_2CO\text{-}SCoA + 14\ NADPH + 14\ H^+ \rightarrow CH_3(CH_2)_{14}CO_2H + 7\ CO_2 + 14\ NADP^+ + 8\ CoA + 6\ H_2O$$

Palmitate is the major product of this system. All other long-chain fatty acids, whether saturated or unsaturated, can be derived from palmitate; the essential fatty acids are exceptions.

The initial step in the biosynthesis of fatty acids is the transfer of an acetyl group from acetyl CoA to the —SH of ACP, forming acetyl ACP, catalyzed by *acetyl CoA–ACP transacylase*. Fig. 12-6 shows that this step occurs at a loading site, B_1. The acetyl group is believed to react at the following intermediate stage:

$$CH_3CO\text{-}SCoA + HS\text{-}ACP \rightarrow CH_3CO\text{-}S\text{-}ACP + CoA$$

with a seryl or threonyl hydroxyl at B_1 before transfer to the ACP-SH. Then the acetyl is transferred from the ACP-SH to a cysteinyl-SH residue (site B_2) of the condensing enzyme (E_{cond}) β-*ketoacyl-ACP synthase* (Fig. 12-6).

$$CH_3CO\text{-}S\text{-}ACP + E_{cond}\text{-}SH \rightarrow CH_3CO\text{-}S\text{-}E_{cond} + ACP\text{-}SH$$

Transfer of the acetyl to site B_2 frees the ACP-SH to accept the first "incoming" malonyl group from the loading site B_1.

$$HO_2CCH_2CO\text{-}SCoA + ACP\text{-}SH \rightarrow HO_2CCH_2CO\text{-}S\text{-}ACP + CoA$$

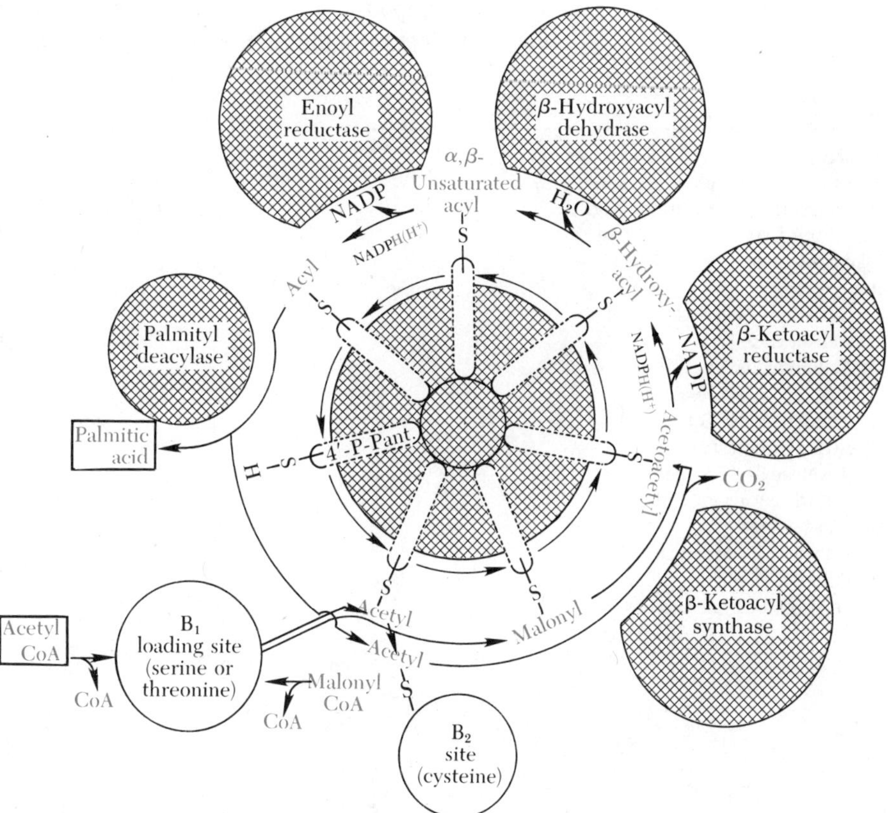

Figure 12-6
Proposed mechanism of fatty acid synthesis. This assembly-line arrangement of enzymes is served by a moving belt (ACP).

From Porter, J.W.: Arch. Biochem. Biophys. **138**:380, 1970.

The enzyme catalyzing this transfer is *malonyl CoA–ACP transacylase.* Next the acetyl group (site B_2) condenses with malonyl ACP; the enzyme is β-*ketoacyl-ACP synthase.*

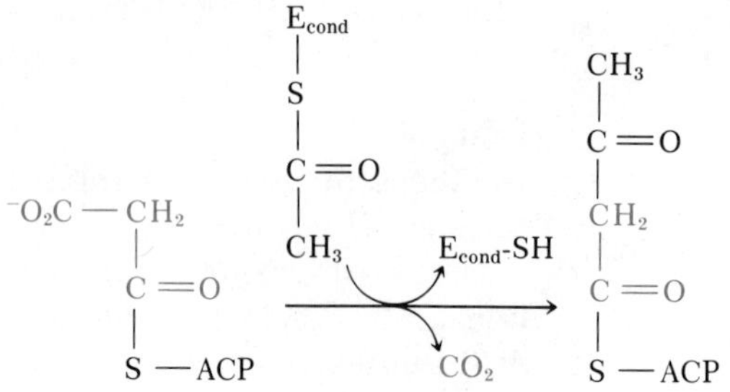

$$^{-}O_2C{-}CH_2{-}C(=O){-}S{-}ACP + E_{cond}{-}S{-}C(=O){-}CH_3 \longrightarrow CH_3{-}C(=O){-}CH_2{-}C(=O){-}S{-}ACP + E_{cond}\text{-}SH + CO_2$$

β-Ketoacyl ACP

The β-ketoacyl ACP is then reduced by β-*ketoacyl-ACP reductase* and NADPH.

$$CH_3COCH_2CO\text{-}S\text{-}ACP + NADPH(H^+) \rightarrow CH_3CHOHCH_2CO\text{-}S\text{-}ACP + NADP^+$$

Dehydration by β-*hydroxyacyl dehydratase* yields an α,β-unsaturated acyl ACP.

$$CH_3CHOHCH_2CO\text{-}S\text{-}ACP \rightarrow H_2O + CH_3CH{=}CHCO\text{-}S\text{-}ACP$$

Reduction with NADPH and *enoyl-ACP reductase* yields the saturated acyl ACP thioester.

$$CH_3CH{=}CHCO\text{-}S\text{-}ACP + NADPH(H^+) \rightarrow CH_3CH_2CH_2CO\text{-}S\text{-}ACP + NADP^+$$

The acyl compound now reacts with the condensing enzyme (site B_2) to form the thioester, liberating the ACP to react with another incoming malonyl group. After

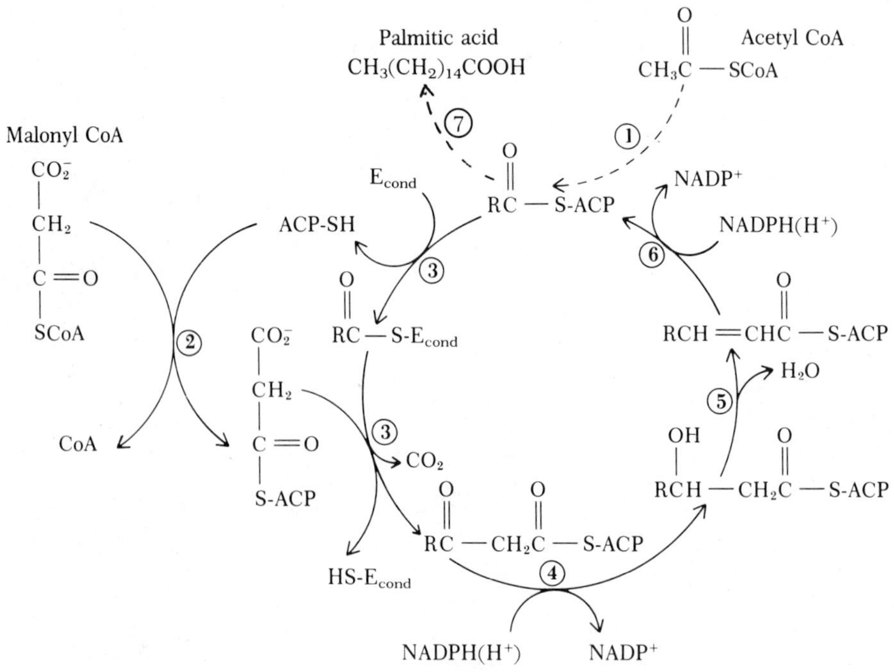

Figure 12-7
Summary of individual steps in the biosynthesis of palmitic acid called the Lynen cycle. The enzymes involved are the following:
1, Initiation: Acetyl CoA–ACP transacylase.
2, Malonyl CoA–ACP transacylase.
3, Condensation: β-ketoacyl–ACP synthetase.
4, β-Ketoacyl–ACP reductase.
5, β-Hydroxyacl–ACP dehydrase.
6, Enoyl-ACP reductase.
7, Palmityl deacylase (after seven turns of cycle).

Palmitic acid $CH_3(CH_2)_{14}COOH$ · Acetyl CoA $CH_3C(=O){-}SCoA$ · Malonyl CoA $CO_2^-{-}CH_2{-}C(=O){-}SCoA$ · CoA · $CO_2^-{-}CH_2{-}C(=O){-}S\text{-}ACP$ · ACP-SH · E_{cond} · $RC(=O){-}S\text{-}ACP$ · $RC(=O){-}S\text{-}E_{cond}$ · $HS\text{-}E_{cond}$ · CO_2 · $RC(=O){-}CH_2C(=O){-}S\text{-}ACP$ · NADPH(H^+) · $NADP^+$ · $RCH(OH){-}CH_2C(=O){-}S\text{-}ACP$ · H_2O · $RCH{=}CHC(=O){-}S\text{-}ACP$ · NADPH(H^+) · $NADP^+$

condensation, the sequence of reduction, dehydration, and reduction is repeated so that each "turn" of the cycle increases the acyl chain by two carbons. After seven turns of the cycle, the final product is palmitic acid, released by the *palmityl deacylase*. It is possible, however, that the palmityl group can be directly incorporated into glycerides without the prior intervention of deacylase. The major events of the fatty acids biosynthesis are summarized in Fig. 12-7.

Sources of reduced NADP$^+$

It is particularly significant that all reduction reactions in the Lynen cycle require reduced NADP$^+$. Since the biosynthesis of fatty acids is cytoplasmic, the supply of NADPH must be likewise. As mentioned previously (Chapter 9), 97% of NADP$^+$ is in the reduced form; most is derived from the pentose phosphate pathway. Glucose metabolism is required to yield the necessary reduced NADP$^+$ for fatty acid synthesis. Another source of NADPH is the reaction involving the enzyme *NADP$^+$–malate dehydrogenase*. This enzyme must not be confused with the malate dehydrogenase of the citric acid cycle. It is often called the "malic enzyme" to avoid possible confusion. Malic enzyme catalyzes the decarboxylation of malate to pyruvate, a reaction requiring NADP$^+$.

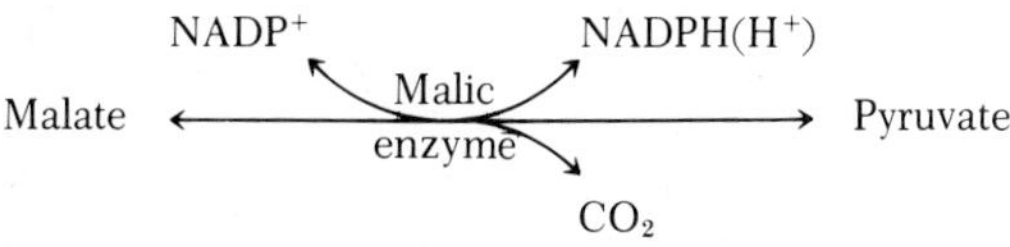

The isocitrate dehydrogenase reaction in the citric acid cycle (Chapter 11) is also involved as a source of NADH in the cytosol.

Elongation of fatty acid chains

As noted above, the Lynen cycle is limited to the formation of palmitic acid. Conversion of palmitate to stearate or other longer chain fatty acids occurs by chain elongation either in microsomal or mitochondrial systems. The microsomal enzyme system is similar to the Lynen cycle except that malonyl CoA is added to an acceptor molecule of acyl CoA without the intervention of the ACP molecule or the other enzymes of the Lynen system. Reduction also requires NADPH as a cofactor. The overall reaction is as follows:

$$\text{R-CO-S-CoA} + \text{Malonyl CoA} + 2\ \text{NADPH} \rightarrow \text{R-(CH}_2)_2\text{CO-S-CoA} + 2\ \text{NADP}^+ + \text{CO}_2 + \text{CoA}$$

Alternatively, palmitic acid may enter the mitochondria by the carnitine-transport system and then participate in an elongation process that is for all intents and purposes a reversal of β-oxidation. Acetyl CoA is added, and reduction occurs by NADH and dehydration, followed again by reduction with NADH.

Unsaturated fatty acids

In man and other mammals the desaturation of fatty acids is limited primarily to the formation of fatty acids having only one double bond, i.e., palmitoleic acid and oleic acid. Linoleic acid (C-18,$\Delta^{9,12}$), when supplied by the diet, can be converted by chain elongation and desaturation to arachidonic acid (C-20,$\Delta^{5,8,11,14}$). Although linoleic acid is the only unsaturated fatty acid that must be supplied by the diet, the three polyenoic acids (linoleic, linolenic, and arachidonic acids) are sometimes considered to be essential fatty acids in man (Chapter 21). The unsaturated fatty acid, arachidonic acid, and several related C_{20} fatty acids are important also because they serve as precursors of a group of biochemically active compounds, the *prostaglandins*. The structures and functions of this group of substances ar described in Chapter 18.

The biosynthesis of oleic acid requires the desaturation of stearoyl CoA. This is a mixed oxygenase system using molecular oxygen and NADPH or NADH.

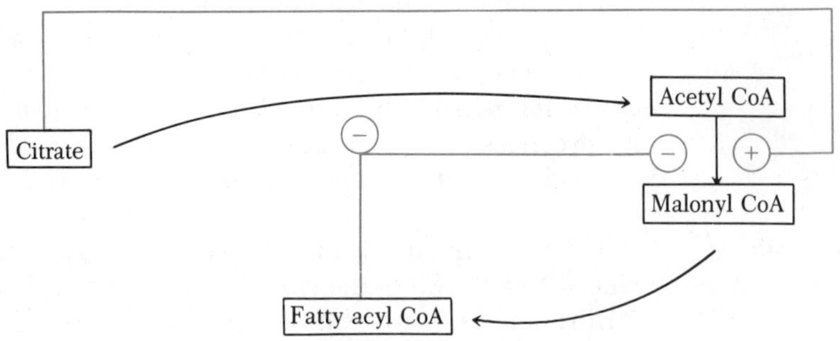

Figure 12-8
Role of citrate and of free fatty acids in the regulation of fatty acid biosynthesis.

$$\text{Stearoyl CoA} \xrightarrow[\text{NADPH(H}^+) \;\rightarrow\; \text{NADP}^+]{O_2 \;\rightarrow\; 2\,H_2O} \text{Oleoyl CoA}$$

The enzyme called *stearoyl CoA oxidase,* or "desaturase," is part of the endoplasmic reticulum of liver (microsomal fraction) cells.

Regulation of fatty acid biosynthesis

A rate-limiting reaction in the biosynthesis of fatty acids is the one catalyzed by *acetyl CoA carboxylase.* This enzyme is stimulated by citrate and several related citric acid cycle components, particularly α-ketoglutarate and isocitrate. Citrate appears to act by favoring the aggregation of an inactive monomer into an active linear polymer of some 10 or more subunits. The production of large amounts of acetyl CoA increases the citrate pool. The latter then leaves the mitochondrion and acts both to stimulate (activate) acetyl CoA carboxylase and to be the substrate for *citrate lyase (cleavage enzyme).* It is clear that a high-carbohydrate diet can lead to an increased synthesis of fatty acids by increasing the available pool of acetyl CoA. Likewise, a high-fat diet, which results in the presence of increased free fatty acids, will inhibit fatty acid biosynthesis by inhibiting the acetyl CoA carboxylase and the citrate lyase (Fig. 12-8).

Control of fatty acid biosynthesis also occurs by way of the fatty acid synthase multienzyme complex. Feeding a fat-free diet to an animal increases the synthase activity; starvation or a high-fat diet decreases the activity. Changes in activity after a fat-free diet are probably related to an increased synthesis of the enzyme system. Rats on a fat-free diet synthesize the hepatic fatty acid synthase at a rate six times greater than do the starved animals.

Several hormones also are involved in the regulation of fatty acid biosynthesis (Chapter 18). Insulin favors lipogenesis, apparently by inhibiting the activation of hormone-sensitive lipase probably, in turn, by reducing the level of cyclic AMP, as is described later in this chapter. Glucagon opposes lipogenesis by inhibiting the hepatic synthesis of fatty acids and by activating hepatic lipase and releasing fatty acids from adipose tissue. Glucagon also increases cyclic AMP levels.

Mobilization of fat (triacylglycerol)

As stressed previously, fatty acids are an important source of energy, stored in the adipose tissue. The key enzyme for the mobilization of fat is *hormone-sensitive triglyceride lipase,* found in adipose tissue cells. Lipolysis, or the hydrolysis of triglycerides, is controlled by hormones such as the catecholamines (epinephrine and

Figure 12-9
Hormonal control of lipolysis mediated by cAMP.

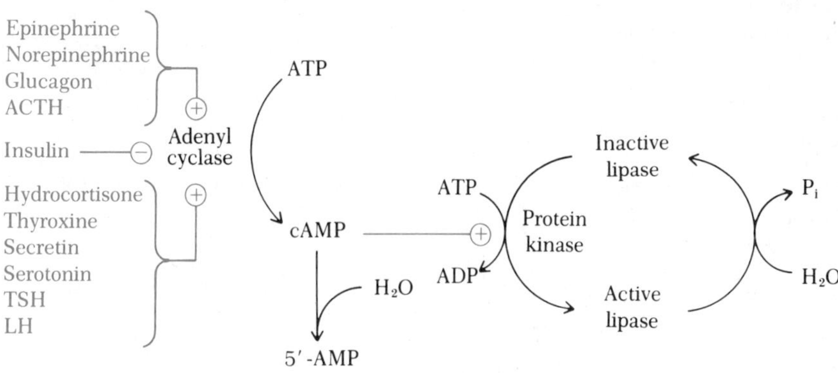

norepinephrine), ACTH, TSH, glucagon, and thyroxine (Fig. 12-9). The action of these hormones is mediated by the biosynthesis of cAMP. The lipase exists in active and inactive forms; activation involves phosphorylation by a *protein kinase*. Cyclic AMP stimulates the action of the protein kinase and therefore the activation of the lipase. Dephosphorylation catalyzed by a phosphatase may inactivate the lipase in the manner of glycogen phosphorylase a.

Hypoglycemia by fasting, for example, stimulates a release of glucagon to restore the blood glucose level. In the absence of insulin, however, glucose transport is reduced and the intracellular supply of carbohydrate in adipose tissue cells is lowered. Because of the low glucose and therefore an inadequate supply of L-glycerol 3-phosphate, lipogenesis is suppressed. On the other hand, lipolysis is increased because of the elevated hormone-sensitive lipase activity stimulated by glucagon (Fig. 12-9). In diabetes the absence of insulin results in an inadequate transport of glucose, and therefore lipogenesis is suppressed. Insulin exerts an antilipolytic effect (Fig. 12-9); its absence favors lipolysis and a flow of free fatty acids into the bloodstream. Consequently the plasma free fatty acids may increase to nearly twice the normal level. Fatty acids mobilized from adipose tissue are carried by the bloodstream (complexed with albumin) to the liver and to the musculature (Fig. 12-10).

Actually free fatty acids can be reesterified to triglycerides in adipose tissue. Whether they are converted to fat or enter the circulation is contingent on the supply of glucose, its metabolism to DHAP, and therefore the availability of L-glycerol 3-phosphate. Glycerol is also freed; it, however, is not reusable in adipose tissues because of the absence of glycerokinase. Rather, the glycerol leaves these cells and enters the bloodstream to be used by the liver for lipogenesis, gluconeogenesis, or both.

When an animal is in a state of fat economy as in hypoglycemia, the fatty acids are the primary source of energy for all tissues except the brain, erythrocytes, and adrenal medulla. Although fatty acids as such cannot supply glucose (Chapter 11) for the brain, the acetyl CoA does stimulate pyruvate carboxylase, thereby promoting gluconeogenesis in the liver. Except for the tissues noted, all others become almost totally incapable of catabolizing glucose.

Hyperglycemia stimulates an increased secretion of insulin, which in turn increases glucose transport into the adipose tissue. Glucose metabolism yields acetyl CoA and DHAP, both of which promote lipogenesis. Insulin, as also prostaglandin E_1 and nicotinic acid, lowers the cAMP, thereby reducing lipolysis. The mechanism of the antilipolytic effect of insulin is presently unknown. It might exert its

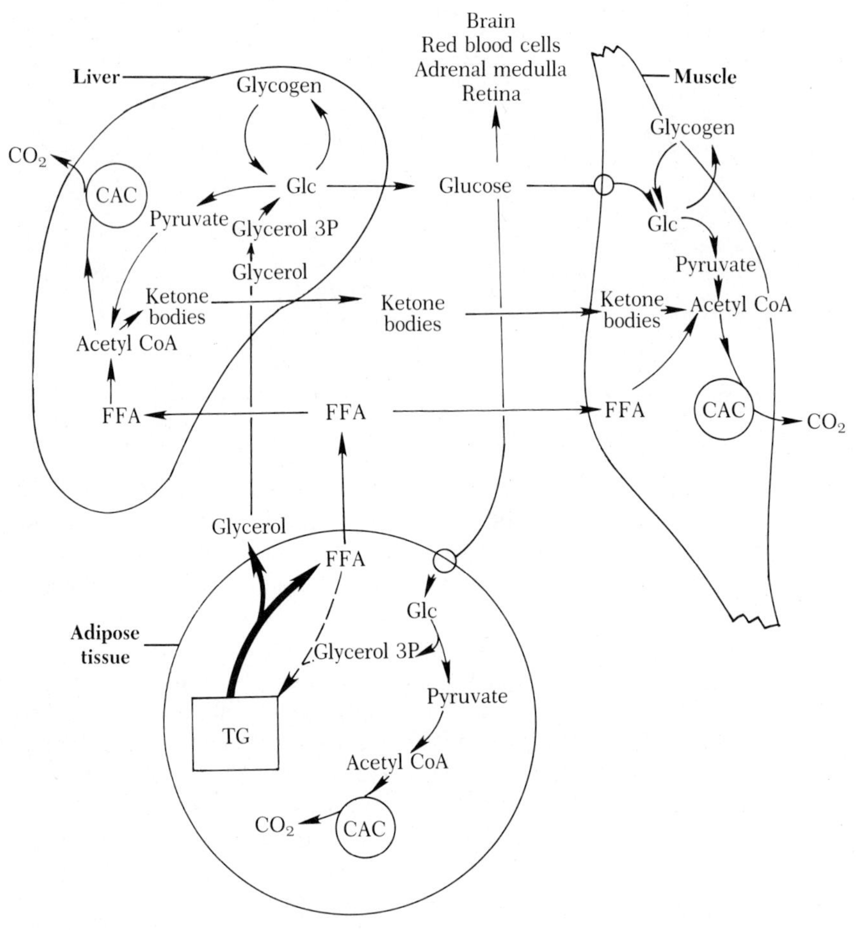

Figure 12-10
Mobilization of fatty acids from the adipose tissue. Also shown is ketogenesis in the liver. Both fatty acids and ketone bodies are used by the musculature as fuel. *CAC*, Citric acid cycle; *Glc*, glucose; *FFA*, free fatty acids; *TG*, triacylglycerol.

action directly by inhibiting adenyl cyclase or it could increase the hydrolysis of cAMP to 5′-AMP by activating a cyclic 3′,5′-nucleotide phosphodiesterase. In summary, during hyperglycemia insulin stimulates lipogenesis by increasing the intracellular pool of glucose in adipose tissues and at the same time inhibits lipolysis by decreasing the cAMP.

Mobilization of fatty acids may lead to a situation in which the liver cannot cope with the acetyl CoA produced. In the subsequent section note that acetyl CoA can be converted to the so-called *ketone bodies*. These enter the bloodstream for distribution and oxidation in the kidneys, muscle, heart, brain, and testes.

Ketogenesis

The term *ketogenesis* means the formation of ketone bodies. These include acetoacetic acid, β-hydroxybutyric acid, and acetone. Acetone, however, is merely a breakdown product of either of the other two, which are the really important substances concerned. *Ketosis* is the production of ketone bodies in excess of the ability of the body to use them. Hence the excess above the renal threshold is excreted in the urine (Chapter 17). Ketosis occurs in severe diabetes, in starvation, in the "acidosis of childhood," and during anesthesia, and it can be precipitated by feeding an unbalanced diet, i.e., one that is high fat and low carbohydrate. The

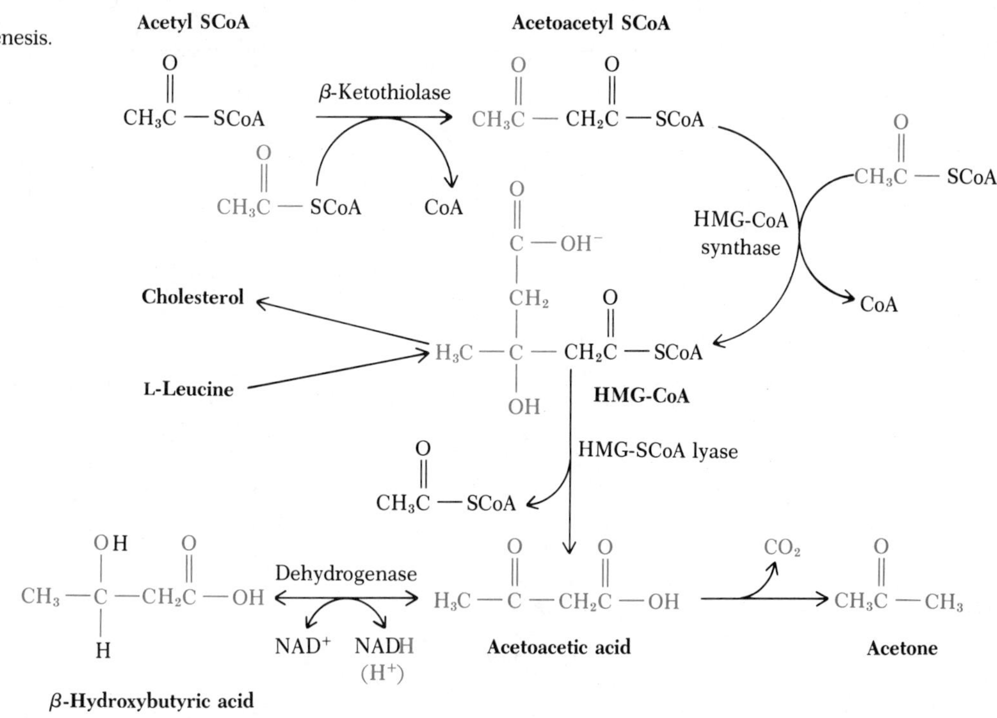

Figure 12-11
Pathway showing ketogenesis.

appearance of these compounds in the urine is a danger sign, indicating an acidosis and warning the clinician of impending coma. As has been seen, they are normal degradation products of the fatty acids. Acetoacetic acid is usually formed from two molecules of acetyl CoA if the latter is produced more rapidly than it can be used.

Ketone body formation in the liver from acetyl CoA and fatty acid oxidation (acetoacetyl CoA) involves β-hydroxy-β-methylglutaryl CoA (HMG-CoA) as an obligatory intermediate. The steps involved are shown in Fig. 12-11. In this process the methyl of acetyl CoA is added to the carbonyl group of acetoacetyl CoA, catalyzed by *HMG-CoA synthase*. The product β-hydroxymethyglutaryl CoA (HMG-CoA) is also an intermediate in the catabolism of leucine (Chapter 13) and an obligatory step in the biosynthesis of cholesterol, which is discussed in a later section of this chapter. In the liver, acetoacetic acid is formed from HMG-CoA by cleavage involving the removal of acetyl CoA. The enzyme catalyzing cleavage (rupture of C—C bond) is called *HMG-CoA lyase*. The acetoacetate can be reduced reversibly in the presence of β-*hydroxybutyrate dehydrogenase* and NADH. However, acetoacetate cannot be reconverted to acetoacetyl CoA because of the absence of the appropriate hepatic thiokinase.

Although other tissues are capable of producing small amounts of these ketones, the liver is by far the chief ketogenic organ. It can, in fact, be regarded as practically the only site of ketone production. The formation of ketone bodies is regulated by one of the anterior pituitary hormones and also, possibly, by glucagon, a hormone secreted by the α-cells of the pancreatic islets of Langerhans (Chapter 18).

As seen in Fig. 12-12, the ketone bodies (including acetone) enter the bloodstream and are transported to extrahepatic tissues. In these, the β-hydroxybutyrate is reconverted to acetoacetate and the latter is then reactivated to acetoacetyl CoA.

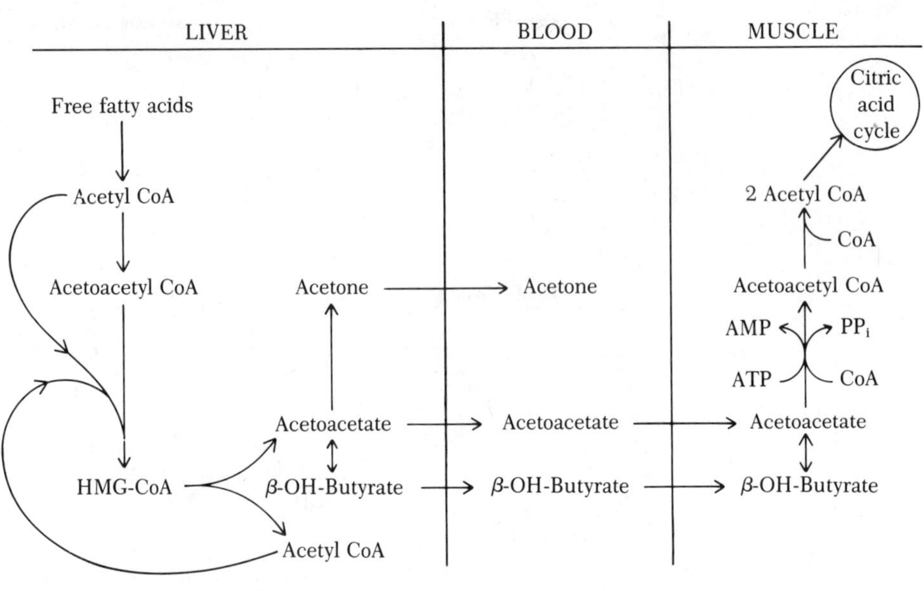

Figure 12-12
Pathway illustrating production and fate of ketone bodies.

Reactivation may occur by two separate paths; one catalyzed by a thiokinase in the presence of CoA and ATP, the other by *CoA transferase*, in the presence of succinyl CoA. In muscle the latter is the predominant process. Acetoacetyl CoA is then cleaved to two acetyl CoAs by β-ketothiolase and CoA. The acetyl CoAs are subsequently oxidized to CO_2 and H_2O by the citric acid cycle and oxidative phosphorylation.

$$CH_3\overset{O}{\overset{\|}{C}}-CH_2-\overset{O}{\overset{\|}{C}}-OH$$

Succinyl CoA → Succinate (CoA transferase); CoA + ATP → AMP + PP_i (Thiokinase)

$$CH_3\overset{O}{\overset{\|}{C}}-CH_2-\overset{O}{\overset{\|}{C}}-SCoA$$

Knowing that the ketone bodies are produced by the liver and used by the tissues, we should see how a ketosis might be caused by (1) increased hepatic production of ketone bodies or (2) decreased use by extrahepatic tissues.

Metabolism of phospholipids

Phospholipids are a class of compound lipids containing phosphate and in most cases a nitrogenous base in addition to the esters of fatty acids with an alcohol (glycerol) (Chapter 23). They are constituents of all cells and play vital roles as constituents of cell membranes and factors regulating cellular permeability. Myelin sheaths of nerve cells and electron-transport particles contain phospholipids. They serve as biologic detergents, aiding the solubilization of the less polar lipids, e.g., triglycerides and cholesterol, in the aqueous fluids of the body. Blood clotting (Chapter 16) is triggered by phospholipid-containing factors. Among the important phospholipids are the derivatives of phosphatidic acid (phosphatidyl choline or lecithin, and phosphatidyl ethanolamine or cephalin) and the sphingomyelins.

Some phospholipids are extremely active metabolically, as evidenced by their rapid turnover rate in tissues, measured by means of ^{32}P labeling. The half-life of hepatic lecithin is less than 24 hours. Phospholipids that serve structural purposes, e.g., brain cephalin, have a much longer half-life, up to several months.

During lipid absorption, phospholipids may enter directly into the systemic circulation; the blood level increases as the blood lipid increases. As a rule, however, high phosphatidyl choline (lecithin) levels persist longer than high total lipid. When radioactive phosphate is administered to animals, it is found in the tissue phospholipid fraction, particularly as phosphatidyl choline and phosphatidyl ethanolamine, with very little as sphingomyelin. The synthesis ratios of phospholipid under these conditions were found to be liver > intestine > kidney > muscle > brain. The slow production (and disappearance) in brain is interesting in view of the high content of phospholipid in this tissue. When radioactive phosphate was injected into hepatectomized dogs, almost no radioactive phospholipid was found in the blood plasma. Therefore most of the plasma phospholipids are hepatic in origin. Normal amounts, however, appeared in the kidneys and small intestine, indicating that these organs can synthesize phospholipids for their own use.

Biosynthesis

Phosphatidic acid serves as a common precursor for the biosynthesis of triglycerides and phospholipids (Fig 12-3). In mammals, phosphatidic acid is the starting point for two separate pathways. In one, phosphatidic acid phosphatase catalyzes the hydrolysis of the phosphate ester to yield the corresponding 1,2-diglyceride. This is the same step as in the biosynthesis of triglycerides (Fig. 12-3). *Choline phosphotransferase* converts the diglyceride to phosphatidyl choline, using the activated form of choline, CDP-choline.

$$(CH_3)_3N^+CH_2CH_2OH \xrightarrow[\text{Choline kinase}]{\text{ATP} \to \text{ADP}} (CH_3)_3N^+CH_2CH_2O-P(=O)(O^-)-O^-$$

Choline → **Phosphorylcholine**

Phosphorylcholine → (CTP → PP_i; Choline phosphate cytidyl transferase) → $(CH_3)_3N^+CH_2CH_2O-P(=O)(O^-)-O-P(=O)(O^-)-CH_2$–ribose–cytosine ($NH_2$, OH, OH)

Cytidine diphosphate choline

CDP-choline then reacts with the diglyceride in the presence of *choline phosphotransferase.*

$$\text{1,2-Diglyceride} \xrightarrow[\text{CDP-choline} \to \text{CMP}]{\text{Choline phosphotransferase}} \text{Phosphatidyl choline}$$

Phosphatidyl ethanolamines are formed from 1,2-diglyceride plus CDP-ethanolamine (Fig. 12-3). Phosphatidyl choline may be derived from phosphatidyl ethanolamine by using S-adenosyl methionine (Chapter 13) and *phosphatidyl ethanolamine methyltransferase*.

A second pathway stemming from L-phosphatidic acid involves a reaction with CTP, catalyzed by *phosphatidic acid transferase*. The product is CDP-diglyceride:

CDP-diglyceride

CDP-diglyceride, in turn, may react with myoinositol (Chapter 15) to yield phosphatidyl inositol and, in bacteria, with serine to form phosphatidyl serine. In mam-

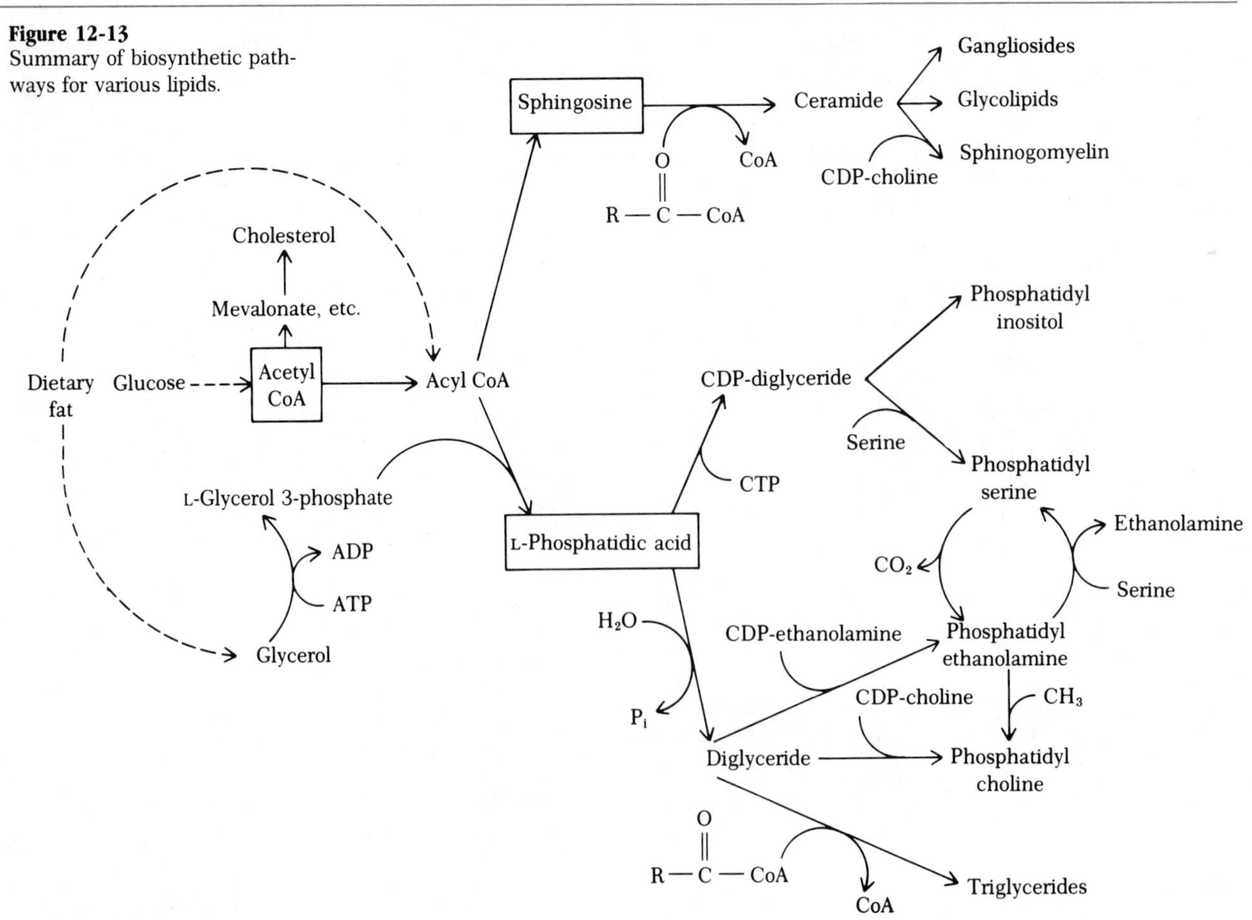

Figure 12-13
Summary of biosynthetic pathways for various lipids.

mals, however, phosphatidyl serine is formed by an exchange reaction of phosphatidyl ethanolamine with serine (Fig. 12-13).

Obviously, many phospholipids are possible, depending on the chain length and unsaturation of the fatty acyl moieties (R_1, R_2, R_3) and their various possible combinations. The distribution of fatty acids in phospholipids is asymmetric; saturated fatty acids are exclusively situated at carbon-1, whereas unsaturated ones are at carbon-2. Linoleic acid is the predominant fatty acid of diphosphatidyl glycerol (*cardiolipin*). This phospholipid is found almost exclusively in mitochondria.

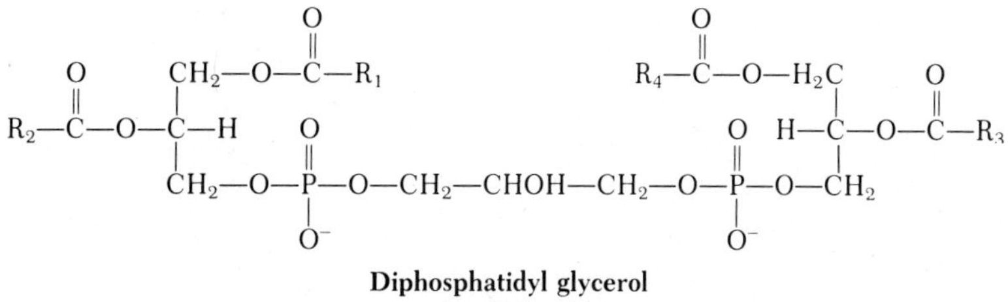

```
                           O                                  O
                           ‖                                  ‖
       O      CH2—O—C—R1                         R4—C—O—H2C        O
       ‖       |                                          |        ‖
R2—C—O—C—H      O                              O   H—C—O—C—R3
               |       ‖                              ‖      |
              CH2—O—P—O—CH2—CHOH—CH2—O—P—O—CH2
                      |                               |
                      O⁻                              O⁻
```

Diphosphatidyl glycerol (cardiolipin)

In another type of phospholipid, the *plasmalogens*, an α,β-unsaturated fatty aldehyde replaces the fatty acid, usually in the α-position (Chapter 23). Stearic and palmitic aldehydes are present in the plasmalogens found in cardiac and skeletal muscle. A large portion of the phospholipids occurring in mitochondria are plasmalogens. Their functions are unclear.

The biosynthetic pathway of the plasmalogens entails the formation of monoacyl dihydroxyacetone phosphate from DHAP and acyl CoA (Fig. 12-13). The acyl group is then exchanged, by an unknown replacement reaction, for a fatty alcohol to form 1-alkyl-DHAP. The latter is now reduced to 1-alkyl glyceryl phosphate with NADPH (or NADH), followed by acylation to form 1-alkyl-2-acyl glyceryl phosphate. This intermediate is dephosphorylated and converted to an alkyl analog of the phosphoglycerides containing choline, ethanolamine, or serine by the reactions described above (see also Fig. 12-13). The alkyl phosphoglyceride is then converted to the plasmalogen by a specific desaturase in the presence of NADPH (or NADH) and O_2.

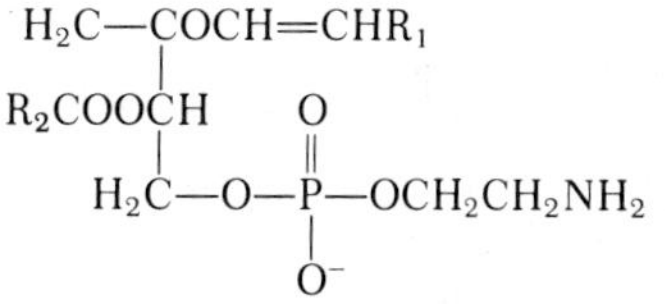

```
H2C—COCH=CHR1
     |
R2COOCH     O
     |      ‖
    H2C—O—P—OCH2CH2NH2
            |
            O⁻
```

A plasmalogen: 1-Alkenyl, 2-acyl glycerol, 3-phosphoethanolamine

Another class of phospholipids of considerable importance that, however, do not have glycerol as the alcohol portion are the *sphingomyelins*. Sphingosine is the alcohol portion of these phospholipids. However, they are not acyl esters but rather acyl amides (ceramides).

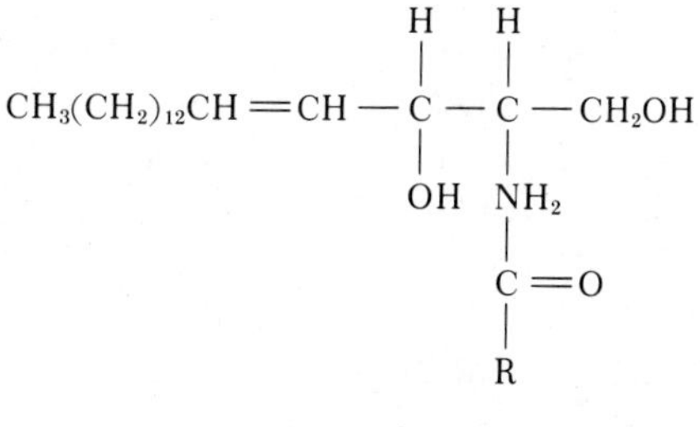

```
                          H    H
                          |    |
CH3(CH2)12CH=CH—C—C—CH2OH
                          |    |
                         OH   NH2
                               |
                               C=O
                               |
                               R
```

Ceramide

The biosynthesis of sphingomyelin is shown below.

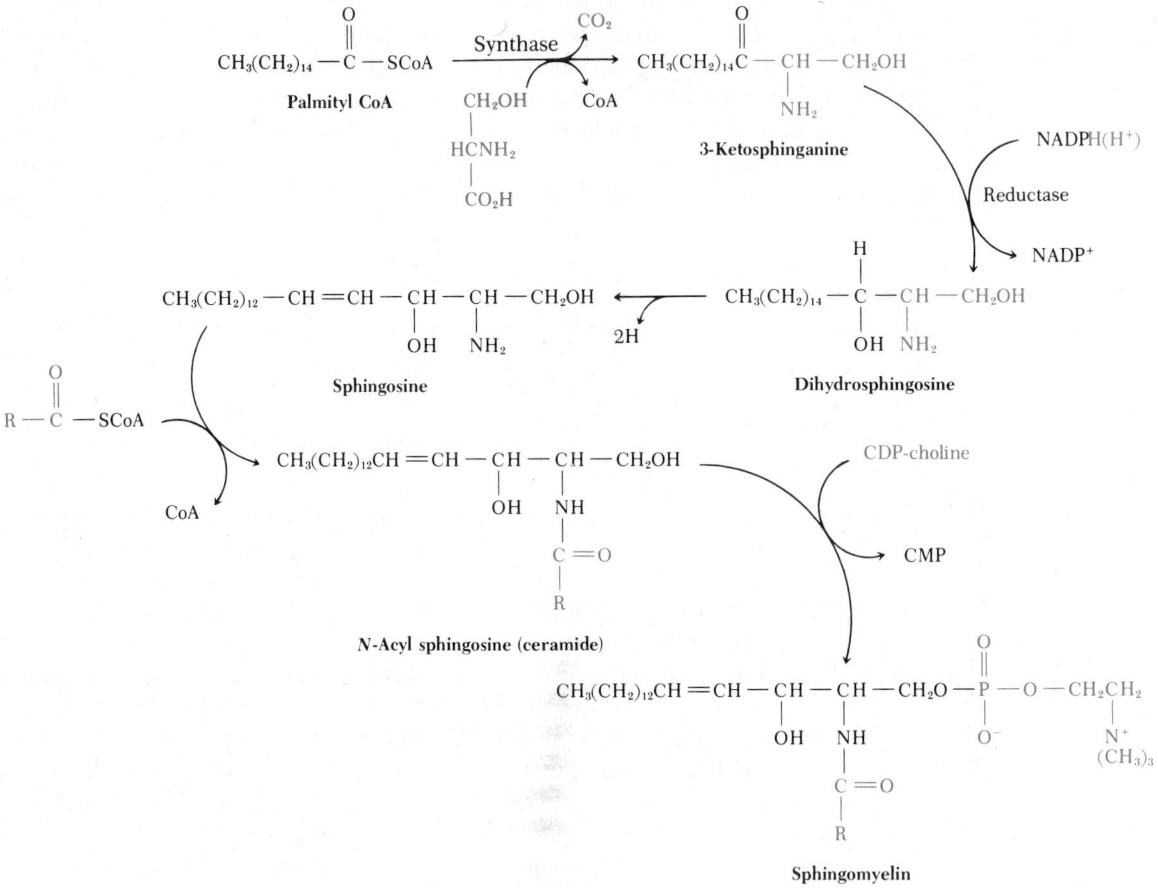

$$CH_3(CH_2)_{14}-\overset{O}{\overset{\|}{C}}-SCoA \text{ (Palmityl CoA)} + CH_2OH-CH(NH_2)-CO_2H \xrightarrow[\text{CoA},\ CO_2]{\text{Synthase}} CH_3(CH_2)_{14}\overset{O}{\overset{\|}{C}}-CH(NH_2)-CH_2OH \text{ (3-Ketosphinganine)}$$

$$\text{3-Ketosphinganine} \xrightarrow[\text{NADPH}(H^+) \to \text{NADP}^+]{\text{Reductase}} CH_3(CH_2)_{14}-CH(OH)-CH(NH_2)-CH_2OH \text{ (Dihydrosphingosine)}$$

$$\text{Dihydrosphingosine} \xrightarrow{-2H} CH_3(CH_2)_{12}-CH=CH-CH(OH)-CH(NH_2)-CH_2OH \text{ (Sphingosine)}$$

$$\text{Sphingosine} + R-\overset{O}{\overset{\|}{C}}-SCoA \xrightarrow{-\text{CoA}} CH_3(CH_2)_{12}CH=CH-CH(OH)-CH(NH-C(=O)-R)-CH_2OH \text{ (N-Acyl sphingosine (ceramide))}$$

$$\text{Ceramide} + \text{CDP-choline} \xrightarrow{-\text{CMP}} CH_3(CH_2)_{12}CH=CH-CH(OH)-CH(NH-C(=O)-R)-CH_2O-\overset{O}{\overset{\|}{P}}(O^-)-O-CH_2CH_2-N^+(CH_3)_3 \text{ (Sphingomyelin)}$$

Palmityl CoA

3-Ketosphinganine

Sphingosine

Dihydrosphingosine

N-Acyl sphingosine (ceramide)

Sphingomyelin

Sphingosine is also an important part of the glycolipids, which, like sphingomyelins, are derivates of ceramide. However, glycolipids are glycosides and not phospholipids; they do not contain phosphate.

Degradation

Phosphatidyl choline can be degraded by four different enzymes as follows:

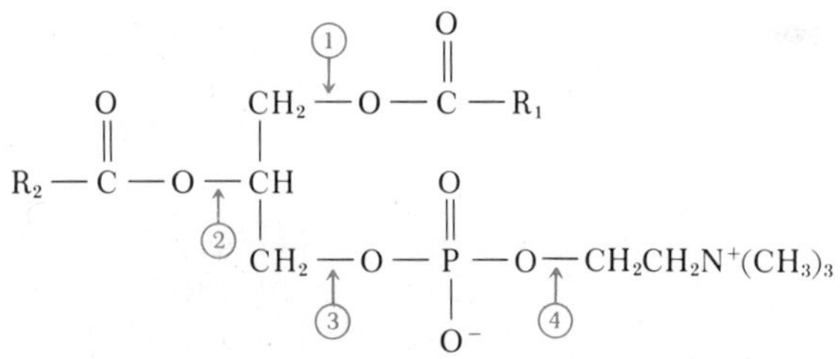

$$\begin{array}{l} CH_2 \overset{(1)}{\downarrow} O-\overset{O}{\overset{\|}{C}}-R_1 \\ R_2-\overset{O}{\overset{\|}{C}}-O\underset{(2)}{\uparrow}CH \\ CH_2 \underset{(3)}{\uparrow} O-\overset{O}{\overset{\|}{P}}(O^-)-O\underset{(4)}{\uparrow}CH_2CH_2N^+(CH_3)_3 \end{array}$$

Phospholipase A attacks at point 2, splitting off one fatty acid and forming lysolecithin, a hemolytic agent. High levels are found in pancreas and small intestine; moderate levels are in testis, spleen, lung, and liver, as well as in some snake venoms. *Phospholipase B* (or lysophospholipase) removes the fatty acid molecules at point 1. It is found generally in animal tissues. *Phospholipase C* acts on the ester linkage between glycerol and the phosphoryl choline (point 3), whereas glycerylphosphoryl choline diesterase hydrolyzes the ester linkge between phosphate and choline at point 4. Such phosphatases are also found in many tissues.

Presumably the other kinds of phospholipids are catabolized similarly by means of appropriate enzymes.

Metabolism of glycolipids

Glycolipids, like the sphingomyelins, are derivatives of acyl sphingosine. Unlike the phospholipids, they do not contain phosphate; rather they are glycosides. These compounds, also called cerebrosides, are present in relatively large amounts in the myelin sheath of brain and nerve cells. The glycolipids are important not only as constituents of cell membranes (Chapter 2) but also as determinants in a number of immunological reactions, including those of the blood group substances (Chapter 16). Smaller amounts are found in liver and other tissues.

Four types of cerebrosides are known (kerasin, phrenasin, nervon, and oxynervon), differing only with respect to the type of fatty acid present.

Glycolipid	Fatty acid	
Kerasin	Lignoceric	24:0
Phrenasin	Cerebronic	α-OH24:0
Nervon	Nervonic	24:1
Oxynervon	Oxynervonic	α-OH24:1

The biosynthetic pathway of the cerebrosides may be shown schematiclly as follows:

$$\text{Sphingosine} \xrightarrow{\text{UDP-galactose} \curvearrowright \text{UDP}} \text{Galactopsychosine}$$

$$\downarrow \; R-\overset{O}{\overset{\|}{C}}-CoA \rightarrow CoA$$

$$CH_3(CH_2)_{12}-CH=CH-\underset{OH}{\overset{H}{C}}-\underset{NH-C(=O)-R}{\overset{H}{C}}-CH_2-O-\text{(galactose)}$$

Galactocerebroside

Gangliosides are like the glycolipids but contain two hexoses, galactosamine and a sialic acid, in addition to sphingosine and a fatty acid in the molecule. The pathway for their biosynthesis may be similar to that of the glycolipids (Fig. 12-14).

The gangliosides are involved in special roles in cell plasma membrane receptors for certain polypeptide hormones (Chapter 18), for viruses, and for a number of drugs. Gangliosides accumulate in certain tissues in several types of lipid storage diseases (Chapter 22).

Another group of glycolipids, the *sulfolipids* or *sulfatides*, are fairly well characterized. They are formed from galactoceramides (cerebrosides) by reaction with "active sulfate," 3′-phosphoadenosine-5′-phosphosulfate (PAPS) (Chapter 6). The

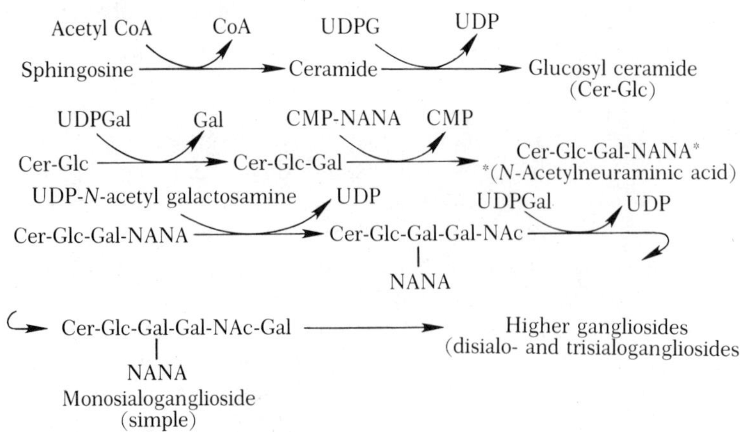

Figure 12-14
Biosynthetic pathway of the gangliosides.

3′-hydroxyl group of the galactosylceramide molecule is sulfated. The pathway is shown schematically below.

$$\text{Cerebroside} \xrightarrow[\text{Sulfokinase}]{\text{PAPS}} \text{Sulfatide}$$

The sulfolipids are found in relatively small amounts in brain and other tissues. Their functions are unclear.

A diagram summarizing schematically the major steps in the biosynthetic pathways of various lipids and their interrelationship is given in Fig. 12-13.

The various glycolipids described in the preceding sections undergo continuous degradation by hydrolytic enzymes generally present in the lysosomes of all cells (Chaper 2). Consequently under normal conditions their biosynthesis is balanced by their degradation in various tissues, resulting in a "steady-state" situation. However, if there is an absence or partial deficiency of any one of the specific hydrolases, accumulation of the lipid may occur in the tissues involved. Nervous tissue is particularly susceptible because of its relatively high glycolipid content. Neurological deterioration and disease result. A number of such diseases, the *lipidoses*, usually hereditary in nature, are discussed in Chapter 22.

Metabolism of cholesterol

Cholesterol is the most important compound of those classified as sterols—those that have cyclopentanoperhydrophenanthrene as a parent structure (Chapter 23). It is found in practically all animal cells, serving as an essential component of the plasma membrane and other membrane structures. It is estimated that a 70-kg adult male has some 75 to 150 gm of cholesterol in his body, about 20% of which is present in a rapidly exchanging cholesterol pool. The remainder is present in various tissues and has a low turnover rate. Cholesterol is a precursor of bile acids (Fig. 12-18), the steroid hormones, and also vitamin D. Cholesterol is not a dietary essential, since it is readily synthesized in the body, as is described later. Dietary sources of cholesterol, amounting to 500 to 750 mg daily in the adult, are primarily animal products, meats, liver, eggs, and milk lipids (Chapter 21). Plant foods contain no or only traces of cholesterol.

Food cholesterol is a variable mixture of free and esterified (with fatty acids) cholesterol. The latter is rapidly hydrolyzed in the intestine by a pancreatic esterase. Free cholesterol is incorporated into lipid micelles, as described earlier, together with conjugated bile acids (Chapter 17) and hydrolytic products of food fats. Absorption occurs mainly in the upper small intestine. On the average, only about 50% of the cholesterol ingested in food and excreted in bile is absorbed. However, the amount absorbed varies directly with the amount of cholesterol ingested.

After absorption, about 80% to 90% of the cholesterol is esterified by acylation with long-chain fatty acids in the presence of an esterase. The free and esterified cholesterol combine with specific apoproteins in the plasma to form chylomicrons and VLDL, described previously. These are then distributed to various tissue cells for metabolic disposal. The serum cholesterol level of a normal adult male ranges from about 150 to 250 mg/100 ml. This exogenously derived cholesterol directly affects by "feedback inhibition" (Chapter 3) the amount synthesized by tissue cells.

In normal human adults cholesterol is found in its largest amounts in the liver (about 0.3%), skin (0.3%), brain and nervous tissue (2.0%), intestine (0.2%), and certain endocrine glands, with the adrenal gland containing some 10%. As much as 50% of the myelin sheath that surrounds nerves is cholesterol. This apparently is related to proper nerve conduction and normal brain function (Chapter 15). The relatively high content of cholesterol in skin may be related to vitamin D formation (Chapter 21), and that in the adrenals and certain other endocrine glands to steroid hormone biosynthesis (Chapter 18).

The various factors that influence the turnover of body cholesterol can be illustrated as follows:

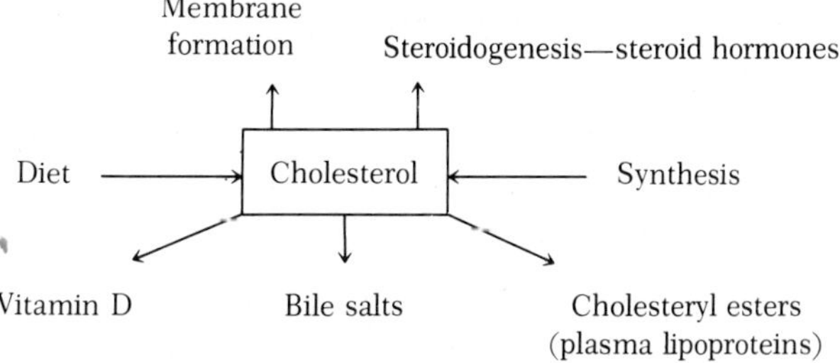

It is estimated that nearly half of the daily cholesterol production is converted to bile acids (such as cholic acid; Chapter 17) and secreted as bile salts in bile. Most of this is reabsorbed and reused by way of "enterohepatic circulation" (Chapter 17). A part of the remainder of an estimated 0.5 to 1.0 gm daily need for cholesterol is used for the formation of steroid hormones (about 40 mg daily), cell membranes, and vitamin D in the skin, as shown in the diagram above.

Any excess cholesterol, amounting to about 750 to 1250 mg daily in the normal adult, is excreted primarily in bile and the intestinal tract. Part of this is reduced by intestinal microorganisms to *coprostanol* and *cholestanol.* These three lipids thus make up the major part of the fecal steroids.

A small amount of cholesterol, approximately 50 mg daily in the adult, is excreted also from the surface of the skin as desquamated cells, sweat, and sebaceous secretions.

Biosynthesis

As indicated above, the de novo biosynthesis of cholesterol provides the major portion of the animal's need for this steroid. Bloch and co-workers showed that the ring structure is completely derived from two-carbon atoms of acetate; both carbon atoms are incorporated into the nucleus and the octyl side chain. We have already seen that acetyl CoAs yield acetoacetyl CoA and then HMG-CoA in the liver, a pathway catalyzed first by β-ketothiolase and then by HMG-CoA synthase. HMG-CoA is then reduced (NADPH) by *HMG-CoA reductase*; CoA is lost in the process

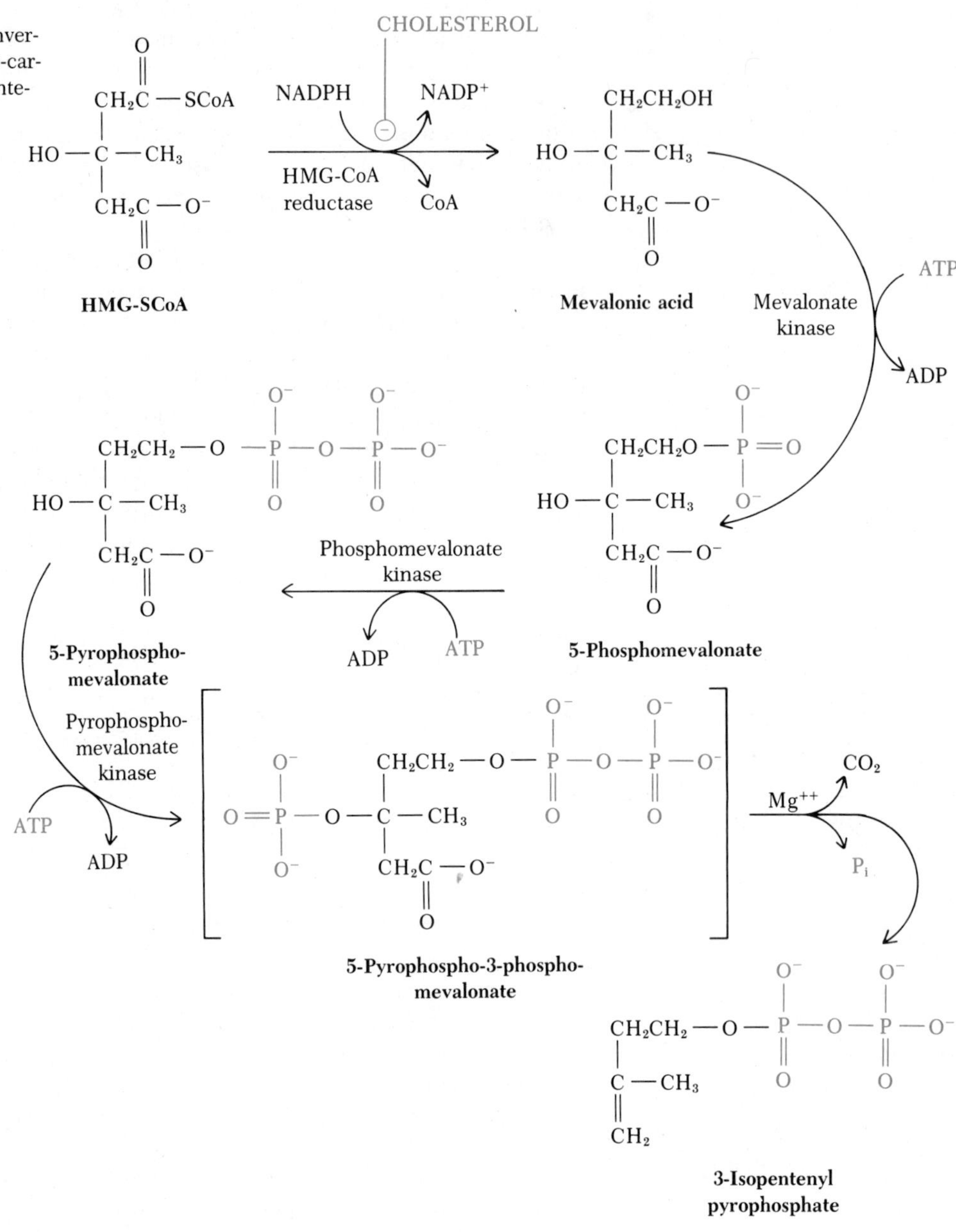

Figure 12-15
Biosynthetic pathway for conversion of HMG-CoA to the five-carbon isoprenoid unit (3-isopentenyl pyrophosphate).

(Fig. 12-15). The product formed is *mevalonic acid.* Mevalonate is phosphorylated consecutively to 5-phosphomevalonate (mevalonate kinase) to produce 5-pyrophosphomevalonate *(phosphomevalonate kinase)* and finally to 5-pyrophospho-3-phosphomevalonate *(pyrophosphorylmevalonate kinase)* (Fig. 12-15). This compound is unstable, loses phosphate and carbon dioxide *(phosphomevalonate decarboxylase)*, and yields 3-isopentenyl pyrophosphate. *Isopentenyl pyrophosphate isomerase* causes an isomerization to 3,3-dimethylallyl pyrophosphate.

It is helpful to consider the steps described in Fig. 12-15 to be the first phase in cholesterol biosynthesis, one that results in the formation of a 5-carbon isoprenoid unit.

Figure 12-16
Summary of pathway from isopentenyl pyrophosphate to lanosterol.

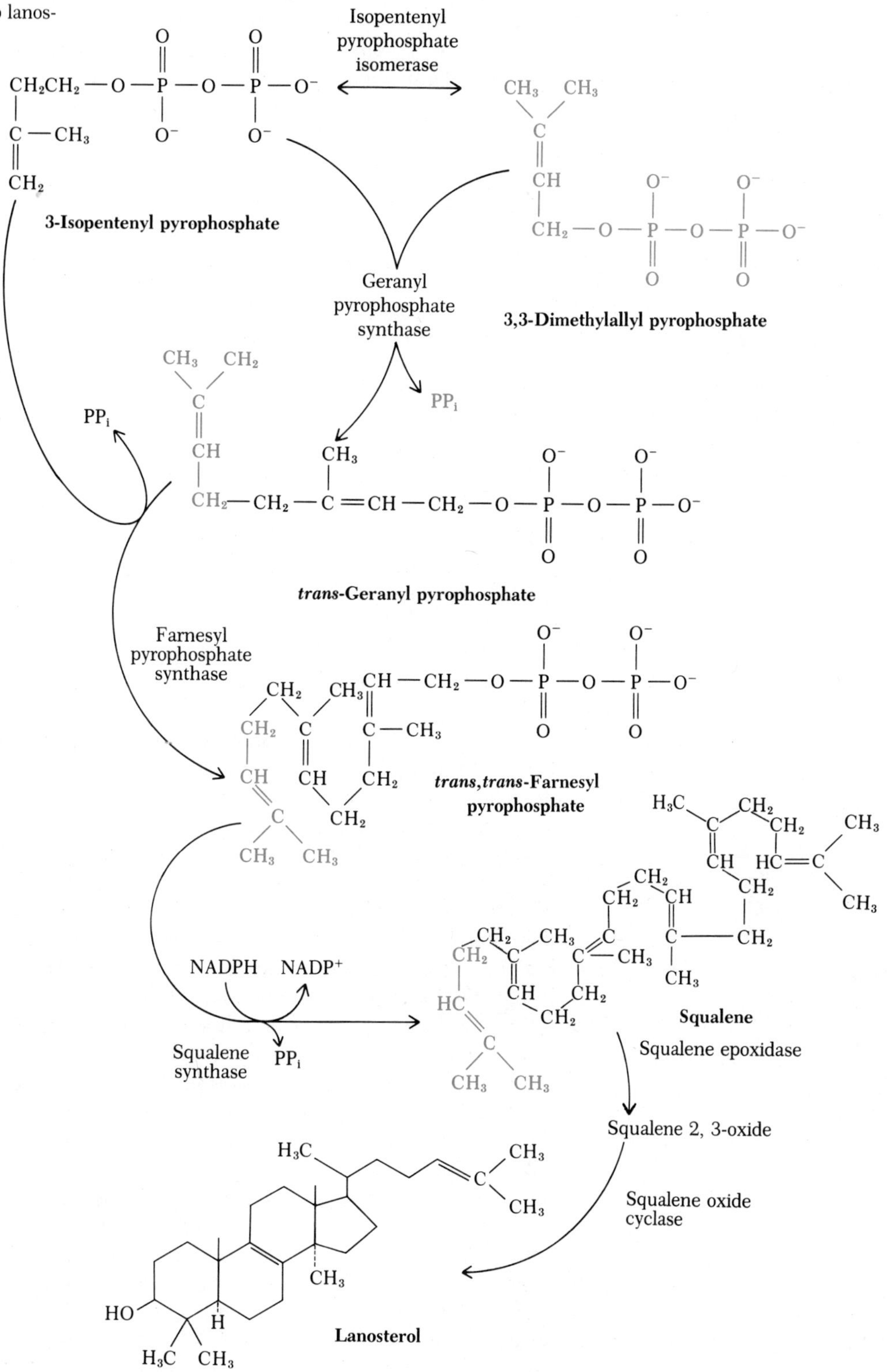

3-Isopentenyl pyrophosphate

Isopentenyl pyrophosphate isomerase

3,3-Dimethylallyl pyrophosphate

Geranyl pyrophosphate synthase

PP_i

PP_i

trans-Geranyl pyrophosphate

Farnesyl pyrophosphate synthase

trans,trans-Farnesyl pyrophosphate

NADPH NADP$^+$

Squalene synthase

PP_i

Squalene

Squalene epoxidase

Squalene 2, 3-oxide

Squalene oxide cyclase

Lanosterol

In the second phase such isoprenoid units condense in a head-to-tail fashion to produce ultimately a 30-carbon structure called *squalene.* First dimethylallyl pyrophosphate condenses with isopentenylpyrophosphate in the head-to-tail manner illustrated below; pyrophosphate is eliminated. This reaction is catalyzed by *geranyl pyrophosphate synthase.*

Dimethylallyl pyrophosphate

```
CH3  CH3
  \  /
   C
   ||
   CH          O         O
   |           ||        ||
   CH2 — O — P — O — P — O⁻
               |         |
               O⁻        O⁻
 +
                                        CH3  CH3
                                          \  /
   CH2                                     C
   ||                                      ||
   C — CH3                                 CH
   |                        PPi            |
   CH2       O        O      ↗             CH2
   |         ||       ||   ———→            |
   CH2 — O — P — O — P — O⁻                CH2
             |        |                    |
             O⁻       O⁻                   C — CH3
                                           ||
                                           CH       O        O
                                           |        ||       ||
                                           CH2 — O — P — O — P — O⁻
                                                    |        |
                                                    O⁻       O⁻
```

Isopentenyl pyrophosphate   ***trans*-Geranyl pyrophosphate**

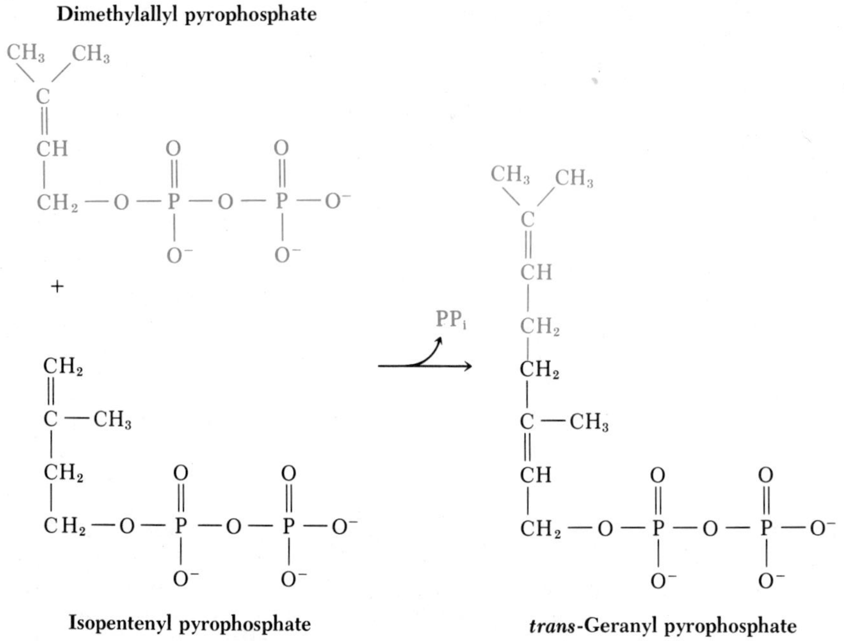

Next another 3-isopentenyl pyrophosphate unit is added in a similar manner; pyrophosphate is eliminated and the product is called *trans,trans*-farnesyl pyrophosphate. The enzyme catalyzing this process is *farnesyl pyrophosphate synthase.* Subsequently two such farnesyl pyrophosphates condense in a tail-to-tail fashion, also with the concomitant elimination of pyrophosphate. The product is called *squalene,* a symmetrical C-30 terpene, and the responsible enzyme is *squalene synthase.* As seen in Fig. 12-16, squalene is then converted into the tetracyclic configuration of lanosterol in two steps. First, the enzyme *squalene epoxidase* catalyzes the conversion of squalene to its 2,3-oxide. The oxygen is derived from molecular O_2 in an as yet uncertain manner. The 2,3-oxide intermediate then undergoes cyclization to lanosterol in a reaction catalyzed by the hepatic enzyme *squalene oxide cyclase.* This reaction apparently is initiated by an attack of a proton on the epoxide ring, followed by electron shifts resulting in ring closure (Fig. 12-16) and the formation of a transient carbonium ion at C-20. Lanosterol thus is derived from squalene 2,3-oxide by a series of concerted hydride and methyl shifts and the elimination of a proton from C-9.

The conversion of lanosterol to zymosterol (Fig. 12-17) entails the oxidation of three methyl groups (positions 4 and 14) to carboxyl groups and elimination as carbon dioxide by decarboxylation. The change from zymosterol to desmosterol involves a shift of the double bond in position 8,9 to position 5,6. The mechanism of this reaction is still uncertain. Saturation of the double bond in the side chain of demosterol by means of NADPH completes the conversion of desmosterol to cholesterol. The enzyme requirements for these later steps are not completely understood.

Alternate pathways and sequences that are tissue dependent exist for the biosyn-

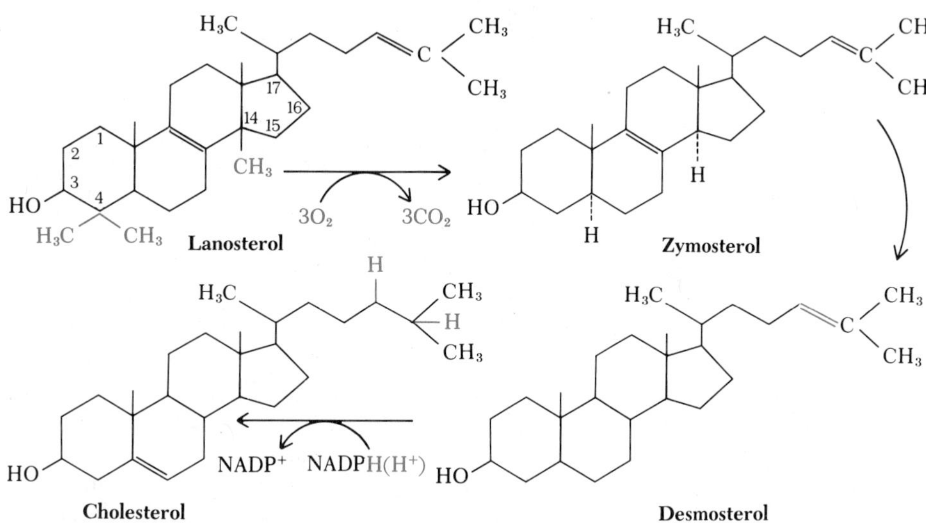

Figure 12-17
Conversion of lanosterol to cholesterol.

thesis of cholesterol from lanosterol. One such pathway is schematically shown as follows:

Lanosterol → 24,25-Dihydrolanosterol → α-Methyl-Δ^8-cholestenol → α-4-Methyl-Δ^7-cholestenol → Δ^7-Cholestenol → 7-Dehydrocholesterol → Cholesterol

Almost all tissues, but especially the liver and intestinal mucosa, are able to synthesize cholesterol. The various enzymes involved are located in the cytoplasm.

Control of cholesterol biosynthesis The rate-limiting step in the biosynthesis of cholesterol is the reduction HGM-CoA to mevalonic acid by HMG-CoA reductase, located in the lipoprotein of the endoplasmic reticulum. Cholesterol inhibits this step, providing an effective product feedback inhibition for controlling its formation. Apparently, first the major transport lipoprotein for cholesterol, LDL, is bound to the surface of cells active in the biosynthesis of cholesterol. Cholesterol is released following lysosomal cleavage of the carrier apoprotein. The released cholesterol accumulates and represses both the biosynthesis and the activity of HMG-CoA reductase, the first and regulatory enzyme of the pathway (Chapter 3). The biosynthesis of additional cholesterol thus is reduced or stopped. Support for the concept of an involvment of LDL in the regulation of cholesterol biosynthesis has been reported from studies of patients with *abetalipoproteinemia* (Chapter 22). The low levels of plasma cholesterol and LDL found in these patients are associated with rates of cholesterol biosynthesis five times higher than that in normal, control subjects.

A second point of control appears to be at the cyclization of squalene into lanosterol. Fasting diverts HMG-CoA to ketone-body formation and hence decreases cholesterol biosynthesis. Higher intakes of saturated fatty acid seem to augment the serum cholesterol level. For a pathway showing this, see Fig. 12-11.

Feedback control of the hepatic biosynthesis is dependent on the total available cholesterol regardless of source. Therefore, dietary cholesterol may also serve to

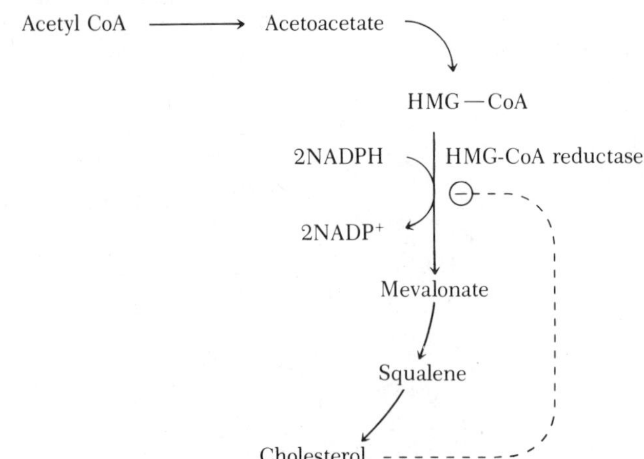

reduce its biosynthesis. Feeding a diet containing 0.5% cholesterol to an animal results in a 95% inhibition of hepatic cholesterol synthesis. Inhibition of hepatic cholesterol genesis has been shown to be true for all vertebrates examined, including birds, rodents, dogs, and primates. It is now clear that feedback control is also operative in the human liver.

Potter first suggested that an impairment of metabolic feedback control might lead to a cellular proliferation and therefore be a cause of cancer. Subsequently, Siperstein demonstrated that cholesterol feedback control was lost in a spontaneous mouse hepatoma. This metabolic lesion was also observed in trout hepatoma caused by aflatoxin, a potent carcinogen. Aflatoxin, a cyclic hydrocarbon produced by the mold *Aspergillus flavus*, has been found in contaminated trout feed and, rarely, also in improperly harvested peanuts. Liver tissue from trout that were fed contaminated food exhibited a functional defect in the hepatic feedback mechanism for the biosynthesis of cholesterol well before the earliest pathologic appearance of malignancy. A loss of feedback control has also been observed in human hepatomas.

Next to the liver, the intestinal mucosa is the most important source of endogenous cholesterol. The biosynthetic pathway is comparable to that in the liver. However, it is not under the feedback control of cholesterol; the HMG-CoA reductase is inhibited by cholic acid instead.

Approximatey 80% of the cholesterol is disposed of by conversion to cholic acid (Chapter 17) and excreted by way of the bile, as the bile salts glycocholate and taurocholate. These bile salts are in turn essential to the intestinal absorption of cholesterol. Regulation of blood cholesterol level is the result of control at the hepatic and intestinal levels. If the amount of bile salts is deficient in the intestinal contents, the absorption of dietary cholesterol to the liver will enhance hepatic synthesis of cholesterol. Furthermore, the reduced level of bile salts will enhance intestinal biosynthesis of cholesterol.

Functions of cholesterol

The esters of cholesterol with fatty acids are normally present in the blood in a more or less definite ratio to free cholesterol. Esterification involves the hydroxyl group on carbon-3, as shown at the top of the next page.

The fatty acids so combined are the most unsaturated in the blood plasma. This suggests that cholesterol acts as a special transport agent for unsaturated fatty acids.

The conversion of cholesterol to cholic acid occurs by a series of hydroxylations, as shown in Fig. 12-18. We know these hydroxylations require vitamin C and O_2.

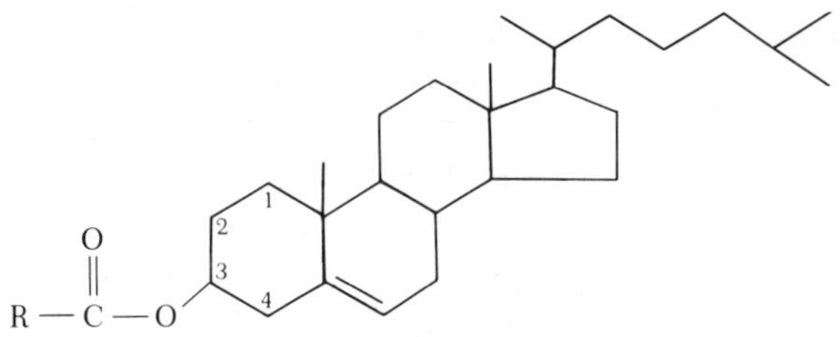

Cholesteryl ester

Both hepatic microsomal fractions and mitochondria are needed for the metabolism of cholesterol to cholic acid. In latent vitamin C deficiency, guinea pigs exhibit a decreased conversion of cholesterol to bile acids. Reduced cholesterol metabolism leads to a hypercholesterolemia and accumulation of cholesterol in the liver. There is a relationship between vitamin C intake and the serum cholesterol level. For example, the hydroxylation at carbon-7 may occur as follows:

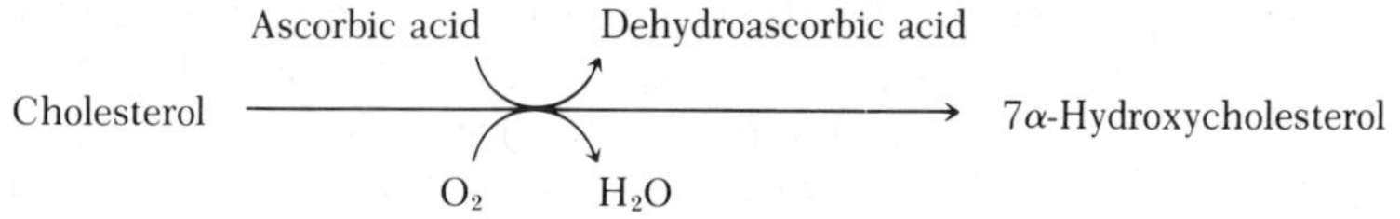

$$\text{Cholesterol} \xrightarrow[O_2 \quad H_2O]{\text{Ascorbic acid} \quad \text{Dehydroascorbic acid}} 7\alpha\text{-Hydroxycholesterol}$$

Figure 12-18
Conversion of cholesterol to cholic acid.

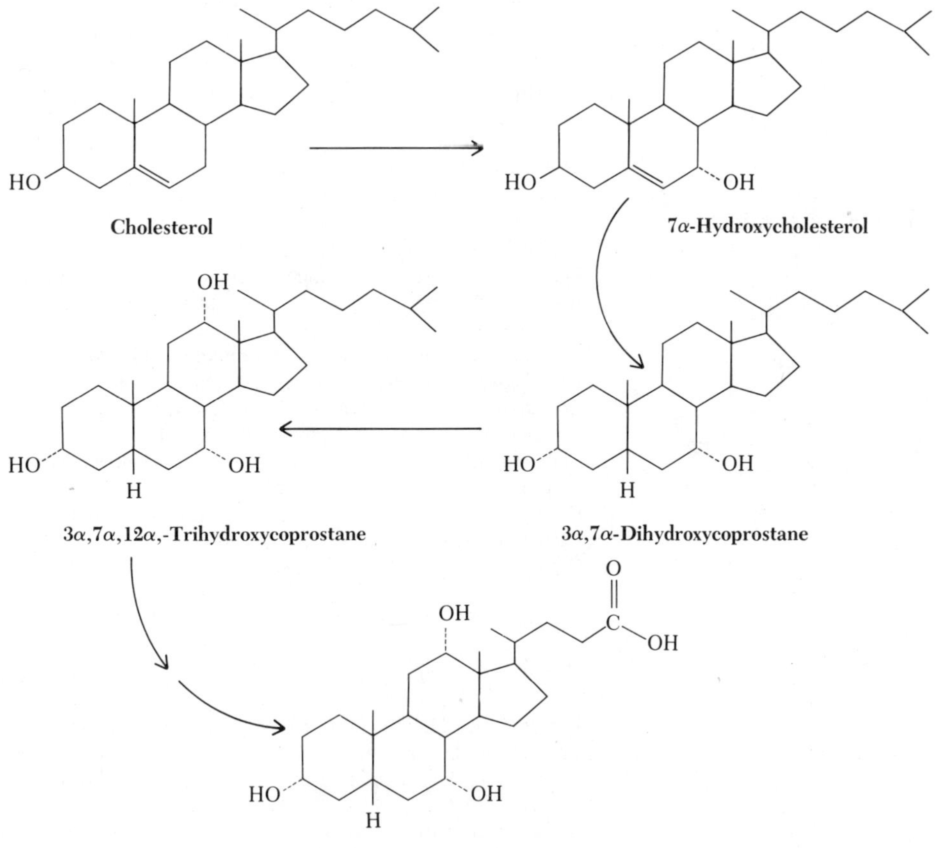

Changes in lipids with aging

Blood cholesterol levels are influenced by both genetic and environmental factors. These include sex, age, diet (Chapter 21), body-weight loss or gain, exercise, stress, and a number of pathological conditions (Chapter 22). During the early years of life females tend to have higher levels of plasma cholesterol than do males. However, during childhood, adolescence, and young adulthood differences between the sexes are minimal. Thereafter, however, at least until 50 years of age men have higher plasma concentrations of cholesterol than do women. During the sixth and seventh decades of life women *may* have higher levels than men. Plasma cholesterol values tend to *increase* with age, except during adolescence, at least to age 69 years. After that a slight decrease may occur.

With increasing age humans normally experience a decreased capacity for fat metabolism. Cholesterol concentration in the tissues increases, e.g., in the aorta by approximately 1 mg of cholesterol per gram per decade. Changes in the nature of the fats may also occur. In neural tissue the average fatty acyl chain length in cerebrosides for a 7- or 8-year-old child is 18 carbons (18:0). ("0" means no double bonds in the carbon chain.) In the elderly person this may increase to more than 26 carbons. Furthermore, there is an increase in odd-numbered carbon chains and in the process of α-hydroxylation.

One of the most common diseases of aging is arteriosclerosis, or "hardening of the arteries." When this ailment is accompanied by an abnormal deposition of cholesterol and other lipids causing a hardening or sclerosis because of calcification, it is called atherosclerosis. Feeding cholesterol to rabbits produces this condition experimentally. It is well known that a high blood cholesterol level usually occurs in patients suffering from atherosclerosis, although many individuals with hypercholesterolemia have no evident arteriosclerosis and not all arteriosclerotic people have elevated blood cholesterol. Biochemical aspects of atherosclerosis, as well as those of other diseases of lipid metabolism, are discussed in Chapter 22.

Suggested readings

Bloch, K., and Vance, D.: Control mechanisms in the synthesis of saturated fatty acids, Annu. Rev. Biochem. **46:**263, 1977.

Brady, R.O.: Hereditary fat-metabolism diseases, Sci. Am. **229:**88, 1973.

Brown, M. S., and Goldstein, J.L.: Receptor-mediated control of cholesterol metabolism, Science **191:**150, 1976.

Eisenberg, S., and Levy, R.I.: Lipoprotein metabolism, Adv. Lipid Res. **13:**1, 1975.

Fredrickson, D.S.: Plasma lipoproteins and apolipoproteins, Harvey Lect. **68:**185, 1972-1973.

Goldstein, J.L., and Brown, M.S.: The low-density lipoprotein pathway and its relation to atherosclerosis, Annu. Rev. Biochem. **46:**897, 1977.

Goodwin, T.W., editor: International review of biochemistry, vol. 4, Biochemistry of lipids I, Baltimore, 1975, University Park Press.

Goodwin, T.W., editor: International review of biochemistry, vol. 14, Biochemistry of lipids II, Baltimore, 1977, University Park Press.

Greenberg, D.M., editor: Metabolic pathways, vol. II, ed. 3, New York, 1967, Academic Press, Inc.

Hancock, W.S., Marshall, G.R., and Vagelos, P.R.: Acyl carrier protein, J. Biol. Chem. **248:**2424, 1973.

Hsai, Y.Y.: Inborn errors of metabolism, part I, ed. 2, Chicago, 1966, Year Book Medical Publishers, Inc.

Iverson, L.L.: The chemistry of the brain, Sci. Am. **241:**134, 1979.

Kritchevsky, D.: How aging affects cholesterol metabolism, Postgrad. Med. **63:**133, 1978.

Krebs, H.A.: The regulation and release of ketone bodies by the liver, Adv. Enzyme Reactions **4:**339, 1966.

Kunau, W.H., and Holman, R.T., editors: Polyunsaturated fatty acids, Champaign, Ill., 1979, American Oil Chemists Society.

Masoro, E.J.: Lipids and lipid metabolism, Annu. Rev. Physiol. **39:**301, 1977.

McMurray, W.C., and Magee, W.L.: Phospholipid metabolism, Annu. Rev. Biochem. **41:**129, 1972.
Morrisett, J.D., Jackson, R.L., and Gotto, A.M.: Lipoproteins: structure and function, Annu. Rev. Biochem. **44:**183, 1975.
Nes, W.R., and McKean, M.L.: Biochemistry of steroids and other isopentendoids, Baltimore, 1977, University Park Press.
Research News: The high-density lipoproteins: the good carriers of cholesterol, Science **105:**677, 1979.
Rommel, K., and Bohmer, R., editors: Lipid absorption: biochemical and clinical aspects, Baltimore, 1976, University Park Press.
Sabine, J.R.: Cholesterol, New York, 1977, Marcel Dekker, Inc.
Stanbury, J.B., Wyngaarden, J.B., and Fredrickson, D.S., editors: The metabolic basis of inherited disease, ed. 4, New York, 1978,McGraw-Hill Book Co.
Van den Bosch, H: Phophoglyceride metabolism, Annu. Rev. Biochem. **43:**243, 1974.
Volpe, J.J., and Vagelos, P.R.: Mechanisms and regulation of biosynthesis of saturated fatty acids, Physiol. Rev. **56:**339, 1976.
Wakil, S.J., editor: Lipid metabolism, New York, 1970, Academic Press, Inc.
Wiegandt, H.: Glycosphingolipids, Adv. Lipid Res. **9:**249, 1971.

Chapter 13

Metabolism of amino acids

The principal objectives of this chapter are to consider both the anabolic and catabolic phases of the metabolism of the some 20 amino acids derived from naturally occurring food proteins and from expended tissue proteins, to describe their interconversions and their conversions into a variety of vital biomolecules and into their excretory products, and to discuss the "essential amino acids."

Background

The metabolism of amino acids is in a dynamic state, as is that of carbohydrates (Chapter 11) and lipids (Chapter 12). Amino acids of the body's "metabolic pool," derived either exogenously from dietary food proteins or endogenously from expended tissue proteins, are in a constant state of flux. They may be used for synthesis into new body protein to replace those lost in the "wear and tear" of life processes or they may be converted into a wide variety of nonprotein nitrogenous substances, some of which perform vital biochemical functions and a few which are excretory waste products. Familiar examples of the former are muscle and other tissue proteins, blood plasma proteins, hemoglobins, and a wide variety of enzymes and protein hormones, such as insulin. Examples of the latter are such important body constituents as creatine, the purines and pyrimidines, thyroxin, serotonin, GABA, epinephrine, and the heme moiety of hemoglobin and cytochromes. The principal excretory products include urea and ammonia.

The metabolism of amino acids and of the proteins from which they are derived is sometimes described by the broader term *nitrogen metabolism*. The concepts of Folin and other biochemists of the early 1900s were that proteins, as well as some other major body constituents, were for structural purposes and were "metabolically inert." This view was changed radically, however, in the 1930s by the classical studies of Schoenheimer and Rittenberg and others, demonstrating by the use of the then new isotopic procedures that body proteins and indeed amino acids and the essential metabolites formed from them are being destroyed and replaced continuously in normal living organisms, including man. The half-life of many of these nitrogenous substances has been found to vary from only a few minutes to several hours or at most up to about 180 days in some instances.

Thus a study of amino acid metabolism must involve a consideration of the biochemical processes by which these nitrogenous substances are formed and degraded in living organisms. The use of amino acids for protein biosynthesis has been discussed in detail in Chapter 10. The metabolism of the some 20 amino acids making up the proteins of living matter (both plant and animal), including their dietary origin, anabolic changes and interconversions, and catabolism is described in this chapter. Their chemical structures and physicochemical properties are discussed in some detail in Chapter 23 for review purposes.

Sources

Food proteins

The prime source of amino acids for metabolic use by man is food proteins, both plant and animal in origin. Since these are supplied to the mammalian organism, with which we are chiefly concerned, as macromolecules, they must be digested into their constituent amino acids before they can be absorbed through the intestinal mucosal epithelium into the blood plasma for transport to cells for metabolism. This process is discussed in a following section of this chapter.

Nitrogen fixation

The ultimate source of the amino acids used by the animal organism, however, is protein foods obtained from plants or certain microorganisms. Their source, in turn, is from amino acids synthesized from ammonia and carbohydrate intermediates formed in photosynthetic processes ("CO_2 fixation"). The ammonia used in this process is formed by "nitrogen fixation" by nitrifying bacteria in root nodules of certain plants or by certain types of other microorganisms, such as blue-gree algae. The enzyme *nitrogenase* and ATP are required as indicated below. Atmospheric nitrogen is reduced to ammonia in the reaction. Nitrogenase is a metalloenzyme containing molybdenum and iron. There is current evidence that molybdenum functions as the catalytic agent. The formation of the amino acid glutamic acid, as an example, is shown schematically as follows:

$$N_2 + 6\ H^+ + 6\ e^- + 12\ ATP + 12\ H_2O \xrightarrow{\text{Nitrogenase}} 2\ NH_3 + 12\ ADP + 12\ P_i$$

$$NH_3 + \alpha\text{-Ketoglutarate} + NADPH(H^+) \underset{\text{Dehydrogenase}}{\overset{\text{Glutamate}}{\rightleftharpoons}} \text{Glutamic acid} + NADP^+ + H_2O$$

Soil nitrates or nitrites also can be converted into ammonia using a nitrate- or nitrite-reductase enzyme system.

Biosynthesis of amino acids

A number of amino acids can be formed from the glutamic acid and the α-keto acid described above by the transamination reaction, which is discussed later. Also, ammonia combined in the form of glutamine or carbamoylphosphate can be used for the formation of certain amino acids and other nitrogenous compounds, to be considered in subsequent sections.

The sulfur atom in the amino acids cystine-cysteine and methionine is "fixed" by a bacterial process analogous to that for nitrogen described above. Either elemental sulfur or soil sulfates are enzymatically reduced by serving as electron acceptors in this type of bacteria, forming sulfide which is then incorporated into the appropriate carbon chain, forming the sulfur-containing amino acids.

The pathways for the biosynthesis of the various amino acids are summarized later in this chapter in the discussions of the metabolism of the individual amino acids.

"Essential" amino acids

Of the 20 amino acids constituting naturally occurring proteins 10 cannot be synthesized in sufficient amounts for growth by humans and most other animal species. Therefore these amino acids must be supplied preformed for this purpose by food proteins. These 10 amino acids are called *essential amino acids*. They include arginine, histidine, isoleucine, leucine, lysine, methionine, phenylalanine, threonine, tryptophan, and valine. All of these with the exceptions of arginine and probably histidine are likewise essential for the maintenance of "nitrogen balance" in most adult animal organisms, including humans. The essential amino acids are discussed further in some detail in Chapter 21.

Although the other 10 amino acids, not listed above, are classed as "nonessential" or, in the case of arginine, cystine, glycine, histidine, tyrosine, and serine, as "semiessential" (Chapter 21), this categorization is true only in a restricted sense. *All* amino acids are essential in the broader sense that they are constituents of essential body proteins, and, further, some serve as precursors of many other essential biomolecules as is discussed later in this chapter.

Digestion of proteins

Gastric digestion

Unlike the carbohydrates and lipids, a significant amount of digestion of proteins occurs in the stomach. This was observed grossly by Beaumont in his classic studies on Alexis St. Martin in the 1820s. Beaumont actually observed pieces of meat and other proteins "liquefy" in the stomach of this subject. Most proteins are acted upon in the stomach by the proteolytic enzyme *pepsin* at the acid pH (approximately 1.0) of gastic juice and converted to smaller peptides. As described previously (Chapter 5), pepsin is not a highly specific proteolytic enzyme; it preferentially hydrolyzes those linkages involving the amino group of the aromatic amino acids, tryptophan, phenylalanine, and tyrosine. In natural protein substrates, pepsin, an endopeptidase, splits other peptide bonds similar to those of aromatic amino acids at a relatively rapid rate, e.g., Leu-Glu, Glu-Asn, Leu-Val, Val-Cys. The peptide bonds attacked by pepsin need not be adjacent to a free carboxyl or an α-amino group.

Although pepsin digests proteins mainly into polypeptides of varying length, some short-chain peptides and even some free amino acids, notably tyrosine and phenylalanine, may be released.

Pepsin is not secreted as such by the "chief" cells of the gastric mucosa; rather, it is secreted as a zymogen form, *pepsinogen* (Chapter 5). Pepsinogen is converted to the active enzyme by the acidity of the gastric juice and also autocatalytically by pepsin itself.

The optimum pH for pepsin activity varies from 1.5 to 2.2, depending on the substrate.

A small amount of pepsinogen apparently is secreted into gastric tissue fluid and is carried by the blood plasma to the kidney, where it is excreted as *uropepsin*. This may be determined clinically and may be of some value in studies of gastric function.

There is some evidence that two forms of pepsin may exist, one secreted by the chief cells, the other by the pyloric cells. The optimum pH values of these forms differ somewhat.

Another proteolytic enzyme, a weak proteinase active at a neutral pH, also has been described in gastric juice. It apparently contributes little to protein digestion in the stomach under ordinary circumstances.

Still another proteolytic enzyme, *rennin*, is secreted in the fourth stomach of the calf and probably other young ruminants. Its action is to clot milk. This is accomplished by the slight hydrolysis of the casein of milk to produce *paracasein*, which coagulates in the presence of calcium ions, resulting in an insoluble calcium paracaseinate curd:

$$\text{Casein} \xrightarrow{\text{Rennin}} \text{Paracasein} \xrightarrow{\text{Ca}^{++}} \underset{\text{(insoluble curd)}}{\text{Ca-paracaseinate}}$$

The purpose of this reaction appears to be to convert milk into a more solid form, thus increasing its retention in the stomach for a longer period of time and hence its gastric digestion.

Apparently pepsin, or perhaps pepsin plus some other proteolytic enzyme, has similar action in older animals since little or no rennin is present in the gastric juice of the adult, at least in humans.

The secretion of gastric juice is stimulated by the hormone *gastrin* (Chapter 18) from the pyloric mucosa. Apparently products of protein digestion, peptides, and other factors in turn cause an increase in gastrin formation.

The gastric digestion of proteins is thus incomplete, producing mainly larger peptides from only a limited number of proteins. The major digestion of proteins is accomplished therefore in the small intestine, where several powerful proteolytic enzymes are secreted.

Pancreatic digestion

Pancreatic juice contains a number of important proteinases, including *trypsin, chymotrypsin, elastase,* and two *carboxypeptidases*. Trypsin and chymotrypsin are secreted in zymogen forms, trypsinogen and chymotrypsinogen respectively, and there are two chymotrypsinogens, A and B (Chapter 5). Chymotrypsinogen A has been crystallized and more intensively studied than chymotrypsinogen B. Trypsinogen is changed to trypsin by *enterokinase* secreted by the intestinal mucosa. This agent, sometimes termed as "activator" of the enzyme, has been definitely shown to be an enzyme. Trypsinogen also can be converted to trypsin autocatalytically; i.e., as soon as some trypsin is formed, it acts on trypsinogen, forming additional trypsin molecules.

The optimum pH for both trypsin and chymotrypsin is in the neighborhood of 8 to 9. Each attacks proteins, splitting off peptide chains of varying length, and amino acids. Trypsin, however, is most effective on partially digested proteins. In fact, it digests collagen, ovalbumin, serum globulins, and hemoglobin very slowly, unless they have been denatured. However, trypsin attacks certain proteins (protamines, histones) that cannot be digested by pepsin. As described in Chapter 5, the peptide linkage hydrolyzed by trypsin is one containing the carboxyl group of either lysine or arginine, i.e., there is a positive charge in the side group immediately adjacent to the point of attack. The net result of trypsin activity is to break down the products of peptic digestion still further and to digest those proteins that pepsin cannot attack. The products are amino acids and various oligopeptides.

Several proteinlike substances that block the action of trypsin and related enzymes have been described. They are called *trypsin inhibitors*. One of them is present in the plasma and also has antithrombin activity (Chapter 16). Apparently it is an α_1-globulin and a glycoprotein. Another substance is present in pancreatic juice and is secreted in zymogen granules along with trypsinogen. It appears to be a peptide containing some 56 amino acid residues, has a molecular weight of 6155, is strongly acidic, and has a relatively high content of glutamic and aspartic acids. It reacts rapidly with trypsin in a stoichiometric manner and apparently protects pancreatic tissue from the proteolytic action of trypsin, which might be activated from trypsinogen. Pancreatic tissue (bovine) contains a similar basic polypeptide with analogous protective properties.

Recent investigations indicate that trypsin binds calcium ions between two carboxyl groups on the surface of the trypsin molecule, near the active site. The bound calcium appears to prevent the autodigestion of trypsin.

A hereditary deficiency of trypsinogen has been reported in several infants. This serious abnormality is discussed in Chapter 22.

Chymotrypsin catalyzes the hydrolysis of the various protein breakdown products resulting from peptic and tryptic action. Again definite linkages are involved. Chymotrypsin preferentially cleaves peptide bonds involving the carboxyl group of the aromatic amino acids (Tyr, Phe, and Trp) (Chapter 5), but it also splits peptide linkages of leucine, methionine, asparagine, and histidine. The products are amino acids and oligopeptides.

There is an enzyme, a collagenase, in pancreatic juice that specifically attacks collagen. It is capable of digesting collagenous fibers present in food. It would also account in part for the tissue necrosis associated with pancreatitis and for the digestion of other tissue that comes in contact with pancreatic juice from a pancreatic fistula.

Other proteolytic enzymes are present in pancreatic juice. Elastase attacks not only elastin but, in addition, a wide variety of proteins. It is secreted by the pancreas as the zymogen, which is converted to the active enzyme by trypsin and enterokinase. Although elastase is capable of hydrolyzing many proteins, it has an unusual ability to digest elastin. The amino acid sequence of elastase has been determined, including that at the active site. Considerable homology exists between the amino acid sequences of elastase, trypsin, and chymotrypsin B, suggestive of a common genetic origin of these proteases.

An elastomucoproteinase and a collagen mucoproteinase hydrolyze the proteins indicated by their names.

Two exopeptidases in pancreatic juice, carboxypeptidases A and B, attack the end-peptide linkage, specifically that having a free carboxyl group (Chapter 5). Thus free amino acids are split off from dipeptides and higher peptides. Their optimum pH is about 7.4.

Carboxypeptidases A and B are secreted in a zymogenic form, *procarboxypeptidases* A and B, and are activated by trypsin. The activation of procarboxypeptidase A, which exists as an aggregate of three large subunits, involves the conversion of one of these subunits by trypsin into carboxypeptidase A. A second unit apparently is converted to a chymotrypsin-like substance. The role of the third unit is unknown. The activation of the pancreatic proteolytic enzymes is depicted schematically in Fig. 13-1.

The secretion of pancreatic enzymes is controlled by the hormone *pancreozymin* (Chapter 18). It is found in the upper intestinal mucosa and is released into the bloodstream by the presence of a variety of substances, including proteins, peptides, and carbohydrates in the duodenum. The hormone *secretin* (Chapter 18) also found in the mucosa of the upper small intestine, likewise stimulates the secretion of pancreatic juice. However, the secretin-stimulated pancreatic juice is rich in bicarbonate rather than enzymes.

Intestinal digestion

The digestion of proteins and of the products of hydrolysis by pepsin, trypsin, chymotrypsin, and the carboxypeptidases is completed by peptidases secreted by the mucosa of the small intestine. These enzymes were formerly believed to be contained in an intestinal juice, or *succus entericus,* analogous to gastric and pan-

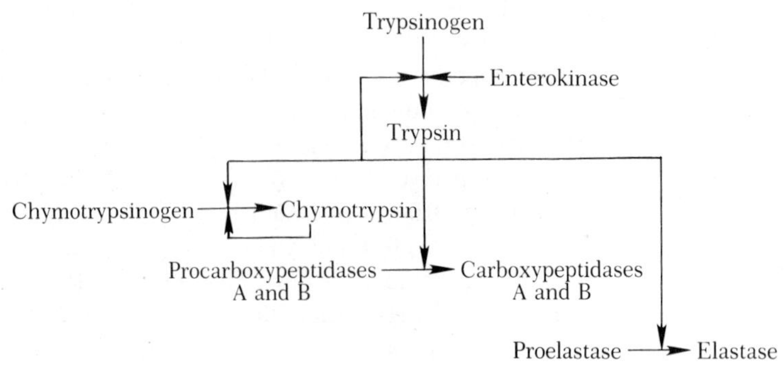

Figure 13-1
Activation of pancreatic proteolytic enzymes. Enterokinase triggers the activating process as shown.

creatic juices. However, they now appear to be primarily intracellular enzymes present in mucosal cells; some of these cells become detached and are released into the lumen of the small intestine. Desquamation and functional replacement of human intestinal mucosal cells are apparently a continuous and rather rapid process, since the half-life of the cells is a period of only a few days. The enzyme content of the intestinal mucosal cells is increased by the hormone *enterocrinin* (Chapter 18), elaborated by the intestinal mucosa. A variety of digestive products stimulate the release of this hormone.

A number of peptidases are present in the intestinal mucosa cells or in the secretions of these cells. Included are a group of *aminopeptidases* and a group of *dipeptidases*. The aminopeptidases split a wide variety of *N*-terminal peptide linkages. Leucine aminopeptidase is an example of this type of enzyme. Aminopeptidases require magnesium or manganese ions for their activity. The dipeptidases are more specific in their action. For example, glycylglycine dipeptidase splits glycylglycine but not glycylglycylglycine. This dipeptidase requires the presence of cobalt or manganese ions for activity.

The hydrolysis of most proteins is thus completed to their constituent amino acids, which are then ready for absorption into the blood plasma. A few proteins, e.g., silk and several other albuminoids, are resistant to digestion. However, the digestion of the majority of proteins is about 95% complete in the normal human subject.

Some individuals apparently lack an enzyme necessary for the hydrolysis of *N*-glutamyl peptides in the small intestine. As a result, such people are intolerant of proteins yielding these peptides on digestion. The chief offending proteins are glutens from wheat, oats, barley, and rye. This syndrome, termed *gluten-sensitive enteropathy* or *celiac disease,* is discussed in Chapter 21 along with other abnormalities involving a deficiency of specific digestive enzymes (Chapter 22).

Absorption of amino acids

Role of transport proteins

As just described, food proteins are almost completely digested to amino acids in the gastrointestinal tract and are absorbed as such from the small intestine. The absorption of most amino acids involves an active transport mechanism (Chapter 19) requiring ATP and specific transport proteins in the intestinal mucosal cells. As is described in Chapter 19, the transport of amino acids is divided typically into 5 groups; the Na^+-dependent "A" and "ASC" systems, the "L" system, and special processes for basic and acidic amino acids. The amino acid transport process involves a two-step, transcellular system. First the amino acids are carried across the membrane from the intestinal lumen into the cytoplasm by a support carrier process. Here a downhill Na^+-gradient provides the energy of transport for the amino acids. The Na^+ gradient is maintained by a Na^+, K^+ATPase. The overall transport process is visualized to occur in a manner comparable to that for glucose (see Fig. 19-7). Carrier proteins, however, are not specific for each amino acid but rather for the groups of amino acids, as described in Table 19-1. The exact nature of the sodium-independent systems is presently unknown. It also appears likely that the intestinal transport includes the glutathione dependent (γ-glutamyl transpeptidase) process.

After active absorption by the intestinal mucosal cell, the amino acids are taken up primarily by the blood capillaries of the mucosa and are transported in the plasma to the liver and other tissues for metabolic use. A significant amount of the absorbed amino acids also appears in lymph.

In some individuals, small amounts of intact proteins may be absorbed through the intestinal mucosa or, in fact, through certain other epithelial barriers. This may lead to the formation of antibodies against the "foreign" protein and anaphylactic reactions or other immunological phenomena after the subsequent absorption of the same intact protein. Thus we have the basis for the common allergies to food proteins.

General pathways of amino acid metabolism

Amino acids derived from the digestion of food proteins, i.e., *exogenous* sources, are absorbed and transported primarily by the blood as free amino acids to the liver, the primary site of amino acid metabolism, and to other organs and tissues of the body for use. In addition, however, amino acids derived from the hydrolysis of worn-out tissue proteins, *endogenous* amino acids, are added to form the "labile amino acid metabolic pool" of the body. The endogenous source of amino acids is by no means insignificant, present estimates being that *two thirds* of the amino acid pool is derived from the turnover of tissue proteins. Thus apparently only about one third of the amino acids used by the body are derived from food proteins.

Amino acids perform many important functions in living matter. As the "building blocks" of proteins (Chapter 4), they are essential constituents of the protoplasm of cells. They are incorporated into cellular structural proteins, e.g., collagen, elastin (Chapter 15); functional proteins, e.g., myosin of muscle, hemoglobin of blood; protective proteins, e.g., keratins of skin, hair, and nervous tissue; catalytic proteins for metabolic reactions in the form of enzymes; transport proteins, e.g., transferrin (Chapter 19); and regulatory proteins in the form of protein hormones. They are involved in the hereditary process as nucleoproteins of genes. As such, amino acids participate in the biosynthesis of many other essential cellular constituents—creatine, choline, purines, pyrimidines, porphyrins, epinephrine, thyroxin, niacin, melanin, bile acids, detoxication products, and even glucose and ketone bodies, to name a few. Some of these substances are discussed in the preceding pages. Others are considered later. Amino acids also may be oxidized in the body, supplying about 4 kcal of energy per gram and some 15% to 20% of the total energy requirement of the average human adult.

Biosynthesis of tissue proteins

The steps involved in the biosynthesis of proteins from amino acids in the animal organism are considered in detail in Chapter 10. This is the *major* route for the use of amino acids, since about 75% of the amino acids metabolized in the normal human adult are used for this purpose. The reason, of course, is the constant destruction of body proteins, the loss in excreta, desquamation of cells, and other minor losses. A large proportion of the "endogenous" amino acids derived from tissue breakdown are recycled, however, as indicated in Fig. 13-2. Present estimates are that at least 140 gm of amino acids are contributed to the amino acid pool in an adult human subject daily by the turnover of tissue proteins. Isotopic studies of the turnover rates of different tissue proteins support these estimates. The half-life ($t_{1/2}$) of liver proteins, for example, is approximately 10 days. Plasma proteins also have a $t_{1/2}$ of about 10 days; muscle proteins, 180 days; and collagen, considerably longer. Some proteins, on the contrary, have much shorter $t_{1/2}$ values. The proteins of the intestinal mucosal cells turn over very rapidly, in a few days, and the $t_{1/2}$ of protein hormones and enzymes is also very short. The $t_{1/2}$ of insulin has been estimated as 6.5 to 9 minutes (Chapter 18). The average $t_{1/2}$ of the total body proteins of man is approximately 80 days. The overall significance of these facts is that, from a quantitative standpoint, the use of amino acids for the formation and replacement of tissue proteins is the major function of amino acids in the body.

There also seems to be a "priority rating" in the biosynthesis of different tissue proteins. For example, when the protein intake is severely restricted in a weanling

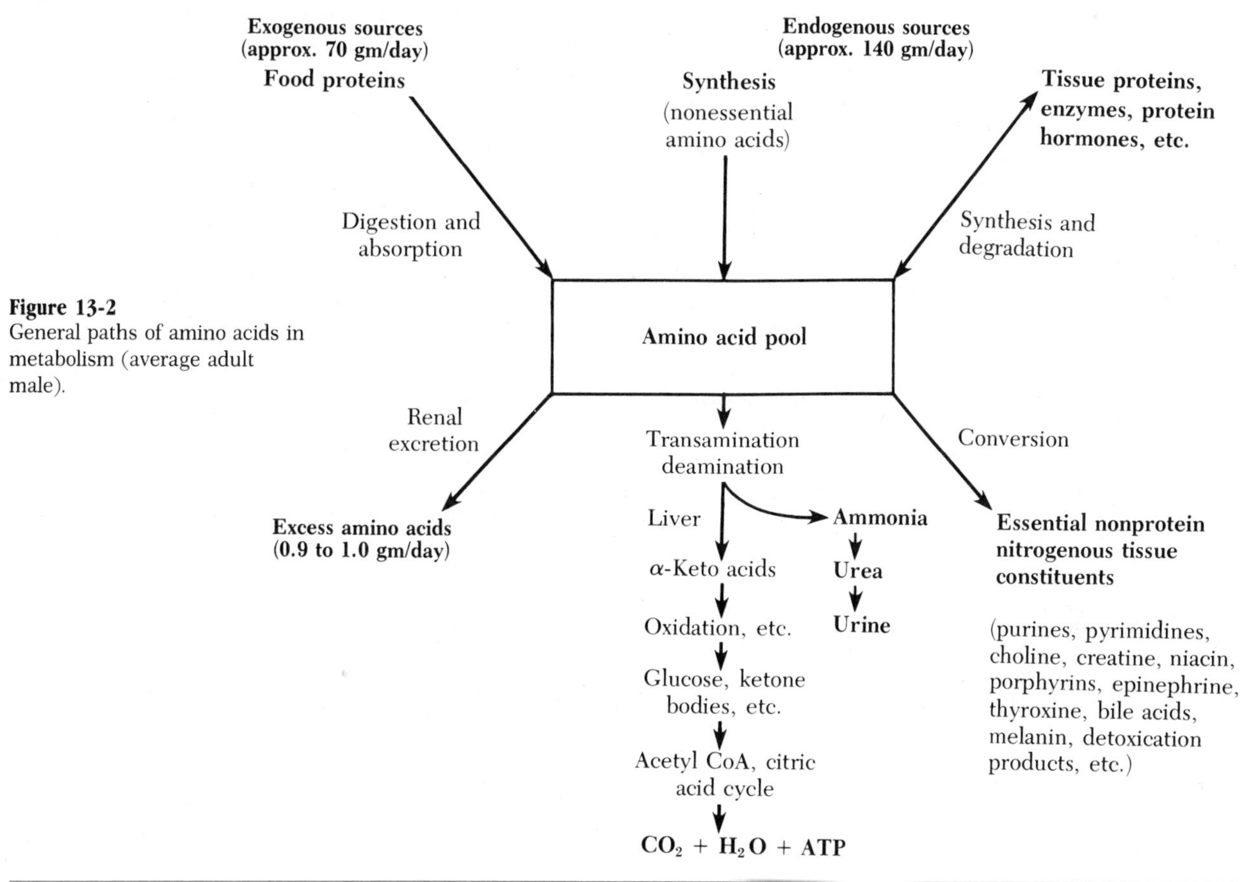

Figure 13-2
General paths of amino acids in metabolism (average adult male).

rat by a low-protein diet, growth virtually ceases, apparently because of a restriction of the formation of the quantitatively major body proteins, collagen, muscle proteins, elastin, etc. On the other hand, the formation of other perhaps more vital proteins, e.g., hemoglobin, shows little change from that of the normal control animals. The "low-protein" rats are only mildly anemic. A satisfactory explanation of this observation may be a concomitant decrease in the formation of the protein hormone erythropoietin, which controls erythrocyte and hemoglobin production (Chapter 16).

The number of different tissue proteins in man is large but difficult to estimate. There are several hundred, perhaps more than 600, different enzymes alone. The total number of tissue proteins in man thus may well be several thousand. If this number is multiplied by the number of species of animals and the number of other forms of life, including plant and microorganisms, the ascertainable total of proteins in living matter becomes enormous indeed. However, such variety is made possible because of the large size and multi–amino acid composition of proteins and the almost limitless capability of nuclear DNA to code for their synthesis. Indeed, it has been estimated that each human cell alone contains enough DNA to code for approximately 7 million different average-sized protein molecules!

Other metabolic uses

In addition to their primary role in the synthesis of tissue proteins, amino acids may be converted to other essential metabolites, as listed previously, or may be oxidized to yield energy after the removal of the amino group by transamination or by deamination, which is discussed next. Small amounts of excess amino acids, some 0.9 to 1 gm per day in the adult male, are excreted in the urine.

General paths for the metabolism of amino acids are shown schematically in Fig. 13-2.

Biochemical transformations of amino acids

Amino acids not used for the formation of the various types of body proteins just discussed may be converted to other essential nitrogenous constituents, as shown in Fig. 13-2, typically by multistep reactions. These are considered later in the discussions of the metabolism of the individual amino acids thus involved.

A third principal route for the metabolic use of amino acids is the degradation to their carbon skeletons, with subsequent conversion to other metabolites (e.g., glucose by gluconeogenesis [Chapter 11], ketone bodies by ketogenesis [Chapter 12], and/or oxidation by way of the citric acid cycle to yield CO_2, H_2O, and ATP). Removal of the amino group from the amino acid is the first step in either case. This is accomplished primarily by *transamination* or by *deamination*, which are described next.

Transamination

First described in 1937, the process of transamination is now recognized as a major reaction involving all of the amino acids, with the possible exception of lysine and threonine. The process is most active in the liver but also occurs in other eukaryotic cells. As the name implies, in the transamination reaction an amino group is transferred reversibly from a donor α-amino acid to a recipient α-keto acid. As a result, the latter becomes an α-amino acid and the former an α-keto acid. This may be depicted in the example:

$$\underset{\textbf{Glutamate}}{^-O-C(=O)-CH_2-CH_2-CH(\overset{+}{N}H_3)-C(=O)-O^-} + \underset{\textbf{Pyruvate}}{CH_3-C(=O)-C(=O)-O^-} \xrightleftharpoons{\text{Transaminase}} \underset{\alpha\textbf{-Ketoglutarate}}{^-O-C(=O)-CH_2-CH_2-C(=O)-C(=O)-O^-} + \underset{\textbf{Alanine}}{CH_3-CH(\overset{+}{N}H_3)-C(=O)-O^-}$$

Three different α-keto acids are known to participate in a number of transamination reactions: pyruvate, α-ketoglutarate, and oxaloacetate. Transamination reactions are catalyzed by specific enzymes called *aminotransferases* or, more commonly, *transaminases*. The preferred specific name is in terms of the two amino acids involved, e.g., *glutamate-alanine aminotransferase* for the above example. However, it is more usually called glutamate transaminase or even glutamate-pyruvate transaminase. Specific aminotransferases for all of the amino acids have been described, with the exception of lysine and threonine. Since only the three keto acids (α-ketoglutarate, pyruvate, and oxaloacetate) appear to serve for the amino-group transfer, the three classes would be termed "glutamate," "alanine," and "aspartate" aminotransferases, respectively, with the name of the amino acid donating the amino group preceding that of the amino acid formed in each instance.

At least several of the aminotransferases, including aspartate, alanine, and glutamate aminotransferases, are present in both cytosol and mitochondria as *isozymic* forms (Chapter 5). These have different physical and catalytic properties and different substrate specificities. The cytosol isozymes are generally more active than the mitochondrial forms.

All the aminotransferases are pyridoxal phosphate dependent, forming pyridoxamine phosphate upon accepting an amino group. (See structures in Chapter 6.)

Transamination reactions are freely reversible, having an equilibrium constant of about 1. The direction of the reaction is therefore determined by the removal of the product. The rate of reaction is determined, however, by the concentrations of the substrates, enzyme, or both. The cellular level of one of the aminotransferases, alanine transaminase, is increased by induction by the glucocorticoid hormones (Chapter 18).

Fig. 13-3 is a schematic representation of the steps involved in a typical transamination reaction. Amino acid 1 reacts with the pyridoxal phosphate–enzyme complex to form a Schiff base with the loss of H_2O. The Schiff base, in an aldimine form, then undergoes molecular rearrangement to a ketimine form, hydrolysis occurs at the double bond, and the α-keto acid 1 forms, with the amino group now attached to the coenzyme to form pyridoxamine phosphate. The reactions are repeated with α-keto acid 2, resulting in the formation of α-amino acid 2 and the regeneration of pyridoxal phosphate to complete the cycle. The catalytic involvement of pyridoxal phosphate in the reaction is evident from its structural changes, shown in Fig. 6-7, Chapter 6.

The transamination reactions appear to play two major roles in amino acid metabolism: (1) to serve as a means for the interconversion of a number of amino acids to increase the amount of one that may be in short supply, and (2) to channel the amino groups of amino acids ultimately to glutamate and aspartate. If the latter are in excess of metabolic needs, amino groups may be released by oxidative deamination to form NH_4^+ for urinary excretion as such or conversion to urea and excretion, as is described next.

It should be added that α-ketoglutarate serves as a *final acceptor* of amino groups, with glutamate being formed as the final product. A second amino group may then be added to form *glutamine*—a temporary nontoxic transport form of ammonia groups, as is discussed later.

Glutamine, as well as asparagine (also an amino acid amide), may participate in transamination reactions. In these cases, the corresponding α-ketoglutaramic and α-ketosuccinamic acids respectively are formed. These are then converted to their α-keto acids and NH_3 by specific ω-amidases for metabolic disposal.

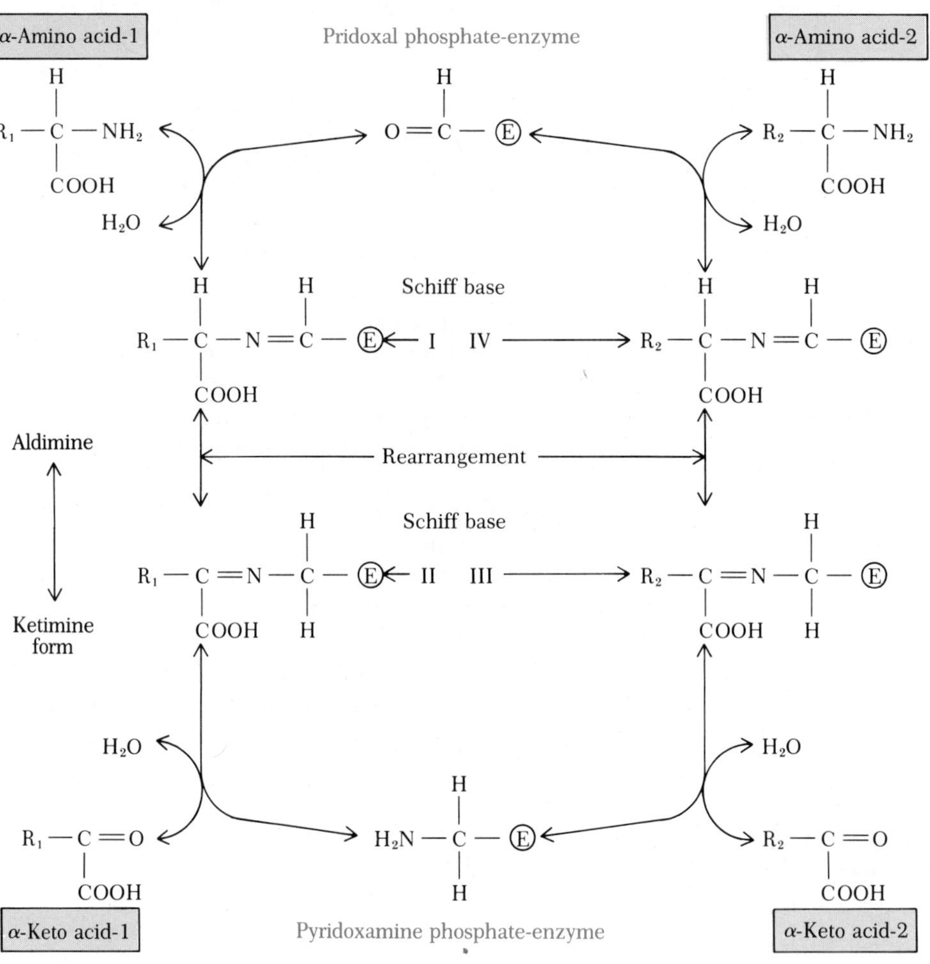

Figure 13-3
Intermediate steps in the transamination of α-amino acid-1 to form α-keto acid-1 and of α-keto acid-2 to form α-amino acid-2. (E) represents a remainder of pyridoxal or pyridoxamine phosphate-enzyme molecule. For the structural changes involved in transamination reactions see Fig. 6-7.

$$\text{L-Glutamine} + \alpha\text{-Keto acid} \rightleftharpoons \alpha\text{-Ketoglutaramic acid} + \text{Amino acid}$$

$$\alpha\text{-Ketoglutaramic acid} \xrightarrow[H_2O]{\omega\text{-Amidase}} \alpha\text{-Ketoglutaric acid} + NH_3$$

Interesting evidence has been reported indicating that α-ketoglutaramate, a normal tissue constituent, is increased in the brain in hepatic coma. Apparently, this is because of the relatively low ω-amidase level in the brain, which limits its capacity to cope with the high blood concentrations of glutamine-α-ketoglutaramic acid and NH_3 found in hepatic dysfunction and coma.

Transaminations are of great importance in the metabolism of many different amino acids and account for numerous transformations, as will be brought out from time to time. Transamination explains why the α-keto acids of analogous essential amino acids are, in certain instances, capable of substituting for these amino acids in the nutrition of rats and, perhaps, also of human beings. Since the transaminases are present in heart and liver tissues in relatively high concentrations, damage to these organs leads to transaminase leakage into the blood serum. As a result, determination of the concentration of different transaminases in blood serum is used to assess the degree of cardiac and hepatic damage.

Oxidative deamination

The process of transamination just described results in the transfer of the amino groups of at least 12 different amino acids (alanine, arginine, aspartic acid, asparagine, cysteine, glutamic acid, glutamine, isoleucine, phenylalanine, tryptophan, tyrosine, and valine) to pyruvate, oxaloacetate, or α-ketoglutarate. Ultimately, however, the amino groups are transferred to the latter and form glutamate, mainly in the cytosol of the hepatic cell. The oxidative deamination of glutamate, catalyzed by the enzyme *glutamate dehydrogenase,* then becomes the major outlet for the ammonia formed in amino acid catabolism. This process may be shown schematically as follows:

$$\text{Glutamate} \xrightleftharpoons[\text{NADP}^+ \;\curvearrowright\; \text{NADPH(H}^+)]{H_2O \;\curvearrowleft\; \text{Glutamate dehydrogenase} \;\curvearrowright\; NH_4^+} \alpha\text{-Ketoglutarate}$$

The reaction is reversible, depending on the concentrations of reactants and their removal. A second type of glutamate dehydrogenase that is NAD^+ dependent is present in hepatic mitochondria. It catalyzes mainly the forward reaction shown just above, i.e., the conversion of glutamate to α-ketoglutarate plus ammonia. It has been estimated that a major portion of the ammonia derived from amino acid degradation is formed by this reaction.

Oxidative deamination by amino acid oxidases Deamination of all naturally occuring amino acids, with the exception of serine, threonine, and the dicarboxylic and basic amino acids, can be catalyzed by amino acid oxidases of liver and kidney, primarily. One group, the L-*amino acid oxidases,* are of rather low activity. This group requires FMN (or, in some microorganisms, FAD) as a coenzyme. The reaction involved is as follows:

$$R-CH(\overset{+}{N}H_3)-COO^- \xrightleftharpoons[\text{FMN} \;\curvearrowright\; \text{FMNH}_2]{H_2O \;\curvearrowleft\; \text{L-Amino acid oxidase} \;\curvearrowright\; NH_3} R-C(=O)-COO^-$$

Liver and kidney also contain an active, FAD-dependent D-amino acid oxidase whose role in the mammalian organism is uncertain. It may provide a means for the oxidative deamination of any D-isomer that might gain entrance from bacterial (cell wall) sources into the animal organism. The enzyme also may be involved in the deamination of glycine to glyoxylic acid, HCOCOOH.

Other amines of biochemical importance, e.g., norepinephrine, epinephrine, and dopamine (Chapter 18), also can be oxidatively deaminated to form ammonia and the corresponding aldehyde by the flavoprotein enzymes monoamine and diamine oxidases.

Nonoxidative deamination The amino groups of serine, homoserine, threonine, cysteine, and possibly homocysteine can be removed nonoxidatively by a group of dehydratases. Pyridoxal phosphate serves as the coenzyme. Ammonia and the corresponding keto acid are formed. For example:

$$\underset{\text{Serine}}{HOCH_2-CH(\overset{+}{N}H_3)-COO^-} \xrightarrow{-H_2O} \underset{\text{Serine dehydratase}}{\left[CH_2=C(\overset{+}{N}H_3)-COO^- \longleftrightarrow CH_3-C(=\overset{+}{N}H_2)-COO^-\right]} \xrightarrow[-NH_3]{+H_2O} \underset{\text{Pyruvate}}{CH_3-C(=O)-COO^-}$$

Metabolic fate of ammonia formed

Since ammonia is extremely toxic to the animal organism in any significant concentration, it is immediately converted to nontoxic metabolites in preparation for reuse or excretion depending on body needs. There are three principal routes for its metabolic disposal—glutamate, glutamine, and urea formation. If any of these processes is impaired, particularly that of urea formation, the level of ammonia, specifically the ammonium ion (NH_4^+), will rise in the blood and tissues and *ammonia intoxication* may result. This condition may occur in patients with hepatomegaly and impaired liver function, since the liver is the site of the conversion of NH_4^+ to urea.

The chief symptoms of ammonia intoxication are a result of impaired function to the brain, and include nausea and vomiting, lethargy, ataxia, convulsions, coma, and even death. The mechanism by which the ammonium ion exerts its toxic effect on the brain is not clear. There may be a decrease in the formation of ATP by way of the citric acid cycle because of a diversion of excessive amounts of α-ketoglutarate to form glutamate and glutamine. The levels of these metabolites, as well as that of NH_4^+, are increased in the brain during ammonia intoxication. Alterations in neural transmission processes, such as by an increased formation of GABA from glutamate (Chapter 15), may also be involved.

Formation of glutamate The formation of glutamate is a major route for the metabolic recycling of ammonia or for its excretion. Glutamate is formed from ammonia and α-ketoglutarate, in the liver primarily, by reversal of the glutamate dehydrogenase reaction as previously described. The amino group could thus become available for the formation of other amino acids by transamination if needed.

Formation of glutamine A second major pathway for the metabolic disposal of ammonia is the formation of glutamine, chiefly in the liver but also in the kidneys and brain. Glutamine may be regarded as a temporary nontoxic storage and transport form. The reaction is:

$$\text{Glutamate } (^{-}OOC\text{-}CH_2\text{-}CH_2\text{-}CH(NH_3^+)\text{-}COO^-) + NH_3 + ATP \xrightarrow[\text{synthetase, } Mg^{++}]{\text{Glutamine}} \text{Glutamine } (H_2N\text{-}CO\text{-}CH_2\text{-}CH_2\text{-}CH(NH_3^+)\text{-}COO^-) + ADP + P_i$$

Glutamate **Glutamine**

The glutamine thus formed is distributed to the liver for conversion of the NH_3 group to urea and to the kidney for excretion of its amide-N as NH_4^+ in the urine. The enzyme *glutaminase* catalyzes the hydrolysis:

$$\text{Glutamine} + H_2O \xrightarrow{\text{Glutaminase}} \text{Glutamate} + NH_4^+ \ (\rightarrow \text{Urine})$$

Formation of urea

The formation of urea by way of carbamoyl phosphate and the "Krebs-Henseleit, ornithine-urea cycle" is a third major route for the metabolic disposition of ammo-

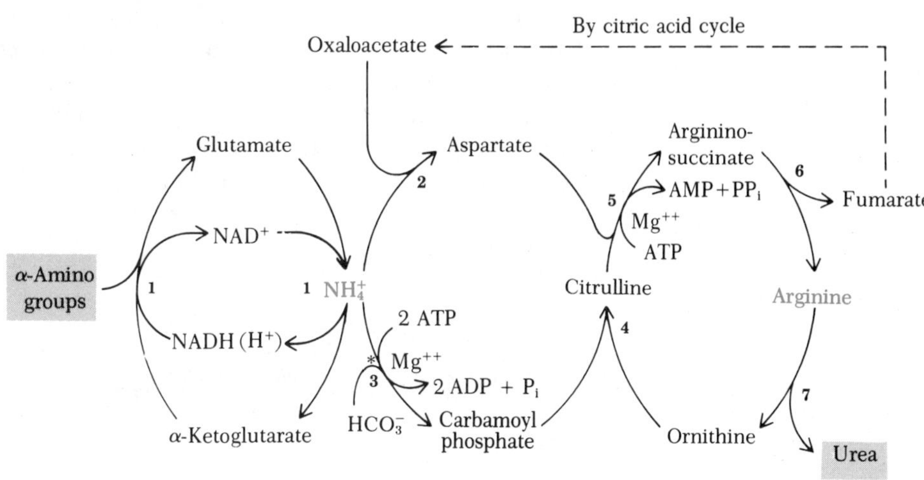

Figure 13-4
Urea cycle. Scheme of principal steps involved in formation of urea in *liver* from ammonia, derived mainly from deamination of amino acids, plus HCO_3^- from metabolic oxidations. Steps to citrulline occur in mitochondria of hepatocytes and the remainder take place in cytosol. *Enzymes involved: 1,* Glutamate dehydrogenase. *2,* Glutamate-oxaloacetate transaminase. *3,* Carbamoyl phosphate synthetase. *4,* Ornithine-transcarbamoylase. *5,* Argininosuccinate synthetase. *6,* Argininosuccinase. *7,* Arginase. *, *AGA* is *N*-acetylglutamate, which acts as a positive allosteric effector of carbamoyl phosphate synthetase.

nia. This series of reactions occurs exclusively in the liver in mammals and is shown schematically in Fig. 13-4.

The sequence of reactions involved in the biosynthesis of urea may be summarized briefly in five steps as follows:

1. Mitochondrial formation of carbamoyl phosphate. *N*-Acetylglutamate acts as a positive allosteric effector for carbamoyl phosphate synthetase in this reaction. Also, arginine appears to serve as a specific stimulator for acetylglutamate synthesis (Chapter 20). In this way, urea formation in the liver is regulated at the committed step in the pathway, as described in Chapter 3.

$$\mathrm{HCO_3^-} \xrightarrow[\mathrm{Mg^{++}},\ 2\,\mathrm{ATP}\ \rightarrow\ 2\,\mathrm{ADP};\ \oplus\ \text{Acetylglutamate}]{\mathrm{NH_4^+}\ \rightarrow\ \mathrm{P_i};\ \text{Carbamoyl phosphate synthetase}} \mathrm{H_2N{-}\underset{\underset{O}{\|}}{C}{-}O{-}\overset{\overset{O^-}{|}}{\underset{\underset{O^-}{|}}{P}}{=}O}$$

Carbamoyl phosphate

2. *Conversion of ornithine to citrulline by carbamoyl phosphate.* This mitochondrial reaction proceeds as follows:

$$\underset{\textbf{Ornithine}}{\mathrm{H_2N{-}(CH_2)_3{-}CH(\overset{+}{N}H_3){-}COO^-}} + \mathrm{H_2N{-}\underset{\underset{O}{\|}}{C}{-}O{-}\overset{\overset{O^-}{|}}{\underset{\underset{O^-}{|}}{P}}{=}O} \xrightarrow{\text{Transcarbamoylase}} \mathrm{P_i} + \underset{\textbf{Citrulline}}{\mathrm{H_2N{-}\underset{\underset{O}{\|}}{C}{-}NH{-}(CH_2)_3{-}CH(\overset{+}{N}H_3){-}COO^-}}$$

The reaction is catalyzed by the hepatic enzyme *ornithine transcarbamoylase,* which may be part of an enzyme complex with carbamoyl phosphate synthetase (step 1 in Fig. 13-4). Citrulline leaves the mitochondrion by means of a *citrulline-ornithine translocase.*

3. *Conversion of citrulline to argininosuccinate.* This reaction, as well as steps 4 and 5, occurs in the cytoplasm. It is catalyzed by the enzyme *argininosuccinate synthetase* and requires ATP and Mg^{++} ions.

$$\underset{\text{Citrulline}}{H_2N{-}C(=O){-}NH{-}CH_2{-}CH_2{-}CH_2{-}CH(\overset{+}{N}H_3){-}COO^-} + \underset{\text{Aspartate}}{^-OOC{-}CH_2{-}CH(\overset{+}{N}H_3){-}COO^-} \xrightarrow[\text{synthetase, } Mg^{++}]{ATP \quad \text{Argininosuccinate} \quad AMP + PP_i,\ H_2O} \underset{\text{Argininosuccinate}}{HN{=}C(NH{-}CH(COO^-){-}CH_2{-}COO^-){-}NH{-}CH_2{-}CH_2{-}CH_2{-}CH(\overset{+}{N}H_3){-}COO^-}$$

The formation of an adenyl-citrulline intermediate is probably involved in this reaction, as indicated by experiments employing citrulline labeled with ^{18}O in the ureido group.

4. *Cleavage of argininosuccinate to form arginine.* This step is catalyzed by *argininosuccinase* and proceeds as follows:

$$\underset{\text{Argininosuccinate}}{HN{=}C(NH{-}CH(COO^-){-}CH_2{-}COO^-){-}NH{-}(CH_2)_3{-}CH(\overset{+}{N}H_3){-}C(=O){-}O^-} \xrightarrow{\text{Argininosuccinase}} \underset{\text{Fumarate}}{H{-}C(COO^-){=}C(H){-}COO^-} + \underset{\text{Arginine}}{H_2N{-}C(=NH){-}NH{-}(CH_2)_3{-}CH(\overset{+}{N}H_3){-}C(=O){-}O^-}$$

The fumarate formed is hydrated to malate and oxidized to oxaloacetate in the citric acid cycle. The oxaloacetate is then converted to aspartate by transamination and the sequence can be repeated.

5. *Hydrolysis of arginine to urea and ornithine.* The conversion of arginine to urea and ornithine is mediated by the hydrolytic enzyme *arginase,* which is found mainly in the liver in the mammalian organism. Less active isozymes of arginase are present in the brain and kidney, but they form little or no urea from arginine. Thus urea formation from arginine is limited almost completely to the liver in mammals. This fact was demonstrated in earlier experiments with hepatectomized and hepatectomized-nephrectomized dogs. Urea did not increase in the blood of the animals in either case; in fact a decrease was found, in contrast to appropriate controls.

$$\underset{\textbf{Arginine}}{H_2N-C(=NH)-NH-(CH_2)_3-CH(\overset{+}{N}H_3)-C(=O)-O^-} + H_2O \xrightarrow{\text{Arginase}} \underset{\textbf{Ornithine}}{H_2N-(CH_2)_3-CH(\overset{+}{N}H_3)-C(=O)-O^-} + \underset{\textbf{Urea}}{H_2N-C(=O)-NH_2}$$

The ornithine formed can now pass back into the mitochondrion by way of the enzyme *ornithine translocase* and react with another molecule of carbamoyl phosphate, and the cycle can be repeated. However, there is evidence that some ornithine may be converted to glutamate and proline, but it can be replaced as required by a reversal of the reaction.

The urea thus formed is excreted in the urine (Chapter 17) and represents the final major end product of nitrogen metabolism in terrestrial mammals and also in amphibians. Birds and reptiles, on the other hand, form uric acid as the final product of amino acid metabolism, whereas in fishes ammonia itself is the terminal product and is excreted directly by the kidney.

The amount of urea excreted in the urine of the normal adult human subject averages 25 to 30 gm per day and represents about 90% of the total nitrogenous substances excreted (Chapter 17). The amount is increased or decreased in direct proportion to the amount of protein ingested, as would be expected. The excretion of ammonia, on the other hand, is quantitatively much less, amounting to only 2.5% to 4.5% of the total urinary nitrogen under ordinary conditions. However, the excretion of ammonia serves the important function of aiding in the maintenance of acid-base balance. Ammonia can be formed in the kidney, as stated previously, for combination with and excretion of excess hydrogen ions as NH_4^+ ions. Thus, the NH_4^+ content of the urine increases in an acidosis and decreases in an alkalosis. These relations are discussed in more detail in Chapter 17.

Since each step of the ornithine-urea cycle is catalyzed by a specific enzyme, one could reasonably expect that an inherited deficiency or functional defect could occur in any of them. Indeed, a deficiency or malfunction of all five enzymes of the cycle has been reported in human patients, mostly children. These are considered again briefly in Chapter 22.

Use of other substituent groups

The metabolism of amino acids is unique not only because of the transfer of the characteristic α-amino group by transamination to form other amino aicds and its removal by deamination and excretion as ammonia-urea, as just described, but also because of the presence of other metabolically reactive groups. These are used in the formation of a wide variety of essential substances, as listed in Fig. 13-2. These groups include the amidine group of arginine used for the synthesis of creatine; the amide group of glutamine and asparagine used for a variety of biosyntheses; the methyl group of methionine used for a number of methylation reactions; and "one-carbon fragments" derived from glycine, serine, histidine, and tryptophan involved in purine, pyrimidine, and other biosyntheses. Since the last two groups involve five different amino acids and are interconvertible, they are considered next. The use of the other groups will be discussed, along with that of the individual amino acid concerned.

Transmethylation and one-carbon transfer

Transmethylation The occurrence of methylated compounds in animal tissues has been known for many years. In fact, Hofmeister, in 1894, postulated that the human organism is unable to synthesize methyl groups and consequently must obtain them preformed in foods. After the discovery of methionine and proof of its essentiality in the diet in the early 1930s, du Vigneaud and his coworkers first demonstrated that its methyl group could be transferred to acceptor molecules by a process termed *transmethylation*. In this way a variety of methylated derivatives

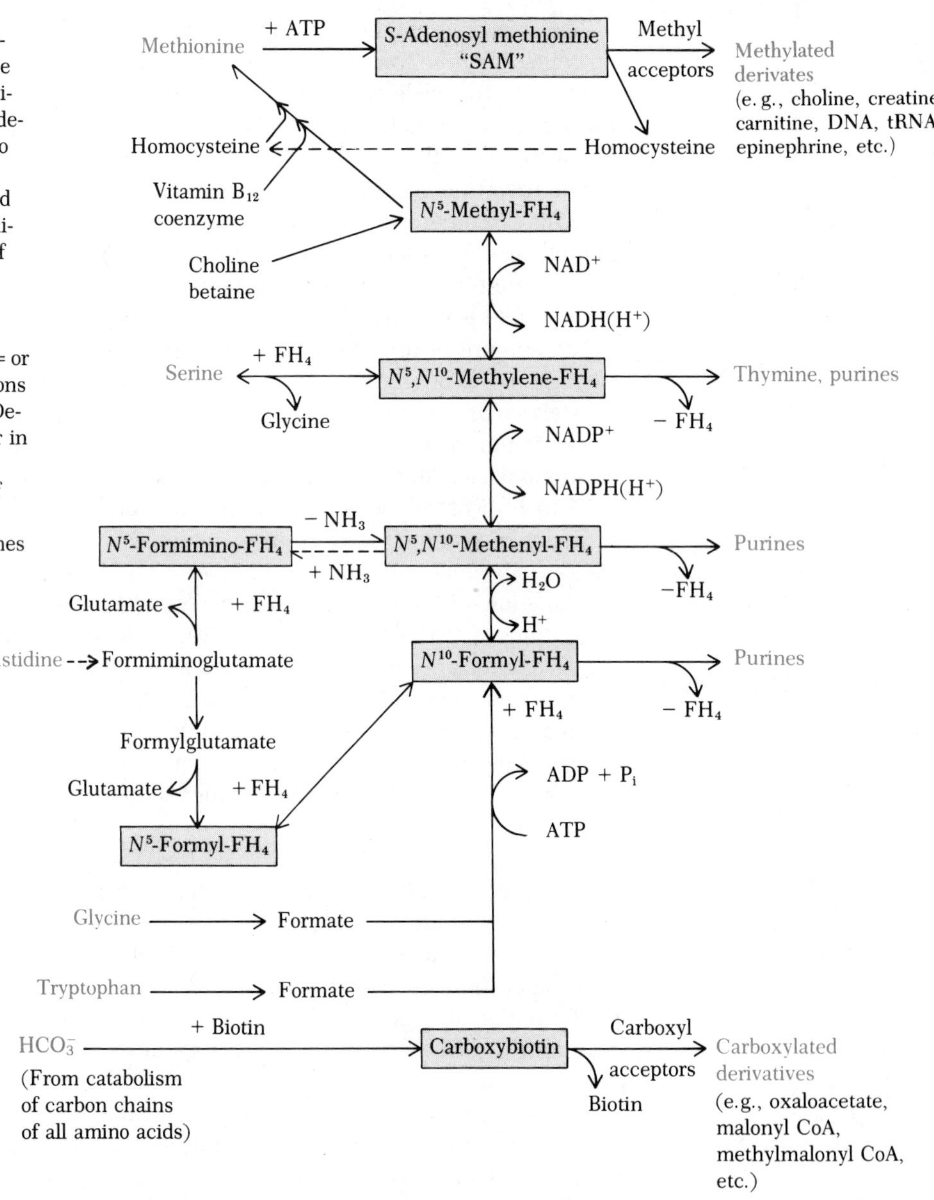

Figure 13-5
Transmethylation and one-carbon fragment transfer. Scheme of interconversions and disposition of one-carbon fragments derived from catabolism of amino acids. *S-Adenosylmethionine, tetrahydrofolic acid (FH_4)*, and *biotin* serve as carriers, as indicated. All levels of oxidation of one-carbon fragments are shown: —CH_3, methyl; —CH_2,—, methylene; =CH—, methenyl; and =C= or O=C=O, carboxy. All reactions are enzymatically catalyzed. Detailed reactions are given later in the text under the individual amino acids. The basic role of tetrahydrofolate (FH_4) with structural changes and enzymes involved is shown in Fig. 6-9.

can be formed, such as choline, creatine, betaine, and a number of others. Homocysteine is formed as a result of the removal of the methyl group of methionine.

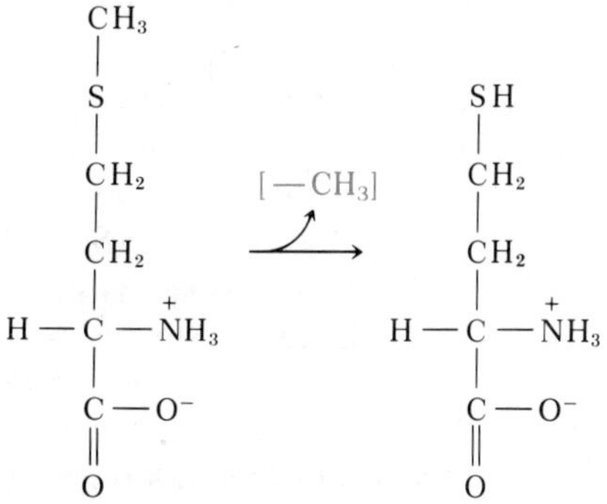

$$\underset{\textbf{Methionine}}{\begin{array}{c} CH_3 \\ | \\ S \\ | \\ CH_2 \\ | \\ CH_2 \\ | \\ H-C-\overset{+}{N}H_3 \\ | \\ C-O^- \\ \| \\ O \end{array}} \xrightarrow{[-CH_3]} \underset{\textbf{Homocysteine}}{\begin{array}{c} SH \\ | \\ CH_2 \\ | \\ CH_2 \\ | \\ H-C-\overset{+}{N}H_3 \\ | \\ C-O^- \\ \| \\ O \end{array}}$$

It was also observed that young rats fed a methionine-free diet can grow satisfactorily if either choline or betaine is administered, along with homocysteine. This observation seemed to indicate that the methyl groups of choline (from phosphatidyl choline, Chapter 12) and betaine could be transferred to an acceptor, homocysteine in this case, like those of methionine. However, subsequent studies demonstrated that this is true only to a limited degree. Homocysteine, for example, is also converted to cysteine, as is described later in this chapter. Therefore, *dietary methionine* from food proteins is the *major source of methyl groups* in the animal organism. This conclusion has been amply confirmed by more recent studies using isotopically labeled methylated compounds.

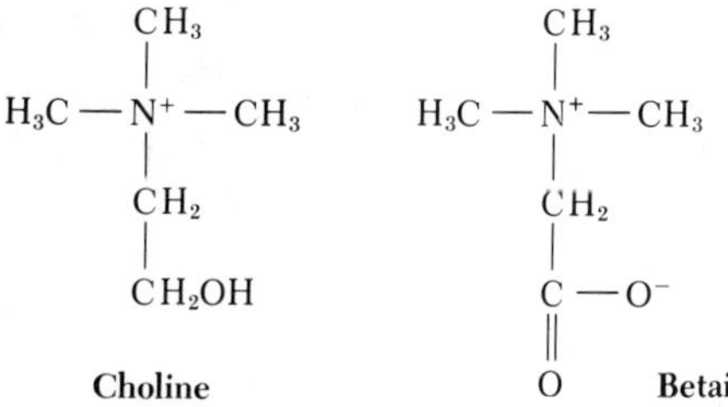

$$\underset{\textbf{Choline}}{\begin{array}{c} CH_3 \\ | \\ H_3C-N^+-CH_3 \\ | \\ CH_2 \\ | \\ CH_2OH \end{array}} \qquad \underset{\textbf{Betaine}}{\begin{array}{c} CH_3 \\ | \\ H_3C-N^+-CH_3 \\ | \\ CH_2 \\ | \\ C-O^- \\ \| \\ O \end{array}}$$

It has been established that the methyl group of methionine becomes available for transmethylation only in an *active form* of methionine, and this has been identified as SAM. ATP and the widely distributed enzyme *methionine adenosyltransferase* are required for its formation, as shown below:

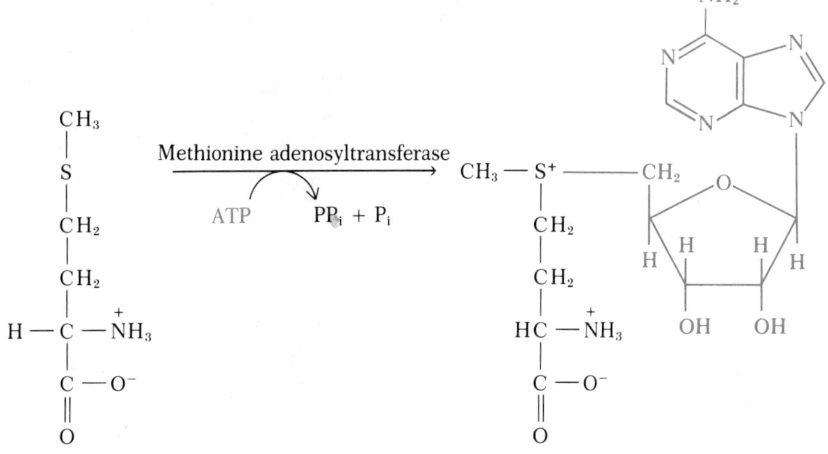

$$\underset{\textbf{L-Methionine}}{\begin{array}{c} CH_3 \\ | \\ S \\ | \\ CH_2 \\ | \\ CH_2 \\ | \\ H-C-\overset{+}{N}H_3 \\ | \\ C-O^- \\ \| \\ O \end{array}} \xrightarrow[\text{ATP} \;\curvearrowright\; PP_i + P_i]{\text{Methionine adenosyltransferase}} \underset{\textbf{S-Adenosylmethionine (SAM)}}{\begin{array}{c} CH_3-S^+-CH_2\text{-(adenosine)} \\ | \\ CH_2 \\ | \\ CH_2 \\ | \\ HC-\overset{+}{N}H_3 \\ | \\ C-O^- \\ \| \\ O \end{array}}$$

The role of methionine in transmethylation reactions by its active form, SAM, to produce methylated derivatives is shown schematically in Fig. 13-5. Specific enzymes, *methyltransferases,* are required.

Note that methionine may be formed from homocysteine and the methyl group from N^5-methyltetrahydrofolic acid (N^5-methyl-FH_4). The methyl group in turn may be derived from choline or betaine. Vitamin B_{12} (cobalamin) coenzyme (Chapter 21) is the prosthetic group of the enzyme *homocysteine methyltransferase* catalyzing the reaction. (See Fig. 13-25.)

One-carbon group transfer "One-carbon fragments" are formed from several amino acids during the course of metabolic reactions. As shown in Fig. 13-5, these include serine, glycine, histidine, and tryptophan. Also the methyl groups of choline and betaine, as well as methionine (perhaps indirectly), can be oxidized to formaldehyde and formate for use.

One-carbon groups are formed in metabolism at all levels of oxidation—from methyl ($—CH_3$), methylene ($—CH_2—$), methenyl ($=CH—$) or formyl ($O=CH—$) to carbon dioxide ($O=C=O$). The first four are used by way of *tetrahydrofolic acid* as a coenzyme "carrier," and the last by way of biotin, as will be discussed later.

Tetrahydrofolic acid (abbreviated FH_4 or THFA) is formed from folic acid (Chapter 21) by the following reaction:

$$\text{Folic acid} \xrightarrow[\text{NADPH(H}^+) \to \text{NADP}^+]{} \text{Dihydrofolic acid} \xrightarrow[\text{NADPH(H}^+) \to \text{NADP}^+]{\text{Dihydrofolate reductase}} \text{Tetrahydrofolic acid}$$

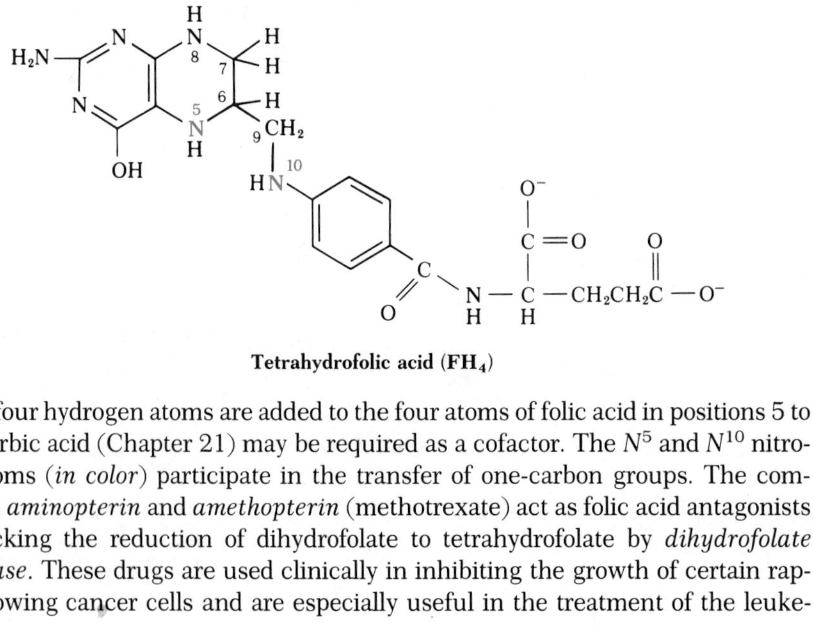

Tetrahydrofolic acid (FH_4)

The four hydrogen atoms are added to the four atoms of folic acid in positions 5 to 8. Ascorbic acid (Chapter 21) may be required as a cofactor. The N^5 and N^{10} nitrogen atoms (*in color*) participate in the transfer of one-carbon groups. The compounds *aminopterin* and *amethopterin* (methotrexate) act as folic acid antagonists by blocking the reduction of dihydrofolate to tetrahydrofolate by *dihydrofolate reductase*. These drugs are used clinically in inhibiting the growth of certain rapidly growing cancer cells and are especially useful in the treatment of the leukemias.

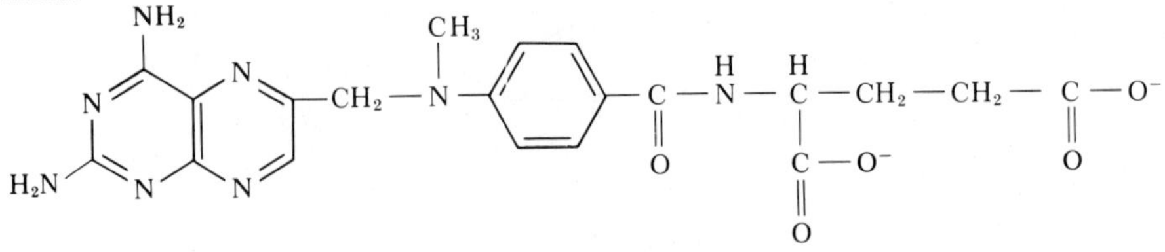

Methotrexate (amethopterin)

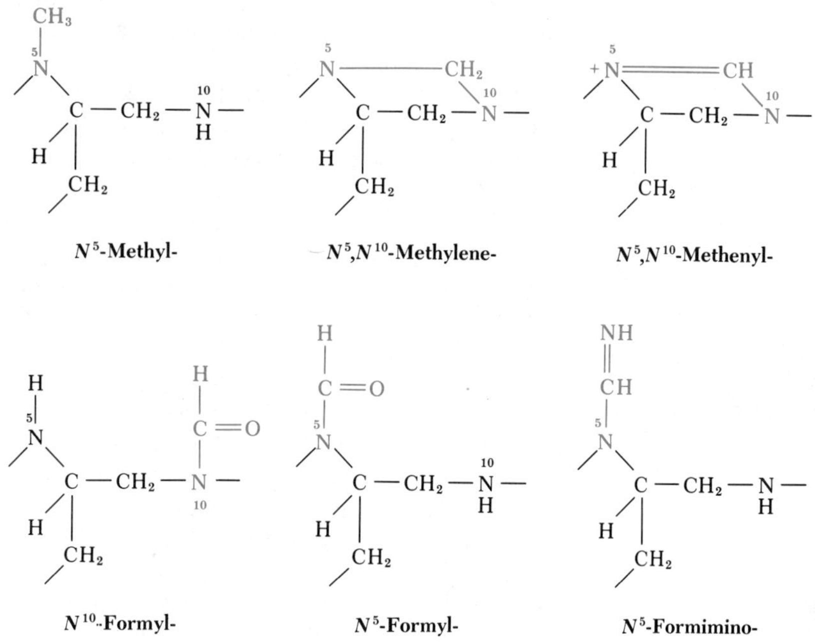

Figure 13-6
Partial structures of one-carbon group derivatives of tetrahydrofolic acid (FH_4).

N^5-Methyl- N^5,N^{10}-Methylene- N^5,N^{10}-Methenyl-

N^{10}-Formyl- N^5-Formyl- N^5-Formimino-

The three different oxidation levels of one-carbon groups listed above are transferred on either or both N^5 and N^{10} nitrogen atoms of FH_4 in Fig. 13-6 (only the reacting portions of the FH_4 molecule are shown).

As shown in Fig. 13-5, the six different one-carbon derivatives of FH_4 are interconvertible, using NAD^+ or $NADP^+$ or their reduced forms as cofactors. Appropriate enzymes are required as catalysts; the four located in the center of the figure require N^5,N^{10}-methylene-FH_4 reductase, N^5,N^{10}-methylene-FH_4 dehydrogenase, and N^5,N^{10}-methenyl-FH_4 cyclohydrase respectively (top to bottom). N^5- and N^{10}-formyl-FH_4 are interconvertible by an isomerase reaction. N^{10}-formyl-FH_4 may be formed also directly from formate in a reaction catalyzed by formyltetrahydrofolate synthetase and requiring ATP.

The conversion of serine to glycine with the formation of N^5,N^{10}-methylene-FH_4 has been studied extensively and, as an example, may be shown schematically as follows:

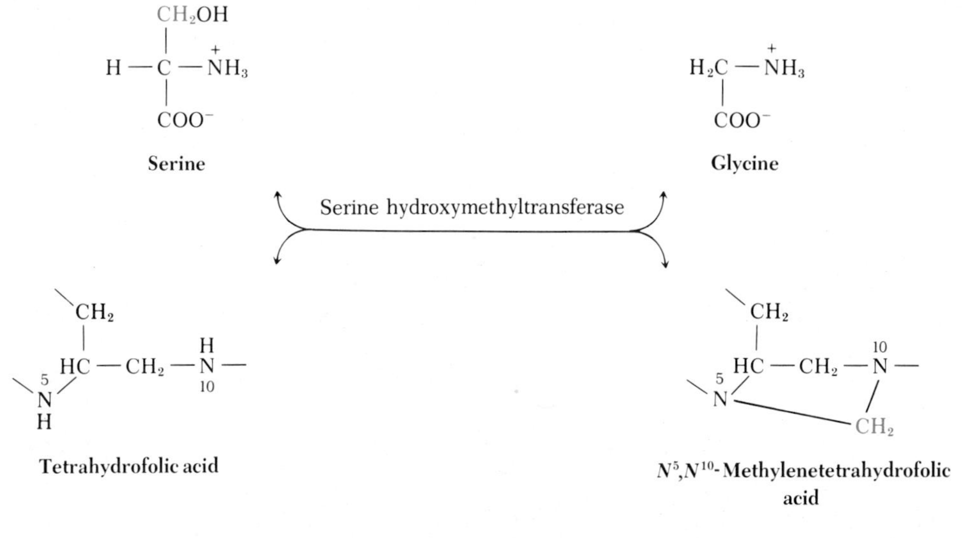

Serine + Tetrahydrofolic acid ⇌ (Serine hydroxymethyltransferase) Glycine + N^5,N^{10}-Methylenetetrahydrofolic acid

The reaction requires the participation of the enzyme *serine hydroxymethyltransferase,* a vitamin B_6–containing enzyme.

Thus one-carbon groups at different levels of oxidation are transferred and made available by way of the tetrahydrofolate and cobalamin coenzymes for use in a wide variety of vital anabolic processes. Fig. 13-5 shows several important examples, their use in the biosynthesis of thymine and purines (Chapter 14), and the interconversion of serine and glycine. Other examples are cited throughout this book. For these reasons, it is understandable why a deficiency of either folic acid or cobalamin (Chapter 21) in man or in experimental animals can lead to metabolic disorders manifested as impaired growth, macrocytic anemia, malabsorption, and other serious pathological conditions.

Transfer of carboxy groups (transcarboxylation)

The transfer of one other type of "one-carbon fragment," carboxy groups, logically may be considered at this point. Carbon dioxide derived from the decarboxylation of amino acids or from the oxidation of their carbon chains by way of the citric acid cycle reacts with biotin (Chapter 6) to form N-carboxybiotin. The fixation of CO_2, as HCO_3^-, in this reaction requires an enzyme, a *carboxylase,* specific for the acceptor to be carboxylated, and ATP (Chapter 6). The reaction apparently proceeds in the following two steps (Fig. 13-7).

As shown schematically in Fig. 13-7, important metabolic intermediates, such as malonyl CoA, methylmalonyl CoA, and oxaloacetate, are formed by transcarboxylation in the above manner. The carboxyl-group acceptors are acetyl CoA, propionyl CoA, and pyruvate respectively. The enzymes required for their carboxylation are acetyle CoA carboxylase, propionyl CoA carboxylase, and pyruvate carboxylase (Chapter 5). Recall that malonyl CoA is a key intermediate in the biosynthesis of fatty acids (Chapter 12), methyl malonyl CoA is involved in the conversion of propionyl CoA to succinyl CoA, and oxaloacetate functions in the citric acid cycle (Chapter 11).

The foregoing transcarboxylation reactions are thus important in recycling carbon atoms released in the degradation of the carbon chains of amino acids, as well as those from other metabolites, into useful metabolic channels.

Transamidination—creatine formation

The process of transamidination involves the transfer of an amidine group from a donor compound to an acceptor molecule, as the name implies. The reaction is catalyzed by the enzyme *transamidinase,* which has been prepared from kidney. Smaller amounts are present also in liver, pancreas, spleen, brain, and other tissues in several species of animals, including humans. The most important example of transamidination reactions is the formation of *creatine,* a constituent of muscle mainly in the form of the high-energy phosphate compound *phosphocreatine* (Chapter 15).

The first step in the formation of creatine is the reversible transfer of the guanidine group of arginine (Fig. 13-8) to glycine, catalyzed by a *transamidinase,* to yield guanidinoacetic acid (also called glycocyamine). This intermediate is then methylated, apparently mainly in the liver, by SAM by an irreversible transmethylation to form creatine. A methyltransferase is required as a catalyst. The foregoing steps involved in the biosynthesis of creatine from the three amino acids, glycine, arginine, and methionine, have been well established by the use of ^{14}C-labeled amino acids. The sequence of reactions may be shown diagrammatically in Fig. 13-8.

The pathway for the biosynthesis of creatine is regulated by feedback repression of enzyme synthesis. Rats and certain other experimental animals fed creatine show a sharp decrease in renal transamidinase activity.

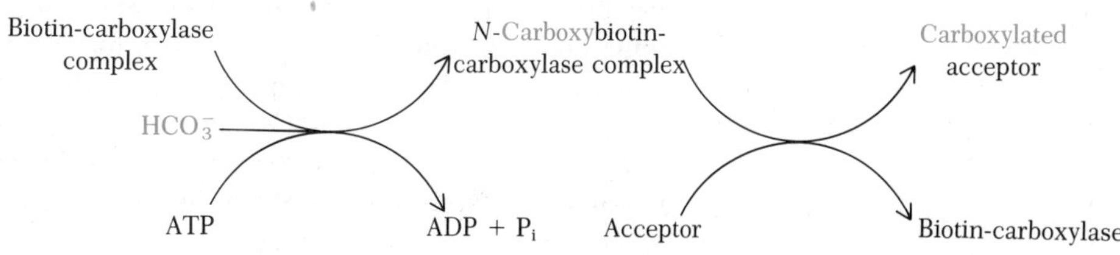

Figure 13-7
Steps involved in transcarboxylation by way of *N*-carboxybiotin. See Chapter 6 for the structural changes involved.

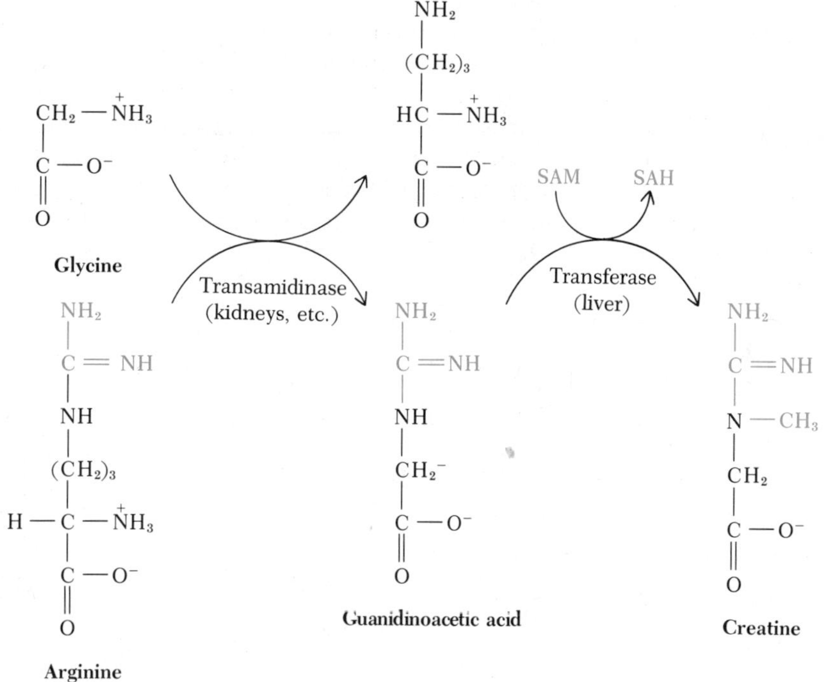

Figure 13-8
Formation of creatine from glycine, arginine, and methionine by transamidination and transmethylation, respectively.

As phosphocreatine is used in muscle (Chapters 7 and 15), the creatine moiety of the molecule is converted to its anhydride, *creatinine,* spontaneously and irreversibly. Creatinine has no known function in the body, diffuses from the muscle, and is excreted in the urine (Chapter 17) as a waste product.

Creatine ⇌ (Creatine kinase; ATP → ADP) Phosphocreatine → (Spontaneously; P_i, H_2O) Creatinine

Creatine | **Phosphocreatine** | **Creatinine**

In muscle | **Excreted in urine**

Decarboxylation

Another important general pathway of amino acid catabolism is their decarboxylation to form the corresponding amines. These reactions are catalyzed by specific amino acid *decarboxylases,* which are widely distributed in tissues. Pyridoxal phosphate is required as a coenzyme for nearly all the decarboxylases, with the possible exception of histidine decarboxylase, as is discussed later. As in the case of transamination, decarboxylation involves the formation of a Schiff base between the amino acid and pyridoxal phosphate, with the carboxyl group being removed as CO_2.

As is described in the discussion of the metabolism of the individual amino acids, a number of extremely important amines are formed as a result of amino acid decarboxylations. These include histamine, tyramine and its related compounds, serotonin (5-hydroxytryptamine), taurine, and GABA in the brain.

Transpeptidation

Although this process was originally studied as a possible means for the biosynthesis of more complex peptides, it has proved to interest more recently in the transfer of amino acids, or even of smaller peptides, from one oligopeptide to another or even to other amino acids. Of particular interest is the γ-glutamyl group that is present in glutathione and glutamine. The enzyme γ-glutamyl transpeptidase, of broad amino acid specificity, for example, is present in mammalian liver, kidney, and brain and transfers this group from glutathione to various amino acids or to other peptides. The following is an example:

$$\underset{\textbf{Glutathione}}{{}^{-}O{-}C({=}O){-}CH(\overset{+}{N}H_3){-}CH_2{-}CH_2{-}C({=}O){-}\text{Cysteinylglycine}} + \underset{\textbf{L-Leucine}}{C_4H_9{-}CH(NH_2){-}C({=}O){-}O^-} \longrightarrow \underset{\textbf{γ-Glutamyl-L-leucine}}{{}^{-}O{-}C({=}O){-}CH(\overset{+}{N}H_3){-}CH_2{-}CH_2{-}C({=}O){-}NH{-}CH(C_4H_9){-}C({=}O){-}O^-} + \textbf{Cysteinylglycine}$$

The synthesis of glutathione has been accomplished by transpeptidation in the presence of *glutathione synthetase* and ATP, as follows:

$$\text{L-Glutamate} + \text{L-Cysteine} + \text{ATP} \xrightarrow{\gamma\text{-Glutamylcysteine synthetase}} \gamma\text{-Glutamylcysteine} + \text{ADP} + P_i$$

$$\gamma\text{-Glutamylcysteine} + \text{Glycine} + \text{ATP} \xrightarrow[\text{synthetase}]{\text{Glutathione}} \text{Glutathione} + \text{ADP} + P_i$$

It has been proposed that γ-glutamyl peptides and a "γ-glutamyl cycle" are involved in amino acid transport (Chapter 19) in the renal tubule cells and possibly in other mammalian tissues. Glutathione appears to serve as the γ-glutamyl donor to the transported amino acid.

Recent investigations also suggest that glutathione may play an important role as a γ-glutamyl donor in fibrin cross-linking to form the insoluble framework of blood clots (Chapter 16).

Disposition of residual carbon chains

Although the conversion of amino acids to body tissue proteins is the major route of use of amino acids, both of exogenous and endogenous origin, as stated before, some 25% of the daily amino acid turnover undergoes oxidative degradation and must be replaced from exogenous proteins. Thus, there appears to be an obligatory requirement for amino acids to meet the normal destruction of various body proteins and of other nitrogenous substances derived from them. Indeed, in human subjects with no exogenous protein intake, i.e., total fasting, a minimum of about 5 gm of nitrogen, equivalent to approximately 30 to 35 gm of endogenous protein turnover, is excreted as urea nitrogen in the urine daily. Even after some 26 days of fasting, 2 to 3 gm of urea nitrogen is excreted per day. Thus, the process of amino acid degradation is continuous and quantitatively significant.

The ultimate pathway for the disposal of the residual carbon chains, "carbon skeletons," of amino acids, after the removal of their amino and certain other substituent groups as just described, is by way of the citric acid cycle (Chapter 11), also called an *amphibolic pathway* (Gr. *amphi,* "from all sides"). From this common, central pathway, where the catabolic and anabolic pathways of amino acids, as well as of carbohydrates (Chapter 11) and lipids (Chapter 12) meet, metabolites may be oxidized or may diverge into other metabolic pathways as the needs of the organism then dictate. Thus the carbon skeletons of some amino acids may be converted into carbohydrate derivatives and are designated by the general term as *glycogenic amino acids* (Gr. *glykys,* "sweet") or even by a somewhat more specific term, *glucogenic amino acids*. The residual carbon chains of two amino acids may be converted into intermediates of fatty acid metabolism (Chapter 12), particularly to ketone bodies, and are therefore called *ketogenic amino acids*. Portions of the residual carbon chains of several other amino acids may be converted into carbohydrate derivatives and other portions into ketone bodies. Thus these amino acids are *both* glycogenic and ketogenic. The amino acids grouped in the three categories are listed in Table 13-1.

Conversion to glucose or to ketone bodies The detailed reactions involved in the conversion of the residual α-keto acids of the various amino acids to glucose or other carbohydrate derivatives, in the case of the *glycogenic* amino acids, or to ketone bodies, in the case of the *ketogenic* amino acids, differ and are described in the considerations of the metabolism of the individual amino acids. Only a general discussion of procedures used and a summary of the results are discussed at this time.

The earlier, classic studies of the glycogenic properties of amino acids were performed on dogs or other animals made diabetic either by pancreatectomy or by the administration of alloxan or phlorizin. Such diabetic animals excrete glucose in the urine even when fasted or when fed a diet containing only protein. When all of their fat and glycogen stores have been depleted, any glucose excreted must have been derived from protein. Since the nitrogen content of proteins averages about 16%, the amount of glucose derived from protein can be calculated from determinations

Table 13-1
Glycogenic and ketogenic amino acids

Glycogenic		Glycogenic and ketogenic	Ketogenic
Alanine	Histidine	Isoleucine	Leucine
Arginine	Hydroxyproline	Phenylalanine	Lysine
Aspartic acid	Methionine	Tyrosine	
Asparagine	Proline	Tryptophan	
Cystine-cysteine	Serine		
Glutamic acid	Threonine		
Glutamine	Valine		
Glycine			

of urinary nitrogen and glucose. This relation was expressed as the glucose-to-nitrogen ratio and was found to reach a maximum of 3.65 in protein-fed, diabetic animals. Thus the amount of glucose that can be derived from 100 gm of protein can be calculated as follows:

$$\textit{Glucose}, \text{ derived from 100 gm protein} = 3.65 \times 16 = 58.4 \text{ gm, or } 58.4\%$$

Thus, approximately 58% of protein is glycogenic. This value is used today for calculating the amount of glucose that can be derived from food proteins when diets for diabetic patients are formulated. About 10% of fat (the glycerol portion) is also glycogenic.

The administration of pure amino acids to diabetic animals has been used extensively to determine those that are glycogenic. Likewise, those that are ketogenic can be indicated by ketone body formation, although the methods available are of limited sensitivity. Much more precise results are being obtained by the use of isotopically labeled amino acids and chromatograpic separation of the glucose or ketone bodies formed. Also, establishment of the catabolic pathways of the individual amino acids will indicate their glucose- or ketone body–forming potentials.

With the above procedures, a classification of amino acids based on their glycogenic or ketogenic properties, as given in Table 13-1, has been fairly well established. Note that several amino acids are glycogenic under some conditions but ketogenic under others. Observe also that lysine has been listed as ketogenic on the basis of the more recent isotopic studies.

Final oxidation by way of the citric acid cycle The points of entry of the various amino acids into the citric acid cycle are indicated in Fig. 13-9. Upon incorporation

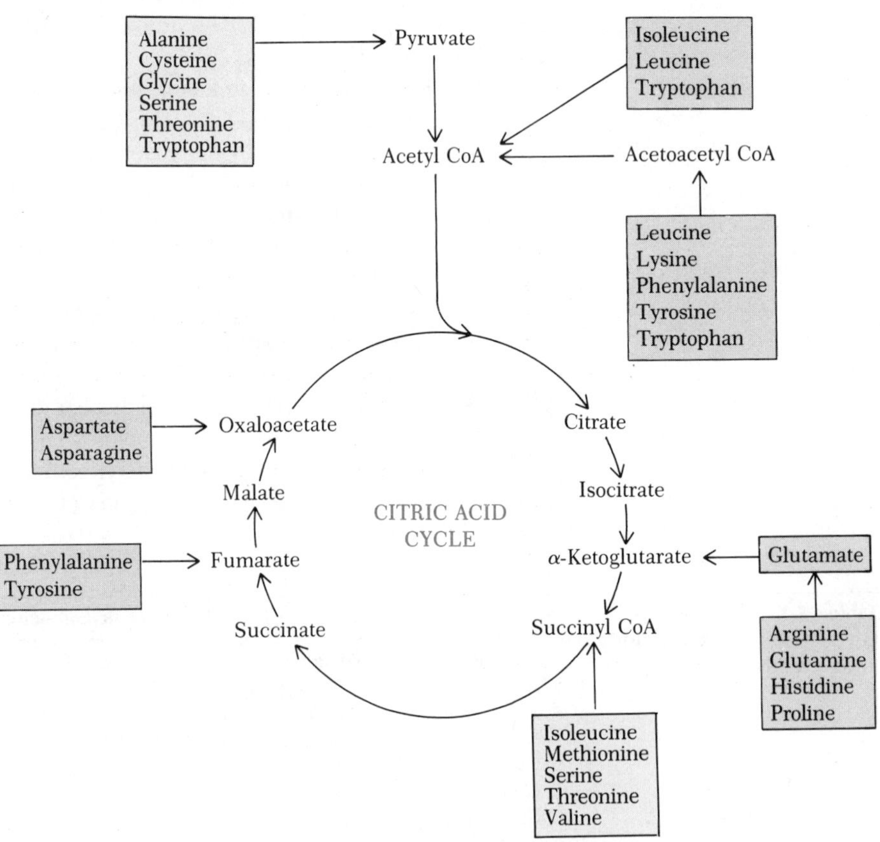

Figure 13-9
Diagram of catabolic disposition of carbon chains, derived from amino acids, by way of citric acid cycle. Some of these reactions are reversible and thus provide general pathways for resynthesis of certain amino acids by transamination, as will be described later under the metabolism of the individual amino acids.

into the cycle, the residual carbon chains of the amino acids are oxidized to CO_2, with the formation of ATP at several points in the cycle (Chapter 11). Amino acids yield approximately 4 calories per gram and thus supply about 15% of the daily caloric requirement of the average normal human adult (see Chapter 21).

Metabolism of individual amino acids

Most of the discussion of amino acid metabolism up to this point has been concerned with amino acids as a class. We now need to consider them individually. They will be taken up arbitrarily in the order of neutral, acidic, and then basic amino acids. The problem to be studied is how the amino acids are used by the liver or other tissues, when they are not being converted into tissue proteins and, further, what their special functions and uses are.

The question of the biosynthesis of individual amino acids is mentioned but is not treated in any detail in this chapter. Only schematic diagrams with the precursors and major intermediates involved are shown. Enzymes and, in some cases, cofactors concerned are not included, since they are uncertain as yet in some instances. Formulas are usually not given, since many of them would be repetitions of those included in the degradative pathways. For further information, refer to the suggested readings at the end of the chapter, particularly those of Greenberg and of Meister. In general, the biosynthesis of the nonessential amino acids may involve a reversal of the catabolic pathways to be described. Usually different enzyme systems are required, however. Transamination of the corresponding α-keto acid with glutamic acid is usually the final step in the biosynthesis of the nonessential acids. The carbon chain "backbone" of the molecule is usually an intermediate of the glycolytic pathway or of the citric acid cycle. The same is true in a general way of the essential amino acids. These, of course, cannot be synthesized in the animal organism, but their biosynthetic pathways and the enzymes and cofactors involved in most instances are fairly well documented in microorganisms.

Since a large number of enzymatically catalyzed reactions are involved in the metabolism of typical amino acids, there are an equal number of possibilities of inherited disorders of amino acid metabolism. These may be *direct* because of a single hereditary enzyme deficiency affecting the metabolism or transport of an amino acid. As a result, an increase in the plasma level and urinary excretion (aminoaciduria) of the amino acid or of one of its metabolites, depending on the site of the deficiency, can occur. Disorders of amino acid metabolism also may be *secondary* (or "indirect") because of some other disease process, e.g., in the liver or kidney, impairing metabolism. Secondary disorders are either hereditary or acquired. Usually they are characterized by a generalized rather than a specific aminoaciduria, as is true of the primary type. Typical examples are the generalized aminoacidurias seen in galactosemia, Wilson's disease, and a number of conditions (burns, poisonings, etc.) in which there is extensive tissue damage, especially of the liver or kidney.

Nearly 100 different primary hereditary disorders of amino acid metabolism have been described in the literature. These are discussed in Chapter 22.

Glycine Glycine may be oxidatively deaminated by glycine oxidase, an enzyme present in liver and kidney tissue.

$$H\!-\!\underset{\underset{\underset{\displaystyle O}{\|}}{\displaystyle C-O^-}}{\overset{\overset{\displaystyle H}{|}}{C}}\!-\!\overset{+}{N}H_3 \xrightarrow[O_2 \;\;\to\;\; \overset{+}{N}H_3]{H_2O \;\;\text{Glycine oxidase}\;\; H_2O_2} \underset{\underset{\underset{\displaystyle O}{\|}}{\displaystyle C-O^-}}{\overset{\overset{\displaystyle O}{\|}}{C}}\!-\!H$$

Glycine — Glyoxylic acid

The enzyme is a flavoprotein and apparently is identical to D-amino acid oxidase. Glyoxylic acid may be decarboxylated to yield formaldehyde and carbon dioxide, both of which take part in many biochemical reactions, as has been described. Glyoxylate may also be converted to malate and then metabolized by the citric acid cycle or may be oxidized to oxalate and excreted. This latter pathway may be involved in the genetic disorder *primary hyperoxaluria,* characterized by a high urinary level of oxalate. By isotope experiments, glycine has also been shown to be converted to acetic acid.

A further important pathway involves the participation of glycine in the synthesis of serine, which then may be converted to pyruvic acid. The enzyme *serine hydroxymethyltransferase* is required for this reaction.

$$\underset{\substack{|\\ C-O^- \\ \| \\ O}}{CH_2-\overset{+}{N}H_3} \xrightleftharpoons[N^5,N^{10}\text{-}CH_2FH_4 \quad FH_4]{H_2O} H-\underset{\substack{|\\ C-O^- \\ \| \\ O}}{\overset{\substack{CH_2OH\\|}}{C}}-\overset{+}{N}H_3$$

^{14}C- or ^{15}N-labeled serine administered to rats gives rise to labeled glycine, showing that the reaction is reversible in vivo. This reversible reaction occurs in the liver and requires as a coenzyme a pteroylglutamic acid derivative and pyridoxal phosphate. In fact it is not as simple a reaction, as indicated above. As described earlier, pyridoxal phosphate forms an addition product, a Schiff's base, with glycine. The folic acid derivative is the N^5,N^{10}-methylene derivative of FH_4, which donates its methylene group to the glycine fraction. Serine is finally split off from this complex.

Glycine, as a nonessential amino acid, may be readily synthesized in the animal organism from serine, as indicated above. By way of serine, it may be formed from 3-phosphoglycerate (Chapter 11).

Although glycine is the simplest of the amino acids chemically and is nonessential, it performs perhaps the most biochemical functions of any amino acid. The following include some of its special functions, which are, of course, in addition to its role as a constituent of various body tissue proteins, protein hormones, and enzymes. It is a prime precursor, along with succinyl CoA, of the porphyrins and heme (Chapter 16). It contributes to the biosynthesis of purines (Chapter 14) and is glycogenic, thus contributing indirectly by glycolytic intermediates to lipid synthesis, carbohydrate formation, and the syntheses of other amino acids. By way of glyoxylate it may form formate, which by complexing with FH_4 may be incorporated into a wide variety of biologically active compounds.

Glycine is one of the substances necessary for the formation of creatine, and creatine is essential in muscle function (Chapter 15). Glycine is also a part of the bile acid glycocholic acid and the tripeptide glutathione, which itself performs numerous important functions.

When benzoic acid or its salts are included in the diet, glycine conjugates with them to form hippuric acid, a detoxication reaction (Chapter 17). It occurs even when no glycine is present in the diet; hence glycine is readily obtainable from other amino acids in vivo.

Several hereditary diseases of glycine metabolism have been described. These are discussed in Chapter 22.

Alanine It was shown previously that, by transamination, alanine can be converted to pyruvic acid and can be metabolized by way of its pathway. As has been seen, alanine fed to diabetic animals is converted to glucose. In normal animals it is similarly converted but is then used. Alanine is thus a glycogenic amino acid. In fact, recent studies in fasting human subjects and animals have demonstrated that alanine plays a key role in hepatic gluconeogenesis. Alanine is released from muscle during fasting and undergoes transamination to form pyruvate, as described earlier. The pyruvate is then converted to glucose by the gluconeogenic pathway given in Chapter 11. Supporting evidence is the fact that the hypoglycemia found in fasting parallels the low blood alanine levels concomitantly observed. Thus alanine now appears to be of major importance in maintaining the fasting blood glucose level by way of gluconeogenesis.

Figure 13-10
Metabolic degradation of valine.

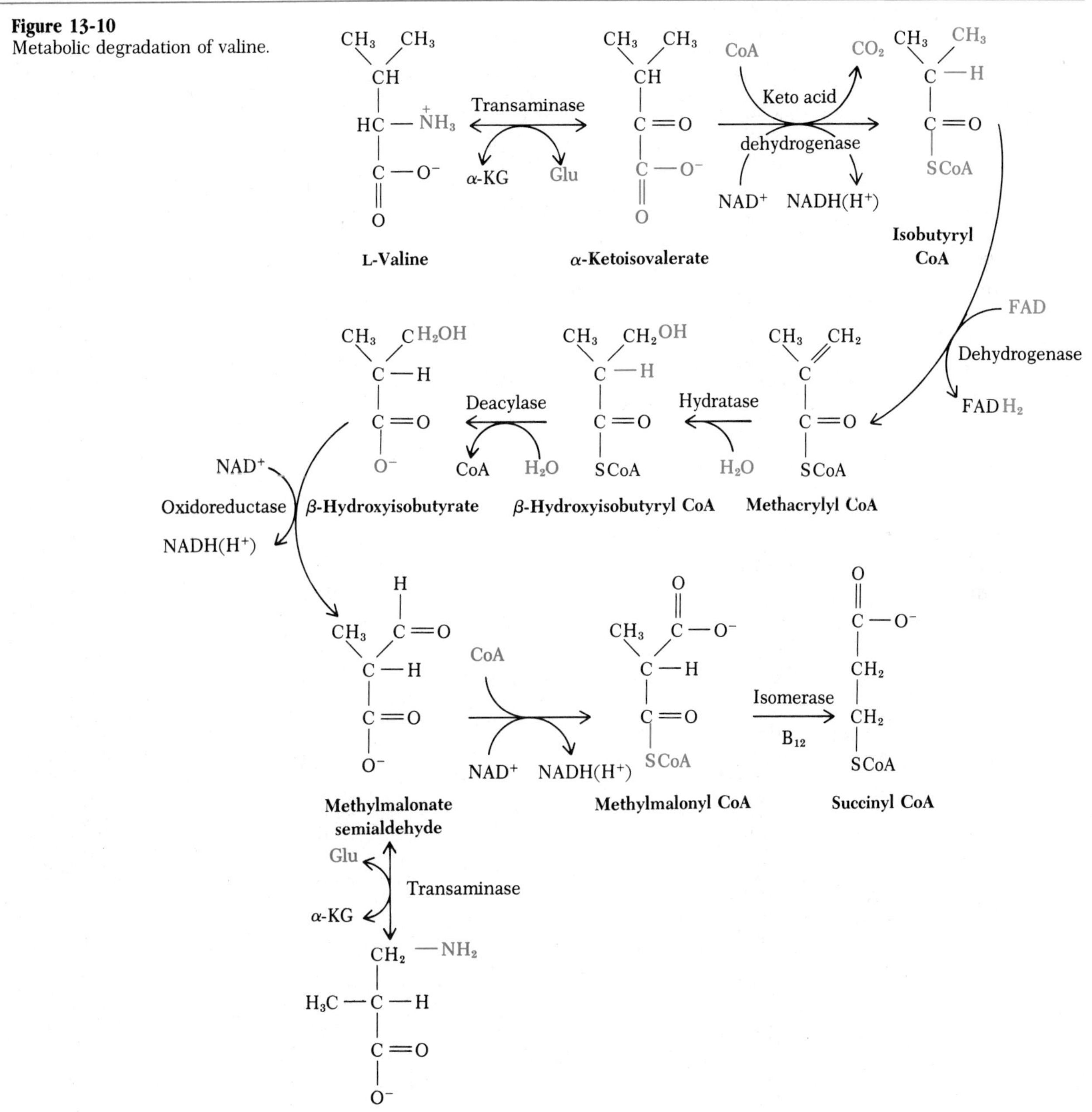

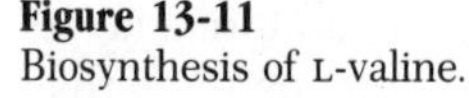

Figure 13-11
Biosynthesis of L-valine.

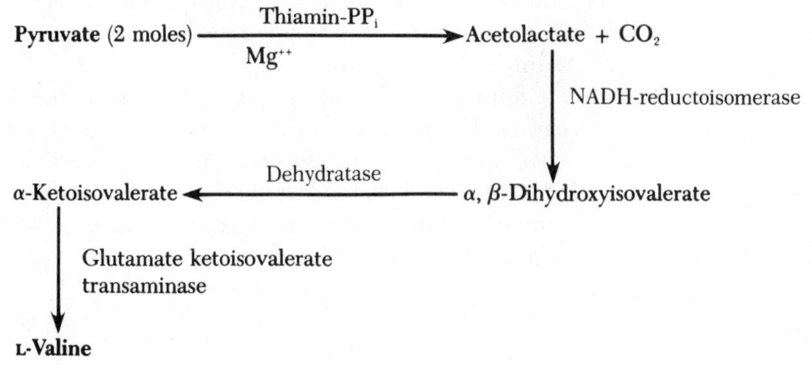

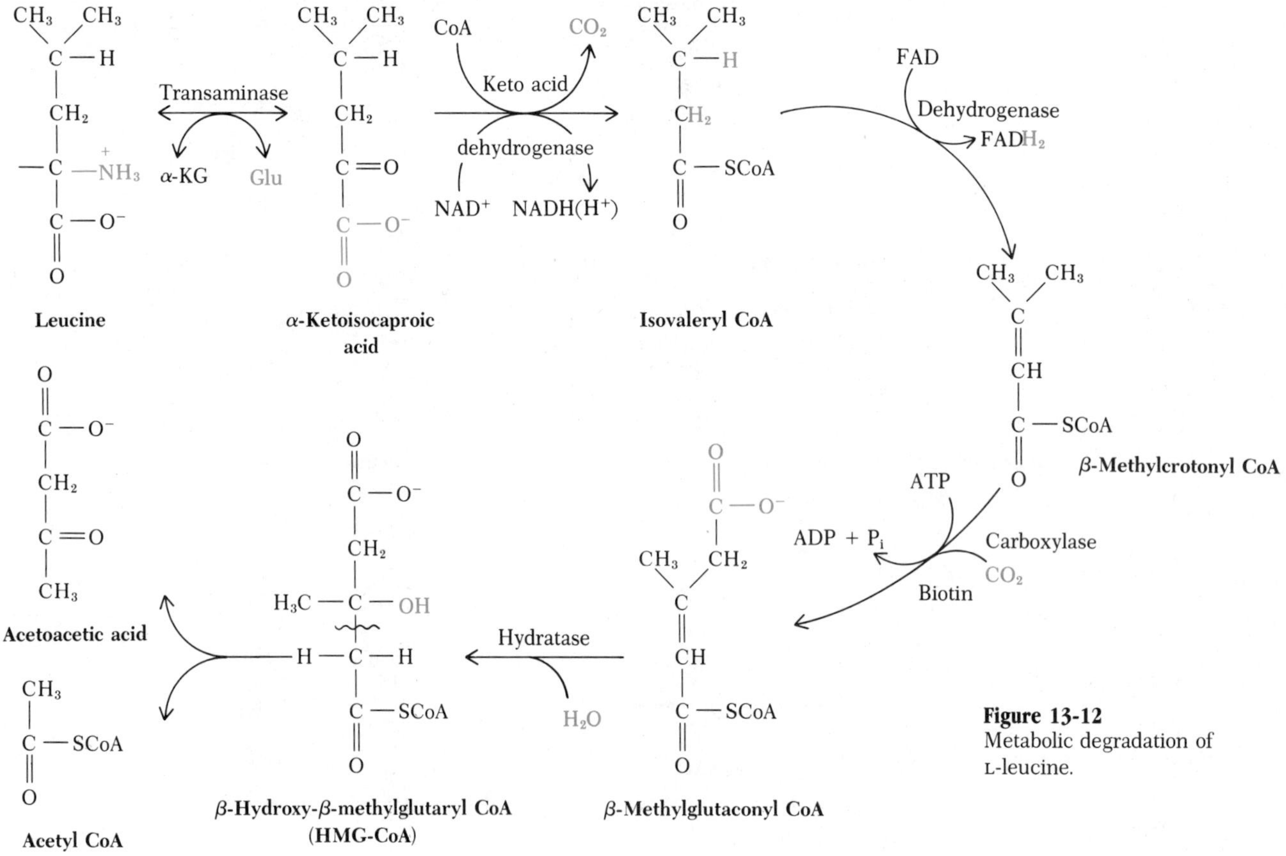

Leucine ⇄ (Transaminase; α-KG → Glu) α-Ketoisocaproic acid → (Keto acid dehydrogenase; CoA, NAD^+ → CO_2, $NADH(H^+)$) Isovaleryl CoA → (FAD → $FADH_2$; Dehydrogenase) β-Methylcrotonyl CoA → (ATP → ADP + P_i; CO_2; Carboxylase; Biotin) β-Methylglutaconyl CoA → (Hydratase; H_2O) β-Hydroxy-β-methylglutaryl CoA (HMG-CoA) → Acetoacetic acid + Acetyl CoA

Leucine: $(CH_3)_2CH-CH_2-CH(NH_3^+)-COO^-$

α-Ketoisocaproic acid: $(CH_3)_2CH-CH_2-C(=O)-COO^-$

Isovaleryl CoA: $(CH_3)_2CH-CH_2-C(=O)-SCoA$

β-Methylcrotonyl CoA: $(CH_3)_2C=CH-C(=O)-SCoA$

β-Methylglutaconyl CoA: $^-OOC-CH_2-C(CH_3)=CH-C(=O)-SCoA$

β-Hydroxy-β-methylglutaryl CoA (HMG-CoA): $^-OOC-CH_2-C(OH)(CH_3)-CH_2-C(=O)-SCoA$

Acetoacetic acid: $^-OOC-CH_2-C(=O)-CH_3$

Acetyl CoA: $CH_3-C(=O)-SCoA$

Figure 13-12
Metabolic degradation of L-leucine.

Figure 13-13
Biosynthesis of L-leucine.

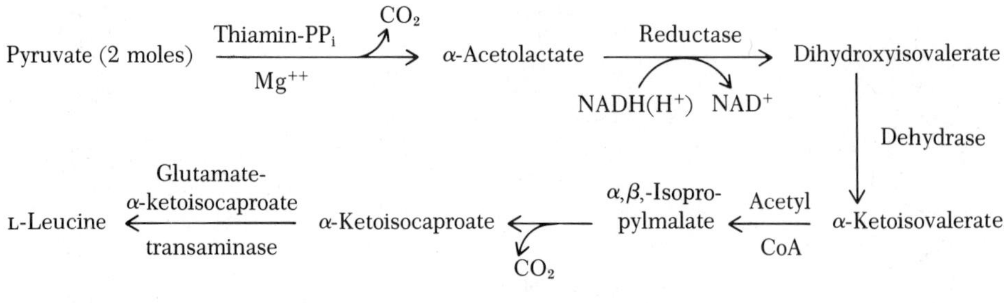

Alanine is a nonessential amino acid, since it is readily synthesized in the body by the transamination of pyruvate by glutamate or other amino acids. It is formed in plants and microorganisms apparently in this same manner.

Valine The metabolic degradation of valine begins with a transamination yielding α-ketoisovaleric acid. The subsequent fate of this is pictured in Fig. 13-10, based on isotope experiments. Note that isobutyryl CoA, rather than isobutyric acid, is formed in the decarboxylation of α-ketoisovaleric acid. This step is analogous to the conversion of pyruvate to acetyl CoA (Chapter 11). Isobutyryl CoA is converted in several steps to methylmalonyl CoA and finally to succinyl CoA by steps discussed in Chapter 12. Succinyl CoA can be metabolized by way of the citric acid cycle and other pathways, as described earlier. It may also be converted to glucose by gluconeogenesis, accounting for the glycogenic effect of valine.

Methylmalonate semialdehyde, an intermediate in valine catabolism, may also be derived from β-aminoisobutyric acid (derived from pyrimidine metabolism, Chapter 14) by transamination with α-ketoglutaric acid. A lack of this transaminase may result in the appearance of large amounts of β-*aminoisobutyric acid* (BAIB) in the urine of affected subjects. However, the condition is benign. Other inborn errors of valine metabolism are considered in Chapter 22.

The biosynthesis of valine, which is an essential amino acid, is limited to plants and microorganisms. The principal steps in its formation, in *Salmonella*, for example, are shown in Fig. 13-11.

Leucine and isoleucine Leucine and isoleucine undergo transamination to form

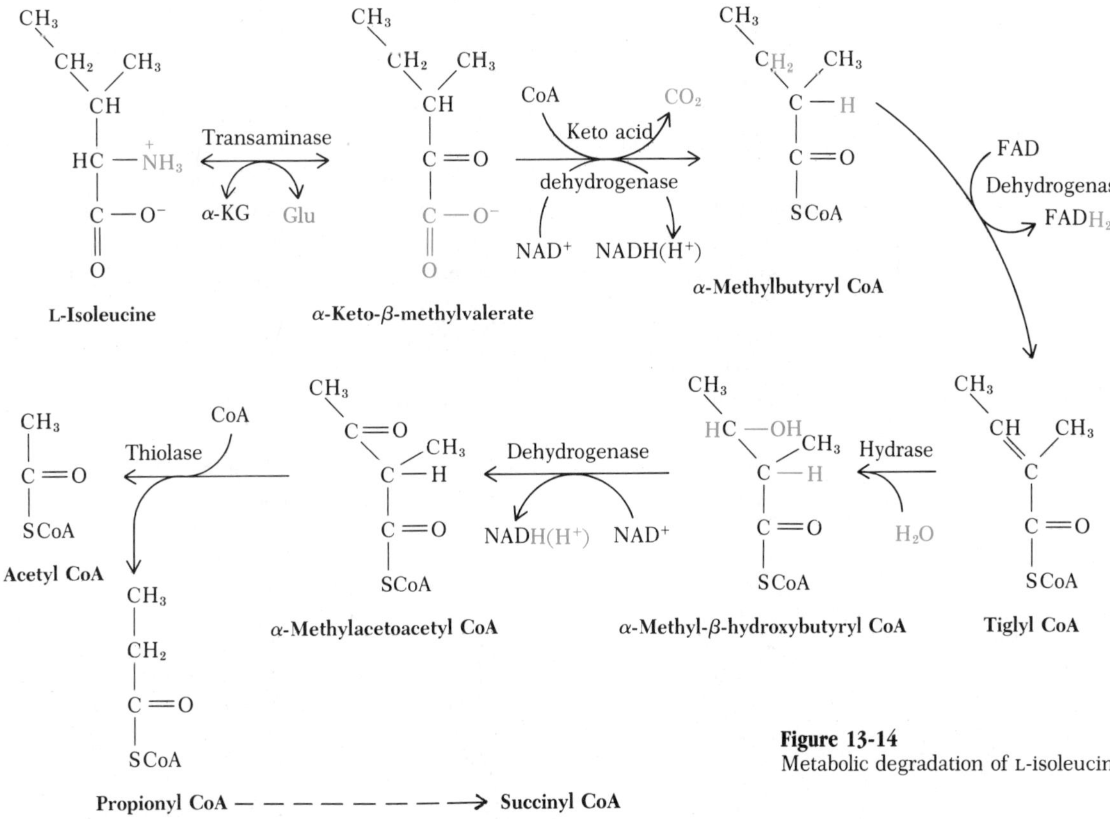

L-Isoleucine
Transaminase
α-KG
Glu
α-Keto-β-methylvalerate
CoA
CO_2
Keto acid dehydrogenase
NAD^+
$NADH(H^+)$
α-Methylbutyryl CoA
FAD
Dehydrogenase
$FADH_2$
Tiglyl CoA
Hydrase
H_2O
α-Methyl-β-hydroxybutyryl CoA
Dehydrogenase
$NADH(H^+)$
NAD^+
α-Methylacetoacetyl CoA
Thiolase
CoA
Acetyl CoA
Propionyl CoA — — — — — — — → Succinyl CoA

Figure 13-14
Metabolic degradation of L-isoleucine.

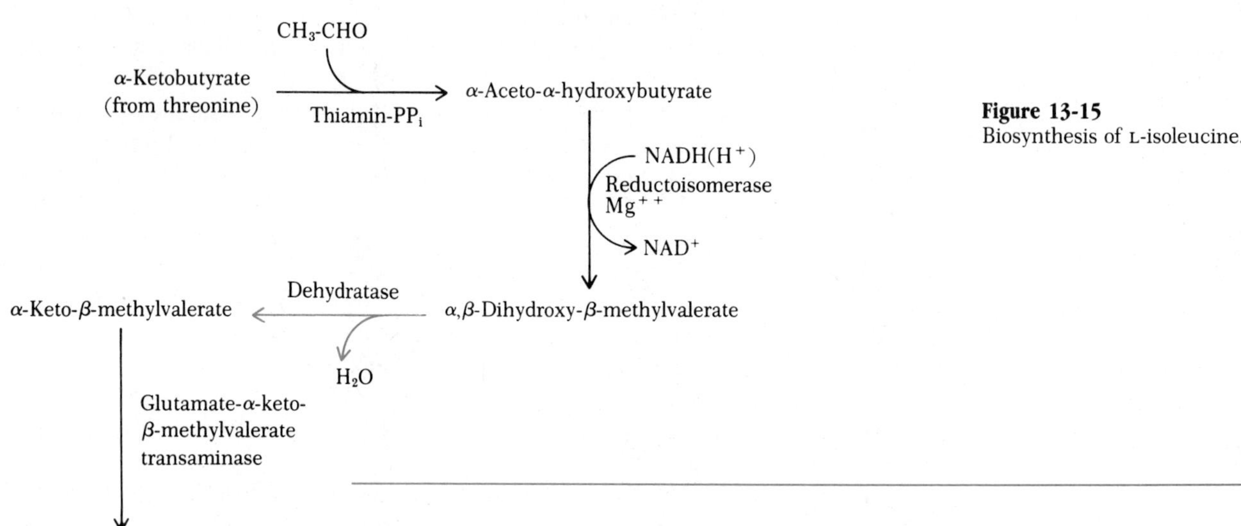

Figure 13-15
Biosynthesis of L-isoleucine.

the corresponding keto acids. It appears that the further degradation of leucine follows a break in the carbon chain between the β- and γ-carbons. Both fragments yield acetoacetate. Investigators determined this by administering leucine containing ^{14}C in various positions in the molecule to animals, or incubating it with liver slices, and finding the exact location of the tagged carbon of the acetoacetate formed in each case (Fig. 13-12). Obviously these reactions explain the origin of acetoacetic acid from L-leucine and reaffirm the fact that it is a ketogenic amino acid. The metabolic fate of acetoacetic acid is discussed in Chapter 12.

Note in Fig. 13-12 that HMG-CoA is an intermediate in the catabolism of L-leucine. HMG-CoA is also an obligatory intermediate in the biosynthesis of cholesterol (Chapter 12) and in ketone body formation from acetyl CoA (Chapter 12).

Leucine is also an essential amino acid. It is synthesized in microorganisms and plants, apparently by the pathway shown in Fig. 13-13.

The metabolism of isoleucine (Fig. 13-14) is analogous to that of leucine. After transamination, the α-keto acid formed, α-keto-β-methylvaleric, is converted in several steps to propionyl CoA and acetyl CoA. These two products account for the weak ketogenic and glycogenic properties of isoleucine, mentioned previously. The propionyl CoA is converted to methylmalonyl CoA and then succinyl CoA in several steps, as indicated previously (Chapter 12).

Isoleucine is an essential amino acid. It is synthesized in microorganims as shown in Fig. 13-15.

A block in the metabolism of leucine and isoleucine results in several inborn errors of metabolism in infants. These are discussed in Chapter 22.

Serine The anaerobic deamination of serine has been demonstrated. In this case, the reaction differs from the general one considered heretofore:

$$\underset{\text{(as pyridoxal complex)}}{\underset{\text{Serine}}{H{-}\overset{\overset{H}{|}}{C}{-}OH \atop H{-}C{-}\overset{+}{N}H_3 \atop {-}C(=O){-}O^-}} \xrightarrow[\text{dehydratase}]{\text{Serine}\ \nearrow H_2O} \left[\begin{matrix} CH_2 \\ \| \\ C{-}\overset{+}{N}H_3 \\ | \\ C{-}O^- \\ \| \\ O \end{matrix} \longleftrightarrow \begin{matrix} CH_3 \\ | \\ C{=}\overset{+}{N}H_2 \\ | \\ C{-}O^- \\ \| \\ O \end{matrix} \right] \xrightarrow[\searrow NH_3]{H_2O} \underset{\text{Pyruvic acid}}{\begin{matrix} CH_3 \\ | \\ C{=}O \\ | \\ C{-}O^- \\ \| \\ O \end{matrix}}$$

It is found to occur in liver and in kidney and is catalyzed by the pyridoxal-dependent enzyme *serine dehydratase*. The possibility that β-hydroxypyruvic acid might be produced by an oxidative deamination is not excluded. Both pyruvic acid and β-hydroxypyruvic acid could easily enter into metabolism, and pyruvic acid likewise would be formed in the transamination of serine. Isotope experiments indicate that the carbon chain of serine can be converted to cysteine, tryptophan, and certain lipids (Chapter 12).

Serine and glycine are interconvertible, as discussed earlier. This relation has been verified by "double-labeling" isotope experiments in rats and guinea pigs.

The activity of serine-threonine dehydratase in liver is increased by the administration of adrenal cortical steroids and glucagon (Chapter 18).

$$\underset{\text{Glycine}}{H-CH(\overset{+}{N}H_3)-{}^{13}C(=O)-O^-} + \underset{\text{Formate (as } FH_4 \text{ derivative)}}{H-{}^{14}C(=O)-O^-} \longrightarrow \underset{\text{Serine}}{H_2{}^{14}C(OH)-CH(\overset{+}{N}H_3)-{}^{13}C(=O)-O^-}$$

Serine may also be formed as indicated in Fig. 13-16. Serine is a constituent of one of the phospholipids found in brain. In addition, it can give rise to ethanolamine, which is one of the constituents of another phospholipid (Chapter 12).

$$\underset{\text{Serine}}{CH_2OH-CH(\overset{+}{N}H_3)-C(=O)-O^-} \xrightarrow{\quad} CO_2 + \underset{\text{Ethanolamine}}{CH_2OH-H_2C-NH_2}$$

Serine also takes part in the biological synthesis of tryptophan in the mold *Neurospora*.

Threonine Threonine is handled similarly to serine, since it also is a hydroxy amino acid. Deamination catalyzed by *threonine dehydratase* occurs first.

The α-ketobutyrate thus formed is converted to propionyl CoA, which forms succinyl CoA as previously shown (Chapter 12). Threonine is thus a glycogenic amino acid.

The aminoacetone formed by the oxidative decarboxylation of the threonine can be transaminated to α-ketopropanol, then oxidized to pyruvate, and thus catabolized. Or it can be converted by an oxidoreductase in the presence of NADH to D-*1-amino-2-propanol*. This substance is used in the synthesis of vitamin B_{12} by *Streptococcus griseus*.

By an indirect method, another path for threonine has been indicated. It is an oxidation of the β-carbon, preliminary to a splitting of the four-carbon chain in half.

The enzyme that catalyzes this reaction requires pyridoxal phosphate as coenzyme:

$$\underset{\textbf{Threonine}}{CH_3-CH(OH)-CH(\overset{+}{N}H_3)-COO^-} \xrightarrow[\text{aldolase}]{\text{Threonine}} \underset{\textbf{Acetaldehyde}}{CH_3-CHO} + \underset{\textbf{Glycine}}{H_2C(\overset{+}{N}H_3)-COO^-}$$

Thus threonine, like serine, participates in many of the reactions of glycine. Threonine was the last of the 10 essential amino acids to be discovered (Chapter 21). It is synthesized in microorganisms, as shown in Fig. 13-18.

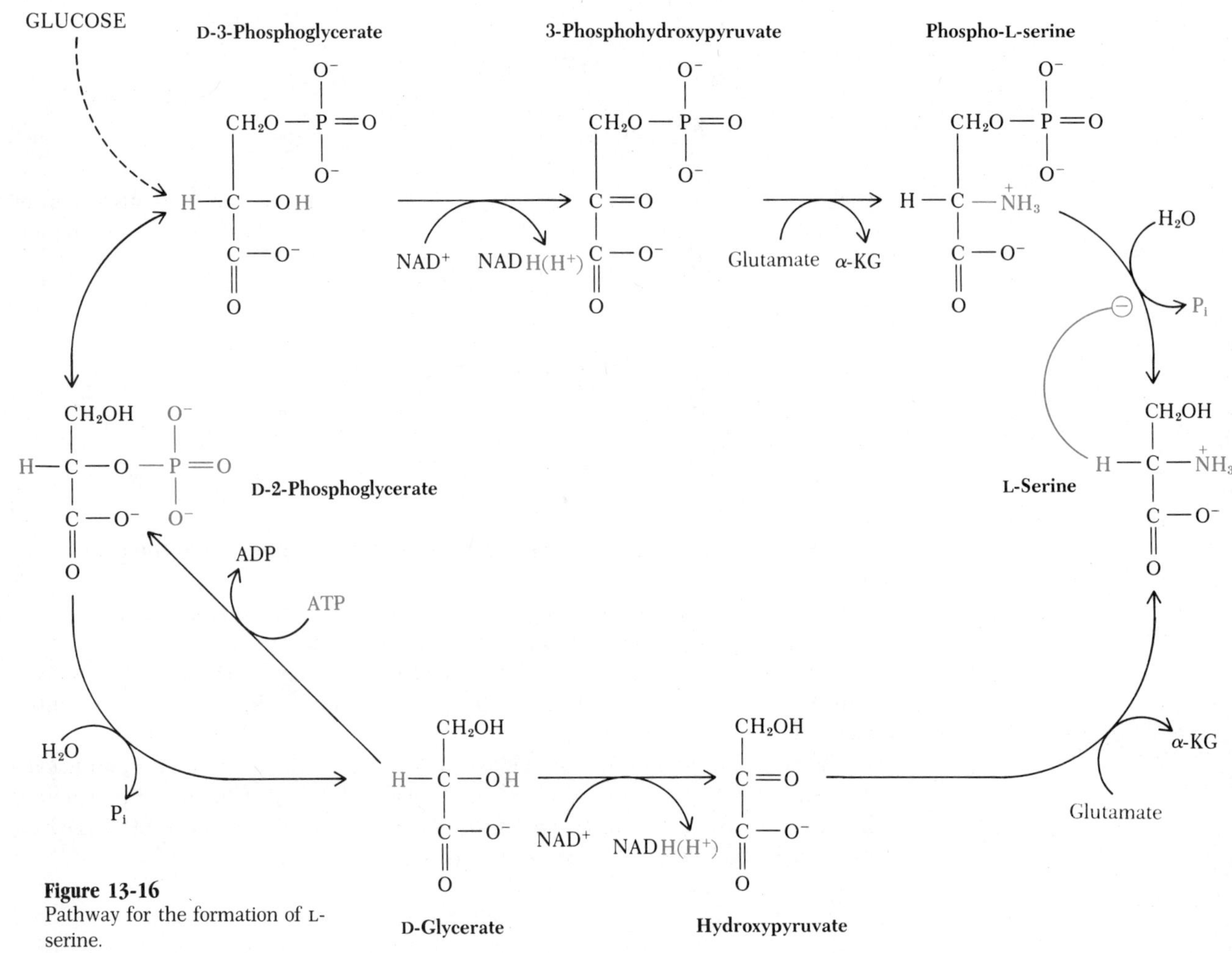

Figure 13-16
Pathway for the formation of L-serine.

Phenylalanine and tyrosine Under usual conditions, the phenylalanine present in the body's amino acid pool, from either exogenous or endogenous sources, that is not required for tissue protein formation is converted primarily to tyrosine (Fig. 13-19). This reaction is catalyzed by the "mixed-function" oxygenase *phenylalanine hydroxylase,* which requires the presence of molecular oxygen and the cofactor *tetrahydrobiopterin* (Chapter 6), a homolog of the pteridine portion of the folic acid molecule (Chapter 21). Tetrahydrobiopterin also serves as a cofactor for *tyrosine hydroxylase,* which catalyzes the formation of L-dopa (3,4-dihydroxyphenylalanine), and *tryptophan hydroxylase*, involved in the formation of 5-hydroxytryptophan and serotonin. These reactions are discussed later.

The oxidized pteridine (dihydrobiopterin) is converted back to the reduced form by an auxiliary enzyme, dihydropteridine reductase, with NADPH as a required cofactor. The reaction is not reversible. Hence, phenylalanine is an *indispensable* amino acid, whereas tyrosine is not, although tyrosine can satisfy part of the requirement for phenylalanine (Chapter 21). The primary reason for the essentiality of phenylalanine is the fact that man and most other animal organisms cannot synthesize the aromatic ring.

Figure 13-17
Metabolic degradation of threonine.

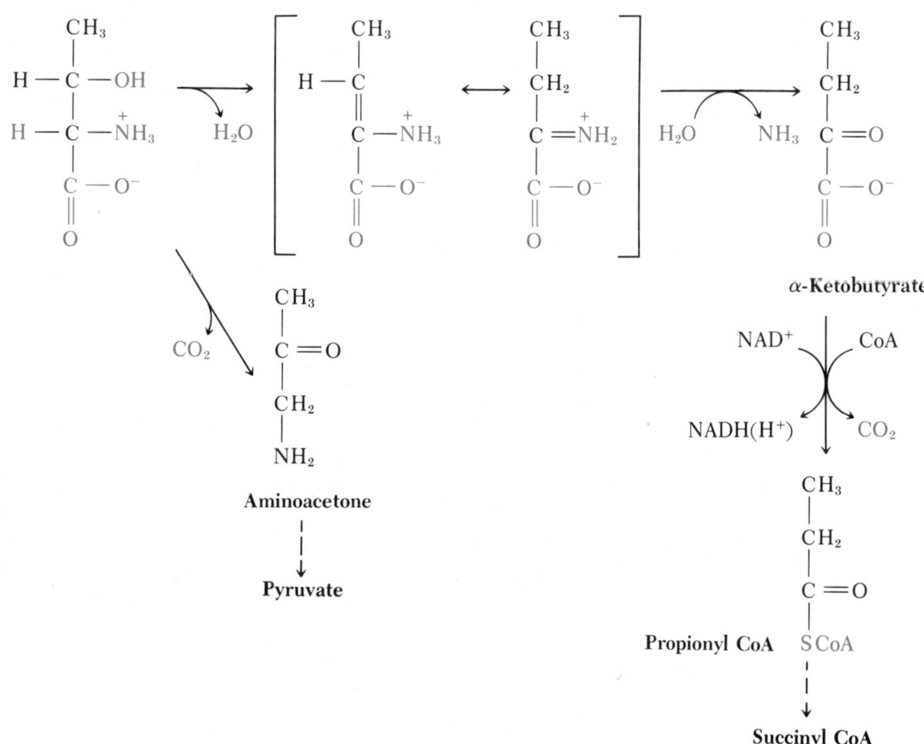

Figure 13-18
Formation of L-threonine in certain microorganisms.

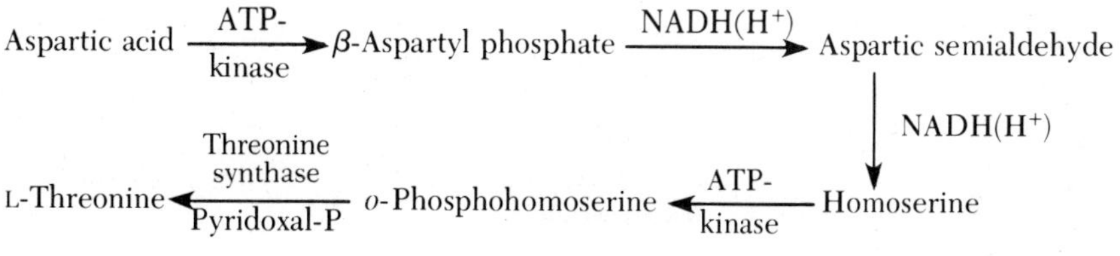

Aspartic acid —(ATP-kinase)→ β-Aspartyl phosphate —($NADH(H^+)$)→ Aspartic semialdehyde —($NADH(H^+)$)→ Homoserine —(ATP-kinase)→ *o*-Phosphohomoserine —(Threonine synthase, Pyridoxal-P)→ L-Threonine

Figure 13-19
Metabolism of phenylalanine and tyrosine: points of blockage in five metabolic disorders are indicated as follows: *, phenylketonuria; †, tyrosinosis; ‡, vitamin C deficiency in infants and guinea pigs; §, alkaptonuria; ‖, albinism.

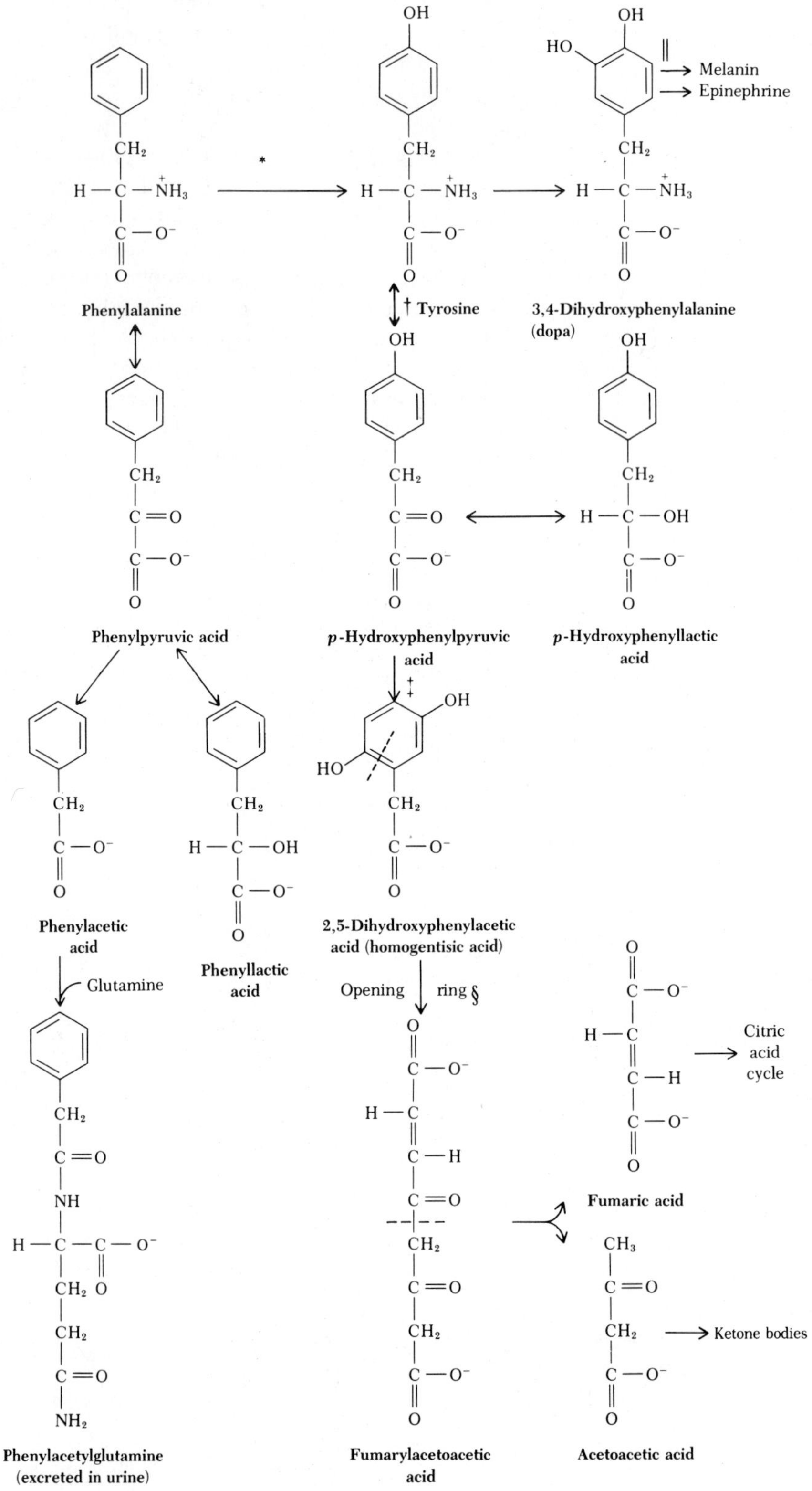

In a minor pathway, some phenylalanine is converted by transamination with either pyruvate or glutamate to phenylpyruvate (Fig. 13-19). This intermediate, in turn, may be reduced to phenyllactate by an aromatic ketoacid reductase, or it may be oxidatively decarboxylated to form phenylacetyl CoA. In man, this substance is conjugated with glutamine to form phenylacetylglutamine, a detoxication product (Chapter 17) that is excreted in the urine.

The first step in the catabolism of tyrosine is its transamination with α-ketoglutarate, catalyzed by the enzyme *tyrosine-glutamate aminotransferase,* to form *p*-hydroxyphenylpyruvic acid, as shown in Fig. 13-19. This intermediate is then converted to 2,5-dihydroxyphenylacetic acid (homogentisic acid) in the presence of the copper-containing enzyme *p-hydroxyphenylpyruvate oxidase.* Ascorbic acid (vitamin C) is apparently required for the activity of this enzyme as indicated by the finding of relatively large amounts of *p*-hydroxyphenylpyruvate in the urine of ascorbic acid–deficient guinea pigs.

Homogentisic acid is oxidized in the presence of *homogentisic acid oxidase,* molecular O_2, Fe^{++} ions, and a sulfhydryl compound (e.g., reduced glutathione) to open the aromatic ring with the formation of fumarylacetoacetic acid, by way of its *cis* isomer, maleylacetoacetic acid. Fumarylacetoacetic acid then splits to yield acetoacetic acid and fumaric acid, which are used by way of the fatty acid metabolic pathway (Chapter 12) or the citric acid cycle (Chapter 11) respectively.

The foregoing major pathways for the metabolism of phenylalanine and tyrosine have been well established by the use of isotopic tracer techniques. For example, the four carbon atoms of acetoacetate have been shown to originate from the four carbon atoms of phenylalanine and tyrosine as dictated by the reactions given in Fig. 13-19. From this pathway it is also clear why phenylalanine and tyrosine are classed as both glycogenic and ketogenic amino acids.

As indicated in Fig. 13-19, phenylalanine and tyrosine also serve as specific precursors for a number of important biolgical compounds. They form the carbon skeleton of the hormones epinephrine and thyroxine, as is described in Chapter 18. L-Dopa is converted into melanin, a brown-black pigment, in the presence of the copper-containing enzyme *tyrosinase.* Melanin is found in skin and hair, in a number of other tissues (including the adrenal medulla, the retina, the choroid, and the ciliary body), and in certain types of tumors, the melanomas. Melanin is formed in skin under the influence of ultraviolet light as a protective mechanism against damage to underlying tissues. Tanning of the skin during exposure to strong sunlight in the summer months is a familiar example.

L-Dopa also is important in brain function as its decarboxylated derivative, *dopamine* (Chapter 18). This compound appears to serve as a "chemical transmitter" at the brain's "dopaminergic synapses" (Chapter 15). Dopamine is deficient in the brains of patients with Parkinson's disease (Chapter 22) and is related in some unclear way to the hand tremors and shuffling gait characteristic of this condition. Dopamine may decrease in the brain with age, resulting in the Parkinson-like symptoms sometimes seen in the elderly.

Another minor catabolic pathway of the metabolism of tyrosine is decarboxylation to form tyramine, which has the property of raising blood pressure. Tyramine is present in some varieties of cheese in significant amounts. This became important when patients who were being treated with monoamine oxidase inhibitors (MAOI), because of the hypotensive action of these drugs in decreasing catecholamines, showed soaring blood pressures when they ingested cheese. The MAOIs prevented the oxidation of tyramine, which, in turn, increased the blood pressure.

At least five inherited disorders of phenylalanine-tyrosine metabolism have been

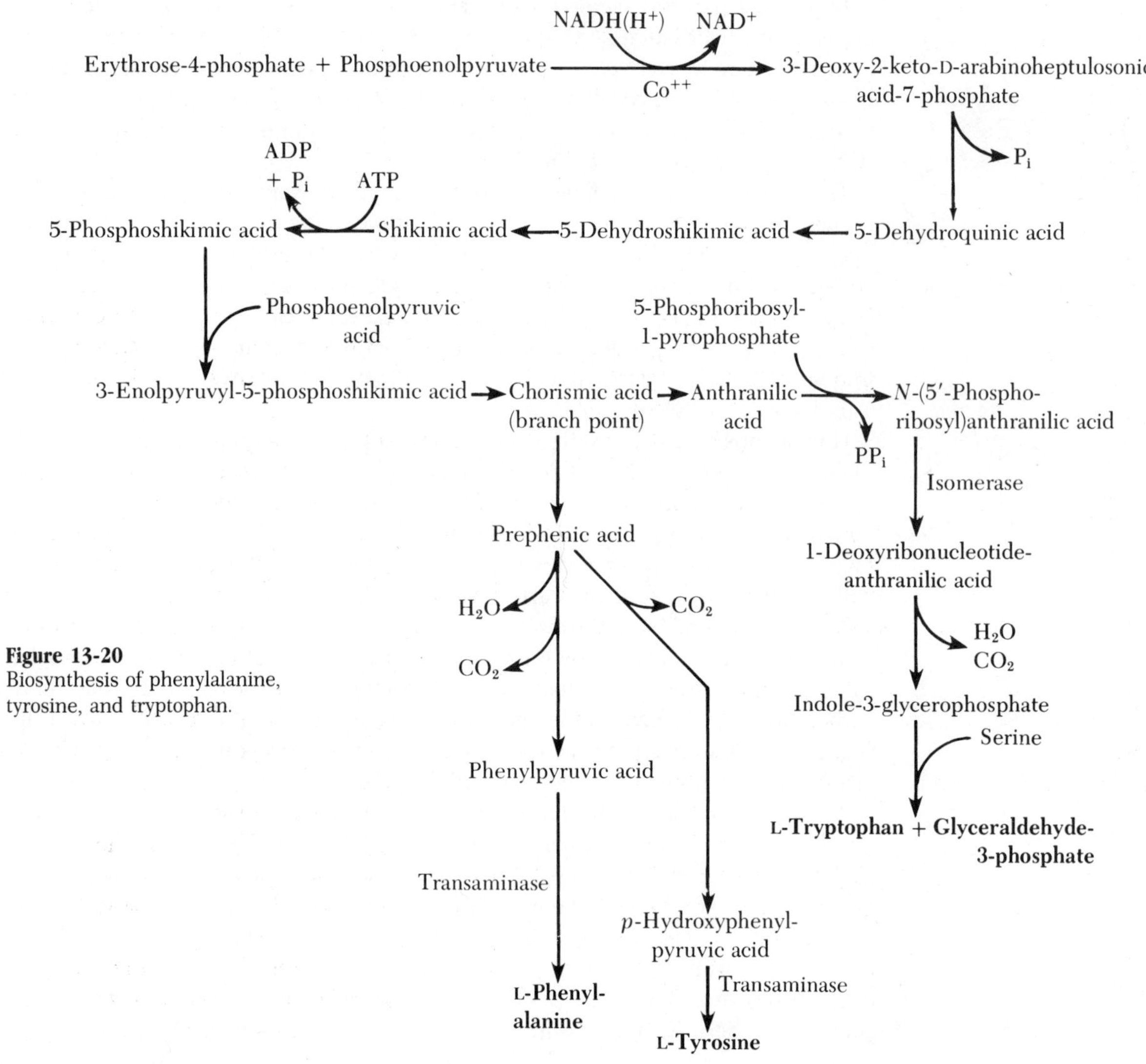

Figure 13-20
Biosynthesis of phenylalanine, tyrosine, and tryptophan.

described in the medical literature, and four have been adequately confirmed. These include albinism, alkaptonuria, phenylketonuria, tyrosinemia, and tyrosinosis. The points at which the metabolic pathways are blocked are indicated in Fig. 13-19. These disorders are discussed briefly in Chapter 22.

The biosynthetic pathway of the two aromatic amino acids phenylalanine and tyrosine, as found in *E. coli,* is summarized briefly in Fig. 13-20, along with the biosynthesis of tryptophan, which shares a portion of the same reactions.

Tryptophan Tryptophan is an essential amino acid and is the only amino acid containing the indole ring. It undergoes reversible deamination in mammals, with the formation of the corresponding keto acid, β-3-indolepyruvic acid. This probably explains why either the D-form or the L-form of tryptophan or its keto acid can maintain nitrogen balance and growth in animals fed a low-tryptophan diet.

Many intermediate products of tryptophan degradation have been identified in the animal organism. Their formation from tryptophan is summarized in Fig. 13-

Figure 13-21
Paths of tryptophan metabolism. Broken lines indicate points blocked by deficiencies of thiamin (*A*), riboflavin (*B*), and pyridoxine (*C*).

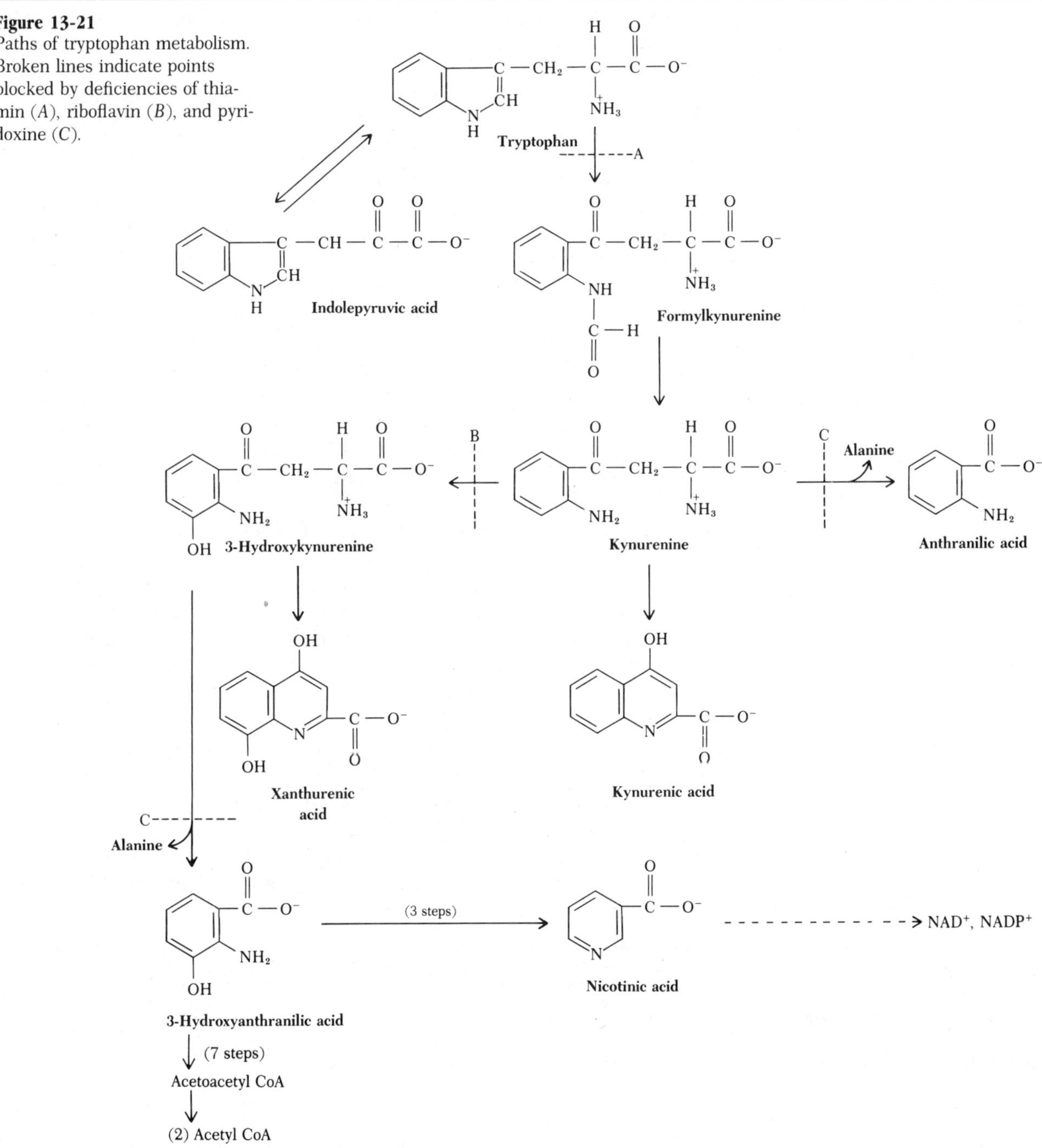

21). One important metabolite of tryptophan is the B-vitamin nicotinic acid. The synthesis of this vitamin from tryptophan supplies about one half the total nicotinic acid requirement of humans (Chapter 21). It has been estimated that 60 mg of tryptophan in food proteins are converted to 1 mg of nicotinic acid in humans. Nicotinic acid (amide) forms the active group of NAD^+ and $NADP^+$ (Chapter 6).

The two major pathways for the degradation of tryptophan are (1) its oxidation to 5-hydroxytryptophan followed by decarboxylation to 5-hydroxytryptamine (serotonin), discussed later, and (2) its oxidation to kynurenine (Fig. 13-21). The oxidation of tryptophan to formylkynurenine involves the opening of the five-membered pyrrole ring and is catalyzed by the inducible heme enzyme *tryptophan dioxygenase,* more commonly called *tryptophan pyrrolase.* Formylkynurenine is converted to kynurenine by removal of the formyl group with the enzyme *formamidase.* Kynurenine can be metabolized by any of three routes, as shown in Fig. 13-21, to form 3-hydroxykynurenine, kynurenic acid, or anthranilic acid. The former is blocked by a deficiency of riboflavin and the latter by a lack of pyridoxine, as indicated in the figure. If pyridoxine is lacking in the diet, nicotinic acid is not formed, but xanthurenic acid results instead. Large amounts are excreted in the urine. Perhaps the most important pathway is its conversion to the 3-hydroxy derivative by *kynureninase* (a hydroxylase) and then to 3-hydroxyanthranilic acid and the B-vitamin *nicotinic acid.* The latter is formed in two steps (Fig. 13-22) from α-amino-β-carboxymuconic-δ-semialdehyde by way of *quinolinic acid,* which loses CO_2, and then reacts with phosphoribose pyrophosphate (PRPP; Chapter 14) to form *nicotinic acid ribonucleotide,* which is used directly for the synthesis of NAD (Chapter 21).

Kynurenine also is converted to a series of intermediates and by-products, some of which are excreted in the urine in relatively small amounts (kynurenic acid, quinaldic acid, xanthurenic acid, 8-hydroxyquinaldic acid, picolinic acid, *o*-aminophenol, anthranilic acid, and 5-hydroxyanthranilic acid). These intermediates apparently have no known function in humans and are excreted in the urine, which fact emphasizes the inefficiency of the use of tryptophan in the body, despite its "essential" nature.

Kynurenine, however, may be metabolized further, as indicated in Fig. 13-22, to yield glutaryl CoA and then acetoacetyl CoA in four steps, involving glutaconyl CoA, crotonyl CoA, and β-hydroxybutryl CoA as intermediates. This accounts for ketogenesis from tryptophan. Finally, two moles of acetyl CoA may be formed, thus apparently accounting for the glycogenic effect of tryptophan.

It is well known that after blood coagulates it possesses vasoconstrictor properties (Chapter 16). The active agent was isolated and crystallized and was called *sero-*

Figure 13-22
Degradation of kynurenine (from tryptophan) to acetoacetyl CoA and acetyl CoA.

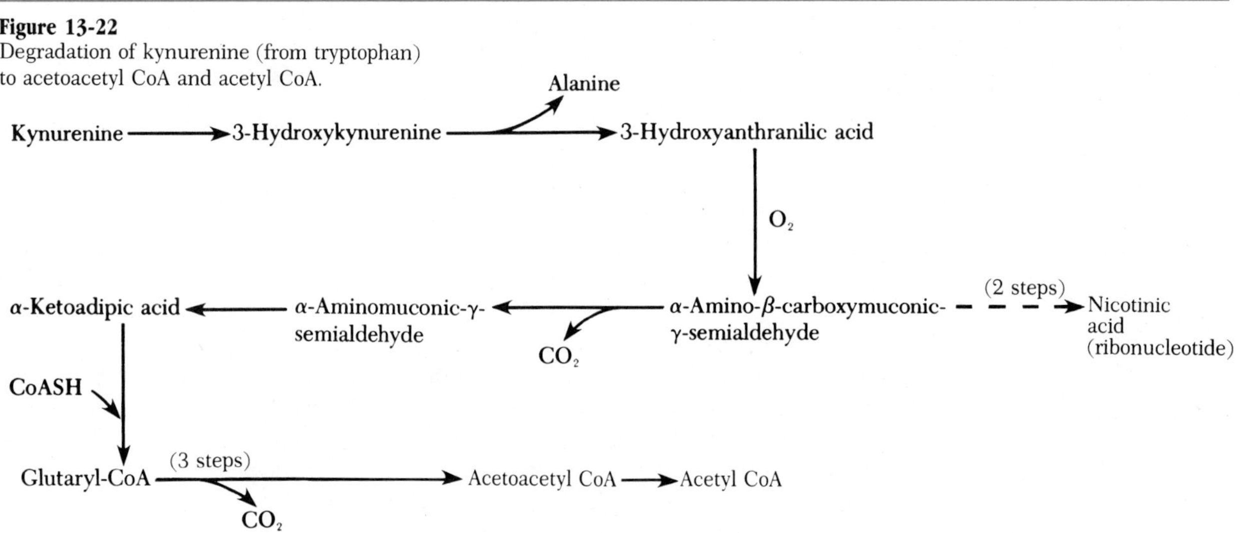

tonin. It was analyzed and found to be a complex of creatinine, sulfate, and 5-hydroxytryptamine. Since the pharmacological properties reside in the 5-hydroxytryptamine, which is separable from the complex, the name *serotonin* was assigned to this compound. Serotonin causes other types of tissue to contract, besides vascular tissue, and it has proved to be an important physiological agent in other respects as well. Antimetabolites of serotonin have been produced. The biosynthesis of serotonin from tryptophan is indicated below. Tryptophan is first oxidized to 5-hydroxytryptophan by *tryptophan hydroxylase,* tetrahydrobiopterin serving as a cofactor, as described earlier in this chapter. This amino acid is then acted on by a specific decarboxylase to yield serotonin.

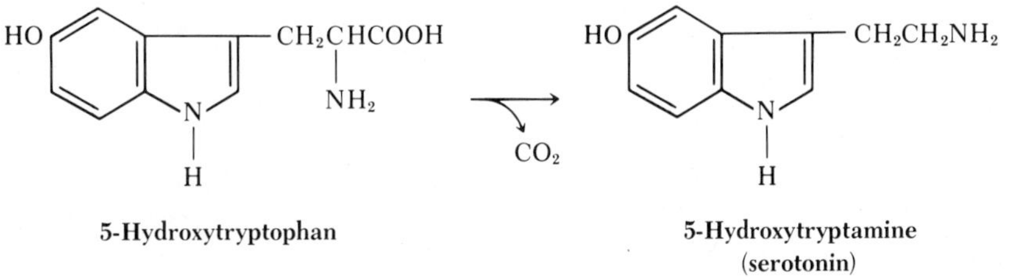

HO—(indole ring, N—H)—$CH_2CHCOOH$ (with NH_2 on CH) $\xrightarrow{-CO_2}$ HO—(indole ring, N—H)—$CH_2CH_2NH_2$

5-Hydroxytryptophan → **5-Hydroxytryptamine (serotonin)**

Serotonin, upon methylation of the hydroxyl group of the aromatic ring and acetylation of its amino group, is converted to *melatonin,* a hormone produced in the pineal gland (see Chapter 18).

Still another pathway of the degradation of tryptophan, important because it occurs as a result of bacterial action in the large intestine, is the following:

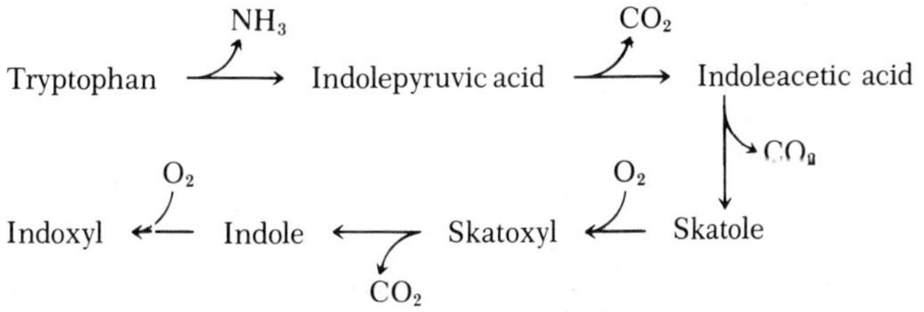

$$\text{Tryptophan} \xrightarrow{-NH_3} \text{Indolepyruvic acid} \xrightarrow{-CO_2} \text{Indoleacetic acid} \xrightarrow{-CO_2} \text{Skatole}$$

$$\text{Skatole} \xrightarrow{O_2} \text{Skatoxyl} \xrightarrow{-CO_2} \text{Indole} \xrightarrow{O_2} \text{Indoxyl}$$

Skatole and indole are partially responsible for the characteristic odor of feces. Some indoxyl may be reabsorbed from the large intestine, conjugated with sulfate in the liver, and excreted in the urine as indican (potassium salt).

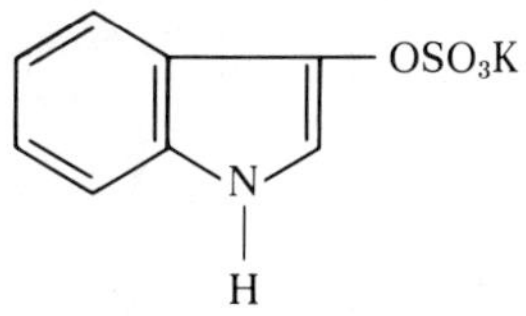

(indole ring, N—H)—OSO_3K

Indican

Tryptophan may be synthesized in certain microorganisms (*E. coli*) from erythrose-4-phosphate and phosphoenolpyruvate as precursors (refer again to Fig. 13-20.

Several inherited disorders of tryptophan metabolism have been described. One of them is called *Hartnup disease,* named after the family in which it was discovered, and results from a deficiency of *tryptophan dioxygenase* (tryptophan pyrrolase). As a result, only a small amount of ingested tryptophan is oxidized and a large amount, at least five to ten times the normal average, is excreted in the urine and feces. There is also a high urinary excretion of indoleacetic acid and indican. A decreased conversion of tryptophan to kynurenine and nicotinic acid occurs, as would be expected (Fig. 13-21). The principal clinical findings are a pellagralike

Figure 13-23
Metabolic pathways of cysteine.

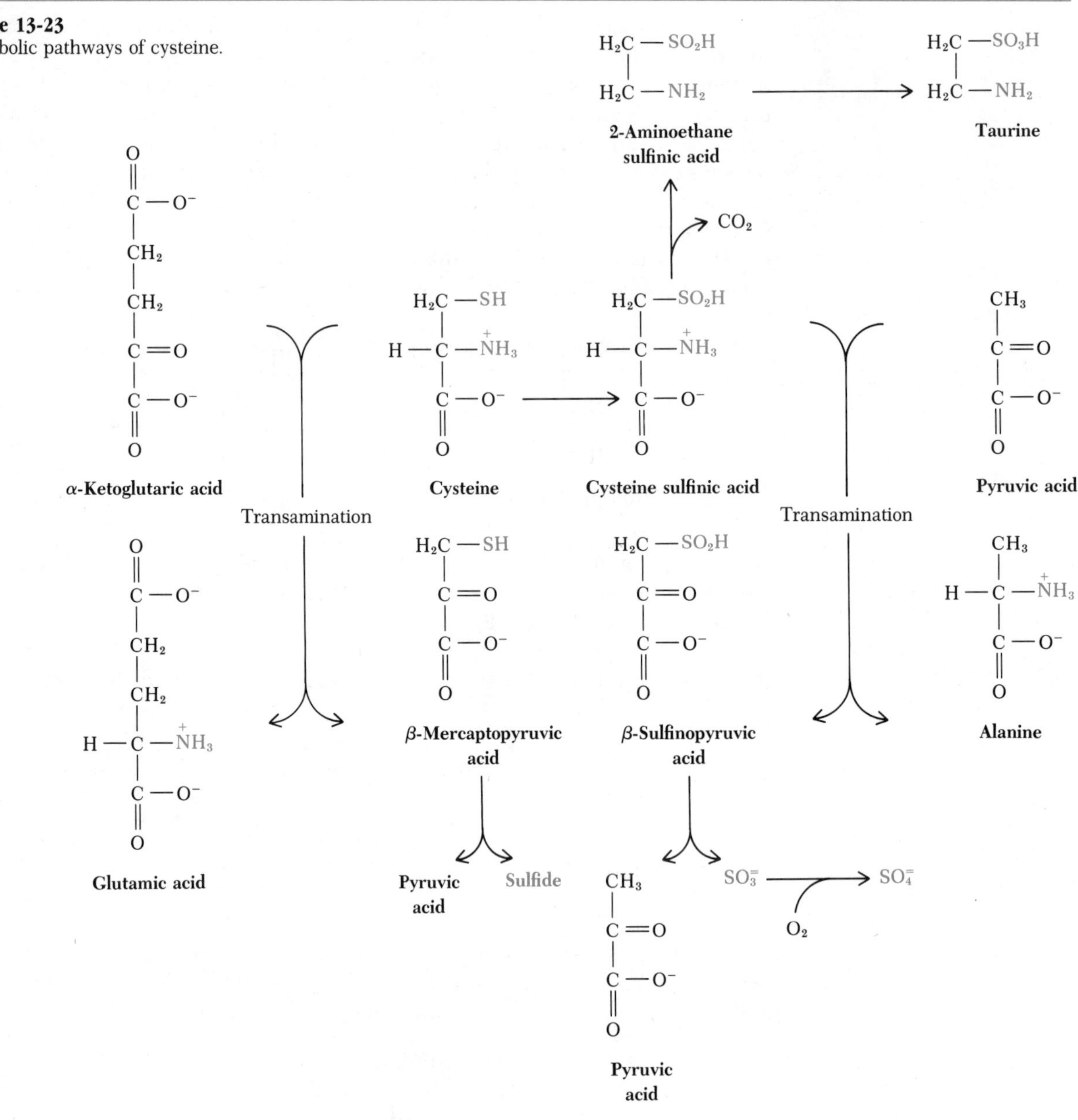

rash and mental retardation, apparently because of the decreased formation of nicotinic acid from tryptophan. This and other hereditary abnormalities of tryptophan metabolism are considered in Chapter 22.

Cystine and cysteine Since cystine and cysteine are readily interconvertible in vitro, they probably undergo similar reactions in the body. The present discussion of the metabolism is therefore concerned primarily with *cysteine*.

$$\underset{\text{Cystine}}{{}^{-}OOC{-}CH(\overset{+}{N}H_3){-}CH_2{-}S{-}S{-}CH_2{-}CH(\overset{+}{N}H_3){-}COO^{-}} \xrightleftharpoons{2H} 2\ \underset{\text{Cysteine}}{HS{-}CH_2{-}CH(\overset{+}{N}H_3){-}COO^{-}}$$

Cysteine may be metabolized by several routes. Three of them are summarized in Fig. 13-23. One principal pathway involves its transamination in the presence of α-ketoglutarate and the enzyme *cysteine-glutamate aminotransferase* to form β-mercaptopyruvate and glutamate, as shown at the bottom of the figure. β-Mercaptopyruvate then undergoes desulfuration with *thiolpyruvate transsulfurase* as the catalyst to form pyruvate and H_2S. The latter may require the presence of a reduced sulfhydryl compound, such as reduced glutathione for sulfide formation. The conversion of cystine-cysteine to pyruvate accounts for the glycogenic effect of these two amino acids. The sulfide formed in the reaction is converted to sulfite by an oxidative step not fully understood and then to sulfate ($SO_4^=$) in the presence of the hepatic enzyme *sulfite oxidase*. The sulfate thus formed is either excreted in the urine or converted to "active sulfate," PAPS, and used for the formation of sulfate esters of various alcohols, phenols, steroids, and polysaccharides. The formation of active sulfate proceeds in two steps as follows:

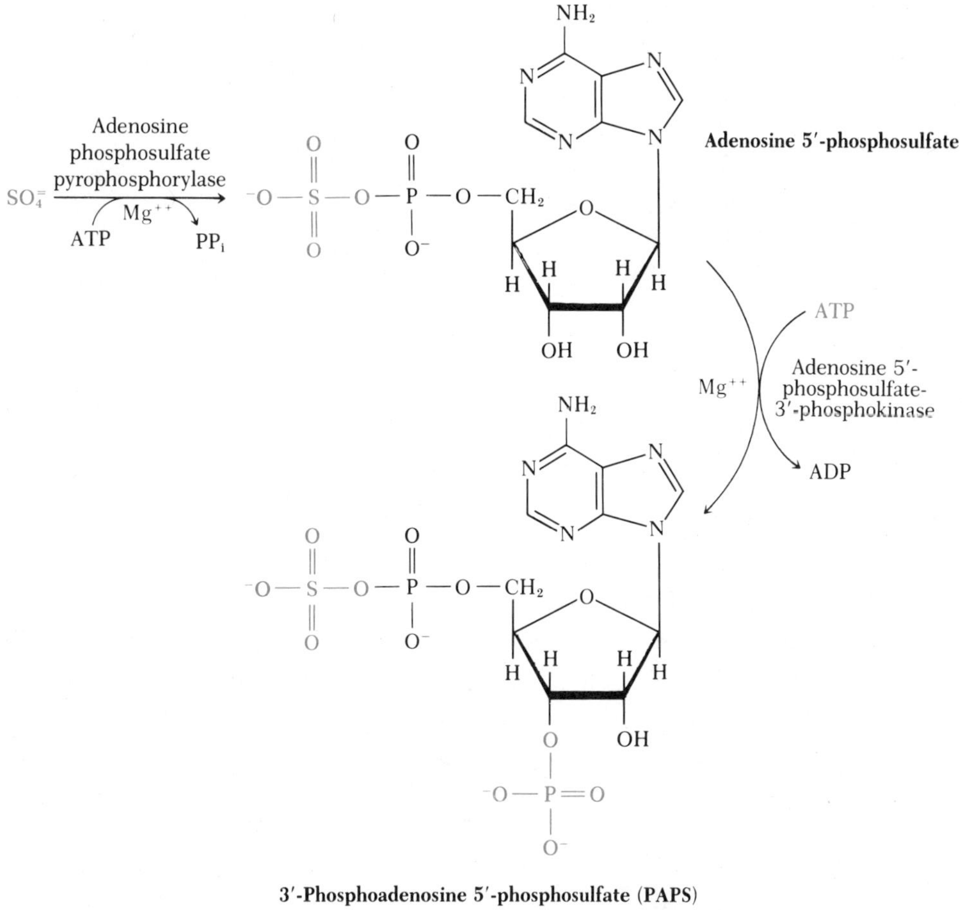

$SO_4^=$ → (Adenosine phosphosulfate pyrophosphorylase; Mg^{++}; ATP → PP_i) **Adenosine 5′-phosphosulfate**

→ (Adenosine 5′-phosphosulfate-3′-phosphokinase; Mg^{++}; ATP → ADP)

3′-Phosphoadenosine 5′-phosphosulfate (PAPS)

Cysteine may also be oxidized to cysteinesulfinic acid, which is then decarboxylated to form 2-aminoethanesulfinic acid, which has been named *hypotaurine*. Hypotaurine may be oxidized to *taurine*. Cysteinesulfinic acid may undergo transamination with pyruvic acid and eventually give rise to sulfate. Thus one may understand how sulfates are formed from cysteine.

A number of other reactions of cystine have been demonstrated. Some of them are shown in Fig. 13-23. Instead of pyruvic acid, α-ketoglutaric acid may enter into the transamination reaction, yielding glutamic acid. Cysteine itself may be transaminated with α-ketoglutaric acid, yielding β-mercaptopyruvic acid, a compound rapidly converted to sulfide and sulfate, as described above, and pyruvic acid, by various tissues. Besides taurine, other sulfur-containing compounds of physiologic importance are also derived from cysteine or cystine. Among them are insulin, CoA, glutathione, and vasopressin. Cysteine is also important in detoxication, as is described in Chapter 17.

Cysteine and cystine are formed in the animal organism from methionine, as is discussed next. Hence these amino acids are not classed as essential; they do, however, "spare" the requirement for methionine.

Several abnormalities in the metabolism of cystine-cysteine have been demonstrated. These are considered in Chapter 22.

L-Cysteine may be synthesized in yeast and certain other microorganisms from L-serine as follows:

$$\text{O-Acetyl-L-Serine} + H_2S \xrightarrow[\text{synthase}]{\text{Cysteine}} \text{L-Cysteine} + H_2O$$

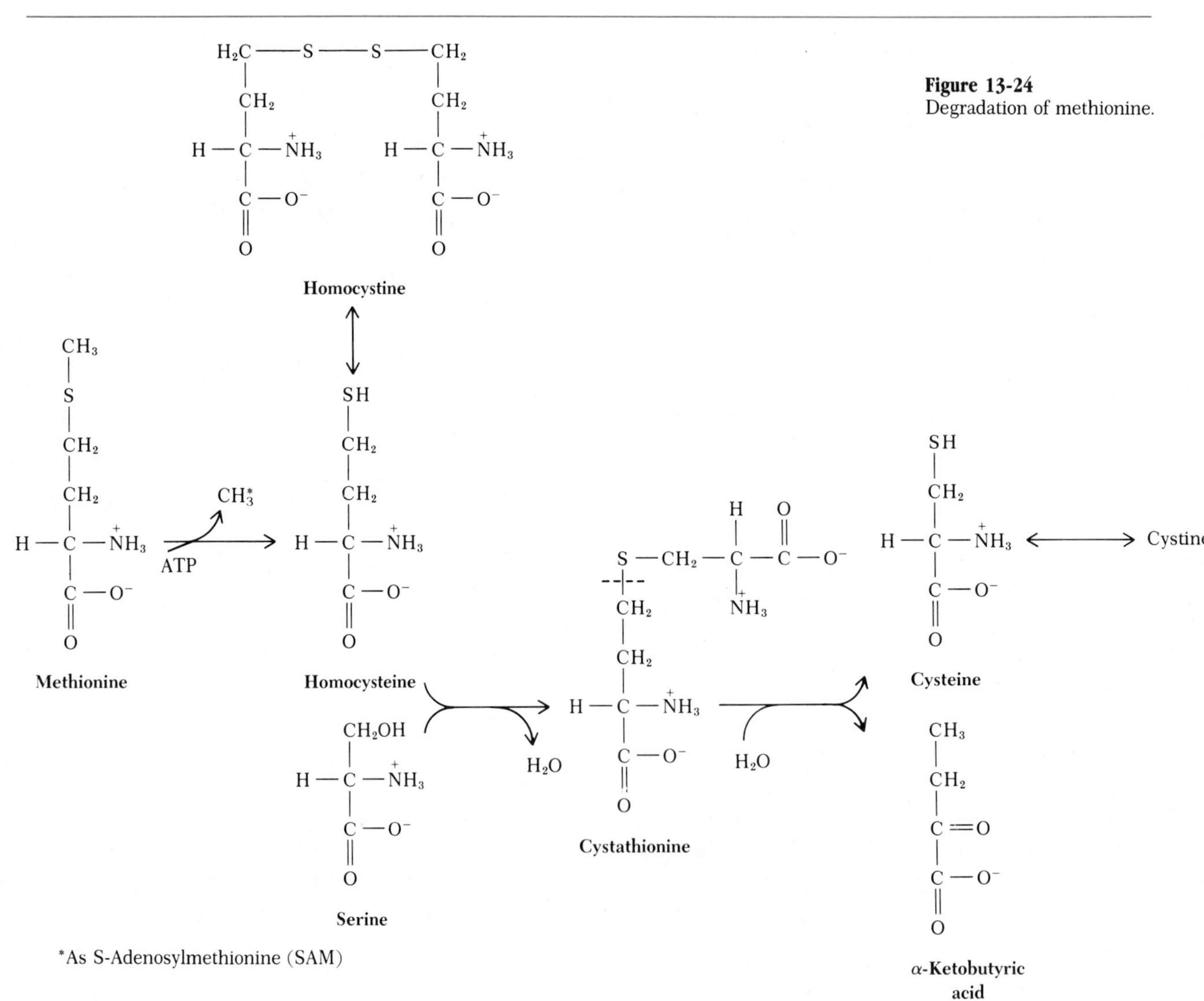

Figure 13-24
Degradation of methionine.

Methionine Methionine has two, perhaps three, principal metabolic pathways in addition to its use in the biosynthesis of tissue proteins. The two primary routes involve transmethylation first, as has been discussed in some detail in the first part of this chapter. The remaining portion is then metabolized as indicated in Figs. 13-24 and 13-25. Transamination to the corresponding keto acid is believed to be a relatively minor pathway.

As is evident from Figs. 13-24 and 13-25, the principal routes of methionine involve the transfer of its methyl group by way of SAM (Fig. 13-5). These have been discussed in more detail earlier. The methyl group may be accepted by a variety of substances to form such important biochemical agents as creatine, epinephrine, choline, DNA, tRNA, and trigonelline. Indeed, methyl groups of over 50 compounds are derived from methionine. The reactions are catalyzed by various specific methyl transferases.

Figure 13-25
Summary of major catabolic pathways for methionine.

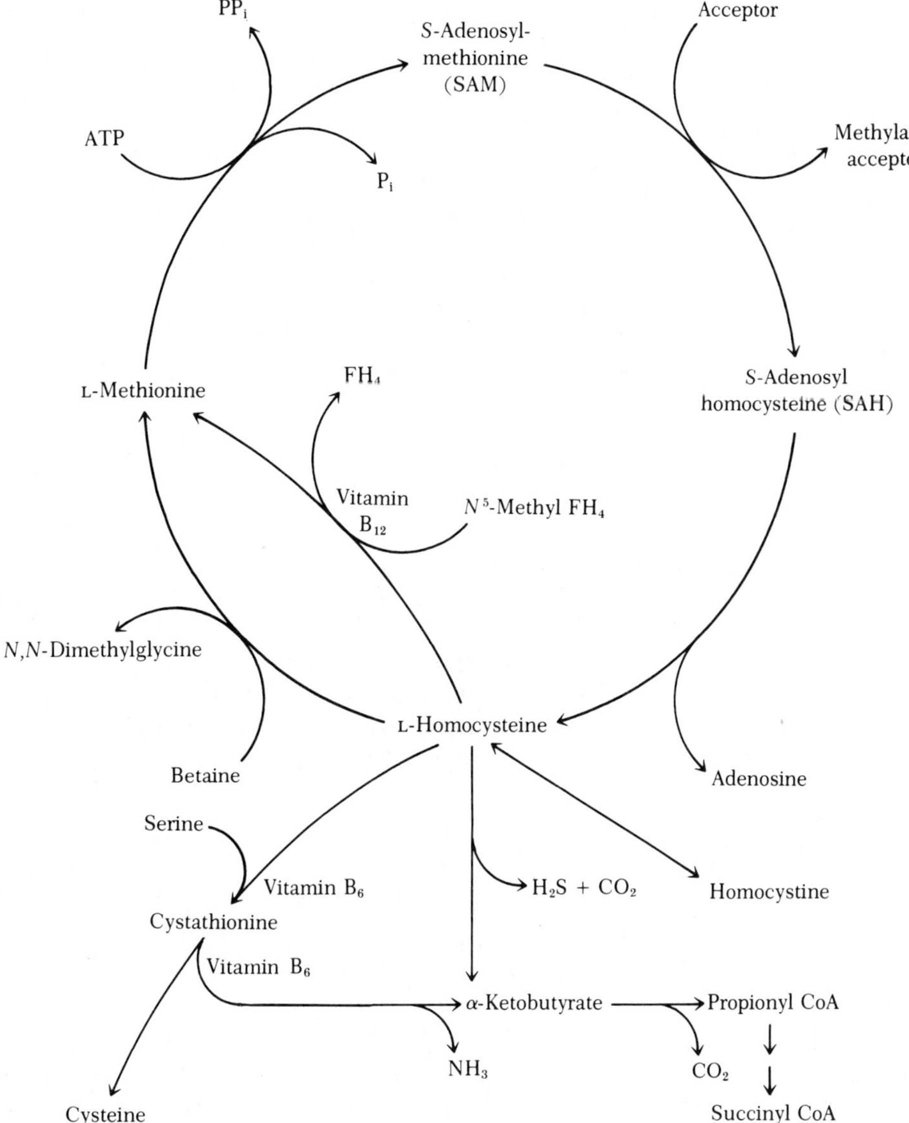

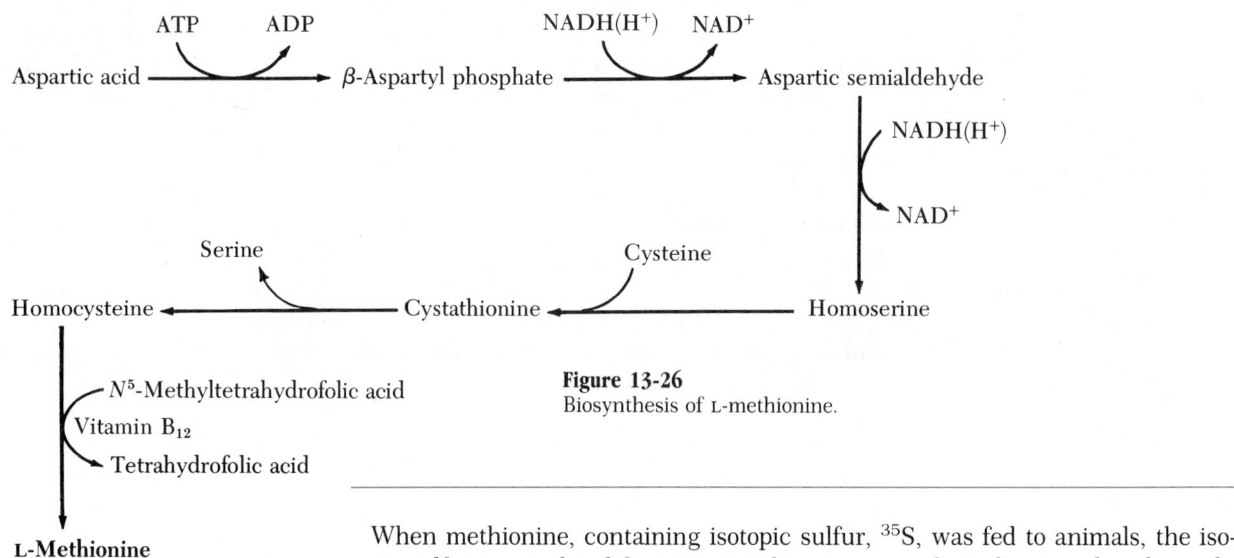

Figure 13-26
Biosynthesis of L-methionine.

When methionine, containing isotopic sulfur, ^{35}S, was fed to animals, the isotopic sulfur was isolated from intermediate compounds, indicating that the pathway was methionine → homocysteine, with the loss of a methyl group. The methyl group may be oxidized to carbon dioxide and water or may enter transmethylation reactions as described previously. Homocysteine then is either oxidized to homocystine or converted to cysteine, after having been coupled with serine to yield cystathionine as an intermediate product. As shown in Figs. 13-24 and 13-25, this is a two-step reaction, requiring the enzymes *cystathionine synthase* and *cystathionelyase,* respectively, with pyridoxal phosphate serving as a coenzyme in each case. The cysteine then may be oxidized to cystine or further catabolized, as described in the discussion of that amino acid. Homocysteine also can be converted to methionine or to α-ketobutyrate, as indicated in Fig. 13-25. α-Ketobutyrate is oxidatively decarboxylated to form propionyl CoA, which is catabolized by way of methylmalonyl CoA and succinyl CoA, as previously described (Chapter 12). Methionine is an essential amino acid, whereas cystine is not; evidently the former can be converted to the latter, but cystine cannot be transformed to methionine.

Two hereditary defects of methionine metabolism in man have been described. These are discussed in Chapter 22.

Since methionine is an essential amino acid and cannot be formed in animal tissues, its synthesis in plants and microorganisms is of interest. Fig. 13-26 is a schematic diagram of the formation of methionine from aspartic acid in *E. coli.*

Aspartic acid, glutamic acid, asparagine, and glutamine Under anaerobic conditions, succinic, fumaric, and malic acids have been found to arise from aspartic acid when it is added to minced muscle.

$$^{-}O-C(=O)-CH_2-CH(\overset{+}{N}H_3)-C(=O)-O^{-}$$

Aspartic acid

$$^{-}O-C(=O)-CH_2-CH(OH)-C(=O)-O^{-}$$

Malic acid

$$^{-}O-C(=O)-CH_2-CH_2-C(=O)-O^{-}$$

Succinic acid

$$^{-}O-C(=O)-CH=CH-C(=O)-O^{-}$$

Fumaric acid

Figure 13-27
Biosynthesis and degradation of glutamine.

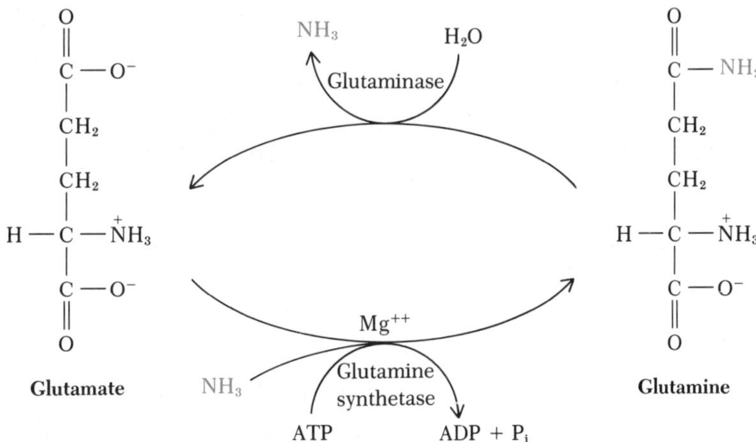

Anaerobically, we would expect oxaloacetic acid to be an intermediate in the catabolism of aspartic acid.

Glutamic acid is transformed by oxidative deamination to α-ketoglutaric acid, as described earlier in this chapter.

$$\underset{\text{Glutamic acid}}{{}^{-}OOC{-}CH_2{-}CH_2{-}CH(\overset{+}{N}H_3){-}COO^{-}} \xrightarrow{\ \ -2H\ \ } \underset{\alpha\text{-Iminoglutaric acid}}{\left[{}^{-}OOC{-}CH_2{-}CH_2{-}C({=}NH){-}COO^{-}\right]} \xrightarrow[\ \ NH_3\ \ ]{H_2O} \underset{\alpha\text{-Ketoglutaric acid}}{{}^{-}OOC{-}CH_2{-}CH_2{-}C({=}O){-}COO^{-}}$$

Transamination also converts glutamic acid to α-ketoglutaric acid and aspartic acid to oxaloacetic acid. Both of these amino acids can form glucose, and in each case three carbons are so used. Both react with ammonia, forming acid amides, glutamine, and asparagine, and because of this both are of importance in ammonia transport and also in ammonia formation (Fig. 13-4).

The conversion of glutamate to glutamine is catalyzed by the enzyme *glutamine synthetase,* which occurs in the liver and other tissues. The reaction is essentially nonreversible, requiring the presence of ATP for the formation of enzyme-bound γ-glutamyl phosphate. Apparently these steps are concerned in the regulation of glutamine synthetase activity by a unique three-stage "cascade type" of allosteric enzyme control (Chapter 5). The synthesis of glutamine from glutamate and its hydrolysis back to glutamine is shown schematically in Fig. 13-27.

The hydrolysis of glutamine is catalyzed by *glutaminases*—there are two of these with different optimum pHs. Thus the enzyme that synthesizes glutamine is different from the one or ones that catabolize it.

Glutamine participates in a number of biosynthetic reactions, usually by supplying amino or ammonia nitrogen. Examples include its involvement in the formation of histidine, tryptophan, arginine, carbamoyl phosphate, cytidine triphosphate, purines, glucosamine, NAD, and of course glutamate and urea. Glutamine is also the principal source of ammonia in the urine (Chapter 17), and in this way plays an important role in the maintenance of acid-base balance.

Another fact of interest in connection with glutamine is that it is capable of transamination with a variety of α-keto acids to yield α-ketoglutaramic acid. The latter is then hydrolyzed by a specific deamidase to produce α-ketoglutaric acid and ammonia. An example follows:

$$\underset{\textbf{Glutamine}}{\mathrm{H_2N{-}C({=}O){-}CH_2{-}CH_2{-}CH(\overset{+}{N}H_3){-}C({=}O){-}O^-}} + \underset{\textbf{Pyruvic acid}}{\mathrm{CH_3{-}C({=}O){-}C({=}O){-}O^-}} \rightleftharpoons \underset{\textbf{α-Ketoglutaramic acid}}{\mathrm{H_2N{-}C({=}O){-}CH_2{-}CH_2{-}C({=}O){-}C({=}O){-}O^-}} + \underset{\textbf{Alanine}}{\mathrm{CH_3{-}CH(\overset{+}{N}H_3){-}C({=}O){-}O^-}}$$

α-Ketoglutaramic acid → α-Ketoglutaric acid + NH_3

α-Ketoglutaramate has become a recent focus of interest as a possible contributory agent in the pathogenesis of the hepatic coma of ammonia intoxication. Threefold to tenfold increases in the concentration of α-ketoglutaramate have been found in the cerebrospinal fluid of patients with hepatic coma. Perfusion of α-ketoglutaramate into the cerebral lateral ventricles of rats depressed the nocturnal locomotor activity and induced circling behavior and myoclonus. Apparently, patients with liver disease and prominent neurological symptoms may have normal or only moderately increased levels of ammonia in the blood and cerebrospinal fluid.

γ-Aminobutyric acid (GABA) is found in brain tissue (Chapter 15). It arises from the decarboxylation of glutamic acid by a specific decarboxylase in brain tissue. It can then be transaminated with α-ketoglutaric to form glutamic acid again and succinic semialdehyde.

The corresponding acid amide of aspartic acid is asparagine. Asparagine apparently undergoes deamidation by asparaginase, and it also undergoes transamination, in much the same manner as described above for glutamine. Extensive studies are in progress on the use of asparagine in the treatment of leukemias and of lymphoid tumors.

Apparently glutamic and aspartic acids are formed in microorganisms and plants by the transamination of α-ketoglutaric and oxaloacetic acids respectively, as they are in animals.

Lysine Lysine, a diaminomonocarboxylic acid, is in a class by itself. It is the only amino acid that, once present in tissues, does not exchange its nitrogen by trans-

$$\underset{\text{Glutamic acid}}{{}^{-}O-C(=O)-CH(\overset{+}{N}H_3)-CH_2-CH_2-C(=O)-O^{-}} \xrightarrow{-CO_2} \underset{\gamma\text{-Aminobutyric acid (GABA)}}{NH_2-CH_2-CH_2-CH_2-C(=O)-O^{-}}$$

$$\text{GABA} + \underset{\alpha\text{-Ketoglutaric acid}}{{}^{-}O-C(=O)-CH_2-CH_2-C(=O)-C(=O)-O^{-}} \rightleftharpoons \underset{\text{Succinic semialdehyde}}{H-C(=O)-CH_2-CH_2-C(=O)-O^{-}} + \underset{\text{Glutamic acid}}{{}^{-}O-C(=O)-CH_2-CH_2-CH(\overset{+}{N}H_3)-C(=O)-O^{-}}$$

amination with other amino acids circulating in the body fluids. When lysine is fed, however, it can give up its nitrogen to those amino acids present in the tissues just as the others do, but if it has lost its nitrogen, it cannot be reaminated.

Using lysine containing an ϵ-labeled carbon, it has been shown that liver converts lysine in vitro to α-aminoadipic acid. When this acid, similarly labeled, was the starting point, it was oxidatively deaminated to α-ketoadipic acid, which was then oxidatively decarboxylated to glutaric acid. (In the formulas below, labeled carbons are indicated by asterisks.)

$$\underset{\text{Lysine}}{H_2{}^{*}C(NH_2)-CH_2-CH_2-CH_2-CH(\overset{+}{N}H_3)-C(=O)-O^{-}} \longrightarrow \underset{\alpha\text{-Aminoadipic acid}}{{}^{-}O-{}^{*}C(=O)-CH_2-CH_2-CH_2-CH(\overset{+}{N}H_3)-C(=O)-O^{-}} \longrightarrow \underset{\alpha\text{-Ketoadipic acid}}{{}^{-}O-{}^{*}C(=O)-CH_2-CH_2-CH_2-C(=O)-C(=O)-O^{-}} \longrightarrow \underset{\text{Glutaric acid}}{{}^{-}O-{}^{*}C(=O)-CH_2-CH_2-CH_2-C(=O)-O^{-}}$$

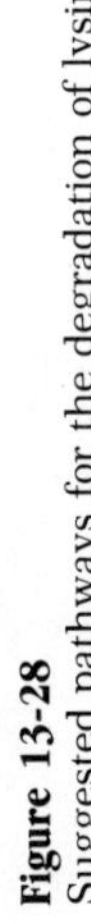

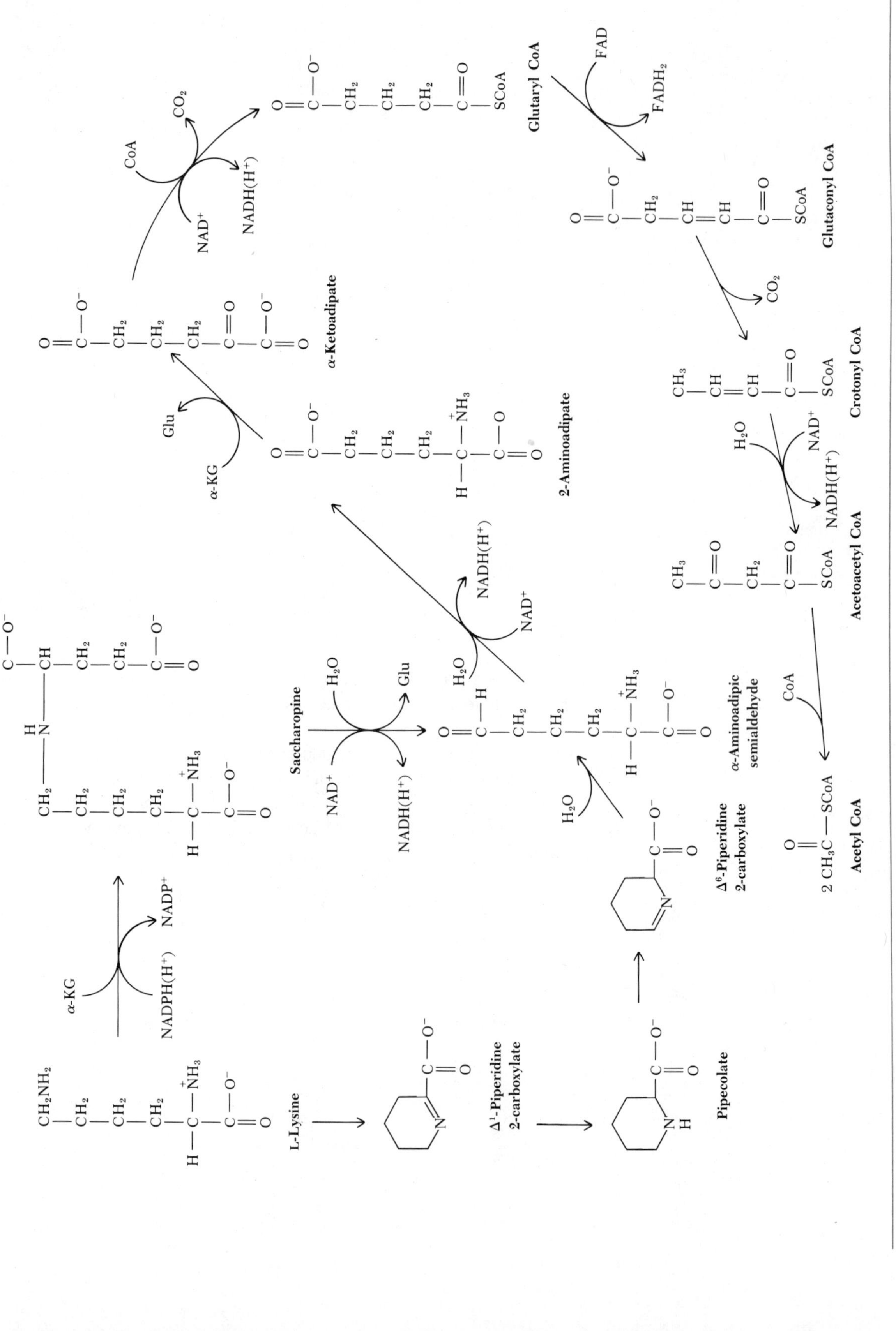

Figure 13-28
Suggested pathways for the degradation of lysine.

Other studies have shown that L-pipecolic acid is involved in this and other pathways of lysine metabolism (Fig. 13-28). However, it appears to be bypassed in one suggested alternate pathway, in which lysine is converted directly to *saccharopine*. This noncyclic intermediate is converted directly to α-aminoadipic-ε-semialdehyde, joining the main pathway at this point.

Still another pathway for the catabolism of lysine is present in mammalian liver, involving ε-*N*-acetyllysine as an intermediate. This route also avoids the formation of cyclic intermediates. The terminal product again is glutaric acid. It undoubtedly is metabolized as shown in Fig. 13-28. However, two other possibilities have been suggested, as shown below.

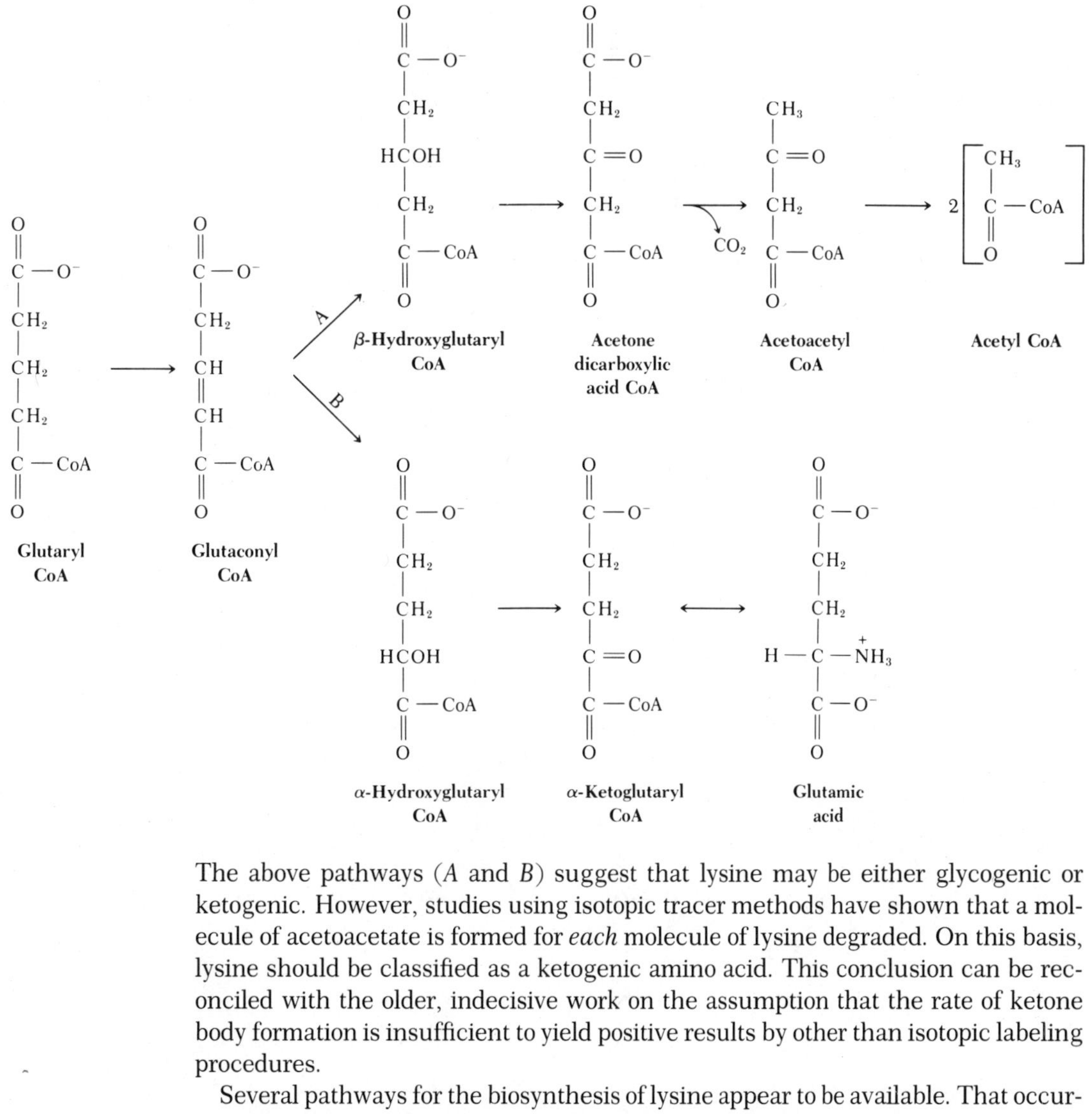

$$^{-}O-C(=O)-CH_2-CH_2-CH_2-C(=O)-CoA \longrightarrow {}^{-}O-C(=O)-CH_2-CH=CH-C(=O)-CoA$$

Glutaryl CoA → Glutaconyl CoA

A:

$$^{-}O-C(=O)-CH_2-HCOH-CH_2-C(=O)-CoA \longrightarrow {}^{-}O-C(=O)-CH_2-C(=O)-CH_2-C(=O)-CoA \xrightarrow{-CO_2} CH_3-C(=O)-CH_2-C(=O)-CoA \longrightarrow 2\,[CH_3-C(=O)-CoA]$$

β-Hydroxyglutaryl CoA → Acetone dicarboxylic acid CoA → Acetoacetyl CoA → Acetyl CoA

B:

$$^{-}O-C(=O)-CH_2-CH_2-HCOH-C(=O)-CoA \longrightarrow {}^{-}O-C(=O)-CH_2-CH_2-C(=O)-C(=O)-CoA \longleftrightarrow {}^{-}O-C(=O)-CH_2-CH_2-CH(\overset{+}{N}H_3)-C(=O)-O^{-}$$

α-Hydroxyglutaryl CoA → α-Ketoglutaryl CoA ⟷ Glutamic acid

The above pathways (*A* and *B*) suggest that lysine may be either glycogenic or ketogenic. However, studies using isotopic tracer methods have shown that a molecule of acetoacetate is formed for *each* molecule of lysine degraded. On this basis, lysine should be classified as a ketogenic amino acid. This conclusion can be reconciled with the older, indecisive work on the assumption that the rate of ketone body formation is insufficient to yield positive results by other than isotopic labeling procedures.

Several pathways for the biosynthesis of lysine appear to be available. That occurring in yeast is shown schematically in Fig. 13-29. Not all the reactions, especially the first, are completely understood at the present time.

Two inherited disorders of lysine metabolism have been reported, *hyperlysinemia* and *saccharopinuria*. These are discussed in Chapter 22.

Arginine Arginine takes part in the formation of urea, as described in the first part of this chapter, yielding ornithine. Not all the ornithine thus formed continues in

Figure 13-29
Biosynthesis of L-lysine.

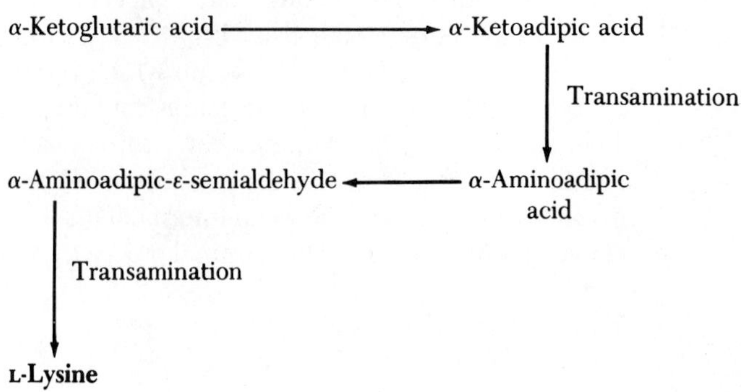

Figure 13-30
Interconversions of ornithine, proline, and glutamic acid.

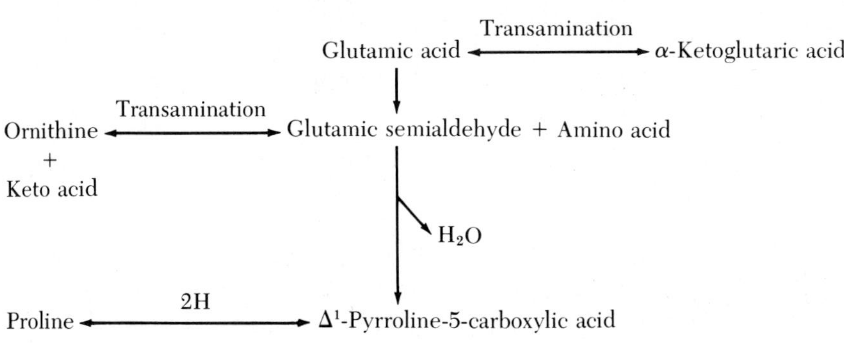

Figure 13-31
Degradative pathway for histidine.

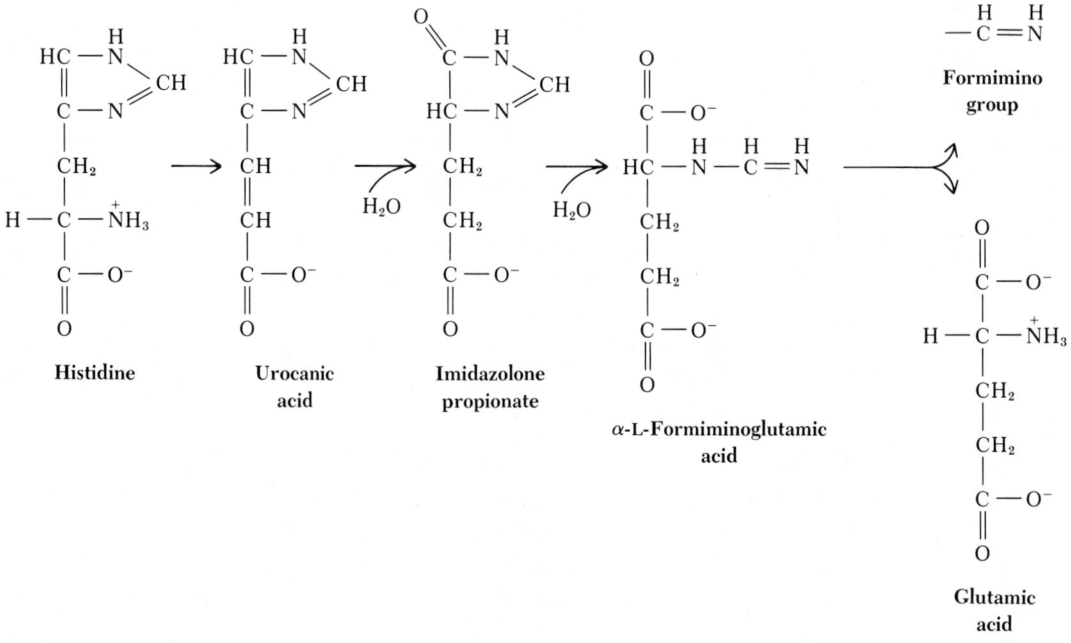

Histidine → Urocanic acid →(H_2O) Imidazolone propionate →(H_2O) α-L-Formiminoglutamic acid → Formimino group + Glutamic acid

that cycle. Some may be converted to proline and glutamic acid. The conversion of ornithine to proline and glutamic acid is by way of glutamic semialdehyde (Fig. 13-30). Thus the ornithine from arginine may be converted to glutamic acid, then by transamination form α-ketoglutaric acid, and eventually by the citric acid cycle be metabolized or form glucose (Chapter 11).

Arginine may be synthesized in the animal from α-ketoglutaric acid or glutamic acid or proline by the reversal of the above process. The ornithine thus produced reacts with carbamoyl phosphate, as in the formation of urea, thus forming arginine.

Arginine is considered to be a semiessential amino acid. It can be synthesized in animal tissues at a rate sufficient for maintenance in the adult but not rapid enough to support growth in the young animal. It is thus an essential amino acid for growth but not for maintenance.

The five hereditary defects in the biosynthesis of urea mentioned earlier (see also Chapter 22) may be considered inherited disorders in the metabolism of arginine, since they share common biosynthetic and catabolic pathways. The same is true, in part at least, of the proline pathways discussed later.

Histidine Histidine, like arginine, is a semiessential amino acid. A dietary protein source of histidine is needed for growth of the young animal (rat) because this amino acid is not synthesized in vivo at a sufficiently rapid rate. However, an amount adequate for the maintenance of nitrogen balance in the fully grown animal, including humans, may be synthesized.

Histidine has several other important functions in addition to the general role of amino acids in tissue protein formation. Upon decarboxylation it forms histamine, which reduces blood pressure, is a vasodilator, and increases the secretion of gastric juice. Allergic reactions appear to stimulate an excessive liberation of histamine. Histamine is converted to β-imidazoleacetic acid by the enzyme histaminase and is excreted in the urine. Histaminase is found in most tissues, with a notable exception being the lungs. Consequently, relatively large quantities of histamine may accumulate in lung tissue. Carnosine and anserine present in muscle extracts (Chapter 15) are β-alanyl dipeptides of histidine and 1-methylhistidine respectively. Ergothionine, present in erythrocytes, liver, and brain, is the betaine of thiolhistidine. The functions of these dipeptides are not understood.

The degradation of histidine in animals occurs mainly in the liver. The major pathway and principal intermediates are shown in Fig. 13-31. Histidine is first converted to urocanic acid by the enzyme *histidase,* with the loss of ammonia from the amino group and adjacent hydrogen atom. The imidazole ring is then hydrolytically cleaved at the nitrogen-3, carbon-4 position by the enzyme *urocanase* to form an intermediate imidazolone propionic acid then formimino-L-glutamic acid. In the presence of FH_4, the formimino group is removed as N^5-formiminotetrahydrofolic acid, or N^5-formyltetrahydrofolic acid plus ammonia, leaving L-glutamic acid, which is converted to α-ketoglutaric acid by transamination or metabolized otherwise, as described previously. N^5-Formiminoglutamic acid (FIGLU) and N^5-formyltetrahydrofolic acid may be used in one-carbon metabolism, discussed earlier in this chapter. Interesting in this connection is the fact that FIGLU is excreted in the urine in folic acid deficiency (Chapter 21).

Another minor pathway for the degradation of histidine by transamination with pyruvate to form imidazolepyruvic acid, which is converted to imidazoleacetic acid and excreted in the urine, is known. During pregnancy, large amounts of histidine are excreted in the urine. This occurs from about the fifth week of pregnancy until a few days postpartum. The absence of histidase from the liver during this period accounts for the phenomenon, and the explanation given is that

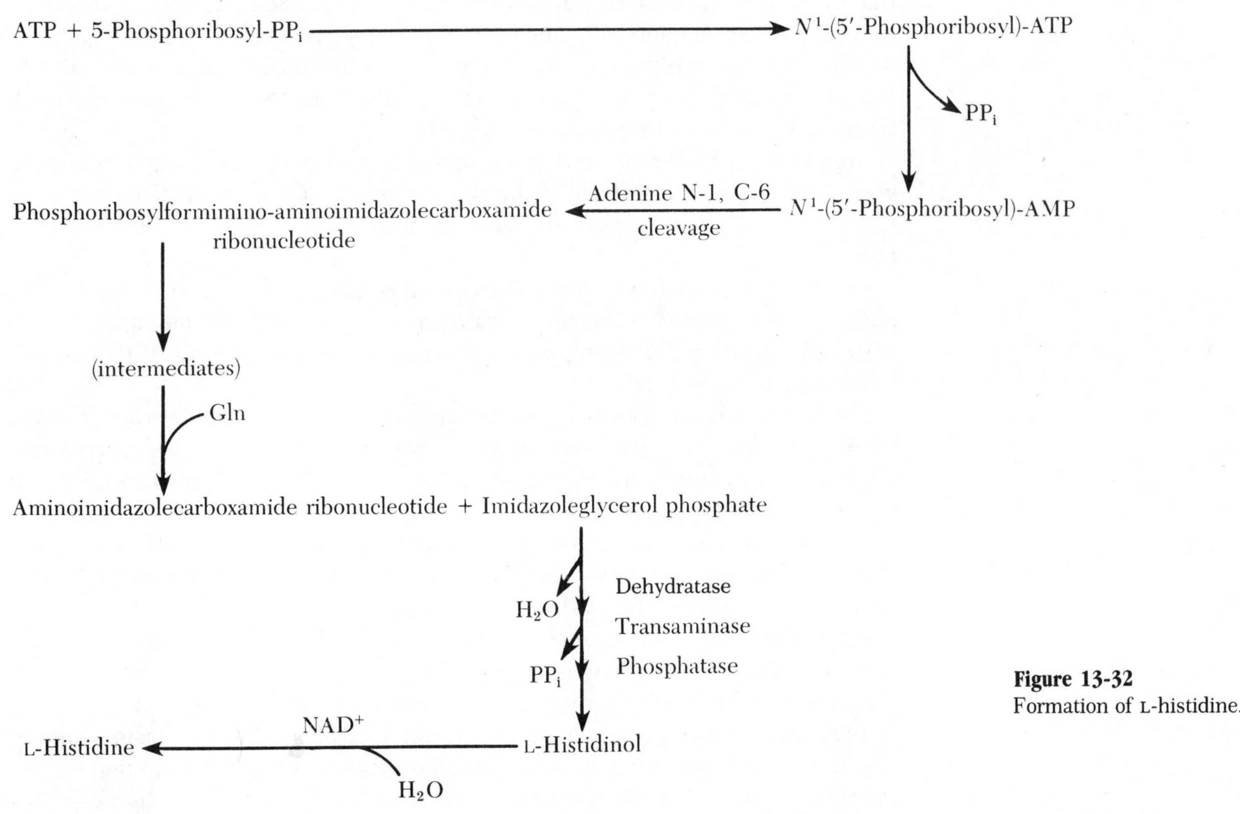

Figure 13-32
Formation of L-histidine.

Figure 13-33
Pathway for the catabolism of proline.

Proline

½ O_2 H_2O

Δ^1-Pyrroline-5-carboxylic acid

Glutamic-γ-semialdehyde

Ornithine

Glutamic acid

nature thus provides the fetus with a superabundance of this indispensable amino acid.

Histidine is clearly a glycogenic amino acid because of its conversion to glutamate.

A hereditary disorder of histidine metabolism, histidinemia, is discussed in Chapter 22.

Histidine is synthesized in microorganisms by a rather complex series of reactions, similar to some of those involved in purine formation (Chapter 14). The pathway may be summarized schematically as in Fig. 13-32.

Proline and hydroxyproline When renal tissue was incubated with proline, α-ketoglutarate, ammonia, and a substance at first thought to be glutamine were found. Later work demonstrated that ornithine was also present, as was glutamate rather than glutamine. Further studies showed that proline was glycogenic. These observations have been confirmed and form the basis for the presently accepted pathway for the catabolism of proline, as presented in Fig. 13-32.

The first reaction is catalyzed by *proline oxidase,* and the second by a dehydrogenase. Both enzymes are found in liver mitochondria. The proline ring can be opened by cleavage at a Δ^5-double bond by proline oxidase, as already mentioned. In this case, α-keto-δ-aminovaleric acid is produced.

Proline and ornithine are readily interconverted in the body (Fig. 13-30). Thus, another metabolic route is seen; i.e., proline may yield ornithine for urea synthesis, or ornithine may be broken down via proline. Proline is a nonessential amino acid and is readily formed in the organism from glutamic acid or ornithine by a reversal of these reactions.

Hydroxyproline is metabolized by routes similar to those of proline (Fig. 13-33). It is metabolized in the liver by a conversion catalyzed by an oxidase into γ-hydroxyglutamic acid through Δ^1-pyrroline-3-hydroxy-5-carboxylic acid as an intermediate. Transamination then occurs with oxaloacetate to form α-hydroxy-γ-ketoglutaric acid and aspartic acid. The former is split by an aldolase type of reaction by way of a Schiff base into glyoxylic acid and pyruvic acid, which are finally metabolized by now familiar routes. The enzyme involved, α-ketoglutaric aldolase, has been purified from bovine liver and found to be a tetrameric protein with a molecular weight of about 125,000. It has no metal ion requirement and is not stereospecific.

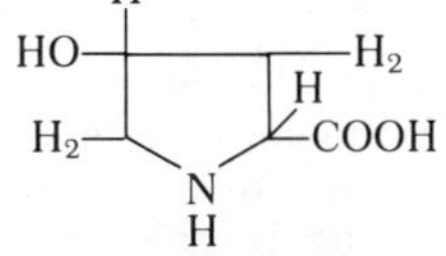

4-Hydroxyproline

4-Hydroxyproline has been isolated from beef tendon collagen, as has a 3-hydroxy form. The 4-hydroxy form apparently predominates. The same enzymes act on both forms, resulting in similar reactions. Both forms also are present in the skeletal proteins of a number of animal species.

Hydroxyproline, like proline, is classed as a glycogenic amino acid. It is an important constituent of collagen (Chapter 15) and is formed from proline by the ascorbic acid–dependent, oxygen-using hydroxylation of proline, apparently after it has been incorporated into a polypeptide in collagen biosynthesis.

Several hereditary disorders or proline and hydroxyproline metabolism have been described (Chapter 22).

Suggested readings

Bondy, P.K., and Rosenberg, L.E., editors: Metabolic control and disease, ed. 8, Philadelphia, 1980, W.B. Saunders Co.

Ciba Foundation Symposia: Protein turnover, No. 9, New York, 1973, Associated Scientific Publishers.

Felig, P.: Amino acid metabolism in man, Annu. Rev. Biochem. **44:**933, 1975.

Frimpter, G.W.: Aminoacidurias due to inherited disorders of metabolism, New Eng. J. Med. **289:**835, 895, 1973.

Garrod, A.E.: Inborn errors of metabolism. Reprinted with supplement by H. Harris. Fair Lawn, N.J., 1963, Oxford University Press.

Greenberg, D.M., editor: Metabolic pathways, vols. 1 and 3, ed. 3, New York, 1969, Academic Press, Inc.

Holmes, F.L.: Hans Krebs and the discovery of the ornithine cycle, Fed. Proc. **39:**216, 1980.

Hsia, D.Y.-Y.: Inborn errors of metabolism, ed. 2, Chicago, 1966, Year Book Medical Publishers, Inc.

Larner, J.: Intermediary metabolism and its regulation, Englewood Cliffs, N.J., 1971, Prentice-Hall, Inc.

Meister, A.: Biochemistry of amino acids, New York, 1965, Academic Press, Inc.

Milunsky, A.: The prenatal diagnosis of hereditary disorders, Springfield, Ill., 1973, Charles C Thomas, Publisher.

Morgan, H.E.: Regulation of protein turnover. Fed. Proc. Symp. **33:**1091, 1974.

Morris, D.R., and Fillingame, R.H.: Regulation of amino acid decarboxylation, Annu. Rev. Biochem. **43:**303, 1974.

Munro, H.N., and Allison, J.B., editors: Mammalian protein metabolism, New York, 1964, Academic Press, Inc.

Neurath, H., and Hill, R.E., editors: The proteins: composition, structure, and function, ed. 3, New York, 1979, Academic Press, Inc.

Nyhan, W.L., editor: Heritable disorders of amino acid metabolism, New York, 1974, John Wiley & Sons, Inc. (Interscience Publishers).

Orten, A.U.: Intestinal phase of amino acid nutrition. Fed. Proc. Symp. **22:**1103, 1963.

Prusnier, S., and Stadtman, E.R., editors: The enzymes of glutamine metabolism, New York, 1973, Academic Press, Inc.

Ratner, S.: A long view of nitrogen metabolism. Annu. Rev. Biochem. **46:**1, 1977.

San Pietro, A., editor: Photosynthesis and nitrogen fixation. In Colowick, S.P., and Kaplan, N.O., editors: Methods in enzymology, vol. 69, part C, New York, 1980, Academic Press, Inc.

Schoenheimer, R.: The dynamic state of body constituents, Cambridge, Mass. 1942, Harvard University Press.

Stanbury, J.B., Wyngaarden, J.B., and Fredrickson, D.S., editors: The metabolic basis of inherited disease, ed. 4, New York, 1978, McGraw-Hill Book Co.

Tabor, H., and Tabor, C.W., editors: Metabolism of amino acids and amines. In Colowick, S.P., and Kaplan, N.O., editors: Methods in enzymology, vol. XVII, New York, 1971, Academic Press, Inc.

Truffi-Bachi, P., and Saari, J.C.: Amino acid metabolism, Annu. Rev. Biochem. **42:**113, 1973.

Umbarger, H.E.: Amino acid biosynthesis and its regulation, Annu. Rev. Biochem. **47:**533, 1978.

Vogel, H.J., editor: Metabolic pathways, New York, 1971, vol. V, Academic Press, Inc.

Weinstein, B., editor: Chemistry and biochemistry of amino acids, peptides, and proteins, vol. 4, New York, 1977, Marcel Dekker, Inc.

Winter, H., and Burris, R.H.: Nitrogenase, Annu. Rev. Biochem. **45:**409, 1976.

Zelitch, I.: Pathways of carbon fixation in green plants, Annu. Rev. Biochem. **44:**123, 1975.

Chapter 14

Metabolism of purines and pyrimidines

Although nucleoproteins are a factor in the diet, they do not provide essential constituents for the biosynthesis of endogenous nucleic acids. In other words, the biosynthesis of cellular nucleic acids is largely dependent on the *de novo* synthesis of purine and pyrimidine nucleotides and is not contingent on a dietary source of bases. Nevertheless, dietary nucleic acids are liberated in the intestinal tract from food nucleoproteins by the action of the proteolytic enzymes (Chapter 13). These polynucleotides in turn are depolymerized by the pancreatic nucleases, ribonuclease and deoxyribonuclease. Intestinal phosphodiesterases assist in converting the nucleic acids into a mixture of nucleotides (Fig. 14-1). These are then attacked by phosphatases or nucleotidases, yielding phosphate and purine and pyrimidine nucleosides. Nucleosides are either absorbed as such or are hydrolyzed by nucleosidases to the constituent bases and pentoses. Absorbed nucleosides are degraded by a phosphorolytic process (nucleoside phosphorylase) in the liver, spleen, and kidneys to free bases and pentose 1-phosphates.

Various factors influence the turnover of purines and pyrimidines (Fig. 14-2). Most of the bases are derived by their complete biosynthesis from simple precur-

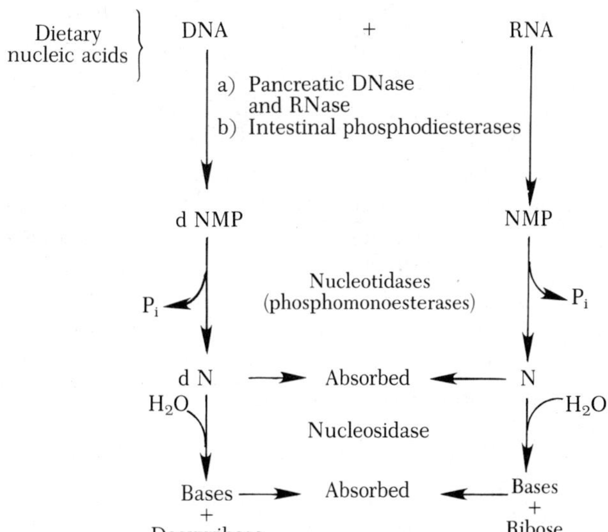

Figure 14-1
Pathway showing the digestion of dietary nucleic acids.

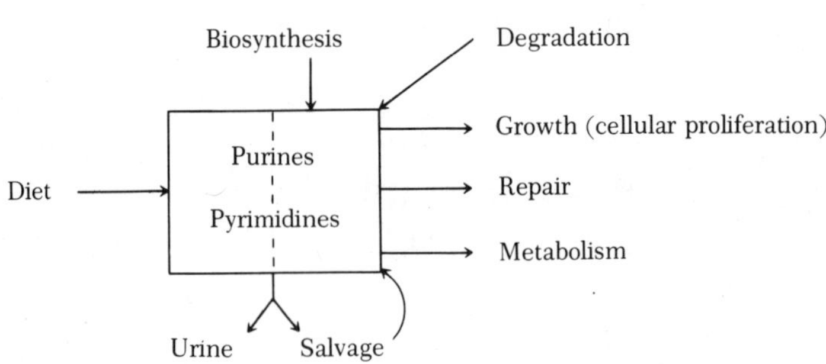

Figure 14-2
Various factors affecting the turnover of purines and pyrimidines. The pool of bases is divided to suggest the existence of a separate handling of the dietary bases. Thus most of the bases derived from dietary nucleic acids are degraded and excreted. Some, especially adenine, are salvaged and reutilized for the formation of endogenous nucleic acids.

sors. The pool of bases is depleted either by degradation to simple metabolites, as in the pyrimidines, or excretion in the urine as uric acid, as with the purines. An active salvage mechanism also exists that serves to recycle the purines to their corresponding nucleotides. In addition, Fig. 14-2 suggests a dietary input of bases. However, as just indicated, this is of minor consequence. Most of the dietary bases are degraded in the intestinal mucosa; adenine, however, may be used in the biosynthesis of nucleic acids.

Metabolic fate of nucleic acids

It is generally presumed that the intestinal digestion of nucleic acids and their catabolism in tissues proceed by relatively similar pathways. In either case the initial process is one of depolymerization by *deoxyribonuclease* or *ribonuclease,* specific for DNA and RNA, respectively. The deoxyribonuclease found in the pancreas, essential for the digestion of dietary DNA, is an enzyme that hydrolyzes the 3′,5′-phosphodiester linkages to produce oligonucleotides, with 5′-phospho terminals; large amounts of mononucleotides are not released. Another deoxyribonuclease is found in the spleen. It hydrolyzes the phosphodiester bond between the phosphorus and the 5′-carbon, thereby releasing 3′-phospho–terminal oligonucleotides. Intestinal mucosa, as well as the pancreas and liver, contains a ribonuclease that acts on the phosphodiester links of RNA between the phosphorus and 5′-carbon, thus releasing 3′-phosphonucleotides. This enzyme requires a pyrimidine base to be part of the nucleoside contributing the 3′-phospho terminus (see diagram at top of p. 377). As indicated, an intermediate stage in the hydrolysis of RNA by ribonuclease is the formation of a 2′,3′-phosphodiester linkage. When this step is prevented, as by the reduction of carbon-2, or blocked, as by methylation, the activity of the enzyme is inhibited.

In the intestine the oligonucleotides are further hydrolyzed to individual nucleotides (phosphodiesterases), which in turn are attacked by nucleotidases or phosphomonoesterases; their products are nucleosides. These steps are summarized in Fig. 14-1. Intestinal absorption may occur at this stage, or the nucleosides may be further degraded to their constituent bases and pentoses by nucleosidases. Absorbed nucleosides are also degraded in the liver by a phosphorolytic process involving nucleoside phosphorylases. These enzymes release purines and pyrimidine bases and pentose phosphates.

$$\text{Purine nucleoside} \xrightarrow[P_i]{\text{Purine nucleoside phosphorylase}} \text{Ribose 1-phosphate} + \text{Purine}$$

As just mentioned, the adenine from dietary nucleic acids may be used, in part at least, for the synthesis of nucleic acids in the tissues. This salvage process, described later in the chapter, is not essential to nucleic acid production; most is derived from nucleotides newly synthesized in the cells.

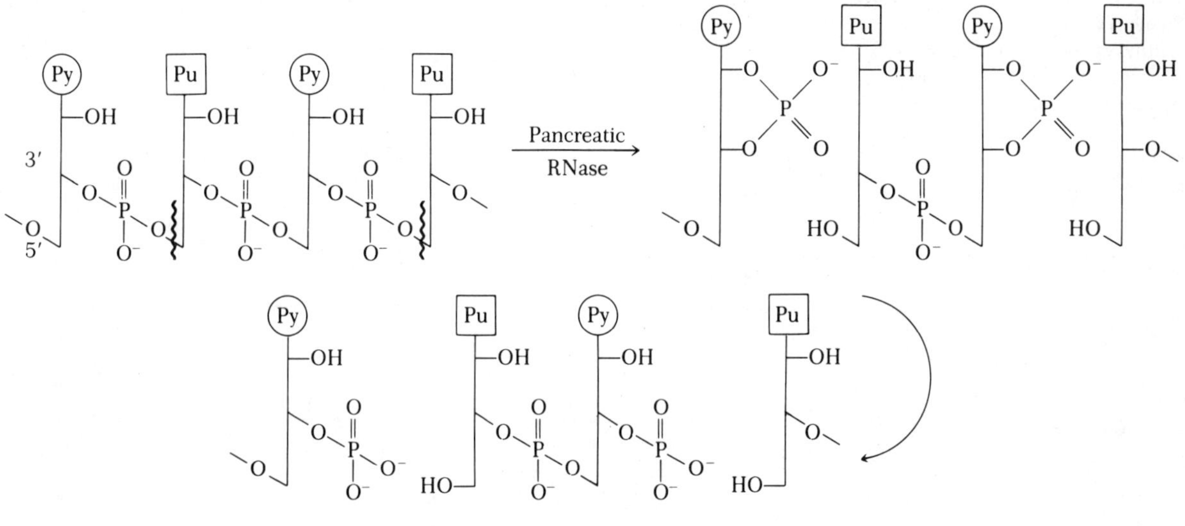

Catabolism of purines

The purine nucleotides derived from dietary or endogenous nucleic acids are degraded (Fig. 14-3). The nucleotides first are converted to their respective nucleosides. Adenosine is hydrolytically converted to inosine by the *adenosine deaminase* found in many tissues. Inosine and guanosine are then converted by a phosphorylase (purine nucleoside phosphorylase) to their constituent purines and pentose 1-phosphates. Guanine is then deaminated by guanase (guanine deaminase) to yield xanthine, whereas hypoxanthine is oxidized to xanthine by *xanthine oxidase,* a process involving molecular oxygen (see following illustration). Xanthine oxidase, found in the liver and intestinal mucosa, is a molybdenum and nonheme iron–containing flavoprotein (Chapter 6). In skeletal muscle the 5′-AMP derived from the degradation of ATP is degraded to IMP (Fig. 14-3) using the enzyme

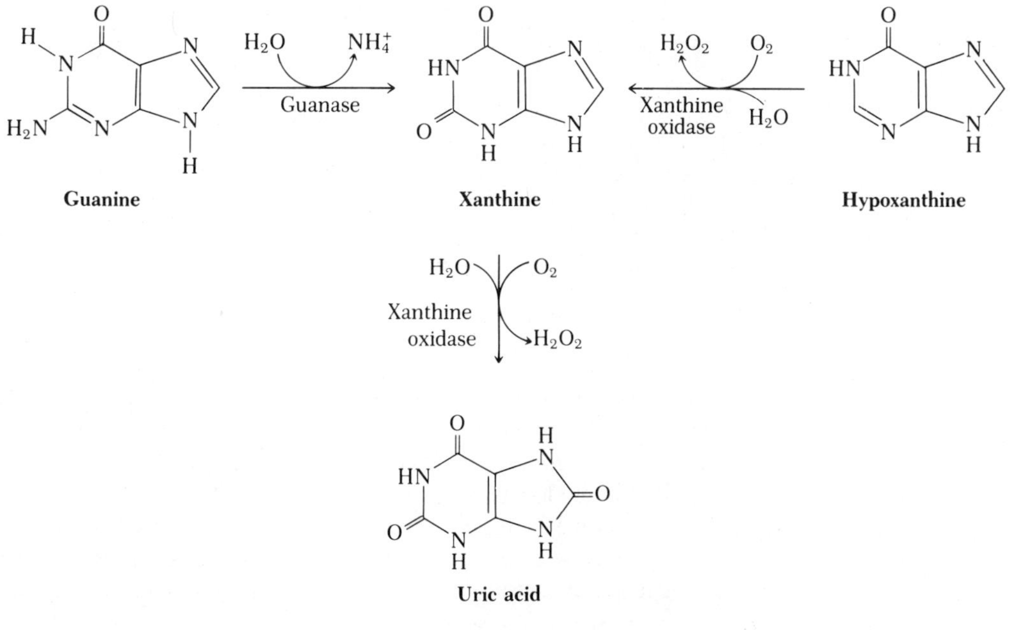

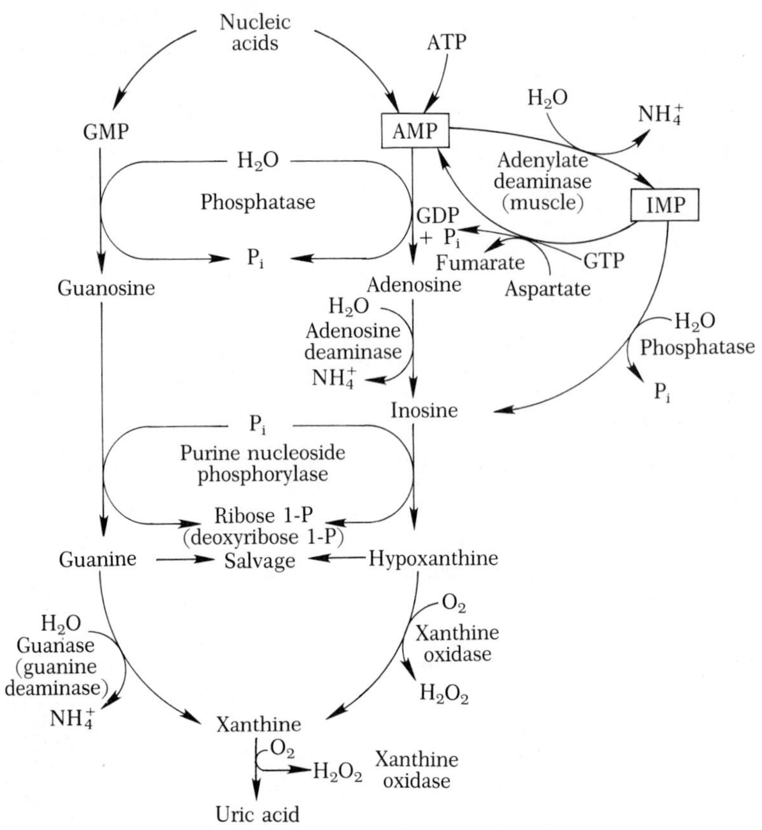

Figure 14-3
Pathways showing the degradation of purine nucleotides ultimately to uric acid.

adenylate deaminase. This results in the release of significant amounts of ammonia. The IMP produced can be restored to AMP at the expense of the amino group from aspartate and energy supplied by GTP.

As seen in Fig. 14-3, the final end product of purine metabolism in humans is uric acid. Again, xanthine oxidase is responsible for the final oxidation of xanthine. As indicated below, uric acid may exist in tautomeric enol and keto forms. The

Xanthine $\xrightarrow{H_2O,\ O_2 \ \rightarrow\ H_2O_2}$ **Uric acid** (**Enol form**) $\longleftrightarrow$ (**Keto form**)

hydroxyl group on carbon-8 has a pK of approximately 5.4 and therefore exists in its anionic state (sodium urate) at physiological pH. Uric acid is excreted at the rate of 0.4 to 0.8 gm/24 hr. This level of excretion varies somewhat according to diet. Foods rich in nucleoproteins result in an elevated excretion. The urinary excretion of uric acid is partly regulated by endocrine hormones, e.g., the administration of 11-hydroxy steroids or ACTH causes an increase. It is not known whether this is caused by a stimulated biosynthesis or by an elevated elimination of uric acid.

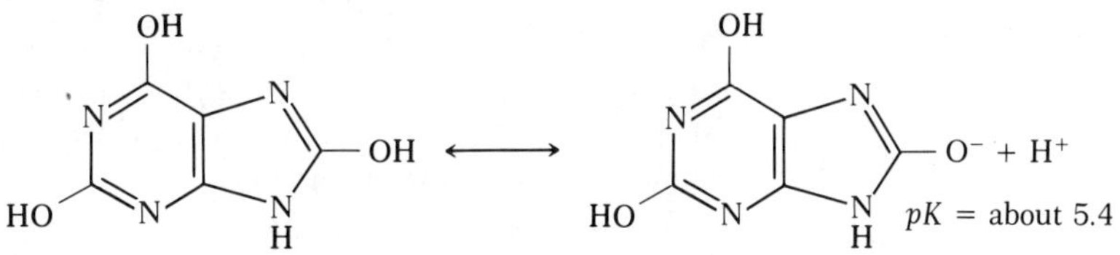

Uric acid (enol form)

pK = about 5.4

Catabolism of pyrimidines

The exact degradative pathway for pyrimidines is not completely clear. However, the amino group of cytosine may be removed either from the free base (Fig. 14-4) or from the nucleoside cytidine. In the latter case the enzyme is *cytidine deaminase*. The product, uridine, is then cleaved by phosphorolysis to release uracil and the appropriate pentose 1-phosphate. The free base cytosine may also be deaminated to uracil (Fig. 14-4, step (1)). The same occurs with thymine and thymidine. The uracil is reduced to dihydrouracil, and thymine is reduced to dihydrothymine. The enzyme involved in step (2) is the $NADP^+$, requiring *dihydropyrimidine dehydrogenase. Dihydropyrimidine amidohydrolase* (step (3)) catalyzes the hydrolysis of these intermediates to the corresponding β-ureido compounds. Uracil (and cytosine) yields β-ureidopropionic acid. Hydrolysis at step (4) by *ureidopropionase*

Figure 14-4
Pathway showing the catabolism of cytosine.

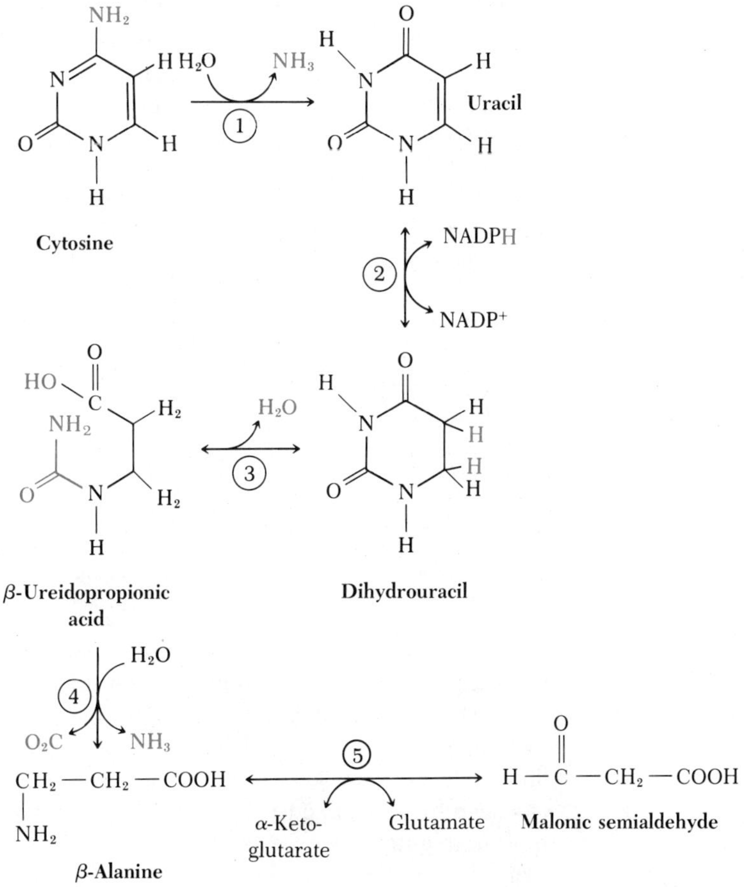

Cytosine

Uracil

Dihydrouracil

β-Ureidopropionic acid

β-Alanine

α-Ketoglutarate

Glutamate

Malonic semialdehyde

then produces β-alanine, beginning with uracil and β-aminoisobutyrate from thymine. The latter is converted to methylmalonylsemialdehyde by transamination (step (5)). Further metabolism occurs by way of its CoA derivative, as shown in the metabolism of valine (Chapter 13).

The catabolism of cytosine and uracil may serve as an important source of β-alanine in the mammalian system. Although this amino acid is a constituent of pantothenic acid and therefore of CoA, synthesis of this vitamin does not occur in humans. However, β-alanine is a component of certain dipeptides, carnosine (β-alanylhistidine) and anserine (β-alanylmethylhistidine), which are synthesized by muscle and brain.

Biosynthesis of purines

Early experiments using isotopically labeled compounds showed that purines are synthesized from simple substances such as amino acids, CO_2, ammonia, and 1-carbon units.

When ^{15}N-labeled glycine was fed to humans, the end product of purine catabolism, uric acid, was found to contain the isotope in position 7. In other words, nitrogen-7 of the purine ring is derived from the amino acid glycine (Fig. 14-5). Other studies showed that nitrogen in position 1 (N-1) is from the amino group of aspartic acid, whereas the amide nitrogen of glutamine becomes N-9 and probably N-3. Glycine contributes its carboxyl carbon to the C-4 of the purine nucleus, its α-carbon becomes C-5, and its amino group becomes N-7. One-carbon units carried as the N^{10}-formyl and N^5,N^{10}-methenyl tetrahydrofolic acid contribute C-2 and C-8, respectively; CO_2 contributes C-6. Thus it appears that the precursors of purines are glycine, aspartate, glutamine, CO_2, 1-carbon units, and various intermediates that yield these compounds in metabolism (Fig. 14-5).

The biosynthesis of the purines begins with ribose 5-phosphate, derived from the pentose phosphate pathway, which is converted to 5-phosphoribosyl 1-pyrophosphate (PRPP) by the transfer of the pyrophosphate group from ATP to C-1 of the ribose. This reaction is catalyzed by the enzyme PRPP synthetase (ATP:D-ribose 5-phosphate pyrophosphotransferase). PRPP is important to both pyrimidine and purine biosynthesis, as emphasized in the following scheme:

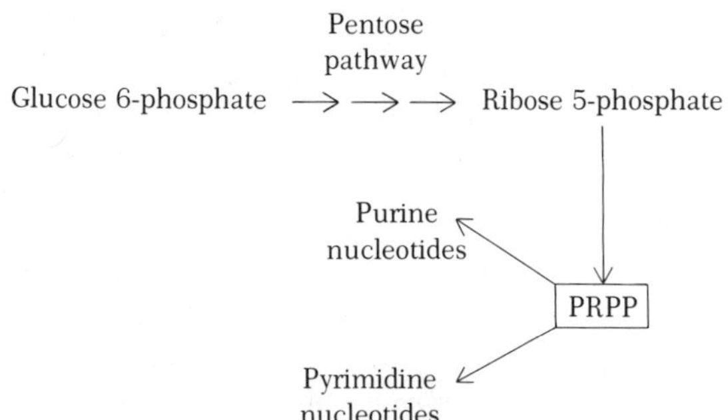

PRPP synthetase has an absolute requirement for inorganic phosphate, whereas purine and pyrimidine nucleotides inhibit the enzyme. An accumulation of phosphate resulting from increased metabolic use of nucleotides will stimulate PRPP production. This is then available for an increased synthesis of purine and pyrimidine nucleotides as well as the salvage process, which is described later.

As seen in step (1) on p. 381, PRPP is aminated by the addition of the amide group from glutamine; the amino sugar 5-phosphoribosyl 1-amine is the product. In the diagram it is apparent that an inversion of configuration has occurred about carbon-1. Thus the 5-phosphoribosylamine has the β-configuration, whereas that

Figure 14-5
Sources of individual atoms in the purine ring.

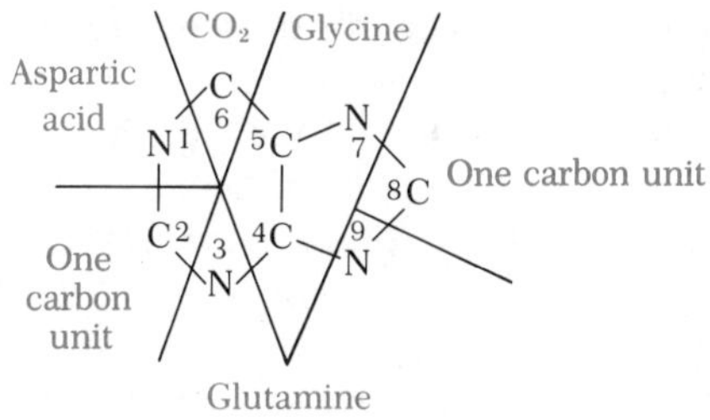

of PRPP has the α-configuration. The enzyme that catalyzes the transfer of the amide nitrogen is called *5-phosphoribosyl 1-pyrophosphate amidotransferase*.

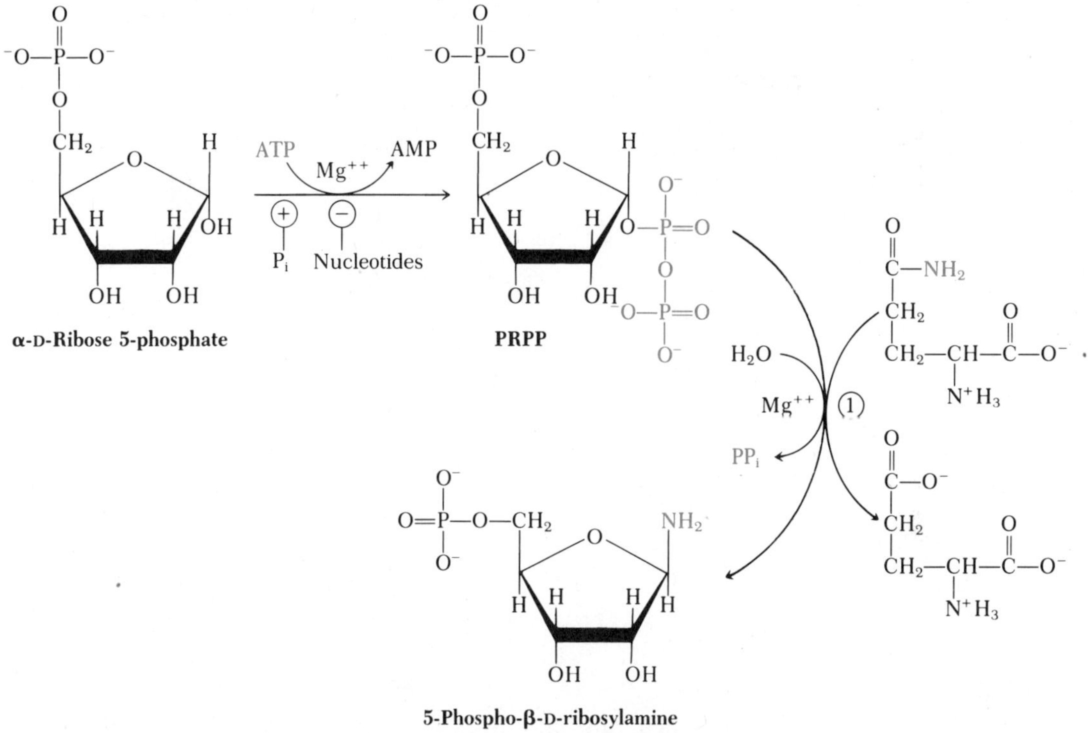

Formation of inosine monophosphate

The synthesis of 5-phosphoribosylamine from PRPP (step (1) in preceding diagram) is the first committed step in the pathway yielding inosine monophosphate (IMP) and is the point at which regulatory control is exerted over this pathway. Purine nucleotides strongly inhibit this reaction. Fig. 14-5 shows that the amino nitrogen of phosphoribosylamine provides N-9 of the completed purine ring. In other words, purine biosynthesis is a building-up process beginning with the sugar phosphate that in turn will be that part of the finished nucleotide, IMP. (In subsequent diagrams the phosphoribose entity of the developing nucleotide is simply abbreviated as RP.)

Carbon-4 and -5, as well as N-7, are next provided by the addition of the amino acid glycine. In step (2) (following diagram) ATP and the enzyme *phosphoribosyl glycinamide synthetase* are required. The name of the product is phosphoribosyl glycinamide or glycinamide ribonucleotide (GAR).

A 1-carbon unit is next transferred to the free amino group of GAR in reaction (3) to form *N*-formylglycinamide ribonucleotide (FGAR). The vitamin folic acid is important to purine biosynthesis because the coenzyme FH_4 (Chapter 6) serves as the carrier of the 1-carbon unit, N^5,N^{10}-methenyl tetrahydrofolic acid. *Phosphoribosyl glycinamide formyl transferase* is the enzyme involved. FGAR then has the required number of atoms to produce the five-membered imidazole portion of the purine structure. Before ring closure occurs, the N-3 of the purine structure is introduced by another amination using glutamine and ATP (reaction (4)). The enzyme is *phosphoribosyl-N-formylglycinamidine synthetase* and, as seen below, the product is 5-phosphoribosyl-*N*-formylglycinamidine (FGAM).

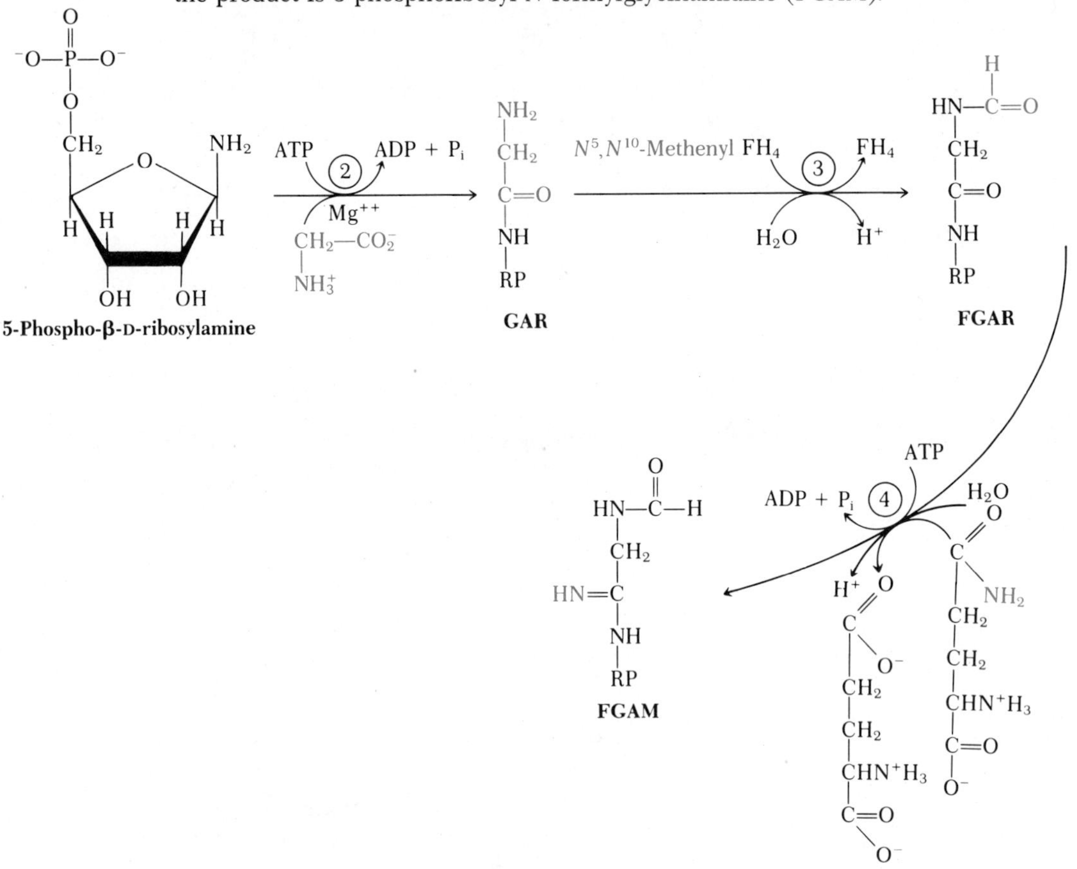

5-Phospho-β-D-ribosylamine **GAR** **FGAR**

FGAM

When ring closure occurs, the product is *5-aminoimidazole ribonucleotide* (AIR). The enzyme for this fifth step (see (5) in following diagram) is *phosphoribosylaminoimidazole synthetase* and requires ATP.

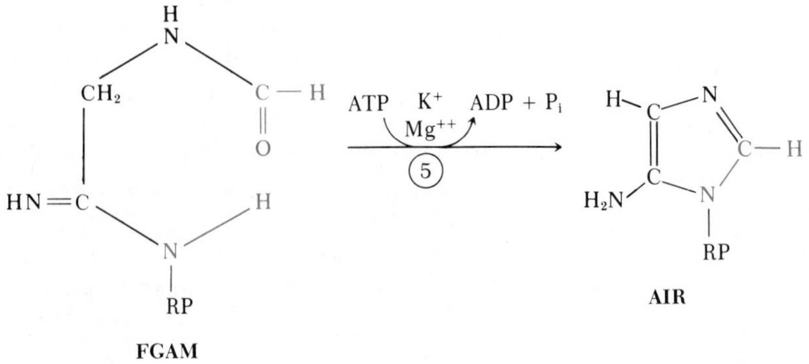

FGAM **AIR**

Carbon-6 of the purine structure (Fig. 14-5) is introduced as CO_2, a reaction catalyzed by *phosphoribosylimidazole carboxylase* (reaction (6)); the product, 5-phosphoribosyl 5-aminoimidazole 4-carboxylate, next accepts an amino group (N-1 of purine) from the amino acid aspartate. This and the following reaction are reminiscent of the biosynthesis of argininosuccinate and arginine with the release of fumarate in the urea cycle (Chapter 13). Thus the enzyme *phosphoribosylaminoimidazolesuccinocarboxamide synthetase* (reaction (7)), which requires ATP, produces 5-phosphoribosyl 5-aminoimidazole 4-(*N*-succino) carboxamide, Fumarate, released in step (8) by *adenylosuccinase,* enters the citric acid cycle. The product of this reaction is phosphoribosyl 5-aminoimidazole 4-carboxamide.

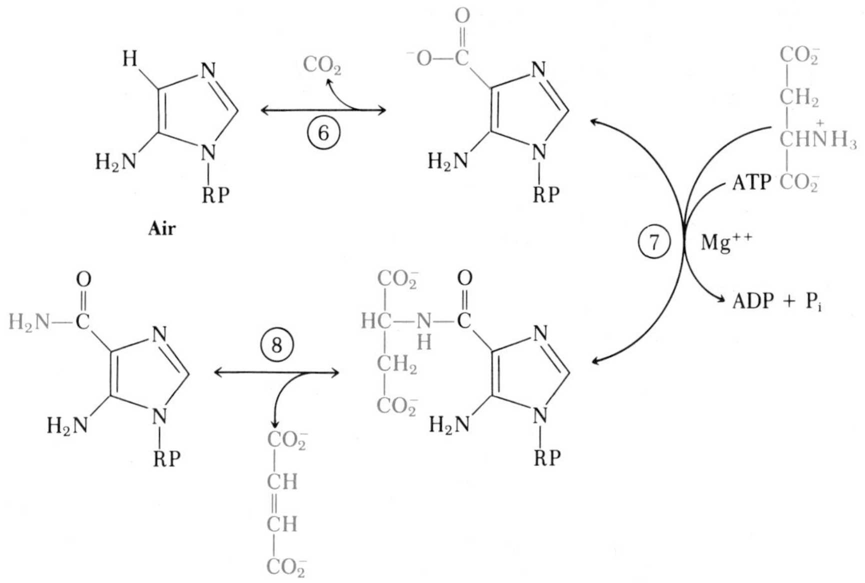

Air

Only carbon-2 (Fig. 14-5) remains to be added before the complete purine ring can be formed. This carbon is contributed in reaction (9) (following) by N^{10}-formyl tetrahydrofolic acid; the enzyme is *phosphoribosylaminoimidazolecarboxamide formyl transferase.* Water is then removed between the carboxamide and formyl groups to yield inosine 5-phosphate (IMP); the enzyme of reaction (10) is called *inosinicase.* Inosinic acid is the first actual purine nucleotide formed.

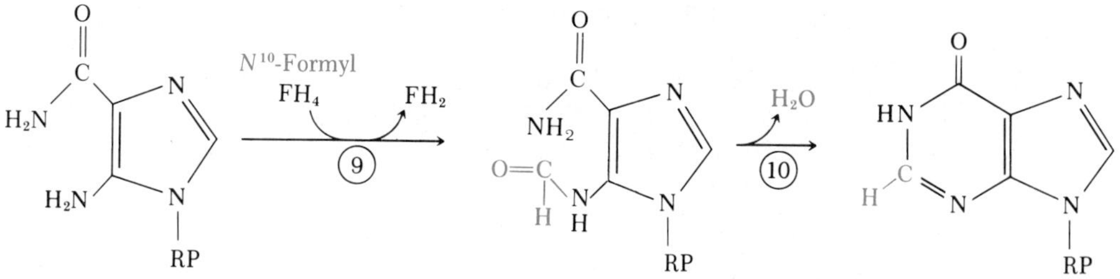

IMP

Conversion of IMP to GMP and AMP

Both adenosine and guanosine monophosphates are produced from inosinic acid using the nitrogens of glutamine and aspartate. As seen in Fig. 14-6, inosinic acid reacts in step (11) with aspartate in the presence of *adenylosuccinate synthetase* and GTP. *Adenylosuccinase* in turn catalyzes in reaction (12) the removal of fumarate to yield AMP. To produce GMP, the inosinic acid must first be oxidized to xanthine monophosphate or xanthylic acid by the NAD^+-requiring enzyme *inosine*

Figure 14-6 Feedback control of biosynthesis of AMP and GMP. Also noteworthy is that ATP is required for the biosynthesis of GMP, whereas GTP is required of AMP.

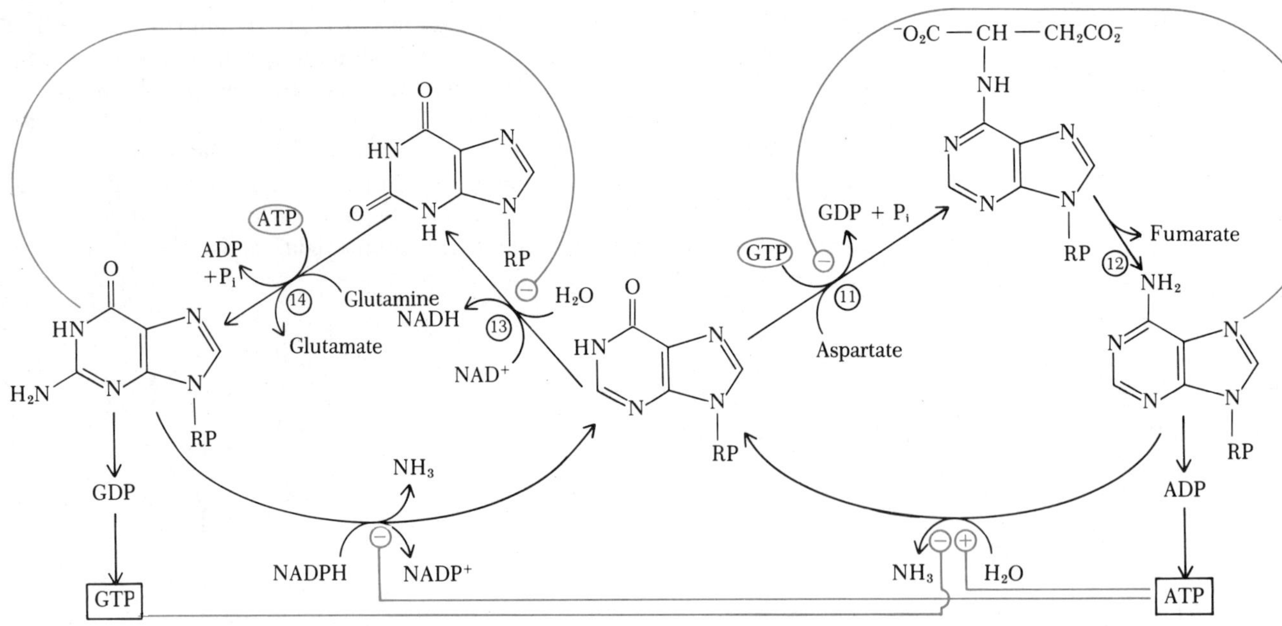

5'-phosphate dehydrogenase (reaction (13)). The latter then accepts an amino group at carbon-2 from glutamine in the presence of ATP in reaction (14); the enzyme is *guanosine 5'-phosphate synthetase.*

Since both nucleotides are needed for the synthesis of nucleic acids, their production must be balanced so that they are available in adequate amounts. This balance is achieved by the feedback system in Fig. 14-6. In bacterial systems it is believed that the ATP level increases GMP production; likewise, GTP increases AMP synthesis. The relative amounts of purine nucleotides are also affected by controlling their catabolism to IMP. ATP inhibits *GMP reductase,* which catalyzes the conversion of GMP to IMP. This reduces both the catabolism of GMP and the possibility that the nucleotide may serve as a precursor for AMP. GTP inhibits the deamination of AMP (adenylic acid deaminase), whereas ATP activates this enzyme.

As mentioned previously, control of the biosynthesis of purines also occurs by negative feedback of the purine nucleotides to the committed step. Thus ATP, ADP, AMP, GTP, GDP, and GMP inhibit allosterically the enzyme PRPP amidotransferase.

Gout

Understanding the biosynthetic pathway of purine formation has helped clarify the nature of certain types of gout, a disease that for centuries was associated with "high" or intemperate living. As mentioned earlier in this chapter, dietary purine contributes little to the overall purine pool. However, a diet rich in nucleic acids (containing meats such as liver, kidneys, or sweetbreads) may raise the blood urate levels sufficiently to promote gout in susceptible persons.

Gout is considered to be derived from both primary and secondary causes. Primary gout may be genetic and caused by defective enzymes such as PRPP synthetase and PRPP amidotransferase. Such defects might lead to a loss of feedback inhibition and a resulting overproduction of uric acid. In a study (Fig. 14-7) the

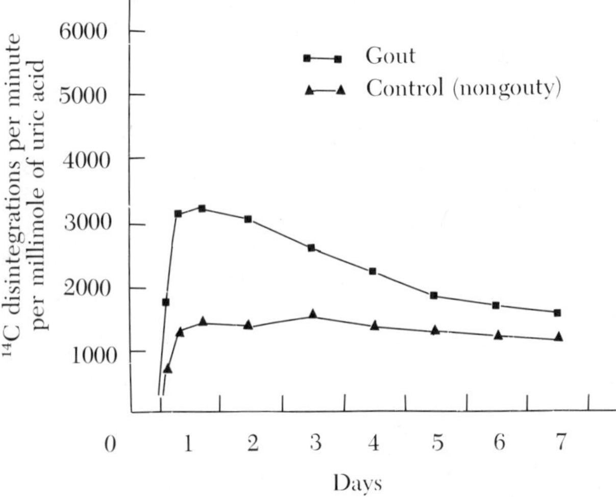

Figure 14-7
Incorporation of glycine-1-^{14}C into urinary uric acid in a control subject and in a patient with gout associated with excessive uric acid production.

Courtesy Seegmiller, J.E.: N. Engl. J. Med. **268:**712, 1963.

incorporation of ^{14}C-glycine into the urinary uric acid of a gouty patient was shown to be elevated compared with that of a normal control patient. Other causes of primary gout are an excessively rapid breakdown of purine nucleotides and a failure in renal function to clear the blood of sodium urate. Secondary gout results from a variety of diseases that cause an elevated destruction of cells, e.g., leukemia and polycythemia. Secondary gout can also be produced from the use of antimetabolites in the treatment of cancer.

Another factor leading to elevated blood urate levels is a genetic defect called Lesch-Nyhan syndrome. This defect occurs in the purine salvage process and leads to an elevated synthesis of IMP. It is described more fully later.

The average normal blood serum level of uric acid is 5 mg and 4 mg/100 ml for males and females, respectively. The upper normal range is 7 mg/100 ml, which is above the solubility of sodium urate in water, i.e., 6.4 mg/100 ml. A hyperuricemic individual has blood serum that apparently is supersaturated with respect to serum urate. Under circumstances leading to hyperuricemia (excessive uric acid in the blood) other avenues of uric acid removal from the blood besides urinary excretion are used by the body. This means that uric acid, as the monosodium salt, deposits in the joints and connective tissues as clusters of crystals called tophi.

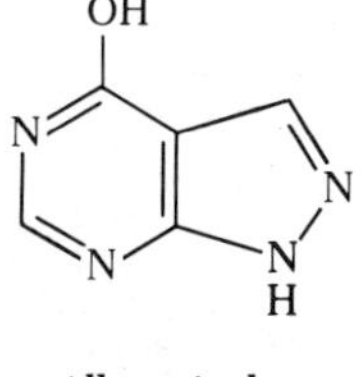

Allopurinol

A valuable drug for the control of gout is allopurinol, an analogue of hypoxanthine (see diagram at left). This compound is a competitive inhibitor of the enzyme xanthine oxidase and therefore reduces the production of uric acid from hypoxanthine and xanthine. Its use results in a pronounced reduction in blood uric acid concentrations accompanied by an elevation in hypoxanthine and xanthine of the serum and urine. Because of the solubility of hypoxanthine, it is excreted with greater facility than is uric acid.

Colchicine is also commonly used in the treatment of gout. Although this drug improves the clinical condition, it does not affect the blood urate levels. Its mode of action is not well understood.

Biosynthesis of pyrimidines

Pyrimidines also are formed from relatively simple precursors: an amino acid (aspartic acid), carbamoyl phosphate, and PRPP (Fig. 14-8). Thus N-1, C-4, C-5, and C-6 are derived from aspartic acid, whereas N-3 and C-2 are from carbamoyl phosphate (ultimately CO_2 and NH_3).

Figure 14-8
Sources of individual atoms in uridine monophosphate.

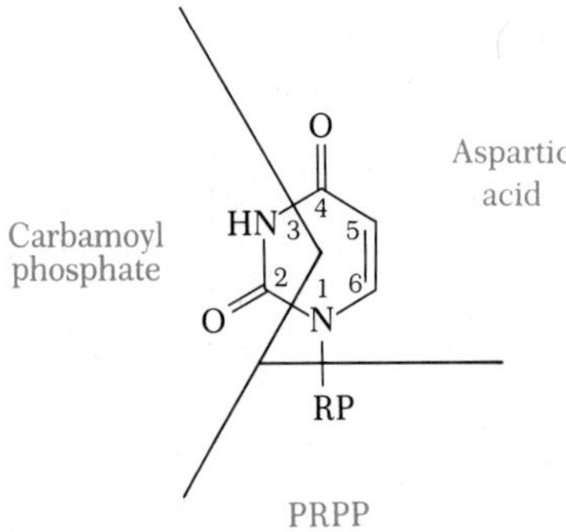

Figure 14-9
Pathway for pyrimidine biosynthesis in mammalian systems showing regulation by negative feedback of UTP and activation by PRPP to carbamoyl phosphate synthetase.

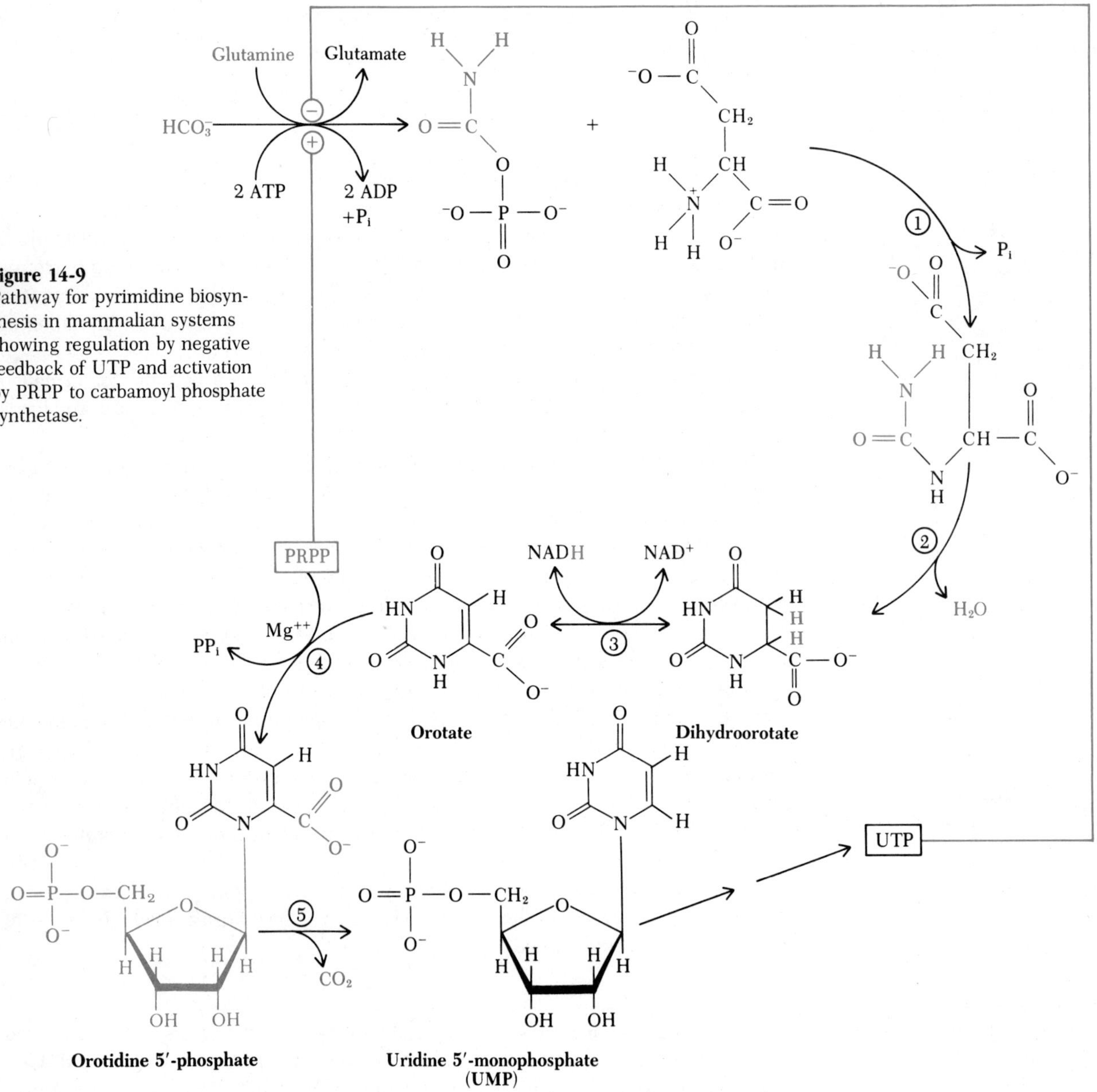

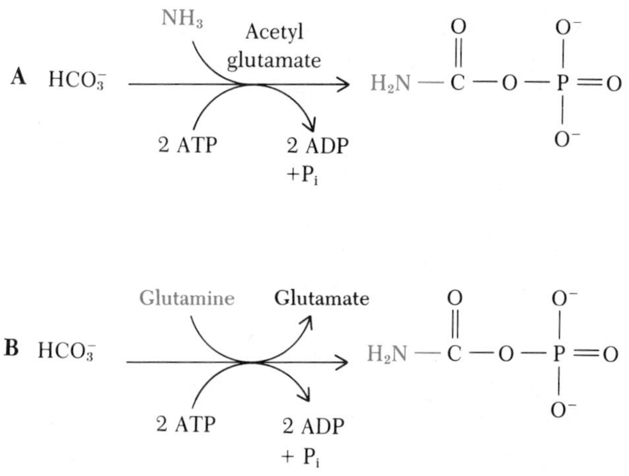

Figure 14-10
Comparison of carbamoyl phosphate synthetase systems found in hepatic mitochondria for urea production **(A)** and in the cytosol for pyrimidine biosynthesis **(B)**.

$$\textbf{A}\quad HCO_3^- \xrightarrow[2\ ATP \to 2\ ADP + P_i]{NH_3,\ \text{Acetyl glutamate}} H_2N-\overset{\displaystyle O}{\overset{\|}{C}}-O-\overset{\displaystyle O^-}{\overset{|}{P}}(=O)-O^-$$

$$\textbf{B}\quad HCO_3^- \xrightarrow[2\ ATP \to 2\ ADP + P_i]{\text{Glutamine} \to \text{Glutamate}} H_2N-\overset{\displaystyle O}{\overset{\|}{C}}-O-\overset{\displaystyle O^-}{\overset{|}{P}}(=O)-O^-$$

Several contrasting features are noteworthy in the biosynthesis of pyrimidines versus purines. In the purine ring the backbone amino acid is glycine; in the pyrimidine ring it is aspartic acid. Carbons are added in the biosynthesis of purines, as CO_2 or 1-carbon units, whereas nitrogen is added as aspartate or glutamine. The remaining carbon of the pyrimidine ring, not contributed by aspartate, is derived from HCO_3^- via carbamoyl phosphate. The purine ring is assembled beginning with the carbohydrate chain; the final product of the pathway is the nucleotide IMP. For the pyrimidine nucleotides the carbohydrate chain is added as PRPP after completion of the ring structure (orotic acid).

In step ① of pyrimidine biosynthesis (Fig. 14-9) the carbamoyl group of carbamoyl phosphate is transferred to the nitrogen of aspartic acid to yield *N*-carbamoylaspartic acid. The enzyme involved is *aspartyltranscarbamoylase (ATCase)*. Recall that carbamoyl phosphate is also an important precursor of citrulline in the urea cycle (Chapter 13). Urea synthesis, however, is a hepatic function, whereas pyrimidine biosynthesis is probably universal. Mammals have separate carbamoyl phosphate synthesizing systems for urea, found in the mitochondria, and pyrimidine, found in the cytosol. These are compared in Fig. 14-10. For urea synthesis (Fig. 14-10, *A*) the enzyme uses NH_3 directly and requires *N*-acetylglutamate as a coenzyme. The pyrimidine pathway (*B*) uses the amide nitrogen of glutamine for its source of nitrogen; it does not require acetylglutamate. Synthetic processes, such as this one, that require a source of nitrogen generally use glutamine rather than the fairly toxic ammonia. The pyrimidine-directed synthetase is a complex enzyme and may also be associated with ATCase.

ATCase of *E. coli* has been extensively studied; it is now the classic example of allosteric control by feedback by an end product of the biosynthetic pathway, i.e., CTP. In bacteria, ATCase is the first committed step in pyrimidine synthesis, unlike the mammalian cells in which the first committed step is actually the pyrimidine-channeled *carbamoyl phosphate synthetase.* This enzyme, or its complex with ATCase, seems to be the actual site of feedback by uridine triphosphate; it is also activated by PRPP (Fig. 14-9).

Carbamoyl phosphate synthetase is composed of two proteins: one synthesizes carbamoyl phosphate from NH_3 but not glutamine, and the other is a glutaminase. The glutaminase breaks down the glutamine, providing the NH_3 needed for the synthesis of carbamoyl phosphate. The regulation of activity occurs by way of specific sites on the synthetase subunit.

Step ② in pyrimidine ring biosynthesis (Fig. 14-9) involves the reversible dehydration by *dihydroorotase* to close the ring and yield *dihydroorotic acid*. Orotic acid is formed by the dehydrogenation of dihydroorotic acid (step ③), a reaction catalyzed by the NAD^+-requiring metalloflavoprotein *dihydroorotic acid dehydrogenase*. In step ④ orotic acid reacts with PRPP catalyzed by *orotate phosphoribosyl transferase* to form the nucleotide orotidine 5′-phosphate. Irreversible decarboxylation shown in step ⑤ is catalyzed by *orotidine 5′-phosphate decarboxylase* and produces uridine 5′-monophosphate (UMP).

A hereditary defect has been described in which the enzymes orotate phosphoribosyl transferase and orotidine 5′-phosphate decarboxylase are deficient. Consequently, orotic acid is excreted in the urine (orotic aciduria), and the resulting deficient synthesis of nucleotides leads to megaloblastic anemia, leukopenia, and retarded growth. In this disease pyrimidine salvage is essential to growth and development; the genetic loss actually produces a nutritional requirement for pyrimidines. Uridine has been used as a treatment. Since there is a reduction in UTP and the inherent loss of feedback control to the pyrimidine-channeled carbamoyl phosphate, pyrimidine synthesis is stimulated. However, because of the absence of the orotate phosphoribosyl transferase, the resultant orotic acid is excreted in the urine.

Orotic aciduria is also produced by two drugs: 6-azauridine (p. 394) and allopurinol (p. 385). The former is converted to 6-azauridylic acid and becomes a competitive inhibitor of orotidine 5′-phosphate decarboxylase. It blocks the metabolism of orotidine 5′-phosphate and stimulates pyrimidine biosynthesis by eliminating feedback control. Allopurinol, used in the treatment of gout, is converted to its ribotide, which is a competitive inhibitor of orotidine 5′-phosphate decarboxylase.

Other reactions involved in the biosynthesis of nucleoside triphosphates

Phosphorylation

As is apparent from the biosynthetic pathways outlined for purines and pyrimidines, nucleotides are the products. These are further phosphorylated at the expense of ATP, a reaction catalyzed by a group of *kinases*. As seen in the following diagram, the phosphorylation of the nucleoside monophosphates is catalyzed by specific nucleoside monophosphate kinases. In the example cited the enzyme is *uridylate kinase* ①. The second phosphorylation ② occurs by the action of a relatively nonspecific *nucleoside diphosphate kinase*.

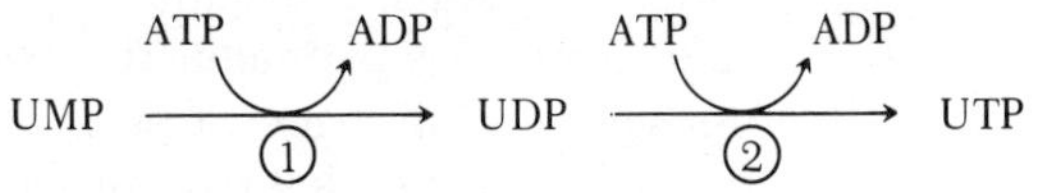

$$\text{UMP} \xrightarrow[\text{①}]{\text{ATP} \;\to\; \text{ADP}} \text{UDP} \xrightarrow[\text{②}]{\text{ATP} \;\to\; \text{ADP}} \text{UTP}$$

Conversion of UTP to CTP

Cytidine triphosphate is produced by the amination of UTP, a reaction catalyzed by *CTP synthetase*. The source of the nitrogen is glutamine; ATP is required probably to form a phosphorylated intermediate; Mg^{++} and GTP are also needed.

It is noteworthy that this enzyme uses ATP and UTP as substrates, produces CTP, and requires GTP as an allosteric effector. GTP increases the V_{max} tenfold and reduces the K_m for glutamine sixfold.

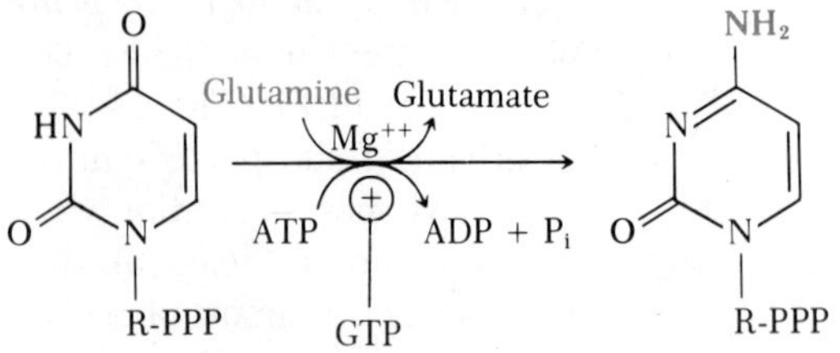

UTP (uracil ring, N–R-PPP) + Glutamine + ATP —(Mg^{++}; GTP ⊕)→ CTP (cytosine ring with NH_2, N–R-PPP) + Glutamate + ADP + P_i

UTP **CTP**

Formation of the deoxyribonucleoside triphosphates

The biosynthesis of DNA requires a ready supply of the four deoxyribonucleoside triphosphates. Reduction of ribose C-2 occurs on the nucleoside diphosphates and is catalyzed by the enzyme *ribonucleoside diphosphate reductase.* The ultimate source of hydrogen atoms is NADPH, but these are actually transmitted directly by a small protein (mol wt=11,700) called *thioredoxin*. Thioredoxin has one cystine residue per 108 amino acids. In the presence of the flavoprotein *thioredoxin reductase* (mol wt=68,000) the disulfide of the cystine is reduced to two sulfhydryl groups, as in reaction ① (following). In this way the hydrogens are passed to the nucleoside diphosphate, thereby reforming the disulfide linkage in thioredoxin. The reduction of the ribonucleoside diphosphates requires two other proteins of the nucleoside diphosphate reductase system, as in reaction ② (following). In mammalian cells the reduction of the purine and pyrimidine diphosphates can be catalyzed by a single enzyme. The resultant dUDP, dCDP, dADP, and dGDP are then phosphorylated at the expense of ATP to yield dUTP, dCTP, dATP, and dGTP.

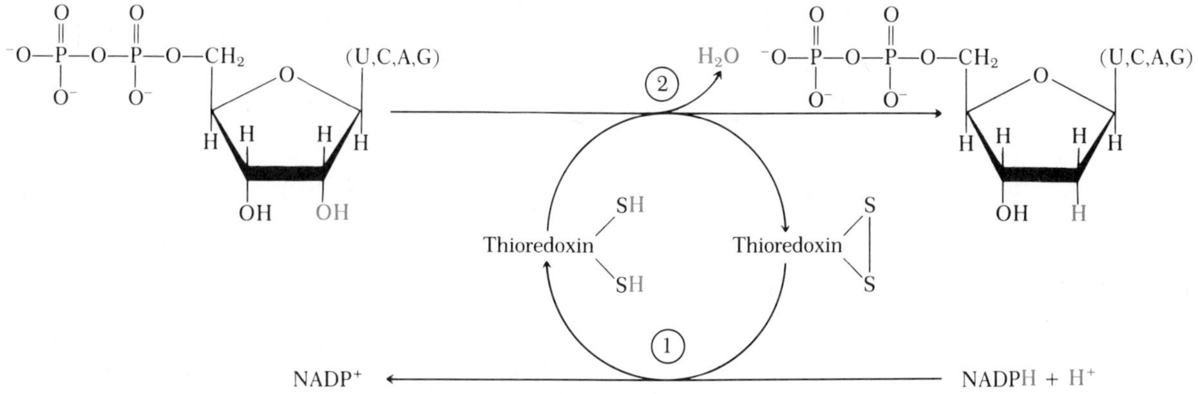

(U,C,A,G) ribonucleoside diphosphate → (② ; H_2O) → (U,C,A,G) deoxyribonucleoside diphosphate
Thioredoxin(SH)(SH) → Thioredoxin(S–S) ; ① $NADPH + H^+$ → $NADP^+$

Although, as just shown, dUDP may be produced, dUTP does not accumulate in the cell and is not available to the DNA polymerases. Rather, any dUTP is rapidly converted to dUMP by a pyrophosphatase (dUTPase). As seen in the following description, dUMP is a precursor for the biosynthesis of deoxyribothymidylate.

Biosynthesis of deoxyribothymidylate

The precursor of dTTP, required for DNA, is dUMP. In mammalian cells this is produced primarily by the deamination of dCTP rather than from dUDP. The sequence of steps involved in the formation of dTMP follow. The potential formation of dUMP from dUTP (dUTPase) is also included.

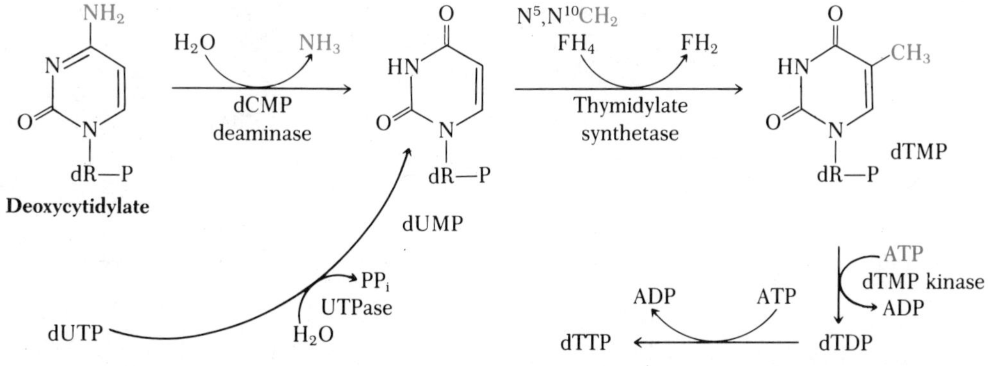

Deoxycytidylate (dR—P) → (H_2O → NH_3; dCMP deaminase) → dUMP (dR—P) → ($N^5,N^{10}CH_2$ FH_4 → FH_2; Thymidylate synthetase) → dTMP (dR—P)

dUTP → (H_2O → PP_i; UTPase) → dUMP

dTMP → (ATP → ADP; dTMP kinase) → dTDP → (ATP → ADP) → dTTP

As seen in the methylation step, catalyzed by *thymidylate synthetase* the N^5, N^{10}-methylene tetrahydrofolate donates the single carbon as well as the necessary hydrogen to produce the methyl group. Consequently, the cofactor is released as the dihydrofolate, which must be restored to the tetrahydro form. The enzyme

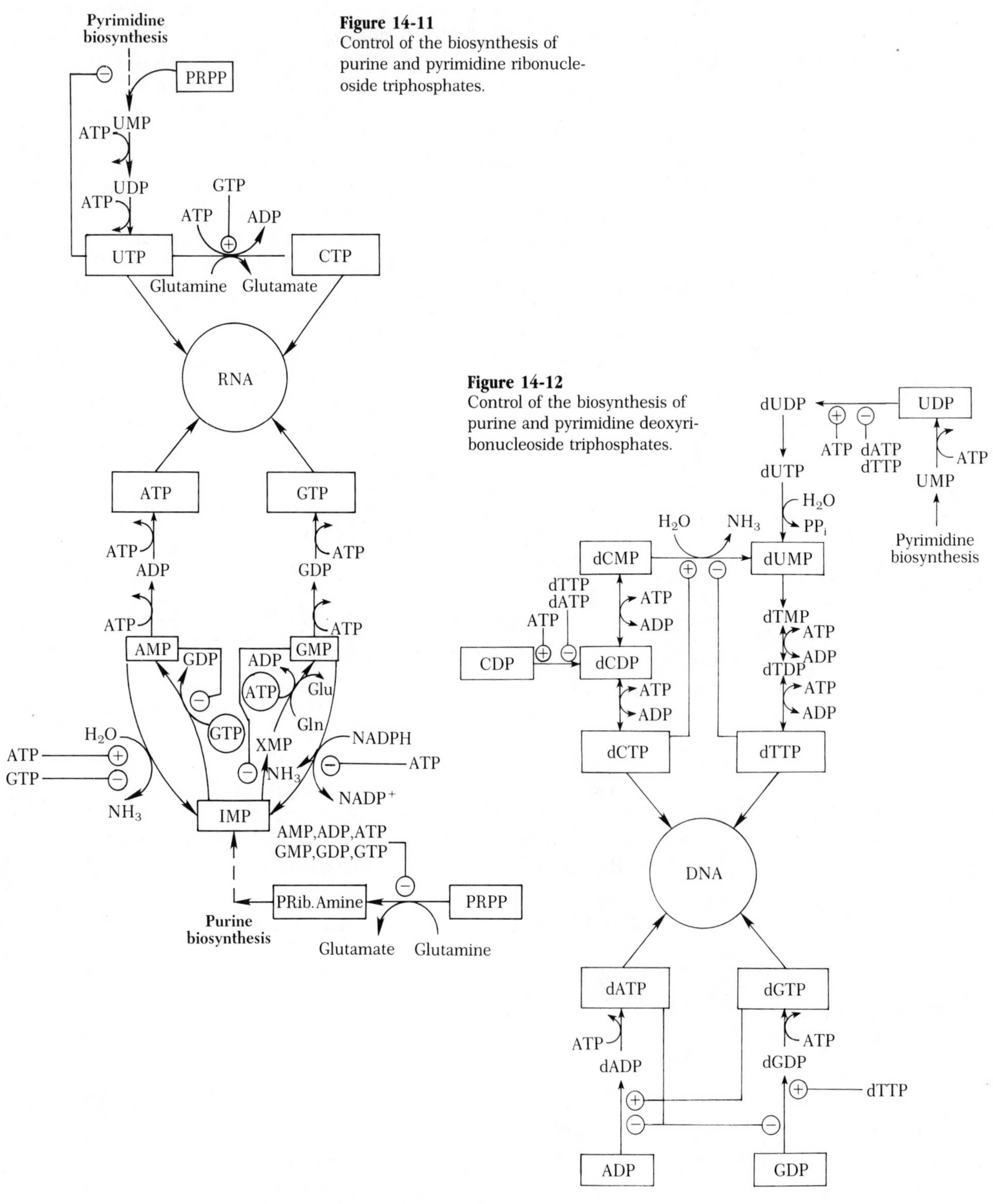

Figure 14-11
Control of the biosynthesis of purine and pyrimidine ribonucleoside triphosphates.

Figure 14-12
Control of the biosynthesis of purine and pyrimidine deoxyribonucleoside triphosphates.

dihydrofolate reductase and reducing equivalents from NADH are required to restore FH_4.

Control of the relative amounts of purine and pyrimidine nucleoside triphosphates

A homeostatic mechanism for balancing the supply of ribonucleoside and deoxyribonucleoside triphosphates is essential to the biosynthesis of RNA and DNA. Three levels of regulation occur: (1) in the biosynthesis of purines, (2) in the biosynthesis of pyrimidines, and (3) in the reduction of the ribonucleotides to deoxyribonucleotides. Both substrate availability and allosteric effects are involved. To illustrate the complex interrelationships that exist, we will consider the ribonucleotides (Fig. 14-11) independently of the deoxyribonucleotides (Fig. 14-12).

Ribonucleoside triphosphates The gross production of IMP is controlled by purine ribonucleotides, which inhibit PRPP transamidinase and regulate the formation of phosphoribosylamine from PRPP (Fig. 14-11).

In the conversion of IMP to GMP the oxidative production of XMP is inhibited by GMP (Fig. 14-6), whereas the amination requires ATP as a substrate. ATP is a negative effector for the enzyme that deaminates GMP. Thus the formation of GMP and therefore GTP is controlled by the level of ATP. GTP in turn is a substrate in the amination of IMP to AMP and a negative effector of the corresponding deaminase (Fig. 14-6). An accumulation of ATP stimulates the deaminase, which depletes AMP. Therefore the level of AMP is regulated by GTP (Fig. 14-11).

The biosynthesis of pyrimidines is controlled by negative feedback of UTP acting on carbamoyl phosphate synthetase. UTP in turn is converted to CTP, a process linked allosterically to GTP (Fig. 14-11).

Deoxyribonucleoside triphosphates In the formation of deoxyribonucleoside triphosphates, a key regulatory step is the nucleoside diphosphate reductase. In the absence of allosteric effectors this enzyme exhibits little activity. The various nucleoside triphosphates function to establish a homeostatic control of the balance of deoxyribonucleoside triphosphates needed in the biosynthesis of DNA. The interrelationships of the allosteric effectors are illustrated in Fig. 14-12. ATP provides the initial stimulus in the reduction of CDP and UDP. The dCDP formed is the primary precursor in the pathway leading to the formation of dTTP. This essential component of DNA then stimulates the reduction of a purine nucleotide, i.e., GDP. The corresponding deoxyribonucleoside triphosphate in turn stimulates the reduction of ADP. In this way the reductions of purine and pyrimidine nucleotides are coupled. An accumulation of dCTP stimulates dCMP deaminase, thereby encouraging the biosynthesis of dTTP. However, when the latter accumulates, the deaminase is inhibited. When dATP accumulates, the activity of the reductase is inhibited and reduction in general is suppressed.

Adenosine triphosphate It is evident from the foregoing discussion that ATP plays a singularly important role in the metabolism of all the nucleotides. ATP is essential as a substrate throughout purine and pyrimidine biosynthesis and also for the various kinases that lead to the ultimate production of the nucleoside triphosphates. Furthermore, it behaves as a negative and a positive effector in many of the enzymes that maintain the proper balance of the purine and pyrimidine ribonucleotides and deoxyribonucleotides.

Salvage of purine and pyrimidine bases

The available pool of purines and pyrimidines depends primarily on the input by biosynthesis. In addition, however, it is maintained by reuse and salvage processes. (Fig. 14-2). Fig. 14-13 illustrates the principal salvage pathways. The nucleotides from catabolized nucleic acids can be reused directly following rephosphorylation (Fig. 14-13, *A*). The nucleotides themselves are not lost from the cell because they

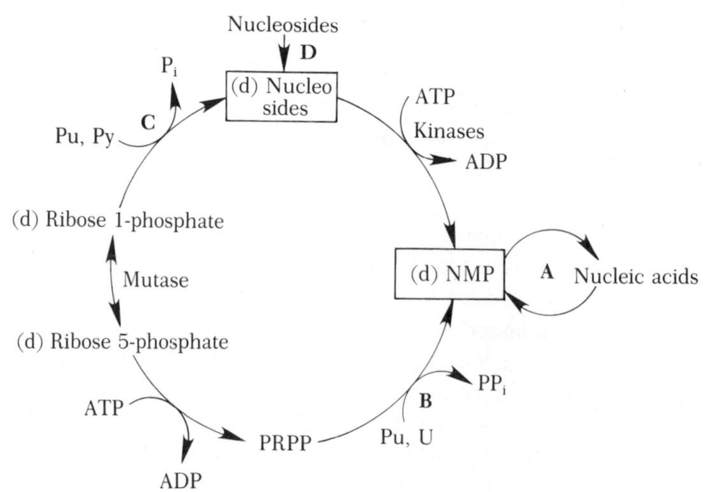

Figure 14-13
A summarizing pathway showing the primary cellular salvaging processes for bases and nucleosides. *Pu,* Purines; *Py,* pyrimidines; *U,* uracil; *(d) NMP,* (deoxy) nucleoside monophosphates.

A Turnover of nucleic acids within the cell. Nucleotides are rephosphorylated (kinases) and reutilized.
B PRPP-dependent salvage involves mainly purines and uracil; enzymes are phosphoribosyl transferases.
C Salvage of purines, uracil and thymine, utilizing nucleoside phosphorylases.
D Kinases may convert some nucleosides directly to nucleotides. Thymidine kinase is ubiquitous. Others such as *(d)* adenosine, *(d)* cytosine, and uridine exist in some cells but not in others.

do not penetrate the plasma membrane unless they are first converted to the nucleoside. The most widely described salvage process is that given in Fig. 14-13, *B*. It involves reactions catalyzed by adenine phosphoribosyltransferase (APRT) and hypoxanthine-guanine phosphoribosyltransferase (HGPRT). In the following diagram HGPRT catalyzes the transfer of the phosphoribosyl group of PRPP to hypoxanthine or guanine to yield IMP or GMP, respectively. In addition, purine analogues such as allopurinol, 6-mercaptopurine, 8-azaguanine, and 6-thioguanine are substrates. APRT requires adenine as substrate and produces AMP.

$$\text{Hypoxanthine} \xrightarrow[\text{HGPRT}]{\text{PRPP} \quad Mg^{++} \quad PP_i} \text{IMP (N-R-P)}$$

Hypoxanthine **IMP**

HGPRT is found in many tissues, including brain, liver, placenta, gonads, erythrocytes, fibroblasts, and leukocytes. The activity appears to be located in the cytosol. A deficiency of this enzyme is found in children afflicted with Lesch-Nyhan syndrome, a genetic disease characterized by aggressive self-mutilation (painful, destructive biting of the fingers and lips), spasticity, growth and mental retardation, lack of speech, inability to sit properly, awkward resting positions, and hyperuricemia.

Hyperuricemia, present also in gout, is always seen in Lesch-Nyhan syndrome. Other manifestations are the accumulation of uric acid such as urinary stones and tophi in the joints. Whereas the blood uric acid in gout reaches 7 to 8 mg/100 ml, it rarely reaches the 10 mg/100 ml seen in the Lesch-Nyhan syndrome. Furthermore, when ^{14}C-glycine is administered to children with the syndrome, a twentyfold increase in the incorporation into uric acid has been observed, compared with a threefold increase in gout. In other words, the absence of this salvage process results in a stimulated biosynthesis of purines, which then leads to an increased production of uric acid.

Of the pyrimidines, only uracil appears to be salvaged by a mechanism involving PRPP (Fig. 14-13, *B*). The following reaction is catalyzed by *uracil phosphoribosyl transferase:*

$$\text{Uracil} + \text{PRPP} \rightleftharpoons \text{Uridine 5'-phosphate} + \text{PP}_i$$

As will be seen later, advantage is taken of the fact that certain derivatives of uracil, e.g., 5-fluorouracil, are converted by this process to the corresponding nucleotide.

A third mode of salvage is process *C* in Fig. 14-13, which is exemplified by the fate of thymine. Thymine is converted to thymidine by a specific phosphorylase. Since the phosphorylases have equilibrium constants of approximately 1, the synthesis of the nucleoside may prevail if there is a supply of the appropriate deoxyribose phosphate. A thymidine kinase is then needed to convert the nucleoside to the deoxyribotide. Thymidine kinase appears to exist in almost all cells, whereas other nucleoside kinases are not commonly found. Therefore, with the exception of thymidine, most pyrimidine nucleosides are not used.

A fourth mode of salvage is illustrated in Fig. 14-13; it represents the direct use of available nucleosides (process *D*). Although there are specific kinases, such as for adenosine-deoxyadenosine, cytidine-deoxycytidine, and uridine, they are not found in all cells. Their usual absence indicates that the direct use of nucleosides is not an important salvage process. As already suggested, the use of thymidine is an important exception.

Treatment of certain cancers

An understanding of the biosynthetic pathways for the formation of purines and pyrimidines has had a number of practical consequences. The entire biosynthetic scheme depends on interrelationships with other metabolic processes, i.e., carbohydrate and protein metabolism. By administering analogues of various intermediates, e.g., of the amino acid glutamine or the vitamin folic acid, there results a competition with the natural compounds that inhibits purine and pyrimidine synthesis and thus also the synthesis of nucleic acids.

Since a characteristic of cancerous tissue is its uncontrolled cell division, an increased synthesis of DNA and RNA is obviously involved. Inhibition of the biosynthesis of purines and pyrimidines would suppress the elevated synthesis of nucleic acids and therefore serve also to restrict the growth of cancerous tissue. Unfortunately the amount of drug needed to block cancer growth often also suppresses normal cell division and therefore can be toxic.

Several analogues that are inhibitors of the biosynthesis of purines and pyrimidines suppress the growth of cancers. For example, azaserine and 6-diazo-5-oxo-L-norleucine (DON) are analogues of glutamine. As has already been seen, glutamine is an important donor of nitrogen to the biosynthetic pathway leading to purines and pyrimidines. It is also involved in the conversions of UTP to CTP and of XMP to GMP.

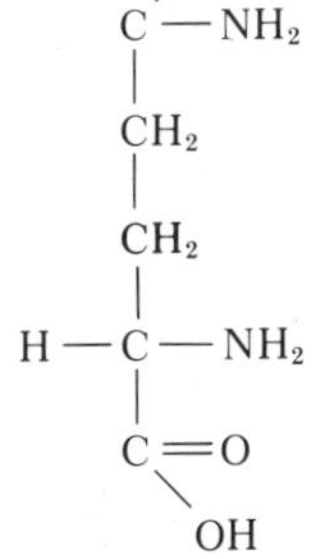

$$\begin{array}{c} \text{O} \\ /\!/ \\ \text{C}-\text{NH}_2 \\ | \\ \text{CH}_2 \\ | \\ \text{CH}_2 \\ | \\ \text{H}-\text{C}-\text{NH}_2 \\ | \\ \text{C}=\text{O} \\ \backslash \\ \text{OH} \end{array} \qquad \begin{array}{c} \text{O} \\ /\!/ \\ \text{C}-\text{CH}=\text{N}^+=\text{N}^- \\ | \\ \text{O} \\ | \\ \text{CH}_2 \\ | \\ \text{H}-\text{C}-\text{NH}_2 \\ | \\ \text{C}=\text{O} \\ \backslash \\ \text{OH} \end{array}$$

Glutamine 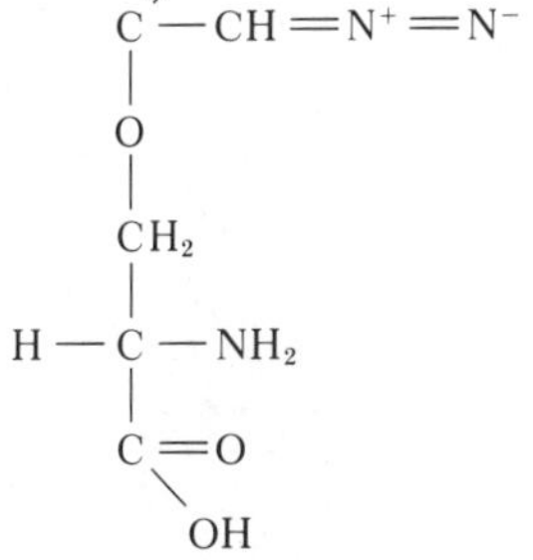 **Azaserine**

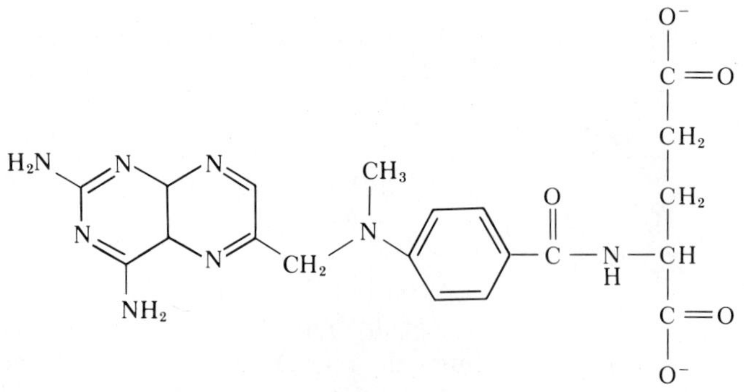

Amethopterin (methotrexate)

Aminopterin and amethopterin (methotrexate) are analogues of folic acid and are used to treat certain forms of leukemia. Methotrexate competitively inhibits the enzyme *dihydrofolate reductase*. Recall that in the biosynthesis of thymidine 5′-phosphate, thymidylate synthetase catalyzes not only the transfer of a 1-carbon unit from FH_4 but also the oxidation of FH_4 to FH_2. Regeneration of FH_4 requires the oxidation of NADPH and dihydrofolate reductase; methotrexate prevents this regeneration and therefore ultimately suppresses DNA synthesis.

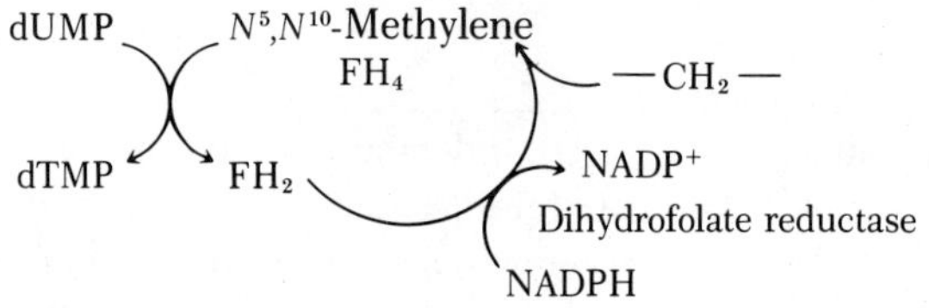

dUMP → dTMP; N^5,N^{10}-Methylene FH_4 → FH_2; —CH_2—; NADPH → $NADP^+$; Dihydrofolate reductase

Other analogues of purine and pyrimidines that interfere with the synthesis of nucleic acids and are used to control cancer are 6-mercaptopurine, 8-azaguanine, 5-fluorouracil, 5-iodouracil, 6-azauracil, and 5-fluoroorotate. The 6-mercaptopurine is converted to the thioguanine and is incorporated into the nucleic acids, thereby interfering with protein synthesis. The thioguanosine ribonucleotide also inhibits the formation of phosphoribosylamine and the synthesis of AMP from GMP. The derivatives of uracil are incorporated into RNA and are also converted into the dUMP derivatives (dFUMP), which act as inhibitors of thymidylate synthetase.

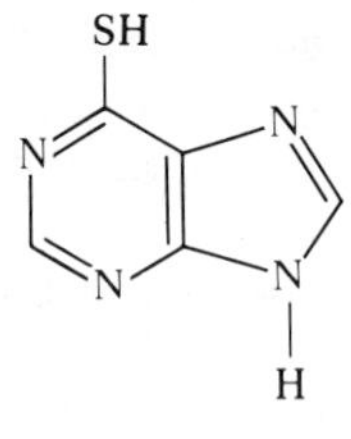

6-Mercaptopurine

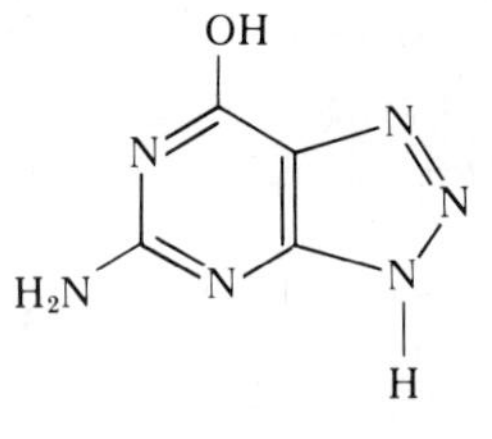

8-Azaguanine

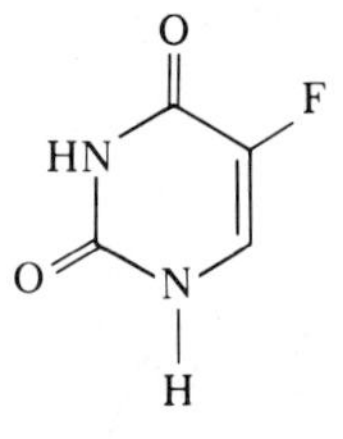

5-Fluorouracil

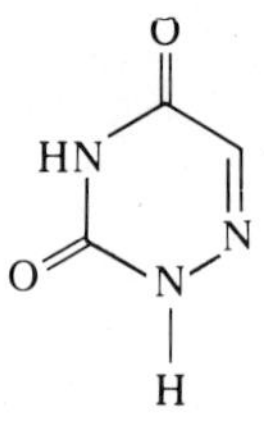

6-Azauracil

Suggested readings

Anderson, E.P.: Nucleoside and nucleotide kinases. In Boyer, P., editor: The enzymes, vol. 9, New York, 1973, Academic Press, Inc.

Blakley, R.L., and Vitols, E.: The control of nucleotide biosynthesis, Annu. Rev. Biochem. **27**:201, 1968.

Boyer, P.D., Lardy, H., and Myrbäck, K., editors: The enzymes, vol. 5, New York, 1961, Academic Press, Inc.

Buchanan, J.M., and Hartman, S.C.: Enzymatic reactions in the synthesis of purines, Adv. Enzym. **21**:199, 1959.

Colowick, S.P., and Kaplan, N.O., editors: Methods in enzymology, vol. 12, New York, 1967, Academic Press, Inc.

Cory, J.G.: Control of ribonucleotide reductase in mammalian cells, Adv. Enzym. Regul. **14**:45, 1976.

Davidson, J.N.: The biochemistry of the nucleic acids, ed. 7, New York, 1972, Academic Press, Inc.

Davidson, J.N., and Cohn, W.E.: Progress in nucleic acid research and molecular biology, New York, 1963-1967, Academic Press, Inc.

Glaser, L.: Biosynthesis of deoxysugars, Physiol. Rev. **43**:215, 1963.

Hartman, S.C.: Purines and pyrimidines. In Greenberg, D.M., editor: Metabolic pathways, vol. 4, New York, 1970, Academic Press, Inc.

Hutchinson, D.W.: Nucleotides and coenzymes, New York, 1964, John Wiley & Sons, Inc.

Kelley, W.N., and Arnold, W.J.: Human hypoxanthine-guanine phosphoribosyl transferase: studies on the normal and mutant forms of the enzyme, Fed. Proc. **32**:1656, 1973.

Kelley, W.N., and Smith, L.H.: Hereditary orotic aciduria. In Stansbury, J.B., Wyngaarden, J.B., and Frederickson, D.S., editors: The metabolic basis of inherited disease, ed. 4, New York, 1978, McGraw-Hill Book Co., Inc.

Kit, S.: Nucleotides and nucleic acids. In Greenberg, D.M., editor: Metabolic pathways, vol. 4, New York, 1970, Academic Press, Inc.

Kornberg, A.: DNA replication, San Francisco, 1980, W.H. Freeman Co.

Murray, A.W.: The biological significance of purine salvage, Annu. Rev. Biochem. **40**:811, 1971.

Reichard, P.: The enzymatic synthesis of pyrimidines, Adv. Enzym. **21**:263, 1959.

Sierakowski, H., and Shugar, D.: Mammalian nucleolytic enzymes, Prog. Nucleic Acid Res. Mol. Biol. **20**:60, 1977.

Smith, L.H., Jr.: Pyrimidine metabolism in man, N. Engl. J. Med. **288**:764, 1973.

Thelander, L.: Mechanism and control of ribonucleoside diphosphate reductase, Biochem. Soc. Trans. **5**:606, 1977.

Thorpe, W.P.: The Lesch-Nyhan syndrome, Enzymes **12**:129, 1971.

Section Four

Specialized tissues and body fluids

Chapter 15

Connective, nerve, muscle, and other tissues

Tissues are an aggregation of similarly specialized cells united in the performance of a particular physiological function. Thus the cells of nerve tissue are specialized for the transmission of nerve impulses, muscle cells for contraction and the performance of mechanical work, etc. With the exception of mature red blood cells, all tissue cells have in common general structural and metabolic properties. However, modified or additional properties adapt them chemically for the performance of their particular tasks. In this chapter we will consider some of the biochemical adaptations of tissue cells that enable them to perform their specialized functions.

All living tissues are in a dynamic state of equilibrium in the adult organism. They are constantly undergoing degradative changes with activity, balanced by anabolic reactions to maintain their structural integrity and function.

A tissue of obvious importance to the human body is the integument. This comprises the skin and its derivatives such as nails and hair. Skin serves a protective function by preventing invasion of injurious substances from the environment and protects the body against desiccation. It aids in the regulation of body temperature and excretion of water, certain lipids, and other substances. The skin is composed of two general parts: the *epidermis,* an epithelial outer layer; and the *dermis,* or inner layer, which contains connective tissues. These layers vary in relative proportion as one considers different parts of the body.

Epidermal tissue

The epidermis consists of a series of layers called the *stratum basale* or *stratum germinativum,* the deepest layer; then the *stratum granulosum;* the *stratum lucidum;* and the outermost layer, the *stratum corneum.* Cells of the stratum basale are the most active physiologically. These cells lose water and are pushed into the next layer, which contains granules consisting of a substance called keratohyaline. It in turn is believed to serve as a precursor for eleidin, a semifluid substance of the stratum lucidum. Eleidin is transformed into keratin, the characteristic constituent of the stratum corneum. The cells of this layer are clear, dead, and scalelike. They are increasingly flattened near the surface from which they are constantly desquamated. In these the nuclei have disappeared and the cytoplasm has been replaced by keratin.

Keratin

Keratin is often described as *soft keratin* of the epidermis and *hard keratin* of nails and hair. Mammalian keratin is classed as an albuminoid, i.e., a protein insoluble in all reagents that do not destroy it. Therefore it makes an ideal protective covering for the body. Its fundamental polypeptide chain is in the right-handed α-helical conformation and is stabilized by hydrogen bonding. Three such α-helices are combined in a left-handed supercoil to form a protofibril. Nine such protofibrils in turn are bundled about two central protofibrils to produce an 11-stranded cable. Such bundles are called microfibrils, and are some 800 nm in diameter. They are in turn cemented together with hundreds of others to form an irregular fibrous bundle or macrofibril. In hair these are packed within the dead cells. In addition to hydrogen bonding the α-helices are cross-linked by disulfide bonds. The soft keratin of skin is poor in sulfur, whereas the hard keratins of hair are relatively rich in sulfur and hence are stabilized by many disulfide bridges. The process of curling hair (permanent) is based on reducing these disulfide bonds and then reforming them after arranging the desired curls. In summary, hair fibers are composed of dead cells containing keratin macrofibrils. These in turn are bundles of microfibrils, protofibrils, and keratin polypeptides in α-helical conformation. Their stability arises first from intrachain hydrogen bonds and in varying degrees to interchain disulfide linkages.

Lipids and other constituents

Subcutaneous tissues contain a considerable quantity of lipid of which about one fifth consists of sterols. One of these, 7-dehydrocholesterol, is converted on irradiation by ultraviolet light to vitamin D (Chapter 21). Fat, fatty acids, phospholipids, and carotene are also present. There is a sex difference in the lipid content of the epidermis of human extremities, with women having a higher percentage of triglycerides and men having a higher percentage of cholesterol and phospholipids. Sebum, the oily secretion of sebaceous glands of the skin, contains a considerable amount of hydrocarbons, about one third of which is squalene. Cholesterol and other lipids are also present; there is little protein. Cerumen, or earwax, is composed of over 40% proteins and about 13% neutral fats, with smaller amounts of phospholipids, cholesterol, and other lipids. Carbohydrates in the skin include pentoses, glucose, and glycogen. It has been claimed that the skin may act as a temporary storage depot for glucose when glucose is present in the blood in large amounts. In psoriasis and certain other scaling skin diseases there is an increased content of carbohydrates. Toxic heavy metals, however absorbed, seem to find their way, in part, to the skin and are deposited there. In silver poisoning (argyria) the skin may become bluish and remain so for years. Significant amounts of lead are deposited in the skin and hair in lead poisoning.

The chief pigment of skin is melanin. Melanin formation is under both hormonal and neurogenic control. It is produced by the melanocytes in the basal layer of the epidermis and occurs in variable amounts as fine granules in the cells and between the cells of the stratum germinativum. The activity of the melanocytes of dark-skinned races is greater than that of the white race, and that of brunettes more than that of blondes. Albinos form no melanin, probably because of the absence of an enzyme that produces melanin from tyrosine by way of 3,4-dihydroxyphenylalanine, or dopa. Skin *tyrosinase* converts dopa to melanin (Chapter 13). This conversion is aided by the tanning action of sunlight, but the mechanism of action, as well as the structure of melanin itself, is unknown. Melanin, a dark brown substance of high molecular weight, is insoluble in all ordinary reagents except alkalies.

As just mentioned, the innermost layer of the skin is the dermis or corium, a region varying in thickness from 0.2 to 4 mm. It is composed of dense, irregularly arranged connective tissue.

Connective tissue

Connective tissue is a system of insoluble protein fibers embedded in a continuous matrix called the ground substance (Fig. 15-1). It is widely distributed in the

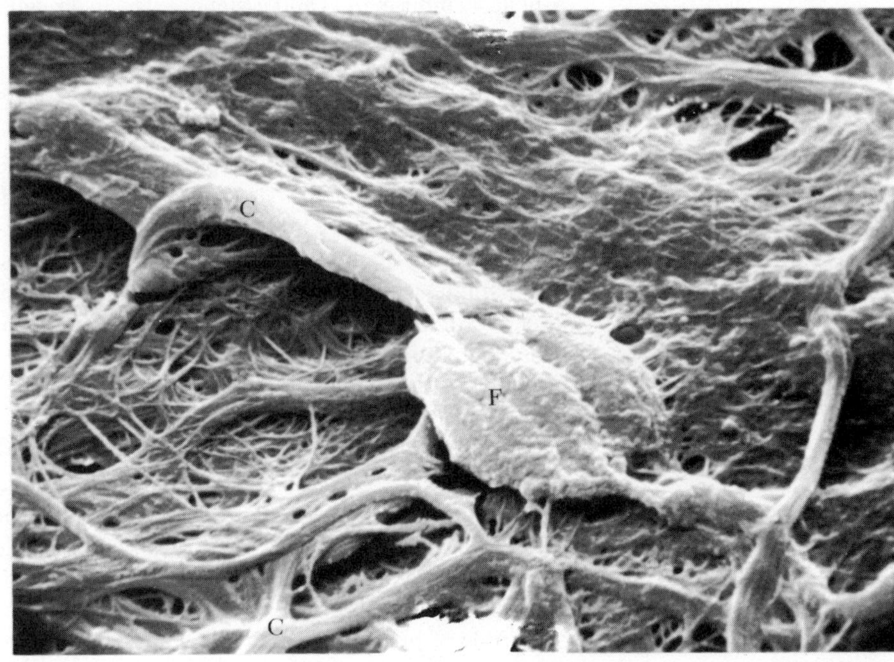

Figure 15-1
Scanning electron micrograph of dorsal loose connective tissue of normal mouse. *C*, Collagen bundles. *F*, Fibroblasts. Note open spaces among fibers; amorphous ground substance fills these spaces.

From Dougherty, T.F., Stevens, W., and Schneebeli, G.L.: Recent Progr. Hormone Res. **29**:287, 1973.

body, composing, in addition to the dermis, the tendons, ligaments, cartilage, and matrix of bone. Its chief function is supportive and is performed by the fibrils of insoluble proteins—collagen and elastin.

The characteristics of connective tissues depend on the proportions of collagen and elastin as well as on the amount of ground substance. The Achilles tendon, for instance, a tissue of great strength, is 32% collagen and 2.6% elastin. The ligamentum nuchae, an elastic tissue, contains 32% elastin and only 7% collagen. Although cartilage is rich in ground substance, the areolar connective tissue contains little of the liquid matrix.

Collagen

Collagen is the most abundant protein in the animal world. It serves as an extracellular framework for all multicellular animals. It is most abundant in fibrous connective tissue (Fig. 15-1) but also is present in some form in practically every tissue. Examples are the "ropes" and "straps" in tendons and ligaments, the woven sheets in skin and fascia, porous membranes (glomeruli of the kidney), the supporting structure of bones and dentin (with embedded calcium salts, etc.), lubricating fluid with proteoglycans in cartilage, bone joints, and intervertebral disks, and special tissues requiring strength yet with unusual properties, such as fatigue-resistant heart valves, and translucency, as in the cornea and lens of the eye.

Extensive investigations, particularly during the past decade, have elucidated many of the details of the molecular structure, covalent cross-linking, and the pathway of the biosynthesis of collagen. As discussed later, at least five different types of collagen molecules have been found in the tissues of higher animals, each apparently adapted for special extracellular functions. Further chemical heterogeneity exists within each of the main types of collagen. There are variations in hydroxylation, glycosylation, and cross-linking. These are related to factors such as tissue types, hormonal factors, and even age. Abnormalities may occur in the structure

Figure 15-2
Electron micrographs of connective tissues, showing reticular fibers, **A,** and collagen, **B.** Both fibers show a characteristic periodicity of 640 Å.

Courtesy J. Gross; from Bevelander, G., and Ramaley, J.: Essentials of histology, ed. 7, St. Louis, 1974, The C.V. Mosby Co.

and biosynthesis of collagen, leading to a variety of human diseases (Chapter 22).

The protein collagen, which represents almost 30% of the total body protein, is formed by fibroblasts embedded in the connective tissues. Electron micrographs of collagen reveal a characteristic banding of 6400 nm (Fig. 15-2). These bands are the result of the regularly spaced binding of an electron-dense material used in electron microscopy. Since this is often an anionic substance, e.g., phosphotungstate, the banding indicates the existence of uniformly spaced regions of polar amino acids.

The formation of insoluble fibers such as collagen by individual cells is difficult to visualize when viewed from the perspective of Fig. 15-1. It is necessary to propose the prior formation of small soluble subunits that can be produced by the cell in accord with the usual processes of protein synthesis (Chapter 10). These subunits presumably leave the cell and aggregate in a regular manner in the extracellular fluid or ground substance. The existence of such subunits was shown by the extraction of connective tissues with cold neutral salt solutions. This removes a readily soluble collagen protein. Warming such solutions results in the reformation of a fibrillar mass showing, by electron microscopy, the characteristically regular spacing of 6400 nm. The solubilized collagen is then called tropocollagen and is considered to be the subunit of the mature fibers.

Tropocollagen is a highly asymmetrical molecule, a relatively rigid rod, with a molecular weight of approximately 300,000 and dimensions of 150 to 28,000 nm. It is composed of three polypeptide chains, each in a left-handed helical conformation having only three residues per turn (Chapter 4). Helices of this type have their amide and carbonyl groups extending perpendicular to the polypeptide chains so that two of every three such groups may participate in interchain hydrogen bonding. The three helices together then wind in a right-handed coil (superhelix) (Fig. 15-3). The interchain hydrogen bonds appear to be the primary linkage holding the structure together.

Collagen is unique in its high concentration of prolines and glycine (see Table 4-2). These prevent the formation of the usual α-helical conformation, which requires the presence of 3.6 amino acid residues per turn. Instead, collagen has 3 amino acid residues per turn. Each of the constituent polypeptides has 1000 residues and a molecular weight of approximately 95,000.

In connective tissues such as bone, mature dermis, tendon, and dentin, collagen has two identical polypeptide chains called the α_1-chains and a third chain of different composition called the α_2-chain. This type of collagen is referred to as $\alpha_1(I)_2\alpha_2$. However, the composition of α-chains varies in other connective tissues. For example, in cartilage the collagen is $\alpha_1(II)_3$, meaning that the three polypeptides are identical but that the α_1 is different from $\alpha_1(I)$ and also the α_2 just mentioned. Young dermis and the cardiovascular tissues contain a third kind of α_1 chain referred to as $\alpha_1(III)$, whereas basement membrane collagen is $\alpha_1(IV)$. In the latter examples the collagen is $\alpha_1(III)_3$ and $\alpha_1(IV)_3$, respectively.

The primary structure of these polypeptides consists of alternate regions of highly polar amino acids separated by regions rich in glycine, proline, and hydroxyproline. The results of amino acid sequencing suggest the separation of polar regions by nonpolar segments that average about 15 residues in length.

$$\text{Polar regions—(Nonpolar regions)}_n\text{—Polar regions}$$

Summarization of information now available shows that the nonpolar regions have the following sequence:

Gly-X-Y-Gly-Pro-Y-Gly-X-Hypro-Gly-Pro-Hypro-

Significant in this sequence are the triads Gly-X-Y, Gly-Pro-X, and Gly-X-Pro, where X and Y are frequently hydroxyproline and alanine. For the left-handed helix to exist, every third residue must by glycine.

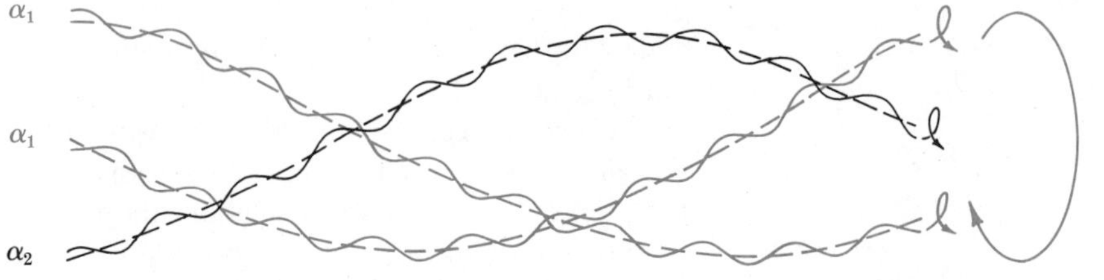

Figure 15-3
Diagram of tropocollagen molecule. Two left-handed α_1-helices (color) are twisted together with an α_2-chain (black) to yield a right-handed supercoil. This would be an example of $\alpha_1(I)_2\alpha_2$.

It is now clear that the α_1- and α_2-chains are not synthesized in their final form as found in tropocollagen but rather are modified in various ways while still in the fibroblasts. For example, neither hydroxyproline nor hydroxylysine is incorporated directly; there are no codons for these two amino acids. Rather, hydroxylation of proline and lysine occurs on the growing polypeptide. Hydroxylation requires a *proline* or *lysine hydroxylase,* ascorbic acid, ferrous iron, α-ketoglutarate, and oxygen. The need for ascorbic acid in these reactions is evidence for the profound effect of scurvy on connective tissue formation and wound healing (Chapter 21).

Hydroxylation of proline and lysine may involve the intermediate formation of a peroxyglutarate, which in turn reacts with proline or lysine (Fig. 15-4). Decarboxylation also occurs in the reaction so that the products are peptidylhydroxyproline (or lysine) and succinate. Ascorbic acid in some way serves as a reducing agent in the system. Proline hydroxylase apparently recognizes those triplets in which proline is the third residue and is followed by glycine.

1 2 3 1
-Gly-X-Pro-Gly
↑
Hydroxylation

Nascent collagen polypeptide chains are also modified by the insertion of carbohydrate. Thus in the α_1-chains lysine-103 is hydroxylated and then glycosylated by the sequential addition of galactose (UDP-galactose) and glucose (UDP-glucose). The side chains are peptide-hydroxylysine-galactose-glucose. Peptide-hydroxylysine-galactose also is present.

A third mode of modification involves a limited proteolytic cleavage to yield the actual tropocollagen. It is now evident that the nascent α_1- and α_2-chains are actually larger than the comparable chains found in tropocollagen. For example, the

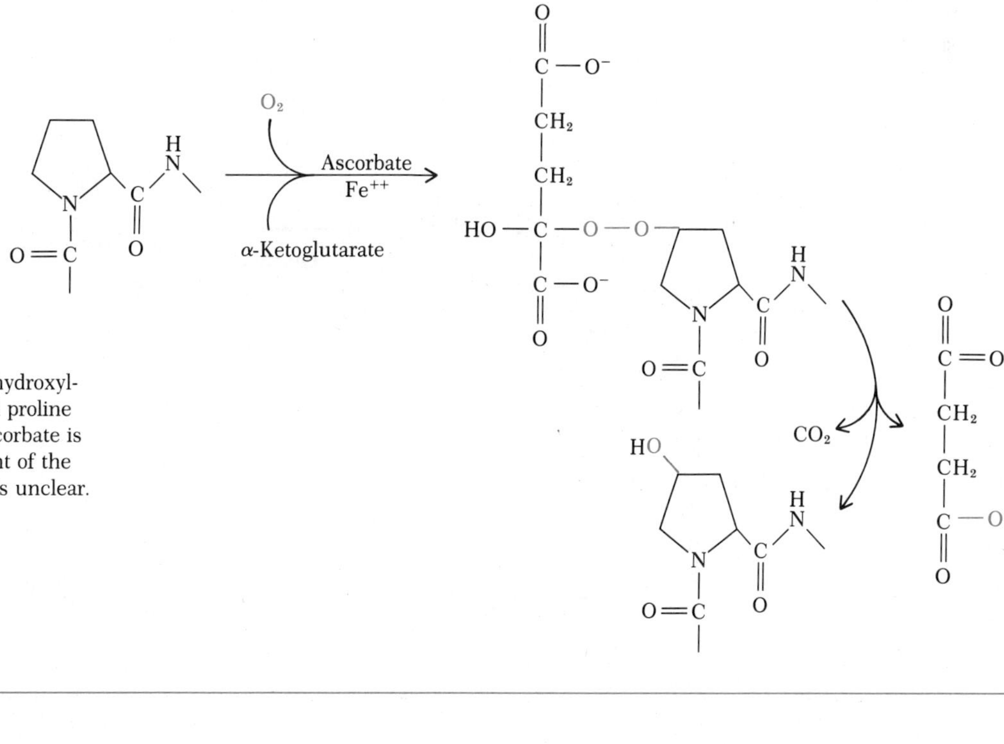

Figure 15-4
Proposed scheme for hydroxylation of peptide-bound proline residues. Although ascorbate is an essential component of the system, its exact role is unclear.

molecular weight of α_1-chains is 95,000; the newly translated chains are 115,000. Therefore these proteins are produced as pro-α_1-chains (procollagen as opposed to tropocollagen) having approximately 1200 amino acid residues and are then converted to α_1-chains of 1000 residues. These removable sequences are at the amino terminal ends of the procollagen chains. It now appears that the N-terminal sequences are needed to assemble the triple-helical supercoil and also to aid in the transport of the protein into the extracellular fluid. Procollagen units do not aggregate into fibers. It is not until the procollagen is cleaved to tropocollagen, apparently in the extracellular fluid, that fiber formation occurs.

The extracellular aggregation of tropocollagen molecules involves an orderly parallel alignment of individual molecules to create a quarter-staggered effect. The

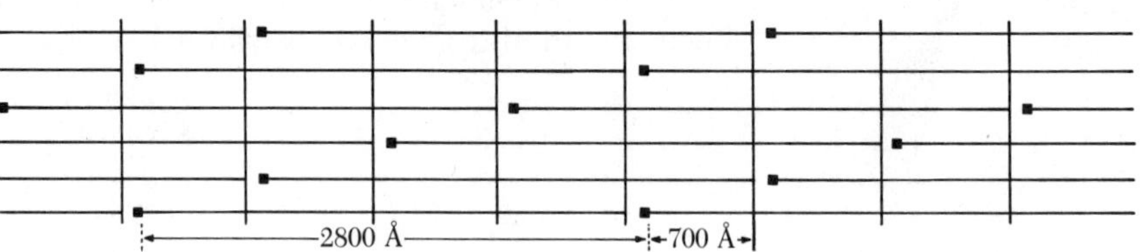

Figure 15-5
Quarter-staggered arrangement of parallel rows of tropocollagen to produce repeated regions of like composition separated by 700 Å.

Figure 15-6
Diagram summarizing various steps involved in biosynthesis and modification of collagen.

Adapted from Miller, E.J.: Fed. Proc. **33:**1197, 1974.

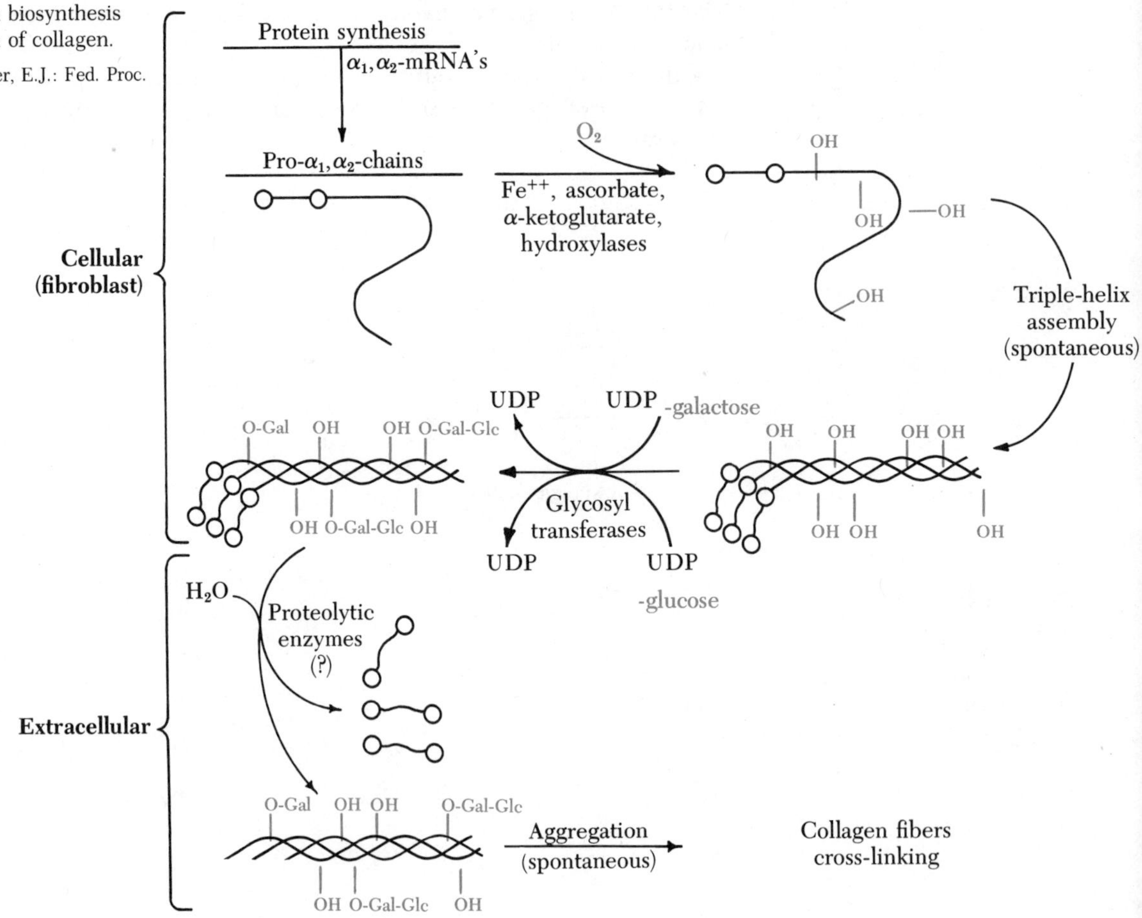

Cellular (fibroblast)
Protein synthesis
α_1, α_2-mRNA's
Pro-α_1, α_2-chains
O_2
Fe^{++}, ascorbate, α-ketoglutarate, hydroxylases
OH
OH
—OH
OH
Triple-helix assembly (spontaneous)
UDP
UDP -galactose
O-Gal OH OH O-Gal-Glc
OH OH OH OH
OH O-Gal-Glc OH
Glycosyl transferases
OH OH OH
UDP
UDP -glucose
Extracellular
H_2O
Proteolytic enzymes (?)
O-Gal OH OH O-Gal-Glc
OH O-Gal-Glc OH
Aggregation (spontaneous)
Collagen fibers cross-linking

asymmetry of the rodlike tropocollagen molecules is suggested in Fig. 15-5 as a line beginning with a small knob. These lines are arranged in a staggered fashion so that they overlap by one fourth of the molecular length, i.e., 7000 nm. Fig. 15-5 shows that this quarter-staggered alignment would yield repeating regions of like ends (knobs). If these knob regions, connected in Fig. 15-5 by black lines, represent sequences rich in basic amino acids (polar regions), they would bind phosphotungstic acid and be the repeating electrondense zones revealed by the electron micrograph.

The process of the biosynthesis, modification, and aggregation of collagen is summarized in Fig. 15-6.

In tissue the newly formed tropocollagen is readily extractable with neutral salt

Figure 15-7
Two kinds of cross-linking found in collagen.

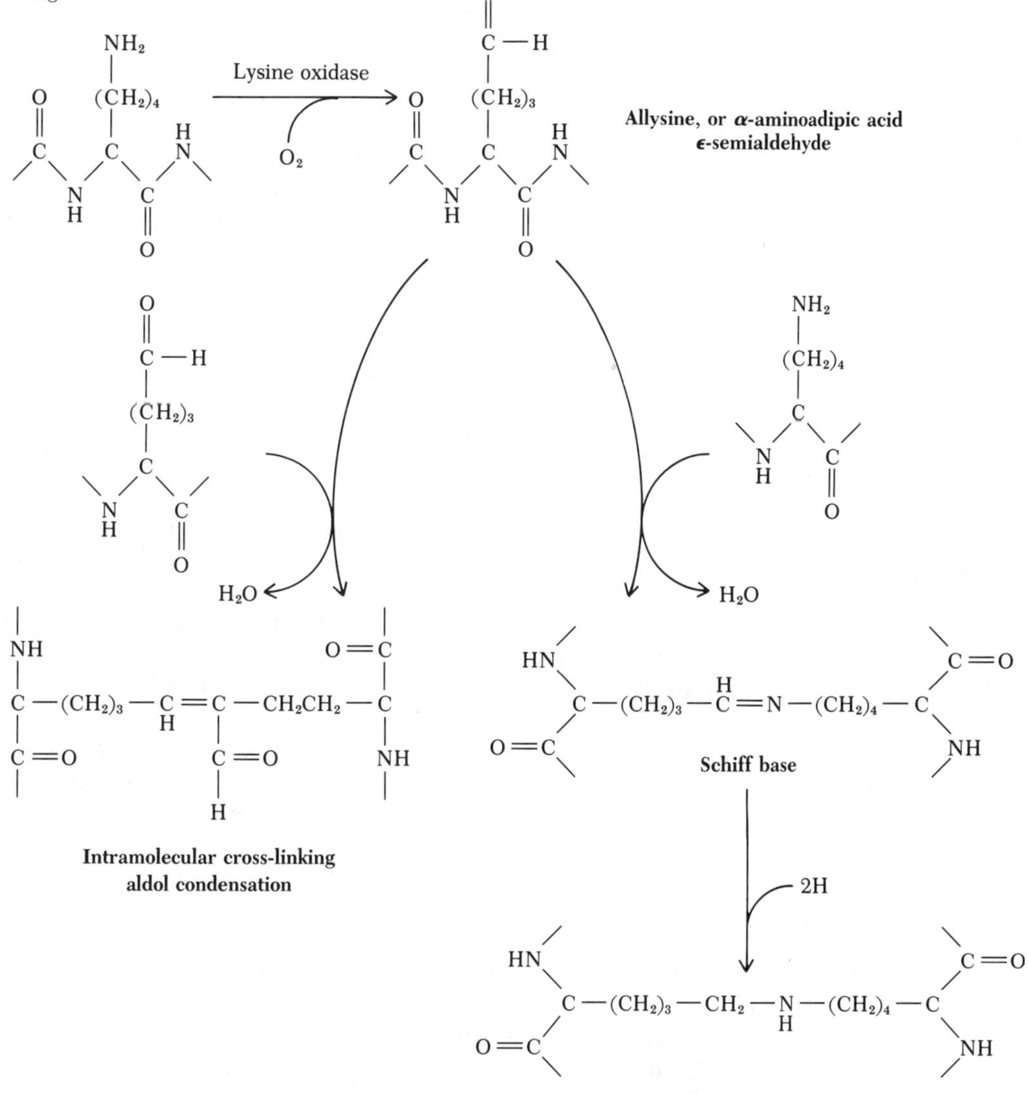

solutions because little or no cross-linking has occurred. On aging the collagen becomes increasingly cross-linked and no longer readily extractable. At one point this collagen is still soluble in a dilute acidic medium, e.g., a solution of citric acid. More extensive cross-linking results in fibers that remain insoluble in dilute alkaline and acidic media.

Cross-linking between tropocollagen units in the collagen fibrils appears to occur in a number of ways. Fig. 15-7 illustrates two, an intramolecular cross-link and an intermolecular cross-link. The abundant occurrence of lysinonorleucine and hydroxylysinonorleucine indicate the importance of the initial formation of aldehydes from lysine and hydroxylysine residues. As seen in Fig. 15-7, lysyl oxidase converts the epsilon carbon of lysine and also hydroxylysine to the corresponding semialdehydes often called allysine and hydroxyallysine, respectively. These in turn react with another allysine in the same polypeptide chain (intramolecular) or with the lysine (or hydroxylysine) residue of a neighboring chain (intermolecular). The former occurs by an aldol condensation, whereas the latter occurs by Schiff base formation. In the case of Schiff base cross-links, some mechanism then brings about a reduction to yield the lysinonorleucine bridge.

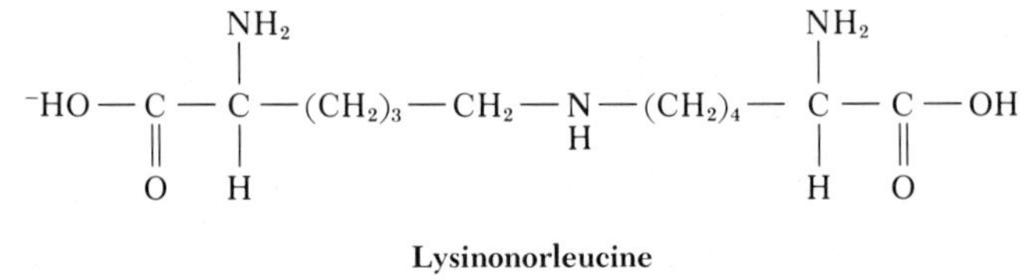

$$\text{HO}-\underset{\text{O}}{\underset{\|}{\text{C}}}-\underset{\text{H}}{\underset{|}{\overset{\text{NH}_2}{\overset{|}{\text{C}}}}}-(\text{CH}_2)_3-\text{CH}_2-\underset{\text{H}}{\text{N}}-(\text{CH}_2)_4-\underset{\text{H}}{\underset{|}{\overset{\text{NH}_2}{\overset{|}{\text{C}}}}}-\underset{\text{O}}{\underset{\|}{\text{C}}}-\text{OH}$$

Lysinonorleucine

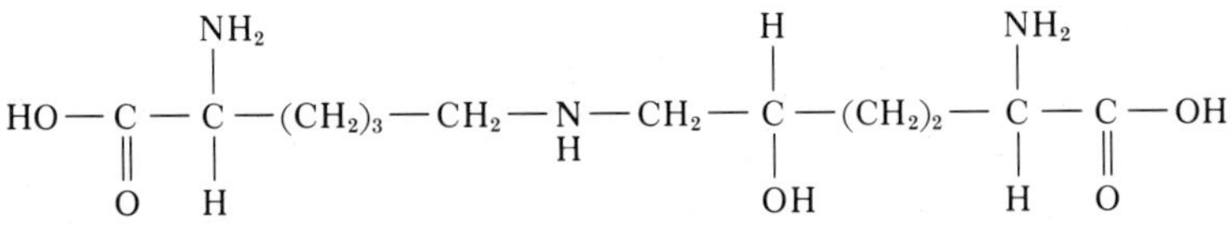

$$\text{HO}-\underset{\text{O}}{\underset{\|}{\text{C}}}-\underset{\text{H}}{\underset{|}{\overset{\text{NH}_2}{\overset{|}{\text{C}}}}}-(\text{CH}_2)_3-\text{CH}_2-\underset{\text{H}}{\text{N}}-\text{CH}_2-\underset{\text{OH}}{\underset{|}{\overset{\text{H}}{\overset{|}{\text{C}}}}}-(\text{CH}_2)_2-\underset{\text{H}}{\underset{|}{\overset{\text{NH}_2}{\overset{|}{\text{C}}}}}-\underset{\text{O}}{\underset{\|}{\text{C}}}-\text{OH}$$

δ-Hydroxylysinonorleucine

Collagen is unique in its transformation into gelatin on heating. This is usually accomplished by boiling in water for a long time or in acid solution for a shorter time. The formation of gelatin appears to result from the untwisting of the triple-strand helix accompanied by a breaking of hydrogen bonds. Although collagen is only slowly digested by pepsin and trypsin, gelatin is readily hydrolyzed. Because of its digestibility, gelatin is a common component in food.

Elastin

Elastin is a protein having distinctly elastic properties as well as high mechanical strength. Compared with that of collagen, our knowledge of elastin remains meager. At first, electron micrographs showed elastin to be purely amorphous masses of material without distinct structure. Now it appears that elastin fibers consist of two components; one is microfibrillar, about 1100 nm in diameter, and the other is amorphous. Elastin has a low content of polar side chains, especially of the basic and acidic amino acids. Like collagen it has a high content of glycine (27%) and proline (13.5%). In contrast, however, there is little hydroxyproline, no hydroxylysine, and a preponderance of nonpolar amino acids, i.e., valine, alanine, leucine, and isoleucine. Elastin is also insoluble in all solvents that do not change its chemical nature. It is not, however, converted to gelatin, as is collagen.

Elastin is probably composed of long-chain precursor molecules (tropoelastin) that are polymerized by a system of cross-links. Distinctive amino acid derivatives have been obtained from elastin and shown to have a cross-linking function. Two of

Figure 15-8
Scheme showing polymerization of four tropoelastin molecules to yield elastin by cross-linking. The linking structure is a pyridine derivative called desmosine. Pyridine nucleus is produced from three allysine residues and one lysine, as shown to the left.

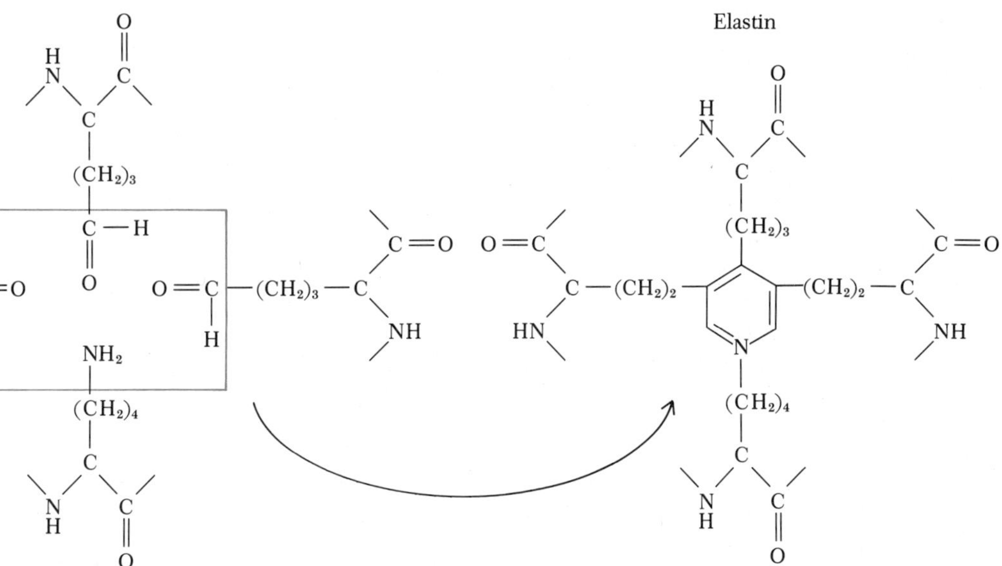

these are called *desmosine* and *isodesmosine*. As shown for the structure of desmosine (Fig. 15-8), four units of lysine are required to produce the compound having a pyridine nucleus. Three lysines are first converted to the α-semialdehydes (allysine), as shown in Fig. 15-7. Then the three allysines condense with an intact peptide-bound lysine to produce the pyridine nucleus (Fig. 15-8). Desmosine and an isomer called isodesmosine are capable of linking together two, three, or four polypeptide chains.

Reticulin

A third fibrous protein found in connective tissues is known as reticulin, which is the component protein of the reticular fibers. This protein resembles collagen in terms of its amino acid composition as well as its 6400-nm axial repeating period, characteristic of the electron micrographs. It appears to be distinctive, however, because of its association with lipids and carbohydrates.

Proteoglycans or ground substance (mucopolysaccharides)

Ground substance, a proteoglycan, may be viewed as a modified dialysate of plasma. It contains some proteins derived from plasma by a certain degree of capillary permeability. The fluid matrix, however, is unique because of its high content of glycosaminoglycans or mucopolysaccharides. In fact, up to as much as 95% carbohydrate may be present. The mucopolysaccharides present are hyaluronic acid and chondroitin sulfates A, B, and C. These polysaccharides are synthesized by the fibroblasts.

A good example of ground substance is the lubricating fluid of all joints, called the synovial fluid. This fluid has the following characteristics:

pH	7.39
Noncollagenous protein, usually serum protein, especially serum albumin	0.75% to 2%
Hyaluronic acid (nonsulfated mucopolysaccharide)	0.15% to 0.2%
NPN, glucose, electrolytes, etc., are similar to those found in blood serum.	

Hyaluronic acid is an unbranched polymer of high molecular weight (1 to 1.5 × 10^6) of a repeating disaccharide consisting of D-glucuronic acid and *N*-acetyl-D-glucosamine. As seen in the following diagram, the monosaccharide subunits are held together by alternating β-1,4- and β-1,3-glycosidic linkages.

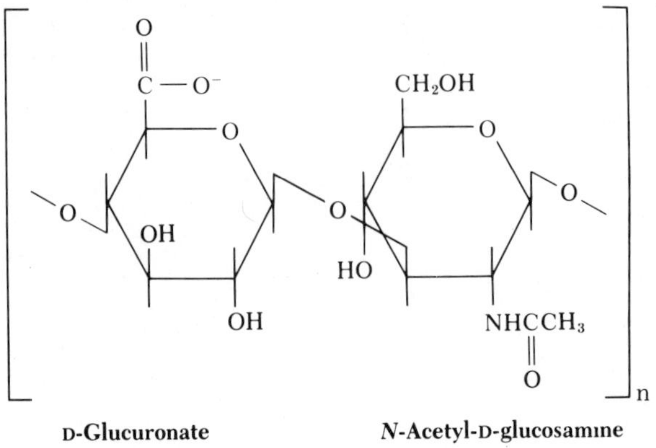

Repeating disaccharide of hyaluronic acid

The chondroitin sulfates are sulfated polysaccharides. Chondroitin sulfate A, now termed *chondroitin-4-sulfate,* is a polysaccharide of glucuronic acid and *N*-acetylgalactosamine bearing a sulfate ester group on carbon-4 of the amino sugar, whereas *chondroitin-6-sulfate* (chondroitin sulfate C) is sulfated on carbon-6 of the amino sugar. Chondroitin sulfate B, now called *dermatan sulfate,* is similar to chondroitin sulfate A except that the uronic acid is of the sugar L-idose rather than D-glucose.

Mucopolysaccharides are believed to be complexed with protein found in the ground substance. Hyaluronic acid is a polysaccharide or glycosaminoglycan that combines loosely with proteins; the nature of the uniting force is unknown. The chondroitin sulfates, on the other hand, are polysaccharide chains, with molecular weights of 15,000 to 30,000, combined covalently with a single protein core (mol wt = 89,000) to produce a total unit of large molecular weight (3 to 6 × 10^6). Therefore the chondroitin sulfates are a special type of glycoproteins and are usually called *proteoglycans.* The oligosaccharide chains are attached to the single polypeptide chain at its seryl residues.

As seen in Fig. 15-9, the innermost carbohydrate is xylose, which is attached to serine by a β-glycosidic bond. Xylose is followed by a sequence of galactose-galactose-glucuronic acid. This oligosaccharide serves to link to the protein the polysaccharide characteristic of the various chondroitin sulfates. For example, a polysaccharide made up of a repeating disaccharide, *N*-acetylgalactosaminyl-D-glucuronic

Figure 15-9
Sugar sequence in the linking of oligosaccharide found in chondroitin sulfates.

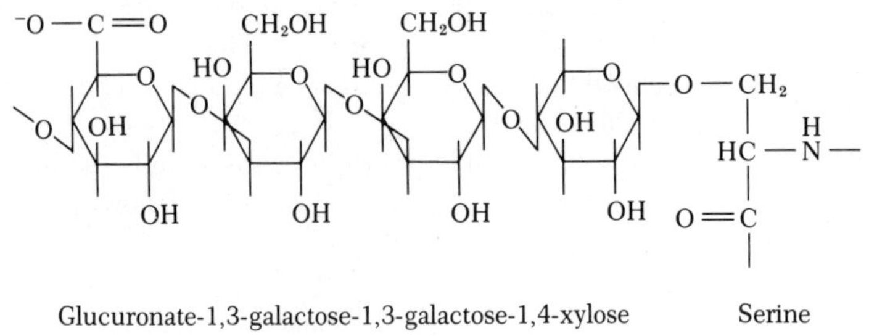

acid, is the constituent of chondroitin sulfates A and C. Chondroitin sulfate B has a repeating disaccharide in which L-iduronic acid replaces D-glucuronic acid. As just mentioned, these polysaccharides are found in connective tissues in their sulfated forms.

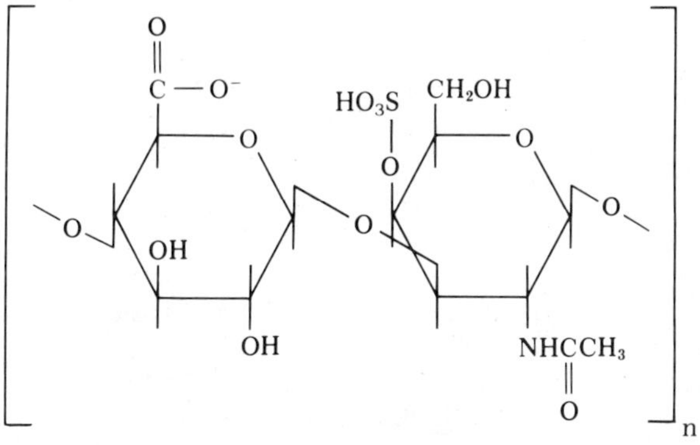

D-Glucuronate **N-Acetyl-D-galactosamine-4-sulfate**

Repeating disaccharide unit of chondroitin-4-sulfate

Synonym	Proteoglycan	Constituents (sugars)	Location of sulfate groups
Chondroitin-4 SO_4	Chondroitin sulfate A	*N*-acetylgalactosamine, D-glucuronic acid	C-4 of *N*-acetyl-D-galactosamine
Dermatan sulfate	Chondroitin sulfate B	*N*-acetylgalactosamine, L-iduronic acid	C-4 of *N*-acetyl-D-galactosamine
Chondroitin-6-SO_4	Chondroitin sulfate C	*N*-acetylgalactosamine, D glucuronic acid	C-6 of *N*-acetyl-D-galactosamine

Two other proteoglycans (mucopolysaccharides) are found in some tissues: *keratosulfate* (keratin) I and II in the cornea, cartilage, spinal disks, and bone; and *heparin* (Chapter 16) in liver, lung, arterial walls, and mast cells. The repeating disaccharide unit in the keratosulfates is mainly D-galactose-*N*-acetyl-D-glucosamine-6-sulfate. In heparin and heparin sulfate the repeating disaccharide unit is D-glucuronic acid–2-sulfate, *N*-acetyl-D-glucosamine-6-sulfate, with variable amounts of *N*-sulfo derivatives of D-glucosamine, rather than *N*-acetyl-D-glucosamine and of iduronic and glucuronic acids.

The biosynthesis of proteoglycans probably occurs by a sequential addition of monosaccharides to the core polypeptide chain. Thus:

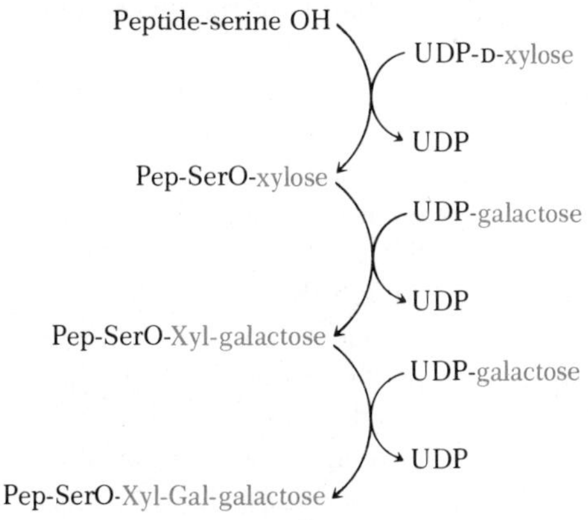

Peptide-serine OH
(UDP-D-xylose → UDP)
Pep-SerO-xylose
(UDP-galactose → UDP)
Pep-SerO-Xyl-galactose
(UDP-galactose → UDP)
Pep-SerO-Xyl-Gal-galactose

Figure 15-10
Pathway summarizing biosynthesis of mucopolysaccharides.

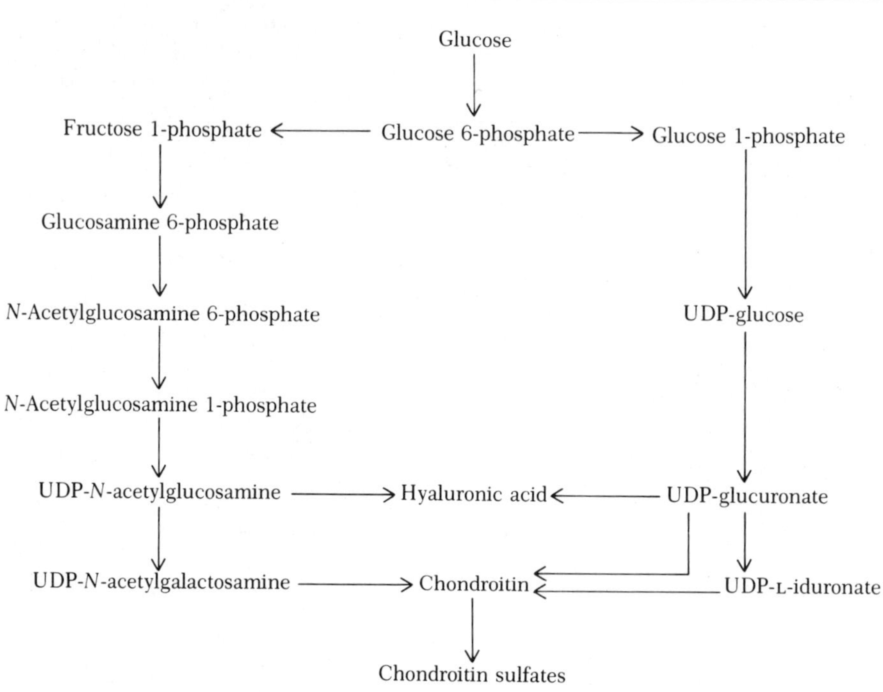

Each monosaccharide is added to the growing polysaccharide chain by a specific glycosyl transferase using the appropriate nucleotide-bound carbohydrate. As previously described (Chapter 4), this process may occur in the smooth endoplasmic reticulum and the Golgi apparatus. The production of proteoglycan probably terminates in an unsulfated glycoprotein such as the chondroitin that is found in corneal tissue.

Control of chondroitin formation is exerted, in part at least, by a feedback inhibition of UDP-glucose dehydrogenase by UDP-xylose (Fig. 15-10). Other feedback control mechanisms of pathways producing precursors of mucopolysaccharides were discussed previously (Chapter 11).

As indicated in Fig. 15-11, the final step in the synthesis of the sulfated proteoglycans is the sulfation process itself. Sulfation requires a mixed phosphosulfate anhydride: 3′-phosphoadenosine-5′-phosphosulfate, or PAPS. PAPS is synthesized as described previously in Chapter 13.

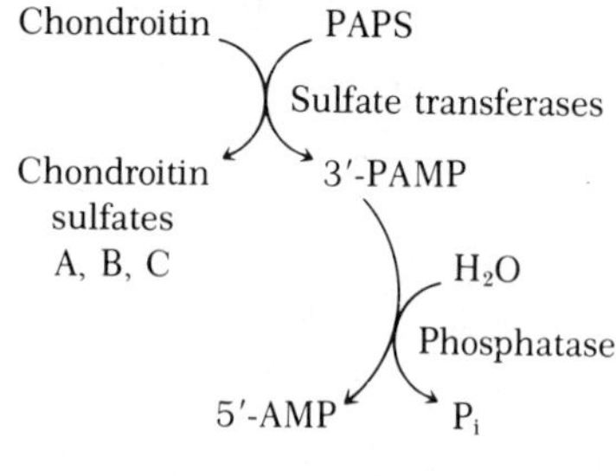

Sulfation of the glycoprotein acceptor proceeds as shown at left.

The proteoglycans (mucopolysaccharide-protein complexes) play numerous roles. They have the particular ability to bind water and cations (even Na^+ and K^+), forming a gel. This is important in structural terms, e.g., maintaining the turgidity of the skin and other tissues. Hyaluronic acid is an important constituent of vitreous humor and the umbilical cord. It imparts the lubricative quality to the synovial fluid of the joints. Mucopolysaccharides help prevent the invasion of the body through the skin by disease-producing microorganisms. Many bacteria, however, produce an enzyme, *hyaluronidase,* that catalyzes the depolymerization of mucopolysaccharides, thereby increasing the invasive capability of the bacteria. Hyaluronidase is a β-glucosaminidase and hydrolyzes the β-1,4-glycosidic bond between *N*-acetylhexasamine and the neighboring uronic acid (Chapter 11). This enzyme is also found in spermatozoa and may be functional in facilitating the

Figure 15-11
Feedback control by xylose of the biosynthesis of uronic acids.

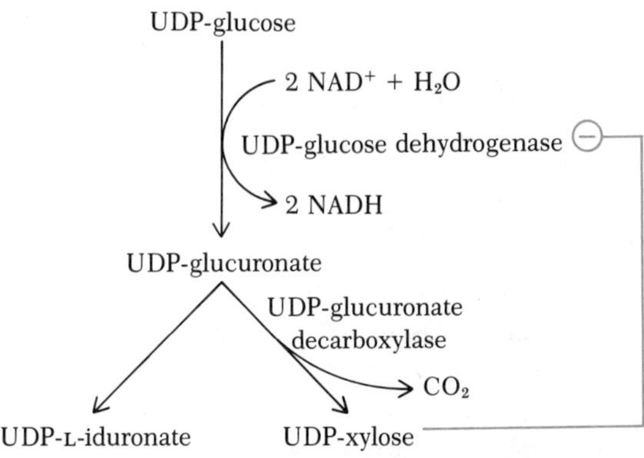

penetration and thus the fertilization of the ovum. Highly purified hyaluronidase is used clinically for the intradermal administration of large volumes of fluid when intravenous injections are contraindicated. The enzyme is given prior to or simultaneously with the fluid and hastens the flow and absorption of the fluid. Hyaluronidase may also be used to enable edematous patients to excrete accumulated tissue fluid more rapidly. It can facilitate the penetration of drugs, e.g., penicillin, into mucous membranes, and it can help spread the effect of local anesthetics over a wider area. Hyaluronidase has been found useful as an adjuvant in infiltration and nerve-block anesthesia, increasing the area and depth but decreasing the duration of anesthesia. Another interesting application of the enzyme is to facilitate the separation of closely adherent structures and thus aid in surgical dissection. A few minutes after the application of a sponge moistened with the warm enzyme solution, adherent tissues are more easily separated. Of course, the use of hyaluronidase in infected areas and in malignancy is not advised, since the enzyme might tend to spread the offending material.

The composition of ground substance changes with aging. In children chondroitin sulfate A predominates, whereas in the adult chondroitin sulfate C is the major component.

Cartilage is also a connective tissue and contains collagen, some noncollagenous protein, and ground substance. The collagen appears to be the same as that in other connective tissues, although the cartilage is devoid of large collagen fibers exhibiting the characteristic 6400-nm periodicity. Fibers of elastin are present in elastic cartilage, tending to give added flexibility to this tissue. Articular cartilage is exceedingly elastic; it recovers quickly and completely from intermittent pressures. This property, which enables cartilage to absorb the shocks to which the body is subjected, is lost on drying but regained when water is restored. Vitamin D is quite possibly related to the development and health of cartilage, in addition to being involved in the normal conversion of cartilage to bone.

Chronic rheumatoid arthritis in humans is characterized by disruption of collagenous structures. Recent investigations have indicated that this may result from an excessive production of the enzyme collagenase by cells of the proliferating synovium. Several investigators reported high collagenase activity in synovial specimens from patients with rheumatoid arthritis, whereas none was demonstrable in specimens from control subjects. The amount of collagenase activity was directly proportional to the degree of local and systemic disease activity. This could account for the destruction of collagen in and about the joints, tendons, capsules, ligaments, cartilage, and bone in patients with this type of arthritis.

A more recent hypothesis regarding the effects of arthritis on joint function suggests a depolymerization of hyaluronic acid by the superoxide radical. It is proposed that superoxide is released by phagocytic cells that have entered the synovial fluid. Superoxide dismutase derived from bovine red blood cells (Chapter 16) is being investigated as a potential therapeutic agent.

Bone

The organic matrix of bone or cartilage is the supporting lattice in which the bone salts are deposited to form a rigid structure. The cartilaginous matrix is flexible and extremely strong, as just described.

As already discussed, collagen is the principal protein present in bone (Table 15-1). Collagenous fibers, formed by the osteoblasts, mature in the mucopolysaccharide-containing ground substance. The ground substance varies in consistency, from interstitial fluid to a thick gel, and forms the communicative medium between the tissue fluid and blood, thereby permitting an exchange of ions and other substances.

Analysis of bone ash reveals a preponderance of calcium, a small amount of sodium, and less magnesium and other cations. As a matter of fact, 99% of the body's calcium is located in bone. The anions are chiefly phosphate, with some carbonate and citrate and small amounts of chloride and fluoride. In general, the mineral of bone is in the form of a *hydroxyapatite* crystal structure having the following formula:

$$3\ Ca_3(PO_4)_2 \cdot Ca(OH)_2$$

This formula accounts for the calcium and phosphate but not for the carbonate, citrate, and small amounts of chloride, fluoride, and other ions that are occasionally found. The belief was formerly held that bone salt was an apatite of changing composition and that these other elements or ions were introduced and withdrawn as they fluctuated in concentration in the blood plasma. Now there is evidence that these other salts are present in an intercrystalline semiliquid medium, which allows the transport of materials from the blood to the bone and vice versa. These two phases of the mineral structure of bone have quite different physical, chemical, and probably physiologic properties. The hydroxyapatite crystals (Fig. 15-12) are relatively stable and are not easily soluble in aqueous fluids yet are subject to rapid ion-exchange reactions at their surfaces. They are extremely small and thus have an enormous surface area. The intercrystalline fraction is far smaller in amount than the crystalline fraction, about 4% as great. However, it is much more soluble, and its calcium and other elements readily exchange with those of the blood.

The hardness and rigidity of bone are attributable chiefly to the hydroxyapatite,

Table 15-1
Composition of bone

	Percent of bovine compact bone
Ash	71.0
Water	8.2
Collagen	18.6
Protein-polysaccharide complex	0.2
Other proteins	1.0
Fat	0.0
Sugars, other than mucopolysaccharides	0.0
TOTAL	99.0

From Eastoe, J.M., and Eastoe, J.M., Jr.: Biochem. J. **57**:453, 1954.

whereas elasticity and toughness are attributable to the proteins. This can be simply illustrated. If a small piece of bone is incinerated, a white ash in the form of the original fragment results. When this is ground in a mortar, it is brittle and can be ground to a fine powder, the hardness of which is apparent. On the other hand, if a bone is subjected to the action of dilute hydrochloric acid for a number of days, the inorganic salts are removed and the material left is again seen to have the exact form of the original bone. However, it no longer is hard and rigid but tough and flexible. Since bone is laid down first as cartilage that becomes increasingly impregnated with inorganic salts, we can easily see why children's bones are less easily broken than are those of adults.

Citrate has been discovered as a constituent of human bone, to the extent of about 1% of the dry weight of bone, representing as much as 70% of the body's citrate content. The function of bone citrate is as yet unknown; it may play a special role in the metabolism of calcium by virtue of its power to bind calcium. The calcium-citrate complex is soluble and diffusible but is un-ionized.

Ossification, the formation of bone salt, requires that the concentration of calcium and phosphate must exceed the saturation point at the site of deposition.

Two enzymes are involved: glycogen phosphorylase and alkaline phosphatase. The latter hydrolyzes the glucose 1-phosphate formed by glycogenolysis and yields high local concentrations of inorganic phosphate. It may also hydrolyze other phosphoric acid esters that may be available, e.g., glycerophosphates and nucleotides. The phosphatase that is found in high concentration wherever bone is formed is probably produced by the osteoblasts; the concentration of phosphate ions is raised locally near these cells as a result of enzyme action. Calcium is present in blood in both ionized and un-ionized form. The un-ionized calcium is partly diffusible and partly not. The nondiffusible calcium is largely that fraction combined with protein, whereas the calcium-citrate complex forms most of the un-ionized diffusible part. Probably the chondroitin sulfates of growing bone unite with calcium to provide a local surplus of available calcium. Normally the concentrations are such that the product of ionic calcium and phosphate is about 36 to 40 mg/100 ml of plasma. Products above 40 mg are found when bone growth or healing is taking place, whereas products below 40 mg generally are seen in active rickets and in other

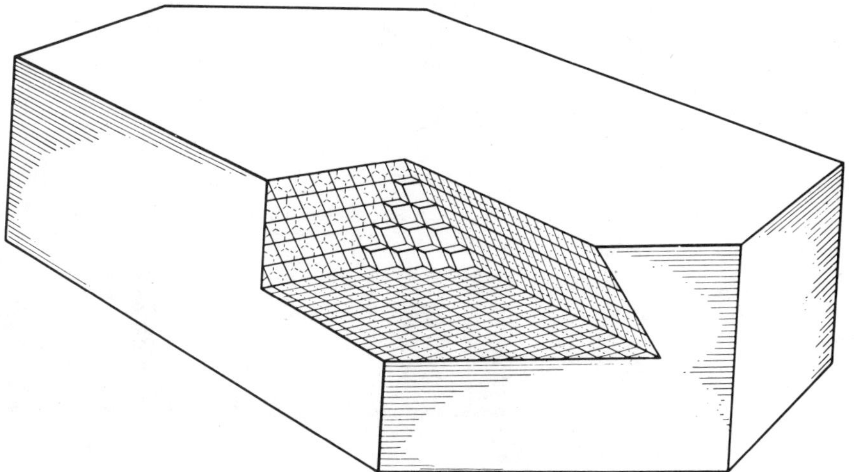

Figure 15-12
Diagram of crystal of hydroxyapatite. This model of a single crystal is cut to show the unit cells of its molecular structure. The model enlarges the crystal about 5 million times.

From McLean, F.C.: Bone, Sci. Am. **192:**84, 1955.

conditions in which bone formation is not occurring properly. When the local concentration of calcium and phosphate ions is increased beyond the saturation point, calcium phosphate is produced. This is changed to hydroxyapatite in a series of steps.

Another enzyme, carbonic anhydrase, may be involved in the deposition of calcium salts in bones and teeth. In vitro experiments showed that carbonate hydroxyapatite is deposited on glass plummets immersed intermittently in either saliva or a synthetic solution containing sodium phosphate and calcium chloride and buffer, provided that carbonic anhydrase is present. This enzyme apparently makes available the carbonate ion for the formation of the bone salt, carbonate hydroxyapatite.

Examination of the ultrastructure of bone by means of the electron microscope reveals an interesting relation between the bone salts and the supporting collagen fibers. The long axes of the bone salt crystals are invariably oriented with the long axes of the collagen fibrils and are arranged around the cross-banding of the fibrils, forming a type of sheath. This is depicted diagramatically in Fig. 15-13.

Note that the bone calcium is in equilibrium with the calcium of the blood. Consequently the blood calcium can be kept at a fairly steady concentration by a slight shift of calcium from the bones to the blood or vice versa. The mechanism whereby this constancy is maintained is dual in character and apparently depends on the two phases of the mineral structure described previously. The concentration of total calcium in the blood is approximately 10 mg/100 ml. If the level of blood calcium ions diminishes, calcium is derived from the intercrystalline material. More calcium is dissolved into the surrounding fluid and enters the blood. This process usually keeps the level at about 7 mg/100 ml. The remaining 3 mg, more or less, are supplied in just the amount needed by means of a feedback mechanism under the control of the parathyroid glands (Chapter 18).

Physical activity has a significant effect on bone composition and bone density. Using a quantitative radiographic technique, recent studies have demonstrated that vigorous physical activity, along with an *adequate calcium intake* and supply of vitamin D (Chapter 21), is required for the development and maintenance of a dense bone in healthy young adult humans. The second phalangeal segment of the small fingers were used for the measurement of bone density.

The incidence of *osteoporosis* (decrease in bone calcium salts) and bone fractures, especially of the hip and distal radius, increase in the elderly, particularly in postmenopausal women. This appears to result primarily from an inadequate intake of dietary calcium (Chapter 21). Increasing the supply of food calcium or even the oral administration of calcium salts, along with estrogen therapy in post-

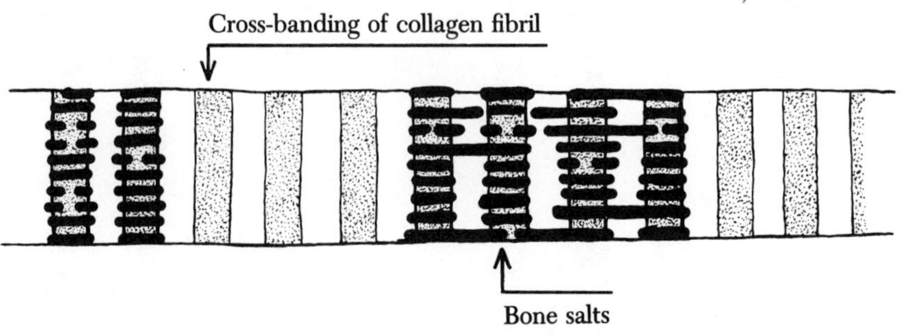

Figure 15-13
Collagen fibril of bone, showing the relation of inorganic crystals to cross-banding.

Adapted from Robinson, R.A., and Watson, M.L.: Ann. N.Y. Acad. Sci. **60**:620, 1955.

menopausal women, appears to reduce the incidence of fractures and to retard the osteoporotic process.

In *rickets* the amount of calcium phosphate in the bones is much below normal. This deficiency is usually the result of inadequate vitamin D, which decreases the absorption and use of calcium and phosphorus. Since the concentration of minerals in the bones is low, the bones become less rigid and consequently bend, resulting in bowlegs or other deformities (Chapter 21). Vitamin A deficiency retards the growth of bone, particularly endochondral bone formation in rats. If the deficiency is established very early in life, skeletal growth is inhibited considerably before the effect on total increase in weight can be observed.

Vitamin C also is essential to bone development, specifically the formation of tropocollagen. In scurvy there are lesions of the epiphyseal junctions of growing bones. Subperiosteal hemorrhages are likely to occur in both growing and adult bone. Rarefaction of the alveolar bone leads to loosening of the teeth; dentine is resorbed, and the gums become spongy.

Bone marrow

Bone marrow is of two kinds: yellow and red. The yellow marrow is composed of connective tissue and large amounts of fat. It has nothing to do with the function of forming red cells. That function belongs to the red marrow; but the yellow marrow may, under some circumstances, be converted to red marrow, which produces the red cells, some of the white cells, and perhaps the platelets (Chapter 16). Red marrow is higher in protein but much lower in fat. Both types contain albumins, globulins, nucleoproteins, fibrinogen, polypeptides, phospholipids, cholesterol, and extractives.

Teeth

The teeth resemble bone chemically to a certain extent. Fig. 15-14 is a diagram of a typical tooth. Over the upper surface of the tooth is the enamel. This is the hardest substance in the body, a property of great value for the masticating and grinding action of the teeth. Only about 5% of enamel is water. The remaining 95% consists of inorganic material chiefly embedded in an organic matrix. This organic matrix of enamel is composed of a protein, resembling keratin but containing no cystine, and a mucopolysaccharide. The inorganic material is hydroxyapatite, a calcium phosphate with the formula $Ca_{10}(PO_4)_6 \cdot (OH)_2$, and possibly also containing fluoroapatite. The greater part of the tooth is dentin, which is identical to bone from a chemical standpoint although different histologically. Dentin protein is largely collagen, and there is chondroitin sulfate present. The inorganic basis is again an apatite, similar to the bone salt.

Administration of labeled phosphorus is followed by rapid uptake of the tracer by developing teeth. Once the teeth are completely formed and calcified, this continuing metabolism is reduced to a minimum. Thus the teeth are not drawn upon for calcium in time of need, as are the bones.

Vitamins A, C, and D are all necessary for proper tooth development and calcification. Lack of vitamins A and C affects the functional activities of the formative cells. Deficiency of vitamin A results in hypoplastic enamel, imperfectly calcified. Lack of ascorbic acid affects the formation of the organic matrix of dentin, as it does in bone. Vitamin D not only aids in the absorption of calcium (Chapter 21) but also apparently promotes the deposition of calcium and phosphorus in teeth.

Dental caries When the enamel breaks and the underlying dentin is exposed, dental caries develops. The cause of this formation of tooth cavities has been a matter of dispute for years and is still unsettled.

Dental caries is one of the most widespread of human diseases, and a tremendous amount of investigation has been instituted to determine the cause and to effect a

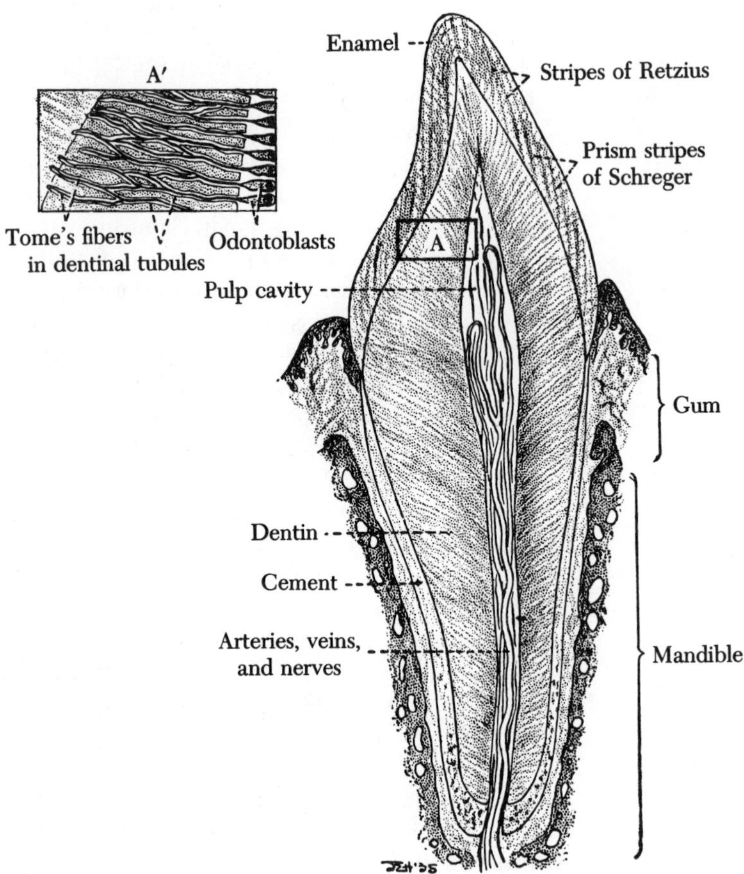

Figure 15-14
Longitudinal section of an incisor tooth. Part *A* is shown at high magnification in *A′*.

cure. The results up to the present time have been rather conflicting. In caries the enamel and dentin are dissolved by chemical action and washed away, thus producing a cavity. The formation of cavities in the teeth is not only a source of pain and discomfort, necessitating dental attention, but also is likely to lead to interference with mastication and thus with proper nutrition. Furthermore, infectious processes occurring in cavities may result in the absorption of toxins or lead to secondary infections in other parts of the body.

In general, there are two schools of thought regarding the initiation of caries; both allow for the possibility that genetic factors, which determine many dental characteristics, are involved also. Advocates of one point of view believe that local bacterial factors are entirely or at least chiefly responsible. Food particles lodged between the teeth or in recesses in the surface of teeth become breeding spots for bacteria, forming bacterial *plaques*. If they are not removed promptly, enough organic (e.g., lactic) acids are produced to dissolve mineral constituents from the enamel and dentin to form a cavity and thus dental caries. Local pH values as low as 4.0 to 5.0 on the tooth surface lend support to this view. Foods particularly rich in easily fermentable sugars (sucrose, glucose, and fructose), especially *sticky sugars* of candies and certain pastries, are most likely to lead to the formation of caries.

Strong supporting evidence for the bacterial fermentation theory of dental caries has been obtained in laboratory experiments on "germ-free" animals. The teeth of such animals developed *no* carious lesions even when they were fed cariogenic high-sugar diets. However, when they were infected with cariogenic microorganisms, dental caries soon developed. *Streptococcus mutans* was found to be highly cariogenic, as were several species of lactobacilli and actinomyces.

Based on the foregoing evidence in support of the bacterial theory of dental caries, predominate current recommendations thus emphasize the importance of lowering the dietary intake of sugars, especially the sticky type, prompt flossing and brushing of the teeth immediately after meals to remove bacterial plaque, and regular dental care to remove "tartar" and deep-seated bacterial deposits.

Current recommendations also stress the importance of proper nutrition, especially with respect to an adequate dietary intake of calcium, fluoride, and vitamin D (Chapter 21), in lessening the occurrence of dental caries. In fact, the second major theory of dental caries proposes that the structure of the tooth itself determines its resistance to caries. An excellent nutritive condition of the individual is essential for caries-resistant tooth structure, beginning with fetal life when the formation of teeth commences. Proper maternal nutrition is thus also an important factor.

The influence of *fluoride* (Chapter 21) must be emphasized. If the amount of fluoride in drinking water is adequate (1.0 to 1.2 ppm), the enamel seems to be *more resistant* to the development of dental caries. Note that fluoride is probably most effective if it is present during the period of tooth development. There are several explanations for the possible inhibiting action of fluoride on caries. One is that fluoride is an essential, or at any rate highly, desirable component of enamel; i.e., fluoride reacts with the tooth substance to form a less soluble complex, less susceptible to the solvent action of oral acids. Using isotope-exchange and ion-competition techniques, it was found that fluoride can replace hydroxyl or bicarbonate ions on the surface of bone, forming an insoluble and resistant fluoroapatite. Perhaps the same phenomenon occurs in the mouth. The fluoride content of the surface layers of enamel is normally approximately 10 times higher than that of the layers near the dentin enamel junction. Another hypothesis is that fluoride acts as an enzyme inhibitor, thus interrupting the chain of fermentative reactions and preventing the formation of organic acids in proximity to the enamel. The use of fluoride in the water supplies of communities (Chapter 21) is advocated as an effective method of preventing dental caries. Fluoridized table salt is also under investigation. In this connection it is interesting to note that the hardness of drinking water may also play a part. The incidence of carious lesions in regions where the drinking water is hard is lower than in those in which it is soft. Whether there is a nutritional effect of the additional calcium or magnesium present or a local effect is not apparent.

A number of other theories of the etiology of dental caries have been proposed. None of these is as well accepted as the two just discussed. However, the possibility of a proteolytic factor has attracted considerable attention. According to this view, caries is a proteolytic process, or perhaps the proteolytic and glycolytic processes go on side by side. It is assumed that bacteria, possessing proteolytic enzymes, multiply at the surface of the teeth and cause disintegration of the lamellae (the flattened bands of organic protein–containing matter extending through the enamel). This permits the easy entrance of fermenting organisms with their production of acid and consequent solution of the inorganic portion of the enamel. The possible formation of amino acid–calcium chelates as a factor in caries development is mentioned in Chapter 23.

Another view is that some component or components of saliva play a role in the prevention of dental caries. Various proteins present in saliva have been alleged to exert anticariogenic effects. Some propose that they act as buffers of the organic acids produced in the mouth by the fermentation of sugars. Others believe that the effect is immunological, certain proteins acting as antibodies. Indeed, there are claims that a group of arginine peptides, termed *sialins,* are the principal agents in saliva conferring resistance to caries. Another cariostatic peptide is alleged to control the incorporation of calcium and phosphorus into teeth. It has been given the name *statherin.*

Thus vigorous biochemical and immunological research continues today in an effort to determine the basic mechanisms involved in the prevention and control of dental caries, one of humankind's oldest and most prevalent chronic diseases.

Adipose tissue

Two types of tissue fat are to be distinguished: protoplasmic fat *(élément constant)* and depot fat *(élément variable).* The former is an essential constituent of protoplasm and includes other lipids besides neutral fat. It is not reduced in amount during starvation. The depot fat, on the other hand, is true adipose tissue and is largely triglyceride, as a reserve energy supply. Depot fat also serves as a protective, cushioning, and heat-insulating material. A cell of adipose tissue, an adipocyte, is literally a droplet of fat contained within a thin membranous living cell. For the droplet of fat to be removed from such tissue, the tissue must be heated, the cell membranes and supporting tissues ruptured, and the fat poured and strained off. When fat is "tried out" or rendered in this way, little protein matter is left behind. Although the distribution of adipose tissue varies in different individuals, a large part of it is in the subcutaneous tissue. Other locations are near the kidneys, in the omentum, and in most other tissues except the brain.

According to isotopic investigations, fat must be deposited in tissues before it can be utilized. Depot triglyceride is thus very active metabolically. Its "half-life" has been estimated by isotopic techniques as some 5 days. Even when given in small amounts to a starving animal, fat apparently is not burned directly as so much fuel but is used only after it has been incorporated into adipose tissue or cellular fat. The mechanisms involved in the deposition and mobilization of storage fat were discussed in Chapter 12.

A *"brown fat"* is deposited in the interscapular glands of the embryo and sometimes of the adult and between the kidneys of hibernating animals. The latter is sometimes called hibernation fat. Brown fat is characterized not only by its color but also by a high content of mitochondria and cytochromes. It has a high rate of oxygen consumption, metabolizing glucose and fatty acids rapidly. However, under cold-stimulated conditions perhaps triggered by the release of free fatty acids via the norepinephrine reaction (Chapter 12), oxidation and phosphorylation apparently are not coupled (Chapter 11) in the mitochondria, so a considerable amount of heat may be released. Little free energy is trapped as ATP. Thus brown fat may serve as a "warming oven" in cold-stimulated, hibernating animals and possibly also in newborn animals.

In obesity there is an abnormal amount of fat laid down in adipose tissue. This is a result of a surplus of food calories over the amount expended. However, there are a number of other factors that enter into the problem (Chapters 12, 21, and 22). Excess obese tissue can be removed surgically—and this is sometimes done—for cosmetic reasons.

Nerve tissue

Nerve tissue, for the purpose of the present discussion, is considered to include the brain, the spinal cord, and the numerous nerves (nerve trunks) that extend to all parts of the body. In brief, the structural unit of nerve tissue is the *neuron,* which consists of a cell body containing the *cell nucleus, dendrites* (extensions in contact with preceding neurons), and a relatively long *axon* in contact via terminal synapses with the dendrites of the next neuron in the chainlike nerve structure. Indeed, axons of the neurons controlling muscles of the foot extend from cell bodies inside the spinal cord to the terminal motor end-plates on the sarcolemma of muscle fibers. The axon is surrounded by a protective *myelin sheath* and a *neurolemma* to form a *nerve fiber*. Bundles of nerve fibers encased in a covering of loose connective tissue, along with blood vessels, fat deposits, etc., make up a nerve or nerve trunk. The brain has a much more complex structure (see textbooks on neuroanatomy for details).

Nerve tissue makes up only about one fortieth of the total weight of the body, yet the brain and nervous system dominate most of its functions. This domination is either directly, by nerve impulses sent to the tissue or organ, or indirectly, by nervous control of the blood supply to the organs. We would therefore expect such a remarkable type of tissue to have a chemical makeup quite different from that of other tissues.

Nerve tissue is characterized by the presence of a large proportion of lipids. Fat, however, is not among the lipids of nerve tissue. Like all other active tissues, there is a large amount of water present—more in embryonic and young nerve tissue and increasingly less with age.

Proteins

Nearly half the dry matter of the human brain consists of proteins (Table 15-2). The gray matter is much richer in proteins than the white matter, as can be seen from the table. *Corpus callosum,* which is composed entirely of white matter, contains 27% proteins, and whole brain, 37%. The proteins of nerve tissue include an albumin, several globulins, a nucleoprotein, and *neurokeratin,* a characteristic albuminoid. Neurokeratin is the material remaining after nerve tissue is subjected to digestion by gastric and pancreatic juices and is then extracted with organic solvents, dilute acid, and alkali. This protein has the physical properties of keratins, to be sure, but contains the amino acids in different proportions from those present in keratins obtained from true epidermal tissue. The amino acid content of neurokeratin and of the combined brain proteins has been determined. The relative amounts of cystine, tryptophan, histidine, tyrosine, lysine, and arginine are about the same in a number of different animals. There is a remarkable constancy in the ratio of lysine to arginine in all species except humans and monkeys. In the other species the ratio lysine: arginine is 100:103-105. In humans and monkeys it is 100:95-96. The percentage of nitrogen in human brain proteins is rather low: 13.4%.

Table 15-2
Solids of the human brain

	Whole brain (child) (percent)	**Whole brain (adult) (percent)**	**Corpus callosum (percent)**
Proteins	46.6	37.1	27.1
Extractives*	12.0	6.7	3.9
Ash	8.3	4.2	2.4
Phospholipids	24.2	27.3	31.0
Cerebrosides	6.9	13.6	18.0
Lipid sulfur	0.1	0.3	0.5
Cholesterol	1.8	10.9	17.1

*The term *extractives* is rather loosely applied to substances that may be extracted from tissues by boiling water. It usually excludes proteins and some inorganic salts.

The myelin sheath of nerve tissue contains several types of protein, mainly in its membranes. These include a *myelin basic protein,* a unique *proteolipid protein* (unusual because of its solubility in organic solvents), several *myelin glycoproteins,* and a number of enzymes. These proteins appear to serve important structural and functional purposes in the myelin sheaths of nerves.

Carbohydrates

Brain and nerve tissue contain very little carbohydrate. Both glucose and glycogen are present in only small amounts; thus there is little reserve supply of carbohydrate. Glucose is extremely important, however, since it is used almost exclusively for energy production in brain and nerve tissue. It is metabolized primarily by way of the glycolytic pathway and citric acid cycle (Chapter 11). A steady and adequate supply of oxygen is essential too, of course. Lack of either glucose or oxygen is followed rapidly by brain dysfunction, coma, and in a few minutes even death. Hypoglycemia (a blood glucose level below 40 mg/100 ml) rapidly is associated with mental confusion, dizziness, convulsions, and even loss of consciousness. The primary need for energy, mainly as ATP, in brain and nerve tissue is for the maintenance of transmembrane potentials and for the biosynthesis of proteins and neurotransmitters, as will be described later.

Abnormal metabolism of carbohydrate may also be related to certain neuroses. It has been shown that patients with anxiety neuroses have an excessively high blood lactate level. This may be related to a chronic overproduction of epinephrine, resulting in excessive lactate production. The insulin shock therapy for schizophrenia is used with some degree of success.

Lipids

As can be seen from Table 15-2, large amounts of lipids are present in brain tissue. In the adult brain they make up more than half the total solids. The phospholipids are present most abundantly (about 40%), with the glycolipids next (about 35%), then cholesterol (about 20%), and finally the sulfolipids. White matter contains more lipids than does gray matter, in general. However, no true fat is present in nerve tissue, and experimental work has shown that the metabolic turnover of fatty acids present is very slow as compared with that in other tissues.

Lipids of the brain and nerve fibers are present primarily in the myelin sheaths and in cellular and intracellular membranes. Most brain lipids are metabolized slowly, as manifested by a relatively low turnover rate. Phospholipids, particularly phosphatidyl choline and phosphatidyl inositides, are exceptions, having a high turnover rate. Cholesterol is readily synthesized (Chapter 12) in the brain, especially by young, growing animals.

Important functions of lipids in the myelin sheath are as an insulating and as a protective layer. The insulation provided by the myelin sheath enables the axon to conduct nerve impulses at a much more rapid rate (usually from 3 to 120 m/sec) than otherwise would be possible. The myelin sheath is separated into short segments by constrictions known as the *nodes of Ranvier.* These appear to serve as gateways through which Na^+ and K^+ ions may flow during the transport of the nerve impulse.

A loss of the myelin sheath of nerve fibers, typically in patchy areas of the brain and/or spinal cord, occurs in *multiple sclerosis,* resulting in marked dysfunction of nerve tissue and in the classic neurological symptoms of this disease (e.g., incoordination, scanning speech, nystagmus). The etiology is uncertain but may result from a viral infection. The nature of the biochemical lesion involved is likewise uncertain. There may be an impaired formation of myelin by the Schwann cells of the neurolemma, an increased degradation of myelin, or some other metabolic dysfunction.

Extractives

The so-called extractives of brain and nerve tissue include a number of substances soluble in hot water. These include small amounts of ATP, phosphocreatine, inositol, and inorganic salts, particularly potassium as the phosphate (basic) and chloride with smaller amounts of sodium salts. The mineral content of gray matter is higher than that of white matter.

Chemical neurotransmitters

The transmission of nerve impulses along the axons of neurons now appears to be well established as an *electrical* phenomenon. However, the mechanism(s) involved in the transmission of nerve impulses from one nerve cell across the synapse to another nerve cell in a nerve tract or to another type of cell (e.g., muscle) is still the subject of extensive investigation. Although electrical transmission may occur in some vertebrate and invertebrate species, the synaptic gap (some 20 nm) between nerve cells is too large in mammals and most other vertebrates to be spanned by the small electrical potentials present. *Chemical* transmission is thus the only form of intercellular transmission in the mammalian nervous system for which there is adequate evidence.

The neurotransmitters apparently are synthesized in the endoplasmic reticulum of the nerve cell body and packaged into vesicles. These then are transported along fast channels in the axon to the synaptic membrane, sometimes a relatively long distance away, and stored there until needed.

The first chemical neurotransmitter identified was *acetylcholine*. It is released at the myoneural junction present at the terminals of all sympathetic and parasympathetic preganglionic fibers and at the endings of postganglionic fibers of the parasympathetic system. Hence the term *cholinergic fibers*.

Subsequently a number of other chemical neurotransmitters have been described—30 or more. Most of them are low–molecular weight compounds containing a positively charged N-atom. Notable exceptions are the more recently discovered *neuropeptides*, as will be discussed later. Each chemical transmitter has a characteristic excitatory or inhibitory effect on a specific type of nerve fiber, although some may be excitatory in one part of the brain or nerve tissue and inhibitory in another.

During the transmission of the nerve impulse the chemical transmitter is ejected by some mechanism from multiple knoblike extensions (called *terminal boutons*) of the axon into the synaptic cleft and then passes into the dendrites of the next nerve cell. The process is extremely rapid, requiring less than 1 msec.

Acetylcholine Acetylcholine is formed in the nerve cell body (Fig. 15-15), transported via the axon in vesicles to the synapse, then stored, bound to the synaptic membrane. On stimulation of the nerve cell acetylcholine is released and, in the

Figure 15-15 Diagram of mechanism of formation of acetylcholine in nerve cell.
Adapted from Nachmansohn, D.: Science **134:**1962, 1961.

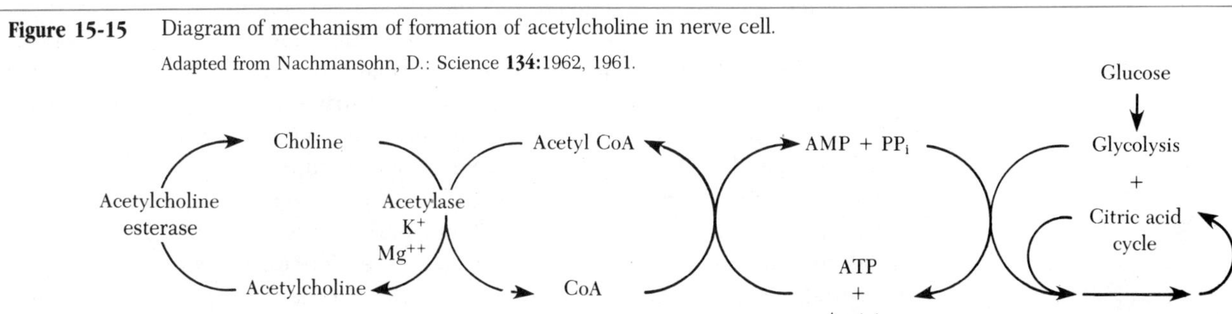

presence of Ca^{++} ions (Mg^{++} ions inhibit), attaches to the synaptic membrane, altering its conformation. Pores develop in the membrane through which sodium ions flow in accord with the $[Na^+]$ gradient and the attraction of a negative charge on the inner surface of the membrane. This shift of sodium ions results in the development of an action potential, with a resultant electrical transmission of the nerve impulse. Some authorities believe that the foregoing process occurs also in the axons of nerve cells as well as the synapse and is thus involved in the propagation of the nerve impulse along the entire nerve cell. Meanwhile, the liberated acetylcholine is destroyed by *acetylcholine esterase* (Fig. 15-15). Sodium ions are pumped out, restoring the original Na^+ and K^+ ion distribution of the resting membrane (Chapter 19), and the process is then ready for repetition. The entire process occurs extremely rapidly, the turnover time being estimated at only 30 to 50 μsec. Thus a nerve can conduct hundreds or perhaps a thousand or more impulses per second.

Acetylcholine has pronounced pharmacological properties when introduced intravenously into an animal. It slows the heartbeat, dilates the arterioles, constricts the bronchi, and has many other effects. There is present in the bloodstream, in the tissues, and especially in the axon and just beneath the cell surface of the ganglion cell an enzyme capable of hydrolyzing acetylcholine into choline and acetic acid, compounds that are much less active than the parent substance. This enzyme has been named *cholinesterase* (or acetylcholine esterase).

$$(CH_3)_3N^+{-}CH_2{-}CH_2{-}O{-}\overset{\overset{O}{\|}}{C}{-}CH_3 \xrightarrow{\text{Cholinesterase}} (CH_3)_3N^+{-}CH_2{-}CH_2{-}OH + CH_3COOH$$

Acetylcholine — **Choline** — **Acetic acid**

There is some evidence of a deficiency of acetylcholine in nerves controlling muscle tissue in cases of *myasthenia gravis*. This could result in the muscle weakness characteristic of this condition. It also could explain the value of certain cholinesterase inhibitors, such as diisopropylfluorophosphate and neostigmine in the treatment of this muscle abnormality. Several other cholinesterase inhibitors are the active ingredients in certain insecticides and in nerve gases proposed for use in gas warfare.

Other neurotransmitters

A growing list of chemical neurotransmitters has been separated in recent years, around 30 different small molecules and perhaps another two dozen or so larger polypeptide molecules. Some serve as excitatory transmitters, some as inhibitory, some as both in different parts of the nervous system. The activity of all chemical transmitters appears to be limited to specific types of nerve fibers.

The list of low–molecular weight neurotransmitters includes a wide variety of compounds in addition to acetylcholine. Examples are norepinephrine, epinephrine, γ-aminobutyric acid, serotonin, dopamine, taurine, glycine, aspartate, histamine, and a number of other putative transmitters for which experimental evidence is not yet complete. Only three of these are considered here.

Norepinephrine Norepinephrine serves as a stimulatory neurotransmitter for most postganglionic fibers of the sympathetic nervous system and for certain areas of the central nervous system. These fibers are therefore termed *adrenergic* fibers, in contrast to cholinergic fibers which use acetylcholine as a mediator. Norepinephrine, like acetylcholine, is stored in vesicles at the end of presynaptic adrenergic fibers. It is released by nerve impulses arriving at the end of the fiber. Its mechanism of action may be similar to that of acetylcholine. Norepinephrine is then destroyed by monoamine oxidase (Chapter 18).

The level of norepinephrine, as measured by the amount of its metabolite (MHPG) (Chapter 18) in urine, has been found to be decreased in patients during certain types of depression. Similar results have been obtained on brain tissue from animals following the withdrawal of certain drugs (amphetamine). These observations may lead to some understanding of the biochemical abnormalities involved in this puzzling mental condition.

γ-Aminobutyric acid (GABA) There is considerable evidence indicating that γ-aminobutyric acid, derived from glutamic acid by decarboxylation (Chapter 13), is an inhibitory neurotransmitter. Apparently its principal sites of action in mammals are the brain and, more specifically, the cerebellum.

Interesting recent studies indicate that *valproic acid,* an 8-C branched–chain fatty acid (an antiepileptic drug), increases the brain and cerebellum levels of GABA. Apparently valproic acid inhibits enzymes that either catabolize GABA or that stimulate its biosynthesis (Chapter 13) or both.

Dopamine Dopamine also appears to function as an inhibitory neurotransmitter. A deficiency of this substance has been found in brain tissue from patients with *Parkinson's disease*. This apparently is responsible for the hand tremors and other neurological manifestions characteristic of this disease. There appears to be an impairment in the ability of the portion of the brain involved to synthesize adequate amounts of dopamine. The administration of L-dopa (L-3,4-dihydroxyphenylalanine) in rather large amounts results in a marked improvement in some patients with Parkinson's disease.

Neuropeptides

A number of polypeptides with profound biochemical activities have been found in brain and central nervous system tissues. The releasing hormones of the hypothalamus (Chapter 18) and somatostatin are important examples. Another group of polypeptides is remarkable because they mimic the effects of morphine. These compounds bind to the same receptors in the brain that bind opiates; hence the term *opioids*. They appear to be *inhibitory neurotransmitters*. The first two discovered were pentapeptides and were given the name *enkephalins* ("in brain"). Their amino acid sequence was determined to be as follows:

Tyr-Gly-Gly-Phe-Met	Tyr-Gly-Gly-Phe-Leu
Methionine enkephalin	**Leucine enkephalin**

The enkephalins are widely distributed in the brain but are in highest concentration in the midbrain and thalamus, where the pain-conducting tracts converge. The enkephalins, like the opiates, inhibit neuronal activity by decreasing sodium ion conductance at Na^+ channels. They also inhibit adenylate cyclase and cAMP formation from ATP, which is necessary for the phosphorylation of membrane and nuclear proteins and in turn for certain functions of the brain.

With the discovery of other opioid peptides a generic name for the groups, *endorphins (endogenous morphine),* has been adopted. Several, including methionine enkephalins, were found to be segments of a 91-residue precursor protein, β-lipotropin, present in the anterior pituitary gland (Chapter 18). Three opioids, α-endorphin, β-endorphin, and γ-endorphin, were found to be composed of residues 61 to 71, 61 to 91, and 61 to 76, respectively, of β-lipotropin. These endorphins are 12 to 100 times more active than the enkephalins.

More recently, additional neuropeptides have been discovered, the list now numbering about 24 compounds. All are candidates for classification as neurotransmitters. One compound, called *substance P,* contains 11 amino acids and has been sequenced. It appears to be involved in the transmission of *pain.* It is degraded rapidly by peptidases (Chapter 5). Another new brain peptide isolated from hog pituitary glands, *dynorphin,* appears to be the most powerful yet found. It is formed from some large brain molecule yet to be identified. However, its structure incorporates that of leucine enkephalin. Its analgesic and behavioral effects are about 50 times that of β-endorphin, 200 times that of morphine, and 700 times that of leucine enkephalin.

Still another neurotransmitter reported is the anxiety-relieving *γ-compound.* It appears to link nerve cells in a special brain pathway. Its composition is yet to be determined, but it has been prepared in purified form from human urine.

The recent explosive progress in the knowledge of the chemical transmission of the nerve impulse is indeed one of the most exciting accomplishments of modern biochemistry. These developments could culminate in the discovery of the long-sought nonaddicting analgesic agent, a substance that could prove of tremendous value in the relief of human pain and suffering. The foregoing discoveries also suggest that the pituitary gland may play important roles in the body in addition to its established function as the master endocrine gland (Chapter 18).

Nerve growth factor Recent studies have shown that a protein with striking similarities to insulin serves as a rather specific *nerve growth factor,* stimulating a rapid growth of neurons. The active factor has been prepared in relatively pure form from embryonic tissue and crystallized in the presence of zinc. It has a molecular weight of approximately 14,000 and consists of a dimer of two identical A and B chains. It contains 118 amino acid residues and has been completely sequenced. Its amino acid sequence has many similarities to that of proinsulin (Chapter 18), suggesting a common ancestral origin and gene. Like insulin, its action is primarily anabolic. Its specificity seems to be related to the fact that it binds to a specific receptor protein on the surface membrane of the nerve cell.

A number of studies have demonstrated that severe malnutrition in infants is associated with decreased growth of the brain and a decrease in the number of brain cells (Chapter 21). Reversible or irreversible impairment of brain function may result, depending on the conditions. An inadequate dietary intake of protein and calories is undoubtedly involved in this serious condition.

Biochemistry of memory Biochemical processes involved in the storage of information (memory) and its retrieval in the brain have been the center of intensive study in the past few years. Controversial claims of the transfer of knowledge from trained animals into naive animals with extracts of brain have also stimulated research in this area. RNA has been suggested as the active constituent of such extracts. Brain RNA also has been suggested as being involved in short-term memory.

Investigations indicate that protein synthesis in the brain is involved in both short-term and long-term memory. The blocking of protein synthesis by puromycin and certain other substances when injected directly into the brain during or very soon after the learning situation does not affect the animal's ability to learn or its short-term memory but interferes dramatically with the formation of long-term memory. If the blocking agent is injected a short time after learning, there is no interference with long-term memory. Protein synthesis in the brain may also be necessary for converting short-term memory to long-term memory. Other work indicates that three key glycoproteins may be involved—two in short-term memory and one in long-term memory.

Further studies along these lines as well as investigations employing memory drugs give promise of elucidating the specific biochemical changes within the nerve cells of the brain that are responsible for the heretofore baffling phenomena of learning and remembering.

Muscle tissue

Muscle forms a large proportion of the active tissue of the body. In normal adults it is fully two fifths of the body weight, but about half the metabolic, or chemical and physical, activity of the body takes place in muscles even during rest. When the muscles are contracting, while doing work, fully three fourths of the total metabolism can be assigned to them.

The three types of muscle—striated (voluntary) skeletal, cardiac, and nonstriated (involuntary) smooth—differ somewhat in their chemistry, but they have the same general characteristics. In skeletal muscle are found the following:

- Water
- Proteins
  - Albumins, globulins, nucleoproteins, myoglobin
  - Albuminoids
- Lipids
  - Cholesterol
  - Phospholipids
  - Triglycerides
- Extractives (soluble in hot water)
  - Nonnitrogenous: glycogen, glucose, inositol, hexose phosphates, lactates
  - Nitrogenous: creatine, creatine phosphate, creatinine, inosinic acid, adenylic acid, adenosine triphosphate, glutathione, purines, pyrimidines, carnosine, anserine, choline, acetylcholine
- Enzymes, hormones, vitamins
- Inorganic salts

Adult muscle is 72% to 78% water. As in the case of nerve tissue, the water content of the muscle of the young and of the fetus is even higher. Here the similarity ends, however. The solids of muscular tissue are largely protein in nature, whereas those of nerve tissue are largely lipid. The total lipid of muscle amounts to only about 3% and the glycogen less than 1%, but the protein content is about 20%.

Structure of skeletal muscle

A typical skeletal muscle consists of many muscle fibers, or *fascicules,* the structural unit of muscle (Fig. 15-16). Numerous fibers are bound together by areolar connective tissue and are arranged parallel to the line of its attachments via tendons to the bones or other structure whose movements they control. Typically, muscle fibers are approximately 0.01 to 0.1 mm in diameter and from 1 to 40 mm in length. Groups of fibers are encased in a thin but tough elastic sheath, known as the *sarcolemma,* beneath which are embedded numerous nuclei and mitochondria. The sarcolemma is an electrically excitable membrane, important in the transport of motor nerve impulses. The sarcolemma also contains numerous longitudinal and transverse *sarcotubules* (transport ducts) and the *sarcoplasmic reticulum.* The network of sacs and tubules of the latter is similar to the endoplasmic reticulum of other kinds of cells and apparently performs the same functions (Chapter 2). Bundles of muscle fibers are surrounded by a semifluid substance, the muscle plasma or *sarcoplasm.* The sarcoplasm contains the red protein myoglobin (muscle hemoglobin), which transports oxygen from adjacent blood capillaries to the sites of oxidation in mitochondria. Myoglobin, like erythrocyte hemoglobin (Chapter 16) can be converted to *metmyoglobin* by various oxidants (e.g., certain drugs) and reconverted to myoglobin by mechanisms similar to those present in red blood cells.

Figure 15-16
Striated muscle dissected in schematic fashion. The muscle is made up of many fibers, which appear cross striated. Single fiber consists of myofibrils, which show alternating dark and light bands. A single sarcomere is the region between two Z lines containing I and A bands. Bands derive from presence of two sets of interdigitating filaments, thick filaments (myosin) with cross bridges and thin filaments (actin).

From Schottelius, B.A., and Schottelius, D.D.: Textbook of physiology, ed. 18, St. Louis, 1978, The C.V. Mosby Co.

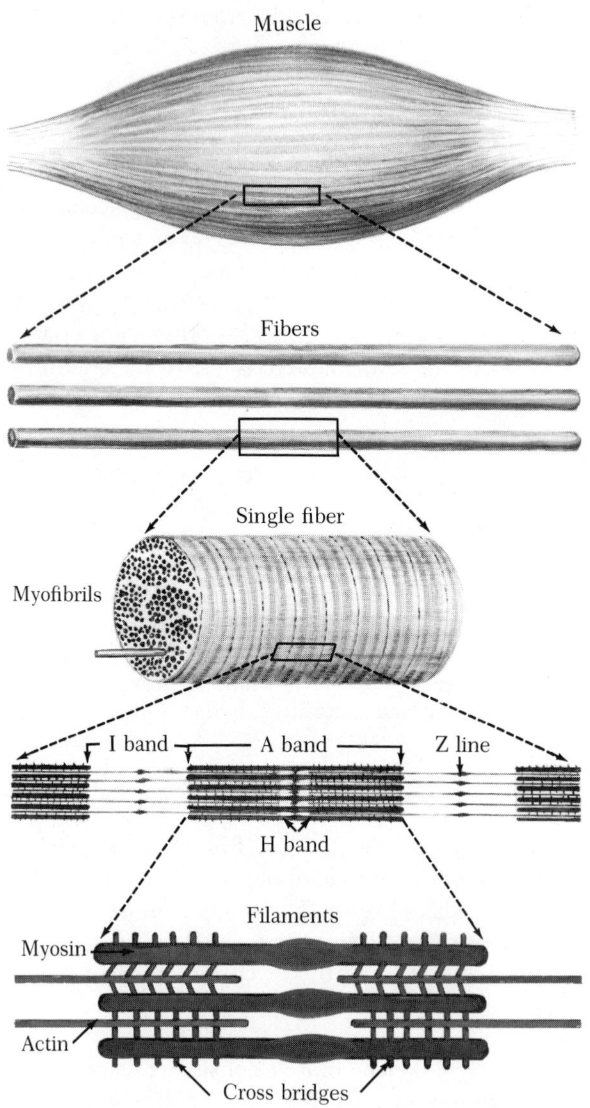

Muscle fibers are made up of numerous smaller structures called *myofibrils.* These contain transverse alternating light (isotopic) I bands bisected by a dark Z *line* (membrane) and dark (aniostropic) *A bands,* having a lighter central *H zone* (Fig. 15-16). These bands are aligned in adjacent myofibrils to appear as continuous bands, or striations, across an entire muscle fiber.

Myofibrils also contain numerous fine longitudinal lines, sometimes termed *myofilaments,* consisting of *actin* (thin) filaments and *myosin* (thick) filaments containing a number of *cross-bridges* (Fig. 15-16). These, as well as other structures in and adjacent to the myofibrils, are clearly visible in the electron micrograph of a skeletal muscle fibril shown in Fig. 15-17. A single *sarcomere,* the *functional* unit of the myofibril bounded by two Z lines, is also indicated in Figs. 15-16 and 15-17.

Fig. 15-16 also shows further details of the structure of a sarcomere. Note the

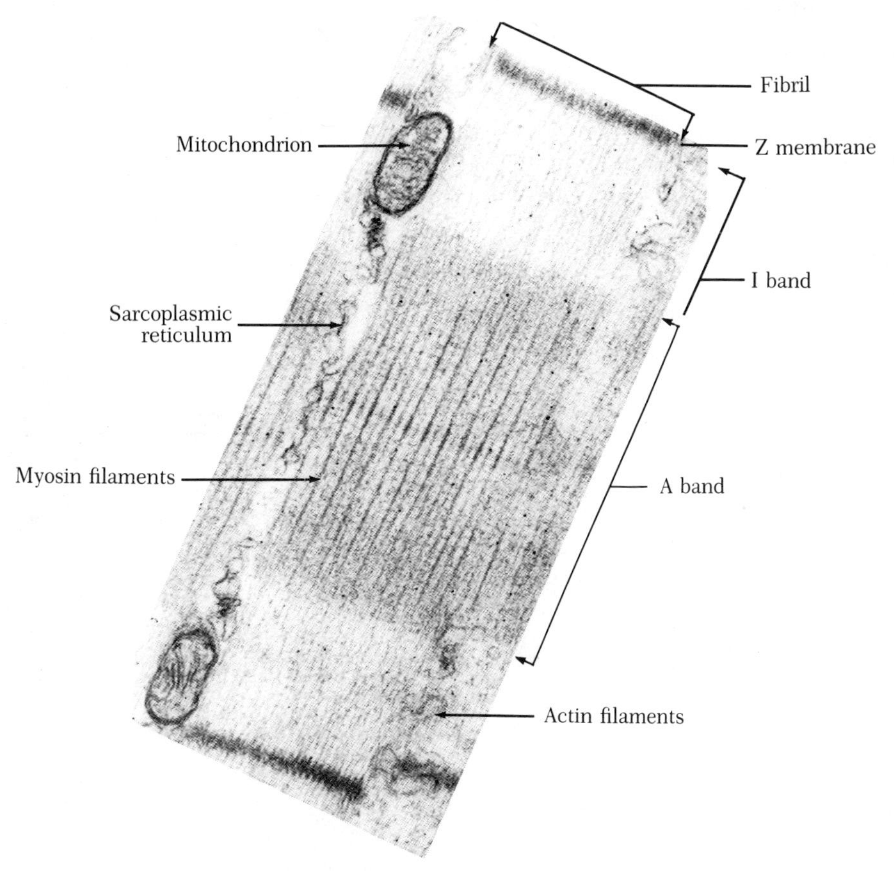

Figure 15-17
Electron micrograph of mouse skeletal muscle fibrils. Actin and myosin filaments are clearly discernible. Mitochondria, sarcoplasm, and sarcoplasmic reticulum also are evident between fibrils. Single sarcomere occurs between Z lines of fibril. Less dense region in center of A band is H zone. (×43,000.)

From Schottelius, B.A., and Schottelius, D.D.: Textbook of physiology, ed. 18, St. Louis, 1978, The C.V. Mosby Co.

arrangement of the two sets of interdigitating myofilaments, the thin actin filaments and thick myosin filaments with cross bridges. This structural arrangement forms the basis of the widely accepted concept of the mechanism of muscle contraction, the "sliding filament" theory, which is discussed later.

Proteins of muscle

As stated before, about 20% of muscle tissue is protein. Over half the muscle proteins are the structural proteins actin, myosin, tropomyosin, and troponin. Other proteins include myoglobin, a number of enzymes involved in the metabolism of muscle tissue, and collagen and other proteins of connective tissue.

Myosin, a globulin, is the most abundant muscle protein. The myosin molecule consists of two globular heads, each joined to a tail-like appendage and twisted into a double-stranded α-helix. The molecular weight is approximately 500,000. The two identical major chains have molecular weights of about 200,000 each, and the four light chains have molecular weights of about 20,000 each. Several hundred myosin molecules bind together tail-to-tail to form the *thick filaments* of the sarcomere A band. Their double heads project at regular intervals (Fig. 15-18, *B* to *D*), forming the cross-bridges of the sarcomere. The tails meet at the central H band of the sarcomere (Fig. 15-16). The head portion of myosin has a marked ability to bind with actin to form *actomyosin,* which has a pronounced ATPase activity.

Actin also is a globulin, with a molecular weight of about 60,000. It is a major constituent of the *thin filaments* of the sarcomere. A lower molecular weight form

Figure 15-18
Organization of thick and thin filaments.

From Schottelius, B.A., and Schottelius, D.D.: Textbook of physiology, ed. 18, St. Louis, 1978, The C.V. Mosby Co.

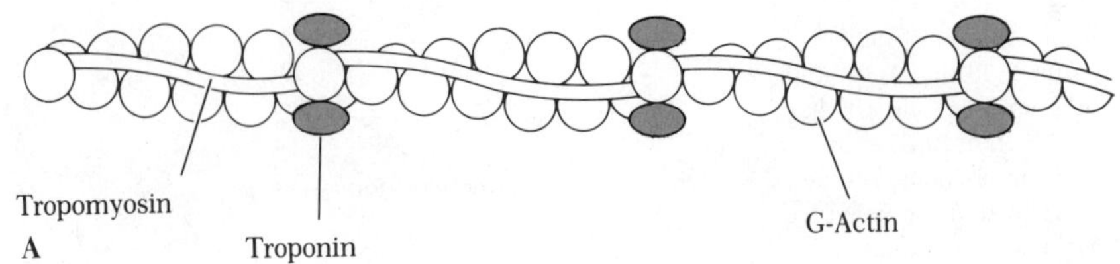

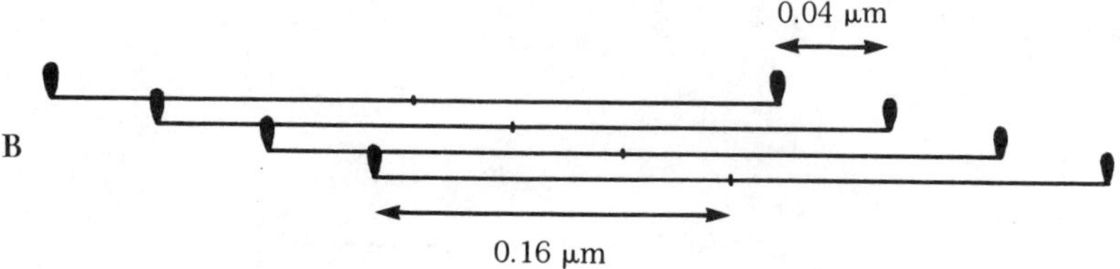

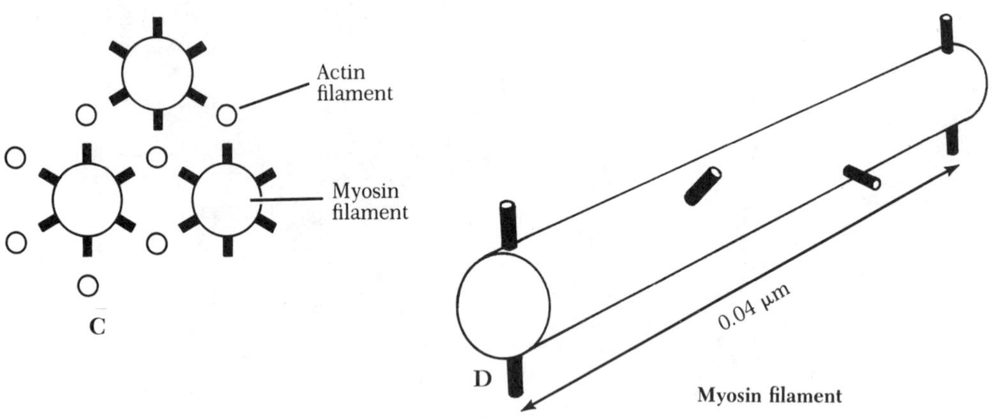

of actin, called *G-actin* (mol wt = 42,000), with a globular configuration, is obtained by extracting muscle tissue with a low–ionic strength medium. If the ionic strength is increased and Mg^{++} ions and ATP are present, G-actin polymerizes into *F-actin,* a fibrous form. It has a double-helical structure (Fig. 15-18, *A*) resembling a double chain of beads.

F-actin combines with myosin to form actomyosin, which has the remarkable property of contraction in the presence of ATP, K^+, and Mg^{++}, as shown in the classic experiments of Szent-Gyorgyi. He correctly concluded that the force of muscular contraction comes from an interaction of actin, myosin, and ATP.

Two other proteins, *tropomyosin* and *troponin,* are now known to be involved in the contraction of muscle. Tropomyosin is a double-stranded α-helical rod with a molecular weight of 70,000. In the thin actin filaments it is intertwined between the two helical strands of F-actin (Fig. 15-18, *A*).

Troponin, which is positioned at regular intervals along actin filaments, is a spherical molecule containing three different types of subunits, with a molecular weight of approximately 70,000. It is the third component of the thin filament (Fig. 15-18, *A*). The three subunits are named according to their functions: *Tn-T* binds

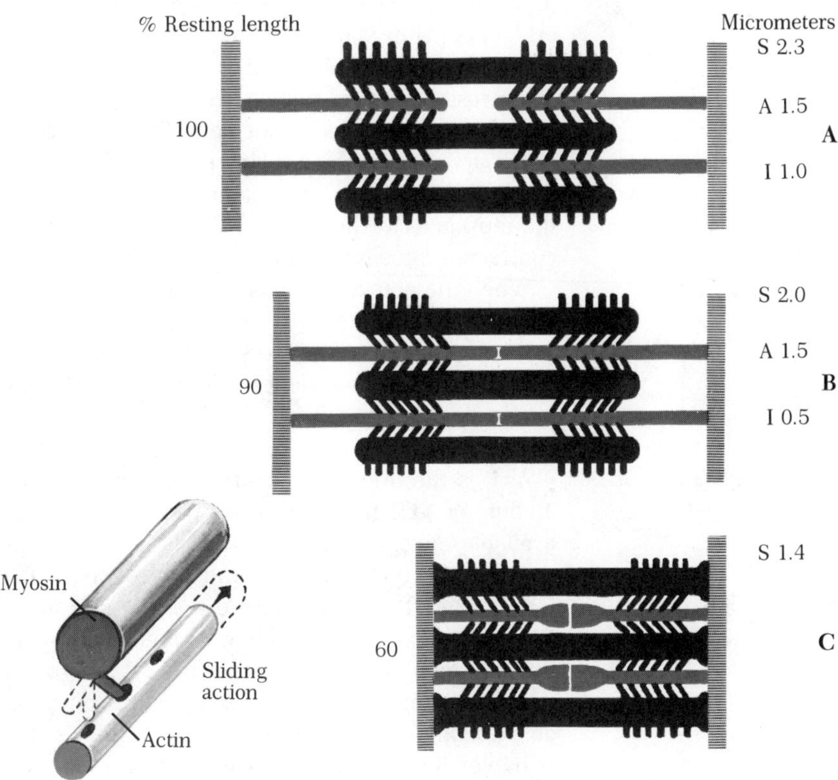

Figure 15-19
A normal sarcomere, **B**, showing changes in arrangement of filaments under stretch **A**, and in contraction, **C**. Relative change in length appears on the left, and approximate length of sarcomere, A band, and I band, in micrometers, is shown on right. Note constancy of A band length except under extreme nonphysiological shortening. Inset in lower left illustrates interaction of myosin cross bridge and actin filament to produce relative movement of filaments in contraction.

From Schottelius, B.A., and Schottelius, D.D.: Textbook of physiology, ed. 18, St. Louis, 1978, The C.V. Mosby Co.

to tropomyosin; Tn-I is the inhibiting unit; and Tn-C is the calcium-binding unit. The Tn-T subunit has a molecular weight of 37,000 and contains 259 amino acid residues, the sequence of which has been determined. It binds the other two subunits to tropomyosin noncovalently. Tn-I contains 179 amino acid residues. Its action appears to be to prevention of the interaction of the head of myosin molecules with actin, thus preventing the binding to actin and ATPase activity. Contraction of the myofibril is thus prevented. Troponin-C, which binds calcium ions, has a molecular weight of about 18,000 and contains 159 residues, a relatively large number being aspartic and glutamic acids. Tn-C also competitively binds Mg^{++} ions.

Sequence of events during muscle contraction

The foregoing sections describe the structural features of muscle and the chemical nature of its contractile proteins. We will now consider the sequence of molecular events that occur during the contraction and relaxation of muscle fibers, a remarkable sequence in which chemical energy is converted into a useful mechanical force.

Contraction of the myofibrils, indeed the sarcomeres of the myofilaments, is initiated by nerve impulses transmitted via motor nerves across the neuromuscular junction to the sarcolemma of muscle fibers. A wave of depolarization spreads from this point along the entire fiber and then inward through openings in the transverse tubules into sarcoplasmic reticulum of the myofibrils (Fig. 15-16). This stimulates a release of Ca^{++} ions by some as yet uncertain mechanism. The Ca^{++} ions bind to troponin-C (of the thin actin filaments), producing a change in its conformation that is followed by a shift of tropomyosin, permitting the attachment of the cross-bridges of the myosin (thick) myofilaments to the actin (thin) myofilaments. Simultaneously, energy is released by the hydrolysis of the ATP by the ATPase action of actomyosin. This energy is used to cause a ratchetlike movement of the myosin cross-bridges attached to the actin filaments, causing the latter to slide along the stationary myosin (thick) filaments toward the center of the sarcomere. This results in the shortening of the sarcomere and in turn the myofibril, the muscle fiber, and the muscle as a whole (Fig. 15-19). The foregoing events occur quite rapidly, the contraction process requiring only a 10 to 20 msec.

When the neurostimulatory phase ends, Ca^{++} ions are drawn back into the sarcoplasmic reticulum network, the cross-bridges of myosin retract, and the sarcomeres and the muscle fibers relax, returning to the resting state. The sequence may be repeated for as long as the supply of ATP and other components essential for muscle contraction permit.

Source of energy for muscle contraction

ATP is the immediate source of energy for muscle contraction. However, the amount of ATP in muscle is relatively small, so a back-up reserve supply of readily available energy is necessary. This is supplied in the form of phosphocreatine ("phosphagen") (Chapters 7 and 13), the high-energy phosphate of which is transferred to ADP by the enzyme *creatine phosphotransferase (creatine kinase* or *CPK)*—the Lohmann reaction. Mammalian skeletal muscle contains about five times as much phosphocreatine as ATP. This still is a fairly limited supply of ATP, so replenishment from other sources is soon necessary. Muscle glycogen is available to a limited extent, forming ATP by means of glycolysis and the citric acid cycle (Chapter 11). Any lactic acid formed in muscle contraction under anaerobic conditions can be converted to glucose-glycogen in the liver (Chapter 11) and then into ATP. Free fatty acids and ketone bodies also can be used as a source of ATP (Chapter 12) in most skeletal muscles. Gluconeogenesis is another source (Chapter 11). ATP also may be derived from ADP by the muscle *myokinase* reaction:

$$2\text{ ADP} \underset{}{\overset{\text{Myokinase}}{\rightleftarrows}} \text{ATP} + \text{AMP}$$

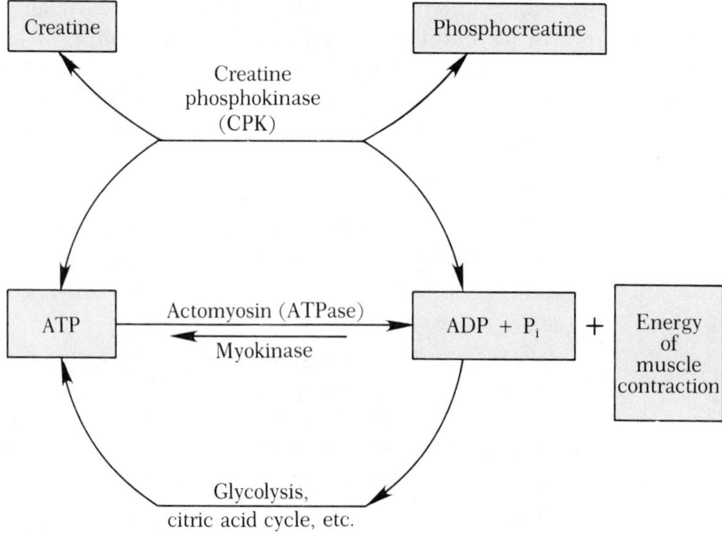

Figure 15-20
Schematic representation of the utilization and resynthesis of ATP and phosphocreatine in skeletal muscle.

The reserve supply of phosphocreatine in muscle is regenerated by the transfer of the high-energy phosphate of ATP to creatine via the reversible action of *creatine phosphokinase*. A schematic representation of the use and resynthesis of ATP and phosphocreatine for the contraction of skeletal muscle is given in Fig. 15-20.

The resynthesis of ATP and phosphocreatine can be blocked by agents, such as iodoacetate (Chapter 11), that inhibit glycolysis. Under these conditions skeletal muscle will continue to contract until the reserve supply of ATP and phosphocreatine is used. Glycolysis thus is the ultimate major source of energy for the contraction of skeletal muscle. Other types of muscle (aerobic), however, can use additional types of fuel efficiently, such as free fatty acids, lactate, and ketone bodies.

Metabolism in other types of muscle

Cardiac muscle, which is characterized by its capability for long, sustained contractions, can readily use fuels requiring aerobic conditions, such as free fatty acids, lactate, and ketone bodies. Indeed, there is some evidence that free fatty acids are a preferred source of energy for the myocardium. The mechanisms of contraction of cardiac muscle are essentially the same as those described for skeletal muscle.

Smooth muscle (involuntary) differs sharply from skeletal muscle in its structure, location in the body, innervation, mode of action, and general functions. However, the chemical processes involved in the contraction of smooth muscle are similar to those found in skeletal and cardiac muscle, e.g., the requirement for calcium ions for initiation and ATP for the force of contraction. For details, consult the pertinent sources at the end of the chapter.

Muscle lipids Besides variable amounts of fat, muscle is found to contain small amounts of cholesterol and larger quantities of phospholipid. Here there are definite differences among the three types of muscle. Smooth muscle has the greatest amount of cholesterol, cardiac muscle next, and striated muscle the least. The ratio of phospholipid to cholesterol is high for skeletal and cardiac muscle and low for smooth muscle. These findings indicate that cholesterol has some relation to the spontaneous muscular activity of cardiac and smooth muscle, and phospholipids are involved in some way with the greater energy production of cardiac and striated muscle.

Extractives If muscle tissue is ground and repeatedly extracted with hot water, a light tan fluid with droplets of fat floating on its surface and particles of coagulated protein suspended is obtained. When this is filtered and concentrated, a dark brown sticky material, commonly known as beef extract, is left. This is composed of all the soluble inorganic salts and all the extractives mentioned earlier. Aside from the small amounts of carbohydrate, amino acids, and peptides present, there is little food value in beef extract.

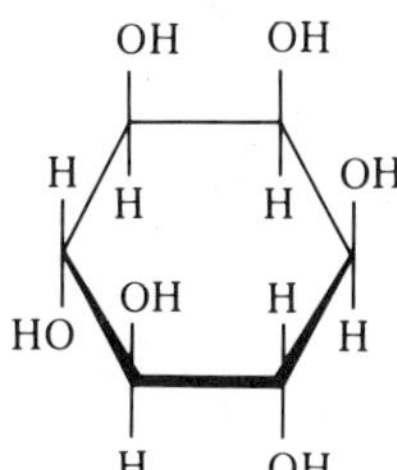

Some of the individual extractives deserve mention at this point. One of them is inositol, $C_6H_{12}O_6$, or better, $C_6H_6(OH)_6$. There are a number of natural isomers of inositol. The most important one was renamed *myoinositol,* instead of *mesoinositol* (see diagram at left). The other isomers differ from this in the arrangement of the hydroxyl groups and hydrogens in space. Although not a sugar, it has a sweet taste. This is a property common to many polyhydroxy alcohols, including glycerol. It is widely distributed in the plant and animal kingdoms and is sometimes considered part of the vitamin B complex (Chapter 21).

At least four peptides have been isolated from muscle extracts. *Carnosine* (β-alanylhistidine), *anserine* (β-alanyl-1-methylhistidine), and β-alanyl-3-methylhistidine are compounds found only in skeletal muscle of vertebrates. They are found alone or in mixtures, depending on the species, at total concentrations ranging from 10 to 60 mmole per kilogram of wet tissue. To date they have no known function. The fourth peptide present in muscle extractives is *glutathione,* a tripeptide, with the composition glutamylcysteinylglycine (Chapter 13). This compound is a hydrogen acceptor and as such plays a role in tissue reactions, such as the reduction of methemoglobin (Chapter 16). It is found in many tissues other than muscle, especially liver, red blood cells, brain, and kidney. It is also present in the lens of the eye and is reduced in amount when cataract occurs. *Carnitine,* a betaine, is also a muscle extractive and is also found in most tissues. Its formula is as follows:

$$(CH_3)_3N^+-CH_2-CH(OH)-CH_2-C(=O)-O^-$$

Carnitine plays a role in the oxidation of fatty acids in muscle tissue, as well as in other tissues, by facilitating their transfer to fatty acid oxidation sites (Chapter 12).

Other constituents Other constituents include glycogen (0.5% to 1%), important as a source of energy for the formation of ATP, traces of free amino acids, a number of enzymes, and inorganic ions, including potassium, magnesium, and sodium as the principal cations in descending amount, and phosphate, sulfate, and bicarbonate as the main anions similarly.

Suggested readings

General

Jeanrenaud, B., and Hepp, D., editors: Adipose tissue regulation and metabolic functions, New York, 1970, Academic Press, Inc.

Epidermal

Bradbury, J.H.: The structure and chemistry of keratin fibers, Adv. Protein Chem. **27:**111, 1973.

Connective

Barnes, M.J., and Kodicek, E.: Biological hydroxylation and ascorbic acid with special regard to collagen metabolism, Vitam. Horm. **30:**1, 1972.

Bornstein, P., and Traub, W.: The chemistry and biology of collagen. In Neurath, H., and Hill, R.L., editors: The proteins, ed. 3, vol. 4, New York, 1979, Academic Press, Inc.

Eyre, D.R.: Collagen: molecular diversity in the body's protein scaffold, Science **207:**1315, 1980.

Fessler, J.H., and Fessler, L.I.: Biosynthesis of procollagen, Annu. Rev. Biochem. **47:**129, 1978.

Gross, J.: Collagen biology: structure, degradation, and disease, Harvey Lect. **68:**351, 1972-1973.

Miller, E.J.: Biochemical characteristics and biological significance of the genetically distinct collagens, Mol. Cell. Biochem. **13:**165, 1976.

Prockop, D.J., et al.: The biosynthesis of collagen and its disorders, N. Engl. J. Med. **301:**13 and 77, 1979.

Ramachandran, G.N., and Reddi, A.H., editors: Biochemistry of collagen, New York, 1976, Plenum Publishing Corp.

Sharon, N., and Lis, H.: Glycoproteins—structure and vital biological roles, Chem. Eng. News **59:**21, 1981.

Spiro, R.G.: Glycoproteins, Adv. Protein Chem. **27:**349, 1973.

Bone

Bourne, G.H., editor: The biochemistry and physiology of bone, ed. 2, vols. 1-3, New York, 1972, Academic Press, Inc.

Jowsey, J.: Metabolic diseases of bone, Philadelphia, 1977, W.B. Saunders Co.

Mandel, I.D.: Dental caries, Am. Sci. **67:**680, 1979.
McLean, F.C., and Urist, M.R.: Bone: fundamentals of the physiology of skeletal tissue, ed. 3, Chicago, 1968, University of Chicago Press.
Rasmussen, H., and Bordier, P.: Bone cells and their functions, Baltimore, 1971, Williams & Wilkins Co.
Shaw, J.H.: Diet regulations for caries prevention, Nutr. News **36:**1, 1973.

Nerve

Bourne, G.H., editor: The structure and function of nervous tissue, vols. 1-5, New York, 1968-1972, Academic Press, Inc.
Flexner, L.B.: Dissection of memory in mice with antibiotics, Am. Sci. **56:**52, 1968.
Glassman, E.: Biochemistry of learning, Annu. Rev. Biochem. **38:**605, 1969.
Goldstein, A.: Opiate peptides (endorphins) in pituitary and brain, Science **143:**1081, 1976.
Guroff, G.: Molecular neurobiology, New York, 1979, Marcel Dekker, Inc.
Iverson, L.L.: The chemistry of the brain, Sci. Am. **241:**134, 1979.
Morell, P., and Norton, W.T.: Myelin, Sci. Am. **242:**88, 1980.
Nachsmansohn, D.: Biochemistry as part of my life, Annu. Rev. Biochem. **41:**1, 1972.
Roberts, E., Chase, T.W., and Tower, D.B., editors: GABA in nervous system function, New York, 1976, Raven Press, Inc.
Rubinstein, E.: Diseases caused by impaired communication between cells, Sci. Am. **242:**102, 1980.
Schwartz, J.H.: The transport of substances in nerve cells, Sci. Am. **242:**152, 1980.
Snyder, S.H.: Opiate receptors and internal opiates, Sci. Am. **236:**44, 1977.
Stevens, C.F.: The neuron, Sci. Am. **241:**54, 1980.

Muscle

Harrington, W.F.: Contractile proteins of muscle. In Neurath, H., and Hill, R.L., editors: The proteins, ed. 3, vol. 4, New York, 1979, Academic Press, Inc.
Huxley, H.E.: The mechanism of muscular contraction, Cold Spring Harbor Symposia on Quantitative Biology, vol. 37, 1972.
Kagen, L.J.: Myoglobin: biochemical, physiological, and clinical aspects, New York, 1973, Columbia University Press.
Koch-Weser, J., and Browne, T.R.: Valproic acid, N. Engl. J. Med. **302:**661, 1980.
Laki, K., editor: Contractile proteins and muscle, New York, 1971, Marcel Dekker, Inc.
Mannherz, H.G., and Goody, R.S.: Proteins of contractile systems, Annu. Rev. Biochem. **45:**427, 1976.
Murray, J.M., and Weber, A.: The cooperative action of muscle proteins, Sci. Am. **230:**59, 1974.
Stephens, N.L.: The biochemistry of smooth muscle, Baltimore, 1977, University Park Press.

Chapter 16

Blood

The evolutionary development of complex multicellular organisms from simpler unicellular forms in direct contact with their environment depended on the simultaneous development of a means of supplying the nutrient needs of the individual cells, removing their waste products, and coordinating and regulating their metabolic activities. Most of these functions are considered in other sections of this book. However, a "circulating tissue," the blood, developed to meet many of these special needs. The biochemical mechanisms by which the blood performs its functions are considered in this chapter.

Functions

Blood has a multitude of functions, all of which are highly necessary to health and to life itself. This is so apparent to the layman that he often becomes unduly alarmed at the loss of even a small quantity. Since blood makes up about one twelfth of the body weight (about 85 ml per kilogram), evidently a person weighing only 54 kg (120 lb) has about 4.5 kg (10 lb) of blood, and the loss of 0.5 L (1 pint; approximately a pound) would not be very serious. Indeed, thousands of persons have donated considerable amounts of blood for whole-blood transfusions and for the production of plasma. However, although single donations do not ordinarily cause more than slight discomfort, they should not be repeated too frequently, and evidence of complete blood regeneration should be required.

Among the chief functions of blood are the following:

1. Transportation of nutrients from the intestine and chyle ducts to the tissues, from the liver to the tissues and back to the liver, and from one tissue to another
2. Exchange of the respiratory gases between the lungs and the tissues
3. Transport of waste products arising in metabolism, such as urea and uric acid, to the kidneys, skin, intestine, and liver for excretion
4. Distribution of hormones and other regulatory substances; the effective agent is brought almost instantaneously to the target organ or tissue to be stimulated or inhibited
5. Protection against microorganisms and various other foreign invaders
6. Aid in the maintenance of acid-base, electrolyte, and water balance
7. Heat regulation of the body, largely by shifting blood to or from the surface of the body
8. Prevention of excessive hemorrhage by coagulation

Some of these functions are discussed in other chapters.

General composition

Circulating blood consists of a fluid portion (plasma) and the formed elements. The plasma constitutes 55% to 60% by volume of whole blood. The formed elements are the red blood cells, or erythrocytes; the white blood cells, or leukocytes; and the blood platelets. The average number of erythrocytes is normally 5,400,000 per cubic millimeter for men and 4,900,000 for women, but higher or lower figures are not uncommon. There is a slight fluctuation in the number of erythrocytes during the day: it is lowest during sleep, rises on awakening, and continues to rise during the rest of the day. Persons living at high altitudes usually have a higher erythrocyte count than do those living at sea level. High erythrocyte counts also occur after muscular exercise, emotional excitement, and increased atmospheric temperature. These are temporary changes, resulting from a flow of a more concentrated blood from the spleen. Any condition that tends to lower the oxygen content of the blood causes a compensatory increase in the number of erythrocytes. On the other hand, any condition that increases the oxygen of the blood causes a decrease in erythrocyte count. High barometric pressure is an example. Pathologically an increase in the erythrocyte count is called *polycythemia; anemia* is a condition in which there is either a lowered count or a subnormal concentration of hemoglobin.

If blood is removed from an artery or vein, it clots or coagulates in a few minutes. The whole mass becomes gelatinous. If it is left undisturbed, a clear straw-colored fluid is gradually squeezed out; this is *serum*. In this case the formed elements have become enmeshed in the clot, which retracts into a semisolid mass. Blood *plasma* can be separated from the formed elements if, immediately after the blood is obtained, an anticoagulant is added, and it is centrifuged. A fluid that is also clear and straw colored is obtained.

$$\text{Whole blood} - \text{Formed elements} = \text{Plasma}$$
$$\text{Whole blood} - (\text{Formed elements} + \text{Clotting factors}) = \text{Serum}$$
$$\text{Plasma} - \text{Clotting factors} = \text{Serum}$$

Clotting, of course, is a process that protects the individual from excessive loss of blood. The mechanisms involved in the coagulation of blood are discussed in a subsequent section of this chapter.

Physical characteristics

Arterial blood is bright crimson; venous blood is a darker red but is not purple or blue. Blood is more viscid than water, the viscosity being attributable to the many corpuscles present and to the high protein content. The pH is approximately 7.4, with a normal range from 7.3 to 7.5. The mechanisms for maintaining this constancy of reaction are varied and are taken up in a later part of this chapter.

The specific gravity of blood ranges from 1.035 to 1.075. This may be determined clinically by Hammerschlag's method, which consists simply of letting drops of blood fall into mixtures of varying proportions of chloroform and benzene, or xylene and bromobenzene, or methyl salicylate and mineral oil. If the drop does not rise or sink in one mixture, it evidently has the same specific gravity as that mixture, which can easily be determined by a hydrometer.

A convenient, practical modification of the falling-drop method for determining the specific gravity of whole blood or plasma, using solutions of copper sulfate of increasing concentrations, has been described. The method also permits a gross estimation of the hemoglobin content of whole blood and the protein concentration in the plasma.

When blood is removed from the circulation, the erythrocytes slowly settle out if clotting is prevented. The rate of settling, or *sedimentation rate,* is frequently determined for clinical purposes. Erythrocytes from men generally sediment more slowly than do those from women. Newborn infant red cells sediment very slowly. The sedimentation rate is greatly increased during menstruation and normal pregnancy. Increased rates are found pathologically in septicemia and pulmonary tuberculosis. The probable explanation for the increased sedimentation rate is a clumping or agglutination of erythrocytes because of an increased globulin and fibrinogen content of the plasma. An increase in the ratio of cholesterol to phospholipid in the plasma also increases the sedimentation rate.

Quantitative composition

Average values and/or usual range of the principal constituents of blood of the normal human adult in the postabsorptive state are given in Table 16-1. The values are given for whole blood, for plasma, and for serum and in the units commonly used by clinical laboratories. The normal samples are usually obtained from healthy adult males unless otherwise indicated. Standard methods of analysis usually automated as used in the clinical laboratory for routine determinations are ordinarily employed. Since many of these values vary considerably after meals, it is desirable that the subject studied be in a postabsorptive state, usually 8 to 12 hours after the last meal. This is the procedure commonly followed in clinical laboratories for diagnostic purposes.

Normal blood values also are being reported now, particularly in European clinical laboratories, in SI units (Système International d'Unités). Most of the SI units are expressed as millimoles (or micromoles) per liter. Values for enzymes are given as microkatals or nanokatals per liter, the katal unit being that amount of an enzyme or other catalytic agent that catalyzes a reaction rate of 1 mole per second in an appropriate assay system. Conversion tables for changing values such as milligrams per 100 ml to SI units may be found in an article by Baron, D.N., et al.: The use of SI units in reporting results obtained in hospital laboratories, J. Clin. Pathol. **27**:590, 1974.

The quantitative composition of the blood of a normal human subject varies with respect not only to the types and amounts of food ingested but also to the age, sex, and activity of the subject and other factors.

The large number of substances present in the blood emphasizes its manifold functions in the body in the transport of nutrients, metabolites, regulatory substances, and waste products. The range of values may deviate sharply from normal in a number of diseases. It has been said that the blood mirrors metabolic and pathologic events in the cells and tissues of the body. This fact is of great practical importance today in modern medical diagnosis and therapy. Examples of the relationship appear repeatedly throughout this text.

Blood plasma

Blood plasma is a light straw-colored fluid with a specific gravity from 1.015 to 1.035. The higher specific gravity of whole blood must be ascribed to the erythrocytes, whose specific gravity is about 1.090. As previously stated, the specific gravity of plasma is related to the protein content, and an approximation of the total protein of plasma may be obtained by determining its specific gravity and applying the following formula:

$$\text{Total plasma protein (grams per 100 ml)} = 373 \text{ (specific gravity, } -1.007)$$

Human plasma contains from 90% to 92% water. Blood owes much of its physiological importance to high water content, for not only is water the medium in

Table 16-1
Composition of normal human blood (postabsorptive state) (values are averages and/or normal ranges)

Whole blood	**Grams percent (gm/100 ml)***
Total solids	19-23
Water	77-81
Hemoglobin	
Adult males	16 (14-18)
Adult females	14 (12-16)
Children	12 (11-17)
Total nitrogen	3.5
	Volumes percent (vol%)
Carbon dioxide content (venous)	50-60
Carbon dioxide content (arterial)	45-55
Oxygen capacity	16-24
Oxygen content (venous)	10-18
Oxygen content (arterial)	15-23
	Milligrams percent (mg/100 ml)
Chloride (as NaCl)	450-500
Cholesterol	150-280
Creatine	3-7
Creatinine	1-2
Glucose	70-120 (depending on method used)
Total combined fatty acids	300-400
Total ketone bodies (as acetone)	1-5
Lecithin (as lipid phosphorus)	5-12
Lecithin (as lecithin)	125-300
Iron	50-70
Lactate	5-20
Pyruvate	0.7-2.0
Nonprotein nitrogen (NPN)	15-35
Urea	20-40
Urea nitrogen (BUN)	8-20
Amino acid nitrogen	3-6
Ammonia nitrogen	0.04-0.1
Undetermined nitrogen	4-10
Uric acid	3-7.5

Plasma	**Milligrams percent (mg/100 ml)**	**Milliequivalents per liter (mEq/L)**
Carbon dioxide capacity	55-75 vol%	20-29
Chloride (as NaCl)	580-630	
(as Na^+)	230-250	99-108
(as Cl^-)	350-380	99-108
Fibrinogen	200-400	
Inorganic phosphorus		
Adults	3-4.5	1.7-2.5
Children	4-7	2.2-3.3
Total lipids	500-600	
Triglycerides	40-160	

Serum	**Grams percent (gm/100 ml)**
Total protein	6-8
Albumin	4-5
Globulin	2-3

Modified from Dotti, L.B., and Orten, J.M.: Laboratory instructions in biochemistry, ed. 8, St. Louis, 1971, The C.V. Mosby Co. See also Normal laboratory values (case records of the Massachusetts General Hospital), N. Eng. J. Med. **290:**39, 1974; and Beck, W.S.: Letter to Editor, N. Engl. J. Med. **290:**695, 1974.
*Although most of the values in this table continue to be expressed in the traditional form—per 100 ml—it should be noted that there is a growing trend, especially in the clinical laboratory fields, to change to the more concise equivalent—*deciliters* (dl). Thus 100 ml = 1 dl.

Continued.

Table 16-1, cont'd
Composition of normal human blood (postabsorptive state) (values are averages and/or normal ranges)

Serum	Milligrams percent (mg/100 ml)	Milliequivalents per liter (mEq/L)
Bilirubin, indirect	0.2-1.5	
Bilirubin, direct	0.1-0.4	
Calcium	9-11	4.5-5.5
Chloride	100-105	
Cholesterol, total	150-280	
Cholesterol, esters	75-200	
Creatine	0.2-0.8	
Creatinine	1.0-1.8	
Iron	50-150 (μg)	
Iron-binding capacity	250-400 (μg)	
Lactate	4-16	
Magnesium	1-3	0.9-2.5
Phosphate	3-4.5	1-2
Phospholipid	150-200	
Potassium	16-22	3.5-5
Sodium	310-333	135-145
Sulfate, inorganic (as S)	0.9-1.1	0.6-1.1
Uric acid	3-5	

Erythrocytes	Percent
Hemoglobin	35
	Milligrams percent (mg 100/ml)
Potassium	420
Sodium	25
Magnesium	6.6
Calcium	Small amount

Enzymes	Units
Amylase, Somogyi	50-150/100 ml serum
Acid phosphatase, Gutman	0.5-2/100 ml plasma
Acid phosphatase, King	1.4-4.5/100 ml plasma
Alkaline phosphatase (Bodansky units)	
Adult	2-4.5/100 ml plasma
Child	5-14/100 ml plasma
Alkaline phosphatase (King units)	
Adult	5-13/100 ml plasma
Child	15-20/100 ml plasma
Cholinesterase	39-51
Creatine phosphokinase (CPK)	10-100 mIU/ml
Lactic dehydrogenase (LDH)	90-200/100 ml serum
Phosphoglucomutase	19-84
Phosphohexoisomerase	14-28
Serum glutamic oxaloacetic transaminase (SGOT)	
Adult	4-40/100 ml serum
Infants	13-120/ml serum
Serum glutamic pyruvic transaminase (SGPT)	
Adults	5-30/100 ml serum
Infants	12-90/100 ml serum

Hormones	Amounts
ACTH	15 μg/100 ml blood
Aldosterone	0.003-0.01 μg/100 ml plasma
Estrogen (as estradiol) during pregnancy	0.2-0.5 μg/100 ml blood
Growth	
Adult male	Below 5 ng/ml plasma
Child	Over 10 ng/ml plasma

Table 16-1, cont'd
Composition of normal human blood (postabsorptive state) (values are averages and/or normal ranges)

Hormones	Amounts
17-Hydroxycorticosteroids	6-44 μg/100 ml plasma
Insulin	6-26 μU/ml plasma (or serum)
Progesterone during pregnancy	530 μg/100 ml plasma
Testosterone	
Adult male	Over 0.30 μg/100 ml plasma
Adult female	Below 0.10 μg/100 ml plasma
Thyroxine	
Total	6-12 μg/100 ml serum
Free	0.8-7.4 μg/100 ml serum
Tri-iodothyronine	70-190 ng/100 ml serum

Vitamins	Amounts
A (as carotene)	15-60 μg/100 ml serum
B_{12}	350-750 μg/100 ml serum
Biotin	1.2 μg/100 ml blood
Ascorbic acid	0.6 mg/100 ml blood
	0.1-1.7 mg/100 ml plasma
D (as D_2)	2.8 μg/100 ml plasma
E	0.9-1.9 mg/100 ml plasma
Niacin	0.6 mg/100 ml blood
Pantothenic acid	30 μg/100 ml blood
Folic acid	3.4 μg/100 ml blood
Riboflavin	2.6-3.7 μg/100 ml plasma
Thiamin	7.6-8.9 μg/100 ml blood
Thiamin + Cocarboxylase	10 μg/100 ml blood

Miscellaneous constituents	Milligrams percent (mg/100 ml)
Bromide (serum)	0.252
Copper (plasma or serum)	100-200 μg
Fluoride (whole blood)	0.28
Iodine (protein bound) (plasma or serum)	0.004-0.008
Lead (whole blood)	50 μg
Manganese (whole blood)	0.005-0.02
Sialic acid (serum)	60 (adults)
	40 (infants)
Silica	
Soluble, whole blood	1.5 (as SiO_3)
Total, whole blood	9 (as SiO_3)
Zinc (plasma or serum)	0.12-0.48

which the water-soluble and water-dispersible substances are carried, it also is needed for maintaining blood pressure, osmotic relations, and heat regulation. As regards heat regulation, water has (1) high specific heat, (2) high heat conductivity, and (3) high latent heat of evaporation. Thus water has great heat-storage properties, i.e., more calories of heat are required to raise the temperature of water a given number of degrees than for most fluids. Its high conductivity results in the rapid removal of heat from the interior of the body by conduction through the water in all the soft tissues and body fluids, as well as in the blood. Finally, a great deal of heat is lost through evaporation from skin and lungs, since the water comes largely from the blood plasma.

Proteins

Proteins form most of the solid matter of plasma. They total between 6% and 8% of the plasma. The major types include albumins and globulins. The approximate concentrations of these in human plasma, as determined by electrophoretic mobil-

Table 16-2
Distribution of major types of human plasma proteins as determined by electrophoretic fractionation

Component	Percent total protein	Concentration (gm/100 ml)
Total protein		6-8
Albumins	50-60	4-5
Prealbumin	0.6	0.1
Globulins	40-50	2-3
α_1	5-13	0.3-0.6
α_2	7-12	0.4-0.9
β	10-14	0.6-1.1
γ	12-19	0.7-1.5
Fibrinogen	5	0.3

Table 16-3
Properties and functions of some purified plasma globulins

Fraction	Component	Approximate amount (mg/100 ml)	Molecular weight	Isoelectric point	Carbohydrate (percent)	Principal functions
	Antitrypsin	200-400	54,000	4.8	12	Inhibits a number of proteolytic enzymes
α_1	Lipoprotein (HDL)	300-800	200,000	5.2	1.5	Transport of cholesterol (from liver)
	Orosomucoid	50-150	44,100	3	42	?
	Retinol-binding protein	3-6	21,000	—	0	Transport of retinol
	Thyroxine-binding protein	1-2	50,000	—	12	Transport of thyroxine
	Transcortin	3-4	50,000	—	25	Transport of cortisol
α_2	Glycoproteins	800	300,000	4.9	16	?
	Mucoproteins	500	—	4.9	13	?
	Haptoglobins (3 types)	400-800	100,000-400,000	4.1	19	Transport of free hemoglobin from destroyed red blood cells
	Lipoprotein (HDL)	200	5-20 million	—	2	Transport of cholesterol and triglycerides
	Ceruloplasmin	20-60	150,000	4.4	8	Transport of copper; increases use of iron as feroxidase
	Prothrombin (bovine)	10	68,900	4.2	11	Proenzyme of thrombin
	Angiotensinogen	—	—	—		Precursor of angiotensin
	Erythropoietin	<5	60,000	3.3	—	Erythropoietic hormone
β_1	Lipoprotein (LDL)	400-1000	Up to 3 million	5.5	2	Transport of triglycerides and other lipids
	Hemopexin	60-100	60,000	—	22	Heme-binding globulin
	Plasminogen	30	143,000	5.6	—	Profibrinolysin
	Fibrinogen	300	341,000	5.8	3	Coagulation factor I
$\beta_1\beta_2$	Complement	1-170	80,000-400,000	—	—	Lysis of foreign cells
	Metal-binding protein (transferrin)	200-400	75,000	5.8	6	Transport of iron
β_2	Unknown	200	—	6.3		?
	Glycoproteins	30	40,000	—	6-19	Unknown
γ	Blood group globulins and immunoglobulins, G, A, M	700-1500	150,000-1 million	6.3-7.3	3-12	Contains various antibodies, blood globulins, complement C_1, C_2, etc.
	Cryoglobulins	10	165,000-600,000	—	—	? (insoluble in cold)

ities, are shown in Table 16-2. At the present time over 100 different plasma proteins have been described. They have been isolated mainly by electrophoretic fractionation. Many have been identified by immunoelectrophoresis, as will be discussed later. Some have been quantitated by immunoprecipitin reactions.

The globulin fraction of the plasma proteins contains a number of components that are of great and varied physiologic importance (Table 16-3). Interesting is the fact that the majority of the plasma proteins appear to be conjugated, e.g., glycoproteins, lipoproteins.

Fibrinogen is the protein that, when blood is shed, is converted to fibrin, the basis of the blood clot. It resembles the globulins in most of the properties but has a few different precipitation reactions. Like globulins, it is precipitated by half saturation with ammonium sulfate, but globulins require full saturation with sodium chloride, whereas fibrinogen requires only half saturation. It is formed in the liver. Animals deprived of their liver by surgical operation (hepatectomy) rapidly lose the fibrinogen of their blood. The fibrinogen content of plasma increases when inflammatory or infectious processes exist and during menstruation and pregnancy.

The liver is also the site of formation of the albumin, prothrombin, and probably more than 80% of the globulins. Normally the plasma proteins are broken down or catabolized continuously, but rather slowly, and the rebuilding process or anabolism keeps pace with this; so the level of these proteins remains constant. Similarly, the fetal plasma proteins, except γ-globulin, are probably derived from the mother, as will be described later.

The fetal serum proteins, like fetal hemoglobin, differ immunochemically from those of the adult. An α_1-fetoglobulin, or *fetuin,* discovered first in calf serum, is an example. Fetuin comprises nearly half the total plasma protein of calf serum. It is a glycoprotein with a molecular weight of about 48,000. It is present in bovine fetal serum until the third trimester of pregnancy, when it gradually decreases, and is almost absent at birth. Its function is unknown, but it promotes growth of the mammalian cells in tissue culture and may increase the adhesion of tissue culture cells to a glass surface.

Separation of plasma proteins

A number of procedures have been developed for separating plasma and other blood proteins on the basis of their differing mobilities in an electric field (Chapter 24). The classic Tiselius procedure has been used extensively. More recently, relatively simple and less expensive methods with filter paper, starch, or thin-layer silica gel plates as a supporting medium have been widely adopted. Also, modified electrophoretic procedures having better separating capabilities, such as *isoelectric focusing* and *immunoelectrophoresis,* have been developed.

Proteins move in an electric field because of the electric charges they carry. In Tiselius apparatus the diluted and dialyzed serum or plasma is overlaid with a buffer solution. During electrophoresis the various proteins at the boundary between the buffer solution and the plasma become separated and move at different rates of speed. As they move, the relative positions and widths of the protein boundaries may be either visibly projected on a ground-glass screen or recorded on a photographic plate. The albumin fraction moves faster than the globulins or fibrinogen, and a study of the patterns obtained reveals six distinct proteins (Fig. 16-1). The area of each hump measures the concentration of the protein moving with that particular mobility. Careful study has shown that fractions formerly designated *pseudoglobulin, euglobulin,* etc., which were obtained by salting-out methods, are really mixtures of several globulins. The older terminology has therefore been abandoned, and the nomenclature of Tiselius is now generally adopted. The proteins shown to be present are albumin, α_1-, α_2-, β_1-, β_2-, and γ-globulins, and

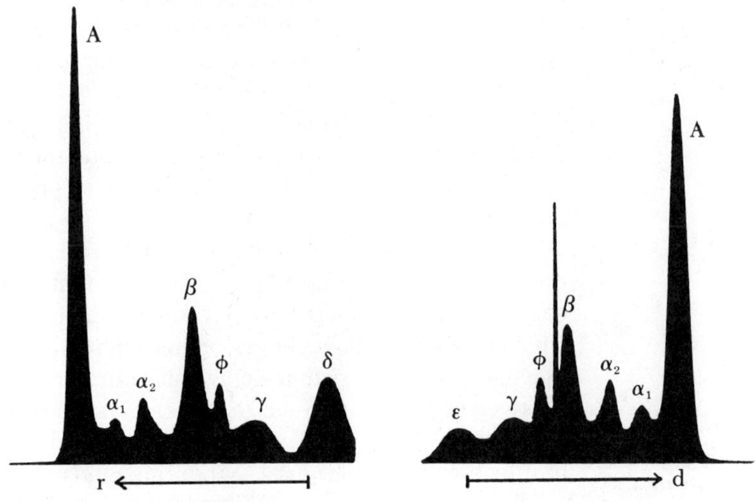

Figure 16-1
Electrophoretic pattern of normal human plasma at pH 8.6. Peaks shown are, from the extreme ends inward, albumin, α_1-, α_2-, and β-globulin, fibrinogen, γ-globulin, and the anomalous peak.

fibrinogen. The patterns do differ with different species and also, even in the case of the same sample, with the buffer used and with the period of electrophoresis. As seen in Fig. 16-2, the albumin peak is the tallest and best defined, whereas the globulin and fibrinogen peaks are lower and often spread over a greater distance. The ascending and descending boundaries are not identical.

Electrophoresis does not separate the protein components into *chemically pure* substances. The albumin component contains bilirubin and the β-globulin cholesterol. All the protein fractions contain carbohydrate and some lipid material. During *chemical* separation most of these combinations are broken; consequently, it is highly probable that the "purified" products ordinarily isolated are really derivatives of the protein complexes as they circulate in the blood and as they are prepared by electrophoresis.

More recently, a modified electrophoretic procedure called *isoelectric focusing* has been developed and is particularly effective in separating some plasma and other blood proteins. In this procedure proteins are separated on the basis of their pI (isoelectric point) using appropriate buffers. Several closely related plasma proteins have been separated by this technique.

Immunoelectrophoresis Valuable as the various modifications of electrophoresis have been, e.g., the Tiselius "moving-boundary" procedure and the several types of "zone" electrophoresis on filter paper or agar plates, these methods do not give a clear separation of closely related mixed proteins such as are present in human plasma. The technique of immunoelectrophoresis appears to be more satisfactory in this respect. Essentially this procedure combines the electrophoretic separation of mixed proteins with the antigen-antibody reaction. The proteins are separated as a series of spots on an agar-covered glass plate by the usual electrophoretic procedure. The current is turned off and the proteins diffuse outward. Meanwhile, an antihuman immune serum from sources such as rabbit or horse, which contains antibodies for the human serum protein under investigation, is placed in a trough at the side of the agar plate, parallel to the electrophoretic migration. The antibody diffuses inward, and, when it meets the antigen, the familiar precipitation occurs, forming an opaque arc. It has been possible thus far to identify more than 25 distinct human serum proteins by this technique, using an antibody-rich horse

Figure 16-2

Immunoelectrophoretic pattern for normal human serum. A sample was introduced into a small round starting well. After electrophoresis in the buffered agar layer, an antiserum was pipetted into the trough (*above*). The specific antibodies diffusing from the trough are precipitated by their corresponding serum protein antigens previously distributed in the gel, according to their electrophoretic mobilities. From left to right (anode to cathode) known constituents are labeled in order of decreasing mobility. **ρ,** Rh_0, a minor component moving in advance of albumin; another ρ component has been identified as a lipoprotein; **A,** albumin; **α_1-A,** α-1-glycoprotein; **α_2-M,** α-2-macroglobulin, molecular weight 900,000; **α_2-H,** haptoglobin, hemoglobulin-binding protein; **β_1-A,** conversion product of β_1-C associated with complement activity; **β_1-B,** β-1-mucoprotein; **T,** transferrin; **β-1,** iron-binding protein; **β_1-C,** active component of the complement system; inactivation involves conversion to β_1-A; **β_2-M,** macroglobulin (molecular weight about 1,000,000) with antibody activity, aberrant in some types of multiple myeloma; **β_2-A,** globulin with some antigenic similarity to α-globulin that is aberrant in some types of multiple myeloma; **γ,** α-globulin.

Courtesy C. A. Williams, Jr., New York.

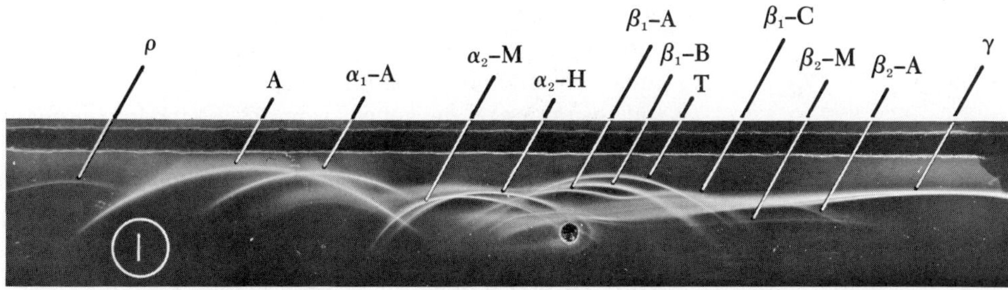

serum. The possible applications appear to be extensive. For example, the technique has proved useful in studies of the serum proteins in various diseases, in studies of the embryologic development of tissues, and in many other types of investigation. The pattern obtained by immunoelectrophoresis of samples of human serum proteins is shown in Fig. 16-2.

Albumin and globulins

Albumin and globulins comprise most of the proteins of blood plasma. The relative concentrations of these two proteins, along with the total protein content, are of great importance because under normal conditions the colloidal osmotic pressure of the blood, attributable almost entirely to the proteins of the plasma, is the force that opposes the hydrostatic pressure in the capillaries. The term *osmotic pressure* is not quite correct from the physicochemical standpoint, since (1) there is no perfect semipermeable membrane and (2) a solution does not exert an osmotic pressure unless it is separated from another solution by such a membrane. In this discussion the term *potential osmotic pressure* or osmotic tendency or quality might be better. With that in mind, we may proceed.

The *total* osmotic pressure is the osmotic pressure from the electrolytes and organic crystalloids present plus that from the colloids. The electrolytes and organic crystalloids are quite diffusible and pass through the capillary walls into the tissue fluids rather freely. Hence the osmotic pressure exerted by them is the same on both sides of the capillary. The plasma proteins, however, are not as freely diffusible and are greater in amount within the capillaries than without. Therefore the osmotic pressure exerted by them, small though it is because of their large molecular size and thus smaller numbers of molecules per unit volume, is greater inside than outside the capillaries. This is a pressure exerted inward, and normally it just balances the hydrostatic pressure (because of factors such as heartbeat and elasticity of the arteries) exerted outward. The total osmotic pressure of plasma is about 6.5 atmospheres or 4940 mm Hg. This is *almost* balanced by a similar osmotic pressure of the tissue fluids bathing the capillaries. The slight difference is due to the difference in protein concentration between the plasma and the tissue fluids.

Osmotic pressure of plasma proteins	= about 28 mm Hg
Osmotic pressure of tissue fluid proteins	= about 10 mm Hg
Difference (oncotic pressure)	18 mm Hg

Therefore the *effective* potential osmotic pressure of the plasma proteins, sometimes called the *oncotic pressure,* is about 18 mm Hg. The hydrostatic pressure inside and outside the capillaries varies in different locations and under different conditions; but assuming that the effective hydrostatic pressure is 32 mm Hg on the arterial side of a given capillary and falls to 12 mm on the venous side, we can see that the following state of affairs would then exist. As a result, on the arterial side the hydrostatic pressure exceeds the colloidal osmotic pressure ($32 - 18 = 14$ mm) and fluid tends to be forced out. On the venous side the reverse is true ($12 - 18 = -6$ mm) and fluid is drawn back. Thus filtration is favored on the arterial side and absorption on the venous side. If the protein content of the plasma were to fall, the effective colloidal osmotic pressure would drop, more water would be forced out, and less would be absorbed. The water of the blood would pass into the tissues and *edema* would result. More than 80% of this colloidal osmotic effect is attributable to the albumin fraction because there is more albumin, with a comparatively low molecular weight (about 70,000) than globulin, most of which have molecular weights over 100,000.

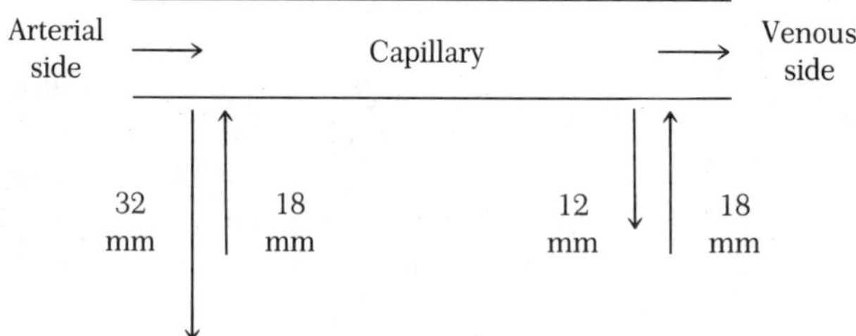

Pathologically, damaged kidneys eliminate proteins in about the same relative proportions as proteins are present in normal plasma. This means a greater loss of albumin than globulin, a particularly unfavorable occurrence if the plasma is already low in albumin. Experimentally such an edema may be produced by *plasmapheresis*. This procedure consists of repeatedly bleeding an animal and reinjecting the blood cells after they have been washed and suspended in a protein-free saline solution. When the protein content of the plasma falls below a critical level, edema occurs. There are a number of mechanisms that might lead to a low plasma protein content, e.g., loss of protein by the kidney (as in nephrosis), inadequate or improper protein intake, inhibition of plasma protein synthesis, and loss of protein as a result of increased permeability of the capillaries.

Edema may result from a number of other causes besides the ones described. The principal causes are related to general or local changes in capillary blood pressure, i.e., if the arterial pressure is relatively lower than the venous pressure, there is a back pressure and a slowing up of capillary flow, with a distension of the capillaries and a consequent forcing out of fluid by this increased capillary pressure. Such a condition frequently results from heart failure or mechanical obstruction of the large veins. Increased permeability of the capillaries may also produce edema, but in this case edema is usually secondary to the changes in capillary pressure, or it may be brought about by avitaminosis (e.g., in beriberi and scurvy), by bacterial or other toxins, or by extreme heat.

As stated previously, two or possibly three albumin fractions are present in adult human plasma. The major fraction is present in amounts varying from 4 to 5 gm/100 ml of plasma. This fraction has a molecular weight of 69,000 and contains approximately 0.8% carbohydrate. A small electrophoretically distinguishable fraction, termed *prealbumin,* is present in an amount of about 0.025%. This fraction has a molecular weight of approximately 61,000 and contains about 0.08% carbohydrate. A third *postalbumin* fraction, actually an α_1-globulin, contains approximately 10% carbohydrate. Specific functions of these minor fractions are unclear.

In addition to their primary function of retaining fluid within the vascular compartment, as just described, plasma albumins are important in the transport of fatty acids, bilirubin, and certain drugs, e.g., the sulfonamides. They are also important for their buffering capacity and for their nutritive effects. The plasma globulins also share in these functions as well as in the transport of certain specific substances (Table 16-3).

Immunoglobulins

Most of substances concerned in immunological reactions are of protein nature. At any rate, the *antibodies* that are present in blood, or are produced there, are modified plasma γ-globulins and are customarily referred to today as *immunoglobulins.* When a foreign protein, an *antigen,* is injected parenterally into an animal, an antibody is formed and appears in the serum of the animal; this may be demonstrated by tests in vitro. The reactions in such tests are termed *precipitin* reactions if the antigen used is of large molecular size. They are called *agglutinin* reactions if the antigen is of cellular size, and *lytic* reactions if the cellular antigen is lysed.

The γ-globulins represent 12% to 19% of the total plasma proteins. Most of the antibodies found in plasma are associated with the γ-globulin fraction, although a minor portion is found in the β-globulin fractions. The γ-globulins are formed extrahepatically, apparently in the plasma cells of lymph tissue.

The γ-globulins were originally so named because they move more slowly than any other plasma protein fraction on electrophoresis at an alkaline pH. They are heterogeneous, comprising a number of distinctly different proteins. At least 25 to 30 different immunoglobulins have been prepared from human γ-globulins. The immunoglobulins may be separated into subfractions by ultracentrifugation and by column chromatography on ion-exchange resins, agar gel, and other adsorbents. These subfractions vary widely in molecular weight (from 150,000 to over 1,000,000) and in isoelectric point (from 6.3 to 7.3) as well as in other physicochemical properties. The immunoglobulins are glycoproteins, containing 3% to 12% carbohydrate, including D-mannose D-galactose, L-fucose, D-glucosamine, and a sialic acid (D-*N*-acetylneuraminic acid) (Chapter 23). They apparently differ only in minor ways with respect to amino acid composition. Their differing physical and immunological properties appear to be attributable mainly to variations in amino acid sequences and conformational differences, particularly at the active site of the molecule. The fact that the immunoglobulins may be partially digested with certain proteolytic enzymes without a loss of immunological properties is significant.

On ultracentrifugation the immunoglobulins separate into at least *three* major fractions and several minor fractions. Several different systems of nomenclature are used for these fractions, creating some confusion. The current preferred nomenclature for the three major fractions, together with synonyms indicated in parentheses, is as follows: IgM (γM, 19S, macroglobulins); IgG (γG, 7S); and IgA

(γA, β_2A, 6.6S to 13S). The *S* in each case refers to the number of Svedberg units (Chapter 25) characteristic of the γ-globulin (1 Svedberg = 1×10^{-13} second and represents the time required for the particle to move a specified distance in an ultracentrifugal field under specific conditions).

Most of the antibodies are in the IgG (7S)-fraction, which represents some 70% of the total γ-globulins. This fraction has a molecular weight of about 150,000 and contains about 3% carbohydrate. It was isolated first from the urine of a myeloma patient with the initials D.O.B. and is therefore sometimes referred to as the *DOB protein*. Its three-dimensional structure has been elucidated by x-ray crystallography using a computerized image display system. It is a T-shaped molecule containing nine sugar molecules that possibly contribute flexibility to the "hinged" T shape. The IgA-fraction has a molecular weight of 180,000 to 500,000. The IgM-, or macroglobulin, fraction has a molecular weight of approximately 1 million and a carbohydrate content of about 10%. A study of a human macroglobulin, purified by agar gel chromatography, showed the molecular weight of the macroglobulin to be 890,000. On treatment with thiol reagents the molecule splits into five subunits, each having a molecular weight of about 185,000.

The different types of immunoglobulins are made up of subunit peptide chains apparently in much the same manner as hemoglobin. Two general sizes of chains are found: a heavy chain, sometimes called *A*, and a light chain, or *B*. The former has a molecular weight of about 40,000; the latter, about 20,000. Three types of heavy chains have been described and termed μ, γ, and α, after the major fractions of γ-globulin from which they are derived. Two types of light chains have been characterized: κ and λ.

Some of the characteristic properties of the major types of human immunoglobulins, together with their subunit compositions, are summarized in Table 16-4. The relatively rapid turnover and synthesis rates of the various immunoglobulins are evident from the data given. Also interesting is that adult blood levels of the immunoglobulins are not attained until 2, 4 to 8, and 1 year of age for G-, A-, and M-types, respectively. The structures and amino acid sequences of the light and heavy chains comprising the immunoglobulins and their evolutionary relationships are the subject of current intensive study. The complete amino acid sequences of several human types IgG and IgM and about half that of an IgA, including (light)

Table 16-4
Some characteristics of the major types of immunoglobulins

Type	Subunit composition	Molecular weight	Carbohydrate content (percent)	Serum level (milligrams per 100 ml)	Synthetic rate (grams per day—70-kg man)	Half-life (days)
IgG (γG)						
Type I	$\gamma_2\kappa_2$	153,000	3.0	0.8-1.6	2.3	25
Type II	$\gamma_2\lambda_2$					
IgA (γA)						
Type I	$(\alpha_2\kappa_2)_n$	180,000-500,000	5-10	0.2-0.4	1.7	6
Type II	$(\alpha_2\lambda_2)_n$ (n = 1 to 4)					
IgM(γM)						
Type I	$(\mu_2\kappa_2)_n$	900,000	10-12	0.05-0.2	0.3	5
Type II	$(\mu_2\lambda_2)_n$ (n = 5, 6, etc.)					

λ-chains and κ-chains have been determined. Both types of light chains contain a segment with a *constant* sequence of amino acids comprising about half the chain, and a *variable* portion in the other half. The heavy chains, μ, γ, and α, also contain a *variable* portion, having a variable amino acid sequence in approximately one fourth of the chain (about 110 amino acids), with the remaining 330 or so residues forming the constant portion of the chain. There are many similarities in the amino acid sequences of the two types of light chains, and indeed of the heavy chains, suggesting that they may have been derived during evolutionary processes by gene duplication from a more primitive "L-chain" comprising about 110 amino acids. The variable portion appears to be related to the specific antibody response of the immunoglobulin, which feature would account for the enormous number of chemically similar but biologically distinct molecules of immunoglobulins that are possible. The light chains, κ or λ, are common to most types of immunoglobulins. Their amino acid sequences and antibody sites have been partially determined. They, as well as the heavy chains, appear to be linked together in the whole immunoglobulin molecule by means of disulfide linkages.

A schematic representation of the gross structure of a typical immunoglobulin M (IgM) is shown in Fig. 16-3. The molecular weight is about 900,000 for the entire pentameric molecule and approximately 180,000 for each monomer. Each monomer consists of two light (κ) chains and two heavy (μ) chains, having molecular weights of about 20,000 and 70,000 each, respectively. The variable portions, consisting of about 110 to 120 amino acids in each, are indicated by the zigzag lines. The constant portion, comprising about 110 amino acids in the κ-chains and 330 in the μ-chains, is indicated by heavy straight lines. The four peptide chains of the monomer are bound together by four disulfide bonds (Fig. 16-3). There are five carbohydrate moieties attached to each μ-chain of each monomer and are either complex oligosaccharides (solid circles) or simple oligosaccharides (open circles). The oligosaccharides are composed of D-galactose, D-mannose, L-fucose, D-glucosamine, and D-*N*-acetylneuraminic acid. The five monomers are held together in the pentameric structure by five disulfide bonds. The circular pentameric structure of IgM shown in Fig. 16-3 is supported by evidence from electron micrograph studies.

The complete amino sequences of both the light (κ) and heavy (μ) chains of several human IgM molecules have been determined by Putnam and his associates, a truly monumental accomplishment. With the aid of electron microscopic studies some progress is being made in elucidating the tertiary conformation of IgM.

The variable portions of the light and heavy chains of the various immunoglobulins, each consisting of about 110 amino acids, contain the *active sites* of the molecule. Therefore each monomer can, theoretically at least, bind two molecules of antigen. Steric hindrance effects apparently reduce this number in some instances, e.g., IgM. The variations in the amino acid sequence in this portion of the molecule are responsible for the enormous number of possible antibodies—optimally one for each invading bacterial, viral, or other toxic antigen. Apparently the plasma cell, which forms antibodies, is "instructed" in some specific manner to synthesize a type of meshing conformation in the variable portions of the two light and heavy chains that will be complementary to the conformation of a portion of the antigen, and this instruction serves as a "lock" for the antigen. The antibody thus combines with the antigen to form an insoluble complex that is taken up by phagocytes and destroyed—a truly remarkable example of the biochemical adaptation of a molecule for the performance of a specific function.

The amino acid composition and sequence for the light-chain component of cer-

Figure 16-3

Diagram of pentameric structure of human IgM immunoglobulin. Each monomeric unit consists of a pair of μ heavy chains and a pair of light chains joined together by disulfide bonds. Each heavy chain has four interchain disulfide bonds: two between the pair of μ-chains in the monomer, one to the light chain, and one intersubunit bridge between the monomers. The N-terminus of each polypeptide chain is depicted by a zigzag line to indicate the variable region, and the C-terminus by a straight line to represent the constant region. The light chains are denoted by the smaller lines and may be of the κ or λ type, but all light chains in any given pentameric molecule are of the same type. The solid circles attached to the heavy chains of one of the monomers (*upper center*) represent complex oligosaccharides; the lower two open circles represent simple oligosaccharides. The other four monomers have the same types of oligosaccharides attached (not shown in the diagram for the sake of simplicity). The antigen-combining site of IgM antibodies is determined by the variable regions of the light and heavy chains. As an antibody, the IgM pentamer is assumed to be decavalent, but it often functions as if it were pentavalent, presumably because of steric factors.

Modified from Putnam, F. W., Shimizu, A., Paul, C., Shinoda, T., and Köhler, H.: Ann. N.Y. Acad. Sci. **190:**83, 1971.

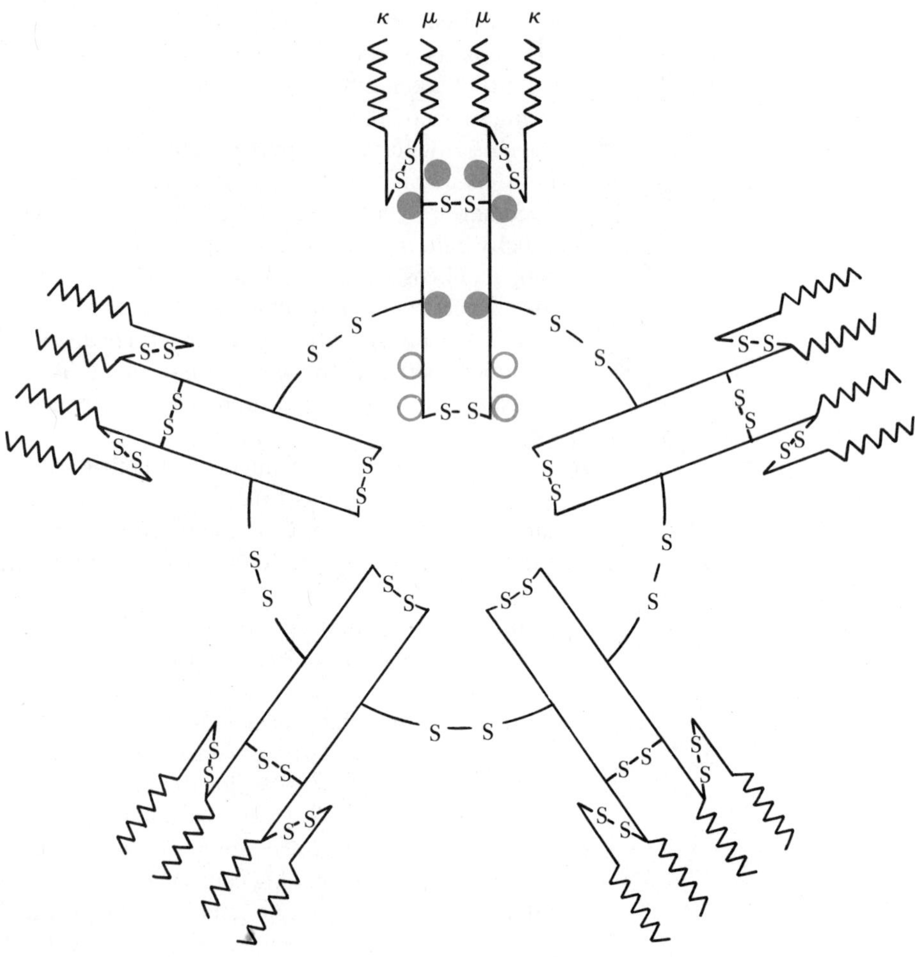

tain myeloma γ-globulins have been determined on preparations of Bence Jones protein made from the plasma or urine of patients with Bence Jones albumosuria. Apparently such proteins are constant in a given patient but vary from one patient to another.

The information about the amino acid sequence of the constant portion of immunoglobulins should aid in clarifying the function of this portion of the molecule. The constant portion is now believed to be involved in attacking the cell wall of invading bacteria, e.g., lysing the wall and thus aiding in the destruction of the bacteria.

Although the amino acid sequences of certain immunoglobulins and the locations of the disulfide bonds appear to be established, full reconstruction of the molecule's three-dimensional conformation must await x-ray analysis of the immunoglobulin crystal. Attempts to crystallize immunoglobulins are now being made at several laboratories as the next step in this direction.

Two other immunoglobulins, *IgD* (γD) and *IgE* (γE), apparently occur in human plasma in very small amounts (1 to 40 mg/100 ml). IgD has a molecular weight of about 150,000. The functions of these two immunoglobulins are uncertain.

Another alleged type of immunoglobulin is the 3S, or microglobulin. This type is also heterogeneous, with molecular weights varying from 10,000 to 40,000. It also contains no carbohydrate. It may be found in small amounts in plasma, in spinal fluid, and in urine. It may consist mainly of free light chains, but its biochemical function is unknown.

A deficiency of dietary protein is now generally believed to result in decreased antibody response. A deficiency of either tryptophan or phenylalanine, but not of methionine, in rats particularly decreases antibody response.

The importance of breast milk, particularly the colostrum (Chapter 17), for increasing the resistance of the young infant to infections has been known for a number of years. Apparently the immunoglobulins derived from breast milk are absorbed unaltered from the intestinal tract of the infant and help "tide him over" until his own immune system becomes operative. The immunoglobulins, of course, pass through the placenta into the vascular system of the fetus, at least to some extent.

A substantial amount of recent evidence indicates that a hormone (Chapter 18) secreted by the thymus gland may stimulate the body's immune system. A substance, termed *thymosin,* has been extracted from calf thymus glands; it mediates the differentiation of precursor blood cells into T cells in the thymus. Thymosin apparently is a mixture of a dozen or so peptides having molecular weights varying from 1000 to 15,000. T cells may either destroy foreign cells or enhance (or suppress) immune functions of the body's own lymphocytes. Thymosin is one agent under extensive investigation for use in diseases that may involve depressed immune responses, such as cancer. In this connection another substance—*interferon,* a protein produced by vertebrates in response to viral infections (Chapter 8)—likewise is being studied intensively for use in the treatment of certain types of cancer.

Quantitative changes in the amounts of immunoglobulins in the plasma and urine are known in several pathological conditions in humans. A deficiency of γ-globulins is found in a rare hereditary disease, *agammaglobulinemia*. Individuals with this disorder lack the ability to synthesize the γ-globulin type of plasma protein, apparently for genetic reasons. As might be expected, such people are particularly susceptible to infectious diseases. Therapy consists of the parenteral administration of γ-globulin preparations.

Abnormally large amounts of certain immunoglobulins may be found in the plasma in several diseases of humans. An increase in IgM, macroglobulins, termed *macroglobulinemia,* may occur in neoplastic diseases, collagen disorders, chronic infections, amyloidosis, and hepatic cirrhosis. The major component found is the 19S fraction, but heavier (24S to 40S) and lighter (14S to 16S) components may also appear. An explanation of this condition at a biochemical level is not now possible.

Increased amounts of a remarkable type of γ-globulin are found in the plasma (also the urine, (Chapter 17) of patients with *multiple myeloma.* This is the so-called Bence Jones protein, named after its discoverer. The sedimentation rate of this globulin is 3.5S, and its molecular weight is about 45,000. It usually appears in the γ-globulin fraction in electrophoresis, although it may be in the β- or even α-fractions. The Bence Jones proteins appear to be either monomers or possibly dimers of the light chains, probably κ_2 or λ_2. They have the remarkable characteristic of precipitating on heating to 50° to 60° C, then redissolving with more heating.

Complement proteins

Normal human plasma contains another group of globulins, known as the *complement proteins* (Table 16-3). These globulins complex with some antigen-antibody systems to mediate certain of their effects, principal of which are the alteration of the structure of cell membranes, leading to lysis of the cell, and the activation of certain cellular immunological functions. These effects are initiated by the binding of an antibody to its antigen, typically on the surface of target cells such as bacteria, erythrocytes, or tumor cells. The complement protein thus generally acts with antibodies, typically IgG or IgM, in defense against infection.

There are about 11 complement proteins in human serum. They move in the various electrophoretic groups α, β, and γ. Their molecular weights vary from about 80,000 to 400,000 and their concentrations in human serum from 0.1 to 170 mg/100 ml. They are usually grouped into the following three functional groups: $C1_{q,r,s}$; C2,3,4; and C5,6,7,8, and 9. The first group contains the *recognition unit,* which binds to an antibody on the target cell surface by means of a Ca^{++} dependent, noncovalent association. The second group, C2,3, and 4, is the *activation unit,* which interacts with a separate site on the cell surface. The third group, C5 to 9, is a *membrane attack* unit that produces a perforation in the cell membrane and thus lysis of the cell. A detailed discussion of the mechanisms involved in the actions of the complement proteins is provided by Putnam (see suggested readings at the end of this chapter).

Blood group globulins

Small amounts of another type of protein (Table 16-3), the *blood group globulins,* are present in the plasma. They are glycoproteins primarily associated with the erythrocyte membrane.

Other plasma proteins

Another unusual type of plasma protein, most commonly associated with the γ-globulins but occasionally with the β- or α-fractions, is the so-called cryoglobulin. This protein precipitates or even gels when plasma or serum in which it is present is cooled, a phenomenon that can be demonstrated in vitro. Traces of cryoglobulin are found in normal serum. Increased amounts, up to 10 gm/100 ml, may be found in a variety of disorders, e.g., rheumatoid arthritis, chronic lymphocytic leukemia, multiple myeloma, hepatic cirrhosis, coronary artery disease, lymphosarcoma. Massive precipitation of the protein may occur at levels of about 1 gm/100 ml. The molecular weight of the cryoglobulins appears to vary widely from about 165,000 to 600,000. The biochemical etiology of *cryoglobulinemia* is unknown.

Included in the various plasma protein fractions are also small amounts of a number of enzymes, e.g., several transaminases, dehydrogenases, peptidases, acid and alkaline phosphatases, aldolase, amylase, an invertase, lipase, catalase, cholinesterase, and β-glucuronidase. These appear to be derived from the tissues in which they are secreted, perhaps representing a leakage at least in some cases. The quantitative determination of certain of these enzymes in plasma is assuming considerable clinical significance.

Biosynthesis of plasma proteins

With the exception of the immunoglobins, which are formed in plasma cells, all plasma proteins are formed in the liver, e.g., fibrinogen, which is synthesized at a remarkably rapid rate. If rabbits are bled until most of the fibrinogen is removed and the defibrinated blood is reinjected, a normal blood level of fibrinogen is almost completely restored in 5 to 6 hours. This does not occur in hepatectomized animals. Other plasma proteins are not replaced as quickly, however. Seven to 14 days are required for their complete regeneration.

More recent studies employing better techniques, e.g., in vitro liver perfusion, immunoelectrophoresis, and fluorescent antibody procedures, confirmed that the liver is the chief site of formation of most of the plasma proteins, including not only fibrinogen but also plasma albumin(s) and α_1-, α_2-, β_1-, and β_2-globulins. The γ-globulins and probably certain generically related β-globulins appear to be formed extrahepatically for the most part, probably in the plasma cells of the reticuloendothelial system. Prothrombin synthesis has been found to occur exclusively in the parenchymal cells of the liver, apparently in the ribosomal particulate fraction of these cells.

The mechanisms involved in the biosynthesis of the plasma proteins appear to be the same as those used for the formation of other body proteins (Chapter 10).

The diet of the animal plays an important role in plasma protein formation; proteins containing a suitable assortment of amino acids are necessary for rapid regeneration. Normally there is a considerable reserve of plasma protein–forming material in the body. This reserve may be reduced by a low-protein diet, by fasting, or by plasma depletion (plasmapheresis or hemorrhage). When such depletion occurs, the animal is much less resistant to infection or stress. However, such states can be remedied readily by feeding adequate protein together with other suitable nutritive factors.

Degradation of plasma proteins

The question of the site of degradation of the plasma proteins is still unsettled. There is some evidence, however, that the lumen of the intestine may be involved. Serum albumin labeled with ^{131}I was found to pass into isolated loops of duodenum (in rabbits). If this transfer occurs in a considerable portion of the intestinal tract, the intestinal enzymatic digestion of plasma albumin so transferred could account for most of the in vivo degradation of this plasma protein. Whether the same site of degradation is also that for the plasma globulins remains to be determined.

Plasma proteins, like other body proteins, are constantly being destroyed and replaced. The half-life of human plasma proteins is about 10 days (Chapter 13). That of the rat is much shorter—about 4 days, as determined by a liver perfusion technique as well as in intact rats. The higher rate of turnover of plasma proteins in the rat may be related to the species' higher rate of metabolism.

Plasma vasoactive peptides—blood pressure regulation

Mechanical interference with the blood flow through the renal arteries of a dog results in the development of a permanent hypertension. Apparently, slowing the circulation causes some substance that produces vasoconstriction to be formed. Such a substance has been demonstrated in the blood coming from such ischemic kidneys. The mechanism seems to be the following: an enzyme, *renin,* is formed in

Figure 16-4
Amino acid sequence of human angiotensin I and angiotensin II.

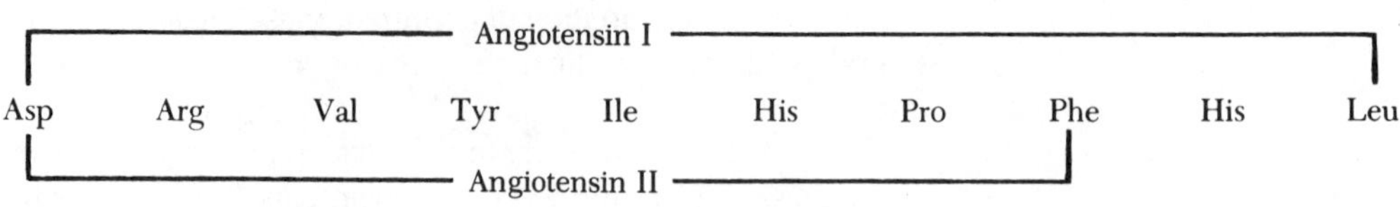

the kidney and is released into the blood; it is a proteinase and acts on an α_2-globulin, present in the systemic blood; a decapeptide, called *angiotensin I*, is split off; this peptide in turn is acted on by a peptidase present in normal serum to form an octapeptide, called *angiotensin II* (Chapter 18). The amino acid sequence of these two peptides has been determined and is shown in Fig. 16-4. Angiotensin I has only a slight effect on blood pressure, whereas angiotensin II is the most powerful pressor agent now known, having an activity about 200 times that of norepinephrine. The pressor activity of angiotensin II appears to be dependent on the presence of (1) the aromatic ring, (2) the free carboxyl group of phenylalanine, (3) the phenolic group of tyrosine, (4) the presence of proline in the seventh position in the peptide sequence, and (5) a hexapeptide structure with specific three-dimensional features. Angiotensin II appears to exert its pressor effect by constricting arterioles and increasing the heartbeat. It is therefore important in regulating the blood pressure in humans, as well as in all other mammals.

Another peptidase, termed *angiotensinase*, produced by the kidney and, to a lesser extent, by other tissues hydrolyzes angiotensin II and thus serves as a balancing antipressor agent.

The foregoing relationships can be shown diagrammatically as follows:

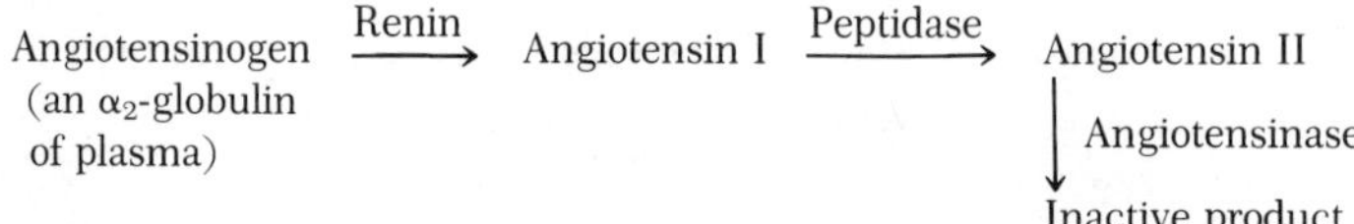

Various other names for the preceding substances have been used in the literature. Angiotensinogen has been called renin activator, hypertensinogen, and renin substrate. Angiotensin II has been termed angiotonin and hypertensin. The inactivating peptidase has been named angiotensinase and hypertensinase. Thus far *renin* has not been found in excess in patients with chronic renal hypertension. According to some investigators the ischemic kidney also forms, in its cortical portion, vasoexcitor material (VEM). This may act as a neutralizing agent for the hepatic vasodepressor material (VDM), which has been claimed to be identical with ferritin (Chapter 21).

Substances that lower blood pressure also have been obtained from kidney extracts. They have not been isolated in pure form but give promise of therapeutic usefulness in cases of hypertension. The use of low-sodium diets in hypertension is discussed in Chapter 21. In this connection it may be remarked that the use of the rauwolfia drugs for hypertension is said to be enhanced by low-sodium diets.

It has long been known that after blood coagulates it possesses vasoconstrictor properties. The active agent has been isolated and crystallized and is called *serotonin* (Chapter 13).

Another group of vasoactive peptides, the kinins, have been studied extensively

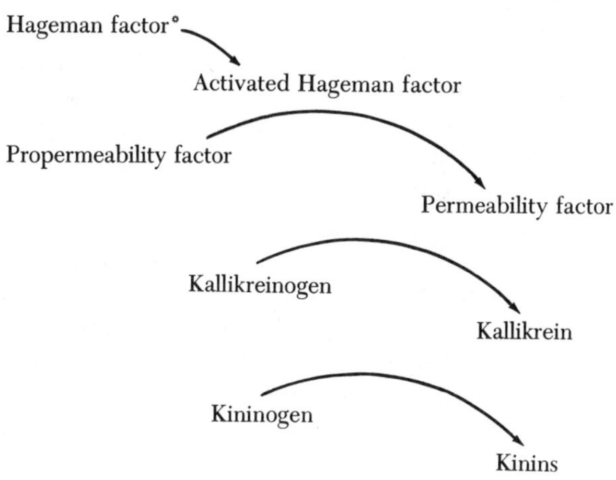

Figure 16-5
Schematic series of interdependent reactions proposed for the in vivo formation of kinins. *A plasma protein is activated by a glass surface or certain chemical agents and initiates blood clotting. This protein was found to be congenitally missing in the blood of Mr. Hageman.

during the past few years. *Kinin* is the generic name for a group of peptides with potent biological activities in causing smooth muscle contraction, vasodilatation, lowering of blood pressure, increasing blood flow and microvascular permeability, and inducing the emigration of granulocytic leukocytes. They thus have some activities similar to and some different from other vasoactive peptides, e.g., oxytocin and vasopressin (Chapter 18), angiotensin, and eledoisin. *Bradykinin* and *kallidin* are typical examples. They are formed from a plasma (or serum) precursor, kininogen, apparently an α_2-globulin containing about 18% carbohydrate, by certain proteolytic enzymes, notably kallikrein and trypsin. Bradykininogen has been found in the tissues of a number of species of animals. Heart contains the most, followed by liver, kidney, and brain. Blood contains about 6 to 8 μg per milliliter. The kinins have strong pharmacological effects, often in nanogram amounts.

Chemically the kinins are small peptides of 9 to 11 amino acids, similar in this respect to angiotensin I, vasopressin, and oxytocin. The active core of the kinins appears to be the following nonapeptide, found in bradykinin:

Arg-Pro-Pro-Gly-Phe-Ser-Pro-Phe-Arg-COOH

Kallidin, a decapeptide, contains lysine in addition on the N-terminal, and plasma kinin contains Met-Lys, in addition, also on the N-terminal.

The kinins appear to be formed in the plasma from the α_2-globulin precursor, kininogen, by a series of interdependent reactions shown schematically in Fig. 16-5.

The kinins are rapidly inactivated by the *kininases* of tissues. The half-life of bradykinin in blood has been estimated as less than 1 minute.

The biological significance of the kinins is not clear at the present time. Kinins appear to be the chemical mediators of inflammation. They may be involved in certain types of shock because of their powerful hypotensive properties. They also may play a role, similar to that of histamine (Chapter 13), as an important mediator of the sequelae of infection, injury, and foreign-body reactions.

Other constituents of plasma

Included in the remaining fraction of plasma solids are amino acids and smaller peptides, glucose, lipids, lactic acid, citric and other organic acids, the ketone bodies, nitrogenous waste products, pigments, inorganic salts, and small amounts of various enzymes, vitamins, and hormones. *Glucose* is present in a concentration of approximately 0.1%. It is most probably α,β-D-glucose. Both the *amino acids* and glucose vary in amount with the state of digestion and nutrition. This is also true of the *lipids*. In general, all three of these components increase after meals, reach a high point, level off, and then fall. Abnormally, all vary to a greater or lesser degree. These fluctuations are considered in other chapters. The lipids of plasma are fats, fatty acids, phospholipids, cholesterol, and cholesterol esters. The phospholipids include lecithin, a smaller amount of sphingomyelin, and a little cephalin. Most of the lipid is bound as the β-lipoprotein, which represents about 5% of the normal plasma proteins.

Lactic acid is present normally in small amounts and increases with exercise and also under pathological conditions. It is a product of carbohydrate metabolism. Traces of the *ketone bodies*—acetone, acetoacetic acid, and β-hydroxybutyric acid—also are normally present. They are derived from fatty acids and are increased when there is increased fat metabolism. The plasma lipids are discussed in some detail in Chapter 12.

The *nitrogenous waste products* result from the breakdown of proteins, purines, and other organic nitrogen-containing substances present in food and tissues. Included are urea, uric acid, creatinine, hippuric acid, and others. Their presence in blood is an indication of the transport of waste products by the bloodstream to the kidneys and other organs of elimination. Creatine is also present in minute amounts. The pigments of normal plasma include the bile pigments and carotene, all in traces, but abnormally these may be increased and others added.

The positive *inorganic* ions present in plasma in appreciable amounts are Na^+, K^+, Ca^{++}, and Mg^{++}. The negative are Cl^-, HCO_3^-, $SO_4^=$, and $HPO_4^=$ (see Fig. 17-2). In addition, lactic, citric, and other organic acids and proteins contribute somewhat to the ionic picture. In plasma and, in fact, in all body fluids the concentration of sodium ions exceeds that of potassium; in the blood cells and other cells the reverse is true. The bicarbonates and the phosphates are quite important as buffers, and the shifting of the chloride ion in and out of the erythrocytes has a definite role in electrolyte balance (Chapter 17). Changes in the concentrations of many of these ions occur under pathological conditions, e.g., Na^+ and Cl^- fall, K^+ rises in Addison's disease, Ca^{++} is diminished when parathyroid function fails. Similarly there are fluctuations of some of the ions present in traces, e.g., iodides, sulfocyanates, copper, and iron (Chapter 21).

Erythrocytes

Formation

The erythrocytes, or red blood cells, are formed in bone marrow. This long-established fact has been reaffirmed in human subjects by the isotopic tracer technique. Injected ^{59}Fe was followed by visualization of various body areas after some 16 hours by means of a scintillation counter. In the normal adult the major uptake of the labeled iron was found to be in the major erythropoietic tissue, the bone marrow of the pelvis, spine, ribs, scapulas, and proximal ends of the bones of the extremities. The liver and spleen did not accumulate sufficient iron to be visualized. In disease processes in which erythropoiesis is considerably increased (e.g., cyanotic congenital heart disease), there is an extension of the erythrocyte-forming marrow into the entire femur, the proximal portion of the tibia and even the tarsal bones, the middle of the forearm, and the wrist. The control of the production of erythrocytes depends on a hematopoietic hormone, *erythropoietin* (Chapter 18), and a *maturation factor,* now known to be vitamin B_{12}. The mechanism of the action of vitamin B_{12} is described in Chapter 21. In addition to these factors, there are required for normal erythropoiesis suitable and adequate dietary protein, folic acid, pyridoxine, niacin, and ascorbic acid, available iron salts, and traces of copper and possibly cobalt (Chapter 21).

Composition

The erythrocytes contain less water than do the cells of most tissues, about 60%. Most of the solid matter is hemoglobin, the conjugated protein that is the red coloring matter of blood. The stroma or meshwork is composed of other proteins and lipids, to which the hemoglobin is probably bound intimately. The lipids are chiefly cholesterol, phosphatidyl choline, and phosphatidyl ethanolamine, and the proteins include an albuminoid, stromatin, and a lipoprotein, elinin. Another protein of the stroma is *hemocuprein,* a bluish copper-containing substance now known to be identical with the enzyme superoxide dismutase (Chapter 21).

Membrane

The erythrocyte is enclosed in a membrane that has been studied extensively during the past few years. It is of special interest biochemically, since human erythrocytes are readily available and the membrane can be rather easily separated and prepared in relatively pure form. Like most cell membranes it consists of a typical phospholipid bilayer interspersed with discrete protein aggregates (Chapters 2 and 19). The overall composition is about 50% protein and 50% lipids. The proteins apparently are of a number of types, including several glycoproteins having molecular weights varying from 60,000 to 80,000 and containing as much as 60% to 70% carbohydrate by weight. There appears to be present a *contractile* protein that in some way may contribute to the mechanical properties of the intact red cell membrane. A relatively high–molecular weight protein, *spectrin,* represents about 40% of the total membrane protein. It appears to be extrinsic to the lipid hydrophobic region. Its function is uncertain at this time.

The lipids of the erythrocyte membrane include cholesterol (25% approximately of the total lipids), phosphatidyl choline (22%), phosphatidyl ethanolamine (20%), sphingomyelin (20%), phosphatidyl serine (10%), and small amounts of phosphatidyl inositol (1%) and phosphatidic acid (2%). The predominate classes of fatty acids of these lipids are the following: 16:0, 18:0, 18:1, 18:2, 20:4, 24:0, and 24:1. (The first number in each case refers to the number of carbon atoms, and the second to the number of unsaturated double bonds of the fatty acids.)

Blood group substances

A group of antigenic substances, the *blood group substances* or *isoagglutinins,* are found on the surface of erythrocytes. These substances are responsible for the major immunological reactions of erythrocytes, the so-called blood types. The antigens of the erythrocyte membrane, designated as the A, B, and H antigens, are associated with the ABO antibody system. Antibodies in the blood of individuals specific for a given antigen will agglutinate erythrocytes containing that antigen. Thus individuals with type A erythrocytes have anti-B antibodies in their blood; those with type B have anti-A antibodies; those with type AB have neither, whereas those with type O (whose erythrocytes contain H-antigen) have both anti-A and anti-B antibodies.

The erythrocyte membrane antigens A, B, and H apparently are glycosphingolipids (Chapter 12), containing significant amounts of L-fucose, 19% in blood group B substance. Other antigens are rather complex glycoproteins, having molecular weights of from 100,000 to 1 million, or are lipoproteins in the important Rh blood group system. Knowledge of the chemical nature of these latter antigens is still fragmentary.

Other constituents

The inorganic cations present include mainly Na^+, K^+, Ca^{++}, and Mg^{++}. They appear to be bound to the membrane, competing for the same binding sites. These cations apparently exert a stabilizing effect on the membrane structure. The accompanying anions include mainly Cl^-, HCO_3^-, $HPO_4^=$.

Although the mature red cell is nonnucleated and in a sense a "dead" cell (i.e., incapable of reproduction, minimal metabolic activities, etc.), the cytoplasmic membrane is very much alive in terms of physiologic capabilities, particularly with respect to the active transport of a variety of substances (Chapter 19).

Various enzymes are present in erythrocytes, including carbonic anhydrase, catalase, peptidases, cholinesterase, and the enzymes of the glycolytic system (Chapter 11). All the glutathione of blood is located in the erythrocytes. Adenosine di- and triphosphates, 2,3-diphosphoglycerate, and di- and triphosphopyridine nucleotides are also important constituents of erythrocytes. Soluble organic crystalloids present include urea, amino acids, creatinine, and glucose. The concentration of glucose in the erythrocyte is about the same as in plasma.

The electrolyte composition of the erythrocytes is qualitatively similar to that of the plasma. It differs quantitatively, however. There is more potassium than sodium—just the reverse of the relation of these two elements in plasma. The osmotic pressure of the interior of the erythrocyte is equal to that of the plasma (i.e., normally equivalent to the osmotic pressure of 0.9% NaCl solution, which is termed *normal* or, better, *physiological* saline, since it is not the same as a chemically "normal" solution). Changes in osmotic pressure of the medium surrounding red blood cells influence the size of the cells. If the medium is hypotonic, water passes into the cell and the size of the cell increases. Not a very great increase occurs before the cell bursts and the hemoglobin is released. This process is called *hemolysis,* or laking, and such blood, which is a clear, transparent crimson fluid, is laked or hemolyzed blood. When erythrocytes are put in a hypertonic solution, i.e., one with a higher osmotic pressure than 0.9% NaCl, they shrink and take on a shriveled appearance. These are described as *crenated* cells.

Hemolysis

Hemolysis may be produced by other means besides the one mentioned. Substances that dissolve or change the physical state of membrane lipids, e.g., ether, chloroform, bile salts, and soaps, accomplish this also. Certain biological toxins, especially those produced by venomous snakes and hemolytic bacteria, also cause the laking of erythrocytes. Some toxins contain enzymes that hydrolyze lecithin, and others act by solution of, or combination with, lipids. Physical forces, e.g., irradiation with ultraviolet rays or alternate freezing and thawing, may so alter the structure of the cell as to cause the release of hemoglobin. Aging also has a similar effect, and this is why whole citrated blood, kept in blood banks, cannot be used after 5 to 7 days. The erythrocytes become more and more fragile. The addition of glucose prolongs their serviceable period to 16 to 30 days, under proper conditions. Hemolysis may occur in the human body under pathological conditions, but it seldom occurs in the body as a result of lowering osmotic pressure. When it does occur, as a result of the action of bacteria, venoms, or other agents, the hemoglobin released into circulation is excreted by the kidney, resulting in *hemoglobinuria.*

Metabolism

Erythrocytes, like other living cells, require energy to maintain their vital processes. Glucose appears to be the major substance oxidized to supply this requirement. Nucleated erythrocytes derive their energy mainly by the oxidation of glucose through the citric acid cycle (Chapter 11). Nonnucleated erythrocytes, on the other hand, including human, appear to metabolize glucose primarily by way of anaerobic glycolysis and by the phosphogluconate shunt.

Recall that a special branch of the glycolytic pathway (Chapter 11), sometimes called the Rapoport-Luebering cycle, produces the metabolite 2,3-diphosphoglycerate, which is important in the transport of oxygen. The latter requires $NADP^+$ with the appropriate dehydrogenase. Regeneration of $NADP^+$ in this pathway is ensured by a special pathway using erythrocyte glutathione, oxidized form, and glutathione reductase, also present in the erythrocyte.

In this connection an interesting congenital hemolytic anemia, called *hereditary spherocytosis,* appears to be the result of a disturbance of glucose oxidation in the patient's erythrocytes. The condition is apparently caused by a congenital defect in intracellular glycolysis, probably from a deficiency of *enolase,* the enzyme that converts 2-phosphoglycerate to phosphoenolpyruvate. A decrease occurs in available energy-rich phosphate bonds (ATP), which seem to be required for the maintenance of the biconcave shape of the normal erythrocyte. Spherocytosis results and the erythrocytes are more susceptible to lysis and destruction. The characteristic anemia follows. Hereditary spherocytosis occurs once in about 20,000 live births. Similarly, a hereditary deficiency of *glucose 6-phosphate dehydrogenase,* which supplies NADPH to the erythrocyte by way of the pentose pathway, may result in increased hemolysis and a severe hemolytic anemia. The anemia may not become evident in some cases unless a drug (e.g., the antimalarial agent *primaquine*) or substances (fava beans) that have a hemolytic effect are ingested. This appears to be attributable to some type of a synergistic effect. The incidence of this enzyme deficiency has been estimated to be about 1% in the white race and 13% in male black Americans. A congenital deficiency of the enzyme *pyruvate kinase* may likewise be responsible for hereditary hemolytic anemia. This enzyme is required for glycolysis, specifically for ATP formation from phosphoenolpyruvate. As many as 80% of the erythrocytes in such cases may appear as bizarre, spiculated forms that are extremely fragile and readily hemolyzed. Deficiencies of other enzymes of the Embden-Myerhof pathway (glycolytic) in the erythrocytes of patients leading to hemolysis and a hemolytic anemia have been reported in recent literature. These include hereditary deficiencies of *pyruvate kinase, hexokinase, glucose 6-phosphate isomerase, triose phosphate isomerase,* and *2,3-diphosphoglycerate mutase.* A genetic deficiency of the *sodium-transport protein* in human erythrocytes leads to an accumulation of Na^+ and water in the affected cells, hemolyses, and a hemolytic anemia. The foregoing hereditary diseases thus emphasize the importance of

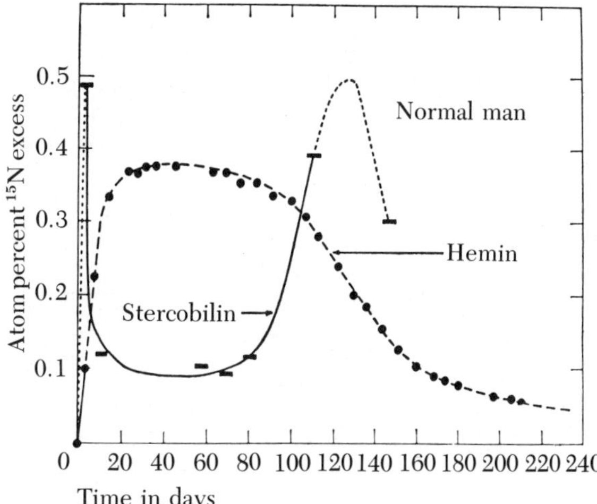

Figure 16-6
Atom percent ^{15}N excess in hemin and stercobilin of a normal man after the start of feeding ^{15}N-labeled glycine for 2 days.

From London, I. M., West, R., Shemin, D., and Rittenberg, D.: J. Biol. Chem. **184**:351, 1950.

energy-producing mechanisms in the erythrocyte for maintaining cellular structure and integrity.

Tracer nitrogen (^{15}N) has been used to study the life span of red cells (Fig. 16-6). After the feeding of labeled glycine is stopped, the concentration of labeled heme does not level off and decrease, as would be expected if the red cells were rapidly being catabolized. It continues to increase for nearly 25 days and then levels off until about the seventieth day. At this time the ^{15}N content begins to diminish. Meanwhile, the excretion of stercobilin, a waste product of hemoglobin destruction (Chapter 17), increases to a maximum at 120 days, then rapidly decreases. The life span of the average erythrocyte thus was found to be about 120 days. Values obtained by the use of different isotopic and other techniques agree closely with the 120-day figure for the life span of normal human red blood cells. The same value has been found for canine erythrocytes, whereas that for the rat is about 100 days.

Destruction of erythrocytes

"Worn-out" senescent erythrocytes are phagocytized primarily by reticuloendothelial cells in the spleen. The free hemoglobin thus released is bound stoichiometrically to the plasma γ-globulin, *haptoglobin*, each molecule combining with two

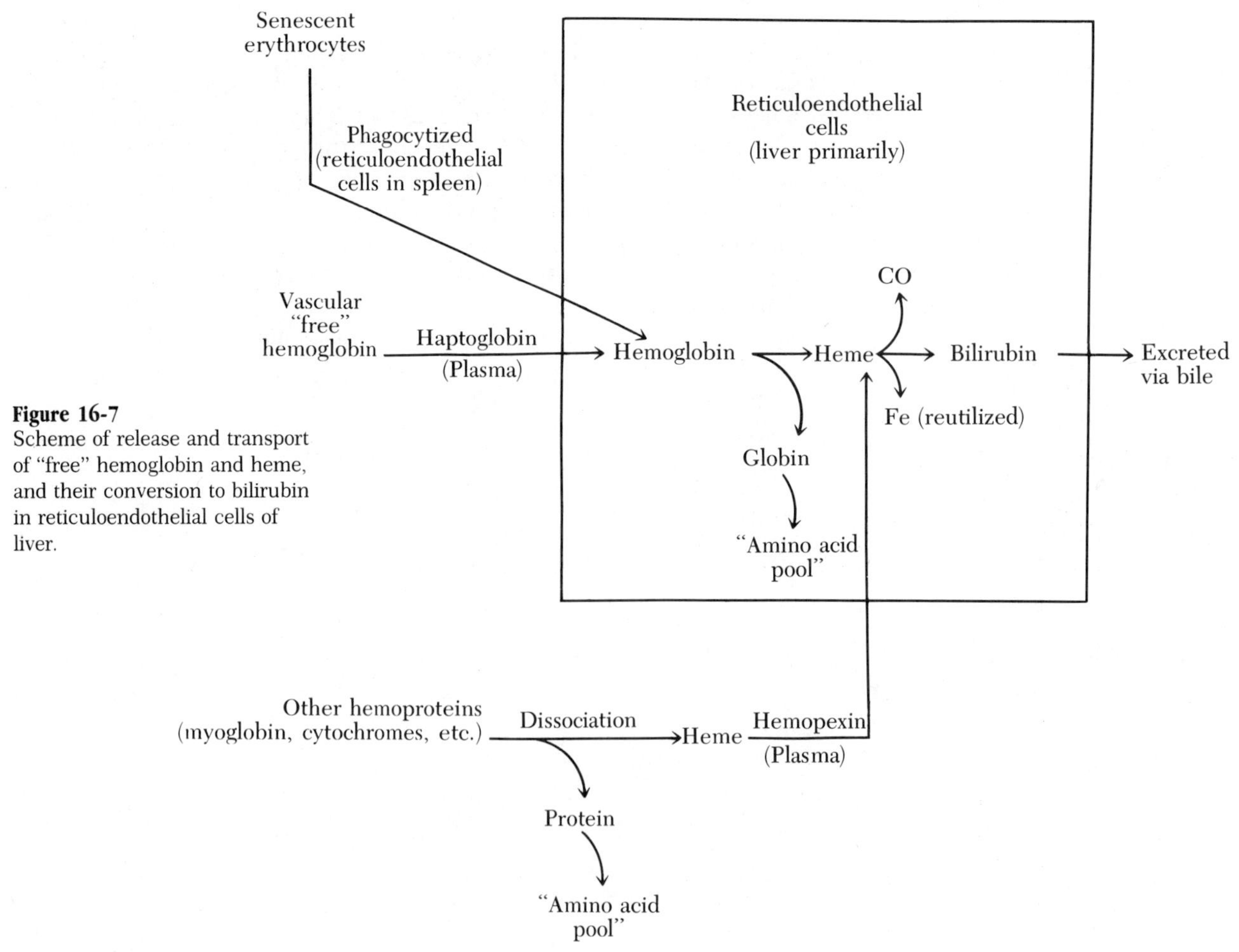

Figure 16-7
Scheme of release and transport of "free" hemoglobin and heme, and their conversion to bilirubin in reticuloendothelial cells of liver.

hemoglobin molecules. This complex is transported in the bloodstream to the liver, where hepatic reticuloendothelial cells split off the hemoglobin portion, converting it into its globin and heme moieties. The former is hydrolyzed into its component amino acids, which are added to the hepatic amino acid pool for reuse. The heme portion, together with any heme derived from other heme proteins (e.g., myoglobin, cytochromes, catalase) is transported to the liver bound to *hemopexin* in the plasma, converted to the bile pigment *bilirubin,* and excreted in the bile. The steps involved in the conversion of heme to bilirubin are described later in this chapter. The iron released from heme in this process is stored in the liver for reuse. The process is shown schematically in Fig. 16-7.

Hemoglobin

All forms of animal life contain a *respiratory substance* capable of transporting oxygen from the environment, usually air or water, to the cells they comprise. In higher animals this substance is the iron-containing pigment *hemoglobin,* our present primary interest. In lower forms of life, however, other types of respiratory substances are used. In the mollusks and arthropods, such as the octopus, lobster, and squid, oxygen is transported by a unique copper-containing protein, *hemocyanin.* Its molecular weight varies between species from about 450,000 up to around 9 million. Oxygen apparently binds to two copper atoms of hemocyanins, possibly as a superoxide. In some lower forms of life, such as certain worms and annelids, a nonheme-iron respiratory protein called *hemerythrin* is the oxygen carrier. These proteins are of relatively low molecular weight, about 13,500, and contain two iron atoms per peptide chain. Oxygen binds to two iron atoms, apparently as a superoxide. These substances are easily crystallized and have been studied extensively by x-ray crystallographic procedures. The amino acid sequences of some are partially known, as are the secondary, tertiary, and quaternary structures (Chapter 4). The oxygen affinities of both the hemocyanins and hemerythrins are pH dependent and show the Bohr effect, to be discussed later.

Human hemoglobins are conjugated proteins in which a prosthetic group, heme, is attached to each of four subunits, two α- and two β-polypeptide chains, in the case of adult hemoglobin A. The α- and β-chains of hemoglobin A are characterized by a relatively high content of histidine and lysine and the absence of isoleucine. By means of isotope tracer technique the α- and β-chains were shown to be formed from amino acids derived, of course, from dietary protein. The peptide chain part of hemoglobin is responsible for the species specificity of the hemoglobin of the species. The pigmentary property and chief respiratory functions are associated with heme, the iron-containing pigment, but the globin fraction plays a role in carrying carbon dioxide, to be discussed later in this chapter. Hemoglobin is a crystallizable protein, and each species has its own characteristic crystalline form. It has the power of uniting in loose combination with atmospheric oxygen, forming oxyhemoglobin. This occurs in the capillaries surrounding the alveoli of the lungs. The oxygen is thus transported in the arterial blood to the tissues, where part of it is released; the venous blood, somewhat depleted of its oxygen supply, returns to the lungs for oxygenation.

Structure of hemoglobin

The structure of the hemoglobin molecule has been extensively studied—probably more than that of any other protein. Most, if not all, mammalian hemoglobins are composed of four subunits (tetramers), consisting of four peptide chains to each of which is attached one heme group. The amino acid composition and sequence of the peptide chains have been determined in a number of species, as will be discussed later. The composite mammalian hemoglobin molecule is nearly spherical because of the remarkable fit of the subunit peptide chains. In horse hemoglobin the molecule is a spheroid measuring 6.4 × 5.5 × 5.0 nm. Most mammalian hemoglobins have a molecular weight of approximately 67,000.

Normal human hemoglobin is of several types, containing four subunits made up of various combinations of four or possibly five different yet related peptide chains. They are designated α, β, γ, δ, and ϵ. Most human hemoglobins contain two α-chains plus two other chains, usually β, γ, or δ. Thus normal adult hemoglobin, commonly called *Hb A*, consists of two α- and two β-chains and is designated therefore $\alpha_2^A\beta_2^A$, or more simply $\alpha_2\beta_2$. Approximately 90% of the hemoglobin of a normal adult is of this type (Hb A). Human fetal hemoglobin is designated *Hb F* and is $\alpha_2\gamma_2$. A minor component of normal adult hemoglobin, present usually to the extent of about 2.5% of the total erythrocyte hemoglobin, is called *Hb A_2* and is designated $\alpha_2\delta_2$. Embryonic hemoglobin is apparently $\alpha_2\epsilon_2$. The four peptide chains of normal human hemoglobins appear to be held together by noncovalent forces (Chapter 4).

The relative amounts and tissue sites of formation of the five types of normal human globin chains synthesized during the development of the human fetus are shown in Fig. 16-8. The reason for the switch from the embryonic to fetal to adult types of hemoglobin during this period is not clear at the present time. Possibly the shifts are related to the higher oxygen affinities of embryonic and fetal types and thus the extraction of oxygen from the maternal circulation.

Apparently multiple forms of adult hemoglobin, along with fetal and embryonic forms, occur in many species of animals as well as in humans. These have been demonstrated in various mammals, including monkeys and chimpanzees, cattle,

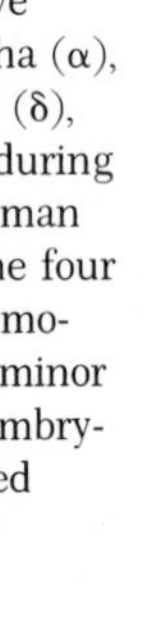
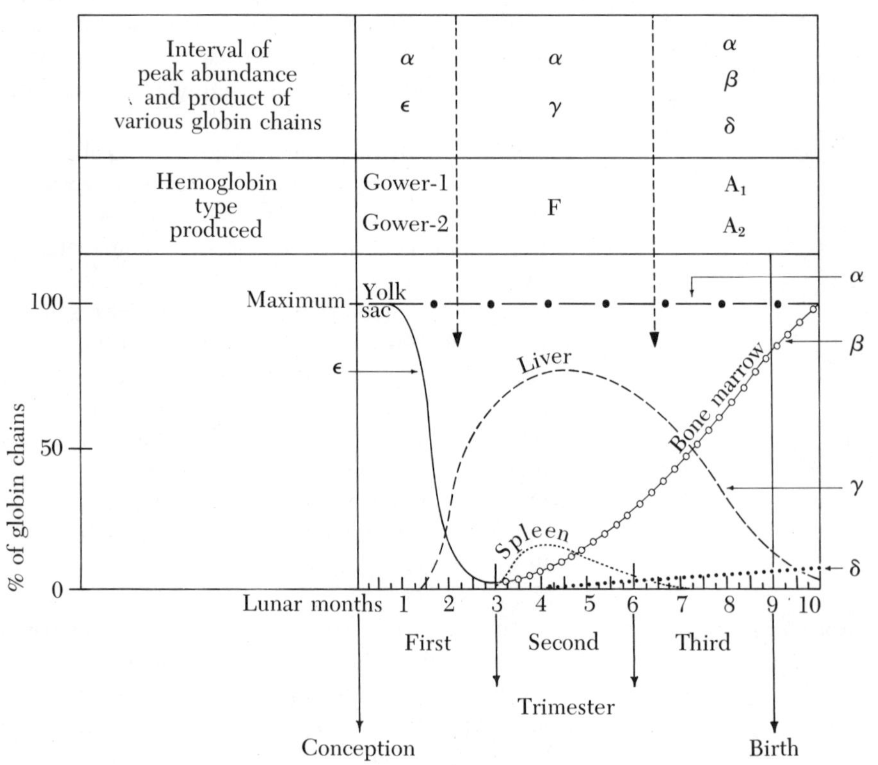

Figure 16-8
Relative amounts and tissue sites of formation of the five types of globin chains, alpha (α), beta (β), gamma (γ), delta (δ), and epsilon (ϵ), produced during the development of the human fetus. Also indicated are the four types of normal human hemoglobin formed, adult (A_1), minor adult (A_2), fetal (F), and embryonic; this last is also termed "Gower-1" and "Gower-2."

pigs, sheep, deer, goats, cats, rats, and mice. Multiple forms of hemoglobin, monomeric in some cases, are also found in lower species of animals, including fish and insects. The value of multiple forms of hemoglobin to the animal organism is uncertain at the present time. These forms may represent fine molecular adaptations to a changing environment with the development and aging of the organism.

In a brilliant series of investigations using x-ray diffraction analysis, Nobel Laureate M.F. Perutz and his colleagues of the University of Cambridge, aided by knowledge of the amino acid sequence of various hemoglobins obtained in other laboratories, succeeded in elucidating the complete tertiary structure of horse hemoglobin. As shown in Fig. 4-12, a photograph of a model constructed in Dr. Perutz's laboratory, the molecule is nearly spherical as a result of the remarkable fit of the two α- and β-chains. The helical nature of the four chains and their folding into the final convoluted subunits are evident.

The conformation of the α- and β-chains is shown more clearly in Fig. 4-12 in separate models of the two chains with superimposed lines showing the course of the central chain. Note that both the α- and β-chains partially enfold and protect the heme groups located in "crevices" near their surface. Also striking is the fact that the folding of the two chains is quite similar. Even more remarkable is the resemblance of the conformation, particularly that of the β-chain, to the tertiary structure of myoglobin (Fig. 4-8), prepared from the sperm whale by Kendrew and his group. The significance of this structural similarity will be mentioned again in a discussion of the evolution of hemoglobin.

The Perutz group has extended their studies on the conformation of horse oxyhemoglobin by much more precise x-ray diffraction analysis and three-dimensional Fourier synthesis at 0.28-nm resolution, rather than at 0.55-nm resolution, used in the earlier work just described. Fifteen months were required to measure the intensities of some 100,000 reflections in the diffractometer. The phase angles of 8000 reflections were determined. The resulting electron-density maps showed the positions of nearly all the amino acid residues of the hemoglobin molecule, the orientations of the side chains, and some details of the heme groups. On the basis of the results, combined with the known stereochemistry and sequence of the amino acids involved, a new atomic model of oxyhemoglobin (horse) was constructed. In construction of this model the x-ray analysis data of the hemoglobin crystals were subjected to a mathematical model building and refinement procedure programed in Fortran for a digital computer. The general conformation of hemoglobin in the newer model confirms that given in the earlier model (Fig. 4-12). However, much more detail is visible. Each polypeptide chain is made up of helical and nonhelical segments, similar to those described earlier (Chapter 4) in sperm whale myoglobin. The heme groups have similar surroundings. The center of the molecule is a cavity filled with water. The contacts between unlike subunits are chiefly nonpolar, whereas those between like subunits, if present, are polar. Thus, when the tetramer $(\alpha\beta)_2$ dissociates into two dimers, it probably breaks at the contacts between $\alpha_1\beta_2$ and $\alpha_2\beta_1$. Dissociation is favored by high concentrations of neutral electrolytes. The structure of the contacts between unlike subunits (α and β) suggests that the *tetramer*, rather than the αβ dimer as some have proposed, is the *functional unit of hemoglobin*.

The general features of horse oxyhemoglobin revealed in the newer model are that polar residues are excluded from the interior of the molecule except for an occasional serine or threonine. Glycine and alanine occur anywhere in the molecule. The larger nonpolar side chains of amino acids are located in the interior or in the surface crevices of the subunits or at the boundaries of unlike subunits. The heme groups lie in nonpolar "pockets" surrounded by the hydrophobic amino acids of the α- and β-chains. There are about 60 interactions between the atoms of the heme groups and those of the surrounding α- and β-chains. All but three are nonpolar.

Two kinds of contacts between the α- and β-chains are present. The majority are nonpolar interactions. Only five probable hydrogen bonds are apparent. The exact locations of the various contacts between the four subunits of hemoglobin and other detailed features are described in this monumental work.

Thus the tertiary and quaternary structures of hemoglobin derive their coherence from weak secondary forces—a few hydrogen bonds and a large number of nonpolar interactions. The ability of the ferrous iron of hemoglobin to combine

Table 16-5
Amino acid sequences of the α-, β-, and γ-chains of human hemoglobin*

	1 2 3 4 5 10 15
α	Val- -Leu-Ser-Pro-Ala-Asp-Lys-Thr-Asn-Val-Lys-Ala-Ala-Trp-Gly-Lys-Val-Gly-Ala-
β	Val-His-Leu-Thr-Pro-Glu-Glu-Lys-Ser-Ala-Val-Thr-Ala-Leu-Trp-Gly-Lys-Val-Asn-
γ	Gly-His-Phe-Thr-Glu-Glu-Asp-Lys-Ala-Thr-Ile-Thr-Ser-Leu-Trp-Gly-Lys-Val-Asn-
	1 2 3 4 5 10 15
	20 25 30 35
α	His-Ala-Gly-Glu-Tyr-Gly-Ala-Glu-Ala-Leu-Glu-Arg-Met-Phe-Leu-Ser-Phe-Pro-Thr-
β	-Val-Asp-Glu-Val-Gly-Gly-Glu-Ala-Leu-Gly-Arg-Leu-Leu-Val-Val-Tyr-Pro-Trp
γ	-Val-Glu-Asp-Ala-Gly-Gly-Glu-Thr-Leu-Gly-Arg-Leu-Leu-Val-Val-Tyr-Pro-Trp
	20 25 29 30 35
	40 45 50
α	Thr-Lys-Thr-Tyr-Phe-Pro-His-Phe- -Asp-Leu-Ser-His- -Gly
β	Thr-Gln-Arg-Phe-Phe-Glu-Ser-Phe-Gly-Asp-Leu-Ser-Thr-Pro-Asp-Ala-Val-Met-Gly
γ	Thr-Gln-Arg-Phe-Phe-Asp-Ser-Phe-Gly-Asn-Leu-Ser-Ser-Ala-Ser-Ala-Ile-Met-Gly
	40 45 50 55
	55 60 65 70
α	Ser-Ala-Gln-Val-Lys-Gly-His-Gly-Lys-Lys-Val-Ala-Asp-Ala-Leu-Thr-Asn-Ala-Val-Ala
β	Asn-Pro-Lys-Val-Lys-Ala-His-Gly-Lys-Lys-Val-Leu-Gly-Ala-Phe-Ser-Asp-Gly-Leu-Ala
γ	Asn-Pro-Lys-Val-Lys-Ala-His-Gly-Lys-Lys-Val-Leu-Thr-Ser-Leu-Gly-Asp-Ala-Ile-Lys
	60 65 70 75
	75 80 85 90
α	His-Val-Asp-Asp-Met-Pro-Asn-Ala-Leu-Ser-Ala-Leu-Ser-Asp-Leu-His-Ala-His-Lys
β	His-Leu-Asp-Asn-Leu-Lys-Gly-Thr-Phe-Ala-Thr-Leu-Ser-Glu-Leu-His-Cys-Asp-Lys
γ	His-Leu-Asp-Asp-Leu-Lys-Gly-Thr-Phe-Ala-Gln-Leu-Ser-Glu-Leu-His-Cys-Asp-Lys
	80 85 90 95
	95 100 105
α	Leu-Arg-Val-Asp-Pro-Val-Asp-Phe-Lys-Leu-Leu-Ser-His-Cys-Leu-Leu-Val-Thr-Leu
β	Leu-His-Val-Asp-Pro-Glu-Asn-Phe-Arg-Leu-Leu-Gly-Asn-Val-Leu-Val-Cys-Val-Leu
γ	Leu-His-Val-Asp-Pro-Glu-Asn-Phe-Lys-Leu-Leu-Gly-Asn-Val-Leu-Val-Thr-Val-Leu
	100 105 110
	110 115 120 125
α	Ala-Ala-His-Leu-Pro-Ala-Glu-Phe-Thr-Pro-Ala-Val-His-Ala-Ser-Leu-Asp-Lys-Phe-Leu
β	Ala-His-His-Phe-Gly-Lys-Glu-Phe-Thr-Pro-Pro-Val-Gln-Ala-Ala-Tyr-Gln-Lys-Val-Val
γ	Ala-Ile-His-Phe-Gly-Lys-Glu-Phe-Thr-Pro-Glu-Val-Gln-Ala-Ser-Trp-Gln-Lys-Met-Val
	115 120 125 130
	130 135 140 141
α	Ala-Ser-Val-Ser-Thr-Val-Leu-Thr-Ser-Lys-Tyr-Arg
β	Ala-Gly-Val-Ala-Asp-Ala-Leu-Ala-His-Lys-Tyr-His
γ	Thr-Gly-Val-Ala-Ser-Ala-Leu-Ser-Ser-Arg-Tyr-His
	135 140 141 142 143 144 145 146

*Gaps have been introduced into the sequences of the peptide chains to show the similarities in the sequences of the chains. These gaps do not actually exist in the hemoglobin molecule.

reversibly with molecular oxygen is apparently dependent on those nonpolar surroundings.

The complete amino acid sequence of the α-, β-, γ-, and δ-chains of normal human hemoglobins have been determined. The procedure used was similar to that developed by Sanger (Chapter 4) for determining the amino acid sequence in insulin. As shown in Table 16-5, the α-chain is made up of 141 amino acids, and the β- and γ-chains are composed of 146 amino acids. The δ-chain, not shown, also is made up of 146 amino acids. It differs from the sequence of the β-chain in only 10 positions. Note the absence of isoleucine in the α- and β-chains.

There are some differences in the amino acid sequences of the three chains. However, more remarkable is the fact that there are many striking *similarities*, e.g., 24 consecutive identical amino acids (positions 88 to 111) in the β- and γ-chains, also 65 identical amino acid placements in the α- and β-chains, 108 in the β- and γ-chains, 55 the same in all three chains. Subsequent work showed that the β- and δ-chains are even more similar, with 138 identical amino acid positions out of a total of 146. Likewise, there is a surprising similarity between the β-chain of human hemoglobin and the amino acid sequence of myoglobin from sperm whale muscle. There are actually 22 identical amino acid positions in these two chains—obtained from such widely different species. The significance of these similarities in amino acid sequence, like those of the tertiary conformations just mentioned, will be discussed later in connection with the evolution of hemoglobins.

Hemoglobin A heterogeneity

Careful studies employing improved chromatographic techniques have demonstrated that there are at least two variants of normal hemoglobin A ($\alpha_2\beta_2$). These hemoglobins are believed to result from a nonenzymatic reaction between glucose 6-phosphate with the NH_2-terminal valine of the β-chain by forming a Schiff base adduct. This form, called HB A_{1b}, then rearranges to form 1-deoxy-1-(*N*-valyl) fructose 6-phosphate, HB A_{1c}, at the N terminal of the β-chain. It is interesting that these variants of normal hemoglobin A are *elevated* in patients with diabetes mellitus up to about 10% to 12% of the total hemoglobin. Current investigations indicate that monitoring the amount of this hemoglobin variant in the blood of diabetic patients is the *best* index now available for assessing the precision of the control of the blood glucose level by insulin. In turn, this may prove invaluable in controlling at least some of the undesirable side effects of a poorly regulated blood glucose level, such as diabetic retinopathy, neuropathy, and possibly even atherosclerosis.

Evolutionary changes in hemoglobin

As mentioned previously, the striking similarities of various hemoglobins led Ingram and others to propose that a genetic, evolutionary relationship existed. He suggested that the α-chain is probably the oldest form, from an evolutionary standpoint, because of its similarity in amino acid sequence to that of lamprey hemoglobin, a monomeric primitive form of hemoglobin. He further postulated that by gene duplication the γ-, β-, and δ-chains evolved in succession in that order. Calculations from the numbers of variant amino acid residues between chains have indicated that the time (in millions of years) of divergence from a proposed common ancestral chain is 600 for α and γ, 565 for α and β, 260 for β and γ, and 44 for β and δ. The divergence of myoglobin and the α-chain from the primitive ancestral chain presumably occurred over 600 million years ago.

Subsequent determinations of amino acid sequences of the α- and β-chains of various primates and of other species supported and extended Ingram's proposal. The following generalizations were made from the data:

1. Amino acid sequence differences vary directly with evolutionary relation-

ships. The closer the species to humans, the more similar is the amino acid sequence to that of human hemoglobin.

2. The β-chain varies more than the α-chain.
3. Sequences *essential* to the function of the molecule show little or no variation. This has been found true in all the mammals studied thus far. The functional portions of the molecule conserved include the basic and hydrophobic clusters, the binding groups to heme, and the C-terminal.
4. Amino acids with desirable but not essential properties show a lesser degree of evolutionary restraint.
5. Other amino acids simply take up space and vary widely.

Another interesting point found in these studies is the striking resemblance between the β-chain of lemurs and shrews and the γ-chain of human fetal hemoglobin.

The foregoing studies have also shown that there is a surprising correlation between the evolutionary times calculated from variant amino acid residues in hemoglobins of various species and paleontological records. For example, paleontological records show that humans and horses separated from a common ancestor about 100 to 150 million years ago. The amino acids of the α-chains of these two species differ in 18 positions. Assuming that there were nine mutations in each species, evidently each mutation occurred at an interval of approximately 12 to 15 million years. Using this figure, application to the number of substitutions in the β-chains of primates gives values agreeing well with paleontological data for these species. Likewise, application of the calculation to the numbers of variant amino acid residues in cytochrome c (a hemeprotein related to hemoglobin) in various species, as obtained in the work of Margoliash, gives values agreeing well with paleontological records. Thus the times in millions of years for the divergence of the following species compared with humans calculate to be approximately as follows: pig, 83; horse, 130; chicken, 160; tuna, 230; and yeast, 500. As was pointed out earlier (Chapter 1), the cytochrome molecule is believed to have existed some 2 billion years. In view of some similarities between the hemoglobin, myoglobin, and cytochrome molecules, many authorities believe that the three are derived from a common primordial ancestor molecule that existed some 1 to 2 billion years ago!

Biosynthesis of hemoglobin

The need for the continual formation of hemoglobin to replace that destroyed by the "wear and tear" in the circulation is obvious. As stated previously, the life span of the average normal human erythrocyte is only some 120 days. It is then lysed and its hemoglobin is converted into products for reuse or excretion. It has been estimated that there are about 30 trillion erythrocytes in the circulating blood of an average adult male and that aproximately 3 million of those are destroyed every second! Assuming that there are an estimated 300 million molecules of hemoglobin in each erythrocyte, we may see that the average adult male must synthesize some *900 trillion* molecules of hemoglobin each second to maintain a normal blood hemoglobin level! This seems a staggering amount, but it is accomplished in the healthy individual. However, in more familiar terms this amounts to only about 8 gm of hemoglobin per day.

The biosynthetic pathway of the heme portion of the hemoglobin molecule will be considered later. The globin moiety is formed from amino acids from the body pool in amounts of about 8 gm per day in the normal adult. Thus about 14% of the amino acids from the average daily protein intake are used for globin formation. Apparently globin biosynthesis has a high priority call on the labile amino acid pool of the body. A number of studies using ^{14}C-labeled amino acids (e.g., leucine, lysine, glycine) demonstrated that the process occurs primarily in the nucleated erythrocyte (normoblast) of the bone marrow and also in the reticulocyte. The mechanism involved in the biosynthesis of globin has probably been studied more extensively than that for any other protein because of the relative ease of obtaining the necessary synthetic system, even in cell-free form, from reticulocytes. Apparently globin synthesis occurs on ribosomes in much the same manner as that of other proteins. This process has been described in some detail (Chapter 10). Ultracentrifugal data and electron micrographic studies indicate that hemoglobin synthesis in the reticulocyte occurs on a multiple ribosomal structure containing five ribosomes. These ribosomes appear to be strung on an RNA strand, apparently messenger RNA. An interesting study reported, however, that hemoglobin synthesis occurs in the nucleus of nucleated avian erythrocytes. The nucleus of human nucleated erythrocytes (bone marrow), in contrast, does not appear to synthesize significant amounts of hemoglobin. The formation of the α- and β-chains of globin does not appear to involve any special mechanisms or genetic factors. The two types of chains are apparently formed independently under the control of different genes although normally at the same rate. There is some evidence that hemin increases the rate of globin synthesis in rabbit reticulocytes while decreasing that of heme synthesis. Thus heme may serve as part of a control mechanism to maintain the synthesis of globin and presumably its constituent α- and β-chains at a near 1:1 molecular ratio. Protoporphyrin IX has also been reported to increase globin synthesis.

Abnormal hemoglobins

Since the α-, β-, γ-, and δ-chains of the globin moiety of hemoglobin are synthesized from amino acids under genetic control, as are other body proteins, occasional errors in their formation are to be expected. However, the occurrence of such resulting *abnormal hemoglobins* in humans (and other species) fortunately is uncommon. Moreover, many of the genetic mutations do not result in the production of a hemoglobin that is *functionally* defective, even though it differs structurally from the normal hemoglobins. Because of their availability, the abnormal hemoglobins represent one of the most completely studied effects of gene mutation on the amino acid composition and structure of protein molecules. Pauling and his associates aptly called these *molecular diseases*.

Since Pauling's classic pioneer studies on hemoglobin S and sickle cell anemia, there has been widespread application of improved electrophoretic and chromatographic techniques and peptide mapping and amino acid analysis to hemoglobins isolated from the blood of patients with diverse abnormalities of erythrocyte function. As a result some 300 different mutant human types of hemoglobin have been described in the literature at the present time. A few selected examples are considered next. For further details the readings at the end of this chapter should be consulted.

The abnormal hemoglobins may be grouped into two general types: (1) those in which there are altered combinations of normal α-, β-, γ- or δ-chains, and (2) hemoglobins in which there is an altered sequence of amino acids usually in one of the constituent chains.

The first type of abnormal hemoglobins apparently results from mutations that affect the *regulator genes* and thus reduce the rate of synthesis of one of the normal globin chains. These mutations result in the *thalassemias*. In the so-called *α-thalassemias* the synthesis of α-chains is repressed, resulting in a compensatory increase in the formation of β-chains that associate into pure β-chain tetramers (β_4A, *hemoglobin H*) or into pure γ-chain tetramers (γ_4F), such as *Bart's hemoglobin*. The excess of β-chains may even form a precipitate that appears as inclusion bodies in erythrocytes. On the other hand, if the formation of β-chains is repressed, an excess of α-chains is formed and a β-*thalassemia* results. The excess of α-chains can combine with γ-chains, causing an increase in hemoglobin F ($\beta_2\gamma_2$), or with δ-chains, resulting in an increase in hemoglobin A_2 ($\alpha_2\delta_2$).

The β-thalassemia gene is allelic with the gene for hemoglobins C or S; thus by interactions with the C or S genes a mixture of hemoglobins may result, making up as much as 80% to 90% of the total hemoglobin. On the other hand, the gene for α-thalassemia is not allelic with that for hemoglobins C or S, so there is no interaction. Then hemoglobin C or S makes up only about half the total hemoglobin. *Thalassemia major* (Mediterranean anemia or Cooley's anemia) occurs as a result of homozygous inheritance of thalassemia genes, whereas *thalassemia minor* is the heterozygous form.

Abnormal hemoglobins composed of only β-, γ-, or δ-chains do not show the Bohr effect and have abnormal oxygen dissociation curves, to be discussed later. Therefore they do not function normally in the transport of oxygen.

The thalassemia hemoglobins are difficult to identify by the usual electrophoretic techniques and may appear as normal hemoglobin or in combination with other hemoglobin variants, C or S.

The second general type of abnormal hemoglobins is characterized by an altered sequence of amino acids, usually involving only one of the constituent globin chains. Usually there is a *single* amino acid substitution resulting from a single point mutation, as originally shown by the classic work of Ingram on sickle cell hemoglobin. However, two-point mutations, deletions, frame shifts, extended sequences, fusions, and perhaps other types of genetic alterations apparently may occur. The number of point mutations in the α-chain now is about 76; 155 in the β-chain; 14 in the γ-chain; and 9 in the δ-chain. A number of additional abnormal hemoglobins are a result of other types of mutations or genetic alterations.

Originally the abnormal hemoglobins were given designations such as C, D, or S on the basis of some characteristic, such as sickling in the case of hemoglobin S, or of their electrophoretic mobility. As the number of new variants discovered increased, some with identical mobilities, the number of letters available was soon exceeded. Thus the abnormal hemoglobins now are named either for the geographical area in which they were first identified, the cultural group involved, or the

Table 16-6
Some selected abnormal human hemoglobins with position of amino acid substitution

Designation*	Substitution From	To	Position	Biochemical alterations and clinical symptoms
Substitution of amino acids in contact with heme groups				
α-Chain				
Torino	Phe	Val	43	Inclusion body anemia
M Boston	His	Tyr	58	Cyanosis, methemoglobinemia (increased O_2 affinity)
β-Chain				
Hammersmith	Phe	Ser	42	Inclusion body anemia, cyanosis (decreased O_2 affinity)
M Zurich	His	Arg	63	Inclusion body anemia (on treatment with sulfonamides) (increased O_2 affinity)

*Abnormal hemoglobins also are designated according to their chain composition and the amino acid substitution with its sequence number, e.g., Hemoglobin S = $\alpha_2^A\beta_2^{6Val}$.

Table 16-6, cont'd
Some selected abnormal human hemoglobins with position of amino acid substitution

Designation*	Substitution From	Substitution To	Position	Biochemical alterations and clinical symptoms
Substitution of amino acids in contact with heme groups, cont'd				
M Hyde Park	His	Tyr	92	Cyanosis, methemoglobinemia
Sabine	Leu	Pro	91	Methemoglobinemia; inclusion body anemia
Substitution of amino acids at contacts between α- and β-chains				
α-Chain				
Chesapeake	Arg	Leu	92	Polycythemia (high O_2 affinity; decreased heme-heme interaction)
J Capetown	Arg	Gln	92	Mild polycythemia
β-Chain				
E	Glu	Lys	26	Mild hemolytic anemia (homozygotes only)
Yakima	Asp	His	99	Polycythemia (increased O_2 affinity)
New York	Val	Glu	113	None reported
Substitution of amino acids in general positions				
α-Chain				
Etobioke	Ser	Arg	84	None reported
Manitoba	Ser	Arg	102	None reported
β-Chain				
Freiburg	Val	?	23	Cyanosis (high O_2 affinity)
Gun Hill	(deletion of 5-amino between 91 and 97)			Hemolytic anemia
Ranier	Tyr	His	145	Polycythemia
External substitutions of amino acids				
α-Chain				
J Oxford	Gly	Asp	15	No clinical symptoms in heterozygotes
Mexico	Gln	Glu	54	No clinical symptoms in heterozygotes
O Indonesia	Glu	Lys	116	No clinical symptoms in heterozygotes
β-Chain				
S	Glu	Val	6	No clinical symptoms in heterozygotes (severe sickling and sickle cell anemia in homozygotes)
C	Glu	Lys	6	No clinical symptoms in heterozygotes (mild anemia in homozygotes)
N Seattle	Lys	Glu	61	No clinical symptoms in heterozygotes
F Alexandra	Thr	Lys	12	Apparently no clinical effects
F Hull	Glu	Lys	121	Apparently no clinical effects
γ-Chain				
F Texas I	Glu	Lys	5	Apparently no clinical effects
F Jamaica	Lys	Glu	61	Apparently no clinical effects
F Malta I	His	Arg	117	Apparently no clinical effects
δ-Chain				
A_2	Gly	Arg	16	Apparently no clinical effects
A_2 Flatbush	Ala	Glu	22	Apparently no clinical effects
A_2 Bahinga	Gly	Asp	136	Apparently no clinical effects

name of the patient in whom the variant was first observed. They also are designated more precisely by indications of the type of chain and the location in its amino acid sequence of the substitution. Thus hemoglobin S is $\alpha_2^A\beta_2^{6\ Val}$, and hemoglobin C (Harlem, a two-point mutation) is $\alpha_2^A\beta_2^{6\ Val+73\ Asn}$.

In general, structural changes in the globin chains of hemoglobin may result in alterations in one or more of the following three properties of the abnormal hemoglobin: (1) solubility, as in the case of hemoglobin S; (2) affinity for oxygen; (3) ease of denaturation and precipitation. These are noted in the following discussion of some of the abnormal hemoglobins and in Table 16-6.

Some substitutions may alter the net charge on the hemoglobin molecule, thus changing the isoelectric point and electrophoretic mobility of the molecule. Thus the substitution of valine for glutamic acid in sickle cell hemoglobin changes the net charge +2 from Hb A. This is of course the basis for the separation of various abnormal hemoglobins by electrophoresis. Amino acid substitutions also may change the solubility of the abnormal form. In sickle cell hemoglobin, for example, the deoxygenated form is much less soluble than that of Hb A and thus tends to form insoluble tactoids, increasing the rate of destruction of the transporting erythrocyte and leading to the hemolytic anemia (the blockage of small blood vessels, the tissue damage, and excruciating pain) characteristic of the "crisis" of this disease.

The life span of sickle erythrocytes is only 30 days—one fourth that of normal red blood cells. Extensive research investigations are in progress to discover effective but nontoxic agents that will prevent tactoid formation and sickling in deoxygenated erythrocytes containing hemoglobin S. Currently the benzyl esters of certain hydrophobic amino acids are giving promising preliminary results.

Most of the known substitutions in amino acid sequence in the abnormal hemoglobins do not significantly impair the oxygen-transport ability of the hemoglobin affected. An interesting exception, however, is hemoglobin M types, in which the histidine residue in position 58 of the α-chain or in position 63 of the β-chain is replaced by tyrosine. Since the histidine is the one that apparently alternates with oxygen in attaching to the iron of heme, impairment of oxygen transport occurs. The ferrous iron of this type of hemoglobin becomes converted to ferric and thus to methemoglobin and is then useless for subsequent oxygen transport. This apparently explains the striking cyanosis seen in individuals with hereditary methemoglobinemia, resulting from the presence of hemoglobin M. Hemoglobin H (β_4) likewise has an altered capacity for oxygen transport as a result of its increased affinity for oxygen.

The description of the new more precise model of three-dimensional conformation of oxyhemoglobin (Chapter 4) has made possible the logical categorization of the new mutants of human hemoglobin and the prediction of their possible functional defects in oxygen transport and the clinical sequelae. The three structural features of the hemoglobin molecule, apparent from studies of the new three-dimensional model, that are of primary functional importance in oxygen transport are (1) the points of contact binding the two symmetrical α,β-dimers of hemoglobin, (2) the heme-binding contacts with α- and β-chains, and (3) the invariant sequence of the nonpolar amino acids in the helical portions of the α- and β-chains. Amino acid substitutions affecting these three vital structural features may seriously affect oxygen transport by the mutant hemoglobin and lead to gross clinical manifestations. Substitutions in other parts of the molecule, especially on the surface, should have little or no effect on oxygen transport and therefore produce no clinical symptoms. This prediction has been supported by studies of abnormal human hemoglobin, to be discussed next.

Substitutions altering the points of contact between the α,β-dimers (1) thus might be expected to produce an increased affinity of the molecule for oxygen, leading to tissue hypoxia and in turn clinically to cyanosis or a compensatory polycythemia or both. Substitutions (or deletions) affecting the binding of the heme groups by the α- and β-chains (2) should result in an unstable hemoglobin that could precipitate in the erythrocyte as inclusion or Heinz bodies and lead to structural weakness of the cell and hemolytic anemia clinically. Substitutions affecting the invariant, nonpolar amino acids in the helical portions of the α- or β-chains (3) might allow internal bonding with ferrous heme, forming ferric heme and methemoglobin, which cannot carry oxygen. This would be manifested clinically by a methemoglobinemia and possibly weakness and other symptoms.

Substitutions involving predominately polar amino acids on the external surface of the molecule may be placed in a fourth category. These form the largest group of abnormal hemoglobins at the present time. For the most part there is no impairment in the oxygen transport of these mutant hemoglobins, hence no clinical symptoms. In the few exceptions, e.g., hemoglobins S and C, hemolytic anemia may occur in *homozygotes only* apparently because of mechanical damage to the erythrocyte transporting the mutant hemoglobin. Heterozygotes usually remain symptom free, i.e., patients with sickle cell trait.

Selected examples of these four different types of abnormal hemoglobins are given in Table 16-6, together with the amino acid substitutions involved and the clinical manifestations resulting.

Since the atypical hemoglobins are usually found in population surveys using electrophoretic methods at pH 8.6 or by examination of patients showing clinical symptoms, it is probable that many mutations involving neutral amino acids are missed. As many as one in every 600 persons may carry a mutant hemoglobin.

The mechanical injury to the erythrocyte produced by certain abnormal hemoglobins was shown clearly in the case of hemoglobin Sabine. The half-life of erythrocytes from one patient was found by the ^{51}Cr method to be only 4 days, in contrast to about 28 days for normal subjects. One explanation for the mechanical damage to the erythrocyte is that the amino acid substitution in the vicinity of heme attachment lessens the avidity of the α- and β-chains for heme, liberating free heme, which is metabolized to dipyrroles and excreted in the urine. The remaining free globin moiety is unstable and precipitates in the erythrocyte, forming the inclusion bodies or Heinz bodies frequently seen clinically in certain blood dyscrasias. The inclusion bodies attach to the erythrocyte membrane and alter its permeability, resulting in osmotic damage to the cell, early destruction, and the characteristic hemolytic anemia. One patient showed severe hemolytic anemia even though only 8% of her total hemoglobin was Hb Sabine, which suggests that clinical symptoms can occur in heterozygotes as well as in homozygotes having this mutant hemoglobin.

Abnormal hemoglobins are known to occur in animals also. Sickling of erythrocytes and a type of hemoglobin S have been described in deer.

Apparently several of the abnormal hemoglobins (F, I, Lepore, A_2) bind to human haptoglobin, an α_2-globulin that transports free hemoglobin from destroyed erythrocytes to the reticuloendothelial system for catabolism (Chapter 17), to the same degree as does Hb A. However, Hb H and Hb Bart's fail to bind to human haptoglobin.

The foregoing monumental expansion of knowledge of the conformation and biosynthesis of normal and abnormal hemoglobins is of far-reaching significance not only in itself but also insofar as it is prophetic of future similar developments for other proteins of equal biological importance (of which there may be an estimated 10^5 and 10^6 in humans alone).

Structure of heme

```
HC —— CH
||    ||
HC    CH
  \  /
   N
   H
```

Pyrrole

Heme, the prosthetic group of hemoglobin, is a chelate of ferrous iron with protoporphyrin, a tetrapyrrole derivative. For this reason the heme of hemoglobin is sometimes called protoheme. Heme may be represented by the structural formula in Fig. 16-9.

The arrangement of the double bond varies in the different pyrrole groups. For each heme group at least one of the coordination valences of the iron is believed to be connected to one of the imidazole nitrogens of a histidine in the peptide chain of the subunit. There is some evidence that each heme iron coordinates to two imidazole nitrogens of histidine, probably at positions 58 and 87 in the α-chain and 63 and 92 in the β-chain. Some investigators believe that one of the imidazole ligands (probably His-58 in the α-chain and His-63 in the β-chain) is reversibly displaced by oxygen rather than by water during oxygen transport. Other investigators, however, believe that neither is possible, since water cannot exist in the heme hydrophobic pocket, and that the distal histidines are too far from the Fe^{++} of heme to permit corrdinate-covalent bonding. They believe, rather, that the two unpaired electrons of Fe^{++} remain unpaired until O_2 is picked again. The other two linkages to the α- or β-chains are postulated to be combined with the two propionic acid groups present in heme.

Several types of heme occur in different hemoproteins. Hemoglobin contains protoheme, as mentioned. Certain cytochromes contain heme C, in which the two vinyl groups appear with sulfhydryl groups linked to cysteine of the protein moiety of the molecule. Heme A has different alkyl groups replacing the two vinyl groups.

Biosynthesis of heme

As stated previously, the normal adult male must synthesize approximately 900 trillion molecules of hemoglobin per second to maintain a normal hemoglobin level in his blood. This is equivalent to 3.6 *quadrillion* molecules of *heme* per second! Obviously an efficient biosynthetic mechanism is essential to maintain this production requirement even though the total amount of heme formation required per day actually is only about 300 mg for a 70-kg adult male.

Although the chemical structure and method of synthesis of heme and several of

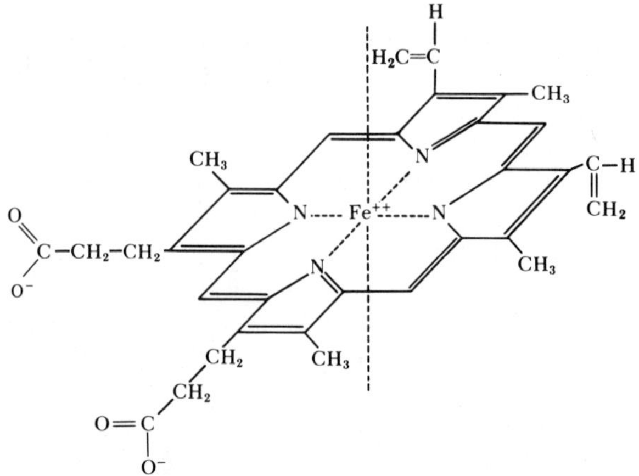

Figure 16-9
Chemical structure of heme moiety of hemoglobin.

the porphyrins (Gr., "purple") have been known for over 60 years, largely from the classic work of Nobel laureate Hans Fischer, definitive knowledge of its formation in vivo has developed only during the past three decades. This has been made possible primarily by the development of more sophisticated techniques for studying metabolic reactions, particularly isotopic labeling and highly efficient chromatographic methods for isolating the minute amounts of metabolic products in biological materials. By 1950 Rittenberg, Shemin, and co-workers were able to show that heme could be synthesized in significant amounts in vitro by mammalian reticulocytes and by avian erythrocytes (nucleated). Also, they demonstrated that the relatively abundant, simple substances glycine and succinate served as the prime precursors of the cyclic tetrapyrrole porphyrin ring of heme.

These fundamental observations were followed by the extensive investigations of a number of workers; thus today a fairly complete knowledge of the porphyrin-heme pathway is available, together with the structures of the intermediate metabolites formed, the chemical nature of the enzymes participating, and in some instances details of the chemical mechanisms of the reactions involved. The isolation and elucidation of the chemical structures of two key intermediates of the pathway, δ-aminolevulinic acid and porphobilinogen, stand out as especially significant accomplishments in our present understanding of porphyrin-heme formation.

Our discussion of the biosynthesis of porphyrins and heme will consider the *three* principal successive steps in the pathway:

1. Biosynthesis of δ-aminolevulinic acid from the prime precursors glycine and succinate as succinyl CoA
2. Formation of porphobilinogen from δ-aminolevulinic acid
3. Conversion of porphobilinogen to the cyclic, tetrapyrrole porphyrin ring and heme

Biosynthesis of δ-aminolevulinic acid (ALA) The formation of δ-aminolevulinic acid from glycine and succinyl CoA, derived from the citric acid cycle, is the first and the committed step (Chapter 3) in the biosynthesis of porphyrins and heme. The overall reaction, catalyzed by the enzyme δ-aminolevulinic acid synthase (ALAS), is as follows:

$$\underset{\textbf{Glycine}}{\begin{array}{c}CH_2NH_2\\|\\COOH\end{array}} + \underset{\textbf{Succinyl-coenzyme A}}{\begin{array}{c}COOH\\|\\CH_2\\|\\CH_2\\|\\O{=}C{-}CoA\end{array}} \xrightarrow[\text{Pyridoxal phosphate, } Mg^{++} \text{ (ALAS)}]{CoA} \underset{\text{(Probable intermediate)}}{\left[\begin{array}{c}COOH\\|\\CH_2\\|\\CH_2\\|\\C{=}O\\|\\CHNH_2\\|\\COOH\end{array}\right]} \xrightarrow{CO_2} \underset{\textbf{δ-Aminolevulinic acid}}{\begin{array}{c}COOH\\|\\CH_2\\|\\CH_2\\|\\C{=}O\\|\\CH_2NH_2\end{array}}$$

α-Amino-β-ketoadipic acid is a probable intermediate, as indicated.

A postulated mechanism of the reaction, shown in Fig. 16-10, is as follows. Glycine forms a Schiff base with pyridoxal-5′-phosphate, bound via a phosphate ester linkage to the enzyme ALAS. Mg^{++} ions, or certain other less active divalent cations, are required as a cofactor. As a result of the Schiff base formation, the α-carbon atom of glycine becomes a better nucleophile, stabilized in an anion form (Fig. 16-10), thus facilitating a nucleophilic attack on the thiol ester carbonyl of succinyl CoA, made more positive by associated Mg^{++} ions. Enzyme-bound α-

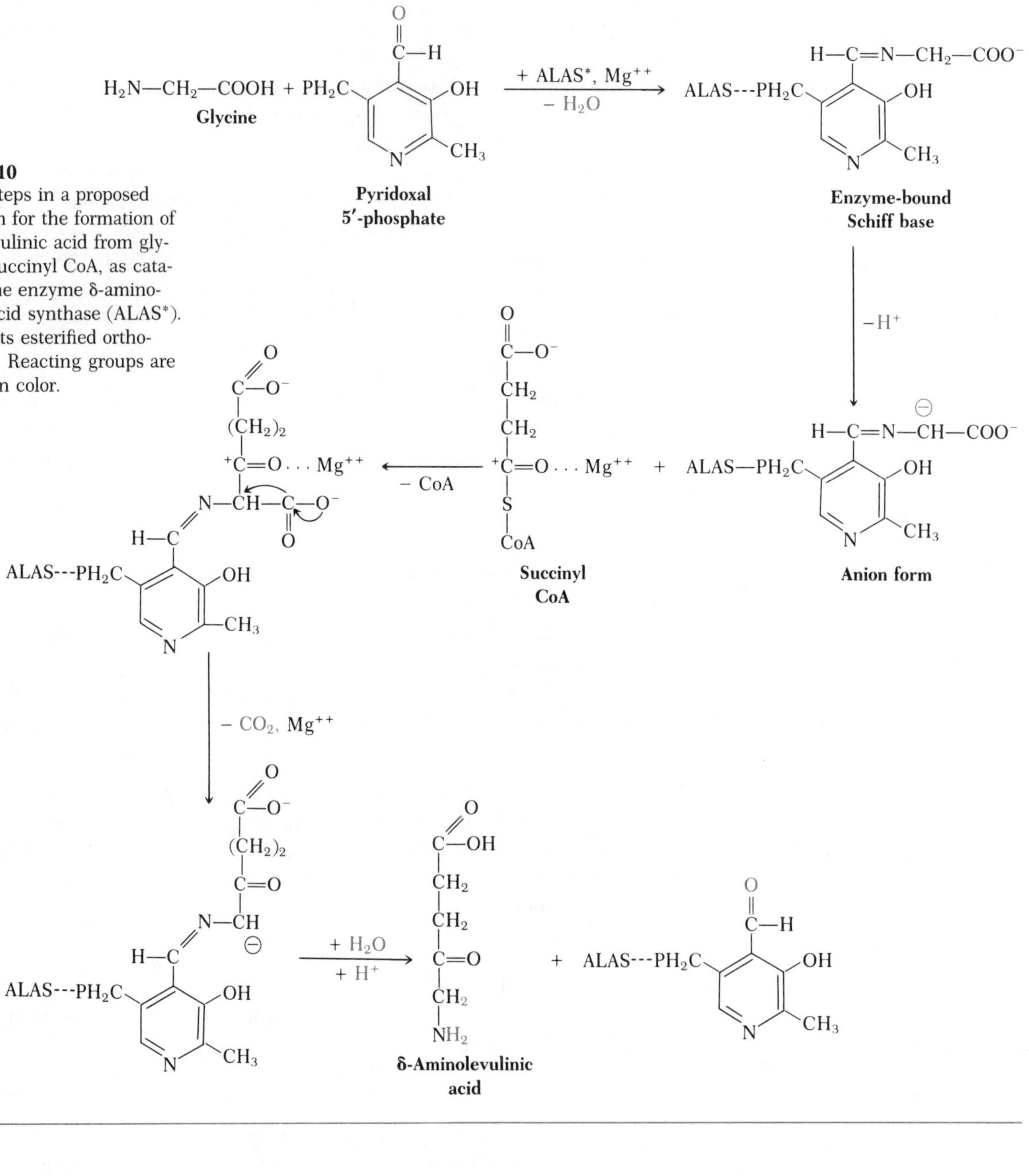

Figure 16-10
Principal steps in a proposed mechanism for the formation of δ-aminolevulinic acid from glycine and succinyl CoA, as catalyzed by the enzyme δ-aminolevulinic acid synthase (ALAS*). *P* represents esterified orthophosphate. Reacting groups are indicated in color.

Glycine

Pyridoxal 5′-phosphate

+ ALAS*, Mg^{++}
− H_2O

Enzyme-bound Schiff base

$-H^+$

Anion form

Succinyl CoA

− CoA

− CO_2, Mg^{++}

+ H_2O
+ H^+

δ-Aminolevulinic acid

amino-β-ketoadipic acid is formed as a labile intermediate, which loses CO_2 to form enzyme-bound ALA. On hydrolysis this yields free ALA plus the enzyme ALAS and bound pyridoxal phosphate.

δ-Aminolevulinate synthase, a mitochondrial allosteric enzyme, has been prepared in crystalline form from beef liver. It has a molecular weight of about 3 to 5 × 10^5. It is the primary rate-controlling enzyme of the heme biosynthetic pathway, being inhibited by the end products of the pathway, heme, hemoglobin, and certain other heme proteins. Inhibition apparently occurs at the transcription or possibly the posttranscription stage of ALAS synthesis. The enzyme is rapidly metabolized, having a half-life of only about 1 hour in mammalian liver.

Several substances have been reported to induce the synthesis of hepatic δ-aminolevulinate synthase, including several 5β-H steroid hormone metabolites, ethanol, barbiturates, and allylisopropyl acetamide. The latter drug produces as much as a fiftyfold increase in the hepatic enzyme level, resulting in an *experimental porphyria* that is similar to the human hereditary disease *acute intermittent porphyria,* to be discussed later in this chapter. There also is some evidence that the renal hormone erythropoietin may induce the formation of ALAS.

Formation of porphobilinogen The discovery of *porphobilinogen* (PBG) in the urine of patients with acute intermittent porphyria a number of years ago and the subsequent determination of its structure were major steps in the elucidation of the biosynthetic pathway of the porphyrins and heme. The compound was immediately postulated as an intermediate in porphyrin formation, as its name implies. The preparation later from beef liver of an enzyme, *porphobilinogen synthase* (PBGS) (also called δ-aminolevulinate dehydratase), which converts two molecules of δ-aminolevulinate to one molecule of PBG, together with supporting evidence from isotopic-labeling studies, established this compound's key position in the porphyrin biosynthetic chain. The overall reaction is:

$$\underbrace{HOOC{-}CH_2{-}CH_2{-}C(=O){-}CH_2{-}NH_2 + HOOC{-}CH_2{-}CH_2{-}C(=O){-}CH_2{-}NH_2}_{\text{δ-Aminolevulinate}} \xrightarrow[\text{(PBGS)}]{2\ H_2O} \text{Porphobilinogen}$$

δ-Aminolevulinate **Porphobilinogen**

PBGS has been prepared in crystalline form from beef liver. It has a molecular weight of about 285,000 and is made up of eight identical polypeptide subunits, each having a molecular weight of approximately 35,000. The active site of the molecule apparently contains a lysyl residue and two sulfhydryl groups that are essential to its activity, as demonstrated by Shemin and his co-workers. The cysteinyl groups are believed to participate in the acid-base catalysis required for the protonation-deprotonation sequences in the biosynthesis of PBG (Fig. 16-11). The beef liver enzyme also contains 4 atoms of zinc per molecule, which is essential for its activity. Recent evidence suggests that Zn^{++} ions interact with the sulfhydryl groups at the active site and may be involved in the protonation-deprotonation reactions just mentioned. It has also been proposed that Zn^{++}ions may act as a Lewis acid in polarizing the carbonyl group on one molecule of ALA, rendering it more susceptible to nucleophilic attack (Fig. 16-11).

The mechanism of action of PBGS (Fig. 16-11) is postulated first to involve the formation of a Schiff base between one molecule of δ-aminolevulinate with an amino group of the enzyme, probably the ϵ-amino group of lysine at the active site, forming an intermediate anion. A molecule of H_2O and a proton are released in the reaction. This is followed by a nucleophilic attack on the carbonyl carbon atom of a second molecule of δ-aminolevulinate and by protonation to link the two ALA molecules. The thiol groups of cysteine and probably an imidazole group of histidine at the active site of the enzyme are believed to mediate the protonation-deprotonation sequences. Protonation of the Schiff base complex then occurs, and one molecule of H_2O is released. The free amino group of the second ALA molecule replaces the

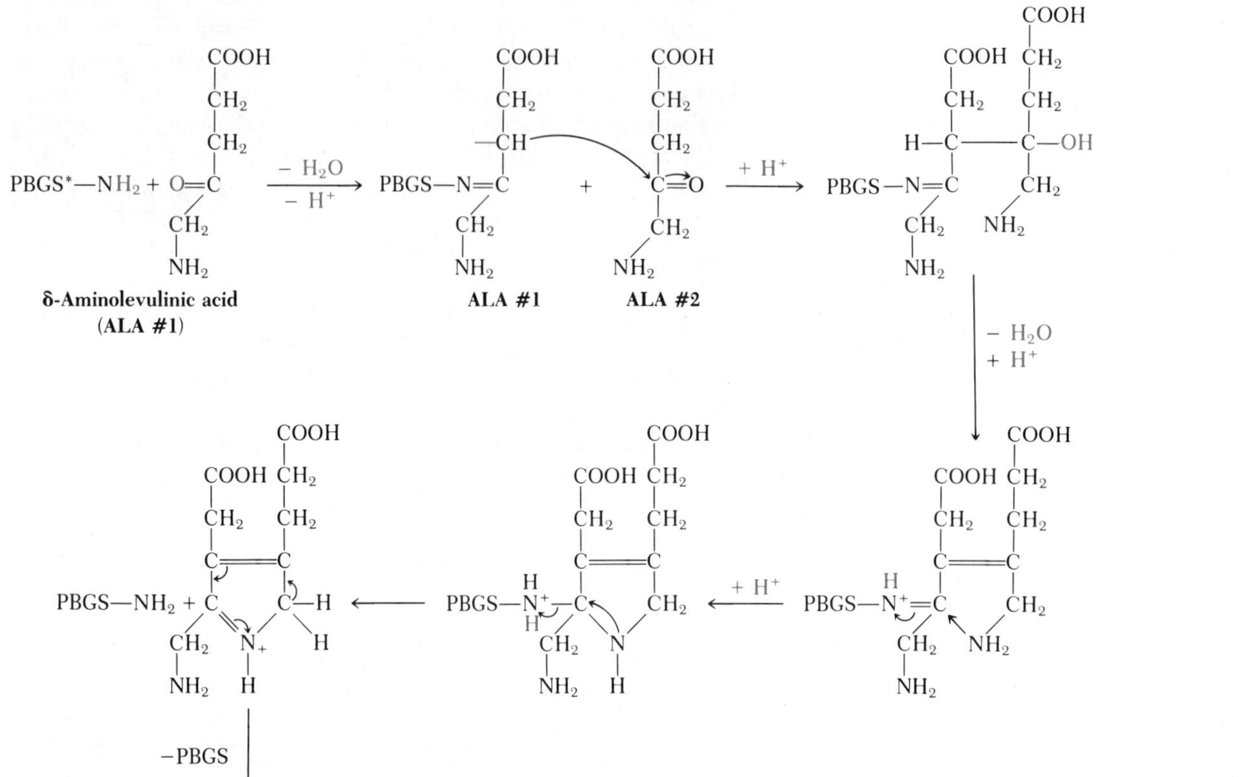

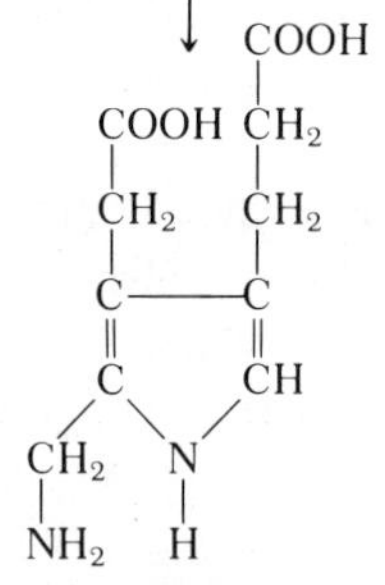

δ-Aminolevulinic acid (ALA #1)

ALA #1 **ALA #2**

Porphobilinogen

Figure 16-11
Proposed mechanism for the biosynthesis of porphobilinogen (PBG) from two molecules of δ-aminolevulinic acid (ALA) as catalyzed by the enzyme porphobilinogen synthase (PBGS*). The $-NH_2$ group of the enzyme shown is probably the $\epsilon-NH_2$ of a lysyl residue in the enzyme molecule. Reacting groups are indicated in color.

amino group of PBGS by a trans-Schiff reaction. Protonation followed by an intramolecular rearrangement occurs, thus forming PBG. Free PBG and free PBGS are released to complete the cycle.

PBGS is subject to feedback inhibition by heme, hemoglobin, and several other heme proteins. The enzyme is inhibited by relatively low concentrations of lead, which apparently reacts with thiol groups essential for the activity of the enzyme. This accounts for the large excretion of δ-aminolevulinate in the urine of individuals with *lead poisoning*. Indeed, the quantitative determination of δ-aminolevulinate in urine is one of the better analytical means of monitoring the severity and control of lead poisoning in human subjects.

Conversion of porphobilinogen to porphyrins and heme The reactions involved in the formation of heme from PBG, as now understood, are shown in Figs. 16-12 and

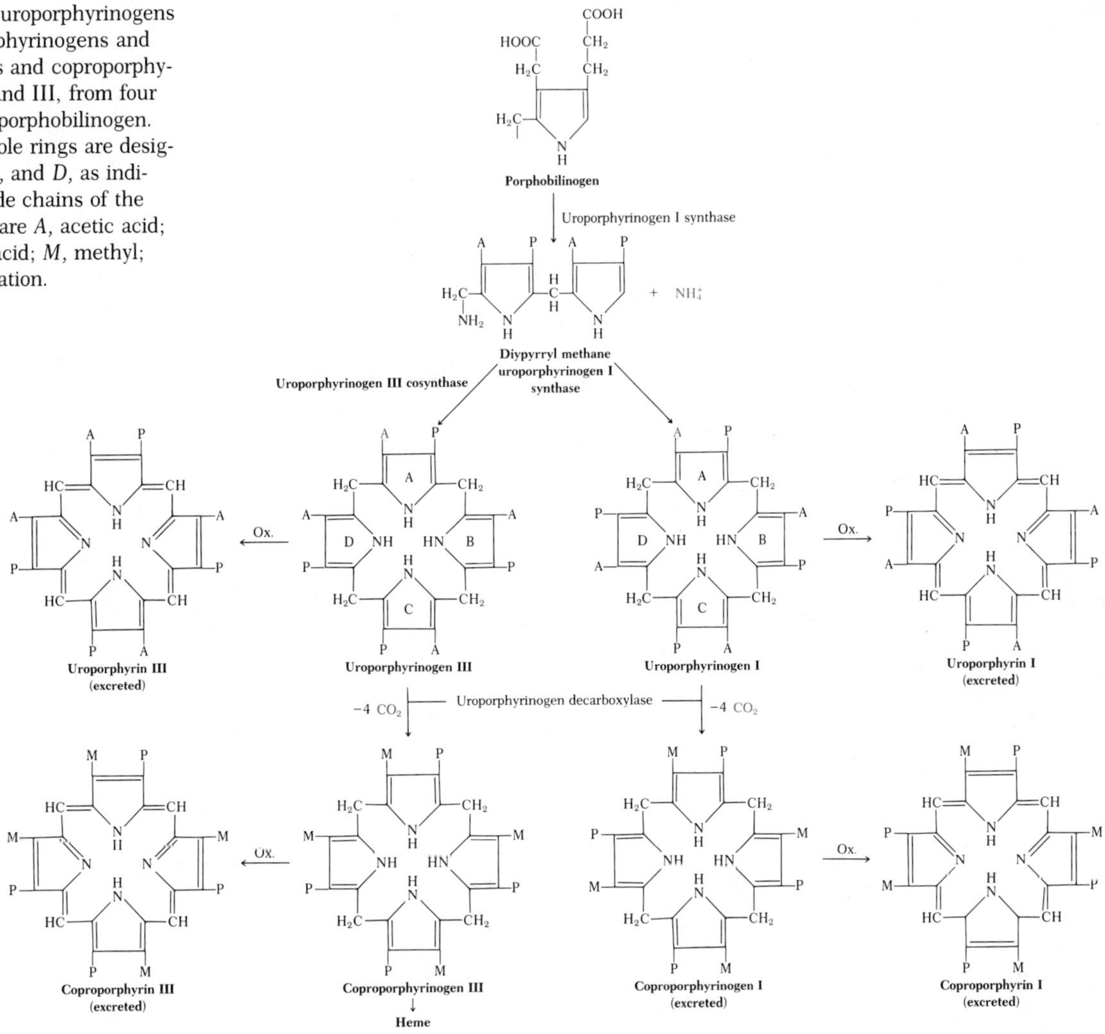

Figure 16-12
Formation of uroporphyrinogens and coproporphyrinogens and uroporphyrins and coproporphyrins, types I and III, from four molecules of porphobilinogen. The tetrapyrrole rings are designated *A, B, C,* and *D*, as indicated. The side chains of the pyrrole rings are *A*, acetic acid; *P*, propionic acid; *M*, methyl; *Ox.*, autooxidation.

16-13. Some details are still incomplete, such as the chemical identity of several of the enzymes required and the exact mechanisms of some of the reactions involved. Overall, four molecules of PBG combine with the loss of four NH_4^+ groups to form cyclic tetrapyrrole, porphyrin ring system. Ferrous iron is then introduced to form heme.

The first step is catalyzed by the enzyme *uroporphyrinogen I synthase* (formerly called porphobilinogen deaminase). The NH_2 group of one molecule is removed together with a hydrogen atom of a second molecule (Fig. 16-12). NH_4^+ is released, and a $—CH_2—$, methene bridge, is formed between the two PBG molecules. The resulting intermediate is a *dipyrryl methane*. According to one theory this substance reacts with two more PBG molecules, possibly as another dipyrrylmethane, with the loss of three more NH_4^+ groups to form a closed tetrapyrrole porphyrin ring,

Figure 16-13
Formation of protoporphyrin III (no. 9) and heme from coproporphyrinogen III and Fe^{++}. Side chains of the pyrrole rings are designated *M*, methyl; *V*, vinyl; *P*, propionic acid.

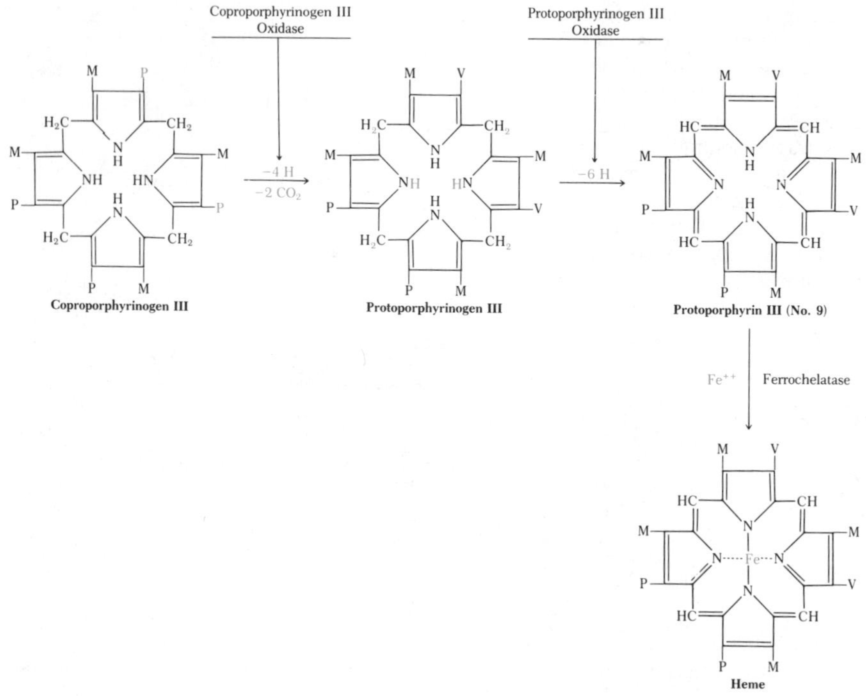

uroporphyrinogen I. Note in Fig. 16-12 that the *A* (acetic acid) and *P* (propionic acid) substituent groups of the four pyrrole rings are symmetrically arranged, i.e., clockwise: A-P, A-P, etc. Uroporphyrinogen I is a *colorless* intermediate, not used by the animal organism. It is therefore either oxidized (by autooxidation) to purple uroporphyrin I* and excreted, or its four acetate (*A*) substituent groups are decarboxylated to methyl (*M*) groups by *uroporphyrinogen decarboxylase* to form *coproporphyrinogen I*. This colorless intermediate also is unusable and undergoes autooxidation to purple *coproporphyrin I*†, which is excreted in the urine and feces as a waste product.

The enzyme *uroporphyrinogen I synthase* is widely distributed in animal and plant tissues and in microorganisms. It has been highly purified by chromatographic and gel electrophoretic procedures. The enzyme in human erythrocytes has been separated by gel filtration techniques into three active protein fractions having molecular weights of between 38,000 and 40,000. The three components are apparently *isoenzymes*. The mechanism involved in the coversion of four PBG molecules into uroporphyrinogen I by uroporphyrinogen I synthase remains unclear.

The major portion of the PBG used by the animal organism, however, is *not* converted to the type I tetrapyrrole isomer uroporphyrinogen I, but rather to a type III isomer, *uroporphyrinogen III*. This isomer differs from the symmetrical form in type I by the *reversal* of the order of the substitutent groups to P-A, on pyrrole ring *D*. This isomerization is catalyzed by the enzyme *uroporphyrinogen III cosynthase* (formerly called isomerase). The enzyme is *not active* in the *absence* of uroporphy-

*So named because it was first isolated from urine.
†So named because it was first isolated from feces.

rinogen I synthase. The type III isomer formed by the combined action of these two enzymes is used for the formation of heme. The two other isomeric forms, types II and IV porphyrins, do *not* occur naturally. They have been synthesized and studied in the laboratory, however.

The mechanism by which four molecules of porphobilinogen are converted into uroporphyrinogen III by the combined efforts of the two enzymes, uroporphyrinogen I synthase and uroporphyrinogen III cosynthase, remains an enigma that has perplexed biosynthetic chemists for several decades. About 20 theories based on over 200 scientific reports have been published. Only one of these explanations, the "switch reaction," i.e., the reversal of the *D* ring, is considered here. Several studies using a ^{13}C-nuclear magnetic resonance procedure have demonstrated that the switch reaction involves an *intramolecular* rearrangement. A recent concept, based on isotopic labeling studies, proposed that the *aminomethyl group* of some dipyrrylmethane intermediates (Fig. 16-12) becomes dissociated from the molecule and chemically attaches to the enzyme uroporphyrinogen III cosynthase. The resulting enzyme complex apparently is leaky, allowing some of the attached aminomethyl groups to recombine to form a rearranged dipyrrylmethane. This could then combine with a symmetrical dipyrrylmethane in a head-to-head and tail-to-tail manner to yield the type III isomer of uroporphyrinogen. Further studies will be required for the complete elucidation of this challenging biosynthetic reaction.

The uroporphyrinogen type III formed is readily converted to *coproporphyrinogen III* by the decarboxylation of its four acetate *(A)* side chains to form methyl groups *(M)*. The reaction apparently is catalyzed by a poorly characterized enzyme, *uroporphyrinogen decarboxylase*. The propionic *(P)* substituent group on pyrrole rings A and B of coproporphyrinogen III are then oxidatively decarboxylated to vinyl *(V)* groups by *coproporphyrinogen III oxidase* to form *protoporphyrinogen III,* which in turn is oxidized by *protoporphyrinogen oxidase* to *protoporphyrin III* (no. 9 of 15 isomers).

As the final step, Fe^{++}, available by way of plasma transferrin, is inserted into the protoporphyrin molecule, catalyzed by the enzyme *ferrochelatase,* to form *heme*—the end product of the biosynthetic pathway. About 85% of the heme thus formed is used to form hemoglobin by combining with α-, β-, or other globin chains. About 10% is used for myoglobin formation and the remaining 5% for cytochromes and other heme proteins.

Very small amounts of the intermediates uroporphyrinogen III and coproporphyrinogen III may become autooxidized in the tissues and converted into purple uroporphyrin III and coproporphyrin III, respectively. These, like those of type I, are not usable by most animal organisms and are therefore excreted in the urine and feces as waste products.

Rather small amounts of uroporphyrins and coproporphyrins are excreted daily, however, by normal adults. Coproporphyrin type III is the principal porphyrin waste product, about 50 to 250 μg being excreted in the urine daily; approximately 10 to 50 μg of the type I isomer is also excreted. Somewhat larger amounts are eliminated daily in the feces (approximately 240 to 400 μg of type III and 60 to 100 μg of type I). Much smaller amounts of the uroporphyrins, mainly type I, are excreted daily in the urine and feces by normal adults (10 to 35 μg).

Considerably larger amounts of both uroporphyrins and coproporphyrins, either types I or III or both, may excreted by patients with various types of porphyria (Table 16-7). As will be discussed later, their qualitative and quantitative determinations are valuable aids in the diagnosis of the type of porphyria involved and thus in indicating appropriate therapeutic measures.

Interesting recent studies indicate that protoporphyrin may have a possible prac-

Table 16-7
The porphyrias*

Type	Enzyme defect	Principal metabolite increased	Principal clinical manifestations
Erythropoietic			
Congenital	Uro'gen III cosynthase	Uro I copro I in blood, feces, and urine markedly increased	Photosensitivity, cutaneous lesion; pink bones and teeth; hemolytic anemia
Protoporphyria	Ferrochelatase	Proto in RBCs, urine, feces; some copro III	Photosensitivity; cirrhosis; solar urticaria
Hepatic			
Acute intermittent	Uro'gen I synthase	ALA and PBG in urine during attacks; urine darkens on standing; copro III and Uro III in urine	Abdominal pains; spasms of smooth muscle; convulsions; neurological abnormalities; confusion; hypertension; hypercholesterolemia
Coproporphyria	Copro'gen III oxidase (?)	Copro III in feces; ALA and PBS in urine	Some similarities to acute intermittent
Cutanea tarda	Uncertain (↓ uro'gen decarboxy) ? (↑ ALA synthase)	ALA, PBG in urine; copro in feces	Cutaneous lesions; some photosensitivity
Variegate	Decreased proto'gen III oxidase	Fecal copro I and III and proto; ALA and PBG in urine during attacks	Cutaneous lesions; abdominal and neuropsychiatric symptoms; attacks precipitated by alcohol and drugs.
"Toxic"	(?) Increased ALA synthase	ALA and PBG in urine during attacks; copro III in urine and feces	Reactions to certain drugs (barbiturates, alcohol, apronalide, sulfonmethane allylisopropyl acetamide); hexachlorobenzene; certain steroids; pesticides; and oral contraceptives

*Uro, Uroporphyrin; *copro*, coproporphyrin; *proto*, protoporphyrin; *uro'gen*, uroporphyrinogen; *copro'gen*, coproporphyrinogen; *proto'gen*, protoporphyrinogen; *ALA*, δ-aminolevulinic acid; *PBG*, porphobilinogen.

tical use as a chemotherapeutic agent in the treatment of certain types of cancer. In experimental animals *localized* tumors of a variety of types can be obliterated completely by injecting protoporphyrin and then radiating the tumor mass by means of a quartz optical device inserted into it. Protoporphyrin is well known to accumulate in tumor tissue. Radiation causes it to fluoresce strongly, thus apparently forming free radicals in the surrounding tumor mass and destroying it. This is analogous to the destruction of cutaneous tissue in certain types of porphyrias on exposure of the skin to strong daylight — a phenomenon known as *photosensitivity*.

Conclusive proof of the preceding pathway for the formation of heme has been obtained by using the isotopically labeled precursors ^{15}N- and ^{14}C-glycine and uniformly labeled ^{14}C-succinate. The individual N- and C-atoms of the labeled heme thus formed were isolated by means of the Fischer degradation procedure. In this way the origin of each carbon and nitrogen atom of heme was traced to the two prime precursors, glycine and succinate (Fig. 16-14). The carboxyl carbon atom of glycine is not used but comes off as CO_2 during the formation of δ-aminolevulinate (Fig. 16-10). The other C- and N-atoms of glycine and succinate are used as indicated in Fig. 16-14.

This pathway for the biosynthesis of heme is the *only* known pathway for the formation of this vitally important compound. Furthermore, the pathway described is the most efficient means of forming heme from a thermodynamic standpoint.

The porphyrias In a biosynthetic pathway such as that for heme, with a number of enzymatically catalyzed reactions, the possibility of partial or complete blockages because of genetically deficient or defective enzymes increases. The metabolite or

Figure 16-14
Diagram of part of heme molecule indicating sources of N and C atoms. The α-C of glycine is shown as the source of the methene bridge C as well as of one pyrrole C, whose exact position in the ring is probably as shown. The carboxyl C atom of glycine is not used.

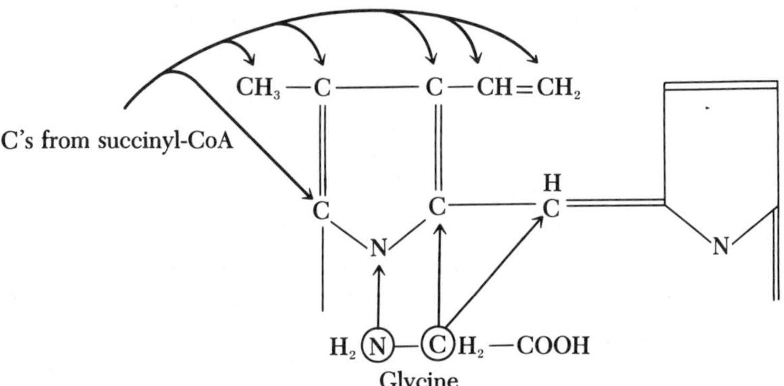

metabolites preceding the impaired reaction then may accumulate in the tissue involved, and usually also in the blood, urine, and/or feces, and may be determined by appropriate quantitative or qualitative procedures. Tissue changes and abnormalities usually follow. *Six* hereditary enzyme deficiencies have been demonstrated in the porphyrin-heme biosynthetic pathway, leading to the categorization of a group of heritable diseases called the *porphyrias*.

The porphyrias are classified on the basis of the principal tissue site of the enzyme defect or biochemical lesion where the affected metabolite(s) accumulates. The two major types of porphyria are *erythropoietic* and *hepatic*. The various subtypes of these, together with the enzyme defect, the principal metabolite(s) accumulating, and the outstanding clinical manifestations, are summarized in Table 16-7.

Since the porphyrias are genetic diseases, no "cure" is currently available. Thus treatment is palliative, designed to relieve any undesirable symptoms.

Two types of porphyria merit further comment. One is the rare *congenital erythropoietic porphyria,* in which relatively large amounts of the type I porphyrin isomers are produced because of a deficiency of the enzyme uroporphyrinogen III cosynthase. The type I isomers are not usable and accumulate in erythropoietic tissues (and then other tissues), urine, and feces, giving them a reddish, Burgundy wine color. These porphyrins fluoresce an intense vermilion in ultraviolet light. Relatively large amounts of uroporphyrin and coproporphyrin I also are deposited in the bones and teeth of these patients, causing a deep red to pink discoloration. Some of the porphyrins are also deposited in the skin, causing extreme photosensitivity, apparently resulting from free radical formation. Blisters and skin lesions develop in even diffuse daylight and, if infection occurs, result in severe disfiguration. Such patients must avoid light or be protected in exposed areas by skin lotions that screen out ultraviolet light.

An interesting type of therapy being studied in patients with this type of porphyria is the intravenous administration of heme. The rationale, of course, is the feedback inhibition of porphyrin-heme formation by heme. This treatment has shown some promise.

Congenital erythropoietic porphyria also occurs in highly inbred strains of domestic animals, cattle, pigs, and the common ground squirrel.

A second type of porphyria that merits further comment is hepatic *acute intermittent porphyria*. It occurs much more frequently than the preceding form because it is autosomal dominant. Its chief manifestations are periodic attacks of intense abdominal pain, mental aberrations and other neurological symptoms, and usually hypertension. Photosensitivity usually is absent. Large amounts of *PBG* and some excess δ-aminolevulinate are excreted in the urine. There is a significant *decrease* in *uroporphyrinogen I synthase* in the erythrocytes of patients with this disease.

Recently heme therapy and high-carbohydrate foods have been employed with some success in the treatment of acute intermittent porphyria. The rationale for increasing carbohydrates is to ensure an adequate supply of glucuronate (Chapter 11) for the detoxication of the 5β-H steroid hormones alleged to be increased in this form of porphyria because of lack of a reductase (hepatic 5β-reductase) required in their normal metabolism (Chapter 12). Hypercholesterolemia is commonly found in this type of porphyria.

Other findings indicate that an altered steroid metabolism may also be present in some patients. Assays indicate that there is a decrease of hepatic 5β-reductase activity; thus steroid metabolites are diverted into the 5β-hydrogen pathway. The 5β-H steroid metabolites, such as etiocholanolone (Chapter 18), are strongly porphyrinogenic, increasing the hepatic level of the enzyme δ-aminolevulinate synthase, as is also found in patients with acute intermittent porphyria. It appears possible that a simultaneous decrease in *both* of these enzymes is necessary for the production of severe symptoms. This could account for the well-known intermittent nature of the disease.

Significant investigations by Granick and his group, using chick embryo liver cultures, demonstrated that certain steroids induce the biosynthesis of porphyrins and heme. Steroids, e.g., pregnanolone, of the 5β-androstane or 5β-pregnane type, with alcohol or ketone substituents at carbons numbered 3, 17, 20, and possibly 11, are active inducers. Steroids such as the androgens, estrogens, cortisol, and aldosterone have little or no activity. Glucuronide conjugates, even of the most potent 5β-H steroids, are likewise inactive. Steroids affect the biosynthesis of heme apparently by serving as an inducer of the formation of ALA synthase, the key enzyme of the biosynthetic chain. Granick believes the steroids block or displace heme from its binding site on the aporepressor protein of the Jacob-Monod model, thus rendering the repressor mechanism inoperative. This leads to increased ALA synthase formation, with the resultant increased production of ALA, porphyrins, and heme.

An interesting article cited evidence indicating that the British royal family of the 1700s, and particularly King George III, who reigned at the time of the American Revolution, suffered from an acute intermittent type of porphyria. His recurring abdominal pain, diagnosed at the time as "biliary concretions in the gall duct," darkening of the urine, neurological symptoms, and delirium are all classic symptoms of this hereditary metabolic disease. He was generally believed to be insane at the time. The same historical study indicates that three royal houses, including the Hanoverian and Prussian royal lines, were affected. Apparently Mary Queen of Scots (1542-1587) was the first known to be afflicted. Modern history records a number of such examples of this hereditary metabolic disease.

Of considerable interest is a recent report that *variegate porphyria* is caused by a deficiency of the enzyme *protoporphyrinogen III oxidase*. The enzyme level was found to be significantly lower in cultures of skin fibroblasts from five patients with typical variegate porphyria, compared with culture specimens from normal control subjects.

Patients with these as well as other types of porphyria are always advised to *strictly avoid* the use of *ethyl alcohol* and *barbiturates*, since these are strongly porphyrinogenic substances, acting as inducers of ALAS formation. Other porphyrinogenic drugs that may produce a toxic porphyria include allylisopropyl acetamide, apronalide (Sedormid), hexachlorobenzene, sulfoxylureas, dicarbethoxydihydrocollidine, certain steroids, certain pesticides, and oral contraceptives.

Degradation of heme

The hemoglobin of senescent erythrocytes is released by phagocytosis in reticuloendothelial cells primarily in the spleen. It is transported to the liver bound to the plasma protein *haptoglobin*. It is mainly in the hepatic reticuloendothelial cells (Kupffer cells) that hemoglobin is cleaved, the globin portion is reused as amino acids, and the heme moiety then is converted in several steps to the bile pigments *biliverdin* and *bilirubin* and excreted in bile (Chapter 17). The iron atom of heme is removed in the process, bound to plasma *transferrin* and either recycled as new hemoglobin or stored in the liver (primarily) as ferritin (Chapter 21). Heme from other hemoproteins is transported in the plasma bound to the protein *hemopexin*, and similarly is converted to biliverdin and bilirubin.

The degradation of heme into bile pigments involves a unique oxidation of the α-methene bridge carbon atom to form *carbon monoxide*. The reaction is catalyzed by the microsomal enzyme *heme oxygenase*, a mixed function oxygenase (Chapter 5), in the presence of molecular oxygen, cytochrome *c* reductase, and NADPH. The tetrapyrrole ring of heme is opened and the ferrous iron atom released, possibly from an intermediate, *verdohemin*, and recycled or stored as stated. *Biliverdin* is thus formed (Fig. 16-15). It is then reduced by the cytoplasmic enzyme *biliverdin reductase* in the presence of NADPH to *bilirubin* by the addition of two hydrogen atoms to the double bond of the central γ-methene bridge (Fig. 16-16).

Some bilirubin is formed in other tissues of the body from various hemoproteins and is transported to the liver bound and loosely associated with plasma *albumin*. Bilirubin itself is insoluble in an aqueous medium.

In the liver, bilirubin is conjugated with glucuronic acid to form *bilirubin diglucuronide*, which is water soluble and is readily excreted by means of the bile (Chapter 17) into the intestine. The formation of bilirubin diglucuronide involves the

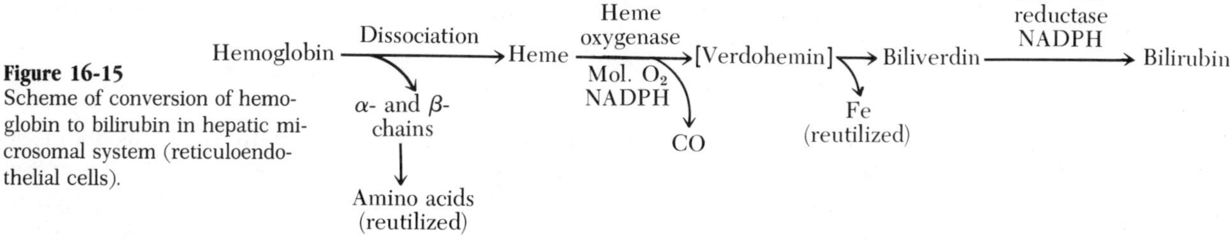

Figure 16-15
Scheme of conversion of hemoglobin to bilirubin in hepatic microsomal system (reticuloendothelial cells).

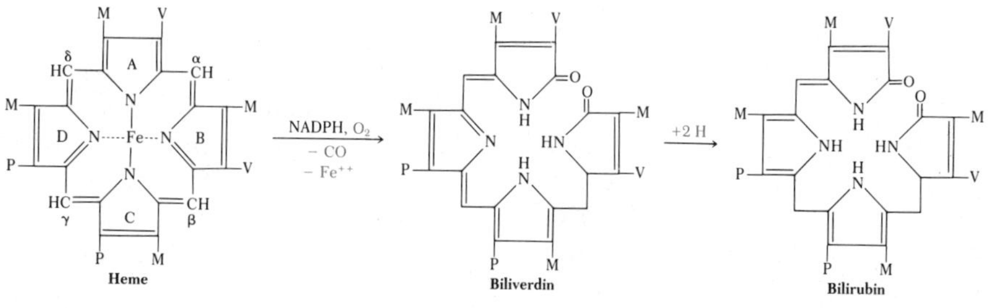

Figure 16-16
Oxidative conversion of heme to biliverdin and bilirubin, with the liberation of carbon monoxide and ferrous iron. ^{14}C-labeled hemoglobin injected into dogs yields ^{14}C-labeled bilirubin. The conversion of heme to biliverdin is catalyzed by the enzyme *heme oxygenase;* the second reaction is catalyzed by *biliverdin reductase.*

Figure 16-17
Scheme of conjugation of bilirubin in endoplasmic reticulum of hepatic parenchymal cell.

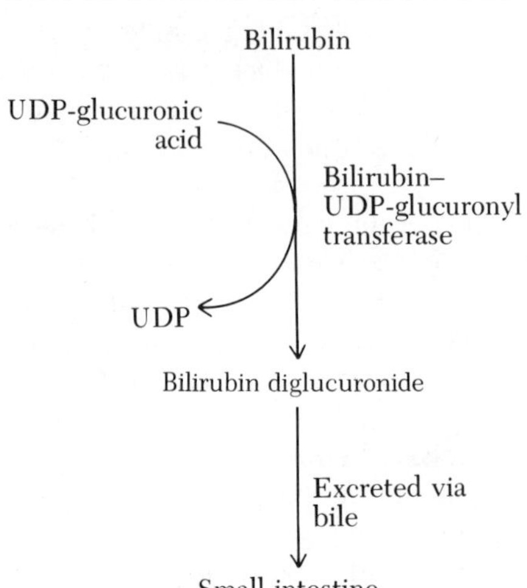

reaction of its two *carboxyl* groups with C-1 hydroxyl groups of two UDP-glucuronic acid molecules (Chapter 11). The reaction is catalyzed by the hepatic microsomal enzyme *bilirubin-UDP-glucuronyl transferase* (Fig. 16-17).

Bilirubin diglucuronide is hydrolyzed in the intestine by a β-glucuronidase. The bilirubin portion is reduced by bacterial floral action to colorless *d-* or *l-urobilinogen,* part of which is reabsorbed and excreted in the urine as an oxidized orange-yellow pigment, *l-urobilin.* The remaining portion of *l*-urobilinogen is reduced in the intestine to *l-stercobilinogen,* which is excreted as an oxidized brown pigment, *l-stercobilin,* in the feces. The structural changes involved in these reactions are discussed further in Chapter 17.

The origin of fecal stercobilin from heme (as hemin) has been demonstrated conclusively in normal human subjects by the administration of ^{15}N-glycine (see Fig. 16-6). The data also confirm the average life span of the human erythrocyte as being about 120 days.

Elevations of the amounts and types of bilirubin in the plasma may occur in certain types of liver jaundice and other diseases. These are discussed in Chapter 17.

Some chemical properties of hemoglobin

The iron in reduced hemoglobin is in the ferrous state. In oxyhemoglobin it is apparently still in the ferrous state (although there is some evidence that it is in the low-spin ferric state in oxyhemoglobin), but oxygen is attached. The oxygen is assumed to be linked loosely to the ferrous iron by a residual valence force, probably by sharing the unpaired electrons of ferrohemoglobin with those of molecular oxygen.

A *hemochromogen* is a compound of heme with any of a number of nitrogenous substances. Thus hemoglobin and most of its derivatives are hemochromogens, as are the respiratory pigments of the invertebrates, as well as the cytochromes. Heme and hemoproteins react with pyridine to form pyridine-hemochromogen. This reaction has been employed in the determination of total hemoproteins in samples.

Heme may be oxidized to hematin, which contains a hydroxyl group. This group may be substituted by a chloride ion, forming hemin. Hemin crystallizes in characteristic brown crystals, which may be easily recognized under the microscope. This procedure is used as a test for blood. Dilute alkali similarly splits off the heme groups to form alkali hematin.

Since each molecule of hemoglobin has four heme groups, it contains four ferrous ions. Each heme group unites with one molecule of oxygen. We may accordingly represent deoxyhemoglobin as globin (ferroheme)$_4$, and the reaction whereby it performs its respiratory function as:

$$\underset{\textbf{Deoxyhemoglobin}}{\text{Globin (ferroheme)}_4} + 4\ O_2 \rightleftharpoons \underset{\textbf{Oxyhemoglobin}}{\text{Globin (ferroheme-}O_2)_4}$$

Usually, however, in discussions concerning this reaction a less exact expression is used:

$$\underset{\textbf{Deoxyhemoglobin}}{Hb} + O_2 \rightleftharpoons \underset{\textbf{Oxyhemoglobin}}{HbO_2}$$

Most of the oxygen present in arterial blood is held in this loose chemical combination with hemoglobin.

Some investigators believe that the oxygenation of hemoglobin involves the migration of an electron from ferrous heme iron to the oxygen molecule. The heme iron atoms of oxyhemoglobin are thus formally in the ferric low-spin state. The ease with which this union can be brought about and broken may be demonstrated in the laboratory. Blood that has been rendered nonclotting may be poured from one vessel to another a few times and it becomes bright crimson (oxyhemoglobin). Addition of a mild reducing agent changes its color to a very dark red (deoxyhemoglobin), after which it may be oxygenated again as before. These changes may be followed spectroscopically if the blood is suitably diluted. Of interest in this connection is the distinct displacement of the β-chains of Hb A when oxygen combines with it. This phenomenon will be considered later in this chapter. In addition to the oxygen combined with hemoglobin, there is a small amount held in solution in the plasma. The tension or pressure of oxygen in the plasma, together with other factors, determines the degree of dissociation of oxyhemoglobin into oxygen and hemoglobin. The partial pressure of the oxygen in atmospheric air at the barometric pressure of 760 mm Hg is 159 mm Hg (i.e., 20.9% $O_2 \times 760$ mm Hg). If blood is placed in contact with oxygen at this pressure, the hemoglobin becomes completely converted to oxyhemoglobin. Increase of oxygen pressure can add no more oxygen to the hemoglobin but can force more oxygen into solution in the plasma. Lowering the partial pressure of oxygen causes dissociation to occur, but even at 102 mm Hg, which is the partial pressure of oxygen in arterial blood, the hemoglobin is 95% saturated. Still lower pressures, such as those obtained in the tissues, cause further dissociation or release of atmospheric oxygen near the site of tissue oxidations. This discussion will be continued when the biochemistry of respiration is taken up.

Absorption spectra of hemoglobin and its derivatives Hemoglobin, its derivatives, and a number of other compounds have characteristic absorption spectra, i.e., if such a solution is interposed between a source of white light and the prism of a spectroscope, the light of certain wavelengths is absorbed and dark bands or shadows appear in the spectrum wherever the light has been taken out. Thus hemoglobin in the deoxygenated state (reduced hemoglobin) has one broad band in the yellow-green section, its center being at 559 nm. Oxyhemoglobin has two narrow bands. One, the narrower of the two, is in the yellow, with its center at 579 nm; the other, the wider, is nearer the green, with its center at 542 nm. On great dilution the wider one disappears first. Dilute solutions of hemoglobin thus may be detected spectroscopically.

The absorption spectrum of carbon monoxide hemoglobin is similar to that of oxyhemoglobin, having two main absorption bands. However, the absorption maxima for carbon monoxide hemoglobin are shifted slightly to the right (blue end of spectrum), being at approximately 570 and 535 nm. Therefore, from the absorption spectrum, carbon monoxide hemoglobin is very difficult to identify in cases of carbon monoxide poisoning, particularly since 20% to 50% of the oxyhemoglobin may still be present in the blood in severe or even fatal cases of carbon monoxide poisoning. The absorption spectrum of methemoglobin, on the other hand, is quite characteristic and is commonly used for the qualitative detection of this pigment. There are *four* absorption bands for methemoglobin with maxima at approximately 634, 575, 540, and 490 nm, respectively. The characteristic absorption band used in the detection of methemoglobin is in the red portion of the spectrum at 634 nm.

This method has various practical applications in the recognition of a number of derivatives of hemoglobin.

Carbon monoxide hemoglobin Carbon monoxide combines with the heme portion of hemoglobin to form carbon monoxide hemoglobin, also called carboxyhemoglobin and carbonylhemoglobin. This is a much firmer combination than the one between oxygen and hemoglobin. The affinity of hemoglobin for carbon monoxide is about 210 times that for oxygen. If, therefore, carbon monoxide is in the inspired air, it forms this firm combination to a greater extent than its proportion in the air would seem to warrant. Consequently, if enough carbon monoxide is present, the blood does not have sufficient oxyhemoglobin for respiratory purposes, and asphyxiation occurs.

Carbon monoxide hemoglobin has a cherry red color that is not changed readily by reducing agents. Carbon monoxide hemoglobin may also be detected by chemical tests. The simplest is to dilute the suspected blood greatly, after treating it with a little sodium hydroxide, and compare the color with normal blood similarly treated. Normal blood shows a greenish hue after such treatment, whereas carbon monoxide blood remains pink.

Poisoning by carbon monoxide is a common danger of modern life. Carbon monoxide is particularly lethal for two reasons: (1) it is odorless and colorless and consequently cannot be readily detected, and (2) its action is insidious and rapid. The victims frequently become unconscious in a few minutes, and death often follows quickly.

This gas is found wherever incomplete combustion of carbonaceous materials occurs—in automobile exhaust gas (4% to 7%), in chimney gases and smoke, and in blasting gases. It is also a constituent of manufactured illuminating gas (derived from coal or oil), in which its presence varies from 4% to 40%, depending on the source materials and the method of manufacture. Natural gas contains no carbon monoxide.

Poisoning may be either acute or chronic. Both are important from the standpoint of public health. Deaths resulting from the inhalation of automobile exhaust gas have increased alarmingly in the past few years. An automobile engine emits 1 cubic foot of carbon monoxide per minute per 20 horsepower. In a small individual garage with no ventilation this amount may be fatal to a person in 5 minutes. Therefore a door or window of a garage must always be open, even in coldest weather, when the engine is running. Vehicular tunnels are also hazardous because of the possible accumulation of carbon monoxide from automobile exhaust gas in the atmosphere. The adequate ventilation of such tunnels is consequently of utmost importance.

National Safety Council figures show that about 850 Americans die annually from automobile exhaust fumes. The actual figure is probably considerably higher if deaths from carbon monoxide from other sources (e.g., faulty stoves, chimneys, heating systems) are included.

Several factors determine the degree of toxicity of carbon monoxide, but all relate to one point, i.e., the rate of absorption of this gas. The chief factors are (1) concentration of carbon monoxide in the air respired, (2) duration of exposure, and (3) rapidity of respiration. Rapidity of respiration depends on the activity of the individual, his age and size, and the temperature and humidity of the atmosphere. The symptoms produced depend on the percentage of hemoglobin combined with carbon monoxide and thus rendered physiologically useless, at least for the time being. Following are the symptoms that occur with the percentages of hemoglobin saturated with carbon monoxide:

Percent	Symptoms
0 to 10	Usually none
10 to 20	Possibly slight headache
20 to 30	Headache, throbbing in temples
30 to 40	Severe headache, weakness and dizziness, dim vision, nausea, vomiting, possibly collapse
40 to 50	Like the above but with greater possiblility of collapse, increased pulse and respiration
50 to 60	Unconsciousness, coma with intermittent convulsions, Cheyne-Stokes respiration (periodic type of respiration)
60 to 70	Like the above, but with depressed heart action and respiration, possibly death
70 to 80	Weak pulse, respiratory failure, death

The proportion of carbon monoxide in the air necessary to produce such saturation figures depends on the factors first mentioned, but in a general way, if the respired air contains the following percentages:

Percentages	
0.01	Symptoms after a few hours
0.04	Perhaps safe for only about an hour
0.10	Uncomfortable and may be dangerous in 2 hours
0.30	Dangerous in 30 minutes
0.60	Dangerous in 10 to 15 minutes
1.30	Dangerous in 1 to 3 minutes

The treatment in cases of carbon monoxide poisoning is (1) rapid removal from the poisoned atmosphere, (2) artificial respiration, using an oxygen and carbon dioxide mixture, if available, and (3) blood transfusion, if necessary.

Methemoglobin Methemoglobin is a derivative in which the iron is in the ferric state. It is produced by the oxidation of hemoglobin, e.g., when potassium ferricyanide is added to blood. It is quite different from *oxygenated* hemoglobin, i.e., oxyhemoglobin, in which the oxygen is united loosely with *ferrous iron*. In methemoglobin oxygen is liberated, leaving the methemoglobin devoid of this gas; but, since there is now an additional positive charge, methemoglobin combines with a negative group, presumably a hydroxyl. Methemoglobin cannot combine with and transport oxygen, since its iron is in the ferric state.

A small amount of methemoglobin is present in the blood of normal individuals. An increased amount is found in clinical methemoglobinemia, which is caused by failure in the normal reconversion of methemoglobin to hemoglobin or by a more rapid production of methemoglobin (by certain drugs, oxidants, etc.) than normal cellular mechanisms can cope with. The normal mechanism for the reconversion of methemoglobin to hemoglobin follows. The amount of methemoglobin present in blood of normal human adults is about 1.7% of the total hemoglobin (approximately 0.3 gm/100 ml blood):

A similar mechanism is operative in muscle tissue, to convert *metmyoglobin* to functional myoglobin.

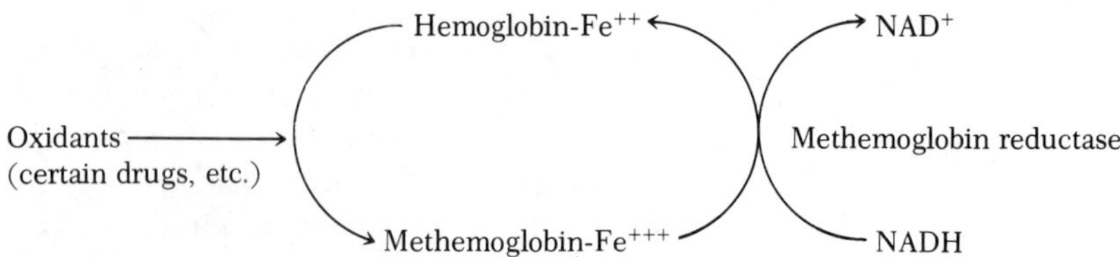

The oxidation of hemoglobin to methemoglobin results in the formation of the superoxide radical, O_2^-, by the transfer of a single electron. The enzyme superoxide dismutase (hemocuprein), present in erythrocytes, catalyzes the conversion of the O_2^- to H_2O_2 and O_2. The H_2O_2 is then decomposed by glutathionine peroxidase system or by catalase, both also present in the erythrocyte. These two systems thus protect the red cell from any damaging effects of an accumulation of the superoxide radical or of hydrogen peroxide. The reactions involved are as follows:

$$2\,O_2^- + 2\,H^+ \xrightarrow[\text{dismutase}]{\text{Superoxide}} H_2O_2 + O_2$$

$$H_2O_2 \xrightarrow{\text{Catalase}} H_2O + \tfrac{1}{2}\,O_2$$

Erythrocyte-reduced glutathione also may aid in the protection from oxidation of sulfhydryl groups of the hemoglobin molecule and thus the formation of so-called inclusion or Heinz bodies.

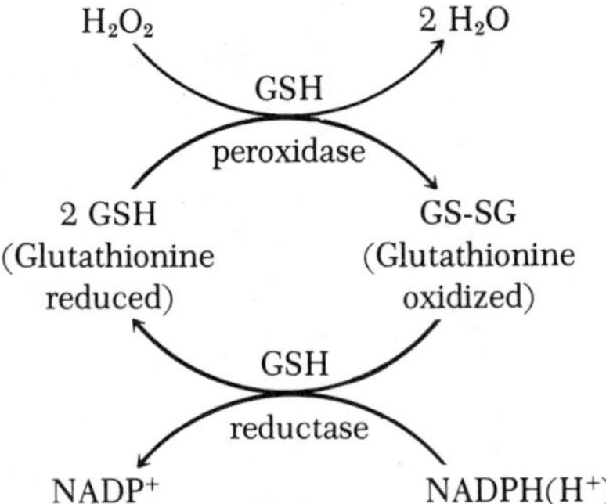

After the administration of certain drugs or exposure to certain poisons, e.g., chlorates, acetanilid, nitrites, nitrobenzene, antipyrine, iodine, phenacetin, sulfonmethane (Sulfonal), sulfonethylmethane (Trional), and, perhaps most important, the sulfonamide drugs, methemoglobin is likely to be present in the circulating blood.

In some industries poisons that cause methemoglobinemia are produced. Nitrobenzene is used in the manufacture of shoe dyes, floor polishes, cosmetics, and explosives. Workers in these industries may be acutely or chronically poisoned if nitrobenzene is absorbed in sufficient amounts, and in such cases methemoglobin is found to be present in the blood. The fumes from carbon arcs contain nitrous oxide, which reacts with atmospheric oxygen to form nitrogen dioxide. If this gas is breathed in high concentrations, methemoglobin may be produced. Motion-picture operators are constantly exposed to such a hazard, but it is believed to be of little danger because the projection booths are usually adequately ventilated. One method of combating this state is to inject glucose or methylene blue intravenously, which helps to reduce methemoglobin (Fe^{+++}) to hemoglobin (Fe^{++}), the pigment thus becoming again available for oxygen transport. Another procedure is to administer ascorbic acid, which also has a strong reducing action.

The use of silver nitrate solution in the treatment of burn patients can cause toxic methemoglobinemia. The mechanism believed to be involved is the conversion of nitrate anion to nitrite by skin bacteria. Nitrite then converts hemoglobin to methemoglobin.

Much work has been done on the toxic effects of methemoglobin. A concentration of 10% to 20% in blood may cause a mild cyanosis as the only symptom. A 20% to 40% level of methemoglobin results in visible cyanosis and mild fatigue and dyspnea with activity. A 40% to 60% blood level produces severe cyanosis, serious cardiopulmonary symptoms, tachycardia, tachypnea, and depression. Levels of methemoglobin above 60% cause ataxia, loss of consciousness, and death.

Patients with a rare hereditary type of *familial methemoglobinemia* have been reported. The condition is caused by a deficiency of the enzyme *methemoglobin reductase* or probably, more specifically, a deficiency of *cytochrome* b_5 *reductase,* a component of the methemoglobin reductase complex, along with FAD. These individuals are usually cyanotic in appearance and may have as much as 30% to 40% of their hemoglobin as methemoglobin. The inherited disorder has also been reported in infants. Cyanosis and respiratory symptoms may be precipitated, e.g., by drinking water containing nitrates over the allowable limit of 5 ppm. Methemoglobinemia is also found in individuals with hemoglobin M and with certain other types of abnormal hemoglobins.

Other derivatives of hemoglobin Hydrogen sulfide reacts with hemoglobin to give a compound having a characteristic absorption spectrum.

Hydrocyanic acid and cyanides, however, do not react directly with hemoglobin but do react with methemoglobin to form cyanmethemoglobin. The principal toxic action of the cyanides lies in their combination with cytochrome oxidase (Chapter 11). Therefore the treatment of cyanide poisoning is based on the production of methemoglobin, to remove the cyanide from this important enzyme. Sodium nitrite and sodium thiosulfate are injected intravenously. The former induces the production of methemoglobin, which quickly combines with the cyanide. Methemoglobin and cyanmethemoglobin, although not useful respiratory pigments, are in themselves nontoxic. Cyanmethemoglobin is slowly converted to hemoglobin and cyanate, which is also nontoxic. The sodium thiosulfate reacts with cyanide, yielding thiocyanate, an innocuous salt that is readily excreted.

A different type of combination is that of hemoglobin with carbon dioxide to form carbaminohemoglobin. In this case the combination is with the globin rather than with the heme. An amino group is responsible, in part at least:

$$\underset{\textbf{Hemoglobin}}{HbNH_2} + CO_2 \rightleftarrows \underset{\textbf{Carbaminohemoglobin}}{HbNHCOOH}$$

This is a normal physiological reaction and accounts for 2% to 10% of the carbon dioxide transported by the blood. Mechanisms for the transport of carbon dioxide are discussed in detail in the next section of this chapter.

Biochemistry of respiration

Respiration is that important physiological function of blood which involves the exchange of gases between the body and the air. In the lungs oxygen passes from the air into the blood, and carbon dioxide from the blood into the air. These are the two ends of the process, but in between occur many surprising phenomena. They include physical and chemical changes. It is often difficult to set forth the stages in consecutive fashion, because the reactions are intricate and closely interrelated and occur simultaneously.

After the blood receives oxygen from the air, it is carried to the tissues, where the oxygen is used in metabolic processes. The end products include inorganic and organic acids, e.g., carbonic, phosphoric, sulfuric, uric, lactic. The acids are largely neutralized by bases in the blood and tissue fluids, chiefly sodium bicarbonate, and thereby salts and carbon dioxide are formed. Most of the salts and acids are excreted by the kidney, and almost all the carbon dioxide by the lungs.

Flow of respiratory gases

At this point it is pertinent to suggest that the gas laws be reviewed. From these laws we know that a given gas tends to flow from a high partial pressure or tension to a lower one, regardless of whether it is in gaseous form or is dissolved in a liquid. First, the composition of the gases present in inspired and expired air and in the air present in the alveoli will be considered (Table 16-8).

The oxygen tensions (or partial pressures of oxygen) of these mixtures of gases are found by multiplying the total pressure by the percentage of oxygen in each case. Consequently the partial pressure of oxygen in inspired or atmospheric air is 20.9% of the total pressure, i.e., 159 mm Hg if the total pressure is assumed to be 760 mm Hg. Similarly, the partial pressure of the oxygen in expired air is 124 mm Hg, and of alveolar air, 108 mm Hg. Thus we see that the direction of the flow of oxygen is toward the alveoli. Why do the alveoli have less oxygen than inspired air? The reason, obviously, is that oxygen must have been removed in the lungs, and the reason for this is that the venous blood brought to the lungs has a low oxygen tension, only 40 or 50 mm Hg. Blood circulates through the capillaries of the lungs at astonishing speed. The combined thickness of the respiratory epithelium and capillary wall, which separates the blood from the air, is not over 0.004 mm. Every corpuscle is thus brought into intimate contact with the alveolar air, and conditions are excellent for the rapid diffusion of gases. There is the added factor of the affinity of hemoglobin for oxygen. Consequently oxygen flows from the partial pressure of 108 mm in the alveoli toward the 40 to 50 mm in the venous blood, building it up almost instantaneously to about 100 mm, the partial pressure of oxygen in *arterial blood,* the state in which it leaves the lungs.

Arterial blood, loaded with oxygen, at a partial pressure of about 100 mm is carried to the muscles, spleen, heart, and other parts of the body. There, in the capillaries, it is separated from the tissue fluids by thin capillary walls and from the cells by their thin walls. The partial pressures of oxygen of tissue fluids are as low as 20 to 50 mm and those of the cell contents about the same or less. Thus the oxygen by physical forces alone would tend to flow out from the blood into the tissues. However, other factors lead in the same direction, factors that serve to dissociate oxyhemoglobin and to combine with the oxygen. As a result the blood comes out of the tissue capillaries and into the veins depleted of much of its oxygen, with a partial pressure of oxygen of 40 to 50 mm, and goes back to the lungs for more oxygen.

To summarize the flow of oxygen in relation to partial pressures: oxygen in *atmospheric air* (159 mm) flows toward *alveolar air* (108 mm) and diffuses into *venous blood* (50 mm); this is rapidly built up with the aid of hemglobin to *arterial*

Table 16-8
Average composition of dry respiratory air reduced to standard temperature and pressure (0° C, 760 mm Hg)

	Oxygen (percent)	Carbon dioxide (percent)	Nitrogen, argon, etc. (percent)
Inspired air	20.94	0.04	79.02
Expired air	16.3	4.0	79.7
Alveolar air	14.2	5.5	80.3

blood (100 mm), which gives oxygen to the *tissues* (50 mm), and thence back to the *alveoli of the lungs*.

Since carbon dioxide forms carbonic acid with water and reacts chemically with bases, its distribution is not entirely a physical matter. However, the direction of its flow is also from higher to lower pressures, as can be seen from the following figures: in venous blood the partial pressure of CO_2 is 46 mm Hg, and in alveolar air it is 40 mm; therefore the tendency is to pass from the venous blood, into the alveoli, to the expired air (20 mm), to the atmospheric air (0.30 mm); after the venous blood has lost CO_2, the partial pressure is down to 40 mm, at which level arterial blood goes to the tissues; here the CO_2 is high, with estimated partial pressures of 50 to 70 mm; consequently CO_2 flows into arterial blood as it courses through the capillaries, bringing the partial pressure up from 40 to 46 at the same time that the partial pressure of O_2 is going down; the venous blood then passes to the lungs again to unload CO_2.

It may be added that there are about 2.5 to 3 vol% of nitrogen in blood. Nitrogen is present in physical solution in the plasma and is ordinarily inert. It may, however, create a problem in individuals who are subject to steep, rapid changes in atmospheric pressure, e.g., high-altitude flyers, deep-sea divers, or workers in pressurized caissons. In the latter case increased amounts of nitrogen dissolve in the plasma of workers subjected to increased pressures in caissons used in deep excavations and tunnels. If decompression when these workers leave the caisson is too rapid, bubbles of nitrogen are released from the plasma and may accumulate in the joints and adipose and other tissues. This can cause intense pain, loss of function, and even intravascular clotting, a condition known as the "bends," or caisson disease. Decompression must be gradual to prevent the occurrence of this condition. The same situation results if ascent to high altitudes is made too rapidly, unless of course the person is protected by a pressurized space suit with oxygen mask or by a pressurized cabin.

An impairment in the flow of respiratory gases occurs in *emphysema*. Apparently this is due to a loss of resiliency of lung tissue (from factors such as excessive smoking), causing a destruction of *elastin* in the alveoli. In turn this appears to be related to a decrease in lung tissue α-1-*antitrypsin*, which prevents the destruction of elastin. Rare instances of a genetic deficiency of α-1-antitrypsin also have been reported.

Transport of oxygen

If arterial whole blood is analyzed for its content of oxygen, it is found to contain from 18 to 20 vol%, when corrected to 0° C and 760 mm Hg. If the plasma is analyzed apart from whole blood, its oxygen content is about 0.3 vol%, i.e., 100 ml of whole blood carry from 18 to 20 ml of oxygen, whereas if blood contained no corpuscles, it could carry only 0.3 ml. The oxygen capacity of whole blood is thus 60 or more times greater than that of plasma because of the presence of erythrocytes with their hemoglobin. If blood contained no hemoglobin, Barcroft says, we would have to have over 150 kg of plasma in our blood system. This would mean the vascular system alone would have to amount to more than twice the weight of the body, and the organism would be unable to cope with the weight of its own blood. Hemoglobin thus is a truly remarkable substance. Its power of combining with oxygen and of releasing the oxygen is not the only role played by this ferroprotein, but it is the most important. The percentage of saturation with oxygen varies with several factors and is shown in dissociation curves, since this is a reversible reaction:

$$\underset{\substack{\textbf{Deoxyhemoglobin} \\ \textbf{(reduced hemoglobin)}}}{\text{Hb}} + O_2 \rightleftarrows \underset{\textbf{Oxyhemoglobin}}{\text{HbO}_2}$$

Figure 16-18
Dissociation curves of human blood exposed to 0, 3, 20, 40, and 90 mm CO_2. *Ordinate:* Percentage saturation with O_2. *Abscissa:* O_2 pressure. *M,* General curve for myoglobin, for comparison.

Adapted from Barcroft, J.: In Bard, P.: Medical physiology, St. Louis, 1956, The C. V. Mosby Co.

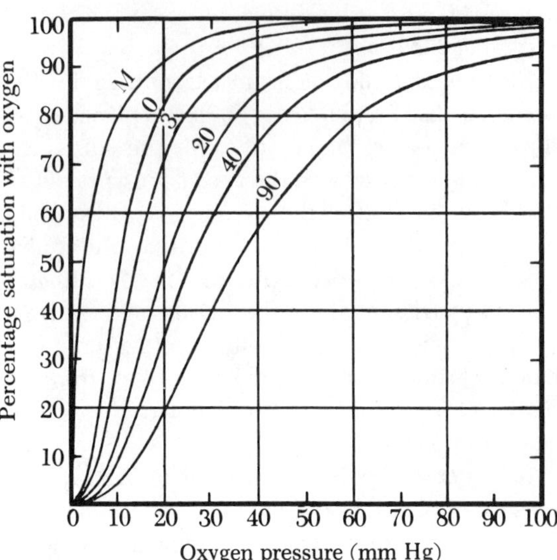

Fig. 16-18 is a series of such curves. The oxygen association-dissociation curves are sigmoid in shape rather than linear. This is believed to be the result of the so-called *heme-heme interaction,* which phenomenon, although not yet fully understood, occurs when oxygen is taken up and the conformations of the four subunits of hemoglobin are altered.

Heme-heme interaction The gross effect of the so-called heme-heme interaction on oxygen transport as manifested by the sigmoidal oxygen-hemoglobin association-dissociation curves (Fig. 16-18) has been known for many years from the classic work of Barcroft and others. There are also attendant changes in color of blood, crystalline forms of oxyhemoglobin and deoxyhemoglobin, and an increase in the acidity of oxyhemoglobin. However, any understanding of the phenomenon at a molecular level was not possible until the elegant work of Perutz and others using x-ray diffraction techniques demonstrated that a change in the molecular conformation of hemoglobin was involved. This was manifested by a movement of the β-chains 0.7 nm closer together in oxyhemoglobin. More recently, sophisticated techniques have demonstrated that there is also a movement of the iron atom into the planar tetrapyrrole moiety of heme by about 0.07 nm on forming a ligand with oxygen. As a result the iron atom changes from its dome-shaped position in the heme ring of deoxyhemoglobin, a "stressed" structure, into the nearly planar, "relaxed" form (Fig. 16-19, *A*) in oxyhemoglobin. This is preceded by a shift in the spin state of the iron atom from *high*-spin to *low*-spin in oxyhemoglobin.

With these basic observations at hand, it became possible for Perutz and others to formulate the following explanation of the heme-heme interaction, perhaps better indicated as the cooperative effects of hemoglobin subunits in the uptake of oxygen.

The basic triggering reaction is the movement of the iron into the planar ring of heme on forming a ligand with oxygen. This is made possible by the prior shift from the high-spin state of the iron atom in deoxyhemoglobin to the low-spin state in oxyhemoglobin. The low-spin iron atom is then small enough to fit into the central cavity of heme (Fig. 16-9). At the same time, the longer ("stretched") bond from the iron atom to the histidine in the α- or β-chain to which it is covalently bound is

Figure 16-19 **A** Scheme of changes in the quaternary structure of hemoglobin as a result of oxygenation and deoxygenation. Note widening of "heme pockets," breaking of salt bridges (*heavy curved lines*), and shifts of subunits associated with liganding with oxygen. As described in the text, the changes in the quaternary structure are apparently triggered by the uptake of oxygen by one of the α-chain subunits, which have slightly wider heme pockets than do the β-chains. Liganding of oxygen by the Fe^{++} atom of heme changes it from the high-spin state to a smaller lower spin state that can fit into the central cavity of heme (Fig. 16-9), thus shortening the covalent bond from the iron atom to an adjacent histidine and "relaxing" the structure as in *right* diagram. Conformational changes follow, with the breaking of salt bridges and widening of heme pockets of the remaining subunits with their oxygenation. This appears to account for the sigmoidal oxygen-uptake curve of hemoglobin (Fig. 16-18). The removal of oxygen from hemoglobin in the tissues reverses these structural changes to the "tense" deoxy form as in *left* diagram.
B Diagram showing relaxation of tension at heme moiety of hemoglobin on going from tertiary deoxy "tense" (*left*) to the "relaxed" oxy structure (*right*). The iron atom (*solid circle*) of the deoxyheme is in the *high*-spin state and lies about 0.7 Å out of the plane of the heme molecule (*left diagram*). Upon oxygenation (*right diagram*), it shifts to the *low*-spin state of oxyheme and nearly into the plane of the heme moiety. Note shortening of bond (*arrow*) from deoxyheme iron atom to histidine (87 in α-chain and 92 in β-chain), upon conversion to the oxy form.
C Concerted changes of tertiary and quaternary structures of hemoglobin on binding of oxygen to ferrous subunits. The "clamp" between the β-subunits in the T (tense) structure (*left*) represents 2,3-diphosphoglycerate (2,3-DPG). The links (*solid lines*) between the α-β and α-α subunits represent salt bridges. 2,3-DPG is expelled and the salt bridges are broken during uptake of oxygen and conversion to oxyhemoglobin. Note relaxed conformation and movement of the β-chains 7 Å closer together upon oxygenation.

A courtesy Dr. M. F. Perutz, Cambridge, England; **B** modified from Perutz, M. F.: Nature **237:**495, 1972; Also Perutz, M. F.: Nature **228:**726, 1970.

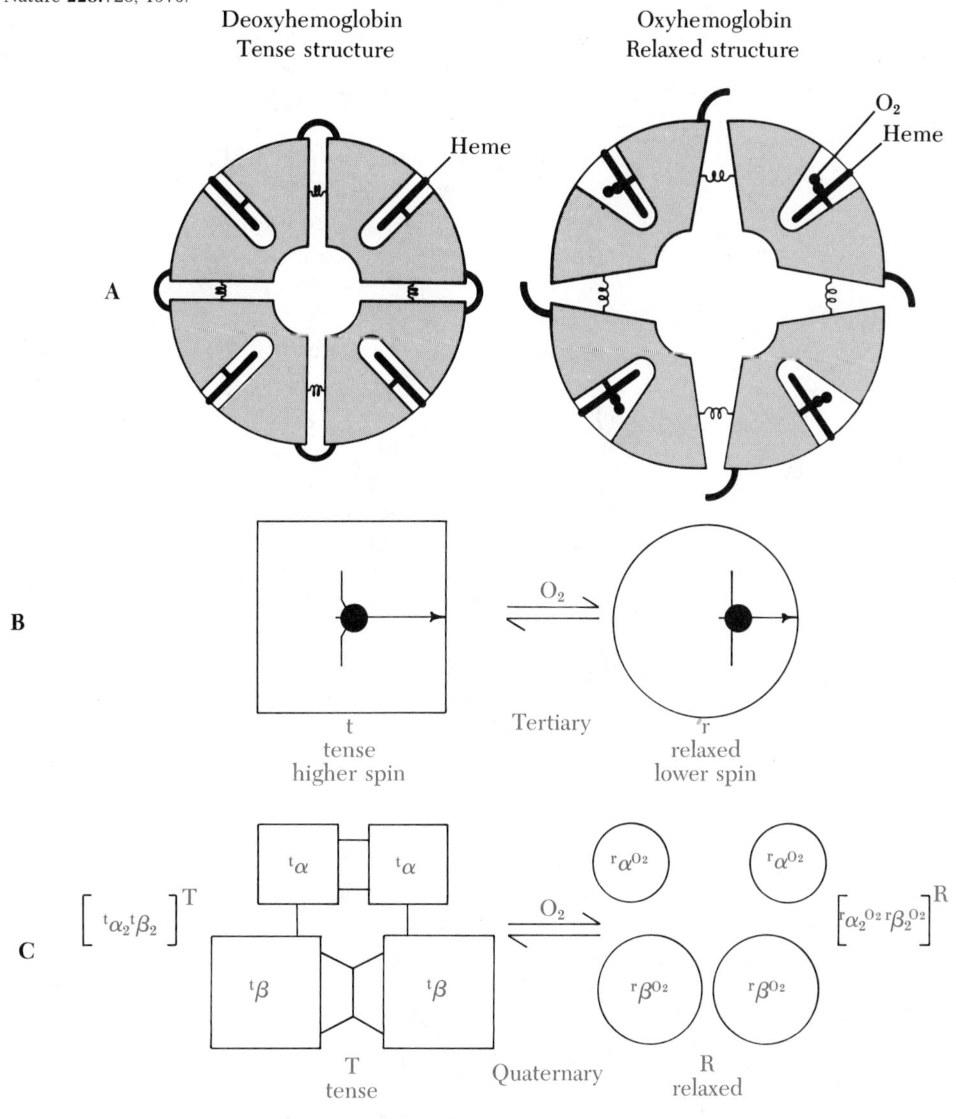

shortened. This shift in the position of the iron atom serves as a "lever," along with the shortening of the Fe-His bond length, to bring about a transition from the tense deoxy structure of hemoglobin to the relaxed form of oxyhemoglobin (Fig. 16-19, *B*). The changes in the quaternary structure are accompanied by the breaking of four salt bridges between the α- and β-chains with the liberation of *Bohr protons;* thus:

$$\text{Val-}\overset{+}{\text{N}}\text{H}_3\text{-}\overset{-}{\text{O}}\text{-}\underset{\underset{\text{O}}{\|}}{\text{C}}\text{-His} \rightarrow \text{Val-NH}_2 + \overset{-}{\text{O}}\text{-}\underset{\underset{\text{O}}{\|}}{\text{C}}\text{-His} + \overset{+}{\text{H}}$$

Simultaneously 2,3-diphosphoglycerate is released by the breaking of the salt bridges attaching it to the β_1,β_2-subunits. Some hydrogen bonds are also opened, apparently. There is a concomitant shift of the β-chains 0.7 nm closer together (Fig. 16-19, *C*), thus making more accessible their heme pockets, and two more molecules of oxygen are taken up. The oxygens liganded first were taken up by the α-chains, whose heme pockets are more readily accessible than are those of the β-chains, when in the deoxy, tense form. Thus the four subunits of hemoglobin are now fully oxygenated (4 moles of O_2 per mole of hemoglobin) with the physicochemical changes observed, including the sigmoid oxygen-uptake curve, the release of Bohr protons, and 2,3-DPG being explained at a molecular level.

The reverse of this process occurs, of course, during the dissociation of oxyhemoglobin in the tissues, by protons being taken up and 2,3-DPG being bound to increase the release of oxygen.

It should be added that the presence of the quaternary structure is essential to the cooperative effects of hemoglobin. Without the change of the quaternary structure of the α- and β-chains just described *no* heme-heme interaction can be detected.

Perutz's work also indicates that the association-dissociation with O_2 in one pair of α- or β-chains has *no significant effect* on the spin state at the other pair of chains unless the reaction is accompanied by a change in quaternary structure, at least in the model studied, of methemoglobin hybrids. In this sense then the term *heme-heme interaction* would appear to be a misnomer. This would account for the fact that the isolated monomers of α- or β-chains or myoglobin, which is a monomer, do not show sigmoidal oxygen association-dissociation curves (Fig. 16-18).

The structural changes in hemoglobin occuring during the respiratory cycle as outlined thus constitute one of the truly most remarkable examples of molecular adaptation to function in all of biochemistry. The heme-heme interaction is of great physiological advantage to the organsim. Much more oxygen is liberated in the tissues as a result than would be the case if the reaction were strictly linear. 2,3-DPG is another favorable factor, as will be discussed later. Of interest, though unexplained, is the fact that hemoglobin H, a hemoglobin containing four β-chains, does not show such an effect or change in structure on oxygenation.

The curves in Fig. 16-18 also show the influence of different carbon dioxide pressures in the dissociation of oxyhemoglobin of human blood. Thus, if curve *0* is followed from right to left, when no carbon dioxide is present, the blood is fully saturated with oxygen at 100 mm oxygen pressure; at 40 mm oxygen pressure it is about 96% saturated; at 20 mm it is 83% saturated; and at 0 mm it contains no oxygen. In other words, when the oxygen tension increases, the above reaction proceeds to the right and more oxyhemoglobin is formed, as in the lungs. When the oxygen tension decreases, as in the tissues, the reaction goes toward the left and more oxygen is liberated. As the carbon dioxide tension, or partial pressure, is increased, the dissociation curves are shifted to the right. This means that, if more carbon dioxide is present, the hemoglobin can hold less oxygen, a phenomenon known as the *Bohr effect,* after its discoverer. A definitive explanation of the Bohr effect is still not available. The decrease in oxygen saturation with an increase in carbon dioxide pressure may result from some change in the conformation of the α- and/or β-chains of the globin moiety of hemoglobin. Such a conformational change may be attributable more directly to an increase in hydrogen ion concentration associated with the increase in carbon dioxide tension. Interesting, but still not satisfactorily explained, is the fact that neither fetal hemoglobin nor myoglobin shows the Bohr effect.

Comparing the same curve (curve *0*) with no carbon dioxide and the one at 40 mm carbon dioxide (curve *40*), we see that at 100 mm oxygen pressure both are practically completely saturated with oxygen, i.e., the hemoglobin is almost all present as oxyhemoglobin. At 90 mm oxygen pressure, which is the pressure in the arteries, they are still nearly the same, with curve *0* being about 99% and curve *40* about 95% saturated. At 40 mm oxygen pressure (oxygen pressure of venous blood) the *0* curve still shows about 95% saturation, whereas curve *40* is down to 72% saturation; i.e., the presence of 40 mm carbon dioxide has caused the oxyhemoglobin to dissociate 23% of its oxygen. Arterial blood has an approximate carbon dioxide tension of 40 mm, and venous blood about 45 mm, as stated. The high venous dioxide pressures of the tissues (50 to 70 mm) would cause the oxyhemoglobin to dissociate still more easily. Other acids have similar effects. Thus the effect of carbon dioxide pressure is just opposite that of oxygen pressure, and both have a desirable physiological effect. In the tissues with low oxygen and high carbon dioxide tensions oxyhemoglobin dissociates more readily, and oxygen is available for tissue needs. In the alveoli the oxygen tension of the air is high and there is no difficulty in forming oxyhemoglobin despite the high carbon dioxide pressure (see tops of all curves). Recall at this point that oxygen is transported by hemoglobin loosely linked to its ferrous ions, probably by sharing unpaired electrons. This combination is easily reversible under the conditions described earlier.

Some other important factors affecting the transport of oxygen by hemoglobin are pH, temperature, and the presence of electrolytes. Slight decreases in pH (more acidic) increase the dissociation of oxyhemoglobin. Thus the slightly more acid pH in tissues by carbon dioxide favors the release of oxygen to the tissues. The slight increase in temperature in the tissues has a similar effect on the dissociation of oxyhemoglobin. The presence of physiological amounts of electrolytes is necessary for the transport of oxygen by hemoglobin.

Certain organic phosphate compounds, mainly diphosphoglyceric acid, have a decided effect on the oxygen-binding power of hemoglobin. The higher the concentration of 2,3-DPG in the erythrocyte, the more readily hemoglobin gives up oxygen. Conversely, when DPG concentrations are low, oxygen is more tightly bound and therefore more slowly released.

This was shown in a classic experiment in which hemoglobin solutions were dialyzed. The oxygen dissociation curve shifted far to the left, meaning that hemoglobin increases its oxygen affinity at a given oxygen tension. Also, it releases its oxygen less readily. The addition of physiological concentrations of DPG to the dialyzed hemoglobin fully restored normal oxygen saturation characteristics. ATP has the same effect, but the amounts required are well above levels normally occurring in erythrocytes. The effects of ATP and DPG were additive, however. Recall that DPG is formed in erythrocytes from glucose and phosphate (Chapter 11). The pathway involved (sometimes called the Rapoport-Luebering cycle) is described in Fig. 16-20.

In further work the Benesches found that DPG is actually bound to hemoglobin,

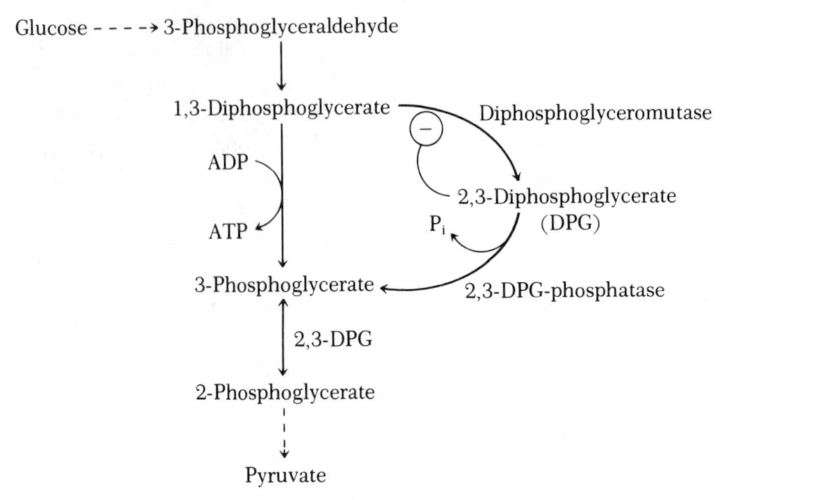

Figure 16-20
Pathway for the formation and degradation of 2,3-diphosphoglycerate, the "Rapoport-Luebering cycle."

but *only* to deoxyhemoglobin and not oxyhemoglobin. The binding is of surprising magnitude, 1 mole of deoxyhemoglobin binding 1 mole of DPG (Fig. 16-13, *C*). Therefore the mechanism of action of DPG appears to be one of shifting the following reaction to the *right*:

$$HbO_2 + DPG \rightleftarrows Hb \cdot DPG + O_2$$

An interesting extension of this work is the fact that fetal erythrocytes have a lower content of DPG than do adult red blood cells and thus a higher oxygen affinity, which may explain how the fetus is able to extract oxygen from the maternal blood supply.

A further application of the *DPG effect* is the inverse relation between the levels of erythrocyte-DPG and hemoglobin in human blood. For example, under identical living and environmental conditions the levels of erythrocyte-DPG in black males are some 10% to 15% higher than in white males, whereas the hemoglobin levels in black males are proportionately lower. Apparently black males are thus able to transport oxygen as efficiently with less hemoglobin.

It is conceivable that some aberrations in oxygen transport may be found to result from lowered erythrocyte-DPG levels caused, in turn, by some impairment in carbohydrate metabolism in erythrocytes and thus in DPG formation.

Studies of the affinities of various hemoproteins for oxygen have shown that important differences exist. The oxygen affinity of cytochrome oxidase is greater than that of myoglobin, which in turn is greater than that of blood hemoglobin. This means, then, that oxygen is removed from the transporting blood hemoglobin first by muscle hemoglobin for storage and second by cellular cytochrome oxidase for biological oxidations in the cell itself, thus ensuring the efficient transfer of oxygen from the lungs to the individual cells of the organism for the ultimate biological function of oxygen, i.e., cellular oxidation for energy.

Transport of carbon dioxide

Carbon dioxide tends to flow from the tissues to the venous blood and from the venous blood in the lungs into the alveoli. But the carriage and elimination of carbon dioxide in the expired air are not entirely a question of pressure. In fact, pressure is one of the least important factors. Of the 50 to 60 volumes of carbon dioxide per 100 ml of blood, only 2 to 3 ml, about 5%, are in solution and exerting a pressure. This is usually written in the hydrated form, H_2CO_3, although over 99% of dissolved carbon dioxide is not in this form. If all of it were in simple solution in an aqueous medium, the pH would be about 4.0, which is far on the acid side, and would mean death to the tissues. Since the pH of the plasma varies only from 7.3 to 7.5 normally (and only a little more abnormally), evidently the major part of the carbon dioxide must be in combined form. Most of it—over 90%—is in the form of bicarbonate, some in the red cells and some in the plasma and tissue fluids. Another fraction, about 3% to 4%, is present as carbamino compounds, formed with proteins, whose free amino groups react with carbon dioxide.

$$\text{Prot-NH}_2 + CO_2 \rightleftarrows \text{Prot-NH} \cdot \text{COOH}$$

By far, the major portion of this fraction is in the red cells because hemoglobin is the most abundant protein in blood. The resulting carbamino compound of hemoglobin is called carbaminohemoglobin.

$$\underset{\textbf{Hemoglobin}}{\text{HbHN}_2} + CO_2 \rightleftarrows \underset{\textbf{Carbaminohemoglobin}}{\text{HbNH} \cdot \text{COOH}}$$

The direction of this reaction is determined almost entirely by the proportion of oxyhemoglobin present in blood, not by the level of the carbon dioxide tension. Oxyhemoglobin is more acidic, due to the release of Bohr protons, than reduced hemoglobin. When more oxyhemoglobin is present, the reaction goes to the left, i.e., more carbon dioxide is released. On the venous side, when hemoglobin is in the less oxygenated, less acid state, more carbon dioxide is combined. Thus the blood can carry more carbon dioxide as carbaminohemoglobin on the venous side. At the instant of oxygenation in the lungs the more acid oxyhemoglobin forces the carbaminohemoglobin to unload some of its carbon dioxide into the alveoli. This, however, is only a small part of the carbon dioxide story.

Chloride shift An important mechanism in the transport of carbon dioxide is the chloride shift. A few factors that play important roles must be mentioned first. Although we know that the reaction $H_2O + CO_2 \rightleftarrows H_2CO_3$ takes place readily in both directions, the rapidity with which this occurs in the body in certain sites has led biochemists to wonder whether it might be catalyzed by some enzyme. This was found to be so. An enzyme that catalyzes the above reaction is present in high concentration in red cells. The enzyme was given the name *carbonic anhydrase*. Thus carbonic acid can be formed with extreme speed; the reaction is reversible equally rapidly by the enzyme under appropriate conditions. Another factor is the permeability of the erythrocyte, which is impermeable to hemoglobin and the plasma proteins but is permeable to water, carbon dioxide, HCO_3^-, Cl^-, OH^-, Na^+, K^+, and H^+. Most of the sodium ions are in the plasma, however, and most of the potassium ions are in the cells. In the erythrocytes a great deal of the hemoglobin is combined with potassium, the amount of which fluctuates in different parts of the respiratory cycle.

With these facts in mind, let us follow the courses of oxygen and carbon dioxide into and out of the erythrocytes and through the various parts of the respiratory cycle.

1. *In the lungs*
   a. Oxygen enters the erythrocyte because of the higher pressure of oxygen in the lungs. Deoxyhemoglobin becomes oxyhemoglobin as shown:

$$\text{HHb} + O_2 \rightleftarrows \text{HHbO}_2; \qquad \text{HHbO}_2 + \text{KHCO}_3 \rightleftarrows \text{KHbO}_2 + H_2CO_3$$

Because oxyhemoglobin *($HHbO_2$)* is a stronger acid than deoxyhemoglobin *(HHb)*, the equilibrium point is shifted toward the right, converting bicarbonate *(HCO_3^-)* to carbonic acid *(H_2CO_3)*. Thus an increased proportion of potassium ions becomes paired with oxyhemoglobin. The increase in acidity of hemoglobin on oxygenation (or the reverse on deoxygenation), without change in blood pH, is called the *isohydric change*.

b. The decrease in bicarbonate (HCO_3^-) concentration in the erythrocyte leads to diffusion of bicarbonate from the plasma, where its concentration is higher, into the erythrocyte.

c. To preserve electroneutrality, i.e., the equality in the number of positive and negative charges, some negative ion must leave the erythrocyte for each bicarbonate ion entering it. Since the cell is readily permeable to chloride ions, which are present in sufficient amount, chloride ions diffuse out of the erythrocyte. This change is called the *chloride shift*. The total cation content (i.e., K^+, Na^+) of the erythrocyte remains essentially constant.

d. The carbonic acid formed is converted quickly in the presence of the carbonic anhydrase (CA) of the erythrocyte to carbon dioxide and water thus:

$$H_2CO_3 \overset{CA}{\rightleftharpoons} H_2O + CO_2$$

e. The low carbon dioxide pressure in the lungs, compared with that of the blood arriving at the lungs, favors the escape of carbon dioxide from the erythrocyte and plasma into the lungs, thereby shifting the above reaction and consequently the first reaction to the right. Because there are fewer osmotically active particles in the erythrocyte after CO_2 escapes, some H_2O leaves the erythrocyte.

2. *In the tissues*

a. Because of the low oxygen pressure of the tissues, as compared with that of the lungs, the oxyhemoglobin of the erythrocyte gives up oxygen to the tissue fluids and becomes deoxyhemoglobin, as shown:

$$KHbO_2 + H_2CO_3 \rightleftharpoons HHbO_2 + KHCO_3$$

$$HHbO_2 \rightleftharpoons HHb + O_2$$

Deoxyhemoglobin (HHb) is a weaker acid than oxyhemoglobin; consequently the equilibrium is shifted toward the right, converting carbonic acid to bicarbonate. As a result an increased proportion of potassium ions now become paired with bicarbonate ($K^+HCO_3^-$).

b. Now the increase in bicarbonate concentration in the erythrocytes leads to diffusion of these ions from the erythrocytes into the plasma.

c. Again a shift of the chloride ions in exchange for the bicarbonate ions occurs, but this time a chloride ion must enter the cell for each bicarbonate that leaves it. The total cation content of the red blood cells continues to remain essentially unchanged.

d. Carbon dioxide diffuses from the tissues, where it is being formed in oxidative processes, into the plasma and then into the erythrocyte. Here in the presence of CA some carbonic acid is formed, as the increase in carbon dioxide shifts the reaction to the left (although the equilibrium constant favors the existence of much more CO_2 than H_2CO_3):

$$H_2CO_3 \overset{CA}{\rightleftharpoons} H_2O + CO_2$$

e. The shifting of the reaction in *d* toward the left results in a shifting of the other reaction (in *a*) toward the right. Because there are more osmotically active particles in the erythrocyte after CO_2 enters it, some water now enters the red cell.

Each phase of this cycle takes place with great rapidity. Therefore a catalyst such as carbonic anhydrase must be present and conditons must be optimal for rapid diffusion of gases. The extremely thin membranes, the small bore of the capillaries (allowing in parts of the system for only a single erythrocyte to pass through at a time), and the temperature all contribute to this end.

The foregoing respiratory cycle is summarized diagrammatically in Fig. 16-21. Carbonic acid (or CO_2 gas) and chloride ions always move in a direction opposite that of bicarbonate in the *Hamburger shift*. The differential distribution of electrolytes between the red blood cells and the plasma, as described here, is also partially explainable by the Gibbs-Donnan equilibrium, but only insofar as the erythrocyte membrane is semipermeable.

The increased acidity of hemoglobin on oxygenation acts as though acid had

Figure 16-21
Diagram of O_2 and CO_2 transport and the chloride shift. *C-A* represents the introerythrocyte enzyme *carbonic anhydrase*.

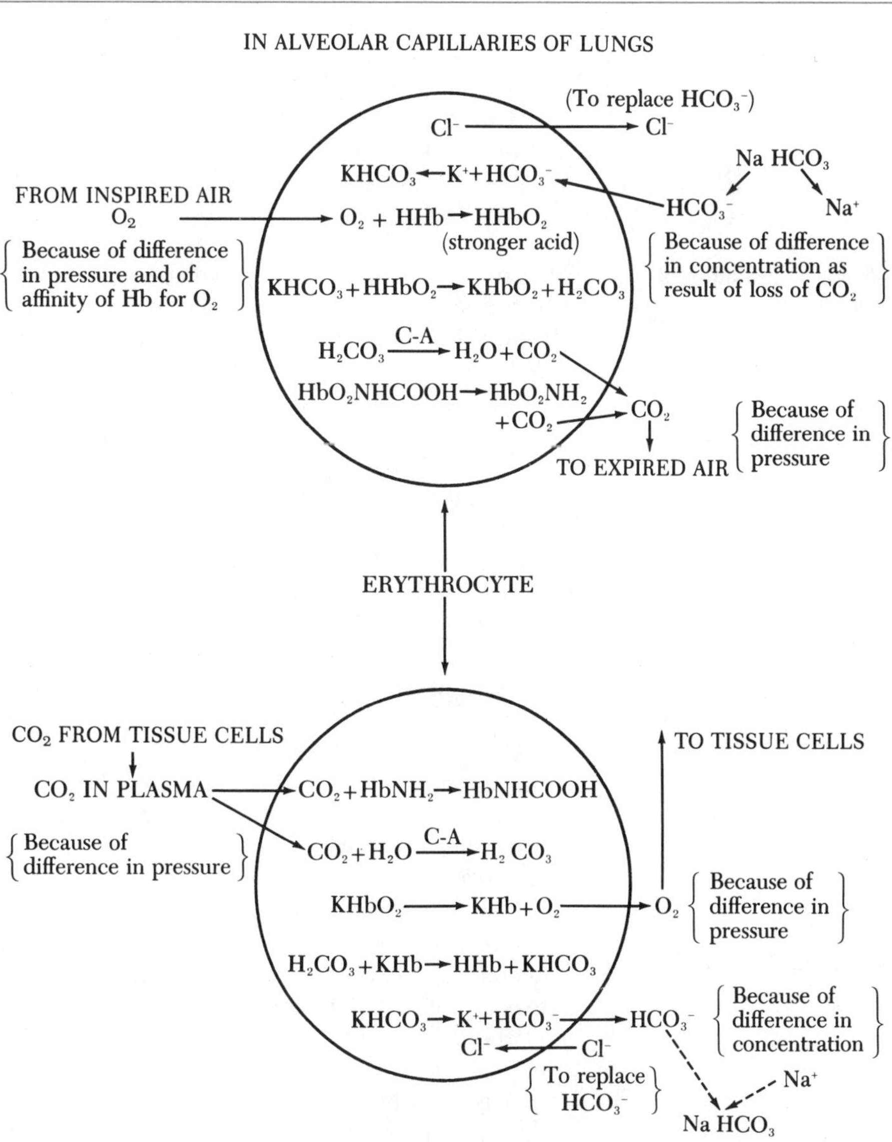

been added to the red cell, liberating carbon dioxide. Conversely, in the tissue capillaries deoxygenation decreases the acidity of hemoglobin, which therefore accepts hydrogen ions from the carbonic acid entering the red cell. This allows most of the carbon dioxide (or H_2CO_3) from the tissues to be carried in the blood as bicarbonate (K^+ or $Na^+HCO_3^-$). The chloride-bicarbonate shift subsequently permits about 60% of the carbon dioxide from the tissues to be carried to the lungs as bicarbonate *in the plasma*. Venous blood has 4% of 10% more total carbon dioxide than does arterial blood, i.e., 2 to 5 more volumes percent, or 1 to 2 mmole more per liter. About three fourths of this extra carbon dioxide in venous blood is carried as K^+ or $Na^+HCO_3^-$. Twice as much K^+ or Na^+ for HCO_3^- formation arises from the isohydric change of hemoglobin and plasma proteins on the invading carbon dioxide (H_2CO_3). The former does not change the pH of the plasma, whereas the latter lowers it slightly. The remaining one fourth of the extra carbon dioxide is carried as carbaminohemoglobin and as physically dissolved carbon dioxide gas. The pH of venous plasma is 0.02 to 0.04 unit lower than that of arterial plasma, but venous plasma has about 1 or 2 mmole more $Na^+HCO_3^-$ per liter than does arterial plasma.

Hemoglobin is the most important buffer against any pH change that would result from carbon dioxide entering the blood, mainly because the isohydric change of oxyhemoglobin to hemoglobin results in the conversion of most of the invading carbon dioxide to bicarbonate.

Chemical regulation of respiration

The control of respiration is considered in detail in textbooks of physiology, and only a few words will be devoted to it here. There is some degree of voluntary control, but the regulation is chiefly involuntary and depends on afferent impulses to the respiratory center, which thereupon sends its impulses to the various muscles involved. Chemical influences have much to do to with some of these mechanisms. Excess carbon dioxide (H_2CO_3) in the blood stimulates the center directly, and, since fixed acids have the same effect, presumably the response is from a slight increase in hydrogen ion concentration (or decrease in OH^- and HCO_3^-) at the site of the center. Increased carbon dioxide of the inspired air raises the rate and depth of respiration, and the result is a remarkable constancy in the percentage of carbon dioxide in the alveolar air (about 5.5%). Air or oxygen given for resuscitation should contain about 5% carbon dioxide to stimulate the respiratory center. Lack of oxygen has little effect unless it is very great. In that case respiration is increased because of the responses of the carotid and aortic bodies, acting as emergency mechanisms.

Under basal conditions, i.e., when the subject is at mental and physical rest, the concentration of the carbon dioxide in the expired air (collected over a period of from 5 to 25 minutes) is constant for normal subjects. The range in a large number of determinations was found to be 18 to 22.5 mm, and an average partial pressure of about 20 mm is an acceptable clinical standard for both sexes and for all ages.

Low concentrations of carbon dioxide in the expired air, indicating respiratory stimulation, would be expected in subjects with circulatory failure, acidosis, severe anemia, and certain forms of pulmonary disease. High concentrations would be expected in alkalosis and in depression of the respiratory center, as occurs after morphine or barbiturate administration. In cardiac patients a definite relationship between degree of cardiac failure and carbon dioxide concentration in the expired air has been found. Patients with slight limitation of physical activity had an average carbon dioxide of 18.55 mm, those with pronounced limitation of physical activity had an average of 15.81 mm, and those who were unable to carry on any physical activity whatever without discomfort averaged 13.44 mm.

Acid-base balance

A major function of blood is the maintenance of the acid-base balance. Although the kidneys and lungs also play vital roles, the blood—specifically the blood buffers—forms the first line of defense against alterations in the pH of body fluids and tissues and in turn their constituent cells.

The oxidation of various metabolites in the living organism results in the formation of a variety of acids and bases, as has been described in previous chapters. Protons, for example, may originate from the ionization of various organic acids that are intermediates in the metabolism of glucose, fatty acids, and amino acids. Uric acid from the catabolism of purines (Chapter 14) is another source of protons (H^+ ions), as is sulfate from the oxidation of cystine-cysteine and methionine sulfur. In general, these metabolic protons are oxidized to water in the process of biological oxidations (Chapter 11) and thus disposed of. Likewise, carbon dioxide, although excreted primarily by the lungs as just described, may be hydrated to form carbonic acid and retained to some extent in the tissues and body fluids. This ionizes to a limited degree and becomes still another source of protons.

Bases, too, can arise during the catabolism of certain metabolites. The formation of ammonia from the deamination of amino acids (Chapter 13) is an important example. The formation of anions such as bicarbonate, biphosphate, and acetate are other examples. The alkalinizing effect of certain basic salts, e.g., citrates of citrus fruits, is still another example.

Thus during normal metabolic processes both acids and bases are formed; usually, however, acidic substances predominate. These substances may enter the blood plasma and other extracellular fluids for metabolic, excretory, or other forms of disposal. In a variety of pathological conditions excessive amounts of metabolic or exogenous acids or bases may accumulate in the cells and tissues of the body, leading to disturbances of acid-base balance. The remaining part of this section will consider mechanisms for the biochemical regulation of acid-base equilibrium in the body.

The blood remains at the remarkably constant slightly basic level of pH 7.3 to 7.5—usually 7.35 to 7.45—during health.* To accomplish this, the body has four lines of defense: (1) the buffer systems of the blood, tissue fluids, and cells, as well as mineral salts of bones; (2) the excretion or retention of carbon dioxide by the lungs; (3) the excretion of an acid or alkaline urine; and (4) the formation and excretion of ammonia or organic acids. Thus the body's internal environment is maintained at a rather constant hydrogen ion concentration.

At this point, a few words are in order concerning the meaning and use of the term *base* as it is ordinarly applied to the subject of acid-base balance. According to current concepts (e.g., the Brønsted) (see also Chapter 24) a base is any substance that combines with protons (H^+ ions). An acid is any substance that yields protons. Accordingly, bases would usually be anions such as OH^-, HCO_3^-, $HPO_4^=$, and CH_3COO^-. Also, ammonia and the amino group would be classed as bases because they can accept protons. Traditionally in the medical and allied sciences, however, the term *base* has been applied to sodium and potassium ions and sometimes even to calcium ions and other cations. Actually, of course, they are not bases in the sense of being proton acceptors. Their salts with weak acids are basic because on hydrolysis the weak acid and the conjugate base are formed, giving a net excess of free hydroxyl ions. Such forms of sodium and potassium ions are probably more accurately termed *alkalies*. In this discussion sodium or potassium may be called bases or alkalies in the sense that they serve as cation carriers or conjugates of

*These pH values are determined at body temperatures 37° to 38° C, at which the neutrality point is 6.8, since $K_w = 10^{-13.6}$ at this temperature.

strongly basic ions such as OH^-, HCO_3^-, $HPO_4^=$, and CH_3COO^-. (All salts, strong acids, and strong bases are virtually completely ionized; thus $KHbO_2$ is $K^+HbO_2^-$, HCl is H^+Cl^-, and NaOH is Na^+OH^-.)

Buffer systems of the blood

In Chapter 24 it is shown that the hydrogen ion concentration of a solution of a weak acid, *HA*, and its salt, B^+A^-, is

$$[H^+] = K \frac{[HA]}{[B^+A^-]}$$

where *K* is the dissociation constant and $[H^+]$, $[HA]$, and $[B^+A^-]$ or $[A^-]$ the concentrations of hydrogen ions, of the acid, and of the salt, respectively. The hydrogen ion concentration of such a buffer pair remains constant if the ratio of the numerator to the demoninator remains constant. Slight additions of acid or base or buffers (or subtractions of either) have very little effect for reasons outlined previously, but large changes, of course, make a decided difference. These relationships should be kept in mind.

In logarithmic form the relationship (Henderson-Hasselbalch equation) is as follows:

$$pH = pK' + \log \frac{[B^+A^-]}{[HA]}$$

For carbonic acid, the pK' is 6.1; for $B^+H_2PO_4^-$, it is 6.8. Therefore, to keep the pH at 7.4, the ratios of these acids to their salts must be kept constant. Substituting in the Henderson-Hasselbalch equation above, we continue as follows:

$$\text{For } H_2CO_3 \; 7.4 = 6.1 + \log \frac{BHCO_3}{H_2CO_3} \text{ or } 1.3 = \log \frac{BHCO_3}{H_2CO_3}$$

and since antilog of 1.3 = 20

$$\frac{B^+HCO_3^-}{H_2CO_3} = \frac{20}{1} \text{ (at pH 7.4)}$$

The ratio 20:1 will be referred to later in this chapter.

$$\text{For } B^+H_2PO_4^-, \frac{B_2^{++}HPO_4^=}{B^+H_2PO_4^-} = \frac{4}{1} \text{ (at pH 7.4)}$$

The principal buffers of the blood are

$$\frac{Na^+HCO_3^-}{H_2CO_3} \quad \frac{Na^{++}HPO_4^=}{Na^+H_2PO_4^-} \quad \frac{Na^+ \text{ Protein}}{H \text{ Protein}} \quad \text{(plasma)}$$

$$\frac{K^+HCO_3^-}{H_2CO_3} \quad \frac{K^+HB^-}{HHb} \quad \frac{K^+HbO_2^-}{HHbO_2} \quad \frac{K^{++}HPO_4^=}{K^+H_2PO_4^-} \quad \text{(red cells)}$$

The buffer pairs in the first line are chiefly in plasma and extracellular fluids, and those of the second are chiefly in the red cells. In the blood, of course, these buffers are all in equilibrium with each other. Therefore the estimation of any one buffer pair would be an index of acid-base equilibrium. Of all the pairs enumerated the $\frac{B^+HCO_3^-}{H_2CO_3}$ is the most important, insofar as action against fixed, i.e., nonvolatile, acids is concerned. The phosphate pair, although more efficient as a buffer, is actually less effective because of its low concentration in plasma. Plasma proteins play a greater buffering role than do phosphates but much less than hemoglobin. The bicarbonates neutralize more than 50% of all acids stronger than carbonic. Finally, in such neutralization carbon dioxide is again formed and is readily eliminated as a gas by the lungs. An increase in $[H^+]$ or $[H_2CO_3]$ stimulates the respiratory center to increase the rate and depth of respiratory ventilation. Similarly, an increase of $[OH^-]$ or $[CO_3^=]$ depresses respiratory ventilation. The lungs thus play a leading role in the minute-to-minute regulation of the pH of the blood and extracellular fluids.

Examples of acids that can alter the acid-base balance include sulfuric, phosphoric, uric, lactic, acetoacetic, and β-hydroxybutyric. Their formation has been discussed in previous chapters.

As acid enters the blood, one of the buffer reactions that occur is as follows:

$$2\ Na^+HCO_3^- + H_2^{++}SO_4^= \rightarrow 2\ H_2CO_3 + Na_2^{++}SO_4^=$$

Here a strongly dissociated acid is transformed into the weakly dissociated acid (H_2CO_3); thus the hydrogen ion concentration is changed but little, i.e., only as much as carbonic acid is dissociated. This slight decrease in the ratio $\frac{BHCO_3}{H_2CO_3}$, caused by an increase in the denominator, can be rapidly brought down to normal because of the easy disposal of carbonic acid by the lungs. Thus any nonvolatile acid stronger than carbonic can be buffered by $BHCO_3$ as long as any bicarbonate is present. Consequently the plasma bicarbonate is a measure of the base remaining after all acids stronger than carbonic have been neutralized. It represents the reserve of alkali available for the neutralization of such strong acids. Thus it has been termed the *alkali reserve.* However, hemoglobinate plays an important role in buffering fixed acids, although not as great a one as bicarbonate. It is not, however, directly measured when the alkali reserve (CO_2 combining power) is determined by the procedure described later, but a decrease in plasma alkali reserve generally parallels a depletion of the reserve of buffering power represented by hemoglobinate.

Again it must be pointed out that the other buffers are in equilibrium with the bicarbonate pair and will react with acids (or bases), but to a lesser extent, because of their lower concentrations. For example:

$$\begin{array}{llll} \text{Na protein} & + HCl & \rightleftharpoons \text{H protein} & + NaCl \\ Na_2HPO_4 & + HCl & \rightleftharpoons NaH_2PO_4 & + NaCl \\ 2\ KHbO_2 & + H_2SO_4 & \rightleftharpoons 2\ HHbO_2 & + K_2SO_4 \\ Na_2CO_3 & + H_2CO_3 & \rightleftharpoons 2\ NaHCO_3 & \\ HHb & + KOH & \rightleftharpoons KHb & + H_2O \end{array}$$

In every case the strong acid or base is transformed to a weak one, and consequently the pH of the blood fluctuates very little. However, the acid formed in largest amounts in the body is carbonic acid or its anhydride carbon dioxide, and this cannot be buffered by bicarbonates. It can be buffered by serum proteins and by phosphates:

$$\begin{array}{l} H_2CO_3 + \text{Na protein} \rightleftharpoons NaHCO_3 + \text{H protein} \\ H_2CO_3 + Na_2HPO_4 \rightleftharpoons NaHCO_3 + NaH_2PO_4 \end{array}$$

Table 16-9
Buffers of body fluids

Estimated percent buffering of invading fixed acid or alkali	Chief buffers present	Location of buffer depots
40	Partly $BHCO_3$ Partly unknown	Tissue cells
30	$BHCO_3$	Extracellular fluid except blood
13	BHb and $BHbO_2$	Blood
17	$BHCO_3$	

Both these factors are of minor consequence. The most important buffer for carbonic acid is hemoglobin. Table 16-9 gives estimates of supplies of buffers in the various compartments of the body.

Role of respiratory system

The respiratory system serves as a mechanism for rapid adjustments of acid-base balance by regulating the concentration of *carbonic acid* in the blood and other body fluids. It thus functions by maintaining one component of the major buffer pair, $HCO_3^-:H_2CO_3$, involved in acid-base equilibrium. The major role of the kidney is in maintaining the proper concentration of the bicarbonate component, to be described later. The respiratory system controls H_2CO_3 concentration in the plasma by the removal of CO_2 in expired air:

$$H_2CO_3 \underset{}{\overset{\text{Carbonic anhydrase}}{\rightleftarrows}} \frac{\text{Expired air}}{CO_2\uparrow + H_2O}$$

Hydrogen ions are also removed in the form of water at the same time.

The respiratory system serves as a rapidly available "emergency" mechanism for maintaining acid-base balance, much as does the blood buffer system. The longer term, complete restoration of acid-base equilibrium, however, is a major function of the *kidney*, as will be described. This relationship is reminiscent of the control of metabolism itself (Chapter 5), an allosteric control serving as a rapid, short-term regulator, whereas longer term, slower adjustments are the responsibility of genetic-biosynthetic mechanisms.

In an acidotic state the respiratory center, located in the medulla of the brain, is stimulated and the respiratory rate is increased. Increased amounts of CO_2 are "blown off" in the expired air, with the residual H^+ being left as H_2O. In an alkalotic state the reverse occurs. Respiration slows and more CO_2 is retained and hence H_2CO_3 is increased. Thus a more normal ratio of $HCO_3^-:H_2CO_3$ is maintained, and the alkalosis (increase in HCO_3^- concentration) is compensated for, at least on a temporary basis until the renal mechanism can effect a permanent, long-term restoration.

Role of the kidney

The kidney contributes to the maintenance of the alkali reserve and to a constant level of blood pH by reabsorbing, secreting, and excreting acidic or basic substances, as the case may be. Moreover, although the lung can help excrete acid, it cannot restore the alkali reserve (HCO_3^-)—something the kidney can do. Although phosphates are present in only small concentrations in the blood, they are concentrated by the kidney and are the principal buffers in urine as excreted. In acid urines there is a relative excess of $H_2PO_4^-$, and in alkaline urines, $HPO_4^=$. There is also a considerable amount of HCO_3^- in alkaline urines—notably potassium bicarbonate from the metabolism of fruits and vegetables. Organic acids, carbonic acid, and salts of organic bases contribute to the urinary pH.

In quite a different way the kidney has another effect on acid-base balance. It is the major site of the formation of ammonia, which is secreted probably as ammonium bicarbonate by the kidney tubules because of the omnipresence of carbon dioxide (H_2CO_3) in the carbonic anhydrase–rich kidney. This results in the conservation and restoration of $B^+HCO_3^-$ or alkali reserve in the following manner: if a strong acid, H^+A^-, has been absorbed into the blood, resulting in the replacement of some of the $B^+HCO_3^-$ by B^+A^-, then the following occurs in the kidney to restore $B^+HCO_3^-$ of the blood:

$$B^+A^- + NH_4^+HCO_3^- \rightleftarrows B^+HCO_3^- + NH_4^+A^-$$

B^+A^-	$NH_4^+HCO_3^-$	$B^+HCO_3^-$	$NH_4^+A^-$
Filtered through glomeruli	Secreted in tubule	Reabsorbed in tubule	Excreted in urine

In acidosis the urinary ammonia rises considerably as a result of increased formation in the kidney; all the ammonium ion produced and excreted in this way takes the place of an equivalent amount of the cations Na^+, K^+, Ca^{++}, or Mg^{++}, which are reabsorbed, paired with the HCO_3^- anion; the undue loss of these cations in the urine is prevented while the level of the bicarbonate anion in the body fluids is restored; the fixed acid anion (A^+), which had disturbed the constancy of the internal environement, is excreted in the urine paired with the NH_4^+ ion.

The mechanism of ammonia formation by the kidney is discussed in Chapter 13. The method whereby an acid urine is formed from a slightly alkaline blood plasma now deserves consideration. Urinary acidification is necessary to provide for the excretion of fixed acids and acid salts and to restore alkali reserve. There are several current theories to account for the phenomenon of a glomerular filtrate of pH 7.4 being converted to a urine having a pH as low as 4.8. One of these concepts is illustrated in Fig. 16-22.

Figure 16-22 Schematic representation of principal reactions involved in renal regulation of acid-base balance. The four major reactions are (1) exchange of H^+ for Na^+ of tubular urine (plasma ultrafiltrate), (2) reabsorption of bicarbonate from tubular urine (as CO_2), (3) formation and excretion of ammonia, and (4) excretion of H^+ as $H_2PO_4^-$ and NH_4^+ in urine.

Modified from Pitts, R.F.: N. Engl. J. Med. **284:**32, 1971.

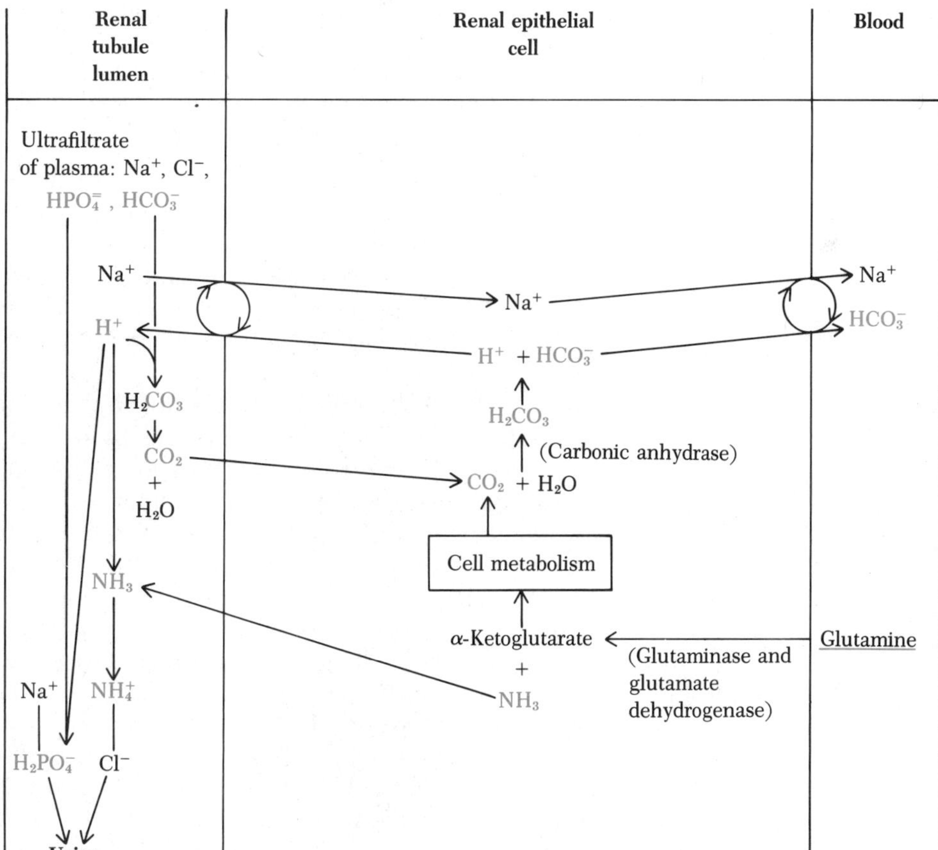

Four key reactions are involved in the renal regulation of acid-base balance and may be summarized as follows:

1. *Sodium ions* present in the tubular urine (ultrafiltrate of plasma) are *exchanged* for *hydrogen ions* formed in the tubular epithelial cells from CO_2 + H_2O. The transport of Na^+ across the tubular-epithelial membrane is apparently mediated by a "sodium pump" mechanism (Chapter 19), indicated by the small circle and arrows. The sodium ion is transported from the epithelial cell into the blood plasma for recycling. The bicarbonate anion formed by the dissociation of H_2CO_3 in the epithelial cell diffuses into the blood plasma as the accompanying ion to Na^+. Base (HCO_3^-) is thus conserved for the blood "alkali reserve."
2. *Bicarbonate* is reabsorbed from the tubular urine, apparently as CO_2. Some hydrogen ions that are exchanged into the tubular urine appear to react with HCO_3^- in the plasma ultrafiltrate to form H_2CO_3, which is dehydrated to $CO_2 + H_2O$. The CO_2 diffuses back into the epithelial cell to add to the pool of CO_2 and thus of HCO_3^-, as well as that of H^+. The epithelial cell has a rich supply of carbonic anhydrase to catalyze the hydration of abundant amounts of CO_2.
3. *Ammonia is formed and excreted.* Ammonia is formed in the tubular epithelial cell from glutamine extracted from the renal blood plasma of the peritubular capillaries. It is deamidated by glutaminase I and deaminated by glutamate dehydrogenase to form 2 moles of ammonia per mole of glutamine. The reactions involved are discussed in detail in Chapter 13. This ammonia diffuses into the tubular urine and binds hydrogen ions to form ammonium ions.
4. *Hydrogen ions are excreted.* The hydrogen ions exchanged for sodium ions in *1* are, for the most part, bound to $HPO_4^=$ to form $H_2PO_4^-$ and to NH_3 to form NH_4^+ and are excreted as such in the urine. The accompanying ions are predominately Na^+ and Cl^-, respectively.

Thus the kidney participates in the regulation of acid-base balance primarily by maintaining the *bicarbonate ion* concentration in the plasma at a level normally varying only from 25 to 28 mEq per liter.

The renal system is effective in restoring acid-base equilibrium from either acidotic or alkalotic states. In the event of an acidosis the excretion in the urine of both $H_2PO_4^-$ (titratable acid) and NH_4^+ salts may be greatly increased until acid-base balance is restored. In an alkalosis the reverse occurs. In a strongly alkaline urine (e.g., pH of 8.0) ammonium salts are virtually zero and nearly all of phosphate is excreted as $HPO_4^=$. A beautiful linearity has been demonstrated between the excretion of NH_4^+ and an increase in the acidity of urine in a pH range of 5 to 8.

Although approximately 90% of the ammonia formed in the kidney is derived from glutamine, small amounts can be formed from the amino-nitrogen of other amino acids, notably alanine and glycine. Also, the kidney clears small amounts of NH_4^+ brought to it by the plasma from other tissues.

Another mechanism, at least partly renal in origin for compensating particularly against an alkalotic state, is the increased urinary excretion of *citric acid*. The administration of alkaline salts (e.g., $NaHCO_3$) to human subjects or to animals is followed by a strong increase in the urinary excretion of citrate. Similar increases have been found during *respiratory alkalosis* as a result of hyperventilation and during the *alkaline tide* after meals. No increase in any other acid of the citric acid cycle was found, nor were there any increases in lactate or pyruvate in the urine under these conditions. Conversely, decreases in urinary citrate occur in acidotic states, as after the administration of acid salts and in severe diabetic acidosis. From these observations it appears that citric acid may serve as a *physiological acid* in the same sense that ammonia serves as a *physiological base.*

Acidosis and alkalosis

An acidosis may be defined as "an abnormal condition caused by the accumulation in the body of excess acid or by the loss from the body of alkali." Similarly, alkalosis is "an abnormal condition caused by the accumulation in the body of excess alkali or by the loss of acid." Ordinary amounts of acid or alkali are taken care of by the mechanisms just considered, i.e., ordinarily the ratio $\frac{B^+HCO_3}{H_2CO_3}$ of the equation $pH = pK + \log \frac{[B^+HCO_3^-]}{[H_2CO_3]}$ is constant at about 20/1, and since this is in equilibrium with all the other sets of buffers, the pH remains at 7.3 to 7.5. In acidosis or alkalosis the ratio may almost be kept constant. If the acidosis is caused by an increase of the denominator, a concomitant and sufficient increase in the numerator holds the ratio almost constant. If it is from a loss of alkali, i.e., diminution of the numerator, a simultaneous decrease in the denominator has the same effect. Similarly, an alkalosis arising from an increased amount of alkali may be compensated for by an increased retention of carbonic acid, and a decrease in the denominator may be followed by a decrease in the numerator. In all these four conditions the pH scarcely changes. Such courses of events result in *compensated acidosis* or *alkalosis*. When the ratio actually changes and the pH is outside of the normal range, the term *uncompensated* is used. However, long before any abnormally great deviation in the pH occurs, the bicarbonate content changes. This is easily detected by determining the carbon dioxide–combining power of blood plasma.

The normal concentration of bicarbonate in plasma is about 0.025 mole or 25 mmole per liter. Plasma carbonic acid (mainly CO_2 gas) is about 1.2 mmole. Plasma alkaline carbonate is 0.1 mmole. The total carbon dioxide (mainly bicarbonate) is thus about 26 mmole. When measured in the laboratory by the addition of acid to plasma, the amount of carbon dioxide liberated from the plasma represents mainly bicarbonate or the alkali reserve and is reported in volumes percent—the number of milliliters of carbon dioxide that would be liberated from 100 ml of plasma. From the gas laws we know that 22.4 ml of a perfect gas represents 1 mmol of the gas. If 1 ml of plasma gave 0.224 ml of carbon dioxide on acidification, it would be reported as 22.4 vol% and would mean that there are 10 mmole of total carbon dioxide in 1 L of plasma. To convert volumes percent to millimoles per liter, divide by 2.24.

Carbon dioxide–combining power of blood plasma

This value is determined by exposing plasma to an atmosphere whose carbon dioxide content is that of alveolar air. The plasma is then acidified, and the carbon dioxide liberated is measured. Normal blood plasma combines with 50 to 70 ml of CO_2 per 100 ml. Lower values are found in an acidosis, higher in an alkalosis (over 70 ml of CO_2 per 100 ml). Values of 40 to 50 are indicative of a mild acidosis, 30 or 40 to 50 are indicative of a mild acidosis, 30 or less of a severe acidosis.

Disturbances in acid-base balance

If you recall that acid-base balance depends on the ratio $\frac{B^+HCO_3^-}{H_2CO_3}$, you can see that there are nine possible states which may occur in the blood. First is a normal relationship. Then there are excesses of either numerator or denominator and deficits of either, i.e., there are four deflections from the normal, and since each may be compensated or uncompensated, they total eight. If the ratio $\frac{B^+HCO_3^-}{H_2CO_3}$ remains within normal limits, i.e., about 16/1 to 25/1, corresponding to pH 7.3 and 7.5, the situation is *compensated.*

Primary alkali deficit In primary deficit of alkali the bicarbonate is diminished as a result of increased production, ingestion, or retention of acid. The increased production occurs in diabetes mellitus, in starvation, and in certain other metabolic disturbances. Such acids as β-hydroxybutyric are not used in the normal manner and they therefore make inroads on the alkali reserve. The ingestion of mineral acids, as might occur from the administration of hydrochloric acid in gastric disturbances, has the same effect. Infantile diarrhea may result in loss of base. In nephritis the kidney may not excrete acids in sufficient amounts, and retention therefore occurs. Except in the case of retention, a primary alkali deficit leads to increased elimination of acid in the urine. There is also a rise in urinary ammonia. Respiration is increased to get rid of carbon dioxide faster. All these compensatory mechanisms tend to reduce carbonic acid (denominator). If the reduction is sufficient to keep the pH in the normal range, a *compensated acidosis* results. If not, the acidosis is *uncompensated*, the pH falls, and the patient may go into coma. Primary alkali deficit has also been called *metabolic acidosis*.

Primary alkali excess The ingestion of excessive amounts of sodium bicarbonate is about the only example of increasing the bicarbonate fraction in an absolute manner. However, removing acid from the body has the same result, relatively. An example of the latter is excessive vomiting as it occurs in pyloric obstruction, with consequent loss of gastric hydrochloric. The physiological mechanisms for combating this are an increased excretion of alkali (HCO_3^-) by the kidney and, at the same time, a diminished formation of ammonia. Respiration is depressed so that loss of carbon dioxide is very low. If these physiological efforts are successful in keeping the pH below 7.5, again there is a compensated condition of alkalosis with few, if any, untoward symptoms. However, if it is uncompensated and the pH rises to an abnormal level, the alkalosis is grave and tetany may occur. Tetany is a condition that may arise from other causes besides severe alkalosis. Neuromuscular excitability is the chief symptom in humans, and even convulsive seizures occur in children as they do in lower animals. *Metabolic alkalosis* is another term applied to this type. It is usually minimally compensated, with the pH being above 7.45 as a rule when the alkaline bicarbonate is above 30 mmole per liter.

Primary carbon dioxide excess A primary excess of carbon dioxide is caused by an obstruction to respiration or depression of it. The former may occur in pneumonia, tracheal obstruction, or emphysema, and the latter from depression of the respiratory center as a result of toxic doses of morphine or other respiratory depressants. Under these conditions usually the lack of oxygen (hypoxia) is more to be feared than the acidosis. However, it is an acidemia, and the compensatory mechanisms are an increase in the renal reabsorption of bicarbonate and a rise in urinary acid and ammonia. This leads to high alkaline bicarbonate with acidemia. Again this may be either compensated or uncompensated. Primary carbon dioxide excess has sometimes been designated *respiratory acidosis* despite the high alkaline bicarbonate.

Primary carbon dioxide deficit A loss of carbon dioxide may occur when respiration is stimulated in some abnormal manner. Examples of this are more common than are usually believed. Fever and hot baths were the two most usual instances formerly cited, but two others have more recently been brought to the attention of observers. One is the lack of oxygen existing at high altitudes. When this is very great, it increases the rate of respiration and carbon dioxide is eliminated more rapidly. A second factor is anxiety or hysteria. Such a mental state results in hyperventilation also, and the two factors may operate together in airplane passengers. In the Army the hyperventilation anxiety syndrome was said to be a rather common condition in hospitalized cases in wartime. It is often difficult to recognize. Primary

Table 16-10 Determination of type of acid-base disturbance from laboratory data

Acid-base state	Urinary pH	Plasma HCO_3^- (mEq per liter)	Plasma H_2CO_3 (mEq per liter)
Normal	6-7	25	1.25
Respiratory acidosis	Decrease	Increase	Increase
Respiratory alkalosis	Increase	Decrease	Decrease
Metabolic acidosis	Decrease	Decrease	Decrease
Metabolic alkalosis	Increase	Increase	Increase

dioxide deficit is, of course, an alkalemia that usually becomes compensated by a reduction of urinary ammonia formation and increased excretion of bicarbonate. A common term for primary carbon dioxide deficit is *respiratory alkalosis,* despite the low alkaline bicarbonate.

From a consideration of these conditions it must be evident that the determination of the alkali reserve alone will not always give a true picture of the condition. Sometimes a pH determination is also needed. For instance, in an uncompensated carbon dioxide deficit there is an alkalemia caused by a loss of volatile acid. In the attempt to compensate for the reduction in carbonic acid there is, as stated, an increased excretion of bicarbonate. Thus we have a lowered blood bicarbonate with an alkemia. On the other hand, in an uncompensated carbon dioxide excess the attempt of the body to compensate is the production and hoarding of biocarbonate. The blood actually is more acid (acidemia) despite the presence of increased bicarbonate. These, of course, are extreme cases, but their implications are important. A good clinical history to determine the cause of the disturbance is most helpful. Acidemia or alkalemia refers to low or high pH respectively.

Determination of the type of acidosis or alkalosis can be made rather easily from laboratory determinations of urinary pH and plasma concentrations of HCO_3^- and H_2CO_3. It is necessary, however, that these determinations be made as soon as possible before compensatory changes have been completed. Typical changes are summarized in Table 16-10.

Leukocytes

The white blood cells, or leukocytes, are much fewer in number than the red cells, and they have a lower specific gravity. Consequently, when whole blood is centrifuged, they form a narrow whitish layer above the red cells. Normally there are from 5000 to 10,000 leukocytes per cubic millimeter. The different types and variations in number cannot be considered here. Of course, in *leukemias* and in many infections and inflammatory conditions leukocytes are greatly increased in number, whereas in typhoid fever and in some other abnormal states a *leukopenia* (i.e., decreased number of white cells) develops. Such a condition is agranulocytosis, which may be caused by an infection or by the use of certain drugs, e.g., amidopyrine, acetophenetidin, dinitrophenol, arsphenamine, or the sulfonamides. Since the leukocytes are typical cells, they contain water, nucleoproteins, albumin, globulin, and other proteins, lipids (especially cholesterol and phospholipids as well as triglycerides), glucose, and other soluble organic substances and inorganic salts. They also possess a variety of enzymes and other substances.

Platelets

Platelets are cytoplasmic fragments of the megakaryocytes found in bone marrow. In the peripheral blood, they are 2 to 3 μ in diameter, and there are 200,000 to 400,000 of them in every cubic millimeter of blood. Platelets disintegrate within a few days, especially in stored blood. They have no nucleus, of course, but mitochondria are present. Thus they have limited metabolism, mainly for energy production.

Platelets contain 86% to 88% water. Nineteen percent of their dry weight is lipid (phospholipids, triglycerides, cholesterol, cholesterol esters), and 57% is protein. Some of the inorganic constituents present are sodium, potassium, manganese, iron, copper, magnesium, and calcium. They contain enzymes, e.g., glucuronidase, catalase, amylase, phosphomonoesterase, cholinesterase, lecithinase, histaminase, trypsin, acid phosphatase, lactic and glutamic dehydrogenases, and others. They also contain agglutinogens apparently comparable to those of the red cells, and they may agglutinate or clump for that reason.

Platelets tend to adhere to foreign surfaces and to each other. Common phrases applied to this phenomenon are platelet "stickiness," platelet adhesion, platelet clumping, and platelet aggregation. These associations can be reversible or proceed further to complete morphological disintegration called *viscous metamorphosis*. The phenomena may occur at a "rough" spot created by injurious substances placed on blood vessel walls, on foreign bodies (e.g., a thread placed in the bloodstream), on glass surfaces, etc. Such clumps have a white appearance and are sometimes called *white thrombi*. Clumping is important for hemostasis because the platelet mass serves as an obstruction to the flow of blood. Platelets also contain serotonin, a vasoconstrictor agent that has many functions, including a role in hemostasis and possibly also in clot retraction, discussed in the next section. Certainly the platelets themselves or their derivatives are necessary for clot retraction, and for that phenomenon plasma factors are also involved. Platelets contain ATP and the contractile proteins actin and myosin, which may account in part for the chemical mechanisms underlying clot retraction.

Most of the foregoing constituents of platelets are probably synthesized in the parent megakaryocyte and are transported to and stored preformed in platelet granules. The so-called α-granules ("light") contain several of the *platelet factors*. The "dense" β-granules contain such substances as ADP and serotonin, which play important roles in the coagulation of blood.

Platelets are the center where blood coagulation begins, if tissue extracts, trypsin, snake venoms, and the like are excluded. Furthermore, they are stimulated by a variety of conditions. The main contribution of platelets to the clotting process seems to be phospholipids. However, many other factors are also found in these formed elements, and it is customary to give the platelet factors arabic numbers. Platelet factor 1 is an accelerator globulin-like activity. Platelet factor 2 is a fibrinoplastic substance, possibly an enzyme, that alters fibrinogen so that the fibrinogen is more readily clotted by thrombin. Platelet factor 3 is very likely a lipoprotein that can be regarded as the main component for coagulation chemistry. Platelet factor 4 is an antiheparin factor. The next number (5) designates a clottable substance, very nearly like fibrinogen, whereas platelet factor 6 is an antifibrinolytic substance. There are different opinions about platelet factors. Some authors believe that they are plasma materials adsorbed on the surface of the platelet; however, extensive washing of platelets does not remove all of them. Platelets contain a transglutaminase that is activated by thrombin and forms cross-linked fibrin. Many other substances are also found in platelets, including ADP, which has much to do with platelet aggregation.

As a general rule stimuli that are followed by platelet aggregation induce a release reaction. It is really a secretory function. This reaction is essential if platelets are to function normally in supplying substances concerned with blood coagulation, fibrinolysis, and hemostatsis. The secretory function is comparable to the release of hormones and other active messengers by specialized cells. Aggregation and the release reaction are separate phenomena that commonly occur at the same time. Below are listed some stimulators of release, some substances that are released, and representative substances that tend to inhibit the release reaction:

Stimuli	Released	Inhibitors
Mechanical	Potassium	Intracellular cAMP
Collagen	ATP	Prostaglandin E_1
Epinephrine	ADP	Aspirin
Serotonin	Platelet factor 3	Adenosine
ADP	Serotonin	Dipyridamole
Thrombin	Hydrolytic enzymes	
	Lysosomal enzymes	
	Platelet factor 4	

Blood coagulation

The clotting of human blood basically involves the conversion of the soluble plasma protein, *fibrinogen,* into an insoluble, tough "hard clot" by the action of a proteolytic enzyme, *thrombin,* in the presence of several other accessory factors. Fibrin, in turn, enmeshes erythrocytes and other formed elements of the blood into a gelatinous mass that adheres to and closes any breaks in the blood vascular system (Fig. 16-23). This process must occur rapidly to prevent excessive blood loss, shock, and even death from exsanguination. Yet the blood otherwise must be maintained in its normal fluid state to continue its numerous vital functions. Unnecessary intravascular clotting must not occur, however, as happens in such serious, often fatal conditions as coronary heart disease and strokes. This *delicate balance* is made possible by the essential clotting factors present in the blood as *inactive forms,* which can be rapidly activated by damage to the blood vascular system or contiguous tissues. The purpose of this section is to describe the chem-

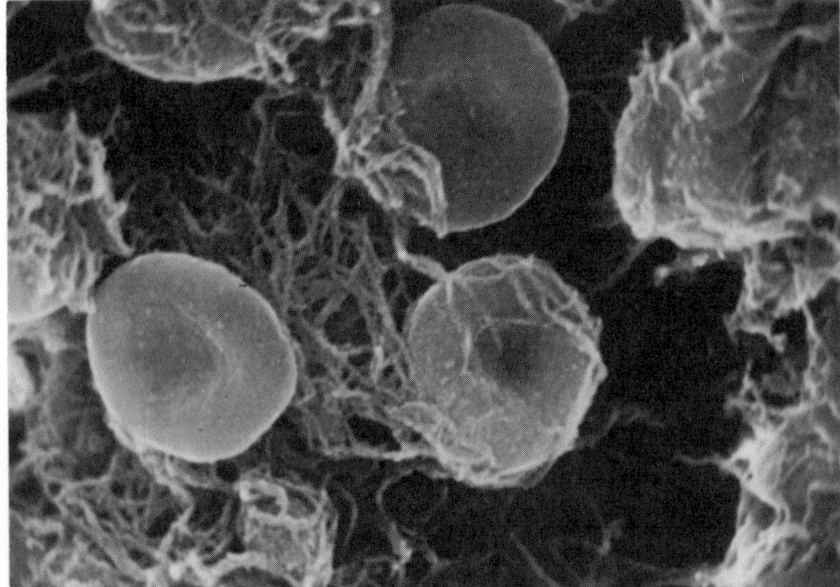

Figure 16-23
Scanning electron micrograph of human erythrocytes enmeshed in a fibrin network clot that formed in vivo and partially occluded a femoral vein. (17,500×.)

Courtesy Dr. M.I. Barnhart, Specialized Center of Thrombosis, Wayne State University School of Medicine, Detroit, Mich.

ical nature of these factors, the reactions involved, and the biochemical functions performed.

Nomenclature

For a number of years the area of blood coagulation was plagued by a multiplicity of terms used to designate individual clotting factors, most investigators using their own preferred names. The problem was compounded by the availability of improved techniques for the separation of new factors—most of them proteins, some with associated oligosaccharides or lipids. The confusion has been minimized, however, in recent years by the general adoption of a system of terminology proposed by an International Committee on the Terminology of Blood Clotting Factors. The Committee, composed of active research investigators in the field, proposed designating the various factors by *Roman numerals* in an approximate order of their discovery. Thus *fibrinogen* was termed factor I. The nomenclature has been expanded to designate the *activated* factors with a subscript *a*. Thus thrombin became factor II_a. The proposed nomenclature together with the common names of the currently recognized factors and some of their chemical properties and biochemical functions is given in Table 16-11.

Several factors derived from platelets are usually designated in Arabic numbers, e.g., PF-3.

Basic reactions

Earlier concepts of the biochemical events that occur during the coagulation of human blood emphasized the view that *three* basic reactions were involved:

1. The presence and release of a *thromboplastin* from damaged blood vessels or tissues and platelets
2. The conversion of an inactive plasma protein proenzyme, *prothrombin,* into an active proteolytic enzyme, *thrombin,* by thromboplastin in the presence of *calcium ions*

Table 16-11
Some properties and functions of the principal clotting factors

Factor (international nomenclature)	Typical common name(s)	Approximate molecular weight	Approximate carbohydrate content (%)	Function of activated factor
I	Fibrinogen	340,000	4	Forms clot framework
II	Prothrombin	68,700	8	Converts fibrinogen to fibrin
III	Thromboplastin (lipoprotein; tissue factor)	—	—	Required by factor VII
IV	Calcium ions	—	—	Cofactor for several clotting reactions
V	Proaccelerin (plasma Ac-globulin; parahemophilia factor)	400,000	—	Accelerates thrombin formation
VI	Not verified	—	—	—
VII	Proconvertin (cothromboplastin)	45,000	—	Accelerates factor X_a formation
VIII	Antihemophilia A-1 factor	100,000	6	Accelerates factor X_a formation
IX	Antihemophilia B (Christmas factor)	55,400	26	Accelerates factor X_a formation
X	Stuart-Prower factor (autoprothrombin III)	55,000	10	Accelerates thrombin formation
XI	Plasma thromboplastin (PTA; antihemophilia factor)	160,000	12	Accelerates IX_a formation
XII	Hageman factor (contact factor)	74,000	15	Accelerates IX_a formation
XIII	Fibrinoligase (fibrin stabilizing; plasma transglutaminase)	146,000* 320,000†	5	Catalyzes cross-linking of fibrin molecules

*In platelets.
†In plasma.

3. The conversion of the inactive plasma globulin, *fibrinogen,* by proteolytic cleavage by thrombin into the insoluble protein *fibrin;* the fibrin network entrapped erythrocytes and other formed elements of the blood to form a *hard clot,* thus closing at least smaller wounds

Subsequent research, although confirming these three basic concepts, has expanded enormously knowledge of the biochemical reactions involved. The operation of several "safeguards" to prevent unnecessary intravascular clotting has been identified, as have "cascade" effects with marked *amplification* to ensure rapid clotting when required. The blood platelets have been found to play a key role in initiating the clotting process as well as in accelerating its rate.

Role of blood platelets

The biochemical reactions that culminate in the formation of a fibrin clot are interlocked with events involving the blood platelets. First, as blood flows out of a severed blood vessel, its platelets come in contact with the subendothelium of the vessel wall, specifically with its contained *collagen* molecules (Chapter 15). Specific glycoproteins on the surface of the platelets adhere to specific acceptor sites on the endothelial collagen in the presence of the *von Willebrand binding factor*. Changes in the shape and structure of the platelets follow, with the release (secretion) of the contents of the platelet granules. The substances released include *ADP,* a powerful stimulant for *platelet aggregation,* and *serotonin* (Chapter 13), a powerful *vasoconstrictor* that narrows the blood vessel wall locally and thus helps to restrict the bleeding area. Serotonin is also a weak platelet aggregating agent. *Fibrinogen,* released from the platelet α-granules, serves as a cofactor, a sort of glue, for platelet aggregation. Other passing platelets adhere to the first layer, building up a loose platelet plug in the wound channel.

Simultaneously factor V and other clotting factors to be described later are released from the α-granules to aid in the local conversion of plasma (and some platelet) *prothrombin* into *thrombin*. Thrombin in turn stimulates further release of platelet granule material, thus providing strong local amplification of the clotting process. Thrombin also converts plasma and platelet fibrinogen into *fibrin,* which enmeshes and reinforces the platelet plug and entraps erythrocytes and other formed elements of the blood. Retraction of the platelet-fibrin clot then occurs as a result of the presence of the contractile protein *actomyosin (thrombosthenin)* in platelets, and a firm mass is formed. Thus smaller, severed blood vessels are closed. Larger wounds may require the application of external pressure.

Table 16-12
Some abnormalities in human platelet function

Disease	Biochemical defect in platelet function
Thrombocytopenia (congenital or acquired)	Deficient platelet production in bone marrow (due to leukemia, aplastic anemia, ionizing radiation, toxic agents, certain drugs)
Idiopathic thrombocytopenic purpura	Unknown (petechial hemorrhages under skin)
Storage pool disease	Impaired aggregation due to congenital absence of dense granules
Thrombasthenia	Impaired aggregation probably due to deficiency of fibrinogen-binding surface glycoprotein
Congenital Afibrinogenemia	Congenital lack of fibrinogen
von Willebrand's disease	Lack of factor required for platelets to adhere to collagen (also lack of antihemophilia A factor)
Bernard-Soulier (giant platelet) syndrome	Probable deficiency of surface glycoprotein that binds von Willebrand factor
Aspirin toxicity	Inhibition of fatty acid cyclooxygenase and thromboxane A_2 formation

Still another substance present in platelet granules is involved in the clotting process. Arachidonic acid is converted to an endoperoxide and *thromboxane* A_2, a metabolite of the prostaglandins (Chapter 18). Thromboxane A_2 stimulates the secretion of ADP from the dense granules, further increasing platelet aggregations and enhancing the formation of the platelet plug. Recent studies have shown that *aspirin* impairs platelet aggregation and thus prolongs bleeding time. Aspirin impairs thromboxane A_2 formation by inhibiting the enzyme fatty acid *cyclooxygenase,* involved in the biosynthesis of thromboxane A_2.

Another metabolite of the endoperoxides, *prostacyclin* (Chapter 18), is an inhibitor of platelet aggregation, thus providing a countercontrol mechanism. Prostacyclin is synthesized in endothelial cells of blood vessels.

The vital role of platelets in the clotting of blood is further emphasized by the marked effects of abnormalities in platelet function on coagulation, especially on bleeding time. The latter is the time required for cessation of bleeding from a small cut in the skin of the forearm. Several of the major abnormalities of human platelets together with the principal function impaired are summarized in Table 16-12.

Figure 16-24 Schematic representation of principal reactions involved in the coagulation of blood. Roman numerals indicate International Commission on Designations of Clotting Factors. A subscript *a* indicates activated factor.

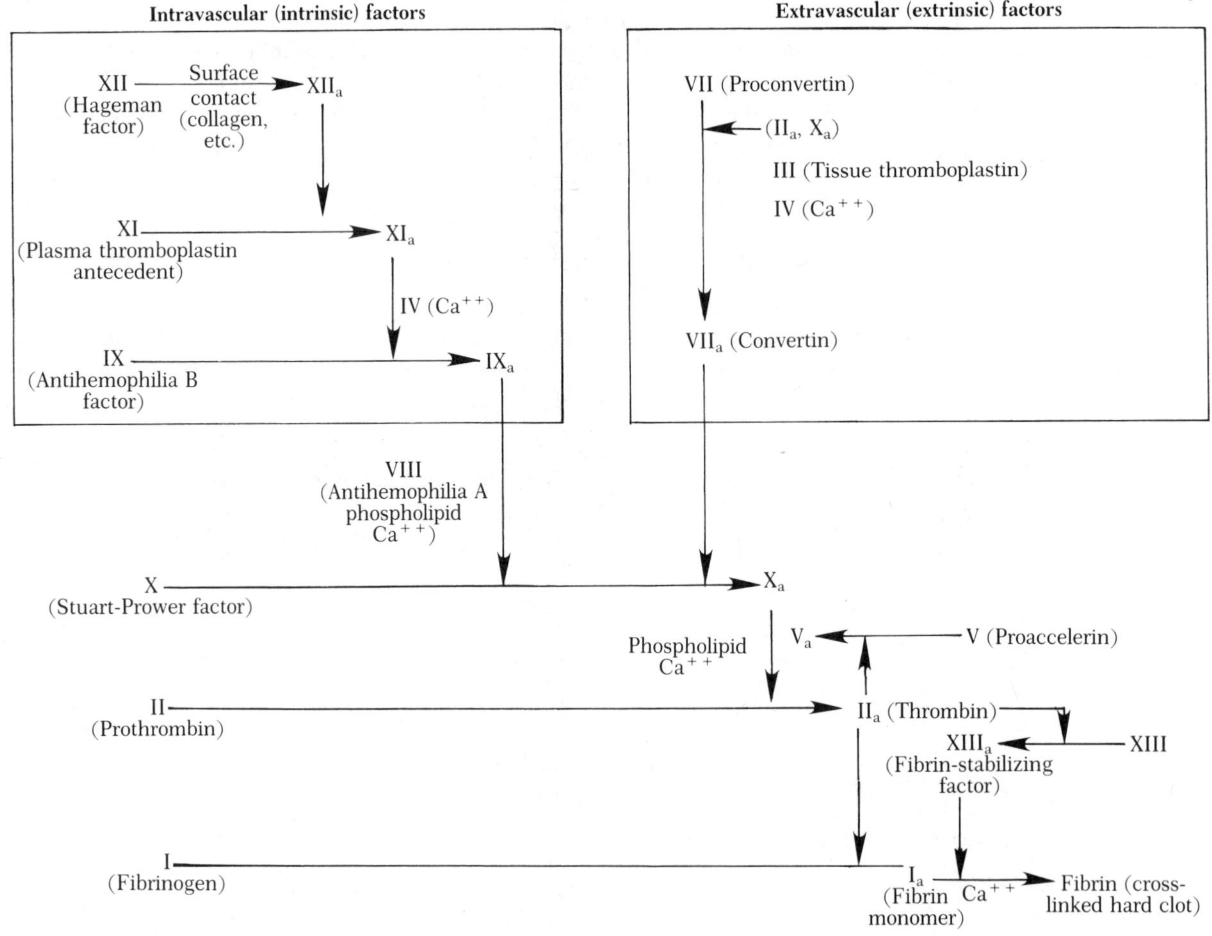

Biochemical reactions in the clotting process

Closely interlinked with the aforementioned events in blood clotting is a sequence of biochemical changes that lead to the conversion of fibrinogen to fibrin and the formation of a firm clot, closing a severed blood vessel. Some of the factors and reactions involved in platelet function are included also in the process now to be described. A current, generally accepted concept of the coagulation of blood is summarized schematically in Fig. 16-24. This figure supplements the summary of the properties and functions of various clotting factors given in Table 16-11.

The following presentation includes current knowledge of the chemical nature of the recognized clotting factors and the mechanisms involved in their reactions. The discussion is arranged in the order of the participation of the factors in the clotting process, in the so-called intrinsic (intravascular) and extrinsic (extravascular) systems.

Stage I

The intravascular system These reactions are so termed because all factors involved are present in the vascular system—the plasma or platelets. The three reactions involved lead to the activation of factor X and in turn to the conversion of prothrombin to thrombin (Fig. 16-24).

Conversion of factor XII to XII_a The common name for this factor is the Hageman, or contact, factor. On contact this factor interacts in some manner with and binds to *collagen* or to a vasoactive peptide such as *kallikrein,* resulting in a change in its conformation to a proteolytically active form, XII_a. The complex of factor XII_a–kallikrein activates more of factor XII, thus amplifying the reaction.

Activation of factor XI to XI_a Factor XII_a, or possibly the factor XII_a–kallikrein complex, activates factor XI apparently by hydrolyzing an internal peptide bond. Factor XI_a also is a proteolytic enzyme.

Activation of factor IX to IX_a The activation of factor IX, antihemophilia B factor or Christmas factor, apparently entails its proteolytic cleavage by factor XI_a in the presence of calcium ions. Since factor IX contains γ-carboxyglutamate residues, its formation in the liver is vitamin K dependent. Factor IX_a has proteolytic activity: cleaving Arg-Gly peptide bonds.

Conversion of factor X to X_a The activation of factor X, called the Stuart-Prower factor, is the final step in both the intravascular and extravascular systems, the latter to be described in the next section. Activation by factor IX_a of the intravascular system requires the presence of factor VIII (hemophilia A factor), platelet phospholipid (PL), and Ca^{++} ions that mediate the binding of factor IX_a to the phospholipid component. Factor VIII may require activation by thrombin. These four substances apparently form a complex that activates factor X.

$$\underset{\text{(Stuart-Prower factor)}}{\text{Factor X}} \xrightarrow[]{\substack{\text{Activation by complex of factors} \\ IX_a \text{ and VIII, phospholipid, and } Ca^{++}}} \underset{\text{Inert peptides}}{\text{Factor } X_a +}$$

Extravascular (extrinsic) system The factors and sequences that make up the intravascular system of blood coagulation are sufficient to promote clotting but at a *slow* rate. The action of the intrinsic pathway is therefore *supplemented* by a second, extrinsic pathway to ensure a *more rapid coagulation* process. The enhancement of clotting is the result of factor VII (proconvertin) being activated to VII_a (convertin) by factor III (tissue thromboplastin) (or by II_a or X_a) in the presence of Ca^{++} ions (Fig. 16-24).

Factor III, sometimes called *tissue factor* or *tissue thromboplastin,* is widely distributed in the vascular endothelial membranes and in leukocytes. It is released after damage to blood vessel walls, and it initiates the rapid clotting sequence characteristic of the extravascular system. Factor III has been partially purified and appears to be a lipoprotein containing phospholipid, either phosphatidyl choline or phosphatidyl ethanolamine, complexed with a specific protein.

Factor VII, *proconvertin,* has been purified from human plasma. Factor VII contains a γ-carboxyglutamate residue; therefore its synthesis is vitamin K dependent. It is activated to *convertin* by either factor III, factor X_a, or thrombin in the presence of Ca^{++}. Factor VII_a has proteolytic activity in the presence of factor III and thus activates factor X.

The reactions in the *extravascular system* may be summarized schematically as follows:

$$\underset{\text{(Stuart-Prower factor)}}{\text{Factor X}} \xrightarrow{\text{Factor VII}_a\text{, factor III, Ca}^{++}} \text{Factor X}_a + \text{Inert peptides}$$

Stage II

Conversion of factor II (prothrombin) to factor II_a (thrombin) The remaining two stages of the clotting process are common to both the intravascular and the extravascular systems (Fig. 16-24). The conversion of prothrombin (factor II), to its active proteolytic form, thrombin (factor II_a), will be considered first. The activation of prothrombin can be accomplished in the laboratory by purified factor X_a alone. However, under physiological conditions it is not present in plasma and platelets in sufficient amounts, and its activity is too low for adequate functioning. Thus in vivo the accessory substances *factor V_a* (accelerin, formed from proaccelerin by thrombin), *platelet factor 3* (phospholipid), and *Ca^{++} ions* present promote the reaction. The reaction may be represented schematically as follows:

$$\underset{\text{(prothrombin)}}{\text{Factor II}} \xrightarrow[\text{platelet phospholipid, Ca}^{++}]{\text{Factor X}_a\text{, factor V}_a\text{ (accelerin),}} \text{Factor II}_a\text{ (thrombin)} + \text{Inactive peptides}$$

Detailed studies of this reaction indicate that it proceeds in the following manner. Prothrombin binds a complex of vesicles of platelet phospholipid–factor X_a–factor

Figure 16-25
Tetragonal thrombin crystal, prepared from bovine serum and crystallized from 2M ammonium sulfate after preliminary purification.

Scanning electron micrograph by J.M. Riddle, L.E. McCoy, D.A. Walz, and D. Tsernoglou; 1260×.

V_a to its NH_2 terminal region mediated by strong binding sites for Ca^{++} on adjacent γ-carboxyglutamic acid residues. Factor X_a then hydrolyzes two peptide bonds, forming proteolytically active *thrombin,* which dissociates from the complex, and an inactive polypeptide with a molecular weight of about 33,500.

Prothrombin (mol wt = 68,700) is synthesized in the liver by a vitamin K–dependent reaction (Chapter 21) required for the formation of γ-carboxyglutamic acid—the essential binding sites for Ca^{++} just described. *Hypoprothrombinemia* thus occurs in vitamin K deficiency and in animals fed the vitamin K antagonist *dicoumarol*. This is used clinically to control blood clotting in certain thrombotic conditions such as strokes and coronary heart disease. Hypoprothrombinemia also occurs in obstructive jaundice (Chapter 17) and in certain other liver disorders.

Thrombin has a molecular weight of about 38,000. It has been crystallized (Fig. 16-25), and its amino acid sequence has been determined. Thrombin is being studied extensively by x-ray crystallographic techniques.

Stage III

Conversion of factor I (fibrinogen) to factor I_a (fibrin) Highly purified fibrinogen prepared from human plasma has a molecular weight of about 340,000. Its amino acid sequence has been determined. The fibrinogen molecule consists of three pairs of polypeptide chains called α-A, β-B, and γ, having molecular weights of 63,000, 56,000 and 47,000, respectively. The *A* and *B* represent fibrinopeptides split off in the conversion to fibrin monomer by thrombin. The chains are bound together by disulfide bonds. Each chain contains oligosaccharide groups, attached by means of asparagine residues. Fibrinogen is synthesized in the liver and has a half-life of about 4 days.

The conversion of fibrinogen to fibrin first involves the cleavage by thrombin of a specific Arg-Gly bond in each of the α-A and β-B chains to release the fibrinopeptides A and B from the NH_2 terminal ends of the chains. The γ-chain is not affected. *Fibrin monomers* are thus formed. These aggregate to fibrin s (soluble fibrin), the soft clot. Fibrin-s in a final step is converted to fibrin-i (insoluble fibrin), the hard clot, by an activated, cross-linking enzyme, factor XIII_a, the fibrin-stabilizing factor *fibrinoligase*. The reaction summarized schematically is:

$$\text{Factor I (Fibrinogen)} \xrightarrow[\text{(Thrombin)}]{\text{Factor II}_a} \underset{\substack{\text{(Fibrin, monomer)} \\ (\alpha\text{-}\beta\text{-}\gamma)_2}}{\text{Factor I}_a} + \underset{(2A + 2B)}{\text{Fibrinopeptides}}$$

$$\downarrow \text{Aggregation}$$

$$\text{Fibrin-s } [(\alpha\text{-}\beta\text{-}\gamma)_2]_n$$

$$\downarrow \text{Factor XIII}_a \text{ (Fibrinoligase)}, Ca^{++}$$

$$\underset{[(\alpha\text{-}\beta\text{-}\gamma)_2]_n^x}{\text{Fibrin-i (hard clot)}} + NH_4^+$$

(*x* represents the cross-linking)

Factor XIII, the fibrin-stabilizing factor (FSF), is present in human platelets and in plasma in two different forms. The plasma factor has a molecular weight of approximately 320,000 and is composed of two A-chains (mol wt = 70,000 each) and two B-chains (mol wt = 90,000 each). It contains about 5% carbohydrate. Platelet FSF apparently does not contain B-chains.

Factor XIII_a is formed from platelet factor XIII by the splitting of an inert peptide by *thrombin* from the NH_2 terminal of each A-chain. A specific Arg-Gly peptide bond is hydrolyzed. Plasma XIII is similarly activated after its B-chains are dissociated in a Ca^{++}-dependent reaction.

Factor XIII_a (fibrinoligase) is a *trans-glutaminase* that catalyzes the formation of cross-linking peptide bonds between the γ-amide of glutamine of one fibrin molecule with the ϵ-amino group of lysine of molecule. The reaction is:

```
                 O            O                                   H            O
                 ||           ||                                  |            ||
             —C—N—CH—C—N—                              —C—N—CH—C—N—
                  |    |          |                       ||        |             |
                  H   CH2         H                       O        CH2            H
                       |                                            |
                      CH2                                          CH2
Lysyl                  |                                            |
donor                 CH2                                          CH2
                       |                                            |
                      CH2                                          CH2
                       |                                            |
                      NH2              Ca++                         NH
                                    Factor XIIIa                    |
                       +           ———————————>                    C=O                + NH4+
                                                                    |
                  H2N    O                                         CH2
                     \  //                                          |
                      C                                            CH2
                      |                                  H          |           O
Glutaminyl           CH2                                 |          |           ||
acceptor              |                              —C—N—CH—C—N—
                     CH2                                ||                      |
                 O    |       O                         O                       H
                 ||   |       ||
             —C—N—CH—C—N—
                  |              |
                  H              H
```

Apparently cross-linking occurs between the α-chains and the γ-chains of fibrin-monomer molecules and not between β-chains. Thus the intermolecularly highly cross-linked polymer of fibrin is formed, giving the typical insoluble hard clot.

Anticoagulants

A number of substances may interfere with the clotting of blood either in vitro or in vivo. Any compound that precipitates Ca^{++} or binds it in nonionized form will prevent coagulation. Soluble salts of oxalate, fluoride, citrate, or EDTA are commonly used for this purpose in vitro. The reason, of course, is that Ca^{++} ions are required for several reactions in the clotting process, as just described. Other anticoagulants include bile salts (inhibitors of thromboplastin), hirudin (extracts of leeches), *dicoumarol,* and *heparin*. The latter two substances are widely used clinically and therefore merit further discussion.

Dicoumarol was first isolated from spoiled sweet clover hay and identified as the causative agent in a hemorrhagic disease in cattle. It was found to be an antagonist of vitamin K and thus inhibits coagulation in vivo by impairing the biosynthesis of the vitamin K–dependent clotting factors II, VII, IX, and X. Dicoumarol (top of p. 517) has been used clinically for years in patients with a tendency to thrombus formation. It is effective orally but only after existing supplies of prothrombin and other vitamin K–dependent clotting factors have been reduced.

Heparin is another powerful anticoagulant drug—the first to be used clinically. It is especially useful in patients during and after various types of surgery to reduce or prevent intravascular clotting and thrombosis. It must be administered *intravenously,* since the molecule may be hydrolyzed by certain digestive enzymes.

Dicoumarol
[3,3′-methylenebis-(4-hydroxy-coumarin)]

Heparin is formed in most cells found principally along blood vessel walls, liver, lung, and other tissues. Chemically it is a complex polysaccharide composed of alternating α-L-iduronic acid or β-D-glucuronic acid and α-D-glucosamine disaccharide units joined by $1 \rightarrow 4$ glycosidic linkages (see Chapter 23 for structure). Most of the glucosamine residues carry O-sulfated amino groups at C-6, although a minor proportion is acetylated. Most, but apparently not all, of the iduronic acid residues are sulfated at C-2. The glucuronic acid residues appear to be nonsulfated. The number of repeating disaccharide units appears to vary per molecule from 6 to 60 or more, with molecular weights ranging proportionally from approximately 3000 to 30,000.

The mechanism of action of heparin as an anticoagulant now appears to be related to its interaction with *antithrombin III,* a plasma protein inhibitor of serine proteases that activate thrombin and several other clotting factors (IX_a, X_a, XI_a, XII_a, and kallikrein). Antithrombin III apparently forms complexes with these activated clotting factors that have no proteolytic activity. Heparin greatly enhances the formation of the inactive complex, possibly by inducing a conformational change in antithrombin III, thus facilitating the interaction between the inhibitor and the activated clotting factor. This would prevent the factor from functioning in the clotting process.

The fibrinolytic system

The dissolution of blood clots, *fibrinolysis,* may be viewed as a final stage of the clotting process (Fig. 16-26). Typically, lysis of clots occurs within a few days of their formation and involves the proteolytic cleavage of fibrin by the enzyme *fibri-*

Figure 16-26
The fibrinolytic system.

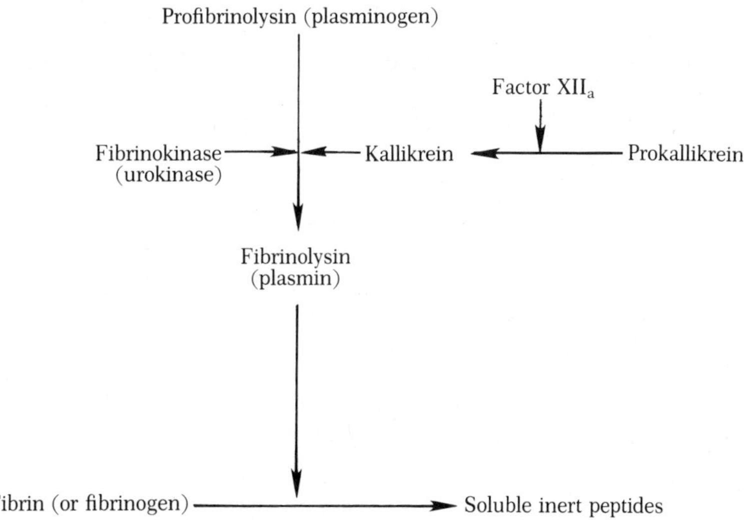

Table 16-13
Some abnormalities of the blood clotting factors

Factor	Abnormality
I	Afibrinogenemia, congenital deficiency, hypofibrinogenemia, low plasma levels, dysfibrinogenemia, abnormal fibrinogen molecule
II	Hypoprothrombinemia, deficient or abnormal prothrombin molecule, vitamin K deficiency, liver disorders
III	Tissue factor deficiency (?), impaired thrombin formation, impaired clotting
IV	Hypocalcemia
V	Congenital parahemophilia, deficiency or abnormal molecule
VII	Congenital absence or abnormal molecule, vitamin K deficiency
VIII	Hemophilia A (classic), deficient, inhibited, or abnormal molecule, sex-linked inheritance, low plasma level in von Willebrand's disease
IX	Hemophilia B, deficient or abnormal molecule, vitamin K deficiency
X	Stuart-Prower factor deficiency, deficient or abnormal molecule, vitamin K deficiency
XI	Hemophilia C, factor absent or not released from platelet
XII	Hageman deficiency, factor absent, or defective molecule
XIII	Congenital deficiency, impaired fibrin polymerization, antithrombin III, thrombophilia, low plasma level, thrombosing tendency, reduced titer in some females using oral contraceptives

nolysin (plasmin). Fibrinolysin is formed by the activation of its zymogen, *profibrinolysin* (plasminogen), which is produced in the kidneys. A number of proteolytic enzymes have the ability to activate the zymogen, including *streptokinase,* obtained from hemolytic streptococci, and several from snake venom. However, the physiological activator is *fibrinokinase* (urokinase), which is present in many tissues and also in normal human urine. Apparently, kallikrein also may activate profibrinolysin. Also, it may be activated in a variety of situations, such as stress and strenuous exercise. Presumably, activation of profibrinolysin under these conditions would serve to remove any small intravascular clots that might form transiently without producing pathological effects. The heparin–antithrombin III complex inhibits fibrinolysin and thus may be a factor in regulating fibrinolysis. The presence of an antifibrinolysin in normal plasma also has been reported.

Promising results have been reported recently on the use of streptokinase in patients following severe heart attacks and in patients at risk of embolism formation. The mortality was reduced 50% in 155 patients. Streptokinase has been used for several years to dissolve clots in veins.

Some abnormalities of the clotting mechanism

Since the coagulation of blood involves a number of interreacting factors, most of which are proteins, dysfunction of the clotting mechanism could result from genetic errors in their formation. These would be manifested by alterations of the blood or tissue level or activity of the factor involved or even by prolonged clotting or bleeding time values, as frequently determined in clinical laboratories. A number of such abnormalities have been described, as well as those involving the blood platelets as described earlier (Table 16-12). Dysfunction of the clotting process may occur also secondary to various pathological conditions, such as liver disorders.

A number of abnormalities of blood coagulation due to dysfunction of the major clotting factors are summarized in Table 16-13. Details of the hematological and clinical manifestations are given in appropriate references at the end of the chapter.

Anemias

Those conditions in which the number of erythrocytes or the amount of hemoglobin is reduced below normal are termed *anemias*. There are a number of types of anemias which can be only briefly considered here. Anemias are caused by (1) loss of blood, (2) increased destruction of blood, or (3) defective formation of blood. The first group includes acute and chronic hemorrhage. Destruction of erythrocytes is brought about by hemolytic agents, which may be of bacterial or metabolic origin, or they may be caused by the absorption of industrial poisons, or by malfunction of the spleen or other erythrocyte-destroying tissues. Anemias of the third group include hypochromic anemias, pernicious anemia, and aplastic anemias.

A hypochromic anemia, i.e., one in which the erythrocytes contain less hemoglobin than normal, may be experimentally induced in animals by feeding them exclusively milk. The lack of iron and copper in this food is the cause of the anemia, for otherwise milk is a superior food. Similarly, in humans an iron-deficient diet may give rise to an anemia (hypochromic anemia) in which the hemoglobin content of the blood is reduced to a greater extent than the number of erythrocytes. The red cells not only contain less hemoglobin but may also be reduced in size. A similar anemia of infants is not uncommon. As stated before, the infant usually comes into the world with a rich store of iron. However, this store is accumulated toward the latter part of gestation. Thus a prematurely born baby may not have enough iron to tide him over the period during which the diet is exclusively milk, and anemia may result. In young women anemias may occur because of a combined effect of malnutrition and menstrual bleeding. This type, called *chlorosis,* is not as common now as it formerly was. Other conditions in which hypochromic anemias sometimes develop are pregnancy and various infectious diseases. Hypochromic anemias are treated mainly by the administration of inorganic iron salts.

Pernicious anemia is caused by an inability to form erythrocytes, not by any difficulty in synthesizing hemoglobin. There results a great diminution in the number of erythrocytes and consequently in the percentage of hemoglobin. The mean corpuscular hemoglobin concentration, however, is high and the blood picture is quite abnormal. The red bone marrow is greatly increased in volume, displacing the yellow marrow and sometimes even invading the true osseous tissue. This greater amount of unused hemoglobin causes a rise in iron and bilirubin in blood plasma, the latter apparently as bilirubin glucuronide, since a direct van den Bergh reaction is observable (Chapter 17). There is invariably a lack of hydrochloric acid in the gastric juice, a fact of great importance in aiding the diagnosis and of interest in explaining the mechanism involved in the production of this type of an anemia.

As might be expected, the administration of iron is to no avail. However, several brilliant series of studies, three by Nobel laureates, demonstrated that pernicious anemia could be treated successfully by large amounts of liver or by liver extracts. The active substance in liver was found to be *vitamin* B_{12} (Chapter 21). Other investigations showed that the gastric juice of patients with pernicious anemia was deficient in a substance called the *intrinsic factor*. This substance was found to be a glycoprotein that binds vitamin B_{12} and enhances its absorption from the lower intestinal tract (ileum). Thus pernicious anemia results from a lack of secretion by the gastric parietal cell (Chapter 17) of an intrinsic factor needed for the absorption of vitamin B_{12}.

Vitamin B_{12} is highly effective in relieving not only the anemic phase but also the lingual and neurological symptoms. This is not true of folic acid (Chapter 21), which benefits only the anemic phase. Folic acid, however, is also efficacious in the treatment of the anemic phase of the macrocytic anemias of pellagra, pregnancy, and sprue. These anemias are nutritional in origin, partially at least, and constitute a group that is not amenable to treatment with iron.

Sickle cell anemia is a hemolytic anemia of a hereditary nature. It occurs mainly in the black race. About 8% of American blacks have the sickle cell trait, but only about one in 40 of the 8% develops the severe chronic anemia. In this disease the erythrocytes undergo reversible changes in shape to crescent and other forms in response to variations in the partial pressure of oxygen. Pauling and associates showed that this is caused by an abnormality of the hemoglobin itself, located in the globin portion of the molecule (hemoglobin S). Those individuals who have the trait, but not the anemia, possess some of this abnormal and some of the normal hemoglobin.

Anemias may also result from toxic inhibition of the bone marrow by certain chemical, bacterial, or physical agents or from replacement of the bone marrow by tumorous or fibrous growths or by leukemia.

Polycythemias

Those conditions in which there is an increase in the number of erythrocytes in the peripheral blood are termed *polycythemias*. A *primary* type of polycythemia of uncertain etiology, called polycythemia vera, occurs infrequently in humans. Erythrocyte counts as high as 12,000,000 cells per cubic millimeter of blood have been recorded. Hemoglobin values are correspondingly increased, up to as much as 20 to 25 gm/100 ml of blood. *Secondary* polycythemias are more common. They are usually compensatory to some other environmental condition or to some disease process. Tissue hypoxia (chronic), as a result of some impairment in oxygen transport, is frequently the underlying cause. Examples are the polycythemia occurring in individuals living at high altitudes and that found in persons having certain types of abnormal hemoglobins, described previously.

Cobalt salts administered in small amounts to experimental animals for several weeks produced a remarkably severe true polycythemia. Plasma erythropoietin (Chapter 18) levels were markedly increased.

Blood transfusion and blood substitution

Loss of blood as a result of hemorrhage or shock is treated by blood transfusion or by injection of a substitute for blood. Blood transfusion, or infusion of a substitute, sometimes preceded by the removal of blood has been used in a number of conditions other than hemorrhage and shock. In general, the purpose is to restore blood volume, to increase the colloidal osmotic pressure, or to provide nutritive or immunological factors.

Whole blood, either citrated or heparinized, is of course the material most approved, being most physiological. Care must be taken that the blood used is not only from a healthy human being but also compatible with that of the recipient. As regards plasma or fractions of plasma, we would expect that serum is the best blood substitute, since no anticoagulant need be added to it. This is not the case, however. Blood serum frequently produces pronounced reactions. Plasma albumin has some value. It makes up about 62% of the blood proteins and exerts about 85% of the colloidal osmotic pressure of the plasma. Because of its great solubility, it may be given in concentrated form.

Blood plasma has proved to be a practical blood substitute and, in fact, is more effective than whole blood in most conditions of loss of blood *fluid*—not, of course, if there is loss of whole blood. Blood of normal human beings known to be *free of viral hepatitis* is collected and citrated, all under rigidly aseptic conditions. It is then centrifuged at 2° to 4° C. It may be preserved in the liquid condition, if properly refrigerated, or it may be frozen or dried. The modern method of drying is the most practical and most widely used. It consists of rapidly freezing the plasma in rotating bottles. This fixes the solid plasma as a "shell." The plasma is then dried, while frozen, under greatly reduced pressure; the procedure is called a lyophilizing process. The proteins are not denatured to any great extent, and the immunological properties are essentially unchanged. The lyophilized, dry, flakey plasma is kept in sterile containers under vacuum until needed. All that is necessary for use is the addition of the required amount of sterile distilled water.

Great progress has been made in fractionating the proteins of plasma so that they might be studied and put to clinical use. There have been separated from human plasma a series of protein products, each a stable white powder responsible for a different natural function. The fractions, with their uses, include the following:

1. Albumin, which is being used instead of dried whole plasma for the reasons mentioned
2. Immunoglobulin, or γ-globulin, which was proved of value in the prevention and treatment of certain infectious diseases; it has been found to be a temporary safeguard against poliomyelitis or to lessen the crippling effects of that disease
3. Agglutinins, for blood typing
4. Fibrinogen, obtained in pure form, which can be made into plastics that have application in surgery
5. Thrombin, which together with fibrinogen, of course, yields fibrin; fibrin films that can substitute for natural membranes have been prepared, and fibrin foams can be used with thrombin to accelerate blood clotting in operative work

The infusion of plasma or plasma proteins has been recommended for many conditions, some of which may be mentioned briefly. It is usually agreed that in shock there is loss of plasma through the capillary walls into the tissue spaces. This leads to a decrease in blood volume, hemoconcentration (i.e., concentration of formed elements), and lowered colloidal osmotic pressure. In severe and extensive burns there is a great loss of proteins from the blood because of transudation of fluid at the site of the burn. At the same time there is believed to be an absorption of toxic substances formed at the burned tissue. These toxins cause an increase in capillary permeability throughout the body, and more plasma is lost. After extensive burns the loss of blood plasma may be even greater than in shock. When hypoproteinemia occurs, as it does in a number of clinical syndromes, plasma may be administered. Loss of blood as a result of hemorrhage, although best replaced by whole blood, may also be replaced by plasma, and this is perhaps the most general use of plasma.

The treatment of shock by the intravenous injection of whole blood, plasma, blood proteins, or other colloids is not accepted by all investigators as the correct method. Some insist that large volumes of physiological saline solution give as good results.

All these procedures are, in a sense, emergency measures. The best and most physiological way to replace blood is by enabling the organism to replenish it in the normal manner. A regimen to accomplish this should follow all transfusion methods.

Recent studies in human subjects indicate that a fluorocarbon fluid may be useful at least on an emergency basis as an "artificial blood." Extensive experiments using animals have given promising results.

Medicolegal tests for blood

Red blood cells, if still fresh, can sometimes be identified microscopically. They can also be identified spectroscopically because the absorption spectrum of hemoglobin is quite specific. The various tests for the catalytic oxidizing effect of heme, e.g., the guaiac, benzidine, and reduced phenolphthalein tests, are helpful but are not specific for blood. Blood reacts after it is heated because of the catalytic action of iron, as well as before because of the enzyme peroxidase. Raw milk, pus, saliva, and other biological materials contain peroxidases, which react similarly, but no reaction is seen after heating. Certain salts also give the guaiac test. If hemin crystals can be prepared, they are indicative of blood; this is Teichmann's test.

However, none of these tests is diagnostic of the species. An immunological technique is necessary if the species is to be determined. The test is based on the fact that blood serum of an animal into which has been injected repeatedly the blood serum of an animal of another species gradually acquires the property of producing a precipitate when mixed with serum of an animal of the species whose serum was injected. This precipitin reaction is frequently used for the detection of human blood.

Electrophoretic procedures using cellulose acetate or other media are particularly valuable in identifying types of human hemoglobin, e.g., A, F, S, and C, even on blood from dried stains. Information on the age, race, and certain other characteristics of an individual may be obtained in this way.

Further forensic identification of human blood samples, fresh or from dried blood stains, usually entails the determination of a variety of genetic markers. Currently these may include a wide range of substances previously discussed in this chapter, such as blood group antigen-antibody systems, hemoglobin and haptoglobin polymorphic protein systems, and the polymorphic enzyme systems, especially of phosphoglucomutase, erythrocyte acid phosphatase, esterase D, adenosine deaminase, 6-phosphogluconate dehydrogenase, and glucose 6-phosphate dehydrogenase. The use of these various genetic markers can narrow down by population frequency data the identity of the individual or individuals, as well as their age and race. Various chromatographic and electrophoretic procedures are used for such analyses (Chapter 25).

Suggested readings

Altman, P.L., and Dittmer, D.S., editors: Biology data book, vol. 3, sections 12 and 13, Blood and other body fluids, Bethesda, Md., 1974, Federation of American Societies for Experimental Biology.

Amzel, L.M., and Poljak, R.J.: Three-dimensional structure of immunoglobulins, Annu. Rev. Biochem. **48:**961, 1979.

Barnard, G.F., et al.: Mechanism of porphobilinogen synthase—possible role of thiol groups, J. Biol. Chem. **252:**8965, 1977.

Bevan, D.R., Bodlaender, P., and Shemin, D.: Mechanism of porphobilinogen synthase: requirement for Zn^{2+} for enzyme activity, J. Biol. Chem. **255:**2030, 1979.

Bogorad, L.: Biosynthesis of porphyrins. In Dolphin, D., editor: The porphyrins, vol. 6, part A, New York, 1979, Academic Press, Inc.

Bradlow, H.L., et al.: Studies in porphyria. II. Evidence for a deficiency of steroid Δ-5α-reductase activity in acute intermittent porphyria, J. Exp. Med. **138:**754, 1973.

Brenner, D.A., and Bloomer, J.R.: The enzymatic defect in variegate porphyria, N. Engl. J. Med. **302:**765, 1980.

Brewer, G.J., editor: Hemoglobin and red cell structure and function, Adv. Exp. Biol. Med. vol. 28, 1972.

Caughey, W.S., editor: Biochemical and clinical aspects of hemoglobin abnormalities, New York, 1978, Academic Press, Inc.

Dolphin, D., editor: The porphyrins, 7 vols., New York, 1978-1979, Academic Press, Inc.

Doss, M., editor: Regulation of porphyrin and heme biosynthesis, Enzyme **16:**1, 1973.

Eales, L.: Clinical chemistry of the porphyrins. In Dolphin, D., editor: The porphyrins, vol. 6, part A, New York, 1979, Academic Press, Inc.

Edelstein, S.J: Cooperative interactions of hemoglobin, Annu. Rev. Biochem. **44:**209, 1975.

Edsall, J.T.: Hemoglobin and the origins of the concept of allosterism, Fed. Proc. **39:**226, 1980.

Finch, C.A., and Lenfant, C.: Oxygen transport in man, N. Engl. J. Med. **286:**407, 1972.

Friedman, M.J., and Trager, W.: The biochemistry of resistance to malaria, Sci. Am. **244:**154, 1981.

Goldwasser, E.: The purification and properties of erythropoietin. In Nakao, R., et al., editors: Erythropoiesis, Baltimore, 1976, University Park Press.

Harris, J.W., and Kellermeyer, R.W.: The red cell: production, metabolism, destruction. Normal and abnormal, Cambridge, Mass., 1970, Harvard University Press.

Henry, R.J., et al., editors: Clinical chemistry, principles, and techniques, Hagerstown, Md., 1974, Harper & Row, Publishers, Inc.

Henry, R.L., and Seegers, W.H.: Drugs related to thrombosis and hemostasis. In Marks, B.H., et al., editors: Principles of pharmacology, New York, 1981, John Wiley & Sons, Inc.

Huisman, T.H.J., and Jonxis, J.H.P.: The hemoglobinopathies: techniques of identification, Clinical and Biochemical Analysis Series, vol. 6, New York, 1977, Marcel Dekker, Inc.

Jackson, C.M., and Nemerson, Y.: Blood coagulation, Annu. Rev. Biochem. **49:**765, 1980.

Kabat, E.A., Wu, T.T., and Bilofsky, H.: Sequence of immunoglobulin chains, Pub. No. 80-2008, Bethesda, Md., 1979, National Institutes of Health.

Litman, G.W., and Good, R.A., editors: Immunoglobulins, vol. 5, New York, 1978, Plenum Publishing Corp.

Marver, H.S., and Schmidt, R.: The porphyrias. In Stanbury, J.B., Wyngaarden, J.B., and Fredrickson, D.S., editors: The metabolic basis of inherited disease, New York, ed. 4, 1978, McGraw-Hill Book Co., Inc.

Masoro, E.J., and Siegel, P.D.: Acid-base regulation: its physiology and pathophysiology, Philadelphia, 1971, W.B. Saunders, Co.

Miyagi, H., et al.: Uroporphyrinogen I synthase from human erythrocytes: separation, purification, and properties of isoenzymes, Proc. Nat. Acad. Sci. **76:**6172, 1979.

Nakao, K., Fisher, J.W., and Takaku, F., editors: Erythropoiesis, Baltimore, 1976, University Park Press.

Pauling, L.: Sickle cell anemia: a molecular disease, Science **110:**543, 1949.

Perutz, M.F.: Regulation of oxygen affinity of hemoglobin: influence of structure of globin on the heme iron, Annu. Rev. Biochem. **48:**327, 1979.

Pitts, R.F.: The role of ammonia production and excretion in regulation of acid-base balance, N. Engl. J. Med. **184:**32, 1971.

Putnam, F.W., editor: The plasma proteins, ed. 2, New York, 1975, Academic Press, Inc.

Reynolds, J.A.: Red cell membranes—fact and fancy, Fed. Proc. **32:**2034, 1973.

Rose, N.R., et al., editors: Principles of immunology, ed. 2, New York, 1979, The Macmillan Co.

Sharon, N., and Lis, H.: Glycoproteins—structure and vital biological roles, Chem. Eng. News **59:**21, 1981.

Smith, L.L., Pownall, H.J., and Gotlo, A.M., Jr.: The plasma lipoproteins: structure and metabolism, Annu. Rev. Biochem. **47:**751, 1978.

Stenflo, J., and Suttie, J.W.: Vitamin K–dependent formation of γ-carboxyglutamic acid, Annu. Rev. Biochem. **46:**157, 1977.

Tietz, N.W., editor: Techniques of clinical chemistry, Philadelphia, 1976, W.B. Saunders Co.

Various authors: Molecular basis of heparin action, Fed. Proc. (Workshop) **36:**10, 1977.

Welt, L.G.: Clinical disorders of hydration and acid-base equilibrium, ed. 3, Boston, 1970, Little, Brown & Co.

West, J.B.: Causes of carbon dioxide retention in lung disease, N. Engl. J. Med. **284:**1232, 1971.

Zucker, M.B.: The functioning of blood platelets, Sci. Am. **242:**86, 1980.

Chapter 17

Body fluids

The major fluid compartments

The total amount of fluid in the human body is approximately 70% of body weight. Most of this is found within three major compartments: intracellular fluid (about 70%), interstitial fluid (e.g., lymph) (about 20%), and blood plasma (7%). The remaining 3% is in the intestinal lumen, cerebrospinal fluid, and other compartments. The distribution of body fluids, as a percentage of body weight, is shown schematically in Fig. 17-1.

Chemical composition

The composition of various body fluids is delicately controlled by homeostatic mechanisms and maintained at relatively constant levels in the normal organism. There are similarities in the chemical composition of the above major compartments of body fluids, yet there are sharp differences. This is shown schematically in Fig. 17-2. For example, the protein content of intracellular fluid is nearly four times that of the blood plasma, which in turn is much greater than that of interstitial fluid. Sharp differences also exist in the electrolyte composition. Potassium and magnesium are the predominant cations in intracellular fluid, whereas sodium predominates in blood plasma and interstitial fluid. Likewise, phosphate and sulfate are the major anions in intracellular fluid, whereas chloride and bicarbonate predominate in plasma and interstitial fluid. Differences in the concentrations of nonelectrolytes, such as glucose and urea, also exist in different compartments. These general differences are attributable to differing membrane permeabilities, active transport mechanisms, and varying metabolic activities in the different areas. Also there are differences in the composition of various body fluids as adaptations to function, as is true of tissues (Chapter 15).

Before we discuss the chemical composition and its regulation of the various specialized body fluids (secretions of the gastrointestinal tract, milk, urine, etc.), we consider the problems of maintaining the water and electrolyte balance of body fluids in general. Biochemical mechanisms involved in the maintenance of acid-base balance and osmotic equilibrium are described in Chapter 16.

Water balance

Water is derived from ingested water and beverages, from the water content of solid foods, and from "metabolic" water. The latter, of course, is derived from the oxidation of metabolites and amounts to some 300 to 350 gm per day in an average adult male. According to composite estimates, 100 gm of fat yields 107 gm of water; 100 gm of starch, 55 gm of water; and 100 gm of protein, 41 gm of water. Water is absorbed in the upper small intestine and distributed by way of the lymph and blood into and from the various tissues and cells of the body. Eventually water is excreted by four channels: kidneys, skin, lungs, and intestine. Various salts accompany the water excreted in the urine, sweat, and intestinal secretions, as is described later.

Figure 17-1 Distribution of body fluids.

Modified from Gamble, J.L.: Chemical anatomy, physiology and pathology of extracellular fluid, Cambridge, Mass., 1958, Harvard University Press.

Figure 17-2 Chemical composition of extracellular fluids and cell fluid. Note that the values are given as milliequivalents per liter of H_2O contained in the fluid. Note also that the patterns of blood plasma and interstitial fluid are almost identical; the greatest single item of difference is in the amounts of protein. This makes necessary adjustment of the concentrations of the diffusible ions, which will preserve the total cation-anion equivalence (Donnan equilibrium). The nonelectrolyte concentration (glucose, urea, etc.) is seen to be very small in comparison with that of the electrolytes, although the total quantity carried to the tissue cells and into the urine over a unit of time is several times larger. Note the predominance of K^+ and the high protein content of cell fluid.

Modified from Gamble, J.L.: Chemical anatomy, physiology and pathology of extracellular fluid, Cambridge, Mass., 1958, Harvard University Press.

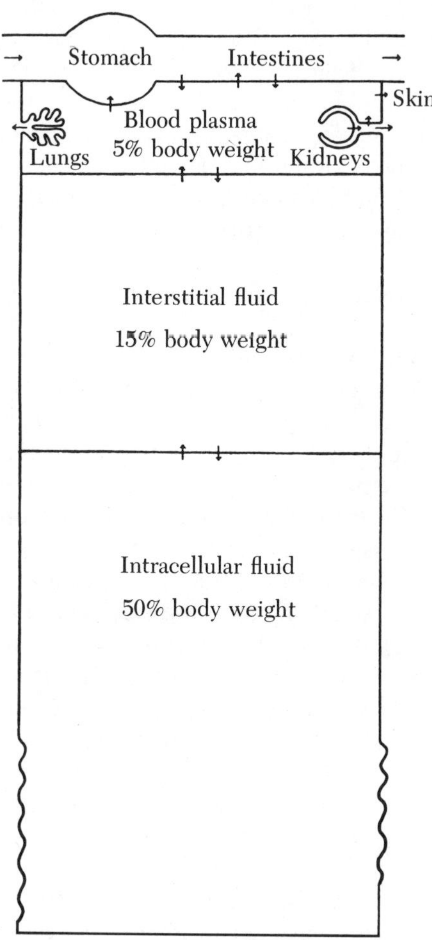

Figure 17-1

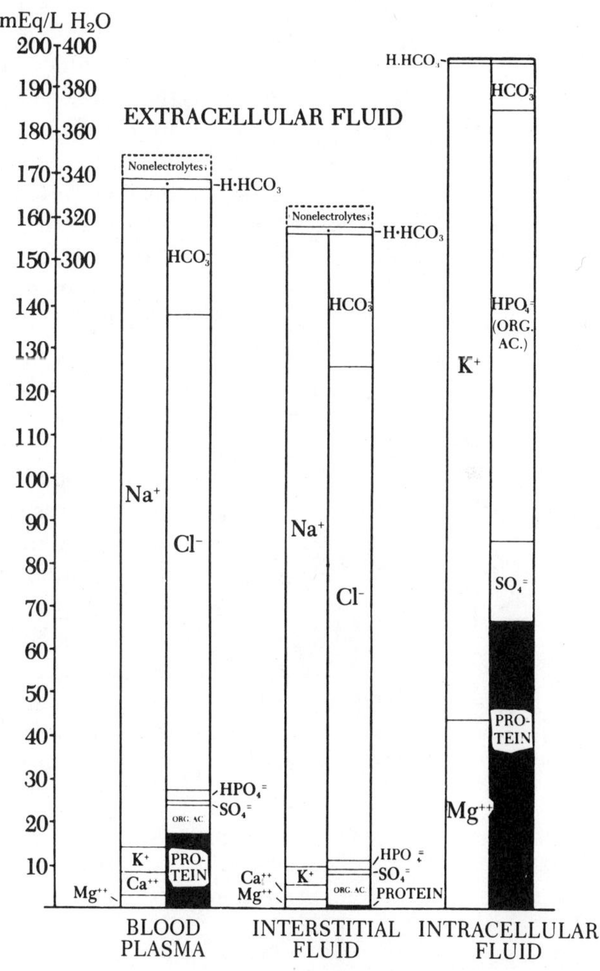

Figure 17-2

Table 17-1
Typical daily water balance—average-sized, normal, adult male

Water intake	Grams	Water output	Grams	
Drinking water	400	Skin	500	
Water in other beverages	580	Expired air	350	
Preformed water in solid foods	720	Urine	1100	
Metabolic water	320	Feces	150	
Total	2020	Total	2100	Balance = −80 gm

A typical daily water balance for an average-sized, normal, adult male is given in Table 17-1. Water ingested as sucn or from beverages will vary widely, from about 700 ml to 1500 ml or more depending on environmental temperature, activity, and other factors. Water output normally will balance the intake, with wide variations for the same reasons.

Distribution of body water The distribution of body water into the various compartments of the body has been determined by a number of procedures. The principle involved is the intravenous injection of a known amount of a substance that will be distributed uniformly in the total body water followed by a determination of the extent of its dilution. Deuterium oxide ("heavy water") and tritiated water, containing the radioactive isotope ^3H, have been successfully employed. These apparently are handled by the body in exactly the same way as is ordinary water. Samples of blood are removed at intervals and the isotope concentration is determined. By this procedure the average total body water in the adult male subject is approximately 60% of the body weight. The half-life of body water in man has been found to be about 9.3 days. Hence there is a constant turnover of water in the body, as is true of other body constituents.

The distribution of body water into the fluids of the compartments of the body is presented in Fig. 17-1.

Maintenance of water balance Under ordinary circumstances water balance is maintained between the cells and fluids of the body at a fairly constant level. Water intake is regulated mainly by "thirst" and water output by excretion by way of the four channels mentioned above. The kidneys play an important part in regulating water balance. If excessive amounts of water (or fluid) are ingested, the kidneys excrete the excess. On the other hand, if the water-fluid intake is low, the kidneys excrete a more concentrated urine so that less water is lost from the body. The skin and lungs also are important channels of water excretion, chiefly for the purpose of heat regulation of the body. "Insensible perspiration" from the skin represents a significant means of water loss (Table 17-1). Water loss in expired air is also appreciable. Indeed expired air is almost saturated with water—easily visible on a cold day! A small amount of water is excreted in the feces (Table 17-1). A decrease in body water, leading to *dehydration,* can result either from an inadequate intake or from excessive excretion, as from excessive sweating, vomiting, or diarrhea. Loss of electrolytes along with water also may occur under the latter conditions, as is discussed later. Further details may be found in the pertinent suggested readings given at the end of the chapter.

Electrolyte balance

Electrolyte content of the major fluid compartments As is true of body water, the concentrations of electrolytes in the three major compartments of body fluids are maintained within narrow ranges during health. Average values, in milliequivalents per liter of H_2O, are given in Fig. 17-2. Note the fact that Na^+ is the major cation in blood plasma and interstitial fluid, whereas K^+ and Mg^{++} predominate in intracellular fluid. Sharp differences in anion concentrations also exist, with Cl^-

Table 17-2
Comparison of the electrolyte content of different body fluids (all values are expressed as mEq per liter)

Ions	Extracellular fluid	Interstitial fluid	Intracellular fluid (muscle cell)
Cations			
Na^+	142 ± 5	145	10
K^+	5 ± 1	4	150
Ca^{++}	5 ± 1	3	2
Mg^{++}	3 ± 1	2	40
Total	155	154	202
Anions			
Cl^-	103 ± 4	114	2
HCO_3^-	27 ± 2	30	10
$HPO_4^=$	2 ± 0.5	2	140
$SO_4^=$	1 ± 0.5	1	5
Organic acids	6 ± 2	6	5
Protein	16 ± 2	1	40
Total	155	154	202

and HCO_3^- predominating in the intracellular fluids and HPO_4^{2-}, SO_4^{2-}, organic acids, and protein predominating in intracellular fluid.

The mean electrolyte content and normal ranges in the three major compartments of body fluids are given in Table 17-2. Note that the values are expressed as milliequivalents per liter of body fluid.

When changes in the electrolyte content of the three major compartments do occur, the blood plasma and interstitial fluid bear the brunt of any fluctuations, the intracellular fluid being the last to suffer any changes. The same is true of water content and acid-base balance (Chapter 16). This fact is reasonable because of the vital importance of these environmental factors in cellular metabolism and hence in cell survival.

Maintenance of electrolyte balance Various mechanisms operate to maintain a normal electrolyte balance in body fluids. Selective transport mechanisms (Chapter 19) are an important factor. The "sodium pump" is an excellent example, maintaining a low intracellular level of Na^+ and a relatively higher extracellular level. Hormonal effects (Chapter 18), Gibbs-Donnan effects (Chapter 24), and cellular metabolic activities are some of the other factors involved.

The intake of electrolytes in amounts sufficient to maintain electrolyte balance is ordinarily met by a well-balanced diet (Chapter 21). Thirst may regulate electrolyte intake to some extent also. Indeed humans apparently cannot discriminate between salt hunger and water hunger, as can some species of animals. Environmental conditions (such as excessive heat) and vigorous exercise can increase the loss of electrolytes, mainly in perspiration. Under these conditions it may be desirable to provide some extra salt with the drinking water to replenish these electrolytes in extracellular fluid. On the other hand, current evidence indicates that the average American is already ingesting excessive amounts of salt, in the form of table salt and certain processed foods (Chapter 21).

The kidney plays an important role in the maintenance of electrolyte balance. Salts are either excreted or conserved by the kidney (by tubular reabsorption, discussed later), depending on the dietary intake and physiological needs.

Dehydration

Dehydration may result from an inadequate intake or excessive loss of water, or both, and is of two types: (1) that from deprivation of water alone and (2) that from pathological loss of water and electrolytes (Fig. 17-3). The output of water may be due to diuresis, to a loss of water from the gastrointestinal tract as a result of

Figure 17-3

Schematic representation of the effect of diabetic acidosis on the volume and composition of extracellular fluid. The vertical dimension represents osmolar concentration, and the horizontal, volume of fluid. For purposes of simplification, only Na^+, $C1^-$, and HCO_3^- are depicted.

A Extracellular fluid of a normal control. The broken lines in **C** and **D** indicate the pattern of normal fluid for comparison.

B The β-hydroxybutyric and acetoacetic acids in excess displace the HCO_3^-. This results in the excretion of some Na^+ as salts of these acids and the elimination of CO_2 by the lungs. For some reason NaCl is also excreted in large amounts. These salts take H_2O along with them and dehydration is accelerated.

C The Cl^- continues to be excreted even after its concentration in the blood serum is greatly diminished. The patient experiences extreme thirst but is unable to retain H_2O by mouth because of nausea and vomiting. In fact, these add to the salt depletion and dehydration. In the attempt to excrete CO_2 through the lungs, overventilation occurs, which takes away more H_2O.

D The final stage. Here the dehydration is caused in part by the high glucose in the blood, which is diuretic. The acetoacetic acid and β-hydroxybutyric acid are buffered by blood HCO_3^-, thus lowering the HCO_3^-. They also must be excreted. Despite an increased formation of NH_4^+ by the kidney to help neutralize these acids so that they can be excreted, the acids take away some Na^+ into the urine. Consequently there is a loss of fixed base from the blood along with a lowered bicarbonate.

From Peters, J. P.: In Duncan, G.G., editor: Diseases of metabolism; detailed methods of diagnosis and treatment, ed. 3, Philadelphia, 1952, W.B. Saunders Co.

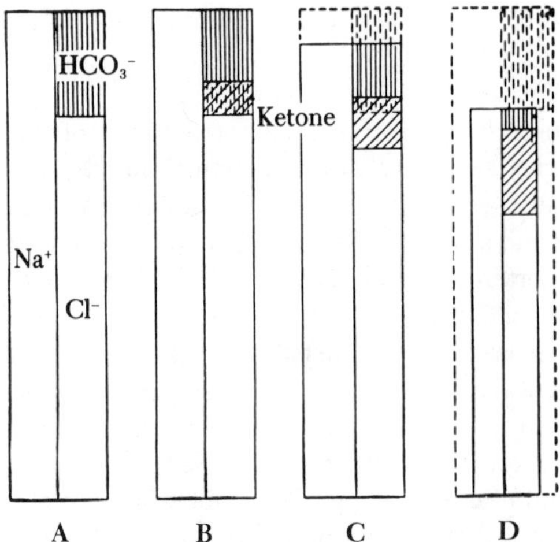

diarrhea, or, more frequently, to persistent vomiting. There are all degrees of dehydration, from a mild state to an exceedingly severe one, which may establish itself more rapidly than one would believe possible. At first the interstitial fluid suffers a shrinkage and not much harm is done; in this situation, if water is taken by mouth, there is a restoration of normal conditions. Losses of the second type require calculated replacement of electrolytes and water.

Pathological alterations of water and electrolyte balance At this point it may be beneficial to repeat the statement that normally the total osmotic effects of the plasma, the interstitial fluids, and the intracellular fluid are all the same. This does not hold for secretions, e.g., sweat and saliva, which are secreted onto relatively impermeable stratified epithelium, but it does hold for all truly internal fluids. The osmotic effect is attributable to nonelectrolytes, glucose, urea in some measure, and to proteins in a very minute degree, but most of the osmotic effect is attributable to the inorganic ions. Consequently, gains or losses of electrolytes, especially sodium or potassium ions, or changes in their concentrations, are usually followed by shifts of fluid to restore osmotic equilibrium.

The volume of blood in an adult's body is roughly 5 liters, of which about 3 liters are plasma. From this blood plasma, of course, all secretions, as well as the interstitial fluid, are derived. As an example of the effect of loss of secreted fluid on water and salt balance, the following illustration may be given: Assume that 500 ml of mixed jejunal and ileal fluids have been secreted and lost from the body. A mixture of equal parts of these two secretions resembles blood plasma in composition except that it contains less protein. Consequently, removing a mixture such as this is like removing 500 ml of protein-free plasma. The results to be expected are as follows:

1. Reduction in plasma volume from 3000 to 2500 ml
2. Reduction in total blood volume of 500 ml, i.e., from 5000 to 4500 ml
3. Rise in erythrocyte count because of blood concentration
4. Increase in the concentration of plasma proteins by 20% with a rise in colloidal osmotic pressure
5. No change in *concentration* of the plasma electrolytes and hence little change in the total osmotic pressure

This would result in no change in the size of the body cells because of the constancy of osmotic pressure. However, additional losses of other body fluids would have other effects. Such losses are caused by longer periods of dehydration, pyloric stenosis, intestinal obstruction, sweating, trauma, and severe burns.

If dehydration and loss of extracellular electrolytes are continued, the volume of the blood plasma decreases, and the plasma is found to have become concentrated. Serum proteins increase in concentration. Blood urea rises, and a negative balance of nitrogen and potassium occurs, which indicates that a generalized tissue disintegration has set in. Since the cells contain potassium, this element thus gets into the interstitial fluid and thence into the plasma and is excreted in the urine. However, prolonged dehydration from any cause has been shown to result in a greater loss of intracellular potassium than can be accounted for by protein catabolism. A continued dehydration, with concentration of blood and loss of cations, eventually leads to death.

Pyloric stenosis or obstruction results in excessive loss of fluid by vomiting. The fluid lost is gastric secretion, which is a varying mixture of sodium chloride, potassium chloride and hydrochloric acid. Therefore a drop in the chloride ion concentration of the plasma occurs. There is a compensatory rise in bicarbonate ion (derived from carbon dioxide) to preserve electroneutrality. Plasma potassium may also be reduced if this ion is not included in the replacement solution.

Diarrhea in infants is accompanied by a decrease in extracellular water from a loss of sodium, chloride, and bicarbonate in the watery stools. These ions are derived from the alkaline intestinal secretions, particularly pancreatic juice and bile (Fig. 17-4). As sodium leaves the plasma and interstitial fluid, potassium salts move out of the cells. As a result, intracellular potassium is lost in tremendous quantities. For this reason potassium salts are added to therapeutic solutions containing sodium salts. Since heart block is produced when potassium rises to a certain level in the plasma, care must be exercised in the intravenous administration of such fluids; hence oral administration is recommended.

Similarly, in any condition involving an excessive loss of sodium chloride and water from the body, e.g., hemorrhage or intestinal obstruction, potassium tends to leave the cells and go into the blood plasma. The potassium salts are excreted in the urine if renal function is satisfactory, but if this function stops, potassium accumulates in the plasma and toxicity is manifested. Potassium also leaves the muscles and other organs in certain disease states and after surgical trauma. Under these circumstances sodium frequently accumulates in excess of the potassium lost.

Figure 17-4

Electrolyte composition of gastrointestinal secretions. The K^+ present in these secretions (included in the black block [see Fig. 17-2] beneath the Na^+) is usually two to five times that of plasma K^+. Gastric vomitus usually is a mixture of gastric juice and gastric mucus, and the Na^+ may be less than, equal to, or greater than the Cl^-.

From Gamble, J.L.: Chemical anatomy, physiology and pathology of extracellular fluid, Cambridge, Mass., 1958, Harvard University Press.

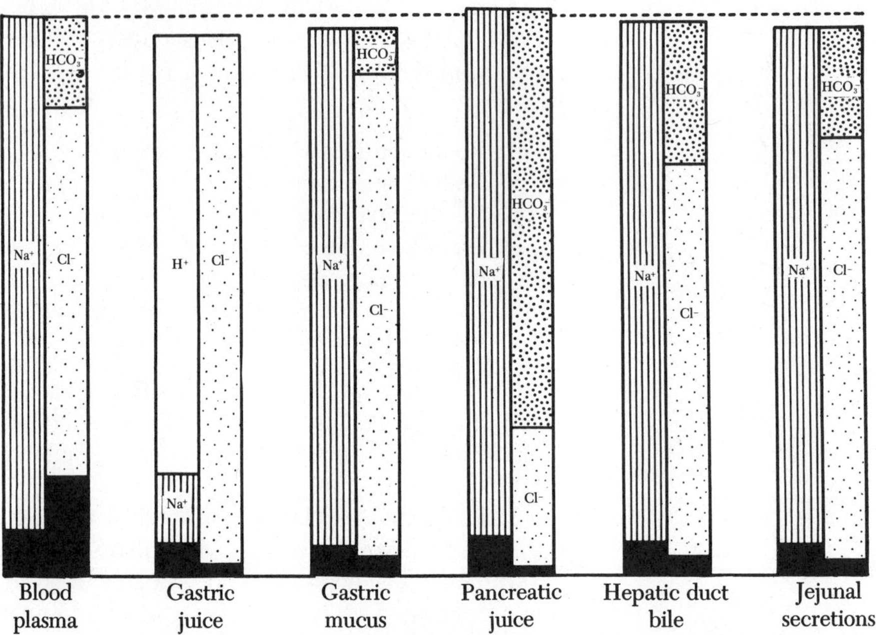

In untreated diabetes there is a loss of water, together with sodium and potassium; these ions are excreted in the urine as salts of the keto acids. When acidosis occurs, such losses are increased. Plasma bicarbonate is diminished. Since vomiting frequently occurs, pronounced dehydration results from losses of electrolytes and water, with effects as pictured in Fig. 17-3. However, the administration of insulin and replacement solutions halts ketosis and further loss of water and electrolytes. The acidoses of childhood and of starvation also are accompanied by similar losses of water and salts.

The adrenal cortex has a profound influence on electrolyte metabolism (Chapter 18). More specifically, it controls the level of sodium ions. In adrenalectomized animals there is a decrease of sodium in the plasma and an increase in the potassium. Attention at first was centered on this rise in the potassium, but the sodium now appears to be the more important factor. In Addison's disease, which is a condition involving adrenal insufficiency, the same relationships are seen. The low plasma sodium is a result of increased excretion of sodium by the kidneys. Sodium is lost not merely from the plasma but also from the interstitial fluid, especially in the muscles, which at the same time gain water, in a manner analogous to the swelling of erythrocytes when placed in hypotonic saline solutions. Treatment with sodium chloride alleviates the symptoms of patients suffering from this disease. Administration of one of the adrenocortical hormones, deoxycorticosterone, cortisone, or aldosterone, in large amounts to animals with acute adrenal insufficiency results in an increased sodium concentration in the blood serum. This probably is caused by a shift of interstitial fluid (tissue fluid) to the blood, resulting in a dilution of the blood with this fluid, which contains sodium salts. A more rapid excretion of water and a diminished excretion of sodium salts then occurs with a consequent improvement of the condition. In edematous states, e.g., nephroses, cortisone frequently causes a diuresis of sodium and water.

In shock from trauma or burns, there is no overall loss of salt from the body, but there is internal loss and a distinct change of the electrolyte pattern. The injured or burned tissues lose potassium, apparently by extrusion from the cells. Sodium passes into the cells in exchange for the potassium. These changes are proportional to the mass of damaged tissue. There is also a considerable gain of extracellular fluid (water, sodium), which is probably the source of the increment of intracellular sodium. The sodium present in the injured cells is really lost from the plasma and interstitial fluid and other uninjured functioning tissue. Tissues remote from the site of injury or burn do not show much change in water content but do show a loss of sodium and a gain of potassium, pointing to extracellular dehydration with intracellular swelling.

The loss of salts and water by sweating may be considerable. When strenuous work is done, especially at high temperatures, as by miners or blast furnace workers, as much as 10 to 15 liters may be lost in 8 hours of work. If each liter contains 3 gm of sodium chloride, this represents a tremendous depletion of the salts of the interstitial fluid. When these stores are gone, first the plasma and then the cells suffer. Violent cramps (stoker's or miner's cramps) and prostration may result from the combined loss of salt and fluid. Replacement of the water alone may make matters worse by diluting the plasma. To guard against this, the drinking water for such workers should contain 0.1% to 0.15% sodium chloride. This does not have an unpleasant flavor and allays thirst quite as well as unsalted water. In fevers, patients may lose large amounts of moisture and electrolytes in perspiration. These should be replaced if ill effects are to be prevented. A moderately increased salt intake during hot weather has also been recommended for most people because of this loss of salt in the persipiration. In cases of renal insufficiency or edema more cautious replacement is necessary.

The foregoing sections of this chapter deal with the intracellular and major extracellular fluids of the body, their distribution, general composition, and balance. The remaining sections consider the specialized fluid secretions and excretions of the body, their chemical composition and biochemical functions.

Secretions of the gastrointestinal tract

The principal constituents of gastrointestinal secretions include enzymes for the hydrolysis of foodstuffs, electrolytes to provide a favorable environment for the activity of the enzymes, and mucus for mechanical lubrication and protection. The chemical properties and catalytic activities of most of the enzymes present in the gastrointestinal tract are discussed in Chapters 5 and 11 to 14. Consideration should now be given to the chemical composition of the secretions themselves.

Under normal conditions, the water and electrolytes of the digestive secretions (Fig. 17-4), as will be discussed, are reabsorbed and little is lost by way of the feces. However, in certain diseases in which either excessive vomiting or diarrhea occurs, fluid and electrolyte loss, particularly potassium, may cause serious derangements of acid-base, electrolyte, or water balance. Knowledge of the composition of the fluids lost therefore is essential to rational therapy.

The secretions of the gastrointestinal tract are produced by the salivary glands, the epithelial cells of gastrointestinal mucosa, the pancreas, and the liver. The relations of these various structures are shown in Fig. 17-5.

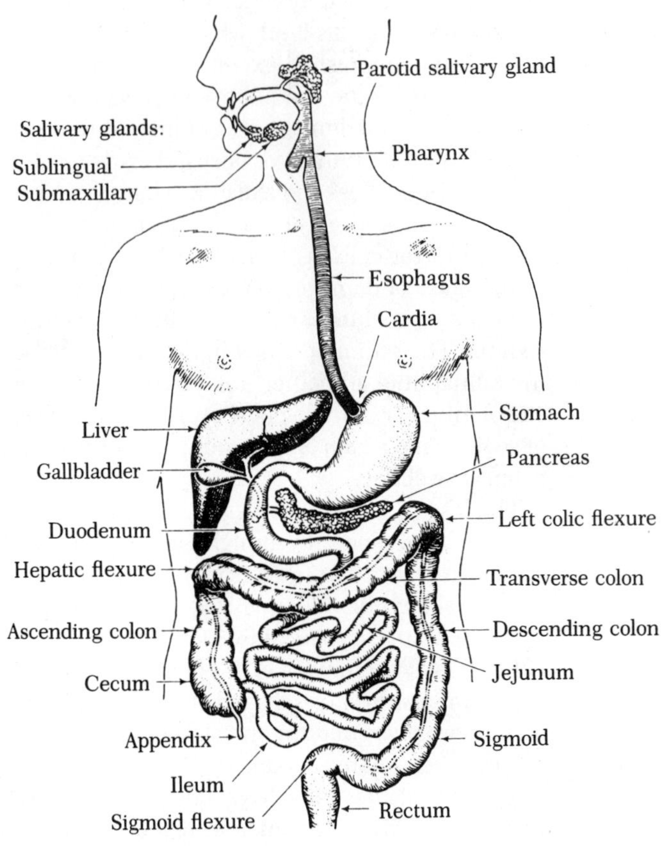

Figure 17-5
Representation of the human gastrointestinal tract with associated structures.

From Schottelius, B.A., and Schottelius, D.D.: Textbook of physiology, ed. 18, St. Louis, 1978, The C.V. Mosby Co.

Saliva

Saliva is the mixed secretion of the parotid, submaxillary, sublingual, and buccal glands. It contains 99.3% to 99.7% water and has a specific gravity of 1.002 to 1.008. Approximately 1500 ml is believed to be the daily secretion in man. The secretion of saliva is entirely under the control of the nervous system. A variety of stimuli cause an increased flow by relex stimulation. This is true whether the stimulus is psychic (sight, smell, or thought of food), mechanical (chewing), or chemical (action of acids, salts, etc. on the taste buds). There seems to be no hormonal control of salivary secretion.

Saliva is almost colorless and rather viscid, and if a quantity of saliva in a vessel is exposed to air, the surface becomes covered with an incrustation consisting of calcium carbonate with a small proportion of organic matter. The reaction of the saliva of a given individual is not constant. Resting saliva is slightly acidic, pH 6.4 to 6.9, whereas saliva obtained during active stimulation of the glands is neutral to slightly alkaline, pH 7.0 to 7.3.

The solid constituents of saliva comprise albumins, globulins, mucins, enzymes, urea, uric acid, and inorganic salts. The salivary mucins are glycoproteins and yield 30% to 45% carbohydrate on hydrolysis. Ovine submaxillary mucin, for example, contains approximately 45% carbohydrate and has about 800 disaccharide units per protein molecule. The disaccharide units are known to be *N*-acetylneuraminyl(2 → 6)*N*-acetylgalactosamine (Chapter 11). The disaccharide units are attached to every sixth amino acid residue of the protein component—a single polypeptide chain. The protein portion is rich in threonine, with a low isoelectric point, about 3.5, probably because of the carbohydrate moiety. This material has a relatively high viscosity and a high degree of hydration, which in part account for the protective and lubricating functions described below. The inorganic components differ greatly in concentration from those of blood serum, but the NPN constituents (urea, uric acid, NH_4^+ salts) appear to bear some relation to these same constituents in the blood. Amino acids and glucose occur in extremely small amounts in the saliva of healthy individuals (11 to 30 mg glucose per 100 ml). Both salivary cholesterol and lipid phosphorus values are very low as compared with blood. The salivary glands therefore appear to be quite selective in secretory action.

The chief inorganic ions present are K^+, $PO_4^{\equiv}$, and Cl^-, with smaller amounts of Na^+, Ca^{++}, and $SO_4^=$ (Fig. 17-4). Some of these may combine to form insoluble precipitates. This may be aided by changes in the pH brought about by decomposing food material left between the teeth or by the loss of carbon dioxide, held in solution in the saliva, as soon as it meets atmospheric conditions. Thus tartar may be formed. This consists chiefly of calcium carbonate and phosphate. Salivary calculi sometimes are formed in the ducts and are similar in composition to tartar (namely, $Ca_3[PO_4]_2$ or $CaCO_3$). It is usually stated that a clump of bacteria or a foreign body establishes a nucleus around which the precipitation of these salts occurs. However, calcium oxalate may be the precipitated salt, which, together with mucin and globulin, may form the calculus. Increased acidity is necessary for oxalate calculus formation.

Functions Saliva has a digestive function because of the enzymes present, but it also has other functions. It moistens and lubricates the food, permitting it to be swallowed easily. Saliva holds the taste-producing substances in solution and so brings them into contact with the taste buds. It dilutes salts, acids, etc., thereby protecting the mucosa and, to some extent, the teeth. It also has a cleansing action on the teeth, gums, and buccal mucosa. It owes its viscous and lubricating property to its content of mucin. This protein is present as an alkaline salt, which is soluble at the pH of saliva but is precipitated on acidification. It is one of the chief buffers present in saliva. A major function of epithelial mucins in general is the protection of the mucosal lining of the mouth, the gastrointestinal tract, and the inner surfaces of other body cavities. They form water-soluble films. In the stomach the acidity probably results in the formation of insoluble gels and although the mucins are not completely resistant to proteolytic enzymes, the action is slow and thus there is considerable protection. Some authorities maintain that saliva has an excretory function, since certain elements and drugs are found in it after administration. Among these are mercury, lead, and potassium iodide. Any part of them lost in expectoration could be considered excreted, but some of the part swallowed may be reabsorbed. Hence it is difficult to see how these elements can be called a true excretion. The same is true of the traces of urea, uric acid, and ammonium salts ordinarily found in saliva.

The parotid and submaxillary salivary glands have been implicated in the deiodination of the hormone thyroxine and hence in the regulation of the thyroxine level of the blood. The iodine thus released is excreted in the saliva and is reabsorbed in the small intestine for reuse, completing an iodine cycle.

Enzymes The principal enzyme of human saliva is an amylase, *ptyalin*. There are also traces of possibly a maltase, catalase, lipase, urease, protease, and others. The role of ptyalin in the digestion of food starches is discussed in Chapter 11.

Some investigators maintain that an important function of salivary amylase is a cleansing agent for the oral cavity; i.e., it digests starch particles or pastes left in or near the teeth.

Gastric juice

Gastric juice consists of water (99.4%), hydrochloric acid, mucins, and the enzymes pepsin, rennin (in infants), carbonic anhydrase, and lipase. The hydrochloric acid is secreted by the parietal cells and the pepsin (as pepsinogen, Chapter 5) by the chief cells. The gastric mucous substances comprise (1) the mucoid of the visible gastric mucus, secreted by the surface epithelium, (2) dissolved mucoproteose, a digestion product of the visible gastric mucus, and (3) "glandular mucoprotein," secreted by the neck mucous cells of the gastric glands. Glandular mucoprotein is considered to be the main carrier of the "intrinsic factor" of human gastric juice (Chapter 21).

Control of secretion The secretion of gastric juice is intermittent in humans, being increased reflexly by psychic stimuli (thought, sight, smell, or taste of food), mechanical stimuli (presence of food in the stomach), certain chemical substances (histamine, alcohol, "meat extractives," etc.), and the hormone *gastrin* (Chapter 18). Secretion may be inhibited by a variety of psychic and chemical agents. Fat has a definite inhibitory effect. The common belief that "greasy foods are hard to digest" rests on a solid basis. Fat causes the release of an inhibitory hormone, *enterogastrone* (Chapter 18), from the intestinal mucosa. Not only is the secretion of gastric juice inhibited, but so is the motility of the intestinal tract. Another inhibiting effect is that of the hydrochloric acid secreted, termed *acid inhibition*.

Inhibitors of carbonic anhydrase have also been shown to have a depressant effect on the secretion of gastric hydrochloric acid. This enzyme plays a role in the formation of hydrochloric acid by the gastric mucosa (see below), and its inhibition would be expected to decrease the secretion of the acid.

Gastric emptying The stomach does not retain its contents until gastric digestion is completed. Soon after food has reached the stomach, some material is ejected into the duodenum, and intestinal digestion and absorption begin. There are a number of factors that influence gastric emptying. The propulsive force lies chiefly in the antrum. Extragastric factors are mainly inhibitory, e.g., enterogastrone. The result is to keep the acidity of the duodenal lumen within certain limits so that damage to the mucosa of the small intestine is minimized and its functions may be maintained.

Hydrochloric acid The secretion of a strong mineral acid by the gastric mucosa is almost unique from a biological standpoint. At the instant of secretion by the parietal cells, the hydrochloric acid has a concentration of about 0.17N and a pH of 0.87. Pure parietal secretion apparently contains no phosphate, neutral chloride, or combined acid. It is practically free of ions other than H^+ and Cl^- and is approximately isotonic with blood plasma and body fluids. It is of remarkably constant composition. How then does the parietal cell manufacture such a strong acid from fluids, e.g., blood plasma and tissue fluid, that are neutral or slightly alkaline (pH 7.3)? There have been a number of theories to account for this phenomenon, but only one is outlined here. It is based on the fact that the parietal cells contain a high concentration of *carbonic anhydrase* (Chapter 16). This hypothesis is illustrated schematically in Fig. 17-6.

The fact that the blood plasma receives HCO_3^- during acid gastric secretion harmonizes with analyses of blood at such times and also with the fact that usually soon after meals the urine excreted is alkaline. The so-called *alkaline tide* is one of the mechanisms for keeping hydrogen ion concentration of the blood constant.

Functions of hydrochloric acid and factors decreasing its strength The hydrochloric acid of gastric juice initiates the conversion of the zymogen pepsinogen to active pepsin and provides a favorable pH for the activity of pepsin. Besides these most important functions, it serves other purposes. It has some physical action on the proteins, swelling some and making them more easily digested. It has a slight hydrolytic action, perhaps more on the disaccharides than on other foodstuffs, but even in the case of disaccharides, it is not of great significance. Another action of hydrochloric acid is to convert the colloidal ferric hydroxide, found in some foodstuffs, to monomolecularly dispersed ferric ions. Then these and any other ferric ions present are more readily reduced to ferrous ions at pH 5 or lower by ascorbic acid, cysteine, or the sulfhydryl groups of proteins, which may be in the food. The strong acid also has a strong antiseptic action. Contrary to popular belief, there is rarely any fermentation in the stomach. Its contents are too acid and the food does not remain long enough for gases to be formed.

Gastric acidity may be decreased by various factors. These include the following:

1. Variations in the rate of parietal secretion (the composition is constant but the rate may vary)
2. Dilution by the secretions of the other cells, especially mucus
3. Dilution and buffering by food
4. Dilution and buffering by the saliva that is swallowed
5. Regurgitation of duodenal fluid and bile

The dilution secretion may be formed by the cuboidal cells. It is possibly produced in order to dilute the stomach contents to proper concentration and consistency. The secretion of mucus is undoubtedly of great value. It has a high buffering power and must aid in slowing the acidification of stomach contents.

Enzymes

Pepsin Pepsin, a powerful proteinase, is present in the chief cells as the zymogen *pepsinogen* (Chapter 5). The conversion of pepsinogen to pepsin and its role in the digestion of proteins are discussed in Chaper 13. Pepsinogen is stable up to pH 9, at which point it is reversibly denatured, but at pH 12 or higher the denaturation of pepsinogen is irreversible. Since pepsin is more sensitive to alkali than is pepsinogen, the fact that pepsin exists in the inactive form can be readily demonstrated (Langley's experiment). A neutral extract of gastric mucosa is divided into two parts. Part A is acidified and can be shown to digest protein; it contains pepsin. Part

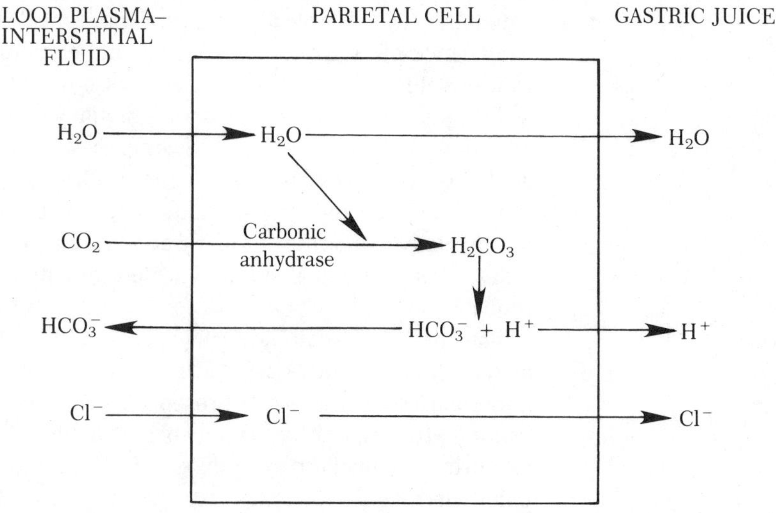

Figure 17-6
Formation of gastric hydrochloric acid by the parietal cell, according to the carbonic anhydrase hypothesis.

B is treated with an equal amount of water; it does not digest protein in neutral solution and presumably contains pepsinogen. Both are now made alkaline (pH 8.3); then they are neutralized and acidified to pH 2 or less. Part A now is incapable of digesting protein, whereas part B, the pepsinogen, has been unaffected by the alkali and now has proteolytic power.

About 99% of the pepsinogen elaborated in the gastric mucosa is secreted into the lumen of the stomach. The 1% remaining is secreted into the tissue fluid, then carried in the bloodstream to the kidney, and excreted in the urine. It is called *uropepsin*. This small amount can be determined in the urine, and since the rate of excretion of uropepsin varies directly with the rate of secretion of pepsinogen into the stomach, this determination may become a useful method in clinical studies of gastric secretion.

There are a number of methods for measuring the amount of pepsin in gastric contents. In patients having peptic ulcers there is usually a high pepsin value, whereas in pernicious anemia, cirrhosis of the liver, and various chronic gastric ailments a low pepsin content is found.

Rennin Some authorities include rennin, another proteolytic enzyme, among the gastric enzymes, but this probably is incorrect. Rennin occurs in the fourth stomach of the calf; it is probably also found in other ruminants and possibly in infants. Its action is to clot milk. This it does by a slight digestive action on casein, as discussed later.

Gastric lipase The lipase of gastric juice is less important than the pancreatic lipase because it is present in very small amounts. It is probably secreted in the active form, not as a zymogen; it is a tributyrase, having almost no action on fats containing long-chain fatty acids. The optimum pH is about 7.8 but in the presence of calcium ions is shifted to the acid side. A further discussion of gastric lipase is given in Chapter 12.

Gastric analysis The clinical procedures for testing various gastric functions include the administration of a test meal or of alcohol or the injection of histamine or insulin and the withdrawal by a stomach tube of the gastric contents for analysis. The routine use of test meals is practically abandoned.

Qualitative tests include those for butyric acid, lactic acid, occult blood, bile, and perhaps trypsin. The presence of the first two acids would point to yeasts or other microorganisms in the gastric secretions and hence a lack of free hydrochloric acid. If blood is present, ulcers, hemorrhages, or other pathological states would be indicated. In testing for blood, a meatless prior meal is imperative. Either bile or trypsin is evidence of regurgitation of intestinal contents; this a frequent normal occurrence. A microscopic examination is also usually made.

The quantitative procedures are gradually being changed to conform to modern chemical ideas. The total acidity comprises the acidity contributed by hydrochloric acid, organic acids, and acid salts, neutralized or buffered by various constituents of the gastric juice and the foodstuffs. With the use of various indicators, a rough idea of the relative amounts of these fractions may be obtained. This is not usually done today. From a clinical standpoint determination of pH or titration of free acid is probably sufficient.

The clinician usually wishes to ascertain whether there is hyperacidity or hypoacidity, high or low pepsin, large or small volume secreted, as well as whether abnormal constituents are present. The absence of hydrochloric acid is termed *achlorhydria*. Achlorhydria may occur in pernicious anemia, in gastric carcinoma, and in a number of other conditions. If hydrochloric acid is not entirely absent but is below normal, the condition is called hypoacidity. Hypoacidity frequently accompanies gastric carcinoma, as well as many gastrointestinal ailments, e.g., gastritis and constipation, secondary anemia, and chronic debilitative diseases. Many normal pregnant women have low gastric acidities. Of the conditions in which the acidity is elevated (hyperacidity), perhaps the most noteworthy are duodenal ulcer and gallbladder disease. The fact should be emphasized that the acidity can never exceed a certain value (pH 0.87), since the parietal cells secrete a fluid of constant composition.

Pancreatic juice

The pancreatic duct joins with the common bile duct to form the ampulla of Vater; thus pancreatic juice and the bile empty into the duodenum at the same point. The total volume of pancreatic juice secreted daily in humans has been estimated at about 500 ml. The solids present amount to about 1.3% to 1.4%; the specific gravity is about 1.007; and the fluid is alkaline, with a pH of about 8. The alkalinity is from bicarbonate (Fig. 17-4). Since carbonic anhydrase is present in pancreatic tissue and since the administration of an inhibitor of this enzyme has been shown to decrease the bicarbonate content of pancreatic juice, most of the bicarbonate ion is probably produced according to the following reactions:

$$CO_2 + H_2O \xrightarrow[\text{anhydrase}]{\text{Carbonic}} H_2CO_3 \rightarrow H^+ + HCO_3^-$$

Enzymes Pancreatic juice, as secreted, contains a number of powerful digestive enzymes, some in their inactive, zymogen forms. The list includes trypsinogen, chymotrypsinogen, elastase (as zymogen), two carboxypeptidases, a lipase (steapsinogen), an amylase (amylopsin), two polynucleotidases (ribonuclease, deoxyribonuclease), lactase, sucrase, possibly a maltase, several esterases, and an alkaline phosphatase. The properties and specific digestive functions of each of these enzymes are discussed in Chapters 5 and 11-14.

Control of secretion The secretion of pancreatic juice is controlled by two principal mechanisms, hormonal and nervous by way of the vagus nerve. Two hormones, *secretin* and *pancreozymin* (which apparently is chemically the same as *cholecystokinin*; see Chapter 18), stimulate the secretion of pancreatic juice. Acid chyme discharged from the stomach into the duodenum plays a role in causing the pancreas to begin its secretory activity. Secretin, present in the duodenal mucosa, is liberated by the action of the hydrochloric acid of the chyme. It is absorbed into the circulation and carried to the pancreas, where it stimulates the secretion of pancreatic juice, which is rich in bicarbonate but not in enzymes. Pancreozymin causes a stimulation of the secretion of enzymes by the pancreas. It is found only in the upper intestinal mucosa, whereas secretin is found also in the gastric mucosa.

Pancreatic function test The fact that secretin stimulates the flow of pancreatic juice has been made use of in a test of external pancreatic function. After the injection of secretin, pancreatic fluid is obtained by intubation of the duodenum. The volume of flow and bicarbonate concentration are measured. Sometimes the enzymes are also determined.

Shortly after the injection of secretin, there occurs an outpouring of pancreatic juice. The duodenal fluid therefore loses its biliary color under normal conditions; but if this bile color remains, a nonfunctioning gallbladder is indicated. The total volume varies normally from 135 to 250 ml in 1 hour, and the bicarbonate, from 90 to 130 mEq.

The test is of value in detecting certain diseases of the pancreas. In pancreatitis with extensive destruction of parenchymal structures there is usually a diminution in the volume of pancreatic juice and bicarbonate output. In less severe pancreatitis about half the cases show these effects. The influence on the enzymes has not been consistent enough to justify their determination. In pancreatic malignancy there is a lowering of the volume response, with less change in the bicarbonate.

Intestinal secretions

The secretion of the intestinal mucosa is at least partly under the control of the nervous system. Mechanical stimuli reflexly cause a flow of this fluid. Secretin probably exerts a hormonal effect for this secretion as well as for pancreatic secretion and for bile secretion. There is also a specific hormone, *enterocrinin* (Chapter 18), secreted by the intestinal mucosa, that stimulates the mucosal glands. Both the volume of fluid and the content of enzymes are increased by enterocrinin. Intestinal juice is not as definite an entity as gastric or pancreatic juice, because it varies at different levels of the intestinal tract and because its composition is not nearly as constant at different periods. Intestinal juice is usually quite turbid because of the presence of leukocytes, epithelial cells, and mucus. The total solids amount to about 1.5%, about half of which is sodium chloride, sodium bicarbonate, and other inorganic salts (Fig. 17-4). It has a pH of about 8.3. The pH of intestinal contents, on the other hand, is slightly acidic. The acid chyme from the stomach is partially neutralized by the alkaline pancreatic juice, bile, and succus entericus so that the pH of the lower duodenal contents in humans range from 4.5 to 5.1. When the ileum is reached, its contents range in pH from 5.9 to 6.5.

The organic material of intestinal juice comprises mucoprotein and a number of different enzymes. Some of these enzymes are undoubtedly not actually secreted but are present in leukocytes and shed mucosal epithelial cells, which disintegrate and liberate their enzymes. Moreover, since intestinal juice is difficult to obtain, most studies have been on extracts of the mucosa. Hence we are not sure whether all the enzymes ascribed to intestinal juice are actually secreted or are in the mucosa or in *desquamated epithelial cells*, where they do their work as intracellular enzymes.

Enzymes An important carbohydrase present in the intestinal mucosa of mammals, is oligo-1,6-glucosidase. This splits the α-1 bonds of the products of digestion of amylopectin, resulting chiefly in maltose (Chapter 11).

The saccharidases maltase, sucrase (or invertase), and lactase split the disaccharides maltose, sucrose, and lactose respectively into their constituent monosaccharides. These enzymes are almost entirely intracellular. Enterokinase is the enzyme that transforms trypsinogen to trypsin. A number of peptidases are present in the intestinal mucosa cells and their secretion. These include a group of *aminopeptidases* and a group of *dipeptidases* (Chapter 13).

A lipase, which is activated by bile salts, is reported to be present in the succus entericus of several species of animals, but it is apparently not of high activity. An amylase also occurs in the intestinal juice.

Three enzymes that hydrolyze nucleic acid to its constituents are present here, also. They are the nucleases, phosphatases, and nucleosidases. The nucleases attack nucleic acids, releasing the mononucleotides in each (Chapter 14). The phosphatases, which are not specific for nucleotides, hydrolyze them to phosphoric acid and purine or pyrimidine nucleosides. The nucleosidases complete the digestion of the nucleosides to purines, pyrimidines, and sugar.

Bile

The bile is secreted, probably continuously, by the liver and passes into the hepatic ducts and into the common duct. It fills the gallbladder, by the way of the cystic duct, and tends to distend all ducts and the gallbladder between digestive periods. The bladder wall, during these intervals, absorbs water from the bile contained in it, thus producing a highly concentrated bladder bile and making more space for the liver secretion. It also absorbs bicarbonate, chloride, and sodium ions and perhaps other inorganic ions. Although bladder bile by this process is four or more times more concentrated than hepatic bile, both have the same osmotic effect as that of blood serum. Gallbladder bile becomes slightly acidic in the process of concentration. The other constituents of bile do not appear to be absorbed to any appreciable extent by the normal gallbladder.

The hepatic secretion is not under nervous control. It can be accelerted by various substances; secretin and bile salts are notable examples, with bile salts being by far the more effective. These are usually referred to as cholagogs. It has been suggested, however, that they be termed *choleretics,* with the word *cholagog* being reserved for substances that stimulate the gallbladder to contract and thus bring about the flow of bile into the duodenum. With regard to hormonal control, an acid extract of intestinal mucosa injected intravenously causes contraction of the gallbladder. The hormone involved is not secretin but is similar to it. It has been named *cholecystokinin* (Chapter 18). A decapeptide called *cerulein,* with striking similarities in amino acid composition to cholecystokinin and gastrin, has been isolated. It is about 16 times more potent than cholecystokinin in causing contraction of the gallbladder. Fatty foods, such as cream and egg yolk, also stimulate contraction of the gallbladder, apparently by causing a release of cholecystokinin from the duodenal mucosa. Protein and carbohydrate have little effect.

Composition Human bile as secreted by the liver is clear, golden, or brownish yellow in color, but sometimes olive green. It has a bitter taste and is a viscid slimy fluid. It is alkaline, with a pH of from 7.8 to 8.6, but bladder bile may be as acid as pH 6.5. The daily volume has been variously estimated at 500 to 1100 ml. In a case of biliary fistula there was an output of 525 ml in 24 hours. Bile contains the following characteristic substances: bile pigments, bile salts, and cholesterol. There are, in addition, variable quantities of proteins, lecithin, inorganic salts (Fig. 17-4), and urea. The proteins include mucin, the usual serum proteins, and a unique protein that migrates electrophoretically more rapidly than the plasma proteins. Hepatic bile contains about 2.5% to 3.5% solid matter, and bladder bile as much as 17%. The high solid content of the latter is the result of the absorption of water; at the same time small quantities of inorganic salts are absorbed. The net result is a fluid with about the same osmotic properties and pH. The specific gravity of the bile in the gallbladder may be as high as 1.040, whereas bile secreted by the liver has a specific gravity of about 1.010.

In Table 17-3 is given the range of various constituents in bile as found by several investigators.

Bile pigments The bile pigments, bilirubin and biliverdin, give bile its color and are excretory products. Bilirubin predominates in human bile. Oxidation and reduction produce a series of varicolored compounds, some of which have received definite names.

$$\underset{\text{(brown)}}{\text{Urobilin}} \xleftarrow{-H_2} \underset{\text{(colorless)}}{\text{Urobilinogen}} \xleftarrow{+4H_2} \underset{\text{(red)}}{\text{Bilirubin}} \xleftarrow{+H_2} \underset{\text{(green)}}{\text{Biliverdin}} \xrightarrow{-H_2} \underset{\text{(blue)}}{\text{Bilicyanin}} \xrightarrow{-H_2} \text{etc. (yellow) etc.}$$

Table 17-3
Composition of bile

Constituent	Bladder bile (percent)	Liver bile (percent)
Water	82.3-89.8	96.5-97.5
Solids	10.2-17.7	2.5-3.5
Bile salts	5.7-10.8	0.9-1.8
Mucus and pigments	1.5-3.0	0.4-0.5
Cholesterol and other lipids	0.5-4.7	0.2-0.4
Inorganic salts	0.6-1.1	0.7-0.8

The Gmelin test is based on this color reaction. Concentrated nitric acid, if overlaid with urine or other fluid that may contain bile pigments, oxidizes them at the junction of the two fluids, producing a rainbow of colors as a positive reaction.

The bile pigments are derived from the heme of hemoglobin from worn-out red blood cells and from other heme proteins, e.g., myoglobin, catalase, cytochromes, peroxidase, and others (see Fig. 16-7). Any "free" hemoglobin that escapes into the plasma is captured by *haptoglobin,* an α_2-globulin. Each molecule of haptoglobin binds two hemoglobin molecules stoichiometrically. Worn-out erythrocytes are removed from the bloodstream by the reticuloendothelial cells, primarily in the spleen, the "graveyard" for senescent red blood cells. The transformation of heme to the bile pigments occurs primarily in the reticuloendothelial cells of the liver, the Kupffer cells, and consequently a major portion of bilirubin originates in the liver itself. Some bile pigment, however, is produced by the reticuloendothelial cells in other parts of the body, whence it is transported to the liver for excretion. Since the bile pigment is water insoluble, it is carried in the plasma in combination with plasma albumin (Fig. 16-7). Involved in the formation of bile pigment are (1) the splitting off of the globin of hemoglobin, (2) oxidation and loss of the α-methene carbon atom as *carbon monoxide* and opening of the tetrapyrrole ring at this point, and (3) removal of iron (see Figs. 16-14 to 16-16). The iron removed is used for the manufacture of new heme or stored temporarily as ferritin mainly in the liver. The *biliverdin* thus formed is next reduced to *bilirubin* by the addition of two hydrogen atoms at the double bond attached to the central γ-methene group.

The conversion of heme to bilirubin is enzymatic, involving a *mixed-function oxygenase* present in liver microsomes. The oxygenase system has an absolute requirement for NADPH and molecular oxygen (Fig. 16-15). The system is inhibited by carbon monoxide, as might be expected. The hepatic microsomal heme oxygenase system also converts other hemoproteins to bilirubin. The latter hemoproteins may be responsible for the so-called early peak in isotopically labeled bile pigments found after the administration of labeled precursors of heme synthesis. Blood hemoglobin is responsible for the late peak, appearing some 120 days after the labeled precursors are given (Fig. 16-6).

The free bilirubin thus formed in the reticuloendothelial system from hemoglobin and other hemoproteins is transported to the liver in the plasma loosely associated with plasma albumin. In the liver the bilirubin is conjugated with glucuronic acid to form bilirubin diglucuronide, which is water soluble and is therefore readily excreted by way of the bile into the intestine and eliminated in the feces (Fig. 16-17). The conjugation of bilirubin with glucuronic acid involves the formation of a diglucuronide with the two carboxyl groups of the propionic acid side chains of bilirubin and the carbon-1 hydroxyl groups of 2 moles of glucuronic acid. The reaction is catalyzed by the hepatic microsomal enzyme, *bilirubin-UDP-glucuronyl transferase*.

A small amount of bilirubin is conjugated with sulfate, as active sulfate (3′-phosphoadenosine-5′-phosphosulfate, or PAPS; Chapter 13) and an appropriate enzyme *(sulfokinase)*. The sulfate ester of blirubin is also water soluble and is excreted in the bile.

After the bile passes into the duodenum, the glucuronic acid (and sulfate) groups are hydrolyzed by β-*glucuronidase* (or a *sulfatase*) and the bilirubin is reduced by the bacterial flora of the intestine (Fig. 17-7). The product is the colorless *d*-urobilinogen, in which four hydrogens are attached at the two double bonds of two methene bridges (β and δ), and two more are used to convert one of the vinyl groups to an ethyl group. The remaining vinyl group of *d*-urobilinogen is then reduced to *l-urobilinogen* (Fig. 17-8). Some of this derivative apparently is absorbed by way of the portal system, reconverted to bilirubin in the liver, and eliminated again in the bile. However, a small amount of the *l-urobilinogen,* 4 mg or less per day, is excreted by the kidney and is autoxidized on standing to orange-red *l-urobilin,* a major urinary pigment (Fig. 17-7).

The major portion of the *l*-urobilinogen, however, is reduced further by intestinal bacterial enzymes to *l-stercobilinogen* (Fig. 17-7). This substance then is apparently converted on standing in the feces by autoxidation to *l-stercobilin,* a brownish-colored pigment, and eliminated in the feces. Approximately 240 mg of *l*-stercobilin are excreted daily in the feces.

The reactions involved in the conversion of biliverdin-bilirubin to the excretory products *l*-urobilin and *l*-stercobilin, as presently understood, are summarized in Fig. 17-8.

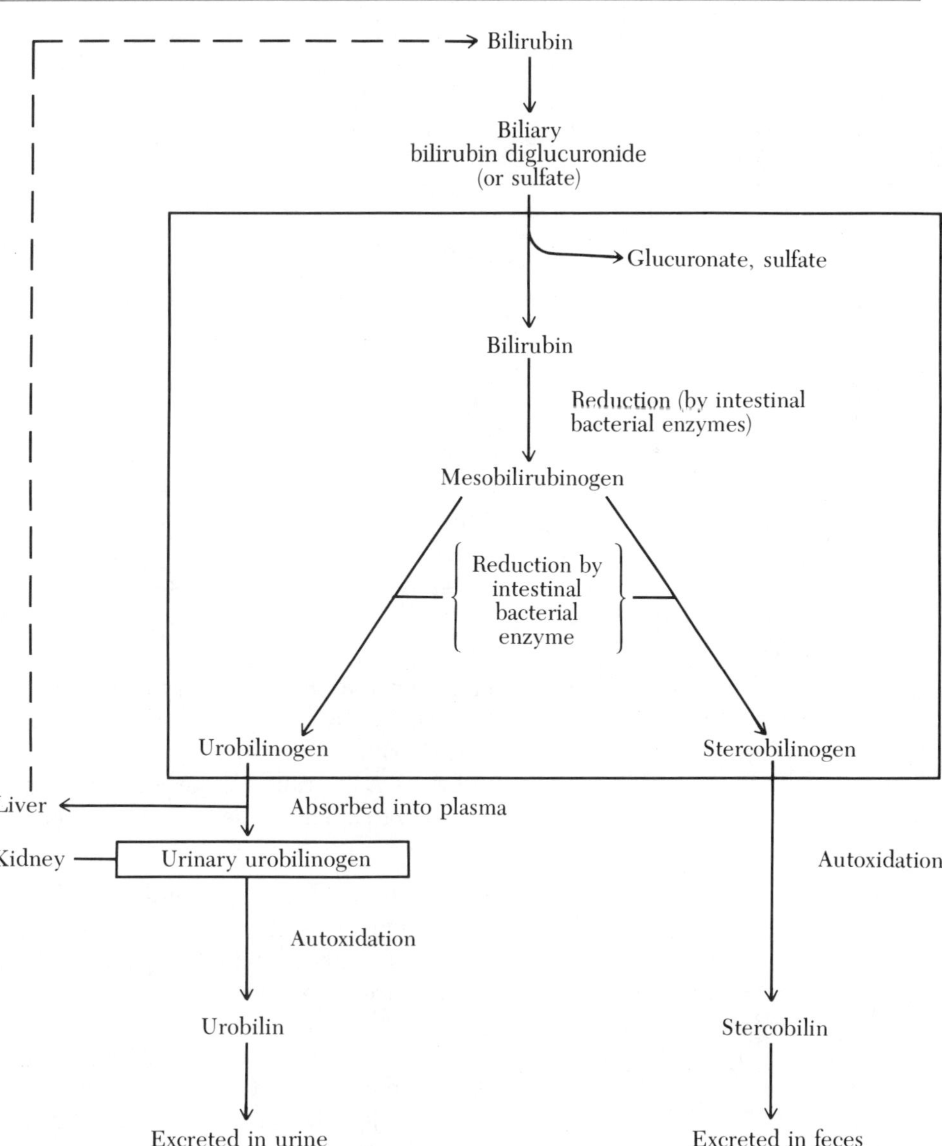

Figure 17-7
Schematic representation of some principal products formed by bacterial enzyme action on bilirubin conjugates in intestine. Excretory fates of end products are shown.

Figure 17-8
Conversion of biliverdin and bilirubin to the principal excretory products *l*-urobilin and *l*-stercobilin. Side groups are designated as follows: *M*, methyl; *E*, ethyl; *V*, vinyl; *P*, propionic acid.

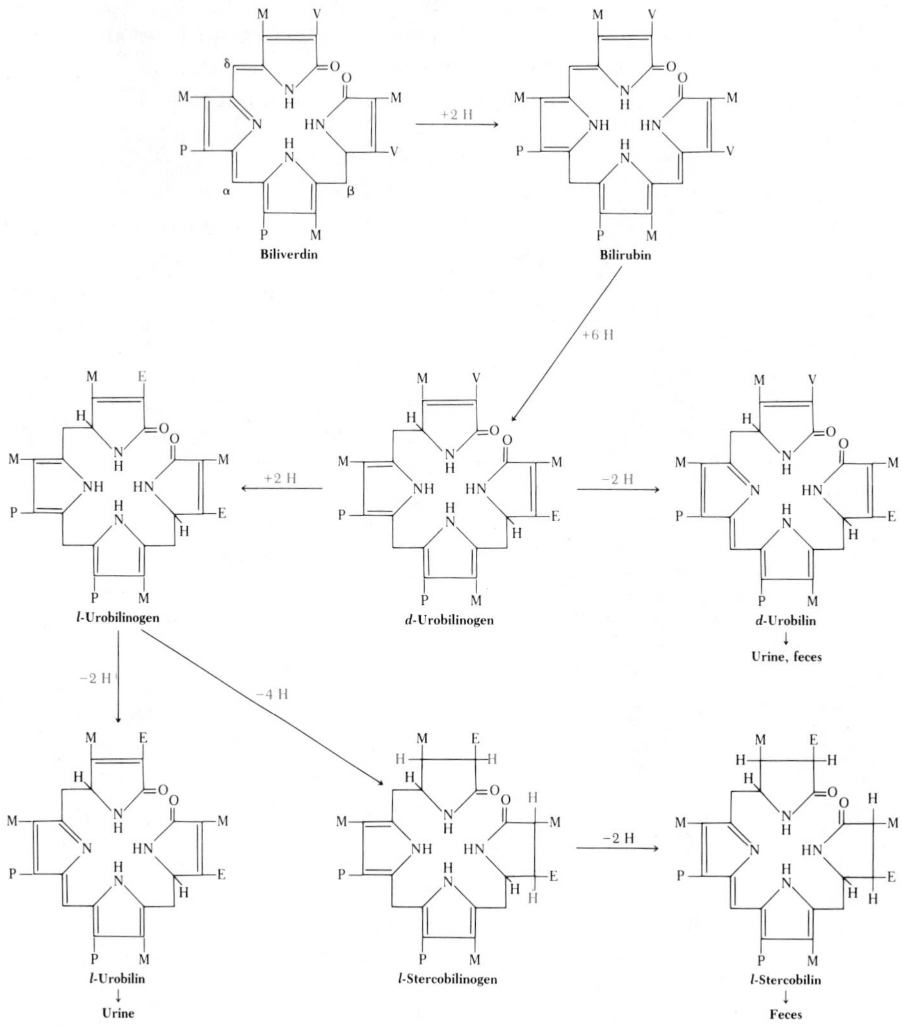

Apparently, any *d-urobilin* that may be formed by the autoxidation of traces of the intermediate, *d-urobilinogen,* would be excreted as a waste product in the urine, feces, or both.

Clinical alterations An elevation of the amount of bilirubin in the plasma may occur in several types of jaundice. There are three principal types, based on the clinical cause:

1. *Hemolytic,* in which excessive erythrocyte destruction results in the formation of bilirubin in amounts exceeding the conjugating ability of the liver and hence its excretion in the bile. This type of jaundice is sometimes termed nonobstructive or retention jaundice. Free bilirubin increases in the plasma as a result.
2. *Obstructive,* caused by the partial or complete blocking of the bile ducts, either inside or outside the liver. This type is also referred to as hepatic or regurgitative jaundice. Conjugated bilirubin is prevented from being excreted into the intestine and consequently appears in increased amounts in the plasma.

3. *Hepatocellular,* in which damage to the liver by toxins, poisons, cardiac failure, or acute or chronic disease impairs the liver's capacity to conjugate circulating bilirubin and hence excrete it. Enlargement of the liver, as in acute hepatitis, may also cause an internal bile duct obstruction, resulting in an increase in both unconjugated and conjugated bilirubin in the plasma of the patient. This type of jaundice may also occur in Wilson's disease.

An elevation of serum bilirubin may also be found in patients in whom there is a hereditary deficiency of the hepatic enzyme bilirubin-UDP-glucuronyl transferase, required for the conjugation of bilirubin *(Crigler-Najjar syndrome).*

Likewise, hyperbilirubinemia is frequently encountered in newborn infants, especially when premature. This may develop into the serious condition *kernicterus* if the serum bilirubin levels become sufficiently high. Brain damage, neurological disorders, and often death may ensue. The elevation in serum bilirubin in this condition is caused by the unconjugated (water-insoluble) form. Crystals of free bilirubin may be deposited in the brain. The accumulation of bilirubin apparently is caused by a relative deficiency of the conjugating enzyme in the liver. Bilirubin-UDP-glucuronyl transferase does not appear to be formed in the liver in sufficient amounts until about the normal time of birth. Hence the problem is particularly acute in the premature infant. Immaturity of the blood-brain barrier may also be a factor. The level of serum bilirubin decreases to normal values soon after birth as the synthesis of the bilirubin-conjugating enzyme increases in the liver.

Interesting recent studies indicate that the hyperbilirubinemia of the premature can be reduced by exposing the infant to fluorescent light—a so-called "bilirubin reduction lamp." Apparently the lamp rays degrade bilirubin to products, possibly di- and monopyrrole derivatives, that are less toxic and can be metabolized or excreted by the infant.

The excretion or nonexcretion of bile pigments in different types of jaundice, as well as the mechanism of the direct and indirect *van den Bergh tests* for serum bilirubin, are explained as follows: Free bilirubin is practically insoluble in water at a pH below 8. It is conjugated in the liver microsomes with glucuronic acid. The diglucuronide is soluble in aqueous solutions and probably circulates in loose combination with a mucoprotein. A small amount may be conjugated with sulfate. The glucuronide and sulfate, being water soluble, give the *direct* van den Bergh test, whereas the free pigment must be solubilized in alcohol first and then is responsible for the *indirect* test. The conjugated water-soluble forms are readily excreted into the bile by the liver. They are also excreted by the kidneys if present in the blood. This occurs when the bile ducts are obstructed and the bile is dammed back into the liver and consequently into the circulation. Hence, in obstructive jaundice, bilirubin diglucuronide is found in the plasma (direct van den Bergh reaction) and in the urine. If for any reason there is an accumulation of free bilirubin, this insoluble molecule is not eliminated by the kidneys but circulates in the blood, attached to blood proteins, chiefly albumin. Such "free" bilirubin in the plasma gives an indirect van den Bergh reaction.

In hemolytic jaundice there is an overproduction of bilirubin, in amounts greater than the ability of the liver to conjugate it. Much of this free bilirubin circulates in the blood but is not excreted by the kidney. There are also excreted into the intestine increased amounts of bilirubin diglucuronide. This is reduced to urobilinogen in the normal way but in abnormally large amounts. Therefore, urobilinogen is increasingly absorbed and excreted in the urine.

There are other types of jaundice in which primarily some damage to the liver occurs. This damage may be caused by toxic agents, e.g., chloroform or arsphenamine, or by some acute or chronic liver disease. In such conditions there is not necessarily an increase in the production of bilirubin, but the damaged or incapacitated hepatic cells cannot conjugate the bilirubin or excrete it in the bile. Hence a considerable amount circulates as unconjugated, "free" bilirubin. No bilirubin appears in the urine, but urobilinogen does because the urobilinogen, absorbed from the intestinal tract, cannot be reeliminated via the bile by the poorly functioning liver.

Bile salts The principal bile salts are chiefly the salts of *glycocholic* and *taurocholic* acids. These acids are conjugates of *cholic acid* with glycine and taurine, respectively, joined together by means of peptide linkages. The salts (chiefly Na^+ and K^+) of the bile acids are water soluble and have a strong detergent action. They have an extremely bitter taste.

The hydrolysis of each of the bile acids is as follows:

$$\underset{\text{Glycocholic acid}}{C_{23}H_{39}O_3{-}C({=}O){-}NH{-}CH_2{-}C({=}O){-}O^-} \xrightarrow{H_2O} \underset{\text{Cholic acid}}{C_{23}H_{39}O_3{-}C({=}O){-}O^-} + \underset{\text{Glycine}}{\overset{+}{N}H_3{-}CH_2{-}C({=}O){-}O^-}$$

$$\underset{\text{Taurocholic acid}}{C_{23}H_{39}O_3{-}C({=}O){-}NH{-}CH_2{-}CH_2{-}SO_2{-}OH} \xrightarrow{H_2O} \underset{\text{Cholic acid}}{C_{23}H_{39}O_3{-}C({=}O){-}O^-} + \underset{\text{Taurine}}{NH_2{-}CH_2{-}CH_2{-}SO_2{-}OH}$$

The structural relations of taurine to both cysteine and methionine are very close. Experiments in which radioactive sulfur, ^{35}S, was introduced in trace amounts into methionine showed that when this amino acid is fed to dogs its sulfur is used in part to form taurine. Cysteine, also, is converted to taurine (Chapter 13). Different species vary in the proportions and even in the nature of the bile salts found in their bile. In humans, besides cholic acid, linked to taurine and glycine, there are also small amounts of *chenodeoxycholic, deoxycholic,* and *lithocholic acids*. Cholic acid is formed from cholesterol in the liver. The formulas show how closely they are related. The steps involved in the conversion of cholesterol to cholic acid (see Fig. 12-18) are first the successive oxidation of rings B and C to form hydroxyl groups in positions 7 and 12. This is followed by the oxidation of one of the terminal methyl groups of the side chain to a carboxyl group, which is converted to its CoA derivative, followed by α,β-dehydrogenation, addition of water, and again dehydrogenation to form a β-carbonyl group. Propionyl CoA is finally split off the side chain, leaving cholyl CoA. This derivative of cholic acid plus taurine or glycine forms taurocholic acid (cholyltaurine) or glycocholic acid (cholylglycine) respectively. Note that the three hydroxyl groups of cholic acid are in the α-position (dotted bond lines) whereas the hydroxyl group of cholesterol is in a β-position (solid bond line). Approximately 80% of the cholesterol in human body fluids and tissues is converted to cholic acid (see top of p. 545). Deoxycholic acid and chenodeoxycholic acid each have one less α-hydroxyl group than does cholic acid, and they are located at positions 3 and 12 and positions 3 and 7 respectively. Lithocholic acid has only one α-hydroxyl group, located on C-3.

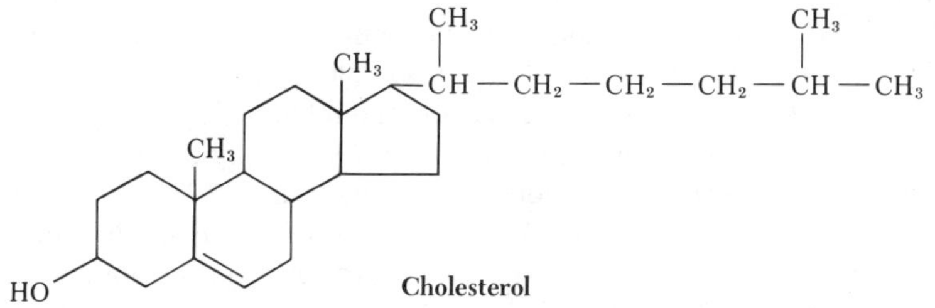

Cholesterol

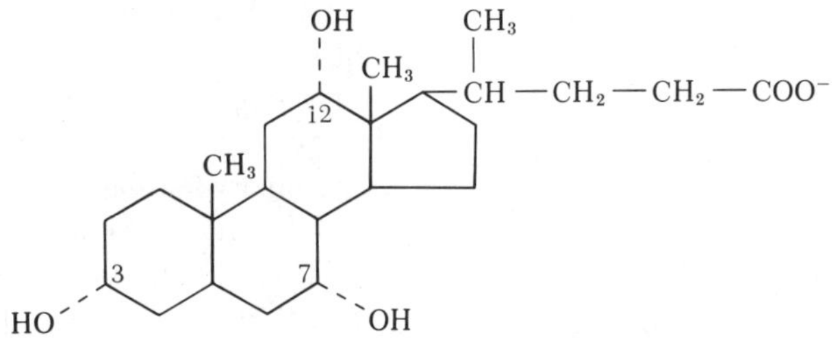

Cholic acid

Functions The bile salts are not excretory products. Some 5 to 15 gm are formed in the liver daily in humans and are the most useful constituents of bile. In human bile, glycocholate predominates over taurocholate in a ratio of about 3:1. After secretion into the intestinal tract, the bile salts are absorbed almost completely and are carried by the portal blood back to the liver for resecretion, the so-called enterohepatic circulation. Apparently, less than 500 mg a day escapes reabsorption and is excreted in the feces.

Several functions of the bile salts have been mentioned previously. They may be summed up as follows:

1. Bile salts accelerate the action of pancreatic lipase. This nonspecific activation transforms a relatively weak enzyme into a quite powerful one.
2. Because of their marked detergent action they aid in the emulsification of fats and tend to stabilize such emulsions. In fact, the bile salts, fatty acids, and lower glycerides are said to form one of the best emulsifying mediums for fats. This may permit the absorption of some emulsified fat and leads to the presentation of a greater amount of surface of the remainder to the lipolytic enzyme, and thus further aids its action.
3. They aid in the absorption of the fat-soluble vitamins. This is particularly important in the case of vitamin K. In the surgery of patients having biliary disease, the administration of bile salts by mouth is necessary to aid in the absorption of vitamin K.
4. It is the bile salts that keep cholesterol in solution.
5. They have great choleretic action. Thus the liver is stimulated to secrete bile as long as bile salts are absorbed. This secretion apparently continues during fat digestion and the absorption of the bile salt–fatty acid complex, i.e., exactly during the period necessary for such secretion.
6. They stimulate intestinal motility.

An unusual mitochondrial defect in the biosynthesis of bile acids from cholesterol has been reported in patients with *Zellweger syndrome*. The exact nature of the defect is uncertain.

Cholesterol The bile seems to be the chief vehicle for the excretion of excess cholesterol. The sources of biliary cholesterol are (1) synthesis by the liver, (2) degradation of red blood cells, and (3) dietary cholesterol. Reabsorption of some of the cholesterol may occur but not after it has been reduced. Reduction occurs by bacterial action, and the product is *coprosterol,* the sterol of the feces. Since cholesterol is not a very soluble substance, it is not surprisingly found to precipitate out of solution from bile and form gallstones.

Functions of bile To sum up the functions of bile, we may say that (1) it tends to neutralize the acid chyme, thus providing a more favorable hydrogen ion concentration for the enzymes secreted by the pancreas and the intestinal mucosa; (2) it aids in fat digestion as an emulsifying agent and an activator of pancreatic lipase; (3) it promotes the absorption of the products of fat digestion, other lipids, and fat-soluble vitamins; (4) it has a choleretic action; and (5) it is an excretory channel for bile pigments, cholesterol, certain drugs, metal ions, etc. Bile has no antiseptic properties; in fact, bacteria grow in bile very rapidly. If bile is diverted to the exterior by a biliary fistula, the feces become clay colored, increased in amount, and greasy, and they have an extremely offensive odor. The color is, of course, due to the lack of stercobilin, and the greasiness and odor, to the undigested fat, which has become rancid. Animals with such fistulas eventually develop abnormalities of the bones, associated with loss of inorganic salts, thus indicating some other function that is vital, since animals with bile fistulas do not survive very long.

Some of the nontoxic emulsifying detergents (Chapter 23) have been tested therapeutically in cases of biliary and pancreatic deficiency in which digestion and emulsification of fat are incomplete. The administration of these agents has been found to be helpful.

Gallstones Gallstones or biliary calculi are composed of material that has precipitated out of bile to form masses of varying size and shape. They usually are found in the gallbladder but may form in the bile ducts. If single, the stone is generally ovoid in shape; but if multiple, they have facets formed by pressing and rubbing against each other. When many are present, the shape of most of them is cuboidal. A gallbladder may contain as many as 2000 calculi. The color, hardness, and inner structure vary with the composition of these stones. When cut in cross section, a central nucleus around which concentric layers of the constituents are deposited may be seen. Gallstones are usually classified, with regard to composition, as (1) cholesterol, (2) pigment, and (3) calcium carbonate stones. As a matter of fact, no gallstones are ever composed entirely of any one constituent. The so-called pure cholesterol stones may contain from 90% to 98% cholesterol, but there is always some bile pigment and some inorganic salt. Human gallstones are almost always of the cholesterol or pigment variety.

The mechanism of the formation of gallstones is not entirely clear. Gallstones apparently form for physical reasons when there is a change in the normal composition of the bile from the following:

1. Change in the relative composition of the major constituents of bile (e.g., excess cholesterol or excess bilirubin as in hemolytic anemias, etc.)
2. Presence of foreign substances or change in trace constituents (e.g., calcium deoxycholate stones), bacterial infection in the gallbladder, etc.

A belief that was held for a long time and still is rather generally accepted is that an infection or injury to the gallbladder mucosa produces a nucleus of microorganisms or a tiny clot around which cholesterol or pigment is deposited. The proteins present in this nucleus are believed to associate with cholesterol, pigments, and calcium carbonate and in that way bring about a precipitation. Since the gallbladder tends to concentrate bile, conditions are favorable for further precipitation. Bladder bile is normally more acid than is liver bile. Infection or other factors that interfere with this acidification of bladder bile could lead to precipitation. A metabolic origin has been put forth as another explanation; i.e., the high concentrations of blood cholesterol (hypercholesterolemia) are often associated with gallstones. It has also been suggested that the protective action of some protein of the bile (together with the solvent action of the bile salts on cholesterol) may be responsible for the usual nonprecipitation of gallstone constituents, and any factor that disturbs the balance is likely to initiate the formation of a concretion. A mucin or a lipoprotein may be the effective agent. Once started, the number and size of the gallstones increase more or less rapidly. Another theory is that the bile salts, in solutions of increasing concentration, show definite alterations in surface tension and conductivity; these alterations denote a change of phase, indicating ionic aggregation or micelle formation. Such concentration can and does occur in the gallbladder. Consequently the cholesterol dissolving power of the bile would be decreased and precipitation of cholesterol would be explained.

Current interesting work indicates that gallstones may be dissolved in asymptomatic patients showing radiolucent gallstones by the oral administration of chenodeoxycholic acid. Reduction in the size and number of the calculi was observed after therapy in 11 of 18 patients for periods up to 2 to 3 years. Only 2 of 13 patients with radiopaque stones responded favorably, however. Control subjects receiving either placebo or cholic acid therapy showed no visible effects. Further studies of this significant, widely occurring problem are being conducted, especially toward the synthesis of less expensive, more effective bile salt derivatives.

Liver function tests Many attempts have been made to devise a test that would measure, with some degree of accuracy, the amount of normal liver tissue actively functioning. Two facts have militated against such an achievement. (1) Although the liver may be quite badly damaged, it may nevertheless perform all its functions, because only a comparatively small proportion of healthy liver tissue is needed for all normal activities; i.e., the liver has a large factor of safety. (2) The functions of the liver are many and diverse in nature. The liver is concerned in protein, carbohydrate, and fat metabolism, in the production of the plasma proteins and heparin, in the secretion of bile, in storage of nutrients, in detoxication and excretion, and in a number of other activities. Often a defect in performing one function is not paralleled by a diminution in others, and the function tested may happen to be one that is not affected.

Nevertheless, if one knows their limitations, certain of these tests have considerable clinical value. Some of the procedures are now presented.

Tests for bile pigments The *icterus index* of blood serum is a crude test, but it has been of considerable service. The modifications of the *van den Bergh test* indicate whether the pigment is combined with glucuronic acid or is free. The former gives the direct test, and the latter the indirect. These were described earlier. They furnish information regarding the concentration of bile pigment in blood serum. These tests, together with the determination of urobilinogen and bilirubin in the urine, aid in the differential diagnosis of obstructive and nonobstructive jaundice. Total serum bilirubin in most healthy individuals is below 1.1 mg per 100 ml of serum. An increase is an unfavorable sign, and if a high total serum bilirubin value decreases, it indicates a remission of liver disease or of biliary obstruction. A stabilized serum bilirubin concentration is considered highly desirable for an operation of biliary obstruction.

Other tests The determination of urinary coproporphyrin is of value as an index of liver function in certain conditions. Other liver function tests include the hippuric acid test for the detoxication function. There are several dye excretion tests in which a nontoxic dye that is excreted almost exclusively by the liver is injected, the amount excreted is measured, and hence the functional capacity of the liver for elimination is determined.

Materials and chemical changes within the large intestine

Although the large intestine does not secrete any significant amount of body fluid but is rather a reabsorptive and an excretory structure, biochemical processes occurring in its lumen can perhaps be appropriately considered at this point.

The biochemical processes that go on in the large intestine are attributable mostly to the activity of the myriads of microorganisms that live and die there. These enter the tract with food and saliva and may survive passage through the stomach, since the hydrochloric acid is not always present in bactericidal concentration. Consequently some living microorganisms pass into the small intestine and begin to multiply as the reaction becomes favorable. However, even near the ileocecal valve the intestinal contents do not contain large numbers of such organisms. At this point are present some undigested food residues, unabsorbed secretions (e.g., bile and pancreatic juice), and cell detritus. The pH is about 5.9 to 6.5.

Feces

Composition In the large intestine, such materials as just described are transformed into feces. A number of enzymes are possibly present in the secretion of the mucosa of the large gut, but digestion by them is generally believed to be of little importance. This secretion is alkaline and viscid and undoubtedly tends to bring the contents over to the alkaline side. The conditions for bacterial growth (particularly anaerobic) are excellent: there is warmth, darkness, little oxygen, an almost neutral medium, and food material in a semisolid condition. The organisms flourish, utilize the food materials, transform them into their own protoplasm, multiply, and die. In fecal material, from one fourth to one half of the dry matter is made up of living and dead bacteria. Water is absorbed by the mucosa and the characteristic consistency results.

In the newborn infant the first fecal discharge is termed *meconium*. This is a dark brownish green semisolid material. It consists of intestinal and biliary secretions that have accumulated in the large intestine from the fourth fetal month on. Meconium continues to be passed for the first 3 or 4 days after birth and accounts for much of the loss of weight that occurs during this period. Usually, with the ingestion of milk, a gradual change to the usual type of infant feces is seen. These are greenish yellow in color and have an acid reaction. The approximate general composition of stools of the infant and of the adult is given in Table 17-4. In the feces of infants, there is very little protein but rather large amounts of fat, fatty acids, and soaps.

Adult fecal material is normally brown, varying in color with fat and water, which lighten the color, and bile pigments, which darken it. About 80 to 170 gm of feces are eliminated per day. The composition varies greatly. Feces contain undigested, indigestible, and unabsorbed food residues, secretions of the gastrointestinal tract, bile constituents, and desquamated epithelial cells. Included in the unabsorbed food may be rather large amounts of iron and calcium complexes. Lipids of various types are also present and constitute about one third of the dry weight.

The bile derivatives in feces are *stercobilin,* a transformed bile pigment that gives the stool its brown color, and *coprosterol.* Coprosterol is a reduced sterol, coming partly from the cholesterol of the bile and partly from any unabsorbed food cholesterol. The pH of the stools of healthy adults on a mixed diet varies from 7.0 to 7.5.

Fecal lipids Normally the main portion of the fecal fatty acids and fatty acid derivatives represents (1) unabsorbed and unaltered fatty acids of dietary fat, e.g., stearic acid, which may be present as esters, soaps, and free fatty acids; (2) saturated fatty

Table 17-4
General composition of stools (in percent)

	Stool of breast-fed infant	Stool of adult
Water	85	75
Organic solids	13	20
Ash	1	5

acids, soaps, or glyceryl esters, derived from unsaturated fatty acids of the diet, apparently by action of the intestinal flora; and (3) fats synthesized by intestinal flora. Increases may be caused by blockage of the bile ducts, the pancreatic ducts, or both; by failure of the pancreas to secrete pancreatic juice; or by incomplete absorption (e.g., when there is increased motility of the upper intestine and the food rushes through too rapidly). Conditions in which the feces contain large amounts of fat, fatty acids, and soaps are called *steatorrheas*. Part of this "fat" may actually be secreted into the intestinal tract.

The normal values for fat and its derivatives in feces vary widely. On a fat-free diet they amount to 0.5 to 1 gm of fatty acid, or fatty acid derivatives, per day. On diets containing the ordinary amounts of food fats, about 5 gm or more will be excreted per day. Most of this increase represents unabsorbed dietary fat, some of which has been made more saturated by action of the intestinal flora. The range of total fatty acid in normal feces is from 7% to 25% of the dry weight. A figure of over 25% is considered abnormal and requires more detailed study. If the neutral fat is high, one should suspect deficient fat digestion; and if the total split fat (i.e., sum of soaps and free fatty acids) is above its usual percentage, probably some abnormality in the absorptive process is occurring. In this way a fractional fecal analysis may aid in diagnosing obscure gastrointestinal conditions.

In steatorrheas, lesions of the bones similar to those seen in rickets and sometimes in dwarfism are found. A low serum calcium is usually present and sometimes tetany results from an excessive loss of calcium in the feces along with the fat. This must be caused by a lowered absorption of calcium from the intestinal tract, which may be accounted for in one of three ways: (1) The excess of fat in the tract holds the fat-soluble vitamin D there and prevents its absorption; vitamin D is involved in the absorption of calcium from the gastrointestinal tract (Chapter 21). (2) The intestinal wall may be impermeable to calcium ions in these conditions. (3) The fatty acids form insoluble soaps with the calcium and these are excreted in the feces as such. There is some evidence for each of these three possibilities.

In addition to neutral fats, fatty acids, and soaps, there are always found various sterols in feces. They are true secretions eliminated by way of the bile or through the intestinal wall. There are also dihydrocholesterol and coprosterol, the reduction products of cholesterol formed by the action of microorganisms, and unabsorbable plant sterols as well as any excess cholesterol of the diet that has escaped absorption or reduction.

Pathological constituents To detect abnormal constituents, one may examine the stools macroscopically, microscopically, and chemically. Each of these methods may yield valuable information. For example, simple observation, after mixing with water and straining through cheesecloth, may enable one to find gallstones, undigested food residues, mucus, epithelial shreds, and, rarely, intestinal concretions. Intestinal concretions are chiefly inorganic, usually ammonium magnesium phosphate, with some admixture of calcium phosphate, calcium carbonate, calcium sulfate, protein, or calcium or magnesium soaps. They always have a nucleus of some indigestible substance, e.g., hair or even a gallstone.

Microscopically one may see crystals that might indicate the presence of salts or organic compounds that are ordinarily absorbed. Undigested food (fat globules, meat and vegetable fibers, starch granules, etc.) is often observed.

Chemically the quantitative estimation of fat, fatty acid, and soaps is sometimes required, but a strict control of intake is then important. A qualitative test for unchanged bile pigments is rarely positive, except in severe diarrhea, when the intestinal contents are rushed through the tract. Ordinarily they are converted to stercobilin and stercobilinogen.

The most important chemical determination is the qualitative one for blood. This aids in the diagnosis of bleeding gastrointestinal lesions such as ulcers and malignancies. A chemical test is often necessary because the colors that blood imparts to feces vary from bright red to black, which in the brown feces may be indistinguishable by the naked eye (*"occult blood"*).

Intestinal gases The volume of gas present in the gastrointestinal tract of the human being is extremely variable but averages about 1 liter daily. The components of the mixed gases vary with the diet. On a high-milk diet the predominant gas is hydrogen; on a vegetable diet, methane; and on a meat or mixed diet, nitrogen. In all cases, these gases, as well as carbon dioxide and usually hydrogen sulfide, are present. The nitrogen is derived from swallowed air, from air dissolved in food and drink, and from that which diffuses out of the blood in the blood vessels of the gut. It is in solution in the blood, having passed into the blood during respiration. A typical analysis of the intestinal gases obtained from a normal man showed nitrogen, 59.4% by volume; methane, 29.6%; carbon dioxide, 10.3%; and oxygen, 0.7%. In intestinal obstructions gases of somewhat similar composition accumulate, except that there seems to be less methane and traces of hydrogen and hydrogen sulfide. The amount of hydrogen sulfide depends on the presence of the sulfur-containing amino acids and the type of decomposition they undergo. The concentrations of carbon dioxide and hydrogen sulfide are low because these gases are quite soluble in the aqueous medium and therefore are absorbed rather easily.

Action of microorganisms on carbohydrates and lipids Bacteria, yeasts, and other organisms in the large intestine act on carbohydrates present, producing butyric, lactic, and perhaps other organic acids, ethyl alcohol, carbon dioxide, methane, and hydrogen. How much absorption, if any, of these takes place in the human being is uncertain.

The action of intestinal organisms on fats is probably simple hydrolysis and saturation of part of the unsaturated fatty acids. There is also synthesis of lipids by microorganisms.

From lecithin and sphingomyelin, choline is split off and is converted to neurine by anaerobic organisms.

$$\underset{\textbf{Choline}}{(CH_3)_3\overset{+}{N}{-}CH_2CH_2OH} \longrightarrow \underset{\textbf{Neurine}}{(CH_3)_3\overset{+}{N}{-}CH{=}CH_2}$$

Both choline and neurine are toxic to animals.

From one cephalin, serine is derived, and this on decarboxylation yields aminoethyl alcohol, which is a derivative of another cephalin (Chapter 23). Aminoethyl alcohol is also called ethanolamine and colamine.

$$\underset{\textbf{Serine}}{HO{-}CH_2{-}\underset{\overset{|}{\overset{+}{N}H_3}}{CH}{-}COO^-} \longrightarrow \underset{\textbf{Colamine}}{HO{-}CH_2{-}CH_2NH_2}$$

Colamine is only slightly toxic, but aminoethyl mercaptan has a pronounced hypotensive effect and this is formed during the bacterial decomposition of cysteine.

Action of microorganisms on proteins The decomposition of proteins by anaerobic organisms is termed putrefaction. The nitrogenous materials that reach the large intestine may be undigested or partly digested food residues, unabsorbed amino acids, or cellular detritus. In addition, there are the proteins of dead bacteria. The action of microorganisms on this varied assortment begins with a digestive action. There may be proteases, e.g., trypsin, that have not been destroyed, proteases from disintegrated epithelial or bacterial cells, or the active enzymes of the living bacteria. Proteolysis results, of course, in the formation of free amino acids, and since little or no absorption takes place in the large intestine, these are attacked by the microorganisms to a varying degree and in two general ways: decarboxylation and deamination. Oxidations, reductions, and hydrolyses also occur. These reactions are all results of the appropriate enzymes. If deamination occurs first, acids are formed, whereas amines result from decarboxylation. The simpler amino acids yield simple organic acids or amines as the case may be. Thus alanine forms propionic acid or ethylamine.

Deamination: Alanine ($CH_3-CH(NH_3^+)-COO^-$) → Propionic acid ($CH_3-CH_2-COO^-$)

Decarboxylation: Alanine → Ethylamine ($CH_3-H_2C-NH_2$)

Alanine **Propionic acid** **Ethylamine**

The amino acid tyrosine may undergo decomposition along two routes also, forming such aromatic compounds as tyramine, *p*-cresol, *p*-hydroxybenzoic acid, and phenol.

Tyrosine ($HO-C_6H_4-CH_2-CH(NH_3^+)-COO^-$) **Tyramine** ($HO-C_6H_4-CH_2-H_2C-NH_2$) ***p*-Cresol** ($HO-C_6H_4-CH_3$) ***p*-Hydroxybenzoic acid** ($HO-C_6H_4-COO^-$) **Phenol** (C_6H_5OH)

Other important compounds formed by the putrefaction of amino acids include: ethyl and methyl mercaptans, hydrogen sulfide, and methane from cysteine; tryptamine, skatole, and indole from tryptophan; cadaverine and putrescine from lysine; and histamine from histidine. Some of these substances exert marked pharmacological effects when administered to animals in appropriate doses. Indole, skatole, and hydrogen sulfide give the characteristic foul odor to feces. The structures of several of these compounds are given on p. 552.

Autointoxication Diarrhea may result from an overabundance of some of the products of intestinal putrefaction, particularly the acidic compounds. The question arises, however, whether the other toxic products, the amines, the choline derivatives, the mercaptans, etc., are harmful in other ways. The idea that the absorption of some of these products of bacterial activity is the cause of many of the

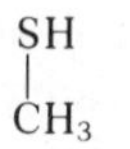

$$\underset{\text{Methyl mercaptan}}{\mathrm{CH_3{-}SH}} \qquad \underset{\text{Ethyl mercaptan}}{\mathrm{CH_3{-}CH_2{-}SH}}$$

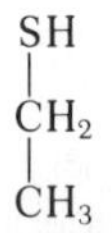

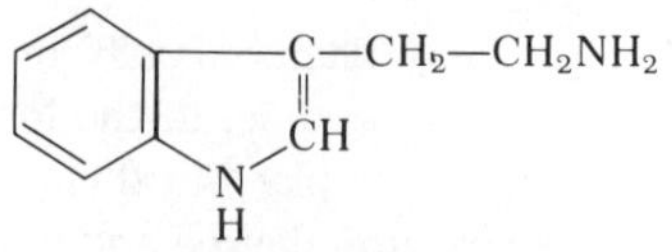

Tryptamine (indole ring with $-C-CH_2-CH_2NH_2$ at C-3; ring CH, NH)

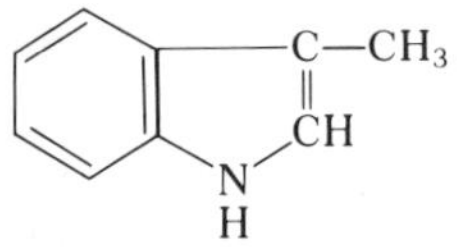

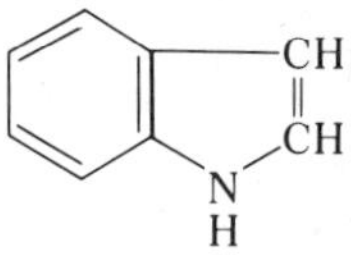

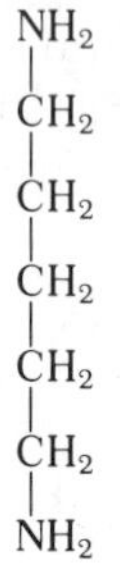

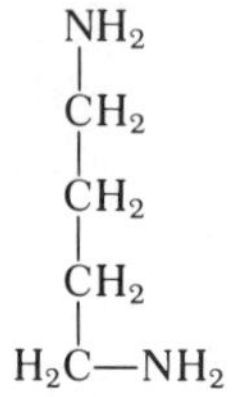

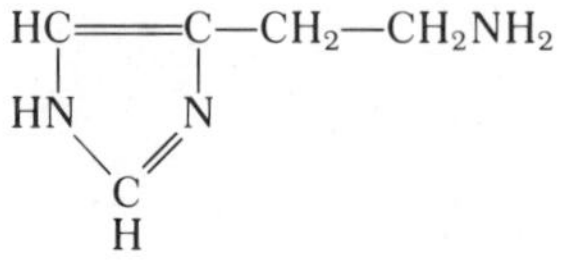

Skatole (indole ring with $-C-CH_3$ at C-3; ring CH, NH)

Indole (benzene ring fused with CH=CH–NH)

$$\underset{\text{Cadaverine}}{\mathrm{NH_2{-}CH_2{-}CH_2{-}CH_2{-}CH_2{-}CH_2{-}NH_2}} \qquad \underset{\text{Putrescine}}{\mathrm{NH_2{-}CH_2{-}CH_2{-}CH_2{-}H_2C{-}NH_2}}$$

Histamine ($HC{=}C{-}CH_2{-}CH_2NH_2$; imidazole ring HN, N, C–H)

ills of mankind has long been prevalent, even among physicians. There is no question that some of the products are somewhat toxic when administered orally and even more so if given parenterally, i.e., by any route other than the mouth. It is doubtful, however, if large enough amounts are ever absorbed to produce harmful effects. For example, the total amount of indole in the feces is seldom over 60 to 70 mg, and yet 1 gm of indole given by mouth produces no ill effects and 2 gm cause only a slight headache and dizziness. Small amounts of indole and skatole are often absorbed, as evidenced by the excretion of their detoxication product, indican, in the urine. They are therefore fully detoxified. The amines produced in putrefaction may be introduced into the gastrointestinal tract in amounts much greater than those that occur in constipation without the appearance of unusual symptoms. Regarding other toxic products formed by bacteria in the large intestine, they are either not absorbed in appreciable quantities or are destroyed in the intestine or the liver.

What is the basis, then, for autointoxication, the symptoms of which often accompany constipation—mental laziness, malaise, headache, dullness, coated tongue, poor appetite, and "biliousness"? Evidently the old idea that these symptoms are caused by absorption of toxic materials from the sluggish intestine is untenable. Current opinion is that most of these symptoms result from mechanical distention and irritation of the rectum by the fecal masses and that their effects are caused by reflex action. Many of the symptoms can be reproduced by simply packing the rectum with cotton.

Milk

Milk is the fluid secreted by the mammary gland for use as food by the young mammal. Consequently milk may be considered from two standpoints: (1) as a secretion and (2) as a food (as discussed in Chapter 21).

Secretion

Milk is secreted by the alveoli of the mammary gland. These are not present in either sex in early childhood. In the female, at puberty, proliferation of the tubules and development of the alveoli occur, and the gland, of course, increases in size. These changes result from the liberation of ovarian and other hormones. During pregnancy the estrogens and progesterone stimulate an increased growth of the breasts and cause functional changes, but the estrogens tend to suppress lactation until after delivery. The actual secretion of milk does not occur until the end of pregnancy. The initiation of lactation may result from a sudden removal of the placenta and from other factors. The chief hormone involved is prolactin, a pituitary factor. One of the posterior pituitary hormones, oxytocin, causes milk ejection. Nervous stimulation induced by suckling is believed to cause the secretion of the hormones that have their continuing secretory effect upon the mammary gland. For further discussion see Chapter 18.

Composition

In addition to being used as the food of the very young, milk, particularly bovine milk, has been adopted as a nutrient for all ages. It is the most complete food found in nature, and for a long period it is the only food of the young mammal. The first secretion of the mammary gland post partum differs a great deal from true milk. It is called *colostrum*. It is a yellowish, alkaline, and slightly viscid fluid. It has a higher content of total solids, the components of which are not exactly the same as those of milk. Colostrum coagulates on heating, whereas milk does not. The lipids present in colostrum have a higher content of cholesterol and lecithins, and the fat has a higher degree of unsaturation. Colostrum seems to have a laxative action and thus may aid in bringing about evacuation of the meconium. The amount of colostrum secreted by the human being is rather small, about 150 to 300 ml in 24 hours. About the third or fourth day, true milk begins to be secreted and the colostrum qualities diminish steadily. For 1 or 2 weeks, however, human milk continues to retain some of the characteristics of colostrum. This is reflected in the changing composition of milk.

Milk is an oil-in-water type of emulsion stabilized by complex phospholipids and proteins absorbed on the surface of fat globules. It contains proteins in colloidal dispersion, lactose in true solution, and a number of minerals, particularly calcium and phosphorus. There are also present some organic acids or their salts, vitamins, enzymes, and some undetermined constituents, including antibodies and substances that possibly have antibacterial properties. In human milk there are present one or more *bifidus factors,* which aid the growth of certain microorganisms found in the breast-fed baby's intestine.

Comparison of human and bovine milk Table 17-5 shows a comparison of the composition of human and bovine milk in two ways: (1) the range of values ordinarily found and (2) averages that represent the usual approximate values. The outstanding differences are a greater concentration of lactose in human than in bovine milk, with lower concentrations of total protein and ash. The percentage of fat is about the same in bovine and human milk, but bovine milk generally comes to a constant composition of about 3.9% at the dairy where the milk is pooled and analyzed. The legal requirement is usually 3.25%.

Table 17-5
Composition of human and bovine milk (values are averages* and ranges† in gm per 100 gm milk)

Constituent	Human milk	Bovine milk
Water	87.5	88.0
Solids	12.5	12.0
Total protein ($N \times 6.38$)	1.1 (1.0-1.8)	3.3 (2.0-6.4)
Caseins	0.44 (40%)	2.64 (80%)
Lactalbumins	0.66 (60%)	0.66 (20%)
Lactose	6.9 (4.6-8.3)	4.7 (2.1-6.1)
Lipid (acylgylcerols 95% to 96%)	4.4 (1.0-8.2)	3.4 (1.7-6.5)
Ash	0.20 (0.18-0.30)	0.72 (0.35-1.2)
Food energy, kcal	70	61

*From National Dairy Council: Newer knowledge of milk, Rosemont, Ill., 1979, the Council.
†From various sources in the literature.

Besides these variations in general composition, several special points should be noted. The distribution of the three milk proteins is quite different. In bovine milk the casein is greater in amount than the albumin-globulin fraction, whereas in human milk the albumin-globulin fraction is slightly greater. The inorganic constituents are about the same qualitatively but not quantitatively. Although human milk contains slightly more iron than bovine milk, this is not sufficient for the baby's day-to-day needs. Nature has provided for this inadequacy by having the infant come into the world with a store of iron in the liver and as extra hemoglobin and erythrocytes—enough to last until the infant can obtain iron from foods other than milk. The calcium and phosphorus are sufficient to provide for bone growth and tissue requirements, but they are not present in excessive amounts.

One unfavorable effect of human milk is the fact that hyperbilirubinemia may occur in very young infants who are breast fed. This seems to be caused by the presence of pregnanediol in the milk. It is known that this hormone occurs in the serum of pregnant women and has an inhibitory (competitive) effect on bilirubin conjugation.

Physical characteristics and reaction Milk is a white to yellowish white fluid having a specific gravity varying from 1.026 to 1.036 (bovine and human). Since the specific gravity of fat is less than 1.000, a high fat (cream) content tends to lower the specific gravity. Watering the milk (diluting with water), of course, has the same effect. Consequently, the determination of specific gravity aids very little in estimating the composition of milk. The reaction is usually faintly acid, with a pH of 6.6 to 6.9. The color is attributable partly to the calcium salt of casein, which is bluish white in solution, and partly to the emulsified fat. The yellowish color often observed is derived from pigments in the food that dissolve in the fat. These are carotene and xanthophyll, chiefly the former. They cannot be synthesized, and, if absent from the diet, similar pigments present in the body fat may be drawn upon. By changing feed, many degrees of color may be observed in butterfat, from almost colorless to a very deep yellow. The yellowish color of whey, the fluid remaining after casein has been precipitated out, is probably caused by the presence of riboflavin.

Lactose Lactose apparently occurs only in milk. Moreover, galactose is not found free in nature but is found in combination. However, not enough galactans or other galactose-containing substances are ingested by the lactating mother to furnish the requisite amount of galactose to form lactose. Therefore lactose must be synthesized in the mammary gland (see Chapter 11).

In human milk there occur growth factors for *Lactobacillus bifidus* var. *Penn*, an organism that grows in the intestinal tract of breast-fed infants. These factors consist of a group of *N*-acetyglucosamine– and sialic acid–containing oligosaccharides and polysaccharides. They are grouped together and are termed the *bifidus factors*. One of them appears to be a tetrasaccharide composed of *N*-acetylglucosamine, D-glucose, D-galactose, and L-fucose, a methylpentose. Another, bifidus factor 2, contains a peptidelike component and hypoxanthine (Chapter 14). The significance of the bifidus factors is not clear, but they may be needed to establish or maintain the growth of a favorable flora in the intestinal tract of infants.

Proteins The proteins of milk are the caseins, the lactalbumins, and the lactoglobulins. These proteins, especially the caseins, are peculiar to milk. They are manufactured by the mammary glands. The sources of the chief milk proteins are the amino acids present in the blood. The fatty acids, glucose, and bicarbonate contribute in greater or smaller amounts to the carbon skeleton of proteins. This was shown by the intravenous injection into lactating rabbits or cows of the precursor, labeled with ^{14}C, and the finding of the labeled carbon in the milk proteins.

Inorganic phosphate, labeled with ^{32}P, when injected into a cow's udder, was incorporated into casein, and there is evidence that casein synthesis in the cow occurs in the udder. Immune globulins and any serum albumin present come directly from blood, although a portion of the globulins may be synthesized in the udder. The loss of protein during lactation has no deleterious effect on the mother's blood proteins, which are maintained at normal levels. Of course, the lactating mother must have a plentiful supply of proteins of good quality in her diet in order to provide the amino acids necessary to build these proteins.

Casein is the principal protein in bovine milk, representing about 80% of the protein nitrogen. In human milk, on the other hand, casein is responsible for only approximately 40% of the total protein nitrogen.

Casein is a phosphoprotein. It is insoluble at its isoelectric point, pH 4.6, but since the pH of milk is nearly 7.0, it is undoubtedly present as a salt, calcium caseinate. On acidification casein precipitates.

$$\text{Ca-caseinate} + 2\ \text{HCl} \rightarrow \underline{\text{Casein}} + \text{CaCl}_2$$

In souring of milk, the same reaction occurs. Rennin (also called *chymosin*) also precipitates casein but at a pH of about 7.0. A partial digestion takes place, some fragment of the protein molecule being split off. It further differs from acid precipitation in that the precipitate contains calcium. In fact in the absence of calcium the precipitation does not occur. Thus:

$$\text{Ca-caseinate} \xrightarrow{\text{Rennin}} \underline{\text{Ca-paracaseinate}} + \text{Peptide}$$

but

$$\text{Casein} \xrightarrow{\text{Rennin}} \text{Paracasein} + \text{Peptide}$$

The paracasein remains in solution until calcium ions are added:

$$\text{Paracasein} + \text{Ca}^{++} \longrightarrow \underline{\text{Ca-paracaseinate}}$$

This can be demonstrated by adding some oxalate to milk and obtaining by filtration the decalcified milk. Rennin, if added to this decalcified milk, does not clot it, whereas it does clot the untreated milk in a few minutes. The subsequent addition of a soluble calcium salt, in excess, brings down a clot. If rennin is added to boiled or evaporated milk, there is no clot formed because the heating has caused the calcium to be precipitated as calcium phosphate. Clotted milk, or junket, is frequently used in the American diet. Under suitable conditions other proteolytic enzymes can cause milk to clot in this way, but rennin, the enzyme present in the fourth stomach of the calf, is especially effective, and apparently its action is limited to this digestion.

Pepsin also clots milk and may serve this purpose in adults, possibly even in infants rather than rennin. The structure of rennin (chymosin) is homologous to that of pepsin, suggesting a common origin of these two proteases. The biochemical significance of the clotting of milk is to ensure its retention in the stomach for a longer period of time and thus to enhance the digestion of its proteins by pepsin (Chapter 13).

Casein is not a single protein but a group of five or more proteins. These have been designated α, β, γ, κ, and λ on the basis of their electrophoretic mobility. κ-Casein is the protein that stabilizes the micelle, preventing the precipitation of the entire complex. An attack by a proteolytic enzyme on this protective colloidal protein renders the protein incapable of protecting the other two, and if calcium ions are present and pH and temperature are favorable, they then precipitate as calcium paracaseinate. The various caseins differ chiefly in their phosphorus content. The α- and β-caseins have a high phosphate content, mainly as *o*-phosphoserine. Bovine β-casein has a molecular weight of about 23,600 and contains 209 amino acid residues. Cystine-cysteine are absent. It has been partially sequenced.

Lactalbumin, like casein, is not a single entity but consists of at least three proteins, β-lactoglobulin,* α-lactalbumin, and blood serum albumin. It is even suggested that these may be multiple in nature. The β-lactoglobulins make up the bulk (about 60%) of the lactalbumin fraction and have some globulin properties; e.g., they are insoluble in water. The lactoglobulins include euglobulin and pseudoglobulin. About 5% of the noncasein proteins are in this group, but colostrum contains much more. They carry the immunological properties of milk and protect the young animal until it can acquire immune systems of its own. Lactalbumins and lactoglobulins are coagulable proteins and coagulate when isolated from milk by suitable separation methods. When present in milk, they do not coagulate on heating because the pH is not favorable, but they are undoubtedly denatured by heat.

As mentioned in Chapter 11, α-lactalbumin forms a complex with the enzyme UDP-galactose : D-glucose-1β-galactosyl transferase to yield *lactose synthase*. The α-lactalbumin appears to act allosterically so that glucose is used as the primary acceptor molecule. The synthesis of lactalbumin appears to be under hormonal control, with the lactogenic hormone and insulin stimulating it and progesterone inhibiting it (Chapter 18). Increased progesterone levels during pregnancy apparently repress the formation of lactalbumin. At parturition, decreased levels of plasma progesterone with increasing prolactin secretion cause an increased α-lactalbumin synthesis.

The sequence of the 123 amino acid residues of α-lactalbumin is partially known. Interesting is the finding that 40 residues in the amino acid sequence are identical with those of the enzyme lysozyme (129 amino acids). This suggests a common genetic ancestry of the two proteins.

The lactalbumins ("whey proteins") represent about 60% of the total protein nitrogen in human milk, whereas in bovine milk they constitute only about 20%.

All the proteins of milk are excellent biologically (Chapter 21), containing a wide assortment of amino acids. The combined milk proteins yield the known amino acids, essential and nonessential. From a consideration of the amino acid content alone, human milk proteins are not nutritionally superior to the proteins of bovine milk. However, the difference in the proportion of albumin to casein explains why bovine milk forms heavy tough curds and human milk soft fine curds that are much more easily and rapidly digested.

Very small amounts of free amino acids are also present in milk. These represent about 4 and 6 mg of nonprotein nitrogen per 100 ml of bovine and human milk respectively. In this connection, it is interesting to note that the objectionable flavor that milk acquires on exposure to sunlight for periods of a half hour or more has been found to be the result of photolysis of methionine, aided by the vitamin riboflavin.

Lipids The fat of milk is in the form of very small globules. There is no appreciable difference in the size of the fat particles in human as compared with bovine milk, although there may be slight variation in the fat globules obtained from dif-

*The apparent confusion in nomenclature here, i.e., placing β-lactoglobulin in the albumin fraction, stems from the decision of milk chemists to designate as albumins those proteins soluble in a saturated solution of magnesium sulfate and as globulins those insoluble in that solution.

ferent individuals. Most of the fat of bovine milk consists of the triglycerides of palmitic, oleic, stearic, myristic, and other higher fatty acids, but a small amount (about 10%) is composed of the triglycerides of butyric, caproic, caprylic, and other fatty acids with short carbon chains. The latter include several volatile fatty acids; in this respect, milk fat differs from other fats formed in the body. Small amounts of cholesterol, phospholipids, and free fatty acids are present. Human milk fat differs from bovine in that few or none of the fatty acids present have chains shorter than 10 carbons. Oleic acid occurs in largest amount, palmitic next, lauric, myristic, stearic, octadecadienoic in smaller amounts, and still smaller quantities of a considerable number of long-chain fatty acids. The fat of human milk resembles human body fat much more than it does typical milk fat of other species. In fact, the milk fat of most mammals except ruminants has a composition similar to that of the remainder of their body fat. The fat of bovine milk contains a variety of triglycerides of fatty acids, of which about two thirds are saturated and one third is unsaturated, and there is a large proportion of short-chain fatty acids. The unsaturated fraction includes small but appreciable amounts of the essential fatty acid linoleic as well as the related linolenic, arachidonic, and vaccenic acids. Furthermore, bovine milk contains traces of fatty acids having odd numbers of carbon atoms, i.e., 7, 9, 11, 13, 15, and 17 carbon atoms. As regards the source of milk fat, isotope experiments indicate that, in the lactating cow, injected acetate gives rise chiefly to the fatty acid fraction of the fat of milk whereas glucose furnishes most of the glycerol. All the ester classes of lipids are readily synthesized in the mammary gland.

Since the fat globules are lighter than water, they rise to the top to form cream. Commercially, the fat of milk is its most valuable constituent, being marketed as cream and butter, entering into the composition of cheese, and being largely responsible for the pleasing flavor of these foods as well as of others into which they enter. The percentage of fat in milk often determines the price that the farmer can get for the milk. It is also an index of the nutritional value of milk. Consequently the analysis of milk for fat is of considerable importance to farmers and dairymen, as well as to food and health authorities and, of course, to the physician. There are many methods of determining fat content in milk, but the quickest, easiest, and almost universally adopted method is the Babcock procedure. This can be used not only for bovine milk but, with modifications, also for cream, skimmed milk, ice cream, human milk, etc.

Table 17-6
Comparison of amounts of inorganic constituents in human and bovine milk (amounts* are averages, as mg/100 gm milk)

Constituent	Human milk	Bovine milk
Total ash	200	720
Calcium	32	119
Chlorine	43	103
Iodine	0.01†	0.004
Iron	0.03†	0.05
Magnesium	3	13
Phosphorus	14	93
Potassium	51	152
Sodium	17	49
Sulfur	14†	25
Zinc	0.17	0.38

Trace elements (less than 1.0 mg each) include: Ag, Al, As, B, Ba, Br, Cd, Co, Cr, Cu, F, Li, Mn, Mo, Ni, Pb, Rb, Se, Si, Sn, Sr, Ti, and Va.

*From National Dairy Council: Newer knowledge of milk, Rosemont, Ill., 1979, the Council.
†From various sources in the literature.

The amount of cholesterol in bovine milk is approximately 11 mg per 100 ml, all of which is free cholesterol. The cholesterol content of the milk does not seem to bear any relation to the level of this lipid in the cow's blood serum.

Inorganic constituents The inorganic salts of milk include chiefly calcium, potassium, and sodium salts of hydrochloric and phosphoric acids. Potassium is present in larger amounts than sodium. Other elements present in milk include magnesium, sulfur, iron, copper, iodine, and zinc (Table 17-6). Although some elements are present only in traces, one must not assume that they are of no value. Only minute amounts of copper, for instance, are needed for hemoglobin formation. Iron, it is true, is present in too small an amount in milk to warrant the exclusive use of this food in later childhood. Iron-containing foods must supplement the diet. Milk is, however, the most practical and adequate source of calcium and phosphorus. Both of these are needed by all cells, but particularly by bones and teeth.

Vitamins Milk contains most of the vitamins in greater or lesser amounts. The fat-soluble vitamins A, D, E, and K and the precursors of A and D are carried in the lipid fraction, whereas the water-soluble vitamins are found in the aqueous fraction. It is quite deficient in ascorbic acid and vitamin D and rather low in the B vitamins, except riboflavin and pantothenic acid, although human milk is richer in ascorbic acid than is bovine milk. To enrich the milk with vitamin D, this factor must be taken just before or during lactation. The content of vitamins A and C also can be increased in human milk if the lactating woman is fed large amounts of these vitamins in a multiple-vitamin supplement. Milk is an excellent source of vitamin A, riboflavin, and pantothenic acid. Table 17-7 gives a comparison of the vitamin contents of human and bovine milk. Since the amounts of the different vitamins vary with diet, exposure to sunlight, and other factors, these figures are to be considered typical rather than absolute values.

Enzymes A catalase, a peroxidase, a phosphatase, xanthine oxidase, and small amounts of several other enzymes are present in milk. It is doubtful whether these have any significance, but the presence of any active enzymes is evidence that the milk has not been pasteurized or sterilized. Indeed, the alkaline phosphatase content of milk is used as a measure of the efficiency of pasteurization.

Urine

The principal function of the kidneys is the excretion of waste products of the body in the urine. However, the kidneys perform other important functions, too—the formation of ammonia to aid in the maintenance of acid-base balance, the conservation by reabsorption of a number of substances essential to the body, and

Table 17-7
Vitamin contents of human and bovine milk (values are averages* or ranges per 100 gm milk)

Vitamin	Human milk	Bovine milk
A (retinol units)	64	31
(Int. units)	241	126
D (Int. units)	0.4-10†	1.14-2.8
E (α-tocopherol) (mg)	0.66†	0.08
K (μg)	—	3.4
Ascorbic acid (mg)	5.0	0.94
Folacin (mg)	0.005	0.005
Niacin (mg)	0.177	0.084
Pantothenic acid (mg)	0.223	0.314
Pyridoxine (mg)	0.011	0.042
Riboflavin (mg)	0.036	0.162
Thiamin (mg)	0.014	0.038
Vitamin B_{12} (μg)	0.045	0.357

*From National Dairy Council: Newer knowledge of milk, Rosemont, Ill., 1979, the Council.
†From various sources in the literature.

the secretion of several physiologically active compounds, such as renin and erythropoietin (Chapter 18).

The functional unit of the kidney is the *nephron,* which consists of a glomerulus, a convoluted and collecting tubule, and blood vessels (Fig. 17-9). The afferent arteriole goes to the glomerulus, carrying blood to the tuft of capillaries that supply it, and then continues on as an efferent arteriole. It breaks up into capillaries again, which are the sole means of blood supply to the tubules. The afferent arteriole enters the glomerulus, and the glomerular filtrate is formed. This is a fluid, very low in protein, formed by the process of ultrafiltration. The concentration of solutes in the glomerular filtrate is similar to the arterial plasma except for the protein. As a result of this filtration, the blood is more concentrated when it leaves the glomerulus in the efferent arteriole. Its protein content is consequently increased, and

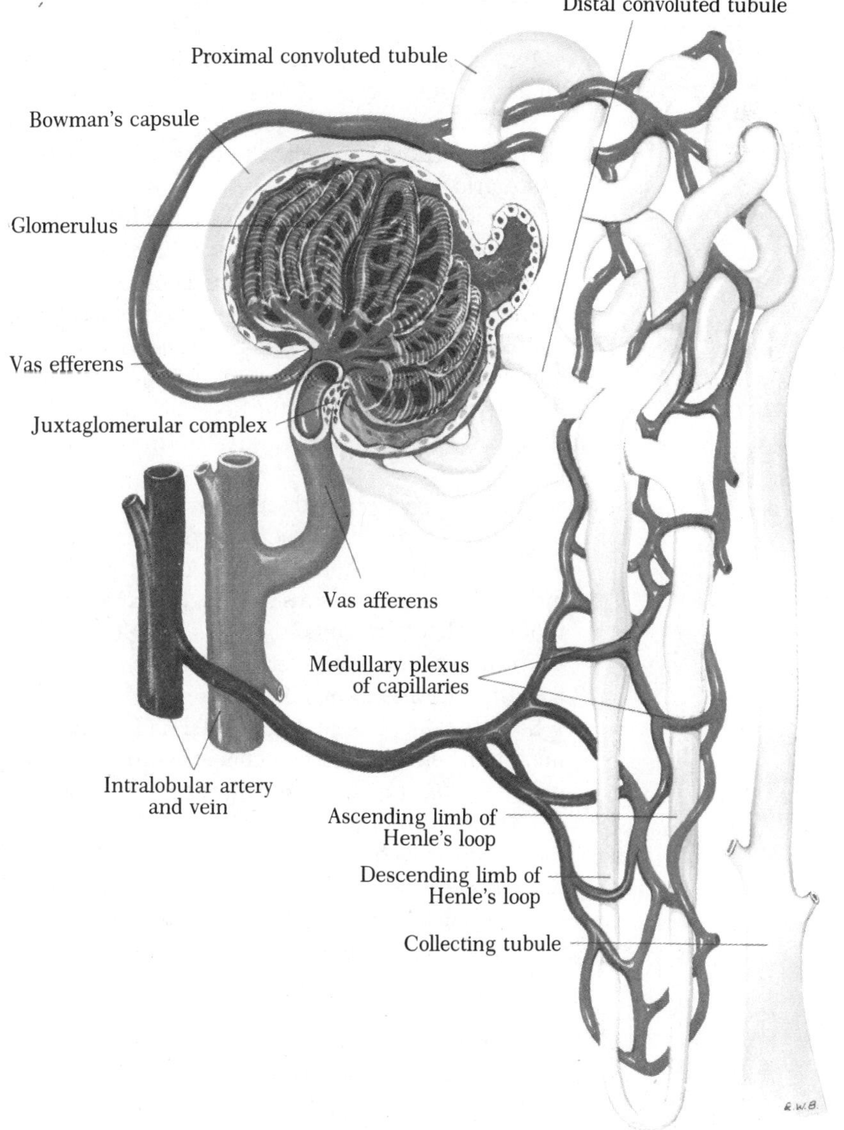

Figure 17-9
Nephron and its associated blood supply. Wall of Bowman's capsule has been cut away to reveal detail of glomerulus.

From Schottelius, B.A., and Schottelius, D.D.: Textbook of physiology, ed. 18, St. Louis, 1978, The C.V. Mosby Co.

therefore its osmotic pressure is greater. This blood surrounds the tubules as the glomerular filtrate flows through them. Most of the filtrate is now reabsorbed through the tubule cells. About 90% of the water is reabsorbed as a result of the difference in pressure. Glucose is reabsorbed after enzymic phosphorylation in the tubular epithelial cells. Sodium, potassium, amino acids, and other substances are also reabsorbed, apparently by specific transport mechanisms. Carbonic anhydrase controls the carbon dioxide–bicarbonate transfer. Urea, creatinine, uric acid, and certain other compounds are not absorbed proportionally to the fluid. At the same time, certain solutes are added to the fluid by an excretory function of the tubules. Among these are hippuric acid, other derivatives of benzoic acid, and still other organic waste products. Of the 10% of water remaining, much is reabsorbed actively by the tubular cells as a result of the action of ADH (Chapter 18) elaborated by the posterior lobe of the pituitary gland, except for 1% or 2%. This small fraction of the glomerular filtrate, necessary as a solute (an "obligatory urine loss") for the excretory waste products (e.g., urea), flows into the urinary bladder as urine. Since an adult excretes from 1000 to 1800 ml in 24 hours, evidently the total glomerular filtrate must be in the neighborhood of 50,000 to 180,000 ml. The end result, urine, is generally a fluid having an osmotic pressure effect greater than that of the body fluids. In this way the kidney conserves water. However, the kidney may also produce a more dilute urine. This occurs, for example, after the ingestion of large quantities of water and leads to *water diuresis,* with the formation of an extremely dilute urine.

The plasma electrolytes pass through the glomerulus, but some potassium leaves the blood by way of the tubules, perhaps by active secretion. Sodium and chloride ions predominate in the glomerular filtrate just as they do in the plasma. As the glomerular filtrate flows through the tubules, the electrolytes are reabsorbed into the blood along with the water. They, like the water, are not absorbed completely, however. Part of the absorption of sodium chloride is brought about through the influence of hormones of the adrenal cortex, aldosterone and deoxycorticosterone (Chapter 18). The tubules are the more particular site of action of these hormones.

Diuresis, the increased secretion of urine, is brought about chiefly by a failure of the tubules to reabsorb their usual quota of the glomerular filtrate. Sodium salts and urea exert diuretic effects. After they filter through the glomerulus, they increase osmotic pressure to such an extent that less filtrate is reabsorbed by the tubules. There are some diuretics, e.g., organic mercurial compounds, that may exert their action on the enzymes in the tubular cells, preventing the absorption of sodium salts and hence of water.

Sodium- and potassium-activated ATP, in the renal medulla and cortex, plays an important role in the urine-concentrating mechanism. If the enzyme is inhibited with *digoxin,* impaired concentrating ability with natriuresis occurs.

The urine thus formed carries off (1) water and salts in such amounts as to maintain the normal equilibria between the extracellular and intracellular fluids, (2) acids or bases to maintain a normal acid-base balance, (3) waste products, (4) toxic and detoxified substances, and (5) other substances that are present in the blood in excessive amounts, if they can be so excreted.

Physical characteristics

Although it is frequently desirable to obtain and analyze casual specimens of urine or the excretory output for short periods, it is usually customary to examine 24-hour specimens in order to have a "yardstick" for comparison and study. The physical characteristics usually noted in examining such a specimen are volume, turbidity, color, odor, specific gravity, and reaction.

Color The color of normal urine is amber yellow. The principal pigment is *urochrome,* which is yellow. This is a compound of uncertain composition urobilin and urobilinogen with a peptide. Small amounts of *uroerythrin, coproporphyrin* and *uroporphyrin* are usually present. Uroerythrin, which is possibly derived from the melanins, is red; uroporphyrin and coproporphyrin are brownish red iron-free pigments arising from heme metabolism. *Riboflavin* and one of its metabolic products, *uroflavin,* may also be present and give a greenish fluorescence to the urine.

On standing, urine usually deepens in color, a result of colorless chromogens changing to colored compounds. Thus urobilinogen and urochromogen on oxidation yield urobilin and additional urochrome respectively.

Abnormally, excessive amounts of some of the normal pigments, notably coproporphyrin and uroporphyrin, may be excreted. The color may also be changed by the appearance of hemoglobin, urobilin, bile pigments, or melanins. Foreign pigments, e.g., dyes, occasionally are found after their administration.

Among the chromogens that are not normal is homogentisic acid, which occurs in that inborn error of metabolism known as alkaptonuria. In this condition the urine, when passed, is of normal color but assumes a smoky or blackish hue on standing. The darkening begins at the top, from exposure to oxygen, and travels downward. Homogentisic acid is a product of the incomplete metabolism of tyrosine and phenylalanine, as shown in Chapter 13. Porphyrinuria occurs in porphyria, an unusual hereditary disorder of heme metabolism involving the cutaneous, gastrointestinal, and nervous systems. This state is to be distinguished from simple porphyrinuria seen in toxic and disease conditions, e.g., lead poisoning, chronic liver disease. The urine may be pink to brown, or at times it may have a normal color when voided and darken only on exposure to light. The specimens exhibit a characteristic pink to red fluorescence when exposed to ultraviolet light. Porphyrins are present in normal urine but in such small amounts that they are not detectable even in ultraviolet light and do not contribute to urine's color.

Volume The rate of secretion of urine is not constant and depends on a number of factors. Normally more urine is secreted during the day than at night, but this is reversed in the case of night workers. The food and fluid intake, the temperature and humidity of the atmosphere, and exercise are the chief influences. Young children excrete more urine in proportion to their weight than do adults. Emotional excitement also increases the volume secreted. The total output in a 24-hour period in the northern part of the United States averages from 1 to 1.5 liters; in the South it is likely to be somewhat less. This difference is, of course, related to climate. In summer a day's output may be as low as 600 ml because of the diversion of water to the skin and lungs. Exercise results in a similiar diminution, for the same reason. Loss of large quanitities of water in diarrheal discharges also lowers the volume of urine.

Foods contain varying amounts of water, and some water arises in the oxidation of foodstuffs. Salty and spicy foods, as a rule, induce diuresis, and certain beverages, e.g., beer, have a decidedly diuretic influence. Among the diuretic drugs is caffeine; so coffee and tea have this property. Urea has the same effect; hence a high-protein diet results in a large output of urine.

Pathologically the volume of urine is increased as a result of injury to the posterior pituitary gland. Diabetes insipidus is a disease in which there is a deficient secretion of this gland. In diabetes insipidus, enormous volumes of urine are eliminated, with resulting intense thirst. An increased urinary output is designated *polyuria*. Polyurias are also seen in diabetes mellitus, because glucose is a diuretic; in malnutrition; in certain endocrine imbalances; and in some renal conditions. In fact, an increased flow of urine at night is frequently one of the earliest symptoms of chronic kidney disease. This is called *nocturia* and is defined as the passage of a volume of over 500 ml of urine having a specific gravity below 1.018 during a 12-hour period. This polyuria of the early stages of chronic glomerulonephritis is believed by some authorities to be an effort on the part of the kidney to compensate for the smaller number of healthy, functioning renal units. Others, however, consider it to be a definite diminution of the ability of the tubules to reabsorb the water from the urine that has filtered through the glomerulus, even if this water is normal in quality and quantity. In later stages, because the involvement of the glomerulus, urine volumes decrease. This is usually also the state of affairs in acute glomerulonephritis. A diminished excretion of urine is termed *oliguria,* and a cessation is known as *anuria*. Oliguria also occurs in fevers, cardiac conditions, and diarrhea. In fevers the explanation is that there is a shift in the water balance, with much of the water of the blood going into the tissues temporarily. From a concentrated blood (anhydremia), the kidney cannot remove water. With a weakened heart action the kidney is not supplied with a sufficient quota of blood and therefore does not secrete urine efficiently. In diarrhea the loss of fluid results in anhydremia, with consequent diminution in the secretion of urine. Other causes of dehydration, e.g., persistent vomiting and excessive sweating, similarly cause oliguria.

Specific gravity The normal range of specific gravity is from 1.008 to 1.030, but usually it is within the limits of 1.015 to 1.025. In a general way, specific gravity varies inversely with the volume of urine excreted. In diabetes insipidus the specific gravity is very low, approaching 1.000, whereas in fevers, in which a small volume is excreted, the urine is concentrated and has a high specific gravity. An exception to the inverse ratio is diabetes mellitus. Here the volume is usually large and the specific gravity is high because of the glucose present.

Normally urine excreted at night has a higher specific gravity than that excreted during the day. A considerable variation is seen from hour to hour throughout the day. In fact, a constancy or fixation of the specific gravity over any appreciable length of time is considered a sign of abnormal renal function. The principle involved is used in several concentration and dilution tests. In each of these, under a fixed set of conditions as regards the kind and amount of food and water taken, the urine is collected at specified intervals and the volume and specific gravity are determined. In the concentration tests, relatively dry food and little drink are given; a highly concentrated urine should be eliminated if the kidneys are normal. On the other hand, normal kidneys are able to secrete the large volume of urine of extremely low specific gravity when a considerable quantity of water is taken, as in the dilution tests. Abnormal kidneys cannot meet the same demands. Moreover, normally the specific gravity is not constant. For example, in the Mosenthal test, specimens are collected every 2 hours. The specific gravities of these samples show a difference of at least 10 points between the lowest and the highest in normal individuals. These tests are excellent criteria of the qualitative detection of renal dysfunction, but they do not give information regarding the extent of damage to the kidneys.

The specific gravity affords a method of estimating the total solids excreted in the urine. Actual determination of the total solids is time consuming and, because of volatile substances present, not very accurate. Consequently a rough method based on a simple calculation is often used. The figure 2.6 (Long's coefficient) multiplied by the last two digits of the specific gravity at 25° C is taken as the total solids in grams *per 1 liter of urine*. The adult usually excretes about 60 gm of urinary solids in 24 hours. About 35 gm of this are organic and 25 gm inorganic.

Acidity Normal human urine may be neutral, acid, or alkaline, having a pH range of from 4.8 to 7.5. It is usually acid, with an average pH of 6.0. The reaction is dependent on the many different inorganic ions as well as organic compounds of acid and basic character present in urine.

Protein diets give rise, in general, to highly acidic urine. The cause is chiefly the sulfur of the sulfur-containing amino acids, which is oxidized to sulfuric acid. The phosphoproteins, in addition, yield phosphoric acid, as do the nucleic acids and the phospholipids. Meats, therefore, are most productive of acid because of their high content of proteins, nucleic acids, and phospholipids. Alkaline urines are excreted when there is a predominance of vegetables and fruits in the diet, since in general these have an alkaline ash (Chapter 21). Thus the proportions of the various foods influence the reaction of the urine. There is another factor that plays an important part. It is the production of ammonia by the kidney, and this modifies the amount of titratable acidity. The total titratable acidity usually is equivalent to 150 to 500 ml of 0.1N acid per day.

Specimens of urine taken at intervals usually vary a great deal in their acidity. Soon after meals the urine excreted is quite alkaline for a while. This "alkaline tide" is explained by the fact that hydrogen ions are excreted in great quantity in the gastric juice. This would result in an alkaline blood if the kidneys did not secrete a preponderance of base at that particular time.

Urine must not be permitted to decompose during or after the collection of a 24-hour sample. If microorganisms begin to grow, they convert the urea to ammonia carbonate and the urine becomes ammoniacal. In addition to having an unpleasant odor, ammoniacal urine indicates a change in the distribution of the nitrogenous constituents, with some loss of nitrogen as volatile ammonia. To avoid this, the urine may be kept cold, or a preservative, e.g., toluene with thymol, may be added to the container at the start of the collection period.

Odor Freshly voided urine has a not unpleasant odor, sometimes described as "aromatic." If urine does not have such an odor soon after it is passed, it may be in a pathological state. A putrid or strongly ammoniacal odor points to decomposition by bacteria, probably occurring in the urinary bladder. Other odors arise from foods eaten, e.g., the unpleasant odor of methyl mercaptan after partaking of asparagus. Oil of sandalwood, cubebs, and other drugs give rise to characteristic odors in the urine. Oil of wintergreen (methyl salicylate) gives rise to a strong odor of evergreens. Perhaps the most important odor, and one that sometimes aids in diagnosis, is the fruity aroma observed when a large amount of acetone is present, especially in uncontrolled diabetese mellitus.

Turbidity Normal urine is almost always perfectly clear and transparent when voided. On standing there is likely to separate out a faintly cloudy flocculence, believed to be nucleoprotein or mucoprotein, which is present only in traces, together with some epithelial cells. Turbidities may be of several types. Ammonium urate may precipitate from alkaline urine, whereas other urates are found only in acid urines. The former dissolve on acidification; the latter, on warming. Calcium phosphate and ammonium magnesium phosphate (triple phosphate) are seen only in alkaline urines, or they may form a cloudy precipitate from an alkaline urine on warming. They dissolve on acidification. It is therefore quite essential to be sure that a urine sample is slightly acidified when a test for heat-coagulable proteins is made. Furthermore, the isoelectric points of most proteins that may appear in the urine are slightly acidic. Hence, adjustment of the pH also facilitates detection of these proteins, especially if only small quantities are present. If a clear urine is acidified slightly without heating and a precipitate is formed, the precipitate is either a mucoprotein or a nucleoprotein, since these proteins come down at an acid pH. A sediment in urine that does not dissolve on adding acid or on heating is most likely made up of cellular matter, i.e., pus, epithelial cells, or microorganisms.

A microscopic study of the sediment present in urine is often of great assistance to the clinician in diagnosis. Both crystalline substances and cells afford many clues that an expert microscopist can use to advantage. A small number of leukocytes are always, or almost always, present. Only an increase is pathological, and the accumulation is then called *pus*. Pus is usually accompanied by protein and may arise from an inflammation of any part of the genitourinary tract. The presence of a small or moderate number of epithelial cells in the urine also is an ordinary occurrence, but a great number is abnormal. However, in the female, catheterized specimens are required for careful diagnostic work because fairly large numbers of erythrocytes and leukocytes and vaginal epithelial cells are likely to be present in uncatheterized urine.

Chemical composition

Normal urine is composed of the following: (1) water, (2) inorganic salts, (3) nitrogenous organic compounds, and (4) nonnitrogenous organic compounds.

The inorganic ions include the following:

Cations—Na^+, K^+, Ca^{++}, Mg^{++}, NH_4^+; traces of Fe^{++}, Fe^{+++}, Cu^{++}, Zn^{++}

Anions—Cl^-, $PO_4^{\equiv}$, $SO_4^=$; traces of NO_3^-, HCO_3^-, SiO_2^-, F^-; also minute traces of many other inorganic compounds

The nitrogenous organic compounds excreted in the urine are, with few exceptions, waste products. The most important ones are urea, uric acid, creatinine, creatine, hippuric acid, indican, purines other than uric acid, peptides, and amino acids.

Nonnitrogenous organic substances are less in amount and include traces of glucose, glucuronic acid, cholesterol, and the ketone bodies; oxalates and salts of other organic acids, including six members of the citric acid cycle, namely, citric, aconitic, α-ketoglutaric, succinic, fumaric, and malic, and organic sulfur compounds. A variety of vitamins and hormones (or their metabolites) and of several enzymes also are present in small amounts.

The quantitative chemical composition of urine is usually expressed as amounts of constituents per a 24-hour sample preserved to prevent bacterial decomposition. Smaller, random samples are not used for this purpose because amounts may vary considerably at different times of the day, following meals, with activity, etc. The ranges of values usually found for the principal urinary constituents for a normal human subject are given in Table 17-8. Values obtained obviously will vary considerably even in a normal healthy adult, depending on the amounts and types of food ingested, body size, activity, sex, and environmental and other factors.

Inorganic constituents

Sulfur Most of the sulfur of our diet is protein sulfur, from the amino acids cystine, cysteine, and methionine. Some enters the body also as chondroitin sulfate, and there are small amounts in other forms. It is excreted as sulfate and as neutral sulfur. Sulfates are both organic and inorganic. The former is called *ethereal sulfate*. There is no definite proportion among the three forms, the inorganic, ethereal, and neutral, but on an ordinary mixed diet from 79% to 84% of the total sulfur is excreted in the form of inorganic sulfate. The remainder is divided between the ethereal sulfates and the neutral sulfur; about 4% to 7% is ethereal sulfate and 16% to 21% is neutral sulfur.

The sulfur of the amino acids is mostly oxidized to sulfate, which forms inorganic sulfate, for the most part, and, to a lesser extent, combines with organic compounds. These may be phenols, cresols, indoxyl, skatoxyl, sterols, or other compounds. Some are toxic, and the formation of the conjugated sulfates transforms them into easily excreted products. The reactions are analogous to that shown for

Table 17-8
Principal constituents of urine of normal adult (amounts per 24 hours)

Type of constituent	Total (gm)	As N (gm)	As mEq
Organic			
Nitrogenous	20-40	10-18	
Urea	18-35	8-16	
Creatinine	1.0-1.8	0.3-0.8	
Creatine	0.06-0.15	—	
Ammonia	1.0-1.8	0.4-1.0	
Amino acids	0.8-2.5	0.1-0.25	
Uric acid	0.3-1.0	0.1-0.3	
Peptides	—	0.3-0.7	
Hippuric acid	—	0.04-0.08	
Indican	0.04-0.15	—	
Protein	< 0.1	—	
Other			
Organic sulfates (as S)	0.06-0.2		
Phenols	0.02-0.05		
Oxalates	0.01-0.02		
Ketone bodies	0.01-0.1		
Inorganic			
Cations			
Sodium	4.0-6.0		170-255
Potassium	1.5-3.0		30-75
Ammonium (NH_4^+)	1.0-1.8		55-100
Calcium	0.2-0.5		10-15
Magnesium	0.1-0.2		8-16
Anions			
Chloride	6.0-9.0		170-255
Inorganic SO_4 (as S)	0.6-1.8		40-100
Phosphate (as P)	0.7-1.6		20-50
Bicarbonate	—		0-50
Other (organic acids, etc.)	—		30-40

the formation of indican from indoxyl, as is described later. Since the sulfates are derived from the sulfur-containing amino acids, the amount of total sulfate excreted is, in a general way, an index of the amount of protein metabolized, although not as accurate an index as the total nitrogen excretion. Hence the total sulfate and the total nitrogen of the urine generally run parallel.

The remaining fraction of urinary sulfur, the neutral sulfur, includes a variety of compounds having the sulfhydryl, sulfide, and thiocyanate groups. It would, accordingly, include sulfur-containing amino acids, e.g., cystine and any peptides containing them. Also in this fraction would be thiosulfates, taurine, ergothioneine, urochrome, and the thiazole part of thiamine. The amount excreted is largely independent of protein intake but is related to cellular protein catabolism.

Pathologically, sulfate excretion is increased when tissue protein catabolism is speeded up, as in acute fevers. Neutral sulfur excretion rises in cases of poisoning by cyanides and nitriles, because of the transformation of these compounds into thiocyanates and their excretion in that form. Chloroform and other anesthetics also increase the excretion of neutral sulfur. In cystinuria, naturally, this fraction is greatly increased.

Phosphates The only phosphorus compounds found in urine in appreciable amounts are the derivatives of phosphoric acid. The total amount of phosphate eliminated in the urine varies with the amount of phosphorus in the food and the amount absorbed. Food phosphorus is contained in the phosphoproteins, phospholipids, nucleoproteins, and preformed phosphate. If calcium or magnesium ions are present in the intestinal tract in abundance at the same time as phosphate ions, insoluble calcium or magnesium phosphate is formed but is poorly absorbed. Other insoluble phosphates are also possible, and consequently the urinary phosphate may represent only 50% to 70% of the food phosphate. Most of the remainder goes through into the feces. The determination of the amount of phosphorus in urine is therefore of little value.

Certain major fluctuations of the urinary phosphorus that are of interest occur, however. In acidosis phosphate excretion may rise (unless the kidney is incapable of secreting it, as may be the case in nephritis). This is a direct effort of the organism to get rid of hydrogen ions, preserve its base as far as possible, and maintain the normal pH. An increased elimination of phosphate is one of the first events in hyperparathyroidism or after the administration of parathyroid hormone. Low urinary phosphate is likely to be associated with diarrhea, because the intestinal contents are hurried through the tract; with acute infections and nephritis, because of failure of the kidneys to function adequately; with pregnancy, as a result of the fetal requirement for phosphate; and with rickets and other bone diseases, in which there is a diminished absorption or increased intestinal elimination of phosphate. When insulin is administered there is an increased requirement of phosphate for the formation of hexose phosphates. This results in a diminished urinary phosphate for a time, often followed by an increase.

Chloride About 10 to 15 gm of chloride (as NaCl) are ingested per day. Normally the amount eliminated in the urine is almost equal to that taken in. Next to urea, the chlorides of the urine are the chief solid constituents. Acid-base balance and water balance are intimately associated with the distribution and elimination of sodium chloride. As has been seen, the sodium may be retained in times of stress, to conserve base. This is evident in the last stages of such a condition as pyloric obstruction.

Remember also that, as the filtrate formed by the glomerulus of the kidney is reabsorbed, much of the sodium chloride is absorbed. Sodium chloride is one of the threshold substances needed by the body and is therefore retained in fairly definite concentrations. Table 17-9 gives a number of the common constituents of blood and urine with their relative concentrations. Sodium, calcium, and chloride ions have about the same concentrations in both urine and blood, which indicates that these are retained normally through the reabsorption mechanism of the kidney. In

Table 17-9
Relative concentrations of constituents of urine and blood*

Substance	Concentration in urine (mg/100 ml)	Concentration in blood (mg/100 ml)	Concentration ratio	Concentration in blood in renal insufficiency
Urea	2000	30	67	Increased
Uric acid	60	2	30	Increased
Creatinine	75	2	37	Increased
Indican	1	0.05	20	Increased
Phosphate	150	3	50	Increased
Sulfate	150	3	50	Increased
Potassium	150	20	7.5	Slightly increased
Chloride	500	350	1.4	Not increased
Sodium	350	335	1	Not increased
Calcium	15	10	1.5	Not increased
Water			1	Not increased

*From Fishberg, A.M.: Hypertension and nephritis, ed. 5, Philadelphia, 1954, Lea & Febiger.

the case of sodium and chloride, the evident result of this retention is a very large contribution to the osmotic pressure of the blood.

Deprivation of salt, as in salt-poor diets or in starvation, leads to a noticeable decrease in the volume of urine. It is the amount of chloride in the interstitial fluid that largely determines chloride and water excretion. Ordinarily the addition of salt to a diet results in the elimination of the excess within 48 hours. After a period of salt deprivation, such an excess will be retained until the volume and salt content of the interstitial fluid have been reconstituted.

An excessive loss of sodium chloride by way of the urine occurs in adrenocortical insufficiency (Addison's disease). A clinical diagnostic test is based on this fact. With a low-sodium and high-potassium diet for 3 days, the excretion of sodium by normal individuals averages 22 mg per 100 ml of urine on the third day, whereas that of patients having Addison's disease averages 206 mg per 100 ml.

During the formation of the exudate in pneumonia, salt is removed from the body fluids. Therefore less salt is available to the interstitial fluid, and both blood and urine chlorides drop. When the exudate is reabsorbed, the condition is reversed and larger quantities of chloride reappear in the urine.

Cations Ammonium, sodium, and potassium ions leave the body chiefly by way of the urine. Calcium and magnesium are excreted both through the intestinal tract and through the urine, chiefly perhaps by the former route. Ammonium salts are considered among the nitrogen compounds.

Nitrogenous constituents

Urea Urea is the diamide of carbonic acid and, as such, is represented by the formula $CO(NH_2)_2$. This simple formula is in accordance with many of the reactions of urea, including its preparation from ammonia and carbonic acid in vitro. The mechanism of urea formation in the body is discussed in Chapter 13.

Urea is a white solid; it is odorless but has a bitter, salty flavor. It crystallizes in long prisms. Although urea is neutral, it reacts with acids as a monobasic amide. Urea nitrate and oxalate form characteristic crystals insoluble in excess of the acid. Urea is soluble in water and alcohol but not in ether or chloroform. The enzyme urease, which occurs in the jack bean and soybean, accelerates its conversion to ammonium carbonate:

$$CO(NH_2)_2 + 2H_2O \rightarrow (NH_4)_2CO_3$$

This is the basis for the quantitative determination of urea in blood, urine, or other fluids. The ammonium carbonate formed is easily measured in a number of different ways.

As has been seen, urea is the chief end product of protein metabolism and is formed in the liver (Chapter 13). It is excreted in the urine in larger amounts than any other substance (about 30 gm per day) and makes up from 85% to 92% of the total nitrogen on a medium- or high-protein diet. On a low-protein diet the proportion (as well as the actual amount) of urea nitrogen is lower. It may be as low as 60% of the total nitrogen. The reason for this is that the urea output parallels protein metabolism whereas the nonurea fraction depends only in part on protein metabolism and in part on other factors. Fig. 17-10 is an indication of how, with a decreasing total nitrogen, the urea nitrogen assumes a smaller and smaller proportional part of the total nitrogen.

Urea is nontoxic even when present in the blood in relatively large amounts. Consequently, high concentrations are regarded with concern not because of any inherent danger from the urea itself but because they indicate inadequate excretory function. Indeed, in neurosurgery, hypertonic solutions of urea are sometimes administered to reduce intracranial pressure and brain volume. This procedure is

Figure 17-10
Relation of urea nitrogen of urine to nonurea nitrogen at various nitrogen levels.

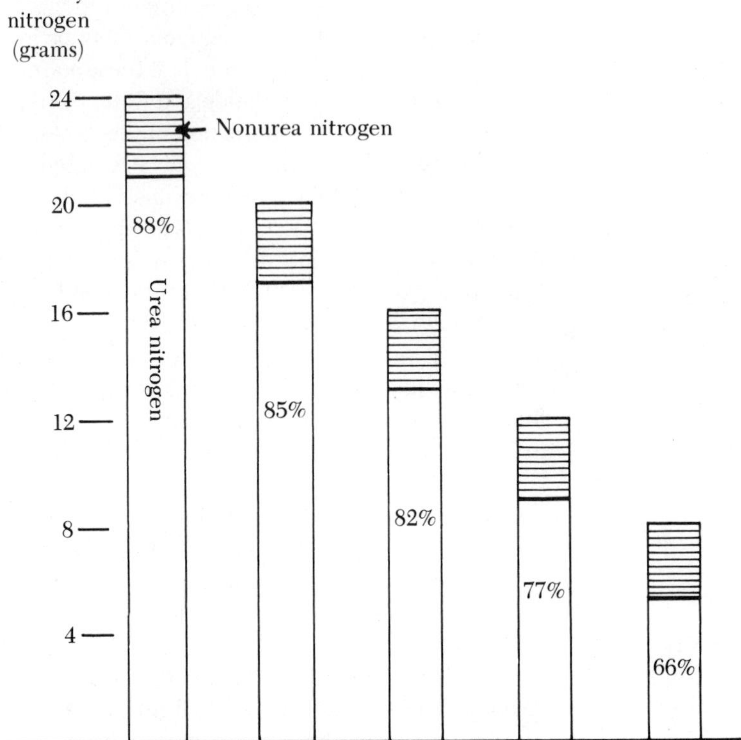

also used in the treatment of glaucoma. Hyaluronidase facilitates absorption. Urea is a diuretic. It freely passes the glomerular filter and is passively reabsorbed to the extent of about 50% under normal conditions. *Passively* indicates by physical rather than by active physiological means. Thus urea diffuses from the relatively higher concentration in the tubules (because of the simultaneous reabsorption of water) to the lower one in the blood plasma.

Ammonia The total amount of urinary ammonia, as ammonium salts, is approximately 0.7 gm per day (0.4 to 1.0 gm). It is formed in the kidney, as discussed in Chapter 13. The production of ammonia aids in the neutralization of acids, with about 67% to 75% of the hydrogen ions excreted in the urine being done so as ammonium ions. By so doing, it conserves sodium and potassium, which are essential for other important physiological activities, e.g., buffer action. Thus when an acid like acetoacetic (H^+A^-) is produced in large amounts, it passes into the blood, where it is buffered by sodium bicarbonate. Thus:

$$\underset{\textbf{Acetoacetic acid}}{H^+A^-} + Na^+HCO_3^- \rightleftharpoons H_2CO_3 + Na^+A^-$$

The H_2CO_3 is excreted in the lungs as CO_2 and H_2O. To restore the lost $Na^+HCO_3^-$, the following occurs in the kidney:

$$\underset{\textbf{Filtered through glomerulus}}{Na^+A^-} + \underset{\textbf{Secreted by tubule}}{NH_4^+HCO_3^-} \rightleftharpoons \underset{\textbf{Reabsorbed into blood}}{Na^+HCO_3^-} + \underset{\textbf{Excreted in urine}}{NH_4^+A^-}$$

Normally the urinary ammonia nitrogen constitutes about 2.5% to 4.5% of the total nitrogen. In acidosis the amount rises about 5%. At one time this was the method of determining the presence and degree of acidosis clinically (Chapter 16). Since urea is easily decomposed and converted to ammonium carbonate, ammonia determinations must be made on fresh or sterile urine.

Uric acid Acidified urine, on standing, shows a reddish crystalline deposit of uric acid. In fact, the normal acidity may be enough to permit this deposition to take place. The crystals assume a variety of shapes, e.g., wedges, prisms, dumbbells, rosettes. Uric acid crystals are odorless, tasteless substances that are insoluble in alcohol and ether, slightly soluble in boiling water, but quite soluble in alkalies and alkali carbonates. Their alkaline solutions have some reducing power on silver salts, copper salts, phosphomolybdates, and phosphotungstates. The reduction of phosphotungstate is used in the Folin method for the determination of uric acid.

When uric acid crystals are oxidized in alkaline solution, allantoin is formed; but in acid solution, alloxan is the product:

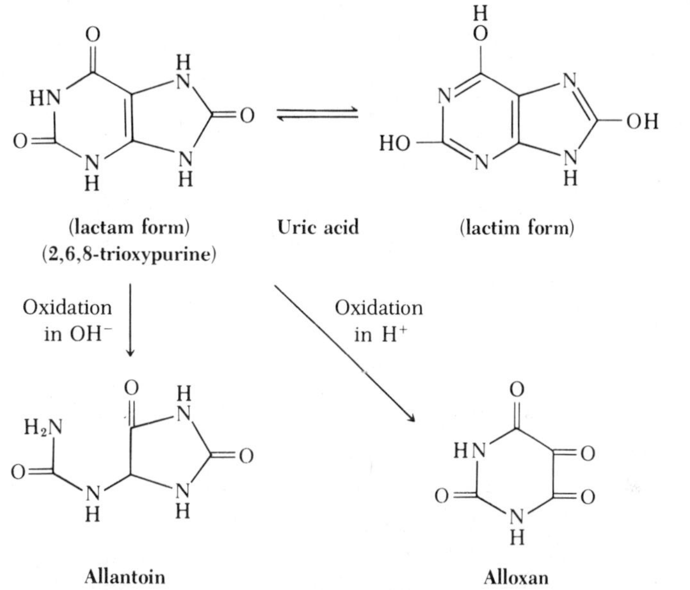

(lactam form)
(2,6,8-trioxypurine)

Uric acid

(lactim form)

Oxidation in OH^-

Oxidation in H^+

Allantoin

Alloxan

Allantoin is an end product of purine metabolism in many animals but not in man. Alloxan is of interest as a substance that can produce diabetes experimentally.

Although uric acid has no carboxyl groups, it acts as a weak, dibasic acid, because of the enolization of the three hydroxyl groups as shown in the above lactim form. This should be tribasic, but apparently the third hydrogen is not dissociated. Thus we have salts of the type $C_5H_3NaN_4O_3$ and $C_5H_2Na_2N_4O_3$. Of the alkali salts, the ammonium one is the least soluble; then, in increasing order of solubility, come sodium, potassium, and lithium. The fact that lithium urate is the most soluble gave rise to the use of "lithia waters" in the treatment of pathological conditions ascribed to an excess of uric acid or uric acid deposits. The lithium was expected to expedite the elimination of uric acid as the more soluble lithium urate.

The total amount of uric acid eliminated per day is approximately 0.7 gm. The determination of uric acid in urine, however, is of little clinical value. Since uric acid is an end product of purine metabolism, it of course normally fluctuates directly with the purine intake. Diets rich in nucleoproteins, e.g., meats, particularly glandular meats, meat extracts, and legumes, lead to an increased excretion of uric acid. Caffeine, theophylline, and theobromine are not converted to uric acid but to methyl purines. On a purine-free diet some of the acid is constantly excreted, amounting to about 0.2 to 0.5 gm per day for an adult. This fraction is referred to as *endogenous* uric acid. If it is determined for a given individual, the excess above this figure, which he or she excretes on a purine-containing diet, is termed *exogenous*. The uric acid excreted on a purine-free diet must arise from synthesis. This is discussed in Chapter 14.

The intensity of nuclear metabolism is reflected in the uric acid output. Thus in leukemia, in which there is a high degree of nuclear catabolism, the uric acid excretion is strongly increased.

Creatine and creatinine In Chapter 13 the biosynthesis of creatine and its relation to creatinine is discussed in detail. Creatinine is a constant normal constituent of urine, whereas creatine is an inconstant one.

Children regularly eliminate creatine in their urine in larger amounts than do adults. It was formerly believed that normal adult males excrete no creatine in their urine but that normal adult women excrete moderate amounts at irregular intervals. This has been the subject of controversy, but now creatine is recognized as a normal component of the urine of healthy men. It constitutes about 6% of the total creatine-creatinine output, i.e., about 60 to 150 mg per day. Most women excrete about twice as much as men and do so much more irregularly. In about 20% of females, the excretion of creatine does not exceed that of the male. During pregnancy the output of creatine increases, and for 2 or 3 weeks post partum it is found in even greater amounts than previously. In pregnancy, a large amount of uterine muscular tissue is formed and is not functioning, and post partum the reduction of this tissue may release creatine stored up there. However, the same increased creatine ouput occurs after cesarean section and removal of the uterus. The ouptut of creatine is also greater in starvation, carbohydrate deprivation, diabetes, hyperthyroidism, fevers, and malnutrition. In all these conditions there is an increased catabolism either of muscular tissue or of tissue proteins in general. Diseases peculiar to the musculature frequently are accompanied by creatinuria. In myasthenia gravis it is not always found, and in myotonia congenita it is seldom found; but in dystrophia myotonia (progressive muscular dystrophy) and in amyotonia congenita, creatine invariably appears in the urine. Creatine elimination is greatly increased in rheumatoid arthritis but not in osteoarthritis.

About 120 gm of creatine and phosphocreatine are present in the human body, mostly in the muscles. Very little creatinine is found there. Neither creatine nor creatinine has been found among the hydrolytic products of any protein.

The relation of creatine to creatinine was investigated at an early date by feeding creatine to animals. Experiments indicate that although creatine is converted to creatinine in the body this is not the only pathway for its catabolism. Isotope experiments have given more direct evidence of this relationship. Bloch and Schoenheimer fed small amounts of creatine containing ^{15}N to adult rats. This ^{15}N was found to be present in the creatine of muscle and internal organs, as well as in urinary creatinine. In a second series the tissue creatine was labeled with isotopic nitrogen by feeding isotopic creatine during a preliminary period. Then, when creatine feeding was discontinued, the isotopic content of the urinary creatinine was identical to that of the body creatine. These facts indicate (1) that creatine in the diet can be absorbed and can replace the creatine of the tissues and (2) that on a creatine-free diet, the tissue creatine is the sole source of urinary creatinine. When isotopic creatinine was fed, however, no isotopic nitrogen was found in the tissue creatine, again emphasizing the fact that creatine → creatinine is biologically irreversible.

Creatinine and creatine are easily determined by spectrophotometric methods. The amount of creatinine eliminated varies chiefly with the weight of the individual, unless he is obese; i.e., it has some relation to the muscular mass. The *creatinine coefficient* for men is about 18 to 32 mg per kg of body weight, and for women, 9 to 26 mg per kg. A round number for the daily excretion is a total of about 1.5 gm.

Amino acids Small amounts of amino acids, both free and combined, are excreted in the urine. Normal adult women excrete more amino acids than do men. The free amino acid amounts to about 1.4 mg per kg for men and 2.3 mg per kg for women. The sex difference does not hold for children, who excrete somewhat less than adults; infants and prematurely born babies excrete about four times more per kilogram than do older children. Growing children excrete higher levels of hydroxyproline than do adults, reflecting an increased metabolism of collagen. Increased excretion of amino acids may occur pathologically. It is frequently observed in wasting diseases and in diseases affecting the parenchyma of the liver, presumably because of the inability of the liver to deaminate the amino acids. Thus in acute yellow atrophy of the liver the amino nitrogen level may rise to 40 mg per 100 ml of urine. Even greater amounts have been found in the urine of a patient having hepatolenticular degeneration, a disease in which there is cerebral degeneration accompanied or followed by cirrhosis of the liver. Alanine, glutamic acid, and aspartic acid have been identified. The aminoaciduria of this condition (Wilson's disease) is now believed to be secondary to an abnormality in copper metabolism (excessive deposition of copper particularly in the liver and kidneys). An increased number of amino acids are excreted by patients suffering from muscular dystrophy and by their mothers and siblings. Among these acids are methionine or valine, isoleucine or leucine, methionine sulfoxide or sarcosine, methylhistidine, and cysteic acid. A generalized hyperaminoaciduria commonly occurs in lead intoxication and indicates damage to the renal tubules. Other abnormalities in which amino acids appear in urine in increased quantities include cystinuria, homocystinuria, histidemia, Hartnup disease, tyrosinosis, and phenylpyruvic oligophrenia. These are discussed in Chapters 13 and 22.

Increased elimination of those amino acids normally present in urine is seen in the *Fanconi syndrome,* a condition in which there is a defect in renal tubular reabsorption. As a result there may be an increased output of any of a variety of substances, including amino acids, other organic acids (e.g., lactic), ketone bodies, glucose, ammonia, and phosphates.

More than 1 gm of peptides is excreted daily by the average normal human adult, which accounts for about half the urinary amino nitrogen and about 2% of the total nitrogen.

Large quantities of β-aminoisobutyric acid are excreted in the urine of a small proportion (about 5%) of otherwise normal people. It has also been found in various pathological states, sometimes with other amino acids. It is apparently caused by a disturbance in the metabolism of thymine and dihydrothymine, which are precursors of this amino acid (Chapter 14).

Hippuric acid Hippuric acid (p. 572) is so called because it was first found in the urine of horses. It is benzoylglycine, a compound of benzoic acid and glycine. This is a physiological metabolic reaction that results in the detoxication of benzoic acid and benzoates. The latter occur in vegetable foods or are derived from the oxidation of aromatic substances. About 0.7 gm of hippuric acid per day is eliminated on the average diet, but deviations from this are to be expected, depending on the amount of precursors in the diet. Glycine is usually present in sufficient amounts to combine with any quantity of benzoates that are likely to be ingested.

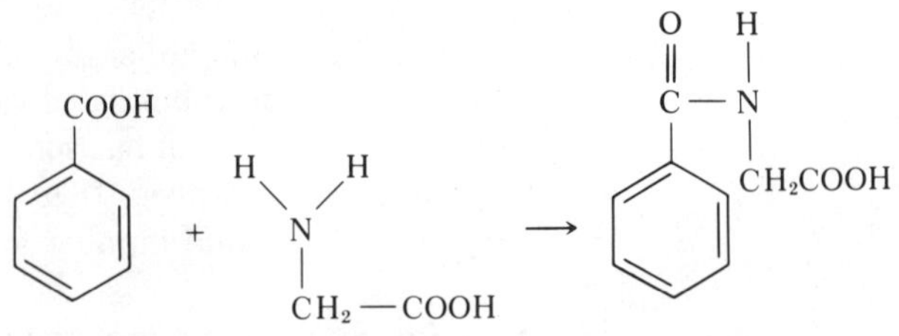

$$C_6H_5COOH + H_2N{-}CH_2{-}COOH \longrightarrow C_6H_5{-}CO{-}NH{-}CH_2COOH$$

Benzoic acid **Glycine** **Hippuric acid**

Cranberries contain 0.05% to 0.09% benzoic acid, and other fruits and berries contain smaller amounts. Some foods, e.g., catsups, are permitted by law to have 0.1% sodium benzoate added as a preservative. This is, of course, also converted to hippuric acid.

In humans the synthesis of hippuric acid takes place almost entirely in the liver. Both ATP and CoA are required for this biosynthesis. The synthetic action of these compounds has been used as a test for liver function but evidently is a test for only this particular liver function, not for all. Nevertheless, subnormal values have been claimed in patients with various types of hepatitis and other liver conditions but not in those with uncomplicated obstruction of the common bile duct; i.e., it may aid in differentiating between hepatic and obstructive jaundice.

Indican Indican is the salt or salts of indoxyl sulfate. It is derived from indole, which in turn arises from the action of putrefying bacteria on tryptophan or on proteins containing it. This occurs in the large intestine. If any indole is absorbed, it undergoes a series of detoxication transformations, probably in the liver, and indoxyl is formed. This is conjugated with sulfate and neutralized to yield a salt. Indican is detected in the urine by oxidizing it to indigo blue. Obermayer's reagent, which is concentrated hydrochloric acid containing a small amount of ferric chloride, is a good oxidizer for this purpose. Chloroform is then added and, after the mixture is shaken gently, the blue dye is taken up by the chloroform. The formation of indican is shown below:

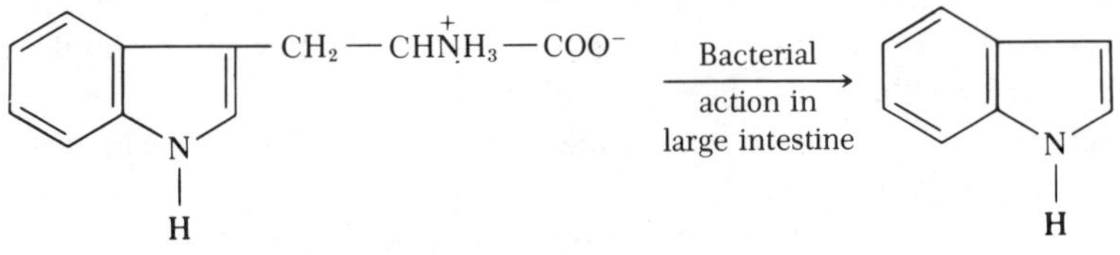

Tryptophan ($-CH_2-CH\overset{+}{N}H_3-COO^-$) —Bacterial action in large intestine→ Indole

—If absorbed, oxidized in liver→ Indoxyl (—OH) —$+ SO_4^=, K^+$→ Indican ($-O-SO_2-O^-$)

Tryptophan **Indole**

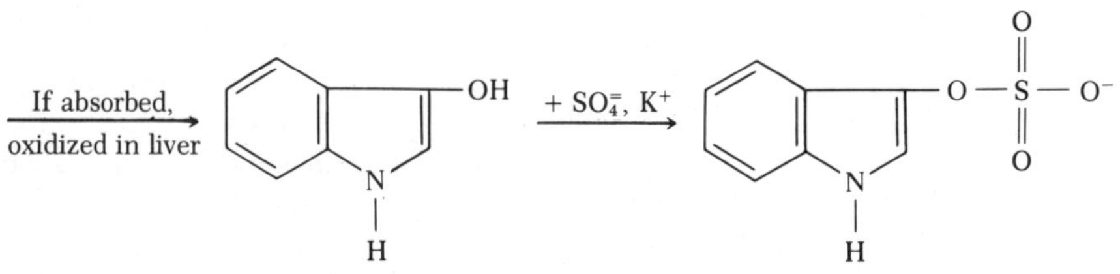

Indoxyl **Indican (potassium indoxyl sulfate)**

The steps between tryptophan and indole are not known, but they are believed to occur in the manner shown in Chapter 13. Remember that indole is not formed in the normal metabolism of tryptophan in the body. It is a product of putrefaction, usually in the intestine, but possibly in other locations. Normally from 4 to 20 mg are excreted daily, but a qualitative test may not be positive for this amount. An increase in urinary indican is found when there is increased putrefaction, provided that the products are absorbed. Putrefaction, you will remember, is the anaerobic bacterial decomposition of proteins. Among the pathological conditions in which this is likely to be observed are hypochlorhydria, because of diminished bactericidal action of the gastric juice; intestinal obstruction and paralytic ileus, because peristaltic movement is inhibited; and obstructive jaundice, because the absence of bile produces voluminous feces with higher nutritive value for the bacteria. Indicanuria is rather rare in simple constipation and rather common in diarrheas. Furthermore, some individuals showing no gastrointestinal symptoms excrete large amounts of indican continually, whereas others, with or without such symptoms, have no indican in their urine. It is thus an index of absorption of putrefactive products but does not necessarily have any other significance.

If putrefaction occurs elsewhere than in the intestinal tract, indole is produced and absorbed and follows the same course. The bacterial decomposition of tissue proteins or of the proteins of body fluids, e.g., exudates, occurs in gangrene, abscesses, empyema, etc. and may lead to a pronounced *indicanuria*.

Nonnitrogenous organic constituents Small but variable amounts of a number of other organic compounds are found in the urine. The range of glucose in the urine of normal individuals is from 0.01% to 0.10%, but the quantity of glucose is in the lower part of that range, and a negative qualitative test is almost invariably observed in normal urines. Glucuronides are formed after the administration of camphor, chloral, menthol, phenol, morphine, aspirin, and other drugs. Some of these, you remember, also combine with sulfate, forming ethereal sulfates. The conjugated glucuronides reduce alkaline copper solutions. The sex hormones are excreted as conjugated glucuronides; they are more soluble in water than are the uncombined hormones.

The ketone bodies or acetone bodies are present in normal urine in very small amounts, less than 0.1 gm in 24 hours. They consist of β-hydroxybutyric acid, acetoacetic acid, and acetone. These substances arise in normal fatty acid catabolism in the liver (Chapter 12) and are carried by the blood to the extrahepatic tissues, where their further degradation occurs. Their concentration in the blood is normally less than 1 mg per 100 ml.

Oxalates are probably excreted constantly in minute amounts. If present in sufficient concentration, they unite with calcium to form insoluble calcium oxalate. This forms characteristic crystals of two types: dumbbell and octahedral. An abnormally great amount of oxalates in urine, *hyperoxaluria,* seems to be an inborn metabolic error involving a failure to catabolize glyoxylate normally, in consequence of which the excess of glyoxylate is converted to oxalate. There is some evidence that tryptophan is a precursor of oxalates, particularly if a deficiency of pyridoxine exists.

Citrate also is a normal constituent of urine. About 0.2 to 1.2 gm are found in adult urine in a 24-hour output under ordinary circumstances. It is increased after the administration of alkali or during a high-carbohydrate, high-fat dietary regimen. This extra citrate does not come from the citrate present in the bones but arises metabolically. The output of citrate also is related to changes in the steroidal reproductive hormones, but this may be an indirect effect. The significance of these facts is not apparent. A number of other organic acids occur in urine in small amounts.

Vitamins, hormones, and enzymes The water-soluble vitamins are excreted by the kidneys in variable amounts. If excessive quantities are present in the diet or are taken as such, some, but not necessarily all, of the excess is eliminated. For certain vitamins the determination in urine has been used as a measure of deficiency; e.g., on an adequate diet the excretion of thiamin amounts to 90 μg or more in males and 60 or more in females; on a deficient diet it is 66 and 43, or less, respectively. Sometimes saturation experiments have proved valuable, i.e., the response to the administration of a large dose of the vitamin. The excess eliminated reflects the degree of saturation previously present. However, the excess is not always eliminated quantitatively. Nicotinamide and its derivatives are found in the urine normally to the extent of 1 to 2 mg together with 5 to 15 mg of the physiologically inert trigonelline, a methylated derivative. Only 27% to 42% of administered nicotinamide is eliminated in urine. On a normal average diet, there is ordinarily excreted from 15 to 28 mg of ascorbic acid in 24 hours. In an avitaminosis, as would be expected, the excretion of the particular vitamin involved is diminished.

As mentioned previously, the steroid hormones are secreted in the urine conjugated with glucuronic acid. A great deal of research has been done regarding the variations in excretion of the individual hormones normally and under various conditions. Improved, more sensitive procedures, including radioimmunoassays (Chapter 25), are now used for their determination.

The enzymes present in urine are small in amount and of little significance. The enzymes of the blood are not likely excreted in appreciable quantities. Those found in urine probably arise from the disintegration of leukocytes and epithelial cells, which always occur in urine. An exception is diastase, which is found in the urine in fairly high concentration in acute pancreatitis. Another exception is urinary pepsinogen and uropepsin, which vary with the secretion of gastric pepsin.

Abnormal constituents

Glucose The term *glycosuria* is normally used for "sugar in the urine," and *glucosuria* for "glucose in the urine." The other kinds of glycosuria include *pentosuria, lactosuria, galactosuria,* and *fructosuria.*

Perhaps at this point mention should be made of the availability of a more specific test for detecting glucose in urine by an enzymatic glucose oxidase procedure. This test is apparently not affected by the presence of other carbohydrates tested. However, glucose oxidase–impregnated test papers may fail to detect glucose in certain types of abnormal urine, e.g., from alkaptonuric or jaundiced patients. Epinephrine and ascorbic acid may also interfere with the reaction.

More than a trace of glucose in a 24-hour specimen of urine is pathological. Easily detectable amounts may be found in specimens voided soon after a high carbohydrate intake, but this alimentary glucosuria seldom gives a positive qualitative test when the entire day's output is pooled and analyzed. In diabetes mellitus the urine is usually light colored, with a higher specific gravity than the color would seem to warrant and a glucose content of a few tenths of 1% up to 12% or 15%. Values as high as 250 gm of glucose per day, nearly the entire carbohydrate intake for a 24-hour period, have been recorded in the literature.

The severity of the condition cannot be gauged by the percentage alone, since this can be modified by varying the volume of fluid ingested. It is the actual amount in grams excreted in 24 hours that is important. If the preformed and potential dietary carbohydrate is calculated and the amount of glucose excreted is subtracted from it, the remainder is the amount in grams of glucose used. This gives the physician a basis for determining what diet to prescribe and whether insulin is necessary. The blood glucose must also be determined and taken into account. Suffice it that if there is a normal renal threshold, i.e., the blood glucose does not rise above 160 mg per 100 ml before glucosuria results, the urinary output is a very good guide for treatment. In certain renal conditions—glomerulonephritis, nephrosclerosis, nephrosis, and renal glucosuria or *"renal diabetes"*—glucosuria may occur. This seems to result from a lowered renal threshold and thus is probably not attributable to any derangement in carbohydrate metabolism.

In addition to diabetes mellitus, glucosuria, as a result of high blood glucose, accompanies about one fourth to one third of the cases of hyperthyroidism. Hyperpituitarism and hyperadrenalism belong in the same category. Ether anesthesia, asphyxia, acidosis, and a variety of other conditions also lead to hyperglycemia and glucosuria.

Lactose Lactose may be found in the urine of a considerable proportion of lactating women. *Lactosuria* seldom occurs during normal pregnancy. However, *glucosuria* is present in from 10% to 15% of all normally pregnant women, with no accompanying hyperglycemia. Since it usually disappears later, it is frequently assumed to be lactosuria. No such assumption is justified, and the urine should be analyzed carefully to determine what sugar is present. Sometimes an early case of diabetes is not diagnosed because no differential analysis or blood sugar determination is made. Lactose may easily be distinguished from glucose by appropriate laboratory tests. The amount of lactose eliminated is usually small.

Pentoses An alimentary *pentosuria,* as a result of ingesting large amounts of prunes, plums, cherries, grapes, or their juices, is likely to be noted in normal individuals. Like alimentary glucosuria, it is temporary and has no significance. The excretion of a pentose in the urine has been reported in cases of morphine addiction. However, the most interesting type of pentosuria is the chronic type, L-*xylulosuria*. This is caused by an inborn error of metabolism, a recessively inherited anomaly that occurs in persons of all age groups. There apparently is a deficiency of L-*xylulose dehydrogenase* (Chapters 11 and 22). However, the use of other carbohydrates is not impaired, the mortality of such individuals is not lowered, and the only danger to the person having this derangement is that it might be mistaken for diabetes mellitus.

The urinary pentose L-xylulose (L-xyloketose) may be detected in urine by several common laboratory methods. L-Xylulose is sometimes present in minute amounts in normal human urine.

Ribosuria has been observed in patients with progressive muscular dystrophies, myotonia congenita, and amyotonia congenita but not with myasthenia gravis or progressive neuropathic atrophy. A provisional test for ribose is a positive qualitative test (in the absence of other sugars) after 45 minutes' heating.

Other sugars *Galactosuria* has been observed in nursing infants suffering from congenital galactosemia (Chapter 11) and also in adults, as well as infants, with hepatic disease. *Fructosuria* is said to occur occasionally in association with glucose, in severe cases of diabetes mellitus. There is also a rare condition, known as *essential fructosuria* or *levulosuria,* in which no other carbohydrate is involved. It may be regarded as another inborn error of metabolism, because persons are afflicted from birth. Insulin does not help the patient to use fructose. The site of the difficulty is believed to be the liver, where fructose normally is stored as glycogen. There appears to be a deficiency of the enzyme *fructokinase* or of *fructose-1-phosphate aldolase* (Chapter 11). No other symptoms are peculiar to this condition, which does not lead to diabetes mellitus or to any change in the use of other carbohydrates. The presence of these sugars in urine may be determined by standard laboratory procedures.

A seven-carbon sugar, D-mannoheptulose, appears in the urine of normal individuals after eating large amounts of avocados. Although some of this sugar is used, enough is excreted to be a possible source of confusion in diagnosis.

Lipids Alimentary *lipuria* may be observed when a large amount of fat has been ingested. The urine is opalescent, or turbid, or even milky when voided. After standing, a peculiar creamy layer is seen at the top in those rare instances in which the fat content of the urine is high. The high blood triglyceride level (lipemia) that sometimes occurs in diabetes mellitus and lipoid nephrosis may lead to lipuria. The same results may be observed after fractures of the long bones with injury to the bone marrow, which is rich in lipids, and after any injuries to the subcutaneous layer of lipids. Other conditions in which lipurias may be seen as pyelitis, pyonephrosis, and alcohol or phosphorus poisoning.

Chyluria is the term applied to the condition resulting from an obstruction to the thoracic duct. This is even more infrequent than lipuria. The lymph vessels of the urinary tract become distended and burst, allowing lymph to pass directly into the urine. The appearance of the urine in chyluria is milky rather than opalescent.

Ketone bodies The ketone bodies are acetoacetic acid, β-hydroxybutyric acid, and acetone. The first two are normal products of fatty acid catabolism (Chapter 12). They are formed in the liver and are destroyed or used by the extrahepatic tissues. If these two activities balance, there is no excess in the blood, i.e., no *ketonemia*. If the formation by the liver is too rapid for the extrahepatic tissues to keep pace with it, ketonemia results, followed by *ketonuria*. Acetone is generally believed to be a secondary product; i.e., it results from the decarboxylation of acetoacetic acid.

$$\underset{\textbf{Acetoacetic acid}}{CH_3 \cdot CO \cdot CH_2 \cdot COO^-} \rightarrow \underset{\textbf{Acetone}}{CH_3 \cdot CO \cdot CH_3} + CO_2$$

Ordinarily a normal person on a mixed diet excretes less than 0.1 gm of ketone bodies in 24 hours. In *ketosis,* as the condition of excessive ketone production is termed, values as high as 100 gm per day, or even higher, have been reported. Ketonuria may be expected to occur in the acidosis of diabetes mellitus and starvation, in normal and toxic pregnancies, after ether anesthesia, and often in alkalosis. It is therefore not necessarily a sign of acidosis, although the most severe ketoses, with strong ketonuria, are seen in severe cases of diabetic acidosis. Acetone and acetoacetic acid are easily detected by qualitative color tests.

Proteins The amount of protein that is excreted in normal urine is insignificant. It primarily consists of serum albumin, serum globulin, and mucoprotein from the blood. A relatively small amount of a glycoprotein is also found. It is secreted by the mucous glands of the normal genitourinary tract. It has a molecular weight of about 300,000. Apparently it serves to protect the epithelial membranes of the tract. Normal urine does *not* give positive reactions with any of the ordinary protein tests.

Abnormally, proteins appear in urine in varying amounts. The condition is commonly known as *albuminuria,* although the albumins seldom are found alone, and consequently the term *proteinuria* is to be preferred. The proteins that are found in the urine in kidney conditions are commonly believed to be plasma proteins that pass the damaged renal epithelium. The albumins, with the smallest molecules, pass most easily, globulins next, and fibrinogen least readily.

Proteinurias may be grouped in two general classes: functional and organic. Functional proteinurias are those that are not related to a diseased organ. The amount of protein excreted is usually small, and the condition is ordinarily temporary. Violent exercise is an example. Soldiers, after long marches, and athletes, after strenuous contests, frequently have proteinuria. Here there may be a slight kidney damage to account for it, but the condition almost always clears up. Cold bathing, leading to constriction of renal blood vessels and anoxia, is another cause, and occasionally an alimentary proteinuria occurs after excessive protein ingestion. In all of these, the subjects may be of any age. Orthostatic, or postural, proteinuria occurs chiefly in children or adolescents, usually 14 to 18 years of age. In these young people, the urine contains protein when they are in the upright position only. When they are lying down, it is free from protein. This is not an evidence of kidney disease but is probably the result of some disturbance in the blood supply to the kidneys, leading to venous stasis and consequent anoxia. These benign proteinurias usually disappear within a few years, but sometimes they continue into adult life. Proteinuria is frequently associated with pregnancy, probably as a result of pressure interfering with the return of blood in the renal veins.

There are many pathological conditions that cause organic proteinuria, which may be classed conveniently as (1) prerenal, (2) postrenal, and (3) renal.

The prerenal conditions causing proteinuria are those that are primarily not related to the kidney. In most cases, however, they affect the kidney in such a way as to render it more permeable to the protein molecule. For example, cardiac disease, by affecting the circulation of the kidney, leads to proteinuria. Any abdominal tumor or mass of fluid in the abdomen does the same, by exerting pressure on the renal veins. Fevers, convulsions, anemias, other blood diseaes, liver diseases, and many other pathological states belong in this category. An increased amount of urinary mucoproteins generally accompanies elevated serum mucoprotein levels. These have been observed in patients with cancer, with highest values when the carcinomatous invasion was most widespread. Collagen disease and inflammatory conditions also have high mucoprotein levels.

Postrenal proteinurias are sometimes called *false* proteinurias, with all others being *true,* because these are conditions in which the protein does not pass through the kidneys. They may be attributable to some inflammatory, degenerative, or traumatic lesion of the pelvis of the kidney, the ureter, bladder, prostate, or urethra. Bleeding into this tract, of course, contributes proteins. Urine containing pus also contains protein, since the exudate that accompanies the pus is rich in protein.

Proteinuria accompanies various types of kidney diseases. These are the *renal* proteinurias. In acute glomerulonephritis, protein is always found in the urine. In the chronic form of this disease, proteinuria is seen in the early stages but may disappear later as the kidney becomes more and more impaired. In nephrosclerosis, albuminuria is frequently but not always found, and the same is true of tuberculosis and carcinoma of the kidney. There are several types of nephrosis—conditions characterized by degenerative lesions of the renal parenchyma. Protein is almost always excreted in nephroses, varying from small to large amounts. Lipoid nephrosis is a form of chronic kidney disease in which lipid deposits occur in the tubules. It is, however, considered to be an affection of the glomerulus. In this disease, large quantities of albumin are lost in the urine, and since albumin is derived from the blood, the concentrations of the blood proteins, particularly serum albumin, fall considerably. A number of investigators believe that lipoid nephrosis is merely a modified glomerulonephritis, not an entirely different condition. This view is supported by evidence that in some cases of definite chronic nephritis the character and quantity of proteins in the urine are the same as those found in the urine of patients with nephrosis.

Polypeptides sometimes are excreted in the urine. This may happen in pneumonia, diphtheria, carcinoma, and other conditions and is caused by some protein-containing material, e.g., an exudate, or a tissue mass or pus undergoing autolysis. If any soluble products of proteolysis with molecules too large for direct use or deamination get into the circulation, they are excreted as foreign bodies.

An unusual protein is eliminated by some patients having multiple myeloma, a tumorlike hyperplasia of the bone marrow, and also in some other diseases of the bone marrow, and sometimes in leukemia. This is the *Bence Jones protein.* It precipitates when the urine is warmed to 40° to 60° C but dissolves almost completely when the temperature is raised to 100° C. It is believed to be a globulin of comparatively low molecular weight, i.e., about 37,000. An increased protein diet does not seem to be followed by a greater output of Bence Jones protein. One study indicated that the Bence Jones protein is really more than one protein, i.e., the proteins excreted by various patients differ from each other in the proportions of amino acids present, electrophoretic mobility, isoelectric point, stability in acid or alkaline solution, etc. Other investigations (Chapter 16) indicate that the Bence Jones protein represents the urinary excretion of an excess of the light-chain moiety of immunoglobulins.

Nucleoproteins are found in the urine in inflammation of the urinary epithelia (e.g., pyelitis, cystitis) and even in nephritis, at times. Since these proteins precipitate in the cold on addition of mineral acids, they have frequently been designated mucoproteins, which have the same property.

Hematuria is the occurrence of blood in urine (i.e., whole blood, including erythrocytes) and is a result of hemorrhage. Sometimes the shade or appearance gives a clue as to the site of bleeding. The hemorrhage may be due to any of a variety of causes, including benign or malignant neoplasms of the kidney or urinary tract, physical injury to the kidney, violent exercise, infection of the urinary tract, or administration of certain drugs (salicylates, methenamine, sulfonamides, barbiturates, anticoagulants). Allergic reactions and low prothrombin levels also cause hematuria, further indicating the difficulty of determining the etiology of this clinical sign. *Hemoglobinuria* refers to the excretion of hemoglobin and occurs after a hemoglobinemia. Excessive hemolysis precedes hemoglobinemia; i.e., the red cells are laked by some hemolytic agent. It appears as pink or light red oxyhemoglobin in alkaline urine and as brown methemoglobin or brownish reduced hemoglobin in acid urine. However, there is a renal threshold for hemoglobin, and normally a reabsorption from the renal tubules. When this threshold is exceeded, hemoglobinuria results. Apparently hemoglobin, when it is in solution in the blood plasma, is treated by the body as a foreign substance and is excreted into the urine. If the kidneys are normal, the threshold is a concentration of about 155 mg per 100 ml of blood plasma, but it may be lower when the kidneys are damaged. When the threshold is exceeded and hemoglobin is excreted into the urine, the hemoglobin may precipitate in the tubules if the urine is acid in reaction. Hence in the treatment of hemoglobinuria, e.g., after incompatible blood transfusions, the administration of alkalies may prove to be helpful. In attempting to differentiate between hematuria and hemoglobinuria, you should remember that the red cells may disintegrate, especially if the reaction is alkaline, and the hemoglobin will dissolve out. Myohemoglobin or myoglobin also may be found in human urine. It is difficult to diagnose spectroscopically, but there is a chemical test for it.

Bile and its derivatives Both bile pigments, or their derivatives, and bile salts may be found in urine in pathological states. If there is a stasis or damming back of bile, bile invariably enters the bloodstream and is excreted in the urine. Both the pigments and the salts are detectable, but only bile pigments are tested for ordinarily. In obstructive jaundice, relatively large amounts of bilirubin diglucuronide may be found in the urine. In hemolytic and in toxic jaundice, on the other hand, little or no bilirubin passes into the urine. The reason is that bilirubin diglucuronide can be excreted readily by the kidney if it is present in the circulating blood, as it is in obstructive jaundice. Free bilirubin is present in the blood in high concentration in hemolytic and toxic jaundice. The kidney cannot excrete it easily, since the renal threshold for this large molecule is relatively high.

Urobilinogen, you will remember, is formed from bilirubin in the intestine by bacterial action (Fig. 17-7). Urobilinogen may be absorbed from the gut and excreted either by way of the bile after having been reconverted to bilirubin or by the kidneys. The presence of urobilinogen in urine therefore is dependent on the passage of bilirubin into the intestine. Thus practically no urobilinogen is found in the urine in obstructive jaundice, but in hemolytic and in toxic jaundice it occurs in appreciable quantities.

Porphyrins Porphyrins are sometimes found in urine, giving it a red color if sufficient amounts are present. The urine may fluoresce a brilliant vermillion color in ultraviolet light. Porphyrins —usually uroporphyrins or coproporphyrins (Types I or III) or both—are found in urine in the *porphyrias,* congenital diseases in the biosynthetic pathway of heme (Chapter 16). Porphyrins may also appear in the urine temporarily after the administration of certain drugs, e.g., sulfonamides and sulfonal, or in alcoholism or lead poisoning, or after excessive exposure to ionizing radiations. The porphyrinuria in these instances usually disappears with the cessation or removal of the causative agent.

Two other intermediates in heme biosynthesis, δ-aminolevulinic acid (ALA) and porphobilinogen (PBG), are also excreted in the urine in certain types of porphyria (Chapter 16) and in *lead poisoning.* Indeed the quantitative determination of ALA in urine is one of the best methods of monitoring and controlling lead poisoning.

Phenylpyruvic acid For early detection of *phenylketonuria* (Chapter 13), it has been suggested that every child's urine be tested for phenylpyruvic acid at the age of 21 days or before. To a fresh specimen of urine is added a little 5% ferric chloride solution. A green coloration, reaching a maximum within 5 minutes and then fading slowly, constitutes a positive result. The urine should not be acidified. If the test is positive, the blood level of phenylalanine should be determined and dietary procedures (Chapter 21) should be instituted.

Homogentisic acid Alkaptonuria is a disturbance in the intermediary metabolism of phenylalanine and tyrosine (Chapter 13). Homogentisic acid is excreted in the urine. Upon exposure to the oxygen of the air, this becomes very dark. Frequently ochronosis and arthritis occur as the patient matures. Ochronosis is a blackening of the cartilages of the individual. *Alkaptonuria* is a rather rare condition.

Cystine Cystinuria occurs in various degrees of severity. Mild cases are not uncommon, but the very severe condition is quite rarely seen. Cystine itself, when fed to patients with *cystinuria,* apparently is metabolized normally, but cysteine and methionine are excreted as cystine (Chapter 13). Persons suffering from cystinuria are likely to have renal cystine calculi, which sometimes must be removed surgically.

Urinary calculi The less soluble constituents of the urine sometimes precipitate out in the urinary tract. They may form minute particles or masses and be passed readily, or they may become larger aggregates, varying in size from "sand" or "gravel" to good-sized "stones." The substances of which they are composed are the same as those that may form sediments in the urine on standing, namely, uric acid and urates, calcium oxalate, calcium phosphate, calcium carbonate, and, very rarely, cystine, xanthine, and others. These substances ordinarily are held in solution at body temperature in urine; some probably are in a "supersaturated state." There is some evidence that certain urinary mucoproteins have a great affinity for calcium, with which they precipitate out of solution. These precipitates may become centers of stone formation as solids are added to them little by little until the concretion becomes macroscopic. The change from a soluble to an insoluble state may well depend on the hydrogen ion concentration of the urine. Calcium phosphate, for example, is far less soluble in neutral urine than in acid and is still less soluble at pH 8.0. Infection and diet may easily tend to change the pH of urine, and stasis may also be a factor in promoting the precipitation of salts.

Investigations have indicated that an *organic matrix* appears to be the one essential component of all urinary calculi. This matrix is a mucoid, containing about 65% protein, 14% carbohydrate, 12% inorganic ash, and 10% "bound" water. The carbohydrate portion may contain glucose, galactose, mannose, rhamnose, fucose, deoxypentose, and hexosamine. The precursor of the matrix is believed to be *uromucoid,* a protein found in small amounts in all human urine. Uromucoid seems to be quantitatively increased in the urine of patients who are actively forming calculi. Its origin is as yet unknown, but it may be derived from depolymerized renal tubular ground substance or possibly from bone matrix. The exact mechanisms by which uromucoid is transformed into the matrix and by which it accretes organic or inorganic crystals to form organized urinary calculi are also unknown at the present time. Certain nutritional deficiencies and a number of pathological states appear to trigger the matrix-forming mechanism and thus the formation of urinary calculi.

The excessive consumption of animal proteins, especially meat, fish, and poultry, also have been suspect for a number of years as a factor in causing urinary calculi. A recent retrospective clinical and dietary study has produced evidence of a positive association between the excessive consumption of animal protein and renal stone disease. Presumably an accompanying increase in the acidity of the urine would be involved.

Certain other factors have been suggested as contributing to the formation of urinary calculi. Among them are hyperparathyroidism, hypervitaminosis D, avitaminosis A, and avitaminosis B_6. Kidney stones are quite prevalent in tropical countries; hence the suggestion that the overproduction of vitamin D by sunlight is a causative factor. The effect of high vitamin D in producing a calcium imbalance, coupled with a possible low vitamin A intake, might induce the formation of calcium stones. An avitaminosis A regularly produces bladder stones in rats. Vitamin B_6 deficiency causes increased urinary oxalate excretion in some animals and in humans. Feeding large amounts of glycine or tryptophan causes increased oxalate excretion that can be corrected only by prolonged parenteral administration of vitamin B_6. Therefore a diet low in oxalates and oxalate precursors and supplemented with this vitamin is recommended in oxaluria. Urinary calculi are often associated with hyperparathyroidism, which results in a removal of calcium salts from bone, with a rise in blood calcium and urinary calcium. However, many types of urinary calculi contain no calcium.

The relative frequency of occurrence of the different varieties of kidney stones is indicated by a series of 510 cases. About 44% were alkaline earth stones, i.e., composed largely of calcium carbonate, calcium phosphate, or ammonium magnesium phosphate. These are "infectional stones," i.e., secondary to an infection, and are whitish gray and may be either rough or smooth. Calcium oxalate stones constituted about 49% of the total. These are associated with a high calcium content of urine and are dark brown to black, exceedingly hard, and usually rough, particularly the larger ones. This roughness is from the protrusion of the sharp octahedral crystals. A small type is termed the "hemp-seed" calculus and may have a smooth surface. The majority of stones found in the kidney at operation are either oxalate or phosphate concretions. These two groups, making up over 90% of all urinary stones, are radiopaque. *Metabolic* stones include uric acid concretions (6%) and cystine calculi (0.8%). Uric acid calculi are always colored, being yellow to reddish brown, and usually, but not always, have a smooth surface. The nuclei or centers of urinary concretions of other types are often composed of uric acid or urates.

Cystine calculi may occur in cases of cystinuria (Chapters 13 and 22). They are white, yellow, or greenish yellow and rather soft. They are very rare. Even rarer are xanthine calculi. These, however, are harder than cystine stones and are brown to red in color.

Other substances A number of other substances may be excreted in the urine as abnormal constituents in individuals with hereditary derangements of metabolism, the so-called inborn errors of metabolism. Such a list would include a variety of metabolites of amino acids, monosaccharides, purines, and pyrimidines not discussed above. Most of these metabolites accumulate in tissues and blood as a result of hereditary defects in the biosynthesis of enzymes or other agents essential for their normal metabolism. Metabolites immediately preceding the defect may accumulate in the affected tissue, pass into the blood plasma, and be excreted in the urine. These hereditary metabolic diseases are discussed in several preceding chapters, mainly in Chapters 11 to 14, and also are summarized briefly in Chapter 22.

A wide variety of drugs, chemicals, and environmental agents, or their "detoxified" products (as are considered next) also may be excreted in the urine and could be classified as abnormal constituents.

Detoxication and its products A variety of toxic substances, such as certain drugs, chemicals, and other foreign agents, frequently may be absorbed from the gastrointestinal tract and thus gain access to the organs and tissues of the body. Some may have extremely deleterious effects. These substances usually undergo a process commonly called detoxication, appropriately considered at this point because the detoxified products are excreted primarily in the urine. Some authorities regard the term *detoxication* as a misnomer, because the detoxified products are sometimes more toxic than the original substance. *Biotransformation* has been suggested as a preferable term.

The primary purpose of the detoxication process is to convert the toxic substance to a more *polar* compound, which is thus less lipid soluble. The object is to decrease the permeability of the toxic compound through lipid cell membranes, thus protecting the cell interior. The object also, if possible, is to increase the water solubility and hence the excretion of the compound in the urine (or bile or intestinal secretions). Most detoxication reactions occur in the liver.

The detoxication mechanisms employed in the liver are ones ordinarily used in normal metabolic processes. They apparently are not specific for any substance simply because of its toxicity. The typical pathways for hepatic detoxication include either oxidation, reduction, hydrolysis, or conjugation, or some combination of these—all familiar metabolic reactions. Likewise, compounds used for conjugation with toxic substances include readily available reactants such as: glycine, glutamine, and ornithine for most acidic substances; glucuronate for some acids, alcohols, and phenols; cysteine for certain aromatic compounds and phenols; sulfate for some steroids and phenols; acetyl CoA for aliphatic and aromatic amines; and methyl group (PAPS) for pyridine derivatives. It should be noted that different species, including humans, may use different compounds for detoxication by conjugation of the same chemical substance. Thus phenylacetic acid is detoxified by conjugation with glutamine in humans, with glycine in dogs and rabbits, and with ornithine in birds.

The following includes a brief consideration of the principal detoxication mechanisms, with examples of each.

Oxidation Oxidation usually occurs first and sometimes is followed by conjugation. Indole is an example (p. 572). It is first oxidized to indoxyl, which is then conjugated with sulfate. Some substances can be completely destroyed by oxidation. Ethyl alcohol, in moderate amounts, can be oxidized by the body to carbon dioxide and water. The fact that methyl alcohol yields intermediate toxic products, formaldehyde and formic acid, in the same kind of oxidation emphasizes the fact that these reactions are general metabolic mechanisms that may fall short of total detoxication.

Aliphatic amines are completely oxidized by the body. An enzyme, *amine oxidase*, that accomplishes this has been found to occur in brain and other tissues. In the case of butylamine, a product of the reaction is acetoacetic acid, which is, of course, a normal metabolite. There is present in liver, intestine, and other tissues a similar enzyme that catalyzes the oxidative deamination of epinephrine and related amines.

Phenyl-substituted fatty acids are oxidized by β-oxidation, losing two carbons at a time in the process. The final products are phenylacetic acid, if the chain contains an even number of carbons, and benzoic acidm if it contains an odd number of carbons. Both phenylacetic acid and benzoic acid are then conjugated with glycine to yield phenaceturic acid and hippuric acid respectively. In humans, however, phenylacetic acid is conjugated with glutamine, as is shown later. Benzene itself is slowly oxidized in the presence of enzymes to phenol and other products, including muconic acid, which involves splitting the ring. Benzene derivatives with a single side chain usually have this side chain oxidized. Thus toluene (or methylbenzene) and benzaldehyde are oxidized to benzoic acid.

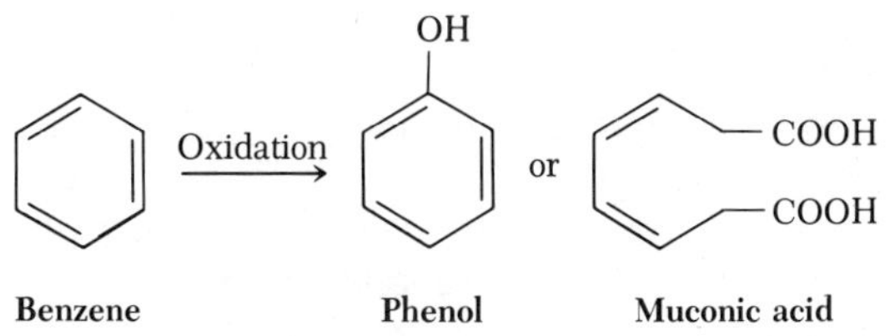

OH

Oxidation → or COOH COOH

Benzene **Phenol** **Muconic acid**

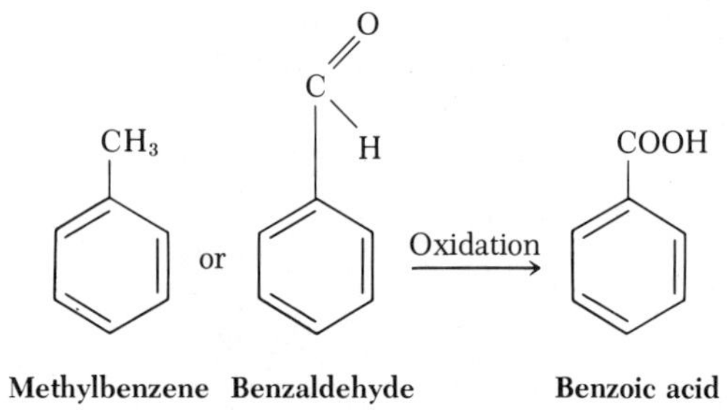

O C H

CH_3 COOH

or Oxidation →

Methylbenzene **Benzaldehyde** **Benzoic acid**

In all these instances the final products are more acidic than the parent substance.

Hydroxylation of certain drugs and steroids The detoxication of a number of drugs and steroids by "oxidation" actually entails only their hydroxylation and subsequent conjugation with glucuronide or sulfate before their excretion in the urine. Hydroxylation is catalyzed by a "mixed-function oxidase" system present in the microsomes of hepatic cells. The hydroxylases introduce a single atom of molecular oxygen into a substrate, as shown by the fact that if labeled oxygen ($^{18}O_2$) is used the label appears both in the hydroxylated substrate and the water formed. The mixed-function oxidase system consists of a stereoselective hydroxylase, a *cytochrome* P_{450} (absorbs at 450 nm), NADPH, usually ferredoxin and FAD, and molecular oxygen. The cytochrome P_{450} differ slightly in their protein carriers, depending on the nature of the drug or steroid inducing them. They belong to the cytochrome b group and have ferriprotoporphyrin as the prosthetic group, which is central to the binding and hydroxylation of the substrate. The protein moiety consists of a single polypeptide chain with a molecular weight of 44,000 to 46,000. The cytochrome P_{450} enzyme induced by the administration of camphor (cytochrome $P_{450_{cam}}$) has been crystallized and its properties described. There may be hundreds or possibly thousands of cytochrome P_{450} enzymes because of the subtle differences in the protein moiety synthesized by the cell in response to the nature of the chemical agent inducing their formation. Thus the cytochrome P_{450} enzymes may rival the immune system of the body in complexity.

The steps involved in the hydroxylation of a variety of steroids and drugs are shown diagrammatically in Fig. 17-11.

Reduction Reduction is less common and apparently less important than oxidation. There are, however, some reductions that are accomplished metabolically. Two probable examples follow.

Picric acid is converted to picramic acid:

OH, O_2N, NO_2, NO_2 (benzene ring) + 6 H → OH, O_2N, NH_2, NO_2 (benzene ring) + 2 H_2O

Picric acid **Picramic acid**

Chloral is reduced to trichloroethyl alcohol:

$$CCl_3 \cdot CHO + 2\,H \rightarrow CCl_3 \cdot CH_2OH$$

Chloral **Trichloroethyl alcohol**

Hydrolysis Aspirin, acetylsalicylic acid, is a good example of hydrolysis within the body. The acetate formed is either oxidized or used for synthesis of physiological compounds, and the salicylic acid is excreted by the kidney, combined partly with glucuronic acid. Glucosides are, in many cases, hydrolyzed to the sugar and the aglycone, each of which is treated by the body according to its particular nature.

Conjugation By conjugation, we mean the chemical combination of one compound with some other. The most common conjugating agents are glycine,

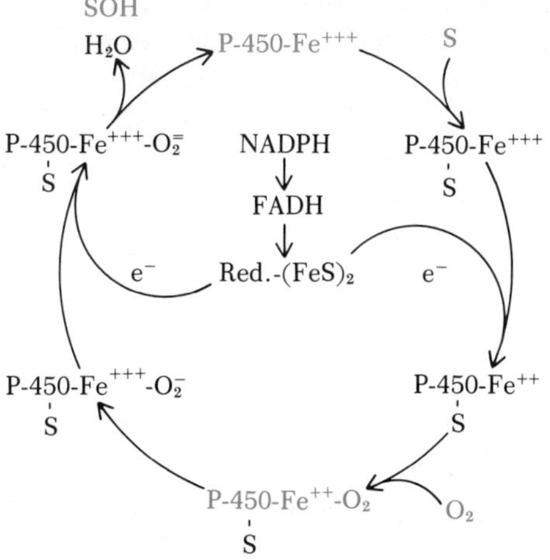

Figure 17-11
Scheme of hydroxylation of certain drugs and steroids by the cytochrome P_{450}–mixed function oxidase system. *S*, Substrate undergoing hydroxylation; *Red.-(Fe-S)$_2$*, a type of redoxin.

cysteine, glutamine, glucuronic acid, and acetate, with sulfate as the best example of an inorganic agent. Methylation is also resorted to in the case of pyridine derivatives.

Glycine The classic example of conjugation with glycine is its union with benzoic acid to form hippuric acid (Chapter 13). Many of the derivatives of benzoic acid are conjugated with glycine in this way, but the presence of a group ortho to the carboxyl inhibits this type of action. Conjugation with glycine is not limited to members of the aromatic series; many other acids are handled in the same way, including, to some extent, niacin. The formation of glycocholic acid, of the bile, by the union of cholic acid with glycine is probably another example of this normal metabolic reaction, which is available for detoxication purposes. There is created a peptide linkage; to accomplish this, the presence of ATP and specific enzymes are required. The type reaction is shown below:

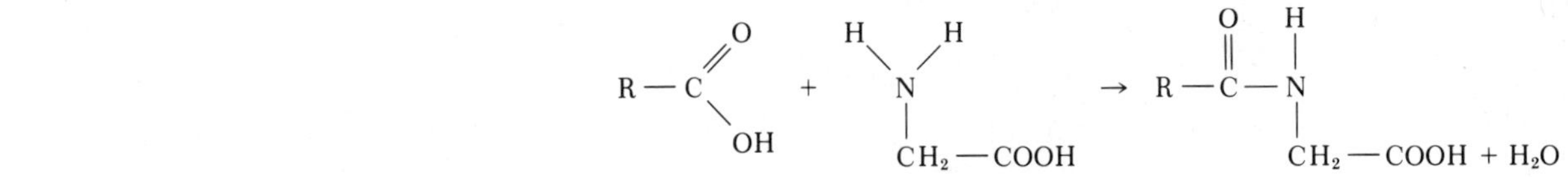

$$R-C(=O)-OH + H_2N-CH_2-COOH \rightarrow R-C(=O)-NH-CH_2-COOH + H_2O$$

Glycine

Glutamine An interesting example of the employment of glutamine in conjugation reactions is the union of phenylacetic acid with this compound. The reaction occurs only in humans and other primates. In other animals, glycine or ornithine is conjugated with it. Here again a peptide linkage is formed.

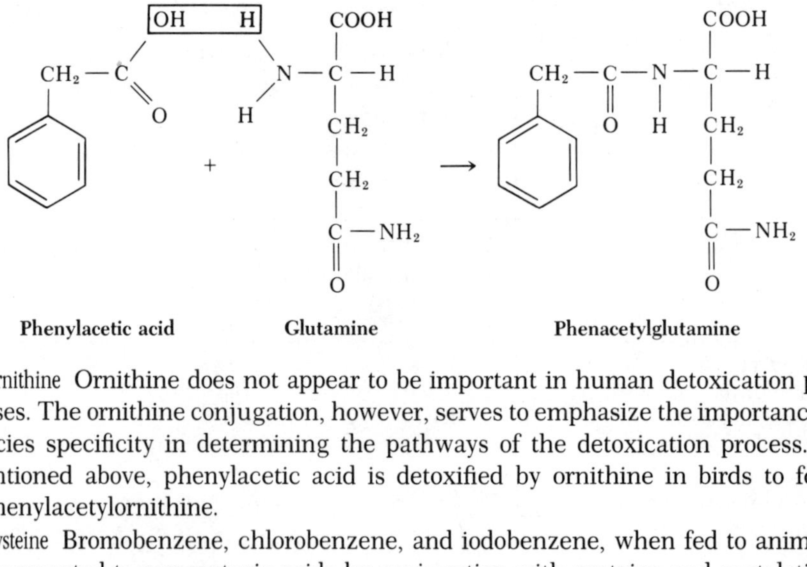

$$C_6H_5-CH_2-C(=O)-OH + H_2N-CH(COOH)-CH_2-CH_2-C(=O)-NH_2 \longrightarrow C_6H_5-CH_2-C(=O)-NH-CH(COOH)-CH_2-CH_2-C(=O)-NH_2$$

Phenylacetic acid **Glutamine** **Phenacetylglutamine**

Ornithine Ornithine does not appear to be important in human detoxication processes. The ornithine conjugation, however, serves to emphasize the importance of species specificity in determining the pathways of the detoxication process. As mentioned above, phenylacetic acid is detoxified by ornithine in birds to form diphenylacetylornithine.

Cysteine Bromobenzene, chlorobenzene, and iodobenzene, when fed to animals, are converted to mercapturic acids by conjugation with cysteine and acetylation.

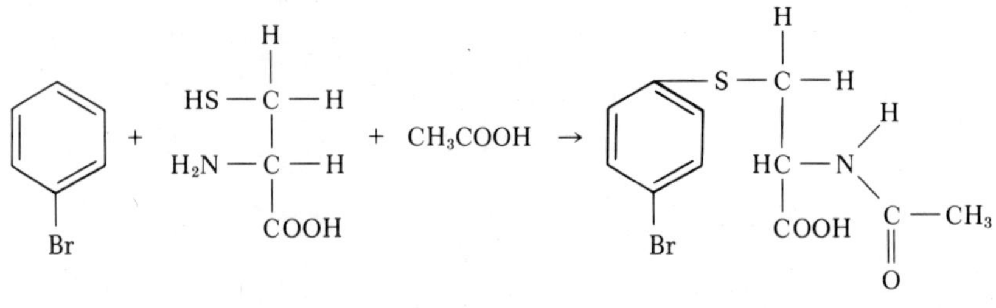

$$C_6H_5Br + HS-CH_2-CH(NH_2)-COOH + CH_3COOH \rightarrow Br-C_6H_4-S-CH_2-CH(COOH)-NH-C(=O)-CH_3$$

Bromobenzene **Cysteine** **Acetic acid** **ρ-Bromphenylmercapturic acid**

Naphthalene, anthracene, benzyl chloride, and a number of other substances are known to be handled similarly by humans to form the corresponding mercapturic acid. The administration of some of these substances to animals results in an inhibition of growth if the protein intake is low. The explanation is that the cysteine required for growth is used in the detoxication process.

Acetate It has just been stated that acetate is used, together with cysteine, in the formation of mercapturic acids. However, conjugation of acetate alone, as acetyl CoA, with other substances having an amino group is a common occurrence. One notable example is the acetylation of the sulfa drugs. This occurs after absorption or parenteral administration, and the efficacy of the drugs as bacteriostatic agents is thereby decreased. CoA is, of course, required for these acetylations.

$$H_2N-C_6H_4-SO_2-NHR \xrightarrow[\text{Condensing enzyme}]{\text{Acetyl CoA +}} CH_3-CO-NH-C_6H_4-SO_2-NHR$$

We would expect *p*-aminobenzoic acid (PABA) to be handled similarly.

$$H_2N-C_6H_4-COOH \xrightarrow[\text{Condensing enzyme}]{\text{Acetyl CoA +}} CH_3-CO-NH-C_6H_4-COOH$$

PABA itself is a detoxicant. Symptoms of hydroquinone poisoning can be overcome by the oral administration of this substance. PABA also can detoxify certain phenylarsonates, which are trypanosomicides, when these are given in toxic doses. Such actions of PABA differ from the others discussed since they are not metabolic reactions of the cells of the body but are brought about by the *administration* of a compound, even though it be a physiological compound.

Sulfate Phenol, cresol, indole, and skatole, formed by the action of intestinal bacteria on some of the amino acids in the large intestine, are transported to the liver, where they are conjugated with sulfate, as discussed previously. The resulting ethereal sulfates appear to be less toxic than their precursors and, because they are more acidic, are more readily excreted by the kidney.

Conjugation with sulfate is accomplished by way of PAPS (Chapters 6 and 13), which is transferred to the substance to be detoxified by the enzyme sulfokinase.

Glucuronic acid Glucuronic acid is an oxidation product of glucose in one of the paths of metabolism of glucose and glycogen (Chapter 11). Perhaps 150 to 200 mg per day is found in the urine of a normal man. This is combined with the number of products of normal metabolism and is increased, sometimes to a considerable extent, after the administration of various drugs.

Glucuronic acid participates in detoxication reactions as its UDP derivative. This is formed from glucose 1-phosphate, as described in Chapter 11. A transfer enzyme, UDP-glucuronyl transferase, is required. The detoxication of bilirubin is an example, forming bilirubin diglucuronide. The detoxication, with glucuronic acid, of the aromatic alcohol phenol and the aromatic acid benzoic acid are shown below.

The products of sex hormone metabolism are, in a number of instances, known to be excreted as glucuronides. Morphine, menthol, camphor, chloral hydrate, borneol, salicylic acid, acetanilid, pyramidon, creosote, vanillin, PABA, and sulfapyridine are representatives of a long list of compounds that are excreted in combination with glucuronic acid.

Note that several of the compounds cited, e.g., benzoic acid, phenol, PABA, and the sulfa drugs, have been shown to be handled by other mechanisms, again an indication that the body uses more than one method of detoxication.

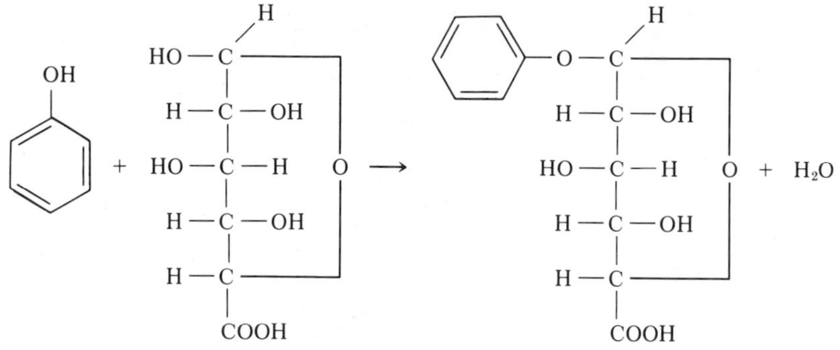

Phenol + β-Glucuronic acid → Phenylglucuronide + H_2O

Phenol **β-Glucuronic acid** **Phenylglucuronide**

Benzoic acid + β-Glucuronic acid → β-Glucuronic acid monobenzoate + H_2O

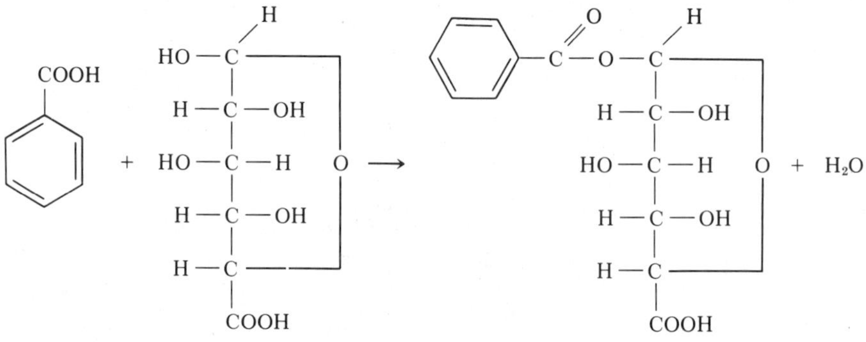

Benzoic acid **β-Glucuronic acid** **β-Glucuronic acid monobenzoate**

The detoxication of cyanide may be another example. The animal organism normally excretes thiocyanates. Human saliva contains an average of 0.01%; normal human blood contains about 1.31 mg potassium thiocyanate per 100 ml. Habitual smokers have values much higher. The formation of thiocyanate is said to be affected by an enzyme, *rhodanese*, which is believed to act on small quantities of cyanide formed during the course of normal metabolism.

$$HCN + S \xrightarrow{\text{Rhodanese}} HSCN$$

In vitro the sulfur can be obtained from thiosulfate, but its source in the body is unknown.

Methylation Methylation and transmethylation are discussed in Chapters 6 and 13, and the importance of methyl group transfer in the formation of methionine and creatine is indicated. Although these are not detoxications in the original sense, they are syntheses, and they indicate that analogous methylations, especially of drugs and other substances that are pyridine derivatives, can occur in the body. Nicotinamide is an important example. It is metabolized in part as follows:

$$\text{Nicotinamide (pyridine-3-CO—NH}_2\text{)} \xrightarrow{\text{Methylation}} N^1\text{-Methylnicotinamide (N}^+\text{—CH}_3\text{ pyridinium-3-CO—NH}_2\text{)}$$

Nicotinamide **N^1-Methylnicotinamide**

Kidney function tests

The body has a considerable factor of safety in renal as well as hepatic tissue. One normal kidney can do the work of two, and if all other organs are functioning properly, less than a whole kidney may suffice. On the other hand, there are extrarenal factors that interfere with kidney function, particularly circulatory disturbances. Therefore methods that appraise the functional capacity of the kidneys are very important. Such tests have been devised but, as in the case of liver function, no single test can measure all the kidney functions, although the kidney is not so versatile an organ as the liver. Consequently more than one test is usually indicated. These procedures throw light on the functional capacity of the kidney as related to the general physiology of the patient, not on the extent of any lesion or pathological process in the patient. Many renal function tests have been proposed and are being used, but only a few can be given space here.

A stepwise increase in three nitrogenous constituents of blood is believed by some authorities to parallel a deteriorating kidney function. Uric acid usually rises first, later urea, and finally creatinine. By determining all three in blood, one can make an estimate of kidney function. However, gout and certain other conditions also result in a high uric acid.

Concentration and dilution tests for renal function are also used. There are a number of variations, but all are based on the principle that the normally functioning kidney is capable of secreting a dilute urine if a large volume of fluid has been ingested and a concentrated urine if the individual has been deprived of fluid.

A number of renal function tests have been described and are used clinically. These include the phenolsulfonphthalein test and the urea, creatinine, inulin, and para-aminohippuric acid clearance tests. Each of these estimates the mass of functionally active renal tissue and its tubular excretory efficiency.

Some other body fluids

Lymph Since the lymphatic capillaries drain the tissue (interstitial) spaces, the fluids present in both are similar. These fluids resemble blood plasma in composition, the chief difference being that blood plasma contains a higher percentage of protein than does lymph and tissue juice. This was mentioned before as the reason for the colloidal osmotic pressure of plasma being higher than that of tissue fluids, whereas the crystalloidal osmotic pressure is about the same. The albumin: globulin ratio is higher in lymph than in plasma. This is so because albumin, with a smaller molecule, diffuses from plasma into lymph more readily than do globulins, although neither diffuses freely. A smaller amount of fibrinogen is present, as well as some prothrombin, and many leukocytes. It clots very slowly. The lymph of the thoracic duct has a protein concentration higher than that of the lymphatic capillaries but lower than that of plasma; in other respects, during the fasting state, it also tends to resemble plasma. Since it drains the abdominal viscera, however, its composition changes with the state of digestion. After a meal, the fat content rises, since more than half the fat absorbed goes by this route. In fact, the lymph or chyle is decidedly milky in appearance if the food contains much fat.

Cerebrospinal fluid Normal cerebrospinal fluid is a clear, colorless fluid, having a specific gravity of from 1.004 to 1.008. It has an extremely low protein content ranging from about 15 to 40 mg per 100 ml, with no fibrinogen and, as already stated, differs considerably from plasma in its concentration of nonelectrolytes. Its electrolyte content and its pH, however, are about the same as that of blood, i.e., pH 7.35 to 7.40. The glucose level, however, is slightly lower (50 to 80 mg per 100 ml) than that of blood.

Pathologically the fluid may be increased in amount and, as a consequence, may be under great pressure. In many of these conditions, the protein content increases appreciably. It is usually referred to in clinical tests as the globulin fraction, since this seems to be the chief constituent to show an increase. Cerebrospinal fluid is frequently analyzed by clinical laboratory procedures for diagnostic purposes.

Semen The study of the composition of semen is of considerable interest because of its possible bearing on the problem of infertility. Most of the work has been done on seminal plasma, the fluid in which the spermatozoa are suspended. The spermatozoa are constituted largely of nucleoproteins, which differ in various species as regards their isoelectric points, amino acid makeup, etc.

Human seminal plasma is a mixture of the secretions of a variety of glands and tubular epithelial linings. This may account for the great differences in analytic figures reported in the literature. The pH is about the same as that of blood plasma, as is the carbon dioxide content. Chloride and cholesterol are much lower, whereas phosphorus and lactic acid are much higher. The high phosphate is undoubtedly of importance in buffering any acid present in the female secretions. Calcium, urea, and sugar are about twice as high in semen as in blood. It is interesting that the sugar present is *fructose* rather than glucose. The proteins present vary considerably, both qualitatively and quantitatively, but most of the work by electrophoretic methods indicates that the protein fractions are qualitatively identical to those of blood serum. From these facts it would appear that seminal plasma is not an ultrafiltrate, and indeed its derivation from so many sources would lend support to this hypothesis.

Transudates and exudates The fluid formed by passage through a membrane is called a transudate. A fluid deposited in or on a tissue is known as an exudate, e.g., nasal and vaginal mucus secretions. Actually the difference between a transudate and an exudate is difficult to define. If inflammation exists, the fluid is an exudate. Thus a transudate may be a normal fluid, e.g., lymph, or it may be a pathological fluid, e.g., some sterile ascitic fluid such as peritoneal fluid. From a physical and chemical standpoint, transudates have a low specific gravity (below 1.015) and a low protein content and clot more slowly than do exudates, if at all. Exudates have a higher specific gravity (above 1.018) and a higher protein content (above 3%) than have transudates and clot rapidly. However, in some rare instances in which these physical and chemical features tend to merge, transudate is difficult to distinguish from exudate.

Electrophoresis is being used to aid in determining the origin of these fluids. Constant and distinct patterns are produced as a result of a different distribution of the proteins; e.g., ascites due to hepatic disease may be differentiated from that due to malignant disease by this procedure.

Synovial fluid As might be predicted, the characteristic constituent of synovial fluid is mucopolysaccharide formed by the cells of the synovium. Synovial fluid contains about 0.9% hyaluronic acid (Chapter 15). This gives synovial fluid its high viscosity, which is essential to its function in the lubrication of joints and other moving parts. The protein content of synovial fluid is much lower than that of blood plasma, about 1 gm per 100 ml. The albumin concentration is relatively higher than in plasma; however, the albumin: globulin ratio is approximately 4:1. Little or no fibrinogen is present, nor are other bloodclotting factors. The glucose concentration is variable, lipids are usually present in only trace amounts, and the amounts of nonprotein nitrogenous (NPN) substances are somewhat less than those in plasma. Electrolytes and other easily diffusible substances apparently exchange readily with those in plasma.

Tears The composition of tears is similar, in general, to that of interstitial fluid. Tears have a lower protein content than plasma, somewhat less than 1 gm per 100 ml. The lipid content is low. The electrolyte and NPN constituents are present in concentrations similar to those of plasma.

A distinctive constituent of tears is the enzyme *lysozyme,* which is also found in nasal and bronchial secretions, milk, and egg white. Its function in tears appears to be to protect the cornea from infection by hydrolyzing the mucopeptide of the polysaccharide cell walls of many microorganisms.

Sweat Sweat is produced continuously by the sweat glands but at a low rate at moderate temperatures so that no visible perspiration is apparent. This *insensible perspiration,* amounting to some 300 to 700 ml per day, plays an important role in body temperature regulation by cooling by evaporation. At higher temperatures or with vigorous activity the production of sweat is increased and visible perspiration occurs. Values as large as 10 to 14 liters per day have been reported. This can create a severe problem in the maintenance of water and electrolyte balance, since the concentrations of the major cations (Na^+, K^+, Mg^{++}, etc.) and anions (Cl^-, HCO_3^-, etc.) are similar to those found in plasma.

The pH of perspiration varies considerably, usually between 5.2 and 7.3. Glucose is present in very small amounts, whereas the concentration of urea is four to six times that of blood plasma. Apparently only traces of protein and lipid are present. Significant amounts of several amino acids have been reported in sweat. The amount of lactic acid in sweat is far greater than that in blood or urine, suggesting that some active transport mechanism may be present.

The excessive loss of sodium chloride in perspiration under extremely hot or humid conditions may result in miners' or stokers' cramps or be a factor in heat prostration. Small amounts of salt should be added to the drinking water under such conditions.

Suggested readings

Beaumont, W.: Experiments and observations on the gastric juice and physiology of digestion, New York, 1959, Dover Publications, Inc. (Originally published in 1833.)

Christensen, H.N.: Body fluids and acid-base balance, Philadelphia, 1964, W.B. Saunders Co.

Davenport, H.W.: Physiology of the digestive tract, ed. 4, Chicago, 1977, Year Book Medical Publishers, Inc.

Editors: Diet and urinary calculi, Nutr. Rev. **38:**74, 1980.

Estabrook, R.W., et al.: Drugs, alcohol and sex hormones. A molecular perspective of the receptivity of cytochrome P-450, Ann. N.Y. Acad. Sci. **212:**27,1973.

Gamble, J.L.: Chemical anatomy, physiology and pathology of extracellular fluid, ed. 6, Cambridge, Mass., 1958, Harvard University Press.

Goldberger, E.E.: Water, electrolyte and acid-base syndromes, ed. 4, Philadelphia, 1975, Lea & Febiger.

Johnson, L.R.: Gastrointestinal hormones and their functions, Ann. Rev. Physiol. **39:**135, 1977.

Kon, S.K., and Cowie, A.T., editors: Milk: its physiology and biochemistry, New York, 1961, Academic Press, Inc.

Lightner, D.A.: Derivatives of bile pigments. In Dolphin, D., editor: The porphyrins, vol. VI, New York, 1979, Academic Press, Inc.

Loeb, J.N.: The hypersomolar state. Engl. J. Med. **290:**1184, 1974.

Muntwyler, E.: Water and electrolyte metabolism and acid-base balance, St. Louis, 1968, The C.V. Mosby Co.

Nair, P.P., and Kritchevsky, D.: The bile acids, vols. 1 and 2, New York, 1971-73, Plenum Publishing Corp.

National Dairy Council: Newer knowledge of milk and other fluid dairy products, Rosemont, Ill., 1979, The Council.

Pesce, A.J., and First, M.R.: Proteinuria: an integrated review. New York, 1979, Marcel Dekker, Inc.

Petryka, Z.J., and Howe, R.B.: Historical and clinical aspects of bile pigments. In Dolphin, D., editor: The porphyrins, vol. VI, New York, 1979, Academic Press, Inc.

Pitts, R.F.: The role of ammonia production and excretion in regulation of acid-base balance. N. Engl. J. Med. **184:**32, 1971.

Pitts, R.F.: Physiology of the kidney and body fluids, ed. 3, Chicago, 1974, Year Book Medical Publishers, Inc.

Share, L., and Claybaugh, J.R.: Regulation of body fluids, Ann. Rev. Physiol. **34:**235, 1972.

Welt, L.G.: Clinical disorders of hydration and acid-base equilibrium, ed. 3, Boston, 1970, Little, Brown & Co.

Williams, H.E.: Nephrolithiasis, N. Engl. J. Med. **290:**33, 1974.

Wright, R., et al.: Liver and biliary disease, Philadelphia, 1979, W.B. Saunders Co.

Section Five

Chemical communication between tissues

Chapter 18

Hormones: chemical regulators of metabolism

Communication is a most important problem in a multiorgan animal. Thus in the human it is essential that various tissues interrelate so that each can play its particular role in the function of the whole body. Communication involves two types of mechanisms: one is neurological and requires the nervous system, the other is chemical and requires the production of regulatory substances called hormones. As we shall see, these two systems interrelate in a carefully tuned regulation of metabolism.

Hormones are produced by special cells or glands, such as the adrenals, ovaries, parathyroids, pituitary, testes, and thyroid. Some tissues associated with nonhormonal functions, such as the pancreas and intestinal mucosa, also contain specialized cells that secrete hormones. These glands secrete their hormones directly into the bloodstream and are known as *endocrine* glands. *Endocrinology* is the study of hormones and their interactions.

Since hormones are secreted into the bloodstream, they are ultimately to be found in the environment of all tissue cells. Important questions are why only certain tissues respond to a given hormone and also why a specific hormone may elicit different effects in different tissues. Because specific tissue cells may respond selectively to a hormone in the circulation, these are called target cells. Important questions arise, such as "How does a target cell know that a hormone molecule to which it must respond is in its environment?" Also, once recognized, "How does the hormone exert its control over cellular metabolism?"

Classification

Hormones are usually classified into three main groups:

1. *Steroids:* derivatives of the cyclopentanoperhydrophenanthrene ring structure and therefore derived from cholesterol
2. *Derivatives of amino acids:* aromatic amino acids such as phenylalanine, tyrosine, and tryptophan serve as parent compounds
3. *Peptides and proteins*

Common characteristics

Hormones do not initiate reactions in cells anew but rather serve to regulate critical or rate-limiting reactions in a metabolic pathway. In other words the necessary equipment for a hormonal response is inherently part of the target cell. This is built in at the time of cell differentiation. There is a limit to what a hormone can do. It is impossible to force the "little toe" to produce glucocorticoids by administering adrenocorticotropin (ACTH) despite the fact that all of the necessary genetic information is present. However, it is possible to force a capon (castrated rooster) to produce certain egg yolk proteins (phosvitin) simply by administering a female sex hormone (in this case diethylstilbestrol, a synthetic hormone). The genetic information to produce phosvitin can be turned on in the "male" liver, since that is the organ which produces this egg yolk protein in the hen.

Hormones are not secreted at a uniform rate

None of the hormones is secreted at a uniform rate. The adrenal hormones are secreted in a diurnal rhythm. The gonadotropins (the gonad-stimulating hormones of the pituitary) are secreted in complex cycles. In the female these coincide with such events as ovulation and menstruation or with pregnancy and lactation. Other hormones are produced in response to environmental factors. For example, insulin and glucagon respond to the glucose level in the blood. Aldosterone responds to the K^+ concentration and parathormone and calcitonin to the Ca^{++} concentration of the blood. Thyroxine responds to the environmental temperature, and enterogastrone responds to the fat in the diet. In very general terms we can say that there are two kinds of hormonal systems in the body: (1) an adaptive system, which responds to the outside environment and (2) an homeostatic system, which responds to the internal environment. Both act to maintain the status quo.

Hormones exert their effects in biocatalytic amounts

All hormones exert their effect in biocatalytic concentrations. Thus glucagon stimulates glycogenolysis in concentrations of only 10^{-10}M. This introduces another concept, that of amplification. A liver cell may respond to single molecules of glucagon yet it releases many molecules of glucose from glycogen. This means that the impact of a single molecule of glucagon can be amplified into an effect involving hundreds or even thousands of glucose molecules.

Turnover of hormone molecules is varied and usually rapid

Hormones are continually lost either by metabolism, conjugation, or excretion. This must be so, because otherwise there would be an uncontrolled accumulation of hormone and the ultimate response would be disastrous. A hormonal response must be transient and carefully controlled. The turnover (half-life) varies widely among different hormones. Thus the half-life for epinephrine is < 30 sec; for insulin, 40 min; and for thyroxine, 6 to 7 days.

Hormones exert multiple actions

If in a given cell a hormone stimulates the biosynthesis of certain enzyme(s), it may simultaneously affect cellular metabolism in a variety of ways. For example, it must stimulate the specific transcription of the requisite mRNA. For this there must be an appropriate supply of the nucleoside triphosphates. An elevated synthesis of protein requires an increased supply of intracellular amino acids. This may depend on a stimulated transport of amino acids across the plasma membrane. An increased input of ATP and GTP is also needed. All of these processes are effects on cellular metabolism that may be stimulated simultaneously by the hormone.

Hormones exhibit a high degree of specificity

Only the target cells will respond to their hormone despite the fact that the hormone is to be found throughout the body and in contact with many different tissues. Thus it is necessary to postulate the existence of a "receptor site" to recognize the presence of the particular hormone in the target cells' environment.

Different tissues may respond differently to a given hormone

A given hormone may have divergent effects on different tissues. Glucocorticoids, for example, will signal the breakdown of proteins in muscle tissue while stimulating the biosynthesis of proteins in the liver.

Mode of action

A particular tissue (such as adipose tissue) possesses a selective sensing mechanism to detect the presence of a hormone (such as insulin) when it is in the environment of the target tissue. At the same time the tissue is exposed to many other hormones to which it simply does not respond. To visualize this selective recognition of hormone molecules, consider the following:

Hormone + Receptor → Hormone-receptor complex

Recognition requires a specific complexing of the hormone molecule to a receptor molecule (a sensor) to form a hormone-receptor (H-R) complex in or on the surface membrane of the adipose cell. Once the sensor has detected the hormone, the resultant H-R complex might lead to a secondary reaction that we could visualize as yielding a "second messenger" called *I*. *I* would be the intracellular communication system and be instrumental in triggering the metabolic responses. This general sequence might be viewed as follows:

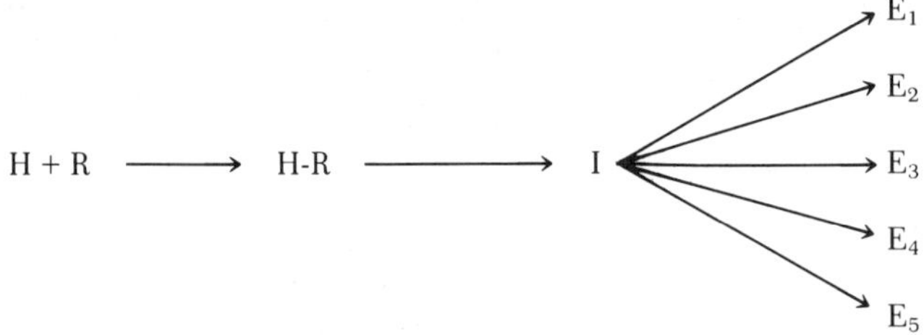

E_1 to E_5 represent metabolic effects of the hormone H. Another possibility would be for the hormone or the complex to penetrate the cell membrane, enter the cytoplasm, and migrate to the nucleus, where the hormone could exert its effect at the genetic level. In other words, two mechanisms may be considered: one a hormone-gene or nuclear-based action and the other an action in which the hormone exerts its effects at the cytoplasmic level. It is likely that a hormone can have both cytoplasmic and nuclear effects. For purposes of instruction the two processes will be considered independently.

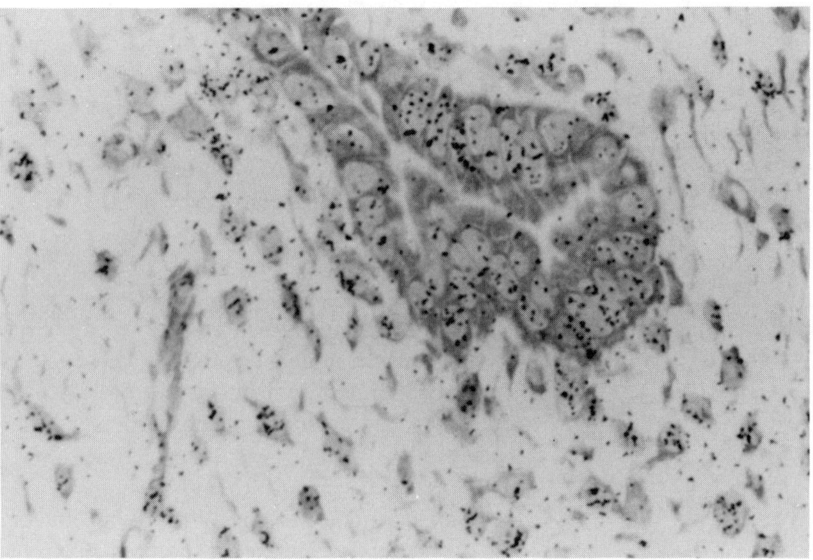

Figure 18-1
Location of radioactively labeled estradiol in the nuclei of uterine tissue cells.

Courtesy W.E. Stumpf: Endocrinology **83**:777, 1968.

Hormone-gene theory

That a hormone may leave its environment, the blood plasma, and traverse the cell membrane to be delivered eventually at the nucleus has been dramatically demonstrated by the location of radioactively labeled estradiol (female sex hormone) in the nuclei of uterine tissue (Fig. 18-1). This hormone will enter its target-organ cells in the uterus, vagina, and anterior (but not posterior) pituitary and accumulate against a concentration gradient with respect to the blood. The process of recognition and transport through the cytoplasm to the nucleus requires a specific cytoplasmic binding protein (Fig. 18-2, *A*). A protein-estradiol complex having a sedimentation constant of 4S is transformed in the cytosol to a 5S species that delivers the steroid to the nucleus. Extradiol then stimulates RNA synthesis, probably by enhancing RNA polymerase activity. In some of its target-organ cells, the male sex hormone (Fig. 18-2, *B*), testosterone, must first be modified to dihydrotestosterone before being bound to a specific binding protein. This complex then acts at the chromosomal level to elicit a selective transcription of mRNA. In both examples the hormonal action elicits the synthesis of the specific proteins.

However, some hormones do not actually enter the cytoplasm of the target cell. Rather, they bind to receptor proteins at the cell membrane and elicit the formation of the cytoplasmic "second messenger," or I molecules. One of the functions of this substance (see cyclic AMP) is to stimulate a selective transcription of mRNA, and therefore an increased biosynthesis of certain proteins.

The genetic induction of protein synthesis, stimulated by a process of hormonal derepression (Chapter 10), would involve the transcription of all types of RNA: mRNA, tRNA, and rRNA. There would follow an increased rate of translation. Some examples of hormones that exert their action at the genetic level are the following:

1. Insulin
2. Glucagon
3. Corticosteroids
4. Estrogens
5. Androgens
6. Erythropoietin

Cytoplasmic actions: the second messenger

A development that contributed much to placing endocrinology in the realm of molecular mechanisms was the discovery of cyclic-3′,5′-adenosine monophosphate (cyclic AMP, or cAMP) and its designation as a second messenger. This concept may be expressed as follows:

$$\text{Gland} \xrightarrow[\text{(Hormone)}]{\text{First messenger}} \text{Target cell} \xrightarrow[\text{(cAMP)}]{\text{Second messenger}} \text{Metabolic effects}$$

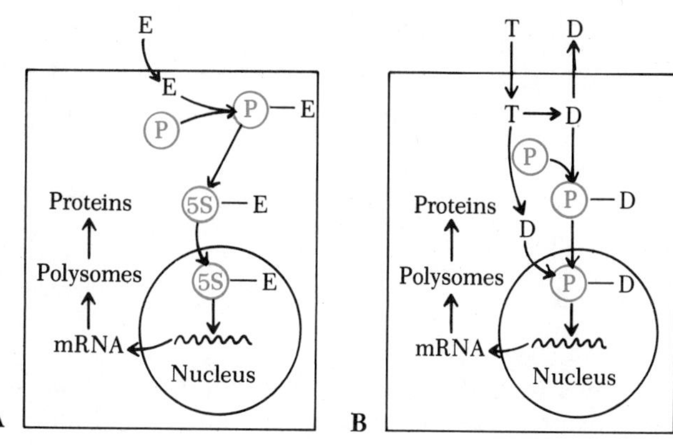

Figure 18-2
Schemes of interactions of estradiol, **A**, and testosterone, **B**, with their target cells. *D*, Dihydrotestosterone; *E*, estradiol; *P*, hormone-binding protein; *T*, testosterone.

A adapted from Jensen and DeSombre, Annu. Rev. Biochem. **41:**203, 1972; **B** adapted from Wilson, N. Engl. J. Med. **287:**1284, 1972.

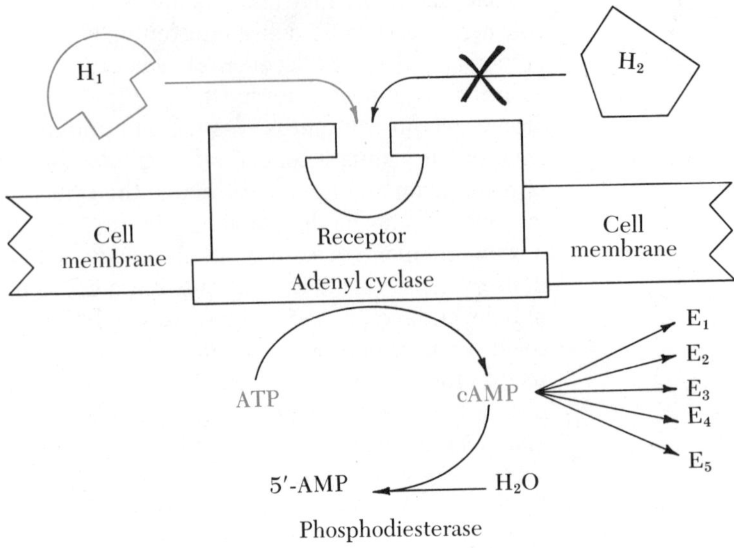

Figure 18-3
Concept of hormone action by way of second messenger, cyclic AMP. H_1 is the hormone recognized by the receptor site, whereas H_2 is a hormone that is not sensed in the environment.

As illustrated in Fig. 18-3, the formation of the H-R complex stimulates the membranal *adenyl cyclase*. This enzyme in turn catalyzes the production of cAMP from ATP. The cAMP level is also determined by its rate of hydrolysis to 5′-AMP by *cAMP phosphodiesterase*. Thus adenyl cyclase controls the formation of cAMP, whereas cAMP phosphodiesterase controls its disposal.

Under experimental conditions the cAMP level may be elevated by inhibiting its hydrolysis. In this way the hormonal effect may be magnified. Compounds that inhibit cAMP phosphodiesterase are pyrophosphate, methyl xanthines, citrate, and several nucleotides. Compounds frequently used are theophylline (1,3-dimethylxanthine) and caffeine (1,3,7-trimethylxanthine).

To illustrate the effect of cAMP on a metabolic system we consider again the process of glycogenesis versus glycogenolysis (Chapter 11). The first messenger, epinephrine, stimulates the formation of cAMP, which then triggers the action of inactive phosphorylase b kinase and the phosphorylation of glycogen synthase I to D. Cyclic AMP therefore simultaneously enhances glycogenolysis and inhibits glycogenesis.

The point has been made that this cytoplasmic mechanism and the genetic process are not self-exclusive. Cyclic AMP is known to be a kinase activator; a nuclear effect could be explained on the basis of an activation of a protein kinase. Phosphorylation of nonhistone proteins, for example, would occur, which then might result in their repulsion from binding sites on DNA, making a gene available for transcription (Chapter 9).

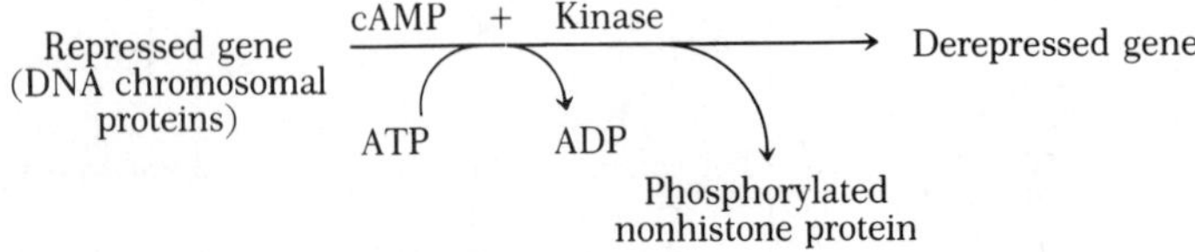

Cyclic AMP is also known to stimulate, in some unknown manner, the transport of certain metabolites, such as amino acids, into the cell. Thus the hormone glucagon stimulates the uptake of amino acids by the liver; the inclusion of theophyl-

Table 18-1
Hormonal actions that are mediated by cyclic AMP

Hormone	Target tissue	Stimulation of metabolism
ACTH	Adrenal cortex	Steroidogenesis
	Fat (rat)	Lipolysis
LH	Corpus luteum, ovary, testis	Steroidogenesis
	Fat	Lipolysis
Catecholamines	Fat	Lipolysis
	Liver	Glycogenolysis
		Gluconeogenesis
	Skeletal muscle	Glycogenolysis
	Salivary gland	Amylase secretion
Glucagon	Liver	Glycogenolysis
		Gluconeogenesis
		Induction of enzymes
	Fat	Lipolysis
	Pancreatic β-cells	Insulin release
Hypothalamic releasing factors	Adenohypophysis	Release of tropic hormones
MSH	Dorsal frog skin	Pigmentation
PTH	Kidney	Phosphaturia
	Bone	Ca^{++} resorption
Prostaglandins	Thyroid	Thyroid hormone release
	Adenohypophysis	Release of tropic hormones
TSH	Thyroid	Thyroid hormone release
	Fat	Lipolysis
Vasopressin	Toad bladder, renal medulla	Permeability

Adapted from Butcher, R.W., Robinson, G.A., and Sutherland, E.W.: In Litwack, G., editor: Cyclic AMP and hormone action, New York, 1972, Academic Press, Inc.

line magnifies this process for reasons noted above. Hormones that influence the permeability of membranes by affecting transport mechanisms are epinephrine, parathormone, ACTH, glucagon, aldosterone, thyroxine, and growth hormone.

Not all hormones, however, stimulate adenyl cyclase, as suggested by Fig. 18-3. Insulin, for example, is bound to a membranal receptor site in adipose and other tissues but is not known to stimulate adenyl cyclase. Those hormones that act by way of cAMP are compiled in Table 18-1.

Hormone receptors

As already mentioned, some hormones enter the cell and are transported to the nucleus to elicit the selective transcription of RNA (Fig. 18-2), whereas others bind at the surface of the cell membrane and promote the release of an informational molecule such as cAMP (Fig. 18-3). The cAMP in turn can elicit a variety of effects both at the cytoplasmic and nuclear levels. A distinction as to whether a hormone may or may not enter a cell cannot be made simply on the basis of molecular size. Thus insulin, ACTH, glucagon, and epinephrine all appear to bind at receptor sites on the membrane. For example, when insulin is chemically attached to beads of agarose it still exhibits the same biochemical effects as the native hormone. Being bound to agarose beads of such dimension as to preclude any possibility of its gaining access to the interior of the cell, the insulin must initiate an action at the cell surface. One might at first conclude that large polypeptide hormones cannot enter the cell because of their size, whereas smaller molecules such as the steroids enter their target cells to exert their hormonal action. However, when norepineph-

Table 18-2
Hormone binding to receptors

Radioactively labeled hormone used for binding study	Tissue from which receptors were prepared
^{125}I-ACTH	Adrenal cortex
^{125}I-angiotensin	Adrenal cortex
^{125}I-growth hormone	Lymphocytes
^{125}I-oxytocin	Fat cells
^{125}I-insulin	Liver, adipose tissue, lymphocytes, fibroblasts
^{125}I-glucagon	Liver
^{125}I-luteinizing hormone	Ovary
^{3}H-vasopressin	Kidney
^{3}H-follicle-stimulating hormone	Testis
^{3}H-prostaglandin E	Adipose tissue
^{3}H-thyrotropin-releasing hormone	Anterior pituitary
^{3}H-norepinephrine	Heart, liver
^{3}H-epinephrine	Erythrocyte, spleen capsule

Adapted from Lefkowitz, R.J.: N. Engl. J. Med. **288:**1061, 1973.

rine is bound to beads of agarose, it still exerts its function. Therefore the presence of an appropriate binding protein on the cell membrane must be the determining factor.

By use of agarose-bound hormones and preparations of solubilized cell membranes, it has been possible to purify numerous receptor sites with the technique of affinity chromatography (Table 18-2). Thus only the norepinephrine-binding proteins would be retained on a chromatographic column made of agarose beads coupled to norepinephrine. Binding proteins for norepinephrine have been obtained from heart, liver, erythrocytes, and spleen. Regardless of the type of tissue, the adrenergic receptor molecules are either closely related or identical.

As has already been noted, steroids enter the cell and are bound to specific cytoplasmic-binding proteins. Using techniques of chromatography, investigators now recognize four receptors simply called binder proteins I to IV. Binder II has been purified and shown to complex with cortisol. Estradiol binds predominantly to binder I. Fig. 18-2, *A*, shows that binding of estradiol to the receptor results in a complex having a sedimentation constant of 4S.

Prostaglandins: modulators of adenyl cyclase

Prostaglandins and the related compound thromboxanes are transient substances that modulate many physiological events. They are considered to be local hormones because their effects appear to be restricted to the cells of origin or neighboring cells. Much of their physiological action seems to depend on modulating the cAMP levels and the intracellular flow of Ca^{++}.

Structures and nomenclature of prostaglandins and thromboxanes

All of the prostaglandins are derivatives of a parent compound known as prostanoic acid. This is a 20-carbon carboxylic acid consisting of a cyclopentane unit and two aliphatic side chains.

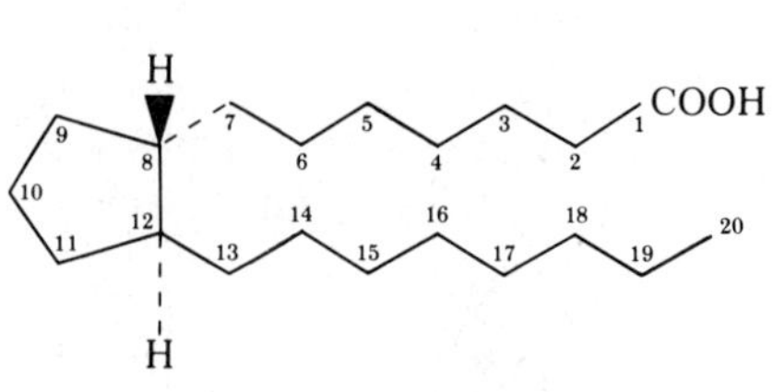

Prostanoic acid

There are five series of prostaglandins, designated PGA, PGB, PGE, PGF, and PGH, of which PGE and PGF are considered to be the primary compounds.

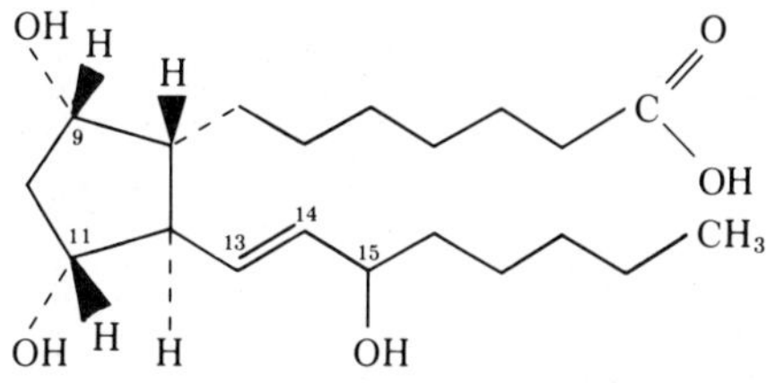

$PGF_{1\alpha}$ 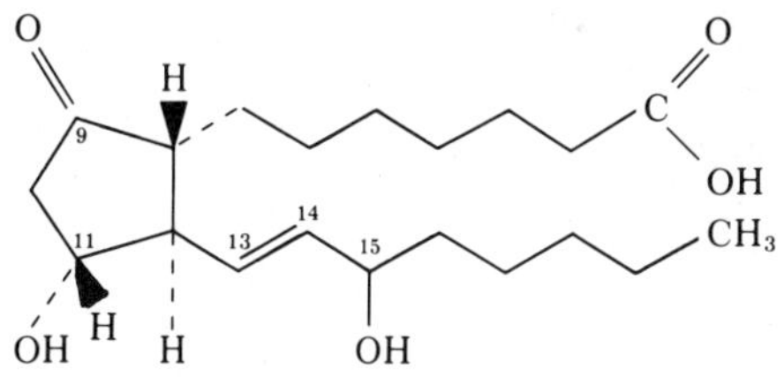 **PGE_1**

As seen above, the cyclopentane ring has either a ketone group at C-9 and hydroxyl at C-11 (PGE) or two hydroxyl groups on the ring (PGF). The α in the nomenclature refers to the configuration of the hydroxyl group on C-9. Although chemical reduction of PGE_1 yields both the $PGF_{1\alpha}$ and $PGF_{1\beta}$, only the PGF compounds with the C-9 hydroxyl in the α position occur naturally. All have a *trans* double bond between C-13 and C-14 and a hydroxyl at C-15. The subscript refers to the number of double bonds in the aliphatic chains. The PGE_3 and $PGF_{3\alpha}$ are found only rarely.

In addition to the prostaglandins there are also two other important and comparable compounds. These are *prostacyclin* and *thromboxane*. As shown below, the prostacyclins are like the prostaglandins in being a derivative of prostanoic acid.

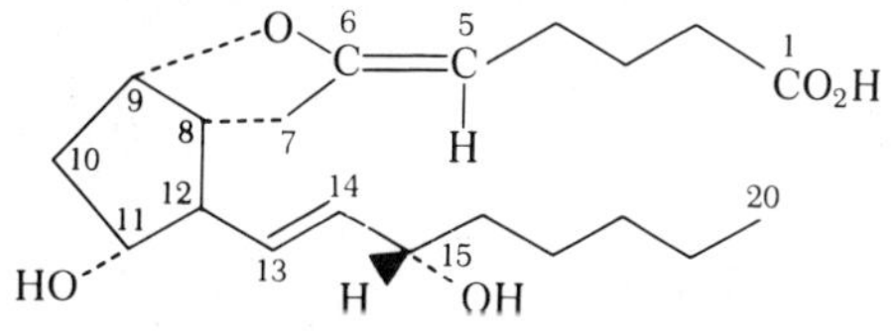

Prostacyclin (PGI_2)

A comparable substance that is not a derivative of prostanoic acid is *thromboxane* A_2.

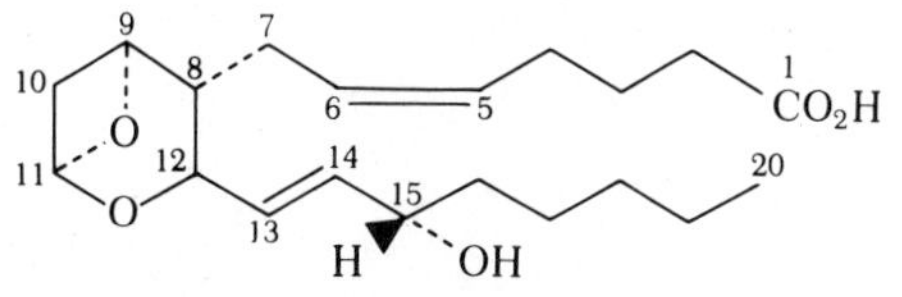

Thromboxane A_2

Biosynthesis

The most prevalent series of prostaglandins are those possessing two double bonds, namely, the PGE_2, $PGF_{2\alpha}$, PGI_2, and thromboxane A_2. All of these are ultimately derived from the precursor, arachidonic acid or 5, 8, 11, 14-eicosatetraenoic acid (Fig. 18-4). As seen from this figure, the pathways are initiated from arachidonic acid by *fatty acyl cyclooxygenase,* which generates the first endoperoxide, PGG_2. PGG_2 may serve as substrate either for *PGI_2 synthetase* to yield PGI_2 or for a *peroxidase* to generate a second endoperoxide, PGH_2. PGH_2 in turn yields thromboxane A_2 (synthetase) or PGE_2 by the action of *PGE_2 isomerase*. PGE_2 in turn may be reduced to the $PGF_{2\alpha}$ by the action of *9-keto reductase*.

Prostaglandins are not stored in the tissues, rather their release appears to be controlled by their rate of synthesis. This is dependent primarily on the availability of the substrate. Arachidonic acid, for example, is stored in the intracellular supply of phospholipids and is selectively released by phospholipase A.

Figure 18-4
Pathways indicating the biosynthetic routes for prostaglandins and thromboxane from arachidonic acid.

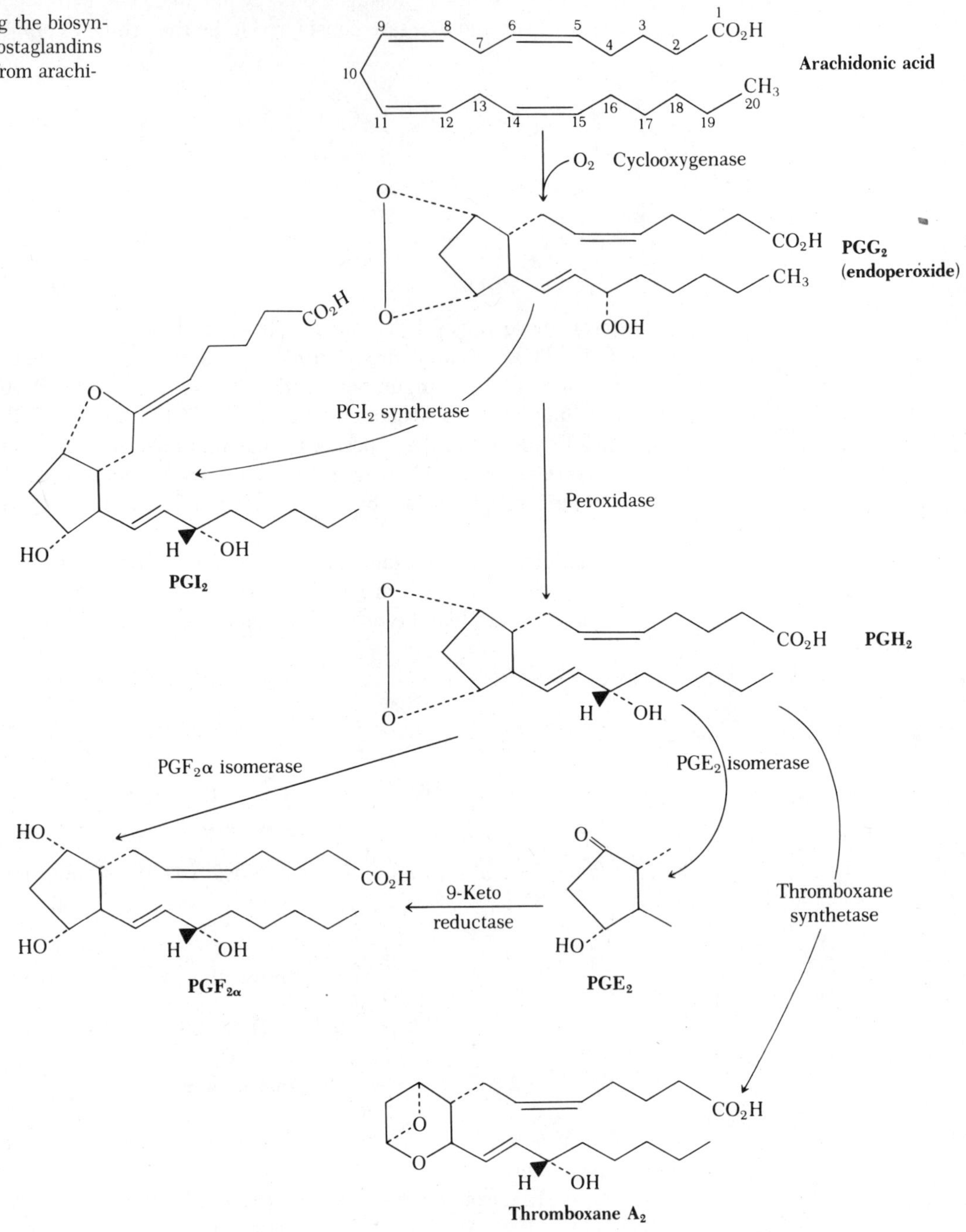

Metabolism

The lungs are a major site for the metabolism of prostaglandin; PGE and PGF are cleared from the bloodstream. The primary mechanism for their inactivation is by way of oxidation and reduction processes. Thus a *15-hydroxyprostaglandin dehydrogenase* (PGDH) converts the 15-hydroxyl to a keto group while a Δ^{13} *reductase* saturates the double bond found at C-13. The compound produced is illustrated at the top of p. 601.

Function

In many instances prostaglandins exert their effects by modulating the production of cAMP. Thus in some endocrine glands prostaglandins act as endogenous stimulators of adenyl cyclase. This appears to be the case in the stimulation of adrenal function by ACTH and the thyroid by thyroid-stimulating hormone (TSH). However, in the case of adipose tissue, in which the adenyl cyclase is stimulated by glucagon, the prostaglandin serves as an endogenous inhibitor of the enzyme. In other words, in the adrenal cortex PGE functions to amplify the response, whereas in the adipose tissue it serves to moderate or act as a "metabolic brake."

Prostacyclin and thromboxane are important factors in blood platelet aggregation (Chapter 16) and therefore in blood clotting. In this function they exert opposing effects. Platelet clumping is prevented by cAMP. PGI_2 is a most potent stimulator of adenyl cyclase, whereas thromboxane A_2 is an inhibitor. The latter compound serves as a "proaggregating" force, whereas PGI_2 is an "antiaggregating" factor. PGI_2, produced primarily in lung tissues, is often viewed as a circulating hormone.

Frequently the prostaglandins oppose one another in their physiological effects. Thus PGE_2 causes a dilation of blood vessels and bronchi, whereas $PGF_{2\alpha}$ constricts them.

Some pharmacological effects

Aspirin, indomethacin, and acetaminophen act by inhibiting cyclooxygenase activity, thereby decreasing the biosynthesis of prostaglandins. An analogue of PGE_2 (namely, the 16,16-dimethyl derivative) and the 15-methyl derivative of $PGF_{2\alpha}$ are finding therapeutic success. These include treatment of hypertension, prevention of conception, induction of abortion, and relief of bronchial asthma.

Neurohypophysial control system

Pituitary gland

The pituitary, or hypophysis, a small gland located at the base of the brain, is attached by a stalk to the hypothalamus. This hypothalamohypophysial system represents an extension of the central nervous system terminating in a hormone-producing gland. In other words, the two systems of communication in the human, the neural and hormonal, act together in concert to control many of the metabolic processes in the body. Neural signals of fright or stress, for example, arrive by the autonomic nervous system at the hypothalamus and result in a dissemination of a variety of responses, some neural, others hormonal. The hormonal responses involve a stimulation of the pituitary and also of the adrenal medulla by autonomic nerves. The pituitary is the master gland; its secretions control a variety of important endocrine glands (Fig. 18-5). The pituitary consists of (1) the *anterior lobe,* or adenohypophysis, which is the largest part and the dominant endocrine structure, (2) the *posterior lobe,* or neurohypophysis, which also possesses endocrine activities, and (3) the very small intermediate region, or *pars intermedia,* which lies between the other two and also has hormonal functions.

In young mammals hypophysectomy results in a cessation of growth and in retardation of both physical and mental development. The animals generally remain immature, their sexual glands do not develop, the epiphyses of bones do not unite, and the first teeth are retained. If the operation is performed on an adult

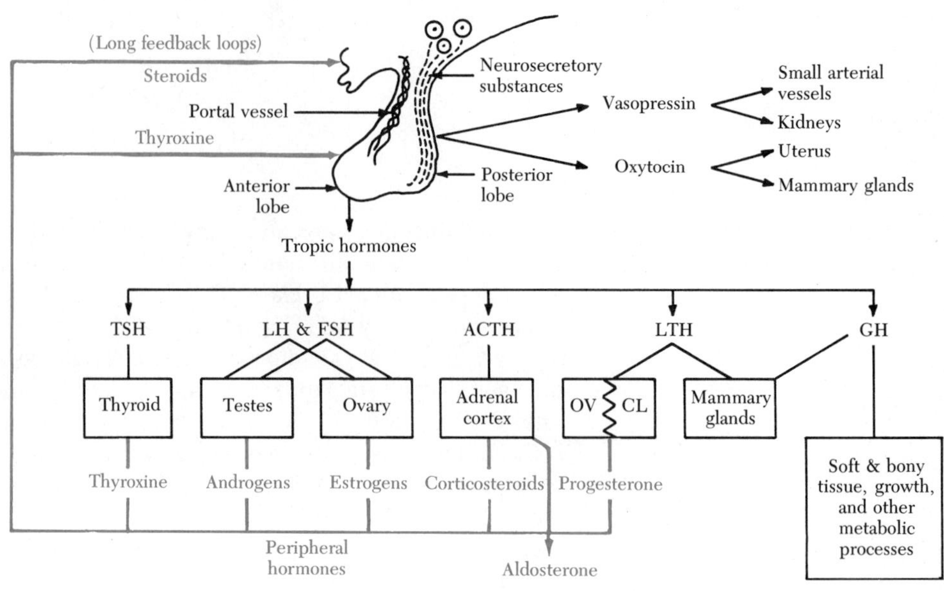

Figure 18-5
Scheme illustrating the relationships between hormones of the anterior and posterior lobes of the pituitary gland and their target structures. *CL*, Corpus luteum; *OV*, ovary.

animal, there is an almost immediate effect on all the other glands of internal secretion. The testes, ovaries, and secondary sexual organs atrophy. The thyroid and adrenal cortex become smaller in size and their activity diminishes. In addition the animal is apathetic, loses its appetite, and becomes emaciated. Metabolism of protein, carbohydrate, and fat becomes deranged.

Posterior pituitary lobe (neurohypophysis) Removal of the neurohypophysis results in a clinical condition in which large volumes of urine are produced, as in *diabetes insipidus,* which is attributed to the lack of a hormone called the *antidiuretic hormone* (ADH). In *diabetes insipidus* as much as 30 liters of water per day are consumed, with an excretion of almost as much urine. ADH is a cyclic nonapeptide called *vasopressin* (arginine vasopressin).

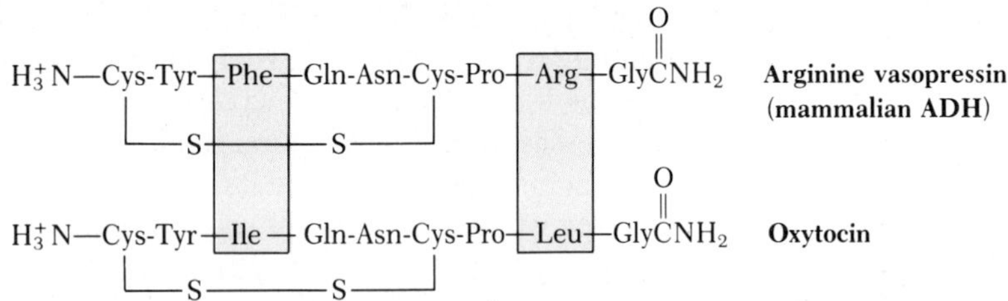

Another nonapeptide found in the posterior pituitary is *oxytocin*. As seen above, its amino acid sequence is like that of vasopressin except at the positions indicated. Characteristic of these structures is the C-terminal glycinamide and the six-residue loop at the N-terminal.

These two polypeptides are synthesized by hypothalamic nerve cells as part of a much larger molecule. Limited proteolysis separates the oxytocin and vasopressin from the remainder, which then serve as carrier proteins called *neurophysins*. As oxytocin-neurophysin I and vasopressin-neurophysin II complexes, they flow down the axons of their parent neurons into the neurohypophysis. Here they accumulate at the nerve endings in storage granules. On stimulation the complexes are discharged into the blood plasma, where the nonapeptide dissociates from the neurophysin carrier molecule.

The primary action of vasopressin is to increase the reabsorption of water by the renal tubules, where about 10% of the glomerular filtrate is actively reabsorbed. Vasopressin appears to exert its effect by the formation of cAMP (Table 18-1). The exact manner in which this alters membrane permeability is unknown. Some substances, e.g., ethyl alcohol, produce diuresis, apparently by suppressing the action of vasopressin.

Changes in the solute concentration of the blood plasma affect the blood's osmotic properties and influence the osmoreceptors of the brain. Increases in concentration stimulate the release or secretion of vasopressin from the posterior pituitary, preventing the loss of more water, whereas a decrease in concentration lessens the secretion of the hormone and permits more water to be lost in the urine.

Oxytocin, on the other hand, causes uterine contraction and acts on mammary glands to stimulate milk ejection. In this gland there is a large amount of modified smooth muscle called myoepithelium. Contraction of this tissue is brought about by minute amounts of oxytocin. Oxytocin-stimulated milk ejection is part of the reflex response to suckling.

Since the structures of vasopressin and oxytocin are similar, differing only in the amino acids at position 3 and 8, their physiological functions overlap. Vasopressin stimulates the pregnant uterus, though not nearly so strongly as oxytocin. Likewise vasopressin promotes milk ejection, but only to a moderate degree compared with oxytocin. The comparative physiological functions of these two hormones illustrate the generality that only slight dissimilarities in structure may alter greatly the physiological properties of the hormones.

Anterior pituitary lobe (adenohypophysis) The adenohypophysis secretes a number of important hormones, all of which are either simple proteins or glycoproteins (Table 18-3). Those that cause other glands to function or increase their activity are termed *tropic* (or trophic) *hormones;* this term is used as a suffix. Fig. 18-5 shows the action of the various tropic hormones on their individual target organs and glands. With the single exception of *growth hormone* (GH), all anterior pituitary hormones are tropic and act on a specific target organ or glands. These various hormones are listed in Table 18-4 with their functions or target organs and abbreviations.

Fig. 18-5 also shows that the adenohypophysis is under the control of the hypothalamus, i.e., controlled by the central nervous system. This regulation is not exerted directly by way of nerve fibers but rather chemically by way of a portal system of blood vessels. The release of pituitary hormones requires first the elaboration by the hypothalamus of special *releasing factors,* which are themselves hormones. Extremely minute quantities of these hormones enter the portal blood

Table 18-3
Hormones of the anterior pituitary

Hormone	Molecular weight	Pituitary content (μg)	Secretion rate (μg per day)	Plasma level (ng per ml)
ACTH	4500	300	10	0.03
GH (human)	21,500	8500	500	1-5
TSH (bovine)	28,300	300	110	1-2
LH (bovine)	27,400	80	30	0.5-1.5
FSH (human)	34,000	35	15	0.5-1

Adapted from Catt, K.J.: The Lancet **1**:827, 1970.

Table 18-4
Hormones of the anterior pituitary and their function

Hormone	Target/function	Abbreviation
Thyrotropic	Thyroid stimulating	TSH
Adrenocorticotropic	Adrenal cortex stimulating	ACTH
Gonadotropins	Follicle stimulating on ovary in female and seminiferous tubules in male	FSH
	Luteinizing hormone on corpus luteum in female and testes in male	LH
	Luteotropic hormone, or prolactin, on mammary gland and ovaries	LTH
Growth hormone	Acts on all tissues	GH

Table 18-5
Hormones of the hypothalamus that release pituitary hormones

Hypothalamic hormone	Abbreviation
ACTH-releasing hormone	CRH or CRF*
TSH-releasing hormone	TRH or TRF
LH-releasing hormone	LH-RH or LH-RF
FSH-releasing hormone	FSH-RH or FSH-RF
GH-releasing hormone	GH-RH or GH-RF
GH release–inhibiting hormone	GH-RIH or GIF
Prolactin release–inhibiting hormone	PRIH-PIF
Prolactin-releasing hormone	PRH or PRF

*H for hormone; F for factor.

vessels between the hypothalamus and pituitary. In the adenohypophysis they elicit the release of specific tropic hormones. As seen from Table 18-5, nine such releasing factors are now reasonably well established.

Thyrotropin releasing hormone, or TRH, is the first for which a structure was determined. It is a tripeptide with the following amino acid sequence:

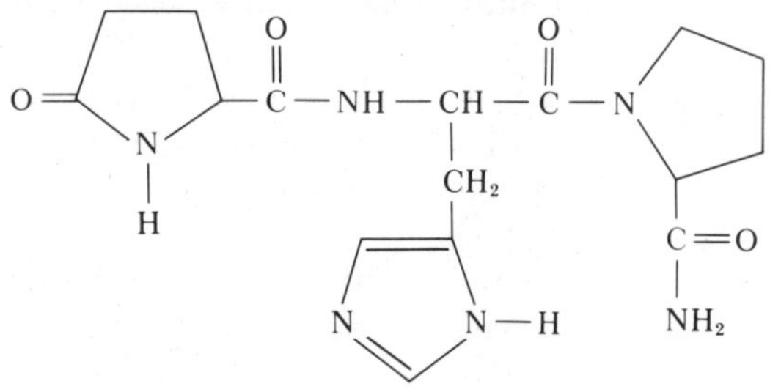

(Pyro)Glu — His — Pro — NH$_2$

Like the polypeptides of the posterior pituitary, TRH possesses an amide C-terminal. The glutamyl residue forming the N-terminal is in the cyclic form, i.e., a pyroglutamyl residue. TRH is rapidly degraded in the blood by the enzymatic cleavage of its amide group from the prolyl end.

Both *luteinizing hormone* (LH) and *follicle-stimulating hormone* (FSH) appear to be released as a result of a single releasing hormone called LH-RH/FSH-RH, which is a decapeptide with the amino acid sequence shown below. Again it is noteworthy that the polypeptide has a C-terminal amide group and a pyroglutamyl residue at the N-terminal.

(Pyro) Glu-His-Trp-Ser-Tyr-Gly-Leu-Arg-Pro-Gly $\overset{O}{\overset{\|}{C}}$—NH$_2$

The release of the adenohypophysial hormones apparently is mediated by cAMP; LH-RH/FSH-RH and TRH have been observed to bind to specific membrane receptor sites and to activate adenyl cyclase. Parallel with the increased accumulation of cAMP there is a release of the corresponding tropic hormones.

Thus when the body is exposed to a stressful stimulus the message is received at the hypothalamus, which initiates the formation of *corticotropin (ACTH)-releasing hormone* (CRH). CRH in turn stimulates the accumulation of cAMP in the appropriate adenohypophysial cells and the release of ACTH into the systemic bloodstream.

In some cases the stimulated release of tropic hormones is counterbalanced by *release-inhibiting hormones,* also produced in the hypothalamus. As seen in Table 18-5, there is a *prolactin release–inhibiting hormone* (PRIH) and a *growth hormone release–inhibiting hormone* (GHRIH). The latter, now called *somatostatin,* is a linear, 14-residue peptide having the following amino acid sequence:

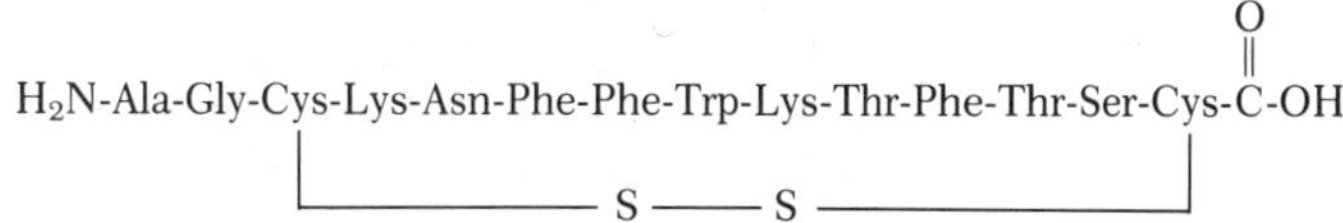

It is effective in concentrations of 1 nM in inhibiting the secretion of GH; it has no effect on the secretion of LH or FSH. Somatostatin is also found in the pancreas, where it inhibits the secretion of glucagon and insulin. In addition it occurs in the gastrointestinal tract, where it functions to inhibit the secretion of the hormones gastrin and secretin.

Control of the production of pituitary hormones also occurs in response to the blood levels of hydrocortisone, gonadal steroids, and thyroxine. As shown in Fig. 18-5, these hormones interact with hypothalamic or pituitary receptors or both to modify the secretion of tropic hormones. For example, thyroxine may act either on the hypothalamus to inhibit the secretion of TRF or directly on the adenohypophysis to depress the release of *thyrotropin.* In some instances there exist short feedback loops whereby the tropic hormones regulate their own release by acting at the hypothalamic level. In these ways the systemic levels of various hormones are maintained by the monitoring action of the hypothalamus and the pituitary.

Hormones of the adenohypophysis The cells of the adenohypophysis are specialized as to the kind of hormones they produce. This gland is composed of three types of cells: chromophobes, acidophils (α-cells), and basophils (β-cells). The chromophobes, which represent about 50% of the total cells, appear to be precursors of the others. The acidophils and basophils in turn are recognized to have subgroups. Histochemical studies show that one kind of acidophil produces GH, whereas another produces prolactin. Gonadotropic hormones are produced by one kind of basophil (gonadotrophs), whereas TSH is produced by a second kind (thyrotrophs). Gonadotrophs may be either FSH or LH producing. ACTH is also produced by basophils.

Thyrotropin One of the effects of removing the hypophysis is a decrease in the size of the thyroid gland. This can be prevented by the injection of extracts of the anterior pituitary. If such extracts are injected into normal animals, the thyroid hypertrophies. The active principle is thyrotropin, or *thyroid-stimulating hormone* (TSH), a glycoprotein of molecular weight 28,300 having two subunits, TSHα (mol wt = 13,600) and TSHβ (mol wt = 14,700). The primary amino acid structure for both chains has been established. Of the two subunits it is the β-chain that exerts the physiological action of TSH. The α-chain, which consists of 96 amino acids and 2 carbohydrate side chains, is identical to the α-chain of LH. A suggestion has been made that the use of the same polypeptide chains in the structure of two hormones (TSH and LH) represents evolutionary conservation.

The action of TSH is on the thyroid gland, stimulating the secretion of the thyroid hormones thyroxine and triiodothyronine. The mechanism by which TSH accomplishes this process is described later (p. 613).

Since TSH acts indirectly on general metabolism through the thyroid, clinical disorders may be attributable either to a primary thyroid disease or to some lesion in the anterior pituitary with a secondary effect on the thyroid. These two glands seem to have a reciprocal relationship, i.e., after thyroidectomy the pituitary gland becomes enlarged.

Adrenocorticotropin One of the most important hormones of the adenohypophysis is adrenocorticotropin (ACTH). It stimulates the synthesis of steroid hormones (steroidogenesis) by the adrenal cortex. ACTH binds to a specific membrane receptor site and stimulates cAMP formation by activating adenyl cyclase. Various enzymatic processes involved in the biosynthesis of steroids are stimulated by the cAMP.

The steroids produced are cortisol, corticosterone, and ketosteroids, with some androgens and estrogens also formed. Aldosterone is also produced, although other factors in addition to ACTH are required for its regulation (p. 622).

ACTH is a polypeptide with a molecular weight of 4500, consisting of 39 amino acid residues in the sequence shown in Fig. 18-6. Bovine, ovine, and porcine ACTH are identical except in the region between residues 25 and 33, a portion rich in acidic amino acids. Of the 39 amino acids, the biological activity resides in the sequence of the first 23 residues. A biologically active sequence of residues 1 to 24 has been synthesized and is used clinically.

It is now thought that ACTH is actually synthesized as a larger (260 residues) prohormone. This same prohormone also may serve as a precursor for other hormones and biologically active peptides (p. 610).

The influence of ACTH on the adrenal cortex can be followed by determining either the cholesterol or the ascorbic acid content of the cortex. Prior treatment with ACTH prevents the usual fall in adrenal ascorbic acid when the animal is exposed to the stress of cold, trauma, etc. The blood level of ACTH influences the rate of secretion of this pituitary hormone, probably by way of the negative-feedback loop exerted by the release of the adrenal steroids (Fig. 18-5). Other factors that regulate the secretion of ACTH are the hypothalamus, epinephrine, or, to a lesser degree, norepinephrine. Catecholamines may stimulate the release of ACTH during acute stress, which activates the sympathetic nervous system.

Figure 18-6
Amino acid sequence of bovine ACTH.

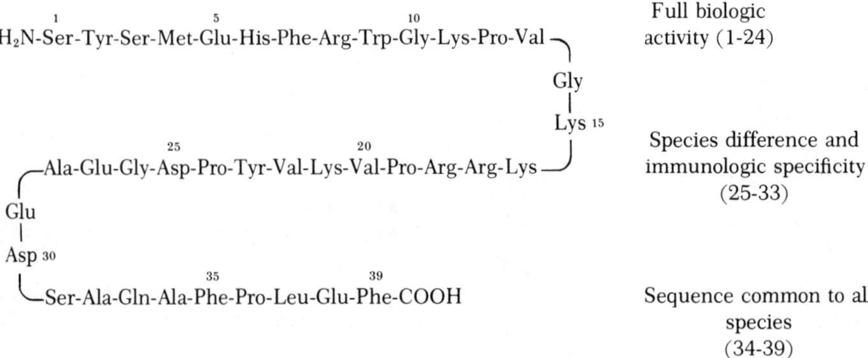

Gonadotropins The primary and secondary sexual glands atrophy after hypophysectomy. These glands may be restored to normal function by the administration of extracts of the anterior lobe or by the implantation of living anterior lobe tissue. Since it makes no difference whether the pituitary glands are derived from males or females, the pituitary hormones involved are identical and the type of effect produced is determined by the sex of the animal affected.

The anterior pituitary hormones involved are recognized as the gonadotropins, i.e., hormones that stimulate the sex glands. All are active in the female, whereas two of the three are functional in the male (Table 18-4). The gonadotropins are: (1) the *follicle-stimulating hormone* (FSH); (2) the *luteinizing hormone* (LH), or *interstitial cell–stimulating hormone* (ICSH); and (3) *prolactin,* also called the *luteotropic hormone* (LTH) or *luteotropin.*

Like TSH, the gonadotropins FSH and LH (or ICSH) are glycoproteins that can be separated into two subunits, namely, an α- and a β-chain. Human FSH has a molecular weight of 34,000 and contains 8% to 13% carbohydrate; bovine LH has a molecular weight of 27,400. Their α-chains are homologous to TSHα. Thus the biological activity of these hormones resides in their β-subunits; their α-chains are interchangeable. LTH is a single polypeptide chain having 198 amino acids (mol wt = 24,000).

In the female, FSH induces growth of a Graafian follicle in the ovary preparatory to the release of the ovum. During this phase of the menstrual cycle the follicle

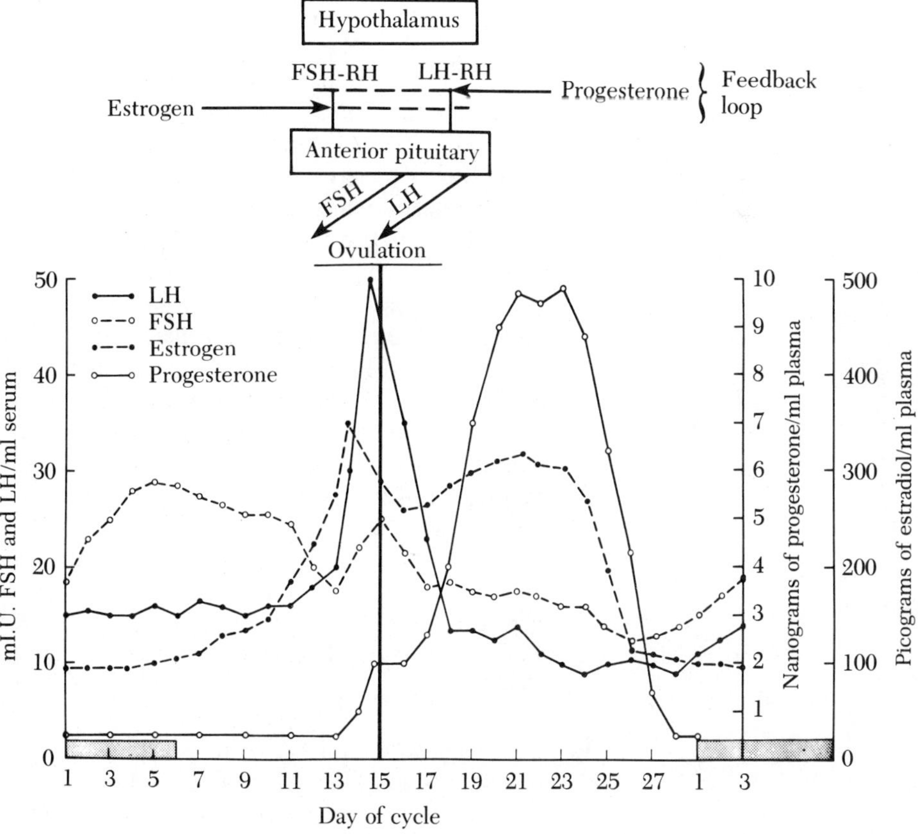

Figure 18-7
Graph showing changes in blood level of various hormones throughout the menstrual cycle. Also shown is the feedback control exerted by estrogens and progesterone.

Modified from Taymor, M. L.: Am. J. Obstet. Gynecol. **114**:445, 1972.

itself excretes a female sex hormone, an estrogen called estradiol. This hormone induces a proliferation and thickening of the endometrium as well as an increase in its vascularity, thus preparing the uterine wall for the reception of the fertilized ovum. LH controls the development of the corpus luteum, which in turn secretes an additional sex hormone, progesterone. Some of these interrelationships are illustrated in Fig. 18-7.

FSH and LH stimulate adenyl cyclase in testicular tissues and the corpus luteum respectively. FSH stimulates testicular protein synthesis, a process inhibited by actinomycin D. LH increases the conversion of cholesterol to progesterone. FSH also stimulates the 20-α-hydroxysteroid dehydrogenase in rat ovaries.

LTH increases RNA and protein synthesis in the mammary glands. Specifically this hormone increases various enzymes involved in mammary carbohydrate and lipid metabolism. Thus UDP-glucose 4-epimerase and UDP-glucose pyrophosphorylase are increased as well as UDP galactosyl transferase and α-lactalbumin (Chapter 11). In addition to stimulating lactose production, LTH also increases glucose 6-phosphate dehydrogenase and 6-phosphogluconate dehydrogenase, which are required for NADPH formation and hence lipid biosynthesis (Chapter 12).

Another gonadotropin characteristically found in primates is the glycoprotein *human chorionic gonadotropin* (HCG). However, it is produced by the placenta and not the pituitary. Since HCG appears in the urine shortly after pregnancy, it serves as the basis for a pregnancy test. An immunoassay using an antiserum against HCG is employed to detect the presence of this hormone in the urine. HCG is LH-like and produces changes in the levels of ovarian enzymes. Thus it increases glucose 6-phosphate dehydrogenase and 6-phosphogluconate dehydrogenase. Lactate and pyruvate dehydrogenases are also stimulated. All of its effects are gonadal.

HCG, a glycoprotein of molecular weight 47,000, also can be separated into α- and β-subunits. The HCGα is similar to and interchangeable with the α-chains of TSH and LH. Once again it is the β-chain that exerts the biological activity. The carbohydrate moiety contains six components, including D-galactose, D-mannose, *N*-acetylglucosamine, *N*-acetylgalactosamine, L-fucose, and *N*-acetylneuraminic acid. The sequence and nature of the carbohydrate moiety are now partially known. *N*-Acetylneuraminic acid is essential to the biological activity of the hormone.

Growth hormone The presence of an anterior pituitary hormone that influences growth is indicated on the one hand by the fact that hypophysectomy inhibits growth and on the other by the occurrence of gigantism as a result of pituitary hyperfunction. Gigantism is caused by pituitary overactivity during childhood, before closure of the epiphyses limits the further growth of the long bones. The opposite is seen in pituitary dwarfism. It is said that if no other adequate cause for retarded growth can be found, e.g., hypothyroidism, the trouble is likely to be hypopituitarism. Children so affected are usually half as tall as they should be for their age. Pituitary hyperactivity in the adult or after closure of the epiphyses leads to a condition known as acromegaly. The bones become misshapen, particularly the bones of the face. There is frequently excessive growth of the hands and feet. Fibrous tissue develops excessively, resulting in thickened nose, lips, eyelids, and broadened fingertips.

Human GH (HGH), also called *somatotropin* (STH), has 191 amino acid residues (mol wt = 21,500); its three-dimensional structure is stabilized by the two disulfide bonds. Only GH from primates is effective in the human; this hormone exhibits a significant species specificity. When bovine GH is digested with trypsin, a polypeptide can be obtained having an amino acid sequence of 37 residues, which is closely homologous with HGH. This polypeptide is biologically active in the human. Perhaps GH from various species have in common a biologically active amino acid sequence.

The level of HGH in anterior pituitary is 5 to 15 mg per gland, compared with only microgram quantities for the various tropic hormones (Table 18-3). Its concentration in plasma is 1 to 5 ng per ml for normal individuals.

As indicated in Fig. 18-5, HGH acts on many tissues, especially muscle, adipose tissue, cartilage, and connective tissues. In the latter, HGH stimulates proteoglycan and collagen synthesis. The action of HGH occurs at the nuclear level of tissues, regulating the biosynthesis of all forms of RNA. An elevated synthesis of RNA may be caused by an increased formation of RNA polymerase.

GH affects the metabolism of protein, carbohydrate, and lipids. Hypophysectomy results in hypoglycemia, whereas supplementation with large doses of GH leads to hyperglycemia. Thus there is an interplay between the pituitary and the pancreas, between GH and insulin. GH exhibits an antiinsulin effect; for example, it blocks the insulin-induced uptake of glucose by muscle tissue. It also stiumulates the mobilization of fatty acids from adipose tissue so that the plasma–free fatty acid level is increased. There is a reduced catabolism and enhanced uptake of amino acids by muscle, liver, adipose, and connective tissues.

Starvation leads to an elevation of GH in humans, rats, and rabbits. This then is manifested in increased fat mobilization and reduced glucose use. GH therefore also has a ketogenic effect.

At the same time that growth hormone has direct effects on metabolism, it also elicits the production, probably from the liver, of a family of polypeptides that are

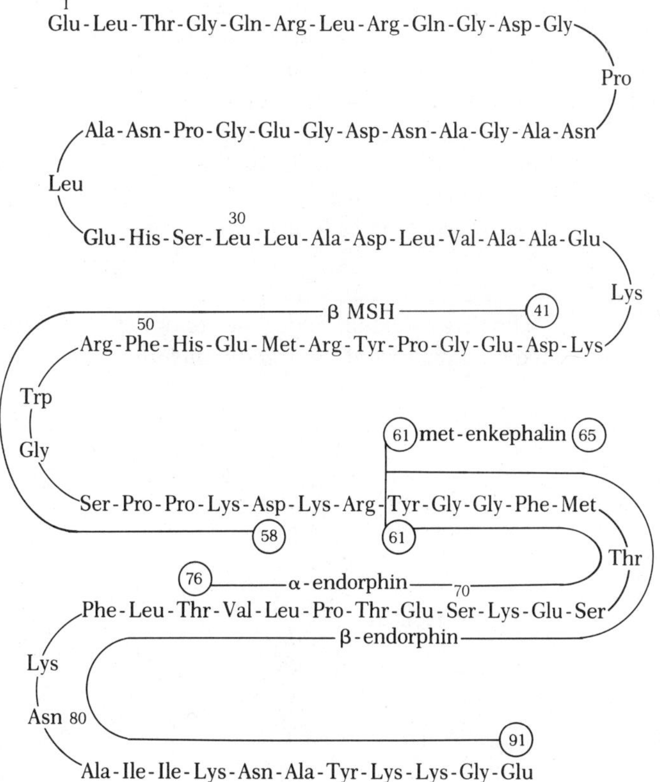

Figure 18-8
Amino acid sequence for β-lipotropin as well as homologous sequences for β-MSH α- and β-endorphins, and methionine-enkephalin.

called *somatomedins*. These polypetides circulate in the blood plasma and likewise affect metabolism. For example, it is likely that the stimulation of protein synthesis and the increased sulfation of proteoglycans in connective tissues are mediated by these factors. One of them, somatomedin C, exerts insulinlike actions; it may be related to a factor called the *nonsuppressible insulinlike activity* (NSILA). This factor increases glycolysis and decreases the blood glucose level and the mobilization of free fatty acids. These effects actually oppose those of the GH itself.

Lipotropins Two polypeptides produced by the adenohypophysis have a lipid-mobilizing effect on adipose tissue. These lipolytic peptides are β-*lipotropin* (β-LPH), with 91 amino acids, and γ-*lipotropin* (γ-LPH), with 58 residues. The amino acid sequence for β-LPH is shown in Fig. 18-8. This polypeptide is derived from the same prohormone as ACTH. As seen from the figure, β-LPH also serves as a precursor for other biologically active polypeptides. These are β-MSH (see below), residues 41-58 of LPH, and a class of peptides, *methionine-enkephalin* (61-65) and β-*endorphin* (61-91), that have morphinelike effects and are thought to be endogenous analgesic compounds.

Hormones of the pars intermedia Between the anterior and posterior lobes lies a small, relatively avascular region called the *pars intermedia,* which is almost absent in the human being by comparison with some lower animals. This portion of the hypophysis was thought to secrete the *melanocyte-stimulating hormone* (MSH), a process apparently controlled by the hypothalamus by way of a melanocyte-releasing hormone (MRH) and a melanocyte releasing–inhibiting hormone (MRIH). Two forms of MSH of differing chain lengths have been described. Both peptides may be derived from the ACTH/LPH prohormone.

However, the human skin does not contain melanocytes such as those found in lower animals (fish and amphibia). Rather skin pigmentation results from the accumulation of melanin in the epidermal cells. Apparently β-LPH itself may elicit skin pigmentation in man; MSH may not be required.

In Addison's disease (p. 622), the adrenal lesion results in a hypersecretion of ACTH. The reason is the absence of circulating adrenocorticoids to control by feedback inhibition the release of this hormone. Since ACTH includes the amino acid sequence of α-MSH (residues 1-13), it stimulates skin pigmentation. Therefore the excessive release of these two hormones in Addison's disease results in a characteristic darkening or bronzing of the skin.

Thyroid hormones

The thyroid gland, under the control of TSH, synthesizes and secretes the iodinated hormones, *thyroxine* and *triiodothyronine*. This gland, consisting of two lobes

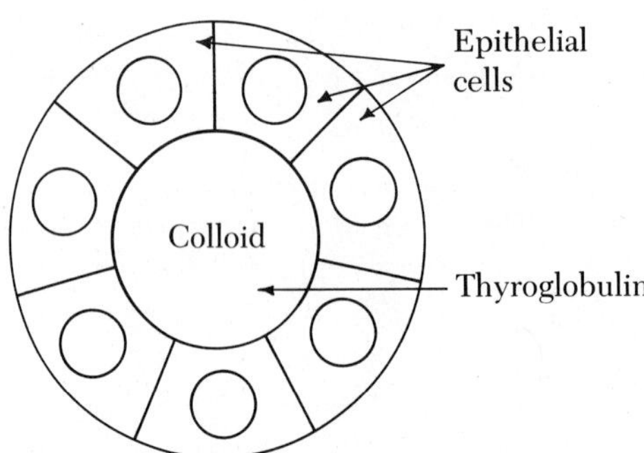

Figure 18-9
Scheme of thyroid follicle.

(one on each side of the trachea just below and anterior to the larynx), is composed of a large number of tiny closed vesicles or follicles. These are approximately 300 μ in diameter and are lined with epithelial cells (one cell in thickness), filled with a material commonly called the *colloid* (Fig. 18-9). The colloid contains the protein *thyroglobulin,* which is the storage form of the actual hormones thyroxine and triiodothyronine.

The thyroid has a powerful ability to concentrate and store the body's iodine supply. There are normally 6 to 7 mg of iodine in the adult body, of which 90% is concentrated in the gland. Practically all is organically bound iodine retained in the form of the storage protein thyroglobulin. Iodine is bound primarily as the hormones thyroxine (35% of total iodine) and triiodothyronine (5%), whereas the remainder exists as diiodothyronine (34%) and monoiodothyronine (24%).

As seen below, thyroxine is the tetraiodo derivative of the iodine-free parent compound called thyronine. Thus thyroxine, or T_4, is 3,5,3′,5′-tetraiodothyronine. Two

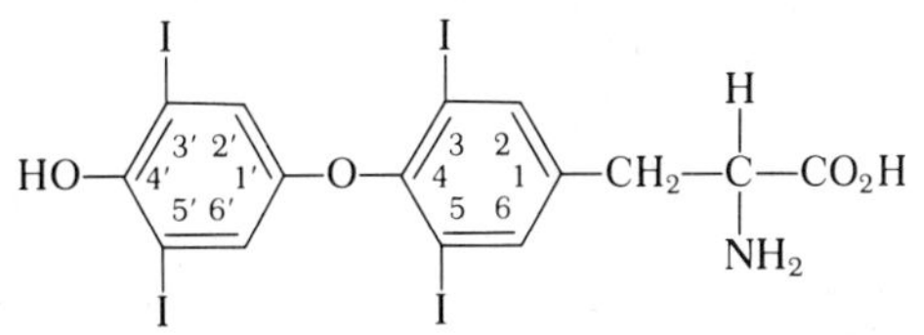

Thyroxine (T_4)

other iodinated thyronines are 3,4,3′-triiodothyronine (T_3) (below left) and 3,3′,5′-triiodothyronine, also known as reverse triiodothyronine (RT_3) (below right).

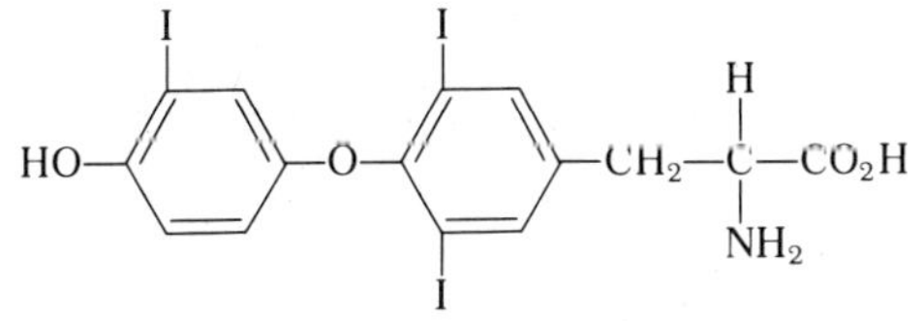

T_3 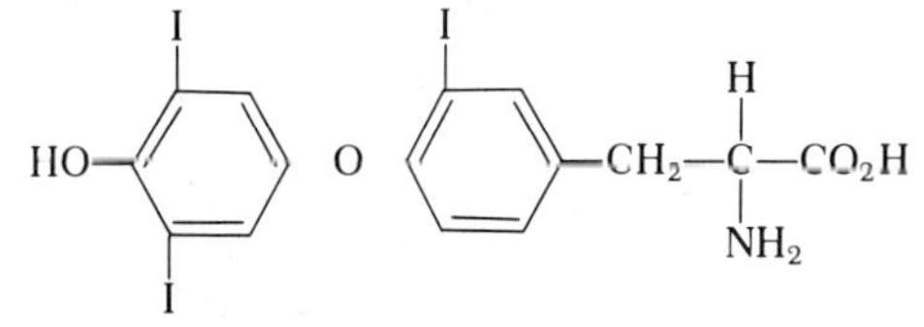 **RT_3**

Biosynthesis, storage, and mobilization of T_4 and T_3

Fig. 18-10 illustrates the various processes involved in the biosynthesis, storage, and mobilization of thyroid hormones. It is apparent from this diagram that the biosynthesis of T_4 and T_3 is contingent upon an adequate supply of iodine. Normally blood contains 0.5 μg of iodide per 100 ml and the thyroid about 10 μg per 100 gm, corresponding to a tissue:blood ratio of 20:1. This means that active transport must pump the iodide ions against a concentration gradient. As indicated in the figure, the transport of iodide (process 1) is linked or coupled in some fashion to an Na^+, K^+ ATPase. In other words, ATP is used in the iodide pump mechanism.

Once inside the cell, the iodide is oxidized by an enzymatic process requiring *thyroid peroxidase* and hydrogen peroxide.

$$2\,I^- + H_2O_2 \xrightarrow{\text{Peroxidase}} I_2 \text{ (active iodine)}$$

This "active" iodine is then involved in the process of organification, i.e., the iodination of tyrosine residues. The tyrosine acceptors are provided by the protein apothyroglobulin (ATG). TG is a glycoprotein having a molecular weight of 660,000 (4 subunits) and 140 tyrosyl residues. As seen in Fig. 18-10 (process 2), the iodination of ATG occurs at the lumenal membrane of the epithelial cells. The enzyme catalyzing the iodination of the phenolic groups is also thyroid peroxidase, which is associated with the lumenal membrane. As seen in Fig. 18-11, iodination yields monoiodotyrosine (MIT) and diiodotyrosine (DIT). Iodination is followed by an unusual transformation or coupling reaction. Because of the globular structure of the TG protein, some iodinated tyrosines are brought into close proximity. This is accompanied by a migration of one iodinated group to another, thereby yielding thyroxine. As indicated in Fig. 18-11, this is an intramolecular process so that the T_4 or T_3 remains bound within the polypeptide chain. The iodinated TG is then stored in the colloid; several weeks' supply of thyroid hormone may be stored in this way.

Figure 18-10 Processes involved in the biosynthesis, storage, and mobilization of thyroid hormones in the epithelial cells of the thyroid gland. The processes indicated are (1) iodide uptake, (2) iodination of ATG, and (3) mobilization via endocytosis and proteolysis. *ATG*, Apo-thyroglobulin; *TG*, iodinated thyroglobulin; *PLys*, primary lysosomes; *PhLys*, phagolysosomes.

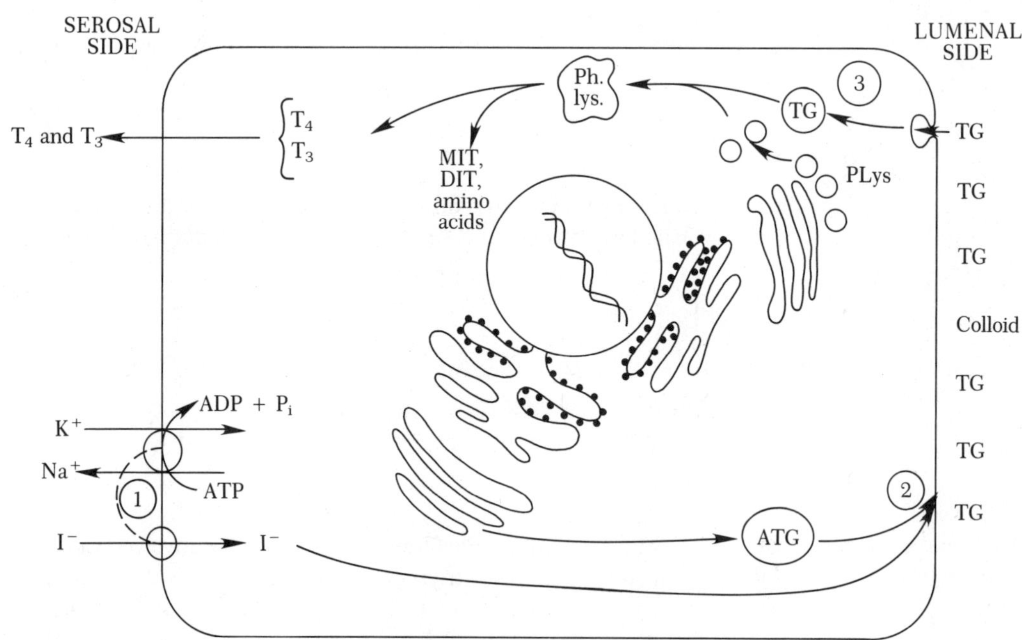

Figure 18-11
Migration of one diiodophenyl group to condense with another to form thyroxine.

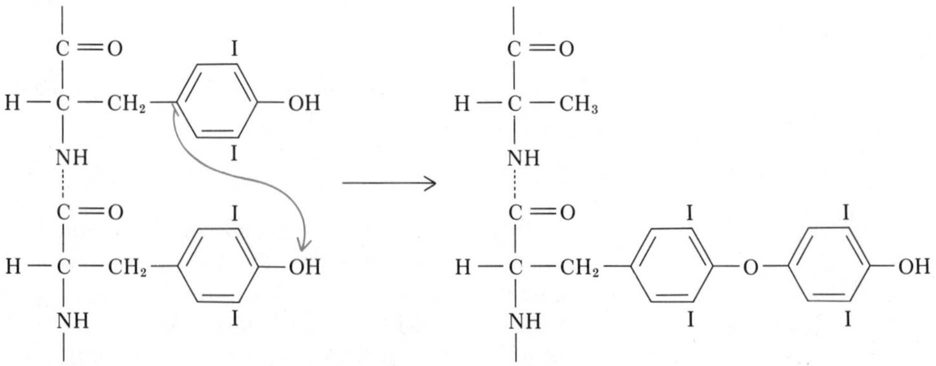

The mobilization of T_4 and T_3, as shown in Fig. 18-10 (process 3), requires first that the storage thyroglobulin be removed from the colloid by way of the process of endocytosis. The endocytotic vacuoles fuse with primary lysosomes containing proteolytic enzymes. These hydrolyze the TG within the phagolysosomes, thereby releasing T_4, T_3, MIT, DIT, and amino acids. Although MIT and DIT are released, they are deiodinated and hence do not leave the cell. T_3 and T_4, on the other hand, are released to the bloodstream. Deiodination of MIT and DIT requires *deiodinase* and a supply of NADPH.

$$\text{Thyroglobulin} \begin{cases} \left.\begin{matrix}\text{DIT}\\ \text{MIT}\end{matrix}\right\} \xrightarrow[\text{deiodinase}]{\text{NADPH}} \text{Tyrosine} + \text{I}^- \\ \left.\begin{matrix}T_3\\ T_4\end{matrix}\right\} \longrightarrow \text{Into bloodstream} \end{cases}$$

Both the iodide and tyrosine are recycled.

Role of TSH in controlling thyroid metabolism

TSH is bound at a membrane-binding site that stimulates adenyl cyclase to produce cAMP. This mediator exerts at least five activities, all of which have a controlling effect on thyroid metabolism. As seen in Fig. 18-12, the five effects are:

1. A stimulation of the iodide pump, thereby increasing the ion-trapping process.
2. A nuclear effect increasing the production of ATG.
3. Stimulation of the oxidation of iodide to iodine, thereby increasing the organification process.
4. Stimulation of glucose metabolism by way of the pentose pathway, thereby increasing the production of NADPH. NADPH may be a factor in the production of H_2O_2 and also in the deiodinase process.
5. Stimulation of endocytosis, or the engulfing of storage thyroglobulin.

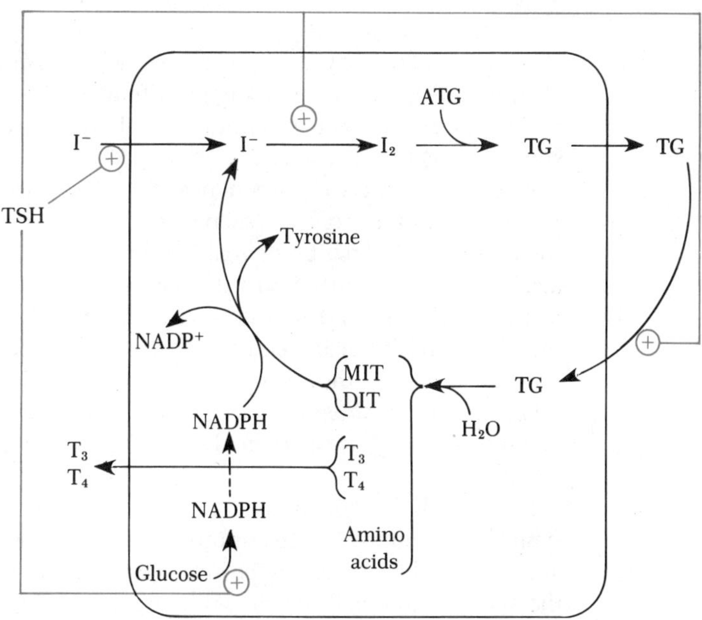

Figure 18-12
Summary of effects of TSH on the production of thyroxine. *ITG*, Iodothyroglobulin; *TG*, thyroglobulin.

The daily output is 70 μg of T_4 and 25 μg of T_3. Despite the lower output of T_3, this hormone is three to eight times as effective as thyroxine itself; the MIT and DIT are hormonally inactive and are not released from the epithelial cells.

Once in the bloodstream, thyroxine does not remain free as part of the circulating pool of amino acids but rather is bound to one or more specific transport proteins. These are *thyroxine-binding globulin* (TBG), which is classified as an inter-α_1-α_2-globulin (Chapter 16), and a prealbumin; if the capacities of these two forms are exhausted, thyroxine will bind with serum albumin. The thyroxine-binding protein is a glycoprotein of molecular weight 40,000 to 50,000. A familial (genetic) deficiency of TBG is known (e.g., in Turner's syndrome and possibly in other diseases involving thyroid-hormone transport). A determination of the iodine-binding capacity of plasma is sometimes used as a diagnostic tool.

In the normal individual (euthyroid) the total blood iodine as T_3 and T_4 is 4 to 8 μg per 100 ml of blood serum, of which only 0.3% is T_3. Both the T_3 and T_4 apparently dissociate from the transport proteins before entering the target cells.

Function of thyroid hormones

Most tissues are affected by the thyroid, for example, the cardiovascular system, gastrointestinal tract functions, and muscular activity. Both T_4 and T_3 enter tissue cells by diffusion. Specific binding proteins have been observed in the cytosol and nuclei of various tissues. These two hormones increase the synthesis of all forms of RNA: messenger, ribosomal, and transfer. This may well be the fundamental biochemical effect of thyroid hormones in affecting the growth and development of the body and in stimulating total metabolism.

Furthermore, thyroid hormones increase the Na^+, K^+–ATPase in all adult cells except those from the brain and retina. The elevation of ATP consumption is offset by an increase in the number and size of mitochondria. These changes are then accompanied by an elevation in oxygen consumption and an increased use of glucose, lipids, and amino acids. In hyperthyroidism the basal metabolic rate (BMR) increases above normal (by 80%) and more food is consumed, yet there is a general loss of weight. The high rate of metabolism demands a more rapid glycogenolysis, resulting in a mild hyperglycemia and sometimes glucosuria. The patient often feels hot and exhibits symptoms such as protrusion of the eyeballs (exophthalmic goiter), dilated pupils, mental excitement, and irritability. Thyroid hormones also increase the glycerol phosphate dehydrogenase activity of rat liver mitochondria. This observation suggests that the increased metabolism of carbohydrate may result from a stimulation of the glycerol phosphate dehydrogenase shuttle (Chapter 7).

Arrested growth is a consequence of the removal of the thyroid in young animals. In humans a similar effect is seen when the thyroid is atrophied at birth. Apparently the fetus can obtain little thyroid hormone in utero and must make its own, using available iodine. If there is insufficient iodide or if the mother is under the influence of some goitrogenic factor, the newborn infant fails to develop normally. The resulting condition is termed cretinism. Cretins are abnormal dwarfs. Although they may grow to adulthood, they do not develop mentally. In cretinism, treatment with thyroid preparations must be started early to have any beneficial effect and must be continued as long as the thyroid gland fails to function. Mental retardation is usually irreversible.

In the adult the clinical condition of hypothyroidism is known as myxedema. The symptoms include changes in the patient's appearance—the skin becomes thick and puffy and there tend to be swellings under the eyes caused by a deposition in the skin of additional protein. Although general intelligence is not impaired, the patient is slower mentally and physically. There is a low BMR, increased deposition of fat, and high blood cholesterol. Myxedema yields dramatically to treatment with thyroid preparations.

The serum of patients with Graves' disease (diffuse goiter, hyperthyroid) contains a thyroid-stimulating factor that is different immunologically from the TSH of the anterior pituitary. When injected intravenously, this hormone exerts its maximal thyroid-stimulating effect more slowly than does TSH. It is called the *long-acting, thyroid-stimulating factor* (LATS).

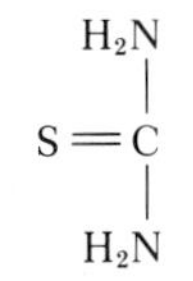

Hyperthyroidism may be treated by surgical removal of some of the overactive tissue. A chemical method of accomplishing the same effect is the administration of thiourea or thiouracil (or large doses of sulfonamides). These antithyroid drugs cause an inhibition of thyroglobulin synthesis, probably by preventing the oxidation of iodide to active iodine or by the coupling of two DIT molecules to form thyroxine (Fig. 18-11). The 6-propyl derivatives of thiouracil are five times more potent than thiouracil itself and are much less toxic. Disulfiram (Antabuse), used in treating alcoholism, has been shown to react with iodine to form a complex that might render iodine unavailable. Antithyroid substances are present in certain foods, excessive ingesting of which may cause simple goiter; hence they are said to be goitrogenic. Among the goitrogenic foods are cabbages, turnips, soybeans, peanuts, and mustard seeds. An antithyroid factor has been isolated from turnips and other goitrogenic vegetables; it is L-5-vinyl-2-thiooxazolidone. When antithyroid substances inhibit the formation of the thyroid hormone, the pituitary gland secretes greater quantities of the TSH and a compensatory thyroidal hypertrophy results.

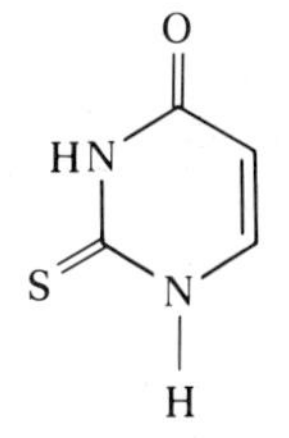

$H_2N-C(=S)-NH_2$

Thiourea

Thiouracil

Endemic goiter is an enlargement of the thyroid caused by an inadequate supply of dietary iodide, which occurs where soil is low in iodide. Approximately 150 μg of iodide should be ingested daily to prevent goiter. The amount of iodide provided by iodized salt is sufficient for this purpose. If an iodine deficiency occurs, the rate of hormone secretion remains constant until the iodine stores present in the gland become depleted. The thyroid removes iodide from the blood more efficiently, and the secretion of the hormone may continue until the iodine deficit becomes acute. Then the pituitary responds by releasing greater amounts of TSH, and the activity of the thyroid gland is further stimulated. Eventually a "work hypertrophy" ensues, resulting in a hypertrophy and hyperplasia of the gland.

Circulating thyroxine and triiodothyronine are metabolized by (1) deiodination, (2) side-chain metabolism, and (3) conjugation. Approximately 80% to 90% of radioactive iodine injected into a human subject as T_4 or T_3 appears in the urine as iodide, provided that thyroidal accumulation is blocked. Deiodination is catalyzed by deiodinases found in the microsomal fraction of liver, kidneys, muscle, brain, and heart. The alanine side chain of T_4 and T_3 may be metabolized by deamination or transamination to yield the pyruvate or lactate analogues, which may be found in the urine and bile. Decarboxylation leads to the formation of tetraiodothyroacetate or triiodothyroacetate. These enzymes are located in the mitochondria of liver and kidneys. T_4 can also be excreted in the bile and urine after conjugation of the phenolic hydroxyl group with glucuronic acid. This requires UDP-glucuronic acid and a microsomal glucuronyl transferase.

Adrenocortical hormones

The pituitary hormone ACTH stimulates the biosynthesis of steroids in the adrenal cortex. When an experimental animal is deprived of both its adrenal glands, the following consequences are observed:

1. *Disturbance of electrolyte and water balance:* Increased excretion of Na^+, Cl^-, and water; retention of K^+, accompanied by an elevation in blood K^+ and decreases in Na^+ and Cl^-; and a loss of water (hemoconcentration)

2. *Elevation in blood urea:* Caused by decreased renal blood flow or decreased kidney function
3. *Muscular weakness:* Secondary to the effects of carbohydrate metabolism and to the salt and water balance
4. *Decreased liver glycogen, with hypoglycemia and sensitivity to insulin:* Results from diminished use of carbohydrate and diminished gluconeogenesis; decrease in quantity of protein catabolized
5. *Reduced ability to withstand stress*

Administration of cortical extracts remedy all of these symptoms, returning the chemical picture practically to normal and extending the life expectancy of the adrenalectomized animals. These effects of adrenalectomy suggest two functions, one on carbohydrate and protein metabolism and the other on electrolyte balance. In other words, we should expect two kinds of adrenal hormones, those controlling metabolism, i.e., the *glucocorticoids,* and those controlling mineral balance, i.e., the *mineralocorticoids.*

Biosynthesis

All of the hormones produced by the adrenal cortex are steroids, i.e., derivatives of *cyclopentanoperhydrophenanthrene* (Chapter 23), and are derived from cholesterol. The first step in *steroidogenesis* (steroid biosynthesis) is the hydroxylation of C-20 and C-22 of the side chain.

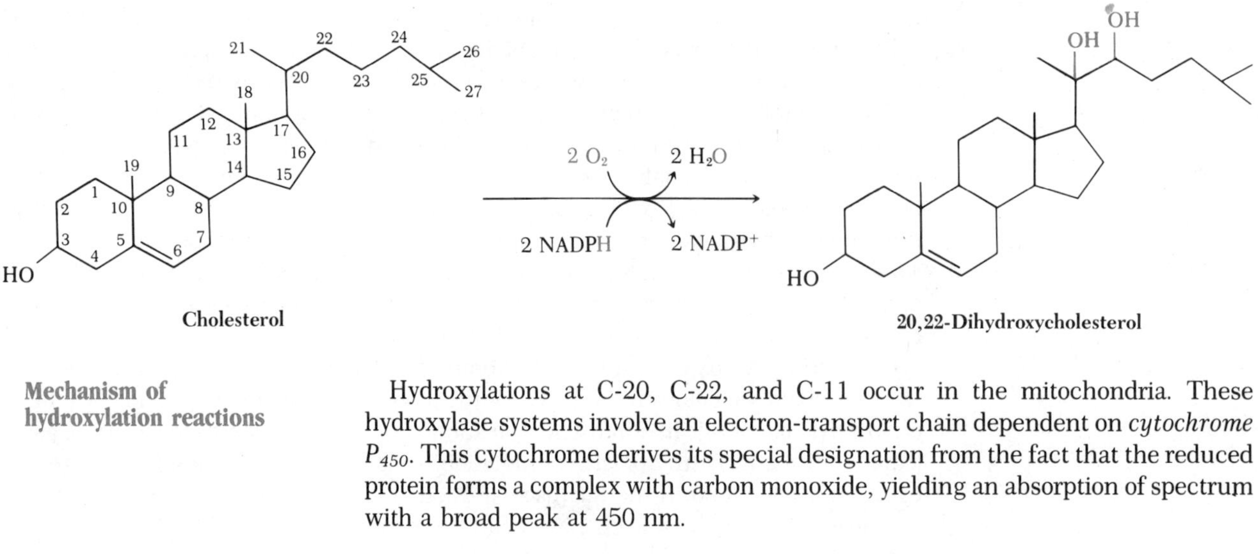

HO

Cholesterol

$2\ O_2$ $2\ H_2O$

2 NADPH $2\ NADP^+$

OH OH

HO

20,22-Dihydroxycholesterol

Mechanism of hydroxylation reactions

Hydroxylations at C-20, C-22, and C-11 occur in the mitochondria. These hydroxylase systems involve an electron-transport chain dependent on *cytochrome P_{450}*. This cytochrome derives its special designation from the fact that the reduced protein forms a complex with carbon monoxide, yielding an absorption of spectrum with a broad peak at 450 nm.

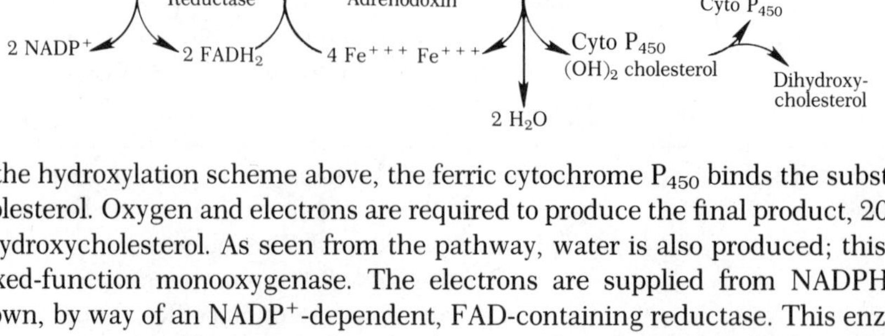

$2\ O_2$

2 NADPH — 2 FAD — $4\ Fe^{+++} \cdot Fe^{++}$ — Cyto P_{450} cholesterol — Cholesterol

Reductase — Adrenodoxin — Cyto P_{450}

$2\ NADP^+$ — $2\ FADH_2$ — $4\ Fe^{+++}\ Fe^{+++}$ — Cyto P_{450} $(OH)_2$ cholesterol — Dihydroxycholesterol

$2\ H_2O$

In the hydroxylation scheme above, the ferric cytochrome P_{450} binds the substrate cholesterol. Oxygen and electrons are required to produce the final product, 20,22-dihydroxycholesterol. As seen from the pathway, water is also produced; this is a mixed-function monooxygenase. The electrons are supplied from NADPH, as shown, by way of an $NADP^+$-dependent, FAD-containing reductase. This enzyme reduces an iron-sulfur protein found only in an adrenal mitochondrial membrane called *adrenodoxin* (mol wt = 22,000). The latter has two atoms each of iron and inorganic sulfur per mole. Each of the two iron atoms is also bound to two cysteine-sulfurs, arranged as shown below. The ferric irons transfer single electrons from the flavoprotein to the cytochrome P_{450}.

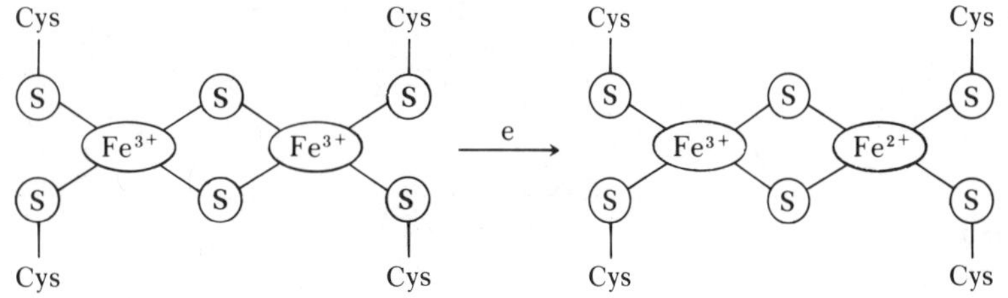

Cys Cys Cys Cys
S S S S S S
Fe^{3+} Fe^{3+} —e→ Fe^{3+} Fe^{2+}
S S S S S S
Cys Cys Cys Cys

The transformation of cholesterol to pregnenolone in gonadal tissues also requires adrenodoxinlike proteins, e.g., *testodoxin,* found in testicular mitochondria.

Conversion of cholesterol to corticosteroids

The cleavage of the C-20 to C-22 bond is catalyzed by a heme-containing enzyme called *cholesterol desmolase* and is found in the mitochondria. Scission yields the products Δ^5-pregnenolone and isocaproic aldehyde.

Pregnenolone leaves the mitochondrion and enters the cytoplasm for further metabolism. As seen below, this involves two enzymes found in the endoplasmic reticulum. These are an NAD^+-requiring 3β-*ol-dehydrogenase* and a $\Delta^{4,5}$-*isomerase* that switches the double bond from Δ^5 to Δ^4. These two steps are shown below.

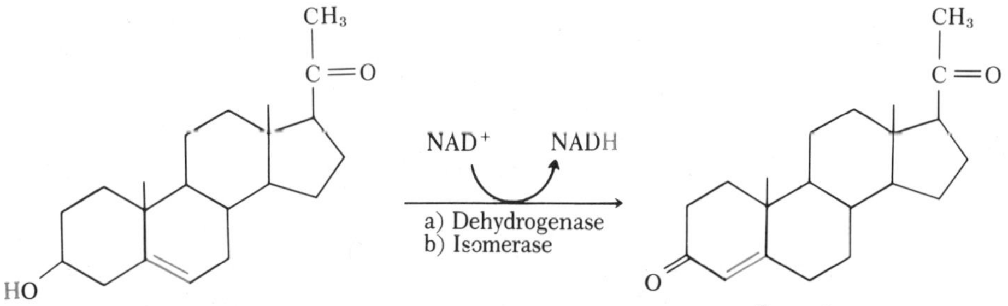

CH_3 C=O HO

NAD^+ NADH

a) Dehydrogenase
b) Isomerase

CH_3 C=O O

Δ^5-Pregnenolone **Progesterone**

As seen, the product *progesterone* has a keto group at C-3 and a double bond between C-4 and C-5.

There are two additional relatively fast-acting hydroxylases located in the endoplasmic reticulum. These are the *17 α-* and *21-hydroxylases.* Both are cytochrome P_{450} monooxygenases and use progesterone as substrate. In the endoplasmic reticulum the flavoprotein NADPH-cytochrome P_{450} reductase catalyzes the transfer of electrons to the cytochrome without an intermediate role for an iron-sulfur protein such as adrenodoxin.

As is seen from Fig. 18-13, the 17 α- and 21-hydroxylases compete for the supply of progesterone and thereby initiate alternative pathways for its metabolism. The 17 α-hydroxylase converts progesterone to the 17 α-hydroxyprogesterone, which is the first step in a pathway leading to the formation of cortisol, the major glucocorticoid hormone. Hydroxylation of progesterone by 21-hydroxylase yields 11-deoxycorticosterone and initiates a pathway leading to aldosterone, the major mineralocorticoid.

The 17 α-hydroxyprogesterone is a substrate for the 21-hydroxylase; the product is 11-deoxycortisol. In the adrenal mitochondria it is converted by 11 β-hydroxylase to cortisol, also frequently called hydrocortisone.

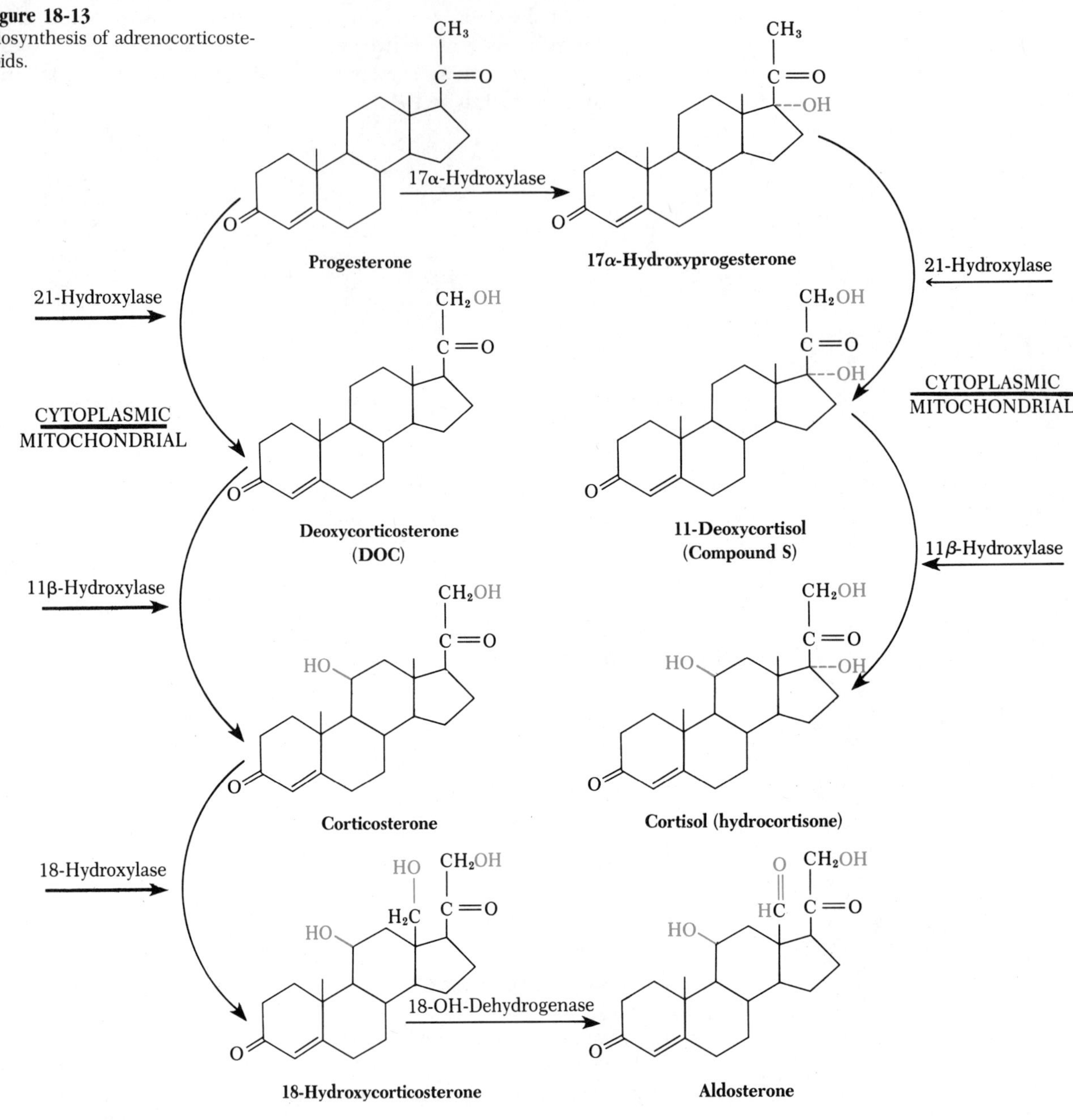

Figure 18-13
Biosynthesis of adrenocorticosteroids.

Deoxycorticosterone is not a substrate for the 21-hydroxylase for reasons of steric hindrance, rather it is converted to corticosterone by the mitochondrial 11 β-hydroxylase. Corticosterone in turn is converted to 18-hydroxycorticosterone by an *18-hydroxylase* and then to aldosterone by an *18-hydroxydehydrogenase*.

The transformation of cholesterol into corticosteroids has been seen to follow two pathways, one terminating in aldosterone and the other in cortisol. Histologically the adrenal cortex consists of two distinct kinds of cell layers. The *zona glomerulosa* lies just beneath the adrenal capsule and functions to produce aldosterone, whereas the *zona fasciculata* is an inner layer of the cortex and serves to produce cortisol.

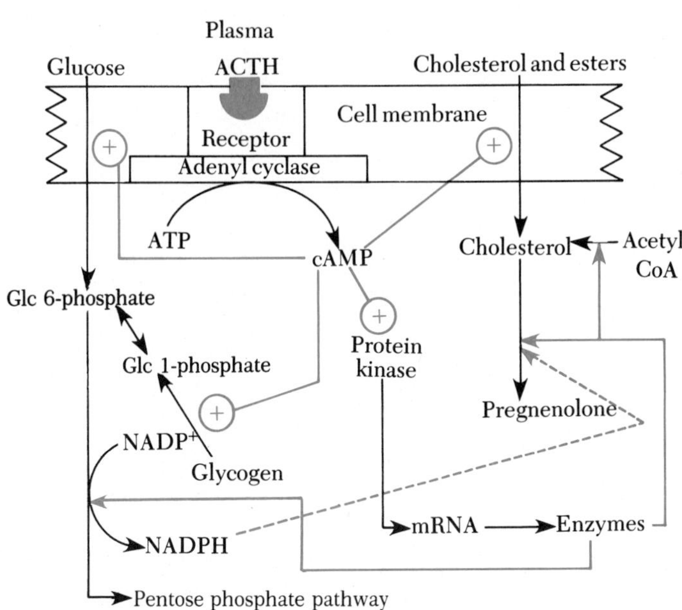

Figure 18-14
Suggested function of cAMP in mediating the hormonal action of ACTH in the adrenal cortex.

It has been estimated that in the normal adult man 10 to 20 mg of cortisol, 3 mg of corticosterone, and 0.3 mg of aldosterone are synthesized daily.

Steroids are only sparingly soluble in water. Therefore the adrenocortical hormones are transported in the plasma to their respective target tissues bound to plasma proteins, mainly α_1-globulins. One of these, *transcortin,* or CBG (corticosteroid-binding globulin), is involved in the transport of cortisol. Normally the total plasma cortisol is 20 μg per 100 ml, of which 19 μg (95%) are bound. The free cortisol (1 μg per 100 ml) represents the metabolically active glucocorticoid; the bound steroid serves as a reservoir.

Metabolism

As noted previously, the biosynthesis of steroids in the adrenal cortex is controlled from the pituitary by way of ACTH. Thus, for example, stress will lead to a liberation of CRH from the hypothalamus, which in turn stimulates the release of ACTH from the hypophysis. The adrenal cells have ACTH-receptor sites; binding of ACTH activates adenyl cyclase. As seen in Fig. 18-14, cAMP then produces an increased uptake of cholesterol and glucose, an increased biosynthesis of hydroxylases, and an increased production of NADPH by way of the pentose pathway. These processes in turn result in an elevated production of pregnenolone. Glucocorticoids are released into the bloodstream; little appears to be stored in the cells.

Adrenal steroids are inactivated primarily in the liver. This occurs at a rate commensurate with their formation, so that a balance with a constant level of the hormones in body fluids is maintained. Under normal circumstances, the biological half-life of cortisols is 90 to 115 minutes. Over 90% of an injected dose of labeled cortisol is found in the urine within 48 hours. The primary mechanism for its metabolism is the reduction the double bond and the ketone group on C-3. This saturated derivative is tetrahydrocortisol (Fig. 18-15). The C-3 hydroxyl is then conjugated with glucuronic acid (Chapter 11) or sulfate; the glucuronides or sulfates are excreted in the urine. If the side chain of tetrahydrocortisol is oxidatively

Figure 18-15
Metabolism of cortisol.

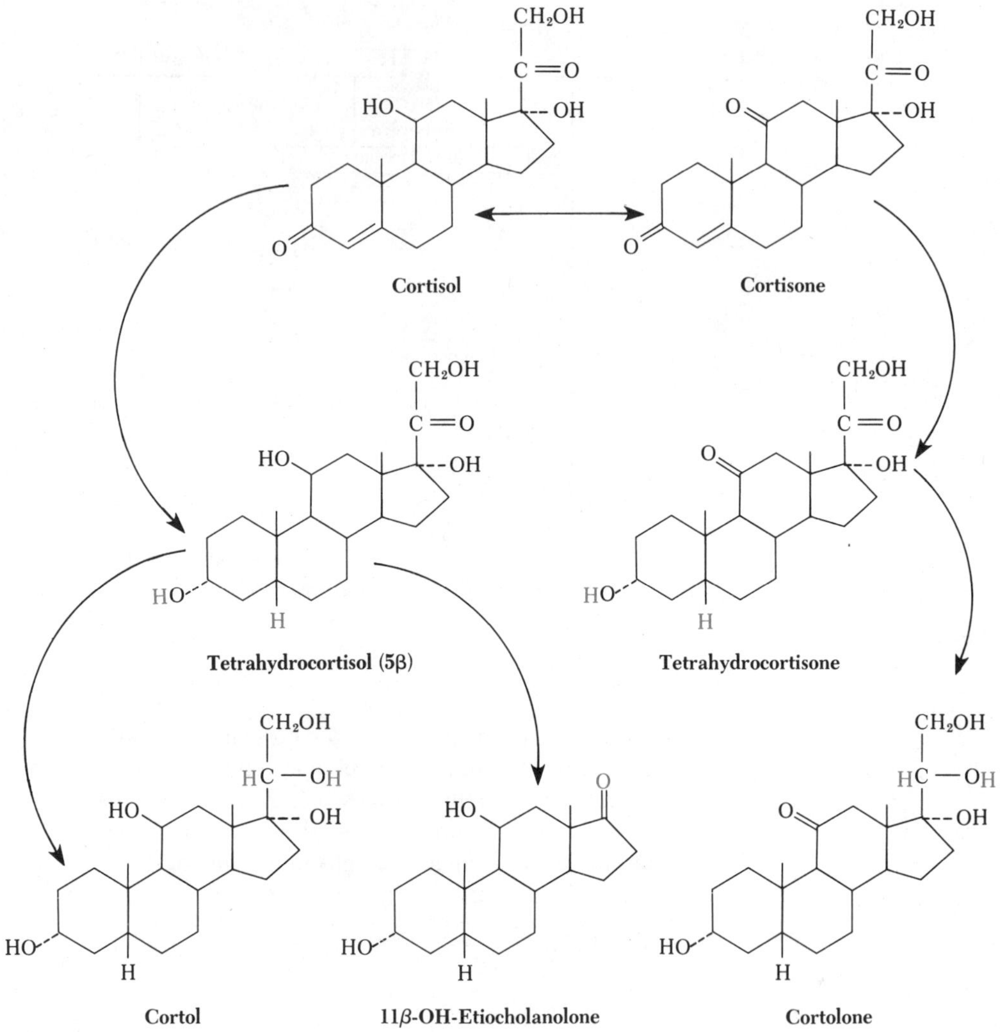

cleaved, a 17-keto steroid called 11 β-hydroxyetiocholanolone is formed (Fig. 18-15). This compound is also excreted as the corresponding glucuronide or sulfate.

Another important metabolite of cortisol is cortisone, which is formed by the reversible oxidation of the C-11 hydroxyl to yield the C-11 ketone (Fig. 18-15). Although cortisone exhibits about two thirds of the glucocorticoid effect of cortisol, it probably owes this action to the widespread distribution of the enzyme *cortisol dehydrogenase*. Thus cortisone is considered to be a metabolite rather than an independent secretion of the adrenal cortex.

Fig. 18-15 illustrates some of the major metabolic pathways of cortisol metabolism. It is apparent from this scheme that both the tetrahydrocortisol and tetrahydrocortisone may undergo further reduction at C-20. The derivative from cortisol is called cortol; that from cortisone is called cortolone. Both cortol and cortolone as well as their tetrahydro precursors are excreted in the urine as the glucuronides.

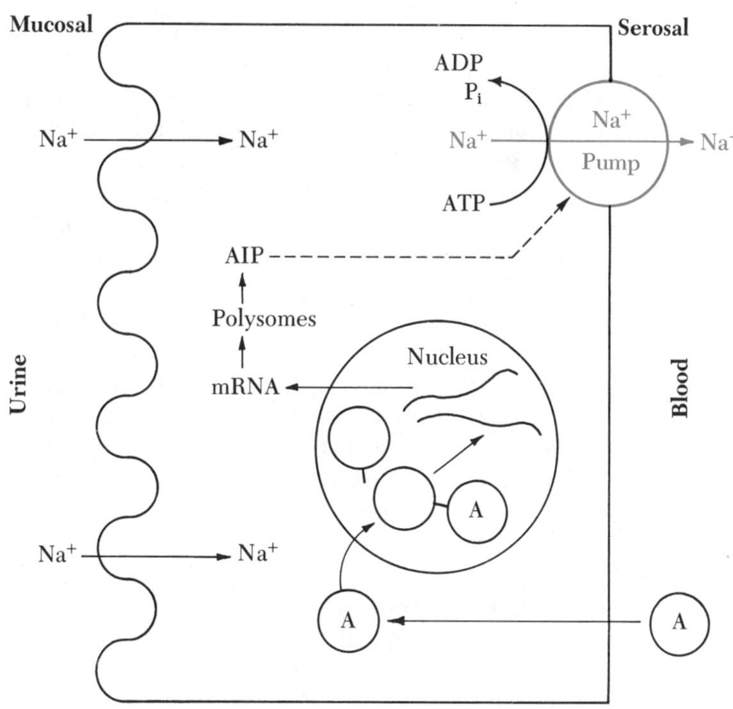

Figure 18-16
Suggested mechanism whereby aldosterone stimulates reabsorption of urinary Na^+ by renal tubule cells. Ⓐ, Aldosterone; —◯, nuclear aldosterone-binding protein; *AIP*, aldosterone-induced protein.

Function of glucocorticoids

The primary metabolic action of glucocorticoids is to stimulate the process of gluconeogenesis (Chapter 11). As indicated earlier, the source of carbon atoms for this process is amino acids. Glucocorticoids promote gluconeogenesis by promoting the catabolism of muscle proteins. In the hepatocyte the steroid hormone is bound by a specific cytoplasmic binding protein. The hormone then increases the transport of amino acids into the cells. It also acts to stimulate the biosynthesis of such gluconeogenic enzymes as PEP carboxykinase, fructose 1,6-bisphosphatase, pyruvate carboxylase, and glucose 6-phosphatase. In addition various transaminases, e.g., tyrosine aminotransferase, are also induced.

While promoting gluconeogenesis, the glucocorticoids inhibit glycolysis but stimulate glycogenolysis by activating glycogen synthase (Chapter 11).

Action of mineralocorticoids

Aldosterone regulates electrolyte balance by stimulating renal sodium reabsorption. The overall effect is to retain Na^+ and excrete K^+. Other target tissues are salivary glands, sweat glands, and the intestinal mucosa.

Kidney tubule cells bind aldosterone to an aldosterone-binding protein (ABP) in the cytosol. Radioautographic studies using labeled aldosterone have shown that the steroid is then concentrated in the cell nucleus. Actinomycin D and puromycin inhibit the regulation of electrolyte balance, demonstrating that the effect of aldosterone is contingent on the biosynthesis of RNA and protein. In other words the reabsorption of Na^+ requires the production of an aldosterone-induced protein (AIP).

Reabsorption of Na^+ from the glomerular filtrate depends on the unique polarization of the distal tubule cells; i.e., their lumenal and serosal cell membranes have different functions (Fig. 18-16). Na^+ diffuses through the lumenal membrane, bathed by the glomerular filtrate, depending on the concentration gradient. Once inside the cell, the Na^+ must be pumped through the serosal membrane against a concentration gradient and into the peritubular circulation. This pump mechanism, which is a constituent of the serosal membrane, requires an input of energy (ATP). AIP may increase the activity of the Na^+ pump either by activating some of its component enzymes or by increasing the production of ATP.

Since aldosterone functions to maintain Na^+ homeostasis, its secretion must be responsive to altered sodium balance. Aldosterone secretion is increased in response to ACTH; however, this appears to be temporary. Other stimuli are an elevated serum K^+, reduction in serum Na^+, and the presence of *angiotensin II*.

A decrease in serum Na^+ triggers a release by the kidney of the enzyme *renin*. Renin acts on a specific plasma protein to release a decapeptide called *angiotensin I*, which is subsequently converted to angiotensin II, an octapeptide (Chapter 16). The latter stimulates aldosterone secretion. Both angiotensin II and ACTH affect the biosynthesis of aldosterone by stimulating the conversion of cholesterol to pregnenolone. Na^+ depletion also appears to stimulate the conversion of corticosterone to aldosterone, whereas the increase in K^+ acts on the transformation of deoxycorticosterone to corticosterone.

Defects of adrenal function

Addison's disease Addison's disease is caused by a hypofunction of the adrenal cortex. The symptoms are low blood pressure, lowered BMR, subnormal temperature, and a disturbance in water and electrolyte balance. This includes loss of sodium and chloride ions, retention of potassium ions, and a loss of body water. There is also a hypoglycemia, which indicates a profound effect on carbohydrate metabolism. The kidneys are affected, resulting in urea retention. Skin pigmentation occurs; frequently the face, neck, and backs of the hands are deeply bronzed. In the absence of sufficient corticosteroids, there is no negative feedback to the pituitary and hence the release of ACTH is increased.

Derivatives of cortisone are used to control the symptoms of Addison's disease and as a replacement therapy for adrenalectomized patients. Δ^1-Hydrocortisone (prednisolone) and Δ^1-cortisone (prednisone) are examples). The Δ^1 double bond in the A-ring increases the glucocorticoid potency and makes the steroid less subject to inactivation.

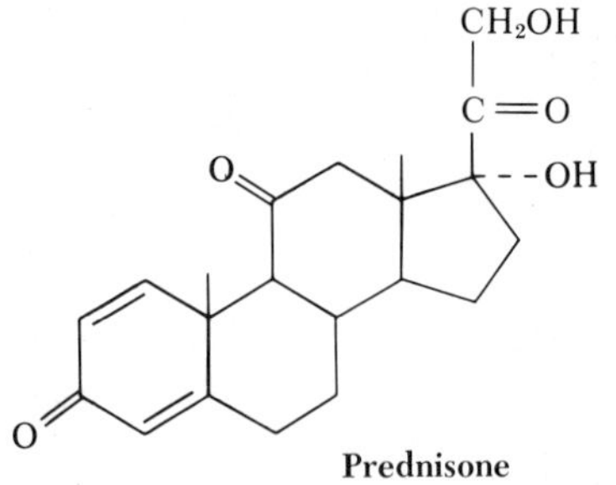

CH_2OH, C=O, OH, O, O

Prednisone

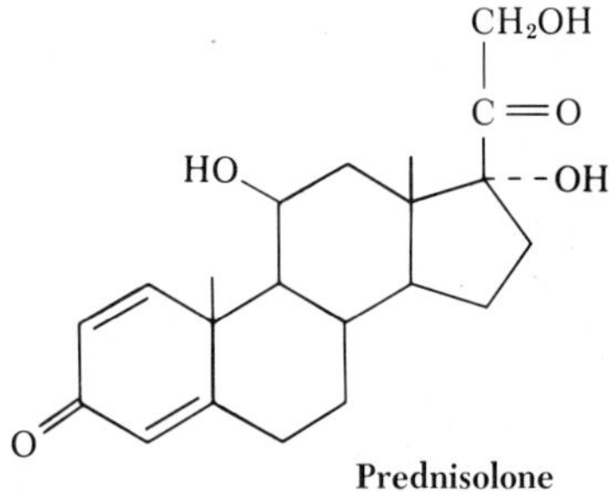

CH_2OH, C=O, OH, HO, O

Prednisolone

Sodium chloride alone is of value in Addison's disease to correct the electrolyte and water balance.

The administration of deoxycorticosterone or aldosterone affects almost exclusively electrolyte and water metabolism. Both restore the blood sodium level to normal and cause a retention of water, thus increasing the volume of the blood plasma and interstitial fluid. They increase the elimination of potassium, resulting in a reduction of serum potassium to normal. Renal function is restored, so that urea and other nonprotein nitrogen (NPN) constituents are excreted, leading to a diminution in the blood NPN. There is also a decrease in the total protein, calcium, and cholesterol concentration of the blood serum, probably a result of the retention of water and dilution of these constituents.

Cushing's syndrome Cushing's syndrome is associated with a hypersecretion of all adrenocortical hormones, the result of a hyperplasia or tumor of the adrenal cortex. The syndrome is characterized by profound disturbances of protein, fat, carbohydrate, and calcium metabolism. There are a rapidly developing, often painful, adiposity of the face, neck, and trunk; rarefaction of the bones, with resulting curvature of the spine, pain, and general body weakness; and often hypertension and polycythemia. In men there is impotence; in women, amenorrhea and masculinization. Cushing's syndrome resembles the adrenogenital syndrome (p. 629). Diabetes mellitus is often present and is usually insulin resistant. Evidently this is caused by the increased secretion of glucocorticoids.

Rheumatoid arthritis and the use of glucocorticoids

Rheumatoid arthritis is a chronic disease affecting the joints. It is characterized by pain, deformity, limitation of motion, and sometimes bony ankylosis, all of which result in debility and weakness. In the past the most useful therapeutic agents have been salicylates. In 1949 the dramatic effects of cortisone and ACTH were discovered. Cortisone decreases the stiffness, tenderness, and pain in the joints in hours or days. Diminution of joint swelling, disappearance of soft tissue deformities, and general improvement in health occur more slowly. ACTH, which of course stimulates the production of cortisol, has similar effects. Cortisone has been used not only for rheumatoid arthritis but also for a number of other ailments, including acute rheumatic fever, acute asthma, and Addison's disease. As has been discussed, glucocorticoids are antagonistic to the action of insulin; therefore prolonged treatment with cortisone may lead to an exhaustion of the insulin-secreting β-cells of the pancreas, causing diabetes.

Hormones of reproduction

Sex hormones responding to gonadotropins

The pituitary gonadotropins FSH and LH (ICSH) stimulate their target gonadal tissues, the ovaries and testes of the female and male respectively. These tissues then produce the sex hormones: androgens in the male and estrogens and progesterone in the female. Each of these is a steroid that controls masculinization, feminization, and gestation.

Male sex hormones Of the male sex hormones, or androgens, *testosterone* is most important. Testosterone is produced by the Leydig cells of the testes. It is a steroid produced from cholesterol, as shown in Fig. 18-17. As in the biosynthesis of adrenocorticosteroids, the first step is the hydroxylation of C-20 and C-22 by the mitochondrial monooxygenase enzymes. The electrons are derived from NADPH by way of a flavoprotein and testodoxin, an iron-sulfur protein. The monooxygenase complex includes cytochrome P_{450}. This step is important in the regulation of the biosynthesis of androgens and is stimulated by the gonadotropin LH(ICSH). The 20,22-dihydroxycholesterol is converted to Δ^5-pregnenolone by the enzyme *cholesterol desmolase*. Pregnenolone must leave the mitochondria and enter the cytosol, where it becomes the substrate either of 17α-hydroxylase to produce 17-α-hydroxypregnenolone or of 3-β-hydroxydehydrogenase and Δ^5-isomerase, in which case progesterone is produced. Progesterone in turn is converted to 17-α-hydroxyprogesterone. These three enzymes are found in the endoplasmic reticulum. In either event, the next step is an oxidative cleavage of the C-17 side chain, yielding a C-17 ketone. The enzyme involved is *C-17, C-20 lyase*. The product formed, as seen at the top of p. 624, is *androstenedione*.

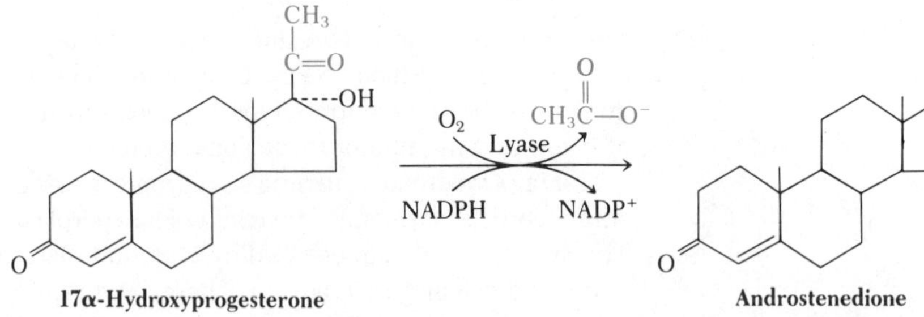

O_2 Lyase CH_3COO^-

NADPH $NADP^+$

17α-Hydroxyprogesterone **Androstenedione**

Alternatively 17-α-hydroxypregnenolone is converted by the lyase to *dehydroepiandrosterone* (DHA) (Fig. 18-17). Both DHA and androstenedione are androgenic. DHA can be converted to androstenedione by the action of 3β-ol-dehydrogenase and $\Delta^{4,5}$-isomerase. A reversible oxidation-reduction of the C-17 keto group results in the formation of testosterone. The optional pathway leading to DHA occurs mainly in the adrenal glands, whereas the pathway by way of progesterone occurs in the testes. Another androgen occurring exclusively in the adrenal is 11-hydroxyl-Δ^4-androstenedione. This steroid is formed from androstenedione by 11β-hydroxylase, an enzyme found only in the mitochondria of the adrenal cortex.

Testosterone is the major circulating androgen and appears to enter the target cell by passive diffusion. Once inside, it may be reduced to *dihydrotestosterone* (DHT) either at the endoplasmic reticulum or at the nuclear membrane (Fig. 18-2, *B*). This reduction, which is biologically irreversible, is catalyzed by a membrane-bound *5 α-reductase* and requires a supply of NADPH. Subsequently DHT is attached to a specific high-affinity protein that enters the nucleus and binds to the chromatin (Fig. 18-2, *B*). The androgen stimulates an increased transcription of RNA either by induction (derepression) or by increasing the RNA polymerase activity or by both.

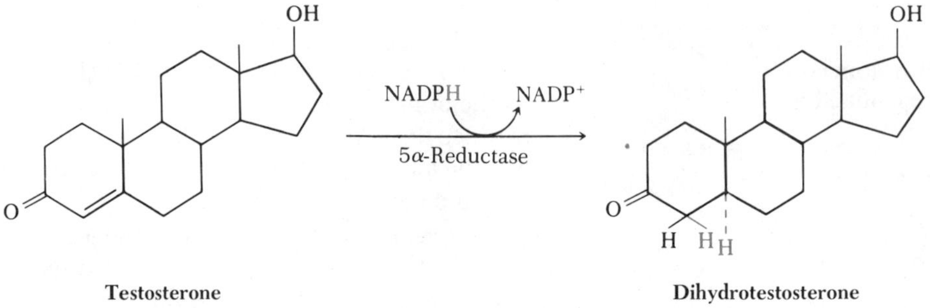

NADPH $NADP^+$

5α-Reductase

Testosterone **Dihydrotestosterone**

DHT is more potent as an androgen than is testosterone itself. Its unique importance may be associated with the fact that testosterone and androstenedione serve as precursors for the biosynthesis of estrone and estradiol by peripheral tissues in the male, whereas DHT does not. Apparently, once the A-ring is totally saturated, as in DHT, it can no longer be aromatized as in the synthesis of estrone (Fig. 18-18). Therefore DHT is entirely androgenic in its action.

At a molecular level the primary action of testosterone is to increase protein synthesis in the target tissues, specifically the prostate and seminal vesicles. Its chief function is to produce normal development of the male reproductive organs and to maintain the secondary male characteristics. Under the influence of the hormone the testes descend into the scrotum. Testosterone is also effective in sustaining spermatogenesis. The secondary male characteristics that develop in its presence are the deep voice, growth and pattern of facial and body hair, and the

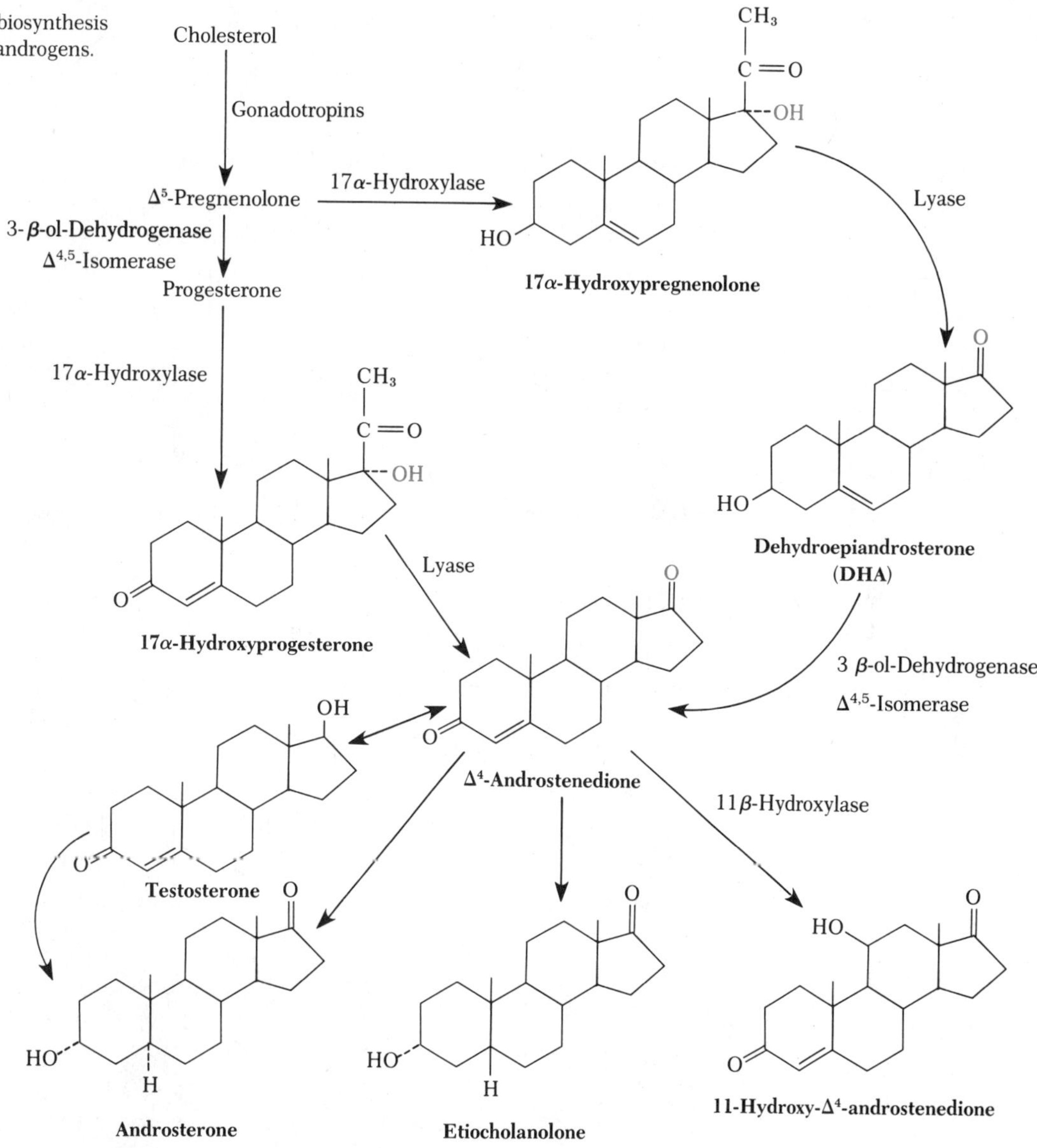

Figure 18-17
Pathways showing biosynthesis and metabolism of androgens.

male type of skeletal muscular development. It inhibits mammary development and function and stimulates libido.

The metabolism of androgens occurs mainly in the liver by way of the reduction of the C-3 keto group and the Δ^4 double bond. This yields α- and β- isomerization in which the C-5 hydrogen may exist as either α or β to the A-B ring system. The two major metabolites are androsterone (α) and etiocholanolone (β) (Fig. 18-17). Of these two, androsterone has androgenic activity, whereas etiocholanolone has little physiological activity. Both are conjugated as glucuronides or sulfates and are excreted in the bile or into the urine.

Testosterone may be used clinically if the testes are absent or nonfunctioning. Such males may have an effeminate appearance—broad hips, prominent breasts,

Figure 18-18
Pathways showing biosynthesis of estrogens.

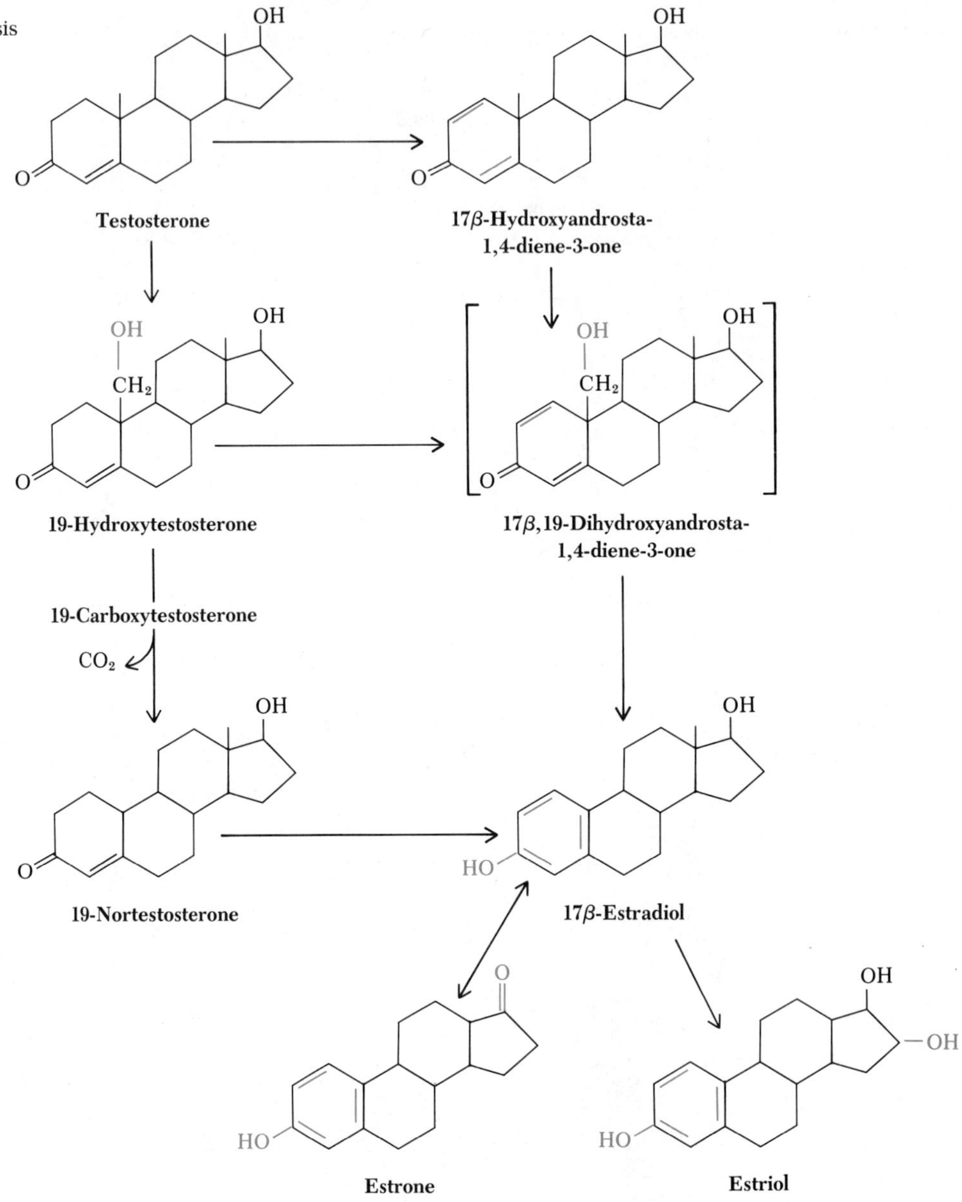

Testosterone

17β-Hydroxyandrosta-1,4-diene-3-one

19-Hydroxytestosterone

17β,19-Dihydroxyandrosta-1,4-diene-3-one

19-Carboxytestosterone

CO_2

19-Nortestosterone

17β-Estradiol

Estrone

Estriol

and a high-pitched voice. They usually are easily fatigued and have a low BMR. Testosterone therapy often brings about enlargement of the sexual organs, prostatic secretion, growth of pubic and axillary hair as well as a beard, and deepening of the voice. The use of gonadotropic hormones is theoretically more sound if there is testicular tissue that can be stimulated. The administration of large doses of testosterone to women may have a masculinizing effect.

Female sex hormones Estrogens are the female sex hormones, the most important of which are estrone, estradiol, and estriol. Although estrone was the first of these to be isolated and was for a long time called the "female sex hormone," it is not as potent as estradiol. The relative potencies of β-*estradiol:* α-*estradiol: estrone: estri-*

ol are approximately 1000:10:100:30. The α and β refer to the relation of the 17-hydroxyl group to the plane of the ring structure.

In the female these hormones are synthesized from cholesterol by way of progesterone and testosterone or androstenedione. The following diagram compares the structure of β-estradiol with that of testosterone.

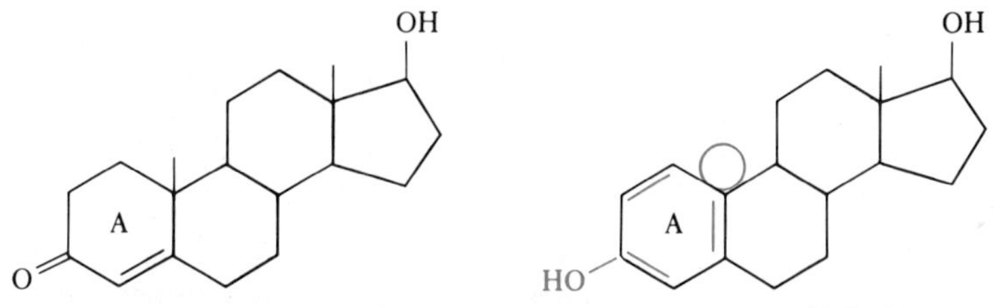

Testosterone **β-Estradiol**

Three differences are noteworthy: (1) the A-ring of estradiol is aromatic, (2) C-3 has a hydroxyl (phenolic), and (3) estradiol has one less $-CH_3$ (colored circle above C-10) and so has 18 carbons, compared with 19 in testosterone.

The biosynthetic pathway in the ovaries is identical to that for testicular tissue, at least to the point of the formation of testosterone. As seen in Fig. 18-18, testosterone is converted to β-estradiol by the oxidation of C-19 to yield 19-hydroxytestosterone, followed by the complete removal of the $-CH_3$ at C-10 and aromatization of the A-ring. This step leads to the formation of estrone. Estrone (Fig. 18-18) is reversibly convertible to estradiol. The third of the estrogens, i.e., estriol, may be produced by the hydroxylation of estradiol at C-16 by the action of *16α-hydroxylase*.

Estradiol is the main estrogen secreted by the ovary. Estrogens are also secreted by the placenta and the adrenal glands. In the male the testes produce small amounts of estrogens, just as the ovaries in females secrete some androgens.

Estriol is conjugated with glucuronic acid to yield the glucuronide, which has little or no physiological action. Likewise estrone forms the glucuronide and also the sulfate and phosphate derivatives. These conjugated forms may be excreted in the urine. The liver is the primary site of estrogen metabolism.

The amount of estrogens in the blood of women shows rhythmic alteration, with peaks at approximately the thirteenth and twentieth days of the menstrual cycle (Fig. 18-7). The amount of estrogen secreted by the ovarian follicle cells is increased by FSH. In turn estrogens control the secretion of FSH by feedback inhibition.

As shown in Fig. 18-2, *A*, estradiol enters the cell by diffusion and is bound by a receptor protein. The estradiol-receptor protein undergoes some modification in the cytosol before entering the nucleus. In the ovariectomized rat, injected estradiol is found bound to chromatin of uterine cells shortly after injection (Fig. 18-1). The primary effect of the hormone is to stimulate the genetic transcription of new RNA. Injections of actinomycin D block the estrogenic stimulation of uterine protein synthesis and the consequent increase in function and growth. Estrogens increase the synthesis of transfer and ribosomal RNA and stimulate the formation of specific enzymes involved in carbohydrate and lipid metabolism.

Corpus luteum hormone The corpus luteum produces *progesterone,* a hormone responsible for the continuing development of the uterus and for the conversion of endometrium to a secretory stage. Progesterone also inhibits ovulation and influences growth of the mammary glands. Its absence, brought about by the removal of the corpus luteum, results in interference with implantation of the ova in humans.

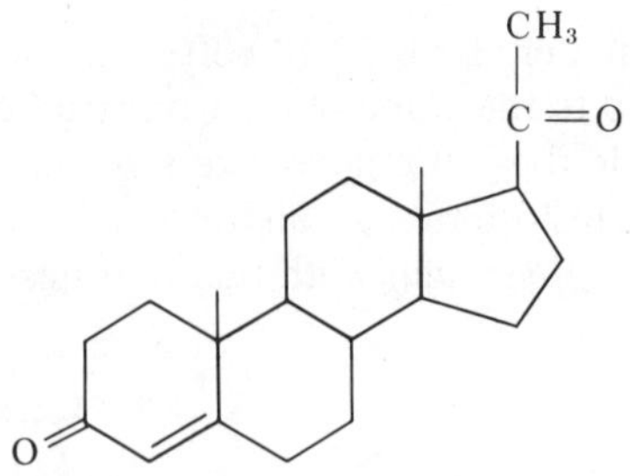

$$CH_3-C(=O)-$$ steroid nucleus with 3-keto, Δ^4

Progesterone

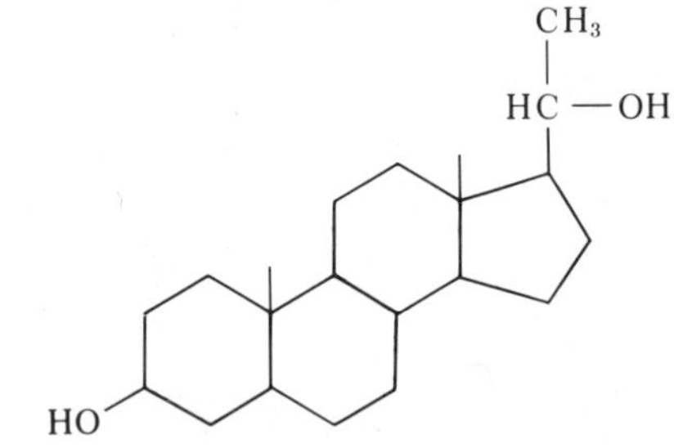

$$CH_3-CH(OH)-$$ steroid nucleus with 3-HO

Pregnanediol

Since progesterone is the immediate precursor of deoxycorticosterone, it is not surprising to find that it has certain adrenocortical properties, especially those influencing salt and water metabolism. Progesterone is a precursor not only of the corticosteroids but also of the androgens and estrogens. The chief metabolic product of progesterone found in the urine is pregnanediol, which is excreted as the glucuronide. Progesterone is used clinically to some extent, particularly in the treatment of amenorrhea (absence of menstruation in young women). Some oral contraceptives are derivatives of progesterone. The two major types are the 19-nor derivatives, which lack the methyl group (carbon-19), and the 17-hydroxyprogesterone diacetate or caproate derivatives.

Progesterone, like other steroid hormones, is biologically active in protein synthesis at the genetic level. Although it induces some of the same enzymes, it is considerably less effective than the estrogens.

True menstruation occurs only in human beings and in members of a closely related group of primates. Under the influence of the pituitary gonadotropic hormones the follicle matures and an increasing amount of estradiol is formed. This occurs during the first 2 weeks of the cycle. Under the influence of estradiol, the endometrium of the uterus increases in thickness and vascularity up to the time of ovulation. The follicle then ruptures and liberates a mature ovum, after which LH causes the ruptured follicle to change to a corpus luteum, which forms progesterone. Progesterone in turn causes the endometrium to assume a turgid secretory condition and to be ready to receive and maintain a fertilized ovum. If the ovum is not fertilized, the corpus luteum regresses, progesterone diminishes in amount, and the endometrium sloughs off, with the occurrence of menstrual bleeding. If the ovum is fertilized, the secretion of progesterone continues, for this hormone is necessary in the maintenance of pregnancy. Prolactin aids in continuing the secretion of progesterone.

In addition to its proliferative action on the endometrium, estradiol seems to maintain the normal size and function of the various parts of the female reproductive organs. It promotes growth of the duct tissue in the breast. Estradiol exerts a controlling action on the secretion of the anterior pituitary, so that the ovary and pituitary seem to have reciprocal effects. Thus indirectly the ovarian hormone influences the other endocrine glands through its feedback to the pituitary. Other effects of estradiol include stimulation of the growth of certain epithelial tissues, such as the mucosa of the vagina.

Estradiol is said to be effective in developing the female organs in sexual infantilism. In juvenile vaginitis from gonorrheal infection and in other forms of vaginitis it is administered to stimulate proliferation of the mucosa. It promotes growth of the breasts under certain conditions but also is said to inhibit the pituitary in the control of excessive milk secretion.

A synthetic estrogenic compound is diethylstilbestrol. Although it does not resemble the estrogens, it exhibits practically all of the physiological effects of estradiol.

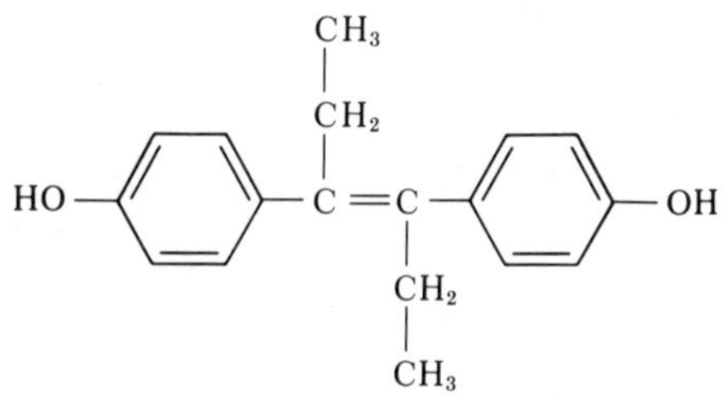

$$HO-C_6H_4-C(CH_2CH_3)=C(CH_2CH_3)-C_6H_4-OH$$

Diethylstilbestrol

Adrenogenital syndrome

Sometimes tumors of the adrenal cortex cause abnormal sexual changes. In the prepubertal female they may produce pseudohermaphroditism, where the external genitals become masculinized and resemble male organs. The breasts atrophy, menstruation ceases, the pubic hair changes to the male pattern, and a beard may grow. Males may show precocious sexual development. Sometimes, both males and females acquire unusual musculature. The prevailing view is that the syndrome is caused by a congenital lack of the 11β-hydroxylase and that therefore the normal hydroxylation of progesterone and 11-deoxycorticosterone to hydrocortisone and corticosterone is defective. This results in a lowered production of these adrenocorticoids and the shunting of precursors to the formation of androgen, e.g., dehydroepiandrosterone (Fig. 18-17). Therefore the sexual changes are generally toward masculinization.

Hormones of the adrenal medulla

All of the hormones discussed so far come under the control of the hypothalamopituitary system and, as already described, are part of neurohormonal processes. Another gland that is a functional part of the nervous system but not of the hypothalamopituitary system is the *adrenal medulla*. In this gland a neural message is converted into a hormonal one; hormones are released for systemic distribution by the bloodstream. The medullary cells have packets of stored hormones that are released on demand. These are amines and derivatives of catechol (i.e., orthodihydroxybenzene) and therefore are called *catecholamines*.

Catecholamines: epinephrine and norepinephrine

The two catecholamines produced by the adrenal medulla are epinephrine (or adrenaline) and norepinephrine (or noradrenaline). These catecholamines are derived from the amino acid phenylalanine. Phenylalanine is converted to tyrosine by the enzyme *phenylalanine hydroxylase* (Fig. 18-19), which in turn is converted to dihydroxyphenylalanine by *tyrosine hydroxylase,* a rate-limiting step. Both of these enzymes are mixed-function oxygenases. As described previously (Chapters 6 and 13), these reactions require oxygen and a supply of electrons provided by NADPH by way of the cofactor tetrahydrobiopterin. The product of tyrosine hydroxylase is dihydroxyphenylalanine, commonly known as DOPA. An aromatic L-amino acid decarboxylase (also DOPA decarboxylase) converts DOPA to dopamine. *Dopamine β-hydroxylase* then catalyzes the conversion of dopamine to norepinephrine. The last step involves *phenylethanolamine-N-methyl transferase* (PNMT), an enzyme that occurs almost exclusively in the adrenal medulla. The methyl donor is SAM. Since epinephrine possesses an asymmetric carbon, two stereoisomers are possible. The natural L form is 15 times more potent than the D form.

Figure 18-19
Biosynthesis of norepinephrine and epinephrine from phenylalanine.

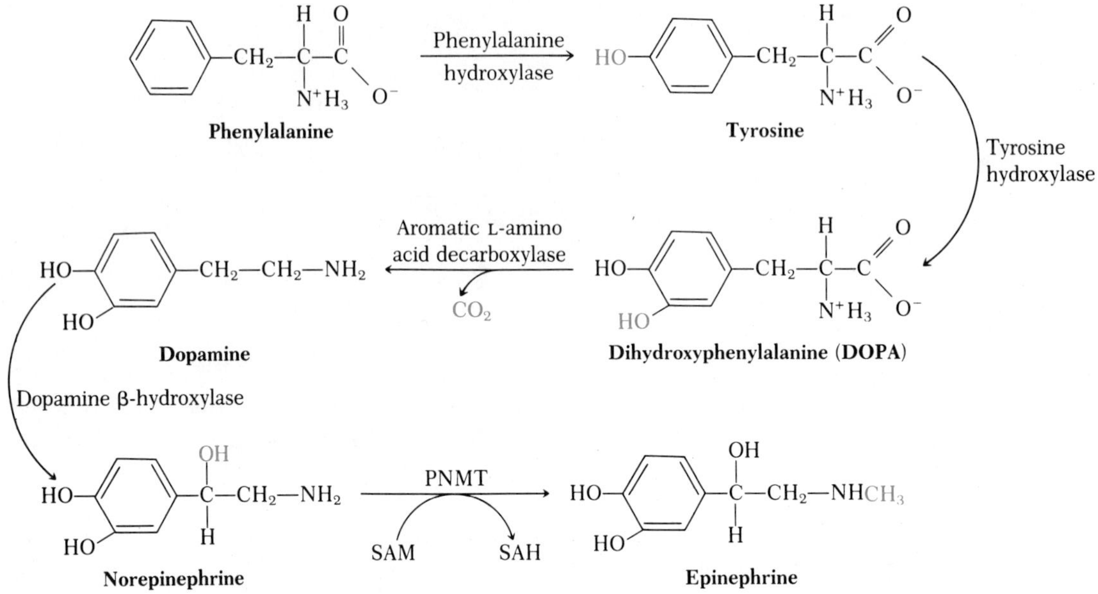

Phenylalanine —(Phenylalanine hydroxylase)→ Tyrosine —(Tyrosine hydroxylase)→ Dihydroxyphenylalanine (DOPA) —(Aromatic L-amino acid decarboxylase; CO_2)→ Dopamine —(Dopamine β-hydroxylase)→ Norepinephrine —(PNMT; SAM → SAH)→ Epinephrine

The catecholamines are found at a constant level in tissues; they are continuously being released, metabolized, and synthesized. The enzymes PNMT and dopamine β-oxidase gradually diminish in hypophysectomized animals. The activity may be restored by administering ACTH. Actually PNMT is induced by glucocorticoids. This suggests that the normal maintenance of the catecholamine biosynthetic enzymes requires ACTH and glucocorticoids.

Injection of epinephrine and, to a lesser degree, norepinephrine causes hyperglycemia and glycosuria. The reason is that epinephrine increases hepatic glycogenolysis by increasing cAMP formation. In this way it can relieve the hypoglycemia produced by insulin. Its effect is greatest if there is a large store of glycogen in the liver. A similar effect on muscle glycogen can also be brought about by epinephrine, resulting in increased lactic acid.

No known clinical syndrome is attributable to a deficiency of the adrenal medulla. Cannon's theory—that epinephrine is secreted in emergencies to raise the blood glucose level to provide ready fuel for the necessary activity (for fight or flight) and, at the same time, to cause a shift of blood from the skin and gastrointestinal tract to the muscles—warrants consideration. If this is the true function of the adrenal medulla, we may have the reason that this gland is not absolutely essential to life although probably useful for optimum physiological performance.

The metabolism of epinephrine and norepinephrine involves the O-methylation of these substances to yield metanephrine and normetanephrine respectively. The enzyme that catalyzes these methylations is *catechol-O-methyltransferase* (COMT), which requires SAM as the methyl donor. Metanephrine either may be oxidatively deaminated by *monoamine oxidase* (MAO) to 3-methoxy-4-hydroxymandelic aldehyde or conjugated with sulfate of glucuronic acid and excreted (see top of p. 631).

HO-C6H3(OH)-CH(OH)-CH2-NH-CH3 → (COMT; SAM → SAH) → HO-C6H3(OCH3)-CH(OH)-CH2-NH-CH3

Epinephrine → Metanephrine

Metanephrine → (MAO) → 3-Methoxy 4-hydroxymandelic aldehyde

Metanephrine → (Conjugation) → Conjugated (R) metanephrine

Control of homeostasis

Claude Bernard proposed that normally the animal organism maintains a constant internal environment. He was impressed by the fact that many constituents of the blood and other fluids remained within fairly narrow limits of concentration. This homeostatic principle reflects an equilibrium between processes that supply the constituents and those that remove them. The balance between them is under the regulatory control of hormonal systems.

Under the term *homeostatic control* are those hormonal mechanisms that respond directly to alterations in the composition of the blood. Thus changes in solute concentration in the blood affect the osmoregulators of the hypothalamus. Even more complex are the controls over the Na^+ concentration. In this section the control of blood glucose level and the concentration of calcium are discussed as specific examples of a homeostatic control mechanism in which the components affect the endocrine gland concerned.

Blood glucose levels

The blood glucose level must be maintained within the limits of 70 to 90 mg per 100 ml. Levels above the normal range are termed *hyperglycemic;* those below are called *hypoglycemic*. Following are factors affecting this delicate balance:

1. *Storage:* As glycogen (glycogenesis versus glycogenolysis)
2. *Supply:* Dietary sugars and gluconeogenesis
3. *Use:* Metabolism
4. *Conversion:* Conversion into fat or other carbohydrates

The regulation of blood glucose is the result of interplay of hormones such as insulin, glucagon, epinephrine, and glucocorticoids.

In addition to its role in the production of digestive enzymes, the pancreas also possesses clusters of cells known as the Islets of Langerhans. The islets, representing about 1% to 2% of the pancreatic tissue, consist of three distinct kinds of endocrine cells, each capable of secreting a specific hormone directly into the bloodstream. As shown in Fig. 18-20, the periphery of the islets is predominantly made up of α-cells, whose function is to secrete the polypeptide hormone glucagon. The α-cells respond to undesirably low blood glucose levels; glucagon serves to elevate blood glucose. Also found within the islets are β-cells (Fig. 18-20); they are the insulin-producing cells. Insulin is required to lower the blood glucose by stimulating glucose transport by the muscle and adipose tissue cells. A third kind of cell, the δ-cells, are distributed as shown in Fig. 18-20; they secrete somatostatin. This hormone inhibits the release of glucagon and insulin. These three hormones function to maintain the blood glucose levels within the normal range of 70 to 90 mg/

Figure 18-20
Diagrammatic representation of the islets of Langerhans showing the distribution of the three cell types. *A*, Glucagon-producing α-cells; *B*, insulin-producing β-cells; *D*, somatostatin-producing δ-cells.

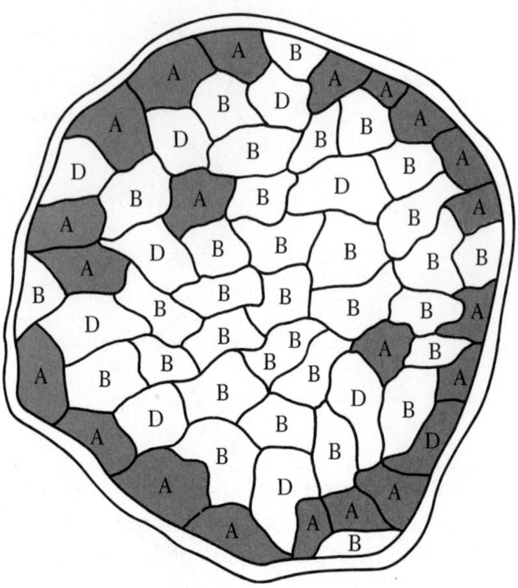

Figure 18-21
Amino acid sequence of human proinsulin. Hydrolysis by trypsin-like enzymes occurs at *1* and 2 and by carboxypeptidase B at *3*. The insulin molecule (A and B chains) is indicated in color.

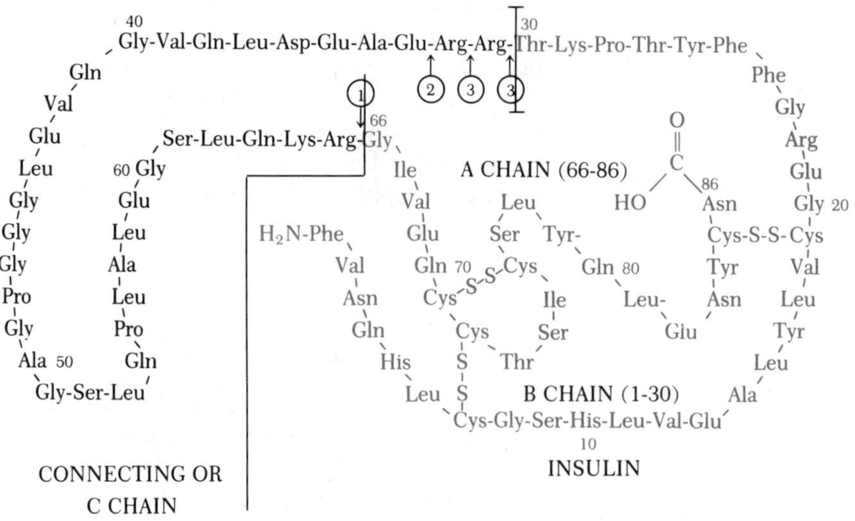

100 ml. Thus the fluctuations resulting from eating, fasting, and exercise are in a sense "buffered" by this finely tuned system.

Insulin As described in Chapter 4, insulin consists of an A-chain having 21 residues joined to a B-chain of 30 residues by two disulfide bonds. It is actually synthesized as a single polypeptide chain, *proinsulin*, possessing 86 amino acid residues (Fig. 18-21). However, proinsulin is produced in the rough endoplasmic reticulum as a *preproinsulin* of 109 residues. The additional 23 residues are the hydrophobic signal peptide, which is attached at the N-terminal of proinsulin, as described in Chapter 10. After removal of these residues the proinsulin folds spontaneously into a conformation of lowest free energy. Oxidation of the free sulfhydryl groups to disulfide bonds occurs to stabilize the functional configuration.

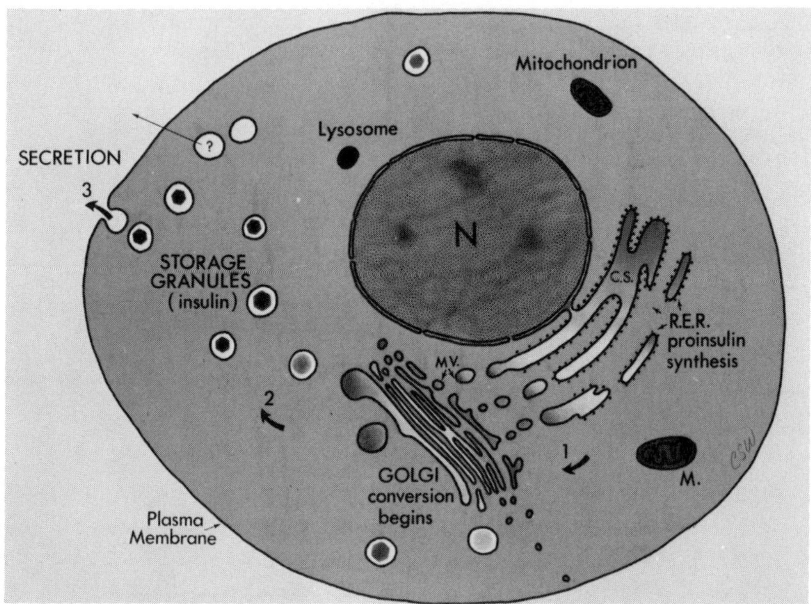

Figure 18-22
Scheme of morphologic organization of biosynthetic and secretory apparatus of pancreatic β-cell. *1,* Proinsulin transferred from endoplasmic reticulum to Golgi apparatus. Removal of C peptide occurs. *2,* Active insulin stored in granules. *3,* Secretion occurs by exocytosis.

From Steiner, D.F.: Diabetes **21** (suppl. 2):572, 1972.

As suggested by Fig. 18-22, the newly completed proinsulin is transferred from the endoplasmic reticulum to the peripheral elements of the Golgi apparatus (process 1). Here it is packaged into secretory granules, where the conversion of proinsulin to active hormone occurs by the proteolytic removal of the "connecting" peptide, or C-peptide. The first step is a scission at points 1 and 2 by a trypsinlike enzyme to produce diarginyl-insulin (Fig. 18-21). A carboxypeptidase B removes arginine residues 31 and 32 from the C-terminal of the B-chain (3). The insulin molecules crystallize with Zn ions and are stored within the granules until they are discharged from the cell by a process called exocytosis (processes 2 and 3, Fig. 18-22). Insulin is discharged into the pancreatic vein, which empties into the portal system. This means that the hormone first traverses the liver before entering the systemic blood supply.

In normal subjects the plasma insulin level increases rapidly from 10-20 μU/ml (0.4-0.8 ng/ml) to 100-140 μU/ml (4-6 ng/ml) after a high-carbohydrate meal (Fig. 18-23). It then decreases to a normal value as the blood glucose level recedes. On a daily basis the pancreas secretes 1 to 2 mg of insulin. The portal blood contains more insulin than peripheral venous blood; the level is approximately three times greater. In a single pass of the liver, 40% to 60% of the insulin of portal blood is removed. It is degraded in this and other tissues by the enzyme *insulinase*. Its half-life in the normal human adult is 6.5 to 9 minutes.

Glucagon Glucagon, a polypeptide consisting of 29 amino acids, is synthesized by the pancreatic α-cells. The primary structure of the hormone is shown in Fig. 18-24. In all likelihood it is synthesized as a "prohormone" and trimmed before secretion, in a manner similar to that described for insulin.

Homeostatic control of blood glucose levels Both insulin and glucagon are essential for the homeostatic control of blood glucose.

A major function of insulin is the control of carbohydrate metabolism. Normally we experience an increased blood glucose level shortly after each meal, a postpran-

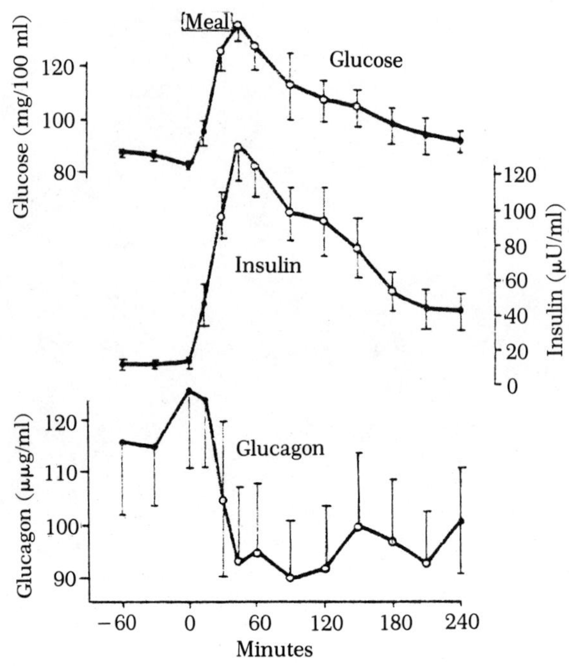

Figure 18-23
Changes in plasma insulin, glucagon and blood glucose following a large carbohydrate meal.

From Muller, W.A., et al.: N. Engl. J. Med. **283:**109, 1970. Reprinted by permission of the New England Journal of Medicine.

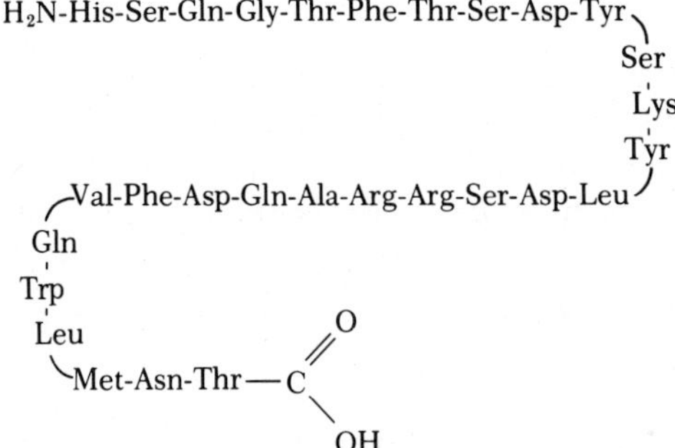

Figure 18-24
Primary structure of bovine glucagon.

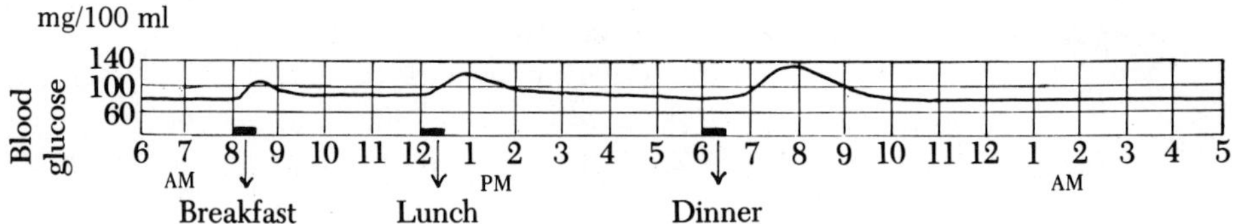

Figure 18-25
Typical blood glucose variations in a normal young man throughout the day.

dial hyperglycemia. This is illustrated in Figs. 18-23 and 18-25, which show typical blood glucose variations throughout the day. In some as yet unknown way the β-cells of the Islets of Langerhans sense the increased levels of circulating glucose and release insulin. This hormone suppresses the blood glucose level in a number of ways. One of the most important is by stimulating the active transport of glucose across cell membranes of muscle and adipose tissues but not of the liver.

As seen below, insulin binds at the cell membrane to a specific receptor protein. The exact mechanism of insulin action is still unclear. A moderating effect of the hormone has been observed on elevated cAMP levels, however, it has no observable influence on normal levels of the cyclic nucleotide. In the liver insulin has effects on both cytoplasmic and nuclear levels. The various metabolic systems affected by insulin in either a stimulatory or an inhibitory manner are shown in the accompanying diagram.

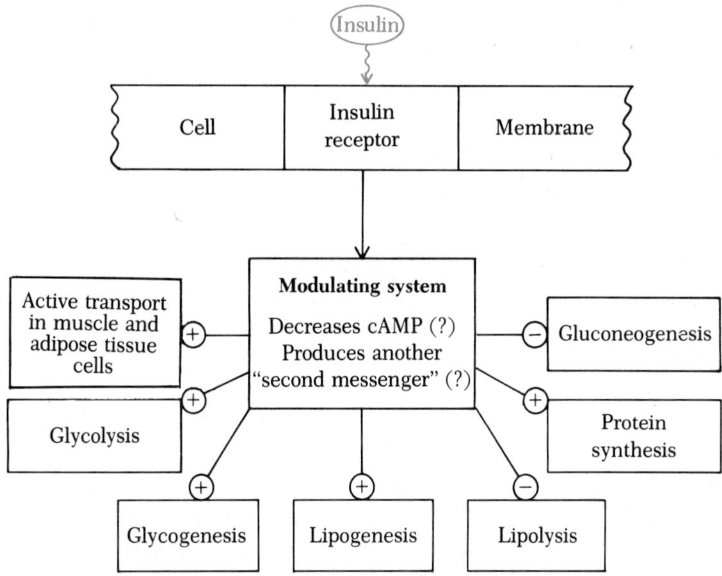

Hypoglycemia, on the other hand, causes a release of glucagon and a mobilization of glucose from the liver. Glucagon also binds to a specific receptor site located in the plasma membrane of liver cells. This complex results in the activation of adenyl cyclase and the production of cAMP. As seen below, glucagon stimulates the processes of glycogenolysis and gluconeogenesis in the liver cell. To provide the carbons necessary for gluconeogenesis, the transport of glucogenic amino acids is also stimulated.

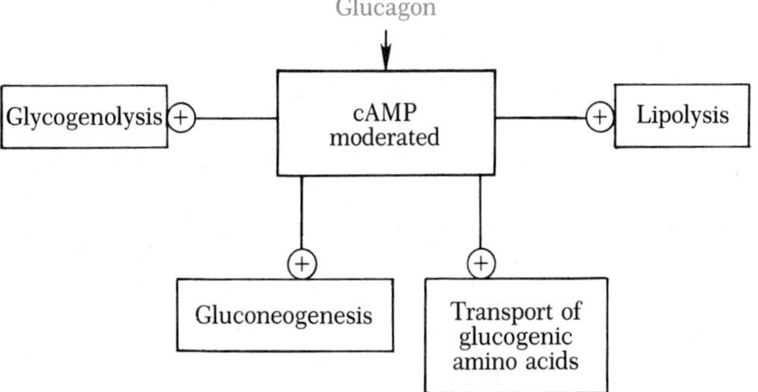

It is apparent therefore that glucose homeostasis is dependent on the ratio of insulin to glucagon (I/G ratio). Thus hypoglycemia lowers I/G; the increased glucagon is needed to mobilize (glycogenolysis) glucose and the insulin level is suppressed. In hyperglycemia (Fig. 18-23) the I/G ratio is increased; glucagon is suppressed to minimize glycogenolysis and insulin is elevated to promote the use and storage of glucose in the tissues.

A reciprocal control of insulin and glucagon is proposed by some investigators to occur at the level of the α- and β-cells. However, the concept of an intercellular communication whereby the glucagon level affects the release of insulin is controversial. The relative actions of the two hormones on the homeostatic control of blood glucose levels may be illustrated as follows:

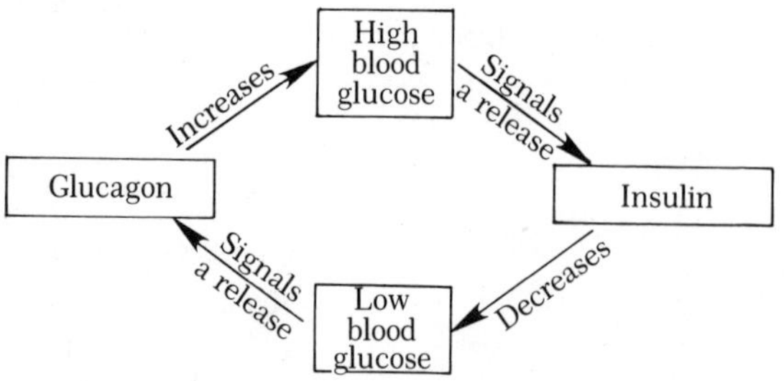

In the liver, insulin increases the use of glucose by inducing the synthesis of key glycolytic enzymes—glucokinase, phosphofructokinase, and pyruvate kinase. It suppresses the action of the enzymes pyruvate carboxylase, PEP carboxykinase, fructose 1,6-bisphosphatase and glucose 6-phosphatase. Furthermore, in liver and muscle tissue insulin exerts a glycogenic effect. This may be caused by reducing elevated cAMP levels and by converting glycogen synthase D to synthase I. Any reduction in cAMP results in an inactivation of phosphorylase, thereby leading to a suppression of glycogenolysis. In adipose tissue glucose yields the glycerol 3-phosphate needed for the formation of triacylglycerols. Likewise insulin opposes the activation of the hormone-sensitive lipase, probably because it reduces the level of cAMP. Thus insulin favors lipogenesis.

In addition to inducing the biosynthesis of specific enzymes, insulin generally increases all protein synthesis. In the liver it produces a positive nitrogen balance by increasing protein synthesis and by decreasing protein catabolism. In muscle it acts at the translational level and on the proper functioning of ribosomes.

Diabetes mellitus In diabetes mellitus either insufficient insulin levels, the production of an incompetent hormone, or a decrease in number and affinity of insulin receptor sites reduces the transport of glucose into muscle and adipose tissue. Since glucose is not rapidly taken up by these tissues in the absence of insulin, the inability to clear the blood of glucose is a typical characteristic of diabetes. The blood glucose level is elevated often to such a point that the renal threshold is exceeded and the glucose spills into the urine (glucosuria). This is accompanied by an increased excretion of urine, and therefore the diabetic has polyuria and an attendant dehydration and hemoconcentration. In diabetes adipose tissue is deficient in glucose (no transport) and therefore has a reduced supply of glycerol 3-phosphate for the synthesis of triacylglycerol. In the absence of insulin there is no control (negative) of the hormone-sensitive lipase, and consequently lipolysis occurs. Lipolysis results in a mobilization of fatty acids and their eventual accumulation in the liver. Hepatic fatty acyl CoA inhibits the citrate-cleavage enzyme and acetyl CoA carboxylase and hence suppresses lipogenesis. The absence of insulin favors glycogenolysis and also gluconeogenesis; thereby the hyperglycemic state is exaggerated. The body attempts to compensate for the lack of intracellular glucose by increasing its availability through gluconeogenesis even though there is an adequate supply of dietary carbohydrate. Since the animal is unable to use glucose for energy, it shifts to a lipid economy. Fatty acids are mobilized and oxidized in the liver to acetyl CoA. The acetyl CoA, however, cannot be oxidized to CO_2 and H_2O because of the inadequate supply of oxaloacetate. Consequently acetyl CoA is converted to acetoacetic and β-hydroxybutyric acids. Muscle tissue may use the ketone bodies for energy metabolism. Usually there is an excessive production

Figure 18-26

General types of blood glucose curves after the administration of regular insulin, protamine zinc insulin, and globin insulin.

Courtesy D.D. Searle, Burroughs Wellcome & Co., Inc., Tuckahoe, N.Y.

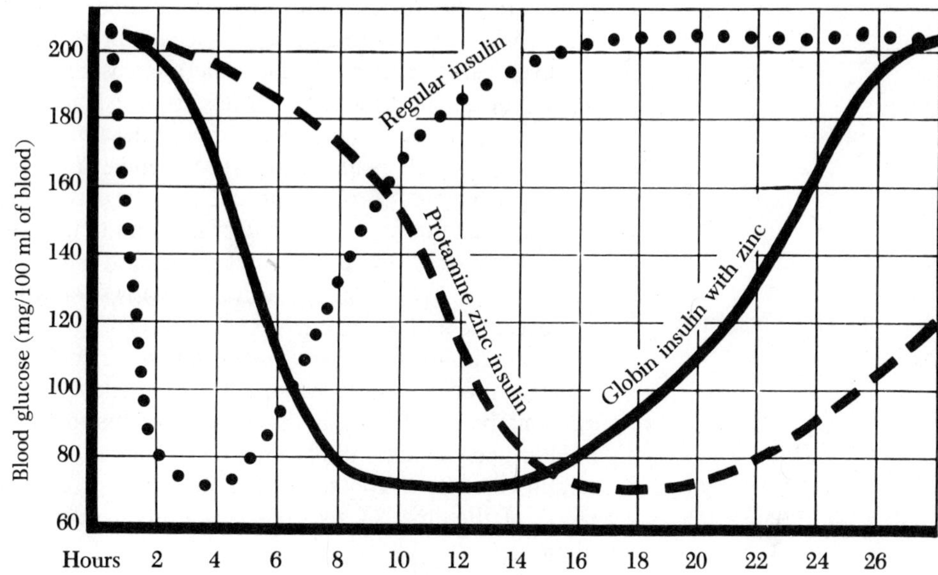

and loss of these substances from the blood by way of the lungs and kidneys. Acetoacetate readily breaks down into acetone, which is exhaled in the breath. Hence a characteristic of the diabetic is his or her so-called acetone breath. Urinary excretion of acetoacetate and β-hydroxybutyrate results in the loss of Na^+, and a state of acidosis occurs. The diabetic attempts to compensate by hyperventilation (loss of CO_2).

The administration of insulin reverses these effects. Fig. 18-26 shows the effects of various preparations of insulin on the blood glucose level. Insulin itself rapidly decreases the blood glucose but its effect is short lived. Protamine zinc insulin, a salt of insulin, has a protracted effect. In the usual treatment of a diabetic patient an adjustment of diet and insulin dosage is required to keep the blood glucose level as near normal as possible. Often a mixture of regular and slow-acting insulins is prescribed—regular insulin for its quick effect and slow-acting insulin to sustain its action. The total caloric and nitrogen requirements are calculated and a diet is carefully planned. In the event of a reduction in the numbers of tissue receptor sites or of a lowering of their affinity for the hormone, an insulin-resistant form of diabetes results. Under these circumstances greater than normal dosages of insulin are required to control the diabetic condition.

Suitable oral hypoglycemic agents have long been sought. One of the best is tolbutamide (1-butyl-3-tolylsulfonylurea). With proper dosage it appears to be nontoxic. Since sulfonylureas have no effect in pancreatectomized animals or in diabetic patients with functionally inactive islets of Langerhans, they probably stimulate the release of endogenous insulin from the pancreatic β-cells.

Blood calcium levels

As in the case of blood glucose, the proper concentration of calcium is extremely important to the well-being of the organism. Of the total calcium in the blood serum, 10 to 11 mg per 100 ml, approximately one half is bound to serum proteins.

Actually, only the free calcium ions are biologically active. They are important for controlling the irritability of nerve cells, blood clotting, contraction and relaxation of muscle, adhesion between cells, and production of milk, in addition to their more obvious function in the mineralization of bones and teeth. Hypocalcemia leads to tetany, whereas hypercalcemia may cause bradycardia, ventricular arrhythmias, and even ventricular fibrillation. The homeostatic control of the calcium level is dependent on the function of three tissues: bone, kidneys, and intestines. This regulatory process is controlled by two if not three hormones: *parathormone* and *vitamin D*, which respond to hypocalcemia, and *calcitonin*, which appears to respond to hypercalcemia. Thus just as in the case of insulin and glucagon we have counterbalancing hormones to decrease and to increase blood calcium levels.

Parathormone If the parathyroid glands are removed, as in the case of a malignancy, tetany may occur (parathyroid tetany). Parathyroidectomy is accompanied by a decrease in the calcium level of the blood. The tetanic symptoms are related to the low serum–calcium ion concentration, since the intravenous injection of calcium salts relieves these symptoms very quickly. *Parathormone* (PTH) is a hypercalcemic hormone; that is to say, it functions to elevate the blood calcium level. PTH is a polypeptide; the human hormone has a molecular weight of 9300 and is composed of 84 amino acids (Fig. 18-27). A biologically active polypeptide composed of the first 34 residues has been synthesized by the solid-phase technique.

Calcitonin Calcitonin, or thyrocalcitonin (TCT), is a polypeptide (32 residues) produced by the parafollicular "C" cells of the thyroid gland. Its molecular weight is

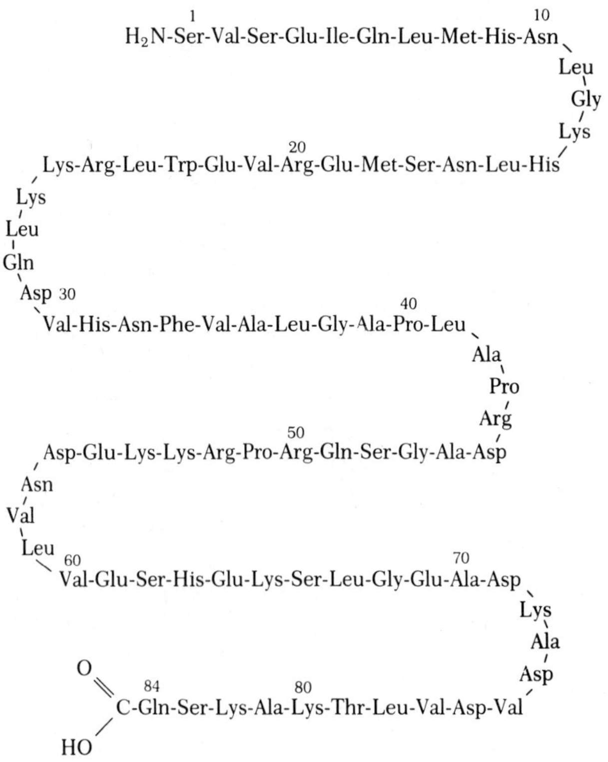

Figure 18-27
The amino acid sequence for human parathormone.

1 10
H_2N-Ser-Val-Ser-Glu-Ile-Gln-Leu-Met-His-Asn
Leu
Gly
Lys
20
Lys-Arg-Leu-Trp-Glu-Val-Arg-Glu-Met-Ser-Asn-Leu-His
Lys
Leu
Gln
Asp 30 40
Val-His-Asn-Phe-Val-Ala-Leu-Gly-Ala-Pro-Leu
Ala
Pro
Arg
50
Asp-Glu-Lys-Lys-Arg-Pro-Arg-Gln-Ser-Gly-Ala-Asp
Asn
Val
Leu 60 70
Val-Glu-Ser-His-Glu-Lys-Ser-Leu-Gly-Glu-Ala-Asp
Lys
Ala
O Asp
84 80
C-Gln-Ser-Lys-Ala-Lys-Thr-Leu-Val-Asp-Val
HO

3600. The primary structure for human calcitonin is show in Fig. 18-28. The secretion of calcitonin is stimulated by an increase in the blood calcium level of as little as 0.15 to 0.5 mg per 100 ml. This hormone increases bone calcification (inhibits bone resorption), thereby decreasing the blood calcium level. However, its physiological significance in the human is uncertain.

Vitamin D: 1,25-dihydroxycholecalciferol Vitamin D has long been recognized as an essential factor in bone ossification and the maintenance of blood calcium levels. However, until recently it had been considered to be exclusively an essential dietary factor. It is now apparent that we may derive sufficient amounts of vitamin D by endogenous synthesis provided that there is adequate exposure of the skin to sunlight. Since the active factor in ossification, namely, 1,25-dihydroxycholecalciferol, can be entirely synthesized, it needs to be considered as a hormone and not as a dietary factor.

As seen from Fig. 18-29, cholecalciferol is synthesized from 7-dehydrocholesterol (Chapter 12). The rupture of the bond between C-6 and C-7 occurs in the skin upon exposure to ultraviolet radiation. Body heat results in the rearrangement shown to yield the vitamin D_3. At 37° C only about one half of the previtamin D_3 is converted within 24 hours. The previtamin remains in the skin, whereas the vitamin enters the bloodstream, where it is bound to a vitamin D_3-plasma–binding protein (transcalciferin). As suggested in Fig. 18-29, dietary vitamin enters the circulating pool of cholecalciferol.

In the liver the vitamin D_3 is hydroxylated by a 25-*hydroxylase* associated with the endoplasmic reticulum. Hydroxylation is catalyzed by a monooxygenase and requires oxygen and a supply of NADH. Complexed to a specific plasma protein, the 25-hydroxycholecalciferol is carried to the kidneys. Here it is further hydroxylated to the hormonally active form, namely, 1, 25-dihydroxycholecalciferol. The 1-hydroxylase involved is located in the mitochondria.

Function of PTH and 1,25-dihydroxycholecalciferol Hypocalcemia stimulates the release of PTH from the parathyroid glands. One of the actions of this hormone is the activation of renal 1-hydroxylase and consequently an elevated production of 1,25-dihydroxycholecalciferol. The active form of vitamin D_3 functions at three sites (Fig. 18-30) to increase: (1) the intestinal absorption of calcium, (2) the renal reabsorption of calcium from the tubular urine, and (3) the mobilization of calcium from bone. PTH acts in concert with the cholecalciferol at the site of intestinal uptake of calcium.

The 1,25-dihydroxycholecalciferol alone functions to stimulate the uptake of calcium by intestinal mucosal cells. The hormone enters the cells by diffusion and is bound by a specific cytoplasmic binding protein. Following translocation to the

Figure 18-28
Primary structure of human calcitonin.

From Dayhoff, M.O.: Atlas of protein sequence and structure, vol. 5, 1972.

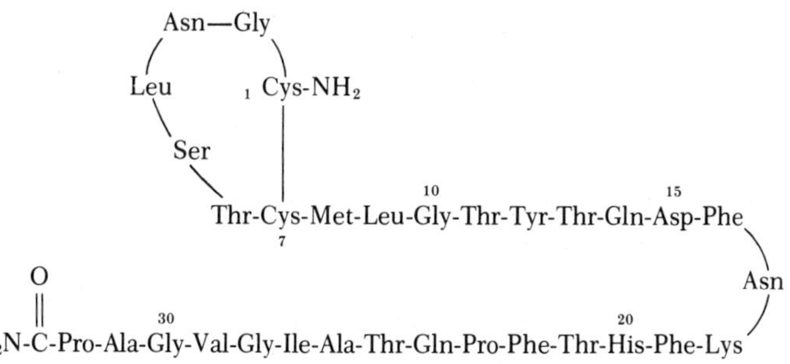

Asn—Gly
Leu 1 Cys-NH_2
Ser
Thr-Cys-Met-Leu-Gly-Thr-Tyr-Thr-Gln-Asp-Phe (7, 10, 15)
Asn
H_2N-C(=O)-Pro-Ala-Gly-Val-Gly-Ile-Ala-Thr-Gln-Pro-Phe-Thr-His-Phe-Lys (30, 20)

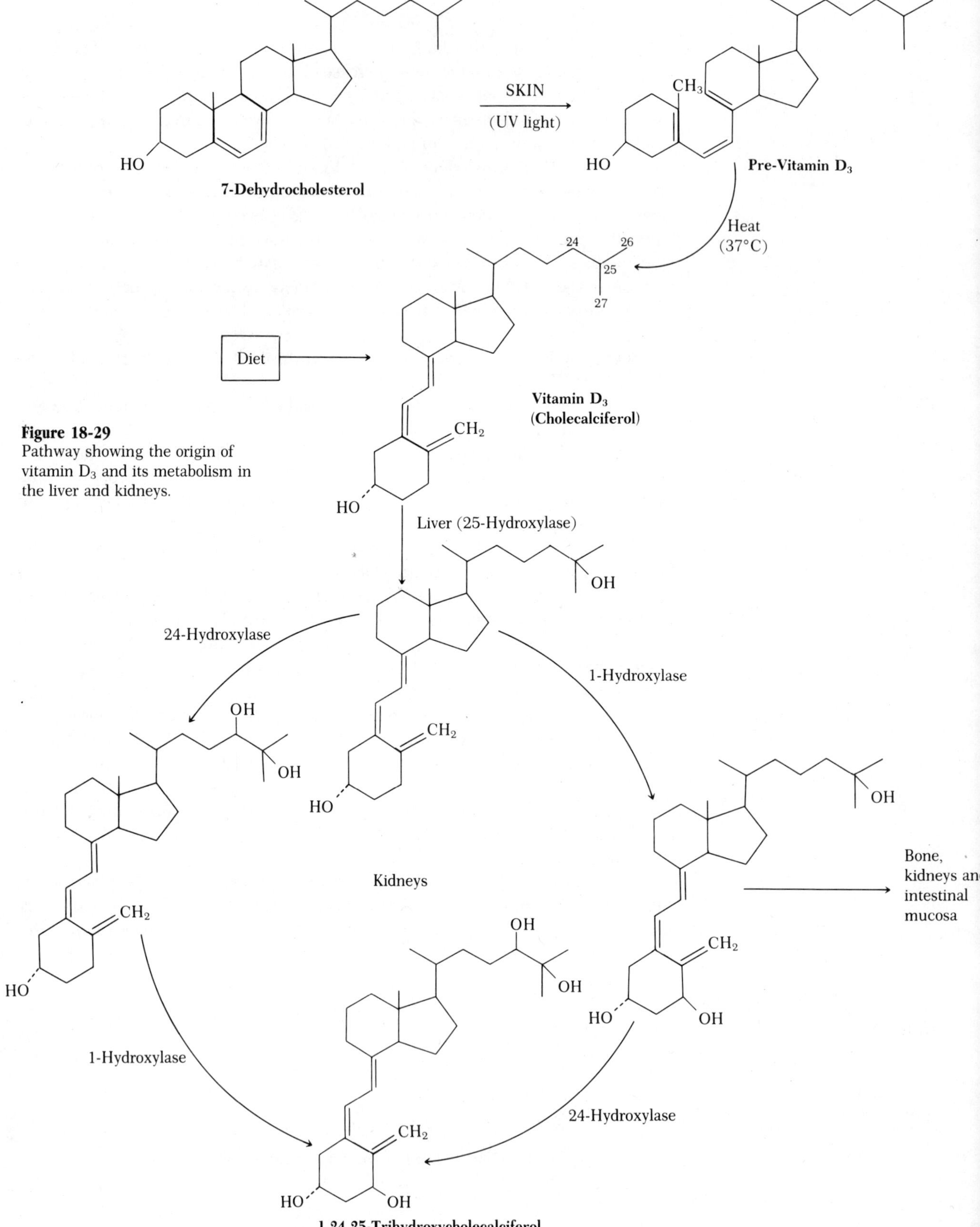

Figure 18-29
Pathway showing the origin of vitamin D_3 and its metabolism in the liver and kidneys.

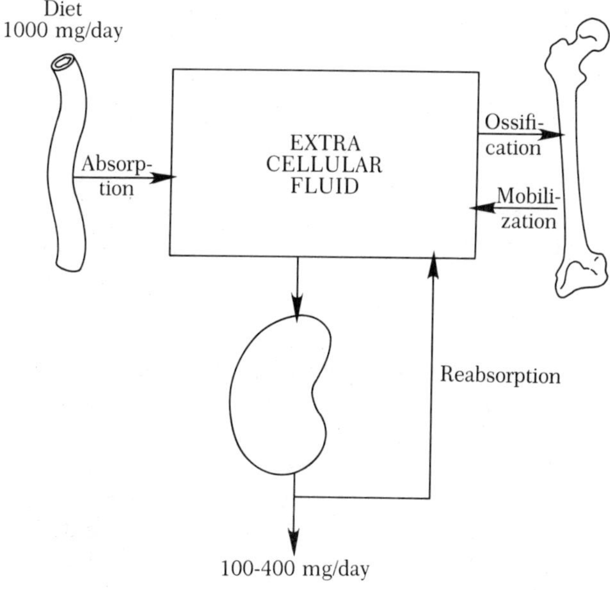

Figure 18-30
Three sites involved in the maintenance of the blood calcium levels. These are absorption of calcium from the intestinal contents, the reabsorption of calcium from the lumenal urine of the renal tubules, and the mobilization of calcium from bone.

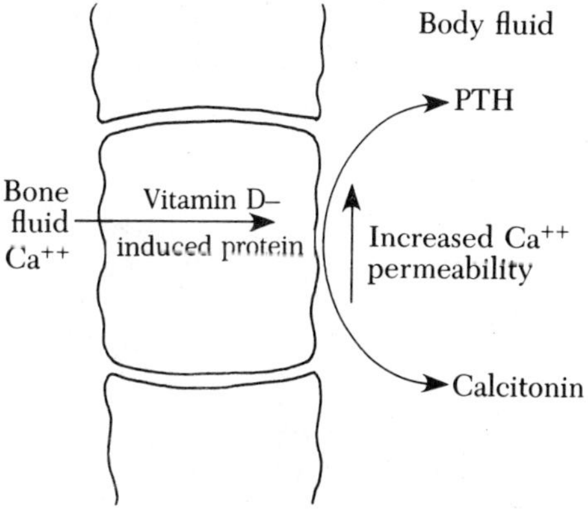

Figure 18-31
Diagram showing role of vitamin D, PTH, and TCT in controlling absorption and resorption of calcium from bone.

From DeLuca, H.: Fed. Proc. **33:**2211, 1974.

nucleus, it stimulates the biosynthesis of a specific "calcium-transport" protein. This protein functions in the microvilli of the mucosal cell as a Ca^{++}-dependent ATPase. In other words, calcium ions are pumped into the cell against a concentration gradient at the expense of ATP.

The mobilization of calcium from the bone requires the concerted action of PTH and 1,25-dihydroxycholecalciferol. As indicated in Fig. 18-31, the bone itself is surrounded by a curtain of cells. PTH is thought to act by stimulating adenyl cyclase and cAMP production. In the osteoclasts, or bone-degrading cells, this in some way leads to an increased solution of calcium ions from the hydroxyapatite crystals. The function of the 1,25-dihydroxycholecalciferol is to get the calcium transported from the bone fluid into the extracellular fluid. Calcitonin may be involved in counteracting the calcium-mobilizing action of PTH (Fig. 18-31).

At the kidneys some 10,000 mg of calcium are filtered daily, yet only about 100 mg are actually lost in the urine. Therefore the difference is actively reabsorbed by the distal tubule cells and restored to the bloodstream (Fig. 18-30). This active transport process is also controlled by the concerted action of PTH and the 1,25-

dihydroxycholecalciferol. Here again protein synthesis may be involved, yielding an increased production of a carrier for the calcium ions (Ca^{++}-dependent ATPase).

Hormones of the gastrointestinal tract

Of the various polypeptide hormones of the gastrointestinal tract, the amino acid sequences of *gastrin, secretin,* and *pancreozymin-cholecystokinin* are known. These hormones are produced by endocrine cells of the gastrointestinal system. Their sites of origin and function are illustrated in Fig. 18-32.

Gastrin

Gastrin is produced by the mucosal cells of the pyloric region of the stomach. Its formation is stimulated by proteins or polypeptides present in or derived from food or possibly by hydrochloric acid. Mechanical stimulation caused by distension of the stomach also results in the production of gastrin. The hormone is absorbed into the bloodstream and carried to the parietal cells of the fundic region of the stomach (Fig. 18-32), causing them to secrete hydrochloric acid.

Gastrin has 17 amino acids in its primary structure (Fig. 18-33). The C-terminal Trp-Met-Asp-Phe-$CONH_2$ sequence is essential for function. Actually two forms (gastrin I and II) have been found; gastrin III is identical to I except that residue 12 (tyrosine) is sulfated.

Gastrin's ability to stimulate the production of hydrochloric acid is shared by a much simpler substance, histamine. It is suggested that the gastrin itself actually produces histamine, which then stimulates the production of hydrochloric acid. In

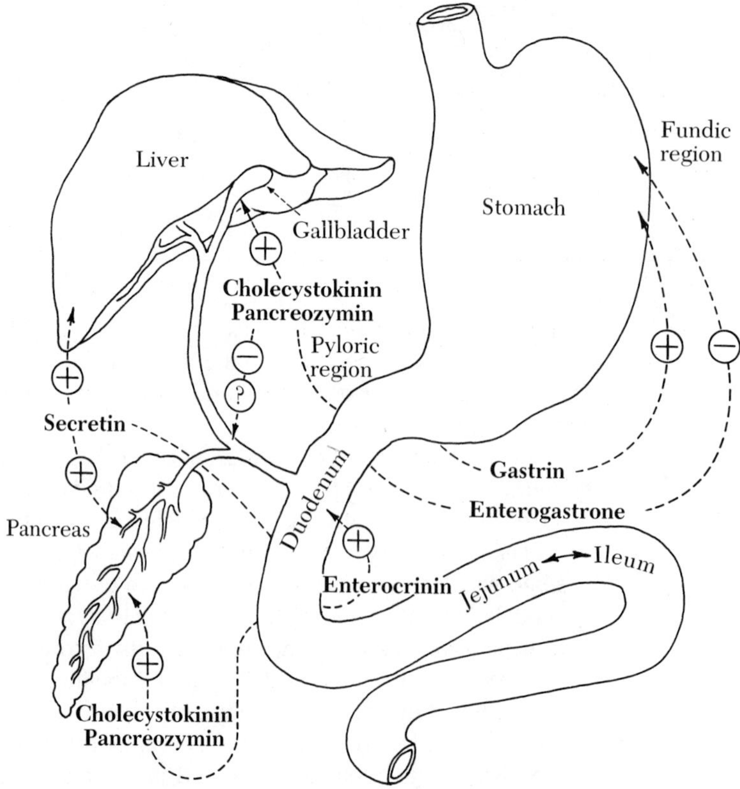

Figure 18-32
Diagram showing place of origin and site of action of five gastrointestinal hormones. Stimulatory action is indicated by plus sign; inhibitory action is indicated by minus sign.

Adapted from Gorbman, A., and Bern, H.A.: A textbook of comparative endocrinology, New York, 1962, John Wiley & Sons, Inc.

support of this concept, gastrin increases the amount of *histidine decarboxylase* in the mucosal tissue by 2 to 12 times. This increase occurs even in the presence of actinomycin D, so that it appears not to be an inductive effect but rather a stimulation of protein synthesis at the translational level. The overall sequence of events can be represented as follows:

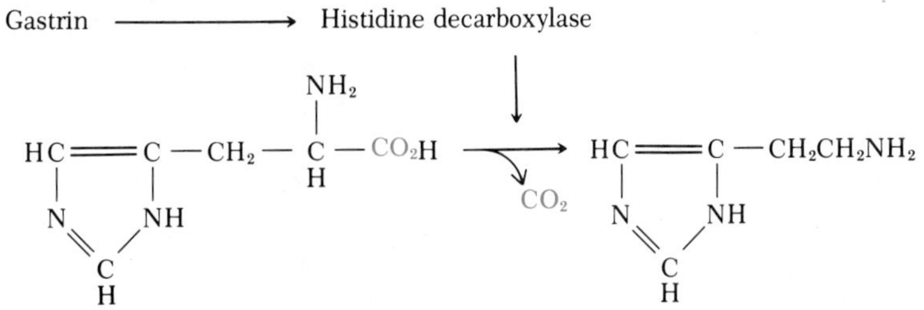

Gastrin ⟶ Histidine decarboxylase

$$\text{Histidine} \xrightarrow{\text{histidine decarboxylase}} \text{Histamine} + CO_2$$

Histidine: $HC{=}C{-}CH_2{-}CH(NH_2){-}CO_2H$ (imidazole ring: HC=C, N, NH, CH)

Histamine: $HC{=}C{-}CH_2CH_2NH_2$ (imidazole ring: HC=C, N, NH, CH)

Histidine **Histamine**

Secretin

Secretin, a polypeptide having 27 amino acid residues, is formed in and liberated from the duodenal mucosal cells by the hydrochloric acid in the acid chyme (Fig. 18-32). It is carried by the bloodstream to the pancreas (see Fig. 12-32), which it stimulates to produce a flow of pancreatic juice rich in bicarbonate but low in enzymes. Like gastrin and vasopressin (other hormones that control the flow of water in biological systems), secretin has a C-terminal amide group. This hormone also stimulates the flow of intestinal juice and is one of the factors that increase the secretion of bile by the liver.

Pancreozymin-cholecystokinin

Although historically this hormone represents two separate physiological functions, it now appears to be a single polypeptide hormone having 33 amino acids. It is produced by the mucosa of the upper small intestinal tract (see Fig. 12-32). Pancreozymin stimulates the pancreas to produce a pancreatic juice rich in enzymes (or their zymogens) as well as in bicarbonate. These include trypsinogen, chymotrypsinogen, lipase, nucleases, and procarboxypeptidase. The cholecystokinin effect is to stimulate the contraction of the gallbladder and to dilate the sphincter of Oddi.

Enterogastrone

Enterogastrone is apparently a polypeptide produced by the duodenal mucosa (Fig. 18-32). Its formation is associated with the presence, in the duodenum, of fat

Figure 18-33
Amino acid sequences for human gastrin (upper) and porcine secretion (lower). *Glx,* Pyroglutamyl residue.

1 10
Glx-Gly-Pro-Trp-Leu-Glu-Glu-Glu-Glu-Glu-Ala-Tyr-Gly-Trp-Met-Asp-
$H_2N{-}C({=}O){-}$Phe 17

1 10
H_2N-His-Ser-Asp-Gly-Thr-Phe-Thr-Ser-Glu-Leu-Ser-Arg-
Leu
Arg
20
Leu-Gly-Gln-Leu-Leu-Arg-Gln-Leu-Arg-Ala-Ser-Asp
27
Val-$CONH_2$

and other substances derived from food; its function is to inhibit gastric secretion and gastric motility. When fat reaches the duodenum, it causes the secretion of enterogastrone, which then slows up gastric digestion and motility, including hunger contractions. There is a diminution in the volume of juice secreted, with a lower concentration of hydrochloric acid and a smaller amount of pepsin. The effect is to permit digestion to be accomplished more completely. A substance of similar effect has been isolated from human urine. It has been called *urogastrone* and may be an excretory product of enterogastrone. In addition to inhibiting gastric secretion, it also seems to inhibit pancreatic secretion.

Other alleged hormones

A protein called *parotin* is elaborated by the salivary gland and is claimed to stimulate calcification of the teeth. It is believed to decrease the calcium and increase the phosphorus level of serum.

The flow of intestinal juice is also under the control of another intestinal hormone called *enterocrinin* or *duocrinin*. It is a polypeptide but distinct from secretin in that it stimulates the secretion of both fluid and enzymes by the intestinal mucosa (Fig. 18-32). *Hepatocrinin* is believed to stimulate the secretion of a dilute, low-salt type of bile; *motilin,* a polypeptide of 22 amino acids, stimulates gastric motility; *vasoactive intestinal peptide,* with 28 amino acids, resembling secretin in structure, inhibits the production of gastric secretin as well as others. The intestinal mucosa also is a source of glucagon (enteroglucagon), which is released by the ingestion of glucose and fat. Intestinal glucagon then stimulates the release of insulin.

Renal hormones

Renal tissue is capable of synthesizing an enzyme called *renin,* which in turn functions to produce the hormone *angiotensin*. In the past this hormone has been associated with the control of blood pressure, a function that is still uncertain. As already stated, the renin-angiotensin system (Fig. 18-34) now appears to play an important role in the control of Na^+ reabsorption by the kidneys. Thus angiotensin II acts to stimulate aldosterone secretion by enhancing the conversion of cholesterol to pregnenolone. There are a number of factors that release renin into the bloodstream. These are arterial pressure, decreased blood Na^+ concentration, and the sympathetic nervous system.

Figure 18-34
Renin-angiotensin system.

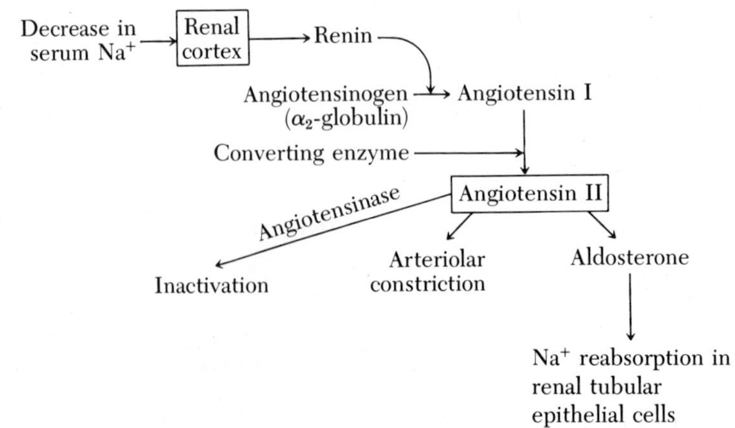

The renin-angiotensin system is unique because the actual production of the hormone occurs in the bloodstream by the hydrolysis of a plasma protein (α_2-globulin) called angiotensinogen. As seen in Fig. 18-34, renin releases a decapeptide, angiotensin I. This polypeptide is only partially active. The actual hormone, an octapeptide, angiotensin II, is produced from angiotensin I (Fig. 18-35) by the action of a "converting enzyme" found in the lungs. This involves the removal of histidine and leucine from angiotensin I. Unlike the other hormones considered, this one is produced in the vascular system rather than in special glands. The level of hormone is controlled by *angiotensinase,* an enzyme found in kidney and other tissues that inactivate it (Fig. 18-34). Angiotensin II also acts to increase the heartbeat and is a vasoconstrictor. Its hypertensive activity is some 40 times that of norepinephrine.

Another renal hormone is *erythropoietin,* a glycoprotein with a molecular weight of 50,000 to 60,000 having about 30% carbohydrate. This hormone stimulates the proliferation of red blood cells in both bone marrow and spleen. The renal secretion of erythropoietin, or the *erythrocyte-stimulating factor* (ESF), is triggered primarily by tissue anoxia and anemia. Androgenic hormones and cobalt also stimulate its secretion or release. One concept is that ESF is secreted as an inactive protein, possibly an enzymelike substance termed the *renal erythropoietic factor* (REF). Like renin, REF then converts a plasma globulin to active erythropoietin.

$$\text{Plasma globulin} \xrightarrow{\text{REF}} \text{Active erythropoietin}$$

Evidence for the renal involvement in erythropoietin formation is that (1) polycythemia is characteristically associated with renal neoplasms and cysts, (2) anemia is commonly found in renal disease, and (3) plasma levels of erythropoietin decrease in nephrectomized animals or animals with bilateral ureteral ligation.

Erythropoietin has been prepared in highly active form from the plasma of anemic sheep and from the urine of human subjects with certain types of anemia. It is an α_1-globulin, a glycoprotein containing equal amounts of glucosamine and sialic acid. Its activity is destroyed by proteolytic enzymes as well as by hyaluronidase and neuraminidase.

The exact mechanism of action is uncertain. The target cell is the stem cell (hemohistioblast) of bone marrow and other erythroid tissue. Erythropoietin stimulates stem-cell differentiation, thereby causing in increase in the number of proerythroblasts, followed by increases in nucleated erythrocytes and finally in reticulocytes and mature erythrocytes in the peripheral circulation.

The action of erythropoietin is inhibited by actinomycin D; therefore its action is

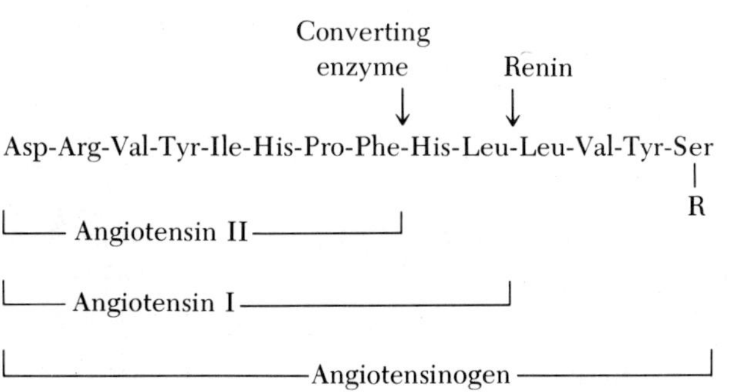

Figure 18-35
Conversion of angiotensinogen to angiotensins I and II.

at the level of transcription. It stimulates the synthesis of a large RNA (150S) by bone marrow cells within a few minutes after the addition of erythropoietin in vitro. Increases in smaller RNA molecules then occur rapidly, probably representing ribosomal, transfer, and messenger RNAs. Erythropoietin enhances the δ-aminolevulinic acid synthetase by 4 to 5 times normal in the livers of polycythemic mice, the δ-aminolevulinic acid dehydrase by 11 times normal, and the iron protoporphyrin–chelating enzyme by 23 times normal.

Pineal gland

The pineal gland synthesizes two substances, serotonin and melatonin, that have biological functions. Melatonin may have a hormonal activity. Both serotonin and melatonin are derived from dietary tryptophan.

As seen in Fig. 18-36, tryptophan is first hydroxylated to the 5-hydroxy derivative by the action of the monooxygenase tryptophan hydroxylase. This step involves oxygen and a supply of tetrahydrobiopterin. An *aromatic amino acid decarboxylase* then converts the 5-hydroxytryptophan to 5-hydroxytryptamine, or serotonin. *Serotonin-N-acetyl transferase* activity is increased when cAMP is present, as a result of the activation of adenyl cyclase by norepinephrine. The catecholamine is released from sympathetic nerve endings in the pineal gland when the experimental animal is kept in the dark. The *N*-acetyl serotonin is then converted by hydroxyindole-*O*-methyl transferase (HIOMT) to melatonin. When light enters the eye, however, the release of norepinephrine is prevented, and therefore the product produced by the pineal is primarily serotonin rather than melatonin.

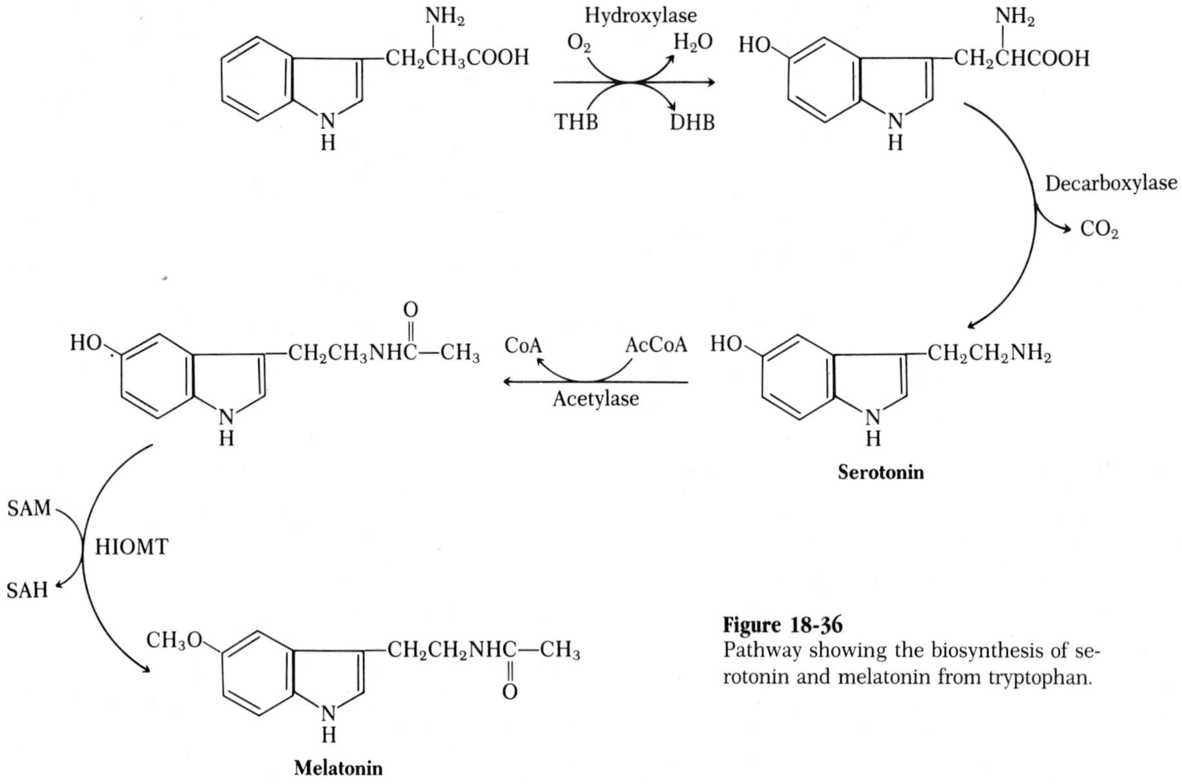

Figure 18-36
Pathway showing the biosynthesis of serotonin and melatonin from tryptophan.

There are also diurnal fluctuations in the production of serotonin and melatonin. This represents a circadian rhythm, or biological clock. Since melatonin inhibits the release of LH, the pineal gland may play an important role in controlling reproductive processes.

Suggested readings

Axelrod, J.: Noradrenaline: fate and control of its biosynthesis, Science, **173:**598, 1971.

Axelrod, J., and Weinshilboum, R.M.: Catecholamines, N. Engl. J. Med. **287:**237, 1972.

Axelrod, J.: The pineal gland: a neurochemical transducer, Science **184:**1341, 1974.

Bethune, J.E.: The adrenal cortex: a Scope monograph, Kalamazoo, Mich., 1974, The Upjohn Co.

Brownstein, M.J., Russell, J.T., and Gaines, H.: Synthesis, transport and release of posterior pituitary hormones, Science **207:**373, 1980.

Burke, G., Chang, L.L., and Szabo, M.: Thyrotropin and cyclic nucleotide effects on prostaglandin levels in isolated thyroid cells, Science **180:**872, 1973.

Cahil, G.F.: Glucagon, N. Engl. J. Med. **288:**157, 1973.

Catt, K.J.: Hormones in general, I, Lancet **1:**763, 1970.

Catt, K.J.: Pituitary function, II, Lancet **1:**827, 1970.

Catt, K.J.: Growth hormone, III, Lancet **1:**933, 1970.

Catt, K.J.: Reproductive endocrinology, IV, Lancet **1:**1097, 1970.

Catt, K.J.: Adrenal cortex, V, Lancet **1:**1275, 1970.

DeLuca, H.F.: The role of vitamin D and its relationship to parathyroid hormone and calcitonin, Recent Progr. Horm. Res. **27:**479, 1972.

DeLuca, H.F.: Vitamin D: the vitamin and the hormone, Fed. Proc. **33:**2211, 1974.

DeLuca, H.F.: The hormonal nature of vitamin D function hormones and cell regulation. In Dumont, J., and Numez, J., editors: Hormones and cell regulation, vol. 2, Amsterdam, 1978, Elsevier/North-Holland Publishing Co., p. 249.

Dorfman, R.I., and Sharma, D.C.: An outline of the biosynthesis of corticosteroids and androgens, Steroids **6:**229,1965.

Dumont, J.E.: The action of thyrotropin on thyroid metabolism, Vitam. Horm. **29:**287, 1971.

Fisher, J.W.: Kidney hormones, New York, 1971, Academic Press, Inc.

Flier, J.S., Kahn, C.R., and Roth, J.: Receptors, antireceptor antibodies and mechanisms of insulin resistance, N. Engl. J. Med. **300:**413, 1979.

Frieden, E., and Lipner, H.: Biochemical endocrinology of the vertebrates, Englewood Cliffs, N.J., 1971, Prentice-Hall, Inc.

Gordon, A.S.: Erythropoietin, Vitam. Horm. **31:**l06, 1973.

Gorman, R.R.: Modulation of human platelet function by prostacyclin and thromboxane A_2, Fed. Proc. **38:**83, 1979.

Guillemin, R., and Burgus, R.: The hormones of the hypothalamus, Sci. Am. **227:**24, 1972.

Habener, J.F., and Kronenberg, H.M.: Parathyroid hormone biosynthesis: structure and function of biosynthesis precursors, Fed. Proc. **77:**2561, 1978.

Halick, M.F., and Clark, M.B.: The photobiogenesis and metabolism of vitamin D, Fed. Proc. **37:**2567, 1978.

Halick, M.F., Garabedian, M., and DeLuca, H.F.: 1,25-Dihydroxycholecalciferol: metabolite of vitamin D_3 active in bone in anephric rats, Science **176:**1146, 1972.

Hinman, J.W.: Prostaglandins, Annu. Rev. Biochem. **41:**161, 1972.

Jensen, E.V., and DeSombre, E.R.: Mechanism of action of the female sex hormones, Annu. Rev. Biochem. **41:**203, 1972.

Jost, J.P., and Rickenberg, H.V.: Cyclic AMP, Annu. Rev. Biochem. **40:**741, 1971.

Kemmler, W., and others: On the biosynthesis, intracellular transport and mechanisms of conversion of proinsulin to insulin and C-peptide, Diabetes **21** (Suppl. 2): 572, 1972.

Labrie, F., and others: Mechanism of action of hypothalamic hormones in the anterior pituitary, TIBS **4:**158, 1979.

Lands, W.: Biosynthesis and metabolism of prostaglandins, Annu. Rev. Physiol. **41:**633, 1979.

Lefkowitz, R.J.: Isolated hormone receptors, N. Engl. J. Med. **288:**1061, 1973.

Litwack, G., editor: Biochemical actions of hormones, vols. 1 and 2, New York, 1970-1972, Academic Press, Inc.

Martin, J.B.: Neural regulation of growth hormone secretion, N. Engl. J. Med. **288:**1384, 1973.

Maurer, A.C.: The therapy of diabetes, Am. Sci. **67:**422, 1979.

McGiff, J.C.: New developments in prostaglandin and thromboxane research (symposium of seven papers), Fed. Proc. **78:**64, 1979.

Moncada, S., and Vane, J.R.: The role of prostacyclin in vascular tissue, Fed. Proc. **38**:66, 1979.

Mulrow, P.J.: The adrenal cortex, Annu. Rev. Physiol. **34**:409, 1972.

Notkins, A.L.: The causes of diabetes, Sci. Am. **241**:62, 1979.

O'Malley, B.W.: Mechanism of action of steroid hormones, N. Engl. J. Med. **284**:370, 1971.

Peach, M.J.: Renin-angiotensin system: biochemistry and mechanisms of action, Physiol. Rev. **57**:313, 1977.

Pierce, J.G.: Eli Lilly lecture: the subunits of pituitary thyrotropin—their relationship to other glycoprotein hormones, Endocrinology **89**:1331, 1971.

Pitot, H.C., and Yatvin, M.B.: Interrelationships of mammalian hormones and enzyme levels in vivo, Phys. Rev. **53**:228, 1973.

Pitt-Rivers, R.V., and Trotter, W.R., editors: The thyroid gland, Washington, D.C., 1964, Butterworth, Inc.

Potts, J.T. Jr., and others: The chemistry of parathyroid hormones and the calcitonins, Vitam. Horm. **29**:41, 1971.

Rasmussen, H.: Cell communication, calcium ion, and cyclic adenosine monophosphate, Science **172**:404, 1970.

Rasmussen, H., and Pechet, M.M.: Calcitonin, Sci. Am. **223**:42, 1970.

Robison, G.A., Butcher, R.W., and Sutherland, E.W.: Cyclic AMP, New York, 1971, Academic Press, Inc.

Samuelsson, B., and others: Prostaglandins and thromboxanes, Annu. Rev. Biochem. **47**:997, 1978.

Schally, A.V., Arimura, A., and Kastin, A.J.: Hypothalamic regulatory hormones, Science **179**:341, 1973.

Schwartz, N.B., and McCormack, C.E.: Reproduction: gonadal function and its regulation, Annu. Rev. Physiol. **34**:425, 1972.

Spanos, E.: New aspects on the regulation of vitamin D metabolism. In Dumont, J., and Numez, J., editors: Hormones and cell regulation, vol. 2, Amsterdam, 1978, Elsevier/North-Holland Publishing Co., P. 271.

Steiner, D.F.: Insulin today (Banting Memorial Lecture 1976) Diabetes **26**:322, 1977.

Sutherland, E.W.: Studies on the mechanism of hormone action, Science **177**:401, 1972.

Tata, J.R.: Hormonal regulation of growth and protein synthesis, Nature **219**:331, 1968.

Unger, R.H., and Orci, L.: The role of glucagon in the endogenous hyperglycemia of diabetes, Annu. Rev. Med. **28**:119, 1977.

Wasserman, R.H., and Corradino, R.A.: Vitamin D, calcium, and protein syntheses, Vitam. Horm. **31**:43, 1973.

Wilson, J.D.: Recent studies on the mechanism of action of testosterone, N. Engl. J. Med. **287**:1284, 1972.

Chapter 19

Molecular aspects of transport

Communication between cells and tissues in a multicellular organism has already been described as a process requiring hormones (Chapter 18); in many instances they control metabolism by regulating the rate of entry of metabolites into tissue cells. An example cited in Chapter 11 is that of insulin, the hormone which regulates the transport of glucose into certain tissues. Thus the lowering of blood glucose immediately following a meal is largely the consequence of an increased transport of sugar into the adipose and muscle tissue cells. Likewise, amino acids are transported into cells (Chapter 13), especially those of the liver and muscles, by processes controlled by insulin and glucagon. Now it appears that there are genetic diseases involving defects in transport mechanisms. For example, there are defects associated with the transport of amino acids, such as Hartnup's disease and cystinuria. Important functions of transport mechanisms are the extraction and concentration by the cells of metabolic fuels and nutrients (Chapter 20). Hexoses such as glucose and fructose and amino acids, all of which are products of the digestion of foods, must be transported from the intestinal lumen into the bloodstream across a layer of epithelial cells in the mucosal lining.

Another function of transport in all cells is the maintenance of ionic concentrations within narrow limits and therefore the regulation of the cell volume. In respiration, oxygen and carbon dioxide dissolved in the blood plasma enter and leave the red blood cells by diffusion; bicarbonate and chloride ions likewise enter and leave these cells to maintain a normal ionic balance. The function of some tissues depends on the actual development of ionic gradients. In the production of gastric juice, for example, the parietal cells must be able to transport H^+ against a large gradient of ions to produce a 0.1M hydrogen ion concentration. Nerve impulses are transmitted because the neural cells are able to maintain an ionic polarization across their membranes by an active Na^+ transport process. Muscle contraction depends on a release of Ca^{++} from the sarcoplasmic reticulum; relaxation requires that the Ca^{++} be transported back across the membrane and sequestered in these membranes.

Understanding transport not only gives us an appreciation of metabolic communication and the function of tissues in the human organism but also provides a rationale for the action of certain kinds of antibiotics capable of disrupting or modifying transport mechanisms.

Transport by passive diffusion

Cell membrane is a selective barrier

Plasma membranes that surround each cell consist of a lipid bilayer (Chapter 2). Molecules of phospholipids are arranged such that the aliphatic chains interact, creating a hydrophobic phase, whereas the polar phosphoglycerol backbone faces outward and is in contact with the aqueous cytoplasm and extracellular medium.

Within this protective lipid layer reside or float a variety of functional and structural proteins. This so-called mosaic model, also described in Chapter 2, for the plasma membrane is an extremely fluid or mobile system, i.e., the proteins are capable of considerable lateral movement.

The plasma membrane therefore defines the limits of the cell and separates the cytoplasm from the environment or outside medium. It represents a highly selective permeability barrier by restricting the molecules entering and leaving the cytoplasm.

Simple diffusion

Many molecules enter the cell by diffusion. Diffusion implies that the molecules are soluble in the lipid phase of the membranes. Fig. 19-1 illustrates this phenomenon by showing the linear relationship between the permeability of a living cell to certain molecules and relative distribution of these molecules between oil and water (partition coefficient). Thus the more lipophilic the substance, the more readily it penetrates the cell membrane. Steroid hormones, for example, enter their target cells (Chapter 18) by simple diffusion. Likewise, oxygen and carbon dioxide diffuse into and out of the red blood cells (Chapter 16) because these gases dissolve in the lipid phase of the membrane.

Simple diffusion occurs from a high concentration of solute to a low one, i.e., it always follows a downhill concentration gradient. If a substance is soluble in the membrane, it will enter the cell as long as the concentration is greater on the outside.

$$\text{Influx: } S_{out} \longrightarrow S_{in} \text{ when } S_{out} > S_{in}$$

When the solute level within the cell increases, then the reverse process occurs and the solute leaves the cell by the same process. Equilibrium is reached when influx equals efflux and the amount of solute entering the cell is exactly counterbalanced by the amount leaving.

$$\text{Efflux: } S_{out} \longleftarrow S_{in}$$

$$\text{At equilibrium: } S_{out} \rightleftarrows S_{in}$$

$$\text{Influx} = \text{Efflux}$$

However, equilibrium may be avoided and influx may continue down the concentration gradient if the solute is altered in some way once inside the cell. Alteration might be a chemical one, as in the technique used to measure the pH within cells or even mitochondria. A dye is used that readily diffuses into the cell (see diagram at top of the next page). However, once inside, esterases remove the acetyl group,

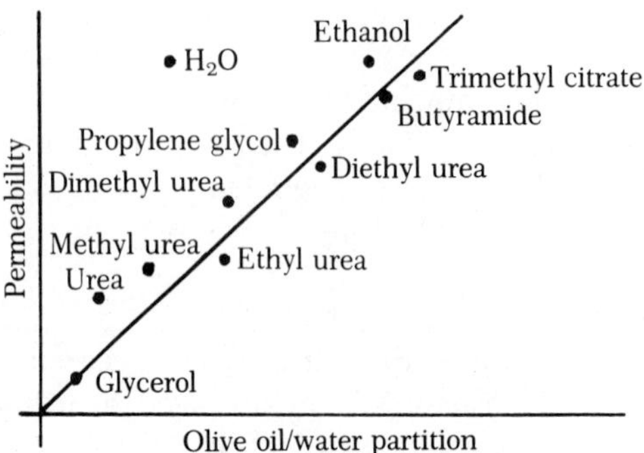

Figure 19-1
Linear relationship between membrane permeability and partition between oil and water phases.

CH_3—C(=O)—O— … —O—C(=O)—CH_3 (CO_2H) → HO— … =O (CO_2H)

producing its free form, which cannot diffuse out of the cell. Therefore the S_{in} has been removed from the reaction and the dye accumulates and is trapped within the cell. Similarly, aspirin with a pK_a of 3.4 exists at the physiological pH of 7.4 predominately in the acetylsalicylate form. However, in the gastric juice, where the pH is about 1.4, the drug is mostly in the uncharged form, which readily penetrates cell membranes. Once inside at a pH of 7.4, the anionic form predominates, which is incapable of passing through the membranes; again, S_{in} has been removed from the reaction, and thus the drug accumulates and is trapped within the cell. This principle is also used for the treatment of aspirin poisoning. Urinary excretion of excess aspirin is encouraged by the consumption of sodium bicarbonate to raise the pH of the urine from its normally acidic value to a distinctly alkaline level. The alkaline urine converts the aspirin predominately to the anionic form and prevents its passage through the cell membranes and entry into the renal tubule cells. In this way renal reabsorption of aspirin is hindered, and urinary excretion is hastened.

COOH, O, O, C, CH_3

Aspirin or acetylsalicylic acid

Diffusion through pores

Fig. 19-1 shows that water does not fit into the linear relationship between oil solubility and membrane permeability; water readily enters the cell. This is dramatically illustrated by the change in shape of red blood cells when exposed to differing osmotic environments. Red blood cells suspended in an isotonic or physiological medium such as 0.9% or 0.15M NaCl retain their normally biconcave shape (Fig. 19-2). This means that the concentration of particles is the same both inside and out; thus there is no net flow of water. However, when suspended in a hypertonic solution of more than 0.15M NaCl, water moves rapidly through the membrane into the more concentrated outer medium. The cells collapse and exhibit the so-called crenated shape. On the other hand, in a hypotonic solution of less than 0.15M NaCl water rapidly enters the cell, which is now more concentrated than the medium causing the spherical shape and eventually leading to hemolysis.

Figure 19-2

Scanning electron micrographs of red blood cells in isotonic **(A)**, hypertonic **(B)**, and hypotonic **(C)** saline solution. All cell shapes shown are typical after exposure to these three concentrations of saline. In isotonic solution (0.15M saline) the cells retain their normal shape; in hypertonic solution (0.15M saline) they shrink (are crenated) as water leaves the cells; in hypotonic solution (0.15M saline) water flows into the cells causing them to swell and burst (hemolysis).

From Singer, S.J.: J. Cell Biol. **70:**193, 1976; micrographs courtesy Robert D. Dyson, M.D., Ph.D., Portland, Ore.

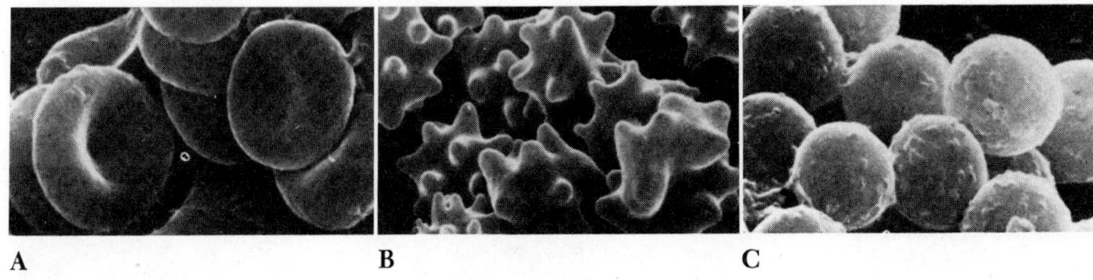

A B C

Figure 19-3
Schematic diagram of a cell membrane showing a pore lined with hydrophilic groups.

Courtesy The Upjohn Co., Kalamazoo, Mich.

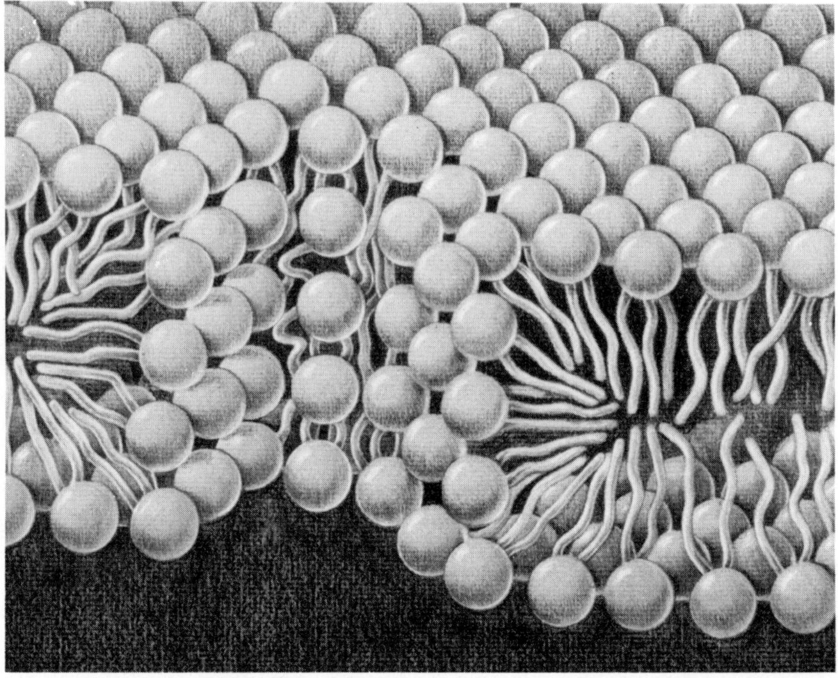

If water does not penetrate the cell membrane by solution in the lipid bilayer, how then may this phenomenon be explained? To accommodate the observed ready passage of water into cells, the existence of pores or channels has long been suggested. It is calculated, for example, that the lipid bilayer possesses pores of 0.40 to 1.0 nm diameter (Fig. 19-3). Such pores would also explain a certain degree of leakage of various ions, such as Na^+ and K^+.

Common characteristics of diffusion

Simple diffusion, whether by lipid solubility or transition via pores or channels, has certain characteristics. It is bidirectional, occurring in either direction across the membrane. Diffusion always occurs in a downhill direction of the gradient. The rate of diffusion or the rate of entry (v) of the solute bears a linear relationship with respect to its difference in concentration across the membrane (Fig. 19-4, *A*).

Figure 19-4
A Linear relationship between the rate of diffusion and the difference in the solute concentration (S_{out}-S_{in}).
B Saturation kinetics exhibited by facilitated diffusion. As the solute is increased, the rate of entry reaches maximum velocity. The curve is a rectangular hyperbola.

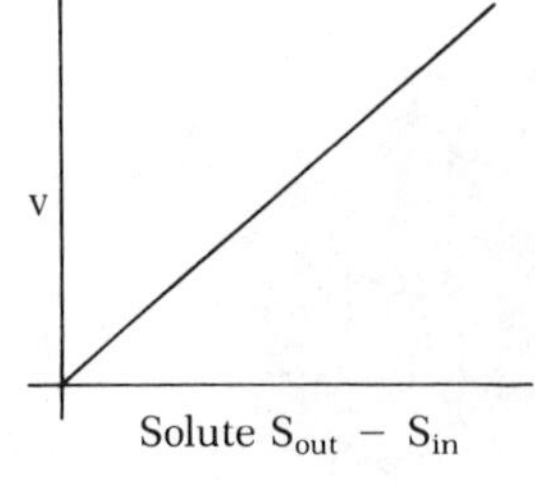

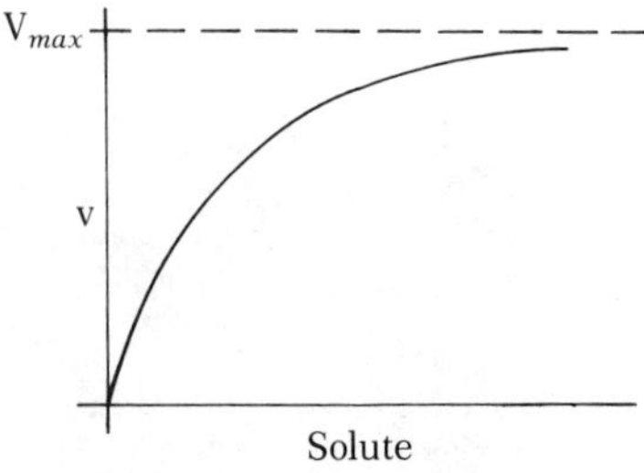

Facilitated diffusion

Carrier-solute complex

$$S_{out} + C \rightleftharpoons CS \rightleftharpoons C + S_{in}$$

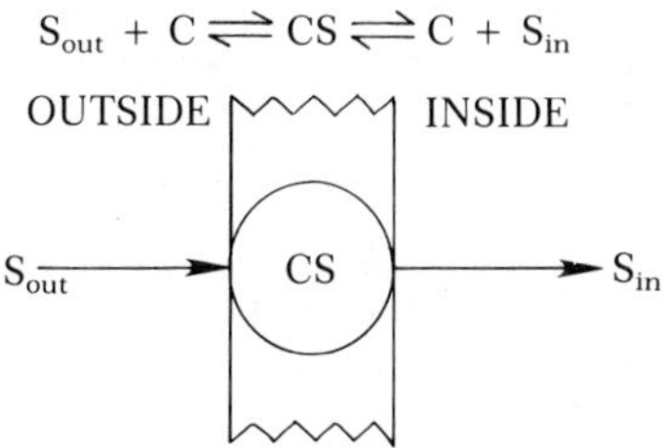

A metabolite such as glucose may enter red blood cells by diffusion but not by either the process of lipid solubility or passage through pores. Rather, glucose transport exhibits quite a different characteristic. For example, the rate of glucose entry into the red blood cells exhibits Michaelis-Menten kinetics. As seen in Fig. 19-4 *B*, the relationship between the rate of accumulation and solute concentration is in the form of a rectangular hyperbola. It exhibits a maximum velocity and can be described by a K_m that is the substrate concentration yielding half maximum velocity. Therefore the process is saturable, suggesting a system not unlike that of an enzyme. Thus the substrate or solute would be conceived as binding to a specific site on a membrane-bound carrier protein (C) in a transient fashion (see left). Although the process of translocation involves the interim formation of a carrier-solute complex (*CS*), the actual movement of solute across the membrane occurs in a down-gradient fashion, i.e., when $S_{out} > S_{in}$. But if $S_{in} > S_{out}$, then the process will be reversed; thus facilitated diffusion, like simple diffusion, is bidirectional. It differs, however, by the presence of the carrier molecule, hence the special designation *mediated* or *facilitated* diffusion.

If it is assumed that the concentration of transient complex CS is constant and that S_{in} is sufficiently low so that efflux can be ignored, then a Michaelis-Menten relationship is valid.

$$v = \frac{V_{max}\, S_{out}}{K_m + S_{out}}$$

or

$$\frac{1}{v} = \frac{K_m}{V_{max}} \frac{1}{S_{out}} + \frac{1}{V_{max}}$$

A Lineweaver-Burk plot may be constructed from which the K_m and V_{max} can be calculated. Furthermore, since these transport systems obey Michaelis-Menten kinetics, it is apparent that there is only a limited number of carriers and/or that each possesses one or a limited number of binding sites.

Specificity

Substrate specificity is another important characteristic of facilitated diffusion. The formation of the intermediate complex implies the existence of a binding site that serves to recognize the appropriate substrate (Fig. 19-5). Thus the K_m for D-glucose entry into red blood cells is 5 mM. Since the blood glucose level is 5 mM,

Figure 19-5 Steps involved in facilitated diffusion. *C*, Carrier-bound protein.

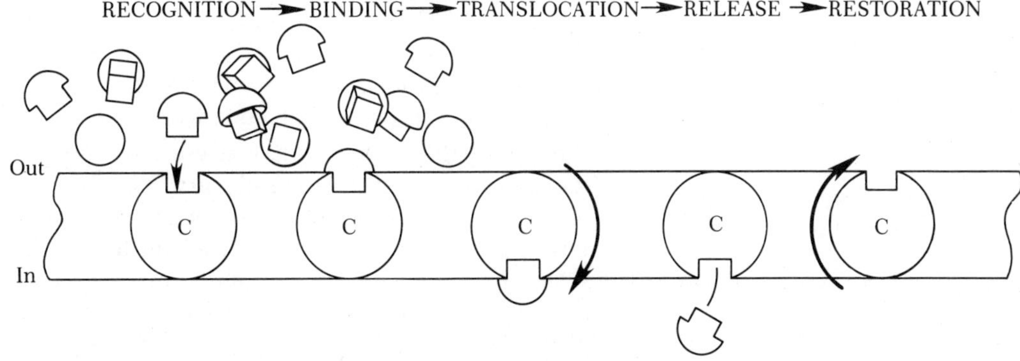

the rate of entry into red blood cells is ½ V_{max}. However, other sugars are also transported; D-mannose (K_m = 20 mM), D-galactose (K_m = 30 mM), and D-xylose (K_m = 60 mM). On the other hand the L-isomer of glucose is not transported; therefore the system is stereospecific. All sugars transported have in common the chair conformation of the pyranose ring (Chapter 23) as well as a number of hydroxyl groups in the equatorial plane of the ring.

Usually kinetic studies are made difficult by the rapid metabolism of the substrate. It is preferable therefore to utilize a nonmetabolizable analogue such as 2-deoxyglucose in place of the readily metabolizable hexose. Then the natural sugars may be employed as competitive inhibitors. Using the transport of 2-deoxyglucose into rat tumor cells (Novikoff hepatoma cells), both glucose and mannose serve effectively as competitive inhibitors.

The important role of the carrier protein molecules in facilitated transport is also illustrated by the inhibitory effects of various group-specific reagents. Compounds such as mercury salts or *N*-ethylmaleimide for -SH groups or fluorodinitrobenzene for amino groups irreversibly inhibit the transport process.

In the facilitated diffusion of glycerol into red blood cells a pH dependency may be demonstrated. An optimum pH of 6 has been observed, which is close to the pK of the imidazolium ring. All the preceding observations lead to the conclusion that the carrier molecules are proteins.

That carrier molecules in facilitated diffusion are in fact proteins was also demonstrated in genetic studies using *E. coli*. As described in Chapter 10, *E. coli* possesses the ability of adapting to the use of lactose as substrate. Adaptation requires the induction of the *lac* operon. In the adaptive process lactose induces the production not only of the enzyme β-galactosidase but also the permease (carrier?) protein needed to transport the β-galactosides into the cell.

Fig. 19-5 illustrates the important steps in facilitated diffusion. First the processes of *recognition* and *binding* are needed to explain specificity, competitive and noncompetitive inhibition, and saturation kinetics. The physical *translocation* of the solute from the outside to the inside requires a downhill concentration gradient. However, the exact manner in which this occurs is unknown. Probably it does not occur by the "turnstile" model or physical rotation or translation of the protein within the lipid bilayer. An alternate suggestion is that the protein(s) creates a channel through which the solute may pass; however, access to the channel must be contingent on a prior binding of the solute with the specific binding proteins.

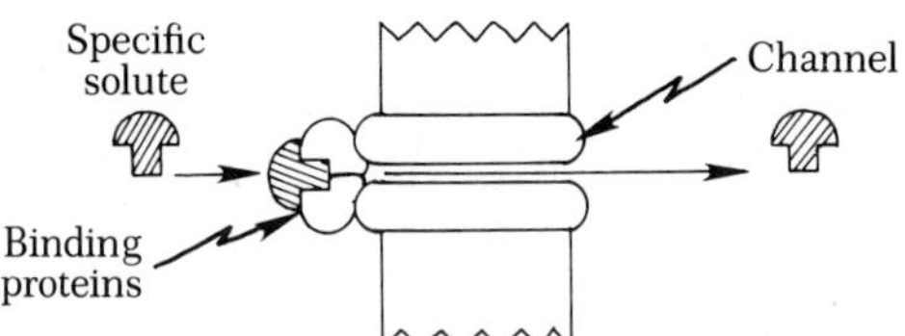

Facilitated diffusion of glucose has been observed not only in red blood cells and leukocytes but also in muscle and adipose tissue cells, as well as in the choroid plexus and the capillaries of the brain. Glucose also enters the liver cells by this process. Of these transport processes insulin increases the rate of glucose transport only into the muscle and adipose tissues. Exactly how insulin affects transport in these tissues in unknown. The downhill gradient for the facilitated diffusion is maintained because the glucose immediately becomes the substrate for hexokinase (glucokinase in the liver) and is converted to glucose 6-phosphate. As mentioned in Chapter 11, the plasma membrane is impermeable to the phosphate, so the glucose is unable to leak out of the cells.

Active transport

In passive diffusion, whether mediated or not, all flow of substrate is in the direction of the gradient (i.e., downhill) and is energetically spontaneous. These energetics may be described for a system consisting of two solutions of a given solute separated by a membrane. In compartment 1 the solute is 5 mM (S_1), whereas in compartment 2 (S_2) it is 0.5 mM. If diffusion occurs in the direction of the gradient (downhill), then the free energy change is:

$$\Delta G' = RTln\ S_2/S_1 = RTln\ 0.5/5 = -1.34\ kcal$$

However, there are many biological processes in which the flow of metabolic substances is in the opposite direction, i.e., against the concentration gradient or uphill. Such a process may be described with the same system as the preceding, the only requirement being that the solute then flows from the compartment of low concentration to the higher one, i.e., from 2 to 1.

$$\Delta G' = RTln\ S_1/S_2 = RTln\ 5/0.5 = +1.34$$

In this instance the process is not spontaneous and would require an equivalent input of energy to make it proceed in the direction needed. In biological systems uphill transport is coupled with the hydrolysis of ATP. An energy requiring uphill transport process is called *active transport*. These concepts are summarized in the following table.

S_1 5 mM	S_2 0.5 mM	$\Delta G'$ kcal	Gradient	Process
⟶	⟶	−1.34	Downhill	Passive
⟵	⟵	+1.34	Uphill	Active

Active transport has the same characteristics as illustrated in Fig. 19-5: recognition, binding, translocation, release, and restoration. It differs, however, in the nature of the translocation process, which is coupled to an input of energy— the hydrolysis of ATP. Because of this, active transport is unidirectional. Nevertheless, it obeys Michaelis-Menten kinetics and exhibits substrate specificity and competitive and noncompetitive inhibition.

Na^+, K^+–ATPase

An important example of active transport found in all animal cells is Na^+, K^+–ATPase. This process is required to maintain the internal ionic environment of the cell. Red blood cells, as all animal cells, have a high internal K^+ and a low Na^+ concentration. This is illustrated by the following table.

	K^+	Na^+
RBC	150 mM	26 mM
Blood plasma	4	144

These ions tend to leak into or out of the cell by simple diffusion, presumably by way of the pores. For example, if red blood cells are kept at low temperatures, e.g. 5° C, then this leakage of Na^+ into and K^+ out of the cells may occur until equilibrium is reached. Now, if these same cells are incubated at 37° C, the original ionic gradients will be reestablished. As long as glucose is present in the medium, an energy-requiring pump will remove the Na^+ from the cells and restore the K^+. This pump uses ATP and therefore is called the Na^+, K^+–ATPase. It is a system composed of two proteins found in the cell membrane; one has a molecular weight of 130,000 and the second, a glycoprotein, has a molecular weight of 45,000. As shown on p. 656, this system transports Na^+ and K^+ simultaneously in opposite directions. Such a coordinated process is described as *antiport* to distinguish it from a *symport* process, in which two substances are carried in the same direction. As seen from the diagram, there is also a definite sidedness about this process; the ATP can only be on the cytoplasmic side of the membrane.

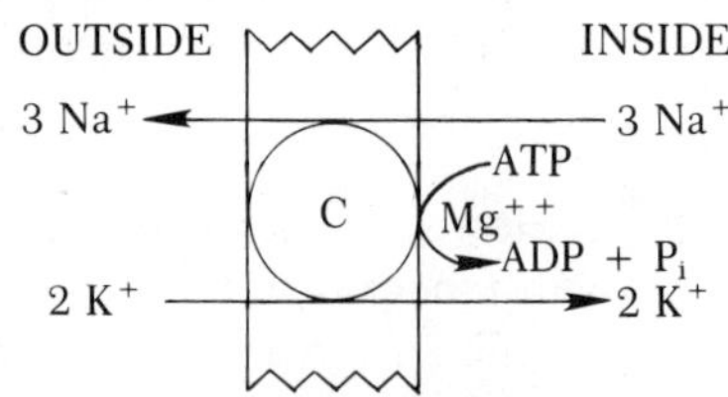

The reactions involved can be separated into two independent reactions:

$$1.\ C + ATP \xrightarrow[Mg^{++}]{Na^+} C \sim P + ADP$$

$$2.\ C \sim P + H_2O \xrightarrow[Mg^{++}]{K^+} C + P_i$$

```
O     O⁻
‖     |
C—O~P=O
|     |
CH₂   O⁻
|
—N—CH—C—
 H     ‖
```

Each reaction, phosphorylation and dephosphorylation, depends on the presence of Na^+ or K^+, as indicated. In the absence of the ions ATP is not hydrolyzed. In other words, the hydrolysis of ATP is tightly coupled to the transport of the ions. For each ATP hydrolyzed, 3 Na^+ are removed while 2 K^+ ions enter the cell. Phosphorylation involves the formation of an intermediate, a mixed phosphoanhydride of an aspartyl residue. Exactly how Na^+ and K^+ are carried is unclear; the overall process may be visualized as in Fig. 19-6.

The Na^+, K^+–ATPase of the cells in the renal distal tubules serves a special function in controlling the excretion and retention of sodium ions. This pump is regulated by aldosterone, a hormone released when sodium retention is required. It communicates to the cells the need for an increased reabsorption of sodium ions from the glomerular filtrate and appears to do so by stimulating the transport of Na^+ out of the cell up-gradient into the bloodstream. This action of the pump establishes and maintains a down-gradient with respect to the glomerular filtrate so that sodium ions then enter the cells possibly by passive diffusion.

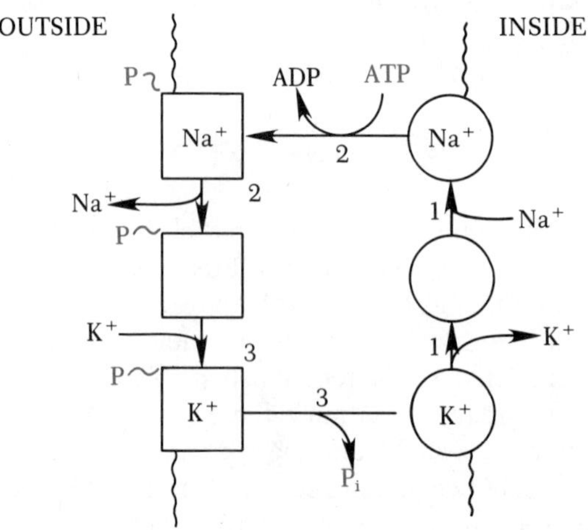

Figure 19-6
Steps involved in the Na^+, K^+–ATPase ion pump. *Step 1:* dephosphorylated carrier has a high affinity for Na^+ and a low affinity for K^+. It binds 3 Na^+ at this stage. *Step 2:* phosphorylation leads to a change in conformation, a process accompanied by translocation and a change in affinity. Na^+ is released outside the cell while 2 K^+ are bound. *Step 3:* potassium binding leads to dephosphorylation, change to original conformation, translocation, and a release of the K^+ inside the cell. The peculiar stoichiometry is not understood.

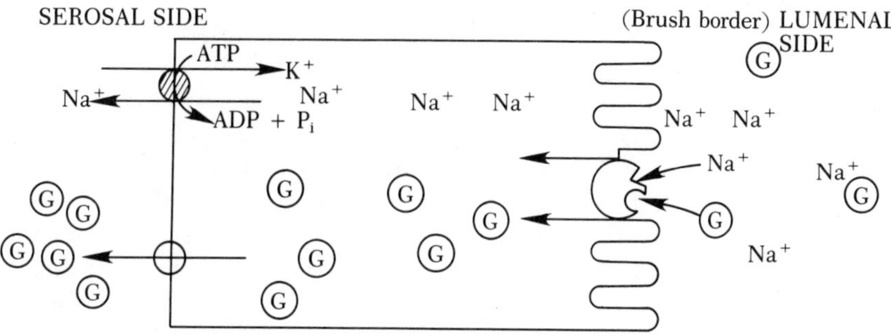

Figure 19-7
Coordinated transport of glucose and sodium ions in epithelial cell layers of the renal tubules and the mucosal lining of the intestine. Glucose is carried into the cytoplasm by cotransport of sodium; the latter is them pumped out of the cell by the Na^+, K^+–ATPase.

Glucose transport

Na^+, K^+–ATPase is also an important functional part of the active transport of glucose across the epithelial cell layers of the intestinal mucosa and the renal proximal tubules. This system is required to move glucose from the intestinal contents and the glomerular filtrate into the bloodstream. The movement of a solute from a compartment of low concentration to one of higher concentration by passage through a layer of epithelial cells may be viewed as a *transcellular process*. The active transcellular transport of glucose across epithelial cells is actually the coordination of three transport processes. The first involves the entrance of glucose into the cell via a cotransport or symport process in conjunction with a downhill sodium ion gradient. Fig. 19-7 suggests the existence of a carrier protein in the brush border membrane having two binding sites, one for glucose and the other for Na^+. The downhill flow of Na^+ "carries" the glucose along with it. Glucose accumulates in the cell until it exceeds that of the extracellular fluid and that of the blood. As suggested by Fig. 19-7, the glucose then leaves the cell at the serosal membrane side, probably by facilitated diffusion. Obviously this process can only continue so long as a favorable downhill Na^+ gradient exists. This is maintained by virtue of the third transport process, a Na^+, K^+–ATPase, which is located in the serosal membrane component. At this point ATP is consumed in the overall transport process.

Table 19-1
Three systems of amino acid transport

Na^+ dependent		Na^+ independent
Alanine or A-system	**Alanine-serine-cysteine or ASC-system**	**Leucine or L-system**
Alanine	Alanine	Glutamine
Glycine	Serine	Methionine
Proline	Cysteine	Histidine
Serine	Threonine	Tyrosine
Threonine	Glutamine	Tryptophan
Glutamine		Valine
Methionine		Leucine
Histidine		Isoleucine
Tyrosine		Phenylalanine
Tryptophan		

Reprinted by permission from Christensen, H.N.: Biological transport, ed. 2, Reading, Mass., 1975, W.A. Benjamin, Inc. (Table 5-1, p. 178).

Transport of amino acids

As with glucose, amino acids enter cells by active transport processes. However, because of the diversity of structures, the problem of specificity is more complex. Much of our present knowledge regarding amino acid transport concerns transmembrane as opposed to transcellular processes and has been studied using only a limited number of tissues: Ehrlich ascites tumor cells, liver, muscle, and avian red blood cells. In the mammalian organism four systems have been defined, three of which are detailed in Table 19-1. The A-, or alanine-preferring, process carries small neutral amino acids, whereas the L-, or leucine-preferring, system transports the larger neutral aliphatic and aromatic amino acids. The basics, lysine and arginine, are carried by a special Ly^+-system, whereas histidine enters cells by way of both A- and L-systems (Table 19-1). Although glutamate and aspartate may have a special transport system in some tissues, especially the intestinal mucosa, the existence of such in Ehrlich cells seems uncertain. Table 19-1 also designates an ASC-process, which serves to carry selectively alanine, serine, and cysteine.

Of the four processes defined for mammalian organisms, the A and ASC are clearly Na^+ dependent, whereas the L-system is Na^+ independent. All processes are considered to be active transport and result in the accumulation of the substrate amino acid. This table suggests that the specificity of amino acid transport is by no means exclusive; rather, it should be apparent that amino acids enter cells by several systems.

Active transport of amino acids also depends on the function of Na^+, K^+–ATPase and is currently believed to function by Na^+ symport, much like glucose transport. Especially important is that glucagon stimulates the A–transport mechanism via mediation with cAMP. It is of interest that the glucagon-stimulated transport involves glucogenic amino acids, which serve as substrates for the process of gluconeogenesis. Insulin also stimulates amino acid transport, but the nature of this or the amino acid systems involved is unknown. Of special significance to the control of metabolism is the fact that insulin counteracts the glucagon stimulation of the A-system. Insulin may increase the entrance of most amino acids into cells so as to provide adequate substrate for protein synthesis.

Once inside the liver cells amino acids may be catabolized, their nitrogen being converted to ammonia and urea. Since this is a mitochondrial process (Chapter 13),

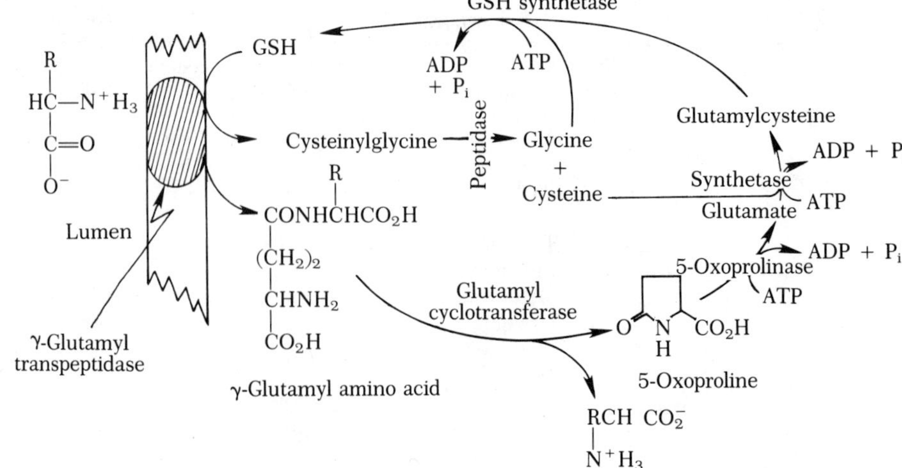

Figure 19-8
Steps in the transport of amino acids across the epithelial cells of the renal proximal tubules. In this mechanism amino acids are transported from the glomerular filtrate into the bloodstream via a recycling process using glutathione.
According to Meister and Tate.

it is likely that some transport systems exist in the membranes of the mitochondria. However, little is known concerning the intracellular transmembrane transport of amino acids.

Transcellular transport of amino acids

To minimize the loss of amino acids in the urine, the proximal tubule cells reabsorb and return them to the bloodstream. The transcellular transport of amino acids in the kidneys appears to occur by a special system in which amino acids are covalently linked to the γ-glutamyl residue of the tripeptide glutathione. Fig. 19-8 illustrates the action of the membrane-bound transpeptidase and the subsequent cyclical nature of the entire system. Three ATPs are consumed in the process.

Genetic defects

A number of inborn errors of metabolism involving the transport of amino acids are known to exist. Cystinuria, for example, is a condition in which the concentration of cystine, lysine, arginine, and ornithine in the urine is elevated because of a failure in the renal reabsorption mechanism. Cystine stones are frequently observed in the ureter or bladder, giving rise to urinary tract obstruction and infection. In Hartnup's disease there is an inability to transport neutral amino acids; in iminoglycinuria the amino acids proline, hydroxyproline, and glycine are not transported.

Other transport processes

In addition to the better known transport system just described, there must be numerous processes for the uptake of metabolites. Fat-soluble vitamins probably cross membranes by diffusion through the lipid bilayer. Bile salts appear to be taken up by an active process involving an Na^+ symport system. The water-soluble vitamins other then B_{12} are thought to be carried by facilitated diffusion. Vitamin B_{12} crosses the intestinal mucosa by way of a special process involving the intrinsic factor secreted by the gastric mucosa.

A number of antiport, facilitated diffusion processes occur in mitochondria. In Chapter 7 several shuttle systems were described, such as the ATP/ADP shuttle, which is specifically inhibited by atractyloside, α-ketoglutarate/L-malate, and L-aspartate/L-glutamate.

The transport and sequestration of Ca^{++} during muscle contraction and relaxation is an important example of intracellular transport. Muscular contraction involves a release of Ca^{++} (Chapter 15) stored at high concentration in the sarcoplasmic reticulum. Relaxation is accompanied by a recapture of Ca^{++}, a process involving an uphill transport of this ion. Ca^{++}-ATPase is found within the lipid bilayer and involves a process of reversible phosphorylation. This reversible phosphorylation is comparable to that already described for Na^+, K^+–ATPase.

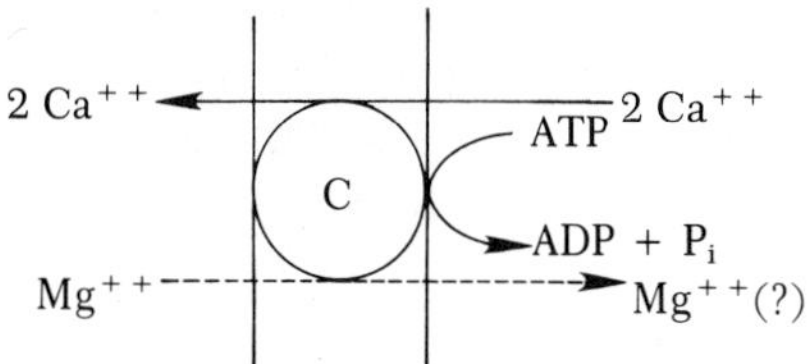

Transport of proteins across membranes

Recently much interest has been displayed in the passage of proteins across membranes. This process involves endocytosis and exocytosis. An important model system is the passage of plasma lipoproteins by endocytosis into cultured fibroblasts. Fig. 19-9 illustrated some of the steps thought to be required. First, intrinsic membrane proteins congregate to produce "caps," or concentrated regions for the recognition and binding of the plasma lipoprotein to be transported. The outer

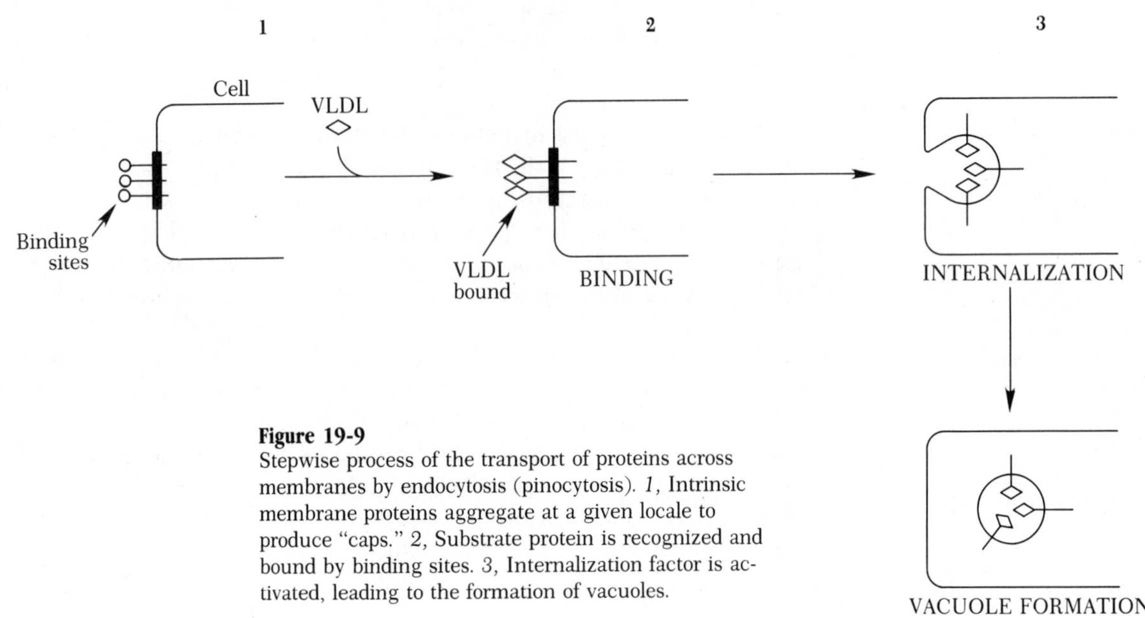

Figure 19-9
Stepwise process of the transport of proteins across membranes by endocytosis (pinocytosis). *1*, Intrinsic membrane proteins aggregate at a given locale to produce "caps." *2*, Substrate protein is recognized and bound by binding sites. *3*, Internalization factor is activated, leading to the formation of vacuoles.

portion of the membrane protein is the binding site. Once binding has occurred, an *internalization* factor is activated, which initiates the endocytotic process. Endocytosis then results in a pinching-off stage in which the bound lipoproteins are totally internalized into vacuoles.

There are many known instances of the uptake of proteins; however, little is known concerning the molecular aspects of this process. In a neonate the intestinal mucosa is permeable to γ-globulins. At this age the infant is incapable of producing antibodies and must derive immunity from the mother through the γ-globulins in the milk (especially the colostrum). Another example of protein passage across lipid bilayer is that of diphtheria toxin (mol wt = 63,000). This polypeptide is composed of two functional regions: the A-segment, which is responsible for its inhibitory effects on protein synthesis, and the B-segment, which binds to specific receptor sites on the cell membrane and initiates the internalization process.

Another process that might be more important than is generally thought is the uptake of protein by the renal tubule cells. Proteins of molecular weight less than 50,000 pass through the glomerular filter to varying degrees. Even serum albumin (mol wt = 69,000) must leak into the glomerular filtrate to some extent. Reabsorption of protein from the glomerular filtrate is 95% to 99% complete; little protein is normally excreted in the urine.

Therapeutic agents affecting membranes

A number of antibiotics owe their actions to their ability to alter membrane permeability, thereby disrupting essential ionic gradients. One is *valinomycin* (Fig. 19-10), a 36-atom ring of alternating hydroxy and amino acids consisting of the sequence D-valine-D-hydroxyisovaleric acid L-valine-L-lactate, repeated three times. Valinomycin assumes a doughnut conformation in the lipid bilayer. It binds K^+ specifically in its cyclic structure and functions as a carrier, shuttling the K^+ through the membrane. Other examples are *nonactin* and *nigericin*. A second type of antibiotic is *gramicidin D*, a 15-residue polypeptide of alternating D- and L-amino

Figure 19-10
A Structure of valinomycin, an antibiotic that disrupts ionic gradients by altering membrane permeability.
B Ouabain, a glycoside that inhibits Na^+, K^+–ATPase.

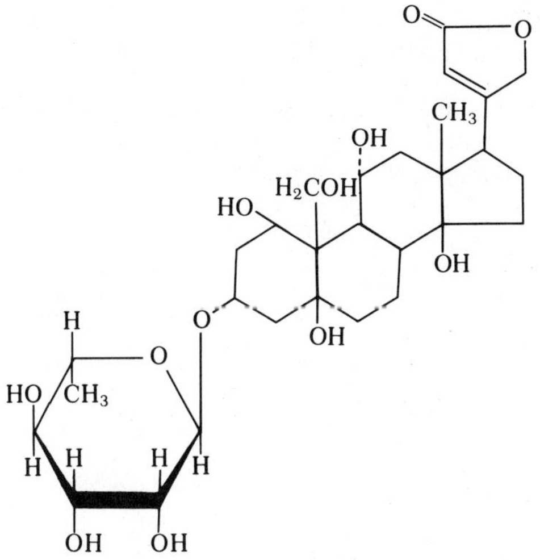

acids. This polypeptide forms a helix of approximately 3½ turns that is stabilized by hydrogen bonds. A dimer of the antibiotic crosses the entire lipid bilayer, producing a channel or pore. Another pore-producing antibiotic is *amphotericin B*.

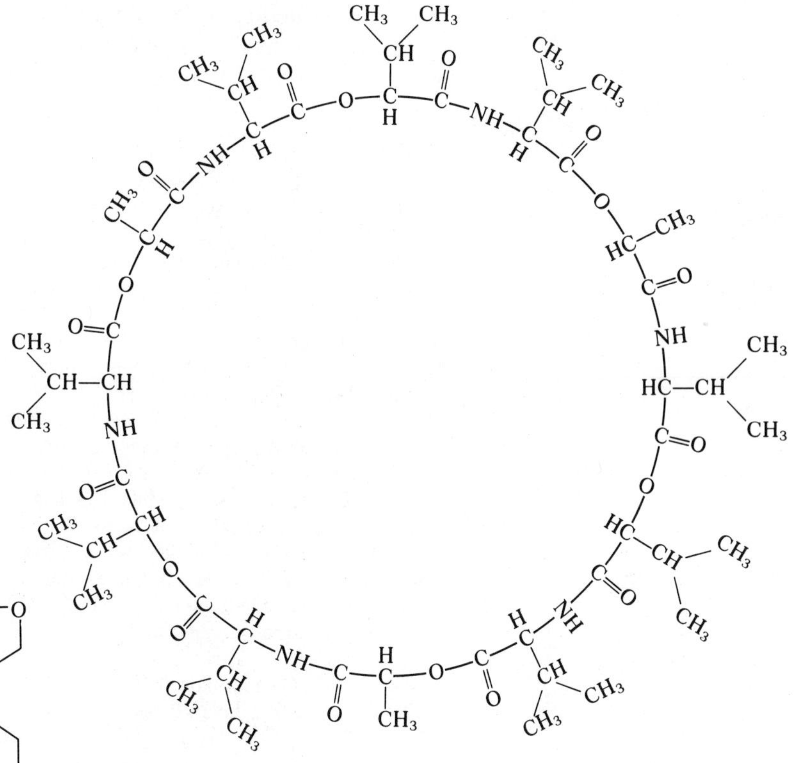

Polymyxin B, colistin, and *tyrocidine* contain lipophilic and lipophobic groups and act as surface-active agents. *Nystatin,* a polyene antibiotic, binds to the membrane and blocks the uptake of glycine by many fungi.

Other drugs known to influence membrane transport processes are the glycosides of strophanthidin, such as *digitalis*. Digitalis is administered in congestive heart failure because it increases the force of contraction of the heart muscle. *Ouabain* is a similar glycoside that has been used frequently in the study of active transport processes. These glycosides inhibit the Na^+, K^+–ATPase apparently by combining with the phosphorylated carrier molecule.

$$C \sim P \xrightarrow{\text{Ouabain}} C \sim P \cdot \text{Ouabain}$$

Inhibition of the ATPase changes the electric properties of the cardiac cells. Perhaps the reduction in Na^{+} efflux in some way reduces Ca^{++} efflux, thereby increasing the available pool of Ca^{++}. Furthermore, the cardiotoxic glycosides might decrease Ca^{++} uptake by the sarcoplasmic reticulum.

Suggested readings

Bittar, E.E.., editor: Membrane: structure and function, vol. 1, New York, 1980, John Wiley-Sons, Inc.

Christensen, N.: Biological transport, ed. 2, Reading, Mass., 1975, W.A. Benjamin, Inc.

Coleman, R.: Membrane-bound enzyme and membrane ultrastructure, Biochim. Acta **300**:1, 1973.

Glynn, I. M., and Karlich, S.J.D.: The sodium pump, Annu. Rev. Physiol. **37**:13, 1975.

Meister, A., and Tate, S.: Glutathione and related γ-glutamyl compounds: biosynthesis and utilization, Annu. Rev. Biochem. **45**:559, 1976.

Mela, L.: Mechanism and physiological significance of calcium transport across mammalian mitochondrial membrane. In Bronner, F., and Kleinzeller, A., editors: Current topics in membrane transport, vol. 9, New York, 1977, Academic Press, Inc.

Oxender, D., and Fox, C.F., editors: Molecular aspects of membrane transport. In Progress in clinical and biological research, vol. 22, New York, 1978, Alan R. Liss, Inc.

Reviews on biomembranes, Biochim. Biophys. Acta, vol. 457, 1976.

Rothstein, A., Cabantchik, Z.I., and Knauf, P.: Mechanism of anion transport in red blood cells: roles of membrane proteins, Fed. Proc. **35**:3, 1976.

Shamoo, A.E., editor: Carriers and channels in biological systems, Ann. N.Y. Acad. Sci., vol. 264, 1975.

Tosteson, D.C., Ovchinnikon, Y.V., and Latorre, R., editors: Membrane transport processes, vol. 2, New York, 1978, Raven Press.

Wallach, D.F.H.: The plasma membrane: dynamic perspectives, genetics and pathology, vol. 18, Heidelberg, 1972, Springer-Verlag.

Wilson, D.B.: Cellular transport mechanisms, Annu. Rev. Biochem. **47**:933, 1978.

Chapter 20

Metabolic interrelationships of tissues

Cellular metabolism was described in Chapter 3 as a state of *dynamic equilibrium* between the various chemical constituents of the cell, resulting in a steady state. It was said that a nutrient such as glucose is constantly being used to meet the energy demands as well as the needs for various synthetic processes of the cell and thus must be replaced. A delicately regulated balance between catabolic and anabolic processes maintains a relatively uniform cellular composition. For the brain and red blood cells to survive, the homeostatic control of blood glucose level is paramount. The maintenance of a constant internal environment was described in the first half of the nineteenth century by Claude Bernard.

The human habit of consuming large amounts of food during a limited number of daily meals results in a cyclical feeding-fasting process. These changes require adaptive or homeostatic processes that involve changes in metabolic patterns as well as in the kinds of fuel used. For example, following a carbohydrate-rich meal the metabolic posture is one of glucose storage to reduce the postprandial hyperglycemia. Glucose is stored first as glycogen and then is converted into fat. In the postabsorptive state these processes are reversed; the glucose stored as glycogen in the liver is mobilized to maintain the blood glucose at normal levels. The fatty acids stored as triacylglycerols are mobilized and used by the musculature as fuel.

The chief fuels used by the tissues for energy are the following:

1. Glucose
2. Fatty acids
3. Ketone bodies
4. Amino acids
5. Lactic acid (from glycolysis)
6. Minor sources such as glycerol, ethanol, and pentoses

The dependency of various tissues on glucose, fatty acids, and ketone bodies as sources of energy is shown in Table 20-1. It is apparent that certain tissues have a complete dependence on glucose, e.g., the red blood cells, the renal medulla, the retina, and intestinal mucosa. The central nervous system usually depends on glucose; however, during extended fasting the brain can adapt to the use of ketone bodies. Table 20-1 also illustrates the versatility of cardiac muscle, which can use all three of the fuels listed. In addition, this organ can use lactic acid as a fuel. The skeletal muscle generally uses fatty acids as a primary fuel. However, as will be seen, it switches to glucose under conditions of exercise and branched chain amino acids during extended fasting.

In Chapter 3 the concept was introduced that the three types of fuels—carbohydrates, lipids, and amino acids—are catabolized by independent pathways. Of

Table 20-1
Dependency of tissues on various fuels as a major source of energy

Tissue	Fuel: Glucose	Free fatty acid	Ketone bodies
Mammalian erythrocytes	+		
Leukocytes and lymphocytes	+		
Renal medulla	+		
Renal cortex	+	+	+
Brain	+		+
Skeletal muscle	+ (in severe exercise)	+ (not in exercise)	+
Cardiac muscle	+	+	+
Liver	+	+	
Retina	+		
Intestinal mucosa	+		

Adapted from Krebs, H.A.: Adv. Enzyme Regul. **10:**397, 1972.

great importance is the appreciation that these three pathways are interrelated and come together at specific points where they share common metabolic intermediates. Fig. 20-1 shows how fatty acids and amino acids serve as fuels by entering the citric acid cycle via acetyl CoA. There are, however, numerous other common intermediates, such as DHAP, glyceraldehyde 3-phosphate, pyruvate, oxaloacetate, α-ketoglutarate, and fumarate.

Metabolic specialization of tissues

Although the liver, skeletal muscle, and adipose tissue each possess the basic metabolic pathways pictured in Fig. 20-1, they have become highly specialized in terms of physiological roles and in metabolism. Specialization of metabolism depends on the selective presence or absence of key enzymes. Table 20-2 compares the activities in the liver and adipose tissue of two gluconeogenic enzymes and two enzymes of the hexose monophosphate pathway. Of the two, only the liver has glucose 6-phosphatase and fructose 1,6-bisphosphatase and thus can function gluconeogenically. Recall from Chapter 11 that the liver is the body's "glucostat"; it is the primary organ for maintaining the blood glucose level. It does this not only by taking up excess dietary glucose and storing it as glycogen but also by remobilizing the glucose as needed. Here glucose 6-phosphatase is a key enzyme; otherwise the liver could not convert the glucose 6-phosphate back to glucose. Furthermore, when the demands are sufficient, glucose can be synthesized *de novo* from noncarbohydrate carbon chains. On the other hand, the epididymal fat pads of the rat (Table 20-2), an example of adipose tissue, are a major site for the storage of lipids and are not programed for gluconeogenesis. In the rat the fat pads are also the important site for the biosynthesis of fatty acids; therefore the first two enzymes in the pentose pathway are very active to supply needed NADPH.

Since the various tissues are metabolically specialized, their activities must be integrated so that the needs of each are met. Metabolic interrelationships of tissues may best be illustrated by comparing the roles of the liver, the skeletal muscle, and the adipose tissues. As viewed in Fig. 20-2, these three may be perceived as a metabolic "three-ring circus" that serves to maintain the blood glucose level. Later it will also be seen that their various roles ensure sufficient fuel and a conservation of glucose so that those organs which are restricted to glucose will always have an adequate supply. It will also be evident that the relative metabolic activities of the three tissues are controlled at two levels—by intracellular and by extracellular (hormonal) signals.

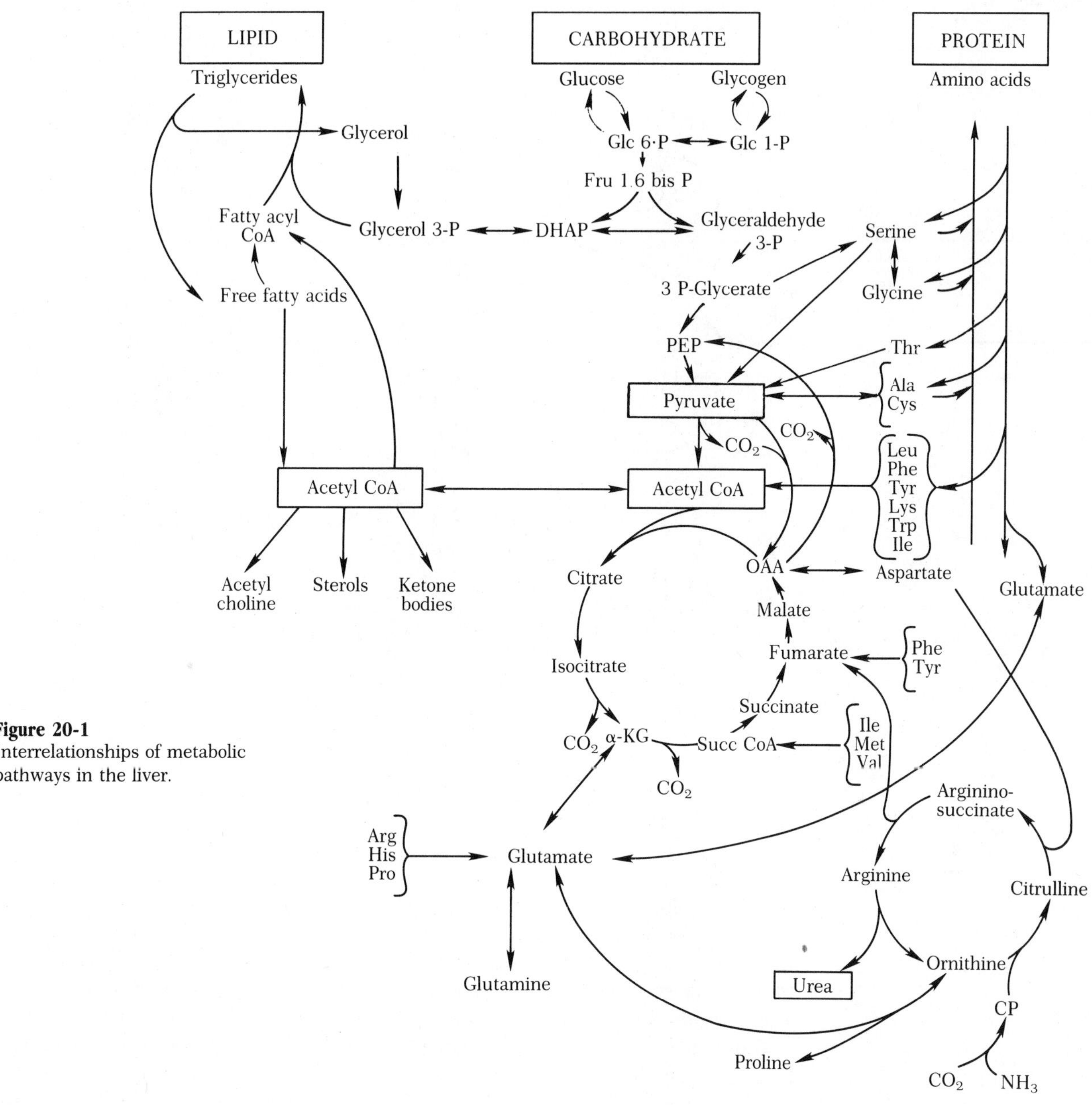

Figure 20-1
Interrelationships of metabolic pathways in the liver.

Table 20-2
Distribution of selected enzymes of carbohydrate metabolism in rat liver and epididymal fat pad

	Activity (moles of substrate per hour per cell × 10^{-7})	
Enzyme	**Liver**	**Fat pad**
Glucose 6-phosphatase	36	0
Fructose 1,6-bisphosphatase	27	0
Glucose 6-phosphate dehydrogenase	4	72
6-Phosphogluconate dehydrogenase	12	49

Adapted from Weber, G.: Adv. Enzyme Regul. **1**:1, 1963.

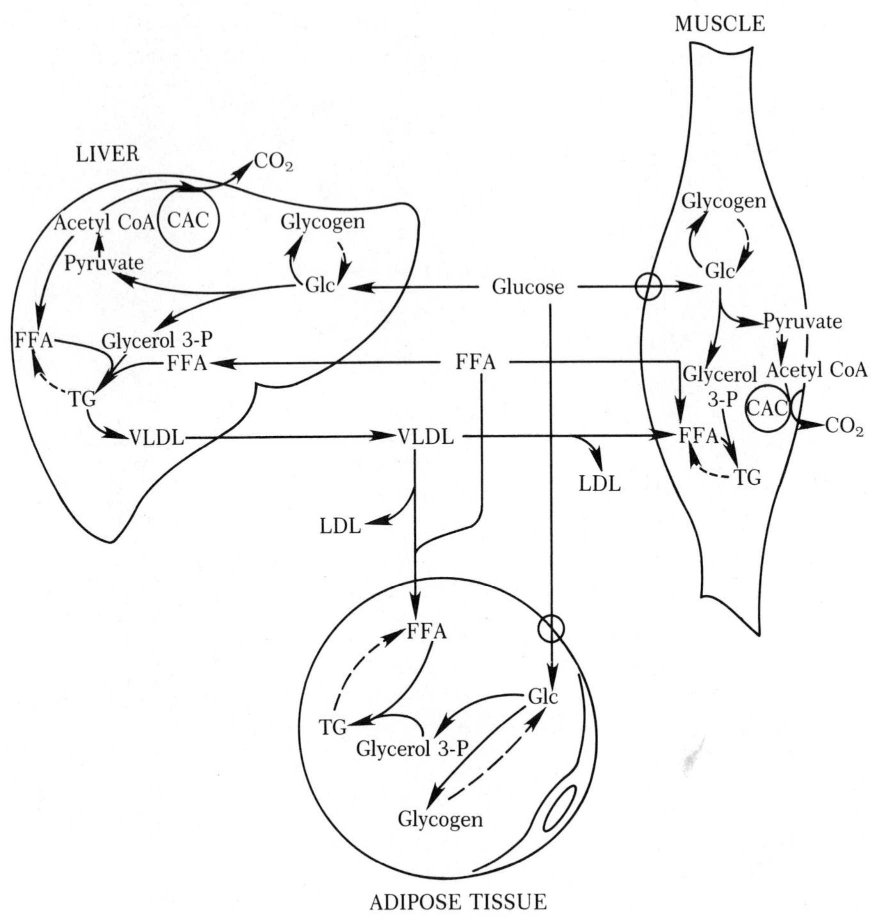

Figure 20-2
The integration of the metabolic activities of the liver, skeletal muscles, and adipose tissues. The major pathways illustrated here are those predominating in the absorptive period or immediately after a meal. *CAC*, Citric acid cycle; *TG*, triacylglycerols; *FFA*, free fatty acids; *Glc*, glucose; *VLDL*, very low–density lipoproteins; *LDL*, low-density lipoproteins.

Liver

The portal system drains the intestinal mucosa and funnels the blood directly into the liver. With the exception of the lipids, most of the products of digestion must pass through the liver before entering the systemic bloodstream. It is the first tissue involved in controlling the level of blood glucose, lipids, and amino acids. Therefore the liver functions as a reprocessing center for these nutrients.

Approximately two thirds of the dietary carbohydrate is removed by this organ. Glucose enters the hepatocytes by facilitated diffusion, a process not affected by insulin. Once inside, it is rapidly converted to glucose 6-phosphate. As described in Chapter 11, the phosphate ester cannot diffuse out and therefore is trapped within the cells. Consequently glucose 6-phosphate accumulates and ensures a continuing downhill gradient of glucose across the hepatocyte membrane. Glucose 6-phosphate in turn is converted to glucose 1-phosphate and then to glycogen (Chapter 11). In this form it is stored as the immediate source of glucose for the maintenance of the blood level. Normally the liver retains approximately 5% of its weight as glycogen (about 70 gm); after a carbohydrate-rich meal this may increase to 10%. Other hexoses, especially fructose and galactose, are converted to glucose as already described. Of these two, fructose is particularly important because of the large amount of sucrose in the human diet. Recall from Chapter 11, and especially Fig. 11-23, that fructose is first phosphorylated to fructose 1-phosphate (hepatic fructokinase). Fructose 1-phophate is cleaved by liver aldolase; the products, dihy-

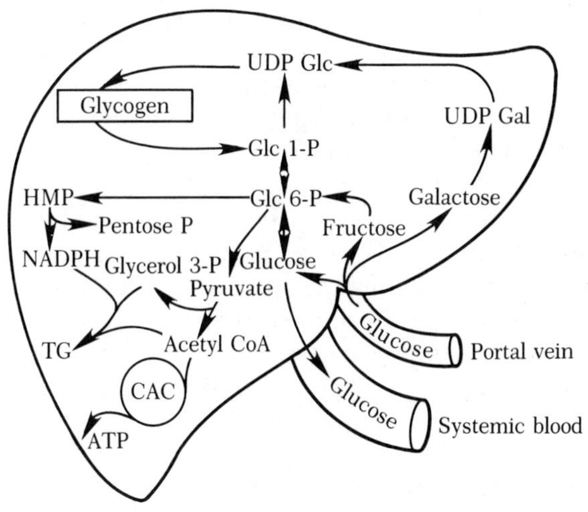

Figure 20-3
Summary of the fate of dietary sugars in the liver. *CAC,* Citric acid cycle; *HMP,* hexose monophosphate pathway; *Glc,* glucose; *Gal,* galactose.

Figure 20-4
Summary of the fate of dietary lipids in the liver. *CAC,* Citric acid cycle; *FFA,* free fatty acids; *TG,* triacylglycerols.

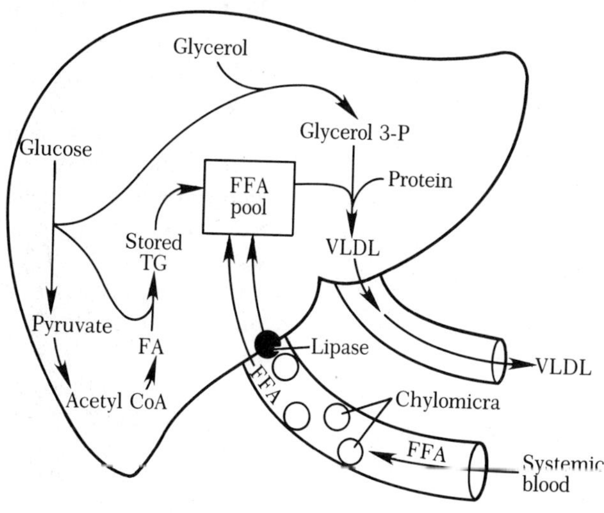

droxyacetone phosphate and glyceraldehyde, are used to produce glucose by way of the gluconeogenic pathway. Excess glucose is converted to fatty acids and then triacylglycerols by the pathways shown in Fig. 20-1. Glucose 6-phosphate is also metabolized by way of the pentose phosphate pathway to provide adequate supplies of NADPH for lipogenesis. The fate of dietary glucose in the liver is summarized in Fig. 20-3.

Dietary lipids enter the liver from two directions. Short-chain free fatty acids are complexed to albumin and arrive at the liver by means of the portal system. Except in infants this is usually a relatively minor process. The major portion of the dietary fatty acids arrive through the hepatic artery as chylomicrons or remnants of chylomicrons. As described in Chapter 12, dietary free fatty acids are converted to triacylglycerols in the intestinal mucosal cells. These, in conjunction with phospholipids and a thin layer of protein (apoprotein), constitute the chylomicrons that enter the lymphatic vessels and the systemic blood. Most of the chylomicrons are removed or cleared from the blood by extrahepatic tissues such as adipose tissue and muscles. In the capillary endothelial cells the lipoprotein lipase catalyzes the release of the fatty acids from the triacylglycerols. The remnants of the chylomicrons enter the liver, where the remaining triacylglycerols are degraded. About 20% of the dietary fatty acids enter the hepatocytes, where they may be used either as fuel or, more commonly, are resynthesized into triacylglycerols (Fig. 20-4).

The liver has three sources of fatty acids to form a single pool (Fig. 20-4). In addition to those from chylomicrons and circulating free fatty acids, the pool is supplied by endogenously synthesized fatty acids derived from acetyl CoA from glucose. These free fatty acids then are used to synthesize triacylglycerols and, in the presence of an apoprotein, form the very low–density plasma lipoprotein, or VLDL (Chapter 12). Therefore, the dietary fatty acids are reprocessed as VLDL and are redistributed to the adipose tissues for storage.

For amino acids there is no storage compound; no one protein can be considered to retain amino acids in the sense that glucose is stored as glycogen. All tissue proteins are functional in some fashion. Dietary amino acids are used to replace the various endogenous proteins and to synthesize the many export proteins, especially the plasma proteins. Once the necessary tissue proteins have been replaced, excess amino acids are promptly catabolized in preference to the degradation of carbohydrate and fat. Thus, under the conditions that prevail immediately following a meal, amino acids also serve as a direct fuel for oxidation. The rate of transamination of dietary amino acids varies in direct proportion to the increased blood levels. This is explained by the high K_m values for the hepatic aminotransferases. For example, the K_m (alanine) for alanine aminotransferase is 34 mM in the rat liver compared with an amino acid concentration in rat blood of only 0.323 mM. Such values are representative; the blood concentrations are generally less than 1 mM, whereas the K_m's are greater than 1.5 mM. Increased blood levels of amino acids also elicit a stimulated synthesis of metabolic enzymes such as alanine aminotransferase, which is elevated 15 times, and threonine dehydratase, which is increased 300 times. Such adaptive changes have been recorded for almost every enzyme initiating amino acid degradation, with the possible exceptions of those for cysteine, glycine, methionine, and proline.

The fate of dietary amino acid in the liver is summarized in Fig. 20-5. The role of the liver in relation to the role of other tissues in amino acid metabolism is illustrated in Fig. 20-6. This figure shows that the non–branched chain amino acids are metabolized by the liver where the carbon chains are used in gluconeogenesis. The branched chain amino acids are used primarily by the skeletal muscle for energy

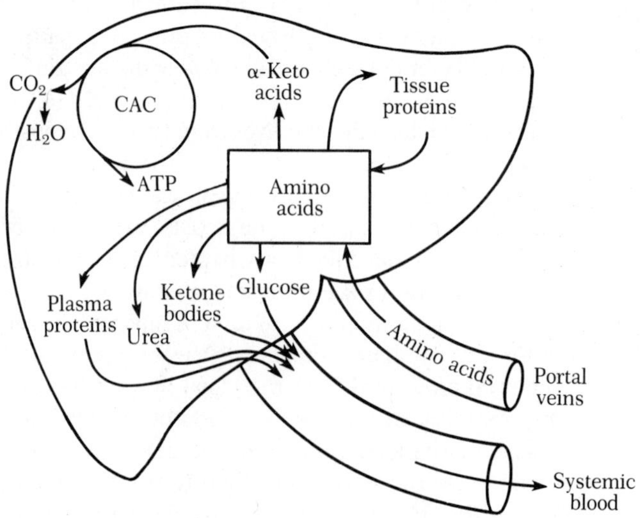

Figure 20-5
Summary of the fate of dietary amino acids in the liver. *CAC,* Citric acid cycle.

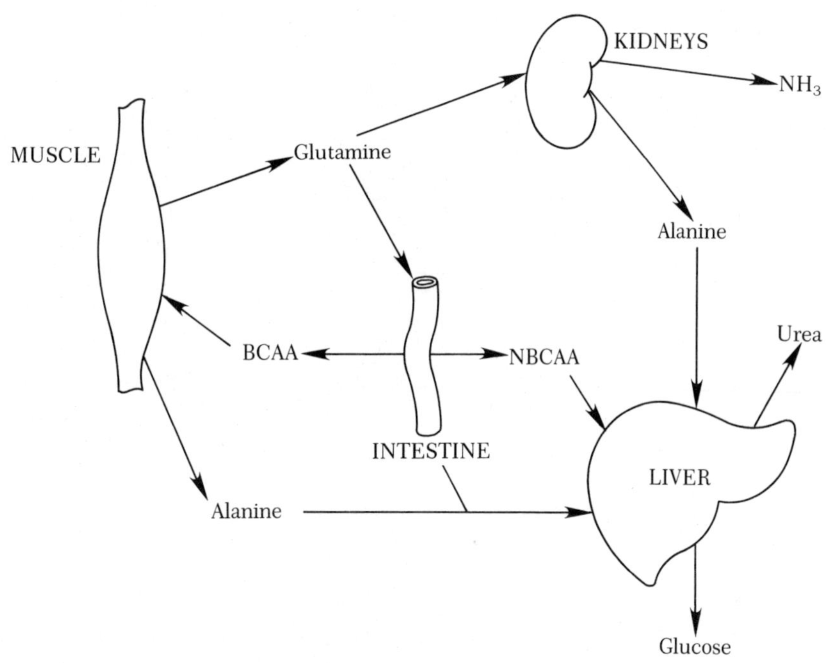

Figure 20-6
A summary of the metabolism of branched chain (BCAA) and non-branched chain (NBCAA) amino acids by various tissues of the body.

Modified from Felig, P.: Recent developments in body fuel metabolism. In Freinkel, N., editor: The year in metabolism, 1975-1976, New York, 1976, Plenum Publishing Corp.

production. In this case the nitrogen is transported through the bloodstream to the liver either as glutamine or as alanine. Glutamine actually goes to the kidneys, where it either yields ammonium ions for excretion or is converted to alanine and returned to the liver. The intestinal mucosa apparently is also involved in the conversion of glutamine to alanine.

In addition to reprocessing of nutrients, as just described, the liver plays specialized roles in the disposal of nitrogen derived from the metabolism of amino acids in other tissues. As described in Chapter 13, the liver is the primary site of urea formation. The hepatocyte mitochondria possess the necessary carbamoyl phosphate synthetase for the initial step in the urea cycle. In addition, the liver is specialized in its ability to convert excess acetyl CoA into ketone bodies and cholesterol. As already described in Chapter 12, the hepatocyte mitochondria contain the necessary HMGCoA synthase and lyase for ketone body and cholesterol biosynthesis.

Skeletal muscle

The muscle cells function to convert chemical energy to mechanical energy. Metabolically they are specialized to degrade the nutrients and to produce the necessary ATP for muscle contraction. Blood glucose enters the cells by way of an insulin-dependent facilitated diffusion. Once inside, glucose is phosphorylated and stored as glycogen. Although skeletal muscle may contain only 0.7% of glycogen, when the total mass of 35 kg is considered, this tissue represents 250 gm of glycogen. During exercise the glucose needed for fuel is derived directly from glycogen. In muscle cells the glycolytic pathway predominates. Since there is no glucose 6-phosphatase, this tissue cannot serve gluconeogenically. As described in Chapter 11 and Fig. 20-7, lactate produced during glycolysis enters the bloodstream and is returned to the liver, where it is converted to glucose (Cori cycle).

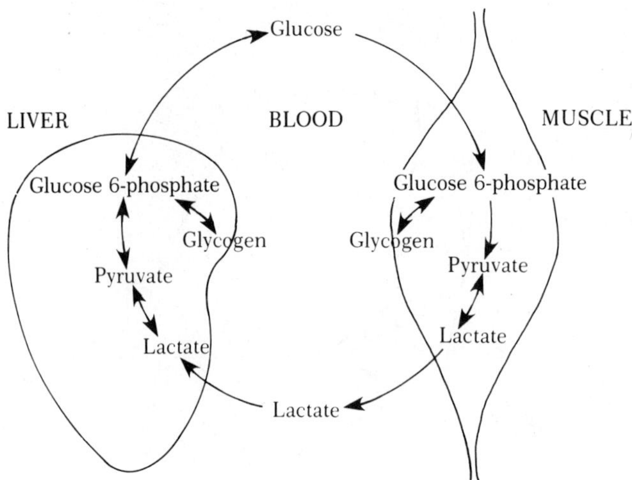

Figure 20-7
Cori cycle: recycling of lactate between muscle and liver through bloodstream. Lactate produced by glycolysis is converted to glucose by gluconeogenesis.

The triacylglycerols of chylomicrons and VLDL are hydrolyzed by lipoprotein lipase. Fatty acids released, as well as circulating free fatty acids (complexed to serum albumin), enter the muscle cells and are used for fuel. Some triacylglycerols may be synthesized and stored.

Amino acids are taken up by the muscle cells from the bloodstream. Following a meal they are used primarily for the restoration of tissue proteins. Insulin stimulates both amino acid transport and a general synthesis of protein.

Under appropriate physiological needs muscle proteins are hydrolyzed to release the amino acids. These in turn are capable of supplying the requirements for energy. Since the muscle cell cannot handle the nitrogen released during amino acid catabolism, alanine and glutamine are produced for the purpose of carrying the nitrogen to the liver in a nontoxic form.

Adipose tissue

The adipocytes are specialized cells that function primarily for the purpose of storing fuel as triacylglycerols. Over 85% of the cell by volume consists of a single large globule of fat.

The fatty acids used for triacylglycerol production are derived mainly from VLDL, chylomicrons, and some lipogenesis (from glucose). As already mentioned, the adipose tissue is the primary site for the removal or clearing of chylomicrons. The free fatty acids from this source, as well as from VLDL, enter the cells. Glucose also enters the adipocytes by an insulin-dependent facilitated transport mechanism. Fatty acids are synthesized from the acetyl CoA derived from glucose metabolism. More important, however, is the dihydroxy acetone phosphate, which is reduced to glycerol 3-phosphate. The triacylglycerols that are produced from the fatty acyl CoAs and the glycerol 3-phosphate are then stored in the cell. Therefore this overall process depends both on the supply of fatty acids (chylomicrons and VLDL) and on carbohydrate metabolism. Insulin affects lipid storage both by stimulating the lipoprotein lipase and by the transport of glucose into the cell. Insulin also suppresses lipolysis by inhibiting the action of the hormone-sensitive lipase.

Because of their low water and high fat content, the adipose tissues have the highest calorie-to-weight ratio (8 kcal per gram of tissue). This makes them the

Table 20-3
Distribution of stored fuels in a 70-kg man

Storage fuels	Weight (kg)	Available energy (kcal)
Fat (adipose)	15	141,000
Protein (muscle)	6	24,000
Glycogen (muscle)	0.12	480
Glycogen (liver)	0.07	280
Glucose (extracellular fluid)	0.02	80
	Total energy	165,842

From Owen, O.E., and Reichard, G.A., Jr.: Progr. Biochem. Pharmacol. **6:**177, 1971.

most efficient tissue for the storage of fuel. As seen in Table 20-3, fully 141,000 kcal, or 85% of the total body energy stores, are found as triacylglycerols in the adipose tissues.

As described in Chapter 12, the fatty acids stored are remobilized by the action of a hormone-sensitive lipase, which is activated by a cAMP-mediated protein kinase. This process is controlled by numerous hormones, especially glucagon and epinephrine.

The metabolic "three-ring circus"

Such specialized tissues as the liver, skeletal muscles, and adipose tissue must function together in a carefully orchestrated fashion for the overall operation of the human body. Together they must respond to the presence of excess nutrients after a meal by initiating storage processes; between meals, when fuel elements must be mobilized, these tissues again must respond by releasing metabolites in a controlled fashion. The overall integration of this metabolic "three-ring circus" requires the existence of two levels of control; these are by intracellular and by extracellular signals.

Intracellular signals

The instant-to-instant control of metabolic pathways depends on the presence and state of a variety of cofactors and nutrients. For example, the ATP/AMP and $NADH/NAD^+$ ratios, the acetyl CoA/CoA ratio, and the levels of Ca^{++}, glucose 6-phosphate, citrate, fatty acyl CoA, and alanine are all important signals for the control of metabolism. Reference to Chapter 11, and especially to Fig. 11-19, will reemphasize where these and other intracellular factors serve to control glycolysis, the citric acid cycle, gluconeogenesis, and lipogenesis.

Extracellular signals

The cellular environment, i.e., the extracellular fluid, provides a constant flow of nutrients as well as hormones. Thus a high level of blood glucose provides the cell with the necessary carbohydrate. The simultaneous presence of insulin stimulates glucose transport into the muscle and adipose tissue, thereby increasing the uptake of glucose by these tissues. In the liver, insulin stimulates the production of glucokinase, phosphofructokinase, and pyruvate kinase, thereby increasing glucose metabolism through glycolysis.

Metabolic adaptation to varying dietary and physiological conditions

As emphasized before, the average American diet consists of approximately 46% carbohydrate and 35% to 40% fat; protein may represent as much as 15% of the diet. After digestion the constituent carbohydrates, lipids, and amino acids are absorbed. The blood glucose level increases sharply (see Fig. 18-23). At that time there also exists a state of hyperlipemia and hyperaminoacidemia. As already described, the human adapts to a cyclical feed-fast process. A crucial aspect of this

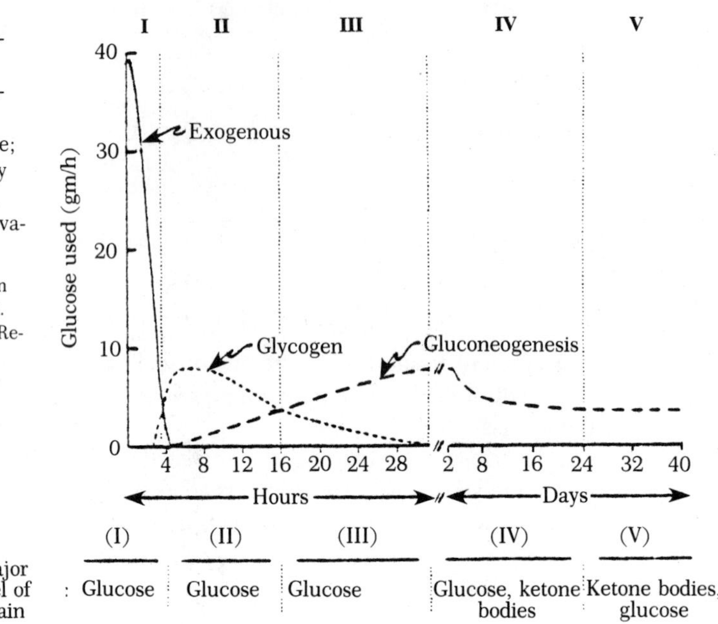

Figure 20-8
Five phases of glucose homeostasis showing the origin of blood glucose following the ingestion of 100 gm of glucose. Phase *I* is the absorptive phase; *II* is postabsorptive; *III* is early starvation; *IV* is intermediate starvation; *V* is prolonged starvation.

Reproduced, with permission, from Ruderman, N.B.: Annu. Rev. Med. **26**:245, 1975. © 1975 by Annual Reviews Inc.

adaptation is the ability to provide the brain continuously with about 140 gm of glucose daily. It is important to appreciate the various phases of metabolic adaptation involved. An insight into the adaptation process is provided by Fig. 20-8; the existence of five phases in glucose homeostasis is suggested. These will be considered sequentially in the following discussion.

Phase I: absorptive state

In this phase the systemic blood glucose is elevated, a condition that is recognized by receptor sites of the β-cells of the islets of Langerhans (see Fig. 18-23). Hyperglycemia (Chapter 18) is a condition that promotes the release of insulin and suppresses the release of glucagon from the α-cells. Under these hormonal conditions (high insulin and low glucagon) glucose is transported into muscle and adipose tissue cells. In the liver, insulin increases the capacity to phosphorylate glucose by increasing the synthesis of glucokinase. In this way the blood glucose levels are reduced to normal levels within 4 hours.

Intracellular signals, especially the ATP/AMP, NADH/NAD^+, the acetyl CoA/CoA ratios, and the citrate, modulate the glycolytic enzymes and the citric acid cycle. Although the liver is normally a gluconeogenic organ, the presence of insulin favors glycolysis and suppresses gluconeogenesis.

Once the needs for energy and glycogen deposition have been met, excess glucose is converted in the liver into triacylglycerol and VLDL. In this way the glucose carbons eventually are stored as fat in the adipose tissues.

Fig. 20-2 illustrates the general metabolic processes that prevail between the liver, skeletal muscle, and adipose tissue during the absorptive phase. In summary, the hormonal status of increased insulin and reduced glucagon have the following effects on metabolism:

1. Elevated glucose transport into adipose and muscle tissues
2. Stimulated glycogenesis and suppressed glycogenolysis

Table 20-4
Alterations in the rate of glucose turnover and oxidation resulting from fasting in a normal man

Duration of fast	Blood glucose (grams per 24 hours)	
	Turnover	**Oxidation**
7-14 hours	152	96
38-74 hours	120	70
5-33 days	87	46

From Owen, O.E., and Reichard, G.A., Jr.: Progr. Biochem. Pharmacol. **6**:177, 1971.

3. Stimulated glycolysis concomitant with increased synthesis of glucokinase, phosphofructokinase, and pyruvate kinase
4. Stimulated lipogenesis in the liver and adipose tissues; a suppression of lipolysis in the adipose tissue
5. Increased formation of triacylglycerols in all tissues, especially in the adipose tissues
6. Reduced oxidation of free fatty acids

The absorptive phase lasts approximately 4 hours; during this time the blood glucose derived from exogenous or dietary sources becomes less and less a factor, and remobilization of glucose from liver glycogen becomes an increasingly important source. After 4 hours the body enters phase II, or the postabsorptive state.

Phase II: postabsorptive state

Once the conditions of glycosemia, hyperlipemia, and hyperaminoacidemia have been normalized, the body is in the second, or postabsorptive, phase (Fig. 20-8). During this time the liver maintains the blood levels for these metabolites according to the needs of the peripheral tissues. This is especially true for blood glucose because, as mentioned earlier, both the brain and the red blood cells are dependent on this fuel. The brain of a normal adult uses 100 to 145 gm of glucose daily. As the circulating blood glucose is consumed, glucagon (Chapter 18) is secreted and glycogenolysis is stimulated. Recall (Chapter 11) that glucagon stimulates hepatic glycogenolysis by releasing cyclic AMP. The inactive phosphorylase b is activated by conversion to the active phosphorylase a. Glucose 1-phosphate, which is derived from glycogen by phosphorolysis, is converted to glucose 6-phosphate by phosphoglucomutase and then to glucose by glucose 6-phosphatase. Since there are only 70 gm of glycogen available for conversion to glucose and since not all is readily mobilizable, with some being retained for emergencies, there is actually not enough liver glycogen to serve the requirements of the brain and red blood cells from the evening meal to breakfast. Therefore the muscle mass must shift to another fuel to conserve glucose. However, even at rest the musculature removes some glucose from the blood and produces pyruvate and lactate, albeit at a reduced level. The continual need for ATP by this and all tissues results in the formation of ADP and a continued operation of glycolysis and the citric acid cycle.

As seen from Fig. 20-8, the importance of liver glycogen as a source of blood glucose begins to decline 8 hours after a meal. At this time blood glucose derived from the amino acid carbon chains by gluconeogenesis becomes an increasingly important factor. Thus in the postabsorptive period the liver of a 70-kg man may synthesize as much as 150 to 160 gm of glucose per day, of which 96 gm are oxidized (Table 20-4). Of the glucose synthesized by gluconeogenesis, 70% is used by the brain, 10% by the heart, and 7% by the musculature.

Accordingly, it is necessary during the postabsorptive state to replenish the glucose supply in the bloodstream. The fuel options available to the body are (1) to release glucose from glycogen, (2) to recycle intermediates derived from glucose, such as lactate, pyruvate, and glycerol, and (3) new synthesis of glucose from amino acids. As already mentioned, the 70 gm of hepatic glycogen hardly suffice to supply the normal rate of glucose use by the brain. The second and third possibilities for glucose production involve hepatic (and renal) gluconeogenesis. Lactate and pyruvate, derived from muscle metabolism, enter the bloodstream, are extracted by the liver, and are recycled through the gluconeogenic process to yield glucose. However, this recycling of 3-carbon units, called the Cori cycle (Fig. 20-7), is dependent on a high rate of anaerobic glucose metabolism, which is usually the consequence of severe exercise and not an important metabolic process in the postabsorptive state. Thus of the options for maintaining the blood glucose level in the postabsorptive state neither glycogen nor the recycling of lactate by the Cori cycle is a prime source; rather, it is gluconeogenesis that is dependent on amino acids derived from muscle tissue.

In the early postabsorptive state glucagon, secreted in increasing amounts as both insulin and blood glucose levels decline, stimulates hepatic glycogenolysis; this hormone also activates the hormone-sensitive lipase of the adipose tissue. Lipolysis leads to the mobilization of free fatty acids and glycerol. The energy of the dietary glucose, stored during lipogenesis as fatty acid, is then released and made

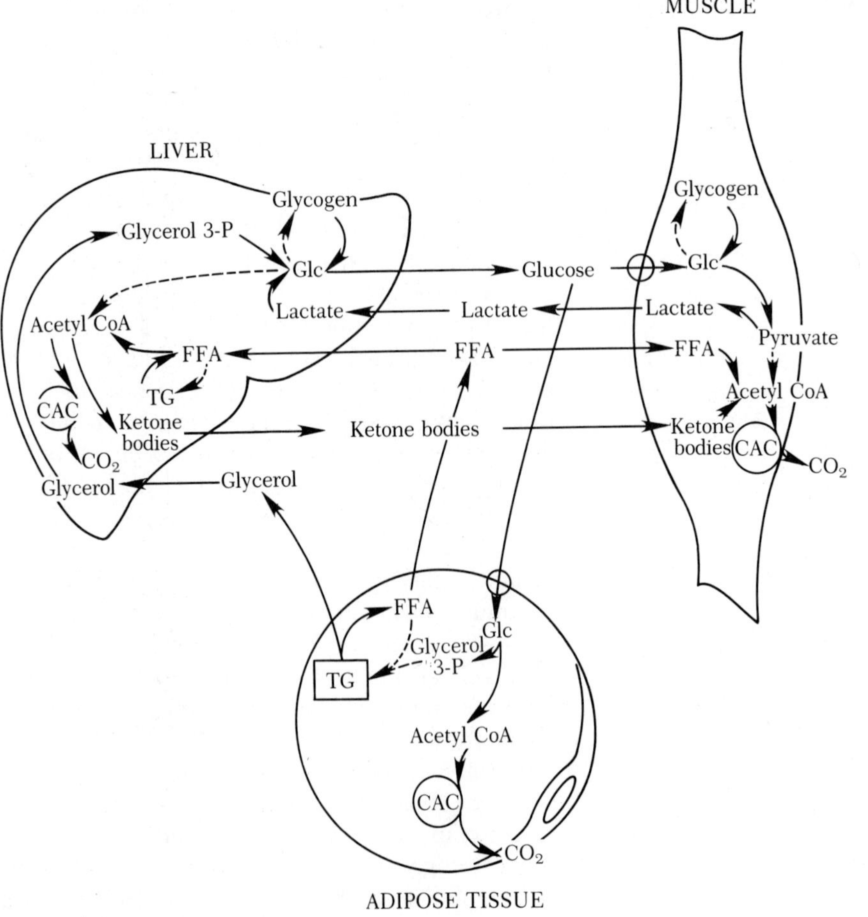

Figure 20-9
Major metabolic postures of the liver, muscle, and adipose tissues during the postabsorptive and fasting periods. *CAC,* Citric acid cycle; *FFA,* free fatty acids; *TG,* triacylglycerols; *Glc,* glucose.

available for ATP formation. The free fatty acids then serve as a major source of energy for all tissues except the brain, red blood cells, and the renal medulla (Table 20-1). Since free fatty acids cannot yield a net synthesis of glucose, they are not even indirectly a source of energy for brain and red blood cells. As described in Chapter 12, even-numbered fatty acids are degraded only to acetyl CoA, which proceeds for terminal metabolism (to CO_2 and H_2O) by the citric acid cycle. Since pyruvate dehydrogenase is physiologically irreversible, acetyl CoA cannot yield pyruvate. Glycerol from triacylglycerols and the terminal propionyl CoA of odd-numbered fatty acids can serve as substrates for gluconeogenesis and thus for the production of blood glucose. Since the fatty acids of most lipids have even-numbered carbon chains, this possibility is of minor importance.

Fig. 20-9 summarizes the metabolism of carbohydrates and lipids that prevails during the early stages of fasting. The hormonal status is an elevation of glucagon and a suppression of insulin. The various effects are as follows:

1. Suppressed transport of glucose
2. Elevated glycogenolysis and a suppressed glycogenesis
3. Suppressed glycolysis
4. Stimulated lipolysis and inhibited lipogenesis
5. Suppressed triacylglycerol formation
6. Elevated ketogenesis
7. Increased oxidation of free fatty acids

As the body progresses from the fed state to the postabsorptive condition, an elevated output of amino acids appears in the bloodstream from the muscle tissue. When the distribution of the individual amino acids in blood plasma was determined, alanine was observed to account for 30% to 40% of the total. Yet less than 10% of the residues in muscle protein are represented by alanine. It is now clear that the circulating alanine is not derived directly from protein but rather is the consequence of the peripheral synthesis of alanine by the transamination of pyruvate. Pyruvate derived from the glycolytic metabolism of glucose in muscle is transaminated to alanine. The amino acid enters the bloodstream, is transported to the liver, and is again converted to pyruvate by transamination. The resultant pyruvate is recycled to glucose by way of gluconeogenesis.

As seen in the following diagram, the increased levels of cellular alanine in the liver have two effects. One is as an allosteric inhibitor of pyruvate kinase and the other as a substrate for alanine aminotransferase.

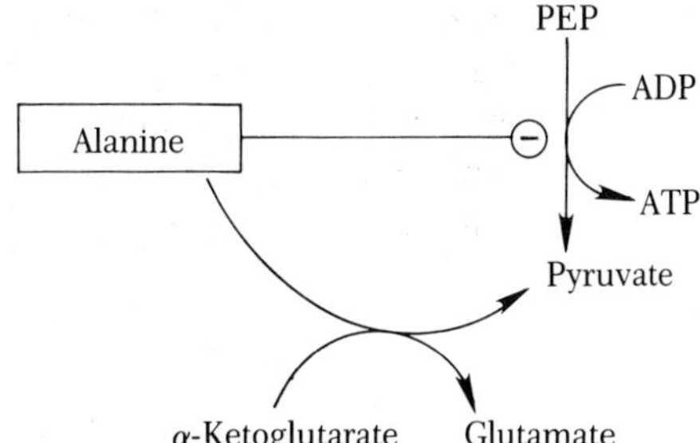

Thus alanine restricts the production of pyruvate from glucose via glycolysis while at the same time providing a source of pyruvate for gluconeogenesis. This overall process, reminiscent of the Cori cycle (Fig. 20-7), is now called the glucose-alanine cycle, as proposed by Cahill and by Felig (Fig. 20-10). In addition to the recycling of

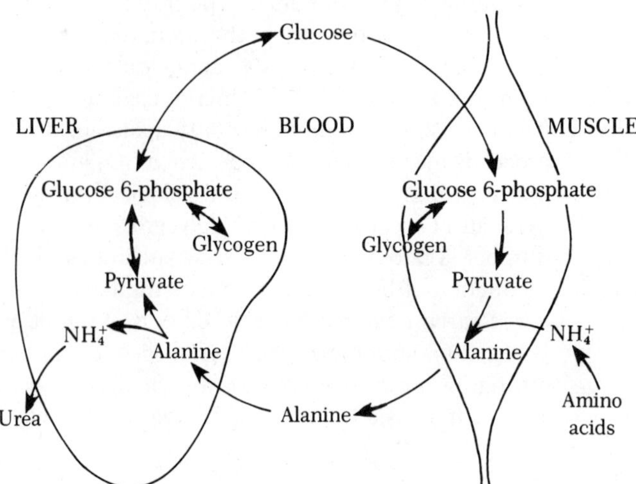

Figure 20-10
Glucose-alanine cycle: recycling of alanine between muscle and liver through bloodstream. L-Alanine produced by transamination of pyruvate provides three carbons for gluconeogenesis and nitrogen for urea formation.

Adapted from Felig, P.: Metabolism **22:**179, 1973; by permission of Grune & Stratton, Inc.

3-carbon units to yield glucose, the glucose-alanine cycle is an efficient mechanism for transporting the nitrogen to the liver from the amino acids liberated by the breakdown of muscle protein. However, this recycling does not produce a net amount of new glucose. Rather, it must be considered to be like the Cori cycle, a carbon-sparing process. The net production of glucose via gluconeogenesis reflects a use of nonbranched and glucogenic amino acids derived from the breakdown of peripheral (skeletal muscle) proteins.

In muscle tissue all the nonessential amino acids are degradable to α-keto acids either by transamination or by deamination. In either event, α-ketoglutarate is the acceptor molecule of the nitrogen (Fig. 20-11). Transamination reactions are catalyzed by the appropriate aminotransferase, whereas NH_4^+ is accepted by α-ketoglutarate via glutamate dehydrogenase. As shown in Figs. 20-1 and 20-11, the carbon chains that result from the catabolism of the nonessential amino acids are metabolized by the citric acid cycle and thus are a source of energy. Of the essential amino acids the branched chain amino acids (isoleucine, leucine, and valine) are degraded in muscle, whereas the remainder are degraded in the liver. Again, the first step is a transamination with α-ketoglutarate acting as the acceptor molecule. The α-keto acids derived from the branched chain amino acids are further degraded to acetyl CoA, succinyl CoA, propionyl CoA, and acetoacetate (see Fig. 20-1 and Chapter 13). All are capable of metabolism by the citric acid cycle for purposes of obtaining energy for muscle activity. As seen below, the three branched chain amino acids may yield approximately 42 ATPs per mole of amino acid.

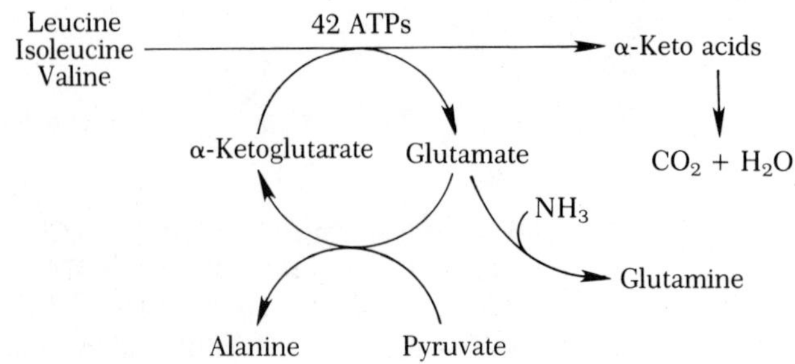

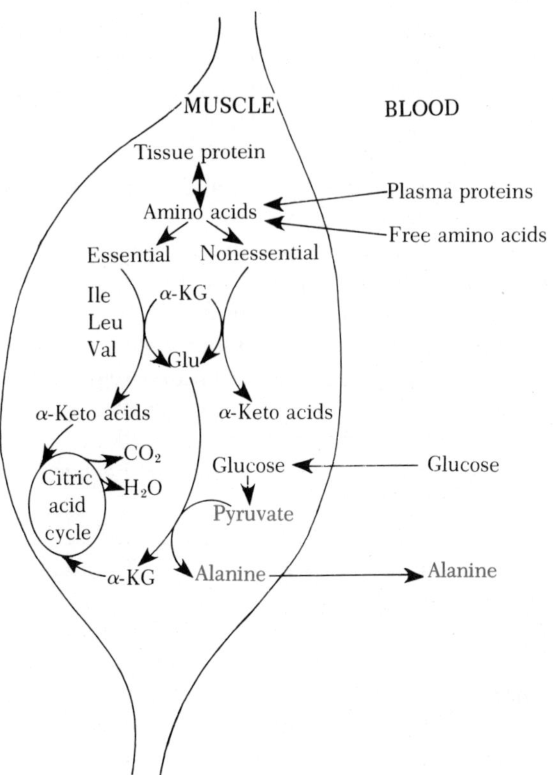

Figure 20-11
Scheme summarizing catabolism of amino acids in muscle tissue.

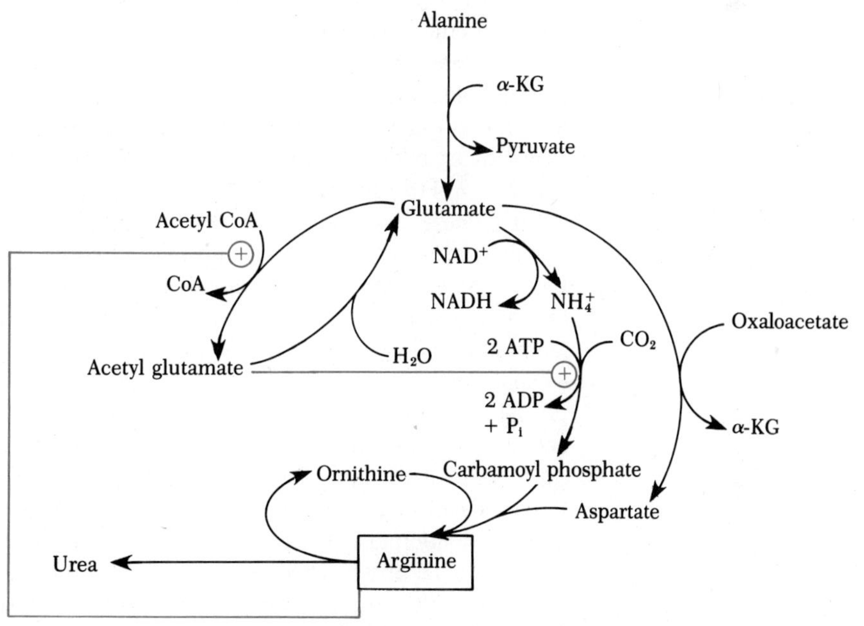

Figure 20-12
Possible system for regulating relative synthesis of aspartate and carbamoyl phosphate en route to urea production.

The glutamic acid produced during the catabolism of the various amino acids donates its nitrogen to pyruvate to yield alanine, which in turn enters the bloodstream and is extracted by the liver. Once the alanine is in the liver, the nitrogen is rapidly transferred back to glutamate. We have already described the kinetic control involving the hepatic alanine aminotransferase. Glutamate in turn yields its nitrogen in two directions (Fig. 20-12). In one, glutamate dehydrogenase catalyzes oxidative deamination; the resulting ammonia is used in the mitochondrial synthesis of carbamoylphosphate. In the other direction aspartate is formed by transamination with oxaloacetate. Stoichiometric amounts of aspartate and carbamoyl phosphate must be produced for the ultimate synthesis of arginine and thus urea (see Fig. 13-4). Although the details are unknown, there appears to be a coordinated formation of carbamoylphosphate and aspartate; they are produced in equal amounts. Perhaps arginine is a positive regulator in this process because it acts as a specific stimulator of acetylglutamate synthesis, which in turn is the regulator for carbamoylphosphate synthetase (Fig. 20-12). The capacity of the urea cycle is normally large enough to remove surplus nitrogen.

Phases III to V: fasting

As seen in Fig. 20-8, the period of fasting or "starvation" may be divided into three phases; early (III), intermediate (IV), and late (V). In phase III, 16 to 24 hours after the meal, gluconeogenesis increasingly provides the source of blood glucose. For this purpose about 155 gm of muscle protein must be degraded daily with a loss of nearly 25 gm of nitrogen. The body has a total of about 1000 gm of nitrogen, of which it cannot lose more than 50% without threat of death. At this rate a starving person might live for only 3 weeks. However, experience shows that an obese person can survive for much longer periods of starvation.

To prevent the dissolution of excessive body protein, survival dictated the evolution of another adaptive process. First, gluconeogenesis, dependent on the glucose-alanine cycle, is reduced considerably. As seen in Table 20-4, the turnover of glucose 7 to 14 hours after absorption is 152 gm/24 hr whereas at 1½ to 3 days after absorption it drops to 120 gm and finally to 87 gm/24 hr after 5 to 33 days of fasting. The arterial alanine concentration in the postabsorptive state and after a 1½ to 2-day fast is approximately the same, whereas after 35 or 42 days of starvation (phase V) the alanine concentration decreases by 70%. Apparently this reflects a severe curtailment in alanine output by muscle tissue. During the fasting period both hypoinsulinemia and hyperglucagonemia persist, so that the control of gluconeogenesis must reside in a reduced production of the amino acid. It seems likely that the degradation of muscle tissue is in some way the controlling factor.

Because of the decline in gluconeogenesis and the threat of a serious reduction in blood glucose, the body must enter a metabolic state in which further glucose metabolism is reduced to a minimum. In starvation, therefore, the metabolism of fat predominates. The increased metabolism of fatty acids by muscle leads to a major production of acetyl CoA and an elevation of the $NADH/NAD^+$ ratio. The accumulation of acetyl CoA and NADH slows the citric acid cycle. Acetyl CoA reacts with oxaloacetate to form citrate, which leaves the mitochondria and serves to inhibit phosphofructokinase; both fructose 6-phosphate and glucose 6-phosphate then accumulate. Hexokinase is inhibited by the accumulation of glucose 6-phosphate. All these changes limit the further use of glucose and favor the breakdown of free fatty acids as the primary fuel. As the glucose turnover diminishes (Table 15-4), the rate of oxidation also decreases; glucose oxidation diminishes from 96 to 46 gm/24 hr once the fasting has been extended to a matter of weeks. Although not apparent from this table, glucose oxidation reaches a minimum value that is not decreased further by extended fasting. These interactions serve to spare the blood glucose for use by the brain and red blood cells.

In the liver, accumulation of acetyl CoA inhibits pyruvate dehydrogenase, thereby preventing its further production from pyruvate, and therefore hepatic glycolysis is also suppressed. Although acetyl CoA stimulates pyruvate carboxylase activity and therefore favors gluconeogenesis, the diminished supply of necessary precursors such as alanine and pyruvate prevents this from being a major avenue of metabolism. Since acetyl CoA must be metabolized by the citric acid cycle, a stoichiometric amount of oxaloacetate is essential. When the acetyl CoA exceeds the oxaloacetate as well as the anaplerotic production of this dicarboxylic acid from a lack of pyruvate, then the accumulating acetyl CoA is directed toward ketone body formation. The biosynthetic pathway, occurring almost entirely in the liver, is described in Fig. 12-11. Acetoacetic acid is reduced to β-hydroxybutyrate by the appropriate dehydrogenase; NADH is obtainable from β-oxidation of free fatty acids. The ketone bodies diffuse into the circulation and, as seen from Table 20-2, are usable for energy metabolism by muscle (cardiac and skeletal), the renal cortex, and the brain. In a normal man weighing 72 kg the ketone body production is approximately 24 gm/24 hr after an overnight fast. After 24 hours the free fatty acids, acetoacetate, and β-hydroxybutyrate have achieved a steady-state concentration (Table 20-5). Use, as well as ketonuria, serves to balance the production of ketone bodies. Starvation of 5 to 6 weeks' duration, however, will elevate the ketone body production to about 92 gm/24 hr. In starvation, therefore, there is a reciprocal relationship between ketogenesis and gluconeogenesis; as ketogenesis increases, gluconeogenesis declines.

Under conditions of starvation the body still cannot do entirely without glucose. The citric acid cycle must be maintained to produce ATP for synthetic reactions. Red blood cells, which derive their ATP only from glycolysis, are totally dependent on blood glucose. The brain, however, can adapt itself to the use of ketone bodies; thereby the need for blood glucose is reduced even further. Apparently the human brain begins to use ketone bodies for its energy needs once the concentration of these metabolites reaches a sufficiently high level in the blood. After 5 to 6 weeks of starvation glucose represents only about 30% of the brain's energy needs. β-hydroxybutyrate supplies 50%; acetoacetate supplies approximately 5%; and amino acids supply perhaps 15%. Even after prolonged starvation there is still some turnover of tissue proteins. This minimal catabolism of protein yields sufficient amino acids to maintain the needs for glucose.

Because of these adaptive mechanisms, the adult human is capable of going without food for incredible periods of time. For example, obese patients have fasted for as long as 236 to 249 days. Obesity is a problem in countries where food supplies are more than adequate. It has been estimated that Americans spend approximately

Table 20-5
Metabolic parameters in fasting man

	Duration of fast		
	Overnight	**3 days**	**28 to 42 days**
Insulin (μU/ml)	37	20	17
Glucagon (pg/ml)	75	140	110
Glucose (mM)	4.8	3.8	3.8
Acetoacetate + β-hydroxybutyrate (mM)	0.1	1.8	7.4
Alanine (μM)	344	321	140
Leucine (μM)	112	152	72
Urine-*N*, gm/day	14	11	5

Reproduced, with permission, from Ruderman, N.B.: Annu. Rev. Med. **26:**245, 1975. © 1975 by Annual Reviews Inc.

\$500,000,000 annually on gimmick diets and food fads in a continuing effort to control weight. The best reducing diet is one balanced in food ingredients and sufficiently low in total calories to produce a loss of weight at a desirable rate. However, there are circumstances involving obesity in which total fasting is the best approach. Since this can be dangerous, total fasting of several weeks' duration is usually conducted under careful medical supervision.

Exercise

An immediate consequence of exercise, during which the rate of muscular contraction is intensified, is the dephosphorylation of ATP to ADP. There are three sources of supply for the rapid replenishment of ATP, all anaerobic (Fig. 20-13). With the rapid accumulation of ADP, oxidative phosphorylation is stimulated and more oxygen is consumed. Within minutes the supply of oxygen to the muscle tissue is increased and the mode of metabolism becomes almost entirely aerobic. This is actually contingent on the intensity and duration of the exercise as well as the physical condition of the individual. An athlete running a 100-yard dash consumes glucose entirely by anaerobic processes, whereas in an athlete participating in a track event requiring sustained efforts, both anaerobic and aerobic mechanisms function. Lactate from anaerobic metabolism may be used as fuel by cardiac muscle and by the kidneys. Muscle glycogen is rapidly depleted; therefore, to maintain exercise for any length of time, the body changes its source of fuel. Glucagon and epinephrine increase the rate of lipolysis (Fig. 12-9), thereby increasing the plasma-free fatty acids. As the supply of glycogen is depleted, pyruvate production decreases and the production of $NADH^+$ also diminishes. Consequently, CoQ will be increasingly subject to reduction by $FADH_2$ of the electron transfer flavoproteins of fatty oxidation. Thus there is a change to a lipid economy; fatty acid oxidation is increased. In moderate exercise performed 2 hours after absorption 25% to 50% of the needed energy is derived from free fatty acids.

The interrelationship of the liver metabolism with that of the skeletal muscle and adipose tissue during exercise is summarized in Fig. 20-14. Here the hormonal status is an elevation of glucagon and epinephrine and a suppressed level of insulin. Following are the effects on carbohydrate and lipid metabolism:

1. Glycogenolysis is stimulated while glycogenesis is inhibited.
2. The production and release into the bloodstream of lactate is elevated; gluconeogenesis in the liver is stimulated.
3. Lipolysis is elevated, whereas lipogenesis is suppressed.
4. The oxidation of free fatty acids is elevated.

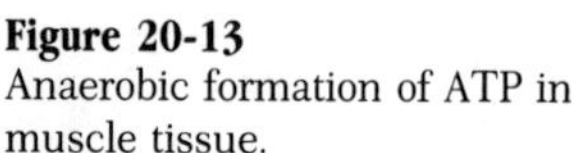

Figure 20-13
Anaerobic formation of ATP in muscle tissue.

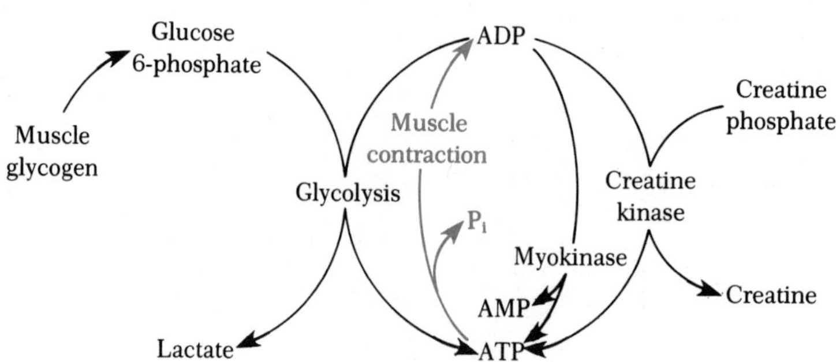

During exercise there is also an elevation in plasma alanine that is proportional to the blood pyruvate and thus the intensity of the exercise. Mild leg exercise, for example, results in a 20% to 25% increase in plasma alanine, whereas intense exercise will increase alanine by 60% to 90%. Most of this alanine is synthesized peripherally from pyruvate by transamination. The nitrogen is derived from the metabolism of the various amino acids released by the catabolism of tissue proteins. Under conditions of heavy exertion the levels of isoleucine, methionine, tyrosine, and phenylalanine are elevated 8% to 35%. As shown by Fig. 20-8, the branched chain amino acids are catabolized by muscle and contribute to the production of ATP. During exercise both the Cori and the glucose-alanine cycles are important mechanisms for maintaining blood glucose levels. After exercise the high ADP level will continue to promote oxygen consumption, stimulate the citric acid cycle by means of isocitrate dehydrogenase, and stimulate glycolysis by means of phosphofructokinase (Fig. 11-19). The oxygen consumption will continue at an elevated level after cessation of physical activity until the ATP pool, creatine phosphate, and glycogen levels have been restored to normal. The amount of oxygen consumed above the resting level during the recovery period, needed to restore these systems, is called the *oxygen debt*.

Figure 20-14
Integration of the major metabolic activities of the liver, skeletal muscles, and adipose tissues during exercise. *CAC*, Citric acid cycle; *FFA*, free fatty acids; *TG*, triacylglycerols; *Glc*, glucose.

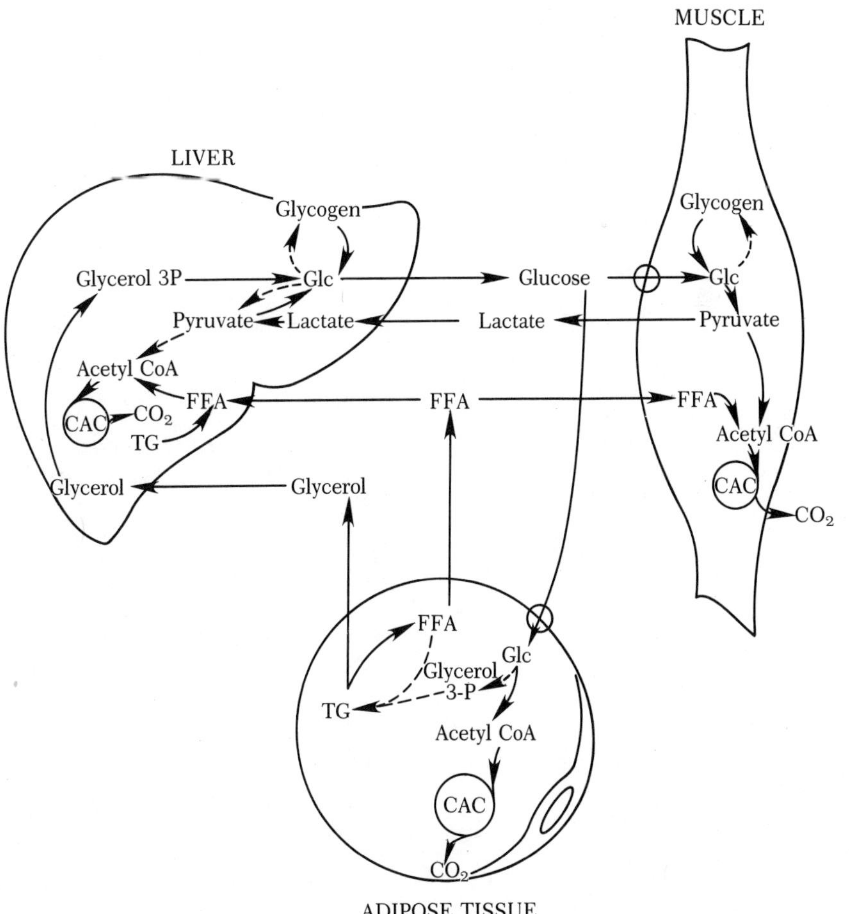

Suggested readings

Felig, P.: The glucose-alanine cycle, Metabolism **22:**179, 1973.

Felig, P., and Wahren, J.: Protein turnover and amino acid metabolism in the regulation of gluconeogenesis, Fed. Proc. **33:**1092, 1974.

Krebs, H.A.: The regulation of the release of ketone bodies by the liver, Adv. Enzyme Regul. **4:**339, 1971.

Krebs, H.A.: Some aspects of the regulation of fuel supply in omnivorous animals, Adv. Enzyme Regul. **10:**397, 1972.

Krebs, H.A., Hems, R., and Lund, P.: Some regulatory mechanisms in the synthesis of urea in the mammalian liver, Adv. Enzyme Regul. **11:**361, 1973.

Masoro, E.J.: Physiological chemistry of lipids in mammals, Philadelphia, 1968, W.B. Saunders Co.

Owen, O.E., and Reichard, G.A., Jr.: Fuels consumed by man: the interplay between carbohydrates and fatty acids, Progr. Biochem. Pharmacol. **6:**177, 1971.

Ruderman, N.B.: Muscle amino acid metabolism and gluconeogenesis, Annu. Rev. Med. **26:**245, 1975.

Snell, K.: Alanine as a gluconeogenic carrier, Trends Biochem. Sci. **4:**124, 1979.

Weber, G.: Study and evaluation of enzyme activity and synthesis in mammalian liver, Adv. Enzyme Regul. **1:**l, 1963.

Young, V.R., and Scrimshaw, N.S.: The physiology of starvation, Sci. Am. **225:**4, 14, 1971.

Section Six

Human nutrition and molecular aspects of disease

Chapter 21

Nutrition: normal and clinical

Today there is an unprecedented interest in human nutrition. The public is becoming increasingly aware of the role of nutrition in the maintenance of health and is more sophisticated in the questions it asks. Further evidence of the current concern with human nutrition is the involvement of the various agencies of the federal government in matters such as establishing recommended dietary allowances through the Food and Nutrition Board of the National Academy of Sciences and dietary goals and guidelines by a U.S. Senate Committee on Food and Nutrition and enacting further legislation on pure food and drug (FDA) requirements and standards. Other national groups (e.g., American Medical Association, American Heart Association, National Cancer Institute) likewise have become more concerned, proposing nutritional guidelines especially for minimizing the risk of developing cancer, coronary heart disease, and certain other pathological conditions. Since the science of nutrition was developed and then nurtured to a large extent by biochemists, this discipline can play an important role in providing definitive, accurate information in this field and reasoned discussion of its controversial issues.

The science of nutrition deals with the processes by which components of food are made available to an organism for meeting energy requirements, for building and maintaining tissues, and in more general terms for maintaining the organism in an optimum functional state. Nutrition is thus concerned with many problems that are traditionally considered to be biochemistry, e.g., the digestion, absorption, transport, metabolism, and biochemical nature and functions performed by the essential nutrients. The close interrelationship between the two fields also is evident from the interests and professional activities of many pioneers and indeed current leaders in the field of nutrition who, by training, also are biochemists. The names of early workers in nutrition trained as biochemists include such eminent scientists as Chittenden, Osborne, Mendel, McCollum, Sherman, W.C. Rose, Steenbock, Elvehjem, King, and du Vigneaud. This is borne out, too, by the strong biochemical roots of the early development of nutrition—the prime importance of energy to the living organism; the essentiality of food proteins for the growth and development of the animal (and human) organism; the discovery of the essential amino acids in the 1910 to 1930 period; the isolation, determination of chemical structure, and biochemical functions of the vitamins in the 1925 to 1945 period; and the role of a number of vitamins and minerals as cofactors in metabolic reactions.

Since these earlier basic studies knowledge of the science of nutrition has increased annually at an almost exponential rate. It is the purpose of this chapter to present current knowledge about nutrition, more specifically, to consider the nutrients now recognized as essential for optimum human nutrition, their chemical nature, sources, requirements, and biochemical functions, and effects of their deficiency. Also, current problems of nutrition, some controversial, will be discussed, such as food additives, food "quackery," desirable dietary adjustments, and diet therapy in a number of pathological conditions afflicting humanity.

Nutrients required

Nutrients are simply those chemical substances essential for the growth and maintenance of normal cells—both animal and plant. The present emphasis, however, is on human cells and tissues.

Some 45 to 50 chemical entities are now known to be "essential" in the sense of being required by humans, either *preformed* in foods or as an appropriate chemical substitute. These may be divided into six main categories: carbohydrate, lipid, protein, inorganic elements, vitamins, and water. Fiber (nondigestible food carbohydrate), although by definition not strictly a nutrient, also is being considered as a necessary food component. The essential nutrients now recognized for humans may be classified into the following broad groups:

1. *Carbohydrate*
   a. Glucose, fiber
2. *Lipid*
   a. Linoleic acid
3. *Protein*
   a. Amino acids
      (1) Leucine
      (2) Isoleucine
      (3) Lysine
      (4) Methionine
      (5) Phenylalanine
      (6) Threonine
      (7) Tryptophan
      (8) Valine
      (9) Histidine
      (10) Arginine (for growth only)
   b. "Nonessential" amino acid nitrogen
4. *Inorganic elements*
   a. Na^+
   b. K^+
   c. Ca^{++}
   d. Mg^{++}
   e. Fe^{++} or Fe^{+++}
   f. P (as phosphate)
   g. S (as Cys, Met)
   h. Cl^-
5. *Trace elements*
   a. Cu^{++}
   b. Co (as vitamin B_{12})
   c. Mn^{++}
   d. Zn^{++}
   e. I
   f. F (possibly essential in humans)
   g. Cr (possibly essential in humans)
   h. Mo (possibly essential in humans)
   i. Se (possibly essential in humans)
   j. Ni (possibly essential in humans)
   k. Si (possibly essential in humans)
   l. Sn (possibly essential in humans)
   m. V (possibly essential in humans)
6. *Vitamins*
   a. Fat soluble
      (1) Retinol (A)
      (2) Cholecalciferol (D)
      (3) α-Tocopherol (E)
      (4) Phylloquinone (K)
   b. Water soluble
      (1) Thiamin (B_1)
      (2) Riboflavin (B_2)
      (3) Niacin (G)
      (4) Pyridoxine (B_6)
      (5) Pantothenic acid
      (6) Biotin
      (7) Folic acid
      (8) Cobalamin (B_{12})
      (9) Ascorbic acid (C)
      (10) Lipoic acid (possibly)
7. *Water*

With dietary foods ingested, digested, and their essential nutrients thus made available for cellular metabolism, the science of nutrition now becomes concerned primarily with the following *four* basic questions:

1. From what foods are the essential nutrients best obtained?
2. How much of each is needed?

3. For what biochemical-nutritional purposes are they required?
4. What are the effects of their deficiency?

The major concerns of traditional biochemistry at this point, on the other hand, are the biochemical pathways (and their regulation) by which these nutrients are metabolized by cells. The foregoing basic questions therefore constitute the *major* thrust of this chapter. In addition, special consideration will be given to the recommended daily dietary allowances, as revised in 1980 by the Food and Nutrition Board of the National Academy of Sciences (USA), and to a practical daily food plan to supply the essential nutrients. Finally, some nutritional problems related to pathological processes in the body will be discussed.

Energy

Since energy is well recognized as a prime requirement for humans and other living organisms, a brief consideration of *energy metabolism* and energy intake and expenditure *(energy balance)* is appropriate at this point. It is common knowledge that if the food energy intake is inadequate to meet the body's energy requirement, loss of body weight occurs, body carbohydrate and fat stores are gradually decreased, and, because of the urgent drive for energy, body protein itself is metabolized to supply energy. Severe emaciation and drastic metabolic alterations (ketosis, acidosis, nitrogen and cation loss, and dehydration) ensue, and if extended, a disastrous, even fatal, outcome may result. Such calorie-protein malnutrition, although rarely seen in the United States, is not uncommon in many other parts of the world today.

A major portion of food energy, approximately 50% to 55% in the United States, is derived from carbohydrate-containing foods. Lesser amounts are obtained from food fats (30% to 40%) and proteins (about 10% to 15%). After these foods are digested and the resulting nutrients are absorbed, they are converted in part to *chemical energy* in the form of ATP by way of oxidative phosphorylation and used as described in Chapters 7 and 11. Part of the nutrients also are used, of course, for the growth and maintenance of body tissues.

The first scientific demonstration that energy is produced in the animal organism by the oxidation of foods was made by Lavoisier in 1794. For this reason Lavoisier is regarded by many as the father of the science of nutrition. Lavoisier discovered that the oxidation of carbon derived from foods was converted to carbon dioxide in the animal body by atmospheric oxygen and that the amount of energy (heat) produced was equivalent to that produced by the combustion of nonliving carbon in a calorimeter. Thus energy production in the animal body from foodstuffs may be determined using this basic principle, by measuring the oxygen consumed and the carbon dioxide produced as well as by measuring the heat evolved directly in suitable *calorimeters*. Thus a carbohydrate, e.g., glucose, is oxidized as follows:

$$C_6H_{12}O_6 + 6\ O_2 \rightarrow 6\ CO_2 + 6\ H_2O + \text{Energy (heat, etc.)}$$

One gram of carbohydrate yields an average of 4.0 kcal of energy and requires 6 moles of O_2 (134 L of O_2 at STP) to produce 6 moles of CO_2 per mole of glucose. The *respiratory quotient* (RQ) for the oxidation of glucose (and carbohydrates in general) is thus $\frac{6\ CO_2}{6\ O_2}$, or 1.0. Similarly, fats have an RQ of approximately 0.7, and proteins about 0.8. The RQ for humans in a resting, postabsorptive state is 0.82, and 1 L of oxygen at standard conditions of temperature and atmospheric pressure (STP) has the equivalence of 4.825 kcal.

Using these principles the *total energy metabolism* and *basal metabolism,* i.e., metabolism in a resting, postabsorptive state, of humans (and experimental animals) can be accurately determined in suitably constructed *respiration calorimeters*. Further information on the methods and conditions employed may be found in appropriate references at the end of this chapter.

In the United States, values for *total metabolism* are usually expressed as *kilocalories* (kcal) per unit time (day, hour, or minute) and/or per unit weight (kilograms or pounds). Basal metabolism is expressed as kilocalories per square meter of body surface per hour for clinical purposes, and also usually as percent above or below the norm for persons of the same age and sex (basal metabolic rate [BMR]). Recall that a *kilocalorie* (large calorie) is the amount of heat required to raise the temperature of 1000 gm of water 1° C, from 15° to 16° C.

At the present time in most European countries and increasingly in the United States, energy values in the nutrition field also are being expressed in the SI unit *megajoules* (MJ) (Chapter 25): 1000 kcal = 4.184 MJ; 1 MJ = 239 kcal (see Table 21-2). Recall also that one *joule* equals 10^7 ergs, or the energy expended when 1 kg is moved 1 m by 1 newton.

The determinations of total and basal metabolism in humans and experimental animals have been extremely useful in the field of nutrition. Basal metabolism, as the term suggests, measures the energy required for sustaining basic life processes, since the subject is at complete physical and mental rest and in a postabsorptive state (14 hours after the last meal, usually the evening meal). About 50% of the energy of basal metabolism is used for ion transport, e.g., Na^+, K^+, and Ca^{++} in or out of cells. The remainder apparently is expended for maintaining functions such as muscle tonus, the heart beat, respiration, and body temperature. In individuals engaged in light to moderate activities basal metabolism amounts to about half the total metabolism. The basal metabolism of an adult man varies from approximately 1600 to 2000 kcal per day and from 1400 to 1670 for adult women, depending on height and weight. Determination of basal metabolism, as compared with the norm for age and sex, has been of considerable value in the diagnosis of a number of diseases, especially those involving endocrine disorders, particularly of the thyroid gland.

Total metabolism values include those for basal metabolism plus energy expenditures for various activities, such as muscular work, mental activities, sleep, emotional states, and others. Some typical values for energy expenditure of adult men

Table 21-1
Typical energy expenditures of mature men and women engaged in various activities

Activity category	Man (70 kg)		Woman (58 kg)	
	kcal/min	kcal/kg/hr	kcal/min	kcal/kg/hr
Sleeping, reclining	1.0-1.2	0.9	0.9-1.1	0.85
Very light Sitting or standing, studying, talking	1.2-2.5	1.5	Up to 2.0	1.3
Light Laboratory work, driving a car, typing, teaching, shop work, walking on level 2.5-3 mph, golf, sailing, table tennis, volleyball	2.5-4.9	2.9	2.0-3.9	2.6
Moderate Walking 3.5-4 mph, gardening, carpentry, cycling, skiing, tennis, dancing	5.0-7.4	4.3	4.0-5.9	4.1
Strenuous Manual labor, forestry work, climbing, basketball, football, swimming, running	7.5-12.0	8.4	6.0-10.0	8.0

*Data taken from various sources.

and women engaged in various activities are given in Table 21-1. As an example, a 70-kg male student might have the following daily energy expenditures:

Activity	Time (hr)	kcal per hr	kcal expended per day
Sleeping	8	63	504
Very light	12	105	1260
Light	2	203	406
Moderate	2	301	602
		Total	2772

To approximately balance this energy expenditure per day, the subject could ingest daily the following:

$$
\begin{aligned}
&400 \text{ gm carbohydrate} \times 4 \text{ kcal/gm} = 1600 \text{ kcal}\\
&100 \text{ gm fat} \times 9 \text{ kcal/gm} = 900 \text{ kcal}\\
&70 \text{ gm protein} \times 4 \text{ kcal/gm} = \underline{280 \text{ kcal}}\\
&\text{Total} = 2780 \text{ kcal}
\end{aligned}
$$

Such a mixture of foodstuffs would be satisfactory from a current nutritional viewpoint, since 57.5% of his energy needs would be satisfied by carbohydrate, 32.3% by fat, and 10.1% by protein.

A plan for supplying this subject's energy needs by actual foods will be presented toward the end of this chapter.

Variations of total metabolism in various pathological conditions also occur. Different foods themselves affect total metabolism, the *specific dynamic action* (SDA) of foods.

Specific dynamic action of foods

The fact that the consumption of foods is associated with an expenditure of energy is well-established. This effect has been termed the *specific dynamic action (SDA)* of foods. A considerable amount of research devoted to this subject has demonstrated that proteins have an SDA of approximately 30%, carbohydrates about 6%, and fats about 4%. Thus from 25 gm of protein only 70 kcal could be available as usable energy, the other 30 kcal being expended as the SDA. However, when proteins are administered as a mixture with carbohydrates and fats, the SDA is only about 6% to 10%. Fats are most involved in reducing the SDA for as yet unclear reasons.

A complete explanation of the SDA effect is not as yet available. Intravenously administered amino acids show the same SDA as that of orally ingested protein, so the effect is not entirely related to activity of the gastrointestinal tract. Hepatectomy is reported to abolish the effect. Thus the SDA of proteins appears to be due primarily to the energy required for the deamination of amino acids and the formation of urea (Chapter 13). The SDA of carbohydrates and fats apparently is due also to the energy expended in their metabolism (Chapters 11 and 12). The energy expenditure for the SDA of foods would be included largely in the individual's basal metabolism.

Recommended energy allowances

Perhaps of greatest importance is the value of total and basal metabolism data in estimating the *caloric requirements* of human subjects. The Food and Nutrition Board of the National Academy of Sciences–National Research Council has relied heavily on such data in the literature in arriving at their recommended energy allowances, as revised in 1980 (Table 21-2). The recommended allowances for each age group listed, male and female, were established at the *lowest value* thought to be consonant with the good health of the *average* person in the group listed. The table also includes updated desirable weights for individuals of different heights.

Note that the recommended energy intakes are expressed in both kilocalories and megajoules.

Energy needs are cited for four adult age groups as well as for infants and children. Each age and sex category for adults shows a rather wide range of ±400 kcal per day of energy needs. This is included to reflect the *large variation* in energy expenditure of grossly similar individuals. The values for the four age groups of adults are given also to emphasize that energy needs decline progressively throughout adult life, due primarily to decreased physical activity. Added energy requirements during pregnancy and lactation also are included.

The recommendation by authorities in the field of nutrition to the American public thus is that the energy intake should balance energy expenditure to *obtain* and *maintain* an appropriate *body weight for height* throughout the entire life cycle. A negative balance caused by an inadequate energy intake to balance caloric expenditure is undesirable, since, if prolonged, it can lead to loss of body weight, increased susceptibility to infectious diseases, ketosis and acidosis, and other pathological manifestations that can be life threatening. On the other hand, a prolonged positive energy balance, more often seen in the United States, can lead to an excessive increase in body weight (fat) and obesity (30% over the mean weight for

Table 21-2
Recommended daily energy intake (revised 1980 by the Food and Nutrition Board of the National Academy of Sciences–National Research Council)*

Category	Age (years)	Weight (kg)	Weight (lb)	Height (cm)	Height (in)	Energy needs (with range) (kcal)	Energy needs (with range) (MJ)
Infants	0.0-0.5	6	13	60	24	kg × 115 (95-145)	kg × .48
	0.5-1.0	9	20	71	28	kg × 105 (80-135)	kg × .44
Children	1-3	13	29	90	35	1300 (900-1800)	5.5
	4-6	20	44	112	44	1700 (1300-2300)	7.1
	7-10	28	62	132	52	2400 (1650-3300)	10.1
Males	11-14	45	99	157	62	2700 (2000-3700)	11.3
	15-18	66	145	176	69	2800 (2100-3900)	11.8
	19-22	70	154	177	70	2900 (2500-3300)	12.2
	23-50	70	154	178	70	2700 (2300-3100)	11.3
	51-75	70	154	178	70	2400 (2000-2800)	10.1
	76+	70	154	178	70	2050 (1650-2450)	8.6
Females	11-14	46	101	157	62	2200 (1500-3000)	9.2
	15-18	55	120	163	64	2100 (1200-3000)	8.8
	19-22	55	120	163	64	2100 (1700-2500)	8.8
	23-50	55	120	163	64	2000 (1600-2400)	8.4
	51-75	55	120	163	64	1800 (1400-2200)	7.6
	76+	55	120	163	64	1600 (1200-2000)	6.7
Pregnancy						+300	
Lactation						+500	

From Recommended dietary allowances, ed. 9, Washington, D.C., 1980, National Academy of Sciences.
*The data in this table have been assembled from the observed median heights and weights of children together with desirable weights for adults given for the mean heights of men (70 inches) and women (64 inches) between the ages of 18 and 34 years as surveyed in the U.S. population (HEW/NCHS data).

The energy allowances for the young adults are for men and women doing light work. The allowances for the two older age groups represent mean energy needs over these age spans, allowing for a 2% decrease in basal (resting) metabolic rate per decade and a reduction in activity of 200 kcal/day for men and women between 51 and 75 years, 500 kcal for men over 75 years, and 400 kcal for women over 75. The customary range of daily energy output is shown for adults in parentheses and is based on a variation in energy needs of ± 400 kcal at any one age, emphasizing the wide range of energy intakes appropriate for any group of people.

Energy allowances for children through age 18 are based on median energy intakes of children these ages followed in longitudinal growth studies. The values in parentheses are 10th and 90th percentiles of energy intake, to indicate the range of energy consumption among children of these ages.

age and sex). Obesity, in turn, is associated with decreased longevity and the development of chronic health disorders such as cardiovascular disease, hypertension, gallbladder disease, maturity-onset diabetes, and a variety of psychological disturbances. Based on measurements of skinfold thickness, as many as 10 to 50 million Americans are in positive energy balance leading to obesity. Approximately 18% of men ages 20 to 74 years and 13% of women in the same age group are 10% to 20% or more above their appropriate weight for height. For those weighing 20% or more above their appropriate weight for height the figures are 14% for men and 24% for women (according to a national survey in 1971-1974 by the National Center for Health Statistics).

Thus obesity, with its concomitant health disorders, and its counterpart, emaciation, are major problems of nutrition in some parts of the world. Today both result from imbalances of energy intake and expenditure. It is now appropriate to consider food sources of energy, primarily from carbohydrates and lipids, although protein foods serve as a minor source of energy. Current opinion among most nutritionists is that food carbohydrates should supply about 55% of total energy needs, no more than 14% coming from *total sugars* (cakes, sweets), not more than 35% from food fats, and *not less than* 10% from food proteins.

Carbohydrates

The principal function of carbohydrates in the body is as a major source of food energy, yielding 4 kcal per gram, about 50% to 60% of the energy requirement coming from this source in the American dietary. In addition, carbohydrates serve a number of other purposes, including increasing the use of proteins for tissue formation and maintenance by decreasing the catabolism of amino acids, enhancing fatty acid metabolism by way of the citric acid cycle, and other functions described in Chapter 11. Dietary lactose increases the absorption of calcium.

The polysaccharides, starches, dextrins, and cellulose along with certain sugars—monosaccharides such as glucose and fructose and disaccharides such as sucrose, lactose, and maltose—are the principal carbohydrates in foods. The starches and dextrins (the "complex carbohydrates"), which supply about 50% to 60% of dietary carbohydrate, are found chiefly in cereal grains, flour, legumes, potatoes, and other root vegetables. The sugars are present mainly as sucrose in jams, jellies, syrups, honey, and some fruits and as lactose in milk and dairy products. About 35% of dietary carbohydrate is supplied by sucrose, 5% by lactose, and the remaining 5% by maltose, glucose, fructose, and other sugars. Pentoses in foods seem to be of limited nutritional value. Moreover, they constitute only a small fraction of the total carbohydrate intake. Cellulose is not used by humans and nonruminant animals, nor are gums, pectins, and other nondigestible residues from the "bulk" or "fiber" of foods.

From a quantitative standpoint the carbohydrates constitute the major foodstuff consumed by humans, representing about 60% by weight of the American dietary, or around 350 to 500 gm per day for the average adult. The amount varies widely, however, depending on physical activity, food preferences, ethnic customs, and environmental and other factors.

There has been no *recommended daily dietary allowance (RDA)* established for carbohydrates as there is for most other essential nutrients and for energy intake (see Tables 21-2 and 21-4). The chief reason for this omission is the fact that carbohydrate can be formed in the body from glycogenic amino acids (Chapter 13) and from glycerol moiety of fats (Chapter 12). However, it is desirable to include sufficient preformed carbohydrate in the diet to *avoid ketosis,* excessive breakdown of body protein, loss of cations, especially sodium, and dehydration. Apparently at least 100 gm of digestible carbohydrate a day will offset these undesirable metabolic responses associated with high-fat diets and fasting.

Nutrition authorities currently are recommending that most Americans *increase* their consumption of complex carbohydrates (starches and dextrins) and fiber while *decreasing* the use of *sugars,* specifically *sucrose* ingested as candies, jams, jellies, syrups, cakes, pastries, beverages, and so on. The principal reason for this recommendation is the mounting evidence incriminating sucrose in the development of obesity and a number of related pathological conditions (e.g., hypertension, cardiovascular disease, maturity-onset diabetes, and decreased longevity) and in dental caries (Chapter 17). Still another reason for decreasing the consumption of sucrose is its crowding out effect on other, less refined foods containing, in addition to carbohydrate, essential vitamins and minerals.

For reasons similar to those cited for sucrose, reduction in the consumption of *alcohol,* which, like that of sucrose, has risen steadily during the past two or three decades, is recommended. Alcohol, because of its high caloric value (6.7 kcal per gram) and refined production is especially likely to crowd out essential food factors.

One of the sugars, fructose, has been promoted recently as having special value in the treatment of obesity and even diabetes. These claims have no scientific basis and have been labeled "rip offs" by some nutrition authorities.

Fiber

The nonutilizable carbohydrates are the indigestible polysaccharides, e.g., cellulose, lignin, agar-agar, and gums. These constitute a large part of the fiber of food, the nondigestible fraction that gives bulk to the feces. The tendency in modern civilization has been toward a refinement of food, with a lessening in the amount of the indigestible parts of grains, fruits, and vegetables. In food tables this fraction is usually termed *fiber*.

The fiber portion is valuable in other respects, as indicated by recent investigations. Epidemiological studies in humans and experiments in animals indicate that pathological conditions such as diverticular disease of the colon, cancer of the colon, and possibly even coronary heart disease may be related to a lack of dietary fiber. Lack of food fiber also has been implicated in such widespread ailments of Western civilization as hernias of the gastrointestinal tract, hemorrhoids, gallbladder disease, appendicitis, and obesity. There is considerable experimental evidence that food fiber has a hypocholesteremic effect. Studies indicate that fiber *binds* bile acids and possibly cholesterol itself in a nonabsorbable complex, increasing the fecal excretion and thus decreasing plasma and tissue cholesterol levels. These investigations support the wisdom of including fiber in the daily diet of normal adults in the form of whole-grain cereals, legumes, fruits, and vegetables.

Lipids

Dietary fats furnish a significant proportion of the body energy needs, averaging from 35% to 45% of the caloric intake in the typical American diet. Fats yield approximately 9 kcal per gram. According to some authorities, about 82% of the energy derived from the metabolism of fatty acids (and glucose) is ultimately converted to *heat* for the maintenance of body temperature. The exact mechanism involved is unclear at the present time. The heat is finally dissipated from the body, mostly through the skin, and to a lesser extent from the lungs, urine, and feces. Fats are rather slowly digested, and large amounts tend to slow the digestion of other foods as well (Chapter 12).

Food triglycerides have several other noncaloric functions: they supply the essential unsaturated fatty acids and they serve as carriers of several fat-soluble essential nutrients, including vitamins A, D, E, and K. Indeed, the human requirement for the fat-soluble vitamins can be met by the consumption of appropriate food fats and oils. Dietary fat has a rather high satiety value, i.e., the ability to satisfy hunger. Complaints about institutional food are sometimes lessened by a moderate increase in the fat content.

Since lipids are poor conductors of heat, in general, their presence in subcutaneous tissues tends to lessen heat loss from the body. The greater the amount of fat, the more effective the heat insulation. This is one very good reason why persons who have an exceptionally thick layer of fat are more comfortable in winter and less comfortable in summer than thinner individuals. Fat deposits in tissues, in addition to serving as a reserve supply of energy, are important in supporting vital organs and protecting them from mechanical stresses.

Requirement

There is no official recommendation for the desirable amount of fat and other lipids in the Food and Nutrition Board's 1980 RDA other than that the amount of fat should be sufficient to supply part of the food energy intake (no more than 35%) and the need for the essential unsaturated fatty acid and the essential fat-soluble vitamins. The former need can be met by 15 to 25 gm of appropriate food fats and vegetable oils in the daily diet. The essential fatty acid (linoleic acid) intake should be 1% to 2% of the total energy intake (3% in infant formulas, according to the American Academy of Pediatrics). Also, for the general population, consuming a relatively low fat intake (less than 29% of calories), 3% of the energy intake as linoleic acid is recommended. This level also will supply significant amounts of the fat-soluble vitamins A, D, E, and K.

Current opinion is that a level of fat supplying over 35% of the energy needs is too high, in view of statistics relating the level of fat intake to the blood cholesterol and the triglyceride level to the incidence of atherosclerosis. Most authorities believe that 20% to 25% of the total caloric intake as triglycerides is a more acceptable level.

Essential fatty acids

If fat is entirely excluded from the diet of rats, there develops a condition characterized chiefly by retarded growth, scaly skin, necrosis of the tail, kidney lesions with bloody urine, and early death. This problem was studied carefully, and it was found that certain unsaturated fatty acids are effective in bringing about a cure of the condition. These are linoleic, linolenic, and arachidonic acids. Strictly speaking, only *linoleic acid* is essential and cannot be synthesized by the body, although the others may replace it or spare it to some extent. Interconversions between several of the polyunsaturated fatty acids may occur. A study with rats demonstrated, for example, that ^{14}C-labeled linoleic acid is converted to arachidonic acid. The pathway is linoleic acid $\rightarrow$ γ-linolenic acid ($\Delta^{6,9,12}$) $\rightarrow$ $\Delta^{8,11,14}$-eicosatrienoic acid $\rightarrow$ arachidonic acid. Arachidonic acid is a 20-carbon chain acid with four unsaturated linkages ($\Delta^{5,8,11,14}$). A related C-20 polyunsaturated fatty acid, *eicosapentenoic acid* with *five* unsaturated bonds ($\Delta^{5,8,11,14,17}$), is found in the marine food chain, especially in mackerel and cod liver (oil). Linoleic and linolenic each have 18 carbons, the former having two and the latter three double bonds.

Vertebrates apparently lack an enzyme required to convert oleic acid to linoleic acid, in contrast to plants and certain microorganisms. Thus, linoleic acid is an essential fatty acid for vertebrates. It is interesting that α-linolenic acid ($\Delta^{6,12,15}$), the plant isomer, is slightly active in promoting growth in animals deficient in linoleic acid but is ineffective in curing the dermal lesions. γ-Linolenic acid ($\Delta^{6,9,12}$), the animal form, is effective in both respects.

Investigation of this condition led to the discovery that, in rats suffering from a lack of essential fatty acids, the serum lipids have a low iodine number. It was soon found that children with eczema likewise have serum lipids with a low content of unsaturated fatty acids. The administration of suitable fats cleared up the skin lesions in many of these cases. Evidently some individuals require a greater than average amount of these essential fatty acids in their diet.

Another deficiency disease with symptoms closely resembling those just described can be produced in rats by withdrawal of pyridoxine from the diet. Apparently there is a relationship between the essential fatty acids and pyridoxine because animals deprived of both the vitamin and the linoleic acid can be relieved by the administration of either. The nature of this relationship is at present uncertain.

It is now known that some of the physiological properties of fats are related to its content of essential fatty acids. These include favorable effects on sex maturation, pregnancy, and lactation. The essential fatty acids protect against the harmful action of x-ray irradiation. Fat-deficient rats have a high capillary permeability and low capillary resistance, which can be remedied by diets containing linseed oil or linoleic acid. The essential fatty acids also are involved in cholesterol metabolism (Chapter 12). The unsaturated fatty acid arachidonic acid, which can be formed from linoleic acid, also serves as a precursor for the prostaglandins PGE_2 and $PGF_{2\alpha}$ (Chapter 18). Eicosapentenoic acid likewise serves as a precursor for prostaglandin I_2 and thromboxane A_3 (Chapter 18).

Of considerable interest is the significant increase in the use of vegetable oils, such as corn, cottonseed, peanut, soybean, safflower, and sunflower oils, during the past few years. This is associated with an increased intake of polyunsaturated fatty acids, of course. Although adequate amounts of these, more specifically of linoleic acid, are needed, excessive amounts may be undesirable. A resulting increased requirement for vitamin E is one possible adverse effect, as are an alleged increase in gallstone formation and an increased incidence of tumor formation in some experimental animals.

Perhaps it should be noted at this point that the hydrogenation of polyunsaturated fatty acid–containing oils not only increases their melting point but also results in the isomerization of their naturally occurring *cis* fatty acids to the *trans*-isomers. The latter have different biological effects than the *cis* forms. Possible undesirable effects of the *trans*-isomers are currently being extensively investigated.

Desirable proportions of energy from fats and carbohydrates

Many nutritionists and physicians believe that the health of a significant proportion of the population of the United States could be improved by several dietary modifications and changes in life-style. Although some of the proposed changes are still controversial, there is sufficient evidence to support several dietary changes that should prove consonant with better health. However, it should be emphasized that most chronic and degenerative diseases have a number of contributing factors, only one of which may be diet. Thus altering the diet only without proper consideration of changing other risk factors undoubtedly would result in minimal desirable effects.

Based on a recent 10-state nutrition survey, up to 25% of adult males and 42% of adult females were classified as obese. Thus reduction of body weight to a desirable level is considered one of the most beneficial measures related to diet that could be implemented in the United States. For much of the population of the United States the maintenance of desirable body weight could be achieved most readily by *controlling the caloric intake* and *increasing physical activity*. This recommendation applies to persons who are *unduly thin*, as emphasized by an eminent nutritionist recently. In this case, of course, the caloric intake probably should be *increased,* probably without significant alteration in physical activity (energy expenditure).

Some further guidelines have been offered by the Food and Nutrition Board of the National Research Council, especially for persons in the high-risk category for certain diseases. The Board emphasizes, however, that these guidelines are not intended to replace therapeutic or modified diets prescribed by a physician for specific medical conditions.

1. Total fat intake, especially in diet of less than 2000 kcal per day, should be reduced so that fat is *no more* than 35% of dietary energy. To achieve this, there should probably be a greater reduction in *saturated* fats (animal fats) than in vegetable fats and oils, containing predominately unsaturated fatty acid. However, an *upper limit* of *10%* of dietary energy as *polyunsaturated fatty acids* is advisable, in the Board's opinion.
2. Intake of refined sugar (sucrose) should be reduced and complex carbohydrates (e.g., starches, dextrins) maintained or even increased. Compelling reasons for this recommendation have been discussed earlier.
3. Alcohol consumption should be decreased for many individuals to assist in achieving proper energy *(and nutritional)* balance.

The Board emphasizes also that the foregoing recommendations for desirable types and amounts of dietary fat and carbohydrate do not entail radical changes in eating habits and could be accomplished with the current American food supply. Furthermore, the alterations just recommended would *increase* the nutrient concentration of the diet and would increase the possibility that all allowances for essential nutrients could be supplied for those whose daily caloric needs are less than 2000 kcal.

Proteins

The importance of proteins as a vital constituent of living matter has been known for at least 150 years. In 1838 Mulder named this group of substances *proteins* (Gr. *protos,* "first"). Indeed, proteins are the most abundant organic constituent of animal cells, constituting about 50% of their dry weight. Since their discovery proteins have become established as being involved in all aspects of cell structure and function. Likewise, proteins have proved to be of prime importance in human and animal nutrition. Questions such as the amounts of protein required, its best sources, and its nutritional significance have concerned nutritionists since the beginning of nutrition as a science. More recently, recognition of the disastrous effects of protein deficiency in humans, afflicting around 700 million people world-wide, and its close interrelationship with caloric deficiency (so-called *protein-calorie malnutrition,* found especially in underdeveloped countries) has led to a perhaps overemphasis of the protein factor in nutrition. This concern was deepened by the discoveries that *kwashiorkor* in several parts of the world was a human protein-deficiency disease and that protein deficiency results in a decreased resistance to infection due in turn to a suppression of several immune systems, particularly to that of the T-cells. In the 1970s the pendulum swung to an overemphasis on *calories* and the deleterious effects of excessive caloric intake on human health and performance. In the 1980s there appears to be a shift in emphasis again, but to a more reasonable intermediate position, with attention focused more on the importance of *balance* between all essential nutrients, including mineral, vitamin protein, and calorie needs, for optimum human health and performance.

This section is concerned primarily with the role of proteins in nutrition—their best food sources, their relative efficiency for tissue formation and maintenance, the daily human requirement, their nutritional significance, and principal effects of their deficiency.

Food sources

Proteins are widely distributed in foods of both animal and plant origin. The principal sources include the red meats, cereal grains, milk and milk products, poultry, eggs, cheese, fish, legumes and a few other vegetables, and nuts. Table 21-3 lists the average amounts of these foods consumed per capita per day in the United States for 1978, along with the grams of protein per day derived from each. Further data from the U.S. Department of Agriculture (1978) show that the percent consumption of the four major red meats and grams of protein per day derived from them are as follows: beef, approximately 65% (15.4 gm); pork, 33% (7.7 gm); veal, 1.2% (0.3 gm); and lamb, nearly 1.0% (0.2 gm). Thus beef, as might be expected, is the principal source of protein in the American diet today, followed closely by milk and milk products, cheese, and then wheat flour. Beef also is a valuable source of other essential nutrients, including especially iron, zinc, and vitamin B_{12}. Note that the American diet (1978) supplied 92 gm protein per capita per day.

Not all food proteins have the same ability to meet the needs of the body for tissue formation, maintenance, and other functions performed by dietary proteins. This is termed the *biological value of proteins* and will be discussed later in this section. In general, animal proteins have a higher biological value than do vegetable proteins, although there is a definite *complementarity* between the two.

Several "instant" dried skimmed milk preparations on the market are excellent, economical sources of protein of high biological value. These preparations may be reconstituted to supply skimmed milk or may be added to milk, soups, cereals, or other foods to increase their protein content. Dehydrated, defatted, and deodorized fish meal also is receiving considerable attention as an economical source of protein of high quality. It may be particularly important in areas where the supply of animal protein is limited, as in parts of Africa and Asia. Soybeans and soybean products, such as meat extenders, also offer an economical source of food protein of high quality.

There is a tremendous food reserve, mainly protein and minerals, in the fish of oceans, lakes, and rivers. Currently approximately 65 million tons are harvested annually, but only half is consumed directly by humans. The other half is converted to fish meal and used for livestock feed. A well-managed world fishery could

Table 21-3
Protein consumption per capita per day in the United States in 1978

Source	Pounds/capita/year	Grams protein/capita/day
Red meats	150	24
Flour	112	17
Milk and milk products	310	14
Poultry	57	9
Eggs	35	6
Cheese	17	5
Potatoes	125	5
Fish	13	4
Vegetables	164	3
Peanuts	7	2
Other	—	3
Total	—	92

From National Food Review—5, Washington, D.C., Dec. 1978, United States Department of Agriculture.

produce four times this amount, or over 200 million tons annually. In North America the average amount of protein obtained from fish per capita per day is only approximately 4 gm, whereas an average of 58 gm daily come from other animal proteins and 28 gm from vegetable proteins. A similar distribution of food proteins is seen in Europe. However, in Asia and Africa the amount of protein from animal sources other than fish is less than in North America, and that from vegetable proteins is much higher. The intake of protein from fish is about the same, although the proportion is higher, since the average daily protein intake is lower.

Requirement

The amount of protein that should be consumed daily to meet human needs is still one of the most popular topics in nutritional biochemistry. The reasons are evident. Food proteins provide the only source of the amino acids required for the biosynthesis of body proteins and the formation of many other vital tissue components (Chapter 13). As has been emphasized repeatedly in the preceding chapters, the body is in a *dynamic state*. Tissue proteins and other equally important nitrogenous compounds derived from amino acids are constantly being degraded and resynthesized. Indeed, some body proteins have a half-life of only *minutes* (Chapter 13).

Earlier views of the daily human protein requirement were based on the *daily intake* of food proteins. Thus in the latter part of the nineteenth century the average daily consumption of food proteins in Germany was found to be *118 gm*. Later, as a result of nitrogen balance studies on students, instructors, soldiers, and other male individuals, Chittenden, in the United States, came to an entirely different conclusion. These human subjects could be maintained in nitrogen balance, i.e., the protein *nitrogen intake equaling the nitrogen excretion* in the urine (as urea, etc., Chapter 17), feces (undigested proteins, etc.) and other secretions-excretions, by from *45 to 53* gm protein daily. Their general health and athletic prowess were usually improved. These results led to the conclusion that 118 gm protein per day was far above the actual need of healthy adult men and that the excess intake might actually lead to various pathological conditions. Recall that the average American per capita intake of protein per day is approximately 92 gm (see Table 21-3). Modern studies using the improved analytical methods and more sophisticated techniques available today have supported, in general, the earlier finding that a lower protein intake can meet the minimal requirement of the normal adult. The value of 0.45 gm protein per kilogram of body weight is now generally accepted as minimal need.

It is important at this point to differentiate between requirement and a recommended dietary intake. Requirement is the *minimum* daily intake needed to meet needs for normal maintenance. A "recommended daily dietary allowance," on the other hand, is that amount needed to meet the minimal requirement plus an added amount to allow for incomplete digestion-absorption and ineffective use of mixed dietary proteins.

Recommended dietary allowance for proteins (1980)

The Food and Nutrition Board of the National Research Council, as stated earlier, establishes the recommended dietary allowance (RDA) for nutrients now recognized as essential for the optimum health and function of normal human subjects. This, of course, includes proteins. In establishing the new 1980 RDA for proteins the Food and Nutrition Board followed three basic steps:

1. Estimation of the minimum requirement of a good quality of protein for the maintenance of nitrogen balance in practically all healthy individuals. Reliable published data in the literature were used for this purpose.

2. Adjustment of this requirement to allow a margin of safety for proper use of the proteins from a mixed diet compared with a diet containing a good quality protein. This amount, then, is the RDA for the maintenance of nearly all healthy adults in this country.
3. Adjustments in protein allowance to meet the added needs of growth, pregnancy, and lactation.

The minimum requirement for protein is based on the obligatory losses of protein nitrogen in healthy young adults receiving protein-free diets. In these subjects the daily urinary nitrogen loss (mainly as urea nitrogen) plateaus at about 37 mg per kilogram of body weight. The daily fecal nitrogen output averages about 12 mg per kilogram of body weight. The amount of nitrogen lost in sweat, secretions-excretions, hair, nails, and sloughed skin is smaller but variable and averages about 5mg per kilogram per day. Thus the total obligatory nitrogen loss in a protein-free diet by all routes of nitrogen elimination averages 54 mg per kilogram of body weight per day. This is equivalent to a daily wastage of 0.34 gm of body protein per kilogram per day, or 24 gm of protein for a 70-kg male. Studies by several investigators have demonstrated that the coefficient of variation of nitrogen loss on the nitrogen-free diet by these subjects is about 15%. Thus, if the mean daily nitrogen loss is adjusted above the mean by twice the coefficient of variation (30%, to cover the nitrogen loss of 97.5% of the population), the obligatory nitrogen loss for healthy young adults becomes 0.45 gm of body protein per kilogram of body weight per day.

Theoretically the consumption of a high-quality protein, such as whole egg protein, equal to the obligatory loss of body protein should result in nitrogen equilibrium. However, even quality proteins that are fully used at submaintenance levels are used less efficiently when consumed in amounts approximating a loss of efficiency of about 30%. Thus the requirement for 97.5% of the adult population is increased from 0.45 to 0.60 gm per kilogram of body weight per day.

Needless to say, this figure is an approximation subject to variations such as technical errors and different energy intakes. It is, however, the most acceptable available today. Accordingly, the Food and Nutrition Board in its 1980 revision of the RDA for proteins has used this value after correcting for a 75% efficiency of use of the protein in a mixed diet, as compared with the reference high-quality proteins of whole egg. Thus the 1980 RDA for adults for the mixed proteins of the American diet is *0.8 gm per kilogram of body weight per day*. This amounts to 56 gm of protein per day for a 70-kg man and 44 gm protein for a 55-kg woman (Table 21-4).

Note that the RDA for protein for infants is quite high (2.0 to 2.2 gm per kilogram of body weight) and also higher for children, although the value decreases gradually to the adult maintenance level at about 18 or 19 years of age. This is to be expected, since infancy and childhood are periods of rapid growth and body tissue formation. There is a definite positive nitrogen balance at this time, i.e., more protein nitrogen consumed from foods than is excreted. Note also that the RDA for persons over 51 years of age is the same as for younger adults, although the energy requirement is somewhat less because of reduced activity (Table 21-2). The increase in the RDA for protein during pregnancy (+30 gm) and lactation (+20 gm) supplies the needs of the growing fetus and then milk formation, respectively.

These RDA values for protein presume an adequate energy intake from other foods. As stated previously, the energy intake has a profound effect on the amount of protein needed and is an important factor in establishing allowances for protein.

There is little evidence that muscular activity increases the need for protein if the energy intake is increased proportionally, unless there is development of new mus-

Table 21-4
Recommended daily dietary allowances (revised 1980 by the Food and Nutrition Board of the National Academy of Sciences–National Research Council)*

	Age	Weight		Height		Protein	Fat-soluble vitamins		
	(years)	(kg)	(lbs)	(cm)	(in)	(gm)	Vitamin A (μg R.E.)†	Vitamin D (μg)‡	Vitamin E (mg α T.E.)§
Infants	0.0-0.5	6	13	60	24	kg × 2.2	420	10	3
	0.5-1.0	9	20	71	28	kg × 2.0	400	10	4
Children	1-3	13	29	90	35	23	400	10	5
	4-6	20	44	112	44	30	500	10	6
	7-10	28	62	132	52	34	700	10	7
Males	11-14	45	99	157	62	45	1000	10	8
	15-18	66	145	176	69	56	1000	10	10
	19-22	70	154	177	70	56	1000	7.5	10
	23-50	70	154	178	70	56	1000	5	10
	51+	70	154	178	70	56	1000	5	10
Females	11-14	46	101	157	62	46	800	10	8
	15-18	55	120	163	64	46	800	10	8
	19-22	55	120	163	64	44	800	7.5	8
	23-50	55	120	163	64	44	800	5	8
	51+	55	120	163	64	44	800	5	8
Pregnant						+30	+200	+5	+2
Lactating						+20	+400	+5	+3

From Recommended dietary allowances, ed. 9, Washington, D.C., 1980, National Research Council.
*The allowances are intended to provide for individual variations among most normal persons as they live in the United States under usual environmental stresses. Diets should be based on a variety of common foods in order to provide other nutrients for which human requirements have been less well defined. The RDA do not represent either a minimal requirement or an optimal level of intake for *every* individual but rather are intakes of nutrients that meet the nutritional needs of the *majority* of healthy persons. Specific requirements would have to be determined on an individual basis by appropriate biochemical and clinical studies.
†Retinol equivalents. 1 retinol equivalent = 1 μg retinol or 6 μg β-carotene. See text for calculation of vitamin A activity of diets as retinol equivalents.
‡As cholecalciferol. 10 μg cholecalciferol = 400 I.U. vitamin D.
§α-Tocopherol equivalents. 1 mg d-α-tocopherol = 1 α T.E. See text for calculation of vitamin E activity of the diet as α tocopherol equivalents.

cle tissue during conditioning. Even this appears to be a relatively small amount.

Extreme stress, either physiological or environmental, increases nitrogen loss. Likewise, infections, fevers, severe burns, and either surgical or accidental trauma can result in substantial urinary nitrogen loss as well as increased energy expenditure. These may require special dietary adjustment, as increased energy and/or protein intake. The same is true during convalescence from wasting diseases.

The allowances for protein recommended for Americans are similar to those recommended in Canada and in India. However, considerably higher protein intakes per kilogram of body weight are recommended for adults in the United Kingdom, the Federal Republic of Germany, Japan, Australia, the Netherlands, and Finland. Apparently these are based on the customary protein intakes in those countries. There appears to be no definite evidence, however, that such higher protein intakes are either beneficial or harmful, except possibly for premature infants. There also is some evidence that calcium metabolism may be affected adversely by the excessive consumption of protein unless accompanied by an increased intake of calcium.

Adjustment to changes in protein intake

Another perplexing observation is that an individual in nitrogen balance on a diet containing a constant level of protein intake can adapt rather quickly to a diet with a higher protein content and again attain nitrogen balance after a day or two at the higher level. An interval of adjustment or "lag" is usually seen (Table 21-5).

Water-soluble vitamins							Minerals					
Vitamin C (mg)	**Thiamin (mg)**	**Riboflavin (mg)**	**Niacin (mg N.E.)‖**	**Vitamin B_6 (mg)**	**Folacin¶ (μg)**	**Vitamin B_{12} (μg)**	**Calcium (mg)**	**Phosphorus (mg)**	**Magnesium (mg)**	**Iron (mg)**	**Zinc (mg)**	**Iodine (μg)**
35	0.3	0.4	6	0.3	30	0.5**	360	240	50	10	3	40
35	0.5	0.6	8	0.6	45	1.5	540	360	70	15	5	50
45	0.7	0.8	9	0.9	100	2.0	800	800	150	15	10	70
45	0.9	1.0	11	1.3	200	2.5	800	800	200	10	10	90
45	1.2	1.4	16	1.6	300	3.0	800	800	250	10	10	120
50	1.4	1.6	18	1.8	400	3.0	1200	1200	350	18	15	150
60	1.4	1.7	18	2.0	400	3.0	1200	1200	400	18	15	150
60	1.5	1.7	19	2.2	400	3.0	800	800	350	10	15	150
60	1.4	1.6	18	2.2	400	3.0	800	800	350	10	15	150
60	1.2	1.4	16	2.2	400	3.0	800	800	350	10	15	150
50	1.1	1.3	15	1.8	400	3.0	1200	1200	300	18	15	150
60	1.1	1.3	14	2.0	400	3.0	1200	1200	300	18	15	150
60	1.1	1.3	14	2.0	400	3.0	800	800	300	18	15	150
60	1.0	1.2	13	2.0	400	3.0	800	800	300	18	15	150
60	1.0	1.2	13	2.0	400	3.0	800	800	300	10	15	150
+20	+0.4	+0.3	+2	+0.6	+400	+1.0	+400	+400	+150	††	+5	+25
+40	+0.5	+0.5	+5	+0.5	+100	+1.0	+400	+400	+150	††	+10	+50

‖1 N.E. (niacin equivalent) is equal to 1 mg of niacin or 60 mg of dietary tryptophan.
¶The folacin allowance refer to dietary sources as determined by *Lactobacillus casei* assay after treatment with enzymes ("conjugases") to make polyglutamyl forms of the vitamin available to the test organism.
**The RDA for vitamin B_{12} in infants is based on average concentration of the vitamin in human milk. The allowances after weaning are based on energy intake (as recommended by the American Academy of Pediatrics) and consideration of other factors such as intestinal absorption.
††The increased requirement during pregnancy cannot be met by the iron content of habitual American diets nor by the existing iron stores of many women; therefore the use of 30-60 mg of supplemental iron is recommended. Iron needs during lactation are not substantially different from those of nonpregnant women, but continued supplementation of the mother for 2-3 months after parturition is advisable in order to replenish stores depleted by pregnancy.

Table 21-5
Example of adjustment to changes in protein intake

Day	Nitrogen in food (grams)	Nitrogen in feces (grams)	Nitrogen absorbed (grams)	Nitrogen in urine (grams)	Nitrogen balance (grams)
Experiment 1					
Before	> 16.96				
1	16.96	0.94	16.02	18.2	−2.18
2	16.96	0.94	16.02	17.0	−0.98
3	16.96	0.94	16.02	15.8	+0.22
4	16.96	0.94	16.02	16.0	+0.02
5	16.96	0.94	16.02	15.7	+0.32
Experiment 2					
1	14.40	0.70	13.70	13.60	+0.10
2	14.40	0.70	13.70	13.80	−0.10
3	14.40	0.70	13.70	13.60	+0.10
4	20.96	0.82	20.14	16.80	+3.34
5	20.96	0.82	20.14	18.20	+1.94
6	20.96	0.82	20.14	19.50	+0.64
7	20.96	0.82	20.14	20.00	+0.14

The subject was a young woman weighing 58 kg, at rest in bed. The first experiment is an example of adjustment to a lowered protein intake because it was known that the subject had previously been on a high-protein diet. Equilibrium occurred after the second day. The second experiment shows the effect of increasing the protein intake. In this case 3 days elapsed before the subject was in nitrogen equilibrium.

Figure 21-1

Determination of indispensability of an amino acid for a human. Initial diet contained 6.7 gm N in the form of a mixture of the 10 amino acids previously found to be essential for the growth of animals. The remaining N came from other components of the ration. The total N content of the diet was kept constant throughout. Valine was removed from the food at the *first arrow* and was returned at the *second arrow*. Horizontal units represent single days.

Courtesy W.C. Rose, Urbana, Ill.

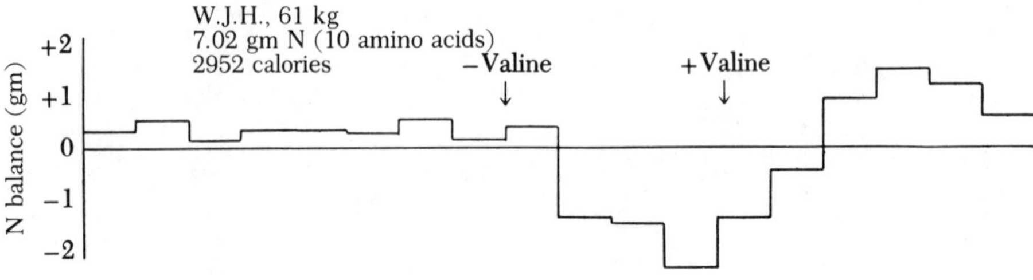

Human essential amino acid requirements

The human requirement for protein just discussed is actually a requirement for the amino acids comprising them, more particularly the *essential* amino acids (Chapter 13). A number of studies have been reported in which the essential amino acid requirements of men, women, children, and infants have been determined by balance methods similar to those used for determining the requirement for protein just described. W.C. Rose, a pioneer investigator in this field, fed mixtures of pure amino acids as the protein-nitrogen source to a group of healthy young adult men, who were also given an adequate mixture of other foods. The results of omitting and then adding one of the essential amino acids, L-valine, on the nitrogen balance of one test subject are given in Fig. 21-1. When L-valine was removed from the amino acid mixture, the subject immediately went into negative nitrogen balance. When L-valine was again added, the subject returned to a positive balance within 2 days. This was repeated in the subjects using the other amino acids found to be essential for growth in young rats. The results demonstrated conclusively that the L-isomer of eight amino acids—lysine, leucine, isoleucine, methionine, phenylalanine, threonine, tryptophan, and valine—were *essential* for young adult healthy men. Subsequent studies demonstrated that histidine also was essential for infants. L-Arginine, a tenth amino acid essential in some species has not been demonstrated as essential for humans. Apparently it can be synthesized in sufficient amounts for human growth and maintenance.

This work has been confirmed by other investigators, and the amounts of the essential amino acids required to maintain nitrogen balance in healthy human subjects have been determined. All subjects were supplied a sufficient amount of nonessential amino acids as well as the other essential nutrients, including carbohydrates, fats, vitamins, and minerals. The human amino acid requirements in milligrams per kilogram of body weight per day as estimated from data in the literature by the Food and Nutrition Board of the National Research Council (1980) are presented in Table 21-6.

The average values shown for adults in the table have been adjusted upward to allow for individual variability in the use of the essential amino acids, as was done for dietary proteins. These values, however, are probably still minimal amounts required, in view of the limited success of investigators to maintain adult subjects in nitrogen balance with diets containing such small amounts of amino acids.

Note the relatively high requirements of infants and children for the essential amino acids as compared with those for adults. This is undoubtedly due to the

Table 21-6
Estimated amino acid requirements of humans

Amino acid	Requirement (per kilogram of body weight) (mg/day)			Amino acid pattern for high-quality proteins (mg/gm of protein)
	Infant (3 to 6 months)	Child (10 to 12 years)	Adult	
Histidine	33	?	?	17
Isoleucine	80	28	12	42
Leucine	135	42	16	70
Lysine	99	44	12	51
Methionine plus cystine	49	22	10	26
Phenylalanine plus tyrosine	141	22	16	73
Threonine	68	28	8	35
Tryptophan	21	4	3	11
Valine	92	25	14	48

From Recommended dietary allowances, ed. 9, Washington, D.C., 1980, Food and Nutrition Board, National Research Council.

increased requirement for growth in the young subjects as compared with the requirement only for maintenance in the adult. Specific amino acid requirements for pregnancy and lactation are unknown at the present time. Note also that cystine can supply part of the requirement for methionine and that tyrosine can supply part of the requirement for phenylalanine.

The amino acid pattern for a high-quality protein is given in the fourth column. Two grams of this protein per kilogram of body weight would supply the amino acid needs of the infant. Also of interest from the data is the fact that approximately 35% of the food protein allowance of infants must be provided as essential amino acids, whereas only 20% is required by adults. The remainder can be supplied by the nonessential amino acids.

Since no one eats purified proteins or mixtures of amino acids, we must ask ourselves how this affects us from a practical standpoint. Are the proteins of any foods deficient in one or more of the essential amino acids? Whole corn is low in tryptophan and lysine, wheat gluten is low in tryptophan, lysine, and threonine, and gelatin is deficient in tryptophan, leucine, isoleucine, threonine, methionine, and valine. Except for gelatin, hemoglobin, and keratin, the animal proteins are, in general, well balanced in their amino acid distribution. Yeast, corn germ, wheat germ, and soybeans yield approximately the same proportions of amino acids as do the animal proteins. The analysis of many food proteins for amino acid content is now available (see Table 4-2 for selected examples). A mixed diet, containing some animal protein—milk, meat, fish, eggs—ensures an adequate mixture of the essential amino acids.

The general effect of a deficiency of one or more essential amino acids, as stated earlier, is to restrict growth and protein synthesis and produce a negative nitrogen balance.

This fact is strikingly illustrated by the growth curves of rats fed different proteins with and without supplementation by missing essential amino acids (Figs. 21-2 and 21-3) and by photographs of experimental rats (Figs. 21-4 and 21-5). Fig. 21-5 also shows the beneficial effect on growth of feeding a mixture of plant and animal proteins. The latter supplement the amino acid deficiences of the plant proteins and thus support improved tissue protein formation and growth. This effect is known as the *complementarity* of proteins.

However, deficiencies of specific amino acids may result in disturbances characteristic of that particular amino acid. The role of tryptophan in niacin formation is

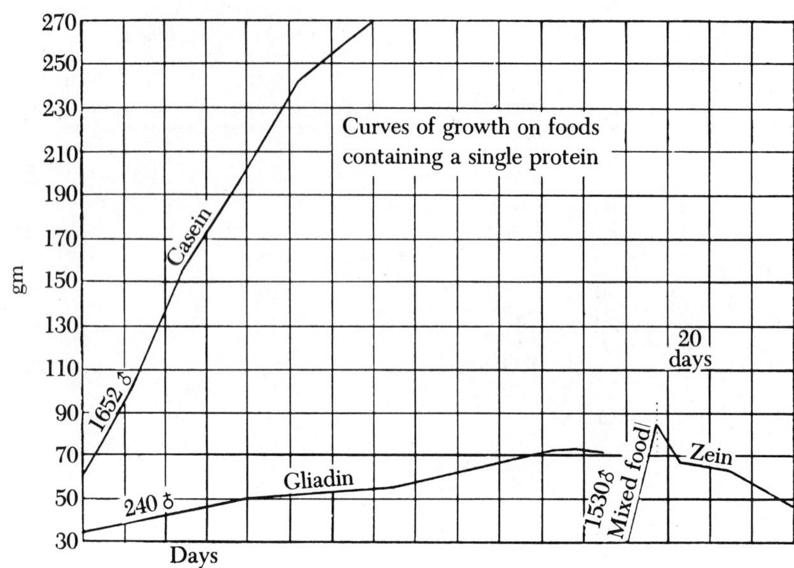

Figure 21-2
Typical curves of growth of rats fed diets containing a single protein. On the casein food (deficient in glycine), satisfactory growth is obtained; on the gliadin food (deficient in lysine), little more than maintenance of body weight is possible; on the zein food (deficient in glycine, lysine, and tryptophan), even maintenance of body weight is impossible.

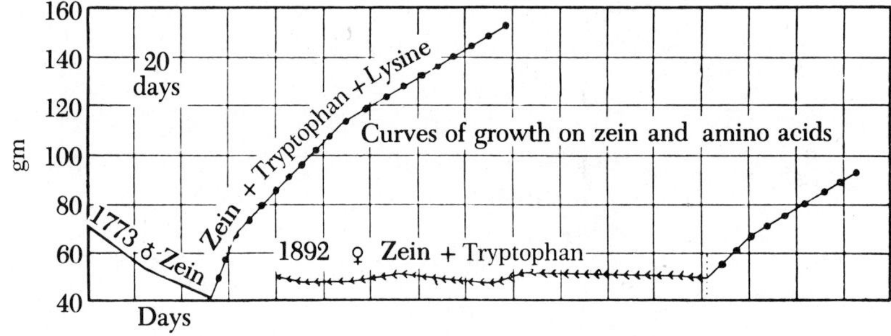

Figure 21-3
Effect of the addition of tryptophan and lysine to zein, which is deficient in them. Addition of tryptophan permits maintenance without growth, whereas further addition of lysine enables the animals to make considerable growth.

an example (Chapter 13). Another is the role of lysine and proline in the formation of hydroxylysine and hydroxyproline in the biosynthesis of collagen (Chapter 16). These amino acids are essential in forming the characteristic tertiary structure of collagen. The participation of ornithine and arginine in urea formation is still another important example. Thus the key roles played by the essential and, likewise, certain nonessential amino acids in metabolic processes emphasize the vital importance of adequate protein nutrition.

Recall that, to be used to the maximum extent, *all* the essential amino acids must be present at the *same* time (Chapter 13). If even one is missing, none will be used for new tissue protein formation. This is logical because nearly all body proteins contain all the essential amino acids, and synthesis cannot occur under these conditions (Chapter 10).

During the earlier studies of the essential amino acids it was found that the keto acids of all the essential amino acids, with the exception of those of lysine and threonine, could be substituted satisfactorily for the corresponding L-amino acid, at least for the growth of rats. The explanation found was that the keto acid was converted to the amino acid by transamination (Chapter 13). This finding thus

Figure 21-4
Essential amino acid deficiency in the growing rat.
A Rat fed a mixture of amino acids lacking valine.
B Same rat after receiving valine added to the amino acid mixture.

Courtesy W.C. Rose, Urbana, Ill.

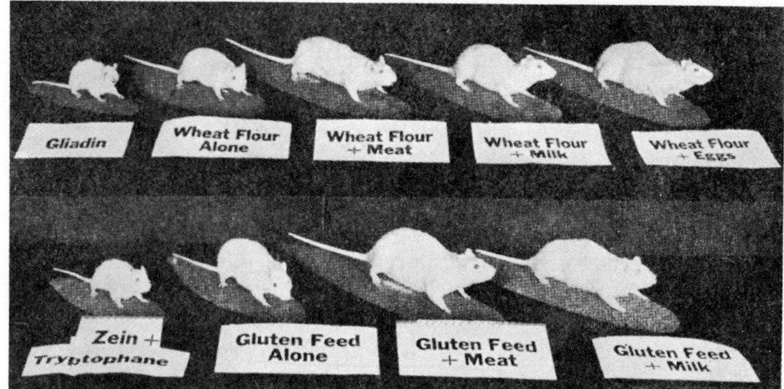

Figure 21-5
These rats were all the same age and fed for the same length of time on diets containing the same proportion of protein. The variation in size is caused by differences in the amino acid composition of the proteins eaten. (From experiments by Osborne and Mendel.)

indicated that the *carbon chain* of those amino acids is the essential part of the molecule, rather than the amino acid itself.

This important fundamental observation is being applied to the treatment of patients with nitrogen-accumulating diseases such as renal failure and certain types of hepatic malfunction. The uremia and hyperammonemia decrease when a portion of the dietary protein or parenteral amino acid mixtures is replaced with mixtures of the α-keto (or α-hydroxy) acids corresponding to leucine, isoleucine, valine, methionine, phenylalanine, and tryptophan. Also, the keto acid mixture can be given *orally* rather than parenterally. Although the results to date are promising, more basic work needs to be done before this type of therapy can be used on a large scale.

The nonessential amino acids must not be considered valueless. They may be considered so important that the body can synthesize them. It might be better to designate them as *synthesizable*. These amino acids contribute to the makeup of body protein, as do the essential ones, but they are more interchangeable, more versatile, and more easily procurable. Glycine, one of the semiessentials, is used in the biosynthesis of creatine (Chapter 13) and heme (Chapter 16) and to detoxify benzoic acid and certain other toxic substances (Chapter 17). Citrulline forms a part of the arginine-urea cycle, and other examples of specific usefulness have been mentioned. However, if any of these semiessential amino acids are lacking, the body can synthesize them from others. Indeed, it has been claimed that "unessential" nitrogen is really an essential dietary factor under certain conditions. If a protein-deficient diet is provided to experimental animals, the addition of glycine, or even urea, may restore some ability to grow.

Biological value of proteins

From the preceding discussion it is evident that the effectiveness of a food protein in supporting growth in the young organism and in tissue maintenance and repair (nitrogen balance) in the adult depends primarily on both the *amount* (quantity) and the *quality* (amino acid composition) of the protein. The *digestibility* of the protein and the efficient absorption of its constituent amino acids are other factors that can affect its utilization. The overall effectiveness of a food protein is termed its *biological value* and is usually expressed in comparison with some arbitrarily designated food protein of high quality, such as whole egg protein, as 100.

Several procedures for determining the biological value of proteins are used, none of which is completely satisfactory. One is the classic "rat growth" method and another the nitrogen balance procedure just discussed. The latter, commonly used in human subjects, entails the determination of the amount of food-protein nitrogen ingested and the nitrogen excreted in the urine and feces. The biological value of the test protein is then expressed as the percentage of the absorbed nitrogen retained (utilized) by the subject. It is calculated as follows:

$$\text{Biological value} = \frac{\text{Food N} - (\text{Fecal N} + \text{Urinary N})}{\text{Food N} - \text{Fecal N}} \times 100$$

One may obtain somewhat more precise values by subtracting from the fecal and urinary nitrogen values the *metabolic* or *endogenous* nitrogen values obtained by analysis of feces and urine when the subject is given an appropriate nitrogen-free diet. The digestibility of the protein can be calculated as follows:

$$\text{Coefficient of digestibility} = \frac{\text{Food N} - (\text{Fecal N} - \text{Fecal N on a protein-free diet})}{\text{Food N}}$$

Two other estimates of the efficiency of protein use are:

$$\text{Net efficiency} = \text{Biological value} \times \text{Coefficient of digestibility}$$

and

$$\text{Protein efficiency ratio} = \frac{\text{Gain in weight of growing animal (10 days)}}{\text{Protein consumed}}$$

As would be expected, the biological values thus obtained for animal proteins (80 to 100) are higher than those for cereal proteins (50 to 70). Legumes give intermediate values.

Another method of expressing the nutritive quality of a protein is based on its amino acid composition or *chemical score*. Essentially, this value is the amount of the most limiting amino acid in the test protein expressed as the percentage of the amount present in whole egg protein. This value does not take into account differences in the digestibility of test proteins nor the intestinal absorption of the amino acids derived from them. However, the chemical scores agree grossly with values as determined biologically and have the advantage over the latter of requiring less time.

In general, the proteins of animal origin, e.g., eggs, milk, meat, fish, poultry, along with soybean protein, have the highest biological value, as stated. Whole eggs are ranked first. In general, proteins of cereals, nuts, and legumes are less adequate, although selected mixtures of grains and legumes serve very well. Cereal grain proteins alone are low in lysine, as is corn protein in tryptophan. However, when cereals are employed as mixtures with other proteins, particularly milk, meat, or eggs, an excellent source of protein results. This is because of the *complementarity effect,* mentioned earlier. The essential amino acids present in one protein balance the deficit or deficits in one or more other proteins of lower biological value. For this reason vegetarians should combine foods such as cereal grains with legumes (soybeans, peas, or beans) or, better still, with milk products or eggs to obtain a more complete protein source.

The enrichment of wheat flour and other cereal grain products with lysine has been suggested as a means of substantially increasing the biological value, making them nearly equivalent to the more expensive animal protein foods. Table 4-2 gives the amino acid composition of some typical animal and plant proteins.

For human nutrition most of the varied dietaries of modern civilization provide adequate mixtures and amounts of the essential amino acids. Even when the cereal grains, especially wheat and rye, furnish most of the protein intake, only one amino acid, lysine, is likely to be lacking. Corn, soybean, and food yeast have been found to be a nutritionally adequate mixture, as has refined wheat flour and yeast. Such plant mixtures might conceivably be substituted for meat, in part at least, or entirely for short periods in emergencies. Another food preparation supplying low-cost protein of good quality is composed of soybean meal fortified with vitamins and minerals. This preparation was developed for the "meals for millions" project and has been studied extensively in the United States and in India and Asia. *Incaparina,* a low-cost mixture of vegetable proteins, has been developed for use in Central America.

Effect of heat Heat processing alters (generally improves) the nutritive value of proteins. Egg white is said to be more easily digested; phaseolin, a protein of the navy bean, has greater nutritive value if cooked; soybeans have a higher nutritional value after cooking, attributable in part to the inactivation of a toxic protein, soyin. As mentioned earlier, there is also present in soybeans a trypsin inhibitor, which is destroyed by heat. *Dry* heat, however, seems to have a deleterious effect, particularly on the proteins of cereals. Apparently lysine becomes less available, although it is not destroyed. A linkage between the ϵ-amino group and a free carboxyl group of a dicarboxylic amino acid may result in the formation of an unnatural peptide linkage that cannot be split by the digestive enzymes. Other explanations have been suggested, e.g., a reaction of free amino or other groups with reducing sugars to form derivatives resistant to enzyme cleavage. This is sometimes called the *browning reaction.* As a result of heating, the protein is thus so altered that the rate of release of the particular amino acids is slowed. The delays result in less efficient amino acid mixtures. The importance of these observations lies in the widespread use of toasted and "puffed" cereals in the American dietary.

Kwashiorkor A nutritional disease chiefly caused by protein deficiency has come into prominence, although undoubtedly it has been known for centuries. Since its victims live in the tropics, the disease has escaped detection in the presence of the great variety of tropical debilitating diseases that occur in that part of the world. Furthermore, its symptoms differ somewhat in different parts of the world. It is called *kwashiorkor* in Africa, and the name is now generally used for it everywhere. The disease occurs in children *after weaning* and is related to a deficiency of protein and apparently other essential nutrients. The diet of such children usually consists of high-starch foods, e.g., yams, bananas, potatoes, maize, cassava. The

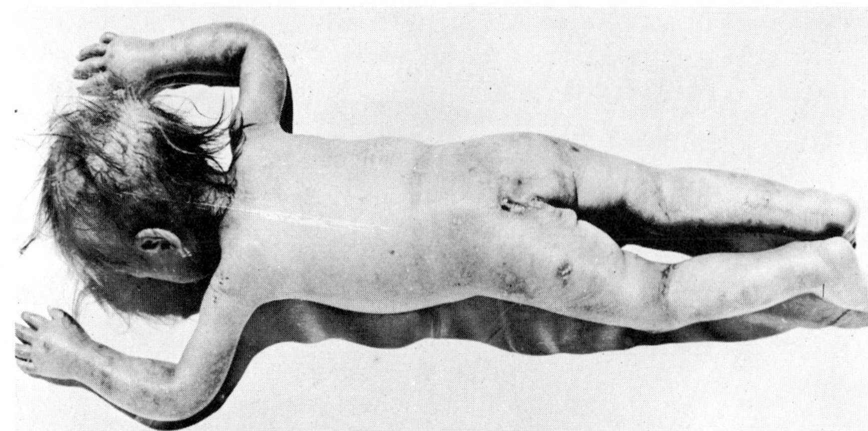

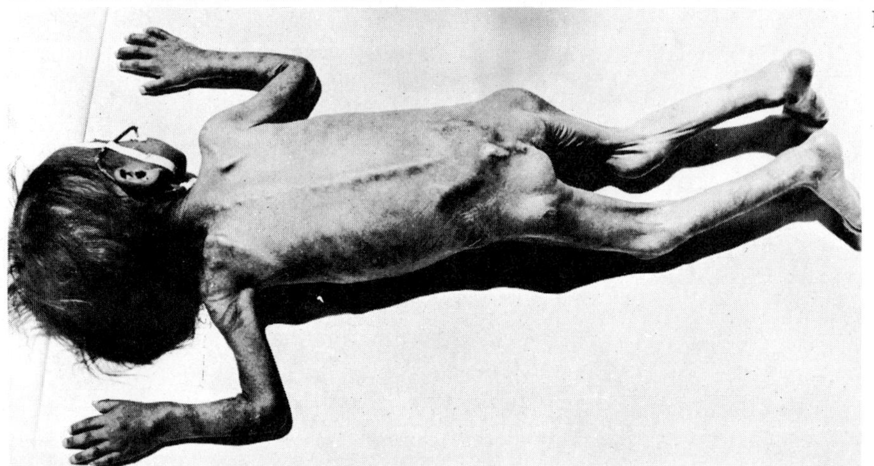

Figure 21-6
Child suffering from kwashiorkor.

A Two years 1 month old on admission. Note the widespread distribution of skin lesions and edema.

B Two weeks after beginning treatment. Note disappearance of edema and improvement in skin lesions. Note also degree of muscular wasting, which had been concealed by the edema.

Courtesy Pan American Sanitary Bureau, Regional Office of WHO.

proteins are poor in quality, being vegetable, and low in amount. A child with this disease often has an enlarged abdomen (Fig. 21-6). A rash occurs in various sites, and the skin cracks behind the ears and at the knees and elbows. Cheilosis, stomatitis, and conjunctivitis may appear. The hair loses its black pigment and looks gray or red. Edema is rather common. The child later loses his appetite and becomes cachectic, and diarrhea occurs. A mild anemia follows, and serum proteins are lowered considerably, as is blood urea. Cholinesterase, alkaline phosphatase, and the enzymes of the digestive tract also are depressed. Many pathological changes are seen, chiefly in the liver and pancreas. β-Aminoisobutyric acid is invariably excreted in the urine, and ethanolamine is usually excreted in this manner, as are some of the nonessential amino acids.

Morphological changes occur in human scalp hair roots during deprivation of protein. Both the bulb and the external root sheath are affected. It is well known that the color and texture of human hair change during protein-calorie deficiency. These observations may be useful in the early diagnosis of protein-calorie malnutrition. The cells of the hair matrix normally proliferate at a rate greater than that of any other tissue, with the possible exception of bone marrow and intestinal mucosal cells. Therefore the cells of the hair matrix have a high protein-synthesizing activity and would be expected to show effects of protein deprivation earlier than other tissues.

The changes in the hair characteristic of kwashiorkor—loss of melanin pigmentation, giving the hair a reddish appearance—are undoubtedly caused by the deficiency of protein. Treatment, of course, is to administer good-quality proteins; but this is not as simple as it seems, partly because of the cost and unavailability of such foods and partly because of feeding and educational problems. Moreover, such treatment is often too late to be of benefit. Prevention is much more important, and the introduction of good nutritional practice should eventually greatly lessen the occurrence of kwashiorkor.

Inorganic elements

Although the inorganic constituents of the body are only a small fraction of the total amount of body tissue, they must not be considered unimportant. They are, in fact, becoming recognized more and more as essential cogs in the human machine. They range in amount from calcium, which makes up about 2% of the average body weight, to cobalt, which is present to the extent of perhaps 0.00004%, and other *trace elements* that may occur in even smaller amounts. The inorganic compounds are required for several purposes. (1) They are needed to provide a suitable medium for protoplasmic activity. The irritability of muscle and nerve cells, the permeability of cell membranes, and the normal functioning of all cells depend on a proper balance of the diverse ions, particularly H^+, Na^+, K^+, Ca^{++}, Mg^{++}, OH^-, HCO_3^-, Cl^-, $HPO_4^=$, and $SO_4^=$. (2) They play a primary role in osmotic phenomena, which have much to do with the flow of tissue fluids, absorption, and secretion. (3) Several anions and cations are also of utmost importance in acid-base equilibria. (4) Certain tissues, especially bones and teeth, have a high mineral content, which accounts for their hardness and rigidity. (5) Some mineral elements become parts of specialized physiological compounds. Hemoglobin's iron and thyroxine's iodine are examples. (6) Other ions are essential to a number of enzyme systems. Copper, manganese, magnesium, potassium, and zinc, as has been seen, are examples of ions needed by enzymes in metabolism. (7) Some 13 trace elements, e.g., Co, Cr, Mo, and Se, appear to be essential as constituents of certain enzymes for other reasons that are discussed later.

Studies in this field have been facilitated in the past few years by improvements in the techniques for determining small amounts of inorganic elements in biological materials, including refinements in flame photometry and the development of other sensitive specific analytic procedures, e.g., atomic absorption spectrophotometry. Atomic absorption devices are particularly useful in determining trace elements as well as Na^+, K^+, Ca^{++}, and Mg^{++} in body fluids and tissues.

Inorganic composition of the body

Seven elements make up 60% to 80% of all inorganic matter in the body. They are calcium, magnesium, sodium, potassium, phosphorus, sulfur, and chloride.

Table 21-7
Inorganic composition of the body

Age	Body weight (kilograms)	Total ash (grams)	Percentage of total ash							
			Ca	Mg	Na	K	P	Cl	S	Total
Fetus, 6 wk	0.88	19	28	0.9	10	7	17	8	8	79
Fetus, 7 mo	1.16	30	23	0.8	8	7	14	10	6	69
Newborn infant	2.9	100	24	0.7	5	5	14	5	6	60
Adult	70	3000	39	0.7	2	5	22	3	4	76

Table 21-8
Inorganic elements in 150 American diets

Elements	Per capita per day Range (grams)	Per capita per day Average (grams)	Per 3000 kcal Range (grams)	Per 3000 kcal Average (grams)
Calcium	0.24-1.87	0.73	0.35-1.47	0.73
Magnesium	0.14-0.67	0.34	0.17-0.53	0.34
Potassium	1.43-6.54	3.39	1.63-5.27	3.40
Sodium*	0.19-4.61	1.94	0.22-4.83	1.95
Phosphorus	0.60-2.79	1.58	0.72-2.30	1.59
Chlorine*	0.88-5.83	2.83	0.83-7.26	2.88
Sulfur	0.51-2.82	1.28	0.80-2.35	1.30
Iron	0.0080-0.0307	0.0173	0.0090-0.0234	0.0174

*Since these dietary records did not show the quantities of table salt used, the figures for sodium and chlorine cover only the amounts in the food as purchased and may be greatly below the actual intake of these elements.

They are the principal inorganic elements in nutrition—at least from a quantitative standpoint. With an increase in age, the total ash, or mineral matter of the body, increases; but a decrease in magnesium, sodium, potassium, chlorine, and sulfur occurs also. This is shown in Table 21-7.

Under average conditions a healthy normal man excretes about 20 to 30 gm of inorganic material daily. This consists chiefly of the chlorides, sulfates, and phosphates of sodium, potassium, calcium, magnesium, and ammonium. Normally the intake should equal the output, except during growth and pregnancy. Table 21-8 shows the amount of inorganic elements in 150 American diets. Study of these figures reveals the great variation in every element determined. Since this might be the result of a divergence in the amounts of food consumed, the figures were also calculated to a uniform basis of 3000 kcal. Even these figures show great differences between the minimum and maximum. The question naturally arises as to whether some of these amounts are too low and others too high for optimal physiological activity or whether only a minimum of each is required and the remainder is an unnecessary excess or a factor of safety.

The effects of a total deficiency of inorganic elements have been studied in young rats fed a low-salt diet, adequate in all other respects. As might be expected, there occurred a failure of growth, the development of a mild anemia, and a decreased mineralization of bone. There also was a distinct hypertrophy of the kidneys, as yet unexplained.

Two factors that concern inorganic metabolism, i.e., ammonia formation and sulfur metabolism, have already been considered (Chapter 13). Therefore these subjects will be taken up only incidentally in this chapter.

The electrolytes: sodium, potassium, and chloride

Sodium is the principal cation in extracellular fluid and is involved primarily with the maintenance of osmotic equilibrium, acid-base balance, and extracellular volume (Chapters 16 and 17). Potassium, on the other hand, is the principal cation in intracellular fluid and is concerned mainly with intracellular homeostasis. Chloride, as a freely diffusible anion, shares in all of the foregoing functions, as well as in the formation of gastric hydrochloride. The principal food sources, functions, estimated requirements, and manifestations of a deficiency of these three electrolytes are summarized in Tables 21-9 and 21-11.

The effects of severe deficiencies of both sodium and potassium have been studied in rats. A lack of sodium results in retarded growth, both somatic and skeletal, atrophy of muscles and testes, and diffuse degenerative changes in many other

Table 21-9
Estimated safe and adequate daily dietary intakes of the major electrolytes (revised 1980 by the Food and Nutrition Board of the National Academy of Sciences–National Research Council)*

	Age (years)	Electrolytes Sodium (mg)	Potassium (mg)	Chloride (mg)
Infants	0-0.5	115-350	350-925	275-700
	0.5-1	250-750	425-1275	400-1200
Children and adolescents	1-3	325-975	550-1650	500-1500
	4-6	450-1350	775-2325	700-2100
	7-10	600-1800	1000-3000	925-2775
	11+	900-2700	1525-4575	1400-4200
Adults		1100-3300	1875-5625	1700-5100

From Recommended dietary allowances, ed. 9, Washington, D.C., 1980, National Research Council.
*Because there is less information on which to base allowances, these figures are not given in the main table of the RDA and are provided here in the form of ranges of recommended intakes.

tissues. A deficiency of potassium likewise produces poor somatic growth, fragility of the bones, sterility in both males and females, renal hypertrophy, paralysis, and a slow heartbeat rate. Chloride deficiency in rats also results in poor growth but apparently few other significant gross changes.

The usual daily intake of sodium chloride is about 10 to 15 gm, or 170 to 256 mEq. This is far greater than is required, but the amount is used chiefly because of its flavor. About 98% is eliminated by way of the urine, and 2% by the feces. The usual amount of potassium in the diet, on the other hand, is only 2 to 4 gm, or 50 to 100 mEq, per day.

Although loss of fluid and loss of salt generally accompany each other, a deficit of sodium chloride alone may be encountered. The symptoms are weakness, fatigue, lack of appetite, nausea, and a diminution of mental acuity. Impairment of renal function with delayed diuresis follows. A thirst that cannot be allayed by drinking develops. Salt, however, does alleviate it.

Chloride is an essential anion. It is closely connected with sodium in foods and in body tissues and fluids and excretions. It is involved in the chloride shift (Chapter 16) and the formation of gastric hydrochloric acid (Chapter 17). Chloride is excreted, mostly as sodium chloride and chiefly by way of the kidney. About 2% is eliminated in the feces and perhaps 4% or 5% in perspiration.

Recently a number of cases of chloride deficiency were reported in infants as a result of the accidental omission of chloride in a certain infant food formula. A metabolic alkalosis ensued (Chapter 16), sometimes with hypovolemia and marked loss of potassium in the urine. Impaired growth, memory defects, and psychomotor disturbances occurred. The administration of chloride remedied the defects.

Ordinary diets contain sufficient sodium, potassium, and chloride, but when there is excessive excretion of any of them, more must be provided. Adrenal insufficiency and acidosis are examples; diarrhea and excessive perspiration are others. Persons working in industries in which they encounter intense heat and perspire freely must have salt supplied with their drinking water to make up for this loss of electrolytes.

No other cation can entirely replace potassium for the performance of a great number of cellular functions. Therefore potassium is an essential element. It can move in and out of most cells more easily than sodium according to the demands of shifting membrane equilibria. Probably changes in acid-base balance influence these shifts considerably. Under normal conditions the respective concentrations of sodium and potassium ions are held within a fairly narrow range, although, as shown by tracer studies with radioactive isotopes, these ions move freely across cell membranes.

In the formation of cells the potassium ions are taken up; this appears to be essential for growth. In infancy and childhood and during pregnancy and lactation there is a comparatively high potassium retention. During muscular contraction there is a loss of potassium from the muscle cells to the extracellular fluid. Subsequently this lost fraction returns to the muscle tissue. The movement of potassium during muscle contraction seems to be related to the contractile process rather than to the neuromuscular transmission of the stimulus. In the steady state the loss from contraction is probably just equal to the gain from recovery. Undoubtedly this is the condition in cardiac contraction, since potassium ions are essential to heart rhythm.

Potassium also is necessary for nerve activity, and the same type of movement of the ion occurs here. Nerve fibers are exceptionally rich in potassium. When the nerve is stimulated, potassium diffuses into the surrounding fluid very rapidly; during rest it diffuses back. This diffusion seems to be associated with a change in potential that occurs during the conduction of the nerve impulse.

In diabetic acidosis the failure to metabolize glucose properly is associated with loss of potassium from the cells. There follows an increased excretion of potassium in the urine if the kidneys are functioning efficiently. Often there is vomiting, with further loss of potassium. However, the plasma level of potassium is usually not below normal because the urinary excretion cannot keep pace with the influx of potassium from the cells. When insulin is administered, the extracellular fluid potassium is shifted into the cells, and a hypopotassemia occurs. This may lead to several alarming symptoms, including paralysis of the respiratory muscles.

In certain types of hypertension rigid restriction of sodium in the diet has been found by some investigators to be beneficial. In hypertensive rats the sodium content of the entire body is elevated, whereas potassium remains unchanged. The data indicate a penetration of the intracellular compartment by sodium. If this occurs, it must be because sodium displaces some other intracellular cation or is in an osmotically inactive state.

There is some evidence in the literature that a relationship between excessive salt consumption and the pathogenesis of human essential hypertension exists. Data from an interesting recent comparison between the sodium intake and blood pressure levels of two Polynesian populations are pertinent in this connection. The subjects, 51 male and 60 female, were inhabitants of two Polynesian islands, Rarotonga and Puka Puka. They were similar ethnically but observed different dietary, social, and economic habits. The mean blood pressure of the Rarotonga group was significantly higher than that of the Puka Puka group and increased with age, especially in females. Only a slight rise with age was seen in the latter group. The blood pressure differences between the two populations showed no correlation with height or weight but correlated significantly with the sodium intake.

Calcium

Functions Calcium is needed by all cells. It is present in blood serum, about half in the ionized form and the rest un-ionized, probably bound to protein for the most part and, to a minor degree, in a calcium-citrate complex. Normally the concentration is about 10 to 11 mg/100 ml of serum. The calcium that is readily available for metabolic use, i.e., the freely moving calcium in soft tissues, extracellular fluid, and blood, is termed the *miscible calcium pool of the body*. The role of the hormones parathormone and calcitonin in regulating the blood level of calcium is discussed in Chapter 18.

Recent investigations have demonstrated that a widely distributed intracellular protein *calmodulin* modulates the intracellular Ca^{++} level to maintain the various calcium-dependent reactions. Any excess calcium ions are "pumped out" of the cell via this regulatory protein.

A particular and important effect of the calcium ion is on nervous tissue. If the ionic calcium of the blood falls, the nervous system becomes hyperirritable. This may lead to tetany. On the other hand, a high calcium content depresses nervous irritability. Thus the administration of calcium salts is indicated in the alleviation of tetany arising from low calcium. Calcium is involved in the contraction and relaxation of muscle (Chapter 15).

Calcium is, of course, required for bone and tooth formation. If the diet is deficient in this element, either or both may suffer. This is also true if the absorption of calcium is inefficient, even in the presence of an adequate amount in the diet. Growing children particularly require an abundance of calcium for teeth and bones as well as for other tissues. During pregnancy and lactation there is likewise a great demand for it in the diet to provide for the growing fetus and for the secretion of the calcium-rich milk. The requirement for calcium ions in blood clotting (Chapter 16) need only be mentioned at this point.

Absorption The absorption of calcium is quite variable. Calcium forms insoluble salts with a number of anions that occur in the intestinal tract. Thus we may find much of the calcium precipitated as the phosphate, carbonate, oxalate, phytate, or sulfate or as calcium soaps, which are also insoluble and therefore poorly absorbed. This depends on the amount of soluble calcium salts present, the negative ions, the pH, and the state of fat digestion and absorption. Calcium salts are more soluble in acid than in basic solutions. Furthermore, all food calcium does not behave in the same way. For example, the calcium of all vegetables is not uniformly absorbed, and in some cases vegetables actually tend to depress the absorption of calcium from other foods. This may be caused by the presence or formation of oxalates or phosphates or by an influence on the pH of the intestinal contents. Insoluble calcium soaps form if fatty acids are present in large amount, resulting of course in diminished calcium absorption. Phytic acid and its compounds also interfere with the absorption of calcium. This substance is abundant in unrefined cereals.

Vitamin D is important in aiding the absorption of calcium and phosphorus. As will be discussed in more detail later, vitamin D appears to induce the biosynthesis of a transport protein for calcium in the intestinal mucosal cells. This calcium-binding protein has been isolated and prepared in a purified form from the intestinal mucosa of several species of animals.

However, at best, normal adults absorb only small amounts of calcium, perhaps 100 to 200 mg per day, this amounting to 10% to 20% of that in the foods eaten.

Lactose exerts a rather striking favorable effect on the absorption of calcium, experimentally in the rat. Investigations using the amount of ^{45}Ca (and ^{85}Sr) deposited in the femur of the rat as an index of absorption demonstrated that lactose significantly increases calcium absorption when given orally but not when administered parenterally. The beneficial effect is proportional to the amount of lactose fed and may be the result of chelation of calcium by lactose, resulting in the formation of a soluble complex of low molecular weight. L-Sorbose, cellobiose, D-xylose, raffinose, melibiose, D-glucosamine, D-mannitol, and D-sorbitol also favor calcium absorption to some extent. D-Glucose, D-galactose, D-fructose, and sucrose are without apparent effect. Of interest in this general connection is the fact that the amino acids lysine and glycine also increase the solubilities of certain calcium salts.

Excretion The excretion of calcium is partly through the kidneys but mostly by way of the small intestine (Chapter 17). Excretion into the feces continues even when the intake is low, and accordingly a negative balance is possible. The intestinal elimination of calcium may be increased by a lack of vitamin D and diminished by a suitable amount in the diet. Measurable amounts of calcium may also be lost in sweat. It is questionable whether an individual on a low-calcium diet ever attains calcium balance under heavy sweating conditions. Since calcium levels have a profound effect on nervous irritability, one can appreciate that a negative calcium balance, if continued long enough, would cause hyperirritability and even tetany as well as decalcification of the skeleton. These are conditions to be guarded against in pregnancy and lactation. In these states the demand for calcium is so great that supplements of calcium and vitamin D should be provided.

Calcium excretion in the urine is increased by the ingestion of excessive amounts of proteins. This may be due to an increase in the acidity of the urine resulting from the metabolism of the proteins (Chapters 13 and 17).

Requirement The amount of calcium retained by the body depends not only on the amount in the diet but also on the efficiency of absorption and on excretion. Therefore it is difficult to set an absolute standard for the calcium requirement. Moreover, it has been shown by several investigations that the need for calcium is flexible. In certain parts of the world where the calcium content of the diet is low the adult population gets along well with little of this element. Apparently, however, such diets must start in childhood and extend throughout life, and adaptation occurs early under such circumstances. This idea is in conflict with the concept that each individual has a definite daily need for calcium, which, if not met by the diet, leads to nutritive disaster in time. The adaptive process must be gradual. An abrupt change from a high- to a low-calcium ration may lead to a negative calcium balance. One should keep in mind the preceding facts when considering the recommended dietary allowances.

Food sources Milk and milk products are the best sources of food calcium. A quart of milk supplies about 1.2 gm of calcium in a readily assimilable form. Consequently, a safe rule to follow is a pint of milk a day for every adult and a quart for every child. However, milk has another virtue in this connection, in addition to its high content of calcium. Milk sugar, lactose, also plays a role in calcium absorption. Other foods relatively high in calcium salts are beans, leafy vegetables, shellfish, and fish of the sardine type (i.e., in which bones are eaten). Vegetables, in general, contain more calcium than do animal foods. Some vegetables, however, have an appreciable amount of oxalic acid present, and this forms insoluble calcium oxalate in the intestinal tract and thus lessens the absorption and utilization of some of the calcium present. Moreover, although calcium is widely distributed, it is not present in most foods in sufficiently high concentrations to prevent a deficiency of this element in our diet. A temporary deficiency or even a long-continued deficiency of calcium in the diet may occur without the appearance of any symptoms attributable to a low calcium intake because the bones act as a storehouse. Under the influence of parathormone, calcium is withdrawn from the bone to maintain the calcium levels of the blood and soft tissues (Chapter 18).

RDA The recommended dietary allowance for calcium established by the Food and Nutrition Board of the National Research Council (1980) is based on the average daily loss of calcium adjusted for the average percent absorption of food calcium in the normal human subject. For example, if urinary excretion is 175 mg per day, endogenous fecal excretion is 125 mg per day, and losses in sweat are 20 mg per day, then a total excretion of 320 mg per day would occur. Assuming a 40% average absorption of food calcium (a figure actually obtained experimentally), a total intake of 800 mg calcium per day would be necessary to replace the excretory loss.

The current recommended dietary allowances for calcium are given for 10 different age groups in Table 21-4. Note that the recommended adult allowance for males and females is *800 mg per day*. An additional 400 mg daily is recommended during pregnancy and lactation to meet the needs of fetus and for milk production, respectively. Likewise, the recommended allowances are higher, on the basis of body weight, for infants and for children during the years of growth.

Bone calcification The calcification of growing bone and the maintenance of mature bone are dependent on an adequate dietary intake of calcium and vitamin D. The latter is discussed later in this chapter. Thus there is a continuing need for both calcium and vitamin D throughout the life cycle, as reflected in the adult RDAs for these nutrients (Table 21-4).

With age the constant withdrawal of calcium from the bone without sufficient replacement results in the gradual development of varying degrees of *osteoporosis* with increased fragility and fractures of bones, particularly of the extremities and pelvis. Postmenopausal women are expecially vulnerable; therefore a greater intake of calcium is desirable for this group. Since it is difficult to obtain regularly more than 800 mg of calcium daily from food sources, supplements of up to 1.5 gm of additional calcium per day as a preventive measure may be prudent. Additional vitamin D and small amounts of fluoride have been reported to have added beneficial effects. There is substantial evidence that an inadequate calcium intake over a period of years contributes to the relatively high prevalence of osteoporosis in older persons.

Contrary to common opinion, the calcium of teeth is not regulated in the same way as the calcium of bone. The adult tooth, already fully formed and calcified, is not subject to decalcification readily when the body requires calcium. This is true in pregnancy and lactation as well, and the old saying "a tooth for every child" has no scientific justification. Disturbances of calcification are of importance only in the *growing* tooth. Therefore children must have an abundance of calcium with vitamin D or sunshine to help them absorb the calcium for tooth as well as skeletal development.

Phosphorus

The vital part played by phosphorus compounds in many phases of metabolism and in acid-base regulation indicates how necessary this element, in sufficient amount, is.

Vitamin D aids in the absorption of phosphates from the gastrointestinal tract, just as it aids in calcium absorption. As a rule, in the absence of vitamin D a low serum phosphorus results. In rickets there is usually a normal serum calcium with a low serum phosphorus. Other cases of rickets occur in which both calcium and phosphorus are low in the serum, or calcium is low and phosphorus is normal. The typical and usual disease, however, is characterized by a low phosphorus and normal calcium. An empirical index for determining whether a child is rachitic is the product of the serum phosphorus and serum calcium (in milligrams per 100 ml). If the index is below 30, rickets is present or will develop, but not if it is above 40.

Phosphorus is found in those food containing phosphoproteins, nucleoproteins, phospholipids, and glycerophosphates, as well as the inorganic phosphates, chiefly calcium and sodium. Since quantitatively the greatest proportion of the phosphorus is used to form the bone salt, which is largely calcium phosphate, evidently the phosphorus intake should bear an optimal relation to the calcium intake. The foods richest in calcium are also richest in phosphorus, i.e., milk, cheese, and beans. Eggs, cereals, fish, and meats are also high in this element.

Phosphorus is present in some food as phytates. In fact, a large proportion of the phosphorus of vegetables is in this form. Although the exact structure of these salts is not known, phytates are quite insoluble mixed calcium and magnesium salts of phytic acid which in turn is a hexaphosphate of inositol. Available evidence indicates that phytic acid and its compounds interfere with the absorption of calcium, zinc, and iron from the intestinal tract. Unrefined cereals also are rich in phytates, but white flour contains little. Hence the phytate problem is not serious in the United States; but in areas of the world where unrefined cereals form a large part of the diet and little calcium is consumed, the interference with calcium absorption may result in serious deficiencies of calcium, including the development of so-called *cereal rickets*. The phosphate and inositol of these substances are, for the most part, unavailable nutritionally. However, slight digestion may be accomplished by gastric juice and somewhat more by intestinal phosphatases.

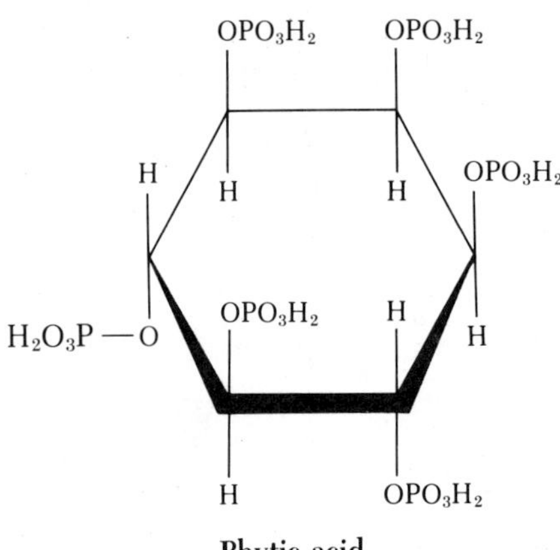

Phytic acid

Requirement The recommended dietary allowance (1980) for phosphorus (Table 21-4) has been established on the basis of an approximate 1:1 relationship with calcium. The recommended daily allowance for both adult men and women is *800 mg*. For infants and children the values range from 240 to 800 mg; for teenagers the value is 1200 mg, depending on age. The amount during pregnancy and lactation should be increased to 1200 mg per day. The recommended 1 pint or more of milk per day for the adult and 1 quart for the child and during pregnancy and lactation to satisfy the calcium requirement will also nearly supply the need for phosphorus.

Magnesium

Magnesium is an essential element. Magnesium-free diets can be prepared experimentally, and animals on such diets have circulatory disturbances, increased irritability, and finally convulsions and death. It is an essential part of chlorophyll and therefore occurs in all green plants.

Magnesium occurs in human bones, muscles, and nervous tissue. Its distribution is uneven, probably because it can replace calcium to some extent, and this depends largely on the amount of calcium available. Human blood serum, however, has a constant magnesium content, 1 to 3.5 mg/100 ml (0.8 to 2.9 mEq per liter). The blood plasma level is about 2 mEq per liter and that of the red blood cells is 5.3 mEq per liter.

The Food and Nutrition Board of the National Research Council (1980) places the recommended dietary allowance of the adult man at 350 mg per day and that of adult women at 300 mg per day. As is true of the other nutrients, more magnesium is required during pregnancy and lactation, 450 mg per day. The recommended dietary allowance for the infant and child varies with age (Table 21-4).

The frequent claim that an adult cannot have a magnesium deficiency on an ordinary diet is incorrect. For instance, there is a large loss of magnesium during diabetic acidosis and in alcoholism, which may cause a negative magnesium balance. Also, there have been other reports of magnesium deficiencies in patients. The clinical signs include neuromuscular disturbances and hyperirritability with tremors and convulsions.

The magnesium ion influences tissue irritability. Thus, when introduced in large amounts parenterally, it is a central depressant, having anesthetic and anticonvulsant effects. These effects are completely antagonized by calcium. Curiously, however, low serum concentrations of either magnesium or calcium lead to the same pharmacological effects, i.e., hyperirritability and convulsions. Magnesium deficiency in rats, produced by the feeding of a magnesium-deficient diet, is likewise characterized by the development of hyperirritability and convulsions. A decrease in growth, a decrease in the efficiency of food utilization, and a striking vasodilatation also occur. The tissues most likely to be damaged by chronic magnesium depletion are cardiovascular, renal, and neuromuscular. Magnesium ions also function as cofactors in a number of enzyme reactions, particularly those requiring ATP.

Toxicity due to the increased use of magnesium-containing laxatives and antacids has been reported recently in the elderly. This group is especially vulnerable because of decreased renal function with age. The chief symptoms are drowsiness, lethargy, and weakness.

Magnesium is excreted by way of the intestine, for the most part. A fraction is eliminated by the kidneys. One of the characteristic crystal forms frequently seen in urinary sediments is the "coffin plate" crystal of ammonium magnesium phosphate ($NH_4MgPO_4 \cdot 6H_2O$).

Iron

The role of iron in the body is closely associated with that of hemoglobin, myoglobin, and other hemoproteins. The great importance of iron is quite out of proportion to the amount present in the entire body, which is the insignificant value of 3 to 5 gm. This small amount is used over and over again in the body. Iron is not like the vitamins or most other organic or even inorganic substances, which are either inactivated or excreted in the course of their physiological functions. Very little iron is lost from the body normally, and, since it is a small part of the hemoglobin molecule (about 0.3%), comparatively little is needed. Iron is also a constituent of many tissues besides blood (e.g., the myoglobin of muscle) and is essential for the composition of such catalysts as the cytochromes, peroxidases, and catalases.

Iron is also a constituent of such nonheme iron-containing proteins as *ferredoxin* and *adrenodoxin*. The former is widely distributed in plant and animal tissues and plays an important role in photosynthesis; it is also involved in methemoglobin reduction, $NADP^+$ reduction, pyruvate metabolism, and nitrogen fixation. It has been crystallized from certain bacteria and from spinach leaf, and its amino acid composition and sequence have been determined. Similarities of the amino acid sequence of spinach ferredoxin to that of bacterial ferredoxin suggest a common evolutionary origin. The ferredoxins have relatively low oxidation-reduction potentials and serve as electron carriers in certain anaerobic bacteria and in photosynthetic organisms. The extremely electronegative character of the ferredoxins suggests that they may have existed as a biological catalyst during the era of the earth's reducing atmosphere before green plants emerged.

Iron has been called a "one-way substance." It may be absorbed in small amounts, but once absorbed, little is excreted. Any excess over and above the amount absorbed is eliminated in the feces. This cannot be considered a true excretion but rather an oversupply, which is thus wasted. However, there is some actual excretion, as determined by following isotopic iron administered intravenously. This is of the order of 0.5 to 1 mg per day and occurs mainly through the small intestine. Almost none is found in the urine, and careful studies of the intake and output have not revealed any appreciable negative balances except in early infancy, and from blood loss.

Hypochromic anemias usually do not result from negative iron balances but from losses of blood, which may be very difficult to detect—so-called *occult blood*. Parasitic infections also may result in blood loss. Hypochromic anemias are those conditions in which there is a greater diminution in the concentration of hemoglobin than in the number of red cells and, accordingly, the red cells are paler than normal. Positive iron balances occur in growing children and in pregnant women. In both instances more iron is absorbed than is excreted; this difference corresponds with the need to synthesize hemoglobin for the expanding blood volume.

Absorption The absorption of iron takes place chiefly in the upper part of the small intestine. Although normally very little is absorbed, an average of approximately 10% of food iron, under certain conditions larger quantities may pass into the body. After a severe hemorrhage, the absorption of iron may be increased 10 to 20 times, but there is usually a delay before this occurs. In hypochromic anemia, iron is absorbed more than normally, and in *hemochromatosis* an astonishing amount may be absorbed and found in the tissues. This is a disorder of iron absorption that is characterized by large deposits in the liver and other organs of two iron proteins, *ferritin* and *hemosiderin*. Hemosiderin is present as brownish granules, apparently consisting of large aggregates of ferritin molecules with a higher iron content. It represents deposits of excess iron. Hemosiderin is derived from hemoglobin and other sources of iron. Its iron content may vary between 9% and 55% and is in the form of ferric hydroxide stabilized by protein. It is also present normally to some extent, but in hemochromatosis the excessive deposition of hemosiderin *(hemosiderosis)* in the liver and other organs may result in liver damage with cirrhosis and diabetes-like symptoms. Significant amounts of hemosiderin also may be deposited in the skin, giving it a bronze pigmentation, sometimes referred to as *bronze diabetes*. A hemosiderosis also has been described in the Bantu peoples of Africa, who cook their food and brew beer in iron kettles. *Iron overloading* thus results. Excessive deposition of iron in the liver and other tissues also may occur in patients receiving repeated transfusions of whole blood.

Ferritin is a protein that may contain as much as 23% iron by weight. The iron is present as micelles or colloidal particles, composed of a ferric hydroxide–ferric phosphate complex, bound rather firmly to the protein. It can be freed from iron without denaturing the protein, and this protein, *apoferritin,* is homogeneous, with a molecular weight of 460,000. Both ferritin and apoferritin can be crystallized with cadmium sulfate. Ferritin has been isolated from bone marrow, spleen, and liver of a number of different animals and has also been found in the gastrointestinal mucosa. Experiments indicate that apoferritin may not always be present in appreciable amounts in the intestinal mucosa but is formed in response to iron feeding, i.e., the feeding of iron in some way brings about the formation of the particular protein that combines with it.

The absorption of iron involves the release of food iron by gastric hydrochloric acid (as Fe^{++}), which then forms a chelate with ascorbic acid and certain sugars and amino acids. These chelates remain soluble in the more alkaline fluids of the duodenum and jejunum, and thus iron absorption is enhanced. It appears that a mucosal receptor in the upper small intestine, *apoferritin,* controls the absorption of iron. Thus, when there is iron need in the body usually as a result of blood loss and resulting increased compensatory erythropoiesis, iron is removed from the intestinal mucosal cell receptor and increased absorption of food iron occurs to replace it. This continues until the body's stores of iron, particularly that of the intestinal mucosal cells, are replenished. The iron-saturated mucosal receptor, ferritin, then once again refuses to absorb available iron. There is also evidence that copper is involved in the intestinal transport of iron. ^{59}Fe fed to copper-deficient swine was absorbed by the intestinal mucosal cell, but its transfer to the plasma was impaired. Injected ^{59}Fe was taken up by cells of the hepatic parenchyma and reticuloendothelial system. The investigators therefore concluded that copper is required for the release of ferrous iron into the plasma for subsequent utilization. This may entail its oxidation to ferric. The copper-containing protein *ceruloplasmin* may promote the incorporation of ferric iron into transferrin, and hence its utilization, by serving as a *ferroxidase*.

The iron in foods is not all equally available. Iron in the heme combination apparently is more available than is nonheme iron. In the therapeutic administration of iron, inorganic iron is probably as useful as organic, and although ferrous iron is preferable, ferric is usually converted to ferrous in the body and is absorbed as such. However, the fact that only small amounts are absorbed should be emphasized. By giving massive doses, one can force slightly larger quantities, but the apoferritin-ferritin regulatory mechanism hinders unlimited absorption no matter how much is available.

Figure 21-7
Absorption, utilization, and storage of iron in normal adult.

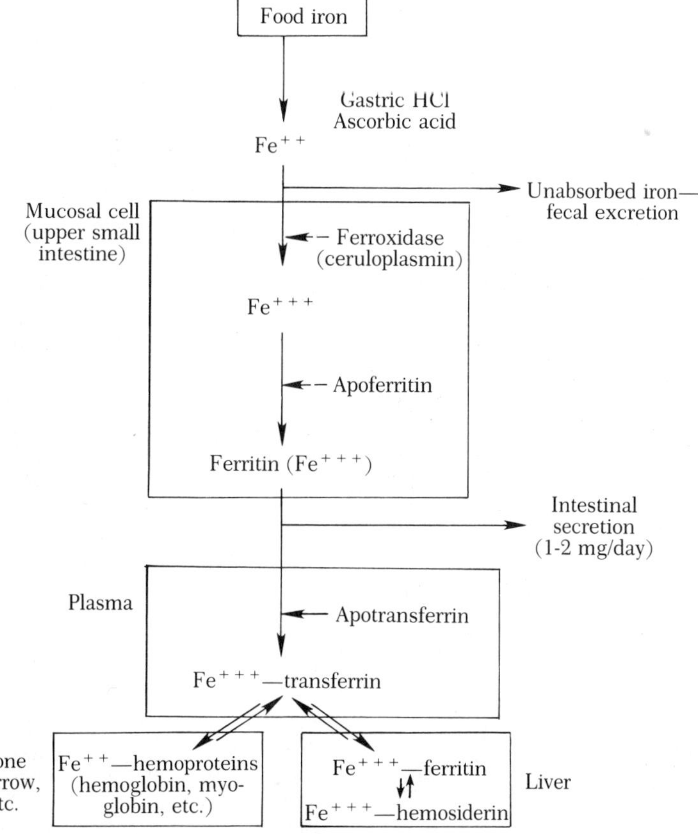

Defective absorption of iron may result from gastrointestinal disturbances, e.g., achlorhydria or diarrhea, leading to anemias that readily yield to large doses of iron. From the mucosal cell the ferrous iron passes into the blood plasma. The ferrous iron is oxidized by *ceruloplasmin* (ferroxidase) and becomes attached to one of the β_1-globulins, called *apotransferrin* (Figs. 21-7 and 21-8). This complex is also in equilibrium with the ferrous iron and ferritin in the liver (the chief storage site for iron) and the spleen and bone marrow. Each apotransferrin molecule binds two atoms of ferric iron. In the bone marrow the ferric ion apparently is reduced again to ferrous iron and is converted to heme by combining with protoporphyrin (Chapter 16).

If there is need for iron by the body, e.g., after hemorrhage or debilitating disease, the bone marrow, liver, and spleen are called upon first to give up their ferritin, for these are the organs with the largest amounts of this protein. Hemosiderin, another form of iron storage, is less soluble than ferritin and probably is not used until ferritin has reached a low level. Only when these major sources have been depleted and the plasma iron concentration diminished is the mucosa called upon for its iron. When this occurs and the physiological saturation of the mucosal cells with respect to ferrous ions is no longer maintained, iron can be absorbed. Thus the ferritin content of the mucosa acts as a valve, permitting the absorption of only enough iron to preserve equilibrium. This is a fortunate provision of nature, since ferric ions are rather toxic.

Fig. 21-8 indicates these relationships in a general way. It also shows how this hypothesis accounts for the saving of iron in the catabolism of hemoglobin. When the erythrocytes have finished their life cycle, the iron is reused. Recent investiga-

Figure 21-8
Metabolism of iron in the human adult.

Adapted from Conrad, M.E.: Borden Rev. Nutr. Res. **28**:49, 1967.

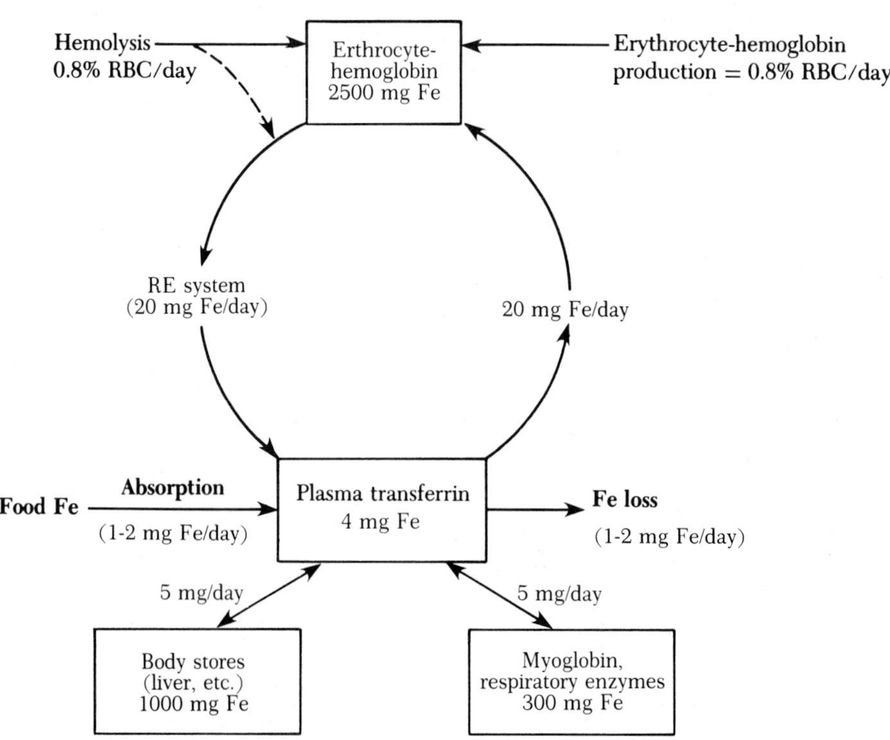

tions using radioactive iron showed that this iron from the "old" erythrocytes is used in preference to storage iron. Thus we see that low absorption of iron is compensated for by the efficient reuse of iron and by the intricate mechanism of storage.

RDA The recommended dietary allowance for iron has been recently revised (1980) by the Food and Nutrition Board of the National Research Council (Table 21-4). The intake for adult men should be 10 mg per day and for adult premenopausal women 18 mg per day. These iron-intake values are considerably higher than the actual amounts of iron required to balance the 1-mg daily loss by healthy adult men and the average 1.8-mg daily loss by adult women, including that lost in menstrual blood. Allowance also is made for the fact that only about 10% of food iron is absorbed by the normal adult. The recommended dietary allowance for boys and girls from 11 to 18 years is also given. Iron needs for women are higher than for men because of menstrual losses and the demands of pregnancy and lactation. The new recommended dietary allowance for iron for women is higher than previous values. This increase is based on findings that iron stores are reduced or absent in about two thirds of menstruating women and in the majority of pregnant women. In view of this the Food and Nutrition Board (1980) recommends the use of supplemental iron (30 to 60 mg daily) during pregnancy. Iron needs during lactation are not substantially different from those of nonpregnant women, but the Board suggests continued supplementation in the maternal diet for 2 to 3 months after parturition to replenish stores of iron depleted by pregnancy.

Food sources The best sources of food iron include liver, meats, egg yolk, green leafy vegetables, whole grains, and enriched breads and cereals. There are two principal types of food iron, classified as *heme iron* and *nonheme iron*. The heme iron group is found principally in meats, poultry, and fish and constitutes about 40% of the total food iron. Vegetables, fruits, legumes, nuts, and breads and cereals contain mainly nonheme iron and comprise about 60% of the total food iron. According to recent investigations about 20% to 23% of heme iron is absorbable, whereas only 3% of nonheme iron is absorbed. However, ascorbic acid increases this value up to about 8%.

A significant amount of iron is obtained from iron-enriched foods, such as breads and cereals. It is estimated that as much as 50% of food iron comes from these sources. The bioavailability of the iron added to foods in the enrichment program (reduced iron) apparently is equal to that of ferrous sulfate. There is some current concern, however, that there may be too much pumping of iron into enriched foods.

Any element that forms insoluble precipitates with the ferric or ferrous ions tends to prevent the absorption of these ions. Thus phosphate or phytate inhibits iron absorption. Conversely, the presence of a sufficient amount of calcium ions to combine with them would increase the possibility of iron absorption. Likewise, the tannic acid of tea and various antacid preparations may decrease the absorption of food iron. The former reacts with iron to produce insoluble iron tannate.

The ingestion of foods (e.g., eggs, meat, and spinach) into which radioactive iron has been incorporated supports the conclusion that, on an average, only 10% of the daily intake of iron is absorbed.

As stated previously, pregnancy demands additional iron for the growing fetus. When the infant is born, he has a considerable store of iron for future use. This is fortunate because milk is extremely low in its content of this element. There is a supply of iron in the infant's spleen and liver, but neither is as great as was formerly believed. The amount of liver iron ranges from a negligible quantity to 60 mg. The chief location of the infant's iron is the hemoglobin of the blood. With a concentration of 22 to 23 gm of hemoglobin per 100 ml, this is higher than at any later period in the individual's life. During the first few weeks, with a constant loss of iron and almost no iron in the milk ingested, there is an appreciable negative balance of iron; but after the second month this balance tends to approach zero. The iron comes, for the most part, from the physiological destruction of hemoglobin. Premature babies or twins may be deficient in iron for obvious reasons, and anemia may result unless iron medication is given. Hypochromic anemia of infants probably is the most common nutritional deficiency in North America.

Iron deficiency is not at all uncommon in the United States; some estimates are as high as 2% of the total population. Nor, indeed, is it uncommon in most of the world in infants and in pregnant women. The deficiency is readily detectable if sensitive measurements such as plasma iron and iron-binding capacity of the plasma (transferrin) are employed. Prophylactic iron administration daily is justified therefore in such infants and during pregnancy. The principal cause of an iron deficiency in humans is blood loss.

Metabolism As indicated in Fig. 21-8, the major amount of iron is present in the hemoglobin of erythrocytes. About 0.8% of the circulating erythrocytes are destroyed daily, having reached their normal life span of about 120 days (Chapter 16), and are hemolyzed. The hemoglobin iron is carefully conserved and reutilized in the biosynthesis of new hemoglobin and erythrocytes in the bone marrow. Body stores of iron in the liver and other tissues, amounting to about 30% of the total body iron, may be called on to replenish plasma iron if needed. The small iron loss daily, 1 to 2 mg, usually in the form of intestinal secretions or desquamated epithelial cells, is compensated for by the absorption of an equal amount of food iron as described. In the adult woman approximately 30 mg of iron are lost during the menstrual period. In pregnancy some 600 mg are transferred to the fetus. These amounts must be replaced by either increased absorption of food iron or therapeutic doses of iron salts.

There is some evidence that vitamin E may play an indirect role in the conservation of body iron stores by increasing the resistance of the erythrocyte membrane to hemolysis and thus decreasing the rate of turnover of hemoglobin. An increased hemolysis of red blood cells is characteristically found in vitamin E deficiency with a concomitant anemia.

Essential trace elements

A number of inorganic elements are present in human, animal, and plant tissues in very small amounts, a few milligrams or less per 100 gm. These are called *trace elements* or sometimes *micronutrients*. They in turn may be divided into *three* distinct categories:

1. Those recognized as essential for human nutrition and for which sufficient information has been obtained to justify the establishment of recommended dietary allowances by the Food and Nutrition Board of the National Research Council. This group includes *zinc* and *iodine*. Some authorities also include *iron*.
2. Those probably essential for human nutrition based on evidence from studies in other mammalian species but for which there is insufficient information on which to base the establishment of dietary allowances. This group includes copper, manganese, fluoride, chromium, cobalt, selenium, and molybdenum.
3. Those which are present in human (and animal) tissues but as yet have not been proved essential. Several elements in this group, however, appear to be

likely candidates for inclusion as essential nutrients: nickel, silicon, tin, and vanadium. A large number of other inorganic elements may be found in minute amounts in human tissues but are probably just accidental constituents or contaminants.

The trace elements are discussed in the order listed.

Zinc

Zinc is an essential element for plants, animals, and humans. It is a constituent of a number of enzymes involved in major metabolic pathways and functions. Carbonic anhydrase, carboxypeptidase, alkaline phosphatase, lactic, malic, and alcohol dehydrogenases, insulin, and most of the enzymes involved in DNA replication, repair, and transcription are examples. Body stores of zinc appear to be small; therefore a human (and animal) dietary deficiency can produce deficiency symptoms promptly. These include loss of appetite, poor growth, skin changes with severe acrodermatitis, impaired wound healing, decreased taste acuity *(hypogeusia)*, and fetal malformations during pregnancy. Pronounced deficiency in humans can also result in hypogonadism and dwarfism. This has been described in the Middle East and probably occurs in other parts of the world (including the United States) where the intake of zinc is low and/or the intake of dietary fiber is high. *Geophagia* (clay eating) is another primary cause of zinc deficiency, particularly in rural areas of Iran and perhaps in other parts of the Middle East. Clay binds food zinc and iron, causing their loss by fecal excretion.

Zinc deficiency has been reported in infants fed formulas low in zinc content. Premature infants are at special risk. Some preparations used for parenteral nutrition also are deficient in zinc (and copper). Human zinc deficiency is characterized by low levels of zinc in the plasma and in hair samples.

There are areas in the United States where the soil is deficient in available zinc, causing zinc deficiency in farm animals and probably also in humans. Zinc enrichment of feeds has been effective in preventing this problem. Phytate and dietary fiber lower the availability of dietary zinc.

Requirement Metabolic studies in human subjects have demonstrated that healthy adults require approximately 12.5 mg of zinc per day when ingesting a mixed diet. Studies using the radioisotope of zinc gave a value of 6 mg per day as the turnover of body zinc. Based on these and other data in the literature and with an allowance of approximately 40% for the absorption of dietary zinc, the Food and Nutrition Board has established the RDA for zinc at 15 mg per day for adults, with an additional 5 mg during pregnancy and 10 mg per day during lactation (Table 21-4). The recommended daily dietary allowance of zinc for infants varies from 3 to 5 mg; for preadolescents it is 10 mg.

Sources Meat, liver, seafoods, and eggs are good food sources of available zinc. The zinc of vegetable and whole grain products (whole wheat or rye, oatmeal, whole corn) is less available largely because of the presence of phytate in the latter. Milk, including breast milk, also is a good source of zinc. The colostrum is an especially rich source.

Toxicity The toxicity of zinc is low in humans, 2 gm or more of zinc sulfate taken orally being required to produce the characteristic symptoms of acute gastrointestinal irritation and vomiting. Excessive intake of zinc may produce evidence of copper deficiency if the intake is marginal.

Iodine

Normally the total iodine content of the body approximates 20 to 50 mg, distributed as follows: muscles, 50%; skin, 10%; skeletal structure, 7%; thyroid, 20%; other endocrine organs plus the central nervous system, 13%. The concentration of iodine in the thyroid gland is more than 1000 times that in muscle and 10,000 times that in blood. Thyroxine and triiodothyronine, the physiologically active substances that are formed by the thyroid, are iodine compounds, but only one fourth or one fifth of the total iodine in the gland is present in those forms (Chapter 18). Probably most of the remainder is in the form of organic precursors of thyroxine, and 1% or less is iodide ion. Little is known of the function of iodide except in the thyroid. The blood plasma normally contains from 4 to 8 μg of protein-bound iodine (PBI) per 100 ml.

A lack of iodine in the food and drinking water is related to the occurrence of simple endemic goiter. Drinking water in different localities in the United States varies in its iodide content from 0.01 to 73.3 parts per billion, with that in the Great Lakes region being especially low. Goiter occurs more frequently in persons living in those regions where the drinking water has a low iodine content. It was also shown that, in general, simple goiter is more prevalent far from the ocean or in sections where ocean winds cannot carry their moisture. The reason is that seawater contains iodine, and, when the sea spray is deposited on coastal regions, it enriches the soil and drinking water with this element. Vegetables grown in these regions take up iodine from the soil, and thus the inhabitants of coastal areas get iodine from drinking water and vegetables, as well as from seafood.

At one time goiter was common in Switzerland. Although not far from the sea, Switzerland is surrounded by high mountains that cause the ocean breezes to deposit their moisture on the outer sides. As a result, Swiss soil and drinking water are low in iodide. Simple goiter can be prevented by an intake of sufficient iodide. This may be accomplished by adding inorganic iodides to the source of water supply or more simply by the addition of iodides to table salt (1:5000 to 1:200,000). The continuous use of iodized salt is necessary in the low-iodine areas to prevent the recurrence of endemic goiter.

Endemic goiter is still prevalent in many parts of the world, especially in mountainous areas remote from the oceans. Frank endemic goiter and cretinism, for example, are present in as much as 54% of the population of rural villages in the Andes regions of Ecuador. The iodine content of the drinking water there is low. Laboratory thyroid function tests (^{131}I uptake, PBI, basal metabolic rate, etc.) showed values typical of endemic goiter. Apparently *coto,* the Indian term for goiter, was noted by the Spanish explorers of this area in the early 1500s. The art and sculpture of that era in Ecuador also clearly show the presence of goiter. Iodine prophylaxis in the form of iodized salt has been instituted in this region.

It is well known that not all goiters are caused by lack of iodine. There is, for example, a goiter from infection. Exophthalmic goiter is a hyperthyroid condition that is not a result of low iodine intake. Experimentally the feeding of agents that block formation of the thyroid hormone (Chapter 18) causes a hyperplasia of the thyroid gland, resembling goiters. Thiourea, thiouracil, and large amounts of certain vegetables, notably cabbage and cauliflower, which contain a goitrogenic agent, have this effect.

Iodine or iodides may be absorbed from mucous surfaces or from the skin. They are excreted chiefly in the urine and to a minor degree in the sweat and feces. If given in large amounts, they are also found in tears, saliva, and bile.

In the 1980 revision of *Recommended Dietary Allowances,* the Food and Nutrition Board of the National Academy of Sciences recommends an intake of iodine for adults of 150 μg per day (Table 21-4). Smaller amounts are recommended for infants and children, depending on size. During pregnancy and lactation an additional 25 and 50 μg daily are recommended, respectively. To ensure adequate dietary iodine, it is desirable to use *iodized salt*. The Food and Nutrition Board has recommended federal legislation requiring such iodization. The current level of enrichment of table salt is 76 μg of iodine per gram. The average use of 3.4 gm of iodized salt per person per day would supply a daily intake of approximately 260 μg iodine. Slightly more than *half* the table salt consumed in the United States is iodized. Its *continuous* use in *all* noncoastal regions of this country is strongly urged by the Food and Nutrition Board. However, although the present intake of iodine in America is considered to be safe, any additional increases (e.g., iodophors, alginates, dyes) in the preparation of foods should be viewed with concern.

Probably essential trace elements

Because of the small amounts of these inorganic substances present in human tissues, direct evidence for their essentiality is usually difficult to obtain. However, it can be predicted reliably from proven essentiality in other mammalian species and from identification of given elements as constituents of normal human enzyme systems or other essential biomolecules.

Requirements for the following trace elements have been established in various animal species, but amounts needed have not been adequately determined. No values have been found on which to base a human requirement. However, the Food and Nutrition Board of the National Academy of Sciences (1980) has recommended an "estimated safe and adequate" *range* of daily intakes for the six selected trace elements listed in Table 21-10.

Copper

Copper is distributed widely in foods but is especially abundant in liver, kidney, egg yolk, oysters, and some legumes. In spite of this, recent evidence shows that copper is deficient in an appreciable number of human diets and is as much as 60% below that now recommended for an adequate diet. Copper also has been found to be deficient in some hospital diets and in some preparations used for parenteral nutrition.

Functions Copper occurs in certain oxidases and possibly in other enzymes. It is now known that at least 11 enzymes, all oxidases, contain copper. These are cytochrome oxidase (a_3), superoxide dismutase (hemocuprein), ferroxidase (ceruloplas-

Table 21-10
Estimated safe and adequate daily dietary intakes of additional selected trace elements*

		Trace elements†					
	Age (years)	Copper (mg)	Manganese (mg)	Fluoride (mg)	Chromium (mg)	Selenium (mg)	Molybdenum (mg)
Infants	0-0.5	0.5-0.7	0.5-0.7	0.1-0.5	0.01-0.04	0.01-0.04	0.03-0.06
	0.5-1	0.7-1.0	0.7-1.0	0.2-1.0	0.02-0.06	0.02-0.06	0.04-0.08
Children and adolescents	1-3	1.0-1.5	1.0-1.5	0.5-1.5	0.02-0.08	0.02-0.08	0.05-0.1
	4-6	1.5-2.0	1.5-2.0	1.0-2.5	0.03-0.12	0.03-0.12	0.06-0.15
	7-10	2.0-2.5	2.0-3.0	1.5-2.5	0.05-0.2	0.05-0.2	0.1-0.3
	11+	2.0-3.0	2.5-5.0	1.5-2.5	0.05-0.2	0.05-0.2	0.15-0.5
Adults		2.0-3.0	2.5-5.0	1.5-4.0	0.05-0.2	0.05-0.2	0.15-0.5

From Recommended dietary allowances, ed. 9, Washington, D.C., 1980, National Research Council.
*Because there is less information on which to base allowances, these figures are not given in the main table of the RDA and are provided here in the form of ranges of recommended intakes.
†Since the toxic levels for many trace elements may be only several times usual intakes, the upper levels for the trace elements given in this table should not be habitually exceeded.

min), tyrosinase (melanomas), uricase (liver, kidney), dopamine β-hydroxylase (adrenal), lysyl oxidase (aorta, cartilage), spermine oxidase, benzylamine oxidase, diamine oxidase, and tryptophan-2,3-dioxygenase (tryptophan pyrrolase). Copper is present to the extent of about 0.34% in one of the plasma α_2-globulins, *ceruloplasmin*. Ceruloplasmin apparently serves as a ferroxidase.

Effects of deficiency It has been known for a number of years that rats fed exclusively whole bovine milk develop an anemia which is not prevented or cured by purified iron preparations. Small amounts of added copper, with the iron, promptly cure or prevent the anemia. It is now known that copper, as ferroxidase, is essential for the intestinal absorption of iron and for its use.

A deficiency of copper in animals is manifested in a number of other ways. For example, in sheep the wool becomes straight instead of curly and black wool loses its pigmentation, becoming grayish white, a steely wool appearance. By adding molybdenum salts, which interfere with copper metabolism, to the diet periodically, a black and white striped wool results. Copper deficiency in lambs and pigs, due to low copper content of the soil and plants, results in ataxia, gross brain abnormalities, swayback, and aneurysms of the aortic arch or abdominal aorta. Anemia may also occur. These abnormalities were attributed to a basic defect from copper deficiency in cytochrome oxidase activity, producing an impairment, particularly in phospholipid synthesis, which in turn results in structural defects of and pathological lesions in the vascular system, brain, and nervous tissue. There was extensive demyelination of neurons.

Similar results were reported in weanling rats maintained on a copper-deficient diet. Neural lesions characterized by hyperirritability, catatonic posture, and convulsive seizures developed. The gross and histological appearance of the brain and nervous tissue was analogous to that produced by severe tissue anoxia. Dissecting aneurysms of the aorta in copper-deficient chicks have been attributed to defective elastin formation. There was a decreased monoamine oxidase activity in the aorta, cartilage, tendons, and skin. Some investigators conclude that copper is essential as a catalyst for the cross-linking of collagen.

Copper deficiency has been described in infants with *Menke's disease*, which is characterized by anemia, loss of hair pigmentation, and aneurysms. The blood and urine copper values are low. The biochemical etiology of the disease is uncertain. It is possible that copper may be trapped in the intestinal mucosal cells by *metallothioneine* and thus not absorbed.

Most of the aforementioned abnormalities found in copper deficiency can be reasonably explained as follows:

1. Achromotrichia—deficiency of tyrosinase
2. Cardiovascular disease and aneurysms—deficiency of lysyl oxidase
3. Central nervous system lesions—deficiency of cytochrome oxidase
4. Anemia—lack of ceruloplasmin, resulting in decreased iron absorption and use
5. "Steely" wool and hair—failure in conversion of SH-groups to form —S—S— bridges and normal tertiary structure; enzyme deficiency uncertain at the present time

Requirement The daily requirement for copper as determined in healthy adult males by balance studies averages approximately 1.24 mg. To allow a margin of safety for losses in the absorption and use of copper, the Food and Nutrition Board (1980) recommends a range of 2 to 3 mg per day as a safe and adequate dietary intake (Table 21-10) for the normal healthy adult, male and female. The recommended intake for infants and children varies between 0.5 and 2.0 mg per day depending on age.

Storage Small amounts of copper are stored in the liver and in other tissues. However, in *hemochromatosis*, a condition in which large amounts of iron are absorbed and retained in the form of hemosiderin in the liver and other tissues, increased amounts of copper also are deposited in the liver and other organs. The cause is uncertain at the present time.

Copper also is found in amounts greater than normal in the liver and brain of patients with *Wilson's disease* (hepatolenticular degeneration), inherited as an autosomal recessive and therefore uncommon disease. Both plasma copper and ceruloplasmin levels are decreased. There is extensive renal damage with hemoglobinuria and amino aciduria. Failure of ceruloplasmin to remove tissue copper may be a biochemical defect in this condition resulting in the accumulation of copper to highly toxic levels.

Manganese

Manganese is an essential element for many animal species. It occurs rather widely in plant and animal tissues. The richest sources are liver, kidney, muscle, lettuce, spinach, tea, and whole-grain cereals. Male rats fed diets deficient in manganese become sterile and have testicular degeneration. Young rats that are born of females on similar diets do not survive long, and the mothers are unable to suckle normal young animals. These symptoms in the female may be cured or prevented by the addition of manganese to the diet. Manganese is also needed by rats for growth. In the chick the presence of manganese in the diet prevents the development of a condition known as *perosis*, an osteodystrophy. The tibiometatarsal joint becomes enlarged, the distal end of the tibia and the proximal end of the tarsometatarsus are twisted and bent, and the gastrocnemius tendon slips from its condyles. An impairment in the sulfation of the bone matrix may be involved. As a result the chicks have shortened leg bones and vertebral columns. Whether a deficiency in humans would have results resembling those observed in the rat or chick cannot be said, since no case of manganese deficiency in humans has been observed.

Recent studies demonstrated that manganese deficiency during the pre- and postnatal periods in young guinea pigs is associated with a reduction in the size of the pancreas. In some animals the pancreas was entirely absent. These animals died at birth or shortly thereafter. The manganese-deficient guinea pigs showed a diabetes-like tolerance curve to glucose administered either orally or intravenously. The administration of manganese resulted in the gradual development of a normal tolerance to glucose.

Another striking effect of manganese deficiency in rats was observed in the offspring of deficient mothers. The young showed a marked ataxia and were unable to right themselves when placed on their backs. The otoliths of the inner ear were deficient or defective. The administration of small amounts of manganese restored normal function. Convulsions and epileptic-like seizures were found in some deficient animals.

Functions Manganese is a cofactor for a number of enzymes. These include several phosphatases, peptidases, succinic dehydrogenase, pyruvate carboxylase, glycosyltransferases, arginase, an adenosine triphosphatase, phosphoglucomutase, cholinesterase, and mitochondrial superoxide dismutase. It may be replaced by certain other ions, e.g., cobalt or magnesium, in some instances.

Requirement The exact human requirement for manganese is not known. Some 2 to 9 mg are ingested daily by the average adult. Variable amounts are excreted in the feces, by way of bile, and almost none in the urine. The estimated safe and adequate range of daily intake recommended by the Food and Nutrition Board (1980) is 2.5 to 5 mg for the adult (Table 21-10). Those for infants and children vary from 0.5 to 2 mg depending on age.

Storage There is little storage of manganese in humans. Small amounts are present in the liver and other organs. Only some 20 mg is present in the entire human body. Excess manganese interferes with the absorption of iron and may produce an iron-deficiency anemia. The anemia is reversed by iron administration.

Fluoride

Fluoride is widely distributed in nature and is found in varying amounts in drinking water and in foods. The intake of fluoride by the average adult has been estimated as 0.7 to 3.4 mg per day. No RDA value has been established as yet. However, the Food and Nutrition Board has proposed (1980) an "estimated safe and adequate" daily dietary intake for fluoride of 1.5 to 4 mg for the adult (Table 21-10). The suggested range of intake for infants, children, and adolescents is proportionally less.

In those localities where the fluoride concentration is relatively high, it usually has deleterious effects on the teeth. If it is ingested in toxic quantities during childhood while the teeth are undergoing calcification, characteristic signs appear. Instead of the normal glistening translucent appearance, the teeth acquire dull white patches, or even the entire surface may look chalky. Pitting, as a result of a breaking off of the ends of the enamel prisms, is a common occurrence. The teeth also may have brown mottling (Fig. 21-9). McCollum demonstrated that the inclusion of fluoride in the diets of experimental animals produces fragility of the teeth and bones, and there is much other evidence that it affects calcium and phosphorus metabolism. It is an inhibitor of various enzymes, notably enolase. Fluoride is sometimes added to blood that is to be analyzed for glucose because it inhibits glycolysis.

The effect of the fluoride ion is not always unfavorable. Smaller amounts, i.e., traces, check the occurrence of dental caries. Enamel of sound teeth contains more fluoride than that of carious teeth. This is the only element known to vary in such a manner, and it was suggested that the increased fluoride may be the effective factor in the prevention of caries. Many other observations point in the same direction. Either the fluoride actually imparts to the tooth structure caries-resistant proper-

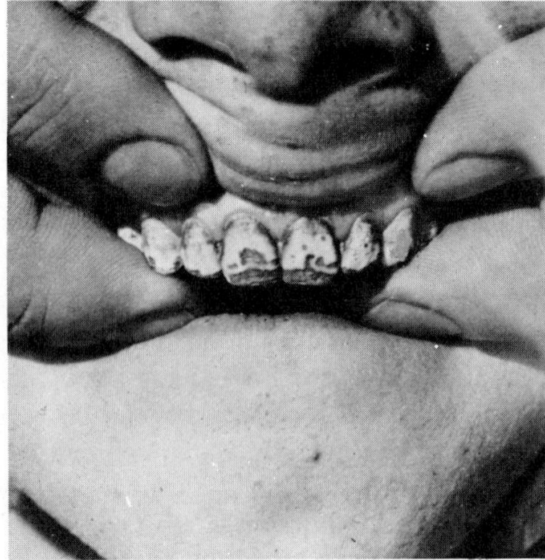

Figure 21-9
Mottled enamel (endemic dental fluorosis) of severe degree. Teeth calcified using water containing 14 ppm of fluoride.

From Dean, H.T., McKay, F.S., and Elvove, E.: Pub. Health Rep. **53**:1736, 1938; courtesy F.A. Arnold, Jr., Washington, D. C.

ties, or it inhibits bacterial action on food particles and on dental tissue. Perhaps both occur. The optimum concentration of fluoride in drinking water that provides a balance between the power to resist caries and the tendency to cause mottling has been calculated to be 0.75 ppm. The presently accepted level is 1 to 2 ppm. A large-scale test began in 1945 in Grand Rapids, Michigan and in two cities in the state of New York. The drinking water of Newburgh had traces of sodium fluoride (F, 1 ppm) added, whereas Kingston, a nearby city of about the same population, served as a control with nothing added to its water supply. The school children in each community had their teeth examined at the beginning of the test and once a year thereafter. At the end of 10 years the Newburgh children, who had been drinking fluoridated water all their lives, had 58% less caries in their permanent teeth than did the Kingston children of the same age group. The 10 to 12-, 13 to 14-, and 16-year-old Newburgh children, who had partaken of fluoridated water for the last 10 years of their lives, had 52%, 48%, and 41% less caries, respectively, than did their Kingston controls. There was no mottled enamel noted in the teeth of the Newburgh children and very little dental fluorosis. Similar results have been obtained in carefully controlled studies in Michigan and in other parts of the United States, Canada, and the United Kingdom.

On the basis of the favorable results obtained in the above pioneer studies, some 4000 communities in the United States now adjust the fluoride content of their public water supplies to the optimal level of about 1 ppm (1 mg per liter). It is estimated that at the present time approximately half the people in the United States drink fluoridated water. Another 10 million live in areas where the drinking water supply contains enough natural fluorides to inhibit tooth decay. The same is true in New Zealand and in several other countries.

Controlled fluoridation of public water supplies has been included by some authorities in the four great mass-preventive health measures of all time, the others being pasteurization of milk, purification of water, and immunization against disease. In each case bitter opposition initially was followed by controversy centered around a set of difficult scientific and political problems. The ultimate fate of fluoridation, however, has been decided largely by public referendum, whereas the fates of the others were decided, more properly, by administrative and legislative means. The fluoridation question has been characterized as the most extensively studied public health problem in the history of humankind.

Recent studies in rural areas of Colombia, South America, where there are no public water supplies, indicate that the use of fluoridated table salt is an *effective* means of reducing the incidence of dental caries by 40% to 50%.

It has been shown that an increased intake of sodium fluoride by elderly patients suffering from osteoporosis (which seems to be more prevalent in low-fluoride vicinities) leads to a diminution of urinary loss of calcium, symptomatic relief, and improvement in bone density. Likewise, fluoride may be of value in protecting against periodontal disease in the elderly. Recent studies also indicate that fluoride administration promotes recalcification of the cochlear capsule of the inner ear, thus preventing the most common cause of progressive loss of hearing in adults.

Chromium

In studies of hepatic necrosis produced by dietary means in the rat a low tolerance to intravenously administered glucose has been found. Selenium salts prevented or cured the degenerative changes in the liver, as will be discussed later, but had no effect on the low tolerance to glucose. In further extensive studies trivalent chromium compounds were found to be extremely effective in correcting the impaired tolerance to glucose. The diet employed in producing the hepatic necrosis syndrome proved to be deficient in chromium as well as selenium and certain other factors involved. Furthermore, minute quantities of chromium, a few micrograms, were found to be essential along with insulin in promoting the utilization of glucose by epididymal tissue for fat synthesis in vitro. Neither chromium nor insulin alone was effective.

Subsequent studies supported the hypothesis that chromium acts as a cofactor for insulin in increasing not only glucose utilization but also the transport of amino acids into cells (heart, liver, and diaphragm). Chromium may also be related to maturity-onset diabetes. Other studies indicated that very small amounts of chromium (0.02 ppm in drinking water) lower the cholesterol level, as well as serum glucose level, in rats fed refined sugar. There also was a significant increase in the plasma level of high-density lipoproteins (HDL) and a decrease in plasma LDL (Chapter 12).

Thus exceedingly small amounts of chromium (perhaps it should be termed an "ultra-trace" element) may play an important role in carbohydrate-lipid metabolism, apparently as a cofactor for insulin. Further clarification of the precise biochemical function of chromium is needed.

Food sources and requirement The best food sources of chromium include brewer's yeast (but *not Torula* yeast), meat products, cheeses, and whole grains. The chromium of vegetables is poorly available. The amount of chromium ingested daily by adults is about 0.05 to 0.12 mg. No RDA intake for chromium has been established, partly because insufficient information on which to base a recommendation is available. Also, analytical methods for determining chromium require improved reliability. However, in its 1980 revision the Food and Nutrition Board has recommended safe and adequate ranges of intake for chromium. For healthy adults the range is 0.05 to 0.20 mg per day (Table 21-10).

Chromium is indeed a *micronutrient*.The entire adult human body contains only some 6 mg. Blood contains about 20 ng/100 dl (about 20 parts per *billion*).

Cobalt

Cobalt is an essential element for some animal species but possibly not for others. For example, cattle and sheep in certain regions develop a peculiar disease characterized by emaciation and anemia. This has been traced to a deficiency of cobalt, and the administration of cobalt is effective in the treatment of the condition. Horses grazing on the same lands remain healthy. A slight excess of cobalt in either metallic or ionic form produces polycythemia in rats and in a number of other species. Rats fed a copper-deficient diet fail to develop this *cobalt polycythemia*. The administration of cobalt to experimental animals significantly increases the level of the hormone erythropoietin (Chapters 16 and 18) in the blood and thus produces a polycythemia.

Since cobalt is a constituent of the vitamin B_{12} molecule, it is evidently necessary for hemoglobin formation and should be regarded as *essential*. This situation is analogous to the role of iron in hemoglobin and other respiratory proteins and of copper in ferroxidase. As mentioned, cobalt may substitute for manganese as an activator of certain enzymes. It is a specific activator for glycylglycine dipeptidase and perhaps for others. Human foods containing over 0.2 ppm include buckwheat, figs, cabbage, lettuce, spinach, beet greens, and watercress, and there are smaller quantities in other vegetable and animal products. It is also a contaminant of many medicinal preparations of iron. The average intake of cobalt in humans is about 0.3 mg per day. The urinary excretion averages 0.26 mg daily.

The Food and Nutrition Board has not as yet established an RDA for cobalt nor presented an estimated safe and adequate daily intake range for humans. More information is needed regarding the nutritional significance and biochemical functions of cobalt.

Selenium

For a number of years it has been known that selenium occurs in relatively large amounts in the soil and vegetation in certain parts of the western United States. Animals grazing in these regions develop *alkali disease,* characterized by emaciation, liver damage, neuromuscular disorders, and other severe toxic manifestations, usually with fatal termination.

More recently, however, accumulated evidence indicates that selenium in smaller amounts has beneficial effects under certain conditions. For example, rats fed a diet low in cystine and vitamin E develop severe hepatic necrosis. Although the addition of cystine and vitamin E offered some protection, another dietary factor was even more potent. This factor was found to contain *selenium.* Inorganic selenium salts proved to be equally effective and far more so than either cystine or vitamin E.

Similar results were obtained in a number of other species of animals. Lambs fed a ration containing raw kidney beans developed a type of muscular dystrophy that could be prevented by selenium plus vitamin E.

Subsequent investigations demonstrated that selenium salts are powerful inhibitors of lipid peroxidation. In this way they are involved in maintaining the integrity of biological membranes, not only of cells but also of mitochondria, microsomes, and lysosomes (Chapter 2). Thus selenium, like vitamin E, serves to protect cells against destruction, such as occurs in selenium deficiency in hepatic necrosis and muscular dystrophy experimentally.

Of considerable recent interest is the fact that the high incidence of myocardial deaths in eastern Finland has been attributed to a deficiency of selenium in the soil and water and thus in the vegetation in that area. Likewise, *Keshan cardiomyopathy* occurring in children in parts of China is due to a low content of selenium in the soil and water in those regions. The condition is prevented by the administration of sodium *selenite.*

Function It has been demonstrated conclusively that selenium as *selenocysteine* is an active part of the enzyme *glutathione peroxidase.* This enzyme protects vital cellular components, e.g., cell membranes, against oxidative damage by H_2O_2 and a variety of hydroperoxides. Selenium, as a constituent of glutathione peroxidase, thus is important in preventing lipid peroxidation and supplementing the action of superoxide dismutase in protecting cells against superoxide (O_2^-) and some other free radicals (Chapters 11 and 16).

Selenium also is a constituent of several bacterial enzymes, including glycine reductase.

Food sources and requirement Liver, kidney, seafoods, and meats are good sources of selenium. Grains have a variable content depending on the region where they are grown. The average dietary intake of selenium is 60 to 70 μg per day but again varies in different areas.

The Food and Nutrition Board (1980) has recommended a safe and adequate range of intake for selenium of 50 to 200 μg per day for normal adults (Table 21-10). This amount is easily obtained from a varied diet. The Board cautions against an excessive intake of selenium, since it may be toxic if continued for a prolonged period.

Molybdenum

The essential role of molybdenum in plants is well established, but not for animals. A deficiency in ruminants maintained on rations containing less than 0.005 μg per gram has been reported to cause a decrease in weight gain, decreased food consumption, decreased life expectancy, and deranged microbiological processes in the rumen. A deficiency in humans has not been described.

Molybdenum has been reported to be a constituent of several enzymes, including

Table 21-11
Summary of essential inorganic elements

Element* and total amount in human body	Best food sources	RDA† 1980	Absorption and metabolism	Principal metabolic functions	Manifestations of deficiency
Sodium (Na^+) 1.8 gm/kg	Table salt, salty foods, animal foods, milk, baking soda, baking powder, some vegetables	1.1-3.3 gm‡	Readily absorbed, extracellular, excreted in urine and sweat; aldosterone increases reabsorption in renal tubules	Buffer constituent, acid-base balance, water balance, osmotic pressure, CO_2 transport, cell membrane permeability, muscle irritability	Dehydration; acidosis; tissue atrophy; excess leads to edema, hypertension
Potassium (K^+) 2.6 gm/kg	Vegetables, fruits, whole grains, meat, milk, legumes	1.9-5.6 gm‡	Readily absorbed, intracellular; secreted by kidney	Buffer constituent, acid-base balance, water balance, CO_2 transport, membrane transport, neuromuscular irritability	Acidosis; renal damage; cardiac arrest
Calcium (Ca^{++}) 22 gm/kg	Milk, milk products, fish bones (cooked)	800 mg	Poorly absorbed (20%-40%) according to body need absorption aided by vitamin D, lactose, acidity; hindered by excess fat, phytate, oxalate; excreted in feces; parathormone mobilizes bone Ca^{++}	Formation of apatite in bones, teeth; blood clotting; cell membrane permeability; neuromuscular irritability	Rickets (child), poor growth; osteoporosis (adult), hyperexcitability
Phosphorus ($PO_4^{\equiv}$) 12 gm/kg	Milk, milk products, egg yolk, meat, whole grains, legumes, nuts	800 mg	Readily absorbed; excreted by kidney	Constituent of bones, teeth; constituent of buffers; constituent of ATP, NAD, FAD, etc.; constituent of metabolic intermediates, nucleoproteins, phospholipids, phosphoproteins	Osteomalacia (rare); renal rickets; cardiac arrhythmia
Chloride (Cl^-) 50 mEq/kg	Animal foods, table salt	1.7-5.1 gm‡	Rapid absorption; excreted in urine; high renal threshold; not stored	Electrolyte, osmotic balance; gastric HCl; acid-base balance	Hypochloremic alkalosis (pernicious vomiting)
Sulfur ($SO_4^{=}$)	Plant and animal proteins as Cys and Met	2-3 gm‡	Derived from metabolism of Cys and Met; excreted in urine	Constituent of proteins, mucopolysaccharides, heparin, thiamine, biotin, lipoic acid; detoxication	Cystinuria; cystine renal calculi; methioninuria (hereditary defects; see Chapter 13)
Magnesium (Mg^{++}) 0.5 gm/kg	Chlorophyll, nuts, legumes, whole grains	350 mg male; 300 mg female	Absorbed readily; competes with Ca^{++} for transport	Cofactor for PO_4-transferring enzymes; constituent of bones, teeth; decreases neuromuscular irritability	Magnesium-conditioned deficiency, muscular tremor, choreiform movements, confusion; vasodilatation, hyperirritability

*The inorganic elements included are those for which evidence exists that they are *essential for humans*. Other elements not included but present in the human body in trace amounts, for which there is fragmentary evidence for some biochemical function, include cadmium, lithium, nickel, vanadium. Other elements present in human tissues in trace amounts as incidental constituents of no known significance include Ag, Au, Al, As, Br, Pb, Rb, Si, Ti, B. The amounts of the element present in the entire human body are averages from the literature (Dairy Council Digest **39**:26, 1968). They are expressed as grams or milligrams per kilogram of body weight *(fat-free basis)* or as milligrams in entire body.

†Recommended dietary allowance per day, established by the Food and Nutrition Board, National Research Council, 1980. The values given are for a normal adult male, 19 to 22 years of age. (See Tables 21-4, 21-9, and 21-10 for other age groups.)

‡An estimated range recommended by the Food and Nutrition Board (1980) as a safe and adequate daily intake for healthy adults. See Tables 21-9 and 21-10 for other age groups (except for those of sulfur and cobalt).

Table 21-11, cont'd
Summary of essential inorganic elements

Element and total amount in human body	Best food sources	RDA 1980	Absorption and metabolism	Principal metabolic functions	Manifestations of deficiency
Iron (Fe^{++} or Fe^{+++}) 75 mg/kg	Liver, meats, egg yolk, green leafy vegetables, whole grains, enriched bread and cereals	10 mg male; 18 mg female	Absorbed according to body need; aided by HCl, ascorbic acid; regulated by apoferritin; stored in liver as ferritin and hemosiderin	Constituent of hemoglobin, myoglobin, catalase, ferredoxin, cytochromes; electron transport, enzyme cofactor	Anemia, hypochromic; pregnancy demands; excess absorption in hemochromatosis
Iodine (I^-) 20-50 mg	Seafoods, iodized salt	150 μg	Concentrates in thyroid; transported as PBI	Constituent of thyroxin, triiodothyronine; regulator of cellular oxidations	Endemic (simple) goiter (hypothyroidism); cretinism
Fluoride (F^-)	Seafoods, some drinking water	1.5-4.0‡ mg (1 ppm in drinkwater)	Easily absorbed; excreted in urine; deposited in bones and teeth	Constituent of fluoroapatite—tooth enamel; strengthens bones and teeth	Dental caries; osteoporosis; excess (5-8 ppm in water) → mottled enamel
Zinc (Zn^{++}) 28 mg/kg	Liver, pancreas, shellfish, most animal tissues, wheat germ, legumes	15 mg	1-2 mg absorbed; phytate decreases absorption	Constituent of insulin, carbonic anhydrase, carboxypeptidase, lactic dehydrogenase, alcohol dehydrogenase, alkaline phosphatase, etc.	Anemia; stunted growth; hypogonadism in male; hypogeusia; decreased protein synthesis and wound healing
Copper (Cu^{++}) 2 mg/kg	Liver, kidney, egg yolk, whole grains	2.0-3.0 mg‡	Limited absorption; transport by ceruloplasmin; stored in liver; excretion via bile	Formation of hemoglobin (increases iron utilization); constituent of 11 oxidase enzymes (tyrosinase, cytochrome oxidase, ascorbic acid oxidase, ferroxidase, etc.)	Hypochromic anemia; aneurysms; CNS lesions; achromotrichia; excessive hepatic storage in Wilson's disease and hemochromatosis
Manganese (Mn^{++}) 20 mg	Liver, kidney, wheat germ, legumes, nuts, tea	2.5-5.0 mg‡	Stored in liver mitochondria and bone; excreted via bile	Cofactor for number of enzymes—arginase, carboxylase, kinases, etc.; biosynthesis of mucopolysaccharides	In animals → sterility, weakness, perosis, congenital ataxia
Cobalt (Co^{++}) 3 mg	Liver, pancreas; vitamin B_{12} in animal proteins	Not established	Limited absorption; stored in liver; excretion via bile	Constituent of vitamin B_{12}	Anemia; deficiency as vitamin B_{12} → pernicious anemia; excess → polycythemia
Chromium (Cr^{+++})	Liver, animal and plant tissue	0.05-0.2 mg‡	Poorly absorbed; traces excreted in urine	Necessary for glucose utilization; cofactor of insulin?	Unknown; deficiency in diabetes claimed; decreased glucose tolerance in rats; possible relation to cardiovascular disease
Selenium (Se)	Liver, kidney, heart; grains, vegetables (varies with Se in soil)	0.05-0.2 mg‡	Trace excreted in urine	Constituent of glutathione peroxidase; inhibits lipid peroxidation	Excess → alkali disease in cattle, sheep; deficiency → liver necrosis and muscular dystrophy in animals; cardiomyopathy in humans
Molybdenum (Mo) 5 mg	Liver, kidney, whole grains, legumes, leafy vegetables	0.15-0.5 mg‡	Readily absorbed; excreted in urine and bile	Constituent of xanthine oxidase, aldehyde oxidase	Decreased growth, food consumption, and life expectancy

xanthine oxidase and aldehyde oxidase. Molybdenum is alleged to be an antagonist to copper, decreasing its utilization.

Food sources and requirement Meats, whole grains, and legumes are the best food sources of molybdenum. There is some variation in amounts in different regions. Investigations in human subjects indicate that balance can be maintained on an intake of 0.10 to 0.15 mg per day. This amount can be supplied by most diets.

Based on balance studies, the Food and Nutrition Board (1980) recommends 0.15 to 0.5 mg as a safe and adequate intake per day for normal adults. They caution against an intake exceeding 0.5 mg molybdenum per day for any protracted period because of its toxicity.

• • •

Brief summaries of the best sources, RDAs, metabolism, principal functions, and deficiency symptoms of the essential major and trace inorganic elements are presented in Table 21-11.

Other trace elements

The following four inorganic elements—nickel, silicon, tin, and vanadium—have been called likely candidates for inclusion in the list of essential nutrients for humans. Deficiencies of each have been produced and confirmed in two or more animal species. They are consistently present in human tissues but as yet have not been proven essential for humans. Some of their unique properties and effects are described briefly. Nonspecific effects, such as impaired growth, occurred in all instances of deficiency.

Nickel

There is some evidence that nickel plays a role as an activator of liver arginase and in maintaining the conformation of membranes. It is consistently found associated with RNA. The average intake of the normal adult varies from 0.17 to 0.7 mg per day.

Silicon

Silicon, as silicates, enters the body chiefly in vegetable foods. Soluble silicates are easily absorbed. Human blood serum ordinarily carries about 1 mg/100 ml. After the ingestion of silicates, this level does not rise because the excess is rapidly excreted by the kidneys. Varying quantities are found in the different organs and tissues. The lungs are highest in silicon because of the inhalation of insoluble particles that lodge there. In industries in which silica dust is produced in large amounts, e.g., stonecutting, the workers inhaling this dust develop *silicosis*. In this condition the lung tissue is replaced by nodular connective tissue overgrowths. Naturally the silicon content of such lung tissue is comparatively high. Similar pathological states result from breathing dusts of other types—coal, steel, etc. Silicon appears to be necessary for calification, growth, and mucopolysaccharide formation as a cross-linking agent. It appears to be essential for chicks and rats. The human intake is variable and uncertain. It has not yet been shown to be essential for humans.

Tin

Tin also occurs in the body; the largest quantities are found in the kidney and skin. In the concentrations usually occurring in foods, as a result of their having been preserved in tin containers, this metal has no deleterious effects. Tin promotes the growth of rats and may function as the active site of several metalloenzymes. It may contribute to the tertiary structure of some proteins. The human intake varies from 1.5 to 3.5 mg per day.

Vanadium

There is some evidence that vanadium may be involved in the mineralization of bones and teeth and in catecholamine and lipid metabolism. This element also appears to increase the resistance of rats to experimentally produced dental caries. The human intake has been estimated as 4 mg per day. The human requirement is considerably less, the estimate being 0.1 to 0.3 mg per day for the adult.

Other elements

Many other elements are present in traces in foods and in body tissues, e.g., boron, rubidium, arsenic, titanium, aluminum, and silver, to list only a few. As far as is now known, these are "incidental" constituents of tissues and have no physiological function or significance. Some inorganic elements, e.g., arsenic, cadmium, lead, and mercury, are highly toxic and are of nutritional concern as possible hazardous contaminants of foods. The accidental poisoning by mercury of some 40,000 persons in the Middle East (with 4000 deaths) from the ingestion of treated wheat intended for use as seed, is an example. Also, of current concern is the contamination of fish by methyl mercury in some industrial areas and of canned fish by lead derived from lead-soldered containers. Several having some special biochemical importance will be considered briefly.

Bromide

Small amounts of bromide sometimes are found in table salt and also in certain vegetables. Normal human serum contains about 1 mg/100 ml. Bromides are absorbed, distributed, and eliminated by the body in almost exactly the same manner as the chlorides, i.e., they are absorbed from the gastrointestinal tract, pass into the various body fluids, penetrate the red cell but not other cell membranes, and are eliminated by the kidney, just as chlorides are. If present in sufficient amounts, bromide tends to replace chloride in the body, doing so in a quantitative manner. It has a sedative effect on nerve tissue, which may be a result of the decrease in concentration of chloride displaced by bromide in the extracellular fluid. Bromide poisoning, known as *bromism,* is fairly common because bromides may be obtained without a physician's prescription. The advanced stages are characterized by mental and neurological disturbances. Bromide has not yet been shown to be an essential trace element.

Cadmium

Cadmium has a hypertensive effect in rats and is present in the kidneys and urine of hypertensive patients in relatively high concentrations. It accumulates in the renal cortex bound tightly to the protein metallothioneine. Severe renal damage occurs. Cadmium has been incriminated as a factor in soft drinking water statistically correlated with cardiovascular mortality. Cadmium apparently is dissolved in increased amounts from pipes by the soft water. There is no definite evidence at the present time that cadmium should be considered as an essential element. The usual intake varies from 13 to 24 μg per day.

Lead

Lead is found in some foods, especially in drinking water. It is stored in the bones and, to a lesser extent, in the liver. In large amounts it is toxic. Its deposition in bones may be explained by postulating that lead, phosphorus, and vitamin D form a system of lead deposition analogous to the deposition of calcium in bones. Lead is a potent inhibitor of certain enzymes dependent on sulfhydryl groups for their activity. δ-Aminolevulinic acid dehydratase, involved in the biosynthesis of heme (Chapter 16), is an example. For this reason the determination of δ-aminolevulinic acid in urine is a sensitive procedure for the early detection of lead poisoning in humans.

Lithium

Lithium salts inhibit the release of norepinephrine and serotonin from brain slices stimulated by an electric current. Lithium salts were found effective in the treatment of mania and related mental disorders in which there is a disturbance of amine metabolism.

The vitamins

The term *vitamine* was introduced into the scientific literature in 1911 by an investigator, Funk, who had isolated a crystalline compound from rice polishings that was effective in curing and preventing a dietary deficiency in pigeons called polyneuritis. The name vitamine was proposed for the compound because of its vital need in this condition and because chemically it was found to be an amine. The term, now *vitamin*, has been adopted universally and applied to a group of biologically essential chemical entities, currently including some 14 compounds, which humans and most animals cannot synthesize. They must therefore be supplied preformed from plant sources, i.e., foods.

Although the term vitamin was not adopted until the early 1900s, a considerable amount of evidence had accumulated long before that time indicating that chemical substances in addition to carbohydrates, lipids, proteins, and inorganic elements were required by animals and humans. In the 1500s explorers on long voyages vividly described scurvy in their sailors and speculated that it was caused by a "lack of something in food." The successful treatment of this condition with citrus fruits was reported in 1753 by James Lind, a physician in the British navy. Since then British sailors have been called "limeys." In the mid-1700s Captain James Cook is said to have found that *sauerkraut* was effective in preventing and curing scurvy in sailors on long voyages while exploring the South Pacific.

Likewise, there were a number of early reports of other diseases in humans and animals caused by a lack of some unknown factor in foods. Such conditions included beriberi, polyneuritis, pellagra, xerophthalmia, and rickets. It became apparent that a number of vitamins were involved, and since their chemical nature was unknown, letter designations were applied, e.g., vitamins A, B, and C. Later, vitamin B was shown to consist of several substances, and subscripts were added, i.e., vitamin B_1, B_2, B_6, etc. These terms continue in use, although they are being gradually replaced or at least supplemented by official chemically descriptive names (e.g., vitamin C is L-ascorbic acid). By the late 1940s, 14 different vitamins had been isolated in crystalline form; their structures had been determined and proved by synthesis.

The vitamins are grouped into two major categories based on their solubilities in the so-called fat solvents and water. The fat-soluble vitamins recognized to be essential for human nutrition are A, D, E, and K. The water-soluble vitamins essential for humans include C (ascorbic acid), B_1 (thiamin), B_2 (riboflavin), niacin, B_6 (pyridoxine), folacin, B_{12} (cobalamin), pantothenic acid, biotin, and possibly lipoic acid. These vitamins and a number of other vitamin-like substances, not proven to be essential for human nutrition as yet but required by some other species, will also be considered in this chapter.

During the past 30 years tremendous progress has been made in elucidating the precise role of the vitamins in many biochemical reactions. These are discussed in preceding chapters, especially in those dealing with metabolism (Chapters 6 and 11-14). The vitamins will now be considered from the standpoint of their role in nutrition.

Avitaminoses (deficiencies of vitamins)

As will be discussed for the individual vitamins, a lack of one or more vitamins (multiple deficiencies being the more usual in humans) produces rather charac-

Figure 21-10
Sequence of events occurring in typical avitaminosis. This sequence applies equally well to many other types of malnutrition, for example, protein-calorie malnutrition, and deficiencies of essential fatty acids, amino acids, and various essential minerals.

Deficient intake	**Secondary conditioning factors**	
(from dietary history) ↘	↙ (from clinical history)	
Gradual decrease in tissue levels		(evaluated by blood, urine, or tissue analysis)
↓ Biochemical lesion		(reduced enzyme levels, altered metabolites, etc.)
↓ Anatomic lesion		(clinical evaluation)
↓ Pathology — disease		(clinical symptoms)

teristic symptoms. These have been studied extensively in experimental animals as well as in humans. The avitaminosis may be *primary* or direct, attributable to a deficient intake resulting from chronic alcoholism, dietary fads, poverty, etc. This is usually discovered from the dietary history of the individual. The avitaminosis may be *secondary,* a "conditioned deficiency" from other factors such as gastrointestinal disorders, poor teeth, anorexia, allergies, malabsorption, increased excretion, imbalance, and others to be discussed later. Avitaminosis from any cause, if prolonged, leads to (1) a gradual decrease in tissue levels of the vitamin or vitamins deficient, (2) a biochemical lesion, and, in time, (3) an anatomic lesion and, finally, cellular disease. This sequence is shown schematically in Fig. 21-10. This sequence of events applies equally well to many other types of malnutrition, e.g., protein-calorie malnutrition and deficiencies of essential fatty acids, amino acids, and the various essential minerals.

Although avitaminoses may not be common in the United States, they occur rather frequently in certain other parts of the world largely for socioeconomic reasons. Therefore consideration must be given to their characteristic features and underlying biochemical causes.

Fat-soluble vitamins

Retinol (vitamin A)

Properties Vitamin A, *retinol,* is soluble in fats and in fat solvents. It is stable at rather high temperatures, except when the conditions are favorable for oxidation. Very little is thus destroyed in ordinary cooking or canning operations. Retinol is destroyed by exposure to ultraviolet light. It is available to the human organism either in the form of the vitamin itself or of a precursor or provitamin, one of a series of carotenoid pigments, commonly called *carotenes*. The carotenes form part of the pigments of many green and yellow vegetables. After absorption they are converted to the vitamin probably in the intestinal mucosal cells. The carotenoids themselves do not possess vitamin A activity and are not all equally potent in their ability to form retinol. It is therefore not surprising that the sources of the vitamin itself are of animal nature. β-Carotene is the most effective provitamin, since, as can be seen from the formula, it is a symmetric compound, each half of which is convertible to a molecule of the vitamin. Other members of this group that lead to the formation of vitamin A, but only half as much as does β-carotene, are α-carotene, γ-carotene, and cryptoxanthine. Retinol has a characteristic absorption spectrum, which in chloroform solution has a maximum (λ_{max}) at 328 nm—quite different from that of carotene, which has an absorption maximum at 335 nm. It also gives a beautiful color reaction, an intense blue (λ_{max} 620 nm) when treated with antimony trichloride. Carotene under the same conditions yields a greenish blue color. This chemical reaction has been used for quantitative determination of the vitamin. Retinol in aqueous dispersions has been recommended for clinical use, since it is claimed to be more rapidly absorbed than its oil solutions.

Structure Vitamin A is a complex primary alcohol, with the empirical formula $C_{20}H_{29}OH$. The terminal hydroxyl is ordinarily esterified. This alcohol was isolated in 1931 and was synthesized in 1946. It contains a β-ionone ring.

β-Carotene has two such rings and no alcohol groups. The other pigments mentioned each have one β-ionone ring and one group similar to it. We can thus see why they yield only half as much vitamin A as β-carotene does. There are actually

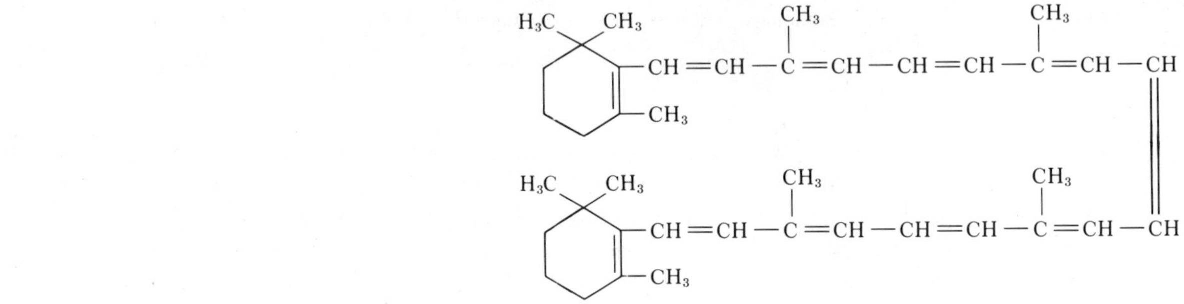

β-Carotene
(provitamin A)

two vitamins A, known as vitamin A_1, or retinol, and vitamin A_2, or 3-dehydroretinol. Vitamin A_1 predominates in the livers of the cod and other saltwater fish, and A_2 in those of freshwater fish. The physiological activity of both seems to be the same qualitatively, and both are justifiably called vitamin A, although A_2 is less active in promoting the growth of rats than is A_1. Vitamin A_2 has the following ring structure, possessing an extra double bond in the 3,4-position:

3-Dehydroretinol

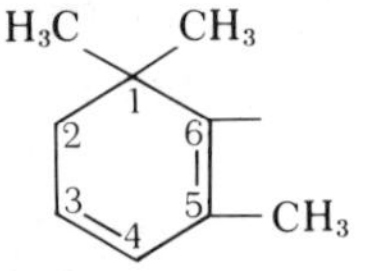

It is thus seen that the β-ionone ring is not absolutely essential for vitamin A action and that certain modifications, but not others, may allow biological activity. Moreover, there are eight possible stereoisomers of retinol, namely, all-*trans*, 9-*cis*, 13-*cis*, 9:13-di*cis*, and four hindered (less probable) configurations, 11-*cis*, 9:11-di*cis*, 11:13-di*cis*, and 9:11:13-tri*cis*. All-*trans* has a straight side chain; all other *cis-trans* isomers are bent at one, two, or three of the double bonds:

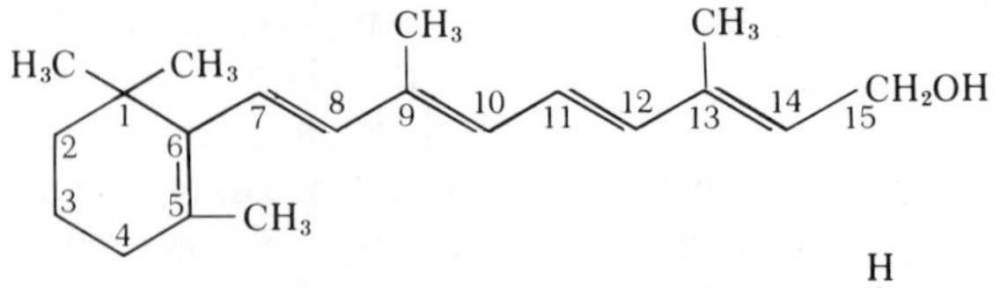

Retinol

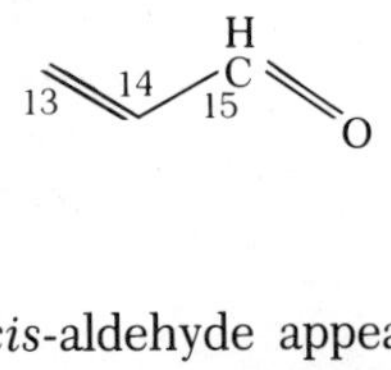

Retinal

All-*trans*

Although a hindered configuration, the 11-*cis*-aldehyde appears to be the one uniquely required for the biosynthesis of the photopigments used in vision:

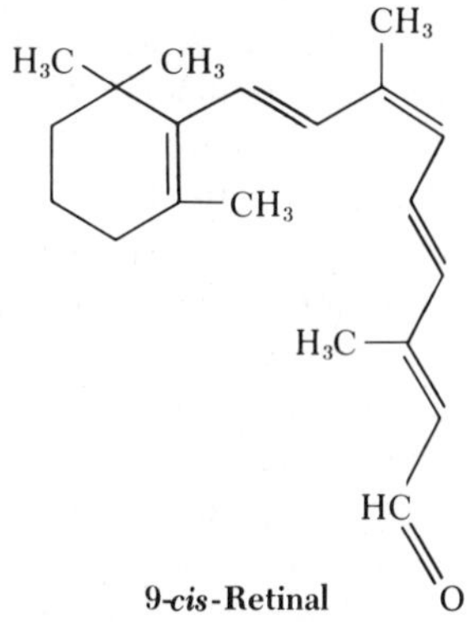

9-*cis*-Retinal

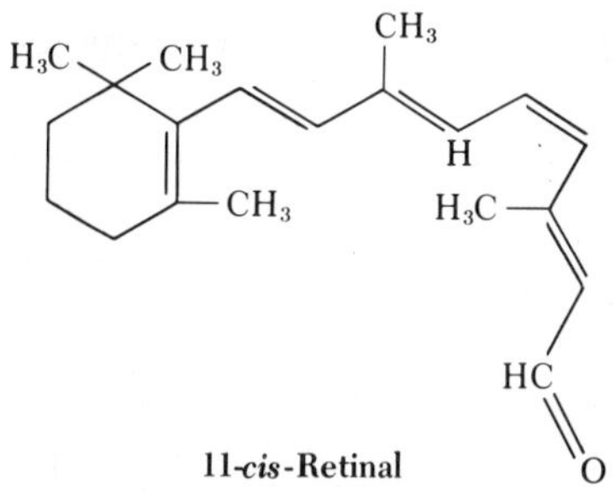

11-*cis*-Retinal

The configuration of 11-*cis*-retinal brings the hydrogen of carbon-10 into proximity with the methyl group of carbon-13, causing *hindrance*. An unhindered isomer, 9-*cis*, appears in retinal extracts as a by-product of photolysis of the visual pigment rhodopsin.

The biologically active form of vitamin A may differ from any of the foregoing forms. An active form has been isolated chromatographically from the liver of rats given ^{14}C-labeled retinoic acid and differs from the known vitamin A compounds. It is highly active in stimulating growth in vitamin A–deficient rats.

Occurrence The best sources of vitamin A are cod liver oil and other fish liver oils, fish roe, the flesh of oily fish, the livers of other animals, butter, eggs, and cheese. The provitamin occurs most abundantly in carrots and other yellow vegetables, e.g., squash, sweet potatoes, and many green vegetables, particularly broccoli, spinach, and beet greens. In fish liver oils retinol is present as esters of fatty acids, chiefly stearic, palmitic, and higher unsaturated acids.

Absorption and metabolism When the liver oil is ingested by humans or animals, the retinol esters are hydrolyzed by pancreatic esterases and the vitamin rapidly absorbed into the intestinal mucosa. In those conditions in which the individual is unable to hydrolyze esters, e.g., celiac disease and tropical sprue, severe deficiencies of vitamin A may occur. The esters cannot be absorbed, unhydrolyzed, in the way that a part of the fats apparently is. The vitamin is recombined with fatty acids *characteristic of the host* (not of the diet) immediately after passage through the gut wall. The vitamin esters are then conveyed by the portal vein to the liver, where they are stored in ester form. From the liver vitamin A is redistributed to the various organs by way of the blood plasma, in the form of a protein complex. The carrier protein *retinol-binding protein*, an α_1-plasma globulin (Chapter 16), has been isolated and characterized. The principal fatty acids present in the retinol esters of the retinas of various animals are palmitic, stearic, and oleic.

For the absorption of the provitamins from the intestinal tract bile salts are necessary. The conversion to the vitamin is not well understood. However, there is evidence to indicate that it occurs primarily in the intestinal mucosal cells and, to some extent, in the liver. The actual process apparently is oxidative. Each molecule of provitamin yields either two molecules or one molecule of retinol, depending on whether the provitamin is β-carotene or one of the other forms. Humans, as well as experimental animals, are inefficient converters of carotene to retinol. On the average, about four times more carotene than retinol is required to maintain normal dark adaptation in adults, and in some cases there are even more remarkable individual variations. We might expect one molecule of β-carotene to yield two retinol molecules, but since it is only about half as active biologically, the β-carotene molecule is apparently split in an unsymmetrical manner. The other carotenoid precursors of vitamin A are even less active biologically, since each molecule can form only one of retinol. This inefficiency of conversion may not be real but may be attributable to the fact that carotene is not absorbed as easily as retinol, and a considerable amount is lost in the feces. As stated, carotene apparently requires the presence of bile salts and fat in the intestine for its absorption, whereas bile does not appear to be necessary for the absorption of vitamin A, although it is helpful. Consequently, in cases where there is a stoppage of bile, bile salts or desiccated bile should be administered to be sure that the provitamin is taken up. Another practical point is that carotene is soluble in mineral oil, which has been shown to remove much of the carotene present in the digesting food and thus cause a deficiency of the vitamin if the mineral oil is taken repeatedly shortly after a meal. This may also be true of retinol. For individuals using mineral oil constantly, such a danger should be recognized. Vitamin E seems to have a sparing action on vitamin A.

Neither retinol nor the provitamins pass the placenta into the fetus very readily, although the vitamin is more easily transferred. Consequently, newborn infants have low stores of both. The milk of well-fed mothers, however, contains ample amounts of this vitamin, mostly in ester form, for the nursing infant's needs.

The principal steps involved in the metabolism of retinol, vitamin A, are shown schematically in Fig. 21-11.

Effect of deficiency In the experimental animal a lack of vitamin A is manifested by a slowing or stopping of growth in the young. This effect, however, is not peculiar to vitamin A, since lack of other vitamins or other essential nutritive factors has similar results. The most manifest specific effect of retinol deficiency is on the eye. This is shown in animals by an avoidance of light *(photophobia)* and by the occurrence of *xerophthalmia* and *keratomalacia*. Xerophthalmia is an eye disease characterized by drying of the eyes. The cells of the lacrimal glands become keratinized and stop secreting tears. The external surfaces thus become dry and have a dull appearance. Ulcers form; bacteria are not washed away; the eyelids swell and become sticky and scaly. Frequently there are bloody exudates and severe eye infections. If not treated in time, blindness results; but in most instances the animals die of respiratory infections before this occurs. The reason for such infections is that retinol deficiency has an effect on other epithelial structures as well as those of the eye. In other words, the eye affection is only one manifestation of the specific influence that this vitamin has on many epithelial structures. This deficit results in

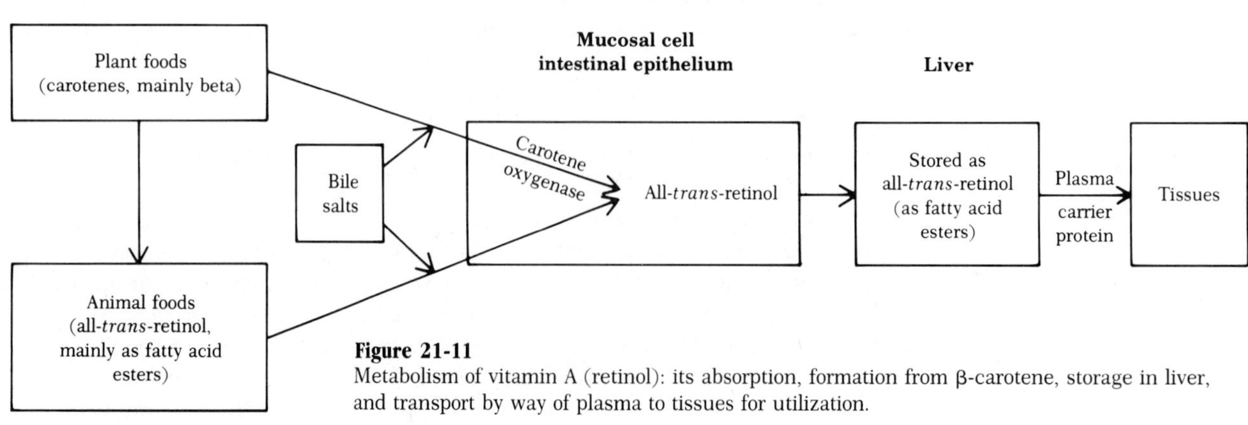

Figure 21-11
Metabolism of vitamin A (retinol): its absorption, formation from β-carotene, storage in liver, and transport by way of plasma to tissues for utilization.

"the substitution of stratified keratinizing epithelium for the normal epithelium in various parts of the respiratory tract, alimentary tract, eyes and paraocular glands, and the genitourinary tract." One of the results of this keratinization is the loss of cilia in the respiratory epithelium. These ordinarily tend to sweep upward bacteria-laden foreign particles and thus combat infection.

Pertinent in this connection is the fact that germ-free rats fed a vitamin A–deficient diet survived as long as 272 days, whereas litter-mate weanling germ-free rats transferred to a conventional animal room and fed the same vitamin A–deficient diet died within 23 to 54 days. This important investigation leaves little doubt that the early death of rats fed a vitamin A–deficient diet must be a consequence of bacterial infection.

In rats vitamin A is definitely necessary for reproduction and lactation. In fact it is just as essential as vitamin E, the "antisterility" vitamin, and must be given in greater amounts than are needed for optimal growth, if normal reproduction and lactation are to occur. Another interesting effect of a deficiency of vitamin A is a loss of normal taste acuity in vitamin A–depleted rats. Both the intensity and the quality of taste are affected. There are distinct histological changes in the taste buds and surrounding tissue.

In humans, deficiencies in vitamin A result in epidermal lesions and ocular changes. The appellation "anti-infective," which was formerly given to this vitamin, may not be justified. As stated, however, a lack of the vitamin may contribute to infection, and there is no doubt that a lowered resistance to bacterial invasion is brought about.

Extreme cases of vitamin A deficiency in humans are rare at the present time in Western civilization, although in Eastern countries they are still seen. During and after World War II many cases of the same condition occurred in Denmark because butterfat was shipped out of that country in large amounts, and substitutes containing no retinol were used. Xerophthalmia, of course, results from a total or nearly total lack of retinol, and seldom does a person today subsist on a diet of this type.

Less serious subclinical symptoms are frequently found in human beings because of a diet containing less retinol than the required minimum. Night blindness, or *nyctalopia,* is often encountered. This is an inability to see in dim light or to adapt to a decrease in intensity of light. Both the rods and the cones of the retina contain substances that depend on retinol for their formation and regeneration. The rods are particularly involved in dark adaptation, and vitamin A, or retinol, is especially needed for this function. Night blindness and slowing down of dark adaptation are frequently associated with cirrhosis of the liver and other hepatic disorders, which may indicate that the liver has something to do with the activity of retinol, in addition to the storage and absorption of the vitamin. The most important chronic disease in which carotene cannot be transformed easily into retinol is diabetes mellitus. If diabetic subjects are on restricted diets without insulin and try to satisfy their hunger with large amounts of green and yellow vegetables, their skin may acquire a yellow tinge because of the deposition of carotene in it. Night blindness is likely to occur under such conditions, but the addition of the vitamin itself to the diet quickly brings a return to normal vision, provided, of course, that no other ocular disease (e.g., cataracts) is present. Perhaps even earlier than night blindness is the occurrence of *xerosis conjunctiva,* minute dry spots, which may be detected with biomicroscopic examination. Both night blindness and the other eye symptoms are treated by administration of carotene or, better, of retinol itself. Fairly large doses are given, but there is a limit to the amount that can be absorbed, or put to work, in healing the damage present. Usually results are noted in a short time.

Skin conditions frequently result from an inadequate retinol intake. Dryness and scaliness of the skin are often seen as early stages of retinol deficiency. Sometimes small pustules, termed *follicular hyperkeratosis,* appear around the hair follicles or extensor surfaces of the upper and lower extremities, on the shoulders, neck, back, lower abdomen, and buttocks. They are hard and pigmented and are surrounded by a zone of pigmentation (Fig. 21-12). In other instances the pimples resemble those of acne except that there is seldom any pus. Large doses of retinol are required over a period of many weeks to cure these conditions. Although the epithelium of the mucous membranes is often keratinized in animals, whether similar pathological changes occur in humans is not certain. Another finding in animals with retinol deficiency is urinary calculi, and here again whether the same result follows in humans is uncertain.

Retinol is an important factor in tooth formation. This is probably related to the fact that the enamel layer is an epidermal structure. As a result of retinol deficiency, there is a defective formation of enamel, with the consequent possible exposure of the dentin. Sound teeth, of course, cannot be expected under such circumstances.

Retinol deficiency has been asserted to result in paralysis and nerve degeneration. The explanation possibly lies in the fact that such a deficiency may retard bone growth and, in particular, the formation of endochondral bone while the central nervous system, as well as other soft tissues, continues to grow at a nearly normal rate. If this occurs at an early age, it has an effect on the nervous system. Because the skull does not grow rapidly enough, there may be overcrowding of the cranial cavity, with distortion of the brain and pressure on the spinal cord and nerve fibers. Therefore the nervous lesions may be entirely mechanical in origin. These results have been seen to occur in laboratory animals, but whether they occur in humans is not certain. The retardation of endochondral bone formation must be specific because bone matrix (osteoid) formation continues.

Figure 21-12
Follicular hyperkeratosis from vitamin A deficiency.

From files of Therapeutic Notes; courtesy O. D. Bird, Ann Arbor, Mich.

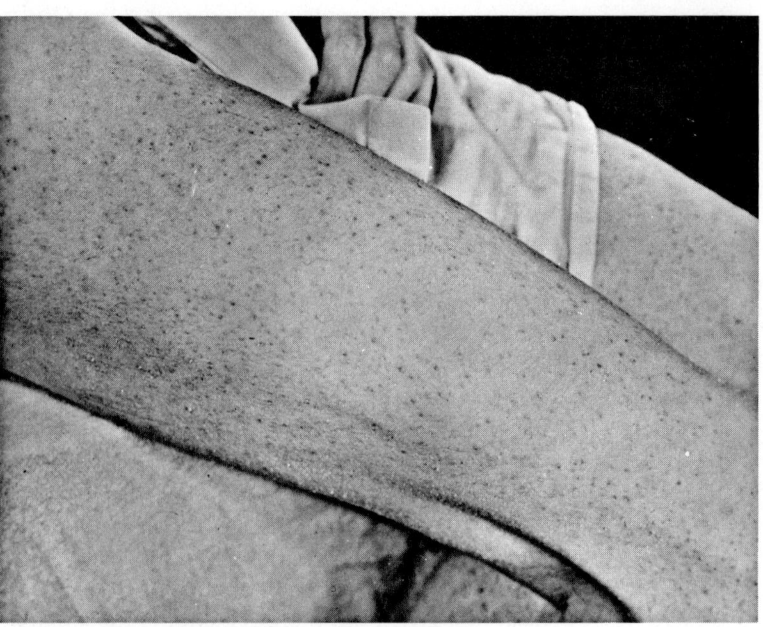

Mode of action Retinol has at least *two* major functions: (1) supporting the growth and/or maintenance of *epithelial tissues* and (2) mediating the visual process as a constituent of the pigment *rhodopsin* and other visual pigments, to be discussed next.

The role of retinol (vitamin A) in the maintenance of epithelial tissue is still unclear. As has been described, there is a general impairment of epithelial tissues in retinol deficiency. There is some evidence that this may be due to an impaired formation of tissue *glycoproteins* and that retinol, as a phosphate, may serve as a mannosyl acceptor for oligosaccharide formation involved in glycoprotein biosynthesis (Chapter 11). Or it may be necessary for chondroitin sulfate synthesis and thus for glycoprotein formation. Also, it appears possible that retinol may play a role in the differentiation of epithelial cells into other specialized cells. For this reason recent work suggests that retinol may in some way exert potential protection from carcinogens (mutagenic agents), particularly in the early, preneoplastic stage. Since tumorigenesis generally involves a loss of differentiation, such a possibility appears plausible. Also, there is evidence that a number of carcinogens are much more potent in animals that have had a long-term deficiency vitamin A.

Effect on vision The light receptors of the eye are the rod and cone cells of the retina. The outer segments of both kinds of receptor contain light-sensitive pigments that require vitamin A for their formation and proper functioning. The rod and cone outer segments are surrounded by pigmented epithelial cells that store vitamin A. The pigment contained in the outer segments of the rods, *visual purple* or *rhodopsin,* is a conjugated protein consisting of a glycoprotein, *opsin,* linked to a prosthetic group, the red aldehyde of 11-*cis*-vitamin A_1, also called *11-cis-retinal.* 11-*cis*-Retinal is bound to opsin by a Schiff base linkage to the ϵ-amino group of lysyl residues and by forces involving sulfhydryl groups. Rhodopsin is extractable from the rods by mild detergents, e.g., bile salts or digitonin, with which it forms a soluble complex; but it is insoluble in the usual protein solvents. It has been estimated as having a molecular weight of about 30,000. Its absorption spectrum is maximal at about 500 nm, corresponding to the wavelength for minimal threshold of the eye in dim light. It is sensitive to light and, when illuminated, changes from red to orange to yellow and, on prolonged exposure, to colorless retinol and opsin. The eye becomes less sensitive to light during the bleaching of rhodopsin (light adaptation). In the dark, rhodopsin is regenerated and the sensitivity of the retina is restored (dark adaptation). The recovery is rapid or slow, according to the wavelength and the intensity and duration of the preexposure to light. If there is a deficiency of retinol in the retina, regeneration is incomplete and the patient is night blind.

The rhodopsin of rod cells on exposure to light is isomerized through a series of several unstable intermediates to form all-*trans*-retinal and, by spontaneous hydrolysis, free opsin (Fig. 21-13). The all-*trans*-retinal is immediately isomerized by *retinal isomerase* in the pigment epithelium of the retina to 11-*cis*-retinal, which spontaneously recombines with opsin to reform rhodopsin and thus complete the visual cycle.

Some of the all-*trans*-retinal, however, may be reduced to all-*trans*-retinol and temporarily stored in the pigment epithelium of the retina (Fig. 21-13). This form of vitamin A, along with additional all-*trans*-retinol derived from food or from that stored in the liver, must be converted to 11-*cis*-retinal before it can combine with opsin to form rhodopsin. This conversion entails the isomerization to 11-*cis*-retinol and its oxidation by NAD^+ (or $NADP^+$) to 11-*cis*-retinal, which is trapped by opsin to form rhodopsin. In this way vitamin A as all-*trans*-retinol from food or hepatic storage or other sources can be utilized in the visual cycle.

Figure 21-13
Visual cycle in *rod* cells of retina, involving the pigment rhodopsin (visual purple). Similar cycle is found in *cone* cells of retina, utilizing *color-sensitive* pigments and resulting in *color vision*. Major color-sensitive pigments are also retinal-opsin complexes and are called "porphyropsin" (red), "iodopsin" (green), and "cyanopsin" (blue).

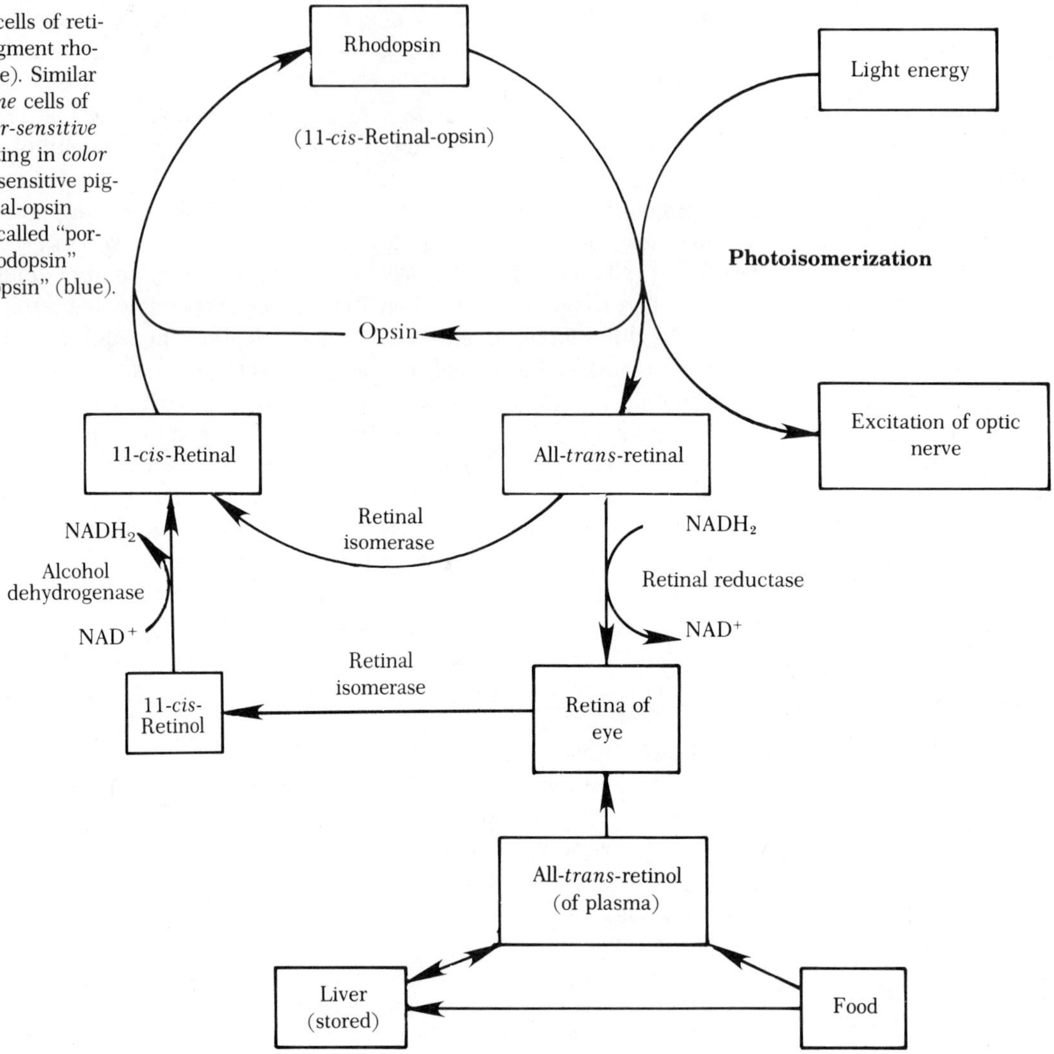

The absorption of light energy by the disclike vesicles in the rod cells results in the conversion of the 11-*cis*-retinal moiety of rhodopsin to all-*trans*-retinal with associated *conformational changes* in structure. These alter the permeability of the vesicle membrane to calcium ions, permitting them to flow out of the vesicle, thus triggering the nerve impulse to the brain, read out as vision. The role of calcium ions as a mediator is reminiscent of its role in the contraction of muscle (Chapter 15).

This, then, is the cycle that operates in the retinal rods in the presence of dim light to give us night vision. If the light intensity is held constant, the entire system enters a *steady state*, rhodopsin being regenerated as fast as it is bleached. As the light intensity is raised or lowered, the rhodopsin concentration is decreased or increased to sustain higher or lower steady states, i.e., when light intensity is raised, the rhodopsin concentration is decreased and night vision sensitivity becomes reduced, and vice versa.

Rhodopsin is the prototype of some 190 visual pigments identified in about 150 animal species. Visual pigments and their derivatives are characterized by the shape of their rather broad (about 200 nm) absorption spectra and particularly by their absorption maxima. Two interconvertible pigments with widely separated absorption maxima are readily converted back and forth by irradiating at appropriate wavelengths.

Color vision While vision in dim light is mediated by rhodopsin of the *rod* cells, color vision is mediated by three different retinal-containing pigments in the *cone* cells. The three pigments are called *porphyropsin, iodopsin,* and *cyanopsin* and are sensitive to the three essential colors: red, green, and blue, respectively. Thus, when light strikes the retina, it bleaches one or more of these pigments, depending on the color quality of the light. The pigments are converted to all-*trans*-retinal, and the protein moiety, opsin, is released as in the case of rhodopsin. This reaction gives rise to the nerve impulse that is read out in the brain as color—red, if porphyropsin is split; green, if iodopsin; or blue, if cyanopsin. If mixtures of the three are converted, the color read out in the brain depends on the proportions of the three split.

The same light-sensitive chromophore, 11-*cis*-retinal, is present in the three kinds of pigments in cone cells responsible for color vision. However, the protein moiety, opsin, of the three pigments (cyanopsin, iodopsin, and porphyropsin) differs and absorbs light at maximums of 440, 535, and 575 nm, respectively. Recent work indicates that the location of charges on the attached opsins are involved in determining the above broad spectrum of absorption maxima and thus enable 11-*cis*-retinal to transmit color signals to the brain. Indeed, this has been demonstrated using a three dimensional model of one opsin constructed from data based on electron density maps, amino acid sequences, and neutron diffraction data. There is a negative charge on the protein molecule near the attachment of retinal on an aspartic acid residue. Retinal is bound to the opsin moiety as a protonated Schiff base with a lysine residue. This negative charge apparently accounts for the shift of retinal's maximum absorption from the ultraviolet range to about 440 nm. The shift to even higher wavelengths appears to result from differing external charges located near retinal's attachment to the respective opsin molecules.

We might ask at what point in the cycle the nerve fibers associated with the rod or cone cells of the retina are stimulated. The answer to this question is not known with certainty, but from our knowledge of the time constants of visual phenomena the stimulation probably takes place immediately after isomerization of rhodopsin by light, when electrically charged groups on opsin have been exposed because of the poor fit of the all-*trans* configuration.

Experiments on patients and normal controls have established the fact that retinol is required by the cone pigments as well as by rhodopsin. In fact, probably the only chemical difference between the rod and cone pigments is in the protein components. Thus, if we substitute cone opsins for rod opsin, the scheme in Fig. 21-13 describes the chemistry of cone vision. The light intensities required, however, are much higher and the absorption maxima are different, in accordance with the requirements of day vision and of color vision.

There is evidence that color blindness is caused by a congenital deficiency of the red- or green-sensitive cone pigments, porphyropsin or iodopsin. Apparently a sex-linked recessive mutation is involved, since the defect is rare in females. The two genes for the red- or green-sensitive pigments probably lie close together on the X chromosome. Blue color blindness, from a lack of cyanopsin, appears to be very rare. However, a small portion of the central part of the normal fovea of the retina lacks cyanopsin in the cone cells. Thus this area is *blue blind*. Trichromatic vision is restricted to a portion of the cone cells in the central part of the retina (20 to 30 degrees). Dichromatic vision, red or green blind, occurs in the next portion of the retina (70 to 80 degrees). The peripheral part of the retina shows only monochromatic vision. Thus there appears to be a definite pattern of distribution of the retinal opsin pigments in the normal retina.

Effect on other processes We have seen that vitamin A has other functions in the body besides those related to vision. Animals on a vitamin A–deficient diet frequently die without any serious eye symptoms.

Retinol appears to be involved in mucopolysaccharide biosynthesis at an enzymatic level. Vitamin A increases the incorporation of ^{35}S- and ^{14}C-glucose into chondroitin sulfate in homogenates of rat colon from vitamin A–deficient rats. The effect may involve sulfate formation or activation, which could explain the damage to mucus-secreting epithelial tissues manifested as xerophthalmia, and dryness and keratinization of the skin and epithelium of the gastrointestinal and genitourinary systems in vitamin A deficiency. Maintenance of epithelial tissue is a second major function of retinol, as emphasized previously. Retinol may have some relation to nucleic acid metabolism, since there is a decrease in the DNA content of several organs during vitamin A deficiency, which is remedied by administration of this vitamin.

Significant investigations indicate that vitamin A or its derivatives counteract the effect of some carcinogenic chemicals by inhibiting tumor formation. A recently discovered vitamin A metabolite, *5,6-epoxyretinoic acid,* has been found to have both vitamin A activity and antipromoter activity against tumor formation.

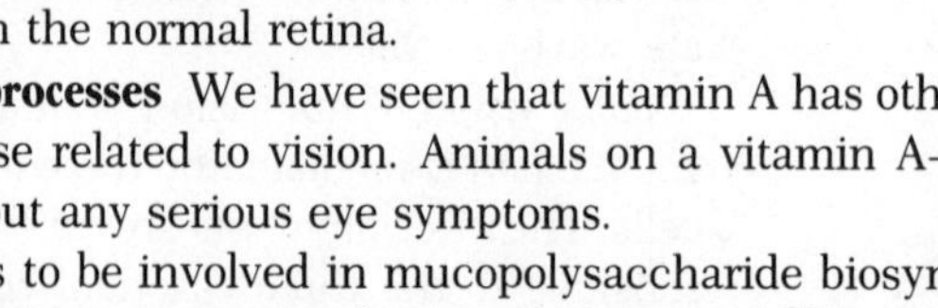

H_3O CH_3 CH_3 CH_3 O OH O

5,6-Epoxyretinoic acid

Other studies have demonstrated that retinoic acid migrates directly into the cell nucleus of vitamin A target tissues, where it becomes bound to a receptor, possibly to the genes themselves. The retinoic acid derivative had antipromoter activity, at least as far as induction of the enzyme ornithine decarboxylase is concerned. These investigators believe that these observations, along with earlier findings in chronic vitamin A deficiency in rats, strengthen the hypothesis that "vitamin A activity and anticarcinogenic activity go hand in hand."

Storage Most of the carotene and retinol absorbed goes to the liver, where it is stored. A sufficient amount of this vitamin can be "hoarded" to last an estimated several months in humans. When a vitamin A–deficient diet is given to rats, the liver stores begin falling immediately. The plasma level remains constant until the liver is almost depleted, with the rhodopsin level in the retina remaining normal until the blood vitamin A is depleted. At this point night blindness begins. When rhodopsin is about 50% depleted, the protein opsin begins to decline and the rod cells of the retina begin to degenerate.

Excretion and secretion Neither vitamin A nor provitamin A is excreted in the urine. Either may appear in the feces but probably as an unabsorbed portion. Even this happens to only a slight extent, the unused material being destroyed by bacteria. Amounts in liver and other tissues in excess of the normal storage capacity or requirement must also be destroyed, but how this happens is still unknown. Both carotene and vitamin A are secreted by the mammary gland. Human colostrum has two or three times more than does human milk, and the latter has from five to 10 times the vitamin A activity of bovine milk.

Human requirements In measuring human requirements of vitamin A, as well as of other vitamins, certain units have been used. These were at first rather arbitrarily fixed and were often based on the amount necessary to prevent avitaminosis in animals under standard conditions. In time, as the vitamins were synthesized, it became possible to base the unitage on the weight of carefully purified and standardized preparations. Thus today the human requirement for vitamin A is expressed in *retinol equivalents* rather than in *international units,* as was formerly done. This change was made because an average of 50% of the vitamin A in the American dietary is supplied as *retinol* and 50% as provitamin A (mainly β-carotene). The value of *1 retinol equivalent* was established as equal to 1 μg *of retinol.* Thus the vitamin A value of foods, expressed as *retinol equivalents,* can be calculated from the following equivalencies:

1 retinol equivalent = 1 μg retinol
= 6 μg β-carotene
= 12 μg other provitamin carotenoid
= 3.33 IU vitamin A activity from retinol
= 10 IU vitamin A activity from β-carotene

Thus the retinol equivalents of a diet can be calculated if the retinol and/or β-carotene content is known in either micrograms or international units.

The new 1980 RDA for vitamin A of the Food and Nutrition Board for the healthy *adult male* is *1000 μg retinol equivalents* per day (see Table 21-4). The basis for this recommendation is that this amount has been found in a number of recent extensive studies on normal adults to *maintain an adequate blood level of retinol* and to *prevent all deficiency symptoms.* The present recommended RDA for the adult *male of 1000 retinol equivalents (RE) equals* the *5000 International Units* formerly recommended.

The 1980 RDA for adult women is *800 RE per day,* based on the smaller average body weight (80% of the adult male).

The 1980 RDA for an infant up to age 6 months is set at 420 μg RE per day, on the basis of the retinol content of human milk at an average daily consumption of 850 ml of milk. After 6 months up to 1 year, as solid foods are fed in addition to milk, the allowance is reduced to 400 μg RE.

During pregnancy the RDA is increased to 1000 RE per day to allow for vitamin A storage in the fetus, and 1200 RE per day during lactation to compensate for the secretion of vitamin A in milk.

The RDA for children and adolescents (Table 21-4) is largely interpolated from the infant and adult male allowances, based on body weight and estimated amounts to satisfy growth needs. As shown in the table, the RDA for vitamin A plateaus at adolescence and remains at this level through adult life.

The well-balanced dietary of most Americans contains the RDA under normal conditions. Foods such as butter, whole milk, fortified margarine, egg yolk, green and yellow vegetables, and yellow fruits are excellent sources of vitamin A. But it is more than possible that the underprivileged may not get the minimum. It is also possible that some individuals require more than the minimum because of either faulty absorption or some other reason. Therefore the addition of supplements may be indicated, especially since a moderate excess seems to be nontoxic.

Toxic effects Overdosage with vitamin A (amounts in excess of 15,000 RE or 50,000 IU per day for long periods) may produce hypervitaminosis and toxic effects, presumably because this vitamin, like vitamin D is not readily excreted, and consequently tissue levels may build up to dangerous concentrations. Excessive amounts of vitamin A, especially in children, may cause loss of appetite, weight loss, irritability, fissuring at the corners of the mouth, and cracking and bleeding of the lips. Later, loss of hair, liver enlargement, and bone and joint pains may occur. Excessive intakes of carotenes may not be as harmful but may result in a yellowish discoloration of the skin *(carotenemia)* that disappears when the carotene intake is reduced.

Cholecalciferol (vitamin D)

Vitamin D is related to rickets and is therefore spoken of as the *antirachitic* vitamin. It is necessary for normal calcium and phosphorus metabolism and consequently for healthy bone and tooth development.

Discovery and properties In the early day of the vitamins, "fat-soluble A" was considered a single substance capable of curing both xerophthalmia and rickets. In 1922 McCollum bubbled oxygen through cod liver oil for several hours at 120° C and found that the resulting oil was ineffective against xerophthalmia but was capable of curing rickets. Vitamin A had been destroyed by this procedure, and the factor remaining was called vitamin D, the antirachitic factor.

Vitamin D is a white crystalline substance, soluble in fats and fat solvents. It is fairly heat resistant and also relatively resistant to oxidation. It is not affected in mild acids or alkalies. When cod liver oil is saponified, the vitamin is found in the nonsaponifiable fraction. It is a sterol derivative.

Relation to radiant energy From the time of Herodotus, the health-giving effects of sunlight have been emphasized, but the particular relation of lack of sunshine to rickets did not begin to be recognized until the last century. In 1822 a Polish physician, Sniadecki, maintained that exposure of the body to direct sunlight had both a preventive and curative effect for rickets. Further studies from different parts of the world revealed the absence of rickets, despite poor sanitary conditions, in those regions where sunlight was most abundant and definitely indicated that rickets was mainly caused by a lack of sunlight. Later it was shown that animals fed a diet deficient in vitamin D but containing a sufficiency of calcium and phosphorus would not acquire rickets if they were irradiated with ultraviolet light. The final chapter in this story was the announcement that cholesterol-containing inactive foods could be given antirachitic properties by irradiating them with ultraviolet light. Dried milk and other foods can be irradiated on a commercial basis. A potent antirachitic agent, ergocalciferol, is produced by the irradiation of ergosterol, a sterol derived from ergot and from yeast.

The explanation of these phenomena is that ultraviolet irradiation causes a change in the molecular structure of certain sterols that, unless so changed, are inactive. If the irradiated sterols are taken by mouth, they act just the same as the naturally occurring vitamin D; if they are produced in or on the skin by the irradiation of the inactive sterols present there, they are absorbed and find their way into the circulation and behave similarly to orally administered vitamin D. These inactive sterols are the provitamins of vitamin D. One of them, 7-dehydrocholesterol, can be synthesized by animals, including humans.

Interesting recent observations suggest that the rate of synthesis of cholecalciferol from 7-dehydrocholesterol in the skin (stratum granulosum) is regulated by the amount of pigmentation and keratinization in the overlying stratum corneum. This regulates the amount of solar ultraviolet radiation, especially at 290 to 320 nm, that penetrates into the stratum granulosum and forms vitamin D. Thus white skin, which contains little pigment and keratin, allows maximal ultraviolet penetration. Yellow skin, which contains more keratin, permits less ultraviolet penetration, whereas black skin, which is more pigmented, permits still less. These differences are apparently a genetic adaptation to climate. Hair or fur in animals is a still further regulatory mechanism. Vitamin D synthesis is thus maintained within physiological limits, estimated to be 0.01 to 2.5 mg of cholecalciferol per day. Skin pigmentation also may correct for seasonal variations, e.g., *tanning* in the summer months in northern latitudes.

Structure At least 10 different compounds are known to have antirachitic properties and are designated D_1, D_2, etc. Five of them are rather well defined as chemical compounds, but only two are of great importance: D_2, of vegetable origin, now called *ergocalciferol;* and D_3, or *cholecalciferol,* of animal origin. The common vitamin D of fish liver oils is D_3. This is probably the form also present in milk and eggs and is produced on irradiation of the skin and when 7-dehydrocholesterol is irradiated. Vitamin D_2 is derived from ergosterol. The formula shows that the effect of irradiation is the opening of ring *B* of the provitamin sterol.

Ergocalciferol (D_2) has the same structure as cholecalciferol (D_3), with the exception of the side chain, which is shown at right:

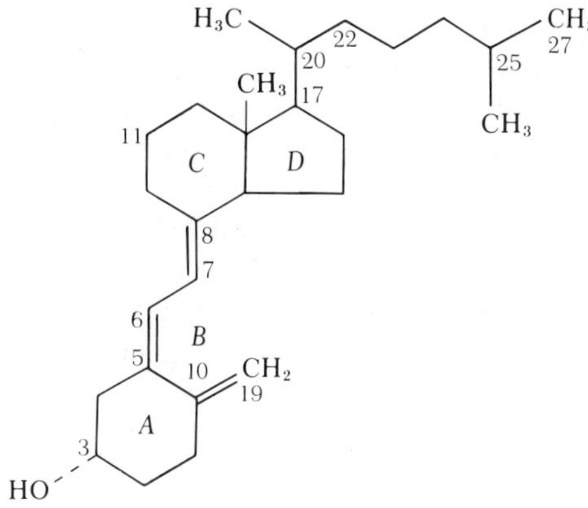

Cholecalciferol (vitamin D_3)

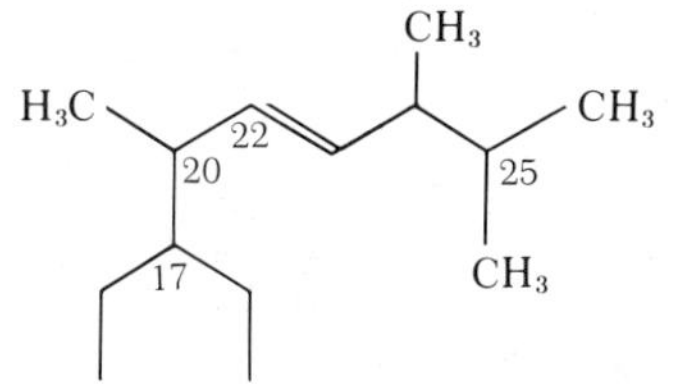

Ergocalciferol

Both forms of the vitamin have about the same degree of activity in the human being. In nature these vitamins occur as esters.

Metabolism During the past few years there have been tremendous advances in knowledge of the metabolism and functions of vitamin D. Recent investigations by DeLuca and his associates indicate, for example, that there are several biologically active forms of cholecalciferol having slightly different structures. They are more polar compounds having additional hydroxyl groups in the carbon-1 and carbon-25 positions. The structure of one of the most active forms, 1,25-dihydroxycholecalciferol, follows:

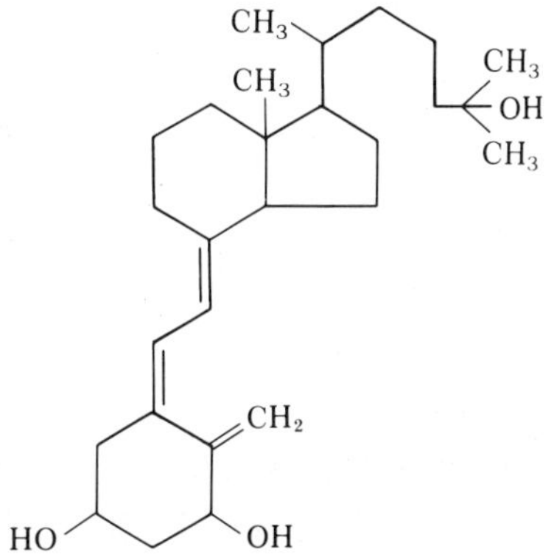

1,25-Dihydroxycholecalciferol
(one of the most active forms of vitamin D_3)

Active forms of cholecalciferol were isolated chromatographically as pure compounds from the plasma of hogs fed large amounts of vitamin D_3. The structures of two were established as 25-hydroxycholecalciferol and 1,25-dihydroxycholecalciferol (as shown on p. 747). They appear to be the major biologically active metabolites of cholecalciferol, although still other forms with high activity have been isolated, e.g., the lα-hydroxy and 24,25-dihydroxy derivatives. 1,25-Dihydroxycholecalciferol has 10 to 15 times more activity than does cholecalciferol in curing rickets in rats, in increasing bone ash in vitamin D–deficient chicks, and in increasing the transport of calcium across the intestinal mucosa of vitamin D–deficient rats. As little as 0.25 μg of 1,25-hydroxycholecalciferol injected intravenously into vitamin D–deficient rats increased calcium absorption from the intestine. The addition of 2.5 μg by means of an arterial blood perfusate to cholecalciferol-deficient rats increased calcium transport to normal levels within 2 hours. 1,25-Hydroxycholecalciferol also produces an earlier rise in serum calcium resulting from bone resorption in vitamin D–deficient rats than does cholecalciferol in a similar dose. These observations, taken as a whole, are strong evidence that 1,25-hydroxycholecalciferol is, indeed, the metabolically active form of vitamin D_3, sometimes aptly termed the "hormone form" of the vitamin (Chapter 18).

Subsequent studies by DeLuca and his associates have shown that 25-hydroxycholecalciferol is formed primarily if not exclusively in the *liver* and that a second hydroxylation to form the 1,25-dihydroxy derivative occurs in the *kidney*. the hydroxylase systems require NADPH and molecular oxygen (Chapter 17). This process is shown diagrammatically in Fig. 21-14. Further intensive investigations are continuing in several laboratories to determine the roles of metabolites of cholecalciferol in maintaining calcium homeostasis.

Recent studies by DeLuca's group demonstrate that 1,25-dihydroxycholecalciferol has *two* primary effects: (1) It acts on the *intestinal mucosal* cells to increase the *absorption of calcium*. Phosphate uptake is simultaneously increased. (2) It

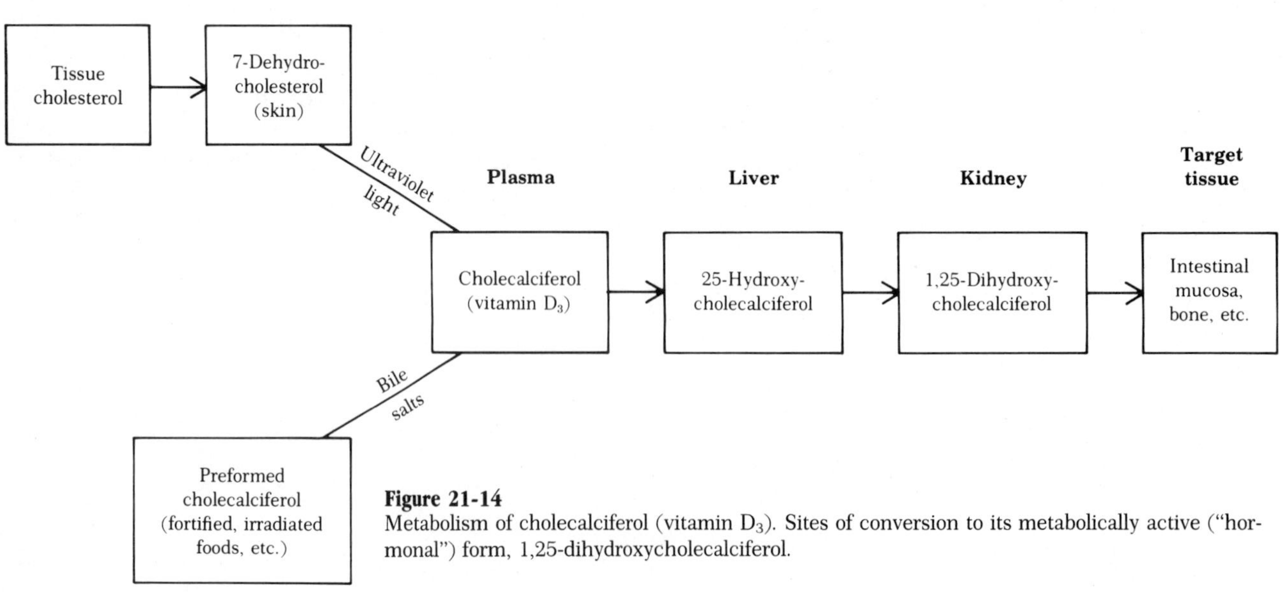

Figure 21-14
Metabolism of cholecalciferol (vitamin D_3). Sites of conversion to its metabolically active ("hormonal") form, 1,25-dihydroxycholecalciferol.

acts on *bone* to increase the *mobilization of calcium* (and phosphate). The biosynthesis of 1,25-dihydroxycholecalciferol in the kidney apparently is stimulated by the parathyroid hormone in response to a decrease in the plasma level of calcium (and phosphate). The 1,25-dihydroxy derivative then acts together with the parathyroid hormone to mobilize calcium from bone. Thus 1,25-dihydroxy D_3 must now be regarded as a calcium-mobilizing hormone (Chapter 18).

Occurrence Cod liver oil and other fish liver oils are the best natural sources of vitamin D. The edible portions of oily fish, e.g., sardines, salmon, herring, are also excellent sources. Egg yolk and liver of the commonly slaughtered animals contain amounts that depend on the food of the animal from which they are derived, but mammalian liver is not very rich in this vitamin. Milk contains little vitamin D unless enriched in one of several ways, and vitamin D milk is now a common article of the American dietary. Many ordinary foods, among them the green plants, contain small quantities, and mushrooms contain slightly greater amounts. In general, this vitamin is not widely distributed, but the fact that it can be provided in three ways should make its deficiency rather uncommon. These three ways of providing vitamin D are (1) by furnishing the vitamin as it occurs naturally in foods or by enriching the food by the addition of vitamin D, (2) by irradiating foods containing precursors of the vitamin, and (3) by irradiating the skin of the individual with ultraviolet light or sunshine.

Absorption Absorption of vitamin D from the intestinal tract requires the presence of bile salts. Here again mineral oil acts as a hindrance because the vitamin is soluble in it and consequently is carried through the intestine into the feces. After absorption, cholecalciferol is apparently transported in the plasma, tightly bound to an α_2-globulin.

Effects of deficiency A deficiency of vitamin D leads to *rickets* in children (Fig. 21-15) and to *osteomalacia* in adults, a condition that might be termed "adult rickets." Rickets usually develops in infancy or early childhood, although juvenile or late rickets is seen in India, with x-ray–detectable rickets being frequently observed up to puberty in that country. Defective ossification is the result of this avitaminosis. The bones become soft and pliable, and a number of different deformities may ensue: bowlegs, knock-knees, enlargement of the ends of bones (epiphyses), rows of beadlike swellings at the rib junctions (the "rachitic rosary"), contracted pelvis, and the development of bosses on the temporal bones. X-ray photographs of the bones reveal that ossification is not normal; the shadows cast are less dense and the ends of the bones less sharply defined (Fig. 21-16). Chemical analysis of the bones reveals a lower content of inorganic constituents and a higher content of organic substances and water. However, the *ratio* of calcium to phosphorus remains constant, indicating that the type of bone salt laid down is normal, although the amount is insufficient. In the blood serum there is usually, but not always, a normal content of calcium, but the phosphate is decreased. If the product of the serum calcium content and the serum phosphate, both expressed as milligrams per 100 ml equals or exceeds 40, rickets does not develop, whereas a product of below 30 always leads to rickets. The plasma level of 1,25-dihydroxycholecalciferol is markedly decreased. There is also a noticeable increase in serum *alkaline phosphatase* in rickets and a decrease when recovery is brought about as a result of vitamin D or 1,25-dihydroxycholecalciferol treatment. The exact significance of this is not known. Some authors believe that the increased serum alkaline phosphatase in rickets (and in other bone diseases) is a result of overproduction in the bone in a vain attempt at bone formation. Others consider it a result of the bone's increased capacity for cellular activities because of the absence of true bone, which is relatively inactive.

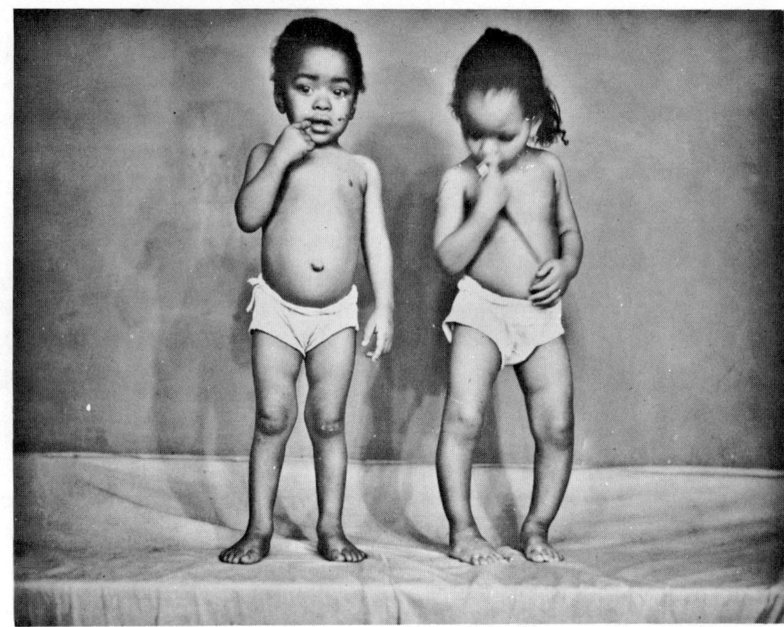

Figure 21-15
Rachitic children, showing knock-knees on left and bowlegs on right.

From files of Therapeutic Notes; courtesy T. D. Spies, Birmingham, Ala., and O. D. Bird, Ann Arbor, Mich.

Vitamin D–resistant forms of rickets in children have been described repeatedly in the literature. This condition is hereditary and characterized by a low serum phosphate level and high serum alkaline phosphatase. The chief clinical manifestation is shortness of the legs, which may be bowed, rather than that of the trunk of the body. Treatment with massive doses of vitamin D does not change the height or deformity or improve the hypophosphatemia, but it does lower the level of alkaline phosphatase. Roentgenographic evidence of improvement in bone structure is usually found. Vitamin D therapy entails the risk of vitamin D toxicity, however. Treatment with phosphates or with human growth hormone have proved of uncertain value.

The pathogenesis of the disease is unclear. There is no demonstrated renal disease and no dietary vitamin D insufficiency. The rickets, however, is clinically similar to the form caused by vitamin D deficiency. The disease may be caused by a renal tubular defect in the reabsorption of phosphate, by a decreased intestinal absorption of calcium, resulting in secondary hyperparathyroidism, or by an abnormal metabolism of or resistance to vitamin D.

It now appears that some patients with familial vitamin D–resistant rickets may lack the ability to convert cholecalciferol to 1,25-dihydroxycholecalciferol, the active form of the vitamin. Some have high plasma levels of 25-OH-vitamin D_3 but low levels of 1,25-dihyroxyvitamin D_3. They have been found to respond satisfactorily to treatment with the 1,25-dihydroxyvitamin D derivative in amounts as small as 1/1700 that of vitamin D_3. This type of vitamin D–dependent rickets therefore appears to result from a genetic defect in *25-hydroxycholecalciferol-1-hydroxylase*, the enzyme in the kidney responsible for the production of 1,25-dihydroxycholecalciferol.

Thus far about *ten* types of vitamin D–refractory rickets have been identified. Some are hereditary. All involve metabolic abnormalities in the absorption or

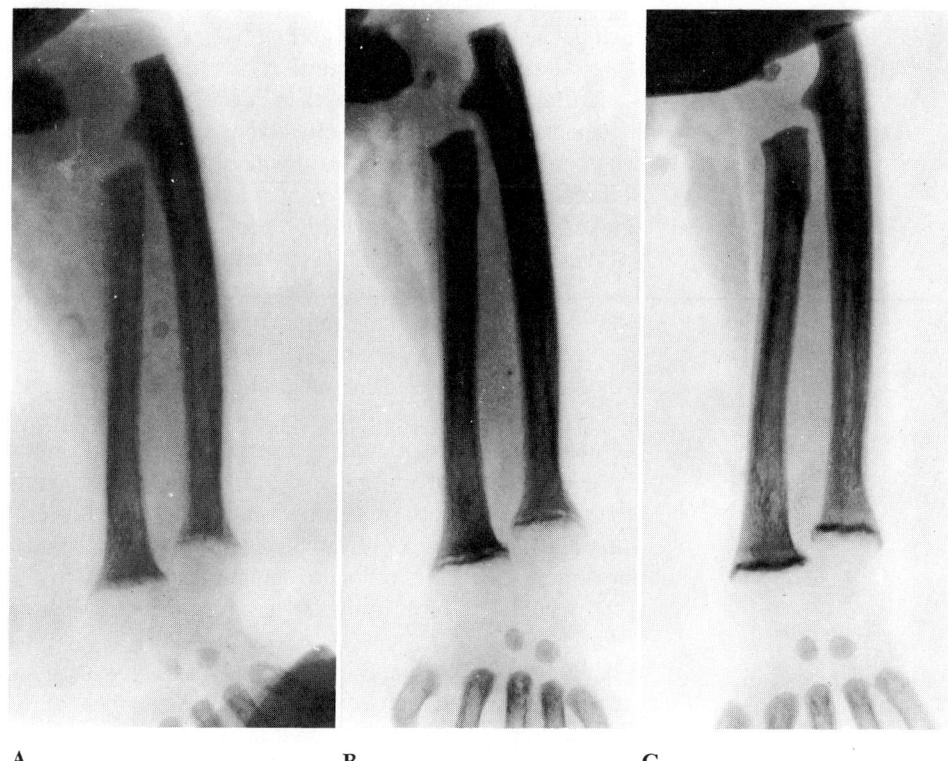

Figure 21-16

X-ray films of forearms of a rachitic child, showing effect of treatment.

A Feb. 11: Pronounced trabeculation of radius and ulna, particularly in the cortical portion of the bone, with slight periosteal thickening. End of bone is slightly mushroomed and cupped and shows distinct fringing. Cartilage is swollen. Distal epiphyses of radius are absent. There are two small centers of ossification in the wrists.

B March 11: New center of ossification has appeared in distal epiphysis of humerus. Centers of ossification at wrists are still two in number but larger and more distinct in outline. There is a fresh line of calcium deposition at the ends of the radius and ulna in the zone of provisional cartilage (line test). Clear area is present between this new calcification and the shaft (submetaphyseal rarefaction). Cupping, fringing, and stippling are present, and spur formation is also noted outlining the swollen cartilage at distal epiphysis of ulna (beginning healing).

C March 24: Healing is now advanced. Shaft shows calcification of subperiosteal osteoid at outer aspect of radius, giving the bone a greater width. Calcification of provisional zone in metaphyses is much more distinct and the submetaphyseal area is filling in.

Case history. Diagnosis: Rickets and infantile tetany. Admitted Feb. 9 because of carpopedal spasm of 6 hours' duration. Never had vitamin D preparations. Had 3 oz. of orange juice daily for 1 month (6 months old upon admission). Physical examination on admission: Carpopedal spasm, separation of the sagittal and lambdoid sutures, with pronounced occipitoparietal craniotabes. Thorax: Flaring of lower ribs with enlargement of costochondral junctions. Chvostek's sign, positive. Laboratory data: Serum CO_2, 41 vol %; Ca, 4.6; P, 5.6 (mg in 100 ml); Ca × P, 25.7; phosphatase, 57.2 units (Bodansky). Treatment: Calcium gluconate, 10 ml of a 10% solution, intramuscularly at 10 P.M. and 12 P.M. $CaCl_2$, 1 dram q4h (1 gm). Orange juice, 2 oz daily. Vitamin D preparation, 10 drops daily, i.e., approximately 2500 IU vitamin D. Feb. 20, $CaCl_2$ discontinued.

metabolism of vitamin D or in the end organ response to its hydroxylated derivatives.

In rickets there is commonly a delay in dentition. The first tooth in rachitic babies seldom appears between the sixth and ninth month, at which time it has appeared in about half the number of normal babies. This would be expected in view of the close relation of bones and teeth. A lack of vitamin D leads to hypoplasia, i.e., poor structural development, of teeth in dogs. This may predispose to dental caries, since a hypoplastic tooth is less effectively protected by enamel. The question as to whether healthy human teeth are likely to be more carious if vitamin D is lacking is still unsettled, as is also the companion question of the effect of a high vitamin D intake in preventing caries. Much work has been done on both these problems, and results indicate that vitamin D probably reduces the incidences of caries indirectly, i.e., by improving the general health and nutrition of the individual.

Osteomalacia presents a somewhat different picture from rickets, although the action on the bones is essentially the same. The bone becomes softer than rachitic bone and the ratio of calcium to phosphorus is changed. The loss of calcium is greater than that of phosphorus, and there is a relative gain in magnesium. This softness of the bones leads to diverse types of deformities. Osteomalacia occurs rarely in America or Europe, except in old age, but is common in India and China, particularly among those confined indoors. Thus they are deprived of exposure to sunshine, and their diet is also deficient in vitamin D and calcium. In osteomalacia the serum calcium is reduced, sometimes to such an extent that tetany ensues. Tetany is a state of muscular twitching that may be brought about by low blood calcium.

Another clinical condition indirectly associated with a lack of cholecalciferol is *celiac disease,* also known as idiopathic steatorrhea, gluten-sensitive enteropathy, and nontropical sprue, discussed later in this chapter. Here, as in osteomalacia, there is an impaired mineralization of the bones, which may result in deformities or dwarfism. Here, too, a low serum calcium and low serum phosphorus are found, with possible manifestations of tetany. Celiac disease is *indirectly* a vitamin D deficiency because the primary abnormality seems to be, in part, a fatty diarrhea. The fatty acids in the intestinal lumen are not absorbed normally and carry with them into the stools calcium soaps and vitamin D.

In all the conditions mentioned the administration of vitamin D in therapeutic doses, or ultraviolet irradiation, or both, produce good results. Deformities cannot be rectified by this means, but further malformation may be checked. Serial x-ray photographs of rachitic bones before and during treatment show this effect in a striking manner (Fig. 21-16).

Mechanism of action Vitamin D has a regulatory influence on calcium and phosphorus metabolism. Both calcium and phosphorus must be present in the diet to have the complex calcium salt deposited in bone. However, no matter how great an amount of these minerals is available, normal calcification does not take place in the absence of this vitamin. On the other hand, even if the supply of calcium and phosphorus is practically at starvation levels, an optimum amount of cholecalciferol can enable them to be utilized and deposited in a nearly normal manner.

In the first place, vitamin D causes an increased absorption of calcium and phosphorus from the intestinal tract. In studies on the transfer of ^{45}Ca across the membrane of everted segments of small intestine of rats, vitamin D greatly increased the rate of passage of ^{45}Ca across the intestinal mucosa. There was an active transport (i.e., against a concentration gradient) in the proximal portion of the small intestine that was dependent on the energy of oxidative metabolism. Vitamin D also increased the rate of facilitated transfer of calcium along the entire length of the small intestine. Using the same technique, investigators have found that ergocalciferol likewise increases the absorption of phosphate.

There is evidence that actinomycin D, which inhibits protein synthesis at the RNA level (Chapter 10), inhibits vitamin D action. The injection of this antibiotic into rats competely prevents both the rise in serum calcium normally induced by vitamin D and the increased transport of calcium by everted sacs of small intestine. Injection of parathyroid hormone did not alter this result, thus eliminating the possibility that the inhibition of vitamin activity was caused by blocked hormone synthesis. Actinomycin D inhibited the action of a subsequent dose of vitamin D_3 in increasing calcium absorption from the intestine. It was suggested that vitamin D may therefore influence the absorption of calcium by inducing the biosynthesis of a protein component of the calcium-transport system in the intestinal mucosal cells.

This suggestion has been strongly supported by other evidence. Vitamin D increases the incorporation of ^{3}H-orotic acid into RNA of the rat intestinal mucosa. Furthermore, a vitamin D–dependent calcium-binding protein from the chick intestinal mucosa has been prepared in purified form by gel filtration and disk electrophoresis. The amount of this protein in the intestinal mucosa decreases in vitamin D deficiency and increases when cholecalciferol is administered. The administration of estrogens and certain other steroids has no effect on its concentration. The amount of the calcium-binding protein is largest in the duodenal mucosa, the primary site of calcium absorption, and is progressively less in the jejunum and ileum. It is absent in mucosa from the colon.

The foregoing current investigations thus strongly indicate that a major function of cholecalciferol, or more probably its active metabolite form, 1,25-dihydroxycholecalciferol, is inducing the biosynthesis of a calcium-binding protein essential to the active transport of calcium and phosphate through the intestinal mucosa. This would increase the plasma level of calcium and phosphate and thus increase their availability for deposition in the bones; these are the major effects of vitamin D in the body.

1,25-Dihydroxycholecalciferol also increases bone resorption in the rat, producing a rise in serum calcium. The mechanism involved is uncertain but may entail an effect on a calcium-transport system in bone analogous to that occurring in the intestinal mucosa. It now appears to be mediated by the parathyroid hormone (Chapter 18). Under conditions of hypocalcemia there is an increased secretion of parathyroid hormone, which in turn stimulates the synthesis of 1,25-dihydroxycholecalciferol in the kidney. The 1,25-dihydroxycholecalciferol then, apparently together with the parathyroid hormone (PTH), mobilizes calcium from previously formed bone. The resulting increase in blood calcium suppresses further PTH secreton and hence further 1,25-dihydroxycholecalciferol synthesis. Apparently 1,25-dihydroxycholecalciferol acts directly in the intestine *without* PTH to increase intestinal calcium absorption.

The effect of 1,25-dihydroxycholecalciferol on plasma and bone phosphate levels may be indirect, resulting from a more direct effect on calcium transport. Further clarification of this relationship is needed.

Tests for vitamin D In experimental animals there are several tests for vitamin D deficiency and recovery. X-ray photographs of the distal ends of the ulna and radius show a characteristic indistinctness. When healing occurs, the entire bone assumes a more homogeneous appearance and the cupped ends fill out (Fig. 21-16). The *line test* is used in experimental studies in ascertaining at autopsy the degree of healing. The split bone is treated with silver nitrate, which stains the provisional zone of recalcification, leaving the uncalcified rachitic tissue unstained.

Another method is to analyze the bone for total ash. In rickets the total ash is low, although the calcium-phosphorus ratio remains normal.

Human requirements The vitamin D requirements of normal infants and children are especially important and depend partly on the amount of ultraviolet light to which they are exposed. Remember that the effective ultraviolet rays do not penetrate ordinary glass. Therefore exposure to sunshine coming through window glass is of little value. Smoke also impedes the penetration of these rays, and consequently city sunshine is not always beneficial. For this and other reasons some vitamin D should be included in the food of younger individuals. The Food and Nutrition Board (1980; Table 21-4) recommends 10 μg cholecalciferol daily for infants, children, and young adults up to the age of 19 years. The same amount is advised for women during pregnancy and lactation. The 1980 RDA for both males and females, ages 19 to 22 years, is 7.5 μg cholecalciferol per day (Table 21-4). Thereafter the RDA is 5 μg cholecalciferol per day for both sexes. The minimal requirement for vitamin D is difficult to determine because of the variability of environmental factors such as sunshine. However, the RDA provides an adequate margin of safety over a minimal value. Ten μg of pure crystalline vitamin D_3 is equivalent to the biological activity of 400 IU formerly used.

Toxic effects After administration of an excess of vitamin D to a mammal the vitamin can be found in the circulating blood for months. Thus the use of enormous doses of vitamin D is not without danger. Severe and even fatal effects have been noted. The toxic manifestations caused by excess dosage include nausea, anorexia, weakness, headache, digestive disturbances, and polyuria. Irreversible damage to the kidneys, as well as calcification of other soft tissues, results. The threshold of toxicity seems to be about 500 to 600 μg cholecalciferol per kilogram of body weight per day. Such doses are not ordinarily employed.

The reason for the toxicity of vitamin D is the difficulty of excretion of this vitamin rather than its storage in the liver. Any excretion is gradual, by way of the bile. Excess cholecalciferol injected into animals remains in the circulation for several months. In contrast, the *water-soluble* vitamins, if given in excess, are excreted promptly in the urine and are therefore relatively nontoxic.

α-Tocopherol (vitamin E)

The possibility that reproductivity might be dependent on a vitamin was first suggested by Mattill in 1920; such a vitamin was discovered 2 years later by Evans and by Sure. They fed rats a diet of purified foodstuffs plus cod liver oil and yeast. At that time such a diet was assumed to contain all the vitamins necessary for the rat. However, although the animals grew at a normal rate, they did not bear young. Addition of a variety of vegetable oils rectified this condition. The factor contained in these foods was termed vitamin E, the *antisterility* vitamin or fertility factor. Numerous subsequent investigations resulted in the isolation of vitamin E in crystalline form, the determination of its structure, and finally its synthesis. It was given the name *tocopherol* (Gr., "child-bearing"; plus *-ol*, "an alcohol").

Properties and structure Vitamin E is a fat-soluble, water-insoluble, light yellow oil, stable to heat and acids, rather unstable to alkalies, and slowly oxidized. It is found in the nonsaponifiable fraction of the vegetable oils.

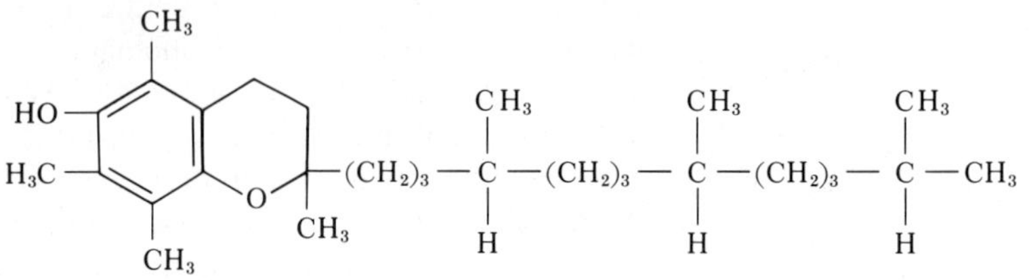

HO, H_3C, CH_3, CH_3, O, CH_3 (chromane ring) —$(CH_2)_3$—CH(CH_3)—$(CH_2)_3$—CH(CH_3)—$(CH_2)_3$—CH(CH_3)—CH_3

α-Tocopherol

Like vitamins A and D, there is more than one form of vitamin E. We now distinguish several different tocopherols. α-Tocopherol is the most potent, has been synthesized, and is commercially available. Its structure is shown above. The other tocopherols, β, γ, etc., differ in the number and position of the methyl groups attached to the benzene ring.

Occurrence The tocopherols are widely distributed in plant and animal tissues and differ from vitamin A in not being concentrated chiefly in the liver. Particularly good sources are cottonseed oil, corn oil, peanut oil, and wheat germ oil, but not olive oil. Green lettuce leaves and orange peels also have a high content, and nearly all green-leaved plants have some of this vitamin. It is also present in meat, butter, milk, eggs, and fish liver oils. Vitamin E activity is displayed by many other organic compounds, some of them quite unrelated structurally to the vitamin. A number of phenols, quinones, and coumarins show some vitamin E action. Slight changes in the structure of the active tocopherols, e.g., shortening the side chain, may reduce or even abolish the physiological effects of these compounds.

Absorption Absorption of the tocopherols is not efficiently accomplished but it is believed to occur similarly to that of the other fat-soluble vitamins. Bile salts and the presence of fats are believed to be useful if not entirely essential. However, rancid fats destroy this vitamin by oxidation (Chapter 23).

If the mother is fed an adequate diet, the fetus absorbs through the placenta sufficient tocopherol for its needs but not enough for storage. This must be supplied to the young animal (and presumably to the infant) by the milk. Storage occurs in various tissues but chiefly in adipose tissue.

Water-soluble forms are available. The water-soluble disodium phosphate ester of α-tocopherol, when administered intramuscularly, has a more rapid and constant effect than the oil-soluble vitamin administered orally.

Most of the plasma vitamin E is transported by lipoproteins (Chapter 12).

Effects of deficiency The effects of a deficiency of vitamin E vary widely in different animal species. The major effects are on the reproductive system, muscle tissue, nervous system, and blood erythrocytes. In rats a lack of α-tocopherol results in damage to the reproductive system of both males and females. There is a degeneration of the germinal epithelium that cannot be remedied, after it is once established, by feeding the vitamin. If the female fed a vitamin E–free diet does become pregnant, the embryo dies and is resorbed.

Besides having effects on the reproductive system, vitamin E is also necessary for the structural and functional maintenance of skeletal muscle, cardiac muscle, smooth muscle, and the peripheral vascular system in a variety of laboratory animals. Indeed, the present opinion is that the effects on muscle are of greater importance than are the effects on fertility. Muscular dystrophy and morphological changes in various tissues are caused by a vitamin E deficit and are accompanied by increased oxygen consumption of the muscle and by alterations in chemical composition and functional behavior. Creatine elimination then is increased (Chapter 17). This latter effect is believed to be the result of an inability of the skeletal muscle to utilize creatine.

A type of muscular dystrophy can be produced in lambs by the feeding of a raw kidney bean–hay ration. The condition can be prevented by administering α-tocopherol *plus selenium*. The unheated beans are believed to contain a heat-labile, antivitamin E or antiselenium factor, or both.

In chicks a vitamin E deficiency results in injury to the nervous system from an impairment of the blood vessels in the brain. In all these conditions there seems to be no comparable effect on human beings.

Vitamin E has a sparing action on vitamin A and carotene, e.g., vitamin A and carotene are more effective in curing their deficiency symptoms if vitamin E is administered at the same time. Ingestion of extra amounts of α-tocopherol increases the storage of vitamin A in the liver of rats, and many other examples of the close connection between these two vitamins could be cited. This biological relationship undoubtedly has a chemical basis. Vitamin E is an antioxidant, i.e., it can prevent the oxidation of various other easily oxidized substances, notably fats (Chapter 23) and vitamin A. For this reason it is often added to foods to prevent oxidation. Possibly this protection is effective even within the cells. The antioxidant properties of E are enhanced by certain other substances, many of which are also antioxidants. Phenols and ascorbic acid are notable examples.

Other work suggests the involvement of vitamin E in combating certain human nutritional anemias. Serum α-tocopherol levels have been found to be low in some types of anemia, particularly in infants. The anemia is a macrocytic type and responds favorably to vitamin E therapy. The erythrocytes appear to be less resistant to hemolysis, resulting in a hemolytic type of anemia. The vitamin E may protect unsaturated fatty acids in the erythrocyte cell membrane from oxidative destruction and thus prevent hemolysis of the cell.

At the present time it cannot be stated definitely that there is a special human requirement for vitamin E for reproduction. Many clinical investigations have been reported in which vitamin E preparations were used to remedy sterility, habitual abortions, and various abnormalities of premature infants. The results are inconclusive.

Possible symptoms of vitamin E deficiency in humans are observed *only* in patients with chronic impairment of fat absorption. These include increased fragility of erythrocytes, increased urinary excretion of creatine (suggesting muscle tissue loss), and the deposition of *ceroid* pigment in the musculature of the small intestine. These symptoms are usually associated with a low blood level of vitamin E.

There have been numerous claims of possible beneficial effects of vitamin E in various human diseases, including myopathies, encephalopathies, erythrocyte abnormalities, certain hereditary conditions, and cardiovascular disease. Again, the results are inconclusive. Vitamin E apparently remains a "vitamin in search of a disease."

Mechanism of action The precise mechanism of action of vitamin E has not been clearly elucidated and remains largely an enigma. However, current evidence indicates that a primary function is as a *biological antioxidant*. Vitamin E, along with glutathione peroxidase (Se), catalase, and superoxide dismutase (hemocuprein), provides a line of defense against peroxidant molecules, such as peroxides, superoxides (O_2^-), and related free radicals. Vitamin E thus may protect cell membranes in general, particularly their unsaturated fatty acid constituents, from lipid peroxidation, which otherwise would result in eventual cell destruction.

Toxicity Tocopherols are considered to be relatively nontoxic in humans. However, according to recent studies large doses are definitely toxic in animals. Therefore, as in the case of vitamins A and D, caution should be exercised in the use of large doses for any prolonged period of time.

Human requirements The basic human requirement for vitamin E is related, in addition to body weight, to the *polyunsaturated fatty acid* content of the body cellular structure. In turn, the latter is variable and is affected by the type of dietary fat. Thus it is difficult to establish a firm requirement and recommended allowance. Thus the requirement values by the Food and Nutrition Board are *average values* based on adequate intakes of vitamin E from *balanced mixed* diets in the United States containing an average amount of polyunsaturated fatty acids. The intake of vitamin E must also maintain a blood level of *total* tocopherols above 0.5 mg/100 ml and an adequate tissue concentration with an appropriate ratio of tocopherols to polyunsaturated fatty acids.

The 1980 RDA (Table 21-4) for vitamin E is expressed as *milligram α-tocopherol* equivalents (mg α-TE). This value in turn is equivalent to 1 mg D-α-tocopherol. In calculating the total milligram α-tocopherol equivalents in a mixed diet the amount of the three principal tocopherols in foods is multiplied by the following factors derived from their relative biological activities. Total mg TE = (1 × mg D-α-tocopherol) × (0.5 mg β-tocopherol) + (0.1 × γ-tocopherol). Thus the 1980 RDAs for normal healthy individuals are:

Adult male	10 mg α-TE
Adult female	8 mg α-TE
During pregnancy	+2 mg α-TE
During lactation	+3 mg α-TE
Adolescents	8 mg α-TE
Children	5-7 mg α-TE
Infants	3-4 mg α-TE

Phylloquinone (vitamin K)

Dam, a Danish investigator, in 1919 discovered a hemorrhagic disease in chicks attributable to the lack of a food factor that he later (1934) called in German *Koagulations-vitamin*. From this came the term *vitamin K*. A deficiency of vitamin K leads to a slowing of the rate of blood clotting. More specifically, there is an increased *prothrombin time* (Chapter 16). This factor therefore appears to be necessary for the production of a normal amount of prothrombin. It is a fat-soluble substance found in various food oils and is of considerable importance from a medical and surgical standpoint.

Properties There are two naturally occurring forms of vitamin K: vitamin K_1, or *phylloquinone*, and vitamins K_2, or *menaquinones*. Both are lipid soluble and stable to heat and to reducing agents. The activity is also abolished by irradiation, alkalies, strong acids, and oxidizing agents.

Structure Vitamin K_1, obtained from the alfalfa leaf, is 2-methyl-3-phytyl-1,4-naphthoquinone. Vitamin K_2, produced by bacterial synthesis, is a group of 2-methyl-3-*N*-isoprenyl-1,4-naphthoquinones, called *menaquinones*. The most common form has seven isoprenyl units in its side chain. Both these natural types have the same general activity. Many synthetic products with similar structures having antihemorrhagic effects have been prepared. Some of these are water soluble, but only one is more potent (weight for weight) than vitamin K_1. This is 2-methyl-1,4-naphthoquinone, which has been given the name *menadione*. It is soluble in oil, sparingly soluble in water, and stable to air when protected from light. Its diphosphate ester is water soluble and is widely used clinically.

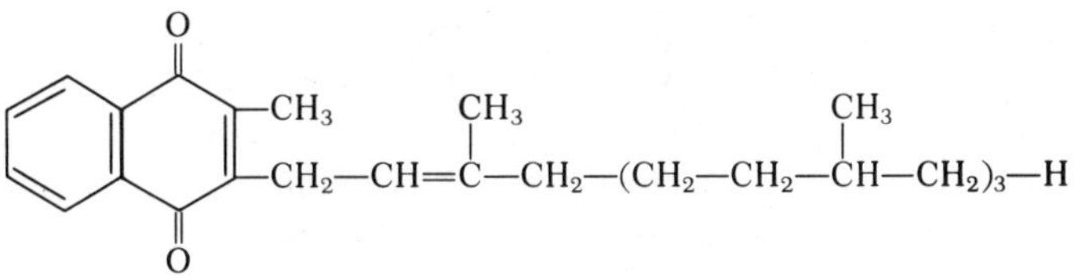

$-CH_3$; $-CH_2-CH=C(CH_3)-CH_2-(CH_2-CH_2-CH(CH_3)-CH_2)_3-H$

Vitamin K_1 (phylloquinone)
(2-methyl-3-phytyl-1,4-naphthoquinone)

$-CH_2$; $-(CH_2-CH=C(CH_3)-CH_2)_n-H$

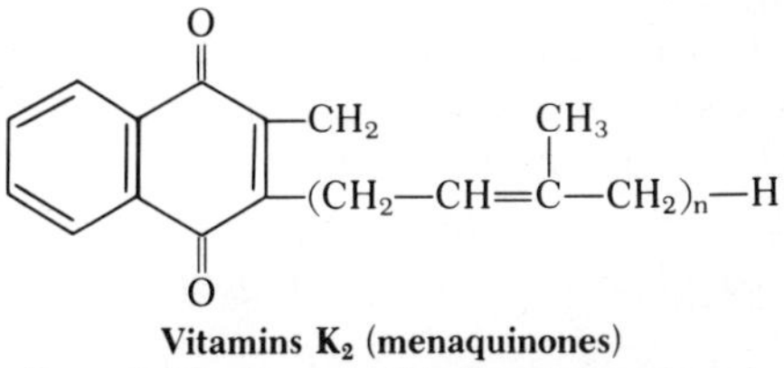

Vitamins K_2 (menaquinones)
(2-methyl-3-n(isoprenyl)-1,4-naphthoquinone)

$-CH_3$

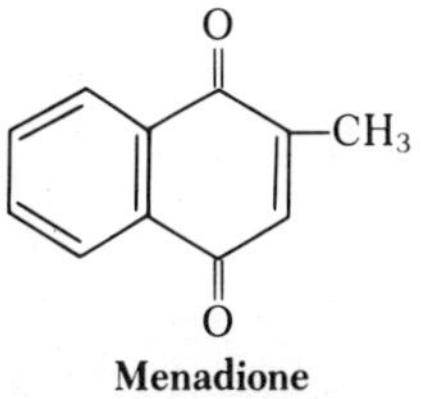

Menadione
(2-methyl-1,4-naphthoquinone)

Occurrence As mentioned, vitamin K_1 is obtained from alfalfa. Other excellent sources are cabbage, cauliflower, kale, spinach, and other green vegetables. Good sources include tomatoes, cheese, dairy products, meat, egg yolk, and liver. The vitamin is also found in a number of bacteria and is synthesized by microorganisms in the intestinal tract and is thus available to the host. A rich source of vitamin K_2 is putrefied fish meal. The average mixed diet provides about 300 to 500 μg of vitamin K per day. About half is as phylloquinone and half is as menaquinones of intestinal origin, as a result of bacterial synthesis.

Absorption The natural vitamins K are fat soluble and require bile to be absorbed. Consequently, absorption occurs in the upper parts of the small intestine where bile salts are present. A vitamin K deficiency is likely to occur whenever bile is prevented from entering the intestinal tract. This is true of most of the fat-soluble vitamins, but it is particularly important in the case of vitamin K because of its bearing on blood clotting. Thus, when there is an obstruction of the bile ducts and jaundice ensues, clotting is delayed. This is not caused by the occurrence of bile in the blood, as was formerly believed, but by a deficiency of vitamin K. As in the case of fatty acids, the bile acids can be ascribed the specific function of absorption of fat-soluble vitamins. Consequently, whenever vitamin K is given orally, the presence of bile is essential, and if there is a deficient bile flow, a bile or bile salt preparation should be administered. The parenteral administration of one of the analogues, e.g., menadione, obviates this necessity. The water-soluble analogues may be given orally without the use of bile or bile salts. Excess vitamin K can be stored to a moderate degree, but in which tissues this occurs is not known.

Effects of deficiency Animals suffering from vitamin K deficiency have a remarkable tendency to bleed profusely from minor wounds, and slight bruises result in extensive subcutaneous hemorrhages. Blood withdrawn from such animal clots very slowly; in some cases it may remain fluid for hours. This is primarily a result of lack of prothrombin. In rats the female is much less susceptible to vitamin K deficiency than the male. Newly hatched chicks on a vitamin K–free diet show a gradual diminution in the concentration of prothrombin in the blood. Their intestinal flora produce some vitamin K_2 but not enough to prevent avitaminosis without additions from the diet. Administration of the vitamin brings the clotting time of their blood up to normal levels within a few hours. The vitamin does not form part of the prothrombin molecule but affects the production of prothrombin by the liver (Chapter 16). Hepatectomized animals show a rapid decline in the blood-clotting power from lowered prothrombin, and administration of vitamin K does not raise it. Vitamin K treatment is ineffective if the liver is so badly damaged that it cannot produce prothrombin or if the intestine is incapable of absorbing the orally administered vitamin.

In humans the same effects are attributable to lack of vitamin K. In normal newborn infants the prothrombin level is low. It continues to fall, reaching the minimum on the third day of life. This is undoubtedly the result of a lack of vitamin K. Apparently the vitamin passes from the mother to the fetus with great difficulty, especially toward the end of pregnancy, and since the intestine of the newborn infant is sterile, there is no opportunity for synthesis by bacteria for a while. However, the prothrombin level may reach normal by the end of the first week, probably as a result of bacterial synthesis of vitamin K, concomitant with the ingestion of milk and the establishment of the normal intestinal flora. Hemorrhagic disease of newborn infants is a frequent cause of infant mortality, and the high incidence of hemorrhage in the newborn infant is thus easily accounted for. Often an intracranial hemorrhage may result in brain injury and, if the infant survives, imbecility or some other mental or neurological condition develops. Prophylactic treatment is recommended by many clinicians. The expectant mother is given vitamin K supplements for several days before delivery is expected. If this has not been done, the infant is given such treatment soon after birth.

In adults *available* vitamin K is seldom lacking. It is either present in the food in sufficient quantity or is produced by bacterial activity in the intestinal tract. A deficiency in the system may usually be referred to one of three fundamental causes: (1) There may be faulty absorption of the vitamin from a lack of bile in the intestine because of an insufficient secretion of bile salts, obstruction of the bile duct, intestinal lesions or obstruction, or surgical procedures in the intestine. (2) This and other fat-soluble vitamins may be swept into the feces, particularly if the intestinal contents are unusually fatty. This has been experienced in diarrheal diseases, e.g., ulcerative colitis, sprue, celiac disease, or after excessive use of mineral oil. (3) The administration of sulfaguanidine, succinylsulfathiazole, or other intestinal antiseptics may cause a deficiency by limiting the production of vitamin K_2 by the intestinal flora. The surgeon about to operate on a patient known to have any of the conditions mentioned gives vitamin K before the operation to avoid excessive hemorrhage during or after the operation. Bile or bile salts may be irritating and produce vomiting; therefore, instead of the natural vitamin, the water-soluble substitutes may be given by mouth, or these or some of the others may be given parenterally. Remember that vitamin K is not always indicated when there is prothrombin deficiency, which may also result from various liver diseases that render the liver incapable of producing prothrombin. Intake of vitamin K cannot restore this function.

Physiological function The role of vitamin K in the synthesis of prothrombin in the liver is well established. Studies employing the fluorescent antibody technique demonstrated that this effect occurs in the liver parenchymal cell and may be observed within a few hours after vitamin K is administered intravenously.

Earlier studies indicated that vitamin K has a genetic action in inducing RNA formation for the synthesis of blood-clotting proteins. The evidence was that actinomycin D inhibits the vitamin K–induced prothrombin formation in chicks deficient in vitamin K. The inhibition involves hepatic RNA formation from ATP, as detected by the use of adenine-8-^{14}C. More recent work, however, indicated that the site of function of vitamin K was not at the transcription level but rather was at a late stage in the translation of prothrombin mRNA to form a functional prothrombin molecule. Extensive subsequent studies indeed have demonstrated that vitamin K is actually involved in a posttranslational modification of prothrombin and at least three other zymogens involved in the coagulation of blood (Chapter 16). More specifically, vitamin K appears to be needed for the carboxylation of the γ-carbon atom of glutamic acid residues in prothrombin to form calcium-binding sites.

$$\cdots C(=O)-CH(CH_2CH_2CO_2^-)-NH\cdots \xrightarrow[CO_2]{\text{Vitamin K}} \cdots C(=O)-CH(CH_2CH(CO_2^-)_2)-NH\cdots$$

Recent investigations indicate that vitamin K also may be involved in the carboxylation of glutamic acid residues in a calcium-binding protein, *osteocalcin,* in bone. Osteocalcin's first appearance in developing chick embryos coincides with

Table 21-12
Estimated safe and adequate daily dietary intakes of additional selected vitamins (revised 1980 by the Food and Nutrition Board, National Academy of Sciences–National Research Council)*

	Age (years)	Vitamins		
		Vitamin K (μg)	Biotin (μg)	Pantothenic acid (mg)
Infants	0-0.5	12	35	2
	0.5-1	10-20	50	3
Children and adolescents	1-3	15-30	65	3
	4-6	20-40	85	3-4
	7-10	30-60	120	4-5
	11+	50-100	100-200	4-7
Adults		70-140	100-200	4-7

From Recommended dietary allowances, ed. 9, Washington, D.C., 1980, National Academy of Sciences.
*Because there is less information on which to base allowances, these figures are not given in the main table of the RDA and are provided here in the form of ranges of recommended intakes.

the beginning of bone mineralization. Vitamin K antagonists markedly reduced the osteocalcin level in chick embryo bones.

Human requirements It is difficult to produce a dietary deficiency of vitamin K because of intestinal bacterial synthesis. However, appropriate patients given antibiotic therapy and intravenous nutrition with vitamin K–deficient fluids showed a decrease of vitamin K–dependent clotting factors to below normal in 4 weeks. The intravenous administration of vitamin K in dosages of 0.03 to 15 μg per kilogram body weight caused a proportional rise in these values to normal. Since the efficiency of absorption of vitamin K varies from 10% to 70%, a total requirement for vitamin K, from both dietary and intestinal bacterial synthesis, has been estimated from these and other data to be about 2 *μg per kilogram body weight.*

Based on these values the Food and Nutrition Board (1980), instead of establishing specific RDA values, has recommended *estimated* safe and adequate daily dietary intake ranges for vitamin K (expressed as μg vitamin K) (Table 21-12).

The lower levels of the indicated ranges are based on the assumption that about half the requirement of 2 μg per kilogram body weight is provided by bacterial intestinal synthesis and half by diet. The upper range levels are calculated by assuming that the entire 2 μg per kilogram body weight requirement is supplied by diet. The suggested intake for adults, 70 to 140 μg per day, is *easily* supplied by diet.

Antagonists The effect of heparin and bishydroxycoumarin in inhibiting blood clotting was discussed in Chapter 16. These anticoagulants act as antagonists to vitamin K, since their action is to diminish the amount of available prothrombin. The salicylates also are antagonistic to vitamin K. Consequently, when any of them are administered over a long period of time, supplements of the vitamin may be required to enable the liver to restore the prothrombin level to normal.

Toxicity Excessive doses of menadione and its derivatives produce a hemolytic anemia in rats and kernicterus (Chapter 17) in infants with low birth weights. This toxic effect appears to result from an increased breakdown of erythrocytes. Vitamin K_1 seems to be free from these side effects, however.

Summary

Table 21-13 summarizes the best food sources, the 1980 recommended dietary allowances, the active cofactor forms, the principal metabolic functions, and the major clinical manifestations of deficiencies of the fat-soluble vitamins just discussed.

Table 21-13
Summary of fat-soluble vitamins

Vitamin	Best food sources	Recommended dietary allowances (1980)*	Active cofactor form	Principal metabolic functions	Major manifestations of deficiency
Retinol (A) Retinal (aldehyde form) (Precursors, carotenes)	Butter, whole milk, fortified margarine, egg yolk, green and yellow vegetables, yellow fruits	1000 μg RE	Uncertain for epithelial tissues Rhodopsin and other visual pigments	Maintenance of epithelial tissues; biosynthesis of certain glycoproteins Constituent of visual pigments; toxic in large amounts	Nyctalopia, xerophthalmia, hyperkeratosis; keratinization of epithelium and other epithelium tissues; faulty tooth formation
Cholecalciferol (D_3) Ergocalciferol (D_2)	Fish liver oils, fortified or irradiated milk	7.5 μg	1,25-Dihydroxycholecalciferol	Induces synthesis of transport protein for Ca in intestinal mucosal cells and (bone?); intestinal and renal absorption of phosphate; toxic in large amounts	Rickets (children) Osteomalacia (adults)
D-α-Tocopherol (E)	Vegetable oils	10 mg (α-TE)	Not identified	Protects cell membranes and various biomolecules against lipid peroxidation and destruction	Hemolytic anemia; degenerative changes in muscle; sterility in rats
Phylloquinone (K_1) Menaquinones (K_2)	Green leafy vegetables, liver, egg yolk, meat, dairy products	70-140 μg (estimated range)	Not identified	Required for carboxylation of glutamic acid residues and formation of calcium-binding sites on prothrombin and other biomolecules	Hemorrhagic disease in newborn and in biliary disease; anemia

*For American men, 19 to 22 years of age, of average activity.

Water-soluble vitamins

The individual water-soluble vitamins bear no closer resemblance to each other chemically than do the fat-soluble vitamins. However, they are conveniently grouped together merely because of their solubility in water. This characteristic was the first basis for their classification. It is also important from the nutritional and clinical aspects, i.e., vitamin deficiencies are likely to be multiple and often may be mixed fat-soluble or mixed water-soluble vitamin deficiencies. The water-soluble vitamins include ascorbic acid, thiamin, riboflavin, niacin, pyridoxine, pantothenic acid, biotin, folic acid, and cobalamin. *p*-Aminobenzoic acid, inositol, α-lipoic acid, and choline are frequently included in this list, but many nutritionists believe that these are not true vitamins, although deficiencies of them in the diet of experimental animals cause characteristic symptoms to develop. None of them, with the exception of α-lipoic acid, appears to be a coenzyme or a part of one.

Ascorbic acid (vitamin C)

Scurvy was probably the first disease to be definitely associated with a food deficiency. It was common in Europe in the fifteenth century and must have been known long before that. It frequently occurred among sailors on long voyages when fresh food was not available. The essential role of vitamin C (ascorbic acid) in the prevention and cure of scurvy both in humans and in certain experimental animals is now well established.

Properties Ascorbic acid is water soluble and insoluble in fats and oils. It is very sensitive to oxidation, particularly in the presence of copper but not of aluminum. Therefore foods prepared in copper vessels lose ascorbic acid quickly. This factor is also rapidly destroyed by alkalies but is fairly stable in weakly acid solutions. Consequently, baking soda has a harmful effect, but cooking in steam has little destructive action on the ascorbic acid of foods, if they are neutral or slightly acid. Drying vegetables usually results in a loss of ascorbic acid, but many attempts have been made to provide desiccated foods containing all the vitamins, including ascorbic acid, unchanged. Freezing has no deleterious effect on the vitamin. Because it is so easily oxidized, ascorbic acid is a strong reducing agent.

Occurrence From a nutritional standpoint the citrus fruits and tomatoes are the best sources of ascorbic acid. Other natural sources may be richer in it, but they are either inedible or are not consumed in considerable amounts. For example, both green peppers and parsley are richer than oranges in this vitamin, but they do not enter into the diet to any great extent. Spinach and other greens are good sources of it also, but they lose their vitamin C content progressively on storage at room temperature. Citrus fruit juices and tomato juice may be canned with slight loss of the antiscorbutic factor. However, they should not be permitted to be in contact with air for a long period of time because of loss by oxidation. Cantaloupes, strawberries, cabbage, and turnips, when raw, are all about equivalent to tomatoes, but the two latter lose some vitamin C in cooking. Potatoes, fresh peas, asparagus, and lettuce are good sources also.

Plant polyphenols apparently play an important part as antioxidants in protecting ascorbic acid from oxidative destruction. A number of polyphenols, especially rutin, quercetin, and related flavonols, have this property. The effect is believed to be indirect because of the chelation of heavy metal ions (Cu^{++}, etc.) that catalyze the oxidative degradation of ascorbic acid. The vitamin C–like action of the so-called bioflavonoids is attributed to this protective action. Bioflavonoids thus decrease oxidative losses of ascorbic acid from foods during storage or in the intestinal tract, especially in individuals in achlorhydria or hypochlorhydria.

Ascorbic acid occurs to some extent in animal tissues. In 1928 Szent-Györgyi found a *hexuronic acid* with high reducing power in the adrenal cortex and later showed that it had antiscorbutic properties. However, the adrenal gland is of no importance as a food source of the vitamin because of the almost insignificant quantity of tissue involved. The same is true of the corpus luteum, which is said to have a high content of ascorbic acid. Most fresh animal tissues have small amounts of vitamin C. Liver is the best animal source, although fish roe and milk are also rather rich in it.

Bovine milk contains small amounts of ascorbic acid, which vary with the cow's fodder. In summer, when the cows are in the pasture, their milk is relatively high in ascorbic acid, but in winter or whenever fresh food is unavailable to the cow the milk has little antiscorbutic value. Human milk has a somewhat higher vitamin C content, but this too is dependent on the quality of the food. Pasteurizing milk is somewhat destructive of vitamin C. Accordingly, babies should have supplements of orange juice or tomato juice at least until they receive a varied diet.

Effects of deficiency The guinea pig is the standard animal for demonstrating vitamin C deficiency, and it has been used in the biological assay of foods for this vitamin. At first there is good growth on the vitamin C–free diet, but in about 2 weeks growth ceases and symptoms begin to appear. The joints become swollen and tender, and the animals show signs of pain when these are present. The animals may lie on their sides or assume a peculiar "scurvy" position, lying flat with hind legs sprawled. They may be excitable at first but soon become very quiet and not easily disturbed. There may also be enlargements of the junctions of the ribs with cartilage, as well as other bone lesions. Hemorrhage of the gums and loosening or breaking of the teeth may occur. Small amounts of orange juice change the picture even at a late stage, and animals may be brought back to an almost normal condition quite rapidly. Pertinent in this connection is the fact that recent work has shown that injected ascorbic acid in the guinea pig tends to concentrate in skin, muscle, and bone, and after injury in the scar tissue of the wound and callus of the fracture.

It has been believed that rats and mice require no ascorbic acid in their diets. However, these animals cannot entirely dispense with the vitamin, and their ability to get along with exceedingly small amounts is attributable to the fact that they synthesize it in their tissues.

In scurvy there occur anemia, pains in the joints, and hemorrhages from the mucous membranes of the mouth and gastrointestinal tract, skin (Fig. 21-17),

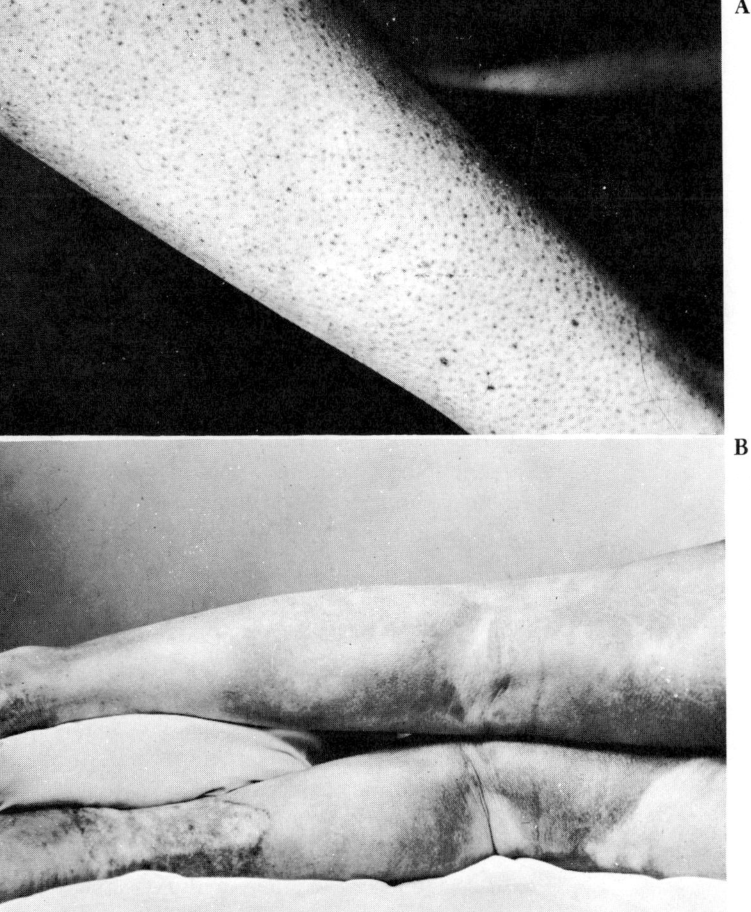

Figure 21-17
A Perifollicular hemorrhages of early scurvy.
B Ecchymosis of scurvy.
From Merck Report, Merck & Co., Inc., Rahway, N.J.

muscles, and subperiosteal tissues. The gums are particularly affected, showing swelling, tenderness, gingivitis, redness, ulceration, and even gangrene. Weakness and emaciation are seen in later stages. There are definite defects in skeletal calcification without much disturbance in mineral metabolism. For example, x-ray examination of the bones in scurvy shows a white line on the outside of the shaft, a line not seen in normal bone. The pathological change leading to all these symptoms is a weakening in the endothelial wall of the capillaries *because of a reduction in the amount of intercellular substance*. The body normally produces intercellular material, absorbs it, and replenishes it continually. This new formation of cementing and supporting material does not occur in the absence of vitamin C. The deficiency in supporting material may extend to the cartilage, bone, muscles, and other tissues and is responsible for the symptoms mentioned. We thus see that ascorbic acid is essential for the production of intercellular material (Chapter 15) and that it is necessary for the healing of wounds and fractures of bones.

In humans an extreme deficiency of vitamin C such as occurs in scurvy is rarely seen in the United States. There do occur, however, deficiencies of various grades, caused either by a subnormal intake of the vitamin or by an increased requirement. These deficiencies may result in slow healing of wounds and decreased ability to combat infections and to metabolize amino acids, especially tyrosine, as well as in the scorbutic symptoms already mentioned.

Investigations have shown that vitamin C is a threshold substance, i.e., it is not excreted by the kidney until the ascorbic acid level in the blood exceeds a certain value, which in turn depends on the degree of saturation of the body tissues. Vitamin C is not stored in the way that vitamins A and D are. The degree of saturation of the tissues with ascorbic acid can be easily estimated by the aid of certain methods. One clinical test is the determination of the concentration of ascorbic acid in blood plasma. The normal range is 0.6 to 2.5 mg/100 ml, but subnormal values have been found in some cases during pregnancy and lactation and in all cases of scurvy.

Some other conditions in which low values have been found are infectious disorders, congestive heart failure, kidney and liver diseases, gastrointestinal disturbances, purpura, endocrine cases, and malignancies. In none of these conditions is the lack of vitamin C a primary causative factor, but the fact that the vitamin C blood level is found to be reduced in many pathological states may be significant. This reduction is probably a result of an increased requirement for the vitamin or a lowered threshold for its excretion, but nevertheless it may contribute to the pathological condition of the patient. After burns, fractures, or extensive surgery, there is also a pronounced diminution of plasma ascorbic acid.

Another clinical test depends on the amount of vitamin C excreted in the urine after a test dose of ascorbic acid has been administered. If the tissues are well supplied with the vitamin, a larger amount is eliminated. If they are in need of it, more is retained. Children on normal diets, but without citrus fruits and tomatoes, quickly develop a lack of vitamin C, as shown by blood analysis, even though no gross clinical symptoms may be present.

Structure Ascorbic acid is a hexose derivative. Its formula, together with the formulas of two closely related compounds, follows. In fact, glucose, labeled at any one or all of its carbons, is converted in the rat to the correspondingly labeled ascorbic acid (see top of next page).

As seen from these formulas, L-ascorbic acid and dehydro-L-ascorbic acid are lactone derivatives of diketogulonic acid. The acidity of ascorbic acid is derived from the dissociation of the enolic hydroxyl groups. When ascorbic acid is oxidized, it loses two hydrogen atoms and becomes the dehydro derivative, which may be reduced to the original ascorbic acid form. The reduced form, L-ascorbic acid, pre-

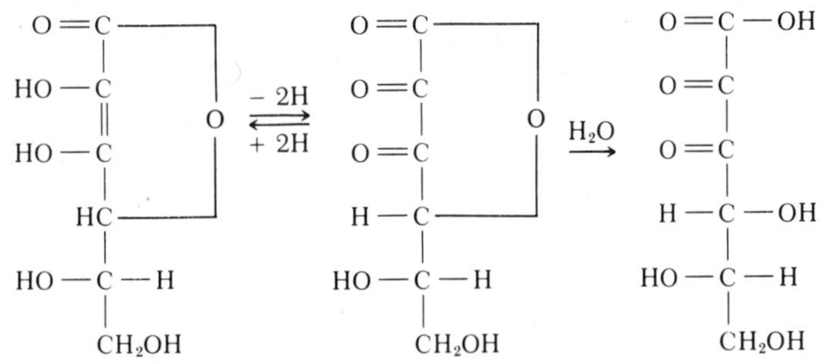

```
 O=C ─────┐                O=C ─────┐                 O=C─OH
   |      |                  |      |                   |
HO─C      |      − 2H       O=C      |                  O=C
   ‖      O     ⇄           |      O      H2O          |
HO─C      |      + 2H       O=C      |      ──→         O=C
   |      |                  |      |                   |
  HC ─────┘                H─C ─────┘                 H─C─OH
   |                         |                          |
HO─C─H                    HO─C─H                    HO─C─H
   |                         |                          |
  CH2OH                     CH2OH                      CH2OH

L-Ascorbic             Dehydro-L-ascorbic        2,3-Diketo-L-gulonic
  acid                        acid                      acid
```

dominates in the plasma and also apparently in tissues at a ratio of about 15(or more):1 of the oxidized form, dehydroascorbic acid. Both of these are biologically active, but the stereoisomer, D-ascorbic acid, is not. When dehydro-L-ascorbic acid is hydrated, it changes to 2,3-diketo-L-gulonic acid, which not only is inactive biologically but also cannot be converted back to either of the active forms in the body. since this hydration takes place spontaneously in neutral or alkaline solution, the oxidation of ascorbic acid frequently means its biological inactivation.

Since L-ascorbic acid is so easily oxidized to its dehydro derivative and vice versa, this is believed to occur in the tissues. Ascorbic acid may be a part of one of the important respiratory enzyme systems. It can be oxidized and reduced by glutathione, which may also be a part of the system. Thus:

$$\text{Ascorbic acid} \underset{\text{2 GSH}}{\overset{\text{GS-SG}}{\rightleftarrows}} \text{Dehydroascorbic acid}\quad (\text{+ Enzyme})$$

The importance of ascorbic acid today is reflected by the fact that at least 12 million pounds of pure crystalline ascorbic acid are used annually in the United States. One manufacturing plant alone capable of producing 17 million pounds annually has been constructed in the United States. In addition to its use in foods and pharmaceutics, ascorbic acid is used in poultry feeds to increase eggshell strength, in plastics to retard yellowing, as an initiation of polymerization in polyvinyl and related types of plastics, and as a spray for fruit trees to facilitate mechanical harvesting.

Mechanism of action From the various effects of vitamin C deficiency seen in humans and experimental animals, evidently a major function of ascorbic acid is the formation of tissue collagen, or *intercellular cement substance*. Ascorbic acid appears to be essential to the activity of the enzymes proline hydroxylase and lysyl oxidase, which catalyze the conversion of proline to hydroxyproline and lysine to hydroxylysine. However, a precise biochemical reaction has not been established as yet. These hydroxylated amino acids are vital in maintaining the tertiary structure of collagen (Chapter 15). The tissue levels of collagen proline hydroxylase have been reported to be lower in scorbutic guinea pigs than in normal animals. Higher concentrations of the enzyme and of ascorbic acid are found in injured tissues in which wound healing and scar tissue or callus formation are occurring.

Ascorbic acid may be involved in biological oxidations in some manner not completely understood at the present time. Thus the oxidation of a metabolite through 1 mole of ascorbic acid would yield 1 mole of ATP by oxidative phosphorylation (Chapter 7). An involvement of ascorbic acid in biological oxidations is a predictable function because of the readily reversible conversion of ascorbic to dehydroascorbic acid.

Ascorbic acid is required in the metabolism of tyrosine (Chapter 13). Alkaptonuria in guinea pigs results when they are fed a diet deficient in vitamin C but containing an excess of tyrosine. This may be due to the fact that ascorbic acid normally protects the enzyme that oxidizes *p*-hydroxyphenylpyruvic acid, a metabolic product of tyrosine. The subsequent administration of ascorbic acid reduces and even abolishes the output of *p*-hydroxyphenylpyruvic acid. Although ascorbic acid does not have the same effect in alkaptonuria in humans, premature infants excrete *p*-hydroxyphenylpyruvic and *p*-hydroxyphenyllactic acids in the urine if ascorbic acid is not present in sufficient amounts in the diet.

Another function of ascorbic acid is its role in the conversion of folic acid to a physiologically active form, tetrahydrofolic acid. Also, there is a need for vitamin C in the hydroxylation of cholesterol to cholic acid (Chapter 12).

Ascorbic acid also appears to act as a regulator of cholesterol metabolism. The rate of conversion of labeled acetate to cholesterol in the adrenal glands was found to be heightened in scorbutic animals as compared with normal ones.

In view of the fact that anemia occurs in scurvy, ascorbic acid not surprisingly enhances the absorption, and possibly the utilization, of iron in the nutritional anemia of infants and children.

In spite of the foregoing considerable amount of information on the biological effects of ascorbic acid, *no precise biochemical reaction* for this vitamin has been established thus far.

Human requirements As a basis for determining the human requirement and from it the RDA for vitamin C, the Food and Nutrition Board has used the extensive data in the literature on the amount necessary to prevent or cure scurvy, the amount metabolized by the body daily (particularly as determined by studies using radioactive ascorbic acid), and the amount necessary to maintain adequate reserves, as measured by plasma levels, leukocyte saturation, and tissue concentration. Ascorbic acid is absorbed at an efficiency of about 80% to 90%.

The basic requirement was found to vary, being increased by acute environmental and emotional stress. Other factors such as age, sex, individual variation, smoking, drugs, and oral contraceptive agents appear to have some effect, although some still are equivocal.

With this above information available, the Food and Nutrition Board (1980) has recommended a dietary allowance of *60 mg* of vitamin C per day for adults of both sexes. The RDA for infants is 35 mg, for children 45 mg, and for adolescents 50 mg (Table 21-4). An increase of 20 mg daily during pregnancy and 40 mg daily during lactation also is recommended.

The 1980 RDA for vitamin C is somewhat increased over that recommended in the 1974 revision. However, the increased recommended allowance is easily obtained from foods available in the United States, as will be discussed later in the chapter. The increased allowance for ascorbic acid is adequate for meeting normal individual needs and for providing a margin of safety. It also may enhance the nutritional status for iron in some groups.

Conditions allegedly requiring larger quantities of ascorbic acid are thyrotoxicosis, achlorhydria, diarrhea, rheumatic fever, rheumatoid arthritis, infections, and after physical trauma and surgery. In such instances 200 mg per day or more may be administered. Even larger amounts have been suggested for use in the prevention of the common cold. Evaluation of such claims, however, by carefully controlled double-blind studies have demonstrated thus far that any possible benefits derived from such large doses of ascorbic acid are too insignificant to justify any routine use. Some caution appears advisable, since one of the possible pathways of ascorbic acid metabolism involves the formation of oxalate.

• • •

B vitamins

When the vitamins were first designated *fat-soluble A* and *water-soluble B,* only one active principle was believed to be present in each. Water-soluble B had growth-promoting properties for the rat and cured the polyneuritis that had been produced in pigeons by feeding them polished rice. In humans the disease beriberi was found to result from a deficiency of vitamin B and to yield to treatment with it. Later pellagra was shown to be attributable to a deficiency of some other factor present in vitamin B preparations. The designation *P-P* (pellagra-preventive) was at first assigned to it. Additional factors were separated, at first on the basis of varying biological reactions, using different sources and different species, and later on the basis of adsorbability by Fuller's earth and other physical or chemical properties.

Two systems of nomenclature were used: one giving each factor a different letter (e.g., vitamin B, vitamin G) and the other naming them vitamin B_1, B_2, B_3, etc. Although the latter system is used more than the former, there is a growing tendency to abandon both in favor of chemical descriptive names of the compounds as soon as they are isolated, in conformity with the official recommendations of the International Union of Pure and Applied Chemistry, the American Society of Biological Chemists, and the American Institute of Nutrition. Thus the group now includes thiamin (B_1), riboflavin (B_2), pantothenic acid (B_3), niacin (B_5), pyridoxine (B_6), biotin (B_7), folic acid (folacin) (B_9), lipoic acid, and cobalamin (B_{12}).

Thiamin (vitamin B_1)

Vitamin B_1, or thiamin, has been called the antineuritic or antiberiberi factor; in Europe it is also designated *aneurin.*

Effects of deficiency A severe deficiency of thiamin in the diet results in the following disorders.

Arrested growth of young animals There is a specific effect on growth not caused by the inhibitory influence on appetite, which this vitamin also exerts.

Polyneuritis in animals Birds develop acute polyneuritis after several weeks and are unable to fly, walk, or even stand. Death occurs unless the vitamin is given. Rats develop, among other symptoms, a bradycardia (slowness of the heartbeat). The curing of both the polyneuritis of pigeons and the bradycardia of rats has been used in methods of biological assay.

There is malfunction of the blood-brain barrier in experimental thiamin deficiency in rats. The evidence is from the labeling patterns of glutamic acid from brain tissue of rats injected with sodium pyruvate-2-^{14}C, which normally does not pass the blood-brain barrier directly but only after conversion to glucose by gluconeogenesis. The results indicate that the alterations in selectivity of the blood-brain barrier permit pyruvic acid to enter the brain directly. Altered metabolic pathways in the brain therefore must occur; these in turn may be responsible for the polyneuritis that develops.

Beriberi in humans This disease is common in the Orient but also occurs in other parts of the world. In the adult it is characterized by polyneuritis, with muscular atrophy, cardiovascular changes, and edema (Fig. 21-18). At first there is weakness and fatigue, followed by headache, insomnia, dizziness, loss of appetite and other gastrointestinal symptoms, and tachycardia. Later the major symptoms may follow chiefly one of the following patterns: (1) nervous symptoms *(dry beriberi);* (2) symptoms associated with edema and serous effusions *(wet beriberi);* (3) symptoms of cardiac involvement *(acute pernicious beriberi).* Often the symptoms are characteristic of more than one of these three classes and are called *mixed beriberi.* Although beriberi is a thiamin deficiency disease, it is almost always accompanied by deficits of other vitamins. This is true of all vitamin B complex–deficiency conditions in humans and perhaps of other vitamin deficiencies as well.

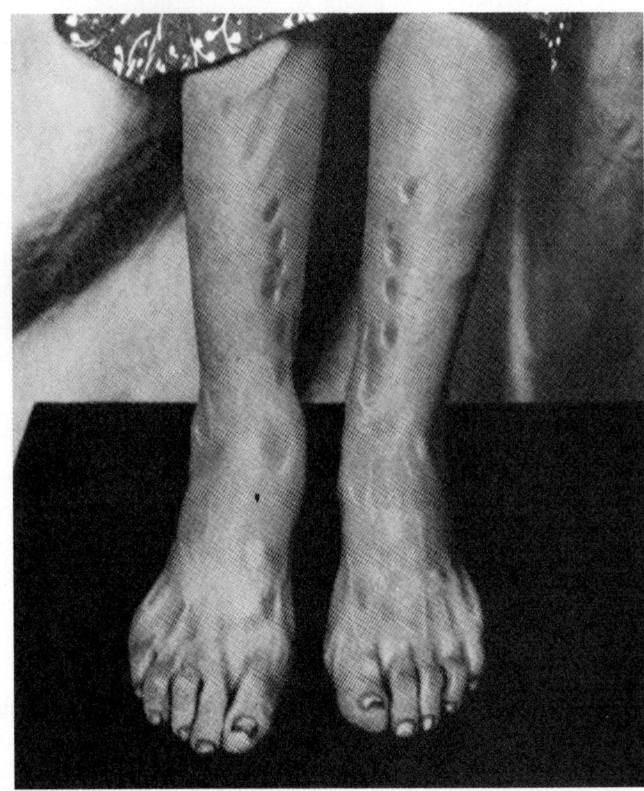

Figure 21-18
Pitting edema of leg in thiamin deficiency.
Courtesy T.D. Spies, Birmingham, Ala.

Beriberi in infants results when their diet is restricted to the milk of mothers suffering from beriberi and undoubtedly is caused by a lack of thiamin in the milk. It occurs suddenly. The symptoms include rigidity of the body, constipation, diminished flow of urine (oliguria), a peculiar whining, weakness, edema, enlargement of the heart, cyanosis, and a rapid, irregular pulse.

Properties Thiamin is a white crystalline compound, readily soluble in water, slightly soluble in ethyl alcohol, but insoluble in ether and chloroform. It has the odor and flavor characteristic of yeast. The aqueous solution has an acid reaction and is optically inactive. In dry heat it is relatively stable up to 100° C but is slowly destroyed by moist heat. Acid retards and alkali hastens this destructive action. In cooking, thiamin is not destroyed to any great extent if the temperature is not much above 100° C, provided that the reaction is not alkaline and heating is not continued for too long a time.

Occurrence Vitamin B_1 is present in many plant and animal foods. Whole grains, legumes, beef, pork, liver, nuts, and yeast are the best sources; fair sources are eggs, fish, and many vegetables. Although the vitamin is widely distributed, many foods have such small amounts present that partial thiamin deficiencies can easily occur. The milling of wheat flour has lowered the thiamin content more than 80%; as a consequence, enrichment of white flour or of white bread with thiamin is widely practiced. Furthermore, because of its solubility in water, much vitamin B_1 may be lost if the water in which foods are cooked is discarded. The desirability of using this water for soups, gravies, and sauces is evident.

Foods may be classed, in general, into three groups as to their thiamin content:

1. Foods highest in thiamin: whole cereals, lean pork, heart, kidney (although none of them is rich to the same degree as are some foods in vitamins A, C, and D)
2. Foods high in thiamin consumed in relatively small amounts, e.g., yeast, and foods low in thiamin but consumed in relatively large amounts: meats other than those mentioned in group 1, milk, fresh fruits, vegetables; the liver and roe of fish are reasonably good sources of thiamin, as well as of the other vitamins of the B complex
3. Foods deficient in thiamin: white flour (not enriched), polished rice, white breakfast cereals, spaghetti, macaroni, refined cane sugar, molasses

Structure The structure of thiamin has been determined to be as follows:

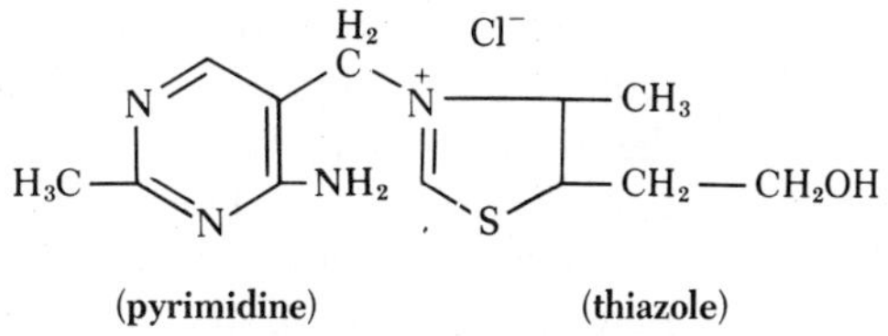

H_2 Cl^- C $\overset{+}{N}$ CH_3 N H_3C NH_2 $CH_2—CH_2OH$ N S

(pyrimidine) (thiazole)

Thiamin (hydrochloride)

It has been synthesized, and the synthetic vitamin is frequently used in medicine. It occurs in nature either as the free vitamin or as the pyrophosphate. Note that the structure of thiamin contains a pyrimidine and a thiazole ring. No other natural compound contains the thiazole group, with the exception of penicillin, which has a hydrothiazole nucleus. The pyrimidine is also unique in that it is the only natural pyrimidine having an alkyl group in position 2. Plants can use a mixture of pyrimidine and thiazole compounds in place of thiamin itself, whereas animals require the complete vitamin. An apparent exception is polyneuritic pigeons, which may be cured by being fed large doses of a mixture of these intermediates.

The three-dimensional conformation of thiamin pyrophosphate is being studied by x-ray analysis. Knowledge of its conformation should provide new insight into the biochemical mechanisms involved in reactions catalyzed by this vitamin.

Mechanism of action Thiamin is involved in the intermediary metabolism of carbohydrates in all the cells of the body. Several oxidative phenomena have been shown to depend on it, and probably they are related to each other. Brain tissue from pigeons having severe thiamin deficiency takes up oxygen at a lower rate than normally. This rate can be increased by adding thiamin to the tissue. Such tissue is also found to have an excess of lactate and pyruvate, and the suggestion has been made that neuritic symptoms may be a result of excess of pyruvate. The vitamin, linked to two molecules of phosphoric acid, becomes thiamin pyrophosphate (TPP) or diphosphothiamin (DPT). This is a coenzyme, *cocarboxylase*. Each coenzyme requires a protein apoenzyme. Carboxylases are present in yeast and other microorganisms, and oxidative carboxylases, e.g., pyruvic dehydrogenase, are found in animal tissues. Thiamin pyrophosphate takes part in the decarboxylation of α-keto acids, notably pyruvic and α-ketoglutaric (Chapter 6). Thiamin is also involved in transketolation (Chapter 11). Usually the content of cocarboxylase in blood parallels that of thiamin. However, in diabetes mellitus there is a high thiamin content with a low cocarboxylase.

Thiamin as the pyrophosphate thus functions primarily as a coenzyme in the decarboxylation of α-keto acids, including pyruvic and α-ketoglutaric. (See Chapter 6 for the reactions involved.)

Human requirements Thiamin is not stored in the tissues to any great extent, and loss of this water-soluble substance continually occurs by way of the urine. Apparently the thiamin content of tissues varies somewhat with the amount in the food, and consequently after a period of overabundance a temporary reserve sufficient to accommodate the needs of the individual for a few weeks is built up. Ordinarily, however, any loss must be made good soon after it occurs. Various factors influence the requirement: (1) age—children require more per kilogram of body weight than do adults; (2) activity—thiamin needs vary with caloric requirements; (3) pregnancy and lactation—greater amounts must be provided for the fetus or the suckling infant. The recommended dietary allowances (1980) (Table 21-4) are 1.2 to 1.5 mg daily for adult men and 1.0 to 1.1 mg for women, increasing by 0.4 mg during pregnancy and 0.5 mg during lactation. This is about 0.5 mg for each 1000 kcal for adults and adolescents on ordinary levels of calorie intake, but when the calorie intake is lower, the thiamin provision should not fall below 1 mg daily. The thiamin requirement of the infant in relation to calories and to carbohydrate content of the diet is similar to that of the adult. In absolute figures it is between 0.3 and 0.5 mg daily. The RDA for children is 0.7 to 1.2 mg.

In view of its limited distribution in foods, thiamin is almost the only vitamin that may be lacking even in a fairly good diet. In patients on a restricted diet or with a diuresis leading to rapid loss, the deficiency may be appreciable. Administration of thiamin under these circumstances is widely practiced.

If raw clams, or certain other raw seafood, are included in the diet, a thiaminase present may destroy enough thiamin to produce a deficiency. The enzyme action is a cleavage between the pyrimidine and thiazole rings. Horses and cattle that consume large amounts of fern sometimes develop "fern poisoning." This has been shown to be another type of thiamin antagonism, not enzymic in nature, since heated ferns are just as toxic. Several possible explanations have been offered, among them the idea that an inhibitory structural analogue, e.g., pyrithiamin, may be present in this plant.

Other clinical applications Clinically the administration of thiamin has met with considerable success in a number of conditions besides frank beriberi. As might be expected, other types of neuritis have been treated with thiamin. Alcoholic neuritis seems to be caused by a lack of this vitamin, and definite improvement is usually seen on treatment with thiamin, preferably together with other constituents of the vitamin B complex. If the neuritis is not associated with a thiamin avitaminosis, the administration of thiamin does not help, nor does it help if permanent destruction of nervous tissue has taken place.

In various forms of nutritional deficiency there are symptoms of cardiovascular disturbance. Vitamin B_1 administration usually ameliorates these symptoms even though the patient may be suffering from a lack of several vitamins. Gastrointestinal disorders also have been ascribed to thiamin deficiency. Lack of appetite and loss of muscular tone of the stomach and intestine were first shown in animals. To correlate a vitamin deficiency with gastrointestinal symptoms in humans is more difficult because of the great number of other factors that influence the condition. However, several investigators believe that a certain group of symptoms are frequently caused by lack of thiamin. These include loss of appetite, low gastric hydrochloric acid, atony of the stomach and intestines, constipation, and a strong tendency toward the development of intestinal inflammatory processes.

Although thiamin is ordinarily nontoxic, it has occasionally produced anaphylactic shock symptoms after repeated intravenous injections.

Riboflavin (vitamin B_2)

No recognized disease is associated with an exclusive deficiency of riboflavin, but in pellagra, which is caused primarily by a lack of niacin, there is usually also a lack of riboflavin. In rats a riboflavin-free diet causes, besides a cessation of growth, vascularization of the cornea, frequently a loss of hair, and scaliness of the skin (with pediculosis); later cataracts may develop. Dogs also develop cataracts if deprived of riboflavin. There may be nervous manifestations in animals suffering from a lack of this vitamin. In the human being, whether cataract is a result of this deficiency is doubtful. In humans there is cheilosis, a condition characterized by inflammation of the lips, fissures at the corners of the mouth, scaliness, greasiness, and fissures in the folds of the ears and nose. Some initial trauma or infection is likely to be followed by a skin lesion if a riboflavin deficiency is present. There may be ocular disturbances like inflammation of the cornea, bloodshot eyes, photophobia, dimness of vision, and itching, burning, and dryness of the eyes with redness of the conjunctivas.

Properties Riboflavin is an orange-yellow crystalline compound. It is water soluble and heat stable, especially in acid solution, but is easily destroyed by exposure to light. Its water solution exhibits a yellow-green fluorescence. It is a pigment consisting of dimethylisoalloxazine attached to a D-ribityl group. In nature it may occur as the free pigment, as riboflavin phosphate, or as a constituent of flavoproteins. Its structural formula is as follows:

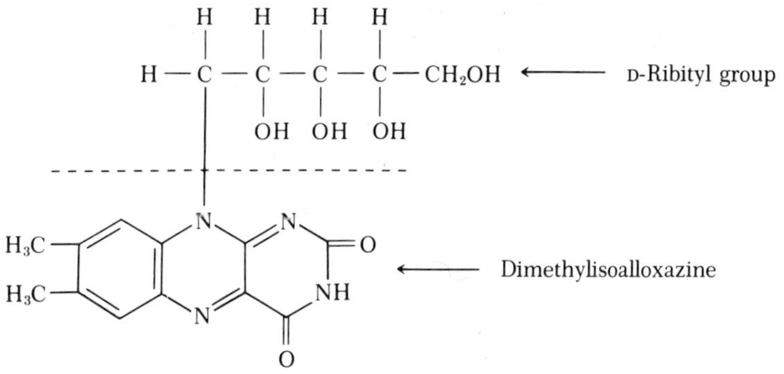

$H-CH-CH(OH)-CH(OH)-CH(OH)-CH_2OH$ ← D-Ribityl group

← Dimethylisoalloxazine

Riboflavin

Distribution Riboflavin occurs widely in nature. Milk is an important source of it. Lactoflavin, one of the pigments of milk, is identical with riboflavin. Other excellent sources are meats, especially liver and kidney, fish, and eggs. Leafy vegetables are richer in riboflavin than they are in thiamin. Fruits and most root vegetables contain moderate quantities. Whole grains, cereals, and milled flour are not good sources.

Mechanism of action The flavoproteins, i.e., combinations of riboflavin with proteins, are enzymes that function in biological oxidations as components of the electron transport system (Chapter 7) and in a number of enzymes, including L- and D-amino oxidases, xanthine oxidase, cytochrome-c reductase, and certain dehydrogenases. The flavin prosthetic group is usually FAD (flavin adenine dinucleotide) (Chapter 6), but in some instances it may be FMN (flavin mononucleotide). The flavoproteins thus form a varied and important group of intracellular enzymes involved in oxidation-reduction reactions. This fact emphasizes the necessity of adequate riboflavin supplies in the diet.

Human requirements The recommended dietary allowances of riboflavin per day vary from 0.4 to 1.7 mg for children and adults (Table 21-4). As is true of other vitamins, additional allowances are recommended during pregnancy and lactation,

0.3 and 0.5 mg per day, respectively. A well-diversified dietary should furnish these amounts, but, as has been pointed out, appetite alone may lead the individual to select food with a distinct deficiency in riboflavin.

Excretion Riboflavin is excreted predominately in the feces and to a lesser extent in the urine. During riboflavin avitaminosis this small urinary fraction diminishes greatly. Riboflavin is excreted mostly in the free form but in varying amounts as the phosphate ester. Another flavin, *uroflavin,* is also found in the urine. It is similar chemically to riboflavin but appears to be more soluble in water. It is probably derived from riboflavin.

Niacin (nicotinamide or niacinamide)

Pellagra is a disease that was long prevalent in southern Europe and in the southern United States. Most cases occurred in the low-income groups, where diet was restricted to a few cheap foodstuffs. Concurrent with monotony of diet was the crowded unsanitary housing of the poor. This coincidence of circumstances lead early investigators to two hypotheses concerning the disease: it was caused by (1) a nutritional defect or (2) an infection. The latter hypothesis had many supporters but was with great difficulty finally set aside in favor of the nutritional nature of the disease. Goldberger and his colleagues in the United States Public Health Service conducted a long series of investigations, the most interesting of which was the prison farm experiment. Twelve convicts were promised pardons if they would agree to subsist on a diet of cornmeal, cornstarch, sweet potatoes, rice, syrup, and pork fat for a year. This was typical of diets consumed by pellagrous families. Before the year was up more than half the subjects showed symptoms of pellagra, whereas no such symptoms appeared among the prisoners on the usual diet under the same sanitary conditions.

At first an amino acid deficiency was believed responsible. Later it was considered purely a vitamin deficiency, and finally, in 1926, Goldberger discovered that yeast, heated to destroy its thiamin, still had curative action on pellagra. This agent was provisionally called the *pellagra-preventive* (P-P) factor. Elvehjem and coworkers proved that *nicotinic acid and its amide* were capable of curing blacktongue, an analogous deficiency disease of dogs, and soon they were shown to be identical with the P-P factor.

Relation to tryptophan However, investigators were confronted with a number of facts about pellagra that could not be explained by a purely vitamin theory. One of these was that for the cure of pellagra not only nicotinamide was needed but also adequate amounts of good-quality protein foods, such as milk, which is low in nicotinamide. Another fact was that a diet composed largely of corn led to the development of pellagra, even though an *apparently* sufficient amount of nicotinamide was present. Animal experiments substantiated this and later led to the observation that addition of casein to the diet has an effect that counteracts the effect of the corn. The amino acid tryptophan was found to be the factor, lacking in corn proteins but present in casein, that simulates nicotinamide formation. The explanation of this nicotinamide-tryptophan relationship is as follows.

Tryptophan can be transformed into nicotinamide by the body tissues and thus contributes to the body's supply of the vitamin. This could occur in the following manner. Tryptophan is normally converted to kynurenine (Chapter 13); this may be converted in the liver and kidney to hydroxyanthranilic acid, which has been shown to substitute in animals for nicotinic acid. For this transformation the presence of pyridoxine seems to be necessary. From extensive studies in human subjects it has been found that approximately 60 mg of tryptophan are equivalent to 1 mg of nicotinamide.

In pellagra there occur patches of dermatitis, soreness and inflammation of the tongue and mouth, alimentary disorders (achlorhydria and diarrhea), and pigmentation and thickening of the skin (Fig. 21-19). There is usually a rash that appears symmetrically on the sides of the body and backs of the hands and arms. The pigmentation of the skin may persist for years after the dermatitis has healed. Nervous disorders and mental disturbances occur, particularly in the later stages. Some of the mental symptoms seen in chronic alcoholism have been ascribed to a nicotinamide avitaminosis, at least in part, since the diet of many alcoholics is deficient in several of the B vitamins. Although these are the typical pellagrous symptoms and the administration of nicotinamide relieves them, generally other symptoms accompany them; also, generally the administration of nicotinamide does not entirely cure the patient. The reason is that, together with the deficiency of nicotinic acid, there is usually a deficiency of riboflavin and thiamin and possibly of other B vitamins as well. Often other food factors are missing so that other vitamins may have to be administered. A well-rounded and complete diet must be insisted on to prevent the recurrence of the symptom complex.

Properties Although nicotinic acid was discovered as a vitamin comparatively recently, the compound was known long before the vitamin era. Since it could be produced by the oxidation of nicotine, it was known as nicotinic acid. This name was considered misleading by some authorities; so the terms niacin and niacinamide, respectively, were coined for the acid and its amide. However, the term *niacin* now is normally used in a *generic* sense for both nicotinic acid and nicotinamide, which are used specifically for the individual chemical compounds. When

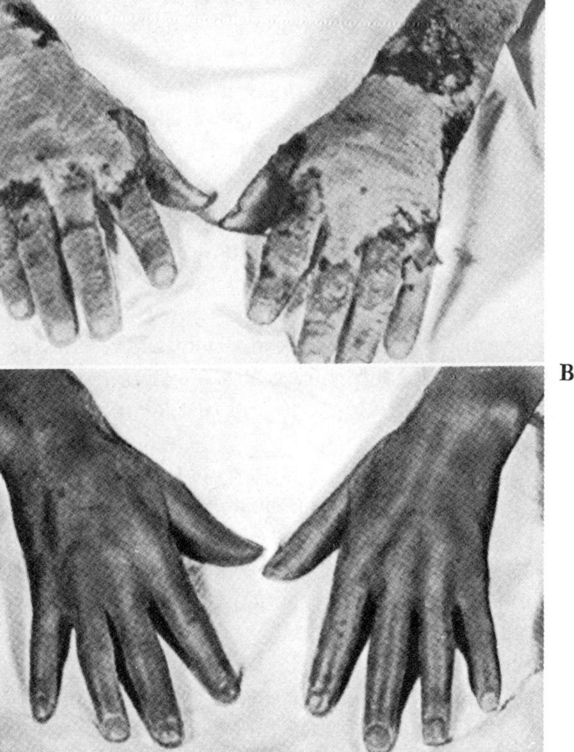

Figure 21-19
Results of severe niacin deficiency.
A Lesions of hands in pellagra.
B Same patient after treatment with nicotinamide.

Courtesy T.D. Spies, Birmingham, Ala.

pure, nicotinic acid occurs as white, needlelike crystals. It is water soluble and stable in air and also in heat. There is little loss in cooking unless the cooking water is discarded.

Structure The structural formulas of the acid and the acid amide are as follows:

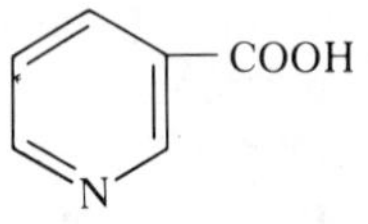

$$\text{(pyridine ring)}\text{—COOH} \qquad \text{(pyridine ring)}\text{—CONH}_2$$

Nicotinic acid 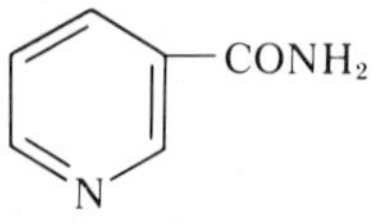   Nicotinamide

Occurrence Niacin is found in largest amounts in meats, especially liver. Fish and eggs are also good sources, as are some cereals and vegetables, notably whole wheat and unpolished rice, and peanuts. However, a number of our staple vegetable articles of diet are not particularly rich in niacin, and it is firmly bound to a protein. Therefore vegetarian diets may be lacking in this vitamin. Although whole wheat is an excellent source of niacin, most of this vitamin, like thiamin, is lost in the milling process. Thus niacin is now one of the substances used to enrich white flour.

Effects of deficiency in animals The dog, pig, and monkey are the only experimental animals that exhibit symptoms as a result of niacin deficiency. Chittenden and Underhill, in 1917, described a condition in dogs known as *blacktongue*. This is characterized by a sudden refusal to eat the deficient diet, apathy, and lesions in the mouth. The inner surfaces of the lips and cheeks become covered with pustules, and the mucous lining comes away in shreds. Intense salivation and bloody diarrhea are additional symptoms, and there may be pustules on the thorax and upper abdomen. In monkeys the chief symptoms are tender bleeding gums, leading to ulceration and necrosis of the gum tissues. Vincent's infection is likely to set in, and, if niacin is not provided, monkeys die usually in 2 to 6 weeks.

Mechanism of action Nicotinamide has been shown to be a part of two important coenzymes, NAD^+ and $NADP^+$, formerly called DPN and TPN and coenzymes I and II, respectively. The Enzyme Commission of the International Union of Biochemistry recommended that they be renamed NAD^+ and $NADP^+$, from their chemical constituents, nicotinamide adenine dinucleotide and its phosphate derivative, respectively. This corresponds with the terminology used for most other coenzymes, e.g., FAD, FMN. These are nucleotides; their structural formulas are shown in Chapter 6. They are members of enzyme systems involved in biological oxidations as components of the electron transport system. As has been seen (Chapter 7), nicotinamide and riboflavin are intimately concerned in these reactions.

The pathway for the synthesis of the active coenzyme forms of nicotinamide and the role of NAD^+ in biological oxidation are discussed in Chapter 6.

The triphospho derivative, $NADP^+$ (Chapter 6), plays an important role in a microsomal system for biological oxidations (Chapter 7) and, as $NADPH(H^+)$, also in many biosynthetic reactions, e.g., the biosyntheses of fatty acids and various steroids (Chapter 12).

Human requirements The RDA (1980) of niacin for infants and children is 6 to 16 mg per day; for adult males it is 16 to 19 mg, and for adult females it is 13 to 15 mg per day. Pregnancy and lactation increase the requirements (Table 21-4). Note that the 1980 RDA values for niacin are given as milligrams of *niacin equilivalents* (NE). One NE = 1 mg niacin or 60 mg tryptophan (see Table 21-4).

Large doses of nicotinic acid or its amide are not toxic. Up to 2 gm per kilogram of body weight have been given to human beings without any toxic effects. Nicotinic acid produces marked vasodilatation and a striking flushing reaction, however.

Pyridoxine

Pyridoxine (vitamin B_6) is another pyridine derivative that belongs to the heat-stable B complex. In pyridoxine deficiency in rats there is a swelling of the ears and dermatitis of the paws and of the nasal region, followed by incrustation. Other species that show symptoms when deprived of pyridoxine are the chick, dog, pig, and rhesus monkey. These symptoms include epileptiform seizures in rats, dogs, and pigs, along with a characteristic anemia in dogs. In the monkey an arteriosclerosis apparently develops, but whether this has any relation to the disease in human beings is uncertain.

Properties Pyridoxine is water and alcohol soluble and is slightly soluble in fat solvents. It is sensitive to light, ultraviolet irradiation, and alkali. Although pyridoxine itself is resistant to heat, its derivatives, pyridoxal and pyridoxamine, are destroyed rapidly at high temperatures.

Structure In addition to pyridoxine, two derivatives, pyridoxal and pyridoxamine, also have B_6 activity. Pyridoxal is the aldehyde and pyridoxamine the amine form. All three are found in foods in varying proportions. Pyridoxine can be converted to either of the other two, but neither of them can be changed to pyridoxine. All three are found in the urine after ingestion, but 4-pyridoxic acid is quantitatively the most important excretion product. The formulas and relationships of these substances follow:

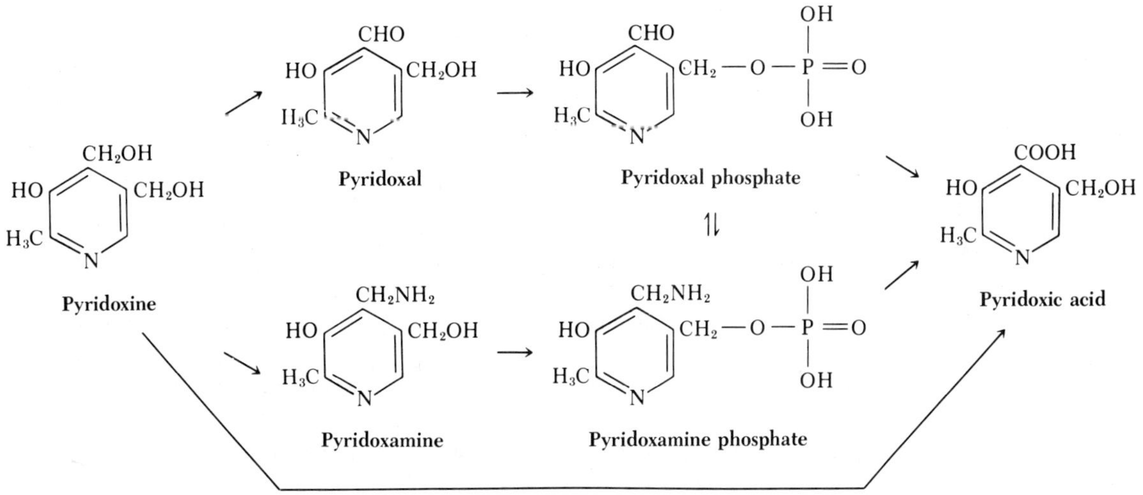

Pyridoxine → Pyridoxal → Pyridoxal phosphate → Pyridoxic acid

Pyridoxine → Pyridoxamine → Pyridoxamine phosphate

Pyridoxal phosphate ⇅ Pyridoxamine phosphate

Pyridoxine → Pyridoxic acid

Pyridoxine and derivatives

Functions A major function of pyridoxine is as a coenzyme for the transaminases (Chapters 6 and 13). The change from pyridoxal to pyridoxamine, and vice versa, is required in transamination reactions. Pyridoxal phosphate is the coenzyme for the decarboxylases that act on a number of amino acids and for two enzyme systems involved in the metabolism of sulfur-containing amino acids (Chapter 13). Vitamin B_6 is essential for the dehydration and desulfhydration of amino acids. It also appears to be essential for the normal metabolism of tryptophan. In its absence large amounts of xanthurenic acid, a product of the incomplete metabolism of tryptophan, are excreted in the urine. For the conversion of tryptophan to nicotinamide, pyridoxine is also needed. Another function of pyridoxine is that in its aldehyde form, pyridoxal, it increases the rate of transport of amino acids and of potassium into cells against a gradient. In rats pyridoxine and magnesium are necessary for the prevention of excessive oxalate formation, but whether this applies to humans has not been determined. Pyridoxine is also believed to be involved in the metabolism of the unsaturated fatty acids.

Occurrence Foods richest in pyridoxine are egg yolk, meat, fish, and milk, among animal sources, and whole grains, cabbage, and legumes.

Effects of human deficiency Although pyridoxine is undoubtedly required by humans, symptoms of pyridoxine deficiency in adult human beings are difficult to produce. Long periods of deprivation are required before any effects are noted. These include a fall in the hemoglobin level and alteration of the leukocyte relationships, depression, and mental confusion. By the use of a structural antagonist, however, skin lesions can be produced in a shorter time. These resemble the ones that occur in riboflavin and nicotinic acid deficiencies.

Since increased amounts of xanthurenic acid are excreted during and shortly after pregnancy, tryptophan metabolism is apparently altered at these times. Extra pyridoxine may be given to women during the second half of pregnancy to correct this abnormality. Clinical reports show that an apparent pyridoxine deficiency occurs in patients receiving the drug isonicotinic acid hydrazide (INH). This is an antitubercular remedy and sometimes produces a peripheral neuritis. This condition results from an impaired vitamin B_6 metabolism and can be prevented or cured by the simultaneous administration of pyridoxine.

Reports of a curious syndrome in infants appeared several years ago in various parts of the United States. There were general irritability, abdominal distention, vomiting and diarrhea, and convulsions. All the babies with these symptoms were being fed the same proprietary liquid food. A change to a different food produced a dramatic cure. Evidently the sterilization procedure had lowered the vitamin B_6 content to a dangerous level. Since pyridoxine added to milk prior to processing is resistant to heat destruction, probably either *pyridoxal* or *pyridoxamine*, or both, had been affected by the sterilization process because both are heat labile.

Human requirements In the cases just described, symptoms developed when the milk contained 60 μg or less per quart, and none developed when the food contained 100 μg per quart. Current pediatric procedure recommends the addition of 2 mg of pyridoxine daily to the diet of an infant. The recommended dietary allowance (Table 21-4) for adult men and women has been set by the Food and Nutrition Board (1980) as 2.2 and 2.0 mg per day, respectively. Since the requirement is closely related to protein metabolism, this amount has been established to allow for a daily intake of 100 gm or more of protein to provide a reasonable margin of safety. The recommended dietary allowance for infants and children is 0.3 to 1.6 mg daily, depending on age. During pregnancy and lactation 2.6 and 2.5 mg daily, respectively, is recommended.

Folacin (folic acid, pteroylglutamic acid)

The discovery of the folic acid group of vitamins is the result of many different investigations. The research proceeded along two chief lines, however. Certain substances were found to be essential to the growth of microorganisms, particularly *Lactobacillus casei* and *Streptococcus lactis* R. Other investigations dealt with factors found to be necessary in the nutrition of chicks, guinea pigs, monkeys, and other species of higher animals. Since these substances were not the same as the known vitamins, they were given new names as their functions became apparent. Not until later was it evident that the chick vitamin and the bacterial growth factor were probably a single substance.

The vitamin to which all of these are related is folic acid (or pteroylglutamic acid [PGA]). This is the *Lactobacillus casei* factor, isolated from liver, first shown to be a factor necessary for the growth of that organism. A number of other compounds, isolated from other sources and having similar or even different biological properties, have been found to be closely resembling substances. In other words, liver *L. casei* factor, fermentation *L. casei* factor, folic acid, vitamin B_c, vitamin M, factors R, S, and U, and yeast norite eluate factor are all vitamins of the same group. All are related to folic acid. The preferred *generic* name of this group of related vitamins now is *folacin*. The structure and synthesis of this compound were soon determined. Folic acid is composed of three main parts: (1) a two-ringed nitrogenous compound called a *pteridine,* which is a yellow pigment first isolated from butterflies' wings, (2) *p*-aminobenzoic acid, and (3) glutamic acid. Its structure follows:

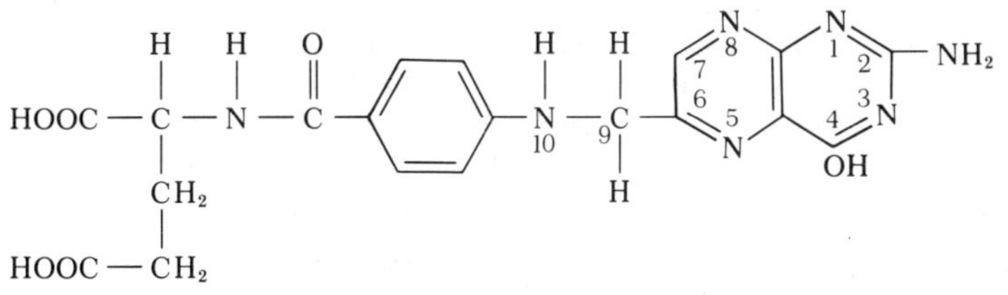

Folic acid
(pteroylglutamic acid)

The related vitamins differ with respect to the number of glutamic acid groups present, the additional glutamic acid molecules being conjugated in peptide linkages. The commonly occurring ones are the monoglutamate, the triglutamate (fermentation factor), and the heptaglutamate. The conjugates, i.e., those compounds having more than one glutamic acid in the molecule, are ineffective for some species that do not possess *conjugase,* the enzyme necessary to release the free vitamin. Normally a conjugase is present in the intestinal mucosal cells.

A derivative of folic acid of considerable interest is the so-called *citrovorum factor,* or *folinic acid*. This derivative is 5-formyl-5,6,7,8-tetrahydrofolic acid (5-formyl-FH_4), to be discussed later. It is a much more active stimulant of erythrocyte formation than is folic acid. It is required to support the growth of the microorganism *Leuconostoc citrovorum;* hence its name. Folacin is ineffective in this respect.

Mechanism of action The major biochemical function of folic acid, as its 5,6,7,8-tetrahydro derivative (FH_4), is as a coenzyme for the transfer of 1-carbon units (Chapter 6), such as formyl, methyl, methenyl, and methylene, for the biosynthesis of a variety of biomolecules. These include serine and methionine (Chapter 13), choline (Chapter 12), and thymine and certain purines and nucleotides (Chapter 14). Folic acid also is involved in tyrosine metabolism (Chapter 13) and in the formation of erythrocytes (Chapter 16).

The formation of the coenzyme form and primary function of folic acid thus may be summarized schematically:

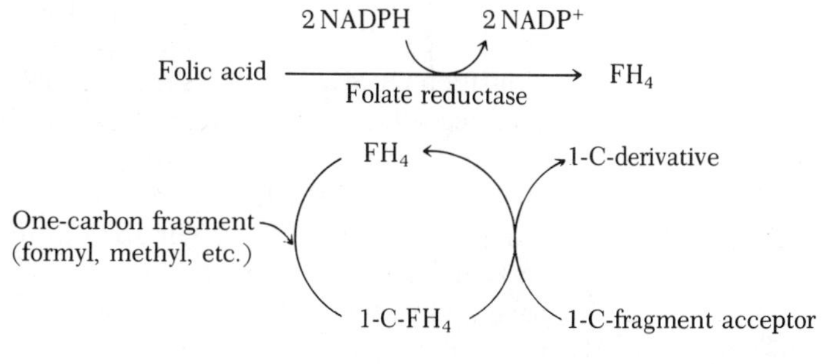

$$\text{Folic acid} \xrightarrow[\text{Folate reductase}]{2\,\text{NADPH} \rightarrow 2\,\text{NADP}^+} FH_4$$

FH_4 → 1-C-derivative
One-carbon fragment (formyl, methyl, etc.)
1-C-FH_4 ← 1-C-fragment acceptor

Properties Folic acid is a yellow substance, only slightly soluble in water; its sodium salt is quite soluble in water, but both the acid and the salt are insoluble in lipid solvents. It is stable to heat in neutral or alkaline solution but is not stable if heated in an acid medium. It is inactivated by sunlight. A considerable loss occurs in foods stored at room temperature.

Occurrence Folacin is widely distributed in nature, particularly in the foliage of plants; hence the name *folic*. Other good sources are yeast, cauliflower, liver, and kidney. Fair sources are beef, veal, and wheat, whereas root vegetables, tomatoes, bananas, rice, corn, sweet potatoes, pork, ham, and lamb contain little.

Most of the folic acid in foods is present as a polyglutamate, containing three to seven γ-linked glutamate residues. These are poorly absorbed. However, as mentioned, human mucosal cells of the duodenum and jejunum contain a *deconjugating* enzyme that splits off the extra glutamate residues. The monoglutamate is then readily absorbed from the normal intestine.

Effects of deficiency In chicks a lack of this factor causes anemia as well as decreased resistance to malarial infection and impairment of the response to estrogens. Rats show achromotrichia (absence of normal pigmentation of the hair) and staining of the fur and whiskers with porphyrin, whereas monkeys respond with a macrocytic anemia, leukopenia, diarrhea, edema, and lesions of the mouth. Similarly, in humans the chief symptoms of folic acid deficiency are megaloblastic anemia, glossitis, and gastrointestinal tract disturbances.

Clinical uses Folic acid is of great value in the treatment of certain macrocytic anemias, i.e., anemias that are characterized by the presence of large red corpuscles in the blood. In fact, folacin deficiency is the most common cause of megaloblastic anemia in the world. The deficiency may result from an inadequate intake, defective absorption, or abnormal metabolism. Among these conditions are sprue, the macrocytic anemias of pregnancy, infancy, and pellagra, and anemias occurring after gastric resection and other intestinal dysfunctions. In such circumstances folic acid has been found to be an effective hematopoietic factor. In sprue not only does it produce satisfactory effects on the blood picture, but it also relieves the gastrointestinal symptoms. It has been suggested as the effective agent in maintaining normal gastrointestinal absorption.

Folic acid also has a favorable effect on hematopoiesis in pernicious anemia. In fact, it has qualitatively the same effect on blood formation as does vitamin B_{12}. However, far less vitamin B_{12} is needed than folic acid, and the latter is unable to check the degenerative changes in the nervous system that take place in pernicious anemia. Because of the efficacy of folic acid in the treatment of macrocytic anemias, its careless use by patients with pernicious anemia, although improving the hematologic picture, may temporarily mask the neurological symptoms that may appear later in an advanced stage refractory to treatment even with vitamin B_{12}. For this reason the Food and Drug Administration restricts the inclusion of folic acid in one-a-day vitamin preparations.

Human requirements Data in the literature indicate that the average mixed dietary in the United States supplies about 690 μg of folacin daily. The bulk of evidence shows that approximately 20% to 50% of dietary folacin is absorbed and is nutritionally available. Present data also indicate that 100 to 200 μg of folacin daily are needed to maintain tissue reserves. Using this information and allowing a margin of safety for individual variation and other factors, the Food and Nutrition Board has established the RDA (1980) for adult men and women and adolescents as 400 μg per day (Table 21-4). The RDA during pregnancy and lactation is increased to 800 and 500 μg per day, respectively. The RDA for infants and children varies with age and body weight (Table 21-4).

The lethal dose for animals ranges from 125 to 600 mg per kilogram. This is many times the therapeutic dosage and indicates the relatively low toxicity of this vitamin.

Cobalamin (vitamin B_{12})

Vitamin B_{12}, called the antipernicious anemia vitamin, was isolated, purified, and extensively studied in 1949. It is the factor of liver extracts responsible for the curative effects of these extracts on pernicious anemia, and it is now believed to be identical with the "extrinsic factor" of Castle. It was isolated in 1948. This vitamin was soon isolated from culture broths of *Streptomyces griseus,* the strain used for the production of the antibiotic streptomycin. Indeed, various related compounds, also members of the vitamin B_{12} group, are present as by-products when other antibiotics are made, and commercial production of these vitamins uses fermentation by special microorganisms. The term *vitamin* B_{12} now is the officially preferred *generic* name for this group of vitamins. *Cobalamin* is used for one specific chemical member of the group.

Structure The structure of vitamin B_{12} was worked out by several teams of scientists, each made up of American and British members. A unique feature of the B_{12} compounds is the presence of the cobalt atom in the trivalent state. No other cobalt-containing organic compound has been found in nature. Cyanide is present in vitamin B_{12a}, but this is replaced by hydroxyl in vitamin B_{12b} and by nitrite in B_{12c}. These latter two may be converted to vitamin B_{12a} by treatment with cyanide. Vitamin B_{12a} is now designated cyanocobalamin; B_{12b} is hydroxycobalamin; and B_{12c} is nitritocobalamin. Several other members of the group also have been isolat-

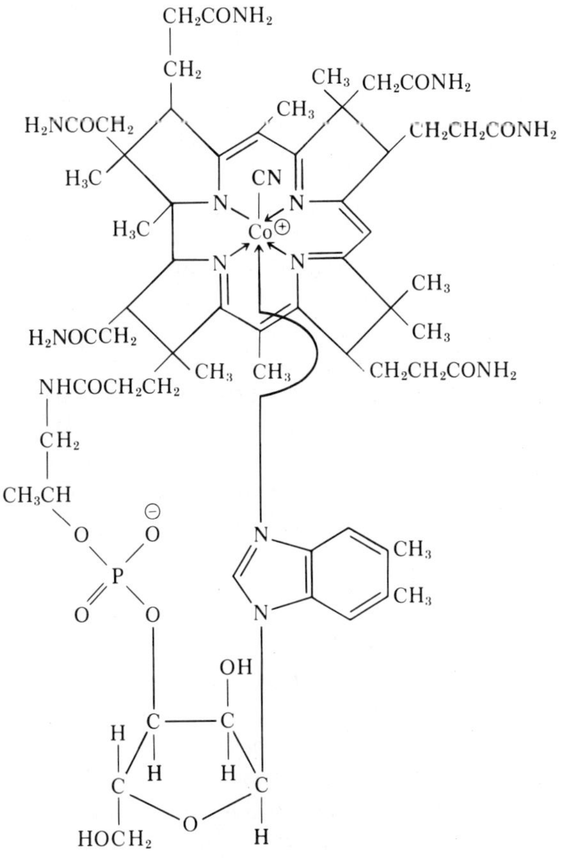

Cobalamin (vitamin B_{12})

ed. The empirical formula of vitamin B_{12a} is $C_{63}H_{90}N_{14}O_{14}PCo$, and the actual formula includes a central *corrin* ring system somewhat similar to the tetrapyrrole ring structure of the porphyrins, e.g., heme and chlorophyll. The single cobalt atom is in the center of the corrinoid ring, like iron in heme. The structure is shown on p. 779.

A series of investigations indicated that cyanocobalamin is a minor component of the total cobalamins actually present in the liver and in certain bacteria. Indeed, it may be an artifact produced from the major component, now shown to be *coenzymes* B_{12}. Several different forms of the cobalamin-containing coenzymes have been isolated: one form crystallized from the bacteria *Clostridium tetanomorphum* and *Propionibacterium shermanii;* two others have been highly purified from the liver of rabbits, chickens, sheep, and humans. These forms are called adenylcobamide, benzimidazolylcobamide, and 5,6-dimethylbenzimidazolylcobamide. In addition, the latter two contain 1 mole of *adenosine,* i.e., adenine plus the sugar deoxyribose. Cyanide is *not* present in any of the three forms. Thus the main form of vitamin B_{12} in human liver is the 5,6-dimethylbenzimidazole form in which the cyanide group is replaced by a 5-deoxyadenosine group to form *coenzyme* B_{12}.

The coenzymes B_{12} are highly active as growth factors for B_{12}-requiring mutants of *E. coli,* for the conversion of glutamate to β-methylaspartate, and for a number of other metabolic reactions. Thus the active forms of the cobalamin vitamin group appear to be the adenine-containing coenzymes B_{12}.

Occurrence The chief source of vitamin B_{12} is liver, although the vitamin is also present in other animal proteins, including milk, meat, eggs, fish, oysters, and clams. Under certain dietary conditions this vitamin may be synthesized by intestinal organisms. In general, it is not present in vegetable foods. Pernicious anemia, which is caused by a deficiency of vitamin B_{12}, must be regarded as the result of a gastric mucosal deficit of the intrinsic factor (Chapter 17) rather than a dietary one, in most cases, since ample quantities of the vitamin (estimated at 3 to 5 μg per day) are present in foods or are provided by the intestinal flora.

Vitamin B_{12} occurs in foods bound to proteins which apparently are split off by proteolytic enzymes.

Human requirements The average mixed diet in the United States supplies approximately 5 to 15 μg of vitamin B_{12} per day. At this level of intake a mean of approximately 16% to 28% is absorbed and available for tissue use. Less is absorbed from higher levels of intake. The daily loss is estimated as about 2.6 μg per day.

The 1980 RDA Using these data, together with amounts of vitamin B_{12} required to initiate a hematological response and to maintain health in patients with pernicious anemia, as well as other related information, the Committee on Dietary Allowances has set the RDA for adults and adolescents at 3 μg of vitamin B_{12} (Table 21-4). This amount will maintain adequate vitamin B_{12} nutrition and a substantial reserve body pool in most normal individuals. The daily recommended allowance for infants varies from 0.5 to 1.5 μg and for children from 2 to 3 μg, varying with age. During pregnancy and lactation the recommended allowance is increased to 4 μg per day.

Properties Vitamin B_{12} is a deep red crystalline compound containing nitrogen, phosphorus, and cobalt but no sulfur. It is soluble in water, alcohol, and acetone but not in chloroform. It is levorotatory and is stable to heat in neutral solutions although destroyed by heat in dilute acid or alkaline solutions.

Physiological effects and clinical uses Vitamin B_{12} is one of the most potent therapeutic agents known. Only 5 μg per day are required for satisfactory hematopoiesis in pernicious anemia. It is a powerful medicament in the parenteral treatment of megaloblastic anemias associated with a deficiency of the intrinsic factor. Apparently the intrinsic factor, present in normal gastric juice, is necessary for the absorption of vitamin B_{12}, the extrinsic factor, from the gastrointestinal tract. Consequently, when administered by mouth, B_{12}, is of little value unless normal gastric juice is present or is given at the same time; or, if administered alone, the dosage required is much greater.

The *intrinsic factor* has been isolated in highly active form, along with another high-capacity B_{12}-binding protein devoid of intrinsic factor activity, from hog pyloric mucosa. Both are glycoproteins and appear to be homogeneous on column chromatography and starch gel electrophoresis. The intrinsic factor glycoprotein is active in binding vitamin B_{12} and facilitating the absorption of B_{12} in amounts as small as 50 μg. The intrinsic factor glycoprotein contains approximately 15% reducing sugars, including galactose and glucosamine, and 33% total carbohydrate. It has a molecular weight of about 50,000 and a sedimentation coefficient of 3.7S. It may form a dimer when complexed with 1 or 2 moles of B_{12}. Apparently the intrinsic factor binds vitamin B_{12} to specific receptor sites in the ileal mucosa, thus mediating its absorption. It is then transported in the plasma bound to the transport protein *transcobalamin*. Maximal absorption of vitamin B_{12} in humans occurs in the lower ileum, as determined by labeling the vitamin with ^{60}Co.

Human vitamin B_{12} deficiency is usually attributable to poor absorption rather than inadequate dietary intake. Vegetarians tend to have low serum B_{12} levels but usually are not anemic.

Hereditary malabsorption of vitamin B_{12} has been reported in a case of pernicious anemia of long standing. A normal amount of active intrinsic factor was present in the gastric juice, and no detectable antibodies to intrinsic factor were found in the gastric contents. The mechanism for the malabsorption of B_{12} in this disorder is obscure.

Thus today it appears that pernicious anemia, long an enigma, may be caused by (1) a chronic dietary deficiency of vitamin B_{12} or (2) malabsorption of B_{12} because of a lack of intrinsic factor in the gastric juice. The latter may be caused by (1) the presence of antibodies to the intrinsic factor in the gastric juice, (2) a lack of secretion of intrinsic factor (from gastric mucosal cell atrophy, etc.), or (3) hereditary malabsorption of uncertain nature. Pernicious anemia thus may frequently be an *autoimmune disorder*.

Vitamin B_{12} may also favorably influence the course of the anemias of sprue, pellagra, and infancy, but in these conditions folic acid may also be needed.

Effect on growth If crude vitamin B_{12} (first called the animal protein factor) is added as a supplement to the food of young animals, a pronounced improvement in the rate of growth is observed. Since the two main constituents of the supplement are vitamin B_{12} and streptomycin, there has been some controversy over which one is the causative agent. Other antibiotics have also been used, often with striking results (e.g., stimulating growth in plants). Although the vitamin alone has growth-promoting properties, the effect is enhanced by the antibiotics. Consequently the effects on animals have been ascribed to (1) improvement of appetite by the vitamin, (2) diminution of the multiplication of intestinal microorganisms by the antibiotics, and (3) specific inhibitory effects of the antibiotics on oxidative phosphorylation. Whatever the mechanism, the influence of such supplements on the growth of fowl and domestic animals may be quite important from an economic standpoint. A somewhat similar effect has been seen in children who have a history of nutritional stress, i.e., not those in normal health. If adequate nutrients and calorie distribution are supplied, the addition of vitamin B_{12} brings these children into conformity with normal height-weight ratios.

Mechanism of action Vitamin B_{12} coenzymes are involved in at least two types of reactions. The first involves a shift of a hydrogen atom to an adjacent carbon atom in exchange for another group (Chapter 6). An example is the conversion of L-methylmalonyl CoA to succinyl CoA. The second involves the transfer of methyl groups in certain methylation reactions, e.g., the conversion of homocysteine to methionine (Chapters 6 and 13). The biosynthesis of choline (Chapter 12) and thymine (Chapter 14) are other examples. Apparently the methyl group is coordinated to the cobalt atom, replacing the 5′-deoxyadenosine group of coenzyme B_{12}, forming methylcobalamin. The methyl group is transferred then to an appropriate acceptor, apparently in the presence of an enzyme, *methyl transferase*. 5-Methyltetrahydrofolic acid (FH_4) probably serves as the source of the methyl group.

The role of vitamin B_{12} (cobalamin) as coenzyme B_{12} in the transfer of methyl groups may be depicted schematically as follows:

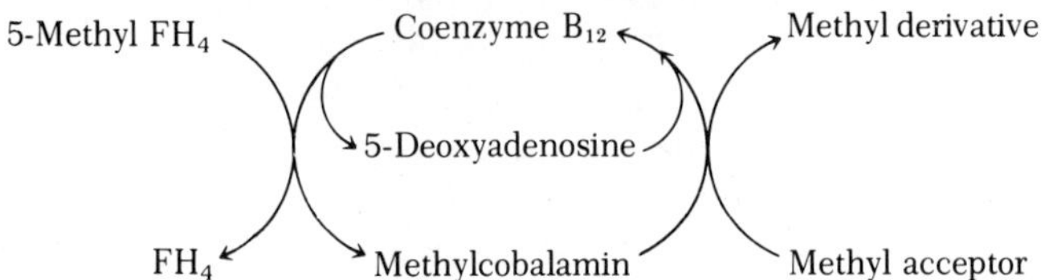

In this way such important biochemical substances as thymine and choline can be formed from the methyl-group acceptors uracil and aminoethanol, respectively. This reaction would be blocked, of course, in vitamin B_{12} deficiency.

Vitamin B_{12} probably as coenzyme B_{12}, appears to be involved in a number of other metabolic reactions. These include the following: reduction of disulfide to sulfhydryl groups; activation of amino acids for protein synthesis; conversion of methylmalonyl CoA to succinyl CoA; conversion of β-methylaspartic to glutamic acid, with coenzyme B_{12} serving as a mutase; reduction of formate to methyl groups; reduction of ribonucleotides to deoxyribonucleotides by a ribonucleotide reductase; dismutation of vicinal diols to their corresponding aldehydes by means of a diol dehydrase reaction; and anaerobic degradation of lysine. With the exceptions of the methylmalonyl CoA reaction and the glutamic mutase reaction, the role of vitamin B_{12} or its coenzyme form in these metabolic conversions is poorly understood.

The methylmalonyl CoA mutase reaction has been the subject of intensive study during the past few years and is a remarkable reaction "without precedent in organic chemistry." Ochoa and his associates discovered that methylmalonic acid plays some role in propionate metabolism in animals. Later the remarkable observation that large quantities, up to 50 to 90 mg daily, of methylmalonic acid appear in the urine of patients with pernicious anemia was reported. This was verified in vitamin B_{12}–deficient patients and rats.

As the result of the work of a number of investigators during the past few years, the relationship between vitamin B_{12} and methylmalonic acid metabolism has been clarified. Vitamin B_{12} coenzyme is a cofactor of the enzyme *methylmalonyl CoA mutase*, which catalyzes the *intramolecular* conversion of methylmalonyl CoA to succinyl CoA and thus facilitates the further metabolism of methylmalonyl CoA presumably by the citric acid cycle. This reaction may be summarized as follows. The remarkable *mutase reaction* given above has been shown, by the use of various isotopically labeled (^{13}C, ^{14}C, ^{3}H) intermediates, to involve the shift of the $-\underset{\underset{O}{\|}}{C}-SCoA$ of L-methylmalonyl CoA from the α-carbon to the CH_3-carbon, with a corresponding shift of the hydrogen atom in the opposite direction. The role of coenzyme B_{12} in this unprecedented type of organic reaction may be illustrated as follows:

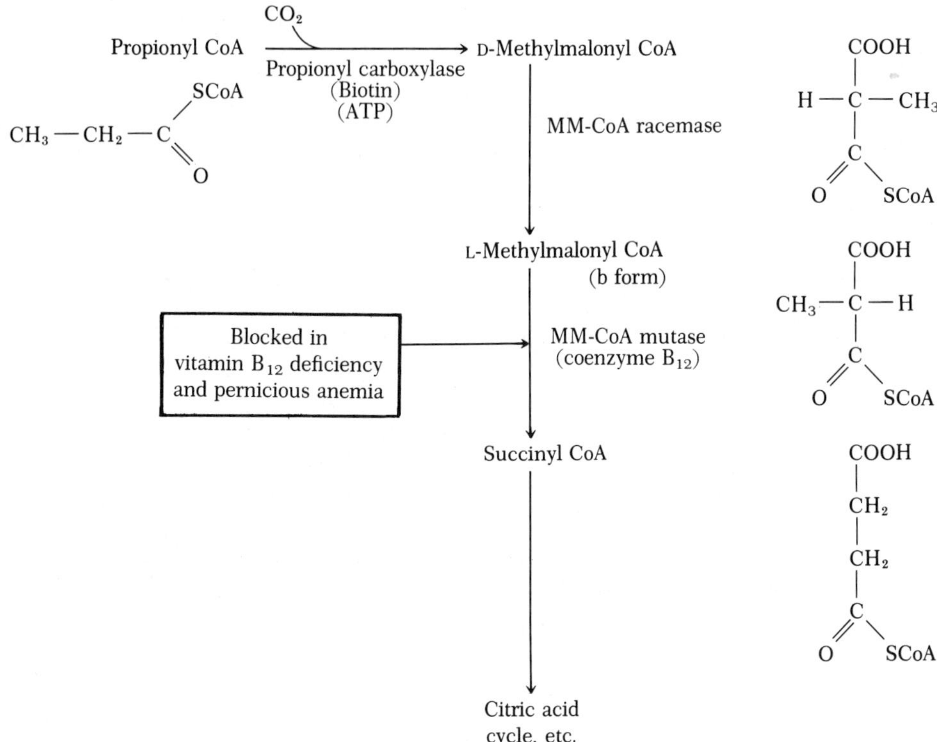

Propionyl CoA and methylmalonyl CoA may be derived from a variety of sources in the animal body, e.g., uracil, β-alanine, the biosynthesis of cholic acid from cholesterol, the metabolism of odd-carbon-atom fatty acids, the metabolism of certain amino acids (Val, Ile, Thr, Met), and the metabolism of thymine. Thus in vitamin B_{12} deficiency, including pernicious anemia in humans, this metabolite appears in relatively large amounts in the urine. Methylmalonic acid recently has been suspect as related to the neurological symptoms of pernicious anemia.

Recent investigations indeed have shown that there is an impairment in fatty acid synthesis in the liver and brain of rats deficient in vitamin B_{12}. This appears to be attributable to a block in the mutase reaction causing an accumulation of propionyl CoA, which in turn competes with acetyl CoA for acetyl CoA carboxylase. Likewise, methylmalonyl CoA accumulates, inhibiting the enzymes of normal fatty acid synthesis. The content of odd-carbon-chain fatty acids increases in these tissues. Thus the inhibition of normal fatty acid synthesis in the brain and nerve tissue (probably glial cells) leading to faulty structure and impaired function, as manifested by neurological symptoms characteristic of cobalamin deficiency and pernicious anemia.

Another recent report describes the occurrence of classic vitamin B_{12} deficiency in the breast-fed infant of a mother who had been a strict vegetarian for 8 years. The infant had a marked methylmalonic aciduria, homocysteinuria, megaloblastic anemia, and typical neurological abnormalities. The plasma level of vitamin B_{12} was low. The mother had a mild megaloblastic anemia, a megaloblastic bone marrow, and depressed values for vitamin B_{12} in the plasma and milk. The abnormalities in the infant were promptly corrected by small doses of vitamin B_{12}.

Biotin

Biotin (B_7) is a food factor that for a long time was known to be necessary for the development of microorganisms. Biotin deficiency cannot be readily induced in animals by feeding biotin-free diets, nor is such a deficiency seen in humans. It is produced by feeding large quantities of raw egg white. If rats are fed such a diet, they develop a characteristic group of symptoms including an extensive dermatitis, with "spectacled eye," hair loss, and involvement of the nervous system. The condition can be prevented by adding yeast or other food rich in biotin. The explanation for these facts is that uncooked egg white contains a protein, called *avidin*, that is responsible for the egg-white injury. Avidin combines with biotin in a firm linkage to form a compound that cannot be hydrolyzed and absorbed and is therefore excreted. Thus an induced biotin deficiency, or egg-white injury, results. The heating of egg white denatures avidin and destroys its ability to bind biotin.

Similarly, biotin deficiency has been produced in humans by a low-biotin diet containing large amounts of raw egg white. This was done experimentally in the case of four volunteers and was also observed in one patient who had for years subsisted on a diet that included 4 to 6 dozen raw eggs a week. The chief symptom that occurred was a fine scaly desquamation of the skin without pruritus. Other symptoms included anemia, anorexia, nausea, lassitude, and muscle pains. Biotin deficiency also has been reported recently in infants due to impaired absorption or binding to a mucosal cell receptor. The chief symptoms included skin rash, lactic acidosis, and hepatomegaly. There was a decrease in pyruvate and other carboxylases, as assayed on the infants' skin fibroblasts. The infants responded promptly to biotin administration.

Properties and structure Biotin crystallizes in long needles and is soluble in water and ethyl alcohol but is insoluble in ether and chloroform. It is heat stable. The structure of this compound was worked out by du Vigneaud as follows:

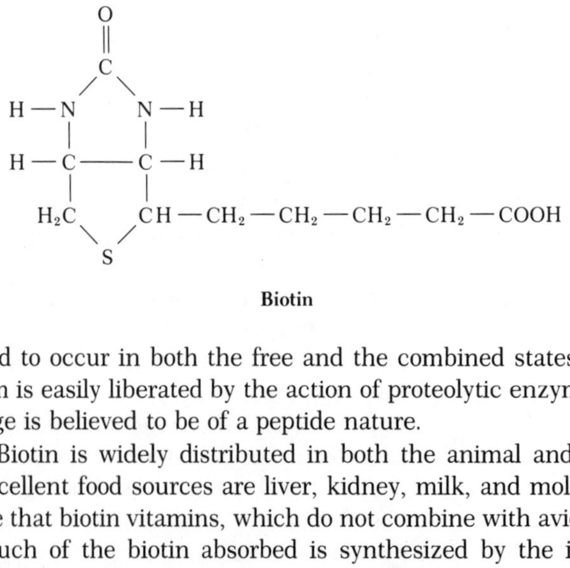

$$
\begin{array}{ccccc}
 & & O & & \\
 & & \| & & \\
 & & C & & \\
 & / & & \backslash & \\
H-N & & & & N-H \\
| & & & & | \\
H-C & - & - & - & C-H \\
| & & & & | \\
H_2C & & & & CH-CH_2-CH_2-CH_2-CH_2-COOH \\
 & \backslash & & / & \\
 & & S & &
\end{array}
$$

Biotin

Biotin is said to occur in both the free and the combined states in foods. The combined form is easily liberated by the action of proteolytic enzymes, and therefore the linkage is believed to be of a peptide nature.

Occurrence Biotin is widely distributed in both the animal and the vegetable kingdoms. Excellent food sources are liver, kidney, milk, and molasses. There is some evidence that biotin vitamins, which do not combine with avidin, also occur. Apparently much of the biotin absorbed is synthesized by the intestinal flora. Therefore a definite requirement for humans is difficult to set, and a deficiency is hardly likely to occur.

Function The role of biotin is primarily of carbon dioxide fixation, or carboxylation, as occurs in the conversion of pyruvic to oxaloacetic acid (Chapters 6 and 11). Biotin is also essential for a number of other carboxylation reactions, including the conversion of acetyl CoA to malonyl CoA in the biosynthesis of fatty acids and of propionyl CoA to methylmalonyl CoA (Chapter 12). The carbon dioxide is carried as a carboxyl group attached to one of the ureidonitrogen atoms of biotin, as described in Chapter 6.

Biotin was found to be a component of crystalline pig-heart propionyl CoA carboxylase. The enzyme has a molecular weight of about 700,000 and contains 4 moles of biotin per mole of protein.

Studies on *E. coli* indicate that biotin is a component of the enzyme carbamoyl phosphate synthetase (Chapter 13). Apparently the biotin-containing enzyme reacts with bicarbonate, glutamine, and ATP in a four-step reaction to form carbamoyl phosphate, glutamate, and ADP. The reaction is inhibited by avidin, and added biotin restores activity. Biotin is present in highly purified enzyme preparations. It probably is also present in mammalian (rat liver) carbamoyl phosphate synthetase. These observations thus implicate biotin in the biosynthesis of arginine and pyrimidines (nucleic acids).

Human requirements Biotin is synthesized by the normal human intestinal flora in amounts that make a significant but difficultly determined contribution to the body pool. However, the Food and Nutrition Board (1980) has recommended a safe and adequate daily dietary intake range of biotin for normal individuals of various age groups (see Table 21-12, p. 760). The values may be summarized as follows: infants up to 1 year, 35 to 50 μg daily; children and adolescents, 65 to 120 μg daily; from 11 years to adulthood, 100 to 200 μg daily.

The average American diet supplies an estimated 150 to 300 μg per day. This amount, plus that uncertain quantity synthesized by the intestinal flora, is more than adequate to supply the recommended allowance under ordinary conditions.

Pantothenic acid

Pantothenic acid (B_3) was at first called the *filtrate factor* and was given its present name by its discoverer, R.J. Williams, in 1938. Pantothenic acid was so called because of its widespread occurrence.

Effects of deficiency A deficiency of pantothenic acid causes dermatitis in the chick and graying of the hair in black rats. This, however, is not the only antigray hair factor. Rusting of the hair and porphyrin caking of the whiskers have been observed in white rats, and in the rat are also seen dermatitis, inflammation of the nasal mucosa, and "spectacled eye" condition (which is more characteristic of biotin deficiency). Atrophy of the adrenal cortex, with necrosis and hemorrhage, occurs in many pantothenic acid–deficient animals, particularly if stress has been experienced. Corneal changes consisting of vascularization, thickening, and opacity are seen. Interference with sexual function and reproduction also is noted in various species, and there are neurological lesions in chicks and pigs.

There seems to be no evidence of a pantothenic acid–deficiency disease in humans. The vitamin is widely distributed, and consequently even on restricted diets no actual deficiency seems to have occurred. However, such a deficiency has been produced experimentally. Human subjects were given a synthetic diet to which was added every known essential mineral and vitamin except pantothenic acid. Although this resulted in lowered blood cholesterol and altered response to corticotropin, no clinical signs of deficiency were evident. To the synthetic diet was added a pantothenic acid antagonist *(thiopanic acid),* designed to still further deplete the system of this vitamin. Then definite signs appeared. The men became easily fatigued, had cardiovascular disturbances, and later had gastrointestinal symptoms. There were numbness and tingling of the extremities and a number of other distressing conditions, including mental depression and upper respiratory tract infections. Administration of fluids intravenously and cortisone, together with dietary supplementation, restored the subjects to normal health.

Occurrence In the first rank of dietary contributors of pantothenic acid may be placed liver, kidney, eggs, lean beef, skimmed milk, buttermilk, molasses, peas, cabbage, cauliflower, broccoli, peanuts, sweet potatoes, kale, and yeast. In the second rank are white potatoes, tomatoes, wheat bran, whole milk, and canned salmon. Many other animal and vegetable foods have a moderately high content of this vitamin.

Properties and structure Pantothenic acid is a water-soluble, yellow viscous oil, stable to moist heat and to oxidizing and reducing agents. It is destroyed by dry heat and by heating in an alkaline or acid medium. It is a β-alanine derivative, possessing a peptide linkage, and has the following formula:

$$\begin{array}{ccccccccccccccc} & OH & & CH_3 & & OH & & O & & H & & H & & H & \\ & | & & | & & | & & \| & & | & & | & & | & \\ H- & C & - & C & - & C & - & C & - & N & - & C & - & C & -COOH \\ & | & & | & & | & & & & & & | & & | & \\ & H & & CH_3 & & H & & & & & & H & & H & \end{array}$$

Pantothenic acid

Functions Pantothenic acid forms a part of *coenzyme A,* the coenzyme in acylation reactions. All cellular pantothenic acid, whether of animal or vegetable origin, is accounted for by the coenzyme. Consequently, coenzyme A (CoA) represents the only known functional form of this vitamin. The structure of CoA has been worked out by a number of investigators. It is an atypical dinucleotide, with the usual mononucleotide being replaced by phosphopantethine. *Pantetheine* is pantothenic

Table 21-14
Summary of water-soluble vitamins essential for human nutrition

Vitamin	**Best food sources**	**Recommended dietary allowances (1980)***	**Active cofactor form**
L-Ascorbic acid (C)	Citrus fruits, tomatoes, strawberries, melon	60 mg	Not identified (cofactor, proline hydroxylase?)
Thiamin (B_1)	Pork, liver, yeast, whole or enriched grains, legumes	1.5 mg	Thiamin pyrophosphate (thiamin diphosphate; cocarboxylase)
Riboflavin (B_2)	Milk, organ meats, animal protein, enriched grains	1.7 mg	FMN (flavin mononucleotide), FAD (flavin adenine dinucleotide)
Niacin (nicotinic acid, niacinamide) (precursor, tryptophan)	Meat, enriched or whole grains, poultry, fish, peanuts	19 mg equiv (1 mg equiv per 60 mg Trp)	NAD^+ $NADP^+$
Pyridoxine (B_6)	Meat, whole grains, poultry, fish	2.2 mg	Pyridoxal phosphate or pyridoxamine phosphate
Folacin (folic acid, pteroylglutamic acid)	Liver, greens, mushrooms, whole grains, legumes	400 μg	Tetrahydrofolic acid
Cobalamin (B_{12})	Animal protein, meats, milk, egg	3 μg	Cobamide coenzymes (with 5-deoxyadenosine)
Biotin	Egg yolk, organ meats, yeast, nuts; widely distributed	100-200 μg†	Biotin
Pantothenic acid	Liver, meat, cereal, milk, legumes; widely distributed	4-7 mg†	Coenzyme A

*For American men, 19 to 22 years of age, of average activity (see Table 21-4, pp. 698-699).
†Estimated range (Food and Nutrition Board) (see Table 21-12, p. 760).

acid, joined to β-mercaptoethylamine through a peptide linkage. At the other end, pantothenic acid is joined by a pyrophosphate bridge to an adenylic acid group. This adenylic acid consists of adenine, D-ribose, and phosphate, but the phosphate is linked onto the carbon-3 of the ribose. The structures of these compounds are given in Chapter 6.

To effect an acylation, CoA must be present in the form of the appropriate acyl CoA. The steps involved together with the various enzymes and cofactors required for a number of acylation reactions have been discussed in Chapters 6 and 11-13.

Human requirements The average amount of pantothenic acid in the daily American diet is about 10 mg. This appears to be adequate for normal human nutrition.

No official value has yet been established as a recommended dietary allowance for humans. However, the Food and Nutrition Board (1980) has recommended a safe and adequate estimated daily dietary intake for normal healthy individuals of different ages (see Table 21-12). The recommended intakes per day follow: up to 1 year, 2 to 3 mg; children and adolescents, 1 to 10 years, 3 to 5 mg; 11 years to adulthood, 4 to 7 mg.

Summary

Table 21-14 presents a brief summary of the best food sources, the 1980 recommended daily allowances, the active cofactor forms, the principal metabolic functions, and the major clinical manifestations of deficiencies of the water-soluble

Principal metabolic functions	Manifestations of deficiency
Intercellular cement substance formation, collagen, capillary walls; metabolism of Tyr, Phe, folic acid; iron absorption	Scurvy, petechial hemorrhages, anemia, delayed wound healing, bone fragility
Decarboxylation of α-keto acids; transketolation of 2-keto sugars	Beriberi (polyneuritis), cardiovascular—vasodilatation, tachycardia, edema; gastrointestinal—anorexia, nausea; neurological—fatigue, apathy, neuritis, paralysis
Coenzyme of electron transfer system	Angular stomatitis (cheilosis), seborrheic dermatitis, conjunctivitis, photophobia, glossitis
Coenzyme of electron transfer system; dehydrogenase reactions; oxidation to produce ATP (NAD^+); biosynthesis of fatty acids, steroids, etc. ($NADP^+$)	Pellagra, diarrhea, scaly dermatitis, dementia, stomatitis
Coenzyme in amino acid metabolism: transamination, decarboxylation, transsulfuration, tryptophan synthetase, amino acid transport	Cheilosis, glossitis, stomatitis, seborrheic dermatitis, convulsions, anemia
Transfer of 1-carbon fragments (formyl); biosynthesis of purines, choline, methionine, etc.	Macrocytic and megaloblastic anemias, sprue, malabsorption, leukopenia, thrombocytopenia
Transfer of 1-carbon fragments (methyl); biosynthesis of purines, choline, methionine, etc.; mutase reactions	Pernicious anemia, neurological lesions; sprue
Carboxylation; transcarboxylation	Dermatitis, alopecia, anemia; experimentally only in humans
Acylation reactions (acetyl group transfers)	Anemia, achromotrichia; human deficiency most unlikely

vitamins, established as essential for human nutrition. They are presented in the order of their discussion in the text.

Water

Water is the most abundant body constituent accounting for 50% to 75% of body weight depending on age and body weight. It is the medium in which most metabolic reactions occur. Deficits or excesses of more than a few percent of total body water are incompatible with health. The turnover rate of water per day in the normal adult is approximately 6% of total body water and more than twice that (15%) in the infant. The body maintains total body water within narrow limits by means of a number of homeostatic mechanisms discussed in Chapter 17.

The major routes of water intake and elimination (without visible sweating) with approximate amounts for a 70-kg adult, resulting in water balance, may be summarized as follows.

	Water intake (ml/day)		**Water loss (ml/day)**
Preformed water in foods	700	Urine	1000
Water from food oxidation	300	Insensible loss (lungs, skin)	850
Drinking water, beverages	1000	Stool	150
Total	2000	Total	2000

Water requirements increase markedly for persons living in hot dry climates or for those engaged in hard work or vigorous exercise. Water losses from the skin and lungs in such individuals may increase tenfold.

The numerous factors determining water loss from the body preclude the setting of any general value for water requirement. The Food and Nutrition Board, however, suggests that a reasonable allowance for adults under ordinary circumstances is 1 ml per kcal energy intake and 1.5 ml per kcal for infants.

Special adjustments may be required for the water needs of patients with fever, polyuria, vomiting, or diarrhea and for those administered diuretics or consuming high-protein diets. The same needs for adjustment applies to infants and to comatose patients, as well as to all persons in hot environments.

Other substances in foods

In addition to the nutrients known to be essential to humans, there are a number of chemical substances occurring naturally in foods. Some of these have been demonstrated to be essential for certain higher animals, but as yet no unequivocal proof exists that they are required by humans. No RDA for these substances has been established, since there is no long-standing and continuing evidence of human need. Some of these substances may be acting pharmacologically. Other substances are known to be growth factors for certain lower forms of life, but, again, no dietary requirement for them has been established for humans or other higher animals. There are still other substances in foods that are claimed (in the lay literature) to be nutrients for humans even though there is no substantial evidence for such claims, not even in other species.

Foods contain many thousands of organic substances as well as a number of inorganic compounds spanning almost the entire gamut of the periodic table of elements. Some of the latter may be accidental constituents from the environment. Some may have as yet undiscovered functions in animal species. Some of the organic substances in foods likewise might be required by higher animal species. Some may have biological functions in the plant or other life form from which they are derived, some may be useful for their pharmacological effects (i.e., as drugs), and some may be simply waste by-products of plant metabolism. Only a very few of these organic substances (about 25 or so) have been established as essential dietary nutrients for humans and for higher animals (the essential fatty acids, the essential amino acids, and the vitamins). A number of other organic substances in foods, although not regarded as essential, nevertheless perform vital roles as sources of energy (the carbohydrates, nonessential fatty acids, glycerol, and the nonessential amino acids) or as precursors of tissue components and biologically active compounds (e.g., hormones, enzymes, respiratory and transport proteins).

Thus it follows that the mere presence of an organic or inorganic substance in foods does not make that substance an essential dietary nutrient. A few such substances present in foods are considered in the following section.

Substances essential for certain higher animals (not including humans)

The list of substances in this category includes choline (or lecithin), glycine, and possibly arginine. Only the first of these will be considered now; the others have been discussed in earlier sections of this chapter.

Choline

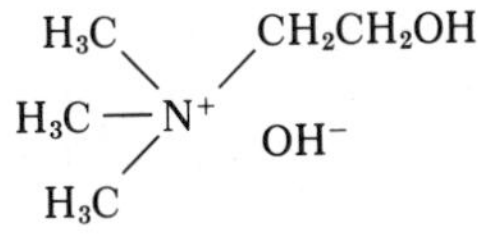

$$(H_3C)_3N^+CH_2CH_2OH \quad OH^-$$

Choline

Choline is not a true vitamin in the strict sense of the term because other naturally occurring substances can substitute for it. However, its importance in nutrition is unquestioned. Choline is a constituent of the lecithins (phosphatidyl choline and sphingomyelins [Chapter 12]), an essential component of cell membranes and lipoproteins. It is a precursor of acetylcholine (Chapter 15) and is a source of labile methyl groups (Chapter 13). It has the formula trimethylhydroxyethylammonium hydroxide.

Effects of deficiency Choline deficiency was discovered in the course of work on diabetic animals. In depancreatized dogs not only does diabetes occur but also a fatty infiltration of the liver. The former condition may be controlled by insulin, but the latter cannot be. Feeding raw pancreas, however, does cure the fatty liver, and the effective agent or one of the effective agents was found to be the choline part of lecithin present in the pancreatic tissue. This so-called *lipotropic* action of choline has been demonstrated in a number of other experimentally produced fatty livers. Choline has been shown to have several other physiological functions.

On a low-choline diet puppies develop a severe anorexia (lack of appetite) and fail to grow; hens do not lay eggs; rats do not have normal lactation. Together with manganese and folic acid, choline prevents perosis (slipped tendon disease) in chicks and young turkeys, and it prevents cirrhosis of the liver in rats. A low-choline diet also produces hemorrhages of the kidneys and eyes in addition to fatty livers in young rats, and if the diet is low in methionine as well, the hemorrhagic condition appears in other organs also. An anemia and hypoproteinemia also develop.

This relation of methionine to choline has been considered previously (Chapter 13). Both compounds contain methyl groups and, with betaine, are sources of these groups in metabolism. If methionine is added to a low-choline diet, it decreases liver fat, probably because it has a methyl group to offer the system, which is lacking in such groups. A shift of the methyl groups is in some way required in fat metabolism. This is called *transmethylation,* and methyl groups are shifted, depending on their need and the dietary supply. Therefore methionine or betaine may replace choline.

Another important effect of choline deficiency has been described. If very young rats are kept on a choline-deficient diet for 6 days and are then given a normal diet, they eventually develop high blood pressure. The severe hypertension may result from the renal damage just described, causing extensive fibrosis of the renal capsule.

As yet, no applications of choline to human nutrition have been definitely established. Possibly human alcoholic cirrhosis of the liver is in large part a result of dietary deficiency of lipotropic agents, of which choline and methionine are important examples. Recent investigations indicate, however, that a deficiency of choline and other *lipotropes* may occur in infants and young children subsisting on protein-deficient foods, rich in highly refined products. A significantly decreased resistance to infection was reported, possibly related to a decrease in immunoglobulin formation.

Occurrence Choline is widely distributed, and no deficiency need ordinarily be expected. The most important sources in our diet are meats, egg yolk, bread, cereals, and various other vegetables, especially beans and peanuts. The average daily intake of choline from foods ordinarily consumed is 400 to 900 mg. This amount, together with an undetermined quantity available from endogenous sources, apparently is ample for the maintenance of human health. No human daily requirement has been established.

Food substances essential for lower forms of life (not as yet for higher animals or humans)

A number of substances in foods have been shown to be growth factors essential for some invertebrates and various types of microorganisms. No evidence exists at the present time for their requirement by higher animals, including humans. Apparently most of these substances, if not all, can be synthesized in the tissues of higher animals, except possibly in rare occurrences of genetic abnormalities.

Food substances in this category include arachidonic acid, asparagine, the Bifidus factor, biopterin, carnitine, certain chelating agents, cholesterol, coenzyme Q (ubiquinones), hematin, inositol, lecithin, linolenic acid, lipoic acid (thioctic acid), nerve growth factor, various nucleotides and nucleic acids, *p*-aminobenzoic acid, a number of peptides and proteins, pimelic acid, certain polyamines, pteridines, and taurine.

Some of these food constitutents are certainly possible candidates for the list of essential nutrients for humans and higher animals. Because of this possibility and their importance in certain metabolic reactions, a few of these will be considered briefly.

Coenzyme Q (ubiquinones)

The coenzyme Q group has been classed as vitamins by some investigators because of the ability of these compounds to cure or protect against vitamin E deficiency in several species of animals. Some of the ubiquinones are also active in electron transport or oxidative phosphorylation or both (Chapter 7).

Inositol

Inositol is hexahydroxycyclohexane. Biologically active inositol—there are nine stereoisomers—is the optically inactive *myoinositol* (see Chapter 17 for formula). It is one of the muscle extractives and is also found in brain, erythrocytes, and tissues of the eye. It occurs widely in the plant kingdom in fruits, vegetables, whole grains, and nuts. Milk and yeast contain considerable amounts.

Inositol is included in the discussion of muscle extractives. It is found in nature in at least four forms: free inositol, phytin, phosphatidylinositol, and a water-soluble, nondialysable complex. Phytin is the mixed calcium and magnesium salt of inositol hexaphosphate (phytic acid) and was formerly believed to be exclusively of vegetable origin. Now it is known to be a constituent of the nucleated erythrocytes of several species of animals, in which it enhances the release of oxygen from oxyhemoglobin. An inositol-containing phosphatide has been called *lipositol* (Chapter 24). It has been isolated in pure form from soybeans and is known to be present in brain and spinal cord.

There have been a number of different types of effects produced in animals fed a diet deficient in inositol. Among them is that inositol deficiency results in retarded growth and a peculiar hairlessness in mice. Inositol has a curative action on the fatty livers produced in rats by the administration of biotin. Its significance in human nutrition is still undetermined.

α-Lipoic acid

α-Lipoic acid is a relatively newly discovered factor. Its formula and that of its reduced form follow:

$$\underset{\text{S}}{CH_2}-CH_2-\underset{\text{S}}{CH}-(CH_2)_4-COOH \;(\text{S—S bonded}) \rightleftharpoons \underset{\text{SH}}{CH_2}-CH_2-\underset{\text{SH}}{CH}-(CH_2)_4-COOH$$

α-Lipoic acid
(6,8-dithiooctanoic acid)

Reduced α-lipoic acid

It appears to be necessary for oxidative decarboxylation of pyruvic acid and α-ketoglutaric acid by certain microorganisms and is probably a coenzyme or part of a coenzyme, sometimes called lipothiamide pyrophosphate (LTPP), for this reaction. It is found in many biologic materials, including yeast and liver, and is also called the pyruvate oxidation factor (POF). α-Lipoic acid has not been shown to be a dietary requirement for mammals.

p-Aminobenzoic acid

In 1941 it was found that failure of lactation occurred in rats whose diet contained all the then known B vitamins, including thiamin, riboflavin, niacin, pyridoxine, pantothenic acid, and choline. The failure to lactate could be cured by yeast and other sources of the B complex. Graying of the hair of black rats also resulted when animals were fed on the deficient diet. Both effects have now been related to *p* aminobenzoic acid (PABA), which had previously been shown to be necessary for growth in the rat and chick and also for bacterial multiplication. As mentioned before, pantothenic acid has also been described as an anti–gray hair factor; however, the possibility that both biotin and PABA have similar properties also exists. Graying of the hair in humans may sometimes be the result of a nutritional deficiency. For instance, Greeley, the Arctic explorer, became gray after a 9-month period of undernutrition. After he had eaten a normal diet for a while, his hair darkened perceptibly. However, neither PABA nor pantothenic acid seems to be the responsible factor in humans.

p-Aminobenzoic acid blocks the bacteriostatic effect of sulfanilamide in vitro. The explanation for this is that there is competition between the two substances in some vitally important enzyme system. PABA is synthesized by the bacteria and is an essential metabolite for the bacterial cell. It takes part in some enzyme reaction necessary for the life of that cell, possibly in a phenolase system. Because of its structural resemblance to PABA, sulfanilamide (*p*-aminobenzene sulfonamide) takes its place in the enzyme system but does not permit the vital reaction to proceed normally. Other sulfa drugs have similar action, although they differ quantitatively. It is interesting to note that the bacteriostatic potency of each sulfa drug is directly proportional to the drug's ability to counteract the antibacteriostatic action of PABA.

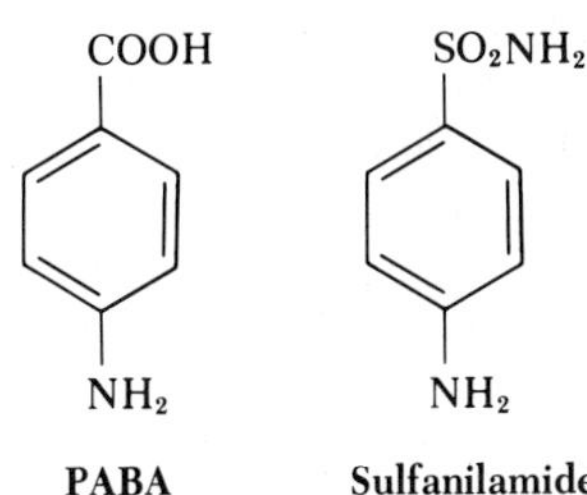

PABA **Sulfanilamide**

PABA forms part of the folic acid molecule. This may be the point of attack by the sulfa drugs. Some investigators have questioned whether PABA itself has catalytic actions; therefore they have not accepted it as a true vitamin. In view of the antagonism between the vitamin and the sulfonamide drugs, the continuous ingestion of extremely large doses of PABA is to be avoided. In itself PABA is relatively nontoxic, but the presence of a high PABA level in blood and tissues might render sulfonamide therapy of little value.

Properties and occurrence *p*-Aminobenzoic acid is a crystalline white compound, slightly soluble in cold water but quite soluble in hot water and in alcohol. It is widely distributed in nature but is more concentrated in liver, yeast, rice bran, and whole wheat.

Food substances that may act pharmacologically

This category of substances present in various natural foods includes compounds that are *not* considered essential nutrients in humans or animals but that do (or may) have some pharmacological effects in humans. The list of such substances includes alcohol, amygdalin or laetrile ("vitamin B_{17}"), bioflavonoids (rutin and hesperidin, vitamin "P" factors), caffeine, chlorophyll, orotic acid, "pangamic acid" (wrongly called vitamin B_{15} but actually an ill-defined mixture of dimethylglycine and sorbitol), "vitamin Q," and "vitamin U."

Other food substances that might conceivably fall in this list include a wide variety of organic compounds, some of which are useful drugs, and a number of trace elements not mentioned elsewhere in this section. Some of these may be simply accidental constituents from the environment having no significant nutritional or biochemical function.

Because of current biochemical-nutritional interest, two of the food substances in this category will be discussed briefly.

Amygdalin ("vitamin B_{17}," "laetrile")

Amygdalin is a nitriloside or β-cyanogenic glucoside known to organic chemists for at least 125 years. It is present in almonds and in peach, apricot, and apple seeds. On hydrolysis it may yield cyanide, cyanate, thiocyanate, benzaldehyde, or aspirin-like aromatic compounds.

Amygdalin has come into prominence recently because of claims for its (as laetrile) effectiveness in the treatment of cancer. These claims have provoked counterclaims and heated controversies, even some attempts at political intervention. Some states have considered legislation permitting its use in certain (terminal) cancer patients. In view of the conflicting clinical claims, the National Cancer Institute is planning controlled clinical trials. The Food and Drug Administration has ruled the substance as "worthless and potentially dangerous." Cyanide poisoning has been reported in dogs administered amygdalin. It is being studied also as an antisickling agent in sickle cell anemia patients.

Bioflavonoids

In 1936 Szent-Györgyi and associates reported the existence in lemon peel of a material that they called *citrin;* it consisted of a mixture of flavonoids and was shown to have physiological activity associated with the maintenance of normal capillary permeability and fragility. The active principle in citrin was found to be hesperidin, which was shown to have similar physiological activity exerted by a number of compounds of like structure, including flavanones, flavones, and flavonols.

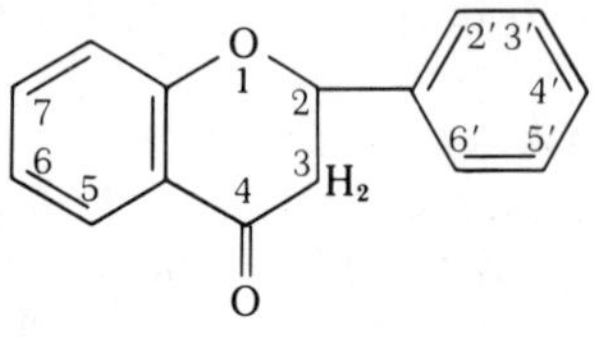

O 1 2 3 4 5 6 7 2′ 3′ 4′ 5′ 6′ H_2 O

Flavanone

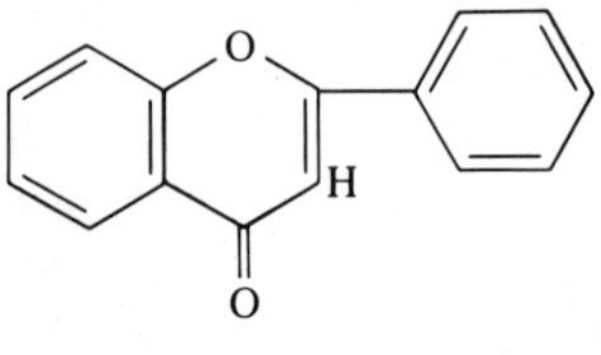

O H O

Flavone

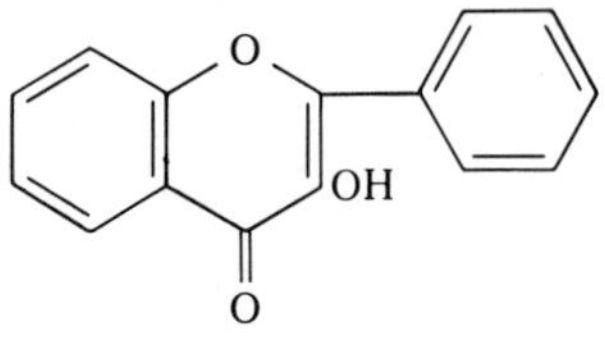

O OH O

Flavonol

Chalconization (opening of the ring system at the 1,2-position) does not affect this biological activity nor, within fairly wide limits, do the types and positions of ring substituents. Thus hesperidin (5,3′-dihydroxy-4′methoxy-7-rhamnoglucosidoflavanone), its aglycone hesperitin, rutin (5,7,3′,4′-tetrahydroxy-3-glucorhamnosidoflavone), and its aglycone, quercetin, all have comparable actions, as do other compounds of the same basic structure. The term *vitamin P* (for permeability) was at first assigned to this group of compounds. They are now more commonly referred to as bioflavonoids to indicate their structural characteristics and the fact that they show physiological activity.

The mechanism by which the flavonoids exert their influence on capillary permeability and fragility has been extensively investigated, and compounds of this type have been found capable of interacting with various metabolites and enzyme systems that can affect the vascular system.

Bioflavonoid deficiency has been produced in animals; it results in a syndrome characterized by increased capillary permeability and fragility. Although the deficiency syndrome has not been observed in humans, the bioflavonoids have been used clinically in the treatment of diseases in which vascular abnormality is a factor.

Investigations mentioned earlier in this chapter indicate that the principal action of the bioflavonoids is as an antioxidant (chelating of heavy metal oxidative ions?), thus protecting ascorbic acid from oxidative destruction. Their effect on the maintenance of normal capillary permeability would thus be indirect, via ascorbic acid.

The various members of this group are widely distributed in nature. They occur in the juice, peel, or pulp of citrus fruits, in tobacco leaves, in buckwheat, and in currants and many other fruits and vegetables. Whether the bioflavonoids can properly be considered to act as vitamins has not been firmly established; daily requirements are not known.

Chemical food additives

In addition to the six major groups of naturally occurring chemical substances that are present in foods—carbohydrates, lipids, proteins, inorganic elements, vitamins, and water—a wide variety of other chemical substances may be present, usually in relatively small amounts. Many of them also are naturally occurring, but some may be added intentionally. The latter group is usually referred to as *chemical food additives*. They may be added during food processing or storage to improve nutritional value, keeping qualities, or other desirable characteristics. In this sense substances used commonly for years, such as salt, sugar, yeast, spices, and other condiments, may be properly regarded as chemical food additives.

Technically, food additives are classed according to the purpose they serve: nutrients, enriching agents, sweeteners, antifoaming agents, bleaches, humectants, sequestrants, emulsifiers, flavoring agents, coloring agents, leavening agents, preservatives, buffers, antimycotics, antioxidants, antibacterial agents, stabilizers, thickeners, maturing agents, and others. Today some 2000 or perhaps 3000 substances are used as direct additives to foods. A look at the list of ingredients on the label of a typical cake mix will illustrate the point.

As stated, some chemical substances are added to improve the nutritional value of foods. This is usually done to replace nutrients lost in the processing of the food. An example is the enriching of flour to replace the vitamins, particularly thiamin, riboflavin, niacin, and minerals (calcium and iron) removed in the milling of wheat. The milling of wheat results in the production of a refined white flour, which has greater keeping qualities than whole-wheat flour. The loss of the germ and bran, however, reduces the vitamin E, thiamin, niacin, riboflavin, pyridoxine, iron, and fiber content of the flour greatly, as well as the calcium, phosphorus, fat, and protein, to some extent. This is partly remedied by the enrichment of white flour, which was accomplished during World War II by federal order and is now carried out by millers under state laws and also voluntarily. Table 21-15 shows that enriched bread compares favorably with whole-wheat bread in selected vitamin and mineral constituents but is still slightly deficient in protein. Since this program went into effect, vitamin B complex deficiencies have decreased considerably among patients in nutrition clinics.

After the introduction of enriched flour, many other foods have had vitamins or minerals, or both, added to them. Among these are cornmeal and corn grits, macaroni and spaghetti, rice, and processed cereals. Hydrogenated fats are devoid of vitamins. For this reason oleomargarine is commonly fortified with at least 9000 USP units of vitamin A per pound. Most of the milk on the market has added vitamin D. Some preparations of nonfat dried milk are now fortified with vitamins A and D.

The purpose of food additives is, as stated, to improve the nutritional value of foods, their keeping qualities, or other desirable characteristics. To quote one authority: "Were it not for food additives, baked goods would go stale and mold overnight, salad oils and dressings would separate and turn rancid, table salt would turn hard and lumpy, canned fruits and vegetables would become discolored or mushy, vitamin potencies would deteriorate, beverages and frozen desserts would lack flavor, and wrappings would stick to the contents."

There are a number of other more general reasons for the use of food additives: (1) the shift of populations from the farms to cities, thus creating the need for the transporting of food, sometimes for rather long distances, and its storage, possibly for prolonged periods of time; (2) population increases, creating the need for larger reserve supplies and their preservation; (3) increased numbers of foods—there are some 3500 items on the shelves of a typical supermarket today and in some the total may reach 8000 items; (4) convenience foods.

The widespread use of chemical additives in today's foods creates the need for their regulation for health and other reasons. Various federal and state legislative acts have been made into law as a result. Familiar examples are the Federal Pure Food and Drug Law, the Federal Food, Drug, and Cosmetic Act, the Pesticide Amendment, Additive Legislation, Color Additives Amendment, Truth in Packaging Law, the Delaney Clause, the recent Truth in Labeling Law, and others. Food products are carefully monitored with chemical analyses to ensure the presence of legal amounts and the absence of contaminants. Rigorous toxicity studies, both "acute" and "chronic," are required in several species of animals and over several

Table 21-15
Comparison of enriched, plain white, and whole-wheat breads

	Plain white bread	Enriched white bread	Whole-wheat bread
Thiamin (milligrams per pound)	0.3	1.1-1.8	1.3
Riboflavin (milligrams per pound)	0.5	0.7-1.6	0.7
Niacin (milligrams per pound)	3.0	10-15	16.0
Iron (milligrams per pound)	3.9	8-12.5	11.8
Calcium (milligrams per pound)	254	254*	272
Protein (grams per pound)	39	39	43

From United States Department of Agriculture, Bull. AIS-39.
*Enriched bread may contain 300 to 800 mg of calcium per pound as well as 150 to 750 USP units of vitamin D.

generations by the Federal Food and Drug Administration before any proposed chemical additive is permitted to be used in foods. At the present time there are only about 600 chemicals on the "generally regarded as safe" (GRAS) list.

The status of ethyl alcohol in this connection is of interest. The question of whether alcohol is a food is difficult to answer. It is rapidly used by the body when ingested in moderate amounts and yields about 7 kcal per gram. This energy is available for heat and work and for sparing other foodstuffs. Therefore the calories derived from alcoholic beverages can be significant and should be taken into account in calculating the caloric content of a diet. Excessive amounts of alcohol have toxic properties and may cause serious pathological changes.

Acid-forming and base-forming properties of foods

When a food is incinerated in a crucible, the ash remaining has an acid, alkaline, or neutral reaction, depending on the proportion and type of anions and cations present. When the same food is consumed by a person, its final products sometimes have the same reaction as the ash, but there are other factors that modify the ash left by metabolic processes. Proteins, phospholipids, and nucleoproteins yield sulfuric, phosphoric, and uric acids. These acids are neutralized by basic elements before excretion and thus tend to diminish the alkaline factors of blood and urine. Fruits and vegetables usually have enough positive ions, e.g., calcium, magnesium, sodium, and potassium, to balance the acid produced from proteins. Organic acids—citric, malic, tartaric, and lactic—present in fruits and vegetables are oxidized to carbon dioxide. Most of this gas is lost by way of the lungs, whereas the potassium salts of the above acids, also occurring in fruits, are oxidized, yielding potassium bicarbonate, which, if present in excess, is excreted in the urine. Thus vegetables, even acid fruits, have an *alkaline effect*. There are some exceptions: benzoic acid, present in cranberries, is not oxidized by the body and is excreted as hippuric acid (after combining with glycine) and thus has an acidic effect on the urine; oxalic acid, found in rhubarb, beet leaves, cocoa, and tea, is oxidized poorly and is neutralized and excreted as oxalate. Table 21-16 lists a number of common

Table 21-16
Potential acidity or alkalinity of foods

Foods having a predominately acidic effect		Foods having a neutral or nearly neutral effect	Foods having a predominately basic effect	
Bacon	Lamb	Butter	Apples	Grapes
Barley	Liver	Buttermilk	Apricots	Lemons
Beef	Lobster	Corn oil	Asparagus	Lettuce
Bread, rye	Macaroni or spaghetti	Cottonseed oil	Bananas	Limes
Bread, whole-grain	Oysters	Cream	Beans, lima	Oranges
Cake, plain	Pastry	Custard	Beans, string	Peaches
Cake, cheese	Plums*	Fudge, chocolate	Broccoli	Pears
Chicken	Pork	Honey	Brussels sprouts	Pineapple
Corn	Prunes*	Ice cream	Cabbage	Potatoes, white
Crab	Sausage	Milk, whole	Cantaloupe	Potatoes, sweet
Crackers	Scallops	Olive oil	Carrots	Raisins
Cranberries*	Shredded wheat	Onions	Cauliflower	Raspberries
Duck	Shrimp	Pie, apple	Celery	Tomatoes
Eggs	Turkey	Sugar	Dates	Walnuts
Fish	Veal	Syrups	Eggplant	
Ham		Tapioca	Figs	
			Grapefruit	

*The ash of these foods is alkaline, but, because of the presence of benzoates, which form hippuric acid, they increase the acidity of the urine in vivo.

foods grouped according to their potential acidity or alkalinity when ingested in average amounts by normal individuals.

Current nutritional problems

Today the science of nutrition, founded on a sound biochemical base, occupies a key position in dealing with many public health problems and their control both in the United States and, through the World Health Organization, in the entire world. Malnutrition, resulting in part from a "population explosion," is currently a problem of grave concern. An estimated 50% or more of the world's population, mainly in the Orient, South America, Central America, and Africa, suffer from hunger and malnutrition. Pockets of undernutrition occur even in the United States. Paradoxically, however, another segment of the American population is overfed and, as a result, subject to a different type of nutritional disorder—obesity, with attendant cardiovascular disease, diabetes, and other related disorders aggravated by overeating.

It is predicted that, unless prompt and adequate remedial steps are taken in time, the world malnutrition problem will become critical by 1990 and catastrophic by 2000, when the projected world population may have more than doubled (to some 6.41 billion).

The major nutritional problems in the underdeveloped countries is an adequate amount of *protein* of good quality and of *calories*. Among the some 300 million children in these areas, protein-calorie malnutrition is found in the majority and causes poor growth, small size, and even a reduction of the head circumference. This may lead to impaired growth of the brain and its development and mental retardation. Malnutrition may likewise lessen resistance to infection, setting up a vicious cycle. These interrelations are represented schematically below:

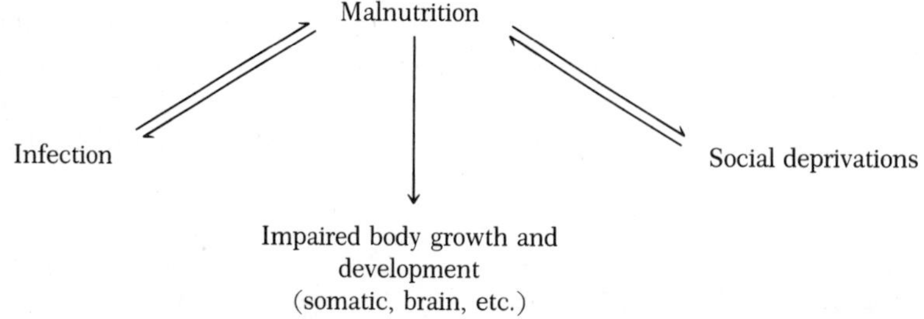

The obvious remedies to the problem recommended by authorities in the field are (1) population (birth) control and (2) increased food production by improved (genetic) varieties, better fertilizers, and better management and distribution practices. Improvements in varieties of rice, wheat, and corn, which are the most important food crops, in that order, are essential. Also, new and increased sources of protein (fish, algae, microorganisms, leaf proteins, petroleum, etc.) are likely to become increasingly important. Education in the use of unfamiliar foods is a necessary accompaniment. Some of these questions are considered in more detail later.

Malnutrition can be caused by factors other than a primary dietary deficiency of one or more essential nutrients. *Secondary*, or *conditioned*, nutritional deficiencies may result from faulty or impaired ingestion, digestion, absorption, and utilization or excessive catabolism caused by pathological processes. A classic example is the magnesium-deficiency syndrome seen in chronic alcoholism, in which the renal reabsorption of magnesium can be impaired. Some therapeutic procedures likewise may, in themselves, precipitate malnutrition. For example, the prolonged use of intravenous glucose-electrolyte solutions, unless appropriately supplemented with essential nutrients, can precipitate iatrogenic malnutrition. Similarly the prolonged use of some drugs without proper nutritional support can lead to a disastrous malnutrition. Emotional disturbance is still another factor just now being recognized as possibly interfering with nutrient utilization and inducing a *negative balance,* i.e., a state in which the output of the nutrient (or its metabolite) in the urine, feces, and sweat exceeds the amount ingested. These examples emphasize the essentiality of nutritional considerations in overall health care.

Dietary goals in the United States

The important problem of adequate nutrition for the population of the United States recently has again become a center of attention and controversy. As a result of a report of a U.S. Senate Committee on dietary goals for the United States, several rather drastic changes in dietary habits were recommended (marked restriction in the intake of animal fats, cholesterol, sugar, and salt) as a means of lessening the risk of several major "killer diseases" (coronary heart disease, strokes,

A practical daily food plan

As a guide for the selection of nutritious foods and for the planning of daily meals, most nutritionists suggest the following general food plan. It is based on the daily use of four *basic food groups* in quantities and variety that will supply the nutrients known to be essential for human nutrition and in balanced amounts that will satisfy the recommended daily allowances desirable for all normal healthy individuals in the United States. One possible exception is the requirement for *iron* in adult women. In this instance supplemental iron may be needed if subnormal hemoglobin levels in the blood are present.

I. *Milk group:* some milk daily:

Children	3 to 4 cups
Teenagers	4 or more cups
Adults	2 or more cups
Pregnant women	4 or more cups
Nursing mothers	6 or more cups

Cheese and ice cream or other dairy products can replace part of the milk

II. *Meat group:* 2 or more servings daily:
Beef, veal, pork, lamb, poultry, fish, eggs, with dry beans and peas and nuts as alternates

III. *Vegetable-fruit group:* 4 or more servings daily, including the following:
A dark green or deep yellow vegetable important for vitamin A—at least every other day
A citrus fruit or other fruit or vegetable important for vitamin C—daily
Other fruits and vegetables, including potatoes

IV. *Bread-cereal group:* 4 or more servings daily:
Whole grain, enriched, restored

Each of the four food groups has been designed to contribute specific nutrients essential for a complete and balanced diet. For example, the milk group is especially important for its contribution of high-quality protein, calcium, and riboflavin. The meat group supplies excellent protein, thiamin, vitamin B_{12}, and minerals. The vegetable-fruit group contributes vitamin A (and β-carotene), ascorbic acid, and other vitamins, minerals, and fiber. The bread-cereal group furnishes carbohydrate, protein, and—if enriched, whole grain, or restored—thiamin, niacin, riboflavin, and iron. Whole-grain cereals also are an important food source of fiber. How these foods and servings are distributed daily is largely a matter of personal preference and convenience. Most preferences are for three meals a day, including a moderate breakfast and lunch and a hearty evening meal. Some could be reserved for a light bedtime snack.

This food plan, along with a variety of selections within each of the food groups and additional carbohydrate or fat as required to meet individual energy needs to maintain a near ideal body weight for age and height, forms a foundation for a well-balanced diet. Most nutritionists agree that it should form the basis not only for the diet of normal individuals but also, with suitable modifications, the diets used for therapeutic purposes.

diabetes mellitus, several types of cancer, hypertension, and obesity). Among the recommendations for decreasing the intake of saturated fats and cholesterol from animal fats was decreasing the consumption of red meats, dairy products, and eggs. This evoked strong protests from the powerful producers of these products, who argued that there was no firm evidence, as yet at least, incriminating cholesterol and saturated fats in coronary heart disease, and perhaps other killer diseases. It also was pointed out that the health and longevity of Americans have steadily improved over the past few decades. Moreover, decrease in the use of these foods would deprive Americans of three of the best sources of several essential nutrients, particularly high-quality protein, iron, zinc, vitamin B_{12}, and calcium.

Other current recommendations (from the Departments of Agriculture and Health and Human Services) have taken a middle-of-the-road position in their *Dietary Guidelines for Americans,* emphasizing the following seven points:

1. Eat a variety of foods; choose daily from fruits, vegetables, whole grains and enriched breads, cereals, and grain products, milk, cheese, milk products, lean meats, poultry, fish, eggs, dried beans and peas (the basic four food groups).
2. Maintain an ideal body weight for height. Obesity is associated with high blood pressure, increased levels of blood lipids, and the common types of diabetes, all of which are associated with increased risks of strokes and heart attacks.
3. Avoid too much fat, saturated fat, and cholesterol. These are associated with increased risk of heart attacks. Moderate the use of eggs, organ meats, and butter, cream, hydrogenated margarines, shortenings, and coconut oil. Trim excess fat from meats and broil, bake, or boil instead of fry.
4. Eat foods with adequate complex carbohydrates (starches, etc.) and fiber, such as whole-grain breads and cereals, beans, peas, nuts, fruits, and vegetables, especially root vegetables. Fiber lessens the incidence of colonic cancer.
5. Avoid too much sugar. Too much sugar can cause tooth decay and may be a factor in obesity. Use less foods containing sugars—candies, soft drinks, ice cream, cakes, cookies, and many baked goods.
6. Avoid too much sodium, as table salt and many processed foods. Too much sodium (salt) is to be avoided, especially by persons with hypertension. Limit the use of salt in cooking and the intake of salty foods, such as potato chips, pretzels, salted nuts, cured meats, pickled foods, and most condiments.
7. If you drink alcohol, do so in moderation. Alcoholic beverages are high in calories and low in nutrients. Heavy drinking can cause cirrhosis of the liver, neurological disorders, and cancer of the throat and neck. One or two drinks a day appear to cause no harm.

Currently, based on epidemiological data, the excessive drinking of coffee has been implicated likewise as a factor in the incidence of pancreatic cancer.

The Food and Nutrition Board of the National Research Council is even more conservative in its recent guidelines:

1. Eat a well-balanced diet selected from a wide variety of foods to enhance an adequate intake of all essential nutrients.
2. Do not overeat. Maintain proper body weight according to height. If overweight, eat less food, particularly sugars, fats and oils, and alcohol. Exercise more but in moderation (with physician's approval).
3. Use salt in moderation.

The FNB, after a 2-year study of the literature, found no conclusive evidence of a link between diet and disease, in particular heart disease, cancer, diabetes mellitus, hypertension, and obesity. Thus the Board, in contradiction to the recommendation of several other groups, found no compelling evidence at this time for any drastic changes in the American diet. However, they realize that at least a decade or more will be required before definitive data can be acquired and final conclusions can be drawn.

Meanwhile, it seems prudent to follow the above recommendations, which agree on most points, in the interest of even better nutrition and better health and performance. The key words are the consumption of a *variety* of healthful foods in *moderation*.

Some clinical applications

Although adequate nutrition is dependent on the availability of the essential nutrients to all cells of the body, a number of factors, both environmental and internal, may affect this process. As in the avitaminoses, discussed earlier in this chapter, general malnutrition may result from *primary* factors attributable to a *deficient dietary intake* or from *secondary* factors attributable to *pathological disorders* (Fig. 21-10). The former would include such interferences with food intake as poverty, floods and famines, food faddism, alcoholism, drug addiction, and certain psychological disorders such as anorexia nervosa. Secondary or *conditioned malnutrition* may result from oral lesions, gastrointestinal disorders (congenital digestive diseases, impaired absorption), metabolic disorders (diabetes mellitus, hypothyroidism or hyperthyroidism), increased excretion (renal or perspiratory), infections, fever, allergy, and some therapeutic diets and procedures. These examples reemphasize the importance of nutrition in overall health care.

A brief consideration of the basic principles followed in making dietary modifications for supportive therapeutic reasons, sometimes called diet therapy, is pertinent at this point. For detailed discussions the appropriate readings given at the end of this chapter should be consulted.

The following is an outline of the general principles involved, with selected examples of the major types of dietary modifications.

1. *General principles*. The foundation of any therapeutic diet is the intelligent and varied selection of foods within the basic four food groups if possible; diet is usually *supportive* to standard therapeutic measures; the patient *requires all essential nutrients,* just as does a normal subject. However, *illness* is *stress*. Therefore the patient's essential nutrient requirements may be considerably altered from the normal. The *principal modifications* of diet for supportive therapeutic reasons include:
   a. *Removal* of offending or low-tolerance food and replacement if possible with a more acceptable substitute of equal nutritional value, e.g.:
      (1) Sugar in diabetes mellitus
      (2) Phenylalanine in phenylketonuria (PKU)
      (3) Allergies
   b. *Replacement* of nutrient poorly absorbed or lost by excretion, increased catabolism, etc., e.g.:
      (1) Protein in certain renal disease; all nutrients in nontropical sprue
      (2) Calcium (and fluoride?) in osteoporosis, osteomalacia
2. *Major modifications*
   a. *Qualitative*. Consistency, flavor, and temperature of diet may be altered.

b. *Quantitative.* Amounts of one or more dietary component may be increased or decreased, e.g.:
   (1) Percentage of protein, fat, or carbohydrate
   (2) Amount of sodium
   (3) Amount of saturated or unsaturated fats
   (4) Amount of cholesterol
   (5) Amount of certain vitamins and minerals

c. *Acceptability.* Diet must be acceptable to the patient; ethnic preferences must be considered.

Obesity

Obesity, an accumulation of excess body fat—the body weight exceeding 20% above the ideal weight, results from an oversupply of calories relative to the total caloric expenditure of the individual. The factors regulating this energy balance are varied and complex. The possible influence of endocrine dysfunction is discussed in Chapter 18, in which the importance of the hypothalamus in the control of food intake is discussed.

Another possible factor involved in obesity has been reported in a recent study, which has demonstrated that obesity is associated with a deficiency of the enzyme Na^+/K^+–adenosine triphosphatase (ATPase). This enzyme not only controls Na^+ and K^+ concentrations in cells but also participates in a number of metabolic reactions that *use energy*. Thus it seems possible that a deficiency of this enzyme could impair normal energy metabolism in obese individuals, thus explaining why they gain more weight even though they ingest an amount of food equivalent to that consumed by individuals of average body weight. Of course, the observed enzyme deficiency could be an effect of obesity rather than a cause.

From a dietary standpoint, a reduction in body fat may be achieved by providing fewer calories than are needed by the individual. An increase in activity with a proportional increase in energy expenditure is helpful, too, of course. A corrective diet that allows for a loss of 1 or possibly 2 lb per week and that supplies a generous amount of protein and all the essential minerals and vitamins should be prescribed.

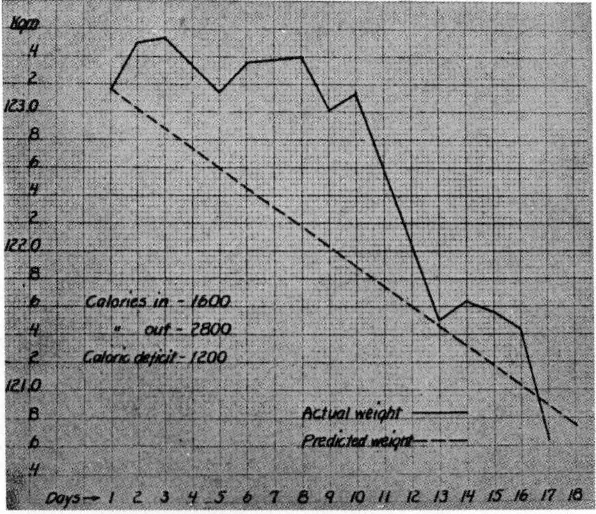

Figure 21-20
Weight chart of an obese patient on a reducing diet. The increases in weight and failure to follow the predicted weight line are ascribed to addition and retention of water by the tissues.

A daily multivitamin preparation, using the recommended dietary allowances as a guide, would provide a margin of safety. The normal diet should be modified based on the basic four groups. Diets containing 1200 kcal daily for women and 1500 to 1800 kcal daily for men are acceptable levels. Since body fat is about one fifth water, a pound is equivalent to 454 gm × ⅘ × 9 kcal/gram, or about 3200 kcal. In other words the maintenance diet for the present weight may be reduced by about 3200 kcal per week, or from 450 to 500 kcal per day. If reduction does not immediately occur, the diet need not necessarily be changed because water is often held temporarily in the tissues in place of fat and is lost later. The serving of smaller meals more frequently, but *without* an increase in the daily total, may be beneficial for some individuals.

Increased physical activity can contribute to weight reduction and is advised, in moderation, to improve physical well-being and to prevent loss of body protein as well as fat.

A puzzling fact is the inability to lose weight frequently experienced by obese individuals when they go on a reducing diet. They frequently do not lose weight for as long as 2 weeks and sometimes even gain weight. This stems from a retention of water by the tissues. After a time, however, the water is eliminated and the weight drops. Fig. 21-20 is a typical graph.

Underweight

Underweight is frequently more difficult to combat than obesity. Increasing the total caloric intake is of *paramount* importance. A good multivitamin supplement is indicated to take care of any unsuspected deficiency and to stimulate appetite. High-fat and high-carbohydrate foods of rather concentrated types are recommended. An adequate protein intake of varied nature should be provided, but an excess is to be avoided because protein has a higher specific dynamic action than do carbohydrates and fats. Frequent small meals are often helpful.

Gastrointestinal disorders

Hyperchlorhydria is a symptom that is frequently encountered. Since hydrochloric acid is secreted at a constant concentration, high acidity simply means an increased volume of gastric juice, too large to be buffered in the normal manner (Chapter 17). One method of combating it is to eliminate highly spiced foods from the dietary, since these may stimulate the flow of gastric juice. Another way is to feed a high-protein diet, since protein exerts a buffering effect.

In gastritis, dietary treatment is usually a bland, smooth, soft diet, whose composition depends on whether hypoacidity or hyperacidity is present. In peptic ulcer, hyperacidity, as well as high pepsin values, is usually present. Many special diets have been devised and recommended. In general, diets for peptic ulcer consist of frequent feedings of small quantities of soft foods that constantly neutralize the secreted acid and prevent its accumulation as free acid; but the amount and kind of food must be regulated to satisfy all the nutritional requirements. An increased fat intake may be of value by decreasing gastric motility and increasing emptying time.

Flatulence, or gas formation in the gastrointestinal tract, occurs in both health and disease. In general, a lack of hydrochloric acid in the stomach may permit more microorganisms than usual to pass into the duodenum, and gas formation ensues as a result of their growth. Some foods are more productive of gas than others. Well-known examples are members of the cabbage family and also turnips, onions, peas, and beans. Melons, cucumbers, and radishes are other offenders in this respect. Various syrups and nondigestible carbohydrate also are gas producers, perhaps serving as media for yeasts in the lower part of the intestinal tract.

Diseases of the liver and gallbladder

In liver diseases a high-carbohydrate, moderately high-protein, and moderately low-fat diet may be indicated. However, the load of protein should not be too great because the liver is the chief site of amino acid degradation and formation of urea, and the excessive amount of protein may be undesirable. Current studies of the substitution of the α-ketoacid derivatives for their corresponding essential amino acids in patients with nitrogen-accumulating conditions was mentioned earlier. Promising results have been obtained thus far in certain renal and hepatic disorders. The restriction of fat is recommended because fat requires bile secretion for its digestion and absorption. Since bile secretion is depressed, the fat-soluble vitamins will not be absorbed in sufficient amounts. For this reason vitamin supplements are advised. If the patient is unable to eat, intravenous feeding of amino acids, together with glucose, vitamins, and salts, may be desirable.

When either cholecystitis or cholelithiasis is present, contraction of the gallbladder may result in severe pain. Because dietary fat stimulates gallbladder contraction, it should be kept at a minimum. During acute phases only 20 to 30 gm of highly emulsified fat daily may be indicated. In chronic cholelithiasis foods rich in cholesterol may be restricted (Table 21-17), since gallstones usually contain significant amounts of cholesterol.

Table 21-17
Cholesterol content of some selected foods

Item,* edible portion, and common serving	Weight (gm)	Cholesterol (mg)
Beef, trimmed, 4 oz	113	103
Brains, 3 oz	85	2700
Butter, 1 pat	7	18
Cheese		
Cheddar, 1 oz	28	28
Cottage, creamed (4% fat), ½ cup	122	24
Cream, 1 oz	28	32
Chicken or turkey with skin, 4 oz	113	99-105
Cream, 20% fat, 1 tbsp	15	11
Egg, 1 large, whole	50	252
Fish, various types, 4 oz	113	70-110
Fruits, vegetables, grains	—	0
Ham, trimmed, 4 oz	113	90
Ice cream, 10% fat, 1 scoop	71	28
Kidney, 3 oz	85	683
Lamb, trimmed, 4 oz	113	113
Lard and other animal fats, 1 tbsp	14	14-17
Liver		
Beef, 3 oz	85	372
Chicken, 3 oz	85	634
Margarine (all-vegetable oils), 1 pat	7	0
Milk		
Whole, 1 cup	244	34
2% Fat, 1 cup	246	22
Skim, 1 cup	246	5
Pork, trimmed, 4 oz	113	100
Shellfish (clams, crab, lobster, shrimp), 4 oz	113	71-170
Sweetbreads, 3 oz	85	396
Veal, trimmed, 4 oz	113	114
Vegetable oils	—	0
Yogurt, from nonfat milk, 1 cup	246	17

*Values for meats are as cooked, without added fat; values for many other foods are given in Feeley, R.M., et al.: Cholesterol content of foods, J. Am. Dietet. Assoc. **61**:134, 1972; and Adams, C.F.: Nutritive value of American foods in common units, Handbook no. 456, Washington, D.C., 1975, U.S. Department of Agriculture.

Celiac syndrome

The celiac syndrome, which occurs chiefly in children, is manifested in *classic celiac disease*. The principal symptoms are *malabsorption* and *generalized malnutrition*. The abdomen is greatly distended, largely becuase of accumulation of intestinal gas. At the same time there are metabolic disturbances; fat stores are used and growth is stunted. Similar conditions are seen in nontropical sprue of adults and in tropical sprue. Fatty stools are common to all these conditions, with a consequent loss also of calcium and vitamin D and other fat-soluble vitamins through the feces. Osteoporosis and osteomalacia may occur. Anemia, either macrocytic or microcytic, is common. Besides an inability to digest and absorb fats, there is frequently also an interference with starch digestion. So-called primary celiac syndrome, or *gluten-sensitive enteropathy* (celiac disease of children, adult celiac disease), appears to result from damage to the intestinal mucosa and resulting malabsorption by the toxic action of undigested *N*-pyrrolidone acidic peptides derived from the protein gliadin, of certain cereal grains. The syndrome is now believed to result either from a lack of enzyme, a peptidase, in the intestinal secretions to digest the toxic peptides or from a severe local immunological reaction to the toxic peptide. The complete elimination of wheat, oats, barley, and rye from the diet has been used with dramatic results.

Cardiovascular disease

Dietary measures may be of considerable value in disorders of the heart and blood vessels (Chapter 12). Obesity is a major obstacle in treating heart diseases. The disadvantages of obesity may be fourfold. (1) Even when the heart is unimpaired, a large body mass puts an abnormally great strain on the heart. (2) In obesity there frequently occur fat deposits on and between the cardiac muscle bundles, decreasing the heart muscle efficiency. (3) Abdominal fat may impede the movement of the diaphragm, and this in turn affects the heart movements. (4) Atherosclerosis, which often accompanies obesity, may involve the coronary vessels and in that way directly affect the blood supply of the heart. Accordingly, cardiac patients are usually advised to reduce their weight to normal or slightly below. A plentiful supply of vitamins, particularly the B complex and C, should be ensured and the protein should be adequate.

The relationship of cholesterol to atherosclerosis was discussed in Chapter 12. The deposition of plaques consisting chiefly of cholesterol in the walls of the coronary arteries and the aorta is well established. Low-cholesterol (Table 21-17), low–saturated fat diets are advised, with the substitution of fats containing unsaturated fatty acids for saturated fatty acids. Although not a cure, such a modification of diet tends to lower the blood cholesterol level. Therefore it is wise to limit severely the intake of animal fats and hydrogenated oils, using polyunsaturated vegetable oils as much as possible. Weight reduction, with the lowering of total fat intake and a shift to the unsaturated type, usually reduces the hypercholesterolemia. In familial hypercholesterolemia, low-cholesterol diets, in addition to the increase in polyunsaturated fats, are useful. The cholesterol concentration of a number of common foods are given in Table 21-17. These are mainly of animal origin; plants contain no cholesterol.

Coronary heart disease is unquestionably one of the most important health problems today. A tremendous amount of research is being conducted, and significant information is being accumulated. However, many unanswered questions remain. The role of diet remains controversial. No single dietary factor appears to be a major etiological factor. Food cholesterol, saturated fatty acids, and even sucrose continue to be suspect. However, several newer nutritional factors currently are receiving increased attention. These include fiber, certain trace elements, xanthine oxidase (Chapter 14), eicosapentenoic acid, and *trans*-fatty acids (Chapter 23).

Diabetes mellitus

Diabetes mellitus is characterized by an impaired ability to use carbohydrate (Chapters 11 and 18). This is true of both major types of diabetes: type 1 (insulin dependent, or juvenile type) and type 2 (insulin independent or maturity onset). Thus control of carbohydrate and caloric intake are important in the management of this, or more accurately, these diseases. Current view emphasizes the prime importance of diet and appropriate exercise, as well as insulin in type 1 diabetes mellitus, in the control of these conditions.

The principal dietary modifications include the control of the caloric intake to maintain as near an ideal body weight as possible and the limitation of the consumption of refined sugars (e.g., sucrose, glucose). Cakes, pies, pastries, and candies are restricted. However, the diet should contain approximately 60% carbohydrate, mainly as complex carbohydrates found in flour and other grain products and root vegetables. The diet otherwise should be well balanced, with about 12% to 20% protein from a variety of meats, fish, poultry, legumes, and dairy products. About 9% to 20% of the diet should comprise food fats, with a saturated to polyunsaturated fatty acid ratio of about 1:1. This, of course, means the use of the polyunsaturated vegetable oils in amounts approximately equal to that of fats from meats and milk and milk products. Fiber in the form of fruits, vegetables, legumes, and whole grain cereals should be consumed in adequate amounts. Indeed, the practical food plan, using the basic four food groups as outlined earlier, is an ideal basic diet for a diabetic with the limitations just described. The individual type 1 diabetic patient, with caloric intake being rigidly regulated and correlated with the time, type, and amount of insulin consumption, must be taught the use of food tables, how to weigh his diet, and the calculations involved.

The practical food plan outlined earlier also is satisfactory for the type 2 diabetic patient. In fact, the hyperglycemia and glycosuria and other symptoms may lessen or even disappear with caloric and refined sugar restriction and the attaining of near ideal body weight. Patients with type 2 diabetes are usually either overweight or frankly obese. Also, with weight reduction their need for oral hypoglycemic agents to control the hyperglycemia and glycosuria frequently disappears.

Alcoholic beverages are not recommended for diabetic patients because of their high caloric density and also the fact that alcohol inhibits gluconeogenesis (Chapter 11) and increases plasma and hepatic lipids. For further details, pertinent readings given at the end of this chapter should be consulted.

Phenylketonuria

The metabolic error in phenylketonuria is the inability of the organism to metabolize phenylalanine (Chapter 13). A diet low in this amino acid therefore is required. The dietary restrictons are difficult because (1) phenylalanine is an essential amino acid and (2) all dietary proteins are about equally rich in it. Fortunately a commercial preparation, Lofenalac, in which most of the phenylalanine has been removed, is available. Multivitamin and mineral supplements are given separately along with milk and a few other foods in moderation. Some phenylalanine is usually metabolized, and tyrosine substitutes for part of it. The prompt administration of a corrective diet to the infant can prevent the brain damage and mental retardation that otherwise result. With the use of this regimen at a sufficiently early age, the mentality of these retarded children improves, as does their general physical condition. The color of the hair also darkens. Later, in many infants synthesis of the missing enzyme, phenylalanine hydroxylase, increases and the children are able to live a reasonably normal life.

Fevers

The total caloric requirements in fever may be calculated by adding 13% of the normal basal metabolic rate for the individual for every degree of fever. To this may be added an additional 10% if there is much extra protein catabolism, as there usually is in most fevers, and from 10% to 30% for the restlessness of the patient. It is thus seen that the caloric needs of a febrile patient may be very high indeed. The diet therefore should be high in carbohydrates to provide for much of this metabolic need, to spare proteins as far as possible, and to aid in combating acidosis. Proteins should be sufficiently high to maintain the patient in nitrogen equilibrium. Fats should be normal in quantity and of a type easily digested. In fevers an alkaline-ash diet is preferred, as a further safeguard against acidosis, and additions of sodium chloride to the diet, to make good the losses of salt in perspiration.

Food allergy

Many individuals are peculiarly sensitive to certain foods, just as others are to pollen or other particles in the air they breathe. The symptoms range from sneezing to vomiting, from headaches to hives, from edema to diarrhea, and many more, some minor and some quite serious. These effects are believed to be due to the release of histamine by an immunological reaction. The symptoms may be caused by more than one food. Proteins have been considered the causative agents, and undoubtedly they are in most cases, but there are some assertions in the literature that fats and even carbohydrates have been found responsible. If the allergen cannot be identified by appropriate skin testing, the patient is put on successive standard elimination diets. These are standardized diets, each one eliminating certain foods or food groups. If the patient is without symptoms on one of these, he is not sensitive to the constituents of that diet. Then foods that are absent from the basic diet are added one at a time until a reaction, peculiar to the patient, is produced by one or more of them. The most usual food allergens are milk, eggs, wheat, and potatoes. During the testing, vitamin or other supplementation may be needed, since the elimination diets may very well be low in one or more essential nutrients. The best treatment, after the offending food or foods is determined, is to plan an adequate diet that does not contain the allergens.

Preoperative and postoperative diets

A good nutritive condition is a great asset before surgery. If an operation can be anticipated, the surgeon should exercise every effort to get the patient into a good nutritive state, correcting any deficiencies that may exist. Fluids and carbohydrates (e.g., sugar candy) are given in fairly large amount on the preceding day, and no food after the evening meal is offered. Postoperative diets depend on the type of operation, condition of patient, etc. In general, a high caloric (2800 kcal minimum to 6000 kcal) intake, therapeutic level vitamin supplementation, and some increase in protein intake are currently recommended. The high caloric intake is to permit maximal use of protein for tissue repair rather than for conversion to energy.

Parenteral feeding and "hyperalimentation"

The intravenous administration of amino acids is in extensive use today. Carbohydrates, vitamins, and salts can also be administered intravenously. During the past few years highly effective mixtures of glucose, amino acids, electrolytes, and vitamins have been developed and found to be effective for parenteral feeding or total parenteral nutrition, sometimes mistakenly called "hyperalimentation." These mixtures are especially useful in maintaining nutrition in patients having various types of gastrointestinal disorders, especially those requiring corrective surgery. Fat suspensions for intravenous feeding have been restricted in the United States because of various toxic symptoms occurring after their administration, e.g., dizziness, dyspnea, low back pain, chills with or without fever, blood clotting abnormalities, hemorrhage, anemia.

It is interesting to note that infants under 2 months of age who have various types of malfunction of the gastrointestinal tract that require long-term total parenteral nutrition have been maintained satisfactorily for periods up to 60 days. Satisfactory weight gain and normal growth were obtained despite repeated surgical procedures, sepsis, and enteric losses during the period. The solution used for intravenous feeding was a mixture containing 20% glucose and 3.3% amino acids (as a fibrin hydrolysate) with appropriate amounts of electrolytes and vitamins added. The solution was delivered continuously with an infusion pump through an indwelling catheter placed in the superior vena cava. The solution was fat free, and the caloric value of infusate was 0.9 kcal per milliliter.

A large series of adult patients with chronic gastrointestinal disease also have been successfully nourished exclusively by intravenous feeding for as long as 210 days, receiving from 2400 to 4500 kcal per day. Positive nitrogen balance was maintained in all cases, with satisfactory wound healing, weight gain, and increased strength and activity occurring. However, extreme care must be taken to ensure that the patients receive *all* essential nutrients during prolonged periods of intravenous feeding. There have been reports of deficiencies of essential unsaturated fatty acids and of trace elements (copper and zinc) in patients administered intravenous fluids lacking these nutrients. Substituting a fluid containing the missing nutrient(s) promptly corrected the deficiency symptoms. Thus, as a result of the remarkable advances in the field of nutrition, tissue maintenance and synthesis in humans can be achieved today by the prolonged, exclusive intravenous administration of essential nutrients. This is indeed a paramount example of the practical value of nutrition to medicine and mankind.

Suggested readings

Alfin-Slater, R.B., and Kritchevsky, D., editors: Human nutrition: a comprehensive treatise, 4 vols., New York, 1979, Plenum Publishing Corp.

American Dietetic Association: Handbook of clinical nutrition, New Haven, 1981, Yale University Press.

Bodwell, C.E., editor: Evaluation of proteins for humans, Westport, Conn., 1977, Avi Publishing Co.

Bray, G.A., editor: Obesity in America, Am. J. Clin. Nutr. **33**(Suppl. 2):376, 1980.

Briggs, G.M., and Calloway, D.H.: Bogert's nutrition and physical fitness, ed. 10, Philadelphia, 1979, W.B. Saunders Co.

Burkitt, T.J., and Trowell, H.C.: Refined carbohydrate food and disease, New York, 1975, Academic Press, Inc.

Cheung, W.Y., editor: Calcium and cell function, vol. 1, Calmodulin, New York, 1980, Academic Press, Inc.

Committee on Recommended Dietary Allowances, Food and Nutrition Board, National Research Council: Recommended dietary allowances, Washington, D.C., 1980, National Academy of Sciences.

Darby, W.J., et al.: Niacin, Nutr. Rev. **33**:289, 1975.

DeLuca, H.F.: the vitamin D system in the regulation of calcium and phosphorus, Nutr. Rev. **37**:161, 1979; Nutr. Rev. **38**:169, 1979.

DeLuca, H.F.: Vitamin D—metabolism and function, New York, 1979, Springer-Verlag.

Food and Nutrition Board, National Research Council: Toward healthful diets, Washington, D.C., 1980, National Academy of Sciences.

Goodhart, R.S., and Shils, M.E., editors: Modern nutrition in health and disease, ed. 6, Philadelphia, 1980, Lea & Febiger.

Goodman, D.S.: Vitamin A and retinoids—recent advances, Fed. Proc. Sympos. **38**:2501, 1979.

Guthrie, H.A.: Introductory nutrition, ed. 3, St. Louis, 1975, The C.V. Mosby Co.

Herbert, V.: Vitamin B_{12}. In Present knowledge in nutrition, ed. 4, Washington, D.C., 1976, The Nutrition Foundation, Inc.

Hodges, R.E.: Nutrition in medical practice, Philadelphia, 1980, W.B. Saunders Co.

Hodges, R.E., et al.: Hemopoietic studies in vitamin A deficiency, Am. J. Clin. Nutr. **31**:871, 1978.

Hoekstra, W.G., et al., editors: Trace element metabolism in animals, Baltimore, 1974, University Park Press.

Horwitt, M.K.: Therapeutic uses of vitamin E in medicine, Nutr. Rev. **38:**105, 1980.
Hurley, L.S.: Developmental nutrition, Englewood Cliffs, N.J., 1980, Prentice-Hall, Inc.
Keys, A.: Overweight, obesity, coronary heart disease and mortality, Nutr. Rev. **38:**297, 1980.
Klevay, L.M., Reck, S.J., and Barcome, D.F.: Evidence of dietary copper and zinc deficiencies, J. A. M. A. **241:**1916, 1979.
Krause, M.Y., and Mahan, L.K.: Food, nutrition, and diet therapy, ed. 6, Philadelphia, 1979, W.B. Saunders Co.
Machlin, L.J.: VItamin E—a comprehensive review, New York, 1980, Marcel Dekker, Inc.
Monsen, E.R.L., et al.: Estimation of available dietary iron, Am. J. Clin. Nutr. **31:**134, 1978.
Multiple biotin-dependent carboxylase deficiencies associated with defects in immunity, Nutr. Rev. **37:**289, 1979.
Neuberger, A., and Jukes, T.H., editors: Biochemistry of nutrition, vols. 1a and 1b, Baltimore, 1979, University Park Press.
Norman, A.W.: Vitamin D—the calcium homeostatis steroid hormone, New York, 1979, Academic Press, Inc.
Nutritional requirements of man, Washington, D.C., 1979, The Nutrition Foundation, Inc.
Olson, R.E., editor: Protein-calorie malnutrition, Washington, D.C., 1975, The Nutrition Foundation, Inc.
Orten, A.U., Burn, C.G., and Smith, A.H.: The effects of prolonged chronic vitamin A deficiency, Proc. Soc. Exp. Biol. Med. **36:**82, 1937; Yale J. Biol. Med. **9:**189, 1936.
Prasad, A.S.: Nutritional zinc today, Nutrition Today **16**(2):4, 1981.
Prasad, A.S., editor: Proceedings of the Conference on the Clinical and Public Health Significance of Trace Minerals, New York, 1981, Alan R. Liss, Publishers.
Present knowledge of nutrition, ed. 4, Washington, D.C., 1979, The Nutrition Foundation, Inc.
Rodriguez, M.S.: A conspectus of research on folacin requirements of man, J. Nutr. **108:**1983, 1978.
Roe, D.A.: Clinical nutrition for the health scientist, Boca Raton, Fla., 1979, C.R.C. Press, Inc.
Shils, M.E.: How to nourish the cancer patient, Nutrition Today **16**(3):4, 1981.
Sipple, H.L., and McNutt, K.W.: Sugars in nutrition, Washington, D.C., 1974, The Nutrition Foundation, Inc.
Stadtman, T.C.: Selenium-dependent enzymes, Annu. Rev. Biochem. **49:**93, 1980.
Stunkard, A.J., editor: Obesity: basic mechanisms and treatment, vol. 1, Philadelphia, 1978, W.B. Saunders Co.
Suttie, J.W., editor: Vitamin K metabolism and vitamin K–dependent proteins, Baltimore, 1979, University Park Press.
U.S. Department of Agriculture and U.S. Department of Health and Human Services: Nutrition and your health—dietary guidelines for Americans, Washington, D.C., 1980.
Van Itallie, T.B.: Obesity, Am. J. Clin. Nutr. **32**(Suppl.):2723, 1979.
Vitamin A: tumor initiation and tumor promotion, Nutr. Rev. **37:**153, 1979.
White, P.L., and Selvey, N., editors: Proceedings: Western Hemisphere Nutrition Congress, Chicago, 1978, American Medical Association.
Williams, S.R.: Nutrition and diet therapy, ed. 4, St. Louis, 1981, The C.V. Mosby Co.

Chapter 22

Molecular aspects of disease

A fundamental premise of biological sciences is that all biological activity is the result of biochemical reactions and that for every *pathology* there is an *underlying biochemical defect*. This broad and rather categorical statement is continuing to be proved valid as further knowledge of the etiology of human diseases becomes available. Perhaps this might be expected, since the metabolism of the organism is in a *dynamic* state and any impairment in its operation, any biochemical defect (as mentioned in Chapter 21), could result in cellular dysfunction, in anatomical and pathological lesions, and unless corrected, in overt disease. Progress in this area during the past decade or so represents one of the most significant achievements of modern biochemistry.

Present evidence has demonstrated that a biochemical lesion or lesions may be responsible for the dysfunction of any tissue or organ in the human body. This is true of diseases involving dietary, metabolic, infectious, or toxic substances, to mention a few. Biochemical dysfunction in turn may involve every major area of biochemistry: the metabolism of carbohydrates, lipids, proteins, purines and pyrimidines, inorganic elements, and the various biocatalysts of metabolic reactions—the enzymes, hormones, and vitamins.

From a *molecular* point of view, human diseases may be grouped into *three* major categories. The first of these categories includes those diseases resulting from *deficiencies* of organic molecules (and certain inorganic elements) that the human organism *cannot synthesize* but requires *preformed*. These substances include the essential nutrients discussed in Chapter 21: the 8 or 10 essential amino acids, linoleic acid, some 13 vitamins, and at least 15 inorganic elements. Human diseases resulting from any one or more of these essential substances are considered in Chapter 21, and those involving inorganic elements and the vitamins are summarized in Tables 21-11, 21-13, and 21-14 respectively.

A second category of "molecular diseases" are those in which there is an impairment or defect in the *biosynthesis* of essential molecules that the human organism *can* form. The defect may be manifested qualitatively or quantitatively, i.e. the essential molecule synthesized (usually an enzyme or other protein) may be altered in its amino acid composition and hence have its structure changed so that its *biological activity* is *reduced* or completely *destroyed,* or the *amount* formed may be deficient. Usually, a *genetic defect* is responsible, as is discussed later. These diseases are termed "diseases of metabolism" and include principally those proteins and enzymes required for special purposes, such a digestion, absorption, transport, and metabolism in a general sense. This group is the major concern of this chapter.

The third category includes molecules that produce disease as a result of a person's excessive intake of or exposure to them in the environment. This group includes diseases resulting from overindulgence in alcoholic beverages or sweet, fatty, and salty foods that may produce damage to the liver and other organs and tissues. These diseases may be manifested as obesity, hypertension, cardiovascular disease, possibly diabetes, and other pathological conditions (Chapter 21). A number of drugs and environmental contaminants also fall into this category, including pesticides, asbestos, a variety of toxic inorganic agents, pathogenic microorganisms, viruses, and fungi. Consideration of the biochemical defects produced by these latter substances more properly belongs in other disciplines, particularly in bacteriology, immunology, and toxicology.

Heritable diseases of metabolism

Heritable diseases of metabolism are a group of molecular diseases that has been of special interest to biochemists for many years. One, cystinuria, was described by Wollaston in 1810. Four others—albinism, alkaptonuria, congenital porphyria, and pentosuria—were known in 1908, when Sir Archibald Garrod introduced the term *inborn errors of metabolism* for them, a term that is still used. Then in 1949 Pauling and his associates introduced the term *molecular disease* for sickle cell anemia as a result of his now classic biochemical studies of sickle cell hemoglobin (Chapter 16).

Today as a result of extensive research in this field over 300 hereditary errors of metabolism have been described. The biochemical defects have been identified in a

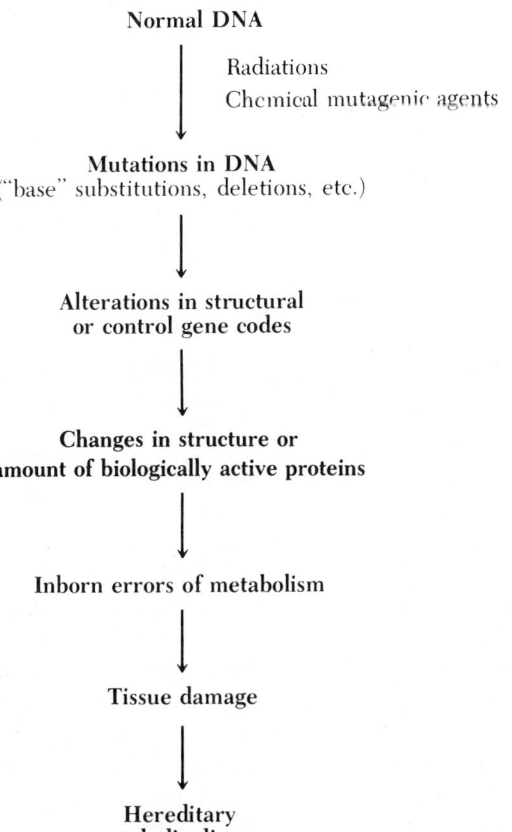

Figure 22-1
Diagram of the effect of mutations in DNA on structure or amount of enzymes or other biologically active proteins formed and the development of heritable metabolic diseases.

number of instances. If the over 300 abnormal hemoglobins, a large number of abnormal plasma proteins, and numerous other proteins are added to this list, the number of genetically altered proteins known in man would be more than doubled, possibly exceeding 1000. It must be added, however, that not all genetic errors of metabolism result in serious pathological conditions. Some are benign and produce no adverse effects whatsoever. The pathogenicity of the molecular defect depends on the degree to which the structure of the protein or enzyme is altered and its effect on the functional or active catalytic site. Thus some proteins have complete biological activity despite amino acid substitutions or deletions, whereas others are less active or totally inactive. Even so, this group of metabolic diseases includes the largest number of diseases afflicting mankind.

The sequence of amino acids and hence the tertiary structure and biological activity of the proteins involved are dictated, of course, by the sequence of purine-pyrimidine bases in the segment of DNA, the gene responsible for their formation (Chapter 10). Fig. 22-1 is a schematic representation of the sequence of events involved in the alterations of DNA and, in turn, the effect of such mutations on the functional activity of the protein thus formed and the possible development of heritable metabolic diseases.

It should now be apparent that any change in the identity of a single purine or pyrimidine base in the gene can affect the composition of the protein ultimately synthesized. An experiment illustrating this point involved treating tobacco mosaic virus–ribonucleic acid (TMV-RNA) with nitrous acid. Nitrous acid replaces the primary amino groups of such bases as adenine and cytosine with a hydroxyl.

$$\text{Cytosine (NH}_2\text{)} \xrightarrow{\text{HONO}} \text{(OH) derivative} + N_2 + H_2O$$

In this way the codon for proline, CCC, may be changed to UUU, CUU, CUC, etc., and may therefore code for another amino acid entirely, e.g., leucine (CUU, CUC). In this example, the *mutant* RNA leads to the formation of viral-coat proteins in which the C-terminal amino acid sequence is slightly altered; i.e., leucine replaces proline.

Pro Ala-Thr-COOH (normal viral protein)
Leu Ala-Thr-COOH (mutant viral protein)

The normal-coat protein is resistant to the action of the protease carboxypeptidase because this enzyme is unable to hydrolyze the proline-containing peptide bond. On the other hand, the mutant coat protein is readily digested so that is is not as effective as the normal one in protecting the all-important genetic material from the destructive action of the host cell. Therefore this experiment also showed, on a molecular basis, how mutations can be detrimental to the organism.

Many mutations anologous to the foregoing example have occurred over the centuries in humans and other living organisms, some resulting in genetic diseases, as pointed out. These have been passed from generation to generation according to the laws of human genetics (Chapters 9 and 10). Examples of some hereditary disorders of human metabolism are summarized in Tables 22-1 through 22-5. Some of these have been mentioned in the earlier chapters on metabolism (Chapters 11-14) and nutrition (Chapter 21). The metabolic disorders listed in the tables are grouped into those affecting: carbohydrates (Table 22-1), lipids (Table

Table 22-1
Some hereditary disorders of carbohydrate metabolism

Disorder*	Molecular deficiency or defect	Principal manifestation†
Diabetes mellitus, type I	Insulin	Glucose (B,U)
Diabetes mellitus, type II	Insulin receptors (?)	Glucose (B,U)
Disaccharide intolerance	Lactase, maltase, or sucrase	Diarrhea, flatulence
Encephalomyelopathy (Leigh's disease)	Pyruvate carboxylase	Pyruvate, lactate (B)
Essential fructosuria	Fructokinase	Fructose (B,U)
Fructosemia	Fructose 1,6-diphosphatase	Fructose (B,U)
Fructose intolerance	Fructose 1-phosphate aldolase	Fructose 1-phosphate (U)
Galactosemia	Galactose 1-phosphate uridyltransferase	Galactose (B,U)
Galactokinase deficiency	Galactokinase	Galactose (B,U)
Gargoylism (Hurler's syndrome)	α-L-Iduronidase	Dermatan sulfate (U,T) Heparan sulfate (U,T)
Glycogen storage diseases		
Type 1 (von Gierke's disease)	Glucose 6-phosphatase	Glycogen (L)
Type 2 (Pompe's disease)	α-1,4-glucosidase	Glycogen (T)
Type 3 (Cori's disease)	Amylo-1,6-glucosidase	Glycogen (T)
Type 4 (Andersen's disease)	Amylo-1,4 → 1,6-trans-glycosylase ("branching enzyme")	Glycogen (T)
Type 5 (McArdle's disease)	Phosphorylase (muscle)	Glycogen (M)
Type 6 (Hers' disease)	Phosphorylase (liver)	Glycogen (L)
Type 7	Phosphofructokinase (muscle)	Glycogen (M)
Type 8	Phosphorylasekinase (liver)	Glycogen (L)
Type 9	Glycogen synthase (liver)	Glycogen (L)
Hunter's syndrome	Sulfoiduronidate sulfatase	Dermatan sulfate (U,T) Heparan sulfate (U,T)
Hyperbilirubinemia	UDP-glucuronyltransferase	Bilirubin (S)
Hyperoxaluria, type I	α-Ketoglutarate-carboligase	Oxalate (U)
Hyperoxaluria, type II	D-Glycerate dehydrogenase	Oxalate (U)
Lactic acidosis	Decreased gluconeogenesis	Lactate (B,U)
Mannosidosis	α-Mannosidase (lysosomal)	Oligosaccharides (T,U)
Pentosuria	L-Xylulose dehydrogenase	L-Xylulose (U)
Primaquine sensitivity (favism)	Glucose 6-phosphate dehydrogenase	Hemolytic anemia
Sanfilippo syndrome A	Heparan sulfate sulfamidase	Heparan sulfate (U,T)
Sanfilippo syndrome B	*N*-Acetyl-α-D-glucosaminidase	Heparan sulfate (U,T)

*See index for page reference to discussion in text.
†Affected metabolite(s) found in increased amounts primarily in: *B*, blood; *F*, feces; *L*, liver; *M*, muscle; *P*, plasma; *S*, serum; *T*, tissues; *U*, urine.

22-2), amino acids (Table 22-3), purines and pyrimidines (Table 22-4), and a miscellaneous group that do not fit into these other categories (Table 22-5). A number of hereditary disorders also are known involving various hormones, vitamins (D, K, B_6, folic acid, and B_{12}), and inorganic elements (Ca, Mg, Fe, Cu, and I). However, these are included in other appropriate groups, since only the protein moiety of the affected enzymes appears to be involved.

For each metabolic disorder listed in the tables the molecular defect involved, as understood at the present time, is given in the second column.

The principal manifestation(s) of the disorder are summarized in the third column. The principal tissue affected or the alteration of the amount of the affected metabolite in the blood, plasma, urine, or feces is included where information permits. If the molecular defect involves a *biosynthetic* enzyme, the metabolite most affected (usually increased) is the one immediately preceding the reaction blocked

Table 22-2
Selected hereditary disorders of lipid metabolism*

Disorder	Molecular deficiency or defect	Principal manifestation
Abetalipoproteinemia	Apoprotein of LDL	Decreased LDL (P)
Acanthocytosis	LDL	Increased LDL (P)
Aldosterone synthesis	18-Hydroxylase	Decreased aldosterone (P)
Cholesteryl-ester deficiency	Lecithin-cholesterol acyltransferase	Decreased cholesterol (P)
Cholesteryl-ester hydrolase deficiency	Specific hydrolase	Increased cholesterol esters (P)
Cholesteryl-ester storage disease	Lysosomal lipase (acid)	Increased cholesterol esters (T)
Fabry's disease	Trihexosylceramide galactosylhydrolase	Trihexosylceramide (T)
Familial hypercholesteremia	Receptors for plasma LDL	Increased cholesterol and LDL (P)
Fucosidosis	α-Fucosidase	Glycolipids (T)
Gangliosidosis (generalized)	β-Galactosidase	Gangliosides (T)
Gaucher's disease	Glucocerebrosidase	Glucosylceramide (L,T)
Hyperlipoproteinemia, type I	Lipoprotein lipase	Chylomicrons (P)
Krabbe's leukodystrophy	Galactocerebrosidase	Galactosylceramide (T)
Lactosyl-ceramidosis	β-Galactosyl hydrolase	Lactosylceramide (T)
Lipase deficiency (pancreatic)	Pancreatic lipase	Triglycerides (F)
Metachromatic leukodystrophy	Arylsulfatase A	Galactosulfatides (T)
Niemann-Pick disease	Sphingomyelinase	Sphingomyelins (L)
Refsum's disease	Phytanate α-oxidase	Phytanate (U)
Sandhoff's disease	*N*-Acetyl-β-hexosaminidase, A and B	Glycosphingolipids (T)
Tangier disease	HDL	Decreased cholesterol (P,T) Increased triglycerides (P,T)
Tay-Sachs disease	Hexosaminidase A	Gangliosides (T)
Testosterone deficiency	Δ^4-5α-Reductase	Decreased testosterone (B)
Vitamin D–resistant rickets	Cholecalciferol 1,25-hydroxylase	Decreased 1,25-dihydroxycholecalciferol (P)
Wohlman's disease	Acid lipase (lysosomal)	Cholesteryl esters (T)
Xanthomatosis	Lipoprotein lipase (etc.?)	Cholesterol and other lipids (T)

*See footnotes to Table 22-1, p. 811.

Table 22-3
Hereditary disorders of amino acid metabolism*

Disorder	Molecular deficiency or defect	Principle manifestation
Albinism	Tyrosine 3-monooxygenase	Tyrosine metabolites (U)
Alkaptonuria	Homogentisic acid oxidase	Homogentisic acid (U)
β-Aminoisobutyrate aciduria	BAIB-Glu transaminase	β-Aminoisobutyrate (U)
Ammonemia, type I	Carbamoyl phosphate synthetase	Ammonia (B,U)
Ammonemia, type II	Ornithine transcarbamoylase	Ammonia (B,U)
Ammonemia, type III	Argininosuccinate lyase	Ammonia (B)
Argininosuccinic aciduria	Argininosuccinase	Argininosuccinic acid (U)
Aspartylglycosaminuria	Specific hydrolase	Aspartylglycosamine (U)
Blue diaper syndrome	Renal and intestinal transport of tryptophan	Indole derivatives (U), tryptophan (F)
Citrullinemia (ammonemia, type IV)	Argininosuccinate synthetase	Citrulline, ammonia (B,U)
Cystathioninuria	Cystathionase	Cystathionine (U)

*See footnotes to Table 22-1, p. 811.

Table 22-3, cont'd
Hereditary disorders of amino acid metabolism

Disorder	Molecular deficiency or defect	Principle manifestation
Cystinosis	Storage-release cystine from lysosomes	Cystine (T-lysomes)
Cystinuria	Renal reabsorption	Cystine (also arginine, lysine, and ornithine) (U)
Fanconi syndrome	Endonuclease (For DNA repair)	Aminoaciduria (generalized) (U)
Glycinemia (ketotic form)	Propionyl CoA carboxylase	Glycine (B,U)
Glycinuria	Renal reabsorption	Glycine (U)
Goiter (familial)	Iodotyrosine dehalogenase	Decreased iodine reuse by thyroid
Hartnup disease	Tryptophan 2,3-dioxygenase	Tryptophan and other neutral amino acids (B,U)
Histidinemia	Histidine ammonia-lyase	Histidine and derivatives (B,U)
Homocystinuria	Cystathionine synthase	Homocystine (B,U)
Hydroxyvaleric aciduria	β-Methylcrotonyl CoA carboxylase	β-Hydroxyisovalerate (U)
Hyperarginemia	Arginase	Arginine (B,U)
Hyperlysinemia	Lysine-NAD-oxidoreductase	Lysine, ammonia (B,U)
Hypermethioninemia	Methionine adenosyltransferase	Methionine (B)
Hyperprolinemia, type I	Proline oxidase	Proline (also glycine and hydroxyproline) (B,U)
Hyperprolinemia, type II	Proline dehydrogenase	Proline (B,U)
Hydroxyprolinemia	Hydroxyproline oxidase	Hydroxyproline (also proline and glycine) (B,U)
Isovaleric acidemia	Isovaleric CoA dehydrogenase	Isovalerate (B)
Leucine intolerance	Unknown	Hypoglycemia (B)
Lysine intolerance	Lysine-NAD-oxidoreductase	Lysine (F)
Maple syrup urine disease	Branched-chain keto acid decarboxylases	Keto acids of leucine, isoleucine, valine, and methionine (U)
Methionine malabsorption	Intestinal transport	Methionine (F)
Methylmalonate aciduria (2 types)	Methylmalonyl CoA mutase and (?)	Methylmalonate (U)
Phenylketonuria	Phenylalanine 4-monooxygenase	Phenylalanine, phenylpyruvate, etc. (B,U)
Pheochromocytoma	Excess decarboxylase, Tyr → norepineprine, epinephrine	Norepinephrine, epinephrine, etc. (B)
Pyroglutamate aciduria	Pyroglutamate hydrolase	Pyrrolidone carboxylic acid (U)
Saccharopinuria	Aminoadipic semialdehyde-glutamate reductase	Saccharopine, lysine (U)
Sarcosinemia	Sarcosine dehydrogenase	Sarcosine (B)
Sulfituria	Sulfite oxidase	Sulfite, thiosulfate (U)
Tyrosinemia, type I	*p*-Hydroxyphenylpyruvate oxidase	Tyrosine, *p*-hydroxyphenylpyruvate (B,U)
Tyrosinemia, type II (tyrosinosis)	Tyrosine aminotransferase	Tyrosine (B), *p*-hydroxyphenylpyruvate (U)
Valinemia	Valine aminotransferase	Valine (B)
Xanthurenic aciduria	Kynureninase	Xanthurenic acid (U)

Table 22-4
Hereditary disorders of purine and pyrimidine metabolism*

Disorder	Molecular deficiency or defect	Principal manifestation
Gout, primary (several types)	Hypoxanthine-guanine phosphoribosyltransferase (in one type)	Uric acid (B,U,T)
Lesch-Nyhan syndrome	Hypoxanthine-guanine phosphoribosyltransferase (nearly complete deficiency)	Uric acid (B,U,T)
Orotic aciduria, type I	Orotate phosphoribosyltransferase *and* orotidine 5-phosphodecarboxylase	Orotic acid (U)
Orotic aciduria, type II	Orotidine 5-phosphodecarboxylase only	Orotic acid (U)
Xanthinuria	Xanthine oxidase	Xanthine (U)
Xeroderma pigmentosum	Endonuclease (for DNA repair)	Skin tumors

*See footnotes to Table 22-1, p. 811.

(Chapter 3) by the defective enzyme. Earlier intermediates also may be increased but usually to a lesser extent. Later metabolites, if any, and the end-product of the pathway usually decrease in amount. Examples are hemoglobin in certain types of anemia and clotting factors in hemophilia and afibrinogenemia. If *degradative* enzymes are involved, such as those for glycogen, mucopolysaccharides, or complex lipids (sphingomyelins, gangliosides, etc.), these substances may accumulate in the tissue involved (e.g. liver, brain, muscle, or sometimes "generalized" tissue) and may produce pathology and disease.

Tables 22-1 through 22-5 are not intended to serve as a complete listing of *all* known hereditary disorders caused by molecular defects. Such a compilation would by itself fill a sizable volume. Rather, the tables include only those well-identified disorders that document the broad impact of molecular defects on the etiology of diseases in humans. The number of disorders listed in the tables also emphasizes the tremendous expansion of knowledge in this area of biochemistry—a decade or two ago these tables would have included only a dozen or so proven hereditary molecular diseases.

Perhaps a few words should be added regarding the current status of the control of hereditary metabolic disorders. There is no "cure" in the usual sense at the present time, since the biochemical defect is an irreversible damage (mutation) in some segment of DNA. Repair of an altered segment of DNA, of course, may be a future possibility. For the present at least, control is palliative. One procedure is decreasing the intake of an offending substance, e.g., milk in the case of lactose intolerance or of phenylalanine in phenylketonuria (discussed later). Lowering the intake of protein, along with its partial replacement with the α-keto acids of certain amino acids, is another example under investigation in hereditary disorders of urea formation and ammonia intoxication. Still another procedure is continued replacement of the defective or deficient enzyme (or protein) with normal plasma (or protein), as in the treatment of certain lipidoses, described later. Prevention by eugenic measures, especially by the avoidance of consanguineous marriage, also may be used.

A procedure for detecting a number of genetic disorders during early pregnancy, now being used more frequently, is *genetic amniocentesis*. Amniotic fluid is examined by high-resolution microscopy for fetal cells with abnormal genes in chromosomes or assayed biochemically for abnormal rates of enzyme activity, abnormal metabolites, or normal metabolites in abnormal concentrations. Hereditary diseases

Table 22-5
Miscellaneous hereditary metabolic disorders*

Disorder	Molecular deficiency or defect	Principal manifestation
Acatalasia	Catalase	Peroxides (T)
Acid phosphatase deficiency	Acid phosphatase (lysosomal)	Organic phosphates (T)
Afibrinogenemia	Fibrinogen deficiency	Impaired blood coagulation
Agammaglobulinemia	γ-Globulin deficiency	Impaired blood coagulation
Analbuminemia	Plasma albumin deficiency	Edema
Anemia (hereditary)	Certain abnormal hemoglobins	Decreased hemoglobin (B)
Atransferrinemia	Transferrin (some 20 variants)	Transferrin (P)
Blood coagulation	Coagulation factors	(See Chapter 16)
Carnosinemia	Carnosinase	Carnosine (B)
Crigler-Najjar syndrome	UDP-glucuronyltransferase	Bilirubin (P)
Dwarfism	Somatotropin	Impaired growth (T)
Formiminotransferase deficiency	Formiminotransferase	Formiminoglutamate (U)
Hemochromatosis	Intestinal iron transport (increases mucosal cell apoferritin?)	Increased iron (L,T)
Hemoglobins (abnormal)	Amino acid substitutions, etc.	(See Chapter 16)
Hemolytic anemias	Erythrocyte enzymes	(See Chapter 16)
Hemophilias	Antihemophilia factors	(See Chapter 16)
Hypomagnesemia	Impaired intestinal transport	Magnesium (B)
Hypophosphatasia	Alkaline phosphatase	Phosphate (S)
Infantile convulsions (vitamin B_6–dependent)	Glutamate decarboxylase	Aminobutyrate (T)
Menke's disease	Impaired intestinal transport of copper	Copper deficiency (B,T)
Methemoglobinemia	NADH-methemoglobin reductase	Methemoglobin (B)
Neonatal jaundice	Glutathione peroxidase	Bilirubin (B,T)
Plasma protein variants	(See Chapter 16)	(See Chapter 16)
Polycythemias, secondary	Certain abnormal hemoglobins	(See Chapter 16)
Porphyrias, 5 types	Enzymes of biosynthetic pathway	Various porphyrins and intermediates (U)
Pulmonary emphysema	α_1-Antitrypsin	Emphysema
Pyridoxine-responsive anemia	δ-Aminolevulinate synthase	Hemoglobin (B)
Trypsinogen deficiency	Trypsinogen	Diarrhea, flatulence
Wilson's disease	Ceruloplasmin (6 variants)	Copper deposition (L)

*See footnotes to Table 22-1, p. 811.

that can be diagnosed by microscopy include sex-linked disorders, e.g., hemophilia and Down's syndrome (mongolism). A number of the disorders listed in the tables can be detected by biochemical assays, e.g. Tay-Sachs disease and others. Sickle cell anemia and thalassemia (Chapter 16) also may be diagnosed by amniocentesis. This valuable procedure, especially when combined with ultrasonography, permits parents to request termination of the pregnancy or to prepare for the birth of an afflicted infant, usually with a short life expectancy.

Disorders of carbohydrate metabolism

At least 40 to 50 heritable diseases of carbohydrate metabolism attributable to molecular defects have been identified. Some of these are listed in Table 22-1. A few should be discussed further because of their special biochemical interest.

Diabetes mellitus Perhaps some explanation should be added regarding the inclusion of diabetes mellitus in Table 22-1. This condition appears to be a group of diseases rather than a single entity. However, a hereditary factor is clearly involved in some types of the disease. Type I, the "juvenile-onset," insulin-dependent form, apparently may have multiple causes, including genetic, environmental (viral infections, nutrition), and autoimmunological factors. In some cases a complex interaction between these several factors may be involved in the etiology. In the "maturity-onset," insulin-independent type II, cellular insulin-binding receptors appear to be defective, again involving possible hereditary factors. In other cases an abnormal type of insulin of lower activity is suspects. In all of these cases genetic factors are clearly involveGMd.

Fructose-related disorders Faulty metabolism of fructose is seen in three inborn errors: hereditary fructose intolerance, essential fructosuria, and fructosemia. In hereditary fructose intolerance there is a deficiency of the *fructose 1-phosphate aldolase;* fructose 1-phosphate cannot be converted to DHAP and glyceraldehyde (see Fig. 11-23), and it therefore accumulates. This results in an inhibition of fructokinase and an impaired clearance of fructose from the blood. At the same time there is a decreased blood glucose level (hypoglycemia). In essential fructosuria a defect at the *fructokinase* stage results in an impaired phosphorylation of fructose. Fructose levels increase in the blood until the renal threshold is exceeded. The sugar then appears in the urine. In fructosemia the defect is in *fructose 1,6-diphosphatase*. Again, fructose (presumably as the diphosphate) accumulates in the blood and is excreted in the urine when the concentration exceeds the renal threshold.

Galactose-related disorders Galactosemia is an inborn error of metabolism caused by a defective enzyme, *galactose 1-phosphate uridyltransferase*. One out of 18,000 infants suffers from this genetic defect, which is characterized by mental retardation and cataracts. Since the problem lies in an inability to metabolize galactose 1-phosphate, the infant patient is simply fed a milk formula in which sucrose replaces lactose as the sugar component.

Cataracts may be caused by faulty galactose metabolism. As already seen, this is one of the consequences of galactosemia. Another inborn error of metabolism that leads to cataract formation is a deficiency of *galactokinase*. Galactose in excess apparently is damaging to the lens proteins of the eye, decreasing their solubility and producing the "cloudiness" that impairs vision.

Glycogen storage diseases Several derangements of glycogen metabolism in liver and muscle have been studied extensively in patients. One of these, von Gierke's glycogen storage disease, has been known for a number of years. The absence of glucose 6-phosphatase has been demonstrated in some cases of this type of glycogen storage disease; in others, there seems to be a lack of the debranching enzyme. This is a rare congenital disorder of carbohydrate metabolism in which the liver in enlarged because of the excessive storage of glycogen. There is usually hypoglycemia, hyperlacticemia, and acidosis. A condition in which the glycogen is stored in the heart is rapidly fatal.

Certain other varieties of glycogen storage disease appear to be caused by the deposition of abnormal forms of glycogen. At the present time, at least nine or possibly ten different types of glycogen storage diseases have been characterized (Table 22-1). Among the treatments recommended is administration of glucagon and synthetic androgen.

Pentosuria A genetic defect or inborn error of metabolism called *idiopathic pentosuria* is caused by a deficiency of *L-xylulose dehydrogenase*. This leads to the excretion of as much as 1 gm per day of L-xylulose. The anomaly is found primarily in Jews and also in American Indians. It is harmless unless it results in an erroneous diagnosis of diabetes mellitus.

Primaquine sensitivity (favism) Individuals with a genetic defect resulting in a deficiency of the enzyme *glucose 6-phosphate dehydrogenase* develop a severe hemolytic anemia from the defective erythrocyte structure and excessive hemolysis. During the Korean War large numbers of American blacks developed acute hemolytic anemia after taking *primaquine* for treatment or prophylaxis of malaria. A similar condition develops in susceptible individuals after ingesting fava beans (favism) or coming in contact with the pollen of the plant, *Vicia fava*. When red blood cells are exposed to the oxidative stress of primaquine, various sulfonamides, analgesics, diuretics, and hypoglycemic agents, hydrogen peroxide is generated. The various reactions for its disposal in red cells are as follows:

1. Red cell + Oxidant $\rightarrow H_2O_2$
2. $H_2O_2 + 2\ GSH \xrightarrow[\text{peroxidase}]{\text{Glutathione}} 2\ H_2O + GSSG$
3. $GSSG + NADPH \xrightarrow[\text{reductase}]{\text{Glutathione}} 2\ GSH + NADP^+$
4. $G6P + NADP^+ \xrightarrow{\text{G6P dehydrogenase}}$ 6 Phosphogluconate + NADPH

Glucose 6-phosphate dehydrogenase has been found to be deficient in 10% to 13% of black males. As a result of this genetic defect the production of NADPH is inadequate both to restore the reduced glutathione level (reaction 3) and to maintain the cell membrane. The consequence is destruction of the red cells and a severe hemolytic anemia unless the condition is checked.

Space does not permit further comment on the 20 or so additional disorders of carbohydrate metabolism listed in Table 22-1. Pertinent readings listed at the end of this chapter should be consulted for further details.

Disorders of lipid metabolism

Heritable disorders of lipid metabolism, like those of carbohydrate metabolism, affect all types of lipids, from the relatively simple triglycerides and sterols to the more complex sphingomyelins and gangliosides (Chapter 23).

Obesity is probably the most common disorder of lipid metabolism and appears to be related primarily to an excess of caloric intake over caloric requirement. Endocrine dysfunction also may be involved in some cases and, according to current studies, perhaps a deficiency of the enzyme ATPase (Chapter 21). Some authorities still maintain that genetic factors also are important, at least in some cases. The same is believed to be true of another widespread disorder involving lipid metabolism, atherosclerosis and its related conditions, which are considered later.

There are probably at least 40 well-documented hereditary abnormalities of lipid metabolism. Some of these are characterized by an excessive accumulation of various lipids in tissues and cells. This type of disorder results from an impaired activity of the enzymes required to catabolize such lipids as sphingomyelins, various cerebrosides, and gangliosides. Table 22-2 lists some representative examples of these types of lipidoses, together with the corresponding molecular defects and principal manifestations. A few merit additional comment.

Gaucher's disease is a hereditary condition that sometimes afflicts several children of the same family. It is characterized by an enlargement of the spleen and liver (splenohepatomegaly) as well as by other symptoms. Although the lipids deposited in the spleen and liver are mixed, the outstanding feature of this abnormality is the presence of a large amount of glucocerebroside. Glucocerebroside is composed of ceramide (sphingosine and a fatty acid) plus glucose in β-glycosidic linkage.

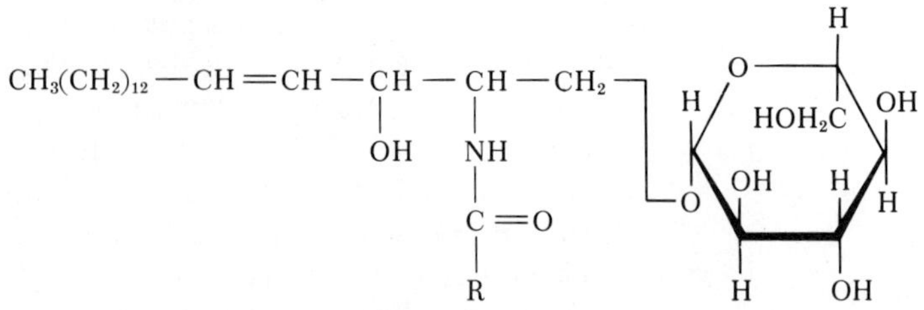

$CH_3(CH_2)_{12}—CH=CH—CH(OH)—CH(NH—C(=O)—R)—CH_2—O$–(glucose: H, HOH_2C, OH, H, OH, H, H, OH)

Normally tissues such as the spleen and liver have an active β-*glucosidase*, which removes the glucose, leaving the ceramide portion. In Gaucher's disease the lack of this enzyme prevents the catabolism of this lipid, resulting in its accumulation. Thus in the normal spleen the amount of glucocerebroside is very low, whereas in Gaucher's disease the concentration may be 4% of the dry weight. The bone and marrow are also involved. Mental retardation is observed in the infantile form of this disease.

Another familial and hereditary disorder is *Niemann-Pick disease*. The liver and spleen are again the site of lipid deposits and are tremendously enlarged, but here the predominant constituent is a mixture of phospholipids, chiefly phosphatidyl choline and sphingomyelin. In this disease the enzyme *sphingomyelinase* is defective, and as a result sphingomyelin is not catabolized and hence accumulates. It occurs in infancy and causes death within a few months.

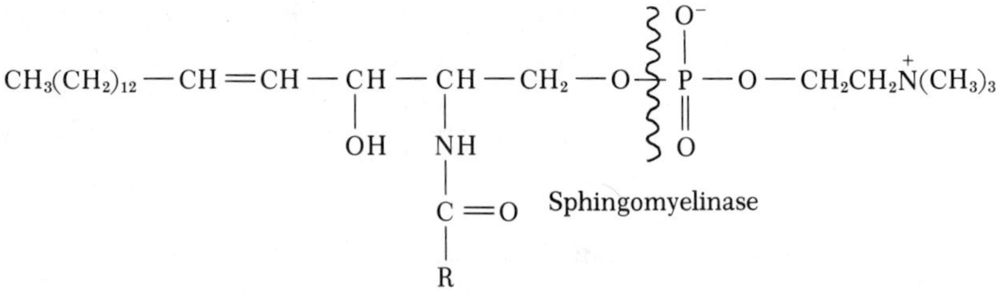

$CH_3(CH_2)_{12}—CH=CH—CH(OH)—CH(NH—C(=O)—R)—CH_2—O—P(=O)(O^-)—O—CH_2CH_2\overset{+}{N}(CH_3)_3$

Sphingomyelinase

Sphingomyelin

Several other hereditary diseases of lipid metabolism have been described, e.g., idiopathic hyperlipemia *(Buerger-Grütz disease)*, essential hypercholesteremic xanthomatosis, amaurotic familial idiocy *(Tay-Sachs disease)*, and lipochondrodystrophy *(gargoylism)*.

In Tay-Sachs disease large amounts of gangliosides are deposited in the brain and nerve tissue. This results from a deficiency in *hexasaminidase A* activity, leading to an inability to catabolize the ganglioside GM_2.

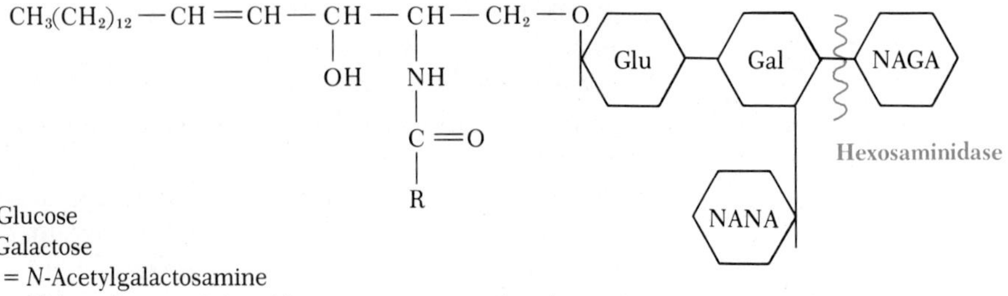

$CH_3(CH_2)_{12}—CH=CH—CH(OH)—CH(NH—C(=O)—R)—CH_2—O$–Glu–Gal(–NANA)–NAGA

Hexosaminidase

Glu = Glucose
Gal = Galactose
NAGA = *N*-Acetylgalactosamine
NANA = *N*-Acetylneuraminic acid

Ganglioside

The accumulated gangliosides apparently interfere with normal cell function and result in the mental defects and the neurological manifestations observed. There may also be a deficiency of fructose 1-phosphate aldolase in children with Tay-Sachs disease. This deficiency may be related to the alteration in the composition of the gangliosides found in these cases.

A pronounced deficiency of β-*galactosidase* has been found in the liver, spleen, kidney, and brain of some patients with *generalized gangliosidosis*. This could result in a decreased catabolism of gangliosides and hence their accumulation in these tissues.

Xanthoma is a disease in which yellow nodules or flat plaques of lipid appear in the skin, especially in the eyelids. The nodules, which vary in size from that of a pinhead to that of a bean, may be found in tendon sheaths, bone, blood vessels, and elsewhere. Xanthomatous deposits are a mixture of various lipids, with cholesterol frequently predominating. The blood lipid level in these cases is often elevated. Sometimes the condition is a complication of diabetes, jaundice, or lipoid nephrosis. *Hand-Schüller-Christian disease,* a rather rare disease of children, is considered to be a form of xanthoma. The nodules are found in cranial and other bones and contain large amounts of cholesterol and cholesteryl esters.

Atherosclerosis is another abnormality of lipid metabolism but is conspicuously absent from the list given in Table 22-2. Heredity is one factor of recognized significance in this condition. However, atherosclerosis was omitted from the table because no specific molecular defect has as yet been identified as an etiological factor. Genetic factors manifested by a family history of coronary heart disease are clearly indicated, along with complex interactions with environmental and perhaps other factors. For example, a current concept of the development of cardiovascular disease stresses the role of the overloading of the lysosomes of the cells of the arterial walls with metabolites that are slowly hydrolyzed, particularly cholesteryl esters and triglycerides. These are transported from the intestine by the plasma low-density lipoproteins (LDL) (Chapter 12). If the amount of these lipids exceeds the capacity of the high-density lipoproteins (HDL) to remove them from use or excretion by the liver, cells in pressure areas such as arterial walls become gorged with lipids, cholesteryl esters in particular. They are familiar to pathologists as "foam cells." The lipid overloading results in impaired cell function and, if continued, eventually in cell death. Continued overloading results in the accumulation of cellular debris, mainly cholesteryl esters, and the formation of a plaque on the vessel wall. Blockage of the artery can follow, with spasms of the muscular layer. Some authorities believe that this series of molecular events is the basic biochemical lesion in cardiovascular disease and is manifested usually as coronary heart disease or strokes.

Disorders of amino acid metabolism

The hereditary disorders of amino acid metabolism include the largest number of metabolic disorders under any of the four major categories of metabolism. This might be expected because of the numerous pathways, both biosynthetic and degradative, involved in amino acid metabolism. Even so, this group does not include the final end product of the biosynthetic pathways, the various body proteins. These form an even larger group of heritable disorders, and are considered separately in the purine-pyrimidine group for that reason.

The well-identified hereditary disorders of amino acid metabolism are listed in Table 22-3 arbitrarily in alphabetical order. The currently accepted molecular defects and the principal manifestations likewise are listed. The principal manifestations in this group refer to the most affected metabolite found in increased amounts in the blood, urine, or both or in a few instances in the feces. Recall that the principal metabolite affected is usually the one immediately preceding the molecular defect in a pathway, both biosynthetic and degradative. Other metabolites preceding the point of blockage also may be increased but usually to a lesser extent.

Additional comments are included in the text for some of the more commonly occurring or biochemically unique disorders of amino acid metabolism. An example is the molecular defects present in the formation of urea in the liver. A marked increase in blood ammonia results, usually in newborn infants afflicted with the enzyme defect. The resulting "ammonia intoxication" may rapidly be fatal. As reflected in the table, infants have been reported with deficiencies of the five enzymes required for the formation of urea from ammonia (NH_4^+).

Ammonemias: inherited disorders of urea formation Since each step in the cycle involved in the biosynthesis of urea is catalyzed by a specific enzyme, one would reasonably expect that an inherited deficiency or functional defect could occur in any one of them. Indeed, this has proved to be the case. A deficiency or malfunction of *all five enzymes* of the cycle has been reported in human patients, usually children. A deficiency or impaired activity of either *carbamoyl phosphate synthetase* or *ornithine-carbamoyl transferase* can result in an increase in blood, urinary, and hepatic ammonia and in ammonia intoxication. The classic clinical symptoms are protein-induced vomiting, coma, progressive spasticity, and cerebral atrophy. A deficiency or defect in any of the other three enzymes, *argininosuccinate synthetase, argininosuccinase,* or *arginase,* is associated with an increase in the blood and urinary levels of the metabolite immediately preceding the affected step, in this case citrulline, argininosuccinate, or arginine respectively. The accompanying clinical symptoms are mental retardation and convulsive seizures. Symptoms of *ammonia intoxication* (vomiting and stupor) may also be seen because of a backup of earlier metabolites, including ammonia. Since urea excretion in the urine, primarily from hepatic synthesis, is normal in these individuals, we assume that a deficiency of the respective isozyme, which can normally form urea in the brain, is responsible for the characteristic symptoms observed. Fortunately, these five inborn errors of metabolism are fairly rare.

Control of the foregoing disorders in urea formation is partly dietary. Decreasing the protein intake of the patient is essential in order to diminish ammonia-urea formation. Interesting current investigations of the therapeutic value of mixtures of the α-keto acids to replace the respective amino acids have been reported.

Other disorders of amino acid metabolism *Albinism* is a result of a hereditary deficiency or defect of the enzyme *tyrosinase* (tyrosine 3-monooxygenase). Lack of melanin pigment formation in the skin, hair, retina, and choroid coat of the eye are the characteristic features. Such individuals (or animals) are called albinos.

Alkaptonuria is a rare genetic disorder resulting from a deficiency of the hepatic enzyme *homogentisic acid oxidase*. As a result, homogentisic acid (once called *alkaptone*) accumulates and is excreted in the urine. Homogentisic acid, on standing in air, is oxidized to a blackish pigment. Hence a dark-colored urine is usually excreted by affected individuals. False tests for glucose in the urine may be obtained when copper-reduction methods are used. The condition is usually benign, although there may be pigmentation of cartilage and other connective tissues *(ochronosis)* in later life.

Cystine occurs in the urine of certain individuals. *Cystinuria* is an inborn error of metabolism and is hereditary. Persons suffering from cystinuria excrete this amino acid in large amounts, even on a protein-free diet. It has been shown that, in contrast to the normal subject, the individual with cystinuria excretes administered cystine rapidly, with only a transient rise in the blood level. Indeed, the renal clearance of cystine in a patient with cystinuria was 30 times that in the normal subject, even though the blood level was somewhat lower. This observation in interpreted as evidence that the metabolic error is in the reabsorption of cystine in the renal tubule and that cystinuria is therefore renal in origin. Such an interpretation would also explain the presence of relatively large amounts of other amino acids, notably arginine, lysine, and ornithine, in the urine of patients with cystinuria. The failure of cystine, when fed, to increase the urinary excretion of cystine in patients with cystinuria in earlier studies apparently is explainable on the basis either of its insolubility or of a defect in its intestinal transport and hence the poor intestinal absorption of cystine.

Cystinuria is usually a rather benign condition unless there is sufficient deposition of cystine in calculi in the kidney to produce renal damage. However, recent studies of a large population indicate that the incidence of homozygous cystinuria is 10 times higher in patients with mental disease than in the general population, now estimated by some authorities to be 1 in 10,000. This adds support to the growing evidence that homozygous cystinuria has a phenotypic effect on the brain, as well as on the kidney and intestine.

A second hereditary abnormality of cystine metabolism, *cystinosis* or *cystine storage disease,* differing from cystinuria, has also been described. There is an excessive deposition of cystine, sometimes as distinct crystals, in various tissues, including the kidney, bone marrow, cornea, and conjunctiva and the peripheral leukocytes. The condition may appear in adults as well as in children. There is a generalized aminoaciduria, glycosuria, and in some cases rickets or osteomalacia. The childhood form of the disease usually leads to early death, apparently from renal damage and uremia. The adult form is usually benign.

Interesting recent studies have demonstrated that the prenatal diagnosis of cystinosis is possible on the basis of an increased content of *nonprotein* ("free") cystine in cultured amniotic-fluid cells. Also, the nonprotein cystine content of the liver, kidney, thymus, spleen, and placenta is 50 to 100 times greater than that of controls.

The accumulation of cystine crystals, recognizable by light microscopy, appears to be related to an intracellular concentration of cystine rather than to a high plasma level. The large intracellular "pool" of cystine appears to be located primarily in the lysosomes.

The exact metabolic defect involved in cystinosis remains unknown. The enzymes concerned in the metabolism of cystine appear to be normal in amount and activity. However, it has been suggested that there may be an impairment in the transport of cystine from affected cells or in the conversion of cystine to cysteine by the enzyme cystine reductase.

Two other inherited diseases of sulfur–amino acid metabolism, *homocystinuria* and *cystathioninuria,* affecting that of cystine-cysteine, have been described. These are commented on later, with disorders of methionine metabolism.

In *glycinemia,* found in newborns (characterized by increased blood levels of glycine), lethargy, mental retardation, severe ketosis, hypogammaglobulinemia, and blood dyscrasias are found clinically. The cause is unknown. Death occurs usually in the first few days of life. More recent studies have shown that there is a pronounced increase, up to a thousandfold, in serum propionate levels and that there is a deficiency of *propionyl CoA carboxylase* in leukocytes and cultured fibroblasts from these patients. This results in a block in the conversion of propionyl CoA to methylmalonyl CoA. The cause of the elevation of blood glycine levels, however, remains obscure. Therapy for glycinemia consists of a low-protein diet and control of the acidosis.

Reports of a nonketotic type of glycinemia in newborns have also appeared. Convulsions occur during the first few days of life, followed by mental retardation in surviving patients. Metabolic studies have indicated that defective decarboxylation of glycine may be responsible.

Another inherited disorder of glycine metabolism, *glycinuria,* is apparently caused by a defect in the renal transport system for glycine. An abnormally high renal excretion of glycine is found. The presence of calcium oxalate renal stones is the major clinical finding.

Hartnup disease is a disorder of tryptophan catabolism resulting from a deficiency of the enzyme *tryptophan 2,3-dioxygenase*. Tryptophan levels in the blood and urine are elevated, and some tryptophan is diverted into other pathways. Hartnup disease is associated with mental retardation.

A hereditary disorder of histidine metabolism has been described. Elevated levels of plasma and urinary histidine are found. The urine gives a greenish color when treated with ferric chloride, sometimes leading to the erroneous diagnosis of phenylketonuria. There is a deficiency of the enzyme *histidase (histidine ammonia-lyase)* so that some histidine is diverted to alternate pathways. The metabolites imidazolepyruvic acid, imidazolelactic acid, and imidazoleacetic acid are also excreted in the urine. The disorder is rather benign, with the principal symptom being speech defects, which may appear during childhood.

Two types of *hyperprolinemia,* called type I and type II, resulting in increased blood and urine levels of proline, are known. An increased urinary excretion of hydroxyproline and glycine may also occur because of the inhibition of their renal reabsorption by the increased level of proline. The deficient enzymes are *proline oxidase* and *Δ^1-pyrroline-5-carboxylic acid dehydrogenase* respectively. Both disorders are characterized clinically by mental retardation. Renal damage is also found in the first type. The interesting occurrence of congenital hyperprolinemia and prolinuria in an inbred strain of mice has been reported.

Hydroxyprolinemia, possibly from a lack of *hydroxyproline oxidase,* has been described. Increased urinary excretion of proline, hydroxyproline, and glycine is reported. Mental retardation and hematuria are prominent clinical features. Impaired renal reabsorption of proline and hydroxyproline, as well as of glycine, has been reported in *glycinuria*. Calcium oxalate kidney stones are a prominent clinical finding in this condition. Hydroxyproline excretion in the urine may also occur in diseases in which there is increased degradation of tissue collagen.

Several hereditary disorders of methionine metabolism have been described. One, *hypermethioninemia,* appears to be caused by a deficiency of the enzyme *methionine adenosyltransferase,* which is involved in the formation of SAM (Chapter 13). As a result, methionine levels in the blood are elevated.

Three other defects in the metabolism of methionine have been reported. One is known as *cystathioninuria*. It appears to be caused by a failure of the cystathionine cleavage enzyme, *cystathionase,* to bind pyridoxal phosphate to its apoenzyme. Mental retardation and a tendency to convulsive seizures are seen in patients with this inherited disorder.

Another defect of methionine metabolism is *homocystinuria,* which involves the catabolism of methionine or more specifically its metabolic intermediates, homocysteine and homocystine. Elevation of homocystine occurs in the plasma, resulting in an "overflow" into the urine amounting to as much as 50 to 100 mg per day. It is associated usually with mental retardation in children or surviving adults. Skeletal deformities involving the spine, thorax, and vertebrae ("codfish vertebrae") may occur. Some affected individuals are extraordinarily tall with long extremities and frequently with flat feet, which "toe out." All have abnormal electroencephalograph results, and some experience seizures. There is a curious dislocation of the lens of the eye, which may be used as a clue to homocystinuria. The disease is apparently caused by an autosomal recessive trait and occurs in approximately 1 in 20,000 live births. The basic biochemical lesion in the disease appears to be a deficiency in, or defect of, connective tissue, perhaps of the intercellular cement substance. It apparently is caused by a genetic deficiency or absence of the enzyme *cystathionine synthase* in the liver. As indicated previously, this enzyme converts homocysteine plus serine to cystathionine. Hence its deficiency results in an increase in the plasma level of homocystine, and homocystinuria results.

A fourth disorder, *methionine malabsorption,* involves a defect in the transport of methionine through the intestinal epithelial cells. The exact nature of the defect is unclear. As a result, methionine from dietary proteins is excreted in the feces.

Hyperlysinemia is characterized by *hyperammonemia* and, as is found in most genetic impairments of amino acid metabolism, mental retardation. There is an elevation of plasma lysine and also plasma arginine levels. The latter is attributed to an inhibition of arginase by lysine, a well-established effect. The precise biochemical defect responsible for hyperlysinemia is as yet uncertain, although there is some evidence that a low activity of the enzyme concerned in the deamination of lysine, *lysine-NAD-oxidoreductase,* is involved.

As stated previously (Chapter 13), saccharopine is an intermediate in an alternate pathway for lysine catabolism. In the two cases of the hereditary disease *saccharopinuria* reported thus far, this substance is excreted in the urine in relatively large amounts, along with lysine. Both patients were mentally retarded. The defect, as indicated by studies using cultured skin fibroblasts, appears to be a deficiency of the saccharopine-catabolizing enzyme, *aminoadipic semialdehyde-glutamate reductase.*

A block in the metabolism of *methylmalonic acid* formed in the catabolism of methionine, threonine, isoleucine, and valine (Chapter 13) has been described. This hereditary condition is characterized by *methylmalonic aciduria,* long-chain ketonuria, intermittent hyperglycinemia, and profound metabolic acidosis. *Methylmalonic aciduria* appears to be caused by a deficiency or defect of *methylmalonyl CoA mutase,* a vitamin B_{12}–dependent enzyme. Some but not all patients with this disorder respond to therapy with vitamin B_{12} (Chapter 21). The defect in the vitamin B_{12}–responsive type appears to be an inability to form 5′-deoxyadenosylcobalamin from vitamin B_{12}. Poor growth and mental retardation are seen in children showing either type of this hereditary disease.

Phenylketonuria (PKU) is apparently the most common hereditary disorder of phenylalanine metabolism and indeed of all amino acid metabolism. The condition is caused by a defect in the formation of the enzyme *phenylalanine 4-monooxygenase* (also called *phenylalanine hydroxylase*). As a result, phenylalanine is diverted into its minor pathway and large amounts of phenylpyruvate, phenyllactate, phenylacetate, phenylacetylglutamine, and even phenylalanine itself are excreted in the urine. The plasma level of phenylalanine is likewise increased above the normal. These patients can metabolize tyrosine, when it is fed to them, just as well as a normal individual. Blood and urine levels of tyrosine are normal. The principal clinical symptoms are mental retardation and convulsions. Special protein hydrolysate preparations of low phenylalanine content are available for therapeutic use in this condition.

Phenylpyruvic acid and its derivatives, phenylacetic and phenyllactic acids, or even phenylalanine itself, are probably responsible for the mental symptoms. The morphological defect in PKU appears to be an impairment of myelination in brain and nerve tissue. The basic biochemical defect is still uncertain. One possible cause is the increase in phenylalanine and its derivatives in the brain, resulting in the inhibition of the glycolytic enzyme *pyruvate kinase*. This explanation has been supported by careful studies in rats administered phenylalanine. The inhibition of pyruvate kinase may be compensated for by an increase in phosphoenolpyruvate, however, since there was no decrease in the amount of ATP or creatine phosphate in the brain. These studies were performed in adult rats however. It is possible that younger animals constantly exposed to high levels of phenylalanine might show the additional changes in the brain characteristic of PKU. This concept is also supported by the fact that if the phenylalanine intake is restricted to infants with PKU, the mental symptoms do not seem to develop. But there is some evidence that a low serotonin content of the brain, attributable to an inhibition of the enzymatic synthesis of serotonin by phenyllactic acid, phenylpyruvic acid, or both, is actually responsible for the mental aberrations of PKU. More recent studies, on the other hand, suggest that a chronic depletion of glutamine, perhaps because of its increased use in forming phenylacetylglutamine and its excretion in the urine, may be more directly involved in the damaging of the brain in PKU.

At the present time, tests for PKU are done routinely in most hospitals on all newborn infants before they are released. Some 30 states now require this early screening procedure before the brain-damaging process has begun. The older ferric chloride method of detection is now considered inadequate, because of insensitivity and nonspecificity. Blood levels of phenylalanine are usually determined by fluorometric or microbiological methods. Recent studies of data from some 600,000 newborn infants indicated an incidence of PKU of about 1 in 14,200, based on plasma phenylalanine levels in excess of 20 mg/100 ml and normal or low plasma tyrosine values. Similar surveys in an adult population of 250,000 showed an incidence of 1 in about 83,000. All were mentally subnormal. Authorities in the field are therefore urging the routine testing of urine for PKU in every woman's first pregnancy and in all mentally ill patients, since this "may avoid illogical treatment for occasional psychotic adults."

Interesting studies have demonstrated that PKU can be produced experimentally in infant monkeys by the feeding of a high-phenylalanine (5%) diet. The feeding of the diet is begun soon after birth. High blood levels of phenylalanine similar to those seen in human PKU patients are produced. Urinary levels of phenylpyruvic and phenyllactic acid are also high. The symptoms, grand mal convulsions and mental retardation, are also similar to those seen in human subjects.

A block in the metabolism of leucine, isoleucine, and valine results in an inborn error of metabolism in infants known as *maple syrup urine disease,* so called because the odor of the urine resembles that of maple sugar. The oxidative decarboxylation of the keto acids does not occur, and hence the branched-chain keto acid derivatives of leucine, isoleucine, and valine accumulate in the urine, since there appears to be no available alternate pathway for their metabolism. The fact that the enzyme or enzymes required for the oxidative decarboxylation of the branched-chain keto acids are deficient or defective has been verified by assays employing either leukocytes or skin fibroblasts from affected infants. It remains to be demonstrated whether the enzyme(s) are controlled by a single mutant or whether closely linked genes are involved. The disorder appears in early infancy and is characterized by severe vomiting, convulsions, and sometimes early death. There appears to be an "intermittent" form of the disease in which the infant may do well unless some stress, such as an infection, occurs. Then ketoaciduria and the classic symptoms follow. The therapeutic management of the condition in surviving infants is primarily dietary. A mixture of synthetic amino acids is used initially, followed by the addition of low-protein foods.

Another genetic disorder of tryptophan metabolism is the "*blue diaper syndrome*" in infants. It appears to be caused by an impaired renal and intestinal transport of tryptophan. The blue color of the diaper apparently is the result of the oxidation of indole compounds excreted in the urine. The biochemical defect is not known, but a synthesis of an intestinal and renal transport protein is suspect.

A hereditary disorder from a lack of the enzyme *sulfite oxidase* has been described. As might be expected, the urine of the case reported, a child, contained large amounts of sulfite and thiosulfate. The clinical symptoms included dislocation of the lenses of the eyes, severe brain damage, and mental retardation. The disease was rapidly fatal.

Tyrosinemia results from a hereditary defect in the formation of *p-hydroxyphenylpyruvate oxidase*. As would be expected, *p*-hydroxyphenylpyruvate is increased in the blood and urine. The plasma level of tyrosine is also increased. Hepatic failure and death occur in most cases during the first 6 months of life. In surviving patients, hepatic cirrhosis and generalized aminoaciduria develop. Treatment with a diet low in tyrosine and phenylalanine has produced satisfactory results in several patients. The hereditary disorder has recently been described in an isolated French-Canadian population. One in 10 to 30 persons in this group was found to be a heterozygous carrier.

Tyrosinosis has been described in only a few cases and is not clearly defined from a biochemical standpoint. There is some evidence from liver biopsy studies that the defect is in hepatic *tyrosine transaminase (tyrosine-glutamate aminotransferase)* rather than in *p*-hydroxyphenylpyruvate oxidase as postulated some 40 years ago. A pronounced increase in plasma tyrosine is the characteristic laboratory finding, along with the excretion of large amounts of *p*-hydroxyphenylpyruvate in the urine. In one case recently described, an infant male, multiple congenital anomalies including microcephaly were reported. Tyrosinosis is an exceedingly rare anomaly in which the aromatic amino acids are eliminated as tyrosine or hydroxyphenylpyruvic acid.

A disorder of valine metabolism, *hypervalinemia,* is caused by a rare recessive defect. These individuals have increased amounts of valine in the blood and urine. α-Ketovaleric acid is metabolized normally. Thus *valine transaminase* would be suspect as the deficient or defective enzyme involved. This has been verified by assays of leukocytes from afflicted subjects. The patients respond well to a low-valine diet. The chief clinical symptoms are vomiting, retarded mental development, and an electroencephalogram showing abnormal findings.

Two other inborn errors of valine metabolism should be mentioned at this point. They are *maple syrup urine disease* and *methylmalonic aciduria*. Both also affect the metabolism of the other two branched-chain amino acids, leucine and isoleucine.

Xanthurenic aciduria has been described and apparently is caused by a deficiency of or modified form of *kynureninase* (see Fig. 13-20).

It is interesting to note that pregnant women suffering from or threatened with eclampsia also have a deranged type of tryptophan metabolism. If given a test dose of tryptophan, they excrete in the urine much larger amounts of xanthurenic acid than do normally pregnant or nonpregnant women under the same conditions. Another derangement of tryptophan metabolism seems to occur in schizophrenia. It has been found that the output of urinary tryptamine in some patients is increased during periods of schizophrenic activity.

Disorders of purine and pyrimidine metabolism

Relatively few hereditary diseases of purine and pyrimidine metabolism per se have been described and established. This may be due in part to the vital role of these two substances in the biosynthesis of living matter. Metabolic defects in their formation could be lethal to the developing organism and to the survival of the species. However, such molecular defects that produce less serious affects on the organism, including gout, have been known for many years. Indeed gout was described and given its name by ancient Greek and Roman physicians. The five principal disorders of purine-pyrimidine metabolism identified at the present time are listed in Table 22-4, together with their enzyme (molecular) defects and principal manifestations. Because of their unique biochemical characteristics and effects, some additional comments will be made on three of these.

Gout, primary type, is one of several forms of this disorder in which uric acid is produced and excreted in excessive amounts. The multiple manifestations of the disease apparently result from several inborn errors of metabolism that are biochemically and genetically heterogeneous. The biochemical defect in the largest subgroup is still undefined. In some forms of gout there is a partial deficiency, to varying degrees, of the enzyme *hypoxanthine-guanine phosphoribosyltransferase* (HGPRT) (Chapter 14). In one form, the Lesch-Nyhan syndrome (discussed later), HGPRT is almost completely absent.

Gout is characterized by persistent *hyperuricemia* and *uric aciduria*. There are recurrent attacks of acute arthritis associated with deposits of monosodium urate as *tophi* in the cartilage of the fingers, big toe, and other locations. Uric acid and urate deposits also may be found as calculi in the kidneys, with resultant renal damage. Males are affected predominately (95%).

Three genetically determined enzymatic defects have been defined that lead to hyperuricemia:

1. HGPRT deficiency, partial or complete (X-linked genetically).
2. Phosphoribosylpyrophosphate synthase variants, which have *increased* catalytic activity (probably X-linked).
3. Glucose 6-phosphatase deficiency (autosomal recessive); recall that this enzyme is deficient in von Gierke's glycogen storage disease.

In all three defects an excessive production of purines and an elevated intracellular level of 5-phosphoribosyl-1-pyrophosphate (PRPP) is found. However, a partial deficiency of the first of the three appears to be of primary importance in some cases of primary gout. An increased availability of cellular L-glutamine (Chapter 14) also may be involved. Other factors, including dietary excesses of calories and excessive consumption of alcohol, also may be related to the hyperuricemia.

It is apparent that what may have seemed originally to be caused by a deficiency of a single enzyme in gout has, on further study and scrutiny, been found to involve multiple molecular defects and to result in a group of metabolic disorders. Treatment with the drug *allopurinol* (Chapter 14), a specific inhibitor of xanthine oxidase, has proven effective in most types of gout.

Lesch-Nyhan syndrome is an X-linked disorder, thus affecting males. It is characterized by a prodigious *overproduction* and *overexcretion* of *uric acid* and by *renal stones,* secondary renal damage, and sometimes gouty arthritis. Patients show constant recurring vermicular movements of the hands and feet, involuntary and irregular jerking movements, spasticity, mental retardation, and a bizarre compulsive self-mutilation. There is almost complete deficiency of HGPRT. The enzyme defect has been demonstrated in patient's skin fibroblasts, erythrocytes, leukocytes, and liver, kidney, and brain tissues.

Allopurinol (Chapter 14) is used effectively to control the symptoms of patients with Lesch-Nyhan syndrome. Preventive control by prenatal diagnosis in suspected cases also is accomplished by amniocentesis. Amniotic fluid cells are cultured and assayed for the level of HGPRT. The diagnosis has been confirmed by enzyme assays of tissue extracts from aborted fetuses.

Orotic aciduria is a unique hereditary disorder of pyrimidine metabolism. It is characterized by an accumulation and urinary excretion of orotic acid (Chapter 14). Affected children fail to grow at a normal rate and develop a *megaloblastic anemia*. There are two types of the disorder. In type I there is a deficiency of *two* enzymes, *orotate phosphoribosyltransferase* and *orotidine 5-phosphodecarboxylase*. In type II only the latter, *orotidine 5-phosphodecarboxylase*, is deficient. Both enzymes are required for the formation of UMP (Chapter 14). The administration of *cytidine* or *uridine* restores normal growth and produces remission of the anemia, as well as reducing the urinary excretion of orotic acid. Orotic aciduria also may result from a deficiency of *ornithine transcarbamoylase* and from other enzyme deficiencies in the urea cycle.

Xanthinuria is an autosomal recessive disorder resulting from a deficiency of hepatic, or possibly also intestinal, *xanthine oxidase*. The disorder is characterized by an increased urinary excretion of *xanthine* and *hypoxanthine* (Chapter 14), and a *hypouricemia*. Xanthine *lithiasis* and secondary renal damage may occur in severe xanthine oxidase deficiency. Xanthine oxidase is unique in that it contains *molybdenum* (Chapter 21).

One other rare hereditary human disease involving purines and pyrimidines is related to "excision repair" (Chapter 8), namely *xeroderma pigmentosum*. Patients with this disorder are extremely *sensitive to sunlight* and tend to develop *skin tumors*. The repair mechanism in skin fibroblasts is defective as a result of a deficiency of an *endonuclease* required for excision repair.

Miscellaneous metabolic disorders

The heading "miscellaneous metabolic disorders" could include a very large number of hereditary disorders not included in the preceding categories, such as those related to the abnormal hemoglobins, variant plasma proteins, certain immunoglobulins, blood clotting factors, and many other proteins performing specialized functions. However, some of these have been considered in earlier chapters and so are only listed as a group in Table 22-5. A few not discussed previously, especially several involving inorganic elements, are considered here in some detail.

Hemochromatosis, primary or idiopathic ("bronze diabetes"), is an hereditary disorder characterized by an *excessive absorption of iron* from the intestine (Chapter 21). As a result, iron slowly accumulates in the tissues, particularly in the liver, pancreas, and heart. After a number of years, usually after age 40, impaired function of these tissues becomes evident as hepatic cirrhosis, diabetes mellitus, and skin pigmentation. Some 30 to 50 gm of iron may have accumulated in the body, at the rate of a few milligrams a day. The condition is some 10 times more frequent in males than in females, who are "protected" to some extent by iron loss in menstrual blood and by pregnancy and lactation. Plasma iron levels are markedly increased, as is the saturation of plasma transferrin. There also may be *hemosiderinuria*. The hepatic parenchymal cells are gorged with iron. The preferred treatment is venesection, which will arrest the condition.

The molecular defect in hemochromatosis may be an increase in *iron-binding* by an intracellular (intestinal mucosal cell) iron carrier (apoferritin?), thus increasing iron absorption. Hemochromatosis also may result from "iron-overloading." This may follow excessive hemolysis in hemolytic diseases, repeated transfusions of whole blood, alcoholic cirrhosis, excessive iron therapy, or the use of iron kettles for cooking and beer making, as was found in the Bantu tribesmen of Africa. For this reason, there is considerable opposition to the overfortification of foods with iron (Chapter 21).

Menke's disease (kinky-hair or steely-hair syndrome) is an X-linked, autosomal recessive disease of infants resulting from *copper deficiency*. Only males are affected, females serving as heterozygote carriers. The molecular defect is an *inability to absorb copper* from the intestine. Recent evidence indicates that the mucosal cells bind copper normally but that the intracellular transport or transport through the serosal part of the cell membrane is defective. Placental transport of copper also may be defective. The incidence is 1:35,000 live births.

The chief symptoms of Menke's disease are changes in the structure and color of the hair, tortuosity of arteries, wormian bone formation in the sutures of the skull, increased susceptibility to infection, and severe neurological deterioration. Death usually occurs by the age of 3 years. *Serum copper* and *ceruloplasmin* levels are low. Treatment with copper salts is effective.

Trypsinogen deficiency has been reported in an 8-week-old infant. The symptoms included severe growth failure, hypoproteinemia, edema, and diarrhea. Specific pancreatic proteolytic enzyme assays disclosed a complete absence of trypsinogen, with a secondary failure in the activation of chymotrypsinogen and procarboxypeptidase. These and other pancreatic enzymes appeared to be present in normal amounts. The disorder responded favorably to dietary management with the parenteral, then oral, administration of protein hydrolysates. Two brothers of the infant had previously died, one at birth, the other at about 1 month of age, with symptoms identical to those just described.

Wilson's disease (hepatolenticular degeneration) is inherited as an autosomal recessive disorder and is characterized by an excessive deposition of *copper* in the *liver, kidneys,* and *brain*. Severe damage to these tissues results in hepatic cirrhosis, extensive renal aminoaciduria, and cerebral degeneration. *Serum copper* and *ceruloplasmin* levels are low, but there is an elevation of urinary copper. The molecular defect apparently is an impairment in the *incorporation* of *copper into ceruloplasmin* in the liver and in its excretion in bile. Current treatment is with *n*-penicillamine, which produces a striking clinical improvement.

The heritable disorders listed in Tables 22-1 through 22-5 thus emphasize the validity of the fundamental premise of the biological sciences stated at the beginning of this chapter—that all biological activity is the result of biochemical reactions—and of its corollary that for every pathology there is an underlying biochemical defect. These hereditary disorders also reemphasize the dynamic state of metabolism in the human organism and demonstrate that these concepts apply to every segment of the discipline of biochemistry. A continuing major goal of biochemistry thus is to further investigate and identify molecular defects of metabolism, to discover means for their control and eventual correction. In these ways, biochemistry is proving to be of immeasurable service to mankind.

Suggested readings

Altman, P.L., and Katz, D.D., editors: Human health and disease, section III, metabolic disorders, Bethesda, Md., 1979, Federation of American Societies for Experimental Biology.

Bondy, P.K., and Rosenberg, L.E., editors: Metabolic control and diseases, ed. 8, Philadelphia, 1980, W.B. Saunders co.

Brady, R.O.: Hereditary disorders of fat metabolism, Sci. Am. **229:**88, 1973.

Brown, D.H., and Brown, B.I.: Some inborn errors of carbohydrate metabolism, Int. Rev. Sci. **5:**391, 1975.

Brown, S.S., Mitchell, F.L., and Young, D.S.: Chemical diagnosis of disease, New York, 1979, Elsevier Press, Inc.

Buttner, J.: From the chemistry of life to chemistry of disease, Clin. Biochem. **12:**232, 1980.

Dickens, F., Randle, P.J., and Whelan, W.J., editors: Carbohydrate metabolism and its disorders, 2 vols., New York, 1968, Academic Press, Inc.

Felig, P.: Amino acid metabolism in man, Annu. Rev. Biochem. **44:**933, 1975.

Frimpter, G.W.: Aminoaciduria due to inherited disorders of metabolism, N. Engl. J. Med. **289:**835, 895, 1973.

Fuchs, F.: Genetic amniocentesis, Sci. Am. **242:**47, 1980.

Garrod, A.E.: Inborn errors of metabolism, Lancet **2:**1, 73, 142, 214, 1980.

Gelehrter, T.D., and Snodgrass, P.J.: Lethal neonatal deficiency of carbamoyl phosphate synthetase, N. Engl. J. Med. **290:**430, 1974.

Goldberg, D.M., and Werner, M., editors: Progress in clinical enzymology, New York, 1980, Masson Publishing USA, Inc.

Goldstein, J.L., and Brown, M.S.: The low-density lipoprotein pathway and its relation to atherosclerosis, Annu. Rev. Biochem. **46:**897, 1977.

Hers, H.G.: The control of glycogen metabolism in the liver, Annu. Rev. Biochem. **45:**167, 1976.

Hsai, D.Y.Y.: Inborn errors of metabolism, ed. 2, parts 1 and 2, Chicago, 1966, Year Book Medical Publishers, Inc.

Milunsky, A.: The prenatal diagnosis of hereditary disorders, Springfield, Ill., 1973, Charles C Thomas, Publisher.

Neufeld, E.F., and others: Inherited disorders of lysosomal metabolism, Annu. Rev. Biochem. **44:**357, 1975.

Nothkins, H.L.: The causes of diabetes, Sci. Am. **241:**62, 1979.

Nyhan, W.L., editor: Heritable disorders of amino acid metabolism, New York, 1974, John Wiley & Sons, Inc.

O'Brien, J.S.: Ganglioside-storage diseases, N. Engl. J. Med. **284:**893, 1971.

Pauling, L.: Sickle cell anemia: a molecular disease, Science **110:**543, 1949.

Ratner, S.: Enzymes of arginine and urea synthesis (and their deficiencies), Adv. Enzymol. **39:**1, 1973.

Ravio, K.O., and Seegmiller, J.E.: Genetic diseases of metabolism, Annu. Rev. Biochem. **41:**543, 1972.

Scanu, A.M., Eisder, R.W., and Getz, G.S., editors: The biochemistry of atherosclerosis, vol. 7 of Biochemistry of disease, New York, 1979, Marcel Dekker, Inc.

Stanbury, J.B., Wyngaarden, J.B., and Fredrickson, D.S., editors: The metabolic basis of inherited disease, ed. 4, New York, 1978, McGraw-Hill Book Co.

Thorpe, W.P.: The Lesch-Nyhan syndrome, Enzymes **12:**129, 1971.

Volk, B., and Aronson, S.M., editors: Sphingolipids, sphingolipidosis, and allied disorders, New York, 1972, Plenum Press, Inc.

Section Seven

Basic chemistry and properties of biological materials

Chapter 23

Chemistry of major cell constituents

Carbohydrates

Carbohydrates, the most abundant compounds found in nature, occur both as simple sugars and as more complex forms. A simple sugar may be attached to a variety of organic or inorganic groups, such as to other sugar molecules to form *polysaccharides* (starches, celluloses, glycogen, etc.); or certain simple sugars may be attached to purines, pyrimidines, and phosphate to form nucleic acids; or they may be attached to proteins to form proteoglycans and glycoproteins; or to lipids to form glycolipids; or to certain alcohols to form glycosides.

Definition The name *carbohydrate* indicates that these compounds are composed of carbon, hydrogen, and oxygen, the hydrogen and oxygen ordinarily being present in the same proportions as they are in water. A proper definition of *carbohydrates* is: polyhydroxy alcohols having potentially active aldehyde or ketone groups, and compounds yielding them on hydrolysis.

The simplest carbohydrates, with three carbons, are known as trioses and are shown below. The suffix *-ose* is used to designate a sugar. Glycerol, a sugar alcohol of the triose, is also shown.

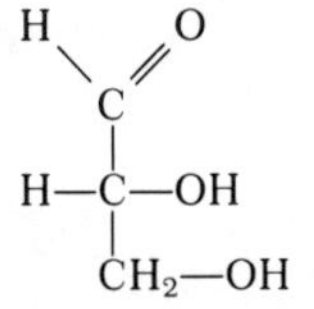

```
H     O
 \   //
   C                 CH2—OH            CH2—OH
   |                 |                 |
H—C—OH              C=O             H—C—OH
   |                 |                 |
   CH2—OH            CH2—OH            CH2—OH
D-Glyceraldehyde   Dihydroxyacetone    Glycerol
```

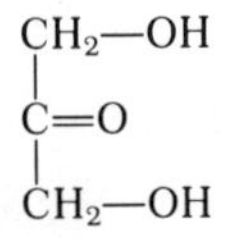

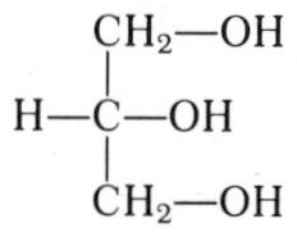

Classification Carbohydrates are generally classified as:

1. *Monosaccharides:* These are sugars that cannot be degraded further into other yet simpler sugars.
2. *Oligosaccharides:* These are small polymers of several monosaccharide units (usually from two to eight or ten) joined together. However, those involving two monosaccharides are specifically termed disaccharides and those having three are called trisaccharides. Anything larger is simply considered an oligosaccharide.
3. *Polysaccharides:* These are a large number of monosaccharides joined together to form a single polymeric molecule. They may form single chains of varying lengths, or they may consist of branched arrangements. Polysaccharides that are polymers of a single monosaccharide are called *homopolysaccharides;* those of different groups are called *heteropolysaccharides.*

Monosaccharides

Simple sugars $(CH_2O)_n$ are further classified according to the number of carbon atoms they contain. Thus trioses have three carbons; tetroses, four; pentoses, five; and hexoses, six. Furthermore, monosaccharides are also described according to the nature of the functional group. Thus there are *aldoses* for aldehyde-containing monosaccharides and ketoses for those having a ketone group. These two classifications may be used together; for example, there are aldohexoses and ketohexoses. The following table serves to illustrate these classifications.

Trioses	$C_3H_6O_3$ Glyceraldehyde (an aldotriose) Dihydroxyacetone (a ketotriose)
Tetroses	$C_4H_8O_4$ Erythrose (an aldotetrose) Erythrulose* (a ketotetrose or tetrulose)
Pentoses	$C_5H_{10}O_5$ Xylose (an aldopentose) Xylulose (a ketopentose or pentulose)
Hexoses	$C_6H_{12}O_6$ Glucose (an aldohexose) Fructose (a ketohexose or hexulose)
Heptoses	$C_7H_{14}O_7$ Sedoheptulose (a ketoheptose or heptulose)

*Ketoses are usually designated by the suffix *-ULOSE*.

Structure Glucose, the sugar of primary physiological significance, is an aldohexose. Its empirical formula is $C_6H_{12}O_6$, and its Fischer projection structure is as shown below. Likewise the structures of mannose and galactose, two other physiologically important aldohexoses, and fructose, a most important ketohexose, are also included.

```
   H   O          H   O          H   O
    \ //           \ //           \ //
     C              C              C            CH2—OH
     |              |              |             |
  H—C—OH        HO—C—H          H—C—OH          C=O
     |              |              |             |
 HO—C—H         HO—C—H         HO—C—H        HO—C—H
     |              |              |             |
  H—C—OH         H—C—OH        HO—C—H         H—C—OH
     |              |              |             |
  H—C—OH         H—C—OH         H—C—OH        H—C—OH
     |              |              |             |
    CH2—OH         CH2—OH         CH2—OH        CH2—OH
  Glucose        Mannose       Galactose      Fructose
```

Asymmetric carbons All sugars possess asymmetric carbon atoms. An asymmetric carbon is one to which is attached four different atoms or groups. If a carbon atom is pictured with four bonds projected into space in four different directions and each is joined to a different atom or group (Q,R,S,T), as shown in Fig. 23-1, then it is possible to construct a similar but not identical figure simply by exchanging the relative positions of any two of the groups. Thus the two models shown in Fig. 23-1 bear the same relation to each other as that of an object to its mirror image. They are not identical, which may be proved by attempting to superimpose one upon the other. Here are two compounds that resemble each other but are different because their carbons are asymmetric. The relationship exhibited by such compounds is called *stereoisomerism*, and the two compounds are called stereoisomers, or enantiomorphs (Gr., "opposite forms").

Figure 23-1
Model of asymmetric carbon atom and its mirror image.

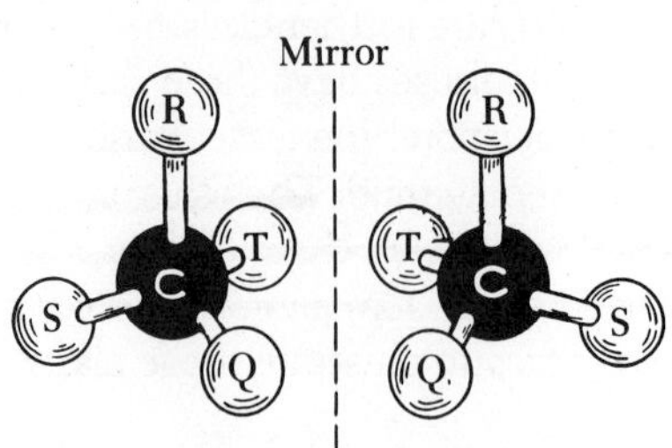

The simplest carbohydrate is the aldotriose glyceraldehyde; its Fischer projection and perspective structures are shown below. Those to the left have their hydroxyl group oriented to the right of the carbon and are known by convention as the D-isomer; the compound is called D-glyceraldehyde. To the right is L-glyceraldehyde, which has its hydroxyl group projecting to the left of the carbon.

D-Glyceraldehyde		L-Glyceraldehyde	
H–C=O	H–C=O	H–C=O	H–C=O
H—C—OH	H▶C◀OH	HO—C—H	HO▶C◀H
CH_2—OH	CH_2—OH	CH_2—OH	CH_2—OH

Since all simple sugars may be regarded as derivatives of these two, glyceraldehyde is considered to be the reference sugar. This means that all sugars with the same configuration as D-glyceraldehyde for the asymmetric carbon farthest from the functional aldehyde or ketone group are given the designation D. Those that are like L-glyceraldehyde in this respect are called L-sugars. The relationship of D-tetroses, D-pentoses, and D-hexoses to D-glyceraldehyde is shown in Fig. 23-2.

Glyceraldehyde has only one asymmetric carbon atom, and there are only two stereoisomers possible. If more then one asymmetric carbon atom is present in a sugar, then the actual number of isomers possible becomes 2^n, where n is the number of asymmetric carbons. From the structure of an aldohexose, it is evident that four asymmetric carbons are present; accordingly, there would be a 2^4, or 16 possible isomers. In Fig. 23-2 is shown the relationship of the eight D-aldohexoses to the reference sugar D-glyceraldehyde. These include D-glucose, D-mannose, D-galactose; the remaining five are uncommon in nature. If we begin with L-glyceraldehyde, then the same eight hexoses will result, but they will have the L-configuration. These will be mirror images, or enantiomorphs, of the corresponding D-sugars. The relationship between D- and L-glucose is shown below.

D-Glucose	L-Glucose
CHO	CHO
H—C—OH	HO—C—H
HO—C—H	H—C—OH
H—C—OH	HO—C—H
H—C—OH	HO—C—H
CH_2OH	CH_2OH

Figure 23-2
Relationships of the D-aldoses. The L-aldoses can be similarly shown, with the hydrogens and hydroxyls of each asymmetric carbon reversed.

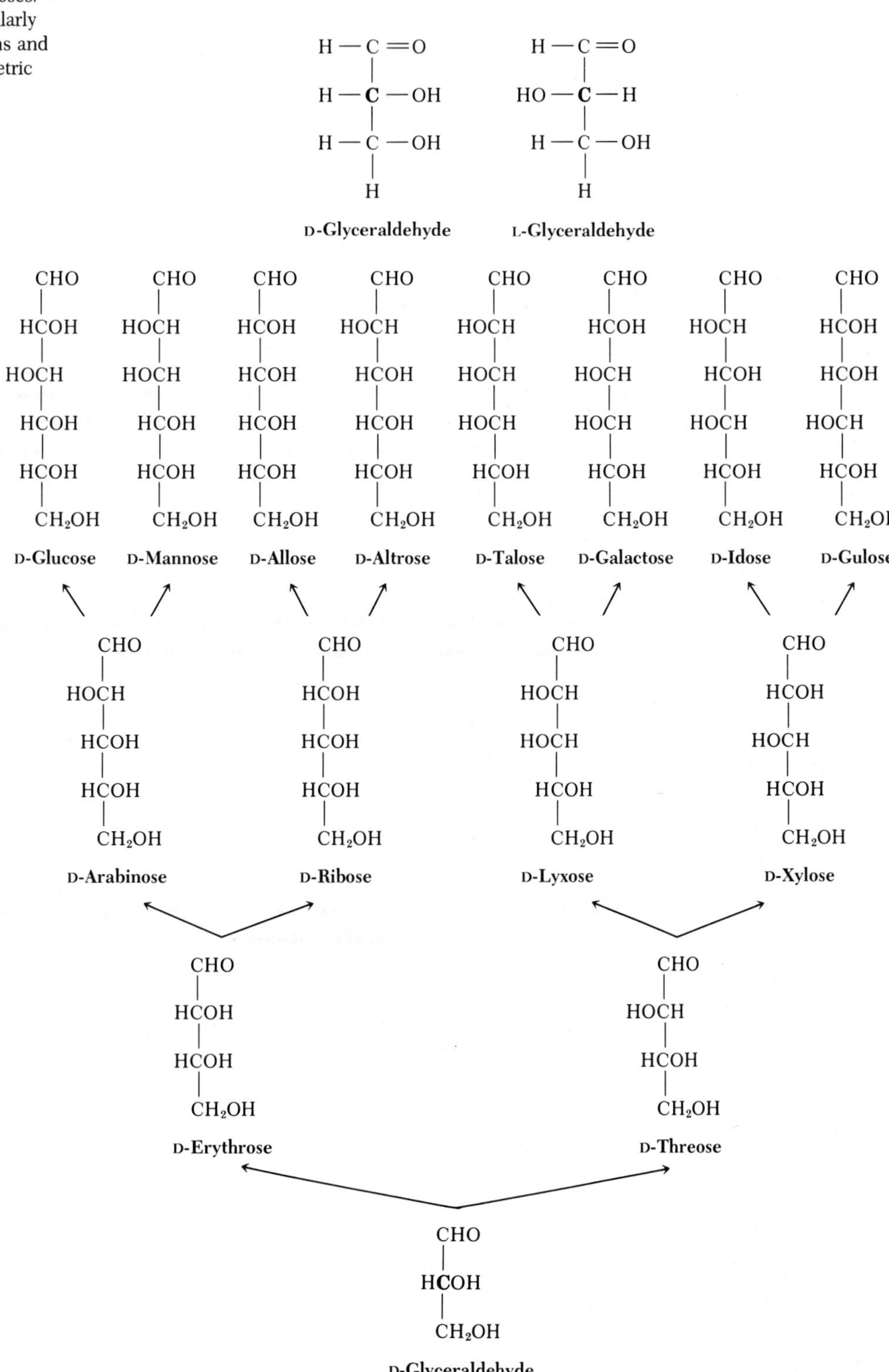

Another kind of stereoisomerism exists between the aldohexoses of a given stereoisomeric family. Consider the relationship of D-galactose and D-mannose to D-glucose. They differ by the orientation of the hydroxyl group only about a single carbon. Such isomers are called *epimers*. However, it should be noted from the following that D-galactose and D-mannose are not epimers of one another.

D-Mannose	**D-Glucose**	**D-Galactose**
CHO	CHO	CHO
HO—C—H	H—C—OH	H—C—OH
HO—C—H	HO—C—H	HO—C—H
H—C—OH	H—C—OH	HO—C—H
H—C—OH	H—C—OH	H—C—OH
CH_2OH	CH_2—OH	CH_2OH

Optical rotation of sugars The differences in structure of sugars are responsible for variations in physical, chemical, and physiological properties, e.g., degree of rotatory power, crystalline form, solubility, reactions, sweetness, nutritive value. Stereoisomers are also frequently optical isomers, because compounds containing asymmetric carbons have the power of turning or rotating the plane of a beam of polarized light. The degree of this optical activity is measured by means of a polarimeter and is expressed in degrees of rotation to the right (dextro) or to the left (levo). Dextro rotation is indicated by a plus sign, levo by a minus sign.

Mutarotation If D-glucose is dissolved in water and observed in a polarimeter it will be found to have a rotation of +52.7 degrees. By crystallizing D-glucose from different solvents two forms were obtained that could be distinguished by their optical rotation; one rotated light +112 degrees and the other +19 degrees. However, when these two forms of D-glucose were kept in solution and continuously observed in the polarimeter, it was found that their rotation changed; one decreased from +112 to +52.7 degrees and the other increased from +19 to +52.7 degrees. This spontaneous change in rotation that reaches a stable, equilibrium value is called *mutarotation*. Mutarotation is explained by the existence of two forms of D-glucose and requires the presence of another asymmetric center. The existence of a fifth asymmetric carbon in D-glucose is explained by the formation of a hemiacetal or ring structure of the hexose. Thus D-glucose, as seen below, forms two ring structures, designated α-D-glucose and β-D-glucose. The ring form includes an oxygen bridge between C-1 and C-5. The hemiacetals readily reopen to yield an equilibrium mixture, with a very small amount remaining in the intermediate aldehyde form.

α-D-Glucose (ring, O bridge C-1 to C-5)	⇌	D-Glucose (aldehyde form)	⇌	β-D-Glucose (ring, O bridge C-1 to C-5)
H—C—OH		H—C=O		HO—C—H
H—C—OH		H—C—OH		H—C—OH
HO—C—H		HO—C—H		HO—C—H
H—C—OH		H—C—OH		H—C—OH
H—C—		H—C—OH		H—C—
CH_2OH		CH_2OH		CH_2OH

α-D-Glucose (α-D-glucopyranose) (sp. rot. +112°)

D-Glucose (aldehyde form)

β-D-Glucose (β-D-glucopyranose) (sp. rot. +19°)

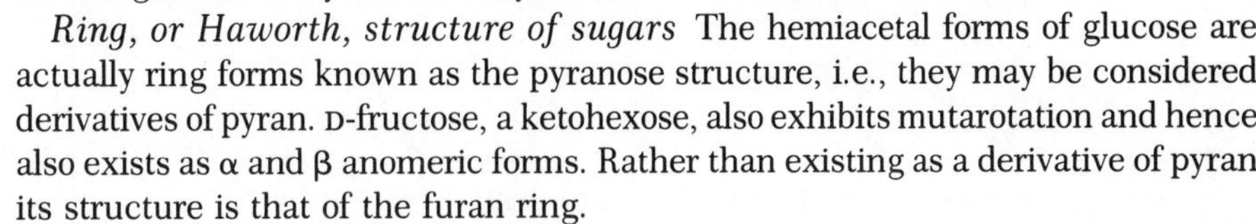

In the ring structure C-1 is asymmetric and is termed an *anomeric* carbon atom, so that glucose really has five asymmetric carbons rather than four.

Ring, or Haworth, structure of sugars The hemiacetal forms of glucose are actually ring forms known as the pyranose structure, i.e., they may be considered derivatives of pyran. D-fructose, a ketohexose, also exhibits mutarotation and hence also exists as α and β anomeric forms. Rather than existing as a derivative of pyran its structure is that of the furan ring.

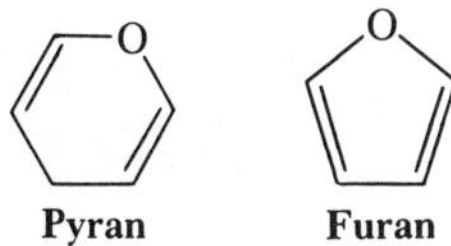

Pyran **Furan**

The pyranose formulas are usually pictured in perspective; the ring of five carbons and one oxygen is represented as a plane, "placed" at right angles to the printed page. The hydrogens, hydroxyls, and one primary alcohol group are oriented either above or below the pyranose ring. In this representation the bonds directed upward and above the plane correspond to the left in the Fischer formulas and the downward orientation corresponds to the right in the Fischer formula. This three-dimensional planar representation is known as the *Haworth projection*.

α-D-Glucose (α-D-glucopyranose) **β-D-Glucose (β-D-glucopyranose)**

D-Fructose ($C_6H_{12}O_6$), a ketohexose, may be represented by the following Fischer and Haworth structures:

α-D-Fructofuranose **Keto-D-fructose** **β-D-Fructofuranose**

α-D-Fructofuranose (perspective formula) **β-D-Fructofuranose (perspective formula)**

Pentoses likewise exist in the hemiacetal ring structure and may be represented in the Haworth formula. As seen on p. 838, the aldopentoses D-ribose and D-deoxyribose, both important constituents of nucleotides and nucleic acids, may be represented in the furanose ring structure.

D-Ribose **β-D-Ribofuranose** **D-2-Deoxyribose** **β-D-2-Deoxyribofuranose**

Xylose, on the other hand, is found in the pyranose form.

D-Xylose **α-D-Xylopyranose**

Conformation of sugars The actual three-dimensional structures, or conformations, of the pyranose rings are not planar as shown above. Rather pyranose rings exist in three-dimensional conformations called the "chair" and "boat" forms, as seen in Fig. 23-3. Of the two, the "chair" conformation is the more stable. The conformation of D-glucopyranose, for example, has been demonstrated to be the chair form.

The bonds to the hydrogen and hydroxyl groups of D-glucopyranose exist either parallel to the axis of symmetry of the ring, in which case they are termed axial bonds, or radiating more or less in the plane of the ring, in which case they are termed equatorial bonds. The conformational representations for α- and β-D-glucopyranose are shown in Fig. 23-3.

Groups such as the hydroxymethyl and hydroxyl occupy the "roomier" equatorial positions. This is probably the reason that β-D-glucopyranose is present in a 2:1 ratio to α-D-glucopyranose in an aqueous equilibrium mixture.

Other common aldohexoses—β-D-mannopyranose, α-D-galactopyranose, and α-D-idopyranose—likewise exist in similar chair conformations. The spatial positions of the hydroxyl groups and hydrogen atoms differ from those of D-glucopyranose. At least two common pentoses, D-xylose and D-arabinose, apparently may also exist in a pyranose ring, chair conformation.

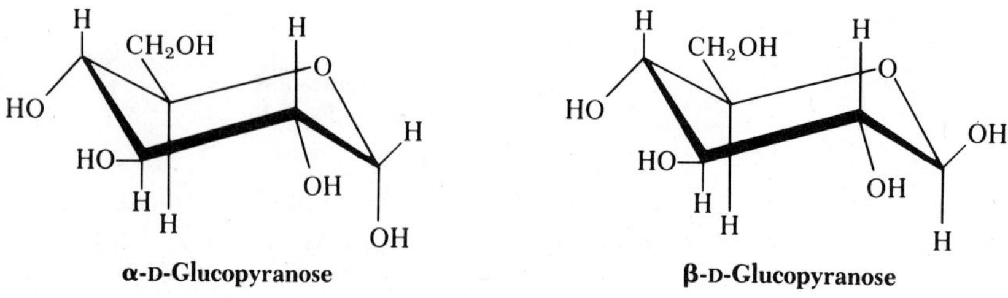

Figure 23-3
Spatial conformation of α- and β-glucopyranose.

Properties of sugars Carbohydrates that have potentially free aldehyde or ketone groups function as reducing agents. This property is usually demonstrated by the reduction of an alkaline copper reagent. Since under these conditions $Cu(OH)_2$ forms and normally precipitates, such reagents are usually prepared in the presence of citrate or tartrate to form soluble complexes. On heating such reagents with a sugar the Cu^{++} is reduced to Cu^{+} and Cu_2O, a red compound precipitates. Sugars with this property are usually called reducing sugars.

When treated with a weak alkali (e.g., barium hydroxide), glucose, fructose, and mannose are each converted to a mixture of the three. Whichever is the starting sugar, the same proportion of all three is finally found at equilibrium. This is the Lobry de Bruyn–van Ekenstein transformation. Remembering that C-4 , C-5, and C-6 have the same configuration in all three sugars, we can show how this reaction proceeds by directing our attention simply to the first three carbon atoms:

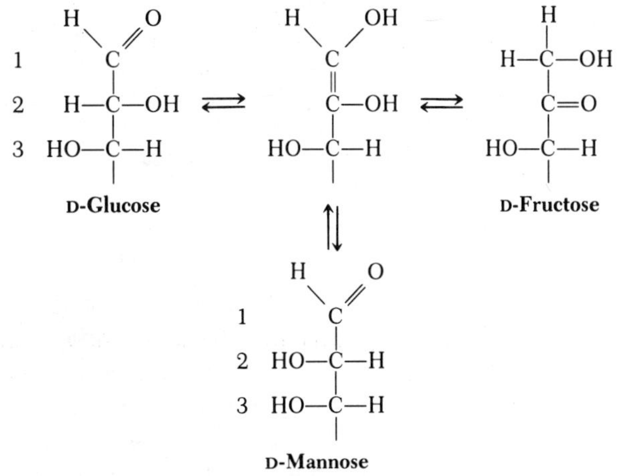

```
     H    O            H    OH              H
      \  //             \  /                |
1      C                 C               H—C—OH
       |                 ||                 |
2   H—C—OH    ⇌        C—OH      ⇌        C=O
       |                 |                  |
3  HO—C—H             HO—C—H             HO—C—H
       |                 |                  |
   D-Glucose             ⇅              D-Fructose

                       H    O
                        \  //
                 1       C
                         |
                 2   HO—C—H
                         |
                 3   HO—C—H
                         |
                     D-Mannose
```

This reaction, which involves the intermediate formation of an enol, is basically the mechanism whereby enzymes convert one hexose phosphate to another in the body. Thus all usable sugars can be converted to D-glucose or D-fructose; the transformation from D-glucose to D-fructose and vice versa occurs continually.

Reduction of monosaccharides to sugar alcohols The reduction of a simple sugar with hydrogen gas under pressure in the presence of a metal catalyst, or with certain other reducing agents, or enzymatically, converts the carbonyl group to an alcoholic hydroxyl group. For example, the reduction of D-glucose yields D-sorbitol (D-glucitol), as shown below. The reduction of L-sorbose also produces both D-sorbitol and L-iditol (alcohol of L-idose). The structural relations between L-sorbose and D-sorbitol will become evident by rotation of the formula of the former through 180 degrees in the plane of the page.

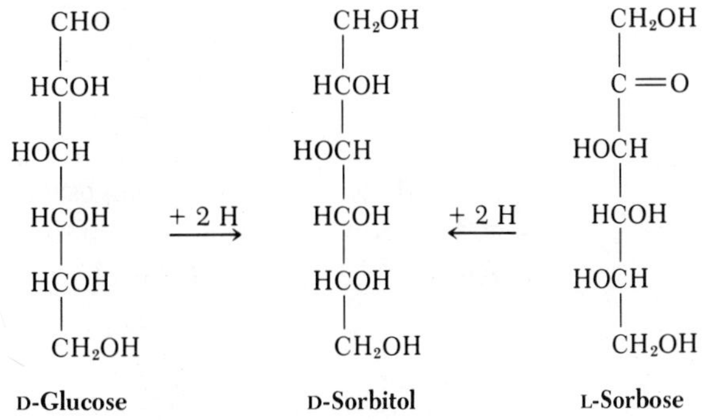

```
   CHO                 CH2OH                 CH2OH
    |                    |                     |
   HCOH                 HCOH                  C=O
    |                    |                     |
  HOCH                 HOCH                  HOCH
    |                    |                     |
   HCOH     + 2 H       HCOH      + 2 H       HCOH
    |       ——→          |        ←——          |
   HCOH                 HCOH                  HOCH
    |                    |                     |
   CH2OH                CH2OH                 CH2OH

 D-Glucose            D-Sorbitol            L-Sorbose
```

The chemical reduction of mannose and galactose yields mannitol and dulcitol respectively. The reduction of the pentoses xylulose and ribose forms xylitol and ribitol respectively.

Deoxy sugars Sugars that have fewer oxygen atoms than carbon atoms are known as deoxy sugars. Several of them are of physiological importance, particularly the pentose 2-deoxy-D-ribose ($C_5H_{10}O_4$), p. 838.

Glycosides When methyl alcohol reacts with the hemiacetal form of D-glucose, a glycosidic linkage is formed with the anomeric hydroxyl group. These compounds are acetals and are called *methyl glycosides* or, in this particular case, methyl glucoside. Here again are five asymmetric carbon atoms instead of four, since the first carbon is asymmetric. The two methyl glucosides, which are named α and β, are represented thus:

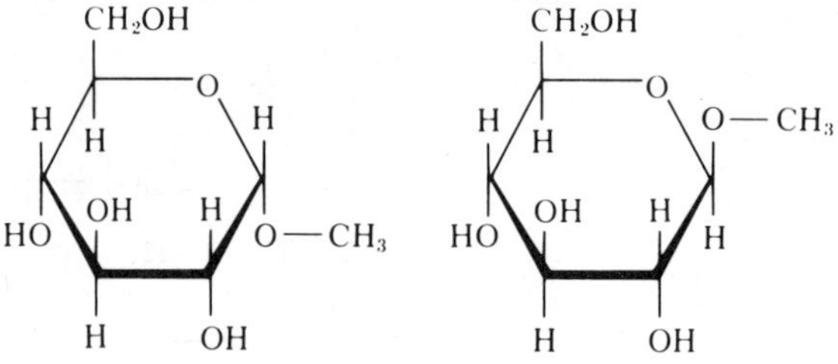

Methyl α-D-glucopyranoside **Methyl β-D-glucopyranoside**

These two differ in optical rotatory power, in solubilities, and in other physical characteristics. They are hydrolyzed by different enzymes: maltase acts on the α-anomer and almond emulsin on the β-anomer.

Glycosides are widely distributed and are present in certain spices, vegetable dyes, and drugs. The cardiac glycosides, which are important drugs, are combinations of steroids and sugar molecules or chains of sugar molecules.

Monosaccharide phosphates Esters of phosphoric acid with sugars are formed in many biological reactions. In fact, the formation of such phosphates appears to be a prerequisite to many physiological reactions. The hydroxyl groups are esterified with phosphate, usually by ATP, in the presence of an appropriate enzyme. The following are two examples of phosphate esters of importance to carbohydrate metabolism.

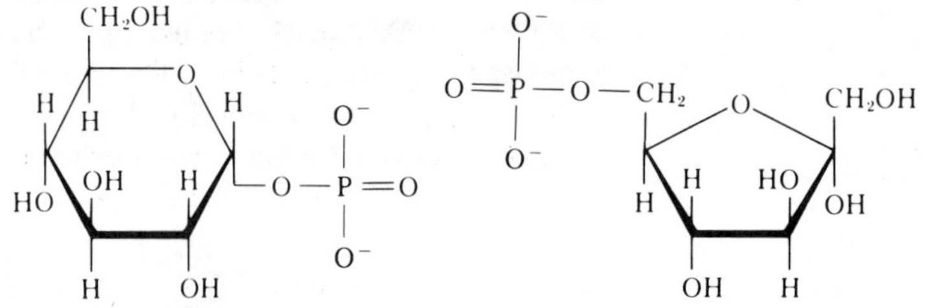

α-D-Glucopyranose 1-phosphoric acid

α-D-Fructofuranose-6-phosphoric acid

Oligosaccharides

The three most common oligosaccharides are the disaccharides maltose, lactose, and sucrose. Each has the empirical formula $C_{12}H_{22}O_{11}$ and is hydrolyzed by hot acids and by appropriate enzymes as follows:

$$C_{12}H_{22}O_{11} + H_2O \rightarrow C_6H_{12}O_6 + C_6H_{12}O_6$$

Maltose → D-Glucose + D-Glucose
Lactose → D-Glucose + D-Galactose
Sucrose → D-Glucose + D-Fructose

Disaccharides are formed by the union of two constituent monosaccharides, with the elimination of a molecule of water. The points of linkage vary, as does the manner of linking, and the properties of the disaccharide depend to a great extent on them. If both of the two potential aldehyde or ketone groups are involved in the linkage, the sugar has no reducing properties. However, if the aldehyde or ketone group of one of the constituent monosaccharide units is free, then the sugar has reducing properties and exhibits mutarotation. Sucrose is formed from D-glucose and D-fructose by union at the aldehyde and ketone carbons. Lactose and maltose both have an unlinked potential aldehyde and consequently are reducing sugars.

A comparison of the formulas of the disaccharides with those for methyl α- and β-glucosides will show that maltose has an α-glucoside linkage and lactose a β-galactoside linkage. The constituents of sucrose are joined by an α-glucoside-β-fructoside linkage.

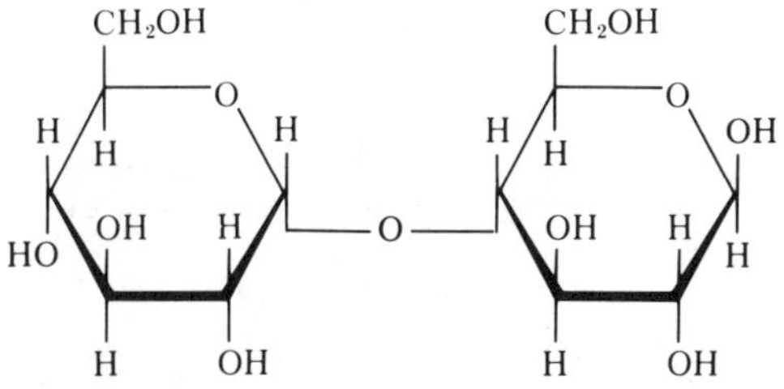

(D-glucose) (D-glucose)
(4-*O*-α-D-glucopyranosyl-D-glucopyranose)
Maltose (β-form)

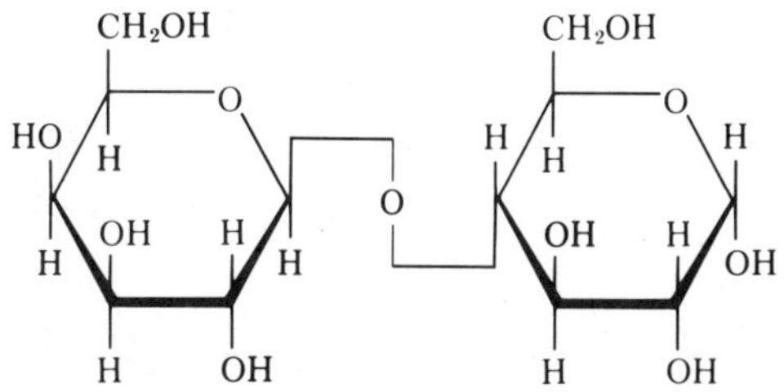

(D-galactose portion) (D-glucose portion)
(4-*O*-β-D-galactopyranosyl-α-D-glucopyranose)
Lactose (α-form)

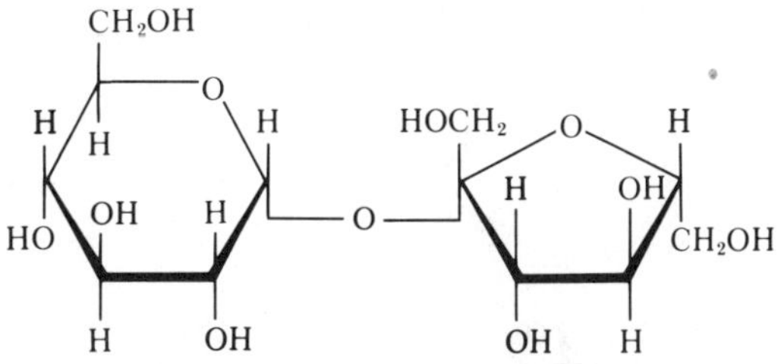

(D-glucose portion) (D-fructose portion)
(α-D-glucopyranosyl-β-D-fructofuranoside)
Sucrose

Polysaccharides

Polysaccharides that are polymers of a single monosaccharide are called homopolysaccharides; those that contain more than one kind of sugar are called heteropolysaccharides. These compounds are usually of very large molecular weight and if not insoluble in water at least form colloidal dispersions. The molecular weight of cellulose ranges from 200,000 to 400,000; that of starches ranges from 10,000 to 1,000,000; and glycogen's molecular weight varies from 1,000,000 to 4,000,000. Usually polysaccharides are either straight-chain or branched-chain compounds.

By convention polysaccharides are given names ending in *-an,* attached to the particular monosaccharide that makes up the polymer. Thus a name for a polysaccharide in general is *glycans,* from *glycose,* a simple sugar. Particular examples are arabans, xylans, mannans, and galactans, which are polymers of arabinose, xylose, mannose, and galactose respectively.

Starch Starch contains two polysaccharides; both are glucans, or polymers of glucose, but they differ in molecular architecture and in certain properties. One is called *amylose* and the other *amylopectin*. There is usually 20% to 28% amylose, and the rest is amylopectin. Amylose consists of 250 to 300 D-glucose units linked by α-1,4-glucosidic bonds (Fig. 23-4, *A*). These tend to twist the chain into a helix. In amylopectin, the majority of the units are similarly connected by α-1,4-glucosidic bonds, but there are occasional α-1,6-glucosidic bonds (Fig. 23-4, *B*). The structure is a branch-on-branch arrangement of perhaps a thousand D-glucopyranose units. Starch granules are insoluble in cold water, but when their suspension is heated, water is taken up and swelling occurs, at first to a slight degree but later to several hundred times their original volume. viscosity increases, and starch gels or pastes result. Both the granules and the colloidal solutions react with iodine to give a blue color. This is chiefly because of the amylose, which forms a deep blue complex, whereas amylopectin solutions are colored blue-violet or purple.

Glycogen Often called animal starch, glycogen occurs in the liver and muscle as the storage form of glucose. Glycogen resembles amylopectin in structure; it is made up of multibranched chains of glucose units. The individual units are

Figure 23-4

A α-1,4-Glucosidic bonds joining the glucose units in the straight chain amylose molecule.
B Branched chain structure of the amylopectin molecule. Shown are examples of both α-1,4- and α-1,6-glucosidic bonds.

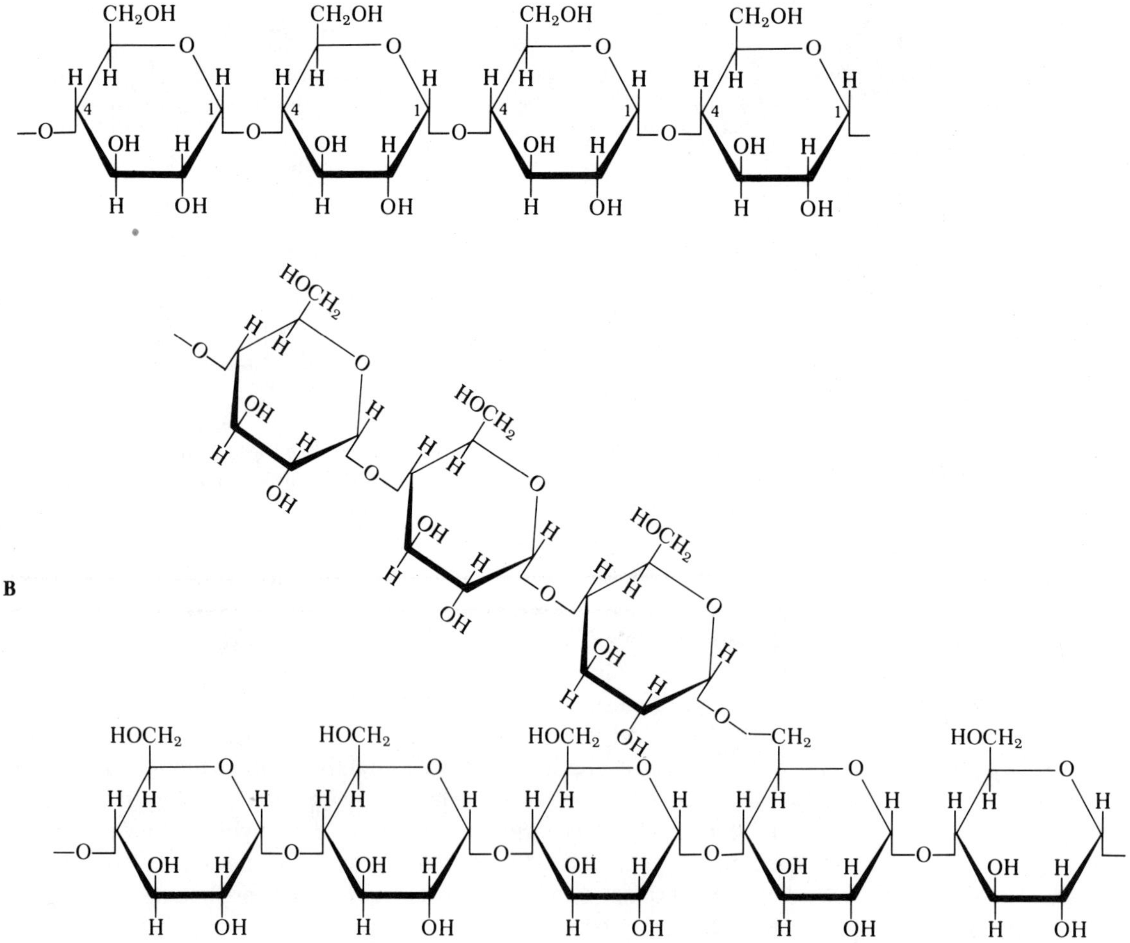

attached to each other by α-1,4-glucosidic linkages, but the chains are somewhat shorter, averaging 12 glucose units. These short chains are joined by way of α-1,6-glucosidic linkages. Glycogen does not exist in the form of homogeneous molecules all having the same molecular weight and possessing the same number of monosaccharide units. Rather glycogen molecules are collections of polymers having the same general branching pattern but a wide range of molecular weights, depending on the animal from which it is derived, and varying even with the tissue of the same animal.

Cellulose Cellulose is a glucan that makes up a large part of plant tissues; it is one of the most plentiful organic substances in nature. Cellulose is a large–molecular weight homopolysaccharide in which the glucose units are held together in the form of a straight chain by β-1,4-glucosidic bonds.

There are no enzymes in the intestinal tract capable of hydrolyzing the β-1,4-glucosidic bonds. Cellulose therefore represents an indigestible part of the human diet.

Proteoglycans Proteoglycans are large, high–molecular weight heteropolysaccharides that are typically composed of amino sugars and uronic acids. These polysaccharides are typically joined to a protein. Proteoglycans, formerly called mucopolysaccharides, are found in connective tissues (Chapter 15) and as a blood coagulant (Chapter 16). Generally these substances are made up largely of heteropolysaccharides linked covalently to a polypeptide core.

One of the most prevalent proteoglycans of connective tissues is called chondroitin (Chapter 15). This polymer is a straight chain composed of alternating units of galactosamine and glucuronic acid.

D-Galactosamine and D-glucosamine represent a class of carbohydrate derivative known as amino sugars, or hexosamines. The structure of these two is given below:

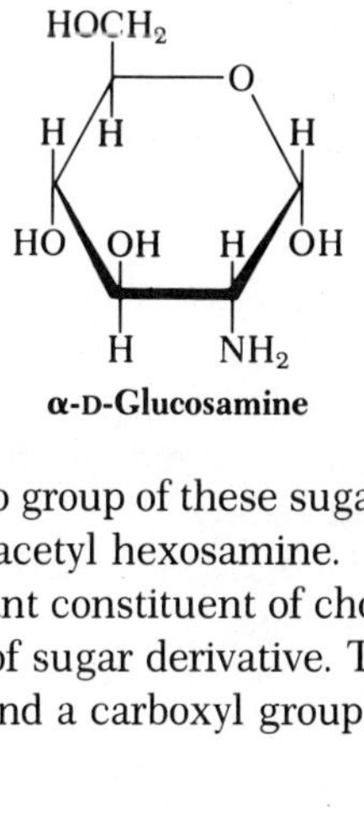

HOCH2, O, H, H, H, HO, OH, H, OH, H, NH2
α-D-Glucosamine

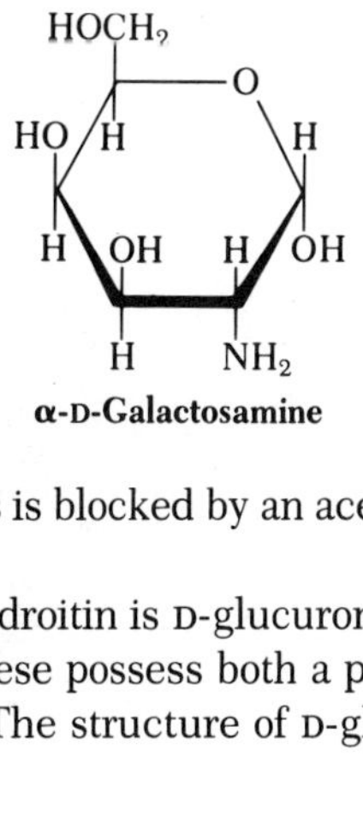

HOCH2, O, HO, H, H, H, OH, H, OH, H, NH2
α-D-Galactosamine

Very often the amino group of these sugars is blocked by an acetyl group, in which case it is termed *N*-acetyl hexosamine.

The other important constituent of chondroitin is D-glucuronic acid. This represents another kind of sugar derivative. These possess both a potentially free aldehyde (hemiacetal) and a carboxyl group. The structure of D-glucuronic acid is as follows:

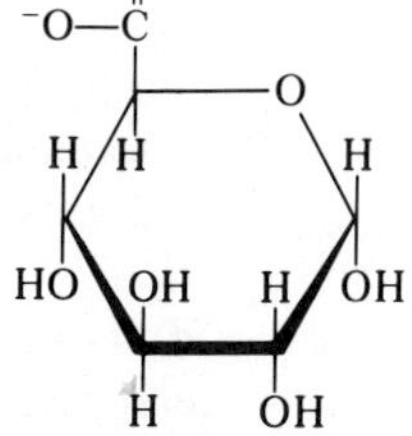

O, ⁻O—C, O, H, H, H, HO, OH, H, OH, H, OH
α-D-Glucuronic acid

In chondroitin the two sugars are linked by alternating β-1,4 and β-1,3 glycosidic bonds. Thus:

$$\beta\text{-GlcUA}(1 \rightarrow 3) - [\beta\text{-GalNAc}(1 \rightarrow 4) - \beta - \text{GlcUA}]_n^-$$

Here GlcUA represents the glucuronic acid and GalNAc is *N*-acetyl galactosamine.

Lipids

Lipids are compounds that have in common: (1) insolubility in water and solubility in one or more organic solvents (ether, chloroform, benzene, acetone—the so-called fat solvents) and (2) some relation to fatty acids as esters, either actual or potential. The following outline shows a classification system for lipids.

- A. Simple lipids: esters of fatty acids with various alcohols
  1. Neutral fats and oils: triesters of fatty acids with glycerol; triacylglycerols or triglycerides
  2. Waxes: esters of fatty acids with monohydroxy aliphatic alcohols higher than glycerol
     - a. True waxes: products of both animal and vegetable origin in which esters are composed of palmitic, stearic, oleic, or other higher fatty acid esters of cetyl alcohol, $CH_3(CH_2)_{14}CH_2OH$, or other higher straight-chain alcohols
     - b. Cholesterol esters: esters of fatty acids with cholesterol
     - c. Vitamin A esters: palmitic or stearic acid esters of vitamin A
     - d. Vitamin D esters
- B. Compound lipids: esters of fatty acids with alcohols plus other groups
  1. Phospholipids: lipids containing phosphoric acid and, in most cases, a nitrogenous base
  2. Glycolipids or cerebrosides: lipids containing a carbohydrate and also nitrogen but no phosphate and no glycerol
  3. Sulfolipids: lipids characterized by possessing sulfate groups
  4. Lipoproteins: lipids attached to plasma or other proteins
  5. Lipopolysaccharides: lipids attached to polysaccharides
- C. Derived lipids: derivatives obtained by hydrolysis of those given in groups A and B that still possess general physical characteristics of lipids
  1. Saturated and unsaturated fatty acids and their derivatives, e.g., the prostaglandins
  2. Monoglycerides and diglycerides
  3. Alcohols
     - a. Straight-chain alcohols: water-insoluble alcohols of higher molecular weight obtained on hydrolysis of waxes
     - b. Sterols and other steroids, including vitamin D
     - c. Alcohols containing a β-ionone ring (vitamin A, certain carotenoids)
- D. Miscellaneous lipids
  1. Aliphatic hydrocarbons: include isooctadecane, found in liver fat, and certain hydrocarbons found in beeswax and plant waxes
  2. Carotenoids, squalene, and other terpenes
  3. Vitamins E and K

Fats and oils: triacylglycerols

Fats and oils are triesters of glycerol and various fatty acids. In general, vegetable oils are triacylglycerols that are liquid at room temperature because of their more highly unsaturated or shorter length carbon fatty acids. Triacylglycerols are the most abundant naturally occurring lipids. The term *fat* represents complex mixtures of triacylglycerols.

A general formula for a triacylglycerol is given below. This diagram shows that the compound is formed by joining three fatty acids (carbons-R_1, -R_2, and -R_3) to a glycerol backbone by ester linkages.

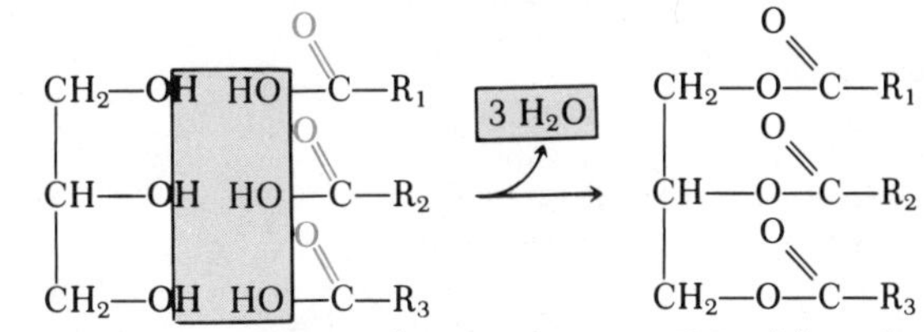

$$\begin{array}{l} CH_2\text{—OH} \\ | \\ CH\text{—OH} \\ | \\ CH_2\text{—OH} \end{array} \quad \begin{array}{l} HO\text{—}C(=O)\text{—}R_1 \\ \\ HO\text{—}C(=O)\text{—}R_2 \\ \\ HO\text{—}C(=O)\text{—}R_3 \end{array} \xrightarrow{3\,H_2O} \begin{array}{l} CH_2\text{—O—}C(=O)\text{—}R_1 \\ | \\ CH\text{—O—}C(=O)\text{—}R_2 \\ | \\ CH_2\text{—O—}C(=O)\text{—}R_3 \end{array}$$

Glycerol **Fatty acids** **Triacylglycerol**

Nomenclature The following structure is that of a monoacylglycerol. It is evident that this compound has an asymmetric carbon atom (C-2) and therefore may exist as two stereoisomers, namely, D- and L-monoacylglycerols. Unfortunately the same monoacylglycerol may be written either as:

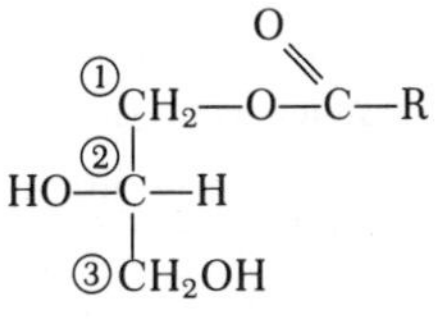

$$\begin{array}{l} ①CH_2\text{—O—}C(=O)\text{—}R \\ \quad | \\ HO\text{—}②C\text{—}H \\ \quad | \\ ③CH_2OH \end{array} \quad \text{or} \quad \begin{array}{l} ①CH_2\text{—OH} \\ \quad | \\ H\text{—}②C\text{—OH} \\ \quad | \\ ③CH_2\text{—O—}C(=O)\text{—}R \end{array}$$

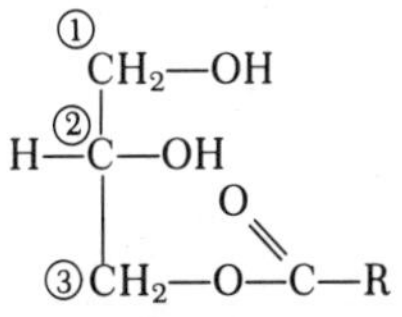

1-acyl-L-glycerol **3-acyl-D-glycerol**

To resolve this ambiguity, a convention of nomenclature has been recommended called *stereospecific numbering*. In this convention glycerol is written as below, with the C-2 hydroxyl group oriented to the left and the carbons numbered 1 through 3, beginning at the top.

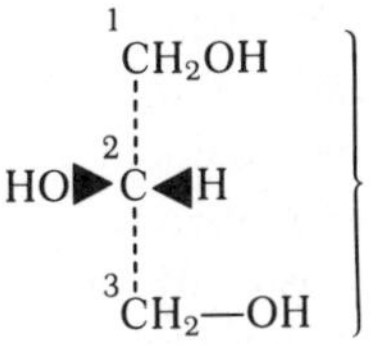

$$\left.\begin{array}{l} {}^{1}CH_2OH \\ \quad \vdots \\ HO\blacktriangleright {}^{2}C \blacktriangleleft H \\ \quad \vdots \\ {}^{3}CH_2\text{—OH} \end{array}\right\} \text{Stereospecific numbering (sn)}$$

Accordingly the following structure would be used to describe monostearoyl glycerol.

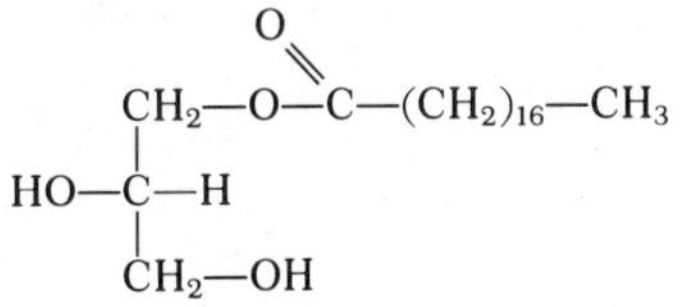

$$\begin{array}{l} CH_2\text{—O—}C(=O)\text{—}(CH_2)_{16}\text{—}CH_3 \\ | \\ HO\text{—}C\text{—}H \\ | \\ CH_2\text{—OH} \end{array}$$

1-Stearoyl sn-glycerol

Likewise a specific example of a triacylglycerol containing palmitic, oleic, and stearic acids would be the following:

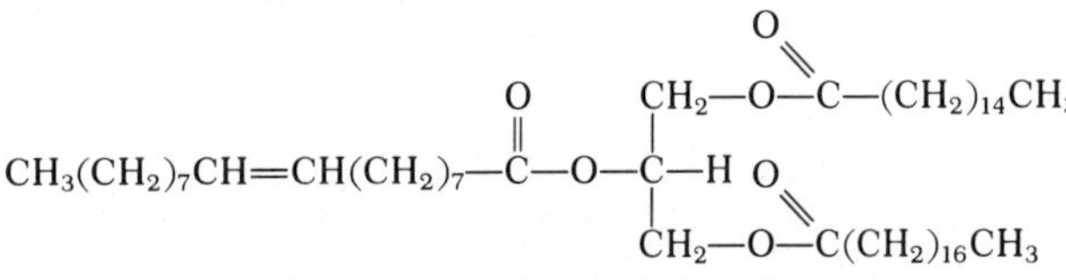

$$\begin{array}{r} \\ CH_3(CH_2)_7CH{=}CH(CH_2)_7\text{—}\overset{O}{\overset{\|}{C}}\text{—O—} \\ \\ \end{array}\begin{array}{l} CH_2\text{—O—}C(=O)\text{—}(CH_2)_{14}CH_3 \\ | \\ C\text{—}H \\ | \\ CH_2\text{—O—}C(=O)(CH_2)_{16}CH_3 \end{array}$$

1-Palmitoyl-2-oleoyl-3-stearoyl-sn-glycerol
(or 1-palmitoyl-2-oleoyl-3-stearoyl-L-glycerol)

Table 23-1
Some common fatty acids

Common name	Systematic name	Structural formula	Melting point (° C)
Saturated series			
Butyric	*n*-Butanoic	$CH_3(CH_2)_2COOH$	−7.9
Caproic	*n*-Hexanoic	$CH_3(CH_2)_4COOH$	−3.4
Lauric	*n*-Dodecanoic	$CH_3(CH_2)_{10}COOH$	44
Myristic	*n*-Tetradecanoic	$CH_3(CH_2)_{12}COOH$	54
Palmitic	*n*-Hexadecanoic	$CH_3(CH_2)_{14}COOH$	63
Stearic	*n*-Octadecanoic	$CH_3(CH_2)_{16}COOH$	70
Arachidic	*n*-Eicosanoic	$CH_3(CH_2)_{18}COOH$	75
Lignoceric	*n*-Tetracosanoic	$CH_3(CH_2)_{22}COOH$	84
Unsaturated series			
Palmitoleic	9-Hexadecenoic (*cis*)	$CH_3(CH_2)_5CH{=}CH(CH_2)_7COOH$	0.5
Oleic	9-Octadecenoic (*cis*)	$CH_3(CH_2)_7CH{=}CH(CH_2)_7COOH$	13.4
Linoleic	9,12-Octadecadienoic (*cis, cis*)	$CH_3(CH_2)_4CH{=}CHCH_2CH{=}CH(CH_2)_7COOH$	−5
γ-Linolenic	6,9,12-Octadecatrienoic (all *cis*)	$CH_3(CH_2)_4CH{=}CHCH_2CH{=}CH$- $CH_2CH{=}CH(CH_2)_4COOH$	−11
Arachidonic	5,8,11,14-Eicosatetraenoic (all *cis*?)	$CH_3(CH_2)_4CH{=}CHCH_2CH{=}CH$- $CH_2CH{=}CHCH_2CH{=}CH(CH_2)_3COOH$	−49.5
Clupanodonic	4,8,12,15,19-Docosapentaenoic (all *cis*?)	$CH_3CH_2CH{=}CH(CH_2)_2CH{=}CHCH_2CH{=}CH$- $(CH_2)_2CH{=}CH(CH_2)_2CH{=}CH(CH_2)_2COOH$	−78

Alternatively each of these may also be described simply as L-isomers, thus 1-stearoyl-L-glycerol and 1-palmitoyl-2-oleoyl-3-stearoyl L-glycerol. In general, naturally occurring triacylglycerols are of the L-stereoisomeric form.

Fatty acids A fatty acid is an aliphatic monocarboxylic acid ranging in chain length from 4 to about 24 carbon atoms. Most of the naturally occurring fatty acids have even-numbered carbon chains. Some are saturated and others unsaturated. There are also fatty acids that contain such substituents as hydroxyl and keto groups. Common fatty acids are listed in Table 23-1. Of these the most abundant saturated fatty acid is palmitic acid; oleic acid is the most abundant unsaturated one.

A frequently employed shorthand nomenclature for fatty acids uses the total number of carbons, the number of double bonds, and their position in the chain. Of the 18-carbon fatty acids, the saturated stearic acid is 18:0 and oleic acid is 18:1(9). Here the number in parentheses indicates that the double bond begins with C-9.

Unsaturated fatty acids and geometric isomerism Unsaturated fatty acids possess one or more double bonds in the carbon chain. Where a double bond exists, there is the opportunity for *cis, trans,* or geometric isomerism. As seen on p. 847, the double bond at C-9 in oleic acid puts a bend in an otherwise straight chain. The *trans* isomer is called elaidic acid.

```
                                   H                            O
                                   |9                          1||
   CH2    CH2    CH2    CH2        C        CH2    CH2    CH2   C
18/   \  /   \  /   \  /   \      //\      /   \  /   \  /   \ / \
CH3    CH2    CH2    CH2    C  10     CH2    CH2    CH2    CH2    OH
                            |
                            H
```

trans or Elaidic acid

$$^{18}CH_3-CH_2-CH_2-CH_2-CH_2-CH_2-CH_2-CH_2-\overset{10}{C}H=\overset{9}{C}H-CH_2-CH_2-CH_2-CH_2-CH_2-CH_2-CH_2-C(=O)-OH$$

***cis* or Oleic acid**

Physical properties of fats The hardness or consistency of acyl glycerols is related to their melting points, which are not sharp, because natural fats are mixtures and not pure substances. The solidification points are considerably lower than the melting points. Triacylglycerols of the lower fatty acids melt at lower temperatures than do those of the higher fatty acids, and triacylglycerols of the unsaturated fatty acids are still lower. These are reflections, in part at least, of the melting points of the constituent fatty acids (Table 23-1). Many triacylglycerols (e.g., triolein) are liquid at room temperatures and are commonly called *oils*. The term *oil* is rather confusing because it is often used for substances having no relation to the lipids (e.g., mineral oil, which is a mixture of hydrocarbons). Here the word *oil* indicates the physical state of a substance, rather than the chemical nature of that substance. The hardness of common acyl glycerols depends largely on the relative amounts of long-chain saturated fatty acids (like palmitic and stearic) and unsaturated fatty acids (like oleic and linoleic) present. The former are solid and the latter liquid at room temperature. There are larger proportions of soft fats in cold-blooded animals than in warm-blooded animals. This facilitates motility at low temperatures. Likewise, within a species of animals, the subcutaneous fat tends to be "softer" than that serving as a protective pad for the internal organs, e.g., the kidneys.

The specific gravity of all acyl glycerols is less than 1.0. Consequently, all fats float on water. They are not soluble in water, at least not to any appreciable extent. Emulsions of acyl glycerols may be made by shaking them vigorously in water, but emulsifying agents such as gums, soaps, and proteins are required to produce more

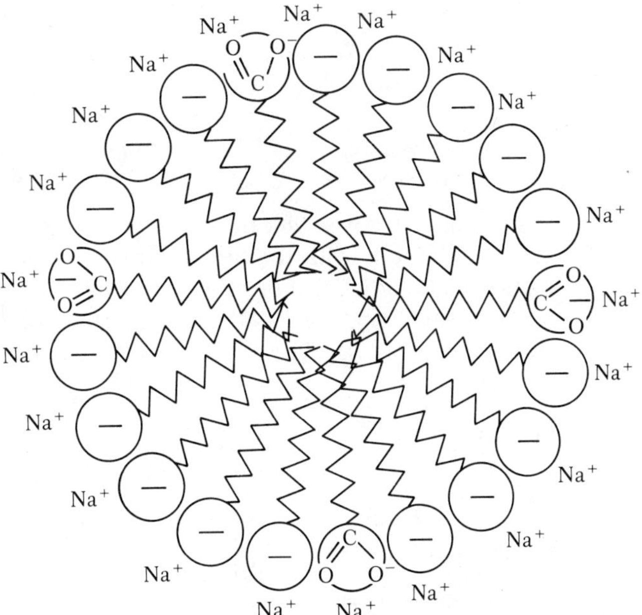

Figure 23-5
Sodium salts of long-chain fatty acids (soaps) form micelles in water. Aliphatic side chains produce a hydrophobic core, whereas the anionic groups produce a polar outer shell.

stable emulsions. The emulsification of fats in the intestinal tract is a prerequisite for digestion and absorption.

Hydrolysis of triacylglycerols Fats can be hydrolyzed by alkalis or by fat-splitting enzymes called lipases. They yield glycerol and the constituent fatty acids. In the case of alkaline hydrolysis, called saponification, the corresponding fatty acid salts or soaps are formed.

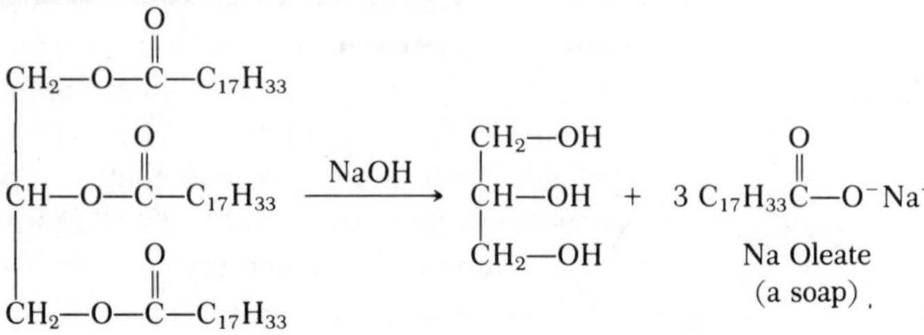

$$\begin{array}{l} CH_2-O-\overset{O}{\overset{\|}{C}}-C_{17}H_{33} \\ | \\ CH-O-\overset{O}{\overset{\|}{C}}-C_{17}H_{33} \\ | \\ CH_2-O-\overset{O}{\overset{\|}{C}}-C_{17}H_{33} \end{array} \xrightarrow{NaOH} \begin{array}{l} CH_2-OH \\ | \\ CH-OH \\ | \\ CH_2-OH \end{array} + 3\ C_{17}H_{33}\overset{O}{\overset{\|}{C}}-O^-Na^+$$

Na Oleate
(a soap).

Soaps function as detergents because of their ability to form micelles. As seen in Fig. 23-5, the hydrocarbon chains arrange themselves to form an hydrophobic core such that the carboxylate groups are at the surface of the micelle. Thus the hydrophilic carboxylate groups are at a lipid-water interface. The micelles are stabilized by the association of these anionic groups with the water molecules. Dirt, fat, and other substances are dissolved in the hydrophobic core.

Phospholipids

Phospholipids are composed of fatty acids, glycerol, phosphoric acid, and in most cases a nitrogenous base. As seen below, they are fatty acyl derivatives of L-glycerol-3-phosphate or *sn*-glycerol-3-phosphate. Since C-2 of glycerol phosphate is asymmetric, phospholipids exhibit stereoisomerism; in nature all are of the L-configuration. The structure seen at the right below is 1-stearoyl-2-oleoyl-*sn*-glycerol-3-phosphate and is usually termed a *phosphatidic acid*. Phosphatidic acids form a

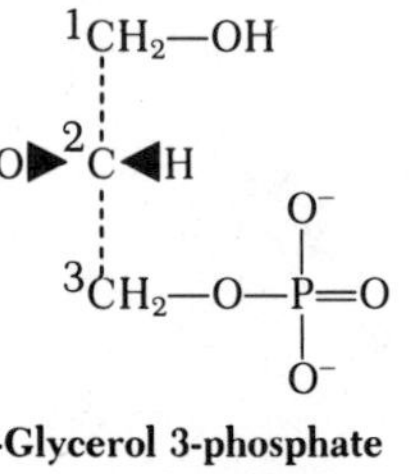

$$\begin{array}{l} {}^1CH_2-OH \\ \vdots \\ HO\blacktriangleright {}^2C \blacktriangleleft H \\ \vdots \\ {}^3CH_2-O-\overset{O^-}{\overset{|}{\underset{O^-}{\underset{|}{P}}}}=O \end{array}$$

sn-Glycerol 3-phosphate
or L-Glycerol 3-phosphate

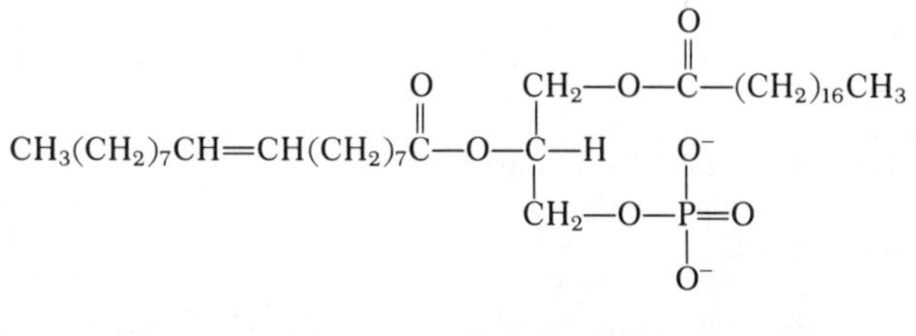

$$CH_3(CH_2)_7CH=CH(CH_2)_7\overset{O}{\overset{\|}{C}}-O-\begin{array}{l} CH_2-O-\overset{O}{\overset{\|}{C}}-(CH_2)_{16}CH_3 \\ | \\ C-H \quad O^- \\ | \qquad\quad | \\ CH_2-O-P=O \\ \qquad\qquad | \\ \qquad\qquad O^- \end{array}$$

1-Stearoyl-2-oleoyl-sn Glycerol 3-phosphate
(a phosphatidic acid)

number of derivatives with nitrogenous bases such as choline, ethanolamine, and serine, as well as with the hexahydroxy cyclohexane, inositol. The structure of L-phosphatidyl choline, formerly called lecithin, is given below. Another derivative of phosphatidic acid involves the nitrogenous base ethanolamine in place of choline (see below); this was formerly called cephalin.

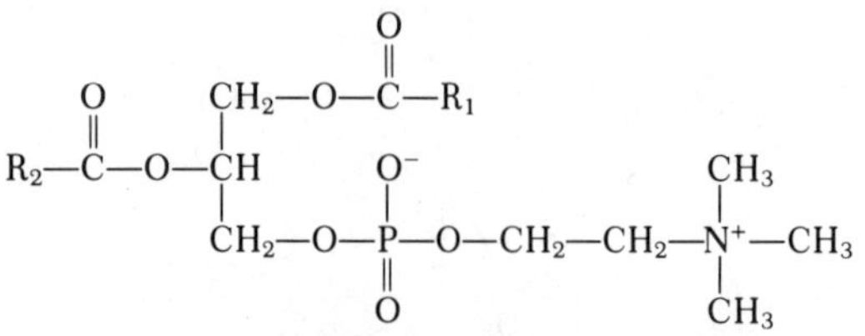

$$R_2-\overset{O}{\overset{\|}{C}}-O-\begin{array}{l} CH_2-O-\overset{O}{\overset{\|}{C}}-R_1 \\ | \\ CH \\ | \\ CH_2-O-\overset{O^-}{\overset{|}{\underset{O}{\underset{\|}{P}}}}-O-CH_2-CH_2-\overset{CH_3}{\overset{|}{\underset{CH_3}{\underset{|}{N^+}}}}-CH_3 \end{array}$$

3-sn-Phosphatidyl choline
or 1,2-Diacyl-sn-glycero-3-phosphoryl choline
or L-Phosphatidyl choline (Lecithin)

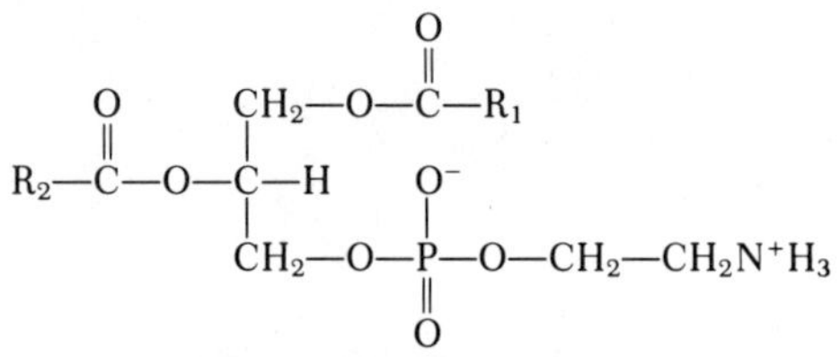

$$R_2-\overset{O}{\overset{\|}{C}}-O-\begin{array}{l} CH_2-O-\overset{O}{\overset{\|}{C}}-R_1 \\ | \\ C-H \\ | \\ CH_2-O-\overset{O^-}{\overset{|}{\underset{O}{\underset{\|}{P}}}}-O-CH_2-CH_2N^+H_3 \end{array}$$

3-sn-Phosphatidyl ethanolamine
or L-Phosphatidyl ethanolamine
(Cephalin)

Usually one of the constituent fatty acids is saturated and one is unsaturated, generally oleic acid.

Phosphatidyl choline and ethanolamine are soluble in fat solvents but are insoluble in acetone. This property is used to separate them from other fats. Although insoluble in water, they readily form micelles (see Chapter 2) because of their amphipathic nature.

Certain enzymes called lecithinases hydrolyze phosphatidyl choline so as to remove the fatty acid from C-2. This forms a compound often called *lysolecithin*, a powerful red blood cell lysing agent (causes hemolysis). The enzyme is a component of cobra venom as well as that of certain poisonous spiders and other stinging insects.

Other important phosphatidates are phosphatidyl serine, which contains the amino acid serine, and phosphatidyl inositol, which has the hexahydroxy cyclohexane inositol as a constituent.

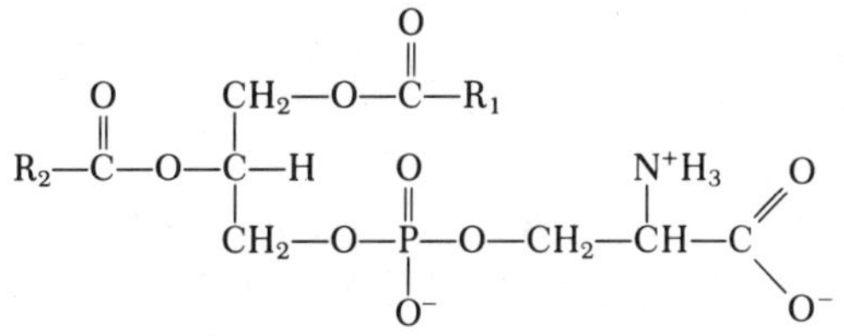

3-sn-Phosphatidyl serine

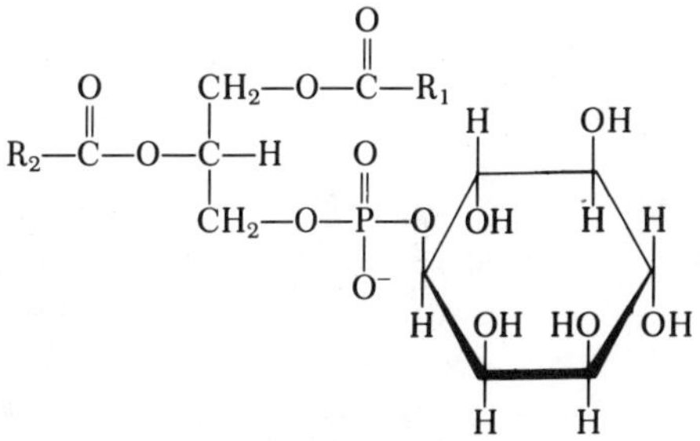

3-sn-Phosphatidyl inositol

An important phospholipid of mitochondrial membrane is cardiolipin. This is a diphosphatidyl glycerol. As seen from the following structure, two phosphatidic acids are joined by a molecule of glycerol.

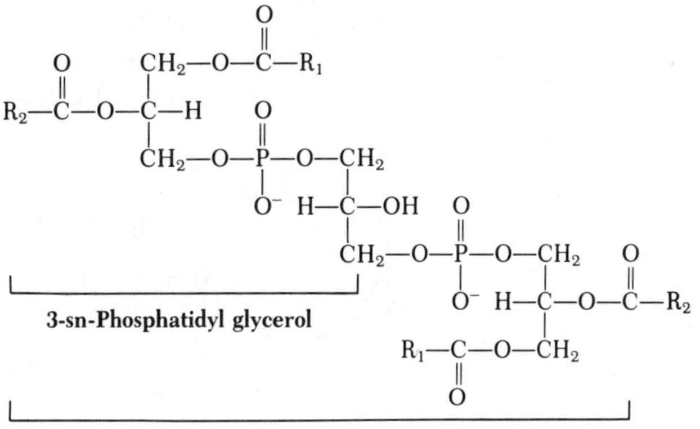

3-sn-Phosphatidyl glycerol

Cardiolipin

These phospholipids are particularly rich in the polyunsaturated fatty acids, especially linoleic acid.

Another kind of phospholipid found frequently in membranes are the *sphingomyelins*. These are not phosphoglycerides but rather have *sphingosine* (see p. 850), an unsaturated aliphatic amino alcohol, as a backbone. Sphingomyelins are unusual in that they are amides of fatty acids rather than esters of fatty acids. The sphingosine–fatty acid amide is given the special designation of *ceramide*. The primary alcohol of the ceramide is phosphorylated; a choline molecule is attached to the phosphate. Sphingomyelins are also unusual in their content of very long fatty

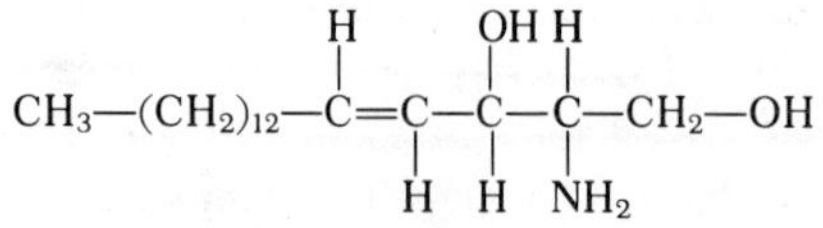

$$CH_3-(CH_2)_{12}-CH{=}CH-CH(OH)-CH(NH_2)-CH_2-OH$$

Sphingosine

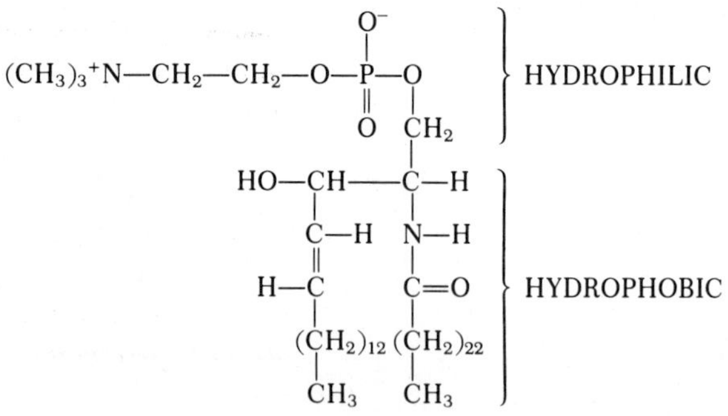

$$(CH_3)_3{}^+N-CH_2-CH_2-O-P(=O)(O^-)-O-CH_2-$$

HYDROPHILIC

HO—CH——C—H; C—H N—H; H—C C=O; $(CH_2)_{12}$ $(CH_2)_{22}$; CH_3 CH_3

HYDROPHOBIC

Sphingomyelin

acids, namely, 24:0 and 24:1 fatty acids. As seen above in the right-hand figure, the sphingomyelin molecule can be folded in such a way that the two aliphatic side chains, one from the sphingosine and the other from the fatty acid, form a hydrophobic region. Like the other phospholipids, the sphingomyelins are amphipathic and have a hydrophilic head composed of the phosphoryl group and the choline.

Glycolipids

On hydrolysis glycolipids yield sphingosine or dihydrosphingosine, a fatty acid, and a sugar (usually galactose or glucose). They contain nitrogen but no phosphorus and therefore are not classed as phospholipids. Like sphingomyelin, they are almost insoluble in ether but are more soluble in acetone than are the phospholipids and are also soluble in hot alcohol, benzene, and chloroform.

Glycolipids occur in the medullary sheaths of nerves and in brain tissue, particularly the white matter. In addition they are constituents of the external surface of cellular plasma membranes.

As in the case of sphingomyelin, the sphingosine moiety provides the backbone for glycolipid structure. Fatty acids are bound to the amino group by an amide bond. As indicated previously, this unit is called a ceramide.

Cerebrosides When the ceramide moiety is linked to a sugar monosaccharide such as glucose or galactose, the compound is called a cerebroside. The bond uniting these two is the β-glycosidic bond, as seen in the structure below.

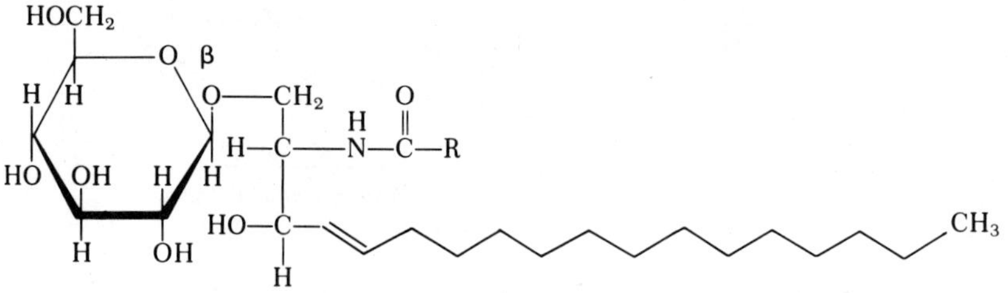

$HOCH_2$; O β; H H; O—CH_2; H—C——N(H)—C(=O)—R; HO OH H H; HO—C—(chain)—CH_3; H OH; H

A glucosyl ceramide, or *glucocerebroside*, has a glucose monosaccharide attached and is often abbreviated βGlc-Cer. Likewise when the sugar is galactose the ceramide is galactocerebroside or βGal-Cer. Cerebrosides are found in the myelin sheath of the brain and peripheral nerves.

More complex sphingolipids have been found having an oligosaccharide chain attached. Thus:

Cer dihexoside	Gal 1 $\xrightarrow{\beta}$ 4 Glc-Cer
Cer trihexoside	Gal 1 $\xrightarrow{\alpha}$ 4 Gal 1 $\xrightarrow{\beta}$ 4 Glc-Cer
Globoside	Gal NAc 1 $\xrightarrow{\beta}$ 3 Gal 1 $\xrightarrow{\alpha}$ 4 Gal 1 $\xrightarrow{\beta}$ 4 Glc-Cer

Gangliosides The gangliosides are a group of glycolipids that were originally found in ganglion cells but that now have been found in the membranes of many cell types. In addition to the ceramide entity, they possess a number of carbohydrate units. As seen below these include *N*-acetyl neuraminic acid (NANA) (Chapter 11).

$$\begin{array}{ccccccc} \text{Gal 1} & \xrightarrow{\beta} & \text{3 Gal NAc 1} & \xrightarrow{\beta} & \text{4 Gal 1} & \xrightarrow{\beta} & \text{4 Glc 1—Cer} \\ \uparrow & & & & \uparrow & & \\ \text{NANA} & & & & \text{NANA} & & \end{array}$$

Terpenes and carotenes

The terpenes are a class of lipids characterized by their containing multiples of five carbon-atom units, related to *isoprene*, i.e., 2-methyl butadiene:

$$\begin{array}{c} \quad CH_3 \\ \quad | \\ CH_2{=}C{-}CH{=}CH_2 \end{array}$$

In Chapter 6 the structure of CoQ serves as an example. In the structure of ubiquinone there is a side chain consisting of 10 repeating isoprenoid units. Other examples include vitamin A (Chapter 21) and squalene, a precursor cholesterol (Chapter 12).

Sterols

Sterols are monohydroxy alcohols that belong to the group of compounds known as cyclopentanoperhydrophenanthrenes, which are steroids. These have a four-ring structure, which is shown below with the rings lettered and the carbon positions numbered. *R* indicates an aliphatic side chain.

Steroid nucleus—conventional form

The steroid nucleus may also be shown as the conformational structure *(chair form)*:

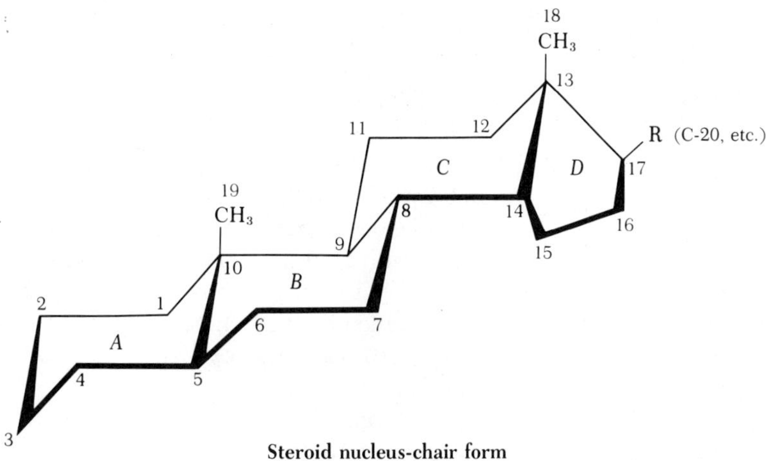

Steroid nucleus-chair form

Those carbons numbered above 17 are not present in every steroid compound. The steroids differ from each other in the arrangement of double bonds in the rings and in the presence of oxygen or of hydroxyl or other groups; in certain cases there may even be a break in one of the rings. This numbering system is frequently referred to in biochemical and clinical literature. Among the steroids are the sterols, bile acids, sex hormones, adrenocortical hormones, cardiac aglycones, and D vitamins.

Nomenclature A number of hydrocarbons serve as the parent substances of different steroids. The ending of their names is *-ane;* the most important ones follow:

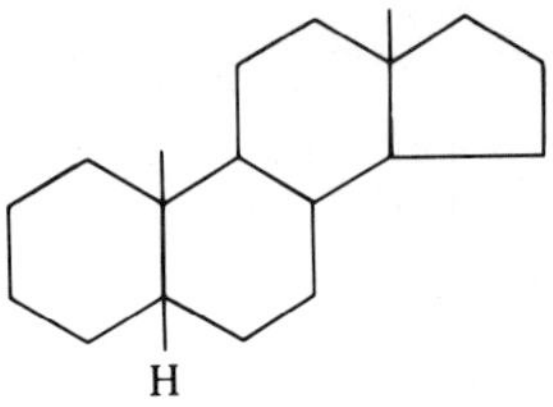

Etiocholane 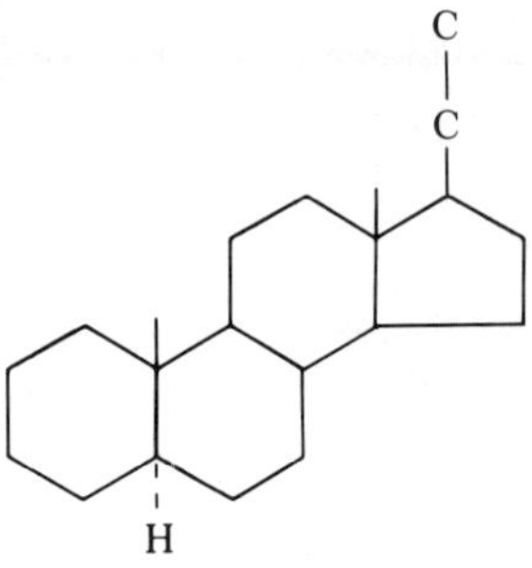 Allopregnane 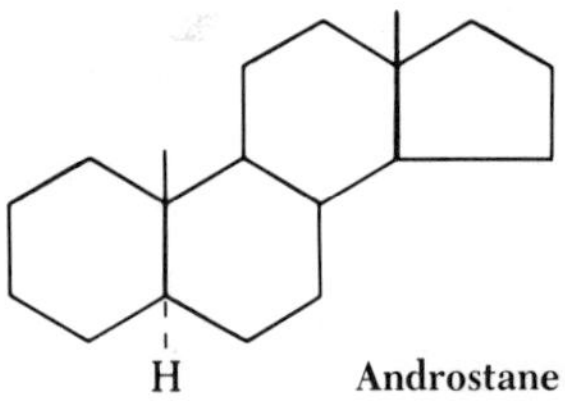 Androstane

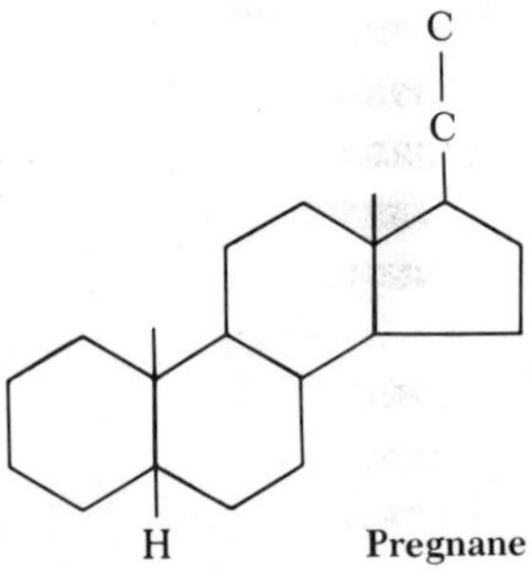

Pregnane 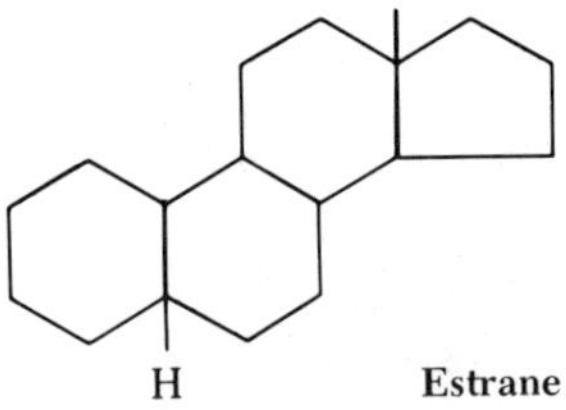 Estrane

The vertical bonds at positions 10 and 13 indicate angular methyl groups with C-19 and C-18 respectively.

A double bond changes the ending of the name of the hydrocarbon to *-ene;* two such bonds make it *-diene,* and so forth. The exact position of the double bond is indicated by Δ with a superscript numeral to indicate where the double bond starts, e.g., Δ^{5}-androstene. The number of the carbon at the end of the double bond is also given if it is not the next higher number (e.g., $\Delta^{7,9:11}$-androstadiene means that the second double bond in this case goes from C-9 to C-11 [not from C-9 to C-10], whereas $\Delta^{7,14}$-androstadiene indicates that the two double bonds go from C-7 to C-8 and from C-14 to C-15).

If a ketone is introduced, the suffix is changed to *-one,* or the syllable *-one-* is added, with a number to indicate its position. Similarly, *-ol* is used for a hydroxyl (e.g., $\Delta^{1,3,5}$-estratriene-3-ol-17-one indicates that the compound has the estrane configuration but with three double bonds, from C-1 to C-2, C-3 to C-4, and C-5 to C-6; it also has a hydroxyl at C-3 and a ketone at C-17). An aldehyde group is indicated by *-al.*

It is obvious that there are many asymmetric carbons in the steroid molecule and that in some cases it might be necessary to show precisely how a given substituent is oriented. Configurations relative to the molecule as a whole are designated β if the orientation of the hydrogen or group corresponds to that of the two methyl groups (C-18 and C-19, which are presumed to be above the plane of the page.) A full line bond is used for such an orientation. If the orientation is opposite the C-18 and C-19 groups, i.e., below the plane of the page, it is called α, and a dotted line bond is used. The terms *trans* and *cis* are also applied to α and β respectively.

The principles just presented will be understood if the following structures are compared with the names beneath them.

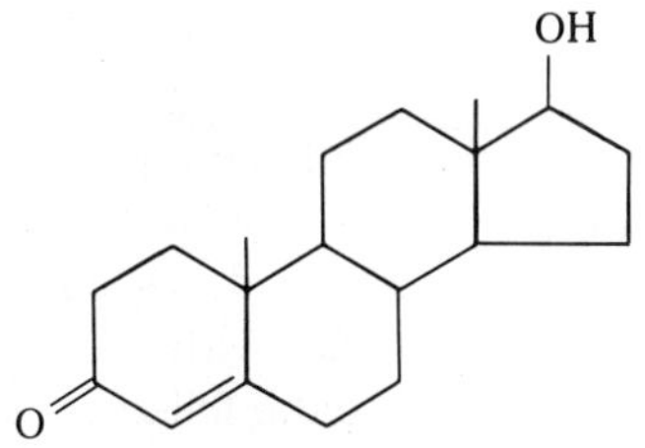

OH

O

Testosterone
Δ^4-Androstene-3-one-17(β)-ol

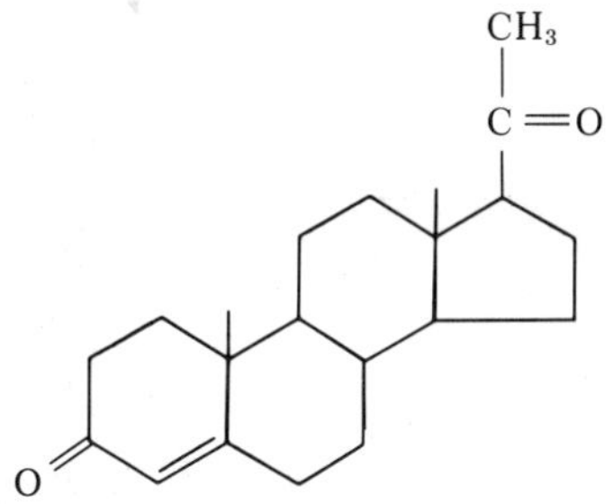

CH_3
C=O

O

Progesterone
Δ^4-Pregnene-3,20-dione

Cholesterol Cholesterol is perhaps the most important sterol; it is widely distributed and has been known and studied for many years. Its structural formula is as follows:

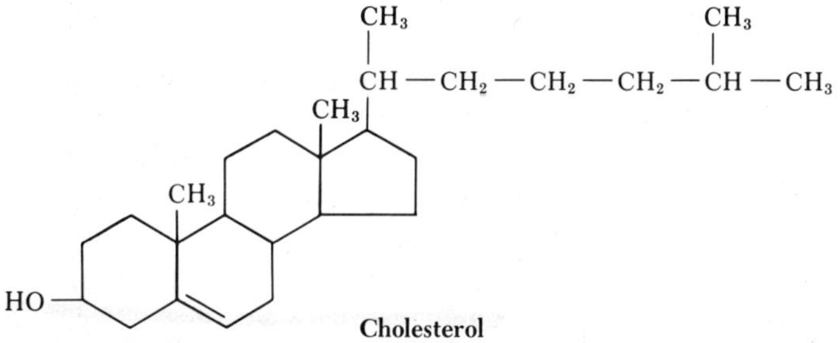

CH_3 CH_3
$CH-CH_2-CH_2-CH_2-CH-CH_3$
CH_3
CH_3
HO

Cholesterol

Since there are 8 asymmetric carbons, theoretically 256 stereoisomers are possible. The hydroxyl group on C-3 is in the β-position; therefore it is shown with a *solid* line bond. Cholesterol is probably a constituent of all animal cells; the corpus luteum and the adrenal cortex are particularly rich in this lipid. The adrenosteroid hormones and probably other steroid hormones and the bile acids are formed from cholesterol in the animal organism. Cholesterol is present in blood and bile and is usually a major constituent of gallstones, from which it was first isolated. It is soluble in many fat solvents, e.g., ether, chloroform, benzene, and hot alcohol, and easily crystallizes from such solutions. Since it has an unsaturated bond, it will take up two halogen atoms. It is not saponifiable.

Cholesterol gives a number of color reactions that form the basis for its quantitative determination. If to a chloroform solution of cholesterol are added acetic anhydride and concentrated sulfuric acid (under as nearly anhydrous conditions as possible), a blue-to-violet color, changing to emerald green, appears. Under carefully controlled conditions the green color produced is proportional to the amount of cholesterol present. Consequently this reaction, known as the Liebermann-Burchard reaction, has become the basis for the quantitative estimation of cholesterol in blood and other biological materials. In clinical work sometimes free cholesterol and cholesterol esters must be determined separately. For this to be accomplished, advantage is taken of the fact that free cholesterol unites with digitonin to form cholesterol digitonide. This is insoluble in petroleum ether, in which the cholesterol esters are freely soluble.

Other important sterols There is present in the skin an important sterol, 7-dehydrocholesterol. This differs from cholesterol only in having a second double bond (between C-7 and C-8) and therefore only one hydrogen at C-7 and none at C-8. It is found in other tissues as well as the skin, probably along with cholesterol, but its special interest lies in the fact that when the skin is irradiated with ultraviolet light this sterol is converted to one of the D vitamins (Chapter 21).

Amino acids

Amino acids are the fundamental constituents of all proteins. The 20 amino acids, commonly found in proteins, may be isolated from hydrolysates of proteins or synthesized by chemical procedures. All amino acids occurring naturally in proteins are α-amino acids and, with the single exception of glycine, all have the L-configuration. They have the following general structure:

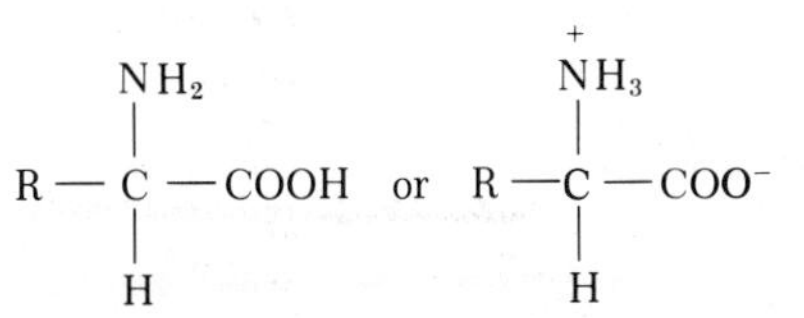

$$R-\overset{\overset{\displaystyle NH_2}{|}}{\underset{\underset{\displaystyle H}{|}}{C}}-COOH \quad \text{or} \quad R-\overset{\overset{\displaystyle \overset{+}{N}H_3}{|}}{\underset{\underset{\displaystyle H}{|}}{C}}-COO^-$$

(Zwitterion form)

In accordance with the nomenclature regarding aliphatic acids, the α-carbon is that carbon nearest the carboxyl; the β-carbon is second; the γ-carbon is third; the δ-carbon is fourth; the ε-carbon is fifth, etc. Amino acids that have their amino groups on other carbon atoms (e.g., β or γ) are also known, but these are not found in proteins.

Classification and structures

Amino acids are obtained from proteins by *hydrolysis*, catalyzed by a proton (HCl, H_2SO_4), a hydroxyl (NaOH), or enzymes such as pepsin, trypsin, and chymotrypsin. If, after hydrolysis, the pH is adjusted to 5.5 and the mixture of amino acids is then separated by paper electrophoresis at the same pH, the amino acids divide into three general groups: (1) A group of nonmigrating, neutral amino acids that have no net charge at pH 5.5 and therefore do not migrate from the origin—these are the *monoamino-monocarboxylic* acids; (2) A second group, which migrates cathodically and consists of those amino acids with a net positive charge—these have more than one nitrogen-containing group and are called *basic* amino acids; (3) A third group, which migrates anodically and consists of amino acids with more than one carboxyl group—these are called *acidic* amino acids.

The various structures of amino acids are considered below in accordance with this classification, i.e., neutral, basic, and acidic. The accepted three-letter abbreviation of each amino acid is indicated in parentheses after the common and chemical names of the amino acid.

Neutral amino acids The neutral amino acids are monoamino-monocarboxylic acids and are characterized by the nature of their side chains.

Aliphatic amino acids Straight and branched chains of carbon atoms as well as of other substituents distinguish these amino acids from each other.

Glycine, aminoacetic acid (Gly)

$$\begin{array}{ccccc} & & \mathrm{NH_2} & & \\ & & | & & \\ \mathrm{H} & - & \mathrm{C} & - & \mathrm{COOH} \\ & & | & & \\ & & \mathrm{H} & & \end{array}$$

Alanine, α-aminopropionic acid (Ala) It is helpful to consider alanine as the basic structure for all the other amino acids. In other words the other amino acids may be considered as derivatives of alanine, substitutions on the β-carbon.

$$\begin{array}{ccccc} & & \mathrm{NH_2} & & \\ & & | & & \\ \mathrm{CH_3} & - & \mathrm{C} & - & \mathrm{COOH} \\ & & | & & \\ & & \mathrm{H} & & \end{array}$$

Valine, α-aminoisovaleric acid (Val) Here two methyl groups have replaced two hydrogens on the β-carbon of alanine.

$$\begin{array}{ccccccc} \mathrm{CH_3} & & & & \mathrm{NH_2} & & \\ & \backslash & & & | & & \\ & & \mathrm{CH} & - & \mathrm{C} & - & \mathrm{COOH} \\ & / & & & | & & \\ \mathrm{CH_3} & & & & \mathrm{H} & & \end{array}$$

Leucine, α-aminoisocaproic acid (Leu) A three-carbon or isopropyl group is attached to alanine.

$$\begin{array}{ccccccccc} \mathrm{CH_3} & & & & & & \mathrm{NH_2} & & \\ & \backslash & & & & & | & & \\ & & \mathrm{CH} & - & \mathrm{CH_2} & - & \mathrm{C} & - & \mathrm{COOH} \\ & / & & & & & | & & \\ \mathrm{CH_3} & & & & & & \mathrm{H} & & \end{array}$$

Isoleucine, α-amino-β-methylvaleric acid (Ile)

$$\begin{array}{ccccccc} \mathrm{CH_3} & & & & \mathrm{NH_2} & & \\ & \backslash & & & | & & \\ & & \mathrm{CH} & - & \mathrm{C} & - & \mathrm{COOH} \\ & / & & & | & & \\ \mathrm{CH_3CH_2} & & & & \mathrm{H} & & \end{array}$$

Serine, α-amino-β-hydroxypropionic acid (Ser) A hydroxyl group is attached to alanine.

$$\begin{array}{ccccccc} & & & & \mathrm{NH_2} & & \\ & & & & | & & \\ \mathrm{HO} & - & \mathrm{CH_2} & - & \mathrm{C} & - & \mathrm{COOH} \\ & & & & | & & \\ & & & & \mathrm{H} & & \end{array}$$

Threonine, α-amino-β-hydroxybutyric acid (Thr) Here both a hydroxyl group and a methyl group have been substituted on alanine.

$$\begin{array}{ccccccc} & & \mathrm{H} & & \mathrm{NH_2} & & \\ & & | & & | & & \\ \mathrm{HO} & - & \mathrm{C} & - & \mathrm{C} & - & \mathrm{COOH} \\ & & | & & | & & \\ & & \mathrm{H_3C} & & \mathrm{H} & & \end{array}$$

Aromatic amino acids The aromatic amino acids consist of amino acids with phenyl, hydroxyphenyl, or indole rings substituted on alanine.

Phenylalanine, α-amino-β-phenylpropionic acid (Phe)

$$C_6H_5-CH_2-\underset{\displaystyle H}{\overset{\displaystyle NH_2}{\overset{|}{\underset{|}{C}}}}-COOH$$

Tyrosine, α-amino-β-hydroxyphenylpropionic acid (Tyr)

$$HO-C_6H_4-CH_2-\underset{\displaystyle H}{\overset{\displaystyle NH_2}{\overset{|}{\underset{|}{C}}}}-COOH$$

Tryptophan, α-amino-β-indolylpropionic acid (Trp)

$$C_8H_6N-CH_2-\underset{\displaystyle H}{\overset{\displaystyle NH_2}{\overset{|}{\underset{|}{C}}}}-COOH$$

Sulfur-containing amino acids Two monoamino-monocarboxylic acids containing sulfur are cysteine and methionine. A third is cystine, which is actually dicysteine and is a diaminodicarboxylic acid.

Cysteine, α-amino-β-mercaptopropionic acid (Cys) A sulfhydryl group is attached to alanine.

$$HS-CH_2-\underset{\displaystyle H}{\overset{\displaystyle NH_2}{\overset{|}{\underset{|}{C}}}}-COOH$$

Methionine, α-amino-γ-methylthiobutyric acid (Met) Here the group $CH_3—S—CH_2$ may be considered to be substituted on alanine.

$$CH_3-S-CH_2-CH_2-\underset{\displaystyle H}{\overset{\displaystyle NH_2}{\overset{|}{\underset{|}{C}}}}-COOH$$

Cystine, di-(α-amino-β-thiopropionic acid) (CyS-SCy)

$$HOOC-\underset{\displaystyle H}{\overset{\displaystyle NH_2}{\overset{|}{\underset{|}{C}}}}-CH_2-S-S-CH_2-\underset{\displaystyle H}{\overset{\displaystyle NH_2}{\overset{|}{\underset{|}{C}}}}-COOH$$

Basic amino acids There are three so-called basic amino acids, all of which have six carbon atoms.

Lysine, α,ε-diaminocaproic acid (Lys)

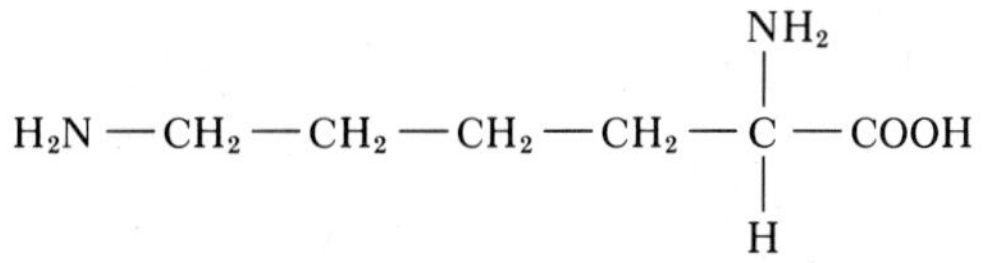

$$H_2N-CH_2-CH_2-CH_2-CH_2-\underset{\displaystyle H}{\overset{\displaystyle NH_2}{\overset{|}{\underset{|}{C}}}}-COOH$$

Arginine, α-amino-δ-guanidinovaleric acid (Arg)

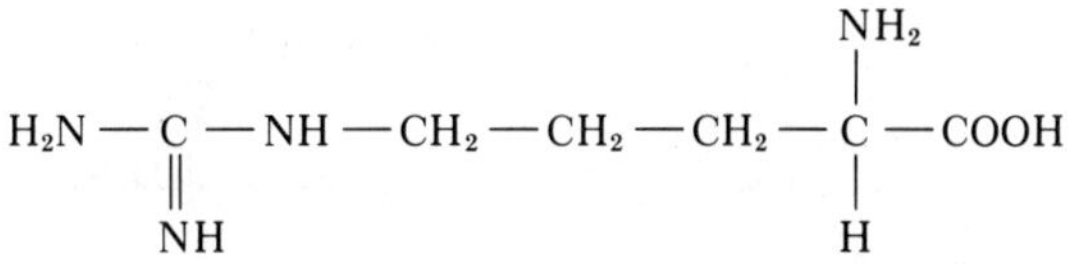

$$H_2N-\underset{\displaystyle NH}{\underset{\|}{C}}-NH-CH_2-CH_2-CH_2-\underset{\displaystyle H}{\overset{\displaystyle NH_2}{\overset{|}{\underset{|}{C}}}}-COOH$$

Histidine, α-amino-β-imidazolylpropionic acid (His) An imidazole ring is substituted on alanine.

$$\text{(imidazole ring: H–C=C–, N, N–H, C–H)}-CH_2-\underset{\displaystyle H}{\overset{\displaystyle NH_2}{\overset{|}{\underset{|}{C}}}}-COOH$$

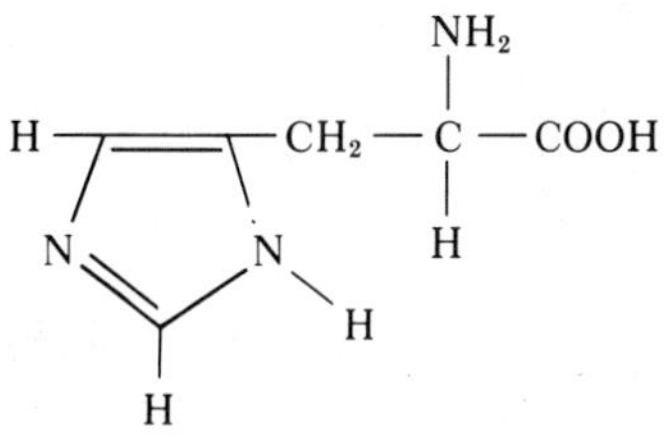

Acidic amino acids There are two acidic amino acids, aspartic acid and glutamic acid; each has two carboxyl groups.

Aspartic acid, aminosuccinic acid (Asp) A carboxyl group may be considered substituted on alanine.

$$HOOC-CH_2-\underset{\displaystyle H}{\overset{\displaystyle NH_2}{\overset{|}{\underset{|}{C}}}}-COOH$$

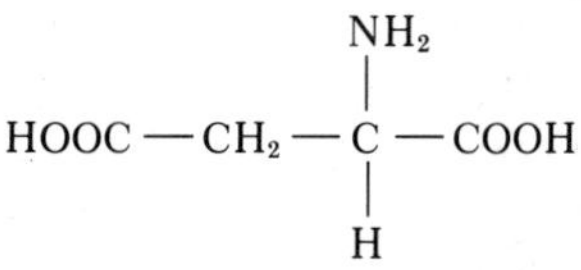

Glutamic acid, α-aminoglutaric acid (Glu) Here the element of acetic acid has been added to alanine.

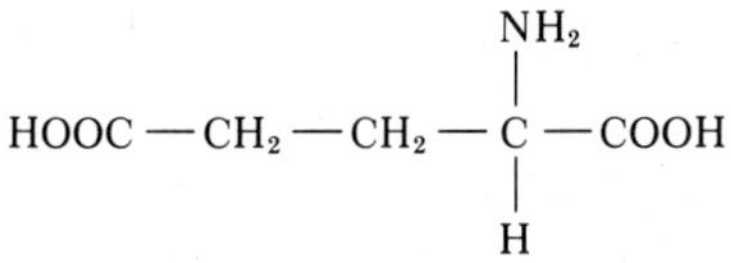

$$HOOC-CH_2-CH_2-\underset{\displaystyle H}{\overset{\displaystyle NH_2}{\overset{|}{\underset{|}{C}}}}-COOH$$

Aspartic acid and glutamic acid occur in proteins largely as the corresponding acid amides, asparagine and glutamine.

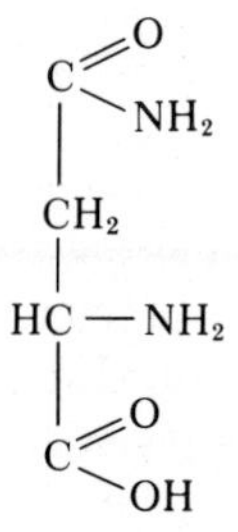

$$\begin{array}{c} C(=O)NH_2 \\ | \\ CH_2 \\ | \\ HC-NH_2 \\ | \\ C(=O)OH \end{array} \qquad \begin{array}{c} C(=O)NH_2 \\ | \\ CH_2 \\ | \\ CH_2 \\ | \\ HC-NH_2 \\ | \\ C(=O)OH \end{array}$$

Asparagine (Asn) 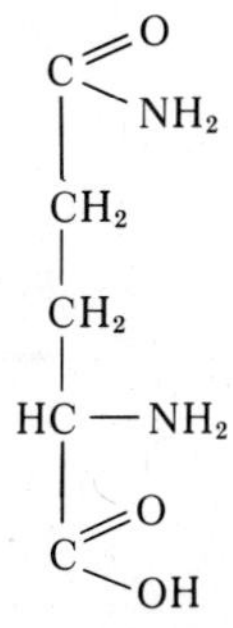 **Glutamine (Gln)**

Imino acids There are two heterocyclic or pyrrolidine amino acids, in which the α-amino nitrogen is part of a ring structure.

Proline, pyrrolidine-2-carboxylic acid (Pro)

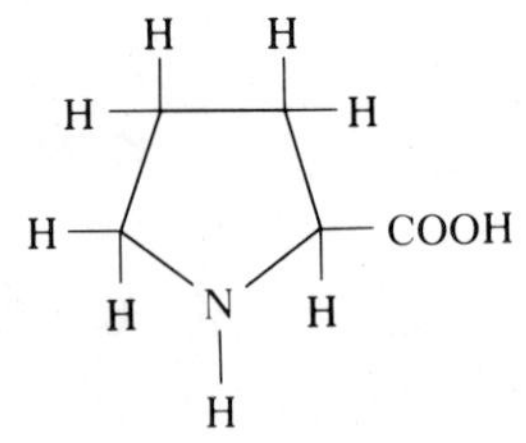

H H
H H
H COOH
H N H
H

Hydroxyproline, 4-hydroxypyrrolidine-2-carboxylic acid (Hyp)

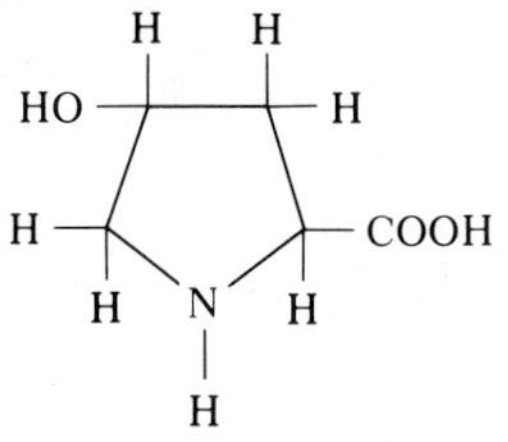

H H
HO H
H COOH
H N H
H

Although the foregoing 20 amino acids are found in most proteins, there are others that either are found free in the cell or participate as substituents in factors other than proteins, e.g., citrulline and ornithine, monoiodotyrosine and diiodotyrosine, and thyroxine.

Stereoisomerism

Examination of the formulas of the amino acids will reveal that all except glycine have an asymmetric carbon atom. It is therefore possible to have a D-form and an L-form of each. Most of the naturally occurring α-amino acids are of the L-form, i.e., their configuration is similar to that of L-glyceraldehyde or more appropriately L-serine.

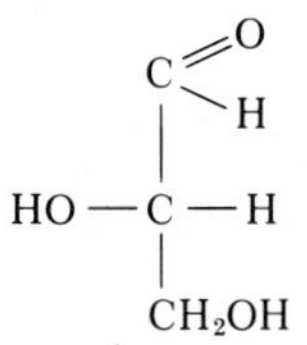

L-Glyceraldehyde	L-Serine
C(=O)H	C(=O)OH
HO—C—H	H_2N—C—H
CH_2OH	CH_2OH

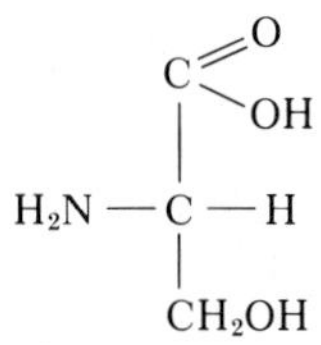

By convention the structure for L-glyceraldehyde is written with the hydroxyl group to the left, provided that the functional group (i.e., the aldehyde) is placed uppermost. Likewise when serine is written with the carboxyl group uppermost, the amino group in the two-dimensional structure appears to the left of the carbon chain.

The D- and L-stereoisomers are mirror images, and the two structures cannot be superimposed. Although all amino acids in most proteins are L-isomers, there are situations where the D-isomers also exist in nature. Thus the antibiotics gramicidin and actinomycin D and the cell walls of certain bacteria contain D-amino acids. The D- and L-nomenclature refers to stereoisomerism, not to dextrooptical or levooptical rotation. To avoid confusing the terms, by convention a dextrorotatory form is indicated by (+) and a levorotatory form by (−). Some amino acids (e.g., histidine) exhibit dextrorotation and levorotation, depending on the pH of the environment.

Note from their structures that two amino acids, isoleucine and threonine, are unique in having *two* asymmetric carbon atoms rather than only one. Hence there should be *four* stereoisomers of each. Indeed all four forms of both are known. The isomer with the configuration on the second carbon atom that does *not* occur in proteins is called the *allo form* or the *diastereoisomer*. The allo form may occur also in either the D- or the L-configuration. Hence these are called D- or L-allo forms, specifically D-alloisoleucine or L-alloisoleucine or D-allothreonine or L-allothreonine. Several other amino acids that may occur in proteins, hydroxylysine, and the two hydroxyprolines, also have two asymmetric centers and accordingly have two allo forms. Since cystine has one asymmetric carbon atom in each half of the molecule, it may exist in an internally compensated or meso form.

α-Amino acids synthesized chemically usually consist of a racemic mixture of the two stereoisomeric forms and are designated by the prefix DL-. Racemization into a DL-mixture can occur when either the L- or the D-isomer is heated in a strongly alkaline solution.

Solubility

Amino acids exhibit a wide range of solubilities. Proline, for example, is soluble to the extent of 162 gm per 100 gm of water (25° C), whereas cystine is soluble only to the extent of 0.011 gm per 100 gm (25° C). In general, amino acids are minimally soluble at their isoelectric points. The composition of the side chains greatly influences solubility, as would be expected. In general, those amino acids with longer aliphatic (hydrophobic) side chains (e.g., Leu, Ile, Val) are less soluble than those with shorter side chains (e.g., Gly, Ala). Polar groups, such as carboxyl and hydroxyl, also tend to increase solubility. They also greatly influence the folding of peptide chains into protein tertiary structure.

Isoelectric point

If, after acid hydrolysis of a protein, the pH of the hydrolysate is adjusted to 5.5 and the mixture of amino acids is then separated by paper electrophoresis at the same pH, the amino acids will divide into the three general groups already described.

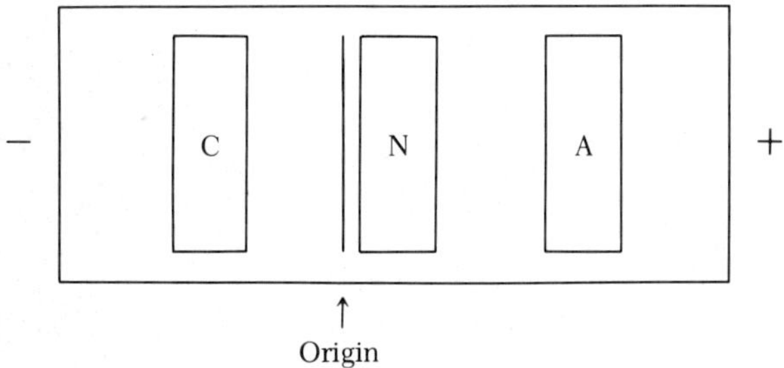

Group *N* represents nonmigrating, neutral amino acids, which have no net charge at pH 5.5 and therefore do not migrate from the origin; group *C* represents the cathodically migrating cationic, or basic, amino acids; and group *A* represents the anodically migrating acidic amino acids. In other words in an electrophoretic system, amino acids migrate to the pole of opposite charge to that of the ionic state or species in solution. Since amino acids have both amino and carboxyl groups, they are ampholytes and may be either positively or negatively charged, depending on the hydrogen ion concentration of the environment. There is a characteristic pH at which an amino acid has no net charge and is not attracted to either electrode of the electrophoretic system. This hydrogen ion concentration is called the *isoelectric*

Table 23-2
Ionization constants and isoelectric points for amino acids

Amino acid (25° C)	pK_1* (COOH)	pK_2	pK_3	pI†
Alanine	2.34	9.69	—	6.00
Arginine	2.17	9.04 (NH_3^+)	12.48 (guanidinium)	10.76
Aspartic acid	1.88	3.65 (COOH)	9.60 (NH_3^+)	2.77
Cysteine (30°)	1.96	8.18 (SH)	10.28 (NH_3^+)	5.07
Glutamic acid	2.19	4.25 (COOH)	9.67 (NH_3^+)	3.22
Glycine	2.34	9.60	—	5.67
Histidine	1.82	6.00 (imidazolium)	9.17 (NH_3^+)	7.59
Leucine	2.36	9.60	—	5.98
Lysine	2.18	8.95 (α-NH_3^+)	10.53 (ϵ-NH_3^+)	9.74
Phenylalanine	1.83	9.13	—	5.48
Proline	1.99	10.60	—	6.30
Serine	2.21	9.15	—	5.68
Tryptophan	2.38	9.39	—	5.89
Tyrosine	2.20	9.11 (NH_3^+)	10.07 (OH)	5.66
Valine	2.32	9.62	—	5.96

From Cohn, E.J., and Edsall, J.T.: Proteins, amino acids and peptides as ions and dipolar ions, New York, 1943, Reinhold Publishing Corp.
*See p. 881.
†pI = isoelectric point.

point. It may or may not be at the neutral pH of 7.0, as seen in Table 23-2. The isoelectric points of the neutral amino acids are in the neighborhood of 5.5 to 6 and hence do not have a net charge in the electrophoretic system described.

Although at its isoelectric point a monoamino-monocarboxylic acid is completely ionized, the charges present on the two ions neutralize each other. This gives, in aqueous solutions, dipolar ions, or *zwitterions*.

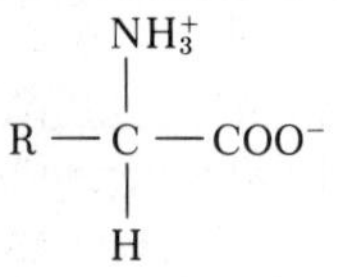

$$\mathrm{R-\overset{\displaystyle NH_3^+}{\underset{\displaystyle H}{C}}-COO^-}$$

If an amino acid solution is titrated with an acid such as hydrochloric acid, the following reaction takes place:

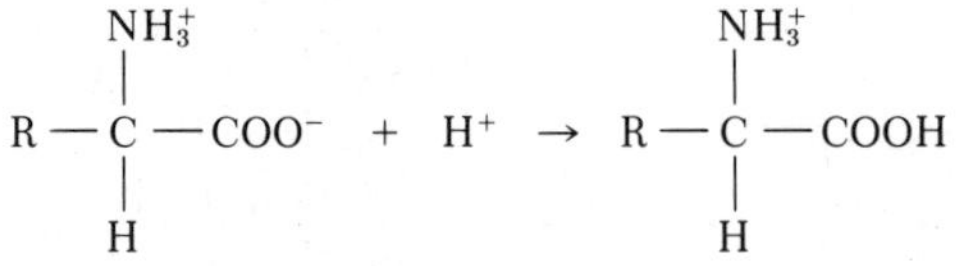

$$\mathrm{R-\overset{\displaystyle NH_3^+}{\underset{\displaystyle H}{C}}-COO^-} + \mathrm{H^+} \rightarrow \mathrm{R-\overset{\displaystyle NH_3^+}{\underset{\displaystyle H}{C}}-COOH}$$

The addition of an acid depresses the ionization of the carboxyl group, and the dipolar ion accepts the proton; thus a net positive charge is placed on the amino acid molecule. In other words, in an electrophoretic system this ionic species migrates toward the cathode. Its structure is more properly written in the salt form, i.e., as the amino acid hydrochloride.

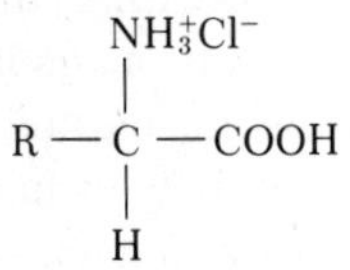

$$\mathrm{R-\overset{\displaystyle NH_3^+Cl^-}{\underset{\displaystyle H}{C}}-COOH}$$

When the amino acid solution at its isoelectric point is titrated with an alkali (e.g., NaOH) the hydroxyl combines with the proton of the ammonium ion to form water.

The amino acid then has a net negative charge from the carboxylate ion. The anionic amino acid species migrates anodically in an electrophoretic system.

Each carboxyl and amino group, as well as all other ionizable groups of an amino acid, has a characteristic pK value (designated as pK_1, pK_2, etc., beginning with the most acidic group titrated). The pK values have been determined for all the common amino acids (Table 23-2).

Titration of amino acids

As shown, each of the functional groups may be titrated either by acid or by alkali. This process follows a typical titration curve of weak acids and bases. Fig. 23-6 shows the curves for three amino acids, glycine, lysine, and aspartic acid. In the curve for glycine the isoelectric point occurs at a pH of 6 *(B)*. Addition of hydrochloric acid equivalent to the amount of amino acid in solution results in the production of the hydrochloride, or the cationic species *(A)*, whereas the addition of an equivalent amount of sodium hydroxide results in the formation of the sodium salt, or anionic species *(C)*. When the carboxyl group is half neutralized by the addition of 0.5 equivalent of hydrochloric acid, the solution will contain an equal concentration of cationic and zwitterion species. In accordance with the Henderson-Hasselbalch equation, the pH at this point will equal the pK. Since this is the pK for the carboxyl group, it is designated pK_1 and is 2.34 (Table 23-2). When alkali is added to the zwitterion species in an amount equal to half the glycine present, anionic glycine is formed equivalent to half the titratable ammonium, or NH_3^+, groups. Since the solution now contains an equal amount of both zwitterion and anionic forms, the pH again equals the pK, but this time of the NH_3^+ group. The pK_2 of the NH_3^+ group is 9.60 (Table 23-2). The titration of each group COOH and NH_3^+ follows the titration curve.

The location of these curves on the coordinates is determined by the various pK values. At each pH the amino acid solution contains a definable proportion of ionic species. If we titrate from pH 1 to 6, the solutions contain proportions of species *B* and *A;* from pH 6 to 11, the species are *C* and *B*. The proportions of species are defined by the Henderson-Hasselbalch equation. One must keep in mind that the titratable groups are defined in accordance with the Brønsted definition of acid and base. In other words a substance contributing protons to a solution, whether it is a COOH or NH_3^+ group, is called an acid; any group accepting protons is a base.

A cationic group that plays an important role in biological functions of many proteins is the imidazolium group of histidine. The pK of this group is 6 and therefore lies near the physiological pH range. The ionization of the various groups in histidine involves the following species of ions:

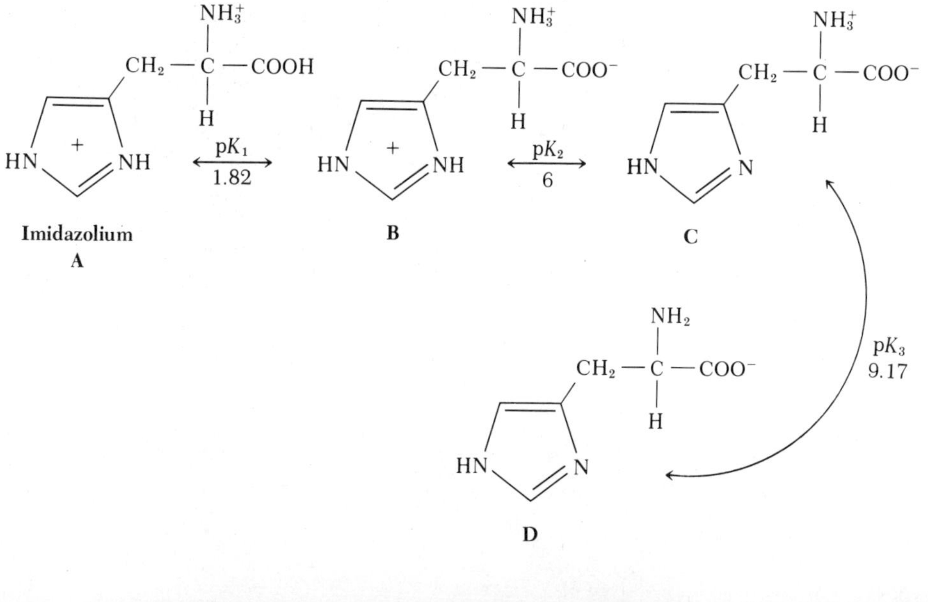

NH_3^+ / $CH_2-C(H)-COOH$ (imidazolium, HN, +, NH) $\underset{1.82}{\overset{pK_1}{\longleftrightarrow}}$ NH_3^+ / $CH_2-C(H)-COO^-$ (HN, +, NH) $\underset{6}{\overset{pK_2}{\longleftrightarrow}}$ NH_3^+ / $CH_2-C(H)-COO^-$ (HN, N)

Imidazolium
A

B

C

$\overset{pK_3}{9.17}$

NH_2 / $CH_2-C(H)-COO^-$ (HN, N)

D

Figure 23-6 Schematic titration curves for glycine, aspartic acid, and lysine. The following amino acid species exist in solutions indicated by lettered points on graph:

Gly:	*A.* NH_3^+ | CH_2COOH	*B.* NH_3^+ | CH_2COO^-	*C.* NH_2 | CH_2COO^-	
Asp:	*A.* $COOH$ | CH_2 | $CH{-}NH_3^+$ | $COOH$	*D.* $COOH$ | CH_2 | $CH{-}NH_3^+$ | COO^-	*E.* COO^- | CH_2 | $CH{-}NH_3^+$ | COO^-	*G.* COO^- | CH_2 | $CH{-}NH_2$ | COO^-
Lys:	*A.* NH_3^+ | $(CH_2)_4$ | $CH{-}NH_3^+$ | $COOH$	*B.* NH_3^+ | $(CH_2)_4$ | $CH{-}NH_3^+$ | COO^-	*F.* NH_3^+ | $(CH_2)_4$ | $CH{-}NH_2$ | COO^-	*H.* NH_2 | $(CH_2)_4$ | $CH{-}NH_2$ | COO^-

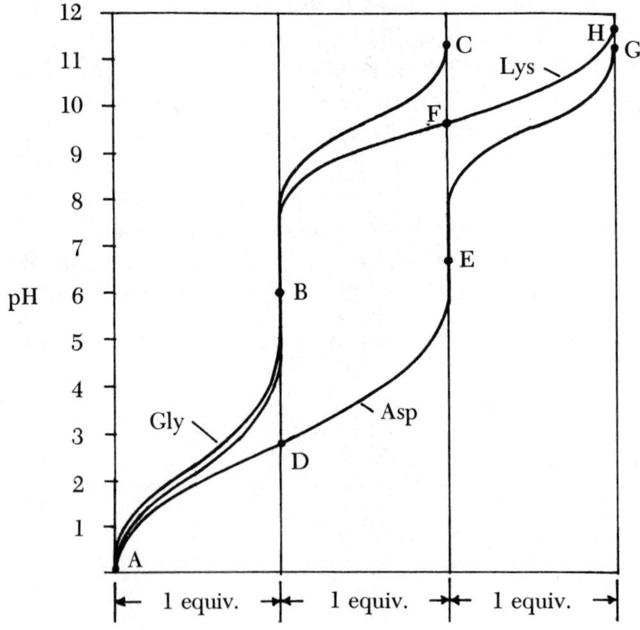

Table 23-3
pK values of acidic and basic groups found in proteins

Group	pK (25° C)
α-Carboxyl	3.0- 3.2
β-Carboxyl (Asp)	3.0- 4.7
γ-Carboxyl (Glu)	About 4.4
Imidazolium	5.6- 7.0
α-Ammonium	7.6- 8.4
ε-Ammonium (Lys)	9.4-10.6
Sulfhydryl	9.1-10.8
Phenolic hydroxyl (Tyr)	9.8-10.4
Guanidinium (Arg)	11.6-12.6

From Cohn, E.J., and Edsall, J.T.: Proteins, amino acids and peptides as ions and dipolar ions, New York, 1943, Reinhold Publishing Corp.

A is the cationic species with a net charge of +2; *B* is cationic with a net charge of +1; *C* is isoelectric with a net charge of 0; *D* is the anionic species with a net charge of −1.

Likewise arginine has a cationic side chain in terms of the guanidinium group. The pK of this ion, however, is 12.48.

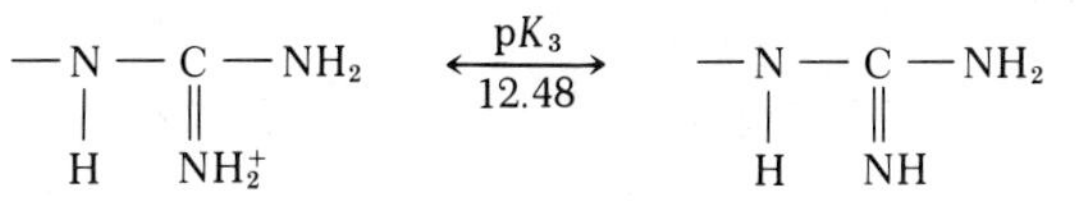

$$-\underset{\substack{|\\ H}}{N}-\underset{\substack{\|\\ NH_2^+}}{C}-NH_2 \underset{12.48}{\overset{pK_3}{\longleftrightarrow}} -\underset{\substack{|\\ H}}{N}-\underset{\substack{\|\\ NH}}{C}-NH_2$$

Guanidinium

In proteins, the pK values of the various functional groups are somewhat modified from the values of free amino acids (compare Table 23-2 with Table 23-3). A consideration of the pK values in Table 23-6 shows that the maximum buffering capacity at a physiological pH of 7.4 would depend on α-ammonium and the imidazolium groups. Since, however, the α-ammonium groups occur only at terminal residues of the polypeptide chain, their quantity is too small to contribute significantly to the buffering action of proteins. The important buffering system (protein) (H^+)/(protein H^+) is attributable largely to the ionization of the histidine (imidazolium) residues.

Reactions of amino acids

Amino acids yield a number of specific color reactions, which have been used traditionally for the purposes of identification. Today they are sometimes used to locate specific amino acids on paper chromatographs. Proteins containing the amino acids in question also yield the reaction. Some of the commonly used reactions are given in Table 23-4.

The following are general reactions of amino acids. In most instances, these are typical of reactions used in either the determination of structure or the synthesis of peptides:

1. All amino acids can be esterified.

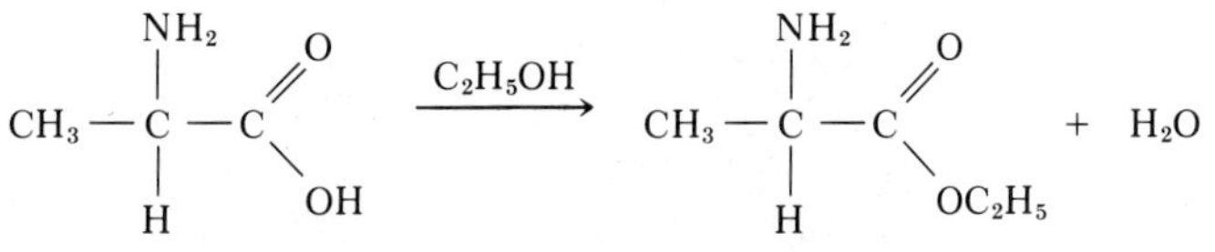

$$CH_3-\underset{\substack{|\\ H}}{\overset{\substack{NH_2\\ |}}{C}}-C\begin{matrix}{\nearrow}O\\ \searrow OH\end{matrix} \xrightarrow{C_2H_5OH} CH_3-\underset{\substack{|\\ H}}{\overset{\substack{NH_2\\ |}}{C}}-C\begin{matrix}{\nearrow}O\\ \searrow OC_2H_5\end{matrix} + H_2O$$

Alanine ethyl ester

This property was used by Emil Fischer for the separation of amino acids by fractional distillation of their volatile esters.

Table 23-4
Amino acid reactions

Test	Conditions, reagents	Color	Amino acid
Xanthoproteic	Concentrated nitric acid	Yellow	Tyr and Trp
Millon	Mercurous nitrate in nitric acid	Red	Tyr
Hopkins-Cole	Glyoxylic acid in concentrated sulfuric acid	Red	Trp
Sakaguchi	α-Naphthol and sodium hypochlorite	Red	Arg
Nitroprusside	Sodium nitroprusside in dilute ammonium hydroxide	Red	Cys
Pauly	Diazotized sulfanilic acid in alkaline medium	Red	His and Tyr

2. Amino acids can be acetylated and benzoylated.

$$CH_3-CH(NH_2)-C(=O)OH \xrightarrow{CH_3C(=O)-Cl} CH_3-CH(NH-C(=O)-CH_3)-C(=O)OH + HCl$$

Acetylation reactions are of general importance in the synthesis of peptides. In vivo, glycine is benzoylated to yield hippuric acid, which is of importance as a detoxication mechanism.

3. Amino acids react with ninhydrin. If a solution of an α-amino acid is boiled with ninhydrin (triketohydrindene hydrate), carbon dioxide is split off and a color is produced. With the exception of proline and hydroxyproline, which give a yellow color, all α-amino acids, peptides, and proteins yield varying shades of purple.

$$\text{Ninhydrin} + R-CH(NH_2)-C(=O)OH \longrightarrow \text{(reduced ninhydrin: C(OH)H)} + R-C(=O)H + NH_3 + CO_2$$

$$\text{Reduced product} + \text{Ninhydrin} + NH_3^+ \longrightarrow \text{(indanedione)}{=}N{-}\text{(hydroxyindenone)}$$

The visualization of amino acids on paper chromatographs and the quantitative determination of amino acids after ion-exchange chromatography usually involve the ninhydrin reaction.

4. Amino acids and certain other primary amines react at room temperatures with *fluorescamine*, to form highly fluorescent fluorophors that are stable for several hours. The fluorescence is directly proportional to the amine concentration. This reaction appears to be very promising for the quantitation of extremely small amounts of amino acids separated from biological materials by various types of chromatographic procedures. The reagent can detect as little as 50 picomoles of an amino acid apparently—a sensitivity approximately 500 times that of ninhydrin reagents.

5. The free amino groups of amino acids or of peptides react with fluorodinitrobenzene to form the dinitrophenyl derivatives, which are yellow dyes. These can readily be separated by suitable procedures of paper chromatography.

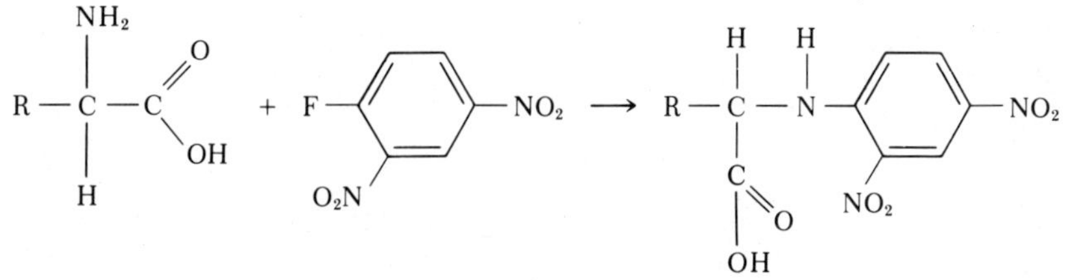

$$R-CH(NH_2)-COOH + F-C_6H_3(NO_2)_2 \rightarrow R-CH(COOH)-NH-C_6H_3(NO_2)_2$$

6. Amino acids also react with certain heavy metals by the process of chelation. Ions such as Cu^{++}, Co^{++}, Mn^{++}, and Ca^{++} react with amino acids to form chelates in the following manner:

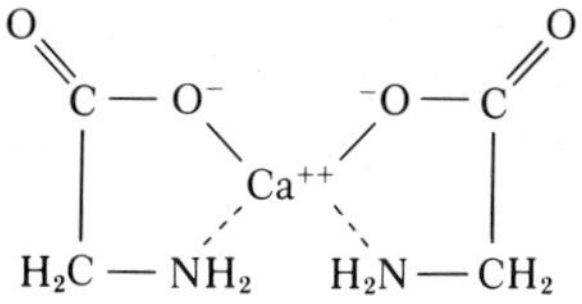

Calcium diglycinate

Chelation is the grasping of secondary valence bonds by the metallic ion. Chelates are nonionic; therefore amino acids and other chelate formers may be used to remove calcium from the bones and teeth. It is possible that dietary amino acids could, in this way, form soluble calcium complexes, causing a loss of calcium and the possible development of caries. Chelating agents have been used for removing calcium from the cornea in ocular lesions.

The primary, secondary, tertiary and quaternary structures of proteins are discussed in Chapter 4.

Purines, pyrimidines, nucleosides, and nucleotides

Structures

Purines Two general kinds of bases are found in all nucleic acids. One type is a derivative of parent compound purine. Principal examples are *guanine* and *adenine*. The structures of these two purines are compared with the parent purine compound below.

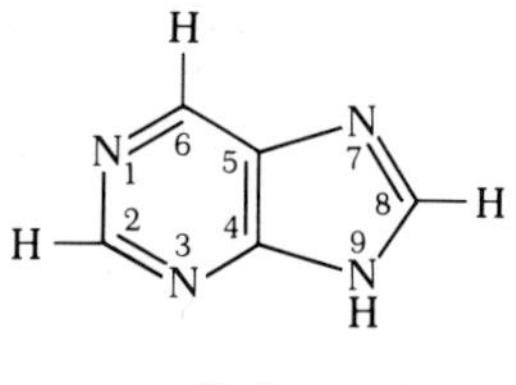

Purine

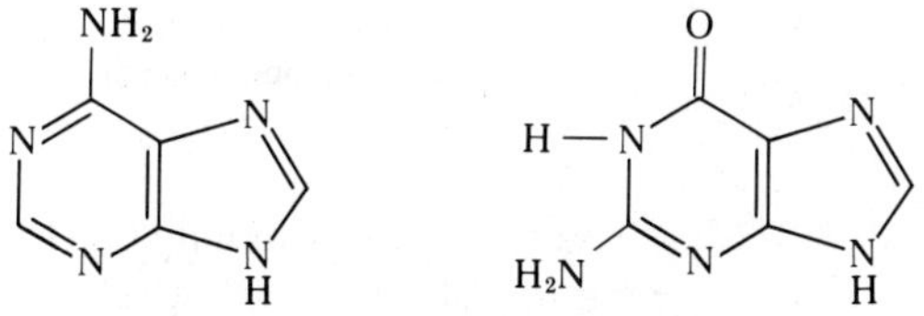

Adenine (6-aminopurine)

Guanine (2-amino-6-oxypurine)

The oxypurines exist in enol-keto (lactim-lactam) forms that are in equilibrium, as shown below for xanthine. The enol (lactim) form for *uric acid* is weakly acidic; it forms monosodium and disodium or potassium salts.

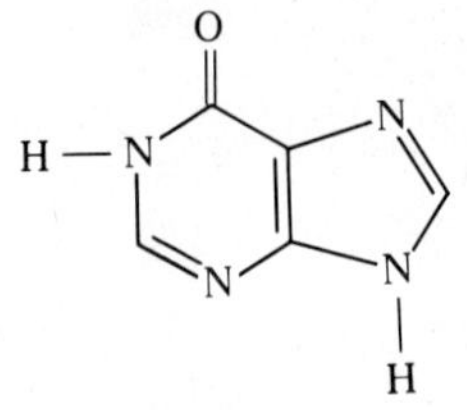

Hypoxanthine (6-oxypurine)

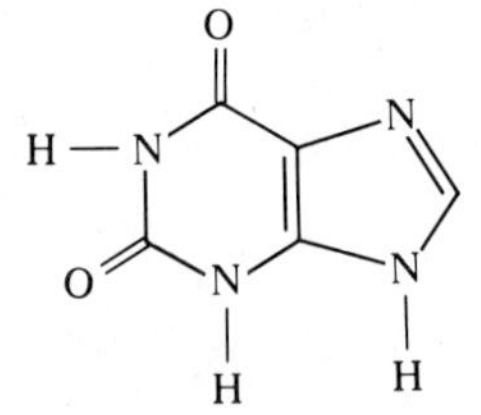

Xanthine (keto)

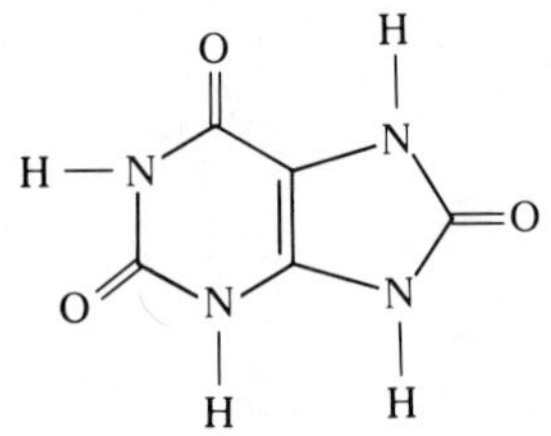

Uric acid (2,6,8-trioxypurine)

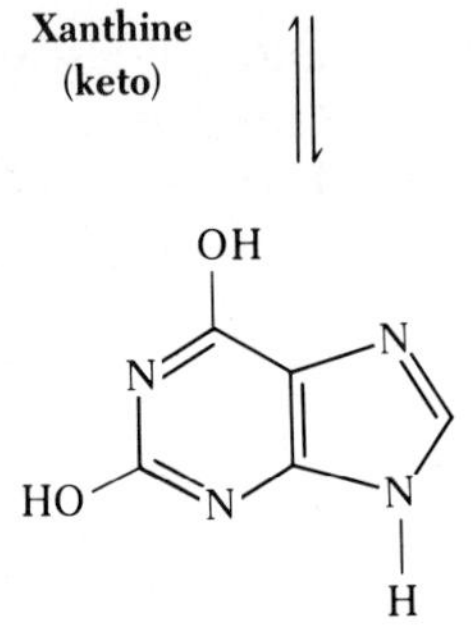

Xanthine (enol) (2,6-dioxypurine)

Pyrimidines The second class of bases found in all nucleic acids is derived from the parent compound pyrimidine. The structures for the principal pyrimidines, i.e., *cytosine, uracil,* and *thymine,* are shown below.

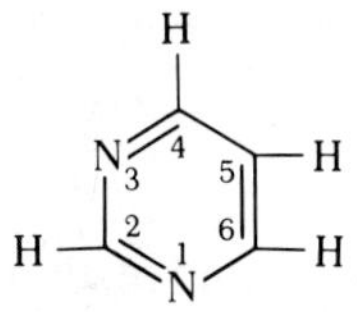

Pyrimidine

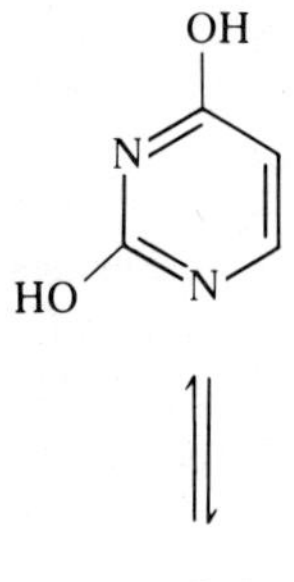

Uracil (lactim or enol form)

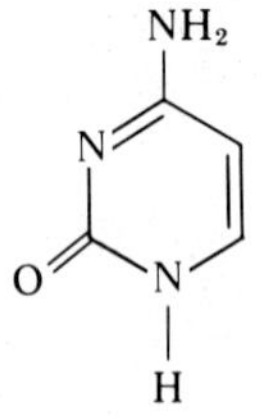

Cytosine (2-oxy-4-aminopyrimidine)

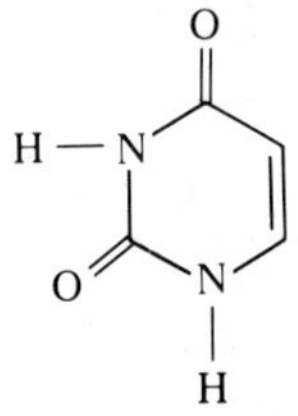

Uracil (lactam or keto form) (2,4-dioxypyrimidine)

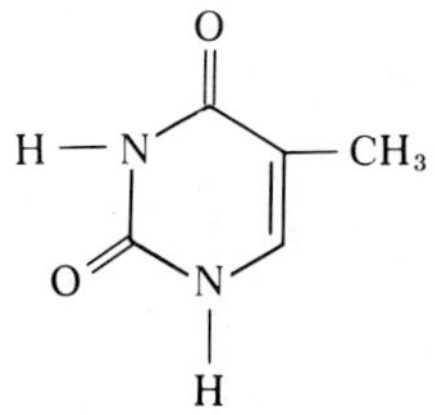

Thymine (2,4-dioxy-5-methylpyrimidine)

The pyrimidines uracil and thymine, which differ only by a methyl group, also serve to distinguish DNA (thymine) from RNA (uracil). Pyrimidines also exhibit tautomerism (see above) and exist in lactim-lactam forms. At physiological pH the lactam, or keto, form predominates.

Modified purines and pyrimidines Although adenine, guanine, cytosine, thymine, and uracil are the purines and pyrimidines most prevalent in DNA and RNA, there are a number of important others. Bacteriophage DNA contains a hydroxymethylcytosine instead of cytosine. Transfer RNA (Chapter 8) is characterized by its content of methylated purines and pyrimidines as well as other derivatives. A number of examples follow:

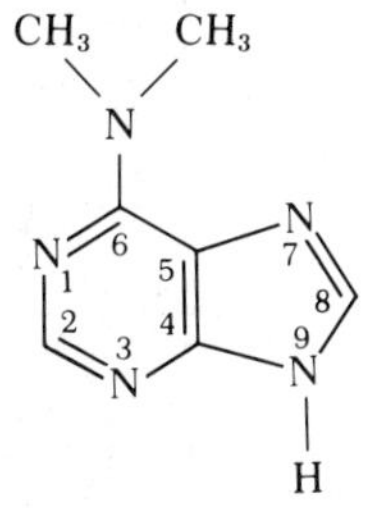

6-*N*-Dimethyladenine

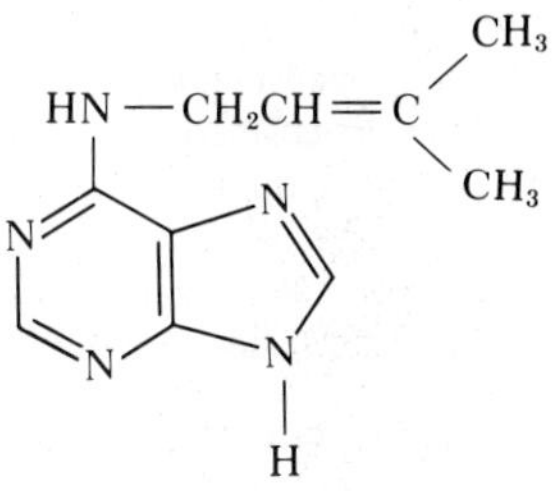

6-*N*-Isopentenyladenine

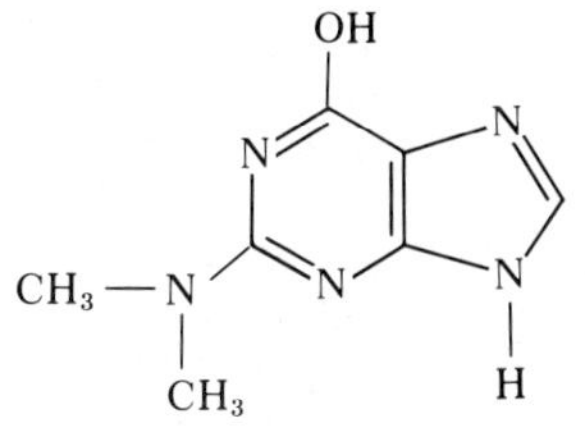

2-*N*-Dimethylguanine

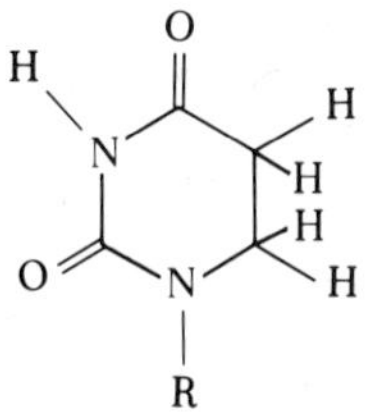

4,5-Dihydrouridine

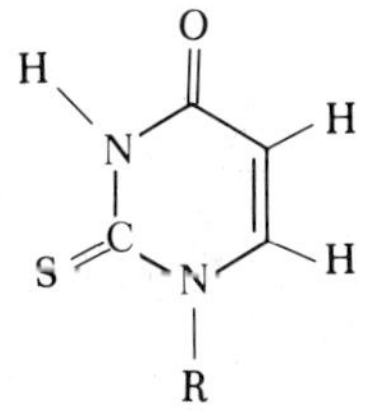

Thiouridine

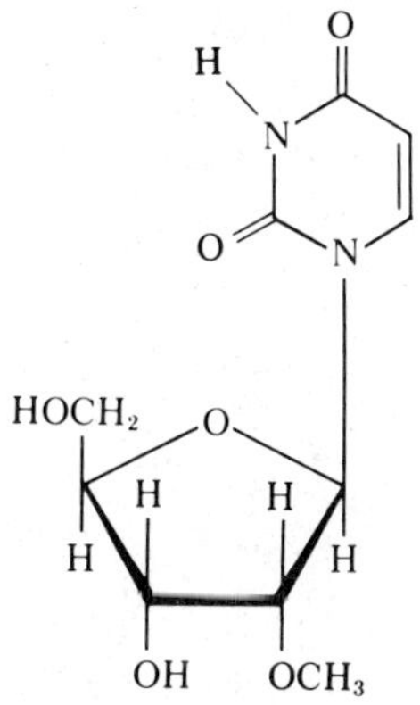

2′-*O*-Methyluridine

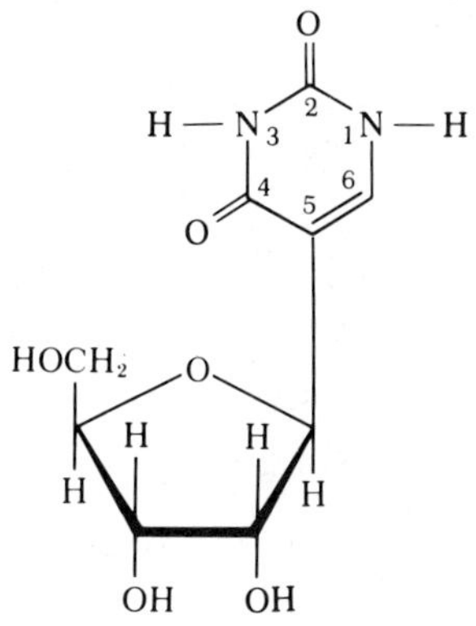

Pseudouridine

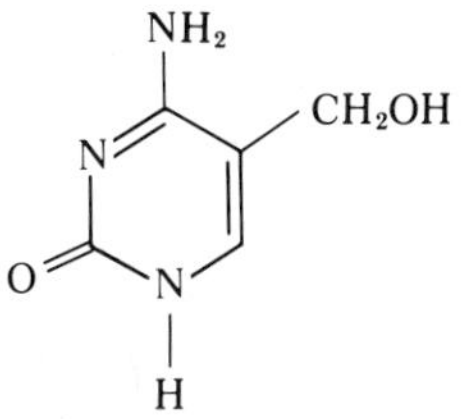

5-Hydroxymethylcytosine

Nucleosides The purine and pyrimidine bases are attached to the appropriate pentose sugar by *N-glycosidic bonds* involving C-1 of the sugar:

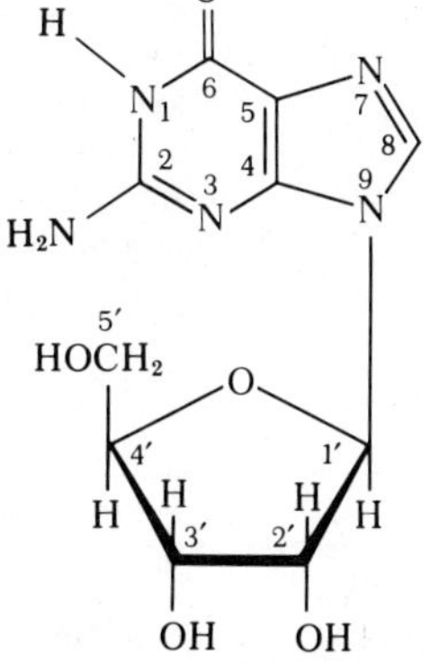

Guanosine

In these glycosidic bonds a molecule of water is eliminated between the hydrogen of N-9 for the purines, or N-1 for the pyrimidines, and the hydroxyl of C-1 of the pentose. The resulting compounds, or glycosides, are called *nucleosides*. In nucleosides, only the β-form of the sugars has been observed; the covalent bond is therefore specifically called a β-*N*-glycosidic linkage.

The atoms of the base in nucleosides are given cardinal numbers, whereas the carbon atoms of the sugars are given primed numbers, as shown in the diagram.

Nucleosides are generally named for the particular purine or pyrimidine present:

Adenine nucleoside:	Adenosine
Guanine nucleoside:	Guanosine
Hypoxanthine nucleoside:	Inosine
Uracil nucleoside:	Uridine
Thymine nucleoside:	Thymidine
Cytosine nucleoside:	Cytidine

Each of these terms is associated with the sugar ribose. For the corresponding deoxyribosides, the prefix *deoxy* is added, e.g., deoxyadenosine, deoxycytidine.

Nucleotides Nucleosides are found in nature primarily as 5′-phosphate esters, called nucleotides. These occur either in the free form or as subunits in nucleic acids.

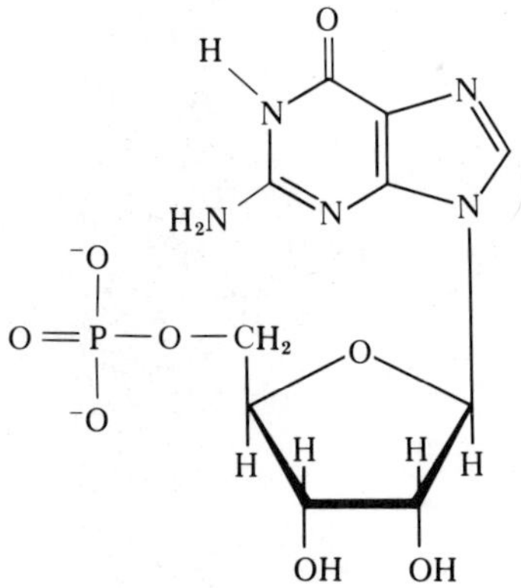

Guanosine monophosphate (guanylic acid)

Nucleotides are named as follows:

Adenine nucleotide, or adenylic acid:	Adenosine monophosphate (AMP)
Guanine nucleotide, or guanylic acid:	Guanosine monophosphate (GMP)
Hypoxanthine nucleotide, or inosinic acid:	Inosine monophosphate (IMP)
Uracil nucleotide, or uridylic acid:	Uridine monophosphate (UMP)
Cytosine nucleotide, or cytidylic acid:	Cytidine monophosphate (CMP)
Thymine nucleotide, or thymidylic acid:	Thymidine monophosphate (TMP)

Table 23-5
Hydrolytic products of RNA and DNA

	Ribonucleic acid	Deoxyribonucleic acid
Acid	Phosphoric acid	Phosphoric acid
Sugar	D-Ribose	D-2-Deoxyribose
Bases		
Purines	Adenine	Adenine
	Guanine	Guanine
Pyrimidines	Cytosine	Cytosine
	Uracil	Thymine

As is apparent from this list, each of the nucleotides is usually referred to by its abbreviation (AMP, UMP, etc.). All these examples are ribotides. When the sugar deoxyribose is involved (deoxyribotides), the prefix *deoxy* must be added to the name, e.g., deoxyadenosine monophosphate, or dAMP.

In nucleoside diphosphates and triphosphates additional units of phosphoric acid are added to the nucleoside monophosphates by means of anhydride bonds.

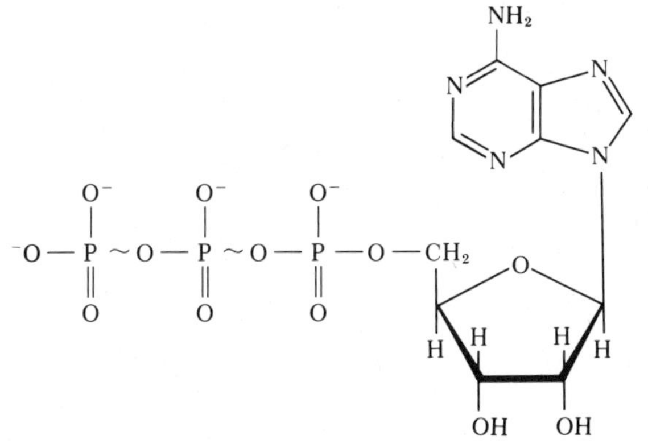

Adenosine triphosphate (ATP)

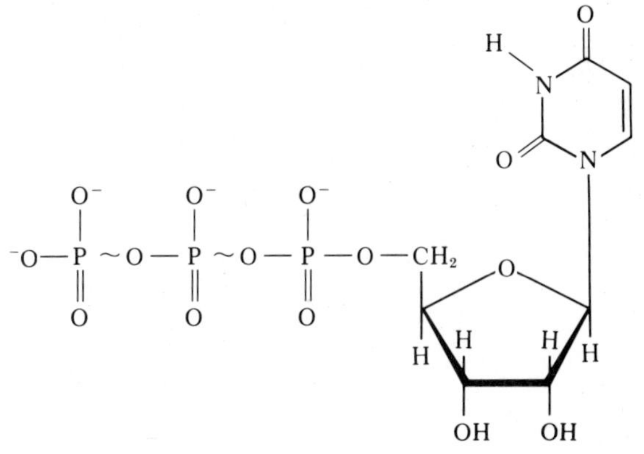

Uridine triphosphate (UTP)

In these structures a tilde is used to designate the high-energy phosphate bonds (Chapter 7).

The 5′-nucleoside phosphates are not the only forms of nucleosides that are important in the cell. The cyclic nucleotide, or the 3′,5′-cyclic phosphate diester of nucleosides, is represented by cyclic AMP.

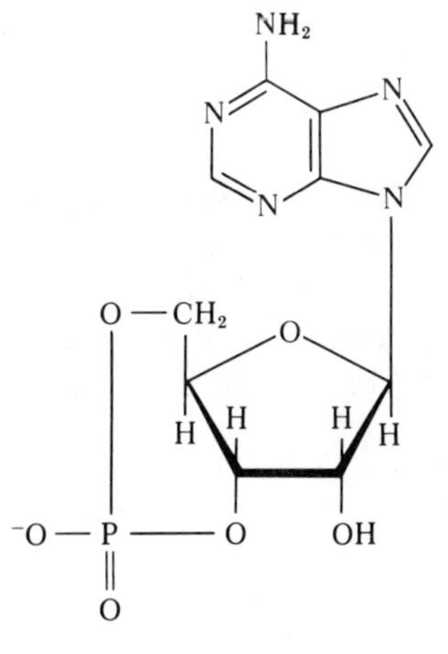

3′,5′-Cyclic adenylic acid (cyclic AMP)

The preceding examples of nucleoside triphosphates are all ribotides. In those nucleotides associated with the biosynthesis of DNA, the presence of deoxyribose must be designated with the prefix *deoxy*. The deoxyribonucleoside triphosphates are abbreviated dATP, dGTP, dCTP, etc. This designation is sometimes omitted in the case of thymidine triphosphate, since the nucleotide is usually found associated with DNA and hence with deoxyribose. However, one must keep in mind that thymine is found in tRNAs (Chapter 8) and hence also exists as the ribotide.

Polynucleotides, or nucleic acids Nucleic acids are of two varieties, ribonucleic (RNA) and deoxyribonucleic (DNA). Upon total hydrolysis the two nucleic acids yield the constituents shown in Table 23-5. They are distinguished primarily on the basis of the five-carbon atom sugar, or pentose, present (p. 838). Nucleic acids are polymers of nucleotide monomers, in other words, polynucleotides.

$$\text{DNA} = (\text{base—deoxyribose—phosphate})_n$$
$$\text{RNA} = (\text{base—ribose—phosphate})_n$$

The constituent units are coupled by means of 3′,5′-phosphodiester bonds.

3′,5′-Phosphodiester bond

The nature, properties, and function of the nucleic acids (DNA, RNA) depend on the exact order of the purines and pyrimidines in the molecule. This sequence of specific bases is called the *primary structure*.

One customarily describes the base sequence of a polynucleotide by the "direction" of its 3′,5′-phosphodiester bonds, i.e., the 5′-phospho terminal is placed to the left and the 3′-hydroxyl to the right, as shown below.

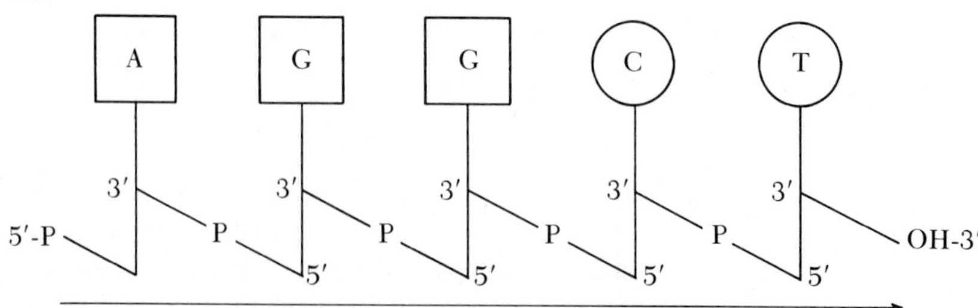

This sequence is more commonly written as pApGpGpCpT or A → G → G → C → T or A-G-G-C-T. The hyphen represents an abbreviated arrow that in turn represents the 3′,5′-phosphodiester linkage.

The structure of polynucleotides and their biological function are discussed in detail in Chapter 8.

Suggested readings

Carbohydrates

Altman, P.L., and Dittmer, D.S., editors: Biology data book, vol. I, section IV, Properties of biologic substances, Bethesda, Md., 1972, Federation of American Societies for Experimental Biology.

Bailey, R.W.: Oligosaccharides, New York, 1965, Macmillan, Inc.

Bentley, R.: Configurational and conformational aspects of carbohydrate biochemistry, Annu. Rev. Biochem. **41:**953, 1972.

Comper, W.D., and Laurent, T.C.: Physiological function of connective tissue polysaccharides, Physiol. Rev. **58:**255, 1978.

Davidson, E.H.: Carbohydrate chemistry, New York, 1967, Holt, Rinehart & Winston.

Ferrier, R.J., and Collins, P.M.: Monosaccharide chemistry, New York, 1972, Penguin Books.

Florkin, M., and Stotz, E., editors: Comprehensive biochemistry, vol. 5, Carbohydrates, New York, 1963, American Elsevier Publishing Co.

Frederick, J.F., editor: Storage polyglucosides, Ann. N.Y. Acad. Sci. **210:**5, 1973.

Heath, E.C.: Complex polysaccharides, Annu. Rev. Biochem. **40:**29, 1971.

Jeanloz, R.W., and Balazs, E.A., editors: The amino sugars: the chemistry and biology of compounds containing amino sugars, New York, 1965-1968, Academic Press, Inc.

Morrison, R.T., and Boyd, R.N.: Organic chemistry, ed. 3, Boston, 1973, Allyn & Bacon, Inc.

Pigman, W., and Horton, D., editors: The carbohydrates: chemistry—biochemistry, vols. IA and IB, New York, 1972, Academic Press, Inc.

Sharon, N.: Complex carbohydrates: their chemistry, biosynthesis and functions, Reading, Mass., 1975, Addison-Wesley Publishing Co., Inc.

Shreeves, W.W.: Physiological chemistry of carbohydrates in mammals, Philadelphia, 1974, W.B. Saunders Co.

Tipson, R.S., and Horton, D., editors: Advances in carbohydrate chemistry and biochemistry, vols. 1-36, New York, 1945-1978, Academic Press, Inc.

Whelan, W.J., editor: Biochemistry of carbohydrates, vol. 5 in MTP International Review of Science, Baltimore, 1975, University Park Press.

Lipids

Altman, P.L., and Dittmer, D.S., editors: Biology data book, vol. I, section IV, Properties of biological substances, Bethesda, Md., 1972, Federation of American Societies for Experimental Biology.

Ansell, G.B., Dawson, R.M.C., and Hawthorne, J.N., editors: Form and function of phospholipids, ed. 2, New York, 1973, American Elsevier Publishing Co.

Burton, R.M., and Guerra, F.C., editors: Fundamentals of lipid chemistry, Webster Groves, Mo., 1974, BI-Science Publishing Div.

Carter, H.E., Johnson, P., and Weber, E.: Glycolipids, Annu. Rev. Biochem. **34:**109, 1965.

Chwala, A., and Anger, V.: Surfactants, Endeavour **30:**97, 1971.

Deuel, H.J., Jr.: The lipids: their chemistry and biochemistry, vol. 1, Chemistry, and vol. 2, Biochemistry, New York, 1951 and 1955, Interscience Publishers, Inc.

Florkin, M., and Stotz, E., editors: Comprehensive biochemistry, vol. 6, Lipids, New York, 1965, American Elsevier Publishing Co.

Goodwin, T.W., editor: Aspects of terpenoid chemistry and biochemistry, New York, 1971, Academic Press, Inc.

Gunstone, F.D.: An introduction to the chemistry and biochemistry of fatty acids and their glycerides, London, 1967, Chapman Hall, Ltd.

Gurr, M.I., and James, A.T.: Lipid biochemistry, ed. 2, New York, 1975, John Wiley & Sons, Inc.

Hanson, J.R.: Introduction to steroid chemistry, Long Island City, N.Y., 1968, Pergamon Press, Inc.

Heftmann, E.: Steroid biochemistry, New York, 1970, Academic Press, Inc.

Jackson, R.L., and Gotto, A.M.: Phospholipids in biology and medicine, N. Engl. J. Med. **290:**24, 87, 1974.

Johnson, A.R., and Davenport, J.B.: Biochemistry and methodology of lipids, New York, 1971, John Wiley & Sons, Inc.

Kates, M., and Wassef, M.K.: Lipid chemistry, Annu. Rev. Biochem. **39:**323, 1970.

King, H.K.: The chemistry of lipids in health and disease, Springfield, Ill., 1960, Charles C Thomas, Publisher.

Morrison, R.T., and Boyd, R.N.: Organic chemistry, ed. 3, Boston, 1973, Allyn & Bacon, Inc.

Nicolaides, N.: Skin lipids: their biochemical uniqueness, Science **186:**19, 1974.

Paoletti, R., and Kritchevsky, D., editors: Advances in lipid research, vols. 1-17, New York, 1963-1979, Academic Press, Inc.
Sober, H.A., editor: Handbook of biochemistry, section on lipids, Cleveland, 1970, Chemical Rubber Co.
Stoffel, W.: Sphingolipids, Annu. Rev. Biochem. **40:**57, 1971.

Amino acids

Barker, R.: Organic chemistry of biological compounds, Englewood Cliffs, N.J., 1971, Prentice-Hall, Inc.
Blackburn, S.: Amino acid determination, New York, 1968, Marcel Dekker, Inc.
Cohn, E.J., and Edsall, J.T.: Proteins, amino acids, and peptides as ions and dipolar ions, New York, 1942, Rheinhold Publishing Corp.
Greenstein, J.P., and Winnitz, M.: Chemistry of the amino acids, New York, 1961, John Wiley & Sons, Inc.
Meister, A.: Biochemistry of the amino acids, ed. 2, New York, 1965, Academic Press, Inc.
Neurath, H., editor: The proteins: composition, structure, and function, ed. 2, New York, 1963-1968, Academic Press, Inc.
Sober, H.A., editor: Handbook of biochemistry, section on amino acids, Cleveland, 1970, Chemical Rubber Co.
Vickery, H.B.: The history of the discovery of the amino acids, II. A review of amino acids described since 1931 as components of native proteins, Adv. Protein Chem. **26:**81, 1972.
Weinstein, D., editor: Chemistry and biochemistry of amino acids, peptides, and proteins, vols. 1-4, New York, 1971-1974, Marcel Dekker, Inc.

Purine and pyrimidine bases, nucleosides and nucleotides

Adams, R.L.P., and others, editors: Davidson's The biochemistry of nucleic acids, ed. 8, New York, 1976, Academic Press, Inc.
Davidson, J.N.: The biochemistry of nucleic acids, ed. 7, New York, 1972, Academic Press, Inc.
Hall, R.H.: The modified nucleosides in nucleic acids, New York, 1971, Columbia University Press.
Henderson, J.F., and Paterson, A.R.P.: Nucleotide metabolism: an introduction, New York, 1973, Academic Press, Inc.
Kornberg, A.: DNA synthesis, 1974, San Francisco, W.H. Freeman & Co., Publishers.
Michelson, A.M.: The chemistry of nucleosides and nucleotides, New York, 1963, Academic Press, Inc.
Tsó, P.O.P.: Basic principles in nucleic acid chemistry, vol. 1, New York, 1974, Academic Press, Inc.
Watson, J.D.: Molecular biology of the gene, Menlo Park, Calif., 1975, W.A. Benjamin, Inc.

Chapter 24

Physicochemical topics of importance in biochemistry

Since the fundamental laws of physics and chemistry apply equally well to biological systems, a brief review of some topics particularly important in biochemistry is presented. More complete discussions may be found in the readings listed at the end of this chapter.

Expressions of concentrations

Most biochemical substances are now expressed in terms of weight per unit of volume. A common example is the expression of the amount of glucose in blood as milligrams per 100 ml or, as is becoming more common now in clinical laboratories, as milligrams per deciliter (dl). Another example is the expression of concentration of reactants in metabolic reactions as moles or as millimoles per liter (mmole/L). The latter is preferable, where possible, because comparisons between reactants are much more precise when the unit of volume is given than when concentrations are expressed on a weight basis only.

Milliequivalents

In dealing with electrolytes it is ordinarily more meaningful to express concentrations in terms of *chemical equivalents* per unit of volume. Recall that 1 equivalent of a chemical reactant is its weight in grams equal to the combining power of 1 gram of hydrogen. In these terms, the amount of one substance as related to that of another permits ready comparison as to their equivalence in many chemical and physical reactions. Since the amounts of most substances involved in fluid and electrolyte reactions are rather small, concentrations are usually expressed as *milliequivalents*, abbreviated *mEq*. Concentrations expressed in ordinary weight-volume terms can be readily converted into milliequivalents per liter, the usual units employed for biologic materials, as follows:

$$\text{Milliequivalents per liter} = \frac{\text{Milligrams per liter} \times \text{Valence}}{\text{Atomic weight}}$$

For example, a serum (human) containing (a) 322 mg of sodium, (b) 10 mg of calcium, and (c) 585 mg of chloride (as NaCl) per 100 ml would contain:

$$\text{(a)}\quad \frac{3220 \times 1}{23} = 144 \text{ mEq of sodium per liter}$$

$$\text{(b)}\quad \frac{100 \times 2}{40} = 5 \text{ mEq of calcium per liter}$$

$$\text{(c)}\quad \frac{5850 \times 1}{58.5} = 100 \text{ mEq of chloride per liter}$$

Milliosmoles and osmotic pressure

Similarly it is often necessary to determine the osmotic pressure that may be exerted by a substance in biological material. This value is expressed as *osmoles* (Osm) or more usually in biological materials as *milliosmoles* (mOsm). An osmolar solution is defined as one exerting the osmotic pressure of a molar solution of a nondissociated solute in 1 liter of solution. Thus the number of osmoles of a non-dissociated substance in a liter of solution would be the weight in grams divided by its molecular weight. The milliosmolar concentration of glucose in a sample of human plasma containing 90 mg per 100 ml therefore would be:

$$\frac{90 \text{ mg per } 100 \text{ ml} \times 10}{180 \text{ (mol wt of glucose)}} = 5 \text{ mOsm per liter}$$

If the solution under study contains a relatively high concentration of some other substance or substances, it is sometimes more desirable to express the value for the substance under study per 1000 gm of *solvent,* usually water in the case of biological materials. In this case, osmol*al* concentration is calculated. Thus, the osmolal concentration of glucose in the above example, if we assume that the plasma contains 93% water, would be:

$$\frac{90 \times 10 \times 100}{180 \times 93} = 5.4 \text{ mOsm per 1 kilogram of } H_2O$$

In the case of electrolytes in solution, each ion exerts an osmotic pressure. Therefore, for the univalent ions, Na^+, K^+, Cl^-, HCO_3^-, etc., 1 mEq = 1 mOsm (per liter of solution). For divalent ions, such as Ca^{++} and $HPO_4^=$, 1 mEq = 0.5 mOsm. Some other examples are as follows:

1 mEq K^+ = 1 mOsm
2 mEq Mg^{++} = 1 mOsm

1 mM NaCl = 2 mOsm (assuming complete dissociation)
1 mM Na_2SO_4 = 3 mOsm (assuming complete dissociation)
1 mM $CaCl_2$ = 3 mOsm (assuming complete dissociation)

The osmolarity of a dilute solution thus approximates the sum of the molarities of the individual solute particles.

The milliosmole concentration, per kilogram of water, as well as osmotic pressure, in atmospheres, of biological fluids such as plasma and urine can also be readily calculated from freezing point depression data. The milliosmole concentration of the fluid would equal the observed freezing point depression in degrees C multiplied by 1000 and divided by the theoretical molal freezing point depression, −1.86° C. This procedure is generally used for the determination of the milliosmole concentrations and osmotic pressures of biological fluids.

The osmolarity of most body fluids, both extracellular and intracellular, is kept remarkably constant by the close regulation of their fluid and electrolyte content, as is described in Chapters 16 and 17. The milliosmolar concentration of human plasma, for example, is maintained between 285 to 295 mOsm per liter under ordinary circumstances. Significant is the fact that all but about 20 mOsm of this total is contributed by Na^+ (150) and its principal accompanying anions, chloride and bicarbonate (140). The plasma proteins contribute only about 0.8 mOsm, i.e., albumins 0.6 and globulins 0.2. However, their osmotic effects ("oncotic pressures") are extremely important, since they are confined to the blood vascular system, as pointed out earlier (Chapter 16). The osmolarity of other extracellular fluids, such as lymph, cerebrospinal fluid, and the digestive secretions (Chapter 17) are about the same as that of plasma, slightly less than 300 mOsm per liter. The osmolarity of urine, on the other hand, varies widely, from 200 to 1400 mOsm per liter, depending on whether it is a "dilute" or a "concentrated" urine (Chapter 17). The osmolarity of *intracellular* fluid (e.g., muscle cells) is also similar to that of plasma, about 300 mOsm per liter. However, the principal contributors in this case are K^+ (150) and phosphates (about 103) (Chapter 16).

The development of relatively simple techniques for the determination of plasma osmolarity has led to increased appreciation of the frequency and implications of clinical states characterized by the hyperosmolarity of body fluids. For example, although glucose and urea normally contribute only small amounts to the osmolarity of the plasma, they may have considerable effect in diabetes mellitus and in certain types of renal disease. Ethanol is perhaps the most common cause of hyperosmolarity of body fluids. However it, like urea, is freely diffusible into cells, which thus become similarly hyperosmolar and create no significant differentials between extracellular and intracellular osmotic pressures.

On the other hand substances that have limited access to intracellular water, such as glucose and sodium, can cause serious differences between extracellular and intracellular osmotic pressures and corresponding compensatory fluid and electrolyte shifts. These can result in the loss of intracellular electrolytes and water and cause true intracellular dehydration and disturbances in acid-base balance. These potentially serious abnormalities of fluid and electrolyte balance are discussed further in Chapters 16 and 17.

SI units (Système International d'Unités)

Now that the metric system, used almost exclusively in the sciences for many years, is gradually replacing our traditional system of weights and measures, further metrication of the fields of science in this country seems likely in the near future. The newest metric system is called the Système International d'Unités (SI) and is being touted as the best of the metric measurement systems. Its terms and symbols are unambiguous and are standard in all languages and disciplines. The

Table 24-1
International System of Units (SI)

Major SI base units

Physical quantity	Name of SI unit	Symbol
length	metre	m
mass	kilogram	kg
time	second	s
electric current	ampere	A
temperature	kelvin	K
luminous intensity	candela	cd
amount of substance	mole	mol

Prefix multiples and fractions

Multiple	Prefix	Symbol
10	deca	da
10^2	hecto	h
10^3	kilo	k
10^6	mega	M
10^9	giga	G
10^{12}	tera	T

Fraction	Prefix	Symbol
10^{-1}	deci	d
10^{-2}	centi	c
10^{-3}	milli	m
10^{-6}	micro	μ
10^{-9}	nano	n
10^{-12}	pico	p
10^{-15}	femto	f
10^{-18}	atto	a

system was devised by the General Conference on Weights and Measures, held in Paris in the early 1960s. It has been adopted widely in many European countries and also in Australia, New Zealand, and South Africa. Some European journals now require that all data be expressed in SI units. It is being seriously considered by the major scientific societies in this country, including the American Society of Biological Chemists, the American Chemical Society, and the American Medical Association. Naturally, any switchover to the SI would be gradual, probably in several stages. The justification for the change would be that the SI would promote better communication and understanding throughout the world, both within the biochemical and medical communities and among the various scientific and engineering disciplines. The SI is summarized in Table 24-1.

The unit of volume (not indicated in the table) in the SI is the *cubic meter* (m^3). However, the *liter* is recognized also and is *exactly* equivalent to *1 cubic decimeter* (i.e., $1 m^3 = 1000 L$). Because of its convenience, the *liter* will probably continue to be used as the unit of volume in biochemistry and in clinical chemistry.

Note in Table 24-1 that *amounts* of substances are expressed as *moles* (mol), millimoles (mmol), etc. In biochemical fluids and reactions, amounts would be expressed as moles per liter (mol/L), millimoles per liter (mmole/L), etc. *Milliequivalents* per liter would become *millimoles* per liter in the SI.

Obviously, if the molecular weight of a substance is not known, its concentration cannot be expressed in terms of moles. This would be true of most if not all proteins, some carbohydrates and lipids, some vitamins, and other substances of biological importance. In these instances, the proposal is to retain the present weight/unit of volume expression. For example, hemoglobin and plasma immunoglobulins would be expressed as deciliter, or per 100 ml, or per liter.

A special SI unit, the *katal* (kat), has been proposed for expressing *enzyme* activity. *One katal* describes *1 mole of substrate transformed per second.*

Law of mass action

The law of mass action applies to the state of equilibrium existing in reversible reactions. This law states that the rate at which a reaction takes place, at constant temperature, is proportional to the product of the concentrations of the reacting substances. The concentrations are expressed as moles per liter, and such concentrations are represented by bracketing the symbols of the substances in question.

The reaction between ethyl alcohol and acetic acid to form ethyl acetate and water is a reversible reaction.

$$C_2H_5OH + CH_3COOH \underset{\text{Reaction II}}{\overset{\text{Reaction I}}{\rightleftharpoons}} CH_3COOC_2C_2H_5 + H_2O$$

According to the law of mass action, the velocity of the reaction proceeding toward the right (Reaction I) depends on the product of the concentrations of alcohol and acetic acid:

$$V_1 \propto [C_2H_5OH] \times [CH_3COOH]$$

where V_1 represents the velocity of Reaction I. Therefore:

$$V_1 = k_1 \times [C_2H_5OH] \times [CH_3COOH]$$

where k_1 is a constant.

In a similar manner Reaction II proceeds at a velocity V_2, which is proportional to the concentrations of ethyl acetate and water:

$$V_2 \propto [CH_3COOC_2H_5] \times [H_2O]$$

and also

$$V_2 = k_2 \times [CH_3COOC_2H_5] \times [H_2O]$$

where k_2 is another constant. Now, at equilibrium, Reaction I must necessarily proceed at the same rate as Reaction II; otherwise it would not be in equilibrium, and

$$V_1 = V_2$$

or

$$k_1 \times [C_2H_5OH] \times [CH_3COOH] = k_2 \times [CH_3COOC_2H_5] \times [H_2O]$$

and by algebraic division we arrive at

$$\frac{[CH_3COOC_2H_5] \times [H_2O]}{[C_2H_5OH] \times [CH_3COOH]} = \frac{k_1}{k_2} = K_{equil}$$

The new constant, K, is the *equilibrium constant* of the reaction. This constant is always the same for a given reaction after equilibrium has been established, no matter what the proportion of the reactants may have been at the start. There is, of course, a different constant for every reaction, and the constant varies with temperature and pressure. Here we are interested mainly in the electrolytic dissociation constants.

If K is always the same for this reaction at equilibrium, then if the equilibrium is upset by adding or removing any of the four reacting substances, the system tends to balance these substances until a new equilibrium is reached and K is reconstituted. For example, if more ethyl alcohol is added, more of it combines with acetic acid to form more ethyl acetate and water.

To make the equation more general, we may say for the reversible reaction:

$$aA + bB \rightleftarrows cC + dD$$

Using the same symbols as above:

$$V_1 = k^1[A]^a \times [B]^b$$
$$V_2 = k_2[C]^c \times [D]^d$$

At equilibrium:

$$k_1[A]^a \times [B]^b = k_2[C]^c \times [D]^d$$

$$\frac{[C]^c[D]^d}{[A]^a[B]^b} = \frac{k^1}{k_2} = K_{equil}$$

Hydrogen ion and hydroxyl ion concentration

Pure water is only slightly dissociated. There is, however, a certain definite concentration of hydrogen ions, which, though small, must be balanced by the same concentration of hydroxyl ions. This state of affairs is reflected in the low but measurable conductivity of water. The dissociation of water may be represented by the following equation:

$$[HOH] \rightleftarrows [H^+] + [OH^-]$$

and therefore

$$\frac{[H^+] \times [OH^-]}{[HOH]} = K_1$$

In this equation the denominator, undissociated water, is extremely large when compared with the numerator and may be considered a constant. An analogy may be drawn between this and a ship that has sprung a leak when in midocean. The amount of water pouring into the ship's hull is of great moment, even though it is an infinitesimal part of the ocean, which, for all intents and purposes, remains constant. Here the numerator is the volume of water passing into the ship and the denominator is the constant ocean. Therefore:

$$\frac{[H^+] \times [OH^-]}{K_2} = K_1$$

$$[H^+] \times [OH^-] = K_1K_2 = K_w$$

K_w, the ionization product for water, is an extremely small value. It has been determined to be 0.00000000000001, or 1/100,000,000,000,000, or 10^{-14} at 25° C.

$$[H^+] \times [OH^-] = K_w = 10^{-14}$$

In pure water the concentration of hydrogen ions must equal to that of hydroxyl ions; therefore since $[H^+] = [OH^-]$, it can be substituted for $[OH^-]$.

$$[H^+] \times [H^+], \text{ or } [H^+]^2 = 10^{-14}$$

Taking the square root of both sides of the equation,

$$[H^+] = 10^{-7}$$

In other words, the hydrogen ion concentration of water is 1×10^{-7} gm per liter, or 1/10,000,000 gm per liter. Either of these methods of expression is unwieldy, and consequently Sørensen suggested that the negative exponent with its sign changed to positive be used and be termed *pH*. Another way of stating this is:

$$pH = -\log_{10}[H^+]$$

The pH of water, or neutrality, then, is 7.0. If acid is added to water, the concentration of hydrogen ions, of course, increases, and instead of 1/10,000,000 gram of H^+ per liter, there is a greater value with a *smaller* denominator. The pH of acidic solutions, accordingly, is less than 7.0 and that of alkaline solutions is greater than 7.0. However, since the product of $[H^+]$ and $[OH^-]$ must remain constant, i.e., 10^{-14}, when $[H^+]$ is greater than 10^{-7} mole per liter, $[OH^-]$ is less than 10^{-7}.

In this connection it is well to define the corresponding term, pOH. The negative logarithm of $[OH^-]$ is pOH, and furthermore we can speak of the negative logarithm of K_w as pK_w. Using HCl and NaOH solutions as examples and assuming complete ionization, we can show the relationship of acidity and alkalinity to pH

Table 24-2
Relation of acidity and alkalinity to pH and pOH

Normality*	pH		pOH
0.1 N HCl	1		13
0.01 N HCl	2		12
0.001 N HCl	3	Acidity	11
0.0001 N HCl	4		10
0.00001 N HCl	5		9
0.000001 N HCl	6		8
0.0000001 N (= water)	7	Neutrality	7
0.000001 N NaOH	8		6
0.00001 N NaOH	9		5
0.0001 N NaOH	10	Alkalinity	4
0.001 N NaOH	11		3
0.01 N NaOH	12		2
0.1 N NaOH	13		1

*A normal solution is one that contains 1 gm equivalent of the substance per liter. For further discussion, see under "Titratable acidity."

and pOH in Table 24-2. (In practice, pOH is seldom referred to.) Notice that the sum of pH and pOH is always 14. This is equivalent to stating:

$$[H^+][OH^-] = 10^{-14}$$

Examples of the conversion of hydrogen ion concentration to pH and vice versa are the following:

1. Given the hydrogen ion concentration $[H^+] = 0.00634$ N, find the pH as follows:

   It is first convenient to express the concentration of hydrogen (or hydronium) ions as a whole number multiplied by 10 raised to the power indicated. Thus:

   $$[H^+] = 0.00634 \text{ N} = 6.34 \times 10^{-3}$$

   Since $\text{pH} = -\log^{10}[H^+]$, and $[H^+]$ is the molar concentration, the logarithm must first be obtained.

   $$\begin{aligned} \log [H^+] &= \log (6.34 \times 10^{-3}) = \log 6.34 + \log 10^{-3} \\ &= 0.8021 + (-3) = -2.1979 \end{aligned}$$

   To get the $-\log$, multiply both sides of the equation by -1.

   $$-\log [H^+] = \text{pH} = 2.20 \text{ (pH is never expressed beyond the second decimal)}$$

2. Given the pH 2.20, determine the $[H^+]$ as follows:

   $$\text{pH} = \log\frac{1}{[H^+]} = \log 1 - \log [H^+]$$

   $$\begin{aligned} \log 1 &= 0 \\ \text{pH} &= 0 - \log [H^+] = -\log [H^+] \\ [H^+] &= 10^{-\text{pH}} \\ [H^+] &= 10^{-2.20} \\ [H^+] &= 10^{-3.0} \times 10^{+0.8} \\ \times &= 10^{0.8} \\ \log \times &= 0.8 \\ \times &= 6.31 \\ [H^+] &= 6.31 \times 10^{-3.0} \text{ or } 0.00631 \text{ N} \end{aligned}$$

   Hydrogen ion concentrations are determined either by electrometric methods or by the use of standard buffers and indicators. A description of the electrometric methods is beyond the scope of this volume. However, it should be pointed out that the instruments have been skillfully developed and simplified to such a degree that pH determinations of accuracy (to ± 0.001 pH unit) may be made in a few minutes. For an understanding of pH and its regulation in body fluids and tissues, a brief discussion of buffers is now pertinent.

Buffers

A buffer solution is one that tends to maintain a constant hydrogen ion concentration when acid or alkali is added to it. A buffer system usually consists of a weakly dissociated acid and the salt of that acid, or a weak base and its salt; e.g., carbonic acid and sodium bicarbonate constitute a buffer system. If acid is added to $NaHCO_3$, the following reaction occurs:

$$NaHCO_3 + HCl \rightarrow NaCl + H_2CO_3$$

A strong acid, HCl, which might be expected to raise the hydrogen ion concentration, reacts with a weak base in such a way as to yield a weak acid, H_2CO_3, and a neutral salt. The hydrogen ion concentration has not been raised appreciably. Also if NaOH is added to NaH_2PO_4,

$$NaOH + NaH_2PO_4 \rightleftarrows Na_2HPO_4 + H_2O$$

the weakly acid sodium dihydrogen phosphate buffers the strong alkali by yielding the weakly alkaline disodium hydrogen phosphate. Again the hydrogen ion concentration has not been changed very much. In the first instance the buffer system or buffer pair H_2CO_3 and $NaHCO_3$ results and the H_2CO_3 is effective in buffering alkalies. In the second instance, NaH_2PO_4 and Na_2HPO_4 become the buffer pair, also effective in buffering in either direction. Both these systems, and several others as well, operate in the body to prevent pronounced changes of hydrogen ion concentration, and they are remarkably efficient.

In such systems it is evident that a common ion effect is operative; i.e., if we add to a weak electrolyte a strong electrolyte having an ion in common with the weak electrolyte, the ionization of the weak electrolyte is diminished and the concentration of the ion not in common is also lessened. For example, if sodium acetate is added to acetic acid (HAc), the ionization of the acid is repressed, resulting in a decreased $[H^+]$.

$$\frac{HAc \rightleftharpoons H^+ + Ac^-}{HAc \rightleftharpoons H^+ + Ac^-}$$

$$\uparrow$$

$$\text{(add) } NaAc \rightleftharpoons NA^+ + Ac^-$$

The addition of the acetate ion (from the sodium acetate), which is in common with the acetate ion from the acetic acid, tends to drive the equilibrium to the left. As a result the concentration of hydrogen ion is decreased by recombination with the acetate ion to form undissociated acetic acid.

The foregoing considerations may now be made quantitative by reviewing some concepts of acid-base chemistry. According to the Brønsted theory, an acid may be defined as any molecule or ion that dissociates to yield hydrogen ion or ions. If the acid *completely* dissociates when in dilute aqueous solution, then it is a *strong acid*, but if only partly dissociates to give hydrogen ions, it is a *weak acid*.

Thus HCl is a strong acid, since it dissociates completely to give hydrogen and chloride ions in a dilute solution.

$$HCl \rightarrow H^+ + Cl^-$$

Note that in this case the reaction is indicated by an arrow in only one direction. But acetic acid dissociated only slightly:

$$HAc \rightleftharpoons H^+ + Ac^-$$

Accordingly, arrows are shown in both directions.

The *Brønsted theory* continues in defining a base as any molecule or ion that reacts with hydrogen ion.

In dilute solution, chloride ion cannot be considered a base, since it does not react with hydrogen ion. (If HCl dissociates completely, then Cl^- apparently has no tendency to remain combined with H^+ ion.) In concentrated solutions, chloride ion associates slightly with hydrogen ion and is therefore a weak base, a fact reflected in incomplete dissociation of HCl in very concentrated solutions.

Acetate ion, however, is a strong base, since it has a ready affinity for hydrogen ion. Chloride ion and acetate ion are called the conjugate bases of their respective acids, HCl and HAc. It is now apparent that the conjugate base of a strong acid is a weak base, whereas that of a weak acid is a strong base.

Returning now to the above reactions, we can see that for:

$$HCl \rightarrow H^+ + Cl^-$$

an equilibrium cannot be reached and an equation of the kind found at the beginning of this chapter cannot be written. On the other hand, a true equilibrium equation can be written for acetic acid:

$$HAc \rightleftharpoons H^+ + Ac^-$$

$$\frac{[H^+]\,[Ac^-]}{HAc} = K$$

It is evident that an increase in acetate ion concentration after equilibrium has been reached tends to increase K. But since K must remain constant, an adjustment that somewhat decreases $[Ac^-]$ and increases HAc just enough to keep the ratio constant must be made. This happens if $[Ac^-]$ associates with $[H^+]$ to form more undissociated acetic acid. A consequence of this is a diminished concentration of $[H^+]$ or a rise in pH. The ions in any solution always act in such a way as to maintain K absolutely constant according to the laws of equilibria.

The hydrogen ion concentration of most body fluids and secretions is on the alkaline side. Urine may be acid, and gastric juice is very acid, but these are exceptions. Many influences tend to change this alkalinity, but the buffers present prevent sharp fluctuations in hydrogen ion concentration. The pH of blood, for example, stays within the limits 7.3 to 7.5 in health. When these limits are exceeded, we have a condition of acidosis or alkalosis with alarming symptoms and, frequently, dire results (Chapter 16). In Table 24-3, are given some of the pH values for various human fluids.

It is quite difficult to keep a solution at constant pH if no buffer is present because of the influence of the CO_2 of the air or the alkali of the glass container or because of other influences. Consequently, buffers are frequently required. Various mixtures, consisting of definite amounts of the acid or base and its respective salt, have been prepared. Since such buffer sets maintain their pH indefinitely, they are used in the indicator method of determining pH.

Henderson-Hasselbalch equation Values of the pH of buffer solutions may be calculated if the composition of the mixture, as well as the ionization constant of the weak electrolyte, is known. For example, in the case of acetic acid:

$$HAc \rightleftharpoons H^+ + Ac^-$$

$$\frac{[H^+] \times [Ac^-]}{[HAc]} = K_{Ac}$$

$$[H^+] = K_{Ac}\frac{[HAc]}{[Ac^-]}$$

In a mixture of the acid and its salt, e.g., NaAc, most of the acid is unionized. Consequently the [HAc] is about the same as the total acid concentration and $[H^+]$ would be extremely small. Also since the salt is completely ionized, the value $[Ac^-]$ is approximately equal to the total salt concentration. Thus, for approximate calculation, the equation may be written as follows:

$$[H^+] = K_{Ac}\frac{\text{Acid}}{\text{Salt}} \text{ or } K\frac{\text{Acid}}{\text{Salt}}$$

Table 24-3
pH values of human body fluids and secretions

Body fluid or secretion	pH values	Body fluid or secretion	pH values
Blood	7.4	Pancreatic juice	8.0
Milk	6.6-6.9	Intestinal juice	7.7
Bile	7.8	Cerebrospinal fluid	7.4
Urine	6.0	Saliva	7.2
Gastric juice (parietal secretion)	0.87	Aqueous humor of eye	7.2

to make the equation applicable to other salts and acids. Taking the negative logarithm of this:

$$-\log [H^+] = -\log \left(K \times \frac{[\text{Acid}]}{[\text{Salt}]}\right)$$

$$= -\log K + \left(-\log \frac{[\text{Acid}]}{[\text{Salt}]}\right)$$

Since $-\log [H^+]$ is called pH, we may call $-\log K$ by the term pK. Therefore:

$$\text{pH} = \text{p}K + \log \frac{[\text{Salt}]}{[\text{Acid}]}$$

In other words, the pH of a buffer solution is determined by the logarithm of the ratio of salt to acid and by the pK (i.e., the negative logarithm of the ionization constant of the acid). This last equation is known as the *Henderson-Hasselbalch equation*. It finds many uses in chemical calculation. Examples follow:

1. To calculate the pH of a buffer solution that contains 0.1 mole of sodium acetate and 0.1 mole of acetic acid at 25° C, $K_a = 1.8 \times 10^{-5}$:

$$\text{pH} = \text{p}K_a + \log \frac{[\text{Salt}]}{[\text{Acid}]}$$

$$\text{P}K_a = -\log K_a = -\log (1.8 \times 10^{-5})$$
$$= -(\log 1.8 + [-5]) = 4.74$$

$$\text{pH} = 4.74 + \log \frac{[0.1]}{[0.1]}$$

$$\log 1 = 0, \text{pH} = 4.74$$

2. To calculate the pH of two solutions, one of which is unbuffered and the other buffered, before and after adding the same amount of a strong base to each:
   a. Given a solution of HCl whose concentration is 0.0001 molar, determine the pH as follows:

$$[H^+] = 0.0001 = 10^{-4}$$
$$\text{pH} = 4.0$$

   Now add 0.0001 mole of NaOH. To determine the pH after this addition: Since the NaOH added exactly neutralizes the acid present,

$$\text{pH} = 7.0$$

   b. Given the buffered solution of Example 1, containing 0.1 mole of sodium acetate and 0.1 mole of acetic acid, with a pH of 4.74, calculate the pH after adding the same amount of NaOH, namely, 0.0001 mole, as follows:

$$\text{pH} = 4.74 + \log \frac{0.1 + 0.0001}{0.1 - 0.0001}$$

   (Because 0.0001 mole of base [numerator] has been added and the same amount of acid has been subtracted [denominator].)

$$\text{pH} = 4.74 + \log \frac{0.1001}{0.0999} = 4.74 + \log 0.1001 - \log 0.0999$$
$$= 4.74 + 0.0008, \text{ or } 4.74$$

These calculations (of Example 2) show also the resistance of a buffer to a change in the pH. A small quantity of base added to an unbuffered solution produces a change of three pH units, whereas an equivalent amount of base added to a buffer solution causes a relatively small change.

Titratable acidity

The total potential acidity or alkalinity of a biological fluid may be determined by titration. On the other hand, the hydrogen ion concentration might be termed the true acidity or alkalinity. Let us consider two different acids, hydrochloric and acetic, having the same normality. By normality, we mean the concentration as related to that of a normal solution, and a normal solution is one containing 1 gm equivalent of a substance per liter of solution. For example, a normal solution of HCl is one that contains 36 gm of HCl per liter, i.e., one that contains 1 gm of H^+ per liter. A normal NaOH solution contains 40 gm of NaOH per liter, i.e., a solution that combines with 1 gm of H^+ per liter. Similarly a normal acetic acid solution has 60 gm of CH_3COOH per liter of solution. In the case of bivalent, trivalent, etc. acids, bases, and salts, we must divide the molecular weight by 2, 3, etc., respectively.

If we wish to combine a gram equivalent weight of NaOH with either HCl or CH_3COOH, we find, of course, that a gram equivalent weight of either of these acids is required.

$$\underset{40\text{ gm}}{NaOH} + \underset{36\text{ gm}}{HCl} \rightarrow \underset{58\text{ gm}}{NaCl} + \underset{18\text{ gm}}{H_2O}$$

$$\underset{40\text{ gm}}{NaOH} + \underset{60\text{ gm}}{CH_3COOH} \rightarrow \underset{82\text{ gm}}{CH_3COONa} + \underset{18\text{ gm}}{H_2O}$$

If, now, the 40 gm of NaOH in each case are dissolved and diluted to 1 liter and the 36 gm of HCl and 60 gm of CH_3COOH are each dissolved in sufficient water to make 1 liter, we have normal solutions of each. One liter of the normal HCl neutralizes 1 liter of normal NaOH; 1 liter of normal CH_3COOH neutralizes 1 liter of normal NaOH. Thus a liter of normal or N/1 HCl is equivalent to the same amount of N/1 CH_3COOH because each is potentially capable of yielding the same amount, namely, 1 gm per liter, of hydrogen ions. The CH_3COOH is not nearly as strong an acid as HCl; i.e., it is not as greatly dissociated. However, if we add NaOH to it (e.g., the N/1 NaOH), little by little the small number of hydrogen ions is neutralized by the base and more and more acid dissociates until finally all the hydrogen ions have been displaced. This process is known as titration. A normal (N/1), tenth normal (N/10), etc. solution of any acid, then, is equivalent to any other N/1, N/10, etc. acid, volume for volume. Again, a N/10 basic solution neutralizes a N/10 acid, volume for volume.

The *stoichiometric point* in an acid-base titration is the point at which an equivalent amount of base has been added to the acid. The end point of such a titration, however, is not necessarily at a pH of 7.0. This will depend on the salt formed by the reaction. Obviously, we are interested in the pH of the solution at the point at which a maximum amount of salt is present. In the titration of HCl with NaOH, it is the exact point at which all the chloride ion is balanced by the sodium ion and we have a solution of NaCl. Since this is a neutral salt with no buffering action, many different indicators may be used to tell the end point, for one additional droplet of alkali above the equivalent amount can cause a great change in the pH. However, in the titration of a weak acid (e.g., CH_3COOH), with a strong base (e.g., NaOH), the salt formed at the stoichiometric point (CH_3COONa) will have an alkaline reaction, as seen from the following reaction generally referred to as hydrolysis:

$$\begin{array}{rcccc} CH_3COONa & \rightleftharpoons & Na^+ & + & CH_3COO^- \\ HOH & \rightleftharpoons & OH^- & + & H^+ \\ & & \uparrow & & \downarrow\uparrow \\ & & NaOH & & CH_3COOH \end{array}$$

In other words, as sodium acetate dissociates in the presence of water, the strong basic reaction overbalances the weak acidic one. Therefore in this titration an indicator that changes at an alkaline pH, e.g., phenolphthalein, must be used. Similarly, in titrating a weak base (e.g., NH_4OH) with a strong acid (e.g., HCl), we finally get NH_4CL at the stoichiometric point. This has an acid reaction and requires an indicator that changes on the acid side, e.g., methyl red.

$$\begin{array}{lcl} NH_4Cl \rightleftharpoons NH_4^+ & + & Cl^- \\ HOH \rightleftharpoons OH^- & + & H^+ \\ \quad\quad\;\; \downarrow\uparrow & & \uparrow \\ \quad\quad\; NH_4OH & & HCl \end{array}$$

Colloidal state

In 1861 Graham classified all substances into two categories, depending on the ability of the substances to pass through parchment and similar membranes. Since those substances that diffused readily were the ones that easily crystallized, e.g., copper sulfate, sucrose, etc., he designated them "crystalloids." Those that did not pass through, e.g., gelatin, starch paste, glue, etc., were considered to be noncrystallizable and were called "colloids," from the Greek word meaning "glue." These terms continue to be used, although we now are able to crystallize many of the colloids and the crystalloids can be converted to a colloidal form. The modern concept of these differences is based on the size of the particles dispersed in the water or other medium. Colloidal particles are large; they cannot pass through the pores of ordinary parchment or collodion membranes. However, they are not large enough to settle out by gravity, as suspensions do, or to float at the top of the medium, as imperfect emulsions do. In true solutions, so-called crystalloidal solutions, the mixture is homogeneous; the constituents are present in the molecular or ionic state and are uniformly distributed throughout and among the molecules of water or other solvent. Colloidal systems are heterogenous; i.e., there are two *phases*—the finely divided particles and the medium in which they are suspended. By "phase," we mean a physically distinct portion of matter. The particles are called the *dispersed phase,* and the medium, usually a fluid, is the *dispersion medium.* Both phases may be solids, liquids, or gases, with a single exception: it is not possible to have a colloidal dispersion of a gas in a gas. Smoke is a solid dispersed in a gas, fog is a liquid dispersed in a gas, and froths and foams are gases dispersed in liquids. We are more concerned with liquids dispersed in liquids, liquids dispersed in solids, and solids dispersed in liquids.

The size of the particles in colloidal systems is generally stated to be from 1 millimicron (mμ) to 100 mμ, but arbitrary limits at either end cannot be set. In fact, the tendency is to place the upper limit somewhat higher, such as at 500 mμ (0.5 μ). A millimicron is one millionth of a millimeter (0.000001 mm). Particles having smaller diameters than 1 mμ are molecular or ionic, and if much above 100 to 500 mμ, they are coarse enough to settle out. The smallest colloidal particles, therefore, are but little larger than crystalloidal molecules, and the largest ones are nearly the size of the particles in a suspension.

Colloidal particles may be removed from the dispersion medium by forcing the fluid, under pressure, through an appropriate membrane. This is termed *ultrafiltration.* By using membranes of varying porosity, it is possible to separate different colloids from each other and to estimate the size of colloid particles. A colloid particle is often termed a *micelle*.

Ultracentrifugation is another method of removing colloid particles. By centrifuging at a very high speed, the dispersed phase may be separated from the dispersion medium. Substances in true solution cannot be separated from their solvents by these two methods. Still another procedure is electrophoresis (Chapter 25).

A simpler method than any of those described is dialysis. This is discussed later in the chapter.

Types of colloids

Colloids may be grouped into two main classes, depending on their ability to take up the dispersion medium. The *lyophilic colloids* (emulsoids) have a great attraction for the dispersion medium; in fact, each particle has a layer of the dispersion medium surrounding it. The *lyophobic colloids* (suspensoids) contain no such layer. The names are quite descriptive, "lyophilic" meaning solvent-loving and "lyophobic" meaning solvent-hating. The lyophilic colloids include starch, egg albumin, blood proteins, soap, and gelatin. This is the more important type physiologically. Examples of the lyophobic colloids are the colloidal metals, e.g., gold, silver, platinum, etc. Both types exist in the fluid state as *sols*. Many lyophilic colloids form semisolid *gels*. A well-known example is gelatin, which, when dilute, is fluid. When a moderately high concentration is allowed to stand, it sets into a jelly or gel. This gel may be converted to its sol by warming, and, on cooling, it may again gel; i.e., the change from a sol to a gel is frequently reversible. However, this is not always the case, as we shall see when we study coagulation and denaturation of proteins. When a gel forms, apparently long chains of molecules of the colloid interlace and entrap the fluid by capillary forces. The gel then is really a liquid dispersed in a solid.

Some sol-gel transformations take place without change of temperature. For instance, if a colloidal iron oxide sol is allowed to stand quietly, it sets into a gel. Upon shaking, a sol is re-formed. This phenomenon is known as *thixotropy*. Protoplasm is said to have thixotropic properties.

Gels possess the tendency to take up and retain water and swell. This is called *imbibition*. A considerable degree of pressure is needed to squeeze out the water that has been taken up by imbibition. This property of lyophilic colloids is of great importance as regards the state of protoplasm in health and disease. For example, in edema, large quantities of water are absorbed and held by certain tissues. Imbibition is also of importance as regards blood volume. Each gram of albumin in blood plasm holds 17 ml of water, which aids in maintaining the normal volume of the circulating blood. If the blood volume is reduced because of shock or hemorrhage, the administration of physiological salt solutions is usually of little value because the fluid leaves the circulation and enters the tissues or is excreted quite promptly. Some material that will hold an amount of water equal to that lost, and thus maintain normal blood volume, must be introduced. If whole blood or plasma is not available, some material that will hold water by imbibition must be used. Gels with this property, called plasma substitutes or *extenders,* are widely used for such purposes clinically (Chapter 16). Imbibition is affected by hydrogen ion concentration and by other electrolytes, which in some instances enhance the swelling produced by pure water. This differs with different ions. Nonelectrolytes such as urea and glucose have little effect.

Preparation of colloidal solutions

Various organic compounds, e.g., gelatin, starch, and soap, form sols or "colloidal solutions," as they are frequently called, simply when added to water. Other colloidal solutions may be formed by chemical reaction; e.g., the reduction of gold chloride yields the lyophobic, colloidal gold. Mechanical grinding in a colloid mill may also be used to reduce a substance to such a fine state that it may readily be dispersed in colloidal form. The dispersal of any solid into the colloidal state is called *peptization*; e.g., the dispersal of gelatin in water is known as the peptization of the gelatin, and the water is the peptizing agent; but of course other liquids or solutes may also be peptizing agents. However, there are substances that may form colloidal solutions in one dispersion medium and true solutions in another; e.g., soap is colloidal in water but crystalloidal in ethyl alcohol.

Electrical charges on colloids

Dispersed colloidal particles carry electrical charges, with the dispersion medium carrying the opposite charge. These charges arise chiefly in two ways: groups on the surface of the colloid may ionize, as is the case for proteins, or ions from the medium may be preferentially adsorbed on the surface of the colloid. Since bodies carrying like charges repel each other, this serves to keep the colloids dispersed and is one of the factors that make for the stability of a colloidal system. If an electric current is sent through a colloidal system, the colloid passes to the anode if it is electronegative or to the cathode if it is electropositive. This is called *cataphoresis* or *electrophoresis*. Furthermore, one may precipitate a colloid by adding a colloid of opposite charge, thereby neutralizing the charge and upsetting the stability of the dispersed substance. A few examples of positively and negatively charged colloids of both the lyophlic and lyophobic classes are given in Table 24-4.

Under certain conditions the phenomenon of *coacervation* may be observed. When two lyophilic colloids of opposite electric charge are mixed, they may not precipitate or flocculate in the ordinary sense but, because of the hydration shell on each, may form microscopic droplets. These droplets, after a while, may coalesce to form a viscous, fluid layer. This is a coacervate. It contains the two colloids, held apart by the hydration shells, and each colloid retains its own electrical charge. Therefore, if the pH is changed or an electrolyte is added to a coacervate, it may again form a sol of the original lyophilic colloids. Many phases of protoplasm are belived to be coacervate in nature. Thus vacuole formation closely resembles a phenomenon seen when complex coacervates are permitted to age.

Tyndall effect A colloidal solution, e.g., dilute starch, appears slightly cloudy or opalescent to the eye. If a beam of light is passed through it, the beam becomes visible as a much cloudier path, particularly if viewed against a dark background. This is the *Tyndall phenomenon* and is the same phenomenon observed when a beam of sunlight enters a darkened room. In this instance the minute (colloidal) particles of dust in the air deflect the light. In the case of the colloidal solution, it is the colloidal particles that partly diffract the light and diffuse it. By means of an ultramicroscope or a dark-field microscope, the Tyndall effect becomes visible in another way. In these instruments, light is sent through a drop of solution in a horizontal direction. Visible particles reflect light to the observer's eye and can be seen as shining objects. Invisible particles, e.g., colloids, may be seen as dancing bright specks. They dance and dart in *Brownian movement*, just as visible particles do. The reason is that they are under constant bombardment from molecules of the dispersion medium. A large particle is likely to be hit by about the same number of bombarding particles from each side at the same instant; therefore it will move less. The smaller particle, with less chance of instantaneous and equal striking from all

Table 24-4
Electrical charges on colloids

Lyophilic	Lyophobic
Proteins in neutral or alkaline solution (−)	Gold (−)
Proteins in acid solution (+)	Platinum (−)
Starch (−)	Stannic oxide (dispersed by HCl) (+)
Soaps (−)	Sols of metallic oxides and hydroxides and basic dyestuffs (usually +)
Aluminum hydroxide (+)	
Gum acacia (−)	

sides, will move whenever it receives an unequal number of blows from different directions.

The nephelometer is an instrument that measures the Tyndall effect. Substances that form extremely fine precipitates in suspension (colloidal particles) may be estimated quantitatively by means of it.

Stability of colloids The constant movement of particles is one of the forces that tend to keep a colloid stable. It keeps the colloidal particles distributed throughout the system, rather than allowing them to settle or rise. The size of the colloidal particle is another factor, since the smaller the size, the more closely the particle approximates molecular dimensions and therefore more nearly resembles a true solution. This is not always the case, however, for the lyophobic colloids may have extremely small dimensions and are in general less stable than the lyophilic colloids. The reason for this is that the lyophobes have little or none of the dispersion medium attached, or adsorbed, to their surfaces; therefore the particles can approach each other closely enough to permit mass attraction to overcome the repulsion because of the electrical charge. The lyophilic colloids have a layer of the fluid adsorbed, which makes them more stable because this keeps them farther apart. The electrical charge present on the colloid, as a stabilizing factor, has already been discussed.

Any procedure that tends to diminish the effect of one or more of the stabilizing factors tends also to precipitate the colloid. Thus, as said before, the addition of another colloid of opposite electric charge has this effect. In the case of the lyophobic colloids, a small amount of electrolyte accomplishes this by also neutralizing the charge. The effect does not occur in the case of lyophilic colloids, although large amounts of electrolytes will precipitate them. Probably this "salting out" occurs as the result of dehydrating the surface layer of fluid. Furthermore, violent agitation, freezing and thawing, or heating, can "break" colloidal solutions by modifying one or more of the stabilizing factors.

Although lyophobic colloids may be precipitated by the addition of small amounts of electrolytes, precipitation may be hindered by the presence of small quantities of a lyophilic colloid. Such "protective" action is believed to be accomplished by the adsorption of the lyophilic colloid onto the surface of the lyophobic colloid, thereby preventing the electrolyte from easily reaching it. The protective action of colloids present in bile and urine may be a major reason for the prevention of the precipitation of almost insoluble constituents of these fluids. Conversely, when the protective colloidal action is not effective, gallstones and kidney stones result. Lange's colloidal gold test is based on this protective colloidal action. Colloidal gold is a bright orange-red sol. The addition of normal cerebrospinal fluid does not precipitate it. Abnormal cerebrospinal fluid may do so, however, with the result showing in various shades, depending on the degree of precipitation. The test is performed under exact quantitative conditions, and curves that are indicative of pathological sates are obtained.

Surface reactions of colloids

The dispersed phase of a colloid differs from a suspension of solid matter in that the colloid is subdivided into much smaller particles. Each suspended particle has a surface. Similarly, each colloidal particle has a surface, and although the colloidal particle is far smaller than the suspension particle, the number of particles is much greater for the same weight and therefore the total area of surface is greater. This can be illustrated by a simple example. A cube of any material 1 cm on a side has a surface area of 6×1 cm^2, or 6 cm^2. If it is divided into eight cubes, with each edge measuring ½ cm, the total area will be ½ cm $\times$ ½ cm $\times$ 6 sides $\times$ 8 cubes $\times$ 12 cm^2. The subdivision of each small cube is continued, and the amount of surface is doubled each time each cube is cut into eight smaller ones. Eventually, when the cubes are down to a size comparable with colloidal dimensions, i.e., 100 mμ on a side, the total of all the tiny cubes will be 600,000 cm^2, i.e., 100,000 times the original surface area, produced by simply subdividing the cube. This indicates the enormous area presented by the surfaces of colloid particles. On these surfaces, substances present in solution may be adsorbed and become concentrated. Adsorption is the phenomenon in which there is condensed on a surface a layer of ions, molecules or aggregates of molecules that are present in the medium with which the surface is in contact. The amount of adsorption depends on the extent of surface exposed and the specific nature of the surface and of the substance adsorbed upon it. Furthermore, the degree of adsorption is increased by a rise in pressure and is diminished by a rise in temperature. Many physiological phenomena are surface reactions. The enzymes, which catalyze so many reactions of the body, probably act through surface forces. They are colloids and the substances they act on are probably adsorbed by them as a first step in the chemical action that is brought about.

Emulsions

Emulsions are dispersions of one liquid in another. If olive oil is shaken vigorously in water, it breaks up into small droplets and a yellow milky fluid results. An emulsion is a heterogeneous system comprising two phases, but the dispersed phase usually consists of particles larger than colloidal particles. An emulsion of olive oil and water does not last very long; the oil droplets coalesce and soon there is a layer of oil floating on the water. Certain substances—among others, soaps, gums, proteins—when added to the system, stabilize the emulsion. The stabilizing substance, an emulsifier, may form a protective layer around the oil droplets and so prevent them from coalescing. This is similar to the *protective colloidal action* just described. With any pair of nonmiscible liquids, two sets of emulsions are possible. Thus, with oil and water we may have (1) oil-in-water and (2) water-in-oil. In a general way, any emulsifying agent that stabilizes an oil-in-water emulsion is unsatisfactory for the corresponding water-in-oil emulsion, and vice versa.

Emulsions occur widely in nature. In fact, protoplasm is probably a mixture of emulsions, containing colloids in one or both phases of many of them. Emulsification, of course, increases the surface of the substance emulsified. This permits biological reactions, which, as said, are frequently surface reactions, to take place more readily. For example, one of the functions of bile is to aid in the emulsification of fats in the small intestine. By doing so, it breaks up these water-insoluble substances into such tiny droplets that the digestive agent can readily attack them. If this were not done, the digestion of the mass of fat would take place extremely slowly.

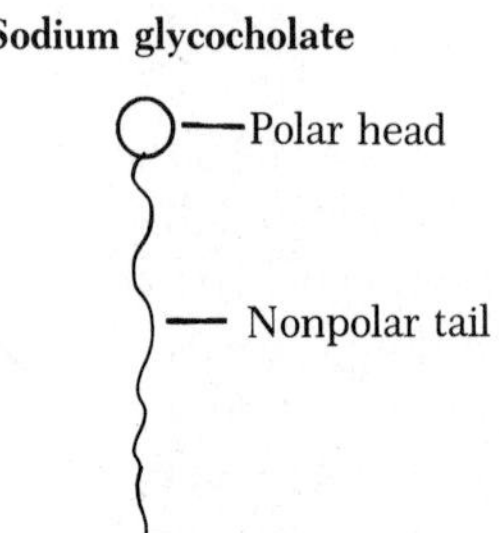

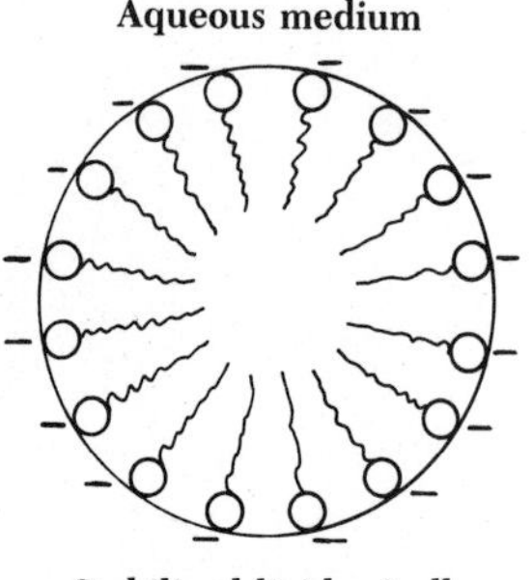

In the formation of oil-in-water emulsions, the emulsifying agent usually is a substance composed of molecules having both strongly polar (hydrophilic) and nonpolar (hydrophobic) groups. Such molecules are called *amphipathic* molecules. Soluble soaps, phospholipids, and bile salts are familiar examples. These molecules are positioned in oil-water emulsions so that the hydrophilic end of the molecule is oriented toward the aqueous medium and the hydrophobic end into the emulsified lipid globule. The surface of lipid particle thus has a charge, usually negative, that will repel like particles and thus stabilize the emulsion. Such minute-sized particles are termed *micelles*. They may be represented schematically as shown on p. 888, using the bile salt (Chapter 17) sodium glycocholate as an example of the emulsifying agent.

Surface tension

The surface of a liquid behaves as if it were a stretched elastic film. This tension is the result of the unbalanced attraction of the molecules to each other. According to Laplace, the molecules of a solution are strongly attracted to each other but only over a very short distance. The attraction is probably greatest at a distance equal to about the diameter of a molecule. Fig. 24-1 illustrates and explains this phenomenon. Molecules *C, D, E,* and *F* are not at the surface and are attracted equally in all directions. They are therefore able to move freely in all directions. Molecules *A* and *B*, however, are at the surface *(XY)* and are not attracted upward because of the absence of molecules of the fluid above the surface. Consequently, they tend to be drawn downward and pulled sideward, and the layer of surface molecules is thereby stretched. The effect of this "film" is seen in the tendency of drops of water and mercury and soap bubbles to assume a spheric form because of the cohesive pull sideward and inward. This phenomenon occurs at any surface or *interface* that separates a liquid from air or other gases or that separates one liquid from another. This explanation is a simplification of the phenomenon; other forces are involved besides that of attraction.

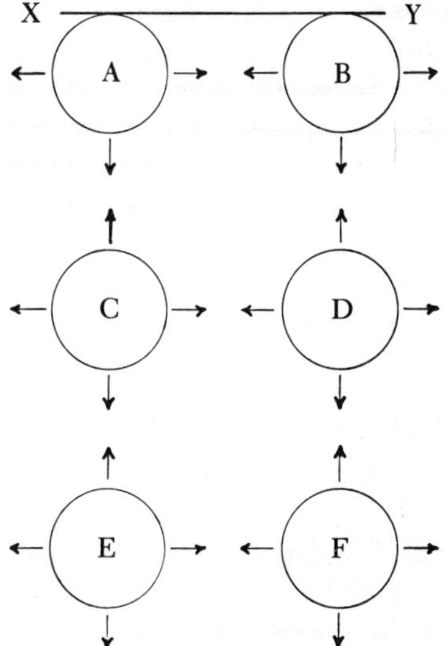

Figure 24-1
Diagram to explain surface tension. Molecules *A* and *B* are at surface, X – Y, and are not attracted upward because of the absence of molecules above them. Consequently, they tend to be drawn downward and sideward, and the layer of molecules at the surface is thus stretched. Molecules *C, D, E,* and *F* are not at the surface and are attracted equally in all directions.

Surface tension may be measured in a number of ways, perhaps most conveniently with a stalagmometer. This is a pipette of special design with a capillary tube ending, permitting a measured amount of fluid to flow out drop by drop. The number of drops depends on the size of the drops, which, in turn, varies with the surface tension. A comparison with the number of drops of pure water permits one to calculate the surface tension of the solution. Surface tension is expressed as ergs per square centimeter or dynes per centimeter. For accurate work, great precautions of cleanliness must be taken, since small amounts of some substances alter the surface tension materially. Soaps, oil, proteins, and salts of the bile acids reduce the surface tension of water, whereas sodium chloride tends to increase it. These and similar effects aid in explaining some physiological actions, e.g., fat digestion and absorption. Substances that reduce surface tension accumulate in the surface film and are said to be adsorbed, whereas the reverse is true of those that increase surface tension. There is a stalagmometric method for the determination of bile acids in bile that has been used as a liver function test, based on the fact that an important function of the liver is the secretion of bile acids. The du Noüy tensiometer is another device for determining surface tension. In this, a light metal ring is set on the surface of the fluid under examination. As the ring is raised, a film of the fluid clings to it. The amount of force required to pull the ring off and break this film is a measure of the surface tension and can be exactly and conveniently measured.

Diffusion, osmosis, and dialysis

Diffusion If a strong solution of a salt, e.g., copper sulfate, is placed in a glass vessel and a layer of distilled water is carefully poured over it, the blue copper sulfate rises gradually into the colorless water until finally the entire body of fluid has the same color. This process is called *diffusion*. The velocity with which it occurs depends on the size of the particles of the substance in solution. Thus Prussian blue, being composed of large particles, diffuses more slowly than does copper sulfate. Higher temperatures also speed up the process. It should be observed that diffusion involves the passage of substances, in true solution or in colloidal solution, through the fluid in which they are suspended. In the many fluids of the body, within cells, during secretory activity, diffusion must be constantly occurring.

Osmosis *Osmosis* is the passage of a solvent through a semipermeable membrane. Such a membrane is permeable only to the solvent, not to the solute, i.e., the substance in solution. The classic experiment of Pfeffer illustrates the point. He precipitated copper ferrocyanide in the walls of an unglazed porcelain jar by filling the jar with potassium ferrocyanide solution after immersing it in a solution of copper sulfate. Such a film of copper ferrocyanide permits water to pass through but does not allow certain soluble substances, e.g., sugars, to do so. Consequently when such a jar, fitted with a glass tube into which the liquid can rise and filled with sugar solution, is placed in distilled water, the water passes through the semipermeable membrane into the sugar solution until the column of diluted sugar solution is no longer increased. The pressure that would have to be exerted on the solution to prevent passage into it of solvent, namely, water, when separated from it by a perfectly semipermeable membrane is called the osmotic pressure. *Osmotic pressure* is evidenced only if there is a semipermeable membrane separating one solution from another. However, in biochemical and physiological literature, it has become customary to allude to the osmotic pressure of a solution even when no membrane is present. Furthermore, in biological systems the membranes are not "perfect" semipermeable membranes, but they permit solutes consisting of small particles or ions to pass through. There is, of course, some osmotic influence even under these circumstances. The student should have these facts in mind when the osmotic pressure of the blood plasma (Chapter 16) is discussed.

The osmotic pressure of a solution is directly proportional to the concentration of the solute. More concentrated solutions give rise to higher osmotic pressures than do weak ones. Therefore, 1 gm of salt dissolved in 100 ml has twice the potential osmotic pressure of 1 gm of salt dissolved in 200 ml. In other words, the osmotic pressure is *inversely* proportional to the volume, showing that Boyle's law is applicable to solutions. Similarly, it has been found that the osmotic pressure increases $1/273$ for each rise of 1° C (Gay-Lussac's law). Avogadro's law also applies to osmotic pressure, since all solutions containing the same number of dissolved particles exert the same osmotic pressure at a constant temperature; i.e., the osmotic pressure depends on the number of particles dispersed in the fluid. This pressure is independent of the nature of the particles. Consequently a given number of ions, undissociated molecules, and aggregates of molecules (colloidal particles) in identical volumes of fluid all exert the same effect. Since each ion has the same effect as a molecule, evidently an electrolyte like sodium chloride, which furnishes two ions, will produce twice as high an osmotic pressure when completely dissociated as will a nonionized substance of the same molecular weight. Also, since large aggregates of molecules have the same effect as ions or small molecules, colloidal "solutions" will exert low osmotic pressures because of the comparatively small *number* of such particles present.

Although osmotic pressure can be determined by an apparatus such as illustrated in Fig. 24-2, in practice it is measured by indirect means. The boiling point of a solvent is raised by the addition of a solute, and the freezing point is lowered similarly. The amount of either change is proportional to the concentration of the particles of the solute and, as we have seen, the same holds true for osmotic pressure.

Solutions that exert the same osmotic pressure are said to be *isosmotic*. If a cell is in contact with a solution having the same osmotic pressure as the cell contents, the amount of water passing into the cell is balanced by that passing out, provided that the cell membrane is impermeable to the solutes. In this case, the solution is not only isosmotic but also *isotonic;* i.e., the cell volume is unchanged; its tone is maintained. If the osmotic pressure of a solution of the same solutes is greater than that of the cell, the solution is hyperosmotic and is said to be *hypertonic,* and water will pass from the cell to the solution. If it is lower, the solution is hyposmotic and is *hypotonic,* and water will flow into the cell. In a hypertonic solution the cell shrinks, and in a hypotonic solution it swells. We have presupposed in each case that the solutes do not permeate the membrane. If, however, one or more of the solutes can pass through the membrane and are not present in the same concentration within the cell, an isosmotic solution may not be isotonic. Osmotic pressure is not the same as tonicity. However, physiologically the terms are almost interchangeable and are found to usually parallel each other rather closely.

Thus 0.9% of NaCl solutions are isotonic to human erythrocytes, and if the two are in contact, the amount of water passing into the cell is balanced by that passing out. A solution of lower salt concentration has lower osmotic pressure than the cell. It is hypotonic. If erythrocytes are immersed in hypotonic solutions, water passes from the more dilute to the more concentrated solution (i.e., from the lower to the higher osmotic pressure) in so great an amount that the erythrocytes swell and may even burst (hemolysis). Hypertonic solutions cause water to pass out of the cells and the cells to become shriveled, or *crenated*. Therefore, solutions used for injecting are made up to be isotonic; otherwise unphysiological effects would be pro-

Figure 24-2
Osmometer.

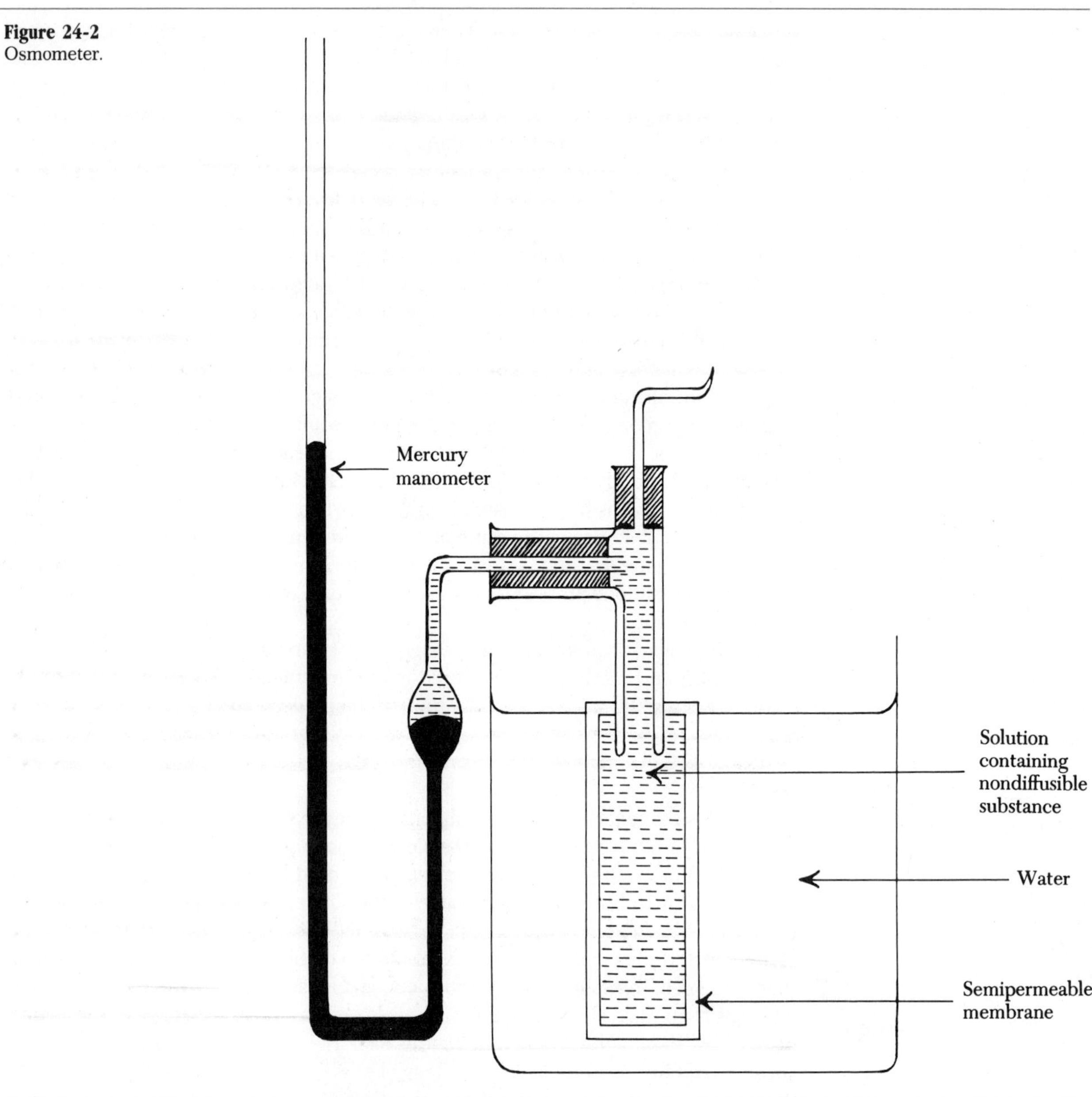

duced on the corpuscles. Similarly, whenever cut tissues or viscera are bathed with fluid, isotonic solutions are used since they cause less change in the cells with which they come in contact. The body adjusts its various fluids to approximately the same osmotic pressure. For example, the osmotic pressures of such extremely divergent types of fluids as blood, hepatic bile, pancreatic juice, and lymph collected simultaneously have been found to be practically the same.

Dialysis As just explained, osmosis requires a membrane that is permeable to the solvent, not to substances dissolved or dispersed in it. Membranes of varying porosity that permit smaller particles of varying sizes to pass through but are impermeable to larger particles exist or can be prepared. In general, membranes such as parchment, collodion, or cellophane allow crystalloids to pass but prevent colloids from doing so. The process involved is termed *dialysis*. Thus, if a mixture of crystalloids and colloids is placed in a dialyzing bag that is then immersed in distilled water, the colloids remain behind while the crystalloids dialyze out. By changing the outside fluid frequently, the colloids may be completely freed of crystalloids. It is easily seen, however, that since colloids themselves vary in size some membranes will permit colloids of smaller dimensions to pass.

Membranes in the animal body

Since all cells, plant and animal, are enclosed by membranes, the question naturally arises: Are these membranes semipermeable or permeable, or do they behave like filter paper? This cannot be answered categorically, since the wall of a living cell cannot be compared with an artificial membrane or even with a dead biological membrane. Animal membranes (see also Chapter 19) do not have the simple structure of artificial membranes. It is probable that cell membranes in different tissues vary in their permeability. Certainly the cells lining different parts of the kidney permit different ions and compounds to pass. Indeed, adjacent portions of the kidney tubules exhibit varying properties of this nature. It is also probable that the same cell wall changes from hour to hour, depending on respiratory and metabolic conditions. Therefore the idea formerly held, that living membranes are impermeable to colloids but are permeable to electrolytes, must be modified.

A remarkable degree of selectivity sometimes is seen in these biological migrations. The erythrocyte membrane, for example, allows cations to pass through less easily but is quite permeable to anions—a fact important in explaining respiratory phenomena. Furthermore, some fat-soluble as well as some water-soluble substances pass through the erythrocyte wall. The membrane is believed to be composed of at least three layers, two outer layers of protein and an inner lipid core, that are permeable to certain fat-soluble and water-soluble materials. However, all the phenomena pertaining to membranes cannot be explained on the basis of porosity, i.e., sieve action, or on the theory of solubility. In some cases, the electrical charge of the membrane must be taken into account. For example, if the charge on erythrocytes is reversed, the cells become impermeable to anions and permeable to cations.

Gibbs-Donnan equilibrium

If a membrane that permits the passage of crystalloids but not colloids is placed between two solutions of a simple electrolyte such as sodium chloride, the sodium and chloride ions pass through the membrane until, at equilibrium, the product of the concentrations of these two ions on one side equals the product of their concentrations on the other side. Concentrations such as gram molecules per liter are again represented by bracketed symbols. The conditions stated may be represented thus:

$$[Na^+]_1 \times [Cl^-]_1 = [Na^+]_2 \times [Cl^-]_2$$

$[Na^+]_1$ means, of course, the concentration of sodium ions on side 1, and $[Na^+]_2$ means the concentration of side 2, etc. Moreover, not only are the products of the concentrations the same, but under these conditions the concentrations of the cations are the same, as are those of anions:

$$[Na^+]_1 = [Na^+]_2$$
$$[Cl^-]_1 = [Cl^-]_2$$

Let us assume that, in addition to a simple electrolyte, we have on side 1 the sodium salt of a colloid, NaR, which can ionize to $[Na^+]$ and $[R^-]$ but the colloid ion is too large to pass through the membrane. Now the additional Na^+ may pass through but not the R^-.

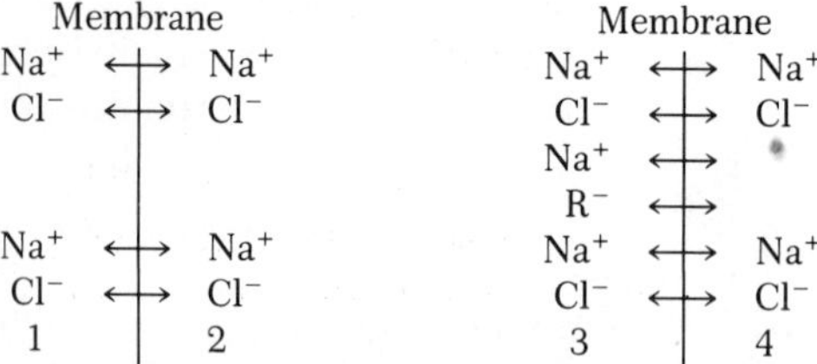

As before, chloride ions may pass back and forth. Finally, equilibrium will occur and the question arises as to how this will affect the distribution of the diffusible ions. As in the first instance, the product of the concentrations of the diffusible cation-anion pair on one side will equal the product of the concentrations of the same pair on the other side. (This is derived from thermodynamics.)

$$(1)\quad [Na^+]_3 \times [Cl^-]_3 = [Na^+]_4 \times [Cl^-]_4$$

But

$$[Na^+]_4 = [Cl^-]_4$$

This is evident, since there is only sodium chloride present on side 4 and there can be no more of one ion than the other. If we remember that on side 3 there is a nondiffusible ion R^- balancing electrically some of the Na^+, it is evident that

$$[Na^+]_3 > [Cl^-]_3$$

Now, referring to the equation (1), if members of the left side are unequal, we can see that the larger value on the left must be larger than either of the two on the right (which are equal to each other) and the smaller value on the left must be smaller than either one on the right, or:

$$[Na^+]_3 > [Na^+]_4$$
$$[Cl^-]_3 < [Cl^-]_4$$

We can therefore see how the presence of a colloidal ion may cause an inequality in the distribution of ions on the opposite sides of a membrane. If there are several colloidal ions and many noncolloidal ions present, the state of affairs becomes exceedingly complex. This is what happens in animal tissues, where different types of cells, bathed by the same or different body fluids under varying conditions, differ fundamentally in the chemical makeup of their contents. This unequal balance of ions also leads to a difference in potential between the solutions on the two sides of the membrane. The phenomenon is known as the *Gibbs-Donnan equilibrium,* after Willard Gibbs and F.G. Donnan, who first studied and explained it.

Membrane potentials One of the most interesting physiochemical phenomena in living matter is the existence of potentials, of as much as 60 or more millivolts, across cell membranes. These are manifested as the clinically important electroencephalogram (EEG), the electrocardiogram (ECG), and the action current in the conduction of nerve impulses.

The phenomenon has been studied extensively for many years, and a number of theoretical explanations have been proposed, none of which is universally accepted. For a complete discussion of these hypotheses, consult the readings listed at the end of the chapter. Only a brief summary is presented here. The fundamental reason for the potential appears to be a different net charge on the two sides of the membrane caused by differing ionic concentrations. The latter may result from a selective permeability of the membrane, a sieve action, a pump mechanism, some sorption or association-induction effect, or a combination of these. The Donnan effect just described is undoubtedly involved in at least some instances. Important in biochemistry is the fact that the membrane potential changes with the transfer of ions across the cell membrane in varying metabolic and physiological situations.

Viscosity

Viscosity is the resistance offered by a fluid to flow. It is the result of the internal friction of the molecules of a liquid. A solvent is almost always less viscid than a solution and considerably less viscid than a colloidal system. Viscosity ordinarily is not expressed in absolute units but is referred to as the viscosity of water. It is measured by allowing a definite amount of the fluid under consideration to flow through a capillary tube at a definite temperature. The time required is compared with that taken by an equal volume of water. With water as unity, the normal viscosity of blood serum is about 1.5 to 2, whereas that of plasma, which has a higher protein content, is about 20% greater, and whole blood has a viscosity of 2.5 to 4. These are approximate values for the viscosity of normal human blood. When dehydration occurs, as it does in some pathological states, the viscosity of whole blood may be three or four times the normal value.

The foregoing are but a few of the many topics that illustrate the dependence of biochemistry today on the laws and the applications of basic chemistry and physics.

Suggested readings

Barron, D.N., and others: The use of S.I. units in reporting results obtained in hospital laboratories, J. Clin. Pathol. **27**:590, 1974.

Bray, H.G., and White, K.: Kinetics and thermodynamics in biochemistry, New York, 1966, Academic Press, Inc.

Clark, W.M.: Topics in physical chemistry, ed. 2, Baltimore, 1952, The Williams & Wilkins Co.

Davenport, H.W.: The ABC of acid-base chemistry, ed. 5, Chicago, 1974, University of Chicago Press.

Dawes, E.A.: Quantitative problems in biochemistry, ed. 5, Baltimore, 1969, The Williams & Wilkins Co.

Edsall, J.T., and Wyman, J.: Biophysical chemistry, vol. I, New York, 1958, Academic Press, Inc.

Klotz, I.M.: Energy changes in biochemical reactions, New York, 1967, Academic Press, Inc.

Montgomery, R., and Swenson, C.A.: Quantitative problems in the biochemical sciences, San Francisco, 1969, W.H. Freeman & Co., Publishers.

Morris, J.G.: A biologist's physical chemistry, Reading, Mass., 1968, Addison-Wesley Publishing Co., Inc.

Muntwyler, E.: Water and electrolyte metabolism and acid-base balance, St. Louis, 1968, The C.V. Mosby Co.

Williams, V.R., and Williams, H.B.: Basic physical chemistry for the life sciences, San Francisco, 1967, W.H. Freeman & Co., Publishers.

Chapter 25

Analytical techniques frequently used in biochemistry

The dependence of modern research on methodology, especially those procedures that permit some degree of quantitation, is an obvious fact. Indeed the tremendous progress made in biochemical and related research during the past several decades was possible because of the development of more sensitive and sophisticated analytical techniques. Two in particular may be mentioned as opening the way, some 30 years ago, to new avenues of research—the development of chromatographic and isotopic labeling techniques. These have been improved and greatly expanded since that time and other equally indispensable techniques have been added and developed.

A detailed, comprehensive discussion of this area of biochemistry is obviously beyond the scope of a general treatise of the field. Yet a brief introductory summary of the principles of some selected important analytical techniques and their applications to biochemical problems seems worthwhile at this time. Accordingly an overview of a few of the techniques specially relevant to current research in biochemistry is included in the following pages. For discussions of other important techniques not included—such as electron-spin resonance, electron paramagnetic resonance, nuclear magnetic resonance, and spin labeling; mass spectrometry; fluorometry; flame and atomic absorption spectroscopy; infrared, Raman, and ultraviolet spectroscopy; circular dichroism; and protein- and nucleic acid–sequencing techniques—the reader is referred to the list of readings at the end of this chapter. Other excellent reference books and monographs are available at many science libraries.

Chromatography

Experience with chromatography goes back to 1850, when Runge described a method for the analysis of mixtures of dyes by applying drops of dye solutions to blotting paper and noting the separation of the different colors. In 1897 Day used a column of limestone for the fractionation of crude petroleum. However, the credit for discovering chromatography is usually given to Tswett, a botanist, who in 1906 effected the separation of plant pigments by filtering a solution of the mixed pigments through a tube containing finely divided, solid-adsorbing material. The individual pigments settled in separate bands and thus formed a *chromatogram,* or pattern of pigments. Each substance adsorbed, called an *adsorbate,* could be removed, or *eluted,* from the *adsorbent*. The term *chromatography* is still used for this type of separation, despite the fact that the process is no longer limited to colored substances. It has been greatly extended and is widely used in biochemistry as well as in other branches of chemistry. Many different adsorbents are used, and they are selected for their affinity for the adsorbates. They include talc, asbestos, clays, charcoal, starches, and filter paper.

Paper chromatography differs somewhat, although the principle is the same. A large sheet of filter paper is treated in one corner with a small amount of the solution to be studied. It is then draped over a glass rod, with the edge containing the sample dipping into a tray of the desired solvent. As the solvent travels up the paper, it takes with it the unknown substances, which are deposited in spots. These spots can be "developed" as colors by suitable chemical treatment. They are identified by running controls in which known substances are used. Sometimes there is a second run, using a different solvent at right angles (a two-dimensional chromatogram), which separates spots placed too closely together by the first run for easy identification.

Within the past few years, many modifications and improvements in chromatographic procedures have been made, increasing the resolving power and applicability of the procedures. For example, *thin-layer chromatography* uses the same principles as those for paper chromatography. Such supports as silica gel or cellulose are dried onto a glass or plastic sheet as very thin layers. In this process the chromatographic separation is quite rapid.

Molecular sieve chromatography takes advantage of differences in molecular size and shape. Protein solutions are passed through columns of synthetic polyacrylamide gels or polysaccharide gels called *dextrans*. These substances function by virtue of differences in pore sizes in the particles. Small molecules pass through the pores and enter the particles, whereas larger ones, or macromolecules, are unable to do so readily. In this way the larger molecules pass through the column, whereas the smaller ones are retained with the particles.

Gas chromatography is also a valuable tool for the separation of volatile substances or volatile derivatives of otherwise nonvolatile substances. The sample molecules are volatilized and separated by being passed, with the help of a carrier gas, through a long column of narrow diameter containing a suitable stationary support.

Ion-exchange chromatography

Ion exchange is a reversible interchange of ions between a liquid and a solid, involving no radical alteration in the structure of the solid. Natural products (e.g., certain sands, peat, and coal), operating on this principle, have been used to soften water for many years; but not until 1935 were synthetic resins employed for this purpose. The English scientists Adams and Holmes reasoned that acidic and basic groups that are not involved in the condensation of the constituents of synthetic resins should be free to ionize. If so, they should permit cation- and anion-exchange processes. Ion exchangers consist, therefore, of insoluble resins or supports, Ⓡ, to which are attached a variety of acidic or basic groups. These substances are often packed into columns, and the solutions containing the exchangeable ions are percolated through the column.

Resins bearing anionic components of the type Ⓡ-$SO_3^-H^+$ or Ⓡ-$CO_2^-H^+$ are called *cation exchangers*. For example the proton may be exchanged by sodium ion so that:

$$\text{Ⓡ-}SO_3^-H^+ + NaCl \rightarrow \text{Ⓡ-}SO_3^-Na^+ + HCl$$

In the separation of amino acids:

$$\text{Ⓡ-}SO_3^-NH_4^+ + \text{Amino acid–}NH_3^+Cl^- \rightarrow \text{Ⓡ-}SO_3^-\text{–amino acid–}NH_3^+ + NH_4^+Cl^-$$

The exchanged cation may then be removed from the column by a regenerating process:

$$\text{Ⓡ-}SO_3^- \text{–amino acid–}NH_3^+ + HCl \rightarrow \text{Ⓡ-}SO_3^-H^+ + \text{Amino acid–}NH_3^+Cl^-$$

Anion exchangers may possess amino groups, Ⓡ-NH_2, that form an ammonium salt with HCl:

$$\text{Ⓡ-}NH_2 + HCl \rightarrow \text{Ⓡ-}NH_3^+Cl^-$$

The chloride may be replaced by another anion in the exchange process:

$$\text{Ⓡ-}NH_3^+Cl^- + \text{Amino acid–}CO_2^- + Na^+ \rightarrow \text{Ⓡ-}NH_3^+\text{–amino acid–}CO_2^- + NaCl$$

The amino acid may be removed by regenerating with HCl.

Proteins are commonly isolated and purified by ion-exchange chromatography. The now standard procedure is to use columns of cellulose derivatives such as diethylaminoethylcellulose and triethylaminoethylcellulose (DEAE and TEAE) as anion exchangers. Carboxymethylcellulose (CM) is often used as a cation exchanger for this purpose.

High-performance (pressure) liquid chromatography (HPLC)

Column chromatography employing a variety of resins or other solid phases for specific purposes and subjected to high pressure to increase flow rates are proving increasingly popular today. Procedures requiring hours of time by usual pressure systems are reduced to minutes by high-pressure techniques. With the appropriate type of solid phase, excellent resolution can be obtained also.

Affinity chromatography

In this type of chromatography, a substance that has a specific affinity for the compound or material to be isolated is attached to an appropriate gel or other solid phase. The attached substance may be a binding protein, an enzyme inhibitor, an antibody, an antigen, or other substance having a specific affinity for the compound to be isolated. A column procedure may be employed in which the solution of the material under investigation is passed through the column of the "affinity gel" to which it is specifically bound. Other substances, ideally at least, are not bound and pass out of the column. The desired substance is then eluted, usually by either increasing the salt concentration or changing the pH or by the use of a specific chaotropic ion, such as isothiocyanate. A chaotropic ion disrupts or destroys the adsorbed complex and elutes the protein desired, usually an enzyme. The eluate should contain a highly purified, homogeneous solution of the substance under study.

In a "batch method" modification of the above column procedure, the sample solution is simply mixed with the "affinity media" for an appropriate period of time. The solid-phase complex is then removed by centrifugation or filtration, washed, and the desired bound substance eluted as above.

During the past few years, the technique of affinity chromatography has been undergoing increasing use in biochemical and related research. This is caused in part by its versatility and applicability in the purification and analysis of biological substances. Affinity chromatography, unlike other forms of chromatography that separate substances on the basis of gross physical and chemical differences, accomplishes separation on the basis of *biospecific interactions* between interacting pairs of substances, e.g., hormone-carrier, enzyme-inhibitor, antigen-antibody, etc. It should prove to be a valuable addition to the analytic techniques of the biochemist and medical scientist.

Electrophoresis

The movement of a charged particle in an external electric field toward the oppositely charged electrode is called electrophoresis. The electrophoresis of biocolloids, for the most part proteins, has been commonly performed in the apparatus developed by Tiselius, in Sweden. Current is applied to the ends of a U-tube, which is filled with a protein solution overlaid by a buffer solution of the same ionic strength, pH, and conductivity. The buffer solution is stratified carefully over the protein layer so that the boundary between the two solutions is sharp. During the electrophoresis period, the various proteins migrate towards the electrode of opposite charge. The degree of migration depends on the magnitude of the protein charges. New boundaries may be formed if more than one protein is present, and their locations in the limbs of the tube (and therefore the distances traveled from the starting position), as well as their relative concentrations, may be determined by an optic process. Light, refracted by the various boundaries, is converted by an optic system into a pattern of peaks, each peak representing a boundary between protein molecules and the buffer.

The Tiselius apparatus is used to determine the isoelectric point and electrophoretic mobility of proteins. The method is also frequently employed to study the homogeneity of protein preparations.

A highly simplified modification of the Tiselius electrophoretic system is called *zone electrophoresis*. In this procedure the electrophoretic separation is performed on an inert support soaked in the buffer to be used. Such supports are usually paper, starch, agar, certain plastics, or cellulose acetate. The paper support, wet with buffer, serves as a bridge between the two electrode vessels. The sample is applied to the support as a narrow strip. Separation of molecules is in the form of zones of proteins that migrate in accordance with their charge densities. After the separation is complete, the paper strip is dried to fix the separate molecules on the paper. The sample molecules are rendered visible by a suitable dye. Their relative concentrations may be estimated by the use of a densitometer.

Isoelectric focusing

Isoelectric focusing is especially useful for the separation and identification of proteins and peptides. It is being used extensively at the present time. The procedure is based on two fundamental principles:

1. Charged particles will migrate in an electric field.
2. These compounds can be immobilized (or precipitated) at a certain pH characteristic of the compound, its *isoelectric point*.

Compounds are thus separated on appropriate chromatographic columns (acrylamide gel, etc.) by use of a suitable eluate with a controlled pH gradient. The gel is then removed and washed, and the separated protein bands are stained with an appropriate dye. Identification may be made by using known standards. The bands may be cut and the stained protein then quantitatively determined by standard spectrophotometric procedures.

Immunoelectrophoresis

Immunoelectrophoresis is an extremely sensitive and useful technique, combining separation by electrophoresis with the specificity of immunological reactions. It is especially useful for separating proteins and other antigenic, charged substances. The technique has been described in some detail in Chapter 16, with an illustration showing the high degree of resolution obtained with plasma proteins.

Ultracentrifugation

The ultracentrifuge produces sufficiently high speeds to cause the sedimentation of macromolecules in solutions. The rate at which sedimentation occurs depends on such parameters as molecular weight and shape. Under carefully controlled conditions, the macromolecules sediment at a characteristic velocity, which can be measured as the movement of the boundary between molecules and solvents.

In the apparatus developed by Svedberg (the analytical ultracentrifuge) an optic system converts the sedimenting boundary of molecules into a peak, which actually represents changes in refractive index. The rate of movement of this peak, dx/dt, is used to calculate the sedimentation coefficient, s.

$$s = \frac{v}{\omega^2 x}$$

In this equation, v is velocity (dx/dt) of sedimentation, ω is angular rotor velocity in radians per second, and x is the distance of the boundary to the center of the rotor in centimeters. The sedimentation constant, s, has the magnitude of 10^{-13} second. The constant, however, is usually given in *Svedberg units, S,* for which the magnitude is 10^{-13} second.

Svedberg then used this constant to develop the equation for molecular weight, M.

$$M = \frac{s \cdot RT}{D\,(1 - \bar{v}\rho)}$$

Here R is the gas constant, T the absolute temperature, and D the diffusion constant. $(1 - \bar{v}\rho)$ is a buoyancy correction in which ρ is the solvent density and $\bar{v}$ the partial specific volume of the solute molecules.

Density-gradient centrifugation

A relatively recent development in the area of ultracentrifugation uses density gradients and a preparative ultracentrifuge. In this method a continuous linear gradient of sucrose, varying, for example, from 20% at the bottom to 5% at the top, is formed in a plastic centrifuge tube. The sucrose gradient is often produced automatically with a controlled mixing device, the gradient maker. The sample is carefully layered on top of the gradient, and the tubes are then spun at high speed in a "swinging-bucket" rotor. During the centrifugation, the molecules migrate through the gradient at characteristic velocities. At the termination of the centrifugation, the tubes are pierced with a hypodermic needle and equal fractions are collected for various assays.

Salts such as CsCl form gradients of high density and low viscosity. Such gradients are used in isopyknic or equilibrium sedimentation because the sample molecules migrate to zones of equal density. This technique was used in the separation of ^{14}N-, ^{15}N-, and ^{14,15}N-DNAs.

Density-gradient methods are popular because they allow both analytical and preparative studies. They may also be used in the determination of molecular weight by the method of Martin and Ames.

X-ray diffraction cyrstallography

When a beam of x rays enters a crystalline substance, the waves are deflected from their course (i.e., diffracted) by the various planes of atoms. In this situation the angle of incidence equals the angle of diffraction. In 1912, Bragg showed that this diffraction of x rays obeys the expression:

$$n\lambda = 2d \sin \theta$$

This equation says, in effect, that when the path difference for waves diffracted by successive sheets of atoms is a whole number of wavelengths, the waves will be in phase, resulting in a summation of intensity and a diffraction maximum. The location of these diffracted beams and their relative intensities are evaluated by a photographic procedure. Since, in a crystalline substance, atoms are arranged in an orderly manner, the characteristic diffraction pattern can be used to determine the relative positions in space of these atoms. In a crystal of protein, the x rays are diffracted by the electron fields of C, O_2, and N_2. This phenomenon is evaluated mathematically, and an electron-density map, called the *Fourier pattern,* is produced. The electron densities of successive sheets of atoms in the crystal are reproduced in such maps. In the final model each successive electron-density contour map is placed on top of the preceding one until a three-dimensional model of the protein is produced. Since a hemoglobin molecule contains 10,000 atoms and many thousands of measurements must be performed, the complexity of the procedure is obvious.

The potentialities of x-ray crystallographic techniques have been greatly increased by combining them with high-speed computer techniques. This combination generates images on TV-like screens, enhancing the visualization of three-dimensional molecular shapes and conformations. Recent improvements also permit color coding and display of three-dimensional structures of molecules in various positions. Images of a number of protein molecules (e.g., IGG immunoglobulins) are being studied by this technique, as well as those of several DNAs and transfer RNAs and even those of certain enzymes *in action!*

Radioimmunoassay and related competitive binding assays

The technique of radioimmunoassay as originally developed in 1959 by Yalow and Berson was designed for the determination of insulin in human serum. It soon became evident that the procedure was equally well adapted for the determination of a number of other protein hormones in plasma and indeed for almost any protein present in biological materials in amounts below the sensitivity range of standard chemical and biological methods. These pioneer investigations stimulated research on the subject during the past decade, as manifested by the publication of some 3000 papers on the subject in the English language alone and the presentation of a number of reviews, symposiums, and workshops. As a result, in turn, considerable improvements in and extensions of the technique were made. Its applicability was also enormously extended by the use of *haptenes,* enabling the induction of specific antibodies to many substances that normally do not elicit an antibody response. Thus the list of substances that may be determined by radioimmunoassay today includes, in addition to many proteins, including hormone-receptor proteins in cells (e.g., insulin-receptor proteins), a number of peptide and steroid hormones and related compounds, drugs, antibiotics, vitamins, nucleic acids, enzymes, cancer antigens, and even structural proteins. The possibilities for the application of the technique in the future appear almost unlimited.

The technique of radioimmunoassay combines the high degree of sensitivity and precision of radiochemistry with the specificity of immunological reactions—an ideal combination for the accurate determination of substances present in extremely small amounts in body fluids and tissues. Indeed, the sensitivity of the procedure extends down to the picogram level for some substances.

The basic principle of the radioimmunoassay technique is the competitive binding of a labeled and an unlabeled antigen by a specific antibody to form labeled and unlabeled antigen-antibody complexes. The unlabeled antigen, of course, is the substance to be determined (insulin, an enzyme, etc.). The antibody to it is produced by injecting the antigen into an animal, usually a rabbit or a goat, etc. The blood serum removed after a suitable period of time contains the antibody. A "double antibody" technique is sometimes employed.

In an actual assay, usually performed in duplicate or triplicate and in several dilutions, the antigen to be determined (Ag), the specific antibody (Ab), and a known amount of the isotopically-labeled (with ^{131}I, etc.) antigen (Ag*) are mixed and incubated for a specific time period. Conditions are so adjusted that there is always an *excess* of antigen. The labeled and unlabeled antigens, Ag^* and Ag, thus *compete* for the limited number of binding sites on the antibody, Ab. Therefore if the amounts of labeled antigen and antibody are held constant, the radioactivity bound to the antibody *must decrease* as the amount of the unlabeled antigen in the sample *increases*. The basic reaction involved may be represented schematically as follows:

$$Ag + Ag^* + Ab \rightarrow Ag \cdot Ab + Ag^* \cdot Ab$$

Therefore, the concentration of $Ag^* \cdot Ab$ varies inversely with the concentration of Ag. The relation is nearly linear.

After the above assay reaction mixture has incubated for a sufficient time, the "free" and "bound" (left side of diagram) fractions are separated, usually by chemical precipitation or adsorption, and the radioactivity of the bound fraction is counted in a scintillation counter or other suitable apparatus. The unbound, soluble fraction is sometimes also counted. Corrections for nonspecific binding, background, etc. may also be necessary. From the amount of radioisotope bound, the amount of the unlabeled antigen in the sample can be determined from a standard calibration curve run simultaneously.

By this technique, which may be *automated* or adapted for *fluorescent immunoassay* techniques, variations in the usually minute amounts of substances such as insulin, other protein hormones, steroid hormones, various drugs, and metabolites can be determined in human plasma or other body fluids or tissues with accuracy and specificity. The applications of this elegant procedure to biochemical and clinical investigation and diagnosis appear to be almost unlimited.

Isotopic labeling with positron emitters

Another development in the use of isotopic labeling techniques, employed with outstanding success in metabolic studies over the past three decades, is that of positron-emitter labeling. Metabolites, pharmaceuticals, and other compounds of biological interest, can be labeled with short-lived positron emitters, such as ^{11}C, ^{15}O, ^{13}N, or ^{18}N, and administered to experimental subjects, human or animal. Used in conjunction with "CAT-scanners" (computerized axial tomography) and scintillation counters, this technique allows the investigator to pinpoint the area in the subject from which the gamma rays are coming and convert the data into a cross-sectional image of the area. The technique shows great promise in studies of normal and abnormal metabolism in various organs of the body, especially the brain.

Suggested readings

Allen, R.C., and Maurer, H.R., editors: Electrophoresis and isoelectric focusing in polyacrylamide gel: advances of methods and theories, biochemical and clinical applications, New York, 1974, Walter De Gruyter, Inc.

Berliner, L.J., and Reuben, J., editors: Biological magnetic resonance, New York, 1980, Plenum Publishing Corp.

Berson, S.H., and others: Immunoassay of protein and peptide hormones, Metabolism **13:**1135, 1964.

Bowie, L.J.: Automated instrumentation for radioimmunoassay, Cleveland, 1980, Chemical Rubber Co. Press.

Bragg, L.: X-ray crystallography, Sci. Am. **219:**58, 1968.

Brown, S.B., editor: An introduction to spectroscopy for biochemists, New York, 1980, Academic Press, Inc.

Colowick, S.P., and Kaplan, N.O., editors: Methods in enzymology, vol. 70, Immunochemical techniques, H. Van Dunakis and J.J. Langone, editors, New York, 1980, Academic Press, Inc.

Cooper, T.G.: The tools of biochemistry, New York, 1977, John Wiley & Sons, Inc.

Cuatrecasas, P., and Anfinsen, C.B.: Affinity chromatography, Annu. Rev. Biochem. **40:**259, 1971.

Curry, R.E., and others: A systems approach to fluorescent immunoassay general principles and representative applications, Clin. Chem. **25:**1591, 1979.

Giddings, J.C., and others, editors: Advances in chromatography, New York, 1960-1980, Marcel Dekker, Inc.

Glick, D., editor: Methods of biochemical analysis, 21 vols., New York, 1954-1972, Interscience Publishers, Inc.

Grob, R.L., editor: Modern practice of gas chromatography, New York, 1977, John Wiley & Sons, Inc.

Gross, E., and Meienhofer, J., editors: The peptides: analysis, synthesis, biology, vol. 3, New York, 1980, Academic Press, Inc.

Horvath, C., editor: High-performance liquid chromatography, vol. 1, New York, 1980, Academic Press, Inc.

Kinney, D.E., and Melville, R.S., editors: Evaluation of uses of automation in the clinical laboratory, Pub. No. (NIH) 79-501, Washington, D.C., 1979, Department of Health, Education, and Welfare.

Lewis, L.A., and Opplt, J.J., editors: Handbook of electrophoresis, Cleveland, 1979, Chemical Rubber Co. Press.

Merritt, C., and McEwen, C.N., editors: Mass spectrometry, parts A and B, New York, 1979-80, Marcel Dekker, Inc.

Needleman, S.B., editor: Protein sequence determination, Berlin, 1975, Springer-Verlag.

Nucleotide sequencing techniques. In Colowick, S.P., and Kaplan, N.O., editors-in-chief: Methods in enzymology, vol. 65, Nucleic acids, part I, New York, 1980, Academic Press, Inc.

Pal, S.B., editor: Enzyme labeled immunoassay of hormones and drugs, New York, 1978, Walter De Gruyter, Inc.

Robinson, J.W., editor: Handbook of spectroscopy, Cleveland, 1974, Chemical Rubber Co. Press.

Schroeder, W.A., and Huisman, T.H.J.: Chromatography of hemoglobin, New York, 1980, Marcel Dekker, Inc.

de Serres, F.J., and Hollaender, A., editors: Chemical mutagens: principles and methods for their detection, New York, 1980, Plenum Publishing Corp.

Strickland, E.H.: Aromatic contributions to circular dichroism spectra of proteins, CRC Crit. Rev. Biochem. **2:**113, 1974.

Ter-Pogossian, M.M., Raichle, M.E., and Sobel, B.E.: Positron-emission tomography, Sci. Am. **243:**170, 1980.

Zweig, G., and Sherma, J., editors: Handbook of chromatography, Cleveland, 1972, Chemical Rubber Co. Press.

Index

D

K

L